Non-Diophantine Arithmetics in Mathematics, Physics and Psychology

Non-Diophantine Arithmetics in Mathematics, Physics and Psychology

Mark Burgin
University of California, Los Angeles, USA
Marek Czachor
Politechnika Gdańska, Poland

World Scientific
NEW JERSEY • LONDON • SINGAPORE • BEIJING • SHANGHAI • HONG KONG • TAIPEI • CHENNAI • TOKYO

Published by

World Scientific Publishing Co. Pte. Ltd.
5 Toh Tuck Link, Singapore 596224
USA office: 27 Warren Street, Suite 401-402, Hackensack, NJ 07601
UK office: 57 Shelton Street, Covent Garden, London WC2H 9HE

Library of Congress Cataloging-in-Publication Data
Names: Burgin, M. S. (Mark Semenovich), author. | Czachor, Marek, author.
Title: Non-diophantine arithmetics in mathematics, physics and psychology / Mark Burgin, University of California, Los Angeles, USA; Marek Czachor, Politechnika Gdańska, Poland.
Description: New Jersey : World Scientific, [2020] | Includes bibliographical references and index.
Identifiers: LCCN 2020032978 | ISBN 9789811214301 (hardcover) | ISBN 9789811214318 (ebook) | ISBN 9789811214325 (ebook other)
Subjects: LCSH: Diophantine analysis. | Arithmetic. | Number theory. | Mathematics--Philosophy. | Mathematical physics.
Classification: LCC QA242 .B886 2020 | DDC 512.7/2--dc23
LC record available at https://lccn.loc.gov/2020032978

British Library Cataloguing-in-Publication Data
A catalogue record for this book is available from the British Library.

For any available supplementary material, please visit
https://www.worldscientific.com/worldscibooks/10.1142/11665#t=suppl

Desk Editors: Ramya Gangadharan/Tan Rok Ting

Typeset by Stallion Press
Email: enquiries@stallionpress.com

Preface

New opinions are always suspected, and usually opposed, without any other reason but because they are not already common.

John Locke

There still remain three studies suitable for free man. Arithmetic is one of them.

Plato

For a long time, mathematicians and other people thought there was only one geometry — Euclidean geometry. Nevertheless, in the 19th century, many non-Euclidean geometries were discovered. It took almost two millennia to do this.

This was the major mathematical discovery and advancement of the 19th century, which changed understanding of mathematics and the work of mathematicians providing innovative insights and tools for mathematical research and applications of mathematics.

A similar event happened in arithmetic in the 20th century.

Even longer than with geometry, mathematicians and other people thought there was only one conventional arithmetic of natural numbers, which can be naturally called the *Diophantine arithmetic*. The reason for this name is that the ancient Greek mathematician Diophantus made important contributions to the development of the classical mathematical field arithmetic transforming it into algebra.

So, almost all people, mathematicians and other, believed that any other arithmetic is impossible and $2 + 2 = 4$ is an eternal law of nature.

Nevertheless, in the 20th century, many non-Diophantine arithmetics were discovered demonstrating that in some of them $2 + 2$ is not equal to 4. It took more than two millennia to do this.

This discovery has even more implications and effects than the discovery of new geometries because the impact of the latter was only on mathematics and physics as the majority of people do not utilize geometry in their life. At the same time, all people use arithmetic of natural numbers. Operation with numbers is one of the most important activities of people in the contemporary society. Every woman and every man perform additions, subtractions and counting many times every day. Mathematicians created arithmetic to provide rigorous and efficient rules and algorithms for operation with numbers. Calculators and computers were invented to help people to operate with numbers. Modern science and technology are impossible without operation with numbers. Political, social and economic applications of mathematics are now multiplying with a great speed and the majority of these applications involve numbers and arithmetic operations.

Moreover, physicists and mathematicians found that geometry forms a foundation of physics, while Descartes demonstrated that geometry can be molded from numbers and relations between them. This shows that arithmetic provides even a more profound foundation of physics.

Non-Diophantine arithmetics contain the same natural or counting numbers but basic arithmetical operations with numbers, such as addition and subtraction, are defined in a different way. For instance, in some non-Diophantine arithmetics two plus two is equal to five while there are other non-Diophantine arithmetics where two plus two is equal to three.

The discovery of non-Diophantine arithmetics demonstrated that there is no a unique universal arithmetic and thus, to correctly apply mathematics and efficiently solve emerging problems, it is crucial to find an appropriate arithmetic for each specific area and each class of problems.

Long before the discovery of non-Diophantine arithmetics, different researchers, such as Helmholtz, Lebesgue, Gasking, Popper, Davis, Rashevsky and Kline, have pointed to various situations in which rules of the conventional Diophantine arithmetic are not valid. Even poets and artists intuitionally felt necessity to go beyond the conventional arithmetic. For instance, the famous poet Lord Byron wrote, "I know that two and two make four — and should be glad to prove it too if I could — though I must say if by any sort of process I could convert two plus two into five it would give me much greater pleasure."

In our times, the expression $1 + 1 = 3$ has become a symbol of synergy (Lawrence, 2011; Jude, 2014; Kress, 2015; Burgin and Meissner, 2017; $1 + 1 = 3$, Urban Dictionary). For instance, now successful mergers in business and industry are metaphorically described by the expression $1 + 1 = 3$. This contradicts the rules of the conventional Diophantine arithmetic but is true in some non-Diophantine arithmetics. The expression "one plus one equals three" is used as a symbolic image of synergy in many areas such as biology and medicine, psychology, human–computer interaction, biochemistry and bioinformatics, optics and photonic, data technology, business and industry, theory and practice of organization, technology, anthropology and sociology, marketing, informatics, agriculture, creativity, education, and politics.

Other expressions from non-Diophantine arithmetics, such as $1+1 = 1$, $1 + 1 = 4$, *one plus one makes more than two*, *one plus one equals three-fourths*, *one plus one equals one and a half*, *negative one plus negative one equals negative three*) and $2 + 2 = 5$, also appear in scientific and popular publications.

Consequently, non-Diophantine arithmetics found applications in different areas. The goal of this book is to present the mathematical theory of non-Diophantine arithmetics as well as their applications in physics and psychology. We analyze the most profound mathematical and philosophical ideas and constructions related to numbers and arithmetic from the very beginning of mathematical sciences to the top achievements of our time. In Introduction, philosophical, methodological and historical issues of numbers and arithmetic are discussed. Chapters 2 and 3 contain mathematics of non-Diophantine arithmetics. In Chapter 4, it is demonstrated how non-Diophantine arithmetics are used for building non-Newtonian calculus. Chapter 5 presents applications of non-Diophantine arithmetics to fractals. Chapter 6 covers utilization of non-Diophantine arithmetics in physics. Section 7 describes employment of non-Diophantine arithmetics in psychology.

In this book, we distinguish three meanings of the term *arithmetic*: arithmetic as a branch of mathematics, arithmetic as a system of numbers with operations and relations in it and arithmetic as the set N of all natural numbers, i.e., numbers $1, 2, 3, 4, 5, \ldots$, with two basic operations —

addition and multiplication. In this context, we delineate two classes of arithmetics — Diophantine arithmetics and non-Diophantine arithmetics. The first class comprises the conventional types of arithmetics, the main of which are:

- the Diophantine arithmetic $\boldsymbol{N}$ of natural numbers,
- the Diophantine arithmetic $\boldsymbol{W}$ of whole numbers,
- the Diophantine arithmetic $\boldsymbol{Z}$ of integer numbers,
- the Diophantine arithmetic $\boldsymbol{Q}$ of rational numbers,
- the Diophantine arithmetic $\boldsymbol{R}$ of real numbers,
- the Diophantine arithmetic $\boldsymbol{R}^+$ of non-negative real numbers,
- the Diophantine arithmetic $\boldsymbol{C}$ of complex numbers.

Non-Diophantine arithmetics are numerical arithmetics, operations in which are intrinsically related to operations in the Diophantine arithmetic (Burgin, 1997). The most fundamental Diophantine and non-Diophantine arithmetics are arithmetics of natural numbers because $\boldsymbol{N}$ is the most innate and primeval arithmetic. However, as it is proved in Chapter 2, all non-Diophantine arithmetics of natural numbers satisfy the famous Peano axioms contrary to the popular opinion that these axioms uniquely characterize the Diophantine arithmetic of natural numbers.

To construct the theory of non-Diophantine arithmetics in a systematic and consistent way, we at first, develop foundations of this theory studying prearithmetics and projectivity relations between them while treating arithmetics as specific types of prearithmetics.

In essence, an abstract prearithmetic is a universal algebra (algebraic system) with two operations and a partial order. Operations are called addition and multiplication but in a general case, there are no restrictions on these operations. It is demonstrated that many important results obtained for Diophantine arithmetics remain true for suitable classes of prearithmetics.

At the same time, abstract prearithmetics contain many important classes of studied before mathematical structures such as rings, fields, semirings, linear algebras, ordered rings, ordered fields, topological rings, topological fields, normed rings, normed algebras, tropical and thermodynamic semirings, normed field Ω-groups, Ω-rings, lattices and Boolean algebras. As a consequence, application of abstract prearithmetics include all applications of these structures as well as of such new areas of mathematics as non-Newtonian calculus (Grossman and Katz, 1972; Pap, 1993; Czachor, 2016), idempotent analysis (Maslov, 1987; Maslov and Samborskii,

1992; Maslov and Kolokoltsov, 1997), tropical analysis (Litvinov, 2007; Speyer and Sturmfels, 2009), non-commutative and non-associative pseudo-analysis (Pap and Vivona, 2000; Pap and Štajner-Papuga, 2001), non-Newtonian functional analysis and idempotent functional analysis (Litvinov *et al.*, 2001).

Some of abstract prearithmetics are numerical, that is, their elements are numbers, e.g., natural numbers or real numbers. A numerical prearithmetic that satisfies additional conditions, in particular, containing all natural numbers and no other elements is called an arithmetic of natural numbers. Everybody knows the conventional Diophantine arithmetic $\boldsymbol{N}$. However, there are also non-Diophantine arithmetics of natural numbers as well as the Diophantine and non-Diophantine arithmetics of whole, integer, real and complex numbers.

In this book, we also try to develop consistent terminology in the area of various arithmetics. By the long-standing tradition, the term arithmetic has been exceptionally used for the arithmetic of natural numbers $\boldsymbol{N}$, which we call the Diophantine arithmetic. However, in the process of the development of mathematics in general and numbers systems in particular, arithmetics of whole, integer, rational, real and complex numbers emerged and became very useful in mathematics and beyond. In some sense, these arithmetics are non-Diophantine because the ancient Greek mathematician Diophantus never used them. However, we call them the Diophantine arithmetics of whole, integer, rational, real and complex numbers, respectively, because they are traditionally accepted as arithmetics. Besides, operations in these arithmetics are induced by the operations in the Diophantine arithmetic N of natural numbers. All other arithmetics are in the same sense non-Diophantine and to encompass all of them, we introduce the general mathematical structure with two operations and one partial order, which we call a prearithmetic. The goal is to develop foundations for the theory of Diophantine and non-Diophantine arithmetics, which are expounded in this book together with its applications. In this book, on the one hand, we derive the basic results of this theory. On the other hand, we apply it to the development of non-Newtonian calculi as well as to scientific disciplines such as physics and psychology.

This book demonstrates how non-Diophantine arithmetics provide innovative means for explorations of the three realms of our world: the physical world, the mental world and the world of structures. Non-Diophantine arithmetics in physics serve for discovering and explanation of different phenomena of the physical universe. As a matter of fact, some researchers, such as Penrose, Zeldovich, Ruzmaikin, Sokoloff, Varadarajan and Volovich, already expressed the opinion that fundamental problems of modern physics are dependent on our ways of counting and calculation while number theory, which is also called higher arithmetic, can be treated as the ultimate physical theory (Penrose, 1972; Volovich, 1987; Zeldovich, *et al.*, 1990; Varadarajan, 2004).

Non-Diophantine arithmetics in psychology make possible more accurate investigation of the mental reality in our world. At the same time, non-Diophantine arithmetics in mathematics bring researchers to new unusual areas in the world of structures, while the world of structures as a whole is the scientific counterpart and interpretation of the world of Plato Ideas.

Reading this book, the reader will see that on the one hand, non-Diophantine arithmetics continue the ancient tradition of operating with numbers, while on the other hand, they introduce extremely novel and pioneering ideas and techniques.

This book is best suitable for researchers and graduate students in mathematics, physics, psychology, and philosophy with knowledge of basic mathematics and interest in innovative developments in mathematics and theoretical physics as well as in foundational issues. At the same time, all intelligent people with curiosity and acquaintance with the basic elements of mathematics will also benefit from reading this book.

It is also possible to use this book for enhancing traditional courses of algebra and calculus for undergraduates, as well as for teaching separate courses for graduate and undergraduate students at colleges and universities. To achieve these goals, exposition in the book goes from simple topics to more and more advanced topics, while proofs of some statements are left as exercises for the students. Special attention is paid to providing a variety of examples illustrating the general theory.

About the Authors

Mark Burgin received his MA and PhD in mathematics from Moscow State University and Doctor of Science (DSc) in logic and philosophy from the National Academy of Sciences of Ukraine. He was a Visiting Professor at UCLA; Professor at the Institute of Education, Kiev, Ukraine; at International Solomon University, Kiev, Ukraine; at Kiev State University, Ukraine; the Head of the Assessment Laboratory of the Research Center of Science at the National Academy of Sciences of Ukraine, and the Chief Scientist at the Institute of Psychology, Kiev. Currently he is affiliated with UCLA, Los Angeles, California, USA. Mark Burgin is a member of the New York Academy of Sciences, a Senior Member of IEEE, of the Society for Computer Modeling and Simulation International, and of the International Society for Computers and their Applications, as well as an Honorary Professor of the Aerospace Academy of Ukraine and Vice-President of the International Society for The Studies of Information. Mark Burgin was the Chair of the IEEE San Fernando Valley Computer and Communication Chapter and the Secretary of the International Society for Computers and their Applications. He was the Editor-in-Chief of the international journals *Integration*, *Information*, and the *International Journal of Swarm Intelligence & Evolutionary Computation* and is an Editor and Member of Editorial Boards of more than 21 journals. He was a member of Program Organization and Scientific Committees in more than 100 international conferences, congresses, and symposia as well as the chief organizer of some of them such as Symposia "Theoretical Information Studies", "Evolutionary Computation and Processes of Life", and "Creativity in Education." Mark Burgin is involved in research and publications, and has taught courses in various areas of mathematics, information sciences, artificial intelligence (AI), computer science, system theory, philosophy, logic, cognitive sciences, pedagogical sciences, and methodology of science. He originated such theories as the general theory of information, theory of named sets, mathematical theory of schemas, theory of super-recursive algorithms, hyperprobability theory, mathematical theory of oracles, theory of non-Diophantine arithmetics, system theory of time, and neoclassical analysis (in mathematics) making essential contributions to such fields as human creativity, theory of knowledge, theory of information, theory of algorithms and computation, theory of intellectual activity, probability theory, and complexity studies. He was the first to discover Non-Diophantine arithmetics; the first to axiomatize and build mathematical foundations for negative probability used in physics, finance and economics; and the first to explicitly overcome the barrier posed by the Church–Turing Thesis in computer science and technology by constructing new powerful classes of automata. Mark Burgin has authored and co-authored more than 500 papers and 21 books, including "*Semitopological Vector Spaces*: *Hypernorms, Hyperseminorms and Operators*" (2017),"*Functional Algebra and Hypercalculus in InfiniteDimensions*: *Hyperintegrals, Hyperfunctionals and Hyperderivatives*"(2017), "*Theory of Knowledge*: *Structures and Processes*"

(2016), "*Structural Reality*" (2012), "*Hypernumbers and Extrafunctions: Extending the Classical Calculus*" (2012), "*Theory of Named Sets*" (2011), "*Measuring Power of Algorithms, Computer Programs, and Information Automata*" (2010), "*Theory of Information: Fundamentality, Diversity and Unification*" (2010), "*Neoclassical Analysis: Calculus Closer to the Real World*" (2008), "*Super-recursive Algorithms*" (2005), "*On the Nature and Essence of Mathematics*" (1998), "*Intellectual Components of Creativity*"(1998), "*Introduction to the Modern Exact Methodology of Science*" (1994), and"*The World of Theories and Power of Mind*"(1992). Mark Burgin is also an Editor of 10 books.

Marek Czachor (1960) is a full professor of theoretical physics at Gdańsk University of Technology. He graduated from the University of Gdańsk, received PhD from the Institute of Physics of Polish Academy of Sciences in Warsaw, and obtained habilitation from the Department of Physics of Warsaw University. He was a recipient of fellowships of Fulbright (Massachusetts Institute of Technology), DAAD and Alexander von Humboldt (both at Arnold-Sommerfeld Institute for Mathematical Physics, Technical University of Clausthal), and NATO (University of Antwerp). For almost two decades he was a close collaborator of Center Leo Apostel for Interdisciplinary Studies at Vrije Universiteit Brussels. His research interests involve foundations of quantum mechanics and theory of relativity, including relativistic quantum information, nonlinear quantum mechanics, and interdisciplinary applications of formally quantum structures. In 2014, he independently discovered non-Newtonian calculus, generalizing the earlier formalisms of Grosmann, Katz, and Pap, and applying it to problems of fractal analysis, dark energy, and Bell's theorem.

Contents

Chapter 1

Introduction: Operation with Numbers as a Base of the Contemporary Culture

It takes a very unusual mind to undertake analysis of the obvious.
Alfred North Whitehead

People started using numbers and counting in prehistoric times. For thousands of years, mathematicians studied numbers and counting learning a lot in this area. Now we know two mathematical disciplines studying numbers, relations between them and operations with them: arithmetic and number theory, which is also called higher arithmetic. At the same time, numbers are used in all areas in general and in all mathematical disciplines in particular.

Numbers and arithmetic, which later developed into mathematics, definitely were among the humanity's greatest inventions or discoveries. They compose a vital part of our culture being inscribed into people's neural architecture and continuing to empower humankind beyond the commonplace thinking, even as the best individuals struggle to fathom its limits. People use numbers for a variety of tasks such as counting, measuring, calculation, computation, characterization, comparison, optimization, representation, description, decision-making and ordering.

Many authors wrote why and how numbers as a central component of mathematics were important for human civilization and culture (cf., for example, Barker, 1967; Boumans, 2005; Burgin, 2001; Dantzig, 1930; Diez, 1997; 1997a; Flegg, 2002; Higgins, 2008; Kneusel, 2015; Kulsariyeva and Zhumashova, 2015; MacNeal, 1995; McLeish, 1994; Miller, 1982;

Porter, 1995; Smeltzer, 1959). For instance, researchers suggest the following view on mathematics:

> "*Mathematics boils down to pattern recognition. We identify patterns in the world around us and use them to navigate its challenges. To do all this, however, we need numbers — or at least the information that our numbers represent.*" (Lamb, 2010)

To more thoroughly analyze this situation, in Section 1.1, we briefly discuss the roles of numbers and arithmetic in the context of the generic stratification of human culture. In Section 1.2, the concept of a number is examined based on opinions of different mathematicians and philosophers presenting diverse approaches and revealing its structure as the named set. In Section 1.3, the concept of arithmetic is discussed with the emphasis on the conventional arithmetic of natural numbers as one of the most (if not the most) basic structures in mathematics. It is natural to call the conventional arithmetic by the name *Diophantine arithmetic* because the ancient Greek mathematician Diophantus made important contributions to the development of arithmetic and transformation it into algebra. In Sections 1.4 and 1.5, the role of numbers in science, philosophy, humanities and technology is analyzed. In Section 1.6, following observations of different philosophers, scientists and mathematicians, we describe and scrutinize various situations in which quantitative descriptions of real situations cannot be correctly modeled using the conventional arithmetics of natural and whole numbers. This brings researchers to the conclusion that people need other arithmetics of natural and whole numbers. Such arithmetics are called non-Diophantine arithmetics and their discovery and informal properties are recounted in Section 1.7. In Sections 1.8, we describe the history of non-Diophantine operations and related constructions in mathematics and its applications. A synopsis of this book is given in Section 1.9.

1.1. Three Pillars of Human Culture

> *Numero pondere et mensura Deus omnia condidit*
> *God created everything by number, weight and measure.*
>
> Isaac Newton

To make their functioning more efficient, people developed various means of operation, many of which have become indispensable for the human civilization forming three pillars of contemporary civilization: language, arithmetic and technology.

Analyzing dynamics of the contemporary society, we can see that its functioning is based on the three pillars and three cornerstones of the contemporary culture:

- Operation/functioning of people manipulating with words is done through language, which is a mental instrument of humankind.
- Operation/functioning of people using numbers is performed employing arithmetic, which is a structural instrument of humankind.
- Operation/functioning of people employing technical devices is based on technology, which is a material instrument of humankind.

In this context, it is possible to treat the dynamic components of culture as pillars and its substantial, static components as cornerstones. Thus, words, numbers and technical devices form static components of culture, while operation with or utilization of words, numbers and technical devices bring into being its dynamic components.

Anthropological, cultural and historical research shows the following regularities:

- *Language* was created (emerged) for information transmission and communication.
- *Arithmetic* was created (emerged) for structuring (counting, comparing and measuring) the reality.
- *Technology* was created (emerged) for making people's functioning more efficient.

We all know that society cannot exist without language.

Numbers are omnipresent when it comes to exploration of nature and society, making decisions and economical activity.

Technology, at first, empowered only material activity of people but then it encompassed all spheres of life. Now electronic devices are everywhere. Computers solve problems and provide services unimaginable decades ago. Technological environment is becoming more important for people than their natural environment. For instance, many people throughout the world cannot live without their cell phones even for a while.

These three pillars and three cornerstones exactly correspond to the global structure of the world, which is described by the *Existential Triad* of the world discovered by Mark Burgin and based on the tradition

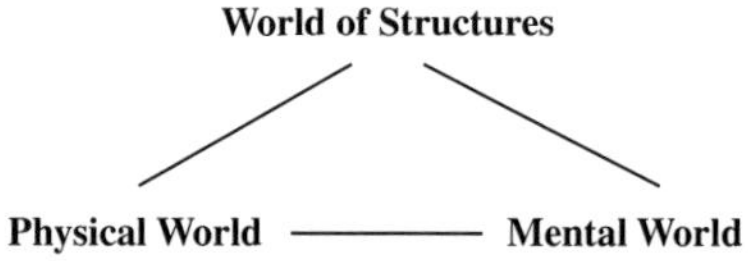

Figure 1.1. The Existential Triad of the world.

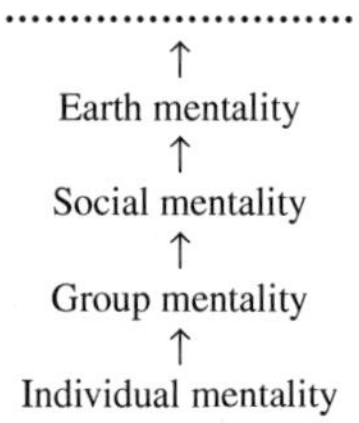

Figure 1.2. The lower levels of the hierarchy of the Mental World (the *basic mental hierarchy*).

coming from Plato and Aristotle (Burgin, 1997b; 2010; 2012a; 2017b). It is presented in Figure 1.1.

In this triad, the Physical (or material) World is interpreted as the physical reality studied by natural sciences, the Mental World encompasses different levels of mentality, and the World of Structures consists of various forms and types of structures.

It is necessary to remark that as physics does not study the physical reality as a whole but explores different parts and aspects of it, psychology also separates and investigates different parts and aspects of the mental reality, such as intelligence, emotions, or unconscious.

For a long time, there existed a very limited understanding of the Mental World reducing to it the individual mentality, which forms only the first level of the Mental World. Now science extended this picture and researchers study the Mental World on the first (lowest) three levels, all of which belong to the Existential Triad (cf. Figure 1.2):

- The *first level* treats mentality of separate individuals and is the subject of psychological studies. As in the case of physical reality, now psychology knows much more about mentality/psyche of people than even the best minds knew at the time of Plato.
- The *second level* deals with group mentality and is the subject of social psychology. In particular, this level includes social conscience, which incorporates collective memory (Durkheim, 1984), collective

wisdom (Suroweicki, 2004), collective intelligence (Brown and Lauder, 2000; Nguen, 2008a), computational collective intelligence (Szuba, 2001; Wolpert and Tumer, 2000) and is projected on the collective unconscious in the sense of Jung (cf. Jung, 1969) by the process of internalization (Atkinson *et al.*, 1990).
- The *third level* encompasses mentality of society as a whole and is the subject of social psychology.

In addition, the Mental World from the Existential Triad comprises higher (than the third) levels of mentality although they are not yet studied by science (Burgin, 1997; 2010). It is possible to relate higher levels of the Mental World to the spiritual worlds described in many religious and esoteric teachings.

The essential difference between the Mental and Physical Worlds is demonstrated by the *thesis of multiple realization*, which is proposed by Putnam and endorses the total irrelevance of physico-material properties for identifying mental states (Putnam, 1960). Such a phenomenon as functional isomorphism when different material objects turn out to be equivalent from the mental perspective employing the same program gives supportive evidence for this thesis.

Finally, there is the World of Structures, which was envisioned by Plato as the world of Ideas/Forms but validated only recently (Burgin, 2011; 2012a). By its essence, the World of Structures is a scientific interpretation and counterpart of the World of Ideas/Forms because Ideas or Forms might be logically interpreted as structures (Burgin, 2017b). Indeed, on the level of ideas and abstract entities, it is possible to link Ideas or Forms of Plato and Aristotle to structures in the same way as the atoms of Democritus may be related to the atoms of modern physics. Only recently, modern science came to a new understanding of Plato Ideas, representing the global world structure as the Existential Triad of the world, in which the World of Structures is much more understandable, exact and explored in comparison with the world of Ideas/Forms of Plato and Aristotle. When Plato and other adherents of the world of Ideas/Forms were asked what an idea or form was, they did not have a satisfactory answer. In contrast to this, many researchers have been analyzing and developing the concept of a structure (Ore, 1935; 1936; Bourbaki, 1948; 1957; 1960; Bucur and Deleanu, 1968; Corry, 1996; Landry, 1999; Burgin, 2012; 2016; 2017a). It is possible to find the most thorough analysis and the most advanced concept of a structure in (Burgin, 2010; 2012). As a result, in contrast to Plato, science has been able to elaborate a sufficiently exact definition of a structure and to prove existence of the World of Structures, demonstrating by means of observations and experiments, that this world constitutes the structural level of the world as the whole. Each system, phenomenon, or process either in nature, technology or society has some structure. These structures exist like material things, such as tables, chairs, or buildings do, and form the *structural level* of the world. When it is necessary to learn or to create a system or to start a process, it is done, as a rule, by means of knowledge of the corresponding structure. Structures determine the essence of things.

If we say that structures exist only embodied in material things, then we have to admit that material things exist only in a structured form, i.e., matter (physical entities) cannot exist independently of structures. For instance, atomic structure influences how the atoms are bonded together, which in turn helps one to categorize materials as metals, ceramics or polymers and permits us to draw some general conclusions concerning the mechanical properties and physical behavior of these classes of materials. Even chaos has its structure and not a single one.

It is also necessary to remark that the World of Structures includes Popper's World 3 from the Pure Popper Triad of the world (Popper, 1959; 1974; Burgin, 2012) because knowledge, as well as the intellectual contents of books, documents, scientific theories, etc., is a kind of structures that are represented in people's mentality (Burgin, 1997; 2004; 2012). This shows that the Existential Triad of the world is a further development of triads of Plato and Popper (cf. Burgin, 2012) describing the world as a whole.

The three worlds from the Existential Triad are not separate realities: they interact, interrelate and interconnect. Individual mentality is based on the brain, which is a material thing, while in the opinion of many physicists mentality influences the Physical World (cf., for example, Herbert, 1987). At the same time, our knowledge of the Physical World to a great extent depends on interaction between Mental and Physical Worlds (cf., for example, von Bayer, 2004).

In addition, there is a projection of the Mental World into the Physical World in the form of material creations of human mentality (creativity), such as books, movies, magazines, newspapers, cars, planes, computers and computer networks. This projection determines the Extended Mental World, which consists of the Mental World and its projection. The Extended Mental World correlates with the World 2 (i.e., mental or psychological world) from the General Popper Triad of the world (cf., Burgin, 2012).

Structural and material worlds are even more intertwined. Actually no material thing exists without a structure. Even chaos has its chaotic structure. Structures determine what things are. For instance, it is possible to make a table from different materials, e.g., from wood, plastics, iron, aluminum, etc. What all these tables have in common is not their material. Their common attribute is specific peculiarities of their structure. In a similar way, the great French mathematician Jules Henri Poincaré (1908) wrote, space is in reality amorphous, and it is only the things that are in it that give it a structure (form). As some physicists argue, physics studies not physical systems as they are but structures of these systems, or physical structures. This idea is the essence of the philosophical direction called structural realism, which maintains that people can learn (cognize) only structures of real things but not real things themselves (Worrall, 1989; Ladyman, 1998; Roberts, 2011). In some sciences, such as chemistry, and some areas of practical activity, such as engineering, structures play a leading role. For instance, the spatial structure of atoms, chemical elements, and molecules determines many properties of these chemical systems.

Contemporary physics treats the Physical World as a net of interacting components (systems), where there is no physical meaning to the state of an isolated object. A physical system (or, more precisely, its contingent state) is represented by the net of relations it entertains with the surrounding objects. As a result, the physical structure of the world is identified with such a global net of system relationships.

As North writes, physics is supposed to be telling us about the nature of the world, while physical theories are formulated in a mathematical language, using mathematical structures, which implies that mathematics is somehow telling us about the physical make-up of the world (North, 2009).

It is necessary to remark that it is possible to see Existential Triads in every person and in each computer. An individual has the physical component — her or his body, the mental component studied by psychologists, and the structural component, which comprises all structures and systems from the other two components underlying their elements and parts. A computer has the physical component — its hardware, the mental component, which consists of everything that is in computer memory, for example, software, and a structural component, which comprises all structures and systems from the other two components. It is also demonstrated that any natural language as a communication system has the structure of the Existential Triad (Burgin and Milov, 1998). The reason is that natural languages emerged (were created) as communication systems for reflecting and describing the world where people lived.

There are also other approaches to the global structure of the World. Many of them have the form of a triangle, which can be called a triangular triad in the context of fundamental triads. For instance, Werner Karl Heisenberg (1901–1976) builds his world stratification based on relations between the states of things and knowledge processes. He writes:

> "*We understand by "regions of reality... an ensemble of nomological* (i.e., representing laws of reality/nature, M.B.) *connections. These regions are generated by groups of relations. They overlap, adjust, cross, always respecting the principle of non-contradiction.... It is clear that the ordering of the regions has to substitute the gross division of world into a subjective reality and an objective one and to stretch itself between these poles of subject and object in such a manner that at its inferior limit are the regions where we can completely objectify. In continuation, one has to join regions where the states of things could not be completely separated from the knowledge process during which we are identifying them. Finally, on the top, have to be the levels of Reality where the states of things are created only in connection with the knowledge process.*" (Heisenberg, 1942/1998)

As knowledge belongs to the World of Structures (Burgin, 2016), Heisenberg's picture of reality reflects his idea of relations and interactions between this world and the Physical World in the context of the Existential Triad.

Naturally, we can see that the considered three pillars and three cornerstones of contemporary culture reflect the Existential Triad.

Technology in its conventional meaning as the diversity of technical devices and operation with them corresponds to the Physical World because technical devices are physical objects.

Language and operation with words correspond to the Mental World as languages are used for mental representation and information exchange.

Arithmetic and operation with numbers correspond to the World of Structures because numbers *per se* are structures.

From all things that people are doing, counting, computing and measuring belong to the most important. Without counting, computing and measuring, people cannot do a lot: cannot develop science and technology, cannot organize mass production, cannot buy and sell, and so on and so forth. Every woman and every man, every boy and every girl perform counting and computing many times a day. Calculators and computers were invented to help people to compute and count. In the old days, people even used to say that computers do not do mathematics but only arithmetic. Later computers began to fulfill much more sophisticated tasks performing much more complex operations than there are in arithmetic. Some of their abilities look miraculous. However, counting lies at the bottom of all computer operations.

As the well-known mathematician Philip Davis (1923–2018) describes, "political, social, and economic applications of mathematics are now multiplying like Fibonacci's rabbits" (Davis, 2000). The majority of these applications involve numbers and arithmetic operations. To mention but one example, we can take the article "Water Arithmetic 'doesn't Adds" (Kirby, 2000), which deals with environmental problems. By standards of the research community of mathematicians, mathematics that people use in everyday life is trivial, "but all those everyday trivialities... made, consciously or unconsciously, philo- or miso-maths of most people."

Now we are accustomed to using technology. Contemporary people cannot live without a multitude of diverse technical devices they use. However, from the very beginning when people became people and not apes, they used primitive technological tools for hunting and for toiling the soil in agriculture.

An interesting peculiarity of technology is that it reflects the configuration of the world exhibiting the structure of the Existential Triad. Namely, the world as a whole is stratified into three components: the Physical World, Mental World, and the World of Structures (or World of Structures). Correspondingly, there are three components (aspects) of technology: technology in the Physical World or material technology comprising physical objects and processes, such as computers, cell phones, cars, planes and radio; technology as a part of the Mental World represented by technological knowledge that people have in their minds, such as technology of car manufacturing or technology of cell phone utilization; and technology as a part of the structural world represented by mathematical and linguistic structures, procedural structures, such as algorithms, and structures of material technological systems.

In this respect, we discern three sides (aspects) of technology, which are represented in Figure 1.3 (Burgin, 1997a; 2003; 2009).

In turn, the technological practice also has three facets represented in Figure 1.4.

The Existential Triad of the world defines three forms of technology: material, mental and structural technologies.

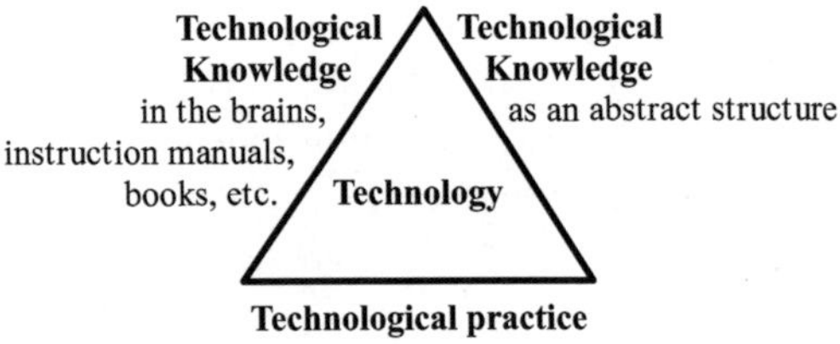

Figure 1.3. The components of technology.

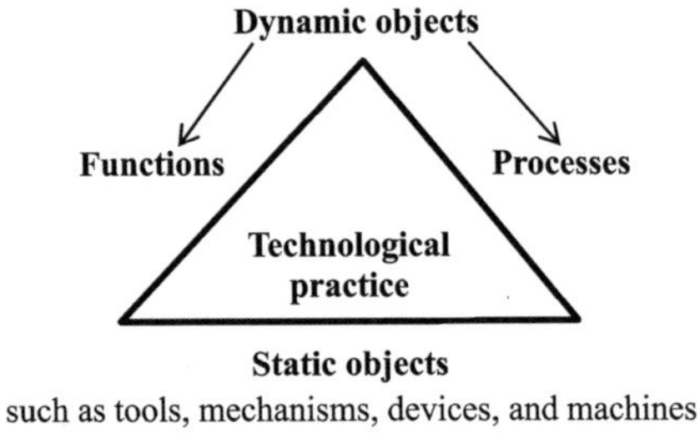

Figure 1.4. The components of the technological practice.

Material technology decreases hardship of physical work and allows performing tasks in the Physical World, which are impossible without technology, e.g., traveling from the Earth to the Moon or from Europe to America.

Mental technology decreases complexity of mental work and allows performing mental tasks, which are impossible without mental technology, e.g., preserving such amount of information that is stored in computers or understanding what is going on in the Universe or in the quantum world.

It is also possible to identify *structural technology* as structures of physical and mental technological systems and processes.

Numbers and other mathematical structures *per se* are structural technological devices. Images of numbers and other mathematical structures in the mind serve as mental technological devices. Psychologists and anthropologists studied mental representations of numbers demonstrating their dependence on culture (Dehaene, 1997; 1997a; 2001; 2002; Dehaene and Cohen, 1995; Dehaene *et al.*, 2008; Goldman, 2010).

Scientific theories also serve as mental technological devices. Indeed, physical technological devices such as telescopes and microscopes allow people to see what they cannot see using only their eyes. In a similar way, theories also allow people to see what they cannot using only their eyes. For instance, physical theories made possible discovery of subatomic particles such as electrons, protons, quarks and neutrino.

The structure of technology reflects the intrinsic feature of the Existential Triad, which is called fractality. We remind that a *fractal* is a complex system displaying self-similarity across different scales (cf., for example, Mandelbrot, 1983; Edgar, 2008). In other words, *fractality* means that the structure of the whole is repeated/reflected in the structure of its parts on many levels (cf., for example, Coleman and Pietronero, 1992; Calcagni, 2010). It is possible to find formalized mathematical definition of fractals in (Lapidus and van Frankenhuijsen, 2013; Lapidus, *et al.*, 2017).

We can trace the idea of fractality to the great German mathematician and philosopher Gottfried Wilhelm Leibniz (1646–1716), who discussed recursive self-similarity. Several examples of fractals were elaborated in the 19th century. At first, the outstanding German mathematician Karl Theodor Wilhelm Weierstrass (1815–1897) presented a definition of an everywhere continuous but nowhere differentiable function, the graph of which was a fractal. Then the great German mathematician Georg Ferdinand Ludwig Philipp Cantor (1845–1918) published examples of subsets of the real line known as Cantor sets, which were also fractals. Later in that century, the outstanding German mathematician Felix Klein (1849–1925) and the great French mathematician Jules Henri Poincaré (1854–1912) studied a class of mathematical structures called self-inverse fractals.

More progress in this direction was achieved in the 20th century. The Swedish mathematician Niels Fabian Helge von Koch (1870–1924), extending ideas of Poincaré, built fractals, which are now called the Koch curves, one of which is the Koch snowflake (von Koch, 1904). Later the outstanding Polish mathematician Wacław Franciszek Sierpiński (1882–1969) constructed several fractals, which were named after him, namely, the Sierpinski triangle, the Sierpinski carpet and the Sierpinski curve.

In 1975, the outstanding French-American mathematician Benoit B. Mandelbrot (1924–2010) solidified all previous work in this area elaborating the general concept of a fractal, introducing new interesting fractals such as the Mandelbrot set and illustrating his mathematical definition with a variety of computer-constructed visualizations (Mandelbrot, 1967; 1983).

Most fractals are geometrical. However, an important property of language, numbers and technology is their fractality. That is, being related to one existential component of the world, each of them is also present in other two components demonstrating fractality of the world.

Language, which is a predominantly mental construction, comes to the Physical World when it is used in speech, printing and writing. At the same time, language incorporates a variety of structures studied in linguistics, and through these structures, language is present in the World of Structures (Burgin and Milov, 1998).

Numbers, which are essentially structural (cf. Section 1.2), are represented in the Physical World by numerals and by written, printed or pronounced words. At the same time, being used by people, numbers have representations in their mentality — either in individual mentality (Núñez, 2009) or in social mentality (cf., Brockman, 1997).

1.2. The Concept of Number

Number is the basis of modern mathematics.

Courant, R. and Robbins, H. "What is Mathematics?"

In this section, we present a theoretical elucidation of the concept *number* explaining its meaning and significance. Pragmatic understanding of this concept is related to theoretical and practical activity of people and is considered in Sections 1.5 and 1.6. The most important philosophical problems related to numbers are: what numbers are, as well as how and where they exist.

Note that there is a diversity of opinions about numbers, especially, about natural numbers, which are the most basic in the variety of all numbers. Here we explain and justify our approach to this basic conception describing and analyzing some of the alternative points of view.

We clarify what numbers are at first explaining the essence of numbers and then describing understanding of numbers by different mathematicians and logicians.

Description 1.2.1. A *number* is an abstract mathematical object — a mathematical structure — represented by a symbol or a system of symbols.

Note that this is not a definition of a number but a general description of the inner structure of a number.

Here we understand symbols not only as individual aggregates of points, such as a or 3, written or printed on the paper or displayed on the screen. From this perspective, symbols are indigenous structures with a material component (Alp, 2010; Burgin, 2012; 2016a; Burgin and Schumann, 2006; Goodman, 1968).

It is also imperative that these representations of numbers (and through them, numbers) are utilized by people and machines for counting, ordering, calculation, measuring and simply for naming or labeling different entities. As we know, people use different representations of numbers. However, to apply these representations in a correct way, it is necessary to assume that, as rule, there are different names (representations) of the same object. For instance, educated people know that the expressions 11, 10+1, 5+6, eleven, five plus six, XI and 1011 and denote the same object. Mathematicians call this object *number* 11 or simply, 11. The difference between the names of this number is that the first three of them are in the decimal positional numerical system, the next two are in English, the next is in the Roman numerical system and the last one is in the binary positional numerical system. This is similar to the situation when we use names Róma in Italian, Rome in English or **Рим** in Russian for the same city in Italy.

Considering numbers from philosophical and methodological points of view, it is necessary to answer three basic questions:

- Do numbers really exist?
- If they exist, then what numbers are?
- If numbers exist, then where or to what reality do numbers belong?

The most fundamental are *natural numbers*, i.e., numbers 1, 2, 3, 4, 5 and so on. They are also called *counting numbers* because they emerged from and are used for counting.

The prominent French mathematician Emil Borel (1871–1956) maintained that "all of mathematics could be deduced from the sole notion of an integer" (Borel, 1903), while integers are constructed from natural numbers.

Similarly Courant and Robbins write that natural number forms the basis of modern mathematics. Then they continue:

> "*While the Greeks chose the geometrical concepts of point and line as the basis of their mathematics, it has become the modern guiding principle that all mathematical statements should be reducible ultimately to the statements about natural numbers.*" (Courant and Robbins, 1960)

This opinion is supported by the result of Hao Wang stating that for each ordinary axiom system S such as the original Zermelo set theory, there are arithmetic predicates expressible in the notation of ordinary number theory such that when they are substituted for the predicates (for example, the membership predicate) in the axioms of S, the resulting assertions are all provable in the system obtained from ordinary number theory by adding the arithmetic statement Cons(S), expressing the consistency of S, as a new axiom (Wang, 1956).

Being ultimate for mathematics, natural numbers caused a lot of controversy about what numbers are (cf., for example, Rosen, 2006; Tubbs, 2009; Dehaene, 2011), what numbers have to be (cf., for example, Dedekind, 1888) and what numbers could not be (cf., for example, Benacerraf, 1965).

As it is written in (National Research Council, 2001), "*Mathematicians like to take a bird's-eye view of the process of developing an understanding of number. Rather than take numbers a pair at a time and worry in detail about the mechanics of adding them or multiplying them, they like to think about whole classes of numbers at once and about the properties of addition (or of multiplication) as a way of combining pairs of numbers in the class.*"

To answer the questions what numbers are and how they exist, people in general and mathematicians, in particular, implicitly or explicitly build models of numbers. These models reflect understanding of numbers by mathematicians in the same way as physical models reflect understanding of nature by physicists.

To build an adequate model of numbers is not a simple problem because as Kauffman writes "the existence of a number is unlike any existence that we call physical or mental" (Kauffman, 1999). Let us consider some philosophical and scientific approaches to this problem.

There are three basic types of models representing natural numbers: operational, analytical and descriptive.

(A) The original *operational model* of natural numbers comes from counting. It is possible to find this approach in the dialogue *Parmenides* by Plato where the existence of numbers is proven from our capacity to count (Blyth, 2000). Note that counting does not only places the counted item in a sequence of other items but also measures sets by assigning to each set the number of elements in it.

According to the contemporary form of the operational model, there are no separate natural numbers but each natural number exists only as a position (place) in the ordered sequence of all natural numbers. For instance, according to Shapiro, "the essence of a natural number is the relations it has with other natural numbers." (Shapiro, 1997)

The general theory of structures explains that the system $\boldsymbol{N}$ of all natural numbers forms an outer structure for each of them (Burgin, 2012a). Namely, the system $\boldsymbol{N}$ has an order structure, in which there are order relations between numbers indicating when one number is larger or smaller than another number. In addition, this system $\boldsymbol{N}$ has its algebraic structure determined by addition and multiplication of numbers.

Another *operational model* of natural numbers comes from the area of computations and algorithms (cf., for example, Turing, 1936). In this approach, it is postulated that 1 is a natural number and then assumed that all natural numbers are obtained by adding 1 to already existing natural numbers. In such a way, we get 2 as $1 + 1$, 3 as $2 + 1$ and so on. However, this representation of natural numbers emerged long before computers were constructed.

In the context of operational model, it is also possible to interpret numbers as electric signals as it is done in computers and calculators. One more way to interpret numbers is presenting them as finger movements (Badets and Pesenti, 2010).

(B) The first *analytical model* of natural numbers was suggested by Gottlob Frege (1848–1925) and later amended by Bertrand Russell (1872–1970). By the definition of number due to Bertrand Russell, "A number is the class of all classes similar to a given class."

Thus, according to this model, any natural number is the collection (class) of *equivalent* (also called *equipotent* or *equipollent* or *equinumerous*) finite sets or classes, i.e., sets between any pair of them there is a one-to-one correspondence (Frege, 1884; Russell, 1903). Such sets are called *similar* by Russell while in the contemporary mathematical terminology, they are called equipotent or equipollent. However, as natural numbers have names, it is more reasonable to consider a natural number n as the named set (X, f, H), in which the support X consists of (all) sets that have n elements, set of names H consists of (all) names of the number n and the naming correspondence f connects each set from X with the names from H (cf. Appendix). As we already mentioned, any (natural) number has many names. For instance, 5, V, five, 5.0, $2 + 3$, 101, $5/1$ and $10/2$ are different names of the same number, which belong to the corresponding set of names H. This correlates with the mathematical custom because according to Tait, the traditional view assumes that a numeral is a common name, under which fall equivalent (equipotent or equipollent) sets, i.e., numerals are treated as names of numbers (cf., Tait, 1996).

In their studies, Lakoff and Núñez come to the mentally based analytical model from a psychological perspective. According to them, the physical truth that people experience in grouping objects can be conceptualized as an arithmetical truth about numbers (Núñez, 2009; Lakoff and Núñez, 2000; Núñez and Lakoff, 2005). For instance, if a number A is greater than a number B, then A plus a number C is greater than B plus C. In this setting,

Lakoff and Núñez use metaphors to describe mathematical structures that are called natural numbers. In particular, they suggest that *numbers are collections of objects*, while *arithmetic is object collection* (Núñez, 2009).

As a result, the analytical model of numbers deals with numbers as patterns for sets. This correlates with the approach to mathematics as a science of patterns (Resnik, 1999).

Another approach to set-theoretical description of natural numbers represents each number by a specific set. For instance, it is possible to treat number 3 as the set with three elements $\{0, 1, 2\}$ or as it is done in set theory, as the set $\{\{\}, \{\{\}\}, \{\{\}, \{\{\}\}\}\}$ where $\{\}$ denotes the empty set. This is a finite part of the von Neumann's approach to set theory where natural numbers are treated as finite cardinal numbers (Fraenkel *et al.*, 1973; von Neumann, 1961).

In essence, the analytical models utilize mathematical structures treating numbers as abstract objects. A common view assumes that this is also true for natural languages where expressions, such as the number of planets, eight, as well as the number eight, are acting as referential terms referring number users, i.e., people or machines, to numbers although often these are named numbers (Moltmann, 2013).

(C) There is one more type of number models. It comprises *descriptive models*, according to which natural numbers are described by axioms. This approach was developed by notable mathematicians such as Hermann Grassmann, Richard Dedekind, Charles Peirce and Giuseppe Peano.

Axioms have the form of propositions and predicates. However, according to the general theory of properties, propositions and predicates are special cases of abstract properties (Burgin, 1985; 1986; 1989a). This makes it possible to represent natural and cardinal numbers by properties, which describe cardinality of sets (Burgin, 1989). In this case, all natural numbers are represented by the named set (C, p, B), in which C is the class, which consists of all finite sets, B consists of abstract mathematical objects called numbers and p is a function, which assigns a number from B to sets from C. Naturally, it is assumed that the same number is assigned to equivalent (equipotent) sets.

Similar view of numbers is suggested by Moltmann, who assumes that "numbers are not primarily treated abstract objects, but rather "aspects" of pluralities of ordinary objects, ... a view that in fact appears to have been the Aristotelian view of numbers" (Moltmann, 2013). In addition, Moltmann argues that natural languages support this view on the ontological status of numbers, on which natural numbers do not act as entities, but rather have the status of plural properties, while the meaning of numerals when acting like adjectives. In addition, this approach correlates with the contemporary philosophy of mathematics where it is explained that "number terms in arithmetical sentences are not terms referring to numbers, but rather make contributions to generalizations about ordinary (and possible) objects. It is only with complex expressions somewhat at the periphery of language such as the number eight that reference to pure numbers is permitted" (Moltmann, 2013).

It is possible to ask the question whether it is possible that the same number is assigned to sets that are not equivalent (equipotent). Even when we are dealing only with finite sets, the answer is positive because when we assign numbers from a modular arithmetic (cf. Section 2.1) to finite sets, one number corresponds to many sets that are not equivalent. For instance, in the modular arithmetic $\boldsymbol{Z}_{10}$, number 1 is assigned to all sets with one element, all sets with eleven elements, all sets with 21 elements and so on. However, for natural numbers, the situation is different because when we assign numbers from the conventional Diophantine arithmetic or even from non-Diophantine arithmetics (cf. Section 1.7), each natural number is assigned only to equivalent (equipotent) sets.

Similar situation exists in transfinite arithmetics where transfinite numbers are assigned to arbitrary sets. If we assign numbers from the Cantor's arithmetic of cardinal numbers, then one number is assigned only to equivalent (equipotent) sets in the standard set theory (Cantor, 1883; 1932; Fraenkel *et al.*, 1973; Kuratowski and Mostowski, 1967). However, if we assign numbers from a cardinal set classification, which is not Cantorian, then one number can be assigned to sets, which are not equivalent (equipotent) in the standard set theory (Buzaglo, 1992; Burgin and Buzaglo, 1994).

The three types of number models considered above are inherently related to the three worlds of mathematics introduced by Tall (2004; 2004a). Namely, he discerns:

- *Conceptual-embodied world* grows out of people's perceptions of the world and consists of people's thinking about things that we perceive and sense, including not only mental perceptions of real-world objects, but also internal conceptions that involve visuospatial imagery.
- *Proceptual-symbolic world* is the world of symbols used for calculation and manipulation in arithmetic, algebra, calculus and so on, enabling

people to switch fluently from processes to do and concepts to think about.

- *Formal-axiomatic world* is based on *properties*, expressed in terms of formal definitions and axioms specifying mathematical structures (such as "group", "field", "vector space", "topological space" and so on).

This shows that the operational models of natural numbers belong to the proceptual-symbolic world, analytical models of natural numbers belong to the conceptual-embodied world and descriptive models of natural numbers belong to the formal-axiomatic world.

Naturally, there is a direct correlation between the three worlds of Tall and the Existential Triad. Indeed, it is easy to see that thinking and perceptions are components of the individual Mental World. Thus, the conceptual-embodied world naturally corresponds to the Mental World. Mathematical structures belong to the World of Structures, i.e., the formal-axiomatic world is logically associated with the World of Structures. The proceptual-symbolic world consists of physical symbols used for calculation and manipulation, while the notion of *procept* builds initially on actions in the embodied world and the initial stages of counting and early arithmetic are basically embodied. It means that signs in the proceptual-symbolic world are material. Thus, the proceptual-symbolic world naturally corresponds to the Physical World. This concludes our demonstration that the structure of mathematics suggested by Tall is globally organized as the Existential Triad.

This naturally implies that the three worlds of Tall correspond to the existential structure of mathematics (Burgin, 2018), which reflects the Existential Structure of the World describing how mathematics exists. It is presented in Figure 1.5.

The proceptual-symbolic world of mathematics exists in the material reality of mathematics because it consists of symbols used for calculation and manipulation. For instance, many comprehend mathematics as the

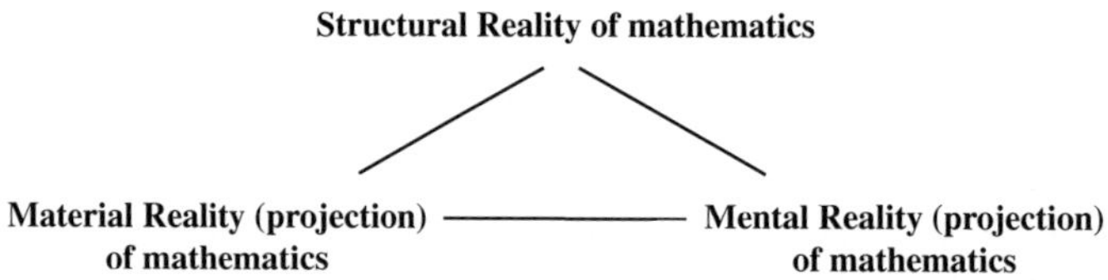

Figure 1.5. The existential structure of mathematics.

contents of mathematical manuscripts, books, papers and lectures, with the increasingly growing net of theorems, definitions, proofs, constructions, conjectures as the results of the activity of mathematicians (Manin, 2007). In the Physical World, numbers are also represented by the states of physical systems such as computers, calculators and cell phones.

The formal-axiomatic world of mathematics exists in the structural reality of mathematics and is formed by comprehension of mathematical objects as abstract structures. Many mathematicians and philosophers understand mathematics as the science of structures (Bourbaki, 1948; 1957; Burgin, 1998; Resnik, 1997).

The conceptual-embodied world of mathematics exists in the mental reality of mathematics. Psychologists and neurophysiologists study the mental reality of mathematics in general and numbers in particular (cf., for example, Piaget, 1964; Baird, 1975; 1975a; Baird and Noma, 1975; Noma and Baird, 1975; Weissman *et al.*, 1975; Dehaene, 1997; 2003; Dehaene and Cohen, 1994). Psychologists conjecture that people have innate comprehension of numbers identifying them in much the same way the brain identifies colors. It is called *number sense*, and the brain comes fully equipped with it from birth. For instance, infants can identify changes in quantity without knowledge about number systems.

Even more, researchers explore representation of and operation with numbers in animals (cf., for example, Dehaene *et al.*, 1998; Biro and Matsuzawa, 2001; Nieder *et al.*, 2002; Nieder and Miller, 2004; Roitman *et al.*, 2007; Sawamura *et al.*, 2010). They exhibit that number sense plays a vital role in the manner animals navigate their environment where objects are numerous and frequently mobile.

A complementary model of mathematics as activity of people is suggested by Brian Rotman. It consists of three components: Person, Agent and Subject (Rotman, 2003). The Person has extra-mathematical physical and cultural presence, is immersed in natural language and the subjectivity it makes available, has insight and hunches, provides motivation for and is the source of intuitions behind concepts and proofs. The Subject corresponds to the domain of the symbols and is the source of the intersubjectivity embedded in mathematical languages, but does not have the Person's capacity of self-reference. The Agent is associated with the domain of procedures functioning as the delegate for the Person through the mediation of the Subject and executing a mathematically idealized version of the actions imagined by the Person. In this framework, it is possible to have the following interpretation: the Person belongs to the Physical World;

the Subject is structural as it belongs to the domain of the symbols; and the Agent is from the Mental World.

All these stratifications of reality have indispensable impact on how mathematicians understand the nature and essence of numbers. It is possible to distinguish four approaches to this problem: the Platonic approach, cultural approach, psychological approach, and pragmatic approach.

Rather popular opinion in the mathematical community is the so-called, the *Platonic approach* when it is assumed that numbers and arithmetic are to be regarded as a part of the realm of objective, self-subsistent, i.e., independently of human thinking, mathematical objects that are timeless, non-spatial, and non-mental (Barker, 1967). Platonism conceives it to be the task of the mathematician to explore this realm of being, which, as we have seen, manifests itself as the World of Structures.

For instance, Frege wrote:

> "*But, it will perhaps be objected, even if the Earth is really not imaginable, it is at any rate an external thing, occupying a definite place; but where is the number 4? It is neither outside us nor within us. And, taking those words in their spatial sense, which is quite correct. To give spatial co-ordinates for the number 4 makes no sense; but the only conclusion to be drawn from that is, that 4 is not a spatial object, not that it is not an object at all. Not every object has a place.*
>
> *For number is no more an object of psychology or a product of mental processes than, let us say, the North Sea is....* (It)*is something objective.*" (Frege, 1984)

In a similar way, describing mathematical Platonism, Øystein Linnebo writes:

> "*Just as electrons and planets exist independently of us, so do numbers and sets.*" (Linnebo, 2018)

It is necessary to remark that in contrast to what Frege writes, in the contemporary ontology, numbers have their place. Only as it was proved later, it is not a physical place somewhere in the material universe but a definite place in the World of Structures (Burgin, 2017c).

An Austrian-born mathematical logician, who studied and worked in Great Britain and America, Georg Kreisel (1923–2015) explains that objectivity of mathematics, which displays itself in agreement on results, supports ontological Platonism. He is treating mathematical systems, such as numbers, as external objects with which mathematicians are in some kind of contact (Kreisel, 1967).

Some researchers think that it is impossible to "get the existence of number without the Platonic realm" (Kauffman, 1999).

Interestingly, the reality of numbers is also supported by some psychologists. For instance, Crollen *et al.*, write:

> *Number, space and time are fundamental properties of the environment constantly used by humans and animals to adapt and regulate their behavior to the external world.* (Crollen *et al.*, 2013)

If we assume, as the mainstream of scientists is doing, that space and time are forms of reality, then this statement implies that *number* also belongs to reality. Only space and time belong to the physical reality while *number* belongs to the structural reality (Burgin, 2012a).

At the same time, mathematical Platonism has not been accepted by many philosophers being among the most hotly debated issues in the philosophy of mathematics over the past several decades.

It is interesting that some philosophers suggest that structuralism is an alternative view to Platonism in the philosophy of mathematics (cf., for example, Bondecka-Krzykowska, 2004). However, contemporary understanding of the Platonic Realm as the Structural Reality posits that ontological structuralism, in essence, is a contemporary form of Platonism (Burgin, 2017b).

Another answer to the question of how numbers exist is given in the *cultural approach*. For instance, for mathematician Reuben Hersh (1927–2020), mathematics has existence or reality only as part of human culture as a social-cultural-historic phenomenon (Brockman, 1997). In essence, numbers and arithmetic evolved over millions of years of evolution (Dehaene, 1997; Dehaene *et al.*, 2003).

In contrast to this, the *psychological approach* places numbers and other mathematical structures in the individual mentality implying that they have been brought forth by specific combinations of everyday cognitive mechanisms that make human imagination and abstraction possible (Núñez, 2009). There is even a more extremist position, according to which the roots of arithmetic reside in single neurons while specific brain circuits perform number processing (Dehaene 1997; 2002; Dehaene *et al.*, 2003).

One more point of view is presented by the *pragmatic approach* implying that everything in mathematics comes from practical activity of people. For instance, the operational model of natural numbers brings us to the situation when objects with different structures are identified, that is, regarded as one and the same object. For instance, the natural number 3 is considered the same as the rational number 3. However, the outer structure of the natural number 3 is the inner structure of the object $\boldsymbol{N}$ of all natural numbers, while the outer structure of the rational number 3 is the inner structure of the object $\boldsymbol{Q}$ of all rational numbers. The inner structure of the natural number 3 is the named set $(X, f, 3)$ where X consists of (all) sets that have three elements and f connects all these sets to the symbol 3, while the inner structure of the rational number 3 is the named set $(Z, r, 3)$ where Z consists of (all) fractions $^3/_1$, $^6/_2$, $^9/_3, \ldots$ and f connects all these fractions to the symbol 3. This shows that mathematical objects are not conventional structures, i.e., structures in that oversimplified sense that is traditionally assigned to the concept of structure.

Thus, we can see that the Platonic approach ascribes numbers to the structural reality, the pragmatic approach puts numbers into the physical reality while cultural and psychological approaches place numbers into the mental reality. Only the psychological approach positions numbers within individual mentality while the cultural approach locates numbers in social mentality.

It is natural to ask a question why different structures are treated in mathematics as the same object. We can answer this question explaining that mathematics is a tool (mechanism) for simplification through a rigorous generalization and unification. That is why mathematics is often comprehended as a science of patterns (Resnik, 1999). Thus, when the natural number 3, the whole number 3, the integer number 3, the rational number 3 and the real number 3 are comprehended as one and the same object *number* 3, it provides a useful unification, simplifying many cases of mathematical reasoning.

From the point of view of structural reality (Burgin, 2012a), the rational number 3, the whole number 3, the integer number 3, the rational number 3 and the real number 3 are structurally different because they have distinct outer structures. For instance, the outer structure of the integer number 3 is the system of all integer numbers, while the outer structure of the real number 3 is the system of all real numbers. At the same time, the natural number 3, the whole number 3, the integer number 3, the rational number 3 and the real number 3 are all instantiations of number 3, which is a structure of a higher level of abstraction than its instantiations.

In addition, this example provides a better understanding of mathematical objects, which usually are not primitive structures but structures of structures, i.e., such structures, in which elements are also structures and which are mathematically represented by nested named sets (Burgin,

2011). While according to the general theory of structures developed in (Burgin, 2012a), conventional structures are structures of the first order, structures of structures are structures of higher orders, i.e., of the second order, third order and so on.

In this setting, number 3 is the structure that includes its analytical representation as its inner structure; its descriptive representation as its intermediate structure; and its outer structures in sets of natural numbers, whole numbers, integer numbers, rational numbers, real numbers, etc. (Burgin, 2012a)

That is why numbers *per se* belong to the World of Structures and are present in other two worlds — physical and mental — through their names or images. As we know, all numbers used by people have names and thus, they are named numbers. People cannot operate with numbers directly — they can only do this, using number names. Some numbers have individual (proper) names, for example, numbers $10, 7, 3, \pi, e$, and many others. At the same time, a great multiplicity of numbers has only common names, such as a *natural number*, a *real number* or a *complex number*.

Rejection of the World of Structures brings even qualified philosophers of mathematics to contradictory conclusions. For instance, Benacerraf after a thorough analysis of the concept *number*, comes to the conclusion that "there are not two kind of things, numbers and number words (number names, MB), but just one words themselves" (Benacerraf, 1965). This means that either there are no numbers or that numbers are "number words". At the same time, Benacerraf writes that different words denote the same number, which contradicts his previous statement demonstrating that numbers do exist while numbers and their names (number words) are different entities.

The approach of Benacerraf is rooted in the nominalistic interpretation of all mathematical objects where numbers are treated only as void names and nothing more (cf., for example, Burgess and Rosen, 1997). Indeed, Benacerraf writes:

> "*Therefore, numbers are not objects at all, because in giving the properties (that is, necessary and sufficient) of numbers you merely characterize an abstract structure — and the distinction lies in the fact that the "elements" of the structure have no properties other than those relating them to other "elements" of the same structure*" (Benacerraf, 1965).

Contrary to this, existence and reality of the World of Structures implies that abstract structures are indeed objects similar to physical objects but abstract structures exist in a different world (Burgin, 2012a).

After we discussed how numbers exist and what models of numbers have been elaborated, it is important to delineate the main roles of numbers. Here we, as before, for the most part, consider natural numbers.

They play three basic roles:

- putting in order some things;
- measuring the quantity of some things;
- naming some things.

It looks reasonable to call natural numbers by the name *counting numbers* when they play the first role, *discretely measuring numbers* when they play the second role, and *nominal numbers* when they play the third role.

This is similar to the situation when we consider the social status of some person, say Antony, we call him a student indicating that he studies at a university. At the same time, we call him a son when we reflect his place in his family.

Thus, when people count, they put things in an ordered sequence, while the last number in the counting is a measure of the counted set. Indeed, as Tal writes, although accounts of measurement varied, the consensus was that measurement is a method of *assigning numbers to magnitudes* (Tal, 2017). For instance, Helmholtz described measurement as the procedure by which one finds the denominate number that expresses the value of a magnitude, where a "denominate number" is a number together with a unit, e.g., 7 kg, and a magnitude is a quality of objects that is amenable to ordering from smaller to greater, e.g., weight (Helmholtz, 1887). In a similar way, Russell explained that measurement is any method by which a unique and reciprocal correspondence is established between all or some of the magnitudes of a kind and all or some of the numbers, integral, rational or real (Russell, 1903). Consequently, assigning natural numbers to sets of objects is a kind of measurement using a discrete measure.

Gödel numbering of arithmetical formulas used in the proof of his famous incompleteness theorems is an example of the nominal role of natural numbers (Gödel, 1931/1932). Natural numbers in the nominal role are also used for coding Turing machines and inductive Turing machines (cf., for example, (Markov and Nagornii, 1984; Sipser, 1997; Burgin, 2005; Hopcroft *et al.*, 2007)).

Interestingly that the basic roles of natural numbers correspond to the basic types of measurement scales used in probability theory, statistics, psychology, sociology and some other fields (cf., for example, Stevens, 1946; 1951; Rozeboom, 1966; Narens, 1981; Burgin, 2015a; 2016):

- nominal scales;
- ordinal scales;
- numerical scales, which include interval and ratio scales.

Some authors regard natural numbers as a separate type of measurement scales (cf., for example, Chrisman, 1998; Mosteller, 1977).

It is necessary to remark that all models of natural numbers considered here are based on set theory. At the same time, there are also category-theoretical models of natural numbers (Parsons, 1984; Jay, 1989).

Utilization of numbers presupposes that numbers must be differentiated from their symbolic representations, which are mathematical names of numbers and are called *numerals*.

There is a conjecture that naming of numbers was developed at least 50,000 years ago for the purpose of counting (Eves, 1990). People used collections of physical objects such as rocks, sticks, bones, clay, stones, wood carvings, and knotted ropes as names of numbers in the process of counting and other operations with numbers. With the development of mathematics, it was discovered that symbolic representations are more efficient and numeral, also called numerical, systems were created. By means of these systems, numbers are methodically named by numerals, which are words used in a language for number representations. In a more exact terms, a symbol or a system of symbols, which represents a number and plays the role of a name of the number, is called a *numeral*. In mathematics, numerals are constructed from *digits*, each of which is a single symbol, e.g., 1, 2, or 3.

Numerals, such as 5, 123, 5.37 or 1,000,000, are mathematical names, while numbers also have names in different natural languages, i.e., English names, Spanish names, French names, etc. For instance, number 10 has the name *ten* in English and the name *diez* in Spanish. It is also possible to name numbers by a predicate (property) in the form of an expression or an algorithm (system of construction rules). For instance, the predicate "the least number, the numeric name of which in the decimal positional numerical system has two digits," is a name of number 10. The expression $3 + 7$ also is a name of number 10.

However, to make sentences not too long and complicated, people identify numbers by their names. For instance, people say or write "number 8" instead of saying or writing "the number with the name 8." In a similar way, people say or write "number 0.5" instead of saying or writing "the number with the name 0.5."

In essence, each number has many names. For instance, number 10 has the following names:

- *ten* in English;
- *diez* in Spanish;
- *dix* in French;
- *zehn* in German;
- दस (Dus) in Hindi;
- עשר in Hebrew;
- *decem* in Latin;
- 10 in the decimal numeral system;
- 1010 in the binary numeral system;
- 101 in the ternary numeral system;
- י in the Hebrew numeral system;
- ∩ in the Egyptian numeral system;
- ι in the Greek numeral system;
- X in the Roman numeral system;
- ═ in the Maya numeral system;
- 丅 in the Common Chinese numeral system;
- 𒌋in the Babylonian numeral system;
- $^{10}/_1$, $^{20}/_2$ and 10.0 are also mathematical names of number 10.

Thus, it is necessary to make a clear distinction between *number systems* and *numeral* or *numeric* or *numerical systems*. Systems of natural numbers, of rational numbers, of integer numbers, of real numbers or of complex numbers are examples of number systems. Systems of decimal numerals, of binary numerals, of duodecimal numerals, of Roman numerals or of sexagesimal numerals are examples of numeral systems.

A *numerical system*, also called a *system of numeration*, *numeric system* or *numeral system*, is a system for mathematical naming of numbers in a chosen number domain, e.g., natural numbers, rational numbers or real numbers. A numerical system uses graphemes or written symbols in a consistent manner and provides rules for number representation, operation and interpretation. Many numerical systems have been created, at first, for natural numbers and then for other kinds of numbers. There are the Hindu-Arabic family of numerical systems, Babylonian, Brahmi, Chinese,

Egyptian, Etruscan, Greek (Ionian), Hebrew, Japanese, Mayan, Roman, Sumerian, and many other numerical systems.

Some researchers (cf., for example, Žarnić, 1999) do not treat numerals as names of numbers because they do not accept numbers as (Platonic) objects assuming that names have to designate some objects but in this case, there are no objects. However, although in terms of named set theory, there are empty or void names, which do not have designated objects (Burgin, 2011), numbers are valid objects because as we discussed above, they exist in the World of Structures.

The most commonly used throughout the world system of numerals is the Hindu-Arabic numerical system, because Indian mathematicians, mostly Aryabhatta of Kusumapura (476–550) and Brahmagupta (598–668), developed it. It is a decimal positional numerical system. Later this system reached Europe in the works of Arabic mathematicians. That is why numbers in this system were called Arabic numbers.

The first version of the decimal positional numerical system was published around 458 C.E. in the Jain astronomical work *Lokavibhāga*, or *Parts of the Universe*. It is interesting that India has produced many numerical systems: Devanagari numerals, Gujarati numerals, Kannada numerals, Tamil numerals, and so on. The first known to us was the Brahmi numerical system, which existed from the third century B.C.E.

In developed numerical systems, numerals are constructed from elementary symbols, which are called *digits*.

All numerical systems are usually divided into positional numerical systems and other types of numerical systems. For instance, the most popular Hindu-Arabic numerical system is a decimal positional system i.e., it has the base 10, while the Roman numerical system is decimal but not directly positional. Positional numerical systems are more efficient in representation of and operation with numbers.

An important characteristic of a positional numerical system is its *base*, i.e., the number of digits used for number representation. For instance, the decimal positional numerical system, which is prevalently used by people now, has the base 10 as there are exactly ten digits 0, 1, 2, 3, 4, 5, 6, 7, 8, and 9. That is why this system is called decimal because in Latin *decem* means ten.

A *positional numerical system*, also called a *place-value notation*, uses digits in such a way that their meaning depends on their position in the number representation. For instance, in the number 111, the first from the left digit 1 means one, the second from the left digit 1 means ten, and the third from the left digit 1 means one hundred. In contrast to this, the Roman numerical system is not positional. That is why in the number III, each digit I means one,

Even in the same numerical system, a number can have different representations (names). For instance, we have

$$7530 = 7 \times 1000 + 5 \times 100 + 3 \times 10 + 0 \times 1. \tag{1.1}$$

In such a way, the meaning of a finite sequence of digits representing a natural number is obtained. The left side of the equality (1.1) is called the *standard representation* and the right side of the equality (1.1) is called the *expanded representation* of a natural number.

In the decimal positional numerical system, each position has a value ten times that of the position to its immediate right neighbor. For instance, in the number 75 the numeral 5 represents five units (ones), and the numeral 7 represents seven tens.

The simplest numerical system called the *unary numerical system* uses only one symbol (digit). In it, every natural number is represented by the corresponding number of symbols. For instance, if the symbol | is chosen for number representation, then the number three would be represented by the numeral |||. Tally marks and fingers, which play the role of digits in a unary numerical system, are still in common use for naming counting numbers (Andres *et al.*, 2012; Burton, 1997). The unary system can be efficient only for small numbers, although it plays an important role in theoretical computer science, e.g., for coding Turing machines (cf., for example, Hilbert and Bernays, 1968; Markov and Nagornii, 1984; Hopcroft *et al.*, 2007).

Binary numerical systems have the base 2, i.e., they use two digits, which usually are 0 and 1 and can be represented by off and on pulses of electric current. They are sufficiently simple for operation with numbers and belong to the most widespread numerical systems because they are used in computers and calculators, which internally work exclusively with binary numerals and display decimal numerals only for people. Actually all data in all computers, cell phones and calculators are represented by binary sequences (words), some of which are interpreted as numbers. In binary positional numerical system, each position has a value twice as great as the position to its immediate right, so that, for example, binary numerals 101 is equal to (represents the same number as) $5 = 1 \cdot 1 + 0 \cdot 2 + 1 \cdot 4$ in the decimal numerical system, and binary numeral 1111 is equal to (represents the same number as) 15 in the decimal numerical system. It is proved that the binary numerical system is more efficient for arithmetical operations, storing and transmitting numbers than the decimal numerical system.

Ternary numerical systems have number 3 as their base and three digits, which usually are 0, 1 and 2. It is proved that the ternary numerical system is the most efficient for storing, transmitting and operating numbers. Ternary numerals are used in ternary computers and in ternary, which are also called three-valued or trivalent, logics, such as the intuitionistic logic. However, logical elements that work with ternary numerals are less reliable than standard logical elements that work with binary numerals. Although mass-produced binary components for computers have reduced ternary computers to a small footnote in the history of computing, the well-known computer scientist Donald Knuth predicts their future prevalence (Knuth, 1997).

Quaternary numerical systems have the base 4 and are used by some Austronesian and Melanesian ethnic groups. Some anthropologists believe that this choice of the numerical system is related to the common village dog, which has four legs.

Quinary numerical systems use the number 5 as the base. There is a hypothesis that quinary numerical system developed from counting by fingers as there are five fingers on each hand. This system appeared very long ago and was utilized in many cultures worldwide being present in the Celtic and Banish numerical systems, as well as in the Inuit languages.

Octal numerical systems are based on the number 8. An octal numerical system is used, for example, in the Pamean languages of Mexico because people there count by using the four spaces between their fingers rather than the fingers themselves.

The vast majority of traditional numerical systems are *decimal* using 10 digits. It is natural to suggest that this prevalence is due to humans having 10 fingers on both hands and using all fingers for counting. At the same time, there are many regional variations of this system. For instance, in the Western system, decimal digits are grouped in triples. In the Indian system, decimal digits are frequently grouped in pairs. Finally, people in East Asia group decimal digits in quadruples. There is historical evidence that the first known decimal numerical system was utilized in ancient Egypt.

Duodecimal numerical systems are based on the number 12 and are employed in Nepal, Indonesia, Melanesia, Polynesia, and Minicoy Island in India. It is also present in some English and American measures, e.g., 12 inches to the foot, 12 Troy ounces to the Troy pound, and in time divisions, e.g., 12 months in a year, the 12-hour clock.

Vigesimal numerical system or *base-20 numerical system* uses the number 20 as the base. The Indians Maya used this system for counting and other arithmetical operations. Anthropologists think this system originated from counting, using all human fingers and toes. Vigesimal numerical systems existed in ancient Mesoamerican cultures and are still in use today in the modern original languages of their descendants, such as the Nahuatl and Mayan languages. Besides, some elements of the vigesimal terminology are also found in some contemporary languages, such as Celtic languages, Basque, French, Danish, and Georgian.

Sexagesimal numerical system uses the number 60 as the base. Very early, in comparison with other systems, it became positional. Sexagesimal numerals were used in Babylonia for commerce, as well as for astronomical and other calculations. Exported from Babylonia sexagesimal numeration became the base for units of measure and counting in many Mediterranean nations, including the Greeks, Romans and Egyptians. It is still used now in measuring time, e.g., 60 minutes per hour and 60 seconds per minute, as well as in angular measurements of arcs and angles, e.g., 60 minutes per degree and 60 seconds per minute, while the measure of a circle is $6 \times 60 = 360$ degrees.

It is interesting that after the French revolution in the 18th century, French scientists transformed the system of measurements making is decimal for many physical quantities, such as length, volume, weight, and mass. However, they were not able to do this with time, measurements of which are sexagesimal.

Having names for whole and integer numbers, mathematicians defined rational numbers as numbers that can be named (represented) by fractions, i.e., by pairs of names of integer numbers. As we know two fractions represent (are the names of) the same rational number if these fractions are equal, e.g., $1/2 = 2/4 = 3/6$. This naming of rational numbers is consistent with the naming of integer numbers and operations with them, such as addition and multiplication, allowing similar operations with rational numbers.

However, there have been more problems with naming real numbers. Some of real numbers, such as π or e, have individual (proper) names. However, it is impossible to give an individual name in the form of a word to all real numbers because according to the Cantor theorem, there are more real numbers than words in any finite alphabet (Fraenkel *et al.*, 1973). To overcome this deficiency, mathematicians invented other theoretical representations for real numbers (cf., for example, Fraenkel *et al.*, 1973). The most popular is the representation of real numbers as classes of equivalent Cauchy sequences (cf., for example, Burgin, 2008).

However, not only mathematical symbols are used as names of numbers. Many numbers have names in natural languages, such as English, Spanish or French. In computers, calculators and cell phones, numbers are represented (named) by states of physical elements and electric signals. Scratches, notches and tokens were also used as names of numbers. In counting, some people still use fingers as names of numbers. The Pythagoreans represented numbers by pebbles or calculi. Counting rods were invented in China around as operational names of numbers. Red rods were used for positive numbers, while black rods were used for negative numbers. In addition, psychologists found that whole numbers have non-symbolic representations (names) in the brain of humans and other primates (cf., for example, Dehaene *et al.*, 1998; Nieder *et al.*, 2002; Roitman *et al.*, 2007; Verguts and Fias, 2004).

As the result, numbers and their names form named sets (Burgin, 1990; 1995; 2011), which are for simplicity called sets of numbers. We remind that a named set is a triad of the form (A, c, B) where A and B are sets while c is a connection (relation) between elements from A and their names from B. For instance, fractions are names of rational numbers and to operate with numbers, people use fractions. Often speaking or writing, people use terms (names) *decimal numbers* and *binary numbers* denoting decimal (correspondingly, binary) representations (names) of numbers. For instance, the decimal numeral 5 and the binary numeral 101 are names of the same natural number.

In common usage and in mathematics education on the lower levels the word *number* is, for simplicity, used for both the abstract object and the numeral. This creates a lot of confusion for the majority of population. The problem is that the students at schools and colleges study "mixed numbers" and "decimal numbers" but what is called "mixed numbers" are a specific notation for rational numbers and what is called "decimal numbers" are a specific notation for real numbers. In other words, "mixed numbers" and "decimal numbers" are numerals and not numbers.

The concept of number has been developing and extending over the years to include more and more mathematical objects. Mathematicians discovered a big variety of numbers and invented various notations for those numbers. As a result, there is no single, all-encompassing definition of number, and the development of the concept of number continues.

In addition to their use in counting and measuring, numerals are often utilized for representation of other objects. Numerals can play the role of labels or names, e.g., telephone numbers, place indicators in ordering, e.g., in the case of serial numbers, and as codes, e.g., as Social Security Numbers in USA.

All this shows that to understand what a number is, it is necessary to take into account three issues.

First, there is a host of different kinds and classes of numbers. Usually researchers have mostly discussed what natural numbers are assuming that all other numbers are generated using natural numbers, as in the cases of integer, rational and real numbers, or generalizing natural numbers, as in the cases of transfinite numbers.

Second, numbers have names called numerals and there is a difference between a number and its name (numeral) in the same way as there is a difference between a person and her/his name. As a result, the concept of a number has the structure of a named set $\mathbf{N} = (X, f, I)$ (cf. Appendix). The support X of it consists of abstract mathematical structures, which represent numbers; the set of names I consists of names of numbers, which include numerals; and the naming correspondence f connects numbers with their names. For instance, the collection of all sets equipotent to the set $\{1, 1, 1\}$ is connected to the set {three, 3, 3.0, $^3/1$, 6/2 , 1+1+1, 1+2, ...}, which consists of all numerals that represent (name) number 3.

Third, numbers acquire their basic operational meaning only because they belong to some arithmetic. Sometimes the same number belongs to several arithmetics. For instance, number 3 belongs to the arithmetic $\boldsymbol{N}$ of all natural numbers, arithmetic $\boldsymbol{W}$ of all whole numbers and arithmetic $\boldsymbol{R}$ of all real numbers.

The most basic system is the class (set) N of all natural numbers. To give an informal definition, it is sufficient to say that natural numbers are numbers 1, 2, 3, 4, 5, and so on. A formal (axiomatic) definition is considered in the Section 2.11.

Natural numbers are used for counting, ordering, calculation and measuring. As it is already explained, a natural number is a measure of sets.

A separate object, for example, a chair, forms a set that consists of this object, or we can say a set that consists of one object. In this case, the (natural) number 1 (one) gives a measure of the considered set.

Adding another object, for example, another chair, to this object, we form a new set, which consists of two objects. In this case, the (natural) number 2 (two) gives a measure of the new set.

Continuing this process, we can add more and more objects obtaining sets with three objects, with four objects and so on. The measure of (objects in) these sets will be natural numbers 3, 4, and so on. In this context, measuring of sets is called counting. This is how natural numbers come into being.

Having natural numbers, people come to a new problem when they need to measure parts of one object, i.e., parts of the whole. This is how positive rational numbers come into being and this development of the concept of a number has continued bringing forth an immense variety of different classes and types of numbers.

The further development of the concept of a number is discussed in Section 1.4 in the context of arithmetic.

It is interesting that although numbers form a foundation of mathematics and are extremely important in society, the working mathematician, as well as the general population, is rarely concerned with the natural question: What is a number? Nevertheless, the efforts in precisely answering this question have motivated much of the work done by mathematicians and philosophers in the foundations of mathematics during the past millennium. In particular, characterization of the natural and whole numbers, integers, rational and real numbers has been the central or one of the most important problems for many outstanding mathematicians and philosophers such as Aristotle (384–322 B.C.E.), Diophantus (around 250 C.E.), Brahmagupta (seventh century C.E.), Simon Stevin (1548–1620), Immanuel Kant (1724–1804), Georg Wilhelm Friedrich Hegel (1770–1831), Carl Friedrich Gauss (1777–1855), Hermann Günther Grassmann (1809–1877), Karl Theodor Wilhelm Weierstrass (1815–1897), Leopold Kronecker (1823–1891), Richard Julius Wilhelm Dedekind (1831–1916), Charles Sanders Peirce (1839–1914), Georg Ferdinand Ludwig Philipp Cantor (1845–1918), Gottlob Frege (1848–1925), Giuseppe Peano (1858–1932), Bertrand Russell (1872–1970), Alfred North Whitehead (1861–1974), Leutzen Egbert Jan Brouwer (1881–1966), and others.

1.3. The Concept of Arithmetic

God ever arithmetizes.

Jacob Jacobi

Arithmetic takes the name from the Greek word for number (*arithmos*). However, the meaning of the word "arithmetic" changed through the ages. As the English mathematician Henry John Stanley Smith (1826–1883) writes, arithmetic (and more exactly, acquaintance with the conventional Diophantine arithmetic) is one of the oldest branches, perhaps the very oldest branch, of human knowledge. According to David Smith, some of the ancient philosophers and mathematicians also used the term "arithmetic" to denote the theory of numbers, which comprised such areas as the study of primes and properties of square numbers (Smith, 1963). In contrast to this, many philosophers used the name *logistic* for the practical use of numbers, including the methods of writing them and calculating with them. This was looked upon by ancient scholars as somewhat a plebeian art, while arithmetic *per se* was considered an intellectual pursuit worthy of the attention of a philosopher. Others treated logistic as a kind of arithmetic but of a lower kind. For instance, Plato in his dialogue *Philebus* maintains: "*Arithmetic is of two kinds, one is popular, and the other philosophical*" (Plato, 1961). Others considered arithmetic as a whole domain, only having different aspects. In modern term, popular arithmetic means applied arithmetic, which all people use for calculations. Theoretical arithmetic means a field of mathematics, which studies properties of numbers and operations with them. Higher arithmetic is usually called number theory (cf., for example, Broadbent, 1971; Davenport, 1992; Hayes, 2009).

Emphasizing primacy and importance of (the conventional Diophantine) arithmetic, the outstanding German mathematician Carl Gustav Jacob Jacobi (1805–1851) said: "*God ever arithmetizes*". The prominent mathematician Leopold Kronecker (1825–1891) wrote: "*God made the integers, all the rest is the work of man.*"

To understand what arithmetic is in the modern culture in general and contemporary mathematics, in particular, it is necessary to know that now the word *arithmetic* has three basic meanings.

According to the **first meaning**, which conveys the most traditional and we may say classical understanding, *arithmetic* is the set N of all natural numbers, i.e., numbers $1, 2, 3, 4, 5, \ldots$, with two basic operations — addition and multiplication — and the order relation on numbers, e.g., $3 < 5$. We denote this arithmetic by $\boldsymbol{N}$ and call it the Diophantine arithmetic.

This point of view is expressed, for example, by Shapiro who writes: "*Arithmetic, then, is about the natural-number structure...*" (Shapiro, 1997).

This traditional conception of arithmetic is still represented in the mathematical field called *number theory*. In spite of the existing diversity of various kinds and types of numbers, number theory studies only natural numbers (cf., for example, Ore, 1948; Weil, 1984). It is even called higher arithmetic (cf., for example, Broadbent, 1971; Davenport, 1992; Hayes, 2009). As the result, it is possible to treat number theory as a subfield (subtheory) of the conventional arithmetic (of natural numbers). This approach is also followed in this book when elements of non-Diophantine number theory are presented.

This understanding of arithmetic is explicitly stated in the book (Courant and Robbins, 1960) where it is written:

> "*The mathematical theory of natural numbers or positive integers is known as arithmetic.*"

Consequently, in many mathematical publications, such as books and papers, the word *arithmetic* means the conventional arithmetic (of natural numbers). Just this arithmetic plays an imperative role in the foundations of mathematics. For instance, the main content of the book (Hilbert and Bernays, 1968) is dedicated to the conventional arithmetic, i.e., to the arithmetic of natural numbers. The famous Gödel's undecidability theorems are about the conventional arithmetic of natural numbers (Gödel, 1931/1932). Trying to build science without numbers, Fields nevertheless treats the arithmetic of natural numbers as one of the most basic fields in mathematics, which cannot be excluded (Fields, 1980).

It is necessary to remark that sometimes 0 is also included in the conventional arithmetic of natural numbers (cf., for example, MacLane and Birkhoff, 1999). This extends the classical meaning of the term *arithmetic* interpreting it as the arithmetic $\boldsymbol{W}$ of all whole numbers.

Interestingly this approach correlates with the suggestion to start counting with 0 instead of 1 (cf. Dijkstra, 1982; Cassani and Conway, 2018). This type of counting is used in several influential programming languages, including C, Java, and Lisp. This provides a simpler implementation of arrays, which are an important computational structure (Richards, 1967; Dijkstra, 1982).

According to the **second meaning**, *arithmetic* in the contemporary understanding is a system of numbers with operations and relations in it. Consequently, each general class of numbers has its own arithmetic. For instance, there is the arithmetic of natural numbers and there is the arithmetic of real numbers. However, all these arithmetics are extensions of the basic arithmetic of natural numbers, which we call the Diophantine arithmetic. Namely, each extension of numbers, for example, extension of rational numbers to real numbers, preserves operations and order relations of the basic subclass, i.e., new operations and order relations are extensions of the existing operations and order relations. For instance, if $2 + 2$ is equal to 4 in the arithmetic of natural numbers, the same remains true in the arithmetic of rational numbers and in the arithmetic of real numbers.

According to the **third meaning**, *arithmetic* is also a branch of mathematics occupied with the study of number systems with operations and relations. In this context, number theory, which is traditionally concerned with the properties of integers or even only of whole numbers, is a part of arithmetic. This is one more reason why number theory is also called higher arithmetic (cf., for example, Broadbent, 1971; Davenport, 1992; Hayes, 2009).

Number theory has several subfields:

- Elementary number theory;
- Analytic number theory;
- Algebraic number theory;
- Geometric number theory;
- Combinatorial number theory;
- Transcendental number theory or transcendental arithmetic;
- Computational number theory or computer arithmetic.

The basic operations of arithmetic as a branch of mathematics are addition, subtraction, multiplication, and division although multiplication can be often reduced to addition and in many number systems, such as integers or real numbers, subtraction can be also reduced to addition. All other operations in arithmetics such as taking powers (for example, squaring or cubing a number), the extraction of roots (for example, taking square roots), and building percentages, fractions, and ratios are constructed from the basic arithmetical operations. Subtraction and especially division are usually partial operations. For instance, in the arithmetic of natural numbers, it is possible to divide 10 by 2 but it is impossible to divide 11 by 2.

In addition, there is *computer arithmetic*, which is a branch of computer engineering dealing with methods of representing integers and real values (e.g., fixed- and floating-point numbers) in digital systems and designing efficient algorithms for manipulating such numbers by means of hardware circuits or software routines (Parhami, 2002; 2010).

There are also metaphoric interpretations of the term *arithmetic*. For instance, Lakoff and Núñez suggest four conceptual metaphors that ground numerical understanding on basic bodily experience (Lakoff and Núñez, 2000):

- Arithmetic is object collection;
- Arithmetic is object construction;
- Arithmetic is the measuring stick;
- Arithmetic is motion along a path.

The great German philosopher Immanuel Kant (1724–1804) made an effort to ground arithmetic in the spatio-temporal structure of reality corresponding arithmetic to time and geometry to space (Kant, 1787).

In this book, we delineate two classes of arithmetics: Diophantine arithmetics and non-Diophantine arithmetics. The first class comprises conventional types of arithmetics, the main of which are:

- The Diophantine arithmetic $\boldsymbol{N}$ of natural numbers;
- The Diophantine arithmetic $\boldsymbol{W}$ of whole numbers;
- The Diophantine arithmetic $\boldsymbol{Z}$ of integer numbers;
- The Diophantine arithmetic $\boldsymbol{Q}$ of rational numbers;
- The Diophantine arithmetic $\boldsymbol{R}$ of real numbers;
- The Diophantine arithmetic $\boldsymbol{R}^+$ of non-negative real numbers;
- The Diophantine arithmetic $\boldsymbol{C}$ of complex numbers.

Non-Diophantine arithmetics are numerical arithmetics, operations in which are intrinsically related to but do not always coincide with operations in Diophantine arithmetics. The most fundamental Diophantine and non-Diophantine arithmetics are arithmetics of natural numbers because $\boldsymbol{N}$ is the most innate and primeval arithmetic.

1.4. The Role of Numbers in Philosophy, Science and Humanities

When I considered what people generally want in calculating,
I found that it always is a number.

al-Khwārizmī

As numbers form one of the foundations of human culture, many philosophers analyzed this concept in their works. The most famous of them was Pythagoras, who tried to define all regularities of the world in terms of numbers.

The great Plato also wrote about numbers but in contrast to Pythagoras, they were on the periphery of his ontology and epistemology, in the center of which were Ideas or Forms. Even in mathematics, Plato makes the main emphasis on geometry defining four basic elements of the world — *fire*, *earth*, *air* and *water* — being made up of particles ("primary bodies" or corpuscles), which are regular geometrical solids: a *tetrahedron*, *cube*, *octahedron* and *icosahedron*, which are now called Platonic bodies. These solids are situated in the empty space, which is "the receptacle of all becoming" (Plato, 1961: *Timaeus*, 49a). However, particles (primary bodies) are not elementary being formed of *Platonic triangles* or *fundamental triangles* as certain geometrical "atoms" (*Timaeus*, 53c–d). Namely, the four particles are like the molecules of matter, which is built of four elements (Platonic bodies), while the triangles are its atoms generating the four particles.

In his metaphysics, Aristotle comprehensively described the differences between Plato and Pythagoras:

> *The name 'participation' was new; for the Pythagoreans say that things exist by 'imitation' of numbers, and Plato says they exist by participation, changing the name. But what the participation or the imitation of the Forms could be they left an open question. Further, besides sensible things and Forms he says there are the objects of mathematics, which occupy an intermediate position, differing from sensible things in being eternal and unchangeable, from Forms in that there are many alike, while the Form itself is in each case unique. ... so is his* [Plato] *view that the Numbers exist apart from sensible things, while they* [Pythagoreans] *say that the things themselves are Numbers, and do not place the objects of mathematics between Forms and sensible things. His divergence from the Pythagoreans in making the One and the Numbers separate from things, and his introduction of the Forms, were due to his inquiries in the region of definitions.*
>
> *The Pythagoreans, also, believe in one kind of number — the mathematical; only they say it is not separate but sensible substances are formed out of it. For they construct the whole universe out of numbers — only not numbers consisting of abstract units; they suppose the units to have spatial magnitude.*

Aristotle paid special attention to these distinctions because *number* was one of the basic categories in the Aristotelian ontology (Aristotle, 1984). Besides, Aristotle distinguished numbers and magnitudes associating numbers with time and magnitudes with space and physical bodies (cf. Annas, 1975).

Based on the opinion of Aristotle, the contemporary Italian philosopher Giuseppe Boscarino explains profound differences between the philosophies of Plato and the Pythagoreans about the number and the concept of reality in the following way (Boscarino, 2018):

1. For the Pythagoreans, numbers are not separated from the corresponding magnitudes. Numbers are logos (knowledge) of magnitudes, which are homogeneous to a magnitude taken as unit of measure. Plato instead

separates numbers from magnitudes and assigns to them a separate and real existence, independent from magnitudes.

2. For the Pythagoreans, there is only one type of number, the mathematical number, which can represent logos, ratio, reason, and law. For Plato, beyond the mathematical number, there is the ideal number. It is between the sensible world and the intelligible world of ideas. It is eternal and immovable, like ideas, but it is multifaceted as sensible things in contrast to ideal forms, which are unique and individual. For instance, over the many 2, which indicate different pairs of objects, e.g., 2 pears or 2 apples, there is the idea $\iota 2$ of 2, which is unique and individual, while the mathematical number 2 is multifaceted as it is predicated of many pairs.
3. For the Pythagoreans number 2 is a magnitude because it is an idealized relationship between physical objects that we count or measure. This number is abstracted from our physical operations of counting and measuring, from which it gets its meaning and its purely logical and nominal existence. For Plato, the idea $\iota 2$ of 2 precedes number 2 while number 2 precedes the corresponding magnitude.

Analyzing dialogues of Plato, it is possible to suggest that mathematical numbers in the sense of Plato are symbolic representations of numbers by numerals. As such, they are placed between the world of ideas and the sensible world.

The concept of numerical characteristic was further developed by Leibniz and Kant. For instance, Leibniz's "principle of continuity" asserts that all natural changes are produced by degrees. Leibniz argued that this principle can be applied not only to changes in extended magnitudes (or numerical characteristics) such as *length* and *duration*, but also to intensities of representational states of consciousness, such as sounds (Jorgensen, 2009; Diehl, 2012).

In this context, it is believed that Kant relied on Leibniz's principle of continuity to formulate his distinction between extensive and intensive magnitudes (numerical characteristics). According to Kant, *extensive magnitudes* are those "in which the representation of the parts makes possible the representation of the whole" (Kant, 1787). In relation to *length*, Kant maintained a line could only be mentally represented by a successive synthesis in which parts of the line joined to form the whole. For Kant, the prospect of such synthesis was grounded in the basic forms of intuition, that is, space and time. In contrast to this, *intensive magnitudes*, such as warmth or colors, also come in continuous degrees, but their comprehension takes place in a moment rather than through a consecutive synthesis of parts. The degrees of intensive magnitudes "can only be represented through approximation to negation," namely, by imagining their gradual diminution until their complete absence (Kant, 1787).

Another outstanding German philosopher Georg Wilhelm Friedrich Hegel (1770–1831) discussed real numbers as the paradigmatic kind of numbers and their relations to the conception of measure. Interestingly, Hegel considered a *quantitative relation* as composed of three components — the two things being related and the relation itself (cf., Kaufmann and Yeomans, 2017). This is exactly a specific form of a named set or fundamental triad, which is the most fundamental structure in mathematics, nature, technology and cognition (Burgin, 2011).

Numbers always have played a crucial role in science because measurement and computing in physics, chemistry and other sciences are done in terms of numbers. As a result, numbers describe and explain the natural world making sense of everything from atoms and molecules to planets and stars.

Moreover, measurement and computing are integral parts not only of contemporary science but also of engineering, business, and daily life. Measurement is often considered a hallmark of the scientific enterprise and a privileged source of knowledge relative to qualitative modes of inquiry. Measurement is usually defined as an assignment of a number to a characteristic of a system, object, process or event.

In his recently published book, Farmelo asserts, following Galileo Galilei (1564–1642), that "the universe speaks in numbers" demonstrating how modern math reveals nature's deepest secrets (Farmelo, 2019).

To efficiently work with numbers, which are so important in science, scientists even elaborated a specific scientific notation. It is a method for writing numbers using powers of 10, which makes representations of very small and very large numbers shorter and thus more comprehensible. Being more efficient in science, scientific notation also conveys the precision of measurement using significant figures.

It is necessary to remark that not only real numbers are important in science. For instance, complex numbers are basic for all predictions in the quantum world made by modern science.

Numbers are prevalent in economics and finance because economical and especially financial processes involve numerical monetary estimation (cf., for example, Berman and Knight, 2008; Boumans, 2005).

As an example of foremost importance of numbers in sociology and economics, we can take an issue of *Science* describing how numerical data allow exploration of the origins, impact, and future of inequality around the world (Chin and Culotta, 2014). In particular, archaeological and ethnographic numerical data are revealing how inequality got its start in our ancestors, while new surveys of emerging economies offer more reliable estimates of people's incomes and how they change as countries develop (Chin and Culotta, 2014).

Numbers are also significant for psychology and especially, for measurement in psychophysics as a tool for measurement (cf., for example, Fechner, 1860; Bachem, 1952; Baird, 1975; 1975a; 1997; Baird and Noma, 1975; 1978; Banks and Coleman, 1981; Heidelberger, 1993; 1993a; Marks, 1974; Noma and Baird, 1975; Schneider *et al.*, 1974; Stevens, 1951; Weissmann *et al.*, 1975; Whalen *et al.*, 1999).

At the same time, as arithmetics is an important area of activity and people are manipulating with numbers constantly, psychologists investigated how people work with numbers exploring number representation in the brain, performance of arithmetical operations, formation of number concept and so on (cf., for example, Arsalidou and Taylor, 2011; Ashcraft, 1995; Ashkenazi, 2008; Badets and Pesenti, 2010; Berteletti and Booth, 2015; Carey, 2001; Cohen Kadosh and Dowker, 2015; Cohen Kadosh and Walsh, 2009; Cohen Kadosh, 2007; Menon, 2015; Dehaene, 1997; 1977a; Dehaene and Changeux, 1993; Dehaene and Cohen, 1995; Dehaene *et al.*, 2003; Eger *et al.*, 2009; Harvey *et al.*, 2015; Nieder and Dehaene, 2009). For instance, psychologists found that arithmetic processing in the brain is shaped by cultures (Tang, 2006). Results of psychologists on numerical skills and arithmetic found applications in learning and education (cf., for example, Menon, 2010; Price *et al.*, 2013). One more interesting result of psychologists implies the existence of a generalized magnitude processing system that would underlie the representation of numerosity, space and time through a common metric system controlled by areas of the parietal cortices (Crollen *et al.*, 2013; Walsh, 2003; 2015).

Moreover, psychologists and biologists studied numerical information processing by insects, fish, birds and animals, such as bees, monkeys, lions, cats and rats (cf., for example, Agrillo *et al.*, 2006; Brannon and Terrace, 1998; Breukelaar and Dalrymple-Alford, 1998; Cantlon and Brannon, 2007; Dehaene *et al.*, 1998; Gibbon, 1977; Hauser *et al.*, 2003; Howard *et al.*, 2018; Merritt *et al.*, 2009; Nieder *et al.*, 2002; Nieder and Miller, 2004; Roberts, 1995; Roberts and Boisvert, 1998; Sawamura *et al.*, 2002; 2010; Sulkowski and Hauser, 2001; Thompson *et al.*, 1970; Vasas and Chittka, 2018; Verguts and Fias, 2004). For instance, psychologists found that generally, animals decide to attack back only when the number of defenders is superior to the number of intruders, which means that animals have number sense (McComb *et al.*, 1994). Evidence was found that nonhuman primates shared three essential numerical processing mechanisms with modern humans: an ability to represent numerical values (Nieder, 2005; Cantlon and Brannon, 2006; 2007a), a general mechanism for mental comparison, and arithmetic algorithms for performing addition and subtraction (Cantlon and Brannon, 2007). Bees, in particular, have been argued to be able to count up to four items and solve complex numerical tasks (Howard *et al.*, 2018; Skorupski *et al.*, 2018).

Some of researchers who study arithmetical skills in humans and animals elaborate neural models for counting (cf., for example, Dehaene and Changeux, 1993; Dehaene and Cohen, 1994; Dehaene *et al.*, 2003; Verguts and Fias, 2004; Prado *et al.*, 2010; Ansari, 2016).

To conclude, it is necessary to mention that in addition to mathematics, science, economy, sociology and other areas where numbers are used in a scientific way, arithmetic and numbers have been frequently used for symbolic purposes in astrology and mystical areas (cf., for example, Losev, 1994; Klotz, 1995; Slaveva-Griffin, 2009). The prominent philosopher, writer and political leader Philo of Alexandria (ca. early first century C.E.), who was called a Pythagorean in antiquity, significantly used numbers as allegorical symbols and methodically compiled a handbook of ancient number symbolism. Plotinus (204/5–270 C.E.), who was one of the most influential philosophers in antiquity after Plato and Aristotle being generally regarded as the founder of Neoplatonism, based his philosophy on numerical symbolism in the style of Pythagoras (Plotinus, 2018; Slaveva-Griffin, 2009).

1.5. The Role of Numbers in Technology

We must not only search for and procure
a greater number of experiments, but also
introduce a completely different method, order,
and progress of continuing and promoting experience.

Francis Bacon

Technology is permeated by mathematics in general and numbers, in particular. Without calculations, technology cannot exist. Every aspect of its design, construction and technical description is based on measurement and reckoning. Technology is impossible without measurements and measurements use numbers as representation of the results of measurements. Nowhere this is more so than for those particular localized implementations of technology we call computing machines, which in extreme cases are little more than fragments of mathematical structures that engineers have found ways to physically realize (Rotman, 2003). Numbers regulate and control design and production of all machines, mechanisms and devices.

To efficiently work with numbers, which are so important in technology, engineers even elaborated specific engineering notation, which is similar to scientific notation but not the same because engineers like exponents in multiples of three. It is also a method for writing numbers using powers of 10 divisible by 3. For instance, the speed of light is approximately equal to 300×10^6 m/s. This makes the wide, dynamic range of numbers that engineers deal with on a regular basis more comprehensible and feasible.

As Weller writes, "To most engineers, mathematics is a means to an end. This end is usually the numerical answer to a problem... In many cases, the numerical answers sought by an average engineer are found by using simple arithmetical formulas" (Beckenbach, 1956). Contemporary engineers use various computer programs but in essence they strive for numbers.

In addition, computers and their precursors, such as the abacus, were spawned by the necessity to calculate, i.e., to work with numbers. The *abacus,* which is also called a *counting frame* in Europe, *soroban* in Japan and *suanpan* in China, is a calculating tool that was exploited for millennia in Europe, China, Japan and Russia. For instance, the Roman abacus was developed from devices used in Babylonia as early as 2400 BC.

The computing devices were created not for entertainment or communication but out of a need to solve crucial number-crunching problems. The first manufactured mechanical calculating device was invented by the great French mathematician, physicist and philosopher Blaise Pascal (1623–1662). When his father Étienne Pascal (1588–1651), who worked as the supervisor of taxes in Rouen, asked Blaise to help him performing necessary calculations, Blaise decided that a machine could do the same work even better inventing and producing an automaton to add and subtract two numbers directly and to perform multiplication and division through repeated addition or subtraction. Pascal began to work on his calculator in 1642, when he was 20 years old, presenting it to the public in 1645.

The great German mathematician, philosopher, logician and diplomat Gottfried Wilhelm Leibniz (1646–1716) further developed Pascal's calculator to make possible one more arithmetical operation — multiplication. After achieving this goal, he improved his device making possible performing additions, subtractions and multiplications automatically and division under operator control. Leibniz struggled for forty years to improve this design elaborating two automata, one in 1694 and one in 1706.

English polymath, mathematician, philosopher, inventor and mechanical engineer, Charles Babbage (1791–1871) developed a project of a steam-driven calculating machine to compute tables of numbers.

By 1880, the U.S. population had grown so large that it would take many years to tabulate the U.S. Census results. The government sought a faster way to get the job done, and Herman Hollerith (1860–1929) designed a punch card system to calculate the 1880 census, accomplishing the task in just three years and saving the government $5 million.

All further history of computers is intrinsically connected to numbers and arithmetic. Now all electronic computers, calculators and cell phones perform their operation using the binary numerical system. With the advent of the digital computer, the entire co-evolutionary dynamics of technology and mathematics underwent a phase shift taking arithmetic to an entirely novel, drastically productive level.

In essence, computers reduce complex mathematics, logic and problem solving to clear-cut arithmetic exponentially increasing its importance. Computers can actually do many kinds of arithmetical procedures with great precision and high speed. This typically means two kinds of arithmetic — one for integers, and one for real numbers. All this brought forth computer arithmetic (Flynn and Oberman, 2001).

The development of the Internet brought researchers to a new type of numbers, which are represented in a new data set *Linked Open Numbers*, which lay the foundation for hitherto unrealized applications for semantic technologies on the Web, and open an avenue for a range of future research topics (Vrandečić *et al.*, 2010).

Algorithms, which control computers and other technical devices, also originated from operation with numbers in arithmetic. The term was derived from the title of the very influential book *Al-jabr wa'l muqabala* written by the great Muslim mathematician and astronomer Muhammad ibn Mūsā al-Khowārizmī (ca. 780–850). His book was brought to Europe and translated into Latin as *Algoritmi de numero Indorum*, which means in English *Al-Khowārizmī* on the Hindu Art of Reckoning, introducing to the European mathematics the Hindu decimal positional system and operations with numbers in this system. In such a way, the name of al-Khowārizmī (*Algoritmi* in Latin) became one of the most popular terms in the 20th and 21st centuries.

Now, numerical algorithms and methods are crucial for solving many mathematical, scientific and practical problems with computers. At the same time, numerical algorithms are almost as old as human civilization. For instance, the Rhind Papyrus (ca. 1650 B.C.E.) from ancient Egypt describes a numerical method for solving a simple equation. An Old Babylonian clay tablet (ca. 1800–1600 B.C.E.) gives a numerical approximation of the square root of 2. Contemporary numerical analysis continues the long tradition of practical mathematical applications. It does not seek exact answers, because exact answers are often impossible to obtain in practice. Instead, it is concerned with obtaining approximate solutions maintaining at the same time reasonable bounds on errors.

1.6. Is the Conventional Diophantine Arithmetic Always Sufficient and Adequate?

The learned is happy, nature to explore.
The fool is happy, that he knows no more.
Alexander Pope

Mathematical establishment treated the arithmetic of natural numbers $\boldsymbol{N}$ as a primordial entity. For example, such a prominent mathematician as Leopold Kronecker (1825–1891) wrote: "*God made the natural numbers, all the rest is the work of man*". It is interesting that Kronecker wrote *natural numbers* but not the *arithmetic of natural numbers* because it was unconditionally assumed that there was only one possible arithmetic of *natural numbers*.

Laymen have been even more persistent on this point of view. Almost all people had and have no doubts that $2+2=4$ is the most evident truth in always and everywhere. Mathematicians supported this attitude. For instance, according to the *American Mathematical Monthly* (April, 1999, p. 375), "Although other sciences and philosophical theories change their 'facts' frequently, $2+2$ remains 4."

People's experience with numbers and, especially, with natural numbers is profound. From ancient times, much longer than they have done this with the Euclidean geometry, people have believed and continue to believe that only one arithmetic of natural numbers — the Diophantine arithmetic — have always existed and no other arithmetic can ever exist.

However, do you think that now people know everything about numbers and counting? The answer is negative because different outstanding thinkers have doubted the absolute character of the conventional Diophantine arithmetic, giving examples when this arithmetic did not correctly describe certain systems and processes. It is possible to find the roots of this fundamental problem of arithmetic relevance in ancient Greece. Long ago, there was a group of philosophers, who were called Sophists and lived from the second half of the fifth century B.C.E. to the first half of the fourth century B.C.E. Sophists asserted relativity of human knowledge and elaborated various brainteasers, explicating complexity and diversity of the real world. One of them, the famous Greek philosopher Zeno of Elea (490–430 B.C.E.), who was said to be a self-taught country boy, invented very impressive paradoxes, in which he challenged the popular knowledge and intuition related to such fundamental essences as time, space, and number (Yanovskaya, 1963). One of them was the *paradox of the heap* or the *Sorites paradox* (as σωρος means a heap in Greek), which is described in the following way.

1. One million grains of sand, for example, make a heap.
2. If we take away one grain, this will be actually the same heap.

Analyzing this situation, Martin Gardner (1914–2010) writes,

> "*Repeated applications of premise 2 (each time starting with one less number of grains), will eventually allow us to arrive at the conclusion that 1 grain of sand makes a heap. On the face of it, there are three ways to avoid that conclusion. Object to the first premise (deny that one million grains makes a heap, or more generally, deny that there are heaps), object to the second premise (it is not true for all collections of grains that removing one grain cannot make the difference between it being a heap or not), or accept the conclusion (1 grain of sand can make a heap). Few, if any, reply by accepting the conclusion. In addition to advocating a response, philosophers who work on this paradox also try to explain why it is that the premise one would have to deny seems so plausible, despite being false.*" (Gardner, 2005)

However, the main problem was not the reasoning but the contradiction with what the conventional arithmetic told people. Namely, it asserts that taking any number and subtracting one, we get a new number. In the same way, taking any big number, say 10,000,000, and adding one, the conventional arithmetic tells that it will be a new number. At the same time, the heap of sand that contains 10,000,000 grains remains actually the same if one grain is added to it.

A reader may ask what for we are interested in puzzles that were suggested thousands years ago and look artificial to the modern reader. However, the paradox of a heap has a direct analogy in our times both in science and everyday life. For example, you are buying a car for $30,000. Then suddenly, when you have to pay, the price is changed and becomes one cent greater. Do you think that the new price is different from the initial one or you consider it practically the same price? It is natural to suppose that any sound person has the second opinion. Consequently, we come to the same paradox: if k is the price of the car in cents, then in the conventional Diophantine arithmetic $k + 1$ is not equal to k, while in reality they are equal as prices because k is much larger than 1.

Moreover, imagine that you are going to receive your salary in cents, the sum of which will be equal to the amount that you receive now but which will be given to you once a year. Do you think it will be the same salary or not if you get one cent less or one cent more? For many people, there is a difference how often they receive their salary, although the sum remains the same.

Here is one more situation from real life. Imagine that you have $100,000,000 and somebody gives you $1. Will you say that now you have $100,000,001? No, you will probably assume that you have the same amount of money. This contradicts the rules of the Diophantine arithmetic where 100,000,000 and 100,000,001 are different numbers. The problems are of a psychological nature, but they have non-trivial implications for economic modeling.

These examples show that in some cases we encounter inconsistency between the real life and the Diophantine arithmetic. There are two basic ways to deal with inconsistencies: one is to elaborate an inconsistent system and try to work with it, and another way is to create new mathematical structures, eliminating inconsistencies. The book *How Mathematicians Think* of William Byers has the subtitle *Using Ambiguity, Contradiction, and Paradox to Create Mathematics* because mathematical reasoning, according to Byers, is not completely algorithmic, computational or based on proof systems. It primarily uses creative ideas to shed new light on mathematical objects and structures, propelling in such a way mathematical progress (Byers, 2007). Ambiguities, contradictions, and paradoxes play the central role in the emergence of creative ideas. Being unsolvable when they appeared, many problems and paradoxes find their solutions on a higher level of cognition.

As we will show in next chapters, non-Diophantine arithmetics solve the paradox of the heap and other paradoxes, which emerge when we compare some real-life situations with the rules of the conventional Diophantine arithmetic.

The same happened with many profound ideas of ancient philosophers. For instance, now all educated people know that material things are built of atoms. However, idea of atoms was introduced much earlier than atoms were really discovered in nature. Outstanding philosophers Democritus (ca. 460 — ca. 370 B.C.E.) and Leucippus (fl. 5th century B.C.E.) from ancient Greece elaborated the idea of atoms as the least particles of matter. For a long time, this idea was considered false due to the fact that scientists were not able to go sufficiently deep into the matter. Nevertheless, the development of scientific instruments and experimental methods made possible to discover such micro-particles that were and are called atoms, although they possessed very few of those properties that were ascribed to them by ancient philosophers.

Another great idea of ancient Greece was the world of ideas whose existence was postulated by Plato. In spite of the attractive character of this idea, the majority of scientists and philosophers believe that the world of ideas does not exist, because nobody had any positive evidence in support of it. The crucial argument of physicists is that the main methods of verification in modern science are observations and experiments, and nobody has been able to find this world by means of observations and experiments. Nevertheless, some modern thinkers, including such outstanding intellectuals as philosopher Karl Raimund Popper (1902–1993), logicians Georg Kreisel (1923–2015) and Kurt Gödel (1906–1978), mathematicians Alain Connes, David Mumford, and René Frédéric Thom (1923–2002), computer scientist Gregory Chaitin, and physicists Werner Karl Heisenberg (1901–1976) and Roger Penrose, continued to believe in the world of ideas giving different interpretations of this world but suggesting no ways for their experimental validation.

However, science is developing, and this development led to the discovery of the World of Structures (Burgin, 2012a). On the level of ideas, this world may be associated with the Platonic world of ideas, in the same way as atoms of the modern physics may be related to the atoms of Democritus. Existence of the World of Structures is proved by means of observations and experiments (Burgin, 2017b). This World of Structures constitutes the structural level of the world as a whole. Each system, phenomenon or process, either in nature or in society, has some structure. These structures exist like tables, chairs, or buildings, and form the structural level of the world. When it is necessary to investigate or to create some system or process, it is possible to do this only by means of knowledge of the corresponding structure. Structures determine the essence of things in the same way as Aristotle ascribed to forms of things. Consequently, structures unite ideas of Plato with forms of Aristotle, eliminating contradictions that existed between their teachings about reality.

Still, while Greek sages and subsequent thinkers posed questions about arithmetic, they suggested no answers that would allow improving the situation. As a result, for more than 2000 years these problems were forgotten and everybody was satisfied with the conventional arithmetic. The reason was that in spite of all the problems and paradoxes, this arithmetic has remained very and very useful in practical activity and theoretical investigations.

The famous German scientist Herman Ludwig Ferdinand von Helmholtz (1821–1894) was may be the first thinker, who in modern times questioned absolute authority of the conventional arithmetic. In his work "Counting and Measuring", Helmholtz considered an important problem of applicability of arithmetic to physical phenomena, although at that time people knew only one arithmetic (Helmholtz, 1887). This was a natural approach of a scientist, who even mathematical statements tested by the main criterion of science — observation and experiment.

Interestingly, this scientific approach to arithmetic essentially differed from the opinion of the great German mathematician Carl Friedrich Gauss (1777–1855), who assumed that arithmetic was purely aprioristic in contrast to geometry, which should be ranked with mechanics.

The first observation of Helmholtz was that because the concept of number was derived from some practice, usual arithmetic had to be applicable to practical experiences. However, it is easy to find many situations when this is not true. To mention but a few situations described by Helmholtz, we give the following examples.

One raindrop added to another raindrop does not make two raindrops but forms only one raindrop. Mathematically, it is described by the equality $1 + 1 = 1$.

Similarly, when one mixes two equal volumes of water, one at 40° Fahrenheit and the other at 50° Fahrenheit, one does not get two volumes at 90° Fahrenheit. This contradicts the conventional arithmetic where $40 + 50 = 90$.

In a similar way, the conventional arithmetic fails to describe correctly the result of combining gases or liquids by volumes. For example (Kline, 1980), one quart of alcohol and one quart of water yield about 1.8 quarts of vodka.

Later the famous French mathematician Henri Léon Lebesgue (1875–1941) facetiously indicated (cf., for example, Kline, 1980) that if one puts a lion and a rabbit in a cage, one will not find two animals in the cage later on. In terms of numbers, it will mean $1 + 1 = 1$.

The famous philosopher Karl Popper also pointed out that the equality $2 + 2 = 4$ is not always true as a physical fact. He wrote:

> *More important is the application in the second sense. In this sense, "$2 + 2 = 4$" may be taken to mean that, if somebody has put two apples in a basket, and then again two, and has not taken any apples out of the basket, there will be four in it. In this interpretation "$2 + 2 = 4$" helps us to calculate, i.e., to describe certain physical facts, and the symbol "+" stands for a physical manipulation — for physically adding certain things to other things.... But in this interpretation "$2+2 = 4$" becomes a physical theory, rather than a logical one; and as a consequence, we cannot be sure whether it remains universally true. As a matter of fact, it does not.... It may hold for apples, but it hardly holds for rabbits. If you put $2 + 2$ rabbits in a basket you may soon find 7 or 8 in it.* (Ryle *et al.*, 1946)

Similar opinion was also expressed by the great French mathematician Jules Henri Poincaré (1854–1912), who believed that arithmetics had to be tested by experiments pointing out that one did not "prove" $2+2 = 4$, one "checked" it (Gonthier, 2008). This is exactly what some researchers did. Performing definite mental experiments, Helmholtz, Lebesgue and Kline demonstrated that there are situations when two plus two is not equal to four.

In contrast to this, another great mathematician Carl Friedrich Gauss had a different opinion about numbers assuming that they had nothing to do with the physical reality. In his 1830 letter to Wilhelm Bessel, he wrote:

> "*We must admit with humility that, while number is purely a product of our minds, space has a reality outside our minds, so that we cannot completely prescribe its properties a priori.*"

Because of this opinion shared by the vast majority of mathematicians, very few (if any) researchers paid attention to the work of Helmholtz on arithmetic as well as to observations of other mathematicians pointing at only partial adequacy of the conventional arithmetic. As a result, because no alternative to the conventional arithmetic was still suggested, the problems with the conventional arithmetic were once more forgotten.

It took almost hundred years to revive observations of Helmholtz on numbers in our times. The most extreme view on the arithmetic was expressed in ultraintuitionism where it was postulated that there was only a finite quantity of natural numbers (Yesenin-Volpin, 1960; 1970). In a similar way, Van Danzig explains why only some of natural numbers may be considered finite (Van Danzig, 1956). Consequently, all other mathematical entities that are traditionally called natural numbers are only some expressions but not numbers. These arguments were later supported and extended in (Blehman *et al.*, 1983). In addition, people and even computers operate only with finite sets of numbers and any computer arithmetic is finite (Flynn and Oberman, 2001).

As a matter of fact, much earlier than non-Diophantine arithmetics were discovered, English mathematician John Edensor Littlewood (1885–1977) considered an example demonstrating how the rules of non-Diophantine arithmetics (in spite of that they were unknown at that time) can be imposed upon the real world (Littlewood, 1953). Several similar and even more lucid examples are given in (Davis and Hersh, 1986) and in (Kline, 1967). For instance, when a cup of milk is added to a cup of popcorn then only one cup of mixture will result because the cup of popcorn will very nearly absorb a whole cup of milk without spillage. So, in this case we also have $1 + 1 = 1$. It is impossible in the conventional arithmetic but it is true in some non-Diophantine arithmetics.

To make the situation, when ordinary addition is inappropriate, more explicit, an absurd but not unrelated question is formulated: If the Mona Lisa painting is valued at $10,000,000, what would be the value of two Mona Lisa paintings?

Let us consider more examples of situations when the Diophantine arithmetic does not work.

1. A market sells a can of tuna fish for $1.05 and two cans for $2.00. So, we have $a + a \neq 2a$.
2. In a similar way, coming to a supermarket, you can buy one gallon of milk for $2.90 while two gallons of the same milk will cost you only $4.40. Once more, we have $a + a \neq 2a$.
3. Even more, coming to a supermarket, you can see an advertisement "Buy one, get one free." It actually means that you can buy two items for the price of one. Such advertisement may refer almost to any product: bread,

milk, juice etc. For example, if one gallon of orange juice costs \$2, then we come to the equality $2 + 2 = 2$. It is impossible in the conventional arithmetic but it true for some non-Diophantine arithmetics.

Another property of the Diophantine arithmetic was also challenged. Some researchers although being moderate in their criticism of the conventional arithmetic, suggested that not all natural numbers are the same in contrast to the presupposition of the conventional arithmetic that the set of natural numbers is uniform (Kolmogorov, 1961; Littlewood, 1953; Birkhoff and Barti, 1970; Rashevsky, 1973; Dummett, 1975; Knuth, 1976). Different types of natural numbers have been introduced, but without changing the conventional arithmetic. For example, one of the greatest mathematicians of the 20th century Andrei Nikolayevich Kolmogorov (1903–1987) suggested that in solving practical problems it is worth to separate *small*, *medium*, *large*, and *super-large* numbers (Kolmogorov, 1961).

A natural number k is called *small* if it is possible in practice to list and work with all combinations and systems such that are built from k elements, each of which has two inlets and two outlets.

A natural number m is called *medium* if it is possible to count to and work directly with this number. However, it is impossible to list and work with all combinations and systems that are built from m elements each of which has two or more inlets and two or more outlets.

A natural number n is called *large* if it is impossible to count a set with this number of elements. However, it is possible to elaborate a system of denotations for these elements.

If even this is impossible, then a number is called *super-large*.

According to this classification, 3, 4, and 5 are small numbers, 100, 120, and 200 are medium numbers, while an example of a large number is given by the quantity of all visible stars. Really, if we invite four people, we can consider all their possible positions at a dinner table. If you come to some place where there are 100 people, you can shake hands with everybody although it might take too much time. What concerns the visible stars, you cannot count them, although, a catalog of such stars exists. Using this catalog, it is possible to find information about any of these stars.

This classification of numbers is based on our counting abilities. Consequently, borders between classes are vague and unstable. Higher counting abilities make borders between classes higher. For example, 10 is a medium number for an ordinary individual, but a small number for a computer. However, some numbers belong to a definite class of this typology in all known situations. For example, 300 is a medium number both for people and computers.

In a similar way to what has been done by Kolmogorov and on the akin grounds, John Edensor Littlewood (1885–1977) separated all natural numbers into an infinite hierarchy of classes (Littlewood, 1953).

To reflect operational properties of natural numbers, Sazonov introduced a formal approach to *feasible numbers*, as well as to *middle* and *small* numbers based on ideas from (Parikh, 1971). According to his approach, the inequality $\log\log N < 10$ is a formal axiom defining feasible numbers (Sazonov, 2002). In a similar way, it is possible to delineate feasible numbers by choosing relevant non-Diophantine arithmetics.

Interestingly, psychologists found that people usually represent and process small and large numbers in a different way (Trick and Pylyshyn 1994; Spelke and Barth, 2003; Cantlon and Brannon, 2006; Goldman, 2010). For instance, reaction time and performance in solving problems with large numbers are determined by their ratio while for small numbers, their size plays the main role.

However, these classifications of natural numbers did not go beyond the conventional Diophantine arithmetic while examples when this arithmetic was not adequate in representing various real-life situations challenged beliefs about uniqueness of the Diophantine arithmetic. Such examples are described in the books of the noted mathematics historian Morris Kline (1908–1992). Let us consider one more example from (Kline, 1967).

If a farmer has two herds consisting of 10 and 25 heads of cows, respectively, he knows by adding 10 and 25 that the total number of cows is 35. That is, he need not count his cows. Suppose, however, he brings the two herds of cows to market where they are selling for $100 apiece. Will a herd of 10 cows which might bring $1000 and a herd of 25 cows which might bring $2500 together bring in $3500? Every businessman knows that when supply exceeds demand, the price may drop, and hence 35 cows may bring in only $3000. In some idealized world the value of the cows may continue to be $3500, but in actual situations this need not be true.

Consequently, continues Kline (1967), mathematicians are, of course, free to introduce the symbols $1, 2, 3, \ldots,$ where 2 means $1+1$, 3 means $2+1$, and so on. We can even deduce from this that $2 + 2 = 4$. But the question is not whether the mathematician can set up definitions and axioms and deduce conclusions. It is necessary to know whether this system necessarily expresses truths about the Physical World.

American mathematician Reuben Hersh (1927–2020) argues that even laws of arithmetic are uncertain by considering a hotel that is missing a thirteenth floor. Take an elevator up eight floors, then go five floors more, and you reach floor fourteen. Hersh apparently thinks this violates the equation $8 + 5 = 13$. What he has done, of course, is a jump from pure arithmetic to applied arithmetic, where applications are often uncertain. Two beans plus two beans make four beans only if you assign to beans what the famous Austrian-American philosopher Rudolf Carnap (1891–1970) called a correspondence rule. In this case, the rule is that each bean corresponds to 1. In the case of Hersh's elevator, if you assume that every floor corresponds to 1, then 8 floors plus 5 floors is sure to make 13 floors. Without correspondence rules, applications of mathematical truths are indeed uncertain.

According to Kline, discovery of non-Euclidean geometries had taught mathematicians that geometry does not offer ultimate truths (Kline, 1967). That was the reason why many turned to the ordinary number system and the developments built upon it and maintained that this part of mathematics still offers unquestionable truths. The same thought is often expressed today by people who, wishing to give an example of an absolute truth, quote $2+2 = 4$. However, examination of the relationship between our ordinary number system and the physical situations to which it is applied vividly demonstrates that it does not always offer truths. For instance, there are biological processes that cannot be correctly described using the conventional Diophantine arithmetic (Cleveland, 2008).

Various examples when $2 + 2$ does not equal 4 are considered by Gershaw, who writes,

> "Believe it or not, sometimes $2+2$ does not equal 4. It depends on what type of measurement scale you are using." (Gershaw, 2015)

Exploring the history of mathematics, it is possible to divide all mathematicians who were wise enough to distrust the complete adequacy of the conventional arithmetic into two groups. Representatives of the first group, such as Helmholtz, Kolmogorov or Littlewood, only explained that in practice natural numbers and operations with them were different from those which were known from mathematics. Others, such as Gasking, Kline or Rashevsky, in addition conjectured that different arithmetics existed but people did not know what they were and how to build them (Gasking, 1940; Kline, 1967; Rashevsky, 1973). This conjecture was proved when non-Diophantine arithmetics were discovered (Burgin, 1977).

It is necessary to note that even before Kline and Rashevsky, Douglass Gasking not only suggested a possibility of existence other arithmetics (of natural numbers) but even described how operations in such arithmetics can correctly describe some situations in people's practice (Gasking, 1940). Analyzing relations between mathematics and the Physical World, he demonstrated how "queer" operations, as he called them, such as $4\times 6 = 12$ or $3\times 4 = 24$, could give useful results when applied to practical tasks of the world. In one of his "queer" arithmetics, Gasking also gave a description of the non-standard multiplication $\otimes$ of even numbers defined by the formula

$$m \otimes n = {}^{1}\!/_{2}(m+2)(n+2).$$

However, Gasking did not build and even did not give a description of any of those arithmetics.

Hector Castaneda further analyzed arguments of Gasking (Castaneda, 1959). Castaneda explains that Gasking uses essentially two arguments:

(1) We could use any mathematics and compensate for deviations from the present one by means of an adequate technique of counting or measuring.
(2) We could use any mathematics different from the present one and modify our physics (Castaneda, 1959).

At the same time, according to Castaneda, Gasking "does not define a queer arithmetic, but it is clear from his discussion that he does not mean one or more changes of labels. In other words, to use the formula '$6\times 4 = 12$' is not merely to use '6' to mean the number 3, and perhaps '3' or 'w' to mean the number 6.... By a 'queer arithmetic' Gasking means something more exciting. Presumably, we are to think of queer arithmetics on the analogy with non-Euclidean geometries. Provisionally, just to fix the sense of the argument, we may say that a queer arithmetic is a system of propositions about natural numbers in which we have a different multiplication table" (Castaneda, 1959).

We see that Castaneda and later Rashevsky suggested that unusual arithmetics would be analogous to non-Euclidean geometries although Castaneda tried to argue that Gasking's claim that "queer" arithmetics describe reality was ungrounded.

An important observation of Castaneda is that "counting is, in fact, the simplest form of measuring, and every measuring includes it or presupposes it" (Castaneda, 1959).

In the same venue, the well-known American mathematician and philosopher Philip Davis (1923–2018) posed a challenging question: Is

One and One Really Two? (Davis, 1972). In his paper, Davis writes that the "ordinary arithmetic is one of the most elementary of the mathematical disciplines" while "among the theorems of arithmetic are various sums" meaning such equalities as $1 + 1 = 2$, $33 + 67 = 100$ or $11111 + 22222 = 33333$. With this in mind, he arrives to the conclusion that "the arithmetic of [meaning *operations with*, MB] excessively large numbers can be carried out only with diminishing fidelity." In essence, it means that while operations with small and medium numbers are performed in the same way as it is done in the conventional Diophantine arithmetic, correct representation of operations with "excessively large" numbers demands new arithmetic.

Boran Berčić also asked, what made it true that $2 + 2 = 4$? (Berčić, 2005). As if to answer this question, Poincaré suggested that arithmetics had to be tested by experiments pointing out that one did not "prove" $2 + 2 = 4$, but one "checked" it (Gonthier, 2008).

Thus, we can see that there are many situations when the conventional Diophantine arithmetic does not correctly describe situations that emerge in science and exist in everyday life. This gives birth to a problem of finding or constructing arithmetics, which correctly represent these situations. This problem implicitly existed almost for two and a half millennia. In the 20th century, it was explicitly formulated by some researchers (cf., for example, Gasking, 1940; Kline, 1967; Rashevsky, 1973; Rotman, 1997). However, the belief in the uniqueness of the Diophantine arithmetic was so strong that it took a lot of time at first to understand the problem and then to solve it near the end of the 20th century. The discovery of non-Diophantine arithmetics took more time than the discovery of non-Euclidean geometries due to the following reasons. First, the conventional Diophantine arithmetic was so solidly ingrained in the mentality of people that it was extremely hard to imagine other possibilities. Second, people in general and the majority of mathematicians in particular even did not understand that there is such a problem. Third, people usually try to minimize their actions, to stay where they are and even attack those who introduce innovations.

Finally, it is necessary to remark that although a quantity of various arithmetics (arithmetic of real numbers, arithmetic of complex numbers, residual arithmetic, arithmetic of algebraic numbers, nonstandard arithmetic, computer arithmetic, arithmetic of computable numbers, etc.) have appeared, followed by the deluge of diverse algebras as the further development of arithmetic, nobody claimed that those mathematical structures solved the problems with natural numbers described above. All previously introduced arithmetics and algebras went beyond natural numbers. Only non-Diophantine arithmetics of natural numbers were able to provide tools for solving these and similar problems in a rigorous mathematical way.

1.7. Discovery of Non-Diophantine Arithmetics

It would be an unsound fancy and self-contradictory to expect that things which have never yet been done can be done except by means which have never yet been tried.

Francis Bacon

Although ancient Greeks discovered some apparent disparities between the standard Diophantine arithmetic and some real-life situations involving (natural) numbers, it took almost 2.5 millennia to discover (construct) the first classes of non-Diophantine arithmetics.

Writing about the discovery of non-Diophantine arithmetics, we allow the reader to choose how to understand innovations in mathematics. Those who believe in the Platonic realm of mathematical structures (cf., for example, Bernays, 1935; Mazur, 2008a; Burgin, 2017c) can assume that non-Diophantine arithmetics were discovered. Those who deem mathematics only as a remarkable creation of the human mind can assume that non-Diophantine arithmetics were invented and constructed. Those who advocate a dualistic approach to mathematics, suggesting that some parts of it exist and are discovered while other parts are created by mathematicians, can apply their approach to non-Diophantine arithmetics, reasoning that non-Diophantine arithmetics were constructed but their properties were discovered.

Note that now the existence of the Platonic realm is scientifically clarified, explained and validated as the world of abstract structures (Burgin, 2012; 2017b). Mathematical structures, as well as other structures, form a part of this world as it is demonstrated in (Burgin, 1994; 2017c; 2018). As a result, mathematics exists as a scientific field, such as physics or biology, having theoretical, experimental and applied components. However, while the domain of physics and biology are fragments of the material world, the domain of mathematics lies in the World of Structures encompassing all mathematical structures.

In any case, in 1973, the renowned Russian mathematician Pyotr Konstatinovich Rashevsky (1907–1983) published a paper, in which he described the paradox of a heap and explicitly formulated the problem of the necessity of building arithmetics of natural numbers different from the conventional one (Rashevsky, 1973).

This paper had an interesting history as one of the authors of the book learned much later. Rashevsky was a member of the Editorial Board of one of the main mathematical journals in the Soviet Union. Naturally, after writing his paper, he brought it to this journal called "*Uspehi Matematicheskih Nauk*," which in English means *Achievements of Mathematics*. However, the other members of the Editorial Board did not want to publish such a radical text. Rashevsky insisted and at last a compromise was achieved. The paper was published outside the main body of the journal's content, in a much smaller font in comparison with other papers in that issue, with a notice that the Editorial Board is not responsible for its content and the paper was published only for a discussion.

Mark Burgin read that paper and started thinking how it might be possible to build arithmetics of natural numbers, which could be different from the conventional Diophantine one. His main contention was that if it had been possible to discover/construct non-Euclidean geometries, then it might be feasible to find/elaborate non-Diophantine arithmetics. Indeed, the first class of non-Diophantine arithmetics of natural numbers called perspective or more exactly, direct perspective arithmetics was discovered (constructed) by Burgin in 1975. However, it was not easy to publish such a revolutionary result. From the history of mathematics, we know that in the case of non-Euclidean geometries, the top mathematician of his time Carl Friedrich Gauss did not want to publish his results on this topic understanding that bigots would attack him, the great Russian mathematician Nikolay Ivanovich Lobachevsky (1792–1856) published his results in the Surveys of the university where he was a Chancellor, and the great Hungarian mathematician János Bolyai (1775–1856) was able to publish his work only as a very short attachment in his father's book.

Naturally, the first paper of Burgin on non-Diophantine arithmetics was published in 1977 in a very compressed form (Burgin, 1977). In that paper, he suggested calling these arithmetics non-Diophantine. It was reasonable to call the conventional arithmetic of natural numbers by the name the *Diophantine arithmetic* due to the foremost contribution of the ancient Greek mathematician Diophantus specifically to arithmetic of natural numbers, making first steps in its transformation into algebra.

It is important to understand that non-Diophantine arithmetics do not change numbers but employ new operations with these numbers. For instance, any non-Diophantine arithmetic of natural numbers contains the same natural numbers as the Diophantine arithmetic but addition and/or multiplication of them can be very different. Moreover, as we show in Section 2.11, non-Diophantine arithmetics of natural numbers satisfy all five Peano axioms for natural numbers.

Note that although a multitude of various arithmetics, such as the arithmetic of real numbers, arithmetic of complex numbers, residual arithmetic, computer arithmetic and arithmetic of integer numbers, as well as the huge amount of algebras have appeared as the further development of the arithmetic of natural numbers, neither of them solved the problems with the Diophantine arithmetic discovered by different researches, such as Helmholtz, Lebesgue, Kline and Rashevsky. A part of these problems was described in the previous section. Only non-Diophantine arithmetics provide efficient means for solving these problems in a rigorous mathematical way (Burgin, 2001; 2001a).

The second class of non-Diophantine arithmetics of natural numbers called dual or, more exactly, dual perspective arithmetics, was discovered (constructed) by Mark Burgin in 1979 and published in 1980, in an even more compressed form (Burgin, 1980). Only much later, these results were included in the book "*Non-Diophantine Arithmetics or what number is* $2+2$" in a more detailed form (Burgin, 1997).

Contradicting the knowledge accumulated by mathematicians through millennia of research as well as going up against the mundane experience of all people, non-Diophantine arithmetics drastically change people's understanding not only of mathematics but also of the whole world because numbers reflect reality. In spite of this (or, maybe, because of this), the discovery of non-Diophantine arithmetics was not noticed by the majority of mathematicians and philosophers of mathematics (not speaking about general public) since it contradicted to such a big extent the conventional knowledge. Usually, instead of changing what they learned from the very young age, people prefer not to see new groundbreaking discoveries.

Being a geometer, Rashevsky, as well as the philosopher Castaneda, predicted that new arithmetics would form parametric families as it had been with non-Euclidean geometries. Namely, he wrote:

> "*It must not be expected that* [such a, MB] *hypothetical theory* [of new arithmetics, MB], *if it would be ever destined to see the light of day, will be unique; on the contrary, it will have to depend on certain "parameters" (with a role distantly reminiscent of the radius of Lobachevsky space when we repudiate Euclidean geometry in favor of non-Euclidean). It may be*

expected that in the limiting case the hypothetical theory should coincide with the existing one." (Rashevsky, 1973)

These predictions of Rashevsky came out true. The first and second discovered classes of non-Diophantine arithmetics formed parametric families (classes) although the parameter was not numerical as in the case of non-Euclidean geometries but functional. It means that in any such arithmetic, properties and laws of its operations depend on a definite function $f(x)$ (cf. Chapter 2). The conventional, Diophantine arithmetic is a member of both parametric families with the parameter equal to the identity function $f(x) = x$. This distinction of arithmetic from geometry shows higher fundamentality of arithmetic in comparison with geometry.

The second prediction of Rashevsky was also validated because in the limiting case, i.e., when the functional parameter was the identity function $f(x) = x$, the member of the parametric family coincided with the conventional Diophantine arithmetic.

It is interesting to know that in contemporary mathematical terms, the technique for building the first families (classes) of non-Diophantine arithmetics is based on mathematical structures called fibered spaces and bundles, which appeared only in the 20th century. These structures are utilized as a tool for operating with numbers in non-Diophantine arithmetics. A similar technique is used for defining operations with hypernumbers when it is impossible to define these operations straightforwardly in the standard fashion (Burgin, 2010a; 2011; 2012c; 2015). This technique is also used by the prominent physicist Paul Benioff for scaling in gauge theory and geometry (Benioff, 2011; 2012; 2012a; 2014; 2015; 2016). This scaling is defined for addition and multiplication by two functions: $h(x) = (1/n)x$ and the identity function $g(x) = x$ from $\boldsymbol{R}$ into $\boldsymbol{R}$ or from $\boldsymbol{C}$ into $\boldsymbol{C}$. This construction of scaling is inherently based on some types of non-Diophantine arithmetics and prearithmetics described in this book. In essence, Benioff's method implicitly utilizes non-Diophantine arithmetics with non-standard multiplication in sets of real and complex numbers without changing addition of these numbers.

The history of non-Diophantine arithmetics clearly resembles what happened with non-Euclidean geometries. Indeed, for quite a while, the discovery of non-Euclidean geometries was not known to the prevalent part of the mathematical community. The same is going on with non-Diophantine arithmetics. For instance, more than 20 years after the discovery of non-Diophantine arithmetics, the American mathematician Brian Rotman put forward the problem to elaborate arithmetics essentially different from the conventional one (Rotman, 1997). He based his suggestion on a series of examples demonstrating that many laws of the conventional arithmetic are not true in different real-world situations. Rotman called those hypothetical structures non-Euclidean arithmetics, although he did not describe them. However, it is more natural to call the conventional arithmetic by the name the *Diophantine arithmetic* than by the name the *Euclidean arithmetic* because Diophantus contributed much more to the development of the arithmetic of natural numbers than Euclid. Consequently, new arithmetics of natural numbers acquired the name *non-Diophantine arithmetics.*

In the same way as it was with the Euclidean geometry, the Diophantine arithmetic was unique and non-challengeable for a very long time when people did not known other arithmetics. Its position in human society has been and is now even more stable and firm than the position of the Euclidean geometry before the discovery of the non-Euclidean geometries. Really, all people use the Diophantine arithmetic for counting. Utilization of numbers and arithmetical operations make all people some kind of consumers of mathematics. At the same time, Euclidean geometry is only studied at school and in real life rather few specialists use it. It is arithmetic, and not geometry, which is considered as a base for the whole mathematics in the intuitionistic approach. As a result, the discovery of non-Diophantine arithmetics changes our understanding of the world similar to the transformation in minds of people caused by the discovery of non-Euclidean geometries. For millennia, people in general and mathematicians, in particular, believed there was only one geometry, absolute and supreme. The discovery of non-Euclidean geometries in the 19th century disproved this misconception demonstrating that the world is much more opulent than it seemed before. In a similar way, for millennia, people in general and mathematicians in particular believed there was only one arithmetic, absolute and ultimate. The discovery of non-Diophantine arithmetics in the 20th century disproved this misconception demonstrating that the world was much more affluent and plentiful than it had seemed before.

At the same time, it is necessary to correctly understand ontological implications of these two discoveries. One may have an impression that the discoveries undermine the existence of the independent mathematical structures, which dwell in the eternal immutable World of Structures or Plato Ideas. For instance, mathematical Platonists argue that "2 plus 2 equals 4" is an eternal truth that would be true even if the Big Bang had never occurred and the universe did not exist. In contrast to this, there are non-Diophantine arithmetics where "2 plus 2 equals 5" (cf., Chapter 2).

However, the eternal truth of "2 plus 2 equals 4" still remains but with a necessary modification. Namely, what is true and unchangeable is

"*There is a mathematical structure called the Diophantine arithmetic, in which* 2 *plus* 2 *equals* 4"

or

"2 *plus* 2 *equals* 4 *in the Diophantine arithmetic*"

It is interesting that when non-Euclidean geometries were discovered, their discoverers, Gauss and Lobachevsky, performed physical experiments trying to find whether such geometries exist in reality (Livanova, 1969). In contrast to this, people in general and mathematicians in particular have already used and are using some kinds of non-Diophantine arithmetics. Moreover, there are many examples demonstrating that various practical calculations are performed according to the rules of non-Diophantine arithmetics and people unconsciously use them. In spite of this evidence, acceptance of non-Diophantine arithmetics is extremely slow due to the powerful bias that "2 plus 2 equals 4" is an eternal truth, which does not depend on any context.

Power of people's stereotypes is vividly demonstrated by the book (Blehman *et al.*, 1983). At first (in Section 1.2.4), the authors of that book explain with many examples and references that our intuition of natural numbers and arithmetic can be very misleading in various situations. After this (in Section 1.2.5), they announce that it is completely impossible that two times two is not equal to four. The authors are even trying to prove this utilizing a probabilistic reasoning. Here we can see their arguments (Blehman *et al.*, 1983: p. 50).

> *Really, the statement that two times two is equal to four may be taken as an example of the most evident truth. Although, nobody doubts that this is a true equality, it is possible to evaluate formally probability that in reality two times two is equal to five, while the standard statement that two times two is equal to four is a result of a constantly repeated arithmetical mistake. Let us suppose that any individual performing multiplication with numbers that are less than ten can decrease the result by one with the probability* 10^{-6}. *This corresponds to several such mistakes during his or her life. If we assume that through the whole history of mankind,* 10^{10} *people performed the multiplication "two times two"* 10^{6} *times during the*

life of each of them, then the probability that they repeated this mistake of decreasing the result is less than $10^{-10^{17}}$. *Thus,* the authors conclude, *the probability is so small that the event is absolutely impossible and we see that two times two is equal to four.*

This is an explicit example of incorrect probabilistic reasoning.

The most famous example of stating that all people saw something and concluding that it is the absolute truth, is attributed to Aristotle. He asserted

All swans are white.

All people known to Aristotle saw only white swans. He saw only white swans. So, Aristotle gave this as an example of an absolute truth. Based on their experience, Europeans had believed in this until they came to Australia where they found black swans and disproved this statement.

The same change happened with the statement

Two plus two is equal to four.

Indeed, all people have known this from their early childhood. So, everybody believed that this is an absolute truth. However, when non-Diophantine arithmetics were discovered, it was found that in some of them two plus two is not equal to four. We will see this in Chapter 2.

When we consider non-Diophantine arithmetics, it is possible to think that they are absolutely formal constructions like many other mathematical objects, which are very far from the real world. However, let us recollect that the discovery of non-Euclidean geometries met similar skepticism and mistrust. Carl Friedrich Gauss understood very well this situation with extreme innovations and in spite of being acknowledged as the greatest mathematician of his time, did not dare to publish his results concerning non-Euclidean geometries. Another reason was that he was not able to find anything that was similar to them in nature. Lobachevsky called his geometry imaginable. Nevertheless, it was demonstrated later that the real physical space fitted non-Euclidean geometries, and that the Euclidean geometry did not have such essential applications as the non-Euclidean ones.

In this respect, the situation with non-Diophantine arithmetics is different. In spite of the relatively short time, which has passed after their discovery, it has been demonstrated that many real phenomena and processes that match the non-Diophantine arithmetics exist (cf., for example, Burgin, 1992; 2001; 2001a; 2007; Czachor, 2015; 2017; 2017a; Burgin and Meissner, 2017; Tolpygo, 1997). Moreover, mathematicians and scientists implicitly used some of non-Diophantine arithmetics without understanding their essence.

Non-Diophantine arithmetics can possess many unusual properties. For instance, recently the expressions $1 + 1 = 3$ and $2 + 2 = 5$ have become a popular metaphor for synergy in a variety of areas: in business and industry (Beechler, 2013; Brown, 2015; Grant and Johnston, 2013; Jude, 2014; Kress, 2015; Ritchie, 2014; Marks and Mirvis, 2010), in economics and finance (Burgin and Meissner, 2017), in anthropology, psychology and sociology (Boksic, 2017; Brodsky, 2004; Bussmann, 2013; Enge, 2017; Frame and Meredith, 2008; Mane, 1952), library studies and informatics (Marie, 2007), creativity (Trott, 2015), biochemistry and bioinformatics (Kroiss *et al.*, 2009), theory and practice of organizations (Klees, 2006), technology (Gottlieb, 2013), computer science (Derboven, 2011; Glyn, 2017; Lea, 2016), networking (Meiert, 2015), physics (Lang, 2014), biology and medicine (Lawrence, 2011; Trabacca *et al.*, 2012; Archibald, 2014; Phillips, 2016), agriculture (Riedell *et al.*, 2002), pedagogy (Nieuwmeijer, 2013) and politics (Van de Voorde, 2017).

Other expressions from non-Diophantine arithmetics also appear in scientific publications such as $1 + 1 = 1$ (Carroll and Mui, 2009; Morris, 2017), $1 + 1 = 4$ (Flegenheimer, 2012), *one plus one makes more than two* (Pascoe, 2017), *one plus one equals three-fourths* (Ries, 2014), *one plus one equals one and a half* (Covey, 2004), *negative one plus negative one equals negative three* (Katsenelson, 2015) and $2 + 2 = 5$ (Cambridge Dictionary).

At the same time, those who use the expressions $1 + 1 = 3$ and $2+2 = 5$ think that it is incorrect mathematics because in the Diophantine arithmetic, $1+1 = 2$ and $2+2 = 4$. They believe synergetic relations defied the laws of mathematics. Some people even call them anti-mathematical formulas. However, the discovery of non-Diophantine arithmetics demonstrates that synergetic relations defy only the laws of the Diophantine arithmetic but not of mathematics because in mathematics there are non-Diophantine arithmetics, in which $1 + 1 = 3$ or $2 + 2 = 5$.

In addition to this, non-Diophantine arithmetics solve some problems that remained unsolved from the time of ancient Greece, just to mention the "paradox of a heap" we have already encountered in the previous section. Indeed, the heap is not changing if we add one grain. Consequently, if we take the number k of the grains in the heap, then adding 1 to

k does not change k. This contradicts the main law of the Diophantine arithmetic stating that for an arbitrary number k, the number $k+1$ is not equal to k, and gives birth to a paradox if we have only one arithmetic. Non-Diophantine arithmetics solve the paradox because in some of them $k+1=k$ for many numbers k.

In a similar way, paradoxes from ancient Greece are sometimes revived in modern physics. For example, theory of chaos encounters many difficult problems caused by insufficiency of modern discrete mathematics to correctly represent chaotic dynamics. Arbitrary small changes in external parameters or/and initial conditions cause essential changes in the behavior of a system, making the classical difference calculus inefficient for simulating a chaotic motion (Gontar and Ilin, 1991). The problem leads us, as emphasizes Vladimir Gontar, to the paradoxes of ancient Greeks: is it mathematically and logically possible to formulate a contradiction-free description of the process of approaching an object when the distance to this object contains an infinite number of segments, involving an infinite number of steps necessary to reach this object? (Gontar, 1993)

Non-Diophantine arithmetics suggest a new understanding of this problem. It is possible that we can approach the object only to a definite distance, a kind of minimal length. It is also feasible to do this in a finite number of steps. All consequent steps cannot make the distance to the object smaller. For instance, it might be impossible to go above 1000 by adding 1. In terms of non-Diophantine arithmetics, this means that 1000 is much greater than 1 (formally $1000 \gg 1$), that is, by adding 1 to 1 to 1 and so on, it is impossible to get number larger than 1,000, and there are non-Diophantine arithmetics where this is true.

In essence, some of non-Diophantine arithmetics possess similar properties to those of transfinite numbers arithmetics built by the great German mathematician Georg Cantor (1845–1918). For example, a non-Diophantine arithmetic may have a sequence of numbers $a_1, a_2, \ldots, a_n, \ldots$ such that for any number b that is less than some a_n the equality $a_n + b = a_n$ is valid. This is an important property of infinity, which is formalized by transfinite (cardinal and ordinal numbers). The equality $a^2 = a$ is another interesting property of some transfinite numbers. This equality may be also true in some non-Diophantine arithmetics. Thus, non-Diophantine arithmetics provide mathematical models in which finite objects — natural numbers — acquire features of infinite objects — transfinite numbers. In such a way, it is possible to model and to describe behavior of infinite entities in finite domains.

Besides, working with numbers, different automata change the Diophantine arithmetic and use non-Diophantine arithmetics. For example, computer arithmetic is a special case of non-Diophantine arithmetics (Parhami, 2010). This is a result of round-off procedures and existence of the largest number in this arithmetic. Consequently, if we want to build better models for numerical computations than we have now, it is necessary to utilize relevant non-Diophantine arithmetics in these models.

Examples also demonstrate that non-Diophantine arithmetics are important for business and economics. Some economical problems and inconsistencies caused by the conventional arithmetic are considered in (Tolpygo, 1997). As some studies of economy show, sometimes finite quantities possess properties of infinite numbers with respect to people's practice (cf., for example, Birkhoff and Barti, 1970). Consequently, when one applies mathematics to solve such problems, the results are often mathematically correct but practically misleading. Utilization of non-Diophantine arithmetics eliminates those problems and inconsistencies.

There are other features of non-Diophantine arithmetics, which are different from the properties of the Diophantine arithmetic. For instance, we know from school that the main laws of the Diophantine arithmetic are:

1. Addition is commutative, i.e., $a + b = b + a$.
2. Multiplication is commutative, i.e., $a \cdot b = b \cdot a$.
3. Addition is associative, i.e., $(a + b) + c = a + (b + c)$.
4. Multiplication is associative, i.e., $(a \cdot b) \cdot c = a \cdot (b \cdot c)$.
5. Multiplication is distributive with respect to addition, i.e., $a \cdot (b + c) = a \cdot b + a \cdot c$.
6. Zero is a neutral element with respect to addition, i.e., $a + 0 = 0 + a = a$.
7. One is a neutral element with respect to multiplication i.e., $a \cdot 1 = 1 \cdot a = a$.

Naturally, we can ask whether these laws are valid for non-Diophantine arithmetics. Exploring them we find that addition and multiplication are always commutative. However, zero is not always a neutral element with respect to addition and one is not always a neutral element with respect to multiplication in non-Diophantine arithmetics. At the same time, the laws of associativity and distributivity fail in the majority of non-Diophantine arithmetics. Only special conditions on the functional parameter of the non-Diophantine arithmetic in question provide validity of these laws (Burgin, 1997).

Besides, the Diophantine arithmetic possesses the so-called Archimedean property, which is important for proofs of many results in arithmetic and number theory. It states that if we take any two natural numbers m and n, in spite that n may be enormously larger than m, it is always possible to add m enough times to itself, i.e., to take the sum $m + m + \cdots + m$, so that the result will be larger than n. This property is also invalid in the majority of non-Diophantine arithmetics. The Archimedean property is important for proving that sets of all natural and prime numbers are infinite. Thus, having in general no Archimedean property in non-Diophantine arithmetics, we encounter such arithmetics that have only a finite number of elements, or such infinite arithmetics that have only a finite set of prime numbers (Burgin, 1997).

One more unusual property of non-Diophantine arithmetics is related to physics. Physicists often use the relation $a \ll b$, which means that a is much smaller than b. However, this relation does not have an exact mathematical meaning and is used informally. In contrast to this, non-Diophantine arithmetics provide rigorous interpretation and formalization for such relations. Namely (cf., Section 2.9), $a \ll b$ if and only if $b + a = b$.

Note that this is impossible in the conventional mathematics because for any number $b > 0$, the sum $b+b$ is larger than b. At the same time, there are non-Diophantine arithmetics, in which $b+a = b$ is true for different numbers a and b when $0 < a < b$, i.e., $a \ll b$. Arithmetics with this property is considered in Sections 2.9 and 2.10.

Interestingly, this property reflects some basic features of nature. Physicists (cf., for example, Penrose, 1972; Zeldovich *et al.*, 1990) emphasize that fundamental problems of modern physics are dependent on our ways of counting and calculation. Mathematicians also call attention to connections between the worlds of quantum physics and number theory expressing even more radical view that number theory can be treated as the ultimate physical theory (cf., for example, Varadarajan, 2002; 2004; Volovich, 2010).

This idea correlates with problems of modern physical theories in which physical systems are described by chaotic processes. Taking into account the fact that chaotic solutions are obtained by computations, physicists ask (Cartwrite and Piro, 1992; Gontar, 1997) whether chaotic solutions of the differential equations, which model different physical systems, reflect the dynamic laws of nature represented by these equations or whether they are solely the result of an extreme sensitivity of these solutions to numerical procedures and computational errors.

It is even clearer that properties of non-Diophantine arithmetics, which reflect the way people count, influence functioning of economy and are important for economical models (cf., for example, Tolpygo, 1997; Burgin and Meissner, 2017). Thus, it would be useful to build models of economical systems and processes employing an appropriate non-Diophantine arithmetic.

One of the interesting properties of projective arithmetics is that they allow us to formalize and make rigorous concepts such as *much smaller* (denoted by $\ll$) and *much larger* (denoted by $\gg$). Namely, we have the following definition:

A number m is *much smaller* than a number n ($m \ll n$) if $n + m = n$.

In this case, the number n is *much larger* than the number m ($n \gg m$).

There are many non-Diophantine arithmetics, which are essentially different from the Diophantine arithmetic $\boldsymbol{N}$. For instance, we know that there are infinitely many prime numbers in $\boldsymbol{N}$. At the same time, it is proved that for any $n > 0$, there is a non-Diophantine arithmetic that has exactly n prime numbers (Theorem 2.9.25) and there are infinitely many non-Diophantine arithmetics with only one prime number (Theorem 2.9.28). We know that in $\boldsymbol{N}$, half of numbers are even and the other half consists of odd numbers. At the same time, it is proved (Theorem 2.9.29) that there are infinitely many non-Diophantine arithmetics with only one odd number.

According to Fermat's Last Theorem, also called Fermat's conjecture, which was proved by English mathematician Andrew Wiles with the help of Richard Taylor (Taylor and Wiles, 1995; Wiles, 1995), the equation $x^n = y^n + z^n$ cannot have positive integer solutions for any natural number n greater than 2. At the same time, it is proved (Theorem 2.9.34) that there are infinitely many non-Diophantine arithmetics of natural numbers, in which for any natural number n, the equation $x^n = y^n + z^n$ has infinitely many solutions.

Non-Diophantine arithmetics also allow elimination of several inconsistencies and misconceptions related to arithmetic. For instance, Rosinger explains

"... we have been doing inconsistent mathematics for more than half a century by now, and in fact, have quite heavily and essentially depended on it in our everyday life. Indeed, electronic digital computers, when considered operating on integers, which is but a part of their operations, act according to the system of axioms given by

- (PA): the usual Peano Axioms for $\boldsymbol{N}$,

plus the *ad-hoc* axiom, according to which

- (MI): there exists M in $\boldsymbol{N}$, $M \gg 1$, such that $M + 1 = M$

Such a number M, called "machine infinity", is usually larger than 10^{100}, however, it is inevitably inherent in every electronic digital computer, due to obvious unavoidable physical limitations, and clearly, the above mix of (PA) + (MI) axioms is inconsistent. Yet we do not mind flying on planes designed and built with the use of such electronic digital computers." (Rosinger, 2008)

In a similar way, Meyer and Mortensen built various inconsistent models of arithmetic (Meyer and Mortensen, 1984), while Priest developed axiomatic systems for inconsistent arithmetics (Priest, 1997; 2000).

Even before Priest, Van Bendegem developed an inconsistent axiomatic arithmetic by changing the Peano axioms so that a number a that is the successor $a + 1$ of itself exists (Van Bendegem, 1994). At the same time, the fourth Peano axiom states that if $x + 1 = y + 1$, then x and y are the same number. In the system of Van Bendegem, starting from some number n, all its successors will be equal to n. Then the statement $n = n + 1$ is considered as both true and false at the same time. This makes the new arithmetic inconsistent. It is possible to rigorously eliminate these inconsistencies using non-Diophantine arithmetics.

In general, there are two basic ways to deal with inconsistencies: one is to elaborate an inconsistent system and try to work with it and another way is create new mathematical structures, eliminating inconsistencies. Ambiguities, contradictions, and paradoxes play the central role in the emergence of creative ideas. According to Byers (2007) one of the main kinds of contradictions is existence of two seemingly contradictory perspectives in a mathematical problem. For instance, Peano axioms imply infiniteness of the arithmetic, while the existence of the largest number implies it finiteness. However, this paradox vanishes with the discovery of non-Diophantine arithmetics and weak arithmetics. Actually, all these inconsistencies and contradictions exist only in the absence of non-Diophantine arithmetics. For instance, the machine arithmetic analyzed by Rosinger (2008) does not satisfy Peano axioms not because it is inconsistent, but since it is non-Diophantine. In essence, the machine arithmetic satisfies the axioms of the corresponding non-Diophantine prearithmetic (cf. Chapter 2), and operating on integers, electronic digital computers perform according to this system of axioms.

The rule $n = n + 1$ of Van Bendegem, which is considered above, is natural for many non-Diophantine arithmetics and causes no inconsistencies and contradictions there. It simply means $1 \ll n$.

It is possible to compare this situation with artificially derived inconsistencies and contradictions with numbers when people knew only natural numbers and positive fractions. Getting information about negative numbers, mathematicians who lived at that time would be able to build an inconsistent formal system by taking two "axioms":

- Only positive numbers exist;
- There are negative numbers.

Naturally these "axioms" give a contradiction. Now, we know that the first "axiom" is valid only for natural numbers and positive fractions if we consider numbers known at that time. Thus, integer numbers combine both positive and negative numbers without any inconsistency.

Mathematicians who lived in the 19th century and earlier were also able to build an inconsistent formal system in geometry, combining together two sets of axioms:

- All postulates of the Euclidean geometry.

and the postulate that is true for the geometry on a sphere, which is considered a geometrical model of the Earth:

- Any two straight lines intersect with one another.

The last axiom is true for the geometry on a sphere, which is considered a geometrical model of the Earth:

Now we know that spherical geometry is non-Euclidean and does not have any contradictions in it.

Kant asserted that the world described by science was a world of sense impressions organized and controlled by the mind in accordance with innate categories of space and time (Kant, 1786). Consequently, continued Kant, there never would be a world's description other than Euclidean geometry and Newtonian mechanics.

One may say that Kant was a philosopher and not an expert in mathematics. However, William Rowan Hamilton (1805–1865), certainly one of the outstanding mathematicians of the 19th century, expressed similar consideration in 1837 when the works of Lobachevsky (1829) and Bolyai (1932) had been already published but were not known to the majority of mathematicians. Hamilton wrote:

"No candid and intelligent person can doubt the truth of the chief properties of *Parallel Lines*, as set forth by Euclid in his *Elements*, two thousand years ago; though he may well desire to see them treated in a clearer and better method. The doctrine involves neither obscurity nor confusion of thought, and leaves in the mind no reasonable ground for doubt, although ingenuity may usefully be exercised in improving the plan of the argument."

Even in 1883, another famous mathematician Arthur Cayley (1821–1895) in his presidential address to the British Association for the Advancement of Science affirmed:

> "*My own view is that Euclid's twelfth axiom* [usually called the fifth or parallel axiom or postulate, MB] *in Playfair's form of it does not need demonstration, but is part of our notion of space, of the physical space of our experience...*"

Another important issue is related to attempts of some mathematicians to build new physical theories based exclusively on numbers suggesting to study physical reality at the Planck scale using number fields of p-adic numbers as the fundamental physical objects (Volovich, 1987). However, it is possible to suggest that non-Diophantine arithmetics, which have functional parameters providing adequate adjustment of the theory to different physical scales, are more relevant for representing quantum objects at the Planck scale than p-adic numbers.

This situation is entirely mirrored in the contemporary mathematics. Mathematicians assume that arithmetic, which comprises mostly representation and operation with numbers, is something completed long ago, trivial and aimed at the beginners, while number theory, which comprises investigation of properties of natural numbers, is a respected field of theoretical mathematics with its highly complicated, "deep" and abstract problems. However, knowing only one arithmetic of natural numbers, mathematicians forget that this arithmetic (in the modern sense) is the base for theory of numbers. If you change arithmetic, you will need new theory of numbers as numbers will change their properties (Burgin, 2018c). For example, the definition of a prime number is completely based on the definition of the operation of multiplication and as it is proved in Section 2, for any $n = 2, 3, 4, 5, \ldots$, there are non-Diophantine arithmetics, in which there are exactly n prime numbers while in the Diophantine arithmetic, there are infinitely many prime numbers.

To conclude, it is necessary to remark that non-Diophantine arithmetics were discussed by Amat Plata (2005) and explicitly applied to physics (Czachor, 2016; 2017; 2017b; Czachor and Posiewnik, 2016), psychology (Czachor, 2017a), fractal theory (Aerts *et al.*, 2016; 2016a; 2018), economics and finance (Burgin and Meissner, 2017). There are also implicit applications of non-Diophantine arithmetics to physics (Benioff, 2011; 2012; 2012a; 2013; 2015; 2016; 2016a) and economics (Tolpygo, 1997).

In addition, researchers used non-Diophantine arithmetics and prearithmetics for the development and application of different kinds of non-Newtonian calculi as it is described in the next section.

Besides, as it is demonstrated in Section 2.6, utilization of logarithmic scales is also an implicit application of non-Diophantine arithmetics and prearithmetics.

1.8. Non-Diophantine Operations and Related Constructions in the History of Mathematics

Arithmetic must be discovered in just the same sense in which Columbus discovered the West Indies, and we no more create numbers than he created the Indians.

Bertrand Russell

In this section, we discuss mathematical constructions related to non-Diophantine arithmetics explicating and exploring their roots in the history of mathematics. As we will see, elements of non-Diophantine arithmetics appeared in ancient times but it took millennia to explicitly discover (construct) these unusual arithmetics.

The most basic class of numbers — natural numbers — came from practical activity of people, which involved counting. Counting, in turn, brought people to arithmetical operations of addition and multiplication, which resulted in creation (discovery) of the Diophantine arithmetic, or more exactly, the Diophantine arithmetic of natural numbers. The subsequent development of arithmetic went by adding new types of numbers.

That is why all these arithmetics naturally bear the common name the Diophantine arithmetic. When 0 was added, the Diophantine arithmetic of whole numbers was created. When that arithmetic was extended by negative numbers, the Diophantine arithmetic of integer numbers emerged. Introduction of rational numbers gave birth to the Diophantine arithmetic of rational numbers. Construction of real numbers from rational numbers brought mathematicians to the Diophantine arithmetic of real numbers while the discovery of complex numbers resulted in the Diophantine arithmetic of complex numbers.

However, while writing this book, the authors found that although Diophantine arithmetics were discovered at the end of the 20th century, elements of non-Diophantine arithmetics appeared millennia ago. Namely, the first appearance of operations with numbers, which were different from operations in the conventional (Diophantine) arithmetic of natural numbers, can be tracked to the Pythagorean theorem, which has the name of the famous Greek mathematician Pythagoras (sixth century B.C.E.) although it was known in ancient Babylon and Egypt between 2000 and 1700 B.C.E. (Neugebauer, 1969), that is, more than 1000 years before Pythagoras was born.

The Pythagorean theorem states that the square of the hypotenuse is equal to the sum of the squares of the other two sides, i.e., $c^2 = a^2 + b^2$ and $c = (a^2 + b^2)^{1/2}$.

At the same time (cf. Section 3.3), in the non-Diophantine arithmetic $\boldsymbol{A}$ that is exactly projective with respect to the Diophantine arithmetic $\boldsymbol{R}^+$ of non-negative real numbers with the generator $f(x) = x^2$, addition is defined by the following formula:

$$a \oplus b = \sqrt{a^2 + b^2} = (a^2 + b^2)^{1/2}.$$

This is exactly the expression of the Pythagorean theorem. It means that, in the context of non-Diophantine arithmetics, we have the following form of the Pythagorean theorem:

The length of the hypotenuse is equal to the non-Diophantine sum of the other two sides.

This shows that implicitly non-Diophantine arithmetics have been used for millennia by mathematicians and engineers.

Later a similar formula based on the non-Diophantine sum was used for defining Euclidean distance on a plane. This explicitly demonstrates that non-Diophantine arithmetics form the foundation of the Euclidean geometry in two and as it is possible to show, in more dimensions.

Interestingly, there is also geometry on a plane, which uses the Manhattan, or taxicab, distance (cf., for example, Krause, 1987) where the distance between two points in a plane is expressed by the formula

$$d((a,b),(c,d)) = |c-a| + |d-b|.$$

Consequently, this geometry is based on the Diophantine arithmetic.

Appearance of other mathematical systems called *arithmetic* did not change the predominant position of the Diophantine arithmetic of natural numbers. The first of those mathematical systems was linked to counting.

While counting different things, people saw that in some situations counting was periodic involving repetition of the same numbers, for example, as it happened in counting of days in a year or hours in a day when after 12 o'clock hours start repeating. This observation resulted in appearance of another type of numbers and operations called *modular arithmetics*. They were the first arithmetics different from the conventional (Diophantine) arithmetic of natural numbers.

A *modular arithmetic*, which is also called *residue arithmetic* or *clock arithmetic*, $\boldsymbol{Z}_n$ is studied in mathematics and it was used in physics and computing. In modular arithmetic, operations of addition and multiplication form a cycle upon reaching a certain value, which is called the *modulus*. Examples of the use of modular arithmetic occur in ancient Chinese, Indian, and Islamic cultures. For instance, written more than 1800 years ago the Chinese text Sun Zi Suan Jing described the solutions to a polynomial equation in modular arithmetics, naturally, without using the term *modular arithmetic*. The rigorous approach to the theory of modular arithmetic was worked out by Carl Friedrich Gauss.

Their construction goes as follows. A natural number n is chosen as the modulus. Then two natural numbers k and m are called congruent modulo n if their difference $k - m$ is divisible by n. This congruence is denoted by the expression

$$k = m \pmod{n}.$$

There are exactly n possible remainders of division by n. These remainders are called numbers or elements of the corresponding modular arithmetic $\boldsymbol{Z}_n$, which contains exactly n elements with the following operations. If numbers k and m belong to $\boldsymbol{Z}_n$, then the sum $k + m$ in

$\boldsymbol{Z}_n$ is equal to the number t from $\boldsymbol{Z}_n$ such that $k + m = t \pmod{n}$ and the product $k \cdot m$ in $\boldsymbol{Z}_n$ is equal to the number q from $\boldsymbol{Z}_n$ such that $k \cdot m = q \pmod{n}$.

In their development, modular arithmetics formed the mathematical base for p-adic numbers, which are useful in mathematics and some applications (cf., for example, Koblitz, 1977; Brekke and Freund, 1993; Gouvêa, 1997).

However, modular arithmetics did not threaten the predominant position of the Diophantine arithmetic of natural numbers because modular arithmetics did not contain all natural numbers. By the same reason, in spite of being useful in various areas in mathematics, modular arithmetics were not able to solve the problems with counting discussed in Section 1.6.

One more constructive feature of non-Diophantine arithmetics also appeared in mathematics long before non-Diophantine arithmetics were discovered (constructed). As we will see in Chapter 2, operations in non-Diophantine arithmetics are usually constructed by functional reduction to the conventional operations in the Diophantine arithmetic. Although non-Diophantine arithmetics were discovered (constructed) only at the end of the 20th century, history of mathematics showed that an important case of such reduction appeared several centuries before this discovery with the invention of logarithms by John Napier (1550–1617) in the 17th century. Logarithms allowed reduction of conventional multiplication to conventional addition and then returning back by exponentiation. Performing addition is much simpler than performing multiplication. Thus, this reduction simplified calculations by means of slide rules and logarithm tables, which were efficiently used by scientists and engineers for several centuries before proliferation of calculators and computers.

Next steps in the direction of non-Diophantine arithmetics were made only in the 20th century. According to contemporary knowledge, in a general form, non-Diophantine arithmetical operations were studied by Andrey Nikolayevich Kolmogorov (1903–1986), who introduced and studied *non-Diophantine average* (Kolmogorov, 1930). Mitio Nagumo (1905–1995) and Bruno de Finetti (1906–1985) also found and investigated this arithmetical concept (Nagumo, 1930; de Finetti, 1931). It was also used in the book (Hardy *et al.*, 1934). The ordinary arithmetical average, or mean, of n numbers $x_1, x_2, \ldots, x_n$ is defined as

$$(x_1 + x_2 + \cdots + x_n)/n.$$

However, mathematicians constructed a variety of other averages (means). In his work, Kolmogorov writes that all known types of averages have the form

$$h((g(x_1) + g(x_2) + \cdots + g(x_n))/n),$$

where g and h are some strictly monotone functions and h is the inverse of g (Kolmogorov, 1930).

This is exactly the form of average in real number prearithmetics and non-Diophantine arithmetics of real numbers with the generator g (cf. Section 3.3). This is also the form of α-average defined by the talented American mathematicians Michael Grossman and Robert Katz (1933–2010) in their non-Newtonian calculus (Grossman and Katz, 1972). Now such operations are often called *quasi-arithmetic means* (*averages*) and are studied by many authors.

Quasi-arithmetic means are efficiently used in decision-making and utility theory (Marichal, 2009).

Non-Diophantine average, which is exactly the Kolmogorov-Nagumo average, is obtained by application of non-Diophantine addition, also called *quasiaddition* and usually denoted by $\oplus$, which was introduced by Karl Menger (1902–1985) in the form of *triangular norms* (*t-norms* for short) in probabilistic metric spaces (Menger, 1942).

By definition, a *t-norm* is a function T: $[0, 1] \times [0, 1] \to [0, 1]$, which in essence, is a binary operation that satisfies the following properties:

- *Commutativity*: $\mathrm{T}(a, b) = \mathrm{T}(b, a)$;
- *Monotonicity*: $\mathrm{T}(a, b) \leq \mathrm{T}(c, d)$ if $a \leq c$ and $b \leq d$;
- *Associativity*: $\mathrm{T}(a, \mathrm{T}(b, c)) = \mathrm{T}(\mathrm{T}(a, b), c)$;
- The number 1 is the *identity element*: $\mathrm{T}(a, 1) = a$.

In the algebraic context, a t-norm is a binary operation on the closed unit interval $[0, 1]$ defining an abelian, totally ordered semigroup with the neutral identity element 1.

Berthold Schweizer (1929–2010) and Abe Sklar provided the axioms of t-norms, as they are used today (Schweizer and Sklar 1960). Many results concerning t-norms were obtained in probabilistic metric spaces.

The main ways of construction of t-norms include using generators, defining parametric classes of t-norms, rotations, or ordinal sums of t-norms. Namely, with the additive generator $f(x)$, a t-norm $T(x, y)$ is defined as

$$T(x, y) = f^{(-1)}(f(x) + f(y)).$$

This is exactly the form of addition in real number prearithmetics and non-Diophantine arithmetics of real numbers with the generator f (cf. Section 3.3).

With the multiplicative generator $h(x)$, a t-norm $T(x, y)$ is defined as

$$T(x, y) = h^{(-1)}(h(x) \cdot h(y)).$$

This is exactly the form of multiplication in real number prearithmetics and non-Diophantine arithmetics of real numbers with the generator h (cf. Section 3.3).

Mathematicians found different conditions when it is possible to represent general t-norms as the non-Diophantine addition (quasiaddition $\oplus$) or non-Diophantine multiplication (quasimultiplication $\otimes$) in the interval $[0, 1]$ (Aczel, 1966; Ling, 1995). Later these operations were extended to arbitrary finite real intervals $[a, b]$ (Maslov and Samborskij, 1992; Pap, 1995; Kolokoltsov and Maslov, 1997) and to the infinite interval $[0, \infty)$ (Marinova, 1986; Kolesarova, 1993; 1996). Note that mathematical systems with quasiaddition, which is usually denoted by $\oplus$, and conventional multiplication or with quasiaddition and quasimultiplication, which is denoted by $\otimes$ or by $\odot$, are special cases of prearithmetics studied in Chapter 2 of this book.

Researchers use t-norms as an important tool for the interpretation of the conjunction in fuzzy logics allowing evaluation of the truth degrees of compound formulas, and are useful for semantics of the intersection of fuzzy sets (Alsina *et al.*, 1983). Besides, t-norms are also utilized in fuzzy control to formulate initial assumptions called antecedents or premises in the form of conjunctions of fuzzy sets, in decision making, in statistics as well as in the theory of measure and game theory (Klement *et al.*, 2000).

The non-Diophantine addition (quasiaddition $\oplus$) was employed for defining new classes of measures and integrals, i.e., $\oplus$-measures and $\oplus$-integrals (Marinova, 1986; Kolesarova, 1993; 1996; Pap, 2002). The special case of $\oplus$-integrals when $\oplus$ is max was studied in Shilkret (1971) and later in such full-sized mathematical theory as tropical analysis (Litvinov, 2007; Speyer and Sturmfels, 2009).

The non-Diophantine addition $\oplus$ has been intensively used in psychophysics (cf., for example, Luce, 1964; 1987; 2002; Narens and Mausfeld, 1992; Czachor, 2017a). Falmagne and Doble analyzed utilization of the non-Diophantine addition and other unconventional, i.e., non-Diophantine, arithmetical operations in scientific theories (Falmagne and Doble, 2015).

The non-Diophantine addition (often called quasiaddition) $\oplus$ or non-Diophantine multiplication (often called quasimultiplication) $\otimes$ were also successfully utilized for defining of *pseudo-integral* (Weber, 1984; Maslov, 1987; Sugeno and Murofushi, 1987; Kolokoltsov and Maslov, 1997; Benvenuti *et al.*, 2002; Pap, 2002) and even more general pan-integral (Yang, 1985; Yang and Song, 1985; Pap, 2002; Wang and Klir, 2009; Pap and Štrboja, 2010; 2013).

Some elements of non-Diophantine arithmetics also implicitly appeared in signal analysis (Oppenheim *et al.*, 1968; 1968a; Oppenheim and Schafer, 1975; Childers *et al.*, 1977). Namely, in the area of nonlinear filtering and cepstral analysis, products are replaced by sums using logarithms for defining logarithmic integral. This approach is called homomorphic filtering (Oppenheim *et al.*, 1968a).

The crucial step in this area was done with the development of non-Newtonian calculus. History of mathematics shows that often important discoveries, inventions and constructions are independently made by different mathematicians. The most spectacular examples are:

- the discovery (construction) of non-Euclidean geometries by Carl Friedrich Gauss (1777–1855), Nikolay Ivanovich Lobachevsky (1792–1856) and János Bolyai (1802–1860);
- the construction (discovery) of the Calculus by Isaac Newton (1642–1727) and Gottfried Wilhelm Leibniz (1646–1716).

One more instructive example is the discovery (construction) of multisets because they were rediscovered multiple times appearing under different names such as an *aggregate*, *heap*, *bunch*, *sample*, *weighted set*, *fireset* (finitely repeated element set) and *occurrence set* (cf. Blizard, 1991; Singh *et al.*, 2007).

In a similar way, non-Newtonian calculus was independently discovered (constructed) in at least three distinct settings by different researchers. The first were Michael Grossman and Robert Katz, who aimed at the further development of the classical calculus. In 1965, they discovered that it was possible to build a variety of non-Newtonian calculi by changing the Diophantine arithmetic of real numbers to some non-Diophantine arithmetic of real numbers (Grossman and Katz, 1972).

Non-Newtonian calculus has roots in the classical mathematics. One of its roots is the product integral introduced by an outstanding Italian mathematician Vito Volterra (1860–1940) in his study of differential equations

(Volterra, 1887; 1887a). Later the concept of the product integral was further developed and utilized by other mathematicians (cf., for example, Rasch, 1934; Birkhoff, 1938; Dollard and Friedman, 1979; Guenther, 1983; Gill and Johansen, 1990). Another root of the non-Newtonian calculus is logarithmic differentiation (cf., for example, Bali, 2005; Bird, 1993; Krantz, 2003) although the logarithmic derivative in the Volterra system is not comprised by non-Newtonian calculi constructed in (Grossman and Katz, 1972).

The non-Newtonian calculus of Grossman and Katz has many applications in different areas including decision making, dynamical systems, differential equations, chaos theory, economics, marketing, finance, fractal geometry, image analysis and electrical engineering.

Almost 30 years later after Grossman and Katz, using the construction of additive t-norms with generators as non-Diophantine addition and extending it to multiplication, Endre Pap independently built and studied a general form of non-Newtonian calculi called *g-calculus* (Pap, 1993). The name *g*-calculus originated from the term "generator", used in (Aczél, 1966). Pap's construction was based on two operations: pseudo-addition $\oplus$ and pseudo-multiplication $\otimes$ in the interval $[0, 1]$, i.e., the non-Diophantine arithmetics of real numbers from the interval $[0, 1]$ formed the foundation for g-calculus (Pap, 1993).

Radko Mesiar and Ján Rybařik modified the initial construction and studied g-calculus, the foundation of which was formed by the non-Diophantine arithmetics of real numbers from the interval $[0, 1]$ with four non-Diophantine operations: addition $\oplus$, subtraction $\ominus$, division $\oslash$ and multiplication $\otimes$ (Mesiar and Rybařik, 1993). As a kind of non-Newtonian calculus, g-calculus has been applied to nonlinear ordinary and partial differential equations, difference equations, generalized functions (distributions), idempotent analysis, measure theory and Laplace transforms.

It is natural to treat the research on $\oplus$-measures and $\oplus$-integrals also as a part of non-Newtonian calculus in a general sense, which, in turn, is a part of the new mathematical area called *pseudoanalysis* (Pap and Vivona, 2000; Štajner-Papuga *et al.*, 2006). However, it is analysis only in a different from the classical domain. Namely, it is analysis in the domain of prearithmetics and non-Diophantine arithmetics.

It is possible to suggest that a better name for this mathematical domain can be *extended analysis* with its two parts: *regular analysis* with Leibniz–Newton's calculus as its part and *irregular analysis*, which would include non-standard analysis, non-Newtonian calculus, idempotent analysis (Maslov and Samborskij, 1992; Kolokoltsov and Maslov, 1997), tropical analysis (Litvinov, 2007; Speyer and Sturmfels, 2009), non-commutative and non-associative pseudo-analysis (Pap and Vivona, 2000; Pap and Štajner-Papuga, 2001), non-Newtonian functional analysis and idempotent functional analysis (Litvinov *et al.*, 2001).

Idempotent analysis was originated by the prominent Russian mathematician Victor Pavlovich Maslov, who introduced several types of non-Diophatine additions $\oplus$ and multiplications $\odot$ in his studies of pseudodifferental equations (Maslov, 1987). In particular, idempotent analysis includes idempotent integration theory, idempotent linear algebra, idempotent spectral theory, and idempotent functional analysis, and is a branch of analysis based on replacing the conventional Diophantine arithmetic by a prearithmetic on an interval of real numbers or a non-Diophantine arithmetic on all real numbers with such operations as maximum or minimum instead of conventional addition (Maslov, 1987; Maslov and Samborskii, 1992; Maslov and Kolokoltsov, 1997). Idempotent analysis studies and applies functions taking values in idempotent semirings, which are special cases of prearithmetics (cf. Section 2.1), while the corresponding function spaces are semimodules over semirings. In turn, semimodules are special cases of quasilinear spaces studied in Section 3.5.

Idempotent semirings, i.e., semi-rings with idempotent addition when $a \oplus a = a$ for all a, were introduced and applied to problems in computer science and discrete mathematics (cf., for example, Kleene, 1956; Pandit, 1961; Vorobjev, 1963; Zimmermann, 1981; Gondran and Minoux, 1979; Cuninghame-Green, 1995; Cohen *et al.*, 1999).

An impetus to the development of idempotent analysis was given by the observation that some problems that are nonlinear in the traditional sense, i.e., over the Diophantine arithmetic of real or complex numbers, turn out to be linear over a suitable prearithmetic or non-Diophantine arithmetic (Maslov, 1987; Maslov and Kolokoltsov, 1997). This helps solving many equations because linearity considerably simplifies the explicit construction of solutions. As a result, idempotent analysis has various applications, which include optimization, optimal design of computer systems and media, optimal organization of data processing, dynamic programming, computer science, discrete mathematics, and mathematical logic.

Almost 20 years later after Pap, coming from problems in physics, the physicist Marek Czachor independently built and studied new kinds of non-Newtonian calculi (Czachor, 2016; 2017; 2019; Aerts *et al.*, 2018).

Physicists often introduced new mathematical structures with the goal of further development of physical theories. For instance, the great physicist James Clerk Maxwell (1831–1879) elaborated a mathematical form of physical fields, the so-called vector fields; two other great physicists Oliver Heaviside (1850–1925) and Paul Adrien Maurice Dirac (1902–1984) invented δ-function: the first generalized function or distribution; one more great physicist Richard Feynman (1918–1988) defined path integrals and mathematicians are still trying to adequately formalize it.

Building non-Newtonian calculus, Czachor introduced two basic innovations in comparison with the previous work in this area. The first one was building non-Diophantine arithmetics and non-Newtonian calculus not only for real numbers but also for complex numbers. The second innovation was elaboration of non-Diophantine arithmetics and non-Newtonian calculus not only for numbers but in essence for arbitrary infinite sets, which were properly connected to real or complex numbers or more exactly, when there was a one-to-one naming of these sets by real or complex numbers.

Actually naming of arbitrary objects by numbers gave birth to mathematics because counting is a process of such naming by natural numbers. Indeed, when counting, people assign one to the first-counted object, two to the second-counted object, three to the third-counted object and so on.

In a similar way, Czachor was naming arbitrary objects by real or complex numbers, which allowed him to construct non-Diophantine arithmetics and non-Newtonian calculus for those objects. This technique allowed Czachor finding diverse applications of Diophantine arithmetics and non-Newtonian calculus in physics and psychology (Czachor, 2016; 2017; 2017a; 2019; Czachor and Posiewnik, 2017). In addition, it made possible developing calculus on fractals and using it for studies of fractals (Aerts *et al.*, 2016; 2016a; 2018).

As the discovery of non-Diophantine arithmetics of natural numbers is discussed in the previous section, here we only make some remarks on these issues explaining similarities and dissimilarities between non-Diophantine arithmetics of natural numbers and non-Diophantine arithmetics used in different forms of non-Newtonian calculus and idempotent calculus.

As the classical calculus in mathematics uses real and complex numbers, the development of non-Newtonian calculus and idempotent calculus brought forth non-Diophantine arithmetics of real numbers and in the version of Czachor, also non-Diophantine arithmetics of complex numbers. Emergence of these arithmetics was caused by the needs of calculus and its applications.

In contrast to this, the discovery of non-Diophantine arithmetics of natural numbers stems from the goal to solve fundamental problems related to counting, which were first formulated in ancient Greece (e.g., the paradox of the heap or the Sorites paradox) and then put forward from time to time by the most insightful researchers (for example, by Herman Helmholtz). Counting makes use of natural numbers, which are the most basic in mathematics according to the opinion of numerous mathematicians. As a result, non-Diophantine arithmetics of natural numbers were discovered (constructed) providing new ways of counting. The construction used for this purpose also allowed building non-Diophantine arithmetics of whole numbers, of integer numbers, of rational numbers, of real numbers and of complex numbers. Besides, the concept of prearithmetic was introduced forming the foundations for non-Diophantine arithmetics and comprising all known kinds of non-Diophantine arithmetics. In particular, specializing prearithmetics to real numbers, it is possible to obtain non-Diophantine arithmetics of real numbers such that go beyond those arithmetics used in conventional (Newtonian) and non-Newtonian calculi. In turn, this would allow building and utilizing new kinds of calculi.

1.9. The Structure of the Book

I know that two and two make four —
and should be glad to prove it too if I could —
though I must say if by any sort of process I could convert
2 plus 2 into five it would give me much greater pleasure.

Lord Byron

To construct the theory of non-Diophantine arithmetics in a systematic and consistent way, we at first develop foundations of this theory studying abstract prearithmetics and projectivity relations between them while treating arithmetics as specific types of prearithmetics. It is demonstrated that many important results obtained for Diophantine arithmetics remain true for suitable classes of prearithmetics.

In Chapter 2, various arithmetics and prearithmetics with two basic operations (addition and multiplication), their characteristics and relations between them are introduced and their properties are studied.

In Section 2.1, we bring in and research abstract prearithmetics, which are mathematical structures with a partial order and two operations: addition and multiplication. Note that in a general case, no other conditions on prearithmetics are assumed. That is why these structures are called abstract prearithmetics.

All known arithmetics, such as conventional arithmetics of natural numbers, of integer numbers of real numbers and of complex numbers, modular arithmetic, and many other mathematical structures, such as rings, lattices or fields, are abstract prearithmetics. In essence, prearithmetics are basic structures for a diversity of mathematical systems.

We explore their properties such as several forms of the Archimedean Property, existence of zeros, successors and predecessors, additive and multiplicative factoring properties, additive and multiplicative cancellation, and existence of subtraction and/or division, investigating relations between arithmetical operations and order. In such a way, we develop a theory of abstract prearithmetics and elements of non-Diophantine number theory analyzing additively and multiplicatively even, odd, prime and composite elements in abstract prearithmetics.

In Sections 2.2–2.4, categorical relations between abstract prearithmetics are studied. In Section 2.2, weak projectivity between abstract prearithmetics is introduced and explored. When two abstract prearithmetics are connected by weak projectivity, they form a fiber bundle and have many similar properties. This allows finding such properties in one of these abstract prearithmetics knowing related properties in the other one. In addition, weak projectivity allows building new prearithmetics and arithmetics from existing prearithmetics and arithmetics. Namely, taking an arbitrary set X and two mappings connecting X with the carrier of some abstract prearithmetic, we induce operations of addition and multiplication on X (Proposition 2.2.1). In particular, as it is done in Sections 2.7 and 2.8, new prearithmetics and arithmetics of whole numbers are constructed using weak projectivity and already known abstract prearithmetics, e.g., the Diophantine arithmetic $\boldsymbol{W}$ of all whole numbers, and deducing many properties of new prearithmetics and arithmetics from properties of abstract prearithmetics used in construction, e.g., properties of the arithmetic $\boldsymbol{W}$. It is also proved (Theorem 2.2.1) that abstract prearithmetics with weak projectivity relations form a category where objects are abstract prearithmetics and morphisms are weak projectivity relations. It would be interesting to compare this category with the category in which objects are abstract prearithmetics and morphisms are their homomorphisms.

Section 2.3 deals with projectivity between abstract prearithmetics, which is a stronger relation in comparison with weak projectivity. As a result, projectivity allows finding more properties of abstract prearithmetics. Besides, prime and composite numbers, subtractability and divisibility of elements as well as prime factorizations and decompositions in abstract prearithmetics are studied in Section 2.3 establishing beginnings of number theory in abstract prearithmetics. In particular, it is proved (Theorem 2.3.1) that abstract prearithmetics with projectivity relations form the category **APAP** where objects are abstract prearithmetics and morphisms are projectivity relations. As projectivity is a particular case of weak projectivity, the category **APAP** is a subcategory of the category **APAWP**, in which objects are abstract prearithmetics and morphisms are weak projectivity relations.

Even a stronger relation between abstract prearithmetics, which is called exact projectivity, is introduced and explored in Section 2.4. In essence, exact projectivity determines isomorphisms between abstract prearithmetics as algebraic systems with two operations and one relation.

Traditionally, the focal relation between algebraic systems is homomorphism with its special types such as monomorphism, epimorphism and isomorphism. The basic property of homomorphisms is that they preserve operations. Systems of algebraic systems such as groups, vector spaces or rings with their homomorphisms form categories.

In the theory of non-Diophantine arithmetics, another basic relation is projectivity with its three basic types — weak projectivity, projectivity *per se* and exact projectivity — and two auxiliary types — direct and inverse skew projectivity. The key property of projectivity relations is that they transfer operations from one prearithmetic to another. Similar to homomorphisms, systems of prearithmetics with their projectivity relations of a fixed type form categories as it is proved in Theorems 2.2.1, 2.3.1 and 2.4.3. Projectivity is intrinsically related to homomorphisms. For instance, exact projectivity between abstract prearithmetics defines isomorphism between the same prearithmetics and *vice versa* as it is proved in Section 2.4.

Section 2.5 studies subprearithmetics of abstract prearithmetics. It is demonstrated that some of the properties of an abstract prearithmetic are preserved for its subprearithmetics, while other properties are not preserved. Subprearithmetics of the Diophantine arithmetics are described.

Skew projectivity of abstract prearithmetics is a special kind of weak projectivity. By definition, weak projectivity is defined by two mapping of abstract prearithmetics. In skew projectivity, one of these mappings is the identity mapping. In Section 2.6, two kinds of skew projectivity between abstract prearithmetics — direct and inverse skew projectivity — are introduced and studied. These types of skew projectivity are interesting as mathematical constructions while their properties are important because some useful mathematical structures, such as the logarithmic scale, relations between percents and their numerical representations or rounding in computations, are formed by particular cases of skew projectivity.

In Section 2.7, prearithmetics of whole numbers and relations between them are introduced and their properties are studied based on the relation of weak projectivity between prearithmetics of whole numbers and the Diophantine arithmetic $\boldsymbol{W}$. In essence, weak projectivity allows building new prearithmetics on whole numbers using the Diophantine arithmetic $\boldsymbol{W}$ and deducing many properties of these prearithmetics from properties of the arithmetic $\boldsymbol{W}$. Similar construction are built and properties are obtained for prearithmetics of natural numbers, which are weakly projective and projective with respect to the Diophantine arithmetic $\boldsymbol{N}$.

In Section 2.8, we consider numerical prearithmetics, which are called weak numerical arithmetics being closer to the Diophantine arithmetic, i.e., having more common properties, than numerical prearithmetics studied in the previous sections. Then we introduce and study projective arithmetics, which are even closer to the Diophantine arithmetic. Finally, when weak arithmetics contain all natural numbers N with the natural order, we come to non-Diophantine arithmetics of natural numbers. Similarly, when a weak arithmetic contains all whole numbers with the natural order, it is a non-Diophantine arithmetic of whole numbers.

In Section 2.9, we introduce and study direct perspective W-prearithmetics, functional direct perspective prearithmetics and functional direct perspective arithmetics with the emphasis on functional direct perspective W-prearithmetics, functional direct perspective prearithmetics and functional direct perspective arithmetics, which are the first discovered class of non-Diophantine arithmetics (Burgin, 1977). Various properties of such arithmetics and prearithmetics are obtained. Some of them are the same as properties of the Diophantine arithmetic. For instance, it is proved that addition and multiplication in these arithmetics are commutative (Proposition 2.9.1). At the same time, new relations, such as "much larger" ($\gg$), "much smaller" ($\ll$), "much much larger" ($\ggg$) and "much much smaller" ($\lll$), are introduced and their properties are studied. In particular, it is proved (Theorems 2.9.2) that relation $\ll$ is compatible from the left with the natural order in functional direct perspective arithmetics. Many functional direct perspective arithmetics are essentially different from the Diophantine arithmetic $\boldsymbol{W}$ or $\boldsymbol{N}$. For instance, it is proved (Theorem 2.9.30) that there are infinitely many functional direct perspective arithmetics in which there is only one prime number. It is also proved (Theorem 2.9.29) that there are infinitely many functional direct perspective arithmetics with only one odd number and (Theorem 2.9.26) there are infinitely many functional direct perspective arithmetics, in which any number is a square.

In Section 2.10, we introduce and study dual perspective W-prearithmetics, functional dual perspective prearithmetics and functional dual perspective arithmetics with the emphasis on functional dual perspective W-prearithmetics, functional dual perspective prearithmetics and functional dual perspective arithmetics. Various properties of such arithmetics and prearithmetics are obtained, some of which are the same as properties of the Diophantine arithmetic. For instance, it is proved that addition and multiplication in these arithmetics are commutative (Proposition 2.10.1). Although construction of dual and direct perspective W-prearithmetics, W-arithmetics, prearithmetics and arithmetics looks very similar, properties of direct perspective W-prearithmetics, W-arithmetics, prearithmetics and arithmetics and of dual perspective W-prearithmetics, W-arithmetics, prearithmetics and arithmetics can be essentially different. For instance, there are more additively Archimedean dual perspective prearithmetics than additively Archimedean direct perspective prearithmetics, i.e., infinitely many functional parameters, which define additively Archimedean dual perspective prearithmetics, define direct perspective prearithmetics that are not additively Archimedean.

In Section 2.11, we study axiomatics of natural-number arithmetics in general and of non-Diophantine arithmetics in particular, starting with the system of Peano axioms and recalling axiom systems suggested for arithmetic by other researchers. We call them postulates to distinguish them from logical axioms. It is possible to admire Peano's intuition, who formulated his postulates only for relations between numbers introducing operations with numbers by definitions. This leaves room for the discovery of non-Diophantine arithmetics, which, as we show (Theorem 2.11.9), satisfy all postulates of Peano.

In Chapter 3, various prearithmetics and non-Diophantine arithmetics of real and complex numbers are introduced. Their properties and relations between them are studied.

In Section 3.1, we introduce and investigate *partially extended abstract prearithmetics* and *arithmetics*, which in addition to two basic operations of abstract prearithmetics — addition and multiplication — have one more operation *subtraction* as the Diophantine arithmetic of integers has. We introduce and explore properties of and relations between partially extended prearithmetics and arithmetics. Although we consider regular types of relations, such as homomorphisms and isomorphisms, the main emphasis is made on new types of relations, such as weak projectivity, projectivity and exact projectivity. In particular, we develop tools for construction of new partially extended (abstract) prearithmetics and arithmetics using already existing partially extended prearithmetics and arithmetics. We also show what properties of new partially extended prearithmetics and arithmetics are deducible from properties of partially extended prearithmetics and arithmetics used for construction.

However, the main emphasis in this chapter is made on *wholly extended abstract prearithmetics* and *arithmetics*, which in addition to two basic operations of abstract prearithmetics, namely, addition and multiplication, have two more operations *subtraction* and *division* as the Diophantine arithmetics of real and complex numbers have. Traditionally, all these operations are treated as *basic arithmetical operations*. Therefore in Section 3.2, we introduce and explore properties of and relations between wholly extended abstract prearithmetics and arithmetics. Although we consider regular types of relations, such as homomorphisms and isomorphisms, the main emphasis is made on new types of relations, such as weak projectivity, projectivity and exact projectivity. In particular, we develop tools of construction of new wholly extended (abstract) prearithmetics and arithmetics using already existing wholly extended prearithmetics and arithmetics. We also show what properties of new wholly extended prearithmetics and arithmetics are deducible from properties of wholly extended prearithmetics and arithmetics used for construction.

Tools developed in Sections 3.1 and 3.2 are used for construction and study of non-Diophantine arithmetics of real (in Section 3.3) and complex (in Section 3.4) numbers. Specifically, in Section 3.3, we introduce and explore properties and relations between wholly extended prearithmetics and arithmetics of real numbers. In this setting, non-Diophantine arithmetics of real numbers are wholly extended prearithmetics of complex numbers that contain all real numbers with the conventional order but have non-conventional arithmetical operations. We show how to build non-Diophantine arithmetics of complex numbers using the Diophantine (conventional) arithmetic of real numbers.

In Section 3.4, we introduce and explore properties of and relations between wholly extended prearithmetics and arithmetics of complex numbers. In this setting, non-Diophantine arithmetics of complex numbers are wholly extended prearithmetics of complex numbers that contain all complex numbers with the conventional order but have non-conventional arithmetical operations. We show how to build non-Diophantine arithmetics of complex numbers using the Diophantine (conventional) arithmetic of complex numbers. To do this, three techniques are developed, which allow building three classes — functional, operational and stratified prearithmetics and non-Diophantine arithmetics of complex numbers.

In Section 3.5, we use prearithmetics and non-Diophantine arithmetics to build and study quasilinear spaces and their important special case — non-Grassmannian linear spaces. Conventional vector (linear) spaces are, in turn, special cases of non-Grassmannian linear spaces. Modules over rings and semimodules over semirings form other subclasses of quasilinear spaces.

Chapter 4 describes construction of non-Newtonian calculi based on non-Diophantine arithmetics of real and complex numbers. The classical calculus, which can be called Newtonian, employs two basic operations with functions — differentiation and integration. In turn, these operations utilize the Diophantine arithmetics of real and complex numbers. The main idea of different non-Newtonian calculi is to use non-Diophantine arithmetics of real and complex numbers instead of the Diophantine ones (Grossman and Katz, 1972; 1984; Pap, 1993; 2008; Czachor, 2016; Aerts *et al.*, 2016a; 2018).

We obtain various properties of the non-Newtonian calculi proving non-Newtonian counterparts of such basic results as linearity of the non-Newtonian derivative (Theorem 4.1.2), the Leibniz rule (Theorem 4.1.1), the chain rule (Theorem 4.1.3), linearity of the non-Newtonian Riemann (Lebesgue) integrals (Theorem 4.1.4), and both fundamental theorems of calculus (Theorems 4.1.6 and 4.1.7). We also introduce and study partial derivatives, integrals along curves and Fourier transforms.

In Chapter 5, the formalism developed in Chapter 4 is applied to two old problems in fractal analysis: (1) Fourier transform on arbitrary Cantor sets and (2) wave equations on fractal space–times. Then the analysis on Cantor sets is generalized to Sierpiński-type sets. The construction of the Cantor and Sierpiński sets is somewhat different from what exists in the typical literature of the subject because the sets considered here do not have to possess any kind of self-similarity. Note that fractals and their history are informally considered in Section 1.1.

In Chapter 6, non-Diophantine arithmetics and non-Newtonian calculi are applied to problems of fractal space–times, Rényi entropies, and dark energy. In particular, non-Diophantine relativity is developed, non-Diophantine forms of Minkowski space–time are constructed and the

Friedman equation, generalized entropies and non-Diophantine probabilities are explored. In addition, a non-Newtonian quantum mechanics is elaborated.

For a long time, it was understood that geometry is not only an abstract mathematical theory but also an inherent characteristic of the Physical World. The results of this chapter demonstrate that arithmetic is not only an abstract mathematical theory but also an inherent characteristic of the Physical World.

One more field where Non-Diophantine arithmetics find essential applications is psychology. Human and animal nervous systems are information channels, creating filters between *us* and the physical reality. Non-Diophantine arithmetics help researchers to better understand regularities of information processing in human organisms. That is why Chapter 7 contains exploration of psychophysics based on non-Diophantine arithmetics.

Several open problems and directions for future research, which are based on the results presented in the previous chapters of the book, are considered in the concluding Chapter 8.

Chapters 1–3 are written by Mark Burgin. Chapters 5–7 are written by Marek Czachor. Chapters 4 and 8 are written together by Mark Burgin and Marek Czachor. Mark Burgin takes responsibility for all epigraphs.

To make the book easier for the reader, notations and definitions of the conventional mathematical concepts and structures used in this book are included in the appendix. Special attention is paid to providing a variety of examples illustrating the general theory.

Acknowledgments

We are grateful to the staff at World Scientific and especially, to Ms. Tan Rok Ting, for their help in bringing about this publication. Our thanks also go to Michael Grossman and Peter Carr for their help in finding various publications related to non-Diophantine arithmetics.

Acknowledgments of Mark Burgin

I would like to thank all my teachers, above all my teachers of mathematics and physics, and especially, my PhD thesis advisor, Alexander Gennadievich Kurosh, who helped shaping his scientific viewpoint and research style. The advice and help of Andrei Nikolayevich Kolmogorov from Moscow State University in the development of the holistic view on mathematics and its connections with physics is also greatly appreciated. I am also grateful to my school teachers and especially, to math teachers Sofya Moiseevna Borshchukova, Lev Demyanovich and physics teacher Raisa Markovna Nesenyuk.

It is also necessary to assert that many discussions and conversations with friends and colleagues were useful in the studies of non-Diophantine arithmetics and their applications. The collaboration with Gunter Meissner from the University of Hawaii contributed to applications of non-Diophantine arithmetics to economics. Credit for the desire to write this book must go to the academic colleagues. I would also like to thank the Departments of Mathematics and Computer Science in the School of Engineering at UCLA for providing space, equipment, and helpful discussions.

Chapter 2

Non-Diophantine Arithmetics of Natural and Whole Numbers

How thoroughly it is ingrained in mathematical science that every real advance goes hand in hand with the invention of sharper tools and simpler methods which, at the same time, assist in understanding earlier theories and in casting aside some more complicated developments.

David Hilbert

To build and explore non-Diophantine arithmetics, we at first describe mathematical structures called prearithmetics (Burgin, 2010), which are algebraic systems more general than Diophantine and non-Diophantine arithmetics but are in some aspects similar to arithmetics. In essence, an abstract prearithmetic is a universal algebra (algebraic system) with two binary operations and a partial order. Operations are called addition and multiplication but in a general case, there are no restrictions on these operations.

In Sections 2.1–2.7, we study properties and relations between abstract prearithmetics, such as the Archimedean property, commutativity, associativity, weak projectivity, projectivity, exact projectivity and skew projectivity, and only then in Sections 2.8–2.11, we come to numerical prearithmetics and non-Diophantine arithmetics by increasing similarity with the conventional Diophantine arithmetic. To develop the foundations of non-Diophantine arithmetics, we start our exploration with a very general concept of a prearithmetic treating at first, abstract prearithmetics and then numerical prearithmetics. In this setting, we elaborate elements of a theory similar to number theory but based on prearithmetics instead of the conventional Diophantine arithmetic $\boldsymbol{N}$, which is the indispensable domain of the contemporary number theory (Davenport, 1992; Friedberg, 1968; Gioia, 2002). In the context of non-Diophantine arithmetics, which form a key subclass of abstract prearithmetics, it is natural to call this theory *non-Diophantine number theory* or *non-Diophantine higher arithmetic*. One of the main goals in this exploration is building efficient tools of construction of new (abstract) prearithmetics and arithmetics using already existing prearithmetics and arithmetics.

As a result, this chapter has the following structure. In Section 2.1, we introduce and study abstract prearithmetics exploring their properties such as several forms of the Archimedean Property, existence of zeros, successors and predecessors, additive and multiplicative factoring properties, existence of subtraction and/or division. Besides, we investigate relations between arithmetical operations and order in abstract prearithmetics such as monotonicity and distributivity. All known arithmetics, such as conventional arithmetics of natural numbers, of integer numbers, of real numbers and of complex numbers, and many other mathematical structures, such as rings, fields, semirings, ordered rings, ordered fields, Ω-groups, Ω-rings, lattices and Boolean algebras, are abstract prearithmetics. In developing fundamentals of number theory in abstract prearithmetics, we pay the main attention to the problems of divisibility and primality, which constitute important areas in the classical number theory (cf., for example, Broadbent, 1971; Davenport, 1992; Hayes, 2009).

In Sections 2.2–2.4 and 2.6, we develop tools for construction of new abstract prearithmetics using already existing prearithmetics. In Section 2.2, weak projectivity between abstract prearithmetics is introduced and studied. When two abstract prearithmetics are connected by weak projectivity, they form a fiber bundle and have many similar properties. This provides the possibility of finding properties of one of these abstract prearithmetics knowing related properties of the other one. In addition, weak projectivity allows building new prearithmetics and arithmetics from existing prearithmetics and arithmetics. Namely, taking an arbitrary set X and two mappings connecting X with the carrier of some abstract prearithmetic, we induce operations of addition and multiplication on X (Proposition 2.2.1).

It is necessary to remark that traditionally the main relation between algebraic systems is homomorphism with its special types such as monomorphism, epimorphism and isomorphism. The basic property of homomorphisms is that they preserve operations. Systems of algebraic systems such as groups, vector spaces or rings with their homomorphisms form categories.

In the theory of non-Diophantine arithmetics, another basic relation is projectivity with its three types: weak projectivity, projectivity *per se* and exact projectivity. These relations form a hierarchy because projectivity is a special case of weak projectivity while exact projectivity is a special case of projectivity. The key property of projectivity relations is that they transfer operations from one prearithmetic to another. Similar to homomorphisms, systems of prearithmetics with their projectivity relations of a fixed type form categories as it is proved in Theorems 2.2.1, 2.3.1 and 2.4.3. Projectivity is intrinsically related to homomorphisms. For instance, exact projectivity between abstract prearithmetics defines isomorphism between the same prearithmetics and *vice versa* as it is proved in Section 2.4.

In Section 2.3, we introduce projectivity as a principal relation between abstract prearithmetics and explore its properties. These properties allow us to build new prearithmetics from existing prearithmetics, deduce properties of one abstract prearithmetic knowing related properties of the other one, and develop elements of number theory in abstract prearithmetics and arithmetics. In Section 2.4, we introduce exact projectivity as an important relation between abstract prearithmetics and explore its properties. We show what properties of new prearithmetics are deducible from properties of prearithmetics used for construction.

In Section 2.5, subprearithmetics are explored. We demonstrate how properties and relations between subprearithmetics are inherited from properties and relations between abstract prearithmetics in which these subprearithmetics are included. In addition, we study projectivity relations between abstract prearithmetics and their subprearithmetics.

In Section 2.6, two important special cases of weak projectivity — direct and inverse skew projectivity are introduced and analyzed.

Prearithmetics, which are studied in Sections 2.1–2.6, serve as the algebraic base for arithmetics allowing inference of various characteristic of arithmetics from previously explored attributes of prearithmetics. In Section 2.7, we study numerical prearithmetics placing an emphasis on whole-number prearithmetics, i.e., prearithmetics in sets of whole numbers, and treating natural-number prearithmetics, i.e., prearithmetics on sets of natural numbers, as subprearithmetics (subalgebras) of prearithmetics of whole numbers. This approach allows deducing many properties of natural-number prearithmetics from properties of whole-number prearithmetics.

We also construct and study different types of non-Diophantine arithmetics of natural and whole numbers. An exceedingly general class of non-Diophantine arithmetics consists of projective arithmetics, which are explored in Section 2.8 providing various properties of these arithmetics. Two other classes of non-Diophantine arithmetics are direct perspective arithmetics, which are studied in Section 2.9, and dual perspective arithmetics, which are studied in Section 2.10.

Axiomatic theory of Diophantine and non-Diophantine arithmetics is presented in Section 2.11. In particular, it is proved (Theorem 2.11.9) that all non-Diophantine arithmetics of natural numbers satisfy all five Peano axioms.

2.1. Abstract Prearithmetics

When a truth is necessary,
the reason for it can be found by analysis,
that is, by resolving it into simpler ideas and truths
until the primary ones are reached.
Gottfried Wilhelm Leibniz

In a similar way, as set theory is a foundation for mathematics, the theory of abstract prearithmetics provides foundations for the theory of non-Diophantine arithmetics and Diophantine arithmetics. In addition, the theory of abstract prearithmetics encompasses as its subtheories theories of various conventional mathematical structures, such as rings, fields, ordered rings, ordered fields, lattices and Boolean algebras. This allows using constructions from the theory of abstract prearithmetics for the further development of its subtheories of conventional mathematical structures. Abstract prearithmetics also provide a unified algebraic context for various other established mathematical constructions, such as the logarithmic scale, modular arithmetics and computer arithmetics, which are used in many applications in mathematics, science and technology. In addition, numerical prearithmetics serve as a tool for constructing new types of the calculus such as idempotent analysis (Maslov and Samborskij, 1992; Kolokoltsov and Maslov, 1997) and tropical analysis (Litvinov, 2007; Speyer and Sturmfels, 2009), which have a variety of applications including optimization, optimal design of computer systems and media, optimal organization of data processing, dynamic programming, computer science, discrete mathematics, and mathematical logic.

2.1.1. *Algebraic structure of abstract prearithmetics*

It is dangerous to be right in matters on which the established authorities are wrong.

Voltaire

An *abstract prearithmetic* is a set (often a set of numbers) A with a partial order $\leq$ and two binary operations $+$ (addition) and $\circ$ (multiplication), which are defined for all its elements. It is denoted by $\boldsymbol{A} = (A; +, \circ, \leq)$. The set A is called either the *set of elements* or the *set of numbers* or the *carrier* of the prearithmetic $\boldsymbol{A}$. As always, if $x \leq y$ and $x \neq y$, then we denote this relation by $x < y$. Operation $+$ is called *addition* and operation $\circ$ is called *multiplication* in the abstract prearithmetic $\boldsymbol{A}$.

Benacerraf (1965) argues that order in the set of natural numbers has to be defined by a recursive definition. It is possible to apply this condition to order in abstract prearithmetics, as well as to their operations, i.e., we can specify the class of *recursive abstract prearithmetics*, which consists of abstract prearithmetics in which the operations addition and multiplication, as well as the order relations are defined by recursive definitions, that is, defined operationally. Note that it is also possible to define operations and order by subrecursive and super-recursive definitions (Burgin, 2005).

Note that an abstract prearithmetic can have more than two operations and more than one order relations.

Example 2.1.1. Naturally, the conventional Diophantine arithmetic $\boldsymbol{N}$ of all natural numbers is an abstract prearithmetic.

Example 2.1.2. Naturally, the conventional Diophantine arithmetic $\boldsymbol{W}$ of all whole numbers is an abstract prearithmetic.

Example 2.1.3. Another example of abstract prearithmetics is *modular arithmetic*, which also known as *residue arithmetic* or *clock arithmetic*. It is studied in mathematics and used in physics and computing. In modular arithmetic, operations of addition and multiplication are defined but in contrast to the conventional arithmetic, its numbers form a cycle upon reaching a certain value, which is called the *modulus*. A rigorous approach to the theory of modular arithmetic was worked out by Carl Friedrich Gauss (1777–1855) in his book *Disquisitiones Arithmeticae*, published when he was 24 years old (Gauss, 1801).

We build a modular arithmetic in the following way. A natural number n is chosen as the modulus. Then two natural numbers k and m are called congruent modulo n if their difference $k - m$ is divisible by n. This congruence is denoted by the expression

$$k = m (\text{mod}\, n).$$

It is possible to show that it is an equivalence relation (cf. Appendix). Indeed, we have

(1) $k = k$ (mod n) for any natural number k;
(2) if $k = m$ (mod n), then $m = k$ (mod n);
(3) if $k = m$ (mod n) and $m = l$ (mod n), then $k = l$ (mod n).

There are exactly n possible remainders of division by n. These remainders are called numbers or elements of the corresponding modular arithmetic $\boldsymbol{Z}_n$, which contains exactly n elements.

The defined congruence preserves addition and multiplication of natural numbers.

Indeed, we have

(4) if $k = m$ (mod n) and $t = l$ (mod n), then $k + t = m + l$ (mod n);
(5) if $k = m$ (mod n) and $t = l$ (mod n), then $k \cdot t = m \cdot l$ (mod n).
This allows us to define operations in the modular arithmetic $\boldsymbol{Z}_n$: If numbers k and m belong to $\boldsymbol{Z}_n$, then the sum $k + m$ in $\boldsymbol{Z}_n$ is equal to the number t from $\boldsymbol{Z}_n$ such that $k + m = t$ (mod n) and the product $k \cdot m$ in $\boldsymbol{Z}_n$ is equal to the number q from $\boldsymbol{Z}_n$ such that $k \cdot m = q$ (mod n). The order in $\boldsymbol{Z}_n$ is partial but the same as the order in $\boldsymbol{N}$.

For instance, when the modulus n is equal to 10, the modular arithmetic $\boldsymbol{Z}_{10}$ contains only ten numbers 0, 1, 2, 3, 4, 5, 6, 7, 8, 9 and when the result of the operation the conventional arithmetic is larger than 10, then it is reduced to these numbers in the modular arithmetic. Readers can find information about modular arithmetics in many books and on the Internet. Here we only give some examples for the modular arithmetic Z_{10}:

$$2 + 2 = 4$$

but

$$5 + 5 = 0$$

and

$$7 + 8 = 5$$

In addition,

$$3 \cdot 3 = 9$$

but

$$5 \cdot 5 = 5$$

and

$$4 \cdot 4 = 6.$$

Naturally, any modular arithmetic $\boldsymbol{Z}_n$ is an abstract prearithmetic.

When n is a prime number, the modular arithmetic $\boldsymbol{Z}_n$ is a field (Kurosh, 1963).

Note that although abstract prearithmetics $\boldsymbol{Z}_n$ have been called by the name *modular arithmetic*, they have never been considered arithmetics of natural numbers. There was only one such arithmetic — the Diophantine arithmetic $\boldsymbol{N}$. However, as we will see in Sections 2.9 and 2.10, non-Diophantine arithmetics are also arithmetics of natural numbers. They even satisfy Peano axioms (cf. Section 2.11).

Example 2.1.4. The conventional arithmetic $\boldsymbol{Q}$ of all rational numbers is an abstract prearithmetic.

Example 2.1.5. The conventional arithmetic $\boldsymbol{Z}$ of all integer numbers is an abstract prearithmetic.

Example 2.1.6. The conventional arithmetic $\boldsymbol{R}$ of all real numbers is an abstract prearithmetic.

Example 2.1.7. The conventional arithmetic $\boldsymbol{C}$ of all complex numbers is an abstract prearithmetic.

All these examples show that conventional arithmetics are abstract prearithmetics. However, there are many unusual abstract prearithmetics.

Example 2.1.8. Let us consider the set N of all natural numbers with the standard order $\leq$ and introduce the following operations:

$$a \oplus b = a \cdot b,$$
$$a \otimes b = a^b.$$

Then the system $\boldsymbol{A} = (N; \oplus, \otimes, \leq)$ is an abstract prearithmetic with addition $\oplus$ and multiplication $\otimes$.

Example 2.1.9. Let us consider the set R^{++} of all positive real numbers is with the standard order $\leq$ and introduce the following operations:

$$a \boxplus b = a + b,$$
$$a \divideontimes b = a \div b.$$

Then the system $\boldsymbol{B} = (R^{++}; \boxplus, \divideontimes, \leq)$ is an abstract prearithmetic with addition $\boxplus$ and multiplication $\divideontimes$.

Example 2.1.10. Many algebraic structures studied in algebra are abstract prearithmetics with a trivial order, i.e., any ring, lattice, Boolean algebra, linear algebra, field, Ω-group, Ω-ring, Ω-algebra (Burgin, 1970; 1972; Burgin and Baranovich, 1975), topological ring, topological field, normed ring, normed algebra, normed field (Gel'fand *et al.*, 1964; Naimark, 1959; 1972), and in essence, any universal algebra with two operations is an abstract prearithmetic with a trivial order (cf Appendix). The same structures with a nontrivial order are also abstract prearithmetics.

Examples are given by ordered rings, ordered linear algebras and ordered fields (cf., for example, Fuchs, 1963; Kurosh, 1963). Besides, it is possible to treat universal algebras with one operation, which is interpreted as addition, as abstract prearithmetics with the trivial order and trivial multiplication, in which the product of any two elements is always the same.

Example 2.1.11. A *semiring* is a commutative semigroup under addition $\oplus$ and a semigroup under multiplication $\odot$ (Kuich, 1986; Golan, 1999) satisfying additional identities of left and right distributivity:

$$r \odot (a \oplus b) = r \odot a \oplus r \odot b$$

and

$$(a \oplus b) \odot r = a \odot r \oplus b \odot r.$$

In some sources, it is assumed that a semiring has zero "0" and multiplicative identity "1". Semirings have diverse applications. Any semiring is a prearithmetic in which addition is commutative and associative, and multiplication is associative. Some authors also demand distributivity of multiplication with respect to addition.

Semirings with multiplicative one (unity) 1 and additive zero "0" are called *semifields*. They are also prearithmetics.

Many researchers utilized *idempotent semirings*, i.e., semi-rings with idempotent addition when $a \oplus a = a$ for all a, and matrices over such semirings for solving various applied problems in computer science and discrete mathematics (cf., for example, Kleene, 1956; Pandit, 1961; Vorobjev, 1963; Zimmermann, 1981; Gondran and Minoux, 1979; Cuninghame-Green, 1995; Cohen *et al.*, 1999). Idempotent semi-rings also have many other applications, in particular, as the basic structure of idempotent analysis (Maslov, 1987; Maslov and Samborskii, 1992; Kolokoltsov and Maslov, 1997) and its special case tropical analysis (Litvinov, 2007; Speyer and Sturmfels, 2009).

Naturally, idempotent semi-rings are also prearithmetics where addition is commutative and associative, multiplication is associative and $x \oplus x = x$.

Example 2.1.12. *Hemirings* (Golan, 1999), also called *pre-semirings* (Gondran and Minoux, 2008), are generalizations of semirings that do not require the existence of a multiplicative identity 1. Any pre-semiring is a prearithmetic.

A further generalization of semirings are *left-pre-semirings* (Gondran and Minoux, 2008), which additionally do not require right-distributivity (or *right-pre-semirings*, which do not require left-distributivity). Any left-pre-semiring or any right-pre-semiring is a prearithmetic.

Example 2.1.13. Let $\mathbf{R}_{\max}$ be the set $A = \boldsymbol{R} \cup \{-\infty\}$ with the operations $\oplus = \max$ and $\odot = +$, which is the usual addition in $\boldsymbol{R}$ and defining $0 = -\infty$ and $1 = 0$. By construction, $\mathbf{R}_{\max}$ is a commutative idempotent semi-ring and thus, a prearithmetic. It is very useful in idempotent analysis (Maslov, 1987; Maslov and Samborskii, 1992; Maslov and Kolokoltsov, 1997).

Example 2.1.14. Let $\mathbf{R}_{\min}$ be the set $A = \boldsymbol{R} \cup \{+\infty\}$ with the operations $\oplus = \min$ and $\odot = +$, which is the usual addition in $\boldsymbol{R}$ and defining $\mathbf{0} = +\infty$ and $1 = 0$. By construction, $\mathbf{R}_{\min}$ is a commutative idempotent semi-ring and thus, a prearithmetic. It is very useful in idempotent analysis (Maslov, 1987; Maslov and Samborskii, 1992; Kolokoltsov and Maslov, 1997).

Example 2.1.15. *Tropical semirings* (Pin, 1998) are prearithmetics. The term "tropical semirings" was introduced in computer science to denote discrete versions of the max-plus algebra $\mathbf{R}_{\max}$ or min-plus algebra $\mathbf{R}_{\min}$ and their subalgebras (Litvinov, 2007).

Example 2.1.16. *Subtropical algebras* with max or min as multiplication (Shiozawa, 1998) are prearithmetics.

Example 2.1.17. A *thermodynamic semiring*, which is related to the Shannon, Tsallis and Rényi entropy functions, non-extensive thermodynamics and multifractals (Marcolli and Thorngren, 2011), is a prearithmetic.

All these examples show that prearithmetics are basic structures for a diversity of mathematical systems.

Note that in a general case, there are no connections between addition and multiplication in prearithmetics and it is even possible that these operations coincide. This allows representation of universal algebras with one binary operation as abstract prearithmetics because it is possible to take the existing operation as, for example, addition, and define multiplication in an arbitrary way. This determines correspondence between a universal algebra $\boldsymbol{A}$ with one binary operation ω and an abstract prearithmetic $\boldsymbol{A}^*$, in which addition coincides with ω. At the same time, it is necessary to be careful because some properties of $\boldsymbol{A}$ and $\boldsymbol{A}^*$ can be different.

In a similar way, it is possible to treat any universal algebra $\boldsymbol{B}$ with more than two binary operation as an abstract prearithmetic $\boldsymbol{B}^*$ having in mind that as in the previous case, some properties of $\boldsymbol{B}$ and $\boldsymbol{B}^*$ can be different.

However, it is necessary to remark that abstract prearithmetics are not partially ordered algebraic system in the sense of Fuchs (1963) and Kurosh (1963) or even in the weaker sense of Matsushita (1951) because the partial order in abstract prearithmetics is independent of operations in the general case, while in partially ordered algebraic system, operations have to be monotone.

It is also possible to assemble different algebraic constructions similar to modules and vector spaces using abstract prearithmetics instead of rings or fields. For instance, taking an abstract prearithmetic $\boldsymbol{A} = (A; +, \circ, \leq)$ and a natural number n, it is possible to build the abstract prearithmetic of n-dimensional A-vectors $\mathrm{V}^n\boldsymbol{A} = (\mathrm{V}^n A; +, \circ, \leq)$, elements of which are vectors in $\boldsymbol{A}$. Namely, elements of n-dimensional A-vector prearithmetic $\mathrm{V}^n\boldsymbol{A} = (\mathrm{V}^n A; +, \circ, \leq)$, i.e., A-vectors, have the form $(a_1, a_2, \ldots, a_n)$ where $a_1, a_2, \ldots, a_n$ are elements from the abstract prearithmetic $\boldsymbol{A}$.

In a similar way, taking an abstract prearithmetic $\boldsymbol{A} = (A; +, \circ, \leq)$ and a pair of natural numbers n and m, it is also possible to build the abstract prearithmetic of $(n \times m)$-dimensional A-matrices $\mathrm{M}^{n\times m}\boldsymbol{A} = (\mathrm{M}^{n\times m}A; +, \circ, \leq)$, elements of which are matrices in $\boldsymbol{A}$. Namely, elements of

$(n \times m)$-dimensional A-matrix prearithmetic $\mathrm{M}^{n\times m}\boldsymbol{A} = (\mathrm{M}^{n\times m}A; +, \circ, \leq)$, i.e., A-matrices, have the form

$$\begin{pmatrix} a_{11} & a_{12} & a_{13} & \cdots & a_{1n} \\ a_{21} & a_{22} & a_{23} & \cdots & a_{2n} \\ \vdots & \vdots & \vdots & \vdots & \vdots \\ a_{m1} & a_{m2} & a_{m3} & \cdots & a_{mn} \end{pmatrix},$$

where all $a_{ij}(i = 1, 2, 3, \ldots, m; j = 1, 2, 3, \ldots, n)$ are elements from the abstract prearithmetic $\boldsymbol{A}$.

Addition and multiplication in these prearithmetics are defined coordinate-wise. For instance, taking two-dimensional Z-vectors (2, 3) and (4, 5) from the prearithmetic $\mathrm{V}^2\boldsymbol{Z}$ of Z-vectors, we define their sum as $(2, 3) + (4, 5) = (6, 8)$ and their product as $(2, 3) \circ (4, 5) = (8, 15)$.

Order in $\mathrm{V}^n\boldsymbol{A}$ is defined by the following condition:

If $(a_1, a_2, \ldots, a_n)$ and $(b_1, b_2, \ldots, b_n)$ are vectors from V^nA, then $(a_1, a_2, \ldots, a_n) \leq (b_1, b_2, \ldots, b_n)$ if and only if $a_j \leq b_j$ for all $j = 1, 2, 3, \ldots, n$.

For matrices, addition, multiplication and order are defined in a similar way.

Note that the defined multiplication is *scalar multiplication* of vectors and matrices, which is different from vector and matrix multiplication.

Prearithmetics $\mathrm{V}^n\boldsymbol{A}$ and $\mathrm{M}^{n\times m}\boldsymbol{A}$ preserve many properties of the abstract prearithmetic $\boldsymbol{A}$. For instance, we have the following results.

Proposition 2.1.1. *If addition is commutative in an abstract prearithmetic $\boldsymbol{A}$, then addition is commutative in the vector prearithmetic $V^n\boldsymbol{A}$ and in the matrix prearithmetic $M^{n\times m}\boldsymbol{A}$.*

Proof is left as an exercise. □

The same is true for multiplication.

Proposition 2.1.2. *If multiplication is commutative in an abstract prearithmetic $\boldsymbol{A}$, then multiplication is commutative in the vector prearithmetic $V^n\boldsymbol{A}$ and in the matrix prearithmetic $M^{n\times m}\boldsymbol{A}$.*

Proof is left as an exercise. □

Proposition 2.1.3. *If addition is associative in an abstract prearithmetic* $\boldsymbol{A}$, *then addition is associative in the vector prearithmetic* $V^n\boldsymbol{A}$ *and in the matrix prearithmetic* $M^{n\times m}\boldsymbol{A}$.

Proof is left as an exercise. □

The same is true for multiplication.

Proposition 2.1.4. *If multiplication is associative in an abstract prearithmetic* $\boldsymbol{A}$, *then multiplication is associative in the vector prearithmetic* $V^n\boldsymbol{A}$ *and in the matrix prearithmetic* $M^{n\times m}\boldsymbol{A}$.

Proof is left as an exercise. □

In the Diophantine arithmetic, multiplication is distributive with respect to addition, i.e., the following identities hold:

$$x \cdot (y + z) = x \cdot y + x \cdot z,$$

$$(y + z) \cdot x = y \cdot x + z \cdot x.$$

However, in abstract prearithmetics, multiplication is not always commutative and we need to discern three kinds of distributivity, namely, *distributivity from the left*

$$x \cdot (y + z) = x \cdot y + x \cdot z$$

and *distributivity from the right*

$$(y + z) \cdot x = y \cdot x + z \cdot x.$$

In addition, multiplication is *distributive* with respect to addition when both identities hold.

Proposition 2.1.5. *If multiplication is distributive (distributive from the left or distributive from the right) with respect to addition in an abstract prearithmetic* $\boldsymbol{A}$, *then multiplication is distributive (distributive from the left or distributive from the right) with respect to addition in the vector prearithmetic* $V^n\boldsymbol{A}$ *and in the matrix prearithmetic* $M^{n\times m}\boldsymbol{A}$.

Proof is left as an exercise. □

Remark 2.1.1. Having an abstract prearithmetic $\boldsymbol{A} = (A; +, \circ, \leq)$, it is possible to build not only abstract prearithmetics of A-vectors and A-matrices but also abstract prearithmetics of multidimensional matrices or arrays in A of arbitrary dimensions, i.e., multidimensional A-matrices or A-arrays, and form their prearithmetics exploring what properties they inherit from the initial abstract prearithmetic $\boldsymbol{A}$.

Now let us explore properties of orders in abstract prearithmetics.

Definition 2.1.1. A partial order $\leq$ in a set X is called *discrete* if it satisfies the following two conditions:

(UD) $\forall x \in X\ [(\exists u \in X\ (x < u)) \Rightarrow (\exists y \in X\ (x < y \leq u)\ \&\ \forall z \in X\ ((x \leq z \leq y) \Rightarrow (z = x \vee z = y))]$;
(LD) $\forall x \in X\ [(\exists v \in X\ (v < x)) \Rightarrow (\exists w \in X\ (v \leq w < x))\ \&\ \forall z \in X\ ((w \leq z \leq x) \Rightarrow (z = x \vee z = w))]$.

Informally, these conditions have the following meaning.

(UD) For any element x from X, if there is an element u from X larger than x, then there is an element y from X such that $x < y \leq u$ and there are no elements z from X, between x and y, i.e., if $x \leq z \leq y$, then z is equal either to x or to y. By the mathematical tradition, we denote such an element y by Sx and call it a *successor* of x.

(LD) For any element x from X, if there is an element v from X less than x, then there is an element w from X such that $v \leq w < x$ and there are no elements z from X, between w and x, i.e., if $w \leq z \leq x$, then z is equal either to x or to w. We denote such an element w by Px and call it a *predecessor* of x.

Thus, a partial order $\leq$ in a set X is discrete if all elements comparable with, at least, one element in X have either a successor or a predecessor or both.

Remark 2.1.2. Often the successor of n is also denoted by n' (cf., for example, Landau, 1966; Hilbert and Bernays, 1968). In contrast to this, Peirce denoted the predecessor of n by n' (Peirce, 1881). That is why we denote the successor of n by Sn and the predecessor of n by Pn.

Note that although the concepts of a successor and a predecessor are dual, it is possible that all elements have successors but not all have predecessors and vice versa. For instance, in the Diophantine arithmetic $\boldsymbol{N}$ all elements have successors but number 1 does not have a predecessor.

Definitions imply the following results.

Lemma 2.1.1. *For any element x from X,* SPx = PSx = x.

Proof is left as an exercise. □

Lemma 2.1.2. *If in an abstract prearithmetic* $\boldsymbol{A}_1 = (A_1; +_1, \circ_1, \leq_1)$ *the order* $\leq_1$ *is discrete, then any element* a *in* $\boldsymbol{A}_1$*, which is not minimal, has a predecessor* Pa.

Proof is left as an exercise. □

Lemma 2.1.3. *If in an abstract prearithmetic* $\boldsymbol{A}_1 = (A_1; +_1, \circ_1, \leq_1)$ *the order* $\leq_1$ *is discrete, then any element* a *in* $\boldsymbol{A}_1$*, which is not maximal, has a successor* Sa.

Proof is left as an exercise. □

Let us consider properties of successors.

Lemma 2.1.4. *If the order* $\leq$ *is linear (total) and an element* x *from* X*, has a successor* Sx*, then* Sx *is unique.*

Proof: If elements a and b are successors of an element x from X, then by condition (UD), $x \leq a$ and $x \leq b$. As the order $\leq$ is linear, we have either $b \leq a$ or $a \leq b$. Consequently, we have either $x \leq b \leq a$ or $x \leq a \leq b$. In both cases by condition (UD), either $a = x$ or $b = x$ or $a = b$.

Lemma is proved. □

Corollary 2.1.1. *If the order* $\leq$ *is linear (total), then* S *is a partial function defined for those elements that have successors.*

Proposition 2.1.6. *If* $|X| > 1$ *and the order* $\leq$ *in* X *is linear (total) and discrete, then* S *is a total function either on the whole set* X *or on* X *less one element.*

Proof: Properties of a linear order imply that in a linearly ordered set X, there is, at most, one element, which is larger than all other elements in X. Indeed, because if we have two elements a and b, which are larger than all other elements in X, then we have $b \leq a$ and $a \leq b$. By the properties of an order relation, $a = b$.

If there is an element d, which is larger than all other elements in X, then for any element x from X, we have $x \leq d$, and by condition (UD), there exists Sx, which is unique by Lemma 2.1.4. Thus, S is a function defined for all elements in X but d.

If the element d larger than all other elements in X does not exist, then for any element x from X, there is a larger element y, i.e., we have $x \leq y$. Consequently, by condition (UD), there exists Sx, which is unique by Lemma 2.1.4. Thus, S is a function defined for all elements in X.

Proposition is proved. □

Both situations considered in Proposition 2.1.6 are possible as the following examples demonstrate.

Example 2.1.18. In the set Z^- of all non-positive integer numbers, all numbers but 0 have successors, i.e., S is a function defined for all elements but 0.

Example 2.1.19. In the set $Z^+ = W$ of all non-negative integer numbers, all numbers have successors, i.e., S is defined for all elements.

Let us consider properties of predecessors.

Lemma 2.1.5. *If the order $\leq$ is linear (total) and an element x from X, has a predecessor* Px, *then it is unique.*

Proof: If elements a and b are predecessors of an element x from X, then by condition (LD), $a \leq x$ and $b \leq x$. As the order $\leq$ is linear, we have either $b \leq a$ or there is, at most, one element, which is larger than all other elements in X. Consequently, we have either $a \leq b \leq x$ or $b \leq a \leq x$. In both cases by condition (LD), either $a = x$ or $b = x$ or $a = b$.

Lemma is proved. □

Corollary 2.1.2. *If the order $\leq$ is linear (total), then P is a partial function defined for those elements that have predecessors.*

Remark 2.1.3. For an arbitrary partial order $\leq$, the statements of Lemmas 2.1.4 and 2.1.5 are not always true as the following examples demonstrate.

Example 2.1.20. Let us take the set $X = \{a, b, c\}$ and define $a \leq b$ and $a \leq c$. Then both are successors of a, i.e., $b = \text{S}a$, $c = \text{S}a$ and $b \neq c$.

Example 2.1.21. Let us take the set $X = \{a, b, c\}$ and define $b \leq a$ and $c \leq a$. Then both are predecessors of a.

Proposition 2.1.7. *If $|X| > 1$ and the order $\leq$ is linear (total) and discrete, then* S *is a total function either on the whole set X or on X less one element.*

Proof: Properties of a linear order imply that in a linearly ordered set X, there is, at most, one element, which is larger than all other elements in X. Indeed, because if we have two elements a and b, which are larger than all other elements in X, then we have $b \leq a$ and $a \leq b$. By the properties of an order relation, $a = b$.

If there is an element d, which is larger than all other elements in X, then for any element x from X, we have $x \leq d$, and by condition (UD), there exists Sx, which is unique by Lemma 2.1.4. Thus, S is a function defined for all elements in X but d.

If the element d larger than all other elements in X does not exist, then for any element x from X, there is a larger element y, i.e., we have $x \leq y$. Consequently, by condition (UD), there exists Sx, which is unique by Lemma 2.1.4. Thus, S is a function defined for all elements in X.

Proposition is proved. □

Corollary 2.1.3. *If X does not have the largest element and the order $\leq$ is linear (total) and discrete, then* S *is a total function.*

In this case, it is possible to treat S as a unary operation in X adding to operations in abstract prearithmetics one more operation.

Let us consider an abstract prearithmetic $\boldsymbol{A} = (A; +, \circ, \leq)$.

Definition 2.1.2.

(a) Addition + *preserves the order* $\leq$, if for any elements a, b, c and d from A, the inequalities $a \leq b$ and $d \leq c$ imply the inequality $a + d \leq b + c$.
(b) Multiplication $\circ$ *preserves the order* $\leq$, if for any elements a, b, c and d from A, the inequalities $a \leq b$ and $d \leq c$ imply the inequality $a \circ d \leq b \circ c$.
(c) A mapping f of A into an ordered set B *preserves the order* $\leq$, if for any elements a and b from A, the inequality $a \leq b$ implies the inequality $f(a) \leq f(b)$.
(d) Addition + *preserves the order* $<$, if for any elements a, b, c and d from A, the inequalities $a < b$ and $d \leq c$ imply the inequality $a + d < b + c$.
(e) Multiplication $\circ$ *preserves the order* $<$, if for any elements a, b, c and d from A, the inequalities $a \leq b$ and $d \leq c$ imply the inequality $a \circ d \leq b \circ c$.
(f) A mapping f of A into an ordered set B *preserves the order* $<$, if for any elements a and b from A, the inequality $a < b$ implies the inequality $f(a) < f(b)$.

Example 2.1.22. In the conventional arithmetic $\boldsymbol{N}$ of all natural numbers, addition and multiplication preserve the natural order $\leq$ and the strict natural order $<$.

Example 2.1.23. In the conventional arithmetic $\boldsymbol{W}$ of all whole numbers, addition and multiplication preserve the natural order $\leq$ and the strict natural order $<$.

Example 2.1.24. In the conventional arithmetic $\boldsymbol{Z}$ of all integer numbers, addition preserves the natural order $\leq$ and the strict natural order $<$, while multiplication does not. Indeed, $3 \leq 5$ but $-5 = (-1) \cdot 5 \leq -3 = (-1) \cdot 3$ or $2 < 7$ but $-7 = (-1) \cdot 7 \leq -2 = (-1) \cdot 2$.

Example 2.1.25. In the conventional arithmetic $\boldsymbol{Q}$ of all rational numbers, addition preserves the natural order $\leq$ and the strict natural order $<$, while multiplication does not.

Example 2.1.26. In the conventional arithmetic $\boldsymbol{R}$ of all real numbers, addition preserves the natural order while multiplication does not.

We remind (cf., for example, Kurosh, 1963) that a binary operation $*$, e.g., addition $+$ or multiplication $\circ$, in an ordered structure is *monotone* if for any elements a, b and d from A, the inequality $a \leq b$ implies the inequalities $a * d \leq b * d$ and $d * a \leq d * b$ (for example, the inequalities $a \circ d \leq b \circ d$ and $d \circ a \leq d \circ b$) and the operation $*$ is *strictly monotone* if for any elements a, b and d from A, the inequality $a < b$ implies the inequalities $a * d < b * d$ and $d * a < d * b$ (for example, the inequalities $a \circ d < b \circ d$ and $d \circ a < d \circ b$).

Let us consider an abstract prearithmetic $\boldsymbol{A} = (A; +, \circ, \leq)$.

Lemma 2.1.6.

(a) *Addition* $+$ *in* $\boldsymbol{A}$ *preserves the order* $\leq$ *if and only if addition* $+$ *is monotone.*

(b) *Addition* $+$ *in* $\boldsymbol{A}$ *preserves the order* $<$ *if and only if addition* $+$ *is strictly monotone.*

Proof: (a) *Necessity* of the condition follows from definitions, which imply that monotonicity is a special case of order preservation.

Sufficiency. Let us suppose that addition $+$ in an abstract prearithmetic $\boldsymbol{A}$ is monotone. Then for any elements a, b, c and d from A, the inequalities $a \leq b$ and $c \leq d$ imply

$$a + c \leq b + c \leq b + d.$$

It means that addition $+$ preserves the order $\leq$.

(b) It is proved in a similar way.

Lemma is proved. □

Lemma 2.1.7. *In an abstract prearithmetic* $\boldsymbol{A} = (A; +, \circ, \leq)$, *multiplication* $\circ$ *preserves the order* $\leq$ *if and only if multiplication* $\circ$ *is monotone.*

Proof is similar to the proof of Lemma 2.1.6. □

Let us consider an ordered set X, in which the order $\leq$ in X is linear (total) and discrete. Then by Lemma 2.1.3, S is a mapping from X into X.

Proposition 2.1.8. *The mapping* S *preserves the order in* X.

Proof: Let us take two elements a and b from X such that $a \leq b$. If $a = b$, then $\mathrm{S}a = \mathrm{S}b$ and S preserves the order because it means that $\mathrm{S}a \leq \mathrm{S}b$.

Let us assume that $a < b$. As the order is linear, we have either $\mathrm{S}a = \mathrm{S}b$ or $\mathrm{S}b \leq \mathrm{S}a$ or $\mathrm{S}a \leq \mathrm{S}b$.

In the first case, we have

$$a < b < \mathrm{S}b = \mathrm{S}a.$$

It is impossible because there are no elements between a and $\mathrm{S}a$ (cf. condition (UD)).

In the second case, we have

$$a < b < \mathrm{S}b \leq \mathrm{S}a.$$

It is impossible because there are no elements between a and $\mathrm{S}a$ (cf. condition (UD)).

Thus, only the third case is possible, i.e., $\mathrm{S}a \leq \mathrm{S}b$. As a is an arbitrary element from X, it means that the mapping S preserves the order in X.

Proposition is proved. □

Corollary 2.1.4. *If* $a < b$, *then* $\mathrm{S}a < \mathrm{S}b$.

Corollary 2.1.5. *If* $a \neq b$, *then* $\mathrm{S}a \neq \mathrm{S}b$.

Because S *is a function, we also have the following result.*

Lemma 2.1.8. *If* $\mathrm{S}a \neq \mathrm{S}b$, *then* $a \neq b$.

Proof is left as an exercise. □

Note that the statement in Lemma 2.1.8 is one of the Peano axioms for arithmetic (cf. Peano, 1908; 1908a; Feferman, 1974; Hilbert and Bernays, 1968; Kleene, 2002).

In the conventional Diophantine arithmetic, addition and multiplication are commutative and associative. For arbitrary abstract prearithmetic, this is not always true. For instance, it is possible that $a_1 + a_2$ is not equal to $a_2 + a_1$. That is why for $n > 2$, we define n-ary operations using specific formulas. Here we consider two operations: n-ary addition and multiplication, which are defined by induction on n.

$$\sum_{i=1}^{1} a_i = a_1,$$

$$\sum_{i=1}^{2} a_i = a_1 + a_2.$$

If $\sum_{i=1}^{n-1} a_i$ is defined, then

$$\sum_{i=1}^{n} a_i = \left(\sum_{i=1}^{n-1} a_i\right) + a_n.$$

In the same way, we have

$$\prod_{i=1}^{1} a_i = a_1,$$

$$\prod_{i=1}^{2} a_i = a_1 \circ a_2.$$

If $\prod_{i=1}^{n-1} a_i$ is defined, then

$$\prod_{i=1}^{n} a_i = \left(\prod_{i=1}^{n-1} a_i\right) \circ a_n.$$

Note that when addition + is associative, we have

$$\sum_{i=1}^{n} a_i = a_1 + a_2 + \cdots + a_n$$

and when multiplication $\circ$ is associative, we have

$$\prod_{i=1}^{n} a_i = a_1 \circ a_2 \circ \cdots \circ a_n.$$

When all a_i are equal to the same element, say a, we use the following notation:

$$\sum_{i=1}^{n} a_i = n[a],$$

$$\prod_{i=1}^{n} a_i = [a]^n.$$

Although in the conventional Diophantine arithmetic, addition and multiplication are commutative and associative, for arbitrary abstract prearithmetic, this is not always true and as a result, it is possible to define other n-ary operations. For instance, it is possible to introduce the n-ary operation $\sum^o$, which equal to $\sum$ for $n = 1, 2$, while for > 2, it is different:

If $\sum\limits_{i=1}^{n-1} {}^\circ a_i$ is defined, then

$$\sum_{i=1}^{n} {}^\circ a_i = a_n + \left(\sum_{i=1}^{n-1} {}^\circ a_i\right).$$

Definitions imply the following result.

Lemma 2.1.9. *For any natural number n and any element a, we have $(n+1)[a] = n[a] + a$ and $[a]^{n+1} = [a]^n \circ a$.*

When addition $+$ is associative, it is possible to remove parentheses and we have

$$\sum_{i=1}^{n} a_i = a_1 + a_2 + \cdots + a_n,$$

$$n[a] = na$$

and when multiplication $\circ$ is associative, it is also possible to remove parentheses and we have

$$\prod_{i=1}^{n} a_i = a_1 \circ a_2 \circ \cdots \circ a_n,$$

$$[a]^n = a^n.$$

Here $\sum$ and $\prod$ are integral operations in the sense of Burgin and Karasik (1976) and Burgin (1982), that is, they can be applied to any finite number of elements from the prearithmetic $\boldsymbol{A}$.

Let us consider an abstract prearithmetic $\boldsymbol{A} = (A; +, \circ, \leq)$.

Proposition 2.1.9. *If the operation + preserves the order $\leq$, then the operation $\sum$ preserves the order $\leq$.*

Proof: As the operation $\sum$ is defined by induction, we prove the statement of the proposition also by induction. Let us take $n = 2$ and four elements a_{11}, a_{12}, a_{21} and a_{22} from A such that $a_{11} \leq a_{12}$ and $a_{21} \leq a_{22}$. Then as the operation + preserves the order $\leq$, we have

$$\sum_{i=1}^{2} a_{i1} = a_{11} + a_{21} \leq a_{12} + a_{22} = \sum_{i=1}^{2} a_{i2}$$

and consequently,

$$\sum_{i=1}^{2} a_{i1} \leq \sum_{i=1}^{2} a_{i2}.$$

Now let us assume that $a_{i1} \leq a_{i2}$ for $i = 1, 2, 3, \ldots, n$. By the induction assumption, this implies the inequality

$$\sum_{i=1}^{n-1} a_{i1} \leq \sum_{i=1}^{n-1} a_{i2}.$$

As the operation + preserves the order $\leq$, we have

$$\sum_{i=1}^{n} a_{i1} = \left(\sum_{i=1}^{n-1} a_{i1}\right) + a_{n1} \leq \left(\sum_{i=1}^{n-1} a_{i2}\right) + a_{n2} = \sum_{i=1}^{n} a_{i2}$$

and consequently,

$$\sum_{i=1}^{n} a_{i1} \leq \sum_{i=1}^{n} a_{i2}.$$

The principle of induction concludes the proof. □

The operation $\prod$ inherits properties of multiplication $\circ$.

Proposition 2.1.10. *If the operation $\circ$ preserves the order $\leq$, then the operation $\prod_{i=1}^{n}$ preserves the order $\leq$.*

Proof: As the operation $\prod_{i=1}^{n}$ is defined by induction, we prove the statement of the proposition also by induction. Let us take $n = 2$ and four elements a_{11}, a_{12}, a_{21} and a_{22} from A such that $a_{11} \leq a_{12}$ and $a_{21} \leq a_{22}$. Then as the operation $\circ$ preserves the order $\leq$, we have

$$\prod_{i=1}^{2} a_{i1} = a_{11} \circ a_{21} \leq a_{12} \circ a_{22} = \prod_{i=1}^{2} a_{i2}$$

and consequently,

$$\prod_{i=1}^{2} a_{i1} \leq \prod_{i=1}^{2} a_{i2}.$$

Now let us assume that $a_{i1} \leq a_{i2}$ for $i = 1, 2, 3, \ldots, n$. By the induction assumption, this implies the inequality

$$\prod_{i=1}^{n-1} a_{i1} \leq \prod_{i=1}^{n-1} a_{i2}.$$

As the operation $\circ$ preserves the order $\leq$, we have

$$\prod_{i=1}^{n} a_{i1} = \left(\prod_{i=1}^{n-1} a_{i1}\right) \circ a_{n1} \leq \left(\prod_{i=1}^{n-1} a_{i2}\right) \circ a_{n2} = \prod_{i=1}^{n} a_{i2}$$

and consequently,

$$\prod_{i=1}^{n} a_{i1} \leq \prod_{i=1}^{n} a_{i2}.$$

The principle of mathematical induction concludes the proof. □

Definition 2.1.3. (a) A mapping f, an abstract prearithmetic $\boldsymbol{A}_1 = (A_1; +_1, \circ_1, \leq_1)$ into an abstract prearithmetic $\boldsymbol{A}_2 = (A_2; +_2, \circ_2, \leq_2)$, *preserves addition*, or is an *additive homomorphism*, or a *homomorphism with respect to addition*, if for any elements a and b from A, we have

$$f(a +_1 b) = f(a) +_2 f(b).$$

(b) A mapping f, an abstract prearithmetic $\boldsymbol{A}_1 = (A_1; +_1, \circ_1, \leq_1)$ into an abstract prearithmetic $\boldsymbol{A}_2 = (A_2; +_2, \circ_2, \leq_2)$, *preserves multiplication*, or is a *multiplicative homomorphism*, or a *homomorphism with respect to multiplication*, if for any elements a and b from A, we have

$$f(a \circ_1 b) = f(a) \circ_2 f(b).$$

Let us study properties of additive and multiplicative homomorphisms.

Lemma 2.1.10. *If a mapping f, an abstract prearithmetic $\boldsymbol{A}_1 = (A_1; +_1, \circ_1, \leq_1)$ into an abstract prearithmetic $\boldsymbol{A}_2 = (A_2; +_2, \circ_2, \leq_2)$, preserves addition, then for all $n > 2$, it preserves the operation $\sum$.*

Proof is done by induction and left as an exercise. □

Lemma 2.1.11. *If a mapping f, an abstract prearithmetic $\boldsymbol{A}_1 = (A_1; +_1, \circ_1, \leq_1)$ into an abstract prearithmetic $\boldsymbol{A}_2 = (A_2; +_2, \circ_2, \leq_2)$, preserves multiplication, then for all $n > 2$, it preserves the operation $\prod$.*

Proof is done by induction and left as an exercise. □

2.1.2. *Archimedean property*

There is nothing more difficult to take in hand, more perilous
to conduct, or more uncertain in its success than
to take the lead in the introduction of a new order of things,
because the innovator has for enemies all those
who have done well under the old condition, and
lukewarm defenders in those who may do well under the new.

Machiavelli

The Diophantine (conventional) arithmetic of natural numbers has the so-called Archimedean property, which is named after the great ancient Greek mathematician Archimedes of Syracuse (ca 287–212 B.C.E.) and is important for proofs of many results in arithmetic and number theory. For instance, the Archimedean property, which is often called the Archimedean axiom, is important for proving that the set of all natural numbers and the set of all prime numbers are infinite. This property (axiom) is also very important for axiomatics in geometry (cf. Veronese, 1889; Hilbert, 1899).

The Austrian mathematician Otto Stolz (1842–1905), who was one of the first to introduce non-Archimedean systems in mathematics (Stolz, 1882; 1883; 1885; 1886), gave the name the Archimedean axiom to this property (Stolz, 1881; 1891) because it appeared in some of the works of Archimedes (cf. also Fisher, 1994). Stolz first stated the importance of the Archimedean axiom (property) for magnitudes in general and number systems, in particular. However, Archimedes credited it to another ancient Greek mathematician Eudoxus of Cnidus (ca 408–355 B.C.E.), who lived before Archimedes and was a student of Plato (Archimedes, 1980).

The Archimedean property (axiom) states that if we take any two (natural) numbers m and n, in spite that n may be enormously larger than m, it is always possible to add m enough times to itself, i.e., to take a sum $m + m + \cdots + m$, so that the result will be larger than n. Note that the arithmetics of whole numbers and of integer numbers do not have this property.

The Archimedean property is used to define special classes of groups, which are called Archimedean groups, special classes of semigroups, which are called Archimedean semigroups, and special classes of rings, which are called Archimedean rings (cf. Fuchs, 1963; Kurosh, 1963). Archimedean semigroups, groups and rings have many additional useful properties. For instance, all linearly ordered Archimedean groups are commutative (Hölder, 1901).

In contrast to the Diophantine arithmetic, the Archimedean property is invalid in many non-Diophantine arithmetics. Other examples of non-Archimedean arithmetics are:

- the arithmetic of cardinal numbers (cf., for example, Abian, 1965; Kuratowski and Mostowski, 1967);
- the non-standard arithmetic of hyperreal numbers (Robinson, 1966);
- the arithmetic of real hypernumbers (Burgin, 2012).

However, these arithmetics include infinitely big (and infinitely small as in the non-standard arithmetic) numbers, thus, going far beyond natural and even real numbers.

Note that all these arithmetics, as well as other non-Archimedean arithmetics (cf., for example, Du Bois-Reymond, 1870/1871; 1875; 1877; Stolz, 1882; 1883; 1885; 1886; Hahn, 1907; Hardy, 1910; Enriques, 1911; Laugwitz, 1961; 1961a; Robinson, 1961; 1966; Conway, 1976; 1994; Tall, 1980; Ehrlich, 1992; 1994; 2012; Burgin, 2002; 2004; 2008a; 2012; 2017) have been either oriented at measurement or elaborated for dealing with infinities. As a result, all of them contained the Diophantine arithmetic of whole numbers without any changes.

In abstract prearithmetics, there are three principal structures — one is relational and two are operational. This shows that for abstract prearithmetics, it is natural to consider five Archimedean properties, which do not coincide in the general case.

Definition 2.1.4. (a) An abstract prearithmetic $\boldsymbol{A} = (A; +, \circ, \leq)$ satisfies the *Successively Archimedean Property* (SAP), or is a *successively Archimedean prearithmetic* (SAPA), if it has the successor function S and for any elements a and b from A, the inequality $a < b$ implies that it is possible to get an element larger than b from the element a by repeating the operation S, i.e., there is a natural number n such that $S^n a$ is larger than or equal to b.

(b) An abstract prearithmetic $\boldsymbol{A} = (A; +, \circ, \leq)$ satisfies the *Additively Archimedean Property* (AAP), or is *additively Archimedean prearithmetic* (AAPA), if for any elements a and b from A, the inequality $a < b$ implies that it is possible to get an element larger than b from the element a by adding a to itself several times, i.e., there is a natural number n such that

$$b \leq n[a]. \tag{2.1}$$

(c) An abstract prearithmetic $\boldsymbol{A} = (A; +, \circ, \leq)$ satisfies the *Multiplicatively Archimedean Property* (MAP), or is *multiplicatively Archimedean prearithmetic* (MAPA), if for any elements a and b from A, the inequality $a < b$ implies that it is possible to get an element larger than b from the element a by multiplying a by itself several times, i.e., there is a natural number n such that

$$b \leq [a]^n. \tag{2.2}$$

(d) An abstract prearithmetic $\boldsymbol{A} = (A; +, \circ, \leq)$ with the additive 0 satisfies the *left Binary Archimedean Property* for addition (BAAPL), or is a *binary for addition Archimedean prearithmetic* (BAAPAL) from the left, if for any elements a and b from A, the inequality $1 < a < b$ implies that there is an element q less than b such that

$$b \leq q + a \tag{2.3}$$

and satisfies the *right Binary Archimedean Property* for addition (BAAPR), or is a *binary for addition Archimedean prearithmetic* (BAAPAR) from the right, if for any elements a and b from A, the inequality $1 < a < b$ implies that there is an element q less than b such that

$$b \leq a + q. \tag{2.4}$$

(e) An abstract prearithmetic $\boldsymbol{A} = (A; +, \circ, \leq)$ with the multiplicative 1 satisfies the *left Binary Archimedean Property for multiplication* (BAMPL), or is a *binary Archimedean prearithmetic* (BAMPAL) from the left, if for any elements a and b from A, the inequality $1 < a < b$ implies that there is an element q less than b such that

$$b \leq q \circ a. \tag{2.5}$$

and satisfies the *right Binary Archimedean Property* for multiplication (BAMPR), or is a *binary for multiplication Archimedean prearithmetic* (BAMPAR) from the right, if for any elements a and b from A, the inequality $1 < a < b$ implies that there is an element q less than b such that

$$b \leq a \circ q. \tag{2.6}$$

Note that if multiplication $\circ$ is commutative, then the right Binary Archimedean Property for multiplication coincides with the left Binary Archimedean Property for multiplication. When an abstract prearithmetic has both the right and left Binary Archimedean Properties for multiplication, then it has the *Binary Archimedean Property for multiplication.*

When addition $+$ is commutative, then the right Binary Archimedean Property for addition coincides with the left Binary Archimedean Property for addition. When an abstract prearithmetic has both the right and left Binary Archimedean Properties for addition, then it has the *Binary Archimedean Property for addition.*

Example 2.1.27. The conventional arithmetic $2\boldsymbol{N}$ of all even numbers has all four properties: the Successively Archimedean Property (SAP), Binary Archimedean Property (BAP) for addition and multiplication, Additively Archimedean Property (AAP) and Multiplicatively Archimedean Property (MAP).

Example 2.1.28. The conventional arithmetic $3\boldsymbol{N}$ of all natural numbers divisible by 3 has all four properties: the Successively Archimedean Property (SAP), Binary Archimedean Property (BAP) for addition and multiplication, Additively Archimedean Property (AAP) and Multiplicatively Archimedean Property (MAP).

However, in general, these properties are independent because there are prearithmetics and arithmetics, which have only one part of the Archimedean Properties.

Example 2.1.29. The conventional Diophantine arithmetic $\boldsymbol{N}$ of all natural numbers has the Successively Archimedean Property (SAP), Additively Archimedean Property (AAP) and Binary Archimedean Property (BAP) for addition and multiplication but does not satisfy the Multiplicatively

Archimedean Property (MAP) because this property does not hold for the number 1.

That is why when the Archimedean Property is defined for multiplicative groups or semigroups, its validity is assumed for all elements but the unit element e (Fuchs, 1963).

Example 2.1.30. The conventional Diophantine arithmetic $\boldsymbol{W}$ of all whole numbers has the Successively Archimedean Property (SAP) and Binary Archimedean Property (BAP) for addition and multiplication but does not have the Multiplicatively Archimedean Property (MAP) and Additively Archimedean Property (AAP) because these properties do not hold for the number 0.

That is why when the Archimedean Property is defined for additive groups or semigroups, its validity is assumed for all elements but the zero "0" (Fuchs, 1963).

Example 2.1.31. The conventional arithmetic $\boldsymbol{R}_1$ of all larger than 1 real numbers does not have the Successively Archimedean Property (SAP) because real numbers do not have successors but has the Binary Archimedean Property (BAP) for addition and multiplication, Multiplicatively Archimedean Property (MAP), and Additively Archimedean Property (AAP).

There are also prearithmetics and arithmetics, which do not have any of the Archimedean Properties.

Example 2.1.32. The arithmetic $\boldsymbol{Ord}$ of all ordinal numbers does not have the Successively Archimedean Property (SAP), Multiplicatively Archimedean Property (MAP), Binary Archimedean Property (BAP) and Additively Archimedean Property (AAP) (Fraenkel and Bar-Hillel, 1958).

Example 2.1.33. The arithmetic $\boldsymbol{NW}$ of all non-standard whole numbers does not have the Successively Archimedean Property (SAP), Multiplicatively Archimedean Property (M AP), Additively Archimedean Property (AAP) and Binary Archimedean Property (BAP) (Robinson, 1966).

Example 2.1.34. The arithmetic $\boldsymbol{NH}$ of all whole hypernumbers does not have the Successively Archimedean Property (SAP), Multiplicatively Archimedean Property (MAP), Additively Archimedean Property (AAP) and Binary Archimedean Property (BAP) (Burgin, 2012).

Proposition 2.1.11. *Any multiplicatively Archimedean prearithmetic (MAPA) is a binary Archimedean prearithmetic (BAPA).*

Proof: Let us consider a multiplicatively Archimedean abstract prearithmetic $\boldsymbol{A} = (A; +, \circ, \leq)$. By Definition 2.1.4, it satisfies condition (2.3), i.e., for some natural number n, we have $b \leq [a]^n = (\cdots(((a \circ a) \circ a) \circ a) \cdots \circ a$. As $a < b$ and $1 \leq n$, we can take the least n such that $b \leq a^n$. It means that $q = a^{n-1} < b$ and $b \leq (a^{n-1}) \circ a = q \circ a$.

Proposition is proved. □

Let us consider an abstract prearithmetic $\boldsymbol{A} = (A; +, \circ, \leq)$ with a discrete order $\leq$ and addition $+$ preserves the order $\leq$.

Lemma 2.1.12. *If* $\mathrm{S}b \leq b + a$ *for any elements* a *and* b *from* A, *then for any natural number* n, *we have* $\mathrm{S}^n a \leq (n+1)[a]$.

Proof: We use induction on n to show that $\mathrm{S}^n a \leq (n+1)[a]$.

For $n = 1$, taking a as b, we have

$$\mathrm{S}a \leq a + a = 2[a].$$

For $n = 2$, taking $\mathrm{S}a$ as b, we have

$$\mathrm{S}^2 a = \mathrm{S}(\mathrm{S}a) \leq \mathrm{S}a + a \leq (a + a) + a = 3[a]$$

as addition $+$ preserves the order $\leq$.

Let us assume that our statement is true for $n - 1$, i.e.,

$$\mathrm{S}^{n-1} a \leq n[a].$$

Then we have

$$\mathrm{S}^n a = \mathrm{S}(\mathrm{S}^{n-1} a) \leq \mathrm{S}^{n-1} a + a \leq n[a] + a = (n+1)[a]$$

as addition $+$ preserves the order $\leq$. The principle (axiom) of the mathematical induction gives us the necessary result.

Lemma is proved. □

Proposition 2.1.12. *If in a successively Archimedean prearithmetic* $\boldsymbol{A} = (A; +, \circ, \leq)$, *we have* $\mathrm{S}b \leq b + a$ *for any elements* a *and* b *from* A *and addition* $+$ *preserves the order* $\leq$, *then* $\boldsymbol{A}$ *is an additively Archimedean prearithmetic.*

Proof: Let us consider elements a and b from A such that $a < b$. As $\boldsymbol{A}$ is a successively Archimedean prearithmetic, there is a natural number n such that $\mathrm{S}^n a$ is larger than or equal to b, i.e., $b \leq \mathrm{S}^n a$. Then by Lemma 2.1.12, we have

$$b \leq \mathrm{S}^n a \leq (n+1)[a].$$

It means that $\boldsymbol{A}$ is an additively Archimedean prearithmetic.

Proposition is proved. □

Elements 0 and 1 have very special properties in the conventional Diophantine arithmetic. Let us explore these properties in the general setting of abstract prearithmetics. To do this, we formalize properties of these numbers.

Let us consider an abstract prearithmetic $\boldsymbol{A} = (A; +, \circ, \leq)$.

Definition 2.1.5.

(a) An element z, which is usually denoted by 0 or 0_A, is called an *additive zero* of $\boldsymbol{A}$ if $a + z = z + a = a$ for any element a from A.
(b) An element z, which is usually denoted by 0 or 0_{mA}, is called a *multiplicative zero* of $\boldsymbol{A}$ if $a \circ z = z \circ a = z$ for any element a from A.
(c) An element b, which is usually denoted by 1 or 1_A, is called a *multiplicative one* of $\boldsymbol{A}$ if $a \circ b = b \circ a = a$ for any element a from A.

Lemma 2.1.13. *An additive zero is unique.*

Indeed, suppose an abstract prearithmetic $\boldsymbol{A}$ has two additive zeros 0_1 and 0_2. Then we have

$$0_1 = 0_1 + 0_2 = 0_2.$$

Lemma 2.1.14. *A multiplicative zero is unique.*

Indeed, suppose an abstract prearithmetic $\boldsymbol{A}$ has two multiplicative zeros 0_1 and 0_2. Then we have

$$0_1 = 0_1 \circ 0_2 = 0_2.$$

Lemma 2.1.15. *A multiplicative one* 1 *is unique.*

Indeed, suppose an abstract prearithmetic $\boldsymbol{A}$ has two multiplicative ones 1_1 and 1_2. Then we have

$$1_1 = 1_1 \circ 1_2 = 1_2.$$

The number 0 in the conventional Diophantine arithmetic is both additive and multiplicative zero while the number 1 is a multiplicative one. However, in a general case of abstract prearithmetics, additive and multiplicative zeros do not coincide as the following examples demonstrate.

Example 2.1.35. Let us define an abstract prearithmetic $\boldsymbol{A} = (N; \oplus, \otimes, \leq)$ where N is the set of all natural numbers by the following rules:

$$m \oplus n = m + n,$$

$$m \otimes n = m \cdot n + 3,$$

where $m+n$ are arbitrary natural numbers, while $+$ is conventional addition and $\cdot$ is conventional multiplication of natural numbers.

We can see that 0 is an additive zero but not a multiplicative zero in $\boldsymbol{A}$.

Example 2.1.36. Let us define an abstract prearithmetic $\boldsymbol{A} = (Z; \oplus, \otimes, \leq)$ where Z is the set of all integer numbers by the following rules:

$$m \oplus n = m + n + 2,$$

$$m \otimes n = m \cdot n,$$

where $m+n$ are arbitrary integer numbers, while $+$ is conventional addition and $\cdot$ is conventional multiplication of integer numbers.

We can see that 0 is a multiplicative zero but not an additive zero in $\boldsymbol{A}$. At the same time, -2 is an additive zero in $\boldsymbol{A}$.

However, in some cases, additive and multiplicative zeros coincide.

As we know, elements of some arithmetics have opposite elements. For instance, number -4 is opposite to number 4. Let us explore opposite elements in abstract prearithmetics.

Definition 2.1.6.

(a) An element c of an abstract prearithmetic $\boldsymbol{A}$ with the additive zero $0_{\boldsymbol{A}}$ is *opposite from the right* to an element d if $d + c = 0_{\boldsymbol{A}}$.
(b) An element c of an abstract prearithmetic $\boldsymbol{A}$ with the additive zero $0_{\boldsymbol{A}}$ is *opposite from the left* to an element d if $c + d = 0_{\boldsymbol{A}}$.
(c) An element c of an abstract prearithmetic $\boldsymbol{A}$ with the additive zero $0_{\boldsymbol{A}}$ is *opposite* to an element d if it is opposite from the right and from the left to d. It is denoted as $-d$.

Lemma 2.1.16.

(a) *If an element c of an abstract prearithmetic* $\boldsymbol{A} = (A; +, \circ, \leq)$ *with the additive zero* $0_{\boldsymbol{A}}$ *is opposite from the right to an element d, then d is opposite from the left to c.*
(b) *For any element c from an abstract prearithmetic* $\boldsymbol{A}$, *we have* $-(-c) = c$.

Proof is left as an exercise. □

Lemma 2.1.17. *For any elements a and c from a distributive abstract prearithmetic* $\boldsymbol{A}$, *we have* $a \circ (-c) = -a \circ c = -(a \circ c)$.

Indeed,

$$(a \circ c) + (a \circ (-c)) = a \circ (c + (-c)) = a \circ 0_{\boldsymbol{A}} = 0_{\boldsymbol{A}}$$

and

$$(a \circ c) + (-a \circ c) = (a + (-a)) \circ c = 0_{\boldsymbol{A}} \circ c = 0_{\boldsymbol{A}}.$$

Consequently, $a \circ (-c) = -a \circ c = -(a \circ c)$.

Existence of opposite elements implies coincidence of additive and multiplicative zeros.

Proposition 2.1.13. *If an abstract prearithmetic* $\boldsymbol{A} = (A; +, \circ, \leq)$ *has an additive zero* 0, *contains an opposite element* $-x$ *for each element* x, *multiplication is distributive with respect to addition and preserves opposite elements, i.e.,* $z \circ (-x) = -(z \circ x)$ *for any elements* z *and* x *from* A, *then* 0 *is also a multiplicative zero.*

Proof: Let us take an abstract prearithmetic $\boldsymbol{A} = (A; +, \circ, \leq)$ that has an additive zero 0, contains an opposite element $-x$ for each element x, multiplication is distributive with respect to addition and preserves opposite elements. Taking arbitrary elements x and z from $\boldsymbol{A}$, we have

$$0 \circ x = (z + (-z)) \circ x = z \circ x + (-z) \circ x = z \circ x + -(z \circ x) = 0.$$

The identity $x \circ 0 = 0$ is proved in a similar way. Consequently, 0 is also a multiplicative zero.

Proposition is proved. □

As a corollary, we obtain a well-known result from the theory of rings (cf. Kurosh, 1963).

Corollary 2.1.6. *In a ring, additive and multiplicative zeros coincide.*

Lemma 2.1.18. *An additively Archimedean prearithmetic cannot have an additive zero* 0.

Indeed, if $0 < b$, then any element $n[0] = 0 + \cdots + 0$ is equal to 0 and still less than b.

Lemma 2.1.19. *A multiplicatively Archimedean prearithmetic cannot have a multiplicative one "1" or a multiplicative zero "0".*

Indeed, if $1 < b$, then any element $[1]^n$ is equal to 1 and still less than b. In a similar way, if $0 < b$, then any element $[0]^n$ is equal to 0 and still is less than b.

Proposition 2.1.14. *If in a successively Archimedean prearithmetic* $\boldsymbol{A} = (A; +, \circ, \leq)$ *with the additive zero* 0, *addition* + *preserves the order* $\leq$ *and* 0 *is the least element in* $\boldsymbol{A}$, *then* $\boldsymbol{A}_{\mathrm{P}} = (A\backslash\{0\}; +, \circ, \leq)$ *is an additively Archimedean prearithmetic.*

Proof: As $0 < a$ for any element a from $\boldsymbol{A}_{\mathrm{P}}$, we have $b = b + 0 \leq b + a$. By Definition 2.1.4, we have

$$b < \mathrm{S}b \leq b + a.$$

Thus, by Proposition 2.1.12, $\boldsymbol{A}_{\mathrm{P}}$ is an additively Archimedean prearithmetic.

Proposition is proved. □

Let us consider an abstract prearithmetic $\boldsymbol{A} = (A; +, \circ, \leq)$ with a discrete order $\leq$ and multiplication $\circ$ preserves the order $\leq$.

Lemma 2.1.20. *If* $\mathrm{S}b \leq b \circ a$ *for any elements* a *and* b *from* A, *then for any natural number* n, *we have* $\mathrm{S}^n a \leq [a]^{n+1}$.

Proof: We use induction on n to show that $\mathrm{S}^n a \leq [a]^{n+1}$.

For $n = 1$, taking a as b, we have

$$\mathrm{S}a \leq a \circ a = [a]^2.$$

For $n = 2$, taking $\mathrm{S}a$ as b, we have

$$\mathrm{S}^2 a = \mathrm{S}(\mathrm{S}a) \leq \mathrm{S}a \circ a \leq (a \circ a) \circ a = [a]^3$$

as multiplication $\circ$ preserves the order $\leq$.

Let us assume that our statement is true for $n-1$, i.e.,

$$\mathrm{S}^{n-1}a \leq [a]^n.$$

Then we have

$$\mathrm{S}^n a = \mathrm{S}(\mathrm{S}^{n-1}a) \leq \mathrm{S}^{n-1}a \circ a \leq [a]^n \circ a = [a]^{n+1}$$

as multiplication $\circ$ preserves the order $\leq$. The principle (axiom) of the mathematical induction gives us the necessary result.

Lemma is proved. □

Proposition 2.1.15. *If in a successively Archimedean prearithmetic* $\boldsymbol{A} = (A; +, \circ, \leq)$ *with a discrete order* $\leq$, *we have* $\mathrm{S}b \leq b \circ a$ *for any elements* a *and* b *from* A *and multiplication* $\circ$ *preserves the order* $\leq$, *then* $\boldsymbol{A}$ *is a multiplicatively Archimedean prearithmetic.*

Proof: Let us consider elements a and b from A such that $a < b$. As $\boldsymbol{A}$ is a successively Archimedean prearithmetic, there is a natural number n such that $\mathrm{S}^n a$ is larger than or equal to b, i.e., $b \leq \mathrm{S}^n a$. Then by Lemma 2.1.20, we have

$$b \leq \mathrm{S}^n a \leq [a]^{n+1}.$$

It means that $\boldsymbol{A}$ is a multiplicatively Archimedean prearithmetic.

Proposition is proved. □

Corollary 2.1.7. *If in a successively Archimedean prearithmetic* $\boldsymbol{A} = (A; +, \circ, \leq)$ *with a discrete order* $\leq$, *we have* $\mathrm{S}b \leq b \circ a$ *for any elements* a *and* b *from* A *and multiplication* $\circ$ *preserves the order* $\leq$, *then* $\boldsymbol{A}$ *is a binary Archimedean prearithmetic.*

Proposition 2.1.16. *If in a successively Archimedean prearithmetic* $\boldsymbol{A} = (A; +, \circ, \leq)$ *with the multiplicative* 1, *multiplication* $\circ$ *preserves the order* $\leq$ *and* 1 *is the smallest element in* $\boldsymbol{A}$, *then* $\boldsymbol{A}_{\mathrm{C}} = (A\backslash\{1\}; +, \circ, \leq)$ *is a multiplicatively Archimedean prearithmetic.*

Proof: As $0 < a$ for any element a from $\boldsymbol{A}_C$, we have $b = b \circ 1 \leq b \circ a$. By Definition 2.1.4, we have

$$b \leq \mathrm{S}a \leq b \circ a.$$

Thus, by Proposition 2.1.15, $\boldsymbol{A}_C$ is a multiplicatively Archimedean prearithmetic.

Proposition is proved. □

Let us study relations between multiplication and addition.

Lemma 2.1.21. *If $b + a \leq b \circ a$ for any elements a and b from A, then for any natural number n, we have $n[a] \leq [a]^n$.*

Proof: We use induction on n to prove the lemma.

For $n = 2$, taking a as b, we have

$$2[a] = a + a \leq a \circ a = [a]^2.$$

Let us assume that our statement is true for $n - 1$, i.e.,

$$n[a] \leq [a]^n.$$

Then we have

$$(n+1)[a] = n[a] + a \leq [a]^n \circ a = [a]^{n+1}.$$

The principle (axiom) of the mathematical induction gives us the necessary result.

Lemma is proved. □

Proposition 2.1.17. *If in an additively Archimedean prearithmetic $\boldsymbol{A} = (A; +, \circ, \leq)$, we have $b + a \leq b \circ a$ for any elements a and b from A, then $\boldsymbol{A}$ is a multiplicatively Archimedean prearithmetic.*

Proof: Let us consider elements a and b from A such that $a < b$. As $\boldsymbol{A}$ is an additively Archimedean prearithmetic, there is a natural number n such that $n[a]$ is larger than or equal to b, i.e., $b \leq n[a]$. Then by Lemma 2.1.18, we have

$$b \leq n[a] \leq [a]^n.$$

Proposition is proved. □

Corollary 2.1.8. *If in an additively Archimedean prearithmetic $\boldsymbol{A} = (A; +, \circ, \leq)$, we have $b + a \leq b \circ a$ for any elements a and b from A, then $\boldsymbol{A}$ is a binary Archimedean prearithmetic.*

Example 2.1.37. In the conventional arithmetic $2\boldsymbol{N}$ of all even numbers, $b + a \leq b \circ a$ for any elements a and b from A. Thus, $2\boldsymbol{N}$ is an additively Archimedean prearithmetic and multiplicatively Archimedean prearithmetic.

Lemma 2.1.22. *If $b \circ a \leq b + a$ for any elements a and b from A, then for any natural number n, we have $[a]^n \leq n[a]$.*

Proof is similar to the proof of Lemma 2.1.21. □

Proposition 2.1.18. *If in a multiplicatively Archimedean prearithmetic* $\boldsymbol{A} = (A; +, \circ, \leq)$, *we have* $b \circ a \leq b + a$ *for any elements* a *and* b *from* A, *then* $\boldsymbol{A}$ *is an additively Archimedean prearithmetic.*

Proof is similar to the proof of Proposition 2.1.17. □

Definition 2.1.7.

(a) Addition in an abstract prearithmetic $\boldsymbol{A} = (A; +, \circ, \leq)$ is *increasing* if $a < b + a$ and $b < b + a$ for any elements a and b from A.
(b) Multiplication in an abstract prearithmetic $\boldsymbol{A} = (A; +, \circ, \leq)$ is *increasing* if $a < b \circ a$ and $b < b \circ a$ for any elements a and b from A.

Example 2.1.38. In the conventional arithmetic $2\boldsymbol{N}$ of all even numbers, multiplication is increasing.

Example 2.1.39. In the conventional Diophantine arithmetic $\boldsymbol{N}$, addition is increasing.

Proposition 2.1.19. *If in an abstract prearithmetic* $\boldsymbol{A} = (A; +, \circ, \leq)$ *with linear order and increasing addition, any bounded chain in* $\boldsymbol{A}$ *is finite, then* $\boldsymbol{A}$ *is an additively Archimedean prearithmetic.*

Proof: As addition is increasing in $\boldsymbol{A}$, we have

$$a < n[a] < (n + 1)[a],$$

where $n = 1, 2, 3, \ldots$. If $a < b$, then $b \leq n[a]$ for some n because order $\leq$ is linear and any bounded chain in $\boldsymbol{A}$ is finite.

Proposition is proved. □

Proposition 2.1.20. *If in an abstract prearithmetic* $\boldsymbol{A} = (A; +, \circ, \leq)$ *with linear order and increasing multiplication, any bounded chain in* $\boldsymbol{A}$ *is finite, then* $\boldsymbol{A}$ *is an multiplicatively Archimedean prearithmetic.*

Proof: As multiplication is increasing in $\boldsymbol{A}$, we have

$$a < [a]^n < [a]^{n+1},$$

where $n = 1, 2, 3, \ldots$. If $a < b$, then $b \leq [a]^n$ for some n because order $\leq$ is linear and any bounded chain in $\boldsymbol{A}$ is finite.

Proposition is proved. □

Proposition 2.1.21. *In an abstract prearithmetic* $\boldsymbol{A} = (A; +, \circ, \leq)$, *which has the additive zero* 0 *as its least element and in which addition of non-zero elements preserves the strict order, addition is increasing.*

Proof is left as an exercise. □

Let us introduce and study stronger than Archimedean properties.

Definition 2.1.8. (a) An abstract prearithmetic $\boldsymbol{A} = (A; +, \circ, \leq)$ is *Exactly Successively Archimedean* (ESAPA) if it has the successor function S and for any elements a and b from A, the inequality $a < b$ implies that it is possible to get an element equal to b from the element a by repeating the operation S, i.e., there is a natural number n such that $\mathrm{S}^n a$ is equal to b.

(b) An abstract prearithmetic $\boldsymbol{A} = (A; +, \circ, \leq)$ is *Exactly Additively Archimedean* (EAAPA) if there is an element d from A, which is called the *additive generator* of $\boldsymbol{A}$, such that for any elements a and b from A, the inequality $a < b$ implies that there is a natural number n such that

$$a + n[d] = b. \tag{2.7}$$

(c) An abstract prearithmetic $\boldsymbol{A} = (A; +, \circ, \leq)$ is *Exactly Multiplicatively Archimedean* (EMAPA) if there is an element d from A, which is called the *multiplicative generator* of $\boldsymbol{A}$, such that for any elements a and b from A, the inequality $a < b$ implies that there is a natural number n such that

$$a \circ [d]^n = b. \tag{2.8}$$

(d) An abstract prearithmetic $\boldsymbol{A} = (A; +, \circ, \leq)$ with the multiplicative 1 satisfies the *left Exactly Binary Archimedean Property* for addition (EBAAPL), or is *exactly additive prearithmetic* (EAPAL) from the left, if for any elements a and b from A, the inequality $1 < a < b$ implies that there is an element q less than b such that

$$b = q + a. \tag{2.9}$$

and satisfies the *right Exactly Binary Archimedean Property* (EBAAPR), or is *exactly additive prearithmetic* (EAPAR) from the right, if for any elements a and b from A, the inequality $1 < a < b$ implies that there is an element q less than b such that

$$b = a + q. \tag{2.10}$$

(e) An abstract prearithmetic $\boldsymbol{A} = (A; +, \circ, \leq)$ with the multiplicative 1 satisfies the *left Exactly Binary Archimedean Property* for multiplication (EBAMPL), or is *exactly multiplicative prearithmetic* (EMPAL) from the left, if for any elements a and b from A, the inequality $1 < a < b$ implies that there is an element q less than b such that

$$b = q \circ a. \tag{2.11}$$

and satisfies the *right Exactly Binary Archimedean Property* for multiplication (BAPR), or is *exactly multiplicative prearithmetic* (EMPAR) from the right, if for any elements a and b from A, the inequality $1 < a < b$ implies that there is an element q less than b such that

$$b = a \circ q. \tag{2.12}$$

Example 2.1.40. The conventional arithmetic $\boldsymbol{N}$ of all natural numbers is Exactly Successively Archimedean (ESAPA) and Exactly Additively Archimedean (EAAPA) because 1 is the additive generator of $\boldsymbol{N}$ but is not Exactly Multiplicatively Archimedean (EMAPA).

Example 2.1.41. The conventional arithmetic $\boldsymbol{W}$ of all whole numbers is also Exactly Successively Archimedean (ESAPA) and Exactly Additively Archimedean (EAAPA) because 1 is the additive generator of $\boldsymbol{W}$ but is not Exactly Multiplicatively Archimedean (EMAPA).

To find an Exactly Multiplicatively Archimedean prearithmetic (EMAPA) is more difficult. However, such prearithmetics exist.

Example 2.1.42. Let us consider the set PT of all powers of 2, i.e., $\{2,\ 2^2, 2^3, \ldots\}$, with the standard order $\leq$ and introduce the following operations:

$$a \oplus b = a \cdot b,$$

$$a \otimes b = a \cdot b.$$

Then the system $\boldsymbol{PT} = (PT; \oplus, \otimes, \leq)$ is an Exactly Multiplicatively Archimedean (EMAPA) and Exactly Additively Archimedean (EAAPA) prearithmetic, in which 2 is both the additive generator and the multiplicative generator.

Remark 2.1.4. Exact Archimedean properties are intrinsically related to the concept of the natural order in partially ordered groupoids, groups and semigroups. We remind that if H is a partially ordered groupoid (semigroup, group), then its order is natural if $a < b$ implies $ax = ya = b$ for some elements a and b from H. Thus, the order in an Exact Archimedean partially ordered groupoid (semigroup, group) is natural.

Lemma 2.1.23. *A prearithmetic is Exactly Successively Archimedean (ESAPA) if and only if it is Successively Archimedean (SAPA).*

Proof: *Necessity.* Any Exactly Successively Archimedean prearithmetic (ESAPA) is Successively Archimedean (SAPA) because for any element a in a partially ordered set, we have $a \leq a$, i.e., if $\mathrm{S}^n a = b$ then $\mathrm{S}^n a \leq b$.

Sufficiency. Let us assume that an abstract prearithmetic $\boldsymbol{A}$ is Successively Archimedean and for some elements a and b from A, we have $a < b$. Then by definition, there is the least natural number n such that $b \leq \mathrm{S}^n a$. It means that we have the inequalities

$$\mathrm{S}^{n-1} a \leq b \leq \mathrm{S}^n a.$$

Because $\mathrm{S}^n a = \mathrm{S}(\mathrm{S}^{n-1} a)$, we have either $b = \mathrm{S}^{n-1} a$ or $b = \mathrm{S}^n a$. Consequently, the abstract prearithmetic $\boldsymbol{A}$ is Exactly Successively Archimedean.

Lemma is proved. □

Remark 2.1.5. For Exactly Additively Archimedean prearithmetics (EAAPA) and Exactly Multiplicatively Archimedean prearithmetics (EMAPA) a similar statement is not always true as the following examples demonstrate.

Example 2.1.43. The arithmetic $\boldsymbol{W}$ of all whole numbers is Exactly Additively Archimedean (EAAPA) because it has a generator 1 but it is not Additively Archimedean (AAPA) because in it, the sum of any numbers of zeros is equal to zero.

Example 2.1.44. The arithmetic $2\boldsymbol{N}$ of all even numbers is Multiplicatively Archimedean (EAAPA) because it has a generator 1 but it is not Exactly Multiplicatively Archimedean (AAPA) because in it, the sum of any number of zeros is equal to zero.

Lemma 2.1.24. *In an Exactly Additively Archimedean prearithmetic* $\boldsymbol{A} = (A; +, \circ, \leq)$ *with linear order and associative commutative addition,* *which strictly preserves order, either* 0 *or the additive generator* d *of* $\boldsymbol{A}$*, which is not maximal, is the least element.*

Proof: At first, we show that any element $b = n[d]$ if $d \leq b$. As $\boldsymbol{A}$ is Exactly Additively Archimedean, in this case, $b = d + n[d]$. As addition is commutative, $b = n[d] + d = (n+1)[d]$.

At the same time, if there is an element b with $d < b$, then assuming $d + d \leq d$, we obtain

$$(d + d) + d \leq d + d \leq d.$$

By induction, we can prove that $n[d] \leq d < b$ for any natural number n and the equality $b = n[d]$ becomes impossible. Consequently, we have

$$d < d + d = 2[d] < \cdots < n[d] < (n+1)[d] < \cdots$$

as addition strictly preserves order and order relation is transitive.

Now let us suppose there is an element a that is less than d.

As $\boldsymbol{A}$ is Exactly Additively Archimedean and d is the additive generator of $\boldsymbol{A}$, in this case, $d = a + n[d]$. As we demonstrated, we have $d < n[d]$. Then by the same token, if $d < a + d$, then $d < a + n[d]$ for any natural number n. If $d > a + d$, then $d > a + n[d]$ for any natural number n. Thus, we come to conclusion that $d = a + d$.

Applying mathematical induction, we see that $a + n[d] = n[d]$ for any natural number n. Thus, for any element b larger than d, we have $b = a + b$, i.e., a is the additive zero for all elements $b \geq d$.

Let us suppose there is an element c that is less than a. Then $a = c + n[d]$ and $d = c + m[d]$ because $c < a < d$ and d is the additive generator of $\boldsymbol{A}$.

At the same time, $c + d < c + n[d] < d$. Thus,

$$d > c + d > c + d + d = 2[d] > \cdots > m[d].$$

This contradict the equality $d = c + m[d]$ demonstrating that a is the least element and the additive zero in $\boldsymbol{A}$.

Lemma is proved. □

The Exactly Additively Archimedean Property allows representing successors Sa using addition as it is done in the conventional Diophantine arithmetic $\boldsymbol{N}$ of natural numbers where S$n = n + 1$. The same is true for many abstract prearithmetics.

Proposition 2.1.22. *If* $\boldsymbol{A} = (A; +, \circ, \leq)$ *is an Exactly Additively Archimedean prearithmetic (ESAPA) with the successor function* S*, an* *additive generator d and strictly monotone associative addition, then* $\mathrm{S}a = a + d$ *for any element a from A.*

Proof: Let us take an Exactly Additively Archimedean prearithmetic $\boldsymbol{A} = (A; +, \circ, \leq)$ with the successor function S, the additive generator d and monotone addition. As by definition $a > \mathrm{S}a$ and the prearithmetic $\boldsymbol{A}$ is Exactly Additively Archimedean, we have

$$\mathrm{S}a = a + n[d]. \tag{2.13}$$

As the order in $\boldsymbol{A}$ is linear, we have three options: $a > a + d, a = a + d$, or $a < a + d$.

If we have the first option, i.e., $a + d < a$, then

$$a + 2[d] = a + d + d < a + d < a,$$

because addition is strictly monotone and associative. By induction, for any natural number n, we have

$$a + n[d] < a.$$

This contradicts the equality (2.13) and shows that the first option is impossible.

If we have the second option, i.e., $a + d = a$, then

$$a + 2[d] = a + d + d = a + d = a,$$

because addition is associative. By induction, for any natural number n, we have

$$a + n[d] = a.$$

This contradicts the equality (2.13) and shows that the second option is impossible.

If we have the third option, i.e., $a + d > a$, then

$$a + 2[d] = a + d + d > a + d > a,$$

because addition is strictly monotone and associative.

By the definition of the successor $\mathrm{S}a$, if $a \leq z \leq \mathrm{S}a$, then z is equal either to a or to $\mathrm{S}a$. Because $a + 2[d] > a + d > a$, it implies that $n = 1$ in the equality (2.13) and $\mathrm{S}a = a + d$.

Proposition is proved. □

Applying Proposition 2.1.19 several times, we obtain the following result.

Proposition 2.1.23. *If* $\boldsymbol{A} = (A; +, \circ, \leq)$ *is an Exactly Additively Archimedean prearithmetic (ESAPA) with the successor function* S, *an additive generator* d *and strictly monotone associative addition, then for any element* a *from* $\boldsymbol{A}$, $\mathrm{S}^n a = a + n[a]$.

Proof is left as an exercise. □

Corollary 2.1.9. $\mathbf{A}_n$ *is an Exactly Additively Archimedean prearithmetic* $\boldsymbol{A} = (A; +, \circ, \leq)$ *with the successor function* S *and monotone multiplication is Exactly Successively Archimedean.*

Proposition 2.1.24. *Any Exactly Additively Archimedean prearithmetic is an exactly additive prearithmetic from the right.*

Proof: Let us consider an additively Archimedean abstract prearithmetic $\boldsymbol{A} = (A; +, \circ, \leq)$ and its elements $a < b$. By Definition 2.1.8, it satisfies condition (2.17), i.e., for some natural number n, we have $b = a + n[d] = a + (\cdots(((d + d) + d) + d)\cdots) + d)$ where d is an additive generator. It means that we can take $q = n[d]$ and $b = a + n[d] = a + q$.

Proposition is proved. □

Corollary 2.1.10. *Any exactly additively Archimedean prearithmetic with commutative addition is an exactly additive prearithmetic from the left and from the right.*

The Exactly Multiplicatively Archimedean Property also allows representing successors Sa using multiplication.

Proposition 2.1.25. *If* $\boldsymbol{A} = (A; +, \circ, \leq)$ *is an exactly multiplicative prearithmetic (ESAPA) with linear order, the successor function* S, *a multiplicative generator* d *and strictly monotone associative multiplication, then* S$a = a \circ d$ *for any element* a *from* A.

Proof: Let us take an exactly multiplicative prearithmetic $\boldsymbol{A} = (A; +, \circ, \leq)$ with the successor function S, the additive generator d and monotone addition. As by definition $a <$ Sa and the prearithmetic $\boldsymbol{A}$ is Exactly Additively Archimedean, we have

$$\mathrm{S}a = a \circ [d]^n. \tag{2.14}$$

As the order in $\boldsymbol{A}$ is linear, we have three options: $a > a \circ d, a = a \circ d$, or $a < a \circ d$.

If we have the first option, i.e., $a \circ d < a$, then

$$a \circ [d]^2 = a \circ d \circ d < a \circ d < a,$$

because addition is strictly monotone and associative. By induction, for any natural number n, we have

$$a \circ [d]^n < a.$$

This contradicts equality (2.14) and shows that the first option is impossible.

If we have the second option, i.e., $a \circ d = a$, then

$$a \circ [d]^2 = a \circ d \circ d = a \circ d = a,$$

because addition is associative. By induction, for any natural number n, we have

$$a \circ [d]^n = a.$$

This contradicts the equality (2.14) and shows that the second option is impossible.

If we have the third option, i.e., $a \circ d > a$, then

$$a \circ [d]^2 = a \circ d \circ d > a \circ d > a,$$

because addition is strictly monotone and associative.

By the definition of the successor $\mathrm{S}a$, if $a \leq z \leq \mathrm{S}a$, then z is equal either to a or to $\mathrm{S}a$. Because $a \circ [d]^2 > a \circ d > a$, it implies $n = 1$ in the equality (2.14) and $\mathrm{S}a = a \circ d$.

Proposition is proved. □

Applying Proposition 2.1.25 several times, we obtain the following result.

Proposition 2.1.26. *If* $\boldsymbol{A} = (A; +, \circ, \leq)$ *is an Exactly Multiplicatively Archimedean prearithmetic (ESAPA) with the successor function* S, *the additive generator* d *and monotone multiplication, then for any element* a *from* A, $\mathrm{S}^n a = a \circ [d]^n$.

Proof is left as an exercise. □

Corollary 2.1.11. *$\boldsymbol{A}_n$ is an Exactly Multiplicatively Archimedean prearithmetic $\boldsymbol{A} = (A; +, \circ, \leq)$ with the successor function* S *and monotone multiplication is Exactly Successively Archimedean.*

We remind that a chain in an ordered set with a partial order $\leq$ has the form

$$\cdots \leq a_m \leq a_{m+1} \leq \cdots \leq a_n \leq \cdots.$$

Then Lemma 2.1.24 implies the following result.

Proposition 2.1.27. *In a Successively Archimedean prearithmetic (SAPA), any chain with distinct elements, which has the beginning and the end, is finite.*

Proof: Let us assume that an abstract prearithmetic $\boldsymbol{A}$ is Successively Archimedean and consider a chain

$$a \leq a_1 \leq a_2 \leq \cdots \leq b \tag{2.15}$$

between elements a and b with distinct elements in it. By Lemma 2.1.23, $\boldsymbol{A}$ is Exactly Successively Archimedean, which implies that there is a natural number n such that $\mathrm{S}^n a = b$. It gives us the maximal chain between a and b because for any number j such that $1 \leq j \leq n$ if

$$\mathrm{S}^{j-1} a \leq c \leq \mathrm{S}^j a$$

then either $c = \mathrm{S}^{j-1} a$ or $c = \mathrm{S}^j a$ where $j = 1, 2, 3, \ldots$ and $\mathrm{S}^0 a = a$.

Besides, we can apply the Exactly Successively Archimedean Property several times obtaining the sequence of natural numbers $k_1, k_2, k_3, \ldots$ such that $a_j = \mathrm{S}^{kj} a$ for all $j = 1, 2, 3, \ldots$. Because all elements a_j in the chain (2.15) are distinct, $k_1, k_2, k_3, \ldots$ are less than n. Consequently, there is only a finite number of numbers $k_1, k_2, k_3, \ldots$, which implies that the chain (2.15) is finite.

Proposition is proved. □

Chain conditions in abstract prearithmetics imply Archimedean properties. Namely, we have the following result.

Proposition 2.1.28. *If in an abstract prearithmetic $\boldsymbol{A}$ any maximal chain with distinct elements, which has the beginning and the end, is finite, then $\boldsymbol{A}$ is Successively Archimedean.*

Indeed, if a chain $a \le a_1 \le a_2 \le \cdots \le a_n \le b$ is maximal, then by the definition of a successor, $\mathrm{S}a = a_1$, $\mathrm{S}a_n = b$ and $\mathrm{S}a_{j-1} = a_j$ for all $j = 2, 3, \ldots, n$. Consequently, $\mathrm{S}^{n+1}a = b$. As a and b are arbitrary elements from the prearithmetic $\boldsymbol{A}$, it means that the prearithmetic $\boldsymbol{A}$ is Successively Archimedean.

Corollary 2.1.12. *If in a prearithmetic* **A** *any maximal chain with distinct elements, which has the beginning and the end, is finite, then* **A** *is Exactly Successively Archimedean.*

Let us consider a set X with a partial order $\le$.

Proposition 2.1.29. *If the order $\le$ is linear (total), then any finite maximal chain in X with distinct elements in it between two elements from X is unique.*

Proof: Let us consider a maximal chain

$$a \le a_1 \le a_2 \le \cdots \le a_n \le b \tag{2.16}$$

between elements a and b from $X$with distinct elements in it. If this chain is not unique for a and b, then there is an element c such that $a \le c \le b$ and c does not belong to the chain (2.16).

As the order $\le$ is linear and $a \le c$, then there is the largest number j such that $a_j \le c$. This gives us the chain

$$a \le a_1 \le a_2 \le \cdots \le a_j \le c \le a_{j+1} \le \cdots \le a_n \le b \tag{2.17}$$

As the chain (2.16) is maximal, either $c = a_j$ or $c = a_{j+1}$. Thus, our suggestion about existence of a different maximal chain is incorrect and the chain (2.16) is unique.

Proposition is proved. □

In the Diophantine arithmetic, the natural order $\le$ has many good properties. Let us study some of these properties for abstract prearithmetics.

Definition 2.1.9. (a) A set X with a partial order $\le$ is called *finitely connected* if for any elements a and b from X, the inequality $a < b$ implies that any chain $a < a_1 < a_2 < \cdots < a_n < \cdots < b$ between elements a and b is finite.

(b) A set X with a partial order $\le$ is called *constrained* if for any elements a and b from X, the inequality $a < b$ implies that the length of any chain $a < a_1 < a_2 < \cdots < a_n < \cdots < b$ between elements a and b is bounded by some number $k(a, b)$.

Example 2.1.45. The set N of all natural numbers is constrained.

The Hausdorff theorem (cf. Kurosh, 1963) implies the following result.

Lemma 2.1.25. *A partially ordered set X, in which any maximal chain that has the beginning and the end is finite, is finitely connected.*

Proof is left as an exercise. □

Definitions imply the following result.

Lemma 2.1.26. *A finitely connected partially ordered set X, any maximal chain, which has the beginning and the end, is finite.*

Proof is left as an exercise. □

Definition 2.1.9 implies the following result.

Lemma 2.1.27. *If a partially ordered set X is constrained, then it is finitely connected.*

Proof is left as an exercise. □

Proposition 2.1.27 implies the following result.

Corollary 2.1.13. *Any Successively Archimedean prearithmetic (SAPA) is finitely connected.*

Remark 2.1.6. The converse is not always true as the following example demonstrates.

Example 2.1.46. Let us consider the interval [0, 1] of real numbers, in which the partial order $\leq$ is defined in the non-standard way. Namely, we have $0 \leq m/p \leq 1$ for any prime number p and any natural number m that is less than p in the standard ordering of natural numbers. In addition, $m/p \leq l/p$ for any prime number p and any natural numbers m and l that are less than p and m is less than l in the standard ordering of natural numbers. The partial order $\leq$ is not defined for all other pairs from the interval [0, 1].

Then for any prime number p, there is a chain $0 \leq a_1 \leq a_2 \leq \cdots \leq a_n \leq \cdots \leq 1$ of the length p. As there are infinitely many prime numbers, the partially ordered set [0, 1] is not constrained but it is finitely connected because it has only finite chains.

However, in some cases, finitely connectedness implies that the set will be constrained.

Let us consider a set X with a partial order $\leq$.

Proposition 2.1.30. *If the order $\leq$ is linear (total) and the set X is finitely connected, then the order $\leq$ is constrained.*

Proof: Let us consider a finitely connected linearly ordered set X and assume that between two of its elements a and b, there are chains of arbitrary big length. Taking a chain l with n elements between a and b, and denoting $n = n_0$, we can find a chain h_1 between a and b with n_1 elements with $n_1 > n_0$. Then we can take an element a_1 from the chain h_1, which is not equal to a, to b or to any element from the chain l. Adding this element to the chain l, we obtain the chain l_1 with $n + 1$ elements because the order $\leq$ is linear.

Now we can find a chain h_2 between a and b with n_2 elements with $n_2 > n+1$. Then we can take an element a_2 from the chain h_2, which is not equal to a, to b or to any element from the chain l_1. Adding this element to the chain l_1, we obtain the chain l_2 with $n + 2$ elements because the order $\leq$ is linear.

As we can continue this process without any limit, by the axiom of induction, we obtain an infinite chain between a and b. This contradicts our assumption that the set X is finitely connected and completes the proof by the principle of excluded middle.

Proposition is proved. □

Proposition 2.1.30 and Corollary 2.1.13 imply the following result.

Corollary 2.1.14. *Any Successively Archimedean prearithmetic (SAPA) with a linear (total) is constrained.*

Finite connectedness also implies discreteness of the order.

Lemma 2.1.28. *If a partially ordered set X is finitely connected, then the order in X is discrete.*

Proof: Let us consider a set X with a partial order $\leq$ and assume that it is finitely connected. To prove our lemma, we need to show that it satisfies condition (UD).

Taking two elements x and u from X such that $x < u$, we see that any chain $x < a_1 < a_2 < \cdots < a_n < \cdots < u$ is finite. If this chain is maximal, i.e., it is impossible to extend it, then $\mathrm{S}x = a_1$.

If this chain is not maximal, then it is possible to extend it. Repeating this process, we either come to a maximal chain that connects x and u or we will continue this process infinitely. In the second case, principle of induction asserts that there is an infinite chain that connects x and u. This means that the set X is not finitely connected. This contradicts the initial conditions demonstrating that it is always possible to build a maximal chain connecting x and u. However, any maximal chain that connects x and u gives a successor $\mathrm{S}x$ of x. It means that the partially ordered set X satisfies condition (UD).

Lemma is proved. □

Lemmas 2.1.27 and 2.1.28 give us the following result.

Corollary 2.1.15. *If a partially ordered set X is constrained, then the order in X is discrete.*

Let us study partial homomorphisms (cf. Appendix) of abstract prearithmetics. Let us take two abstract prearithmetics $\boldsymbol{A}_1 = (A_1; +_1, \circ_1, \leq_1)$ and $\boldsymbol{A}_2 = (A_2; +_2, \circ_2, \leq_2)$.

Definition 2.1.10.

(a) A mapping $f\colon A_1 \to A_2$ is called an *additive homomorphism* of $\boldsymbol{A}_1$ into $\boldsymbol{A}_2$ if for any elements a and b from A_1, we have

$$f(a +_1 b) = f(a) +_2 f(b).$$

(b) A mapping $f\colon A_1 \to A_2$ is called a *multiplicative homomorphism* of $\boldsymbol{A}_1$ into $\boldsymbol{A}_2$ if for any elements a and b from A_1, we have

$$f(a \circ_1 b) = f(a) \circ_2 f(b).$$

(c) A mapping $f\colon A_1 \to A_2$, which is both an additive homomorphism and a multiplicative homomorphism, is called a *homomorphism* of $\boldsymbol{A}_1$ into $\boldsymbol{A}_2$.

Example 2.1.47. The mapping $f\colon N \to N$ defined as $f(n) = kn$ where k is a fixed natural number is an additive homomorphism of $\boldsymbol{N}$ into $\boldsymbol{N}$ but not a multiplicative homomorphism and not a homomorphism.

Indeed, we have

$$f(n+m) = k(n+m) = kn + km = f(n) + f(m)$$

while

$$f(n \cdot m) = k(n \cdot m) = knm \neq f(n) \cdot f(m) = kn \cdot km = k^2nm.$$

Example 2.1.48. The mapping f: $N \to N$ defined as $f(n) = n^2$ is a multiplicative homomorphism of $\boldsymbol{N}$ into $\boldsymbol{N}$ but not an additive homomorphism and not a homomorphism.

Indeed, we have

$$f(n \cdot m) = (n \cdot m)^2 = f(n) \cdot f(m) = n^2 \cdot m^2$$

while

$$f(n+m) = (n+m)^2 \neq n^2 + m^2 = f(n) + f(m).$$

Theorem 2.1.1. *Abstract prearithmetics with additive homomorphisms form a category where objects are abstract prearithmetics and morphisms are additive homomorphisms.*

Proof is left as an exercise. □

In particular, it means that the sequential composition of additive homomorphisms is an additive homomorphism.

Theorem 2.1.2. *Abstract prearithmetics with multiplicative homomorphisms form a category where objects are abstract prearithmetics and morphisms are multiplicative homomorphisms.*

Proof is left as an exercise. □

In particular, the sequential composition of multiplicative homomorphisms is a multiplicative homomorphism.

There are also other categories of abstract prearithmetics. Some of them are studied in next sections.

2.1.3. *Elements of number theory in abstract prearithmetics*

The soft-minded man always fears change.
He feels security in the status quo, and
he has an almost morbid fear of the new.
For him, the greatest pain is the pain of a new idea.

Martin Luther King Jr.

Obtained properties of abstract prearithmetics allow building non-Diophantine number theory, which is also called non-Diophantine higher arithmetic. Here we develop only the very beginning of this theory.

Number theory begins with classification of numbers and studying their properties. An important class of numbers in the Diophantine arithmetic $\boldsymbol{N}$ consists of prime numbers, which are extensively studied in number theory (cf., for example, Dickson, 1932; Friedberg, 1968; Davenport, 1992; Landau, 1999; Crandall and Pomerance, 2001).

In abstract prearithmetics in general and in non-Diophantine arithmetics in particular, there are two classes of prime numbers — additively prime numbers and multiplicatively prime numbers. There are also additively composite numbers and multiplicatively composite numbers. They are counterparts of the well-known concepts of prime and composite numbers in the Diophantine arithmetic $\boldsymbol{N}$. Here we define these classes in abstract prearithmetics.

Definition 2.1.11.

(a) An element p from an abstract prearithmetic $\boldsymbol{A} = (A; +, \circ, \leq)$ with the additive zero 0_A is *additively prime* in $\boldsymbol{A}$ if there are no elements $a, b \neq 0_A$ in $\boldsymbol{A}$ such that $p = a + b$.
(b) An element p from an abstract prearithmetic $\boldsymbol{A} = (A; +, \circ, \leq)$ with the multiplicative one 1_A is *multiplicatively prime* in $\boldsymbol{A}$ if $p \neq 0_A, p \neq 1_A$ and there are no elements $a, b \neq 1_A$ in $\boldsymbol{A}$ such that $p = a \circ b$.
(c) An element a from an abstract prearithmetic $\boldsymbol{A} = (A; +, \circ, \leq)$ is *additively composite* in $\boldsymbol{A}$ if it is not additively prime.
(d) An element a from an abstract prearithmetic $\boldsymbol{A} = (A; +, \circ, \leq)$ is *multiplicatively composite* in $\boldsymbol{A}$ if it is not multiplicatively prime.

Example 2.1.49. In the Diophantine arithmetic $\boldsymbol{N}$, there is only one additively prime number 1 and infinitely many multiplicatively prime numbers. That is why in the conventional (Diophantine) number theory, additively prime numbers are not even introduced but multiplicatively prime numbers, which are simply called prime numbers, are studied with great interest by many mathematicians.

Remark 2.1.7. It is interesting that the great Greek philosopher Aristotle defined additively prime numbers and found two additively prime numbers 2 and 3 because at that time, Greek mathematicians did not consider 1 as a number (Aristotle, 1984).

Note that there are many abstract prearithmetics that do not have additively prime numbers, i.e., all numbers are composite.

Example 2.1.50. For any modular arithmetic $\boldsymbol{Z}_n$, there are no additively prime numbers because any number in $\boldsymbol{Z}_n$ is a sum of two non-zero numbers.

This is a particular case of the following result.

Lemma 2.1.29. *If an abstract prearithmetic $\boldsymbol{A} = (A; +, \circ, \leq)$ is a group with respect to addition, then it does not have additively prime elements.*

Indeed, any element a in $\boldsymbol{A}$ has the opposite element $-a$ and if $a \neq b$ in $\boldsymbol{A}$, then $b = (b + (-a)) + a$ where $b + (-a) \neq 0_A$.

At the same time, there are many abstract prearithmetics that have infinitely many additively prime numbers as the following example demonstrates.

Example 2.1.51. Let us consider an abstract prearithmetic $\boldsymbol{A} = (N; \oplus, \circ, \leq)$ which contains the set of all natural numbers N and in which operations are defined by the following formula:

$$n \oplus m = (n + m)^2,$$

$$n \circ m = (n \cdot m)^2,$$

where $+$ is the standard addition and $\cdot$ is the standard multiplication of natural numbers.

Then all natural numbers that are not squares in $\boldsymbol{N}$ will be additively prime numbers in this prearithmetic $\boldsymbol{A}$.

Proposition 2.1.31. *There is continuum of abstract prearithmetics that have infinitely many additively prime numbers.*

Proof: Abstract prearithmetics are different when they have different multiplication. Because multiplication in abstract prearithmetics is defined independently from addition, we can take the abstract prearithmetic $\boldsymbol{A}_f$, in which addition is defined as in the abstract prearithmetic $\boldsymbol{A}$ from Example 2.1.46 while multiplication is defined by an arbitrary function f from $N \times N$ into N. By construction, all natural numbers that are not squares in $\boldsymbol{N}$ will be additively prime numbers in this prearithmetic $\boldsymbol{A}_f$. As there is continuum of such functions f (Abian, 1965), there is also continuum of abstract prearithmetics that have infinitely many additively prime numbers.

Proposition is proved. □

However, the following result shows that for an arbitrary abstract prearithmetic, the situation can be essentially different.

Theorem 2.1.3. *There are infinite abstract prearithmetics, in which for any natural number $n > 1$, there are exactly n additively prime elements.*

Proof: Let us consider the set W of all whole numbers, take a natural number n and define the following functions:

$$g(m) = \begin{cases} 0 & \text{when } m = 0, \\ n + m & \text{when } m > 0 \end{cases}$$

and

$$h(q) = \begin{cases} q & \text{when } 0 \leq q < n + 1, \\ q - n & \text{when } q > n. \end{cases}$$

We can build an abstract prearithmetic $\boldsymbol{A}_n = (W; \oplus, \otimes, \leq)$ with addition defined for whole numbers m and n larger than 0 by the following formula:

$$m \oplus k = h(g(m) + g(k)) = (m + n) + (k + n) - n = m + k + n.$$

Besides,

$$0 \oplus m = m \oplus 0 = h(g(m) + 0) = (m + n) - n = m$$

for all $m > 0$.

Then the least additively composite number is $n + 2 = 1 \oplus 1$. At the same time, any larger number $r = 2 + n + k$ is also additively composite because $2 \oplus k = h(g(2) + g(k)) = 2 + k + n = r$. Consequently, there are exactly $n + 1$ additively prime elements in $\boldsymbol{A}_n$ and it is possible to build such a prearithmetic $\boldsymbol{A}_n$ for all $n = 1, 2, 3, \ldots$.

Theorem is proved. □

Considering multiplicatively prime numbers, we see that there are also many abstract prearithmetics that do not have multiplicatively prime numbers.

Example 2.1.52. In the modular arithmetic $\boldsymbol{Z}_p$ where p is a prime number, there are no multiplicatively prime numbers because any non-zero number in $\boldsymbol{Z}_p$ is a product of two non-zero numbers.

This is a particular case of the following result.

Lemma 2.1.30. *If an abstract prearithmetic $\boldsymbol{A} = (A; +, \circ, \leq)$ is a group with respect to multiplication, then it does not have multiplicatively prime elements because any non-zero element a in $\boldsymbol{A}$ has the inverse element a^{-1}.*

Indeed, any element a in $\boldsymbol{A}$ has the inverse element a^{-1} and if $a \neq b$ in $\boldsymbol{A}$, then $b = (b \circ a^{-1}) \circ a$ where $b \circ a^{-1} \neq 1_A$.

At the same time, there are many abstract prearithmetics that have infinitely many multiplicatively prime numbers.

Proposition 2.1.32. *There is continuum of abstract prearithmetics that have infinitely many multiplicatively prime numbers.*

Proof: Abstract prearithmetics are different when they have different addition. Because multiplication in abstract prearithmetics is defined independently from addition, we can take the abstract prearithmetic $\boldsymbol{N}_f$, in which multiplication is defined as in the Diophantine arithmetic $\boldsymbol{N}$ while addition is defined by an arbitrary function f from $N \times N$ into N. As it is proved (cf., for example, Davenport, 1992) that $\boldsymbol{N}$ has infinitely many multiplicatively prime numbers. Thus, there are infinitely many multiplicatively prime elements in this prearithmetic $\boldsymbol{N}_f$. As there is continuum of such functions f (Abian, 1965), there is also continuum of abstract prearithmetics that have infinitely many multiplicatively prime numbers.

Proposition is proved. □

However, the following result shows that for an arbitrary abstract prearithmetic, the situation can be essentially different.

Theorem 2.1.4. *There are infinite abstract prearithmetics, in which for any natural number $n > 3$, there are exactly n multiplicatively prime elements.*

Proof: Let us consider the set N of all natural numbers, take a natural number n and define the following functions:

$$g(m) = \begin{cases} 1 & \text{when } m = 1, \\ 2^{n+m} & \text{when } m > 1 \end{cases}$$

and

$$h(q) = \begin{cases} q & \text{when } 1 \leq q < n+1, \\ \lfloor \log_2 q \rfloor & \text{when } q > 2^n. \end{cases}$$

Note that if $q = 2^m$, then $h(q) = m - n$.

We can build an abstract prearithmetic $\boldsymbol{B}_n = (N; \oplus, \otimes, \leq)$ with multiplication defined for whole numbers m and n larger than 1 by the following formula:

$$\begin{aligned} m \otimes k = h(g(m) + g(k)) &= h(2^{n+m} \cdot 2^{n+k}) = h(2^{2n+m+k}) \\ &= \lfloor \log_2(2^{2n+m+k}) \rfloor - n = \log_2(2^{2n+m+k}) - n \\ &= m + k + n. \end{aligned}$$

Besides,

$$1 \otimes m = m \otimes 1 = h(g(m) \cdot 1) = \log_2(2^{n+m}) - n = (m+n) - n = m$$

for all $m > 0$.

Then the least composite number is $n + 4 = 2 \otimes 2$. At the same time, any larger number $r = n + 4 + k$ with $k = 1, 2, 3, \ldots$ is also composite as it is divisible by 2 because $2 \otimes (2 + k) = h(g(2) + g(k)) = 2 + k + 2 + n = r$. Consequently, there are exactly $n+3$ prime elements in $\boldsymbol{B}_n$ and it is possible to build such a prearithmetic $\boldsymbol{B}_n$ for all $n = 1, 2, 3, \ldots$.

Theorem is proved. □

Let us also consider other traditional classes of natural numbers, for example, even and odd numbers. It is also natural to define even and odd elements in abstract prearithmetics.

Definition 2.1.12.

(a) An element a from an abstract prearithmetic $\boldsymbol{A} = (A; +, \circ, \leq)$ is *additively even* in $\boldsymbol{A}$ (with respect to an element b) if there is an element c in A such that $a = 2 + c$ $(a = b + c)$.

(b) An element a from an abstract prearithmetic $\boldsymbol{A} = (A; +, \circ, \leq)$ is *multiplicatively even* in $\boldsymbol{A}$ (with respect to an element b) if there is an element c in A such that $a = 2 \circ c$ $(a = b \circ c)$.

(c) An element a from an abstract prearithmetic $\boldsymbol{A} = (A; +, \circ, \leq)$ is *additively odd* (*with respect to* an element b in $\boldsymbol{A}$) if it is not additively even (with respect to the element b).

(d) An element a from an abstract prearithmetic $\boldsymbol{A} = (A; +, \circ, \leq)$ is *multiplicatively odd* (*with respect to* an element b in $\boldsymbol{A}$) if it is not multiplicatively even (with respect to the element b).

Note that when an abstract prearithmetic $\boldsymbol{A} = (A; +, \circ, \leq)$ has the number 2, then multiplicatively even in $\boldsymbol{A}$ with respect to two elements are simply *even elements* (*numbers*) although they might be essentially different from even numbers in the Diophantine arithmetic $\boldsymbol{N}$. Naturally, all elements, which are not multiplicatively even in $\boldsymbol{A}$ with respect to 2, are *odd elements* (*numbers*).

Example 2.1.53. In the arithmetic $2\boldsymbol{N}$, of all even numbers with conventional addition and multiplication, all numbers larger than 2 are additively and multiplicatively even.

However, there are many abstract prearithmetics that have only one additively (multiplicatively) even number as the following example demonstrates.

Example 2.1.54. Let us consider an abstract prearithmetic $\boldsymbol{A} = (N; \oplus, \circ, \leq)$ which contains the set all natural numbers N and in which addition is defined by the following formulas:

$$n \oplus 2 = 2 \oplus n = 2,$$

$$n \oplus m = n + m \quad \text{if } n, m \neq 2,$$

where $+$ is the standard addition of natural numbers.

We see that in $\boldsymbol{A}$, only 2 is an additively even number.

Example 2.1.55. Let us consider an abstract prearithmetic $\boldsymbol{A} = (N; \oplus, \circ, \leq)$ which contains the set all natural numbers N and in which multiplication is defined by the following formulas:

$$n \circ 2 = 2 \circ n = 2,$$

$$n \circ m = n \cdot m \text{ if } n,\ m \neq 2,$$

where $\cdot$ is the standard multiplication of natural numbers.

We see that in $\boldsymbol{A}$, only 2 is a multiplicatively even number.

In some abstract prearithmetics, even numbers have usual properties.

Proposition 2.1.33. *If addition in an abstract prearithmetic* $\boldsymbol{A} = (A; +, \circ, \leq)$ *is associative, then the sum of any additively even in* $\boldsymbol{A}$ *element with respect to an element* b *with any element* c *is additively even in* $\boldsymbol{A}$ *with respect to the element* b.

Indeed, if a is an additively even in $\boldsymbol{A}$ element with respect to an element b, then $a = b + d$. Consequently, as addition in $\boldsymbol{A}$ is associative, we have

$$a + c = (b + d) + c = b + (d + c).$$

It means that $a + c$ is additively even in $\boldsymbol{A}$ with respect to the element b.

Proposition 2.1.34. *If multiplication in an abstract prearithmetic* $\boldsymbol{A} = (A; +, \circ, \leq)$ *is distributive with respect to addition, then the sum of two multiplicatively even with respect to an element* b *in* $\boldsymbol{A}$ *elements is multiplicatively even with respect to the element* b.

Indeed, if a and d are multiplicatively even in $\boldsymbol{A}$ elements with respect to an element b, then $a = b \circ u$ and $d = b \circ w$. Consequently, as multiplication in $\boldsymbol{A}$ is distributive with respect to addition, we have

$$a + d = (b \circ u) + (b \circ w) = b \circ (u + w).$$

It means that $a + d$ is multiplicatively even in $\boldsymbol{A}$ with respect to an element b.

Proposition 2.1.35. *If multiplication in an abstract prearithmetic* $\boldsymbol{A} = (A; +, \circ, \leq)$ *is associative, then the product of any multiplicatively even in* $\boldsymbol{A}$ *element with respect to an element* b *with any element* c *is additively even in* $\boldsymbol{A}$ *with respect to the element* b.

Indeed, if a is an additively even in $\boldsymbol{A}$ element with respect to an element b, then $a = b \circ d$. Consequently, as multiplication in $\boldsymbol{A}$ is associative, we have

$$a \circ c = (b \circ d) \circ c = b \circ (d \circ c).$$

It means that $a \circ c$ is multiplicatively even in $\boldsymbol{A}$ with respect to the element b.

One of the basic results of the conventional number theory is the prime decomposition theorem, proofs of which it is possible to find in many books (cf., for example, Landau, 1966; Feferman, 1974; Davenport, 1992; Gioia, 2001).

Prime Decomposition Theorem. For any natural number larger than 1 in the conventional Diophantine arithmetic $\boldsymbol{N}$, there is a unique up to the order of factors decomposition (factoring) of this number into the product of prime numbers.

It is also called the Fundamental Theorem of Arithmetic. According to (Davenport, 1992), the first clear statement and proof of this theorem seem to have been given by Gauss in 1801 (Gauss, 1801).

An equivalent form of the Fundamental Theorem of Arithmetic states that any factoring of a natural number can be extended to a unique up to the order of factors prime factorization.

It is interesting that the same result is true for additively prime numbers. Namely, we have the following result.

Proposition 2.1.36. *For any natural number in the conventional Diophantine arithmetic* $\boldsymbol{N}$*, there is a unique up to the order of factors decomposition of this number into the sum of additively prime numbers.*

Indeed, 1 is an additively prime number and any natural number is the sum of some number of 1s.

Note that in the Diophantine arithmetics $\boldsymbol{N}$ and $\boldsymbol{W}$, there is only one additively prime number. At the same time, as it is demonstrated in Section 2.7, there are prearithmetics that have an infinite set of additively prime numbers.

An equivalent form of Proposition 2.1.36 states that any decomposition (factoring) of a natural number into a sum can be extended to a unique up to the order of factors decomposition (factorization) of this number into the sum of additively prime numbers.

These results bring us to the following concepts.

Definition 2.1.13.

(a) An abstract prearithmetic $\boldsymbol{A}$ has the *additive factoring property* if for any of its non-zero elements, any factorization (additive decomposition) of this element into a sum can be extended to a factorization (additive decomposition) of this number into the sum of additively prime elements.
(b) An abstract prearithmetic $\boldsymbol{A}$ has the *strong additive factoring property* if for any of its non-zero elements, any factorization (additive decomposition) of this element into a sum can be extended to a unique up

to the order of factors factorization (additive decomposition) of this number into the sum of additively prime elements.

(c) An abstract prearithmetic $\boldsymbol{A}$ has the *multiplicative factoring property* if for any of its elements but zero and multiplicative one, any factorization (multiplicative decomposition) of this element into a product can be extended to a factorization (multiplicative decomposition) of this number into the product of multiplicatively prime elements.

(d) An abstract prearithmetic $\boldsymbol{A}$ has the *strong multiplicative factoring property* if for any of its elements but zero and multiplicative one, any factorization (multiplicative decomposition) of this element into a product can be extended to a unique up to the order of factors factorization (multiplicative decomposition) of this number into the product of multiplicatively prime elements.

For abstract prearithmetics and even for whole-number and natural-number prearithmetics (cf. Section 2.7), the additive factoring property is not true in a general case as the following example demonstrates.

Example 2.1.56. Let us consider the arithmetic $\boldsymbol{R}^{++} = (R^{++}; +, \circ, \leq)$ of all positive real numbers with standard addition, multiplication and order. The prearithmetic $\boldsymbol{R}^{++}$ does not have additively prime numbers because any positive real number a is equal to $a/2$ plus $a/2$. Consequently, this prearithmetic does not have the additive factoring property and Proposition 2.1.36 is not true for this prearithmetic.

Lemma 2.1.31. *The strong additive factoring property implies the additive factoring property.*

Proof is left as an exercise. □

The inverse implication is not true as the following example demonstrates.

Example 2.1.57. Let us consider the set $F = \{0, 1, {}^1\!/_3, {}^1\!/_2\}$ and the set P of all expressions of the form $a_1 + a_2({}^1\!/_2) + a_3({}^1\!/_3)$ where a_1, a_2 and a_3 are natural numbers. We see that the sum of these expressions have the same form. The multiplication is defined by the following formula:

$$(a_1 + a_2({}^1\!/_2) + a_3({}^1\!/_3)) \circ (b_1 + b_2({}^1\!/_2) + b_3({}^1\!/_3))$$
$$= (a_1 + b_1) + (a_2 + b_2)({}^1\!/_2) + (a_3 + b_3)({}^1\!/_3).$$

Now we can define the set A_2 of numbers that can be represented as expressions from P. Naturally, some of these polynomials define the same number. For instance, $1 = 2({}^1\!/_2)$ or $({}^1\!/_2) \circ ({}^1\!/_3) = 0$.

This gives us the abstract prearithmetic $\boldsymbol{A} = (A; +, \circ, \leq)$, in which order is the same as in the arithmetic $\boldsymbol{R}$ of all real numbers, while addition and multiplication are defined above. In it, $1/2$ and $1/3$ are additively prime elements, while number 1 has two additive prime decompositions (factorizations)

$$1 = 2(1/2) = 3(1/3).$$

In the same way as for the additive factoring property, for abstract prearithmetics and even for whole number and natural number prearithmetics, the multiplicative factoring property is not true in a general case as the following example demonstrates.

Example 2.1.58. Let us consider the arithmetic $\boldsymbol{R}^{++} = (R^{++}; +, \circ, \leq)$ of all positive real numbers with standard addition, multiplication and order. The prearithmetic $\boldsymbol{R}^{++}$ does not have multiplicatively prime numbers because any positive real number a is equal to $a^{1/2}$ times $a^{1/2}$. Consequently, the Fundamental Theorem of Arithmetic is not true for this prearithmetic and it does not have the multiplicative factoring property.

There are also modular arithmetics, which do not have multiplicatively prime numbers.

Example 2.1.59. Let us consider the modular arithmetic $\boldsymbol{Z}_5$. It has five elements 0, 1, 2, 3, and 4. There are the following multiplicative decompositions in $\boldsymbol{Z}_5$:

$$2 \cdot 3 = 1, 3 \cdot 4 = 2, 2 \cdot 4 = 3, 2 \cdot 2 = 4.$$

This shows that all numbers in $\boldsymbol{Z}_5$ are composite.

As a result, we can build multiplicative decompositions of an arbitrary length. For instance, we have

$$2 = 3 \cdot 4 = (2 \cdot 4) \cdot 4 = ((3 \cdot 4) \cdot 4) \cdot 4 = \cdots .$$

At the same time, some modular arithmetics have prime numbers. For instance, 3 is a prime number in $\boldsymbol{Z}_4$.

Lemma 2.1.32. *The strong additive factoring property implies the additive factoring property.*

Proof is left as an exercise. □

The inverse implication is not true as the following example demonstrates.

Example 2.1.60. Let us consider the set $F = \{1, 2, 2^{1/3}, 2^{1/2}\}$ and the set P of all expressions of the form $2^{a_1 + a_2^{(1/2)} + a_3^{(1/3)}}$ where a_1, a_2 and a_3 are natural numbers. We see that the product of these expressions have the same form. Now we can define the set A_2 of numbers that can be represented as expressions from P and their arbitrary sums. Naturally, some of these polynomials define the same number. For instance, $2 = 2^{2(1/2)} = 2^{3(1/3)}$.

This gives us the abstract prearithmetic $\boldsymbol{A} = (A; +, \circ, \leq)$, in which order is the same as in the arithmetic $\boldsymbol{R}$ of all real numbers, while addition and multiplication are defined above. In it, $2^{1/2}$ and $2^{1/3}$ are multiplicatively prime elements, while as we have demonstrated, number 2 has two multiplicative prime decompositions (factorizations).

Let us consider an abstract prearithmetic $\boldsymbol{A} = (A; +, \circ, \leq)$, which is totally ordered, additively Archimedean and exactly additive and in which addition preserves the order.

Theorem 2.1.5.

(a) *If for some elements a and b from $\boldsymbol{A}$, we have $a < b$, then for some natural number n either*

$$b = n[a] \tag{2.18}$$

or

$$b = n[a] + r, \tag{2.19}$$

where $r < a$.

(b) *If in addition, the abstract prearithmetic $\boldsymbol{A}$ is with additive cancellation from the left, then the representation* (2.19) *is unique.*

Proof: (a) As the abstract prearithmetic $\boldsymbol{A}$ is additively Archimedean, for some natural number n, we have

$$b \leq (n+1)[a]. \tag{2.20}$$

It is possible to assume that n is the least number for which the inequality (2.20) is true. If in the inequality (2.20), we have equality, then b satisfies formula (2.18) and the statement (a) is proved.

If the inequality (2.20) is strict and n is the least natural number for which the inequality (2.20) is valid, then we obtain

$$n[a] \leq b < (n+1)[a] = n[a] + a. \tag{2.21}$$

As the abstract prearithmetic $\boldsymbol{A}$ is exactly additively Archimedean with the additive generator d, then for some natural number k, we have $b = n[a] + m[d]$. Now we can take $r = m[d]$.

By construction, $0 \leq r < a$. Indeed, if this is not true, then $a \leq r$ because the order $\leq$ is total. As addition preserves the order, we have

$$n[a] + a = (n+1)[a] \leq n[a] + r = b.$$

Because it is assumed $b < (n+1)[a]$, we come to a contradiction, which by the principle of excluded middle, concludes the proof of the part (a).

(b) By construction, the part $n[a]$ in the representation (2.19) is unique because n is the largest natural number for which $n[a] \leq b$. Now let us suppose

$$b = n[a] + r = n[a] + q.$$

Because the abstract prearithmetic $\boldsymbol{A}$ is with additive cancellation from the left, $r = q$.

Theorem is proved. □

When the abstract prearithmetic $\boldsymbol{A}$ has the additive zero 0 and is additively Archimedean for all non-zero elements, then it is possible to reduce formulas (2.19) and (2.20) to one formula. Namely, we have the following result.

Corollary 2.1.16.

(a) *If for some elements a and b from $\boldsymbol{A}$, we have $a < b$, then for some natural number n, we have*

$$b = n[a] + r$$

where $0 \leq r < a$.

(b) *If in addition, the abstract prearithmetic $\boldsymbol{A}$ is with additive cancellation from the left for all non-zero elements, then this representation is unique.*

Theorem 2.1.5 implies a well-known important result from number theory.

Corollary 2.1.17 (Landau, 1999: Theorem 7). *If for natural numbers a and b, we have $a < b$, then there is a natural number n such that*

$$b = na + r$$

where $0 \leq r < a$.

Let us consider an abstract prearithmetic $\boldsymbol{A} = (A; +, \circ, \leq)$, which is totally ordered, multiplicatively Archimedean and exactly multiplicative and in which multiplication preserves the order.

Theorem 2.1.6. *If for some elements a and b from $\boldsymbol{A}$, we have $a < b$, then*

$$b = [a]^n \circ r$$

where $r < a$.

Proof is similar to the proof of Theorem 2.1.5. □

This result is a multiplicative counterpart of Theorem 2.1.5. There are abstract prearithmetics that have this property. For instance, let us consider an abstract prearithmetic $\boldsymbol{A}_{\text{pow}} = (A; +, \circ, \leq)$, in which A consists of powers of some natural number m, i.e., $A = \{m^n; n = 1, 2, 3, \ldots\}$, multiplication is the same as the conventional multiplication of natural numbers and addition is trivial, i.e., the sum of any two numbers from A is equal to m. In this arithmetic, Theorem 2.1.6 is valid.

Corollary 2.1.18. *If for natural numbers a and b from $\boldsymbol{A}_{\text{pow}}$, we have $a < b$, then there is a natural number n such that*

$$b = a^n r,$$

where $1 \leq r < a$.

Let us consider an abstract prearithmetic $\boldsymbol{A} = (A; +, \circ, \leq)$, which is totally ordered, additively and multiplicatively Archimedean and exactly additive and in which addition is associative and preserves the order.

Theorem 2.1.7. (*a*) *For any elements a and b from $\boldsymbol{A}$, the following property is valid:*

$$b = k_n[a]^n + k_{n-1}[a]^{n-1} + \cdots + k_1[a] + k_0[r], \tag{2.22}$$

where $r < a$, the element k_0 is either 1 or the symbol $\varnothing$, and for $i = 1, 2, 3, \ldots, n$, the element k_i is either a natural number or the symbol $\varnothing$, which means that the corresponding element $[a]^i$ is absent in the left part of (2.22).

(*b*) *If the abstract prearithmetic $\boldsymbol{A}$ is with additive cancellation from the left, then the representation* (2.22) *is unique.*

Proof: (a) To prove existence, we use mathematical induction on n.

Given two elements a and b from $\boldsymbol{A}$, we have either $a > b$ or $a = b$ or $a < b$ because the order in $\boldsymbol{A}$ is total. In the first two cases, the statement (a) is evident. Indeed, if $a > b$, we can take

$$b = k_0[r] = r.$$

If $a = b$, we can take

$$b = 1[a] = a.$$

In the case when $a < b$, we suppose that for all elements d from $\boldsymbol{A}$, such that $a \leq d < b$ the statement (a) is true and prove the equality (2.22). As the abstract prearithmetic $\boldsymbol{A}$ is multiplicatively Archimedean, there is a natural number n for which

$$b < [a]^{n+1} \tag{2.23}$$

because by Corollary 2.1.7, $[a]^n < [a]^{n+1}$ for all $n = 1, 2, 3, \ldots$. Taking the least n for which the inequality is valid, we obtain

$$[a]^n \leq b < [a]^{n+1},$$

where by our supposition, $n > 1$. If $b = [a]^n$, then the statement (a) is proved because

$$b = [a]^n = 1[a]^n.$$

If $b > [a]^n$, then by Theorem 2.1.5, we have

$$b = k_n[a]^n + c, \tag{2.24}$$

where $c < [a]^n$. As by our supposition, formula (2.22) is true for c, we obtain the following equality:

$$c = k_m[a]^m + k_{m-1}[a]^{m-1} + \cdots + k_1[a] + k_0[r] \tag{2.25}$$

with $m < n$ and $r < a$. If we substitute c in the equality (2.18) by the right side of the equality (2.25) and add the necessary number of expressions when $m < n - 1$, we obtain the equality (2.22) as addition is associative. The principle of mathematical induction implies that the statement (a) is true for all elements a and b from $\boldsymbol{A}$.

(b) By Theorem 2.1.5, decomposition (2.24) is unique, while uniqueness of decomposition (2.25) is assumed according to the proof by induction. Uniqueness of decompositions (2.24) and (2.25) implies uniqueness of decomposition (2.22) for the chosen element b. Then the principle of mathematical induction allows us to conclude that the decomposition (2.22) is also unique for any element from A, which is larger than a where a is an arbitrary element from A.

Theorem is proved. □

Corollary 2.1.19 (Landau, 1999: Theorem 8). *If a number a is larger than 1, then a natural number b can be expressed in one and only one way in the form*

$$b = k_n a^n + k_{n-1} a^{n-1} + \cdots + k_1 a + k_0,$$

where $n \geq 0$, $k_0 > 0$ and $0 \leq k_j < a$ for all $j = 1, 2, \ldots, n$.

Abstract prearithmetics allow efficient formalization of the concept *difference.* However, because addition in abstract prearithmetics is not always commutative, we have three kinds of differences: the difference from the right $b \rightharpoondown a$, difference from the left $b \leftharpoondown a$ and full difference or simply, difference $b - a$.

Definition 2.1.14.

(a) The *difference from the right* of b and a is $d = b \rightharpoondown a$ if $d + a = b$ and the equality $c + a = b$ implies $c = d$.
(b) The *difference from the left* of b and a is $d = b \leftharpoondown a$ if $a + d = b$ and the equality $a + c = b$ implies $c = d$.
(c) The *difference* of b and a is $d = b - a$ if $a = b \leftharpoondown a$ and $d = b \leftharpoondown a$.

Let us study properties of differences.

Definition 2.1.14 implies the following results.

Lemma 2.1.33. *For any elements b and d from an abstract prearithmetic $\boldsymbol{A}$, if the difference $b - d$ is defined, then differences $b \rightharpoondown d$ and $b \leftharpoondown d$ are defined.*

Proof is left as an exercise. □

Lemma 2.1.34. *For any elements b and d from an abstract prearithmetic $\boldsymbol{A}$, we have the following properties:*

(a) *If the difference from the right $b \rightharpoondown d$ is defined, then $b = (b \rightharpoondown d) + d$.*
(b) *If the difference from the left $b \leftharpoondown d$ is defined, then $b = d + (b \rightharpoondown d)$.*
(c) *If the difference $b - d$ is defined, then $b = (b - d) + d = d + (b - d)$.*

Proof is left as an exercise. □

These results show that as in the case of the Diophantine arithmetic, difference is the *inverse* operation to sum.

Proposition 2.1.37. *If addition + is associative in an abstract prearithmetic $\boldsymbol{A} = (A; +, \circ, \leq)$, then for any elements a, b and c from $\boldsymbol{A}$, we have the following properties*:

(1) *If differences from the right $b \rightharpoondown a$ and $a \rightharpoondown d$ are defined, then the difference from the right $b \rightharpoondown d$ is defined.*
(2) *If differences from the left $b \leftharpoondown a$ and $a \leftharpoondown d$ are defined, then the difference from the left $b \leftharpoondown d$ is defined.*
(3) *If differences $b - a$ and $a - d$ are defined, then the difference $b - d$ is defined.*

Proof: (1) If the difference from the right $b \rightharpoondown a$ is defined, then $b = e + a$ for a unique element e from $\boldsymbol{A}$. If the difference from the right $a \rightharpoondown d$ is defined, then $a = c + d$ for a unique element c from $\boldsymbol{A}$. Consequently,

$$b = e + a = e + c + d = (e + c) + d$$

for a unique element $e + c$ from $\boldsymbol{A}$. It means that the difference from the right $b \rightharpoondown d$ is defined.

Statements (2) and (3) are proved in a similar way.

Proposition is proved. □

Lemma 2.1.35. *If addition + is commutative in an abstract prearithmetic $\boldsymbol{A} = (A; +, \circ, \leq)$, then for any elements a and b from $\boldsymbol{A}$, the difference $b - a$ is defined if and only if the difference from the right $b \rightharpoondown a$ is defined and/or the difference from the left $b \leftharpoondown a$ is defined, i.e., difference, difference from the right and difference from the left coincide.*

Proof is left as an exercise. □

Proposition 2.1.38. *If addition + is associative in an abstract prearithmetic $\boldsymbol{A} = (A; +, \circ, \leq)$, then for any elements a, b and c from $\boldsymbol{A}$, we have the following properties*:

(1) *If the difference from the right $b \rightharpoondown a$ is defined, then the difference from the right $(b + c) \rightharpoondown (a + c)$ is defined.*
(2) *If the difference from the left $b \leftharpoondown a$ is defined, then the difference from the left $(c + b) \leftharpoondown (c + a)$ is defined.*
(3) *If the difference $b - a$ is defined and addition + is commutative, then the difference $(b + c) - (a + c)$ is defined.*

Proof: (1) If the difference from the right $b \rightharpoondown a$ is defined, then $b = e + a$ for a unique element e from $\boldsymbol{A}$. Consequently,

$$b + c = e + a + c = e + (a + c)$$

for a unique element e from $\boldsymbol{A}$. It means that the difference from the right $(b + c) \rightharpoondown (a + c)$ is defined.

(2) If the difference from the left $b \leftharpoondown a$ is defined, then $b = a + e$ for a unique element e from $\boldsymbol{A}$. Consequently,

$$c + b = c + a + e = (c + a) + e$$

for a unique element e from $\boldsymbol{A}$. It means that the difference from the left $(c + b) \leftharpoondown (c + a)$ is defined.

By Lemma 2.1.35, the statement (3) follows from statements (1) and (2) when addition + is commutative.

Proposition is proved. □

Proposition 2.1.39.

(a) *An element d of an abstract prearithmetic $\boldsymbol{A}$ with the additive zero $0_{\boldsymbol{A}}$ has the unique opposite from the left element c if and only if the difference from the right $0_{\boldsymbol{A}} \rightharpoondown d$ is defined.*

(b) *An element d of an abstract prearithmetic $\boldsymbol{A}$ with the additive zero $0_{\boldsymbol{A}}$ has the unique opposite from the right element c if and only if the difference from the left $0_{\boldsymbol{A}} \leftharpoondown d$ is defined.*

(c) *An element d of an abstract prearithmetic $\boldsymbol{A}$ with the additive zero $0_{\boldsymbol{A}}$ has the unique opposite element c if and only if the difference $0_{\boldsymbol{A}} - d$ is defined.*

Proof: (a) *Sufficiency.* If the difference from the right $0_{\boldsymbol{A}} \rightharpoondown d$ is defined, then $0_{\boldsymbol{A}} = c + d$ for a unique element $c = 0_{\boldsymbol{A}} \rightharpoondown d$ from $\boldsymbol{A}$. By Definition 2.1.6, the element c is opposite from the left to the element d.

Necessity. Let us assume that an element d has the unique opposite from the left element c. By Definition 2.1.14, $c + d = 0_{\boldsymbol{A}}$. By Definition 2.1.14, this means that $c = 0_{\boldsymbol{A}} \rightharpoondown d$.

Statements (b) and (c) are proved in a similar way.

Proposition is proved. □

Proposition 2.1.39 shows that the element $0_{\boldsymbol{A}} \rightharpoondown d$ is the opposite from the left to the element d, the element $0_{\boldsymbol{A}} \leftharpoondown d$ is the opposite from the right to the element d and the element $0_{\boldsymbol{A}} - d$ is the opposite to the element d.

In the same way, as it is done in the Diophantine arithmetic $\boldsymbol{Z}$ of all integers, it is possible to define operation *subtraction* using addition in abstract prearithmetics. However, because addition in abstract prearithmetics is not always commutative, we have three operations: *subtraction from the right* $\rightharpoondown$, *subtraction from the left* $\leftharpoondown$ and *full subtraction* or simply, *subtraction* $-$.

Definition 2.1.15. (a) *Subtraction from the right* in an abstract prearithmetic $\boldsymbol{A}$ is a binary operation that assigns a unique element $b \rightharpoondown d$ called the *right difference* of b and d to the pair (b, d) of elements from $\boldsymbol{A}$ so that the following identities are valid in $\boldsymbol{A}$:

$$(b \rightharpoondown d) + d = b, \tag{2.26}$$

$$(b + d) \rightharpoondown d = b. \tag{2.27}$$

(b) *Subtraction from the left* in an abstract prearithmetic $\boldsymbol{A}$ is a binary operation that assigns a unique element $b \leftharpoondown d$ called the *left difference* of b and d to the pair (b, d) of elements from $\boldsymbol{A}$ so that the following identities are valid in $\boldsymbol{A}$:

$$d + (b \leftharpoondown d) = b, \tag{2.28}$$

$$(d + b) \leftharpoondown d = b. \tag{2.29}$$

(c) *Subtraction* in an abstract prearithmetic $\boldsymbol{A}$ is a binary operation that assigns a unique element $b - d$ called the *full difference* of b and d to the pair (b, d) of elements from $\boldsymbol{A}$ so that all identities (2.26)–(2.29) are valid in $\boldsymbol{A}$, i.e., we have in $\boldsymbol{A}$:

$$d + (b - d) = b,$$

$$(d + b) - d = b,$$

$$(b - d) + d = b,$$

$$(b + d) - d = b.$$

Lemma 2.1.36. *The element $b \rightharpoondown d$ is the right difference of b and d, the element $b \leftharpoondown d$ is the right difference of b and d and the element $b - d$ is the difference of b and d.*

Proof is left as an exercise. □

This shows that as in the case of the Diophantine arithmetic, subtraction is the inverse operation to addition.

Although in a general case, all subtraction operations are partial, there are many prearithmetics and arithmetics where subtraction is a total operation. Examples of such arithmetics are:

- the arithmetic $\boldsymbol{Z}$ of all integer numbers;
- the arithmetic $\boldsymbol{Q}$ of all rational numbers;
- the arithmetic $\boldsymbol{R}$ of all real numbers;
- the arithmetic $\boldsymbol{C}$ of all complex numbers.

Let us study properties of subtraction.

Definition 2.1.15 implies the following results.

Lemma 2.1.37. *In an abstract prearithmetic $\boldsymbol{A}$, if subtraction from the right and subtraction from the left are defined and $b \rightharpoondown d = b \leftharpoondown d$, then subtraction is defined and $b - d$.*

Proof is left as an exercise. □

Lemma 2.1.38. *If addition $+$ is commutative in an abstract prearithmetic $\boldsymbol{A} = (A; +, \circ, \leq)$, then for any elements a and b from $\boldsymbol{A}$, operation $b - a$ is defined if and only if operation $b \rightharpoondown a$ is defined and/or operation $b \leftharpoondown a$ is defined, i.e., subtraction, subtraction from the right and subtraction from the left coincide.*

Proof is left as an exercise. □

Proposition 2.1.40. (*a*) *An abstract prearithmetic $\boldsymbol{A} = (A; +, \circ, \leq)$ with subtraction from the right satisfies additive cancellation from the right, i.e., $a + c = b + c$ implies $a = b$ for any elements a, b and c from $\boldsymbol{A}$.*

(*b*) *An abstract prearithmetic $\boldsymbol{A} = (A; +, \circ, \leq)$ with subtraction from the left satisfies additive cancellation from the left, i.e., $c + a = c + b$ implies $a = b$ for any elements a, b and c from $\boldsymbol{A}$.*

Proof: (a) By definition we have

$$a = (a + c) \rightharpoondown c = (b + c) \rightharpoondown c = b,$$

i.e., $a = b$.

The proof for (b) is similar.

Proposition is proved. □

Corollary 2.1.20. *An abstract prearithmetic $\boldsymbol{A} = (A; +, \circ, \leq)$ with subtraction satisfies additive cancellation, i.e., $a + c = b + c$ or $c + a = c + b$ implies $a = b$ for any elements a, b and c from $\boldsymbol{A}$.*

Proposition 2.1.41. *In an abstract prearithmetic* $\boldsymbol{A} = (A; +, \circ, \leq)$ *with subtraction and commutative and associative addition, for any elements* a, b *and* c *from* $\boldsymbol{A}$, *we have*

$$(b - a) - d = b - (a + d).$$

Proof: By definition, we have

$$(b - (a + d)) + (a + d) = b.$$

At the same time,

$$\begin{aligned}((b - a) - d) + (a + d) &= ((b - a) - d) + (d + a) \\ &= ((((b - a) - d) + d) + a) = ((b - a) + a) = b.\end{aligned}$$

It means that

$$(b - (a + d)) + (a + d) = ((b - a) - d) + (a + d).$$

Then by Corollary 2.1.20, we can cancel $(a+d)$ in both sides of this equality and obtain

$$b - (a + d) = (b - a) - d.$$

Proposition is proved. □

Lemma 2.1.39. *If an abstract prearithmetic* $\boldsymbol{A} = (A; +, \circ, \leq)$ *has the additive zero* $0_{\boldsymbol{A}}$, *then for any element* a *from* $\boldsymbol{A}$, *we have*

$$a - 0_{\boldsymbol{A}} = a.$$

Indeed, $a - 0_{\boldsymbol{A}} = a$ because $a + 0_{\boldsymbol{A}} = 0_{\boldsymbol{A}} + a = a$.

Lemma 2.1.40. *If an abstract prearithmetic* $\boldsymbol{A} = (A; +, \circ, \leq)$ *with subtraction has the additive zero* $0_{\boldsymbol{A}}$, *then for any element* a *from* $\boldsymbol{A}$, *we have* $a - a = 0_{\boldsymbol{A}}$.

Indeed, by Definition 2.1.14, $a + (a - a) = a$, while by Definition 2.1.5, $a + 0_{\boldsymbol{A}} = a$. As by Definition 2.1.14, $a - a$ is the unique element in $\boldsymbol{A}$, for which $a + (a - a) = a$, we have $a - a = 0_{\boldsymbol{A}}$.

As in an abstract prearithmetic $\boldsymbol{A}$ with subtraction (from the right or/and from the left), the difference (from the right or/and from the left) is defined for all elements, Proposition 2.1.39 implies the following result.

Proposition 2.1.42.

(a) *Any element d of an abstract prearithmetic $\boldsymbol{A}$ with the additive zero $0_{\boldsymbol{A}}$ and subtraction from the right has the unique opposite from the left element $0_{\boldsymbol{A}} \rightharpoondown d$.*
(b) *Any element d of an abstract prearithmetic $\boldsymbol{A}$ with the additive zero $0_{\boldsymbol{A}}$ and subtraction from the left has the unique opposite from the right element $0_{\boldsymbol{A}} \leftharpoondown d$.*
(c) *Any element d of an abstract prearithmetic $\boldsymbol{A}$ with the additive zero $0_{\boldsymbol{A}}$ and subtraction has the unique opposite element $0_{\boldsymbol{A}} - d$.*

Proof is left as an exercise. □

Axioms of groups (cf. Appendix) and Proposition 2.1.42 imply the following result.

Proposition 2.1.43. *If addition $+$ is associative in an abstract prearithmetic $\boldsymbol{A} = (A; +, \circ, \leq)$ with subtraction and the additive zero $0_{\boldsymbol{A}}$, then $\boldsymbol{A}$ is a group with respect to addition.*

Proof is left as an exercise. □

Axioms of Abelian groups (cf. Appendix) and Proposition 2.1.43 imply the following result.

Proposition 2.1.44. *If addition $+$ is associative and commutative in an abstract prearithmetic $\boldsymbol{A} = (A; +, \circ, \leq)$ with subtraction and the additive zero $0_{\boldsymbol{A}}$, then $\boldsymbol{A}$ is an Abelian group with respect to addition.*

Proof is left as an exercise. □

Axioms of rings (cf. Appendix) and Proposition 2.1.44 imply the following result.

Proposition 2.1.45. *If addition $+$ is associative and commutative and multiplication is distributive with respect to addition in an abstract prearithmetic $\boldsymbol{A} = (A; +, \circ, \leq)$ with subtraction and the additive zero $0_{\boldsymbol{A}}$, then $\boldsymbol{A}$ is a ring.*

Proof is left as an exercise. □

Distributivity (from the right or/and from the left) with respect to addition implies corresponding distributivity with respect to subtraction.

Proposition 2.1.46. *If multiplication $\circ$ is distributive from the left with respect to addition $+$ in an abstract prearithmetic $\boldsymbol{A} = (A; +, \circ, \leq)$*

with subtraction from the right (from the left), then multiplication $\circ$ *is distributive from the left with respect to subtraction from the right (from the left).*

Proof: (a) Let us assume that multiplication $\circ$ is distributive from the left with respect to addition $+$ in an abstract prearithmetic $\boldsymbol{A} = (A; +, \circ, \leq)$ with subtraction from the right and $a = b \rightharpoondown d$. It means that $b = a + d$. As multiplication $\circ$ is distributive from the left with respect to addition $+$, we have $c \circ b = c \circ (a + d) = c \circ a + c \circ d$. Consequently, $c \circ a = c \circ (b \rightharpoondown d) = c \circ b \rightharpoondown c \circ d$, i.e., multiplication $\circ$ is distributive from the left with respect to subtraction from the right.

(b) Let us assume that multiplication $\circ$ is distributive from the left with respect to addition $+$ in an abstract prearithmetic $\boldsymbol{A} = (A; +, \circ, \leq)$ with subtraction from the left and $a = b \leftharpoondown d$. It means that $b = d + a$. As multiplication $\circ$ is distributive from the left with respect to addition $+$, we have $c \circ b = c \circ (d + a) = c \circ d + c \circ a$. Consequently, $c \circ a = c \circ (b \leftharpoondown d) = c \circ b \leftharpoondown c \circ d.$, i.e., multiplication $\circ$ is distributive from the left with respect to subtraction from the left.

Proposition is proved. □

Proposition 2.1.47. *If multiplication* $\circ$ *is distributive from the right with respect to addition* $+$ *in an abstract prearithmetic* $\boldsymbol{A} = (A; +, \circ, \leq)$ *with subtraction from the right (from the left), then multiplication* $\circ$ *is distributive from the right with respect to subtraction from the right (from the left).*

Proof is similar to the proof of Proposition 2.1.38. □

Propositions 2.1.41 and 2.1.47 imply the following results.

Corollary 2.1.21. *If multiplication* $\circ$ *is distributive from the right (from the left) with respect to addition* $+$ *in an abstract prearithmetic* $\boldsymbol{A} = (A; +, \circ, \leq)$ *with subtraction, then multiplication* $\circ$ *is distributive from the right (from the left) with respect to subtraction.*

Corollary 2.1.22. *If multiplication* $\circ$ *is distributive with respect to addition* $+$ *in an abstract prearithmetic* $\boldsymbol{A} = (A; +, \circ, \leq)$ *with subtraction from the right (from the left), then multiplication* $\circ$ *is distributive with respect to subtraction from the right (from the left).*

Corollary 2.1.23. *If multiplication* $\circ$ *is distributive with respect to addition* $+$ *in an abstract prearithmetic* $\boldsymbol{A} = (A; +, \circ, \leq)$ *with subtraction, then multiplication* $\circ$ *is distributive with respect to subtraction.*

Equalities between differences can be deduced from equalities between sums.

Proposition 2.1.48. *In an abstract prearithmetic* $\boldsymbol{A} = (A; +, \circ, \leq)$ *with subtraction and commutative and associative addition, for any elements* a, b, c *and* d *from* $\boldsymbol{A}$*, we have* $b - a = c - d$ *if and only if* $b + d = c + a$.

Proof: *Necessity.* Let us suppose

$$b - a = c - d.$$

Then by definition, we have

$$b = (b - a) + a = (c - d) + a.$$

Consequently,

$$\begin{aligned} b + d &= ((b - a) + a) + d = ((c - d) + a) + d \\ &= ((c - d) + d) + a = c + a. \end{aligned}$$

Sufficiency. Let us suppose

$$b + d = c + a.$$

Then we have

$$\begin{aligned} b - a &= ((b + d) - d) - a = ((c + a) - d) - a \\ &= ((c + a) - a) - d = c - d. \end{aligned}$$

Proposition is proved. □

Proposition 2.1.49. *In an abstract prearithmetic* $\boldsymbol{A} = (A; +, \circ, \leq)$ *with subtraction and commutative and associative addition, for any elements* a, b *and* d *from* $\boldsymbol{A}$*, we have*

$$(b - a) + d = b - (a - d).$$

Proof: By definition, we have

$$(b - (a - d) + (a - d) = b.$$

At the same time,

$$((b - a) + d) + (a - d) = (b - a) + (d + (a - d))$$
$$= (b - a) + ((a - d) + d) = (b - a) + a = b.$$

It means that

$$(b - (a - d) + (a - d) = ((b - a) + d) + (a - d).$$

Then by Corollary 2.1.20, we can cancel $(a - d)$ in both sides of this equality and obtain

$$(b - a) + d = b - (a - d).$$

Proposition is proved. □

Proposition 2.1.50. *In an abstract prearithmetic* $\boldsymbol{A} = (A; +, \circ, \leq)$ *with subtraction and commutative and associative addition, for any elements* a *and* d *from* $\boldsymbol{A}$*, we have*

$$a - (a - d) = d.$$

Indeed, by Proposition 2.1.48, we have

$$a - (a - d) = (a - a) + d = d$$

because by Lemma 2.1.40, $a - a = 0_{\boldsymbol{A}}$.

Let us consider relations between subtraction and order.

Proposition 2.1.51. *In an abstract prearithmetic* $\boldsymbol{A} = (A; +, \circ, \leq)$ *with linear order* $\leq$*, subtraction and strictly monotone, commutative and associative addition, the operation subtraction is monotone from the left, i.e., for any elements* a, b *and* c *from* $\boldsymbol{A}$*, the inequality* $a \leq b$ *implies inequality* $a - c \leq b - c$.

Proof: Let us suppose

$$a \leq b.$$

Let us check whether it is possible that

$$a - c > b - c.$$

As addition is strictly monotone, we have in this case

$$b = (b - c) + c < (a - c) + c = a.$$

This contradicts initial conditions and shows that

$$a - c \leq b - c$$

because the order $\leq$ is linear (total).

Proposition is proved. □

Sums allow comparing differences.

Proposition 2.1.52. *In an abstract prearithmetic* $\boldsymbol{A} = (A; +, \circ, \leq)$ *with subtraction and monotone, commutative and associative addition, for any elements* a, b *and* c *from* $\boldsymbol{A}$, *we have* $b - a \leq c - d$ *if and only if* $b + d \leq c + a$.

Proof: *Necessity.* Let us suppose

$$b - a \leq c - d.$$

As addition is monotone, we have

$$b = (b - a) + a \leq (c - d) + a.$$

Consequently, transitivity of order relation gives us

$$\begin{aligned} b + d &= ((b - a) + a) + d \leq ((c - d) + a) + d \\ &= ((c - d) + d) + a = c + a. \end{aligned}$$

It means that

$$b + d \leq c + a.$$

Sufficiency. Let us suppose

$$b + d \leq c + a.$$

Then by Proposition 2.1.51, we have

$$\begin{aligned} b - a &= ((b + d) - d) - a \leq ((c + a) - d) - a \\ &= ((c + a) - a) - d = c - d. \end{aligned}$$

Proposition is proved. □

Proposition 2.1.53. *In an abstract prearithmetic* $\boldsymbol{A} = (A; +, \circ, \leq)$ *with subtraction and strictly monotone, commutative and associative addition, for any elements* a, b *and* c *from* $\boldsymbol{A}$, *we have* $b - a < c - d$ *if and only if* $b + d < c + a$.

Proof is similar to the proof of Proposition 2.1.52. □

Let us consider relations between subtraction and order. We can see that subtraction of unequal elements changes the order.

Proposition 2.1.54. *In an abstract prearithmetic* $\boldsymbol{A} = (A; +, \circ, \leq)$ *with linear order* $\leq$, *subtraction and strictly monotone, commutative and associative addition, the operation subtraction is antitone from the right, i.e., for any elements* a, b *and* c *from* $\boldsymbol{A}$, *the inequality* $a \leq b$ *implies inequality* $c - b \leq c - a$.

Proof: Let us suppose

$$a \leq b.$$

Let us check whether it is possible that

$$c - b > c - a.$$

In this case, by Proposition 2.1.53, we have

$$c + b < c + a.$$

At the same time, as addition is strictly monotone, our assumption $a \leq b$ implies

$$c + a \leq c + b.$$

This contradiction shows that

$$c - b \leq c - a.$$

Proposition is proved. □

Propositions 2.1.51 and 2.1.54 imply the following result.

Corollary 2.1.24. *In an abstract prearithmetic* $\boldsymbol{A} = (A; +, \circ, \leq)$ *with linear order* $\leq$, *subtraction and strictly monotone, commutative and associative addition, the inequalities* $a \leq b$ *and* $c \leq d$ *imply the inequality* $a - d \leq b - c$.

In the same way, as it is done in the Diophantine arithmetic $\boldsymbol{N}$, it is possible to define the partial operation *division* using multiplication in abstract prearithmetics. However, because multiplication in abstract prearithmetics is not always commutative, we have three partial operations:

- *division from the right* $\nearrow$, the result of which is the quotient from the right $b \nearrow d$;
- *division from the left* $\swarrow$, the result of which is the quotient from the left $b \swarrow d$;
- *full division* $\div$, the result of which is the full quotient $b \div d$.

Definition 2.1.16.

(a) The *quotient from the right* of b and d is $q = b \nearrow d$ if $q \circ d = b$ and the equality $c \circ d = b$ implies $c = q$.
(b) The *quotient from the left* of b and d is $q = b \swarrow d$ if $d \circ q = b$ and the equality $d \circ c = b$ implies $c = q$.
(c) The *quotient* of b and d is $q = b \div d$ if $q = b \swarrow d$ and $q = b \nearrow d$.

Let us study properties of quotients.
Definition 2.1.16 implies the following results.

Lemma 2.1.41. *For any elements b and d from an abstract prearithmetic $\boldsymbol{A}$, if the quotient $b \div d$ is defined, then quotients $b \nearrow d$ and $b \swarrow d$ are also defined.*

Proof is left as an exercise. □

Lemma 2.1.42. *For any elements b and d from an abstract prearithmetic $\boldsymbol{A}$, we have the following properties:*

(a) *If the quotient from the right $b \nearrow d$ is defined, then $b = (b \nearrow d) \circ d$.*
(b) *If the quotient from the left $b \swarrow d$ is defined, then $b = d \circ (b \swarrow d)$.*
(c) *If the quotient $b \div d$ is defined, then $b = (b \div d) \circ d = d \circ (b \div d)$.*

Proof is left as an exercise. □

These results show that as in the case of the Diophantine arithmetic, division is the inverse operation to multiplication.

Proposition 2.1.55. *If multiplication $\circ$ is associative in an abstract prearithmetic $\boldsymbol{A} = (A; +, \circ, \leq)$, then for any elements a, b and c from $\boldsymbol{A}$,* *we have the following properties:*

(1) *If the quotients from the right* $b \nearrow a$ *and* $a \nearrow d$ *are defined, then the quotient from the right* $b \nearrow d$ *is defined.*
(2) *If the quotients from the left* $b \swarrow a$ *and* $a \swarrow d$ *are defined, then the quotient from the left* $b \swarrow d$ *is defined.*
(3) *If the quotients* $b \div a$ *and* $a \div d$ *are defined, then the quotient* $b \div d$ *is defined.*

Proof: (1) If the quotient from the right $b \nearrow a$ is defined, then $b = e \circ a$ for a unique element e from $\boldsymbol{A}$. If $a \nearrow d$ is defined, then $a = c \circ d$ for a unique element c from $\boldsymbol{A}$. Consequently,

$$b = e \circ a = e \circ c \circ d = (e \circ c) \circ d$$

for a unique element $e \circ c$ from $\boldsymbol{A}$. It means that the quotient from the right $b \nearrow d$ is defined.

Statements (2) and (3) are proved in a similar way.

Proposition is proved. □

Lemma 2.1.43. *If multiplication* $\circ$ *is commutative in an abstract prearithmetic* $\boldsymbol{A} = (A; +, \circ, \leq)$, *then for any elements* a *and* b *from* $\boldsymbol{A}$, *the quotient* $b \div a$ *is defined if and only if the quotient from the right* $b \nearrow a$ *is defined and/or the quotient from the left* $b \swarrow a$ *is defined.*

Proof is left as an exercise. □

Proposition 2.1.56. *If multiplication* $\circ$ *is associative in an abstract prearithmetic* $\boldsymbol{A} = (A; +, \circ, \leq)$, *then for any elements* a, b *and* c *from* $\boldsymbol{A}$, *we have the following properties:*

(1) *If the quotient from the right* $b \nearrow a$ *is defined, then the quotient from the right* $(b \circ c) \nearrow (a \circ c)$ *is defined.*
(2) *If the quotient from the left* $b \swarrow a$ *is defined, then the quotient from the left* $(c \circ b) \swarrow (c \circ a)$ *is defined.*
(3) *If the quotient* $b \div a$ *is defined and multiplication* $\circ$ *is commutative, then the quotient* $(b \circ c) \div (a \circ c)$ *is defined.*

Proof: (1) If the quotient from the right $b \nearrow a$ is defined, $b = e \circ a$ for a unique element e from $\boldsymbol{A}$. Consequently,

$$b \circ c = e \circ a \circ c = e \circ (a \circ c).$$

for a unique element e from $\boldsymbol{A}$. It means that the quotient from the right $(b \circ c) \nearrow (a \circ c)$ is defined.

(2) If the quotient from the left $b \swarrow a$ is defined, $b = a \circ e$ for a unique element e from $\boldsymbol{A}$. Consequently,

$$c \circ b = c \circ a \circ e = (c \circ a) \circ e$$

for a unique element e from $\boldsymbol{A}$. It means that the quotient from the left $(c \circ b) \swarrow (c \circ a)$ is defined.

By Lemma 2.1.43, the statement (3) follows from statements (1) and (2) when multiplication $\circ$ is commutative.

Proposition is proved. □

Proposition 2.1.57. *In an Exactly Multiplicatively Archimedean abstract prearithmetic $\boldsymbol{A} = (A; +, \circ, \leq)$, for any elements a, b and c from $\boldsymbol{A}$, if $a < b$, then the quotient from the left $b \swarrow a$ is defined.*

Proof is left as an exercise. □

Lemma 2.1.44. *If an abstract prearithmetic $\boldsymbol{A} = (A; +, \circ, \leq)$ has the multiplicative one $1_{\boldsymbol{A}}$, then for any element a from $\boldsymbol{A}$, the quotient $a \div 1_{\boldsymbol{A}}$ is defined.*

Indeed, $a \div 1_{\boldsymbol{A}} = a$ because $a \circ 1_{\boldsymbol{A}} = 1_{\boldsymbol{A}} \circ a = a$.

As we know, in some arithmetics, elements have inverse. For instance, number 0.2 is inverse to number 5. Let us explore inverse elements in abstract prearithmetics.

Definition 2.1.17.

(a) An element c of an abstract prearithmetic $\boldsymbol{A}$ with the multiplicative one $1_{\boldsymbol{A}}$ is *inverse from the right* to an element d if $d \circ c = 1_{\boldsymbol{A}}$.
(b) An element c of an abstract prearithmetic $\boldsymbol{A}$ with the multiplicative one $1_{\boldsymbol{A}}$ is *inverse from the left* to an element d if $c \circ d = 1_{\boldsymbol{A}}$.
(c) An element c of an abstract prearithmetic $\boldsymbol{A}$ with the multiplicative one $1_{\boldsymbol{A}}$ is *inverse* to an element d if it is inverse from the right (from the left) to d.

Proposition 2.1.58.

(a) *An element d of an abstract prearithmetic $\boldsymbol{A}$ with the multiplicative one $1_{\boldsymbol{A}}$ has the unique inverse from the left element c if and only if the quotient from the right $1_{\boldsymbol{A}} \nearrow d$ is defined.*

(b) *An element d of an abstract prearithmetic* $\boldsymbol{A}$ *with the multiplicative one* $1_{\boldsymbol{A}}$ *has the unique inverse from the right element c if and only if the quotient from the left* $1_{\boldsymbol{A}} \swarrow d$ *is defined.*

(c) *An element d of an abstract prearithmetic* $\boldsymbol{A}$ *with the multiplicative one* $1_{\boldsymbol{A}}$ *has the unique inverse element c if and only if the quotient* $1_{\boldsymbol{A}} \div d$ *is defined.*

Proof: (a) *Sufficiency.* If the quotient from the right $1_{\boldsymbol{A}} \nearrow d$ is defined, then $1_{\boldsymbol{A}} = c \circ d$ for a unique element $c = 1_{\boldsymbol{A}} \nearrow d$ from $\boldsymbol{A}$. By Definition 2.1.17, the element c is inverse from the left to the element d.

Necessity. Let us assume that an element d has the unique inverse from the left element c. By Definition 2.1.12, $c \circ d = 1_{\boldsymbol{A}}$. By Definition 2.1.16, this means that $c = 1_{\boldsymbol{A}} \nearrow d$.

Statements (b) and (c) are proved in a similar way.

Proposition is proved. □

Proposition 2.1.58 shows that the element $1_{\boldsymbol{A}} \nearrow d$ is inverse from the left to the element d, the element $1_{\boldsymbol{A}} \swarrow d$ is inverse from the right to the element d and the element $1_{\boldsymbol{A}} \div d$ is inverse to the element d.

Although in a general case, all division operations are partial, still there are prearithmetics and arithmetics where division is a total operation. Examples of such arithmetics are as follows:

- the arithmetic $\boldsymbol{Q}^{++}$ of all positive rational numbers;
- the arithmetic $\boldsymbol{R}^{++}$ of all positive real numbers.

In the same way, as it is done in the Diophantine arithmetic $\boldsymbol{Q}$, it is possible to define operation *division* using multiplication in abstract prearithmetics. However, because multiplication in abstract prearithmetics is not always commutative, we have three partial operations: division from the right $\nearrow$, division from the left $\swarrow$ and full division or simply, division $\div$.

Definition 2.1.18.

(a) *Division from the right* in an abstract prearithmetic $\boldsymbol{A}$ is a binary operation that assigns a unique element $b \nearrow d$ to the pair (b, d) of elements from $\boldsymbol{A}$ so that the following identities are valid in $\boldsymbol{A}$:

$$(b \nearrow d) \circ d = b, \tag{2.30}$$

$$(b \circ d) \nearrow d = b. \tag{2.31}$$

(b) *Division from the left* in an abstract prearithmetic $\boldsymbol{A}$ is a binary operation that assigns a unique element $b \swarrow d$ to the pair (b, d) of elements from $\boldsymbol{A}$ so that the following identities are valid in $\boldsymbol{A}$:

$$d \circ (b \swarrow d) = b, \tag{2.32}$$

$$(d \circ b) \swarrow d = b. \tag{2.33}$$

(c) *Division* in an abstract prearithmetic $\boldsymbol{A}$ is a binary operation that assigns a unique element $b \div d$ to the pair (b, d) of elements from $\boldsymbol{A}'$ so that all identities (2.30)–(2.33) are valid in $\boldsymbol{A}$, i.e., we have in $\boldsymbol{A}$:

$$d \circ (b \div d) = b,$$

$$(d \circ b) \div d = b,$$

$$(b \div d) \circ d = b,$$

$$(b \circ d) \div d = b.$$

In other words, the operation of division assigns the quotient to two numbers. However, in general, a quotient to two numbers is not unique while the operation of division assigns the unique quotient to two numbers.

Lemma 2.1.45. *If an abstract prearithmetic* $\boldsymbol{A} = (A; +, \circ, \leq)$ *with division has the multiplicative one* $1_{\boldsymbol{A}}$, *then for any element* a *from* $\boldsymbol{A}$, *we have* $a \div a = 1_{\boldsymbol{A}}$.

Indeed, by Lemma 2.1.35, $a \circ (a \div a) = a$, while by Definition 2.1.8, $a \circ 1_{\boldsymbol{A}} = a$. As by Definition 2.1.5 and Lemma 2.1.15, $1_{\boldsymbol{A}}$ is the unique element in $\boldsymbol{A}$, for which $a \circ 1_{\boldsymbol{A}} = a$, we have the necessary equality $a \div a = 1_{\boldsymbol{A}}$.

Proposition 2.1.59. *If multiplication* $\circ$ *is associative in an abstract prearithmetic* $\boldsymbol{A} = (A; +, \circ, \leq)$ *with division and the multiplicative one* $1_{\boldsymbol{A}}$, *then* $\boldsymbol{A}$ *is a group with respect to multiplication.*

Proof is left as an exercise. □

Proposition 2.1.60. *If multiplication* $\circ$ *is associative and commutative in an abstract prearithmetic* $\boldsymbol{A} = (A; +, \circ, \leq)$ *with division and the multiplicative one* $1_{\boldsymbol{A}}$, *then* $\boldsymbol{A}$ *is an Abelian group with respect to multiplication.*

Proof is left as an exercise. □

Definition of division allows deducing many other properties of division, which are well known for rational and real numbers (cf., for example, Feferman, 1974), but are also valid in a much more general context of abstract prearithmetics.

Proposition 2.1.60 implies the following result.

Proposition 2.1.61.

(a) *Any element d of an abstract prearithmetic $\boldsymbol{A}$ with the multiplicative one $1_{\boldsymbol{A}}$ and division from the right has the unique inverse from the left element.*
(b) *Any element d of an abstract prearithmetic $\boldsymbol{A}$ with the multiplicative one $1_{\boldsymbol{A}}$ and division from the left has the unique inverse from the right element.*
(c) *Any element d of an abstract prearithmetic $\boldsymbol{A}$ with the multiplicative one $1_{\boldsymbol{A}}$ and division has the unique inverse element.*

Proof is left as an exercise. □

In Diophantine arithmetics, it is prohibited to divide by zero. This remains true for abstract prearithmetics.

Lemma 2.1.46. *If an abstract prearithmetic $\boldsymbol{A} = (A; +, \circ, \leq)$ has the multiplicative zero $0^{\mathrm{m}}_{\boldsymbol{A}}$, then for any element a from $\boldsymbol{A}$, the division operations $a \div 0^{\mathrm{m}}_{\boldsymbol{A}}$, $a \nearrow 0^{\mathrm{m}}_{\boldsymbol{A}}$ and $a \swarrow 0^{\mathrm{m}}_{\boldsymbol{A}}$ are not defined.*

Proof: Let us take a non-zero element a from $\boldsymbol{A}$ and assume that $a \nearrow 0^{\mathrm{m}}_{\boldsymbol{A}}$ is defined and equal to c. Then by definition, $a = c \circ 0^{\mathrm{m}}_{\boldsymbol{A}}$. However, any element multiplied by the multiplicative zero becomes zero, i.e., $a = 0^{\mathrm{m}}_{\boldsymbol{A}}$. It means that for a non-zero element a from $\boldsymbol{A}$, division from the right by the multiplicative zero $0^{\mathrm{m}}_{\boldsymbol{A}}$ is not defined.

At the same time, the element $0^{\mathrm{m}}_{\boldsymbol{A}} \nearrow 0^{\mathrm{m}}_{\boldsymbol{A}}$ can be equal to any element from $\boldsymbol{A}$ because $c \circ 0$ is always equal to 0. So, $0^{\mathrm{m}}_{\boldsymbol{A}} \nearrow 0^{\mathrm{m}}_{\boldsymbol{A}}$ also is not defined.

Cases, $a \swarrow 0^{\mathrm{m}}_{\boldsymbol{A}}$ and $a \div 0^{\mathrm{m}}_{\boldsymbol{A}}$ are treated in a similar way.

Lemma is proved. □

Lemma 2.1.46 shows that it is reasonable to consider division only by non-zero elements.

Definition 2.1.19.

(a) *Non-zero division from the right* in an abstract prearithmetic $\boldsymbol{A}$ is a binary operation that assigns a unique element $b \nearrow d$ to the pair (b, d)

of elements from $\boldsymbol{A}$ with $d \neq 0$ so that the following identities are valid in $\boldsymbol{A}$:

$$(b \nearrow d) \circ d = b,$$

$$(b \circ d) \nearrow d = b.$$

(b) *Non-zero division from the left* in an abstract prearithmetic $\boldsymbol{A}$ is a partial binary operation that assigns a unique element $b \swarrow d$ to the pair (b, d) of elements from $\boldsymbol{A}$ with $d \neq 0$ so that the following identities are valid in $\boldsymbol{A}$:

$$d \circ (b \swarrow d) = b,$$

$$(d \circ b) \swarrow d = b.$$

(c) *Non-zero division* in an abstract prearithmetic $\boldsymbol{A}$ is a partial binary operation that assigns a unique element $b \div d$ to the pair (b, d) of elements from $\boldsymbol{A}$ with $d \neq 0$ so that the following identities are valid in $\boldsymbol{A}$:

$$d \circ (b \div d) = b,$$

$$(d \circ b) \div d = b,$$

$$(b \div d) \circ d = b,$$

$$(b \circ d) \div d = b.$$

Lemma 2.1.47. *If an abstract prearithmetic $\boldsymbol{A} = (A; +, \circ, \leq)$ with non-zero division has the multiplicative one $1_{\boldsymbol{A}}$, then for any non-zero element a from $\boldsymbol{A}$, we have $a \div a = 1_{\boldsymbol{A}}$.*

Indeed, by Lemma 2.1.38, $a \circ (a \div a) = a$, while by Definition 2.1.9, $a \circ 1_{\boldsymbol{A}} = a$. As by Definition 2.1.19, a is the unique element in $\boldsymbol{A}$, for which $a \circ (a \div a) = a$, we have $a \div a = 1_{\boldsymbol{A}}$.

Axioms of groups (cf. Appendix) and Proposition 2.1.60 imply the following result.

Proposition 2.1.62. *If multiplication $\circ$ is associative in an abstract prearithmetic $\boldsymbol{A} = (A; +, \circ, \leq)$ with non-zero division and the multiplicative one $1_{\boldsymbol{A}}$, then the set of all non-zero elements from $\boldsymbol{A}$ is a group with respect to multiplication.*

Proof is left as an exercise. □

Axioms of Abelian groups (cf. Appendix) and Proposition 2.1.60 imply the following result.

Proposition 2.1.63. *If multiplication* $\circ$ *is associative and commutative in an abstract prearithmetic* $\mathbf{A} = (A; +, \circ, \leq)$ *with non-zero division and the multiplicative one* $1_{\mathbf{A}}$, *the set of all non-zero elements from* $\mathbf{A}$ *is an Abelian group with respect to multiplication.*

Proof is left as an exercise. □

Axioms of fields (cf. Appendix) and Proposition 2.1.63 imply the following result.

Proposition 2.1.64. *If addition* $+$ *is associative and commutative and multiplication is associative, commutative and distributive with respect to addition in an abstract prearithmetic* $\mathbf{A} = (A; +, \circ, \leq)$ *with subtraction, non-zero division and the additive zero* $0_{\mathbf{A}}$, *then* $\mathbf{A}$ *is a field.*

Proof is left as an exercise. □

Proposition 2.1.65.

(a) *An abstract prearithmetic* $\mathbf{A} = (A; +, \circ, \leq)$ *with non-zero division from the right satisfies multiplicative cancellation from the right, i.e.,* $a \circ c = b \circ c$ *and* $c \neq 0$ *implies* $a = b$ *for any elements* a, b *and* c *from* $\mathbf{A}$.
(b) *An abstract prearithmetic* $\mathbf{A} = (A; +, \circ, \leq)$ *with non-zero division from the left satisfies multiplicative cancellation from the left, i.e.,* $c \circ a = c \circ b$ *and* $c \neq 0$ *implies* $a = b$ *for any elements* a, b *and* c *from* $\mathbf{A}$.

Proof: (a) By definition, we have

$$a = (a \circ c) \nearrow c = (b \circ c) \nearrow c = b,$$

i.e., $a = b$.

The proof for (b) is similar.

Proposition is proved. □

Corollary 2.1.25. *An abstract prearithmetic* $\mathbf{A} = (A; +, \circ, \leq)$ *with non-zero division satisfies multiplicative cancellation, i.e.,* $a \circ c = b \circ c$ *and* $c \neq 0$ *or* $c \circ a = c \circ b$ *and* $c \neq 0$ *implies* $a = b$ *for any elements* a, b *and* c *from* $\mathbf{A}$.

Multiplicative cancellation allows obtaining other relations between division and multiplication.

Proposition 2.1.66. *In an abstract prearithmetic* $\boldsymbol{A} = (A; +, \circ, \leq)$ *with [non-zero] division and commutative and associative multiplication, for any elements* a, b *and* d *from* $\boldsymbol{A}$ *[with* $a \neq 0$ *and* $d \neq 0$*], we have*

$$(b \div a) \div d = b \div (a \circ d).$$

Proof: By definition, we have

$$(b \div (a \circ d)) \circ (a \circ d) = b.$$

At the same time,

$$((b \div a) \div d) \circ (a \circ d) = ((b \div a) \div d) \circ (d \circ a)$$
$$= (((b \div a) \div d) \circ d) \circ a) = (((b \div a) \circ a) = b.$$

It means that

$$(b \div (a \circ d)) \circ (a \circ d) = ((b \div a) \div d) \circ (a \circ d).$$

Then by Corollary 2.1.25, we have

$$(b \div a) \div d = b \div (a \circ d).$$

Proposition is proved. □

Lemma 2.1.48. *If an abstract prearithmetic* $\boldsymbol{A} = (A; +, \circ, \leq)$ *with non-zero division* $\div$ *has the multiplicative one* $1_{\boldsymbol{A}}$*, then for any element* a *from* $\boldsymbol{A}$*, we have* $a \div 1_{\boldsymbol{A}} = a$.

Indeed, $a \div 1_{\boldsymbol{A}} = a$ because $a \circ 1_{\boldsymbol{A}} = 1_{\boldsymbol{A}} \circ a = a$.

Lemma 2.1.49. *If an abstract prearithmetic* $\boldsymbol{A} = (A; +, \circ, \leq)$ *with non-zero division* $\div$ *has the multiplicative one* $1_{\boldsymbol{A}}$*, then for any element* a *from* $\boldsymbol{A}$*, we have* $a \div a = 1_{\boldsymbol{A}}$.

Indeed, by Lemma 2.1.34, $a \circ (a \div a) = a$, while by Definition 2.1.9, $a \circ 1_{\boldsymbol{A}} = a$. As by Definition 2.1.19, $a \div a$ is the unique element in $\boldsymbol{A}$, for which $a \circ (a \div a) = a$, we have $a \div a = 1_{\boldsymbol{A}}$.

Proposition 2.1.67. *If multiplication* $\circ$ *is associative in an abstract prearithmetic* $\boldsymbol{A} = (A; +, \circ, \leq)$ *with non-zero division and the multiplicative one* 1_A*, then all non-zero elements from* $\boldsymbol{A}$ *form a group with respect to multiplication.*

Proof is left as an exercise. □

Division is also distributive with respect to addition.

Proposition 2.1.68. *In an abstract prearithmetic* $\boldsymbol{A} = (A; +, \circ, \leq)$ *with [non-zero] division and commutative and associative multiplication, which is distributive with respect to addition, for any elements* a, b *and* d *from* $\boldsymbol{A}$ *[with* $d \neq 0$;*], we have*

$$(b + a) \div d = b \div d + a \div d.$$

Proof: Assuming $d \neq 0$, by definition we have

$$((b + a) \div d) \circ d = b + a.$$

At the same time,

$$(b \div d + a \div d) \circ d = ((b \div d) \circ d) + ((a \div d) \circ d) = b + a.$$

It means that

$$((b + a) \div d) \circ d = (b \div d + a \div d) \circ d.$$

Then by Corollary 2.1.25, we have

$$(b + a) \div d = b \div d + a \div d.$$

Proposition is proved. □

Under definite conditions, relations between addition and division in abstract prearithmetics are the same as in the conventional arithmetic $\boldsymbol{Q}$ of all rational numbers.

Proposition 2.1.69. *In an abstract prearithmetic* $\boldsymbol{A} = (A; +, \circ, \leq)$ *with [non-zero] division and commutative and associative multiplication, which is distributive with respect to addition, for any elements* a, b, c *and* d *from* $\boldsymbol{A}$ *[with* $c \neq 0$ *and* $d \neq 0$*], we have*

$$b \div c + a \div d = (b \circ d + a \circ c) \div c \circ d.$$

Proof: Assuming $c \neq 0$ and $d \neq 0$, by definition, we have

$$((b \circ d + a \circ c) \div c \circ d) \circ (c \circ d) = b \circ d + a \circ c.$$

At the same time, as multiplication is distributive with respect to addition,

$$
\begin{aligned}
(b \div c + a \div d) \circ (c \circ d) &= (b \div c) \circ (c \circ d) + (a \div d) \circ (c \circ d) \\
&= (((b \div c) \circ c) \circ d) + (((a \div d) \circ d) \circ c) \\
&= b \circ d + a \circ c.
\end{aligned}
$$

Consequently,

$$((b \circ d + a \circ c) \div c \circ d) \circ (c \circ d) = (b \div c + a \div d) \circ (c \circ d).$$

Then by Corollary 2.1.25, we have

$$(b + a) \div d = b \div d + a \div d.$$

Proposition is proved. □

Equalities between quotients can be deduced from equalities between products.

Proposition 2.1.70. *In an abstract prearithmetic* $\boldsymbol{A} = (A; +, \circ, \leq)$ *with* [*non-zero*] *division and commutative and associative multiplication, for any elements* a, b, c *and* d *from* $\boldsymbol{A}$ [*with* $a \neq 0$ *and* $d \neq 0$], *we have* $b \div a = c \div d$ *if and only if* $b \circ d = c \circ a$.

Proof: *Necessity.* Let us suppose that $a \neq 0$ and $d \neq 0$ and

$$b \div a = c \div d.$$

Then by definition, we have

$$b = (b \div a) \circ a = (c \div d) \circ a.$$

Consequently,

$$
\begin{aligned}
b \circ d &= ((b \div a) \circ a) \circ d = ((c \div d) \circ a) \circ d \\
&= ((c \div d) \circ d) \circ a = c \circ a.
\end{aligned}
$$

Sufficiency. Let us suppose

$$b \circ d = c \circ a.$$

Then we have

$$b \div a = ((b \circ d) \div d) \div a = ((c \circ a) \div d) \div a$$
$$= ((c \circ a) \div a) \div d = c \div d.$$

Proposition is proved. □

Proposition 2.1.71. *In an abstract prearithmetic* $\boldsymbol{A} = (A; +, \circ, \leq)$ *with non-zero division and commutative and associative multiplication, for any non-zero elements* a, b *and* d *from* $\boldsymbol{A}$, *we have*

$$(b \div a) \circ d = b \div (a \div d).$$

Proof: By definition, we have

$$(b \div (a \div d)) \circ (a \div d) = b.$$

At the same time,

$$((b \div a) \circ d) \circ (a \div d) = (b \div a) \circ (d \circ (a \div d))$$
$$= (b \div a) \circ ((a \div d) \circ d) = (b \div a) \circ a = b.$$

It means that

$$(b \div (a \div d) \circ (a \div d) = ((b \div a) \circ d) \circ (a \div d).$$

Then by Corollary 2.1.25, we can cancel $(a \div d)$ in both sides of this equality and obtain

$$(b \div a) \circ d = b \div (a \div d).$$

Proposition is proved. □

Proposition 2.1.72. *In an abstract prearithmetic* $\boldsymbol{A} = (A; +, \circ, \leq)$ *with non-zero division and commutative and associative multiplication, for any non-zero elements* a *and* d *from* $\boldsymbol{A}$, *we have*

$$a \div (a \div d) = d.$$

Indeed, by Proposition 2.1.70, we have

$$d \div (a \div d) = (a \div a) \circ d = d.$$

because by Lemma 2.1.40, $a \div a = 1_{\boldsymbol{A}}$.

Proposition 2.1.58 implies the following result.

Proposition 2.1.73.

(a) *Any element $d \neq 0_{\boldsymbol{A}}$ of an abstract prearithmetic $\boldsymbol{A}$ with the multiplicative one $1_{\boldsymbol{A}}$ and non-zero division from the right has the unique inverse from the left element.*
(b) *Any element $d \neq 0_{\boldsymbol{A}}$ of an abstract prearithmetic $\boldsymbol{A}$ with the multiplicative one $1_{\boldsymbol{A}}$ and non-zero division from the left has the unique inverse from the right element.*
(c) *Any element $d \neq 0_{\boldsymbol{A}}$ of an abstract prearithmetic $\boldsymbol{A}$ with the multiplicative one $1_{\boldsymbol{A}}$ and non-zero division has the unique inverse element.*

Proof is left as an exercise. □

Proposition 2.1.74. *In an abstract prearithmetic $\boldsymbol{A} = (A; +, \circ, \leq)$ with [non-zero] division and strictly monotone, commutative and associative multiplication, for any elements a, b, c and d from $\boldsymbol{A}$ [with $a \neq 0$ and $d \neq 0$], the operation division is strictly monotone from the left, i.e., for any elements a, b and c from $\boldsymbol{A}$, the inequality $a \leq b$ implies inequality $a \div c \leq b \div c$.*

Proof: Let us suppose

$$a \leq b.$$

Let us check whether it is possible that

$$a \div c > b \div c.$$

As multiplication is strictly monotone, we have in this case

$$b = (b \div c) \circ c < (a \div c) \circ c = a.$$

This contradicts initial conditions and shows that

$$a \div c \leq b \div c$$

because the order $\leq$ is linear (total).

Proposition is proved. □

Inequalities between quotients can be deduced from inequalities between products.

Proposition 2.1.75. *In an abstract prearithmetic $\boldsymbol{A} = (A; +, \circ, \leq)$ with [non-zero] division and monotone, commutative and associative* *multiplication, for any [non-zero] elements* a, b *and* c *from* $\mathbf{A}$*, we have* $b \div a \leq c \div d$ *if and only if* $b \circ d \leq c \circ a$.

Proof: *Necessity.* Let us suppose

$$b \div a \leq c \div d.$$

As multiplication is monotone, we have

$$b = (b \div a) \circ a \leq (c \div d) \circ a.$$

Consequently, transitivity of order relation gives us

$$b \circ d = ((b \div a) \circ a) \circ d \leq ((c \div d) \circ a) \circ d$$
$$= ((c \div d) \circ d) \circ a = c \circ a.$$

It means that

$$b \circ d \leq c \circ a.$$

Sufficiency. Let us suppose

$$b \circ d \leq c \circ a.$$

Then by Proposition 2.1.73, we have

$$b \div a = ((b \circ d) \div d) \div a \leq ((c \circ a) \div d) \div a$$
$$= ((c \circ a) \div a) \div d = c \div d.$$

Proposition is proved. □

Proposition 2.1.76. *In an abstract prearithmetic* $\mathbf{A} = (A; +, \circ, \leq)$ *with [non-zero] division and strictly monotone, commutative and associative multiplication, for any [non-zero] elements* a, b *and* c *from* $\mathbf{A}$*, we have* $b \div a < c \div d$ *if and only if* $b \circ d < c \circ a$.

Proof is similar to the proof of Proposition 2.1.75. □

Division by unequal elements changes the order.

Proposition 2.1.77. *In an abstract prearithmetic* $\mathbf{A} = (A; +, \circ, \leq)$ *with linear order* $\leq$*, division and strictly monotone, commutative and associative multiplication, the operation division is antitone from the right, i.e., for* *any elements* a, b *and* c *from* $\boldsymbol{A}$, *the inequality* $a \leq b$ *implies inequality* $c \div b \leq c \div a$.

Proof: Let us suppose

$$a \leq b.$$

Let us check whether it is possible that

$$c \div b > c \div a.$$

In this case, by Proposition 2.1.70, we have

$$c \circ b < c \circ a.$$

At the same time, as multiplication is strictly monotone, our assumption $a \leq b$ implies

$$c \circ a \leq c \circ b.$$

This contradiction shows that

$$c \div b \leq c \div a.$$

Proposition is proved. □

Propositions 2.1.72 and 2.1.77 imply the following result.

Corollary 2.1.26. *In an abstract prearithmetic* $\boldsymbol{A} = (A; +, \circ, \leq)$ *with linear order* $\leq$, *subtraction and strictly monotone, commutative and associative addition, the inequalities* $a \leq b$ *and* $c \leq d$ *imply the inequality* $a \div d \leq b \div c$.

Here we studied properties of abstract prearithmetics, while in the next section, we explore relations between abstract prearithmetics and show how these relations influence properties of abstract prearithmetics making possible to define new abstract prearithmetics using known abstract prearithmetics.

2.2. Weak Projectivity of Abstract Prearithmetics

There are three classes of people:
- *those who see,*
- *those who see when they are shown,*
- *those who do not see.*

Leonardo da Vinci

There are different relations between abstract prearithmetics and arithmetics. One of the most important is weak projectivity, which we study in this section.

Let us take two abstract prearithmetics $\boldsymbol{A}_1 = (A_1; +_1, \circ_1, \leq_1)$ and $\boldsymbol{A}_2 = (A_2; +_2, \circ_2, \leq_2)$ and consider two mappings $g\colon A_1 \to A_2$ and $h\colon A_2 \to A_1$.

Definition 2.2.1. (a) An abstract prearithmetic $\boldsymbol{A}_1 = (A_1; +_1, \circ_1, \leq_1)$ is called *weakly projective* with respect to an abstract prearithmetic $\boldsymbol{A}_2 = (A_2; +_2, \circ_2, \leq_2)$ if there are following relations between orders and operations in $\boldsymbol{A}_1$ and in $\boldsymbol{A}_2$:

$$\begin{aligned} a +_1 b &= h(g(a) +_2 g(b)), \\ a \circ_1 b &= h(g(a) \circ_2 g(b)), \\ & a \leq_1 b \text{ only if } g(a) \leq_2 g(b). \end{aligned}$$

(b) The mapping g is called the *projector* and the mapping h is called the *coprojector* for the pair $(\boldsymbol{A}_1, \boldsymbol{A}_2)$.

In this case, we will say that there is a *weak projectivity* between the prearithmetic $\boldsymbol{A}_1$ and the prearithmetic $\boldsymbol{A}_2$ and there is an *inverse weak projectivity* between the prearithmetic $\boldsymbol{A}_2$ and the prearithmetic $\boldsymbol{A}_1$.

When an abstract prearithmetic $\boldsymbol{B}$ is weakly projective with respect to an abstract prearithmetic $\boldsymbol{A}$ with the projector g and the coprojector h, we denote $\boldsymbol{B}$ by $\boldsymbol{A}_{g,h}$ to show this relation between abstract prearithmetics $\boldsymbol{A}$ and $\boldsymbol{B}$.

Let us consider some examples of weak projectivity of abstract prearithmetics.

Example 2.2.1. It is interesting that mathematicians started implicitly using a special case of weak projectivity and projectivity (cf. Section 2.3) of abstract prearithmetics and arithmetics several centuries ago when Scottish mathematician John Napier (1550–1617) introduced logarithms in the early 17th century as means to simplify calculations. He publicly put forward logarithms and described their utilization for calculations in the book *Mirifici Logarithmorum Canonis Descriptio*, the title of which is translated into English as *Description of the Wonderful Rule of Logarithms* (Napier, 1614). After introduction, logarithms were quickly adopted by engineers, scientists, navigators, and many others to carry out calculations more easily by means of slide rules and logarithm tables (Pierce, 1977). Techniques based on logarithms provided means to replace tiresome multi-digit multiplication by table look-ups and much less complicated addition because in the logarithmic scale the product of numbers is changed to their sum. This possibility is based on the following property of logarithms:

$$\log_b (x \cdot y) = \log_b x + \log_b y. \tag{2.34}$$

This equality is true for all positive numbers b, x and y when the *base* of the logarithms $b \neq 1$. Usually such bases as 2, 10 or e are used.

The contemporary structure of logarithms as a real function was developed by the great mathematician Leonhard Euler (1707–1783), who connected them to the exponential function in the 18th century.

One more useful property of logarithms is the possibility to reduce wide-ranging quantities to much smaller numbers using logarithmic scales, which have been utilized in physics, chemistry, and other sciences.

Besides, logarithms are commonplace in scientific formulae, which are used as measures, describe properties and express various laws in physics, biology, information theory, economic and psychological models, complexity of algorithms and the theory of fractals.

However, it is necessary to remark that utilization of logarithms by Indian mathematicians precedes Napier's invention by more than 1000 years. For instance, the *Sthananga Sutra*, an Indian religious text dated back to the second century, contains descriptions of operations with logarithms with the base 2.

Utilization of logarithms is based on the following procedure of multiplication. To multiply two real numbers x and y, they are converted to logarithms $\log_b x$ and $\log_b y$. Then these logarithms are added obtaining $\log_b(x \cdot y)$ based on the formula (2.34) and finally the result $x \cdot y$ is obtained by applying the exponential function b^z to the value $\log_b(x \cdot y)$. By the formula (2.34), it gives the product $x \cdot y$.

This procedure determines the weak projectivity of the arithmetic $\boldsymbol{A_R}$ of all positive real numbers, in which multiplication is defined by the projector $g(x) = \log_b x$ and coprojector $h(z) = b^z$, with the arithmetic $\boldsymbol{R}^{++}$ of all positive real numbers with the conventional addition and multiplication. We see that the described weak projectivity essentially simplifies multiplication of numbers reducing to addition and has other useful properties.

Weak projectivity between abstract prearithmetics is inherently related to other mathematical structures as it is demonstrated below.

Remark 2.2.1. Weak projectivity, as well as projectivity studied in Section 2.3 and exact projectivity studied in Section 2.4, is a structure that includes two sets A_1 and A_2 connected by two mappings g: $A_1 \to A_2$ and h: $A_2 \to A_1$. As a result, weak projectivity, as well as projectivity and exact projectivity, defines (cf. Figure 2.1) a bidirectional named set in the sense of Burgin (2011, 2017). In the case, when there are (arithmetical) operations in both sets A_1 and A_2, this named set is used to derive properties of one of these sets properties of another one. In the case, when there are (arithmetical) operations in one of these sets, this named set is used to define the same operations in another one. In particular, this technique is used for construction of non-Diophantine arithmetics (Burgin, 1977; 1997).

However, Figure 2.1 does not reflect relations between operations in abstract prearithmetics and arithmetics connected by weak projectivity and projectivity. To represent these relations, we expand Figure 2.1 to Figures 2.2 and 2.3.

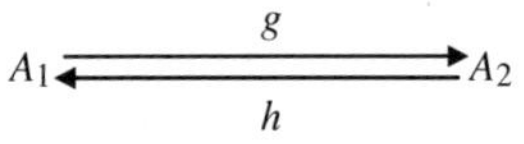

Figure 2.1. Weak projectivity and projectivity of abstract prearithmetics and arithmetics in the form of a bidirectional named set (Burgin, 2018b).

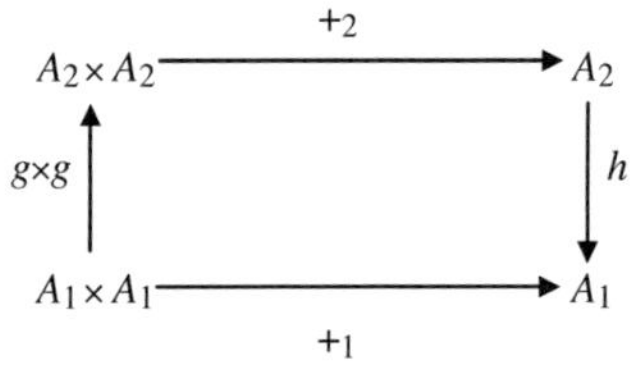

Figure 2.2. Weak projectivity and projectivity of addition.

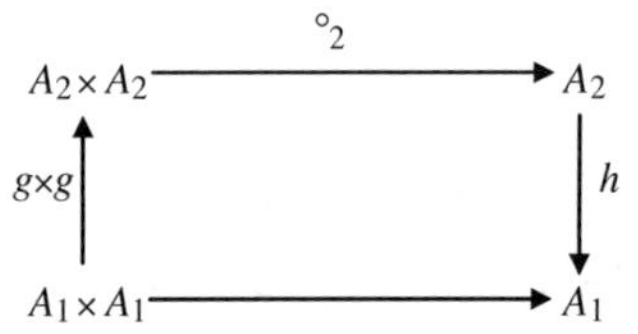

Figure 2.3. Weak projectivity and projectivity of multiplication.

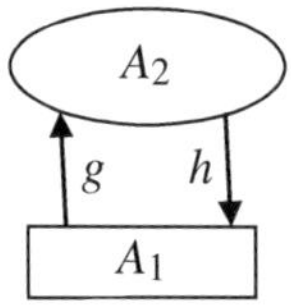

Figure 2.4. Projectivity of abstract arithmetics in the form of a fiber bundle.

Remark 2.2.2. Weak projectivity and projectivity of abstract prearithmetics and arithmetics is also related to basic topological constructions. When h is a surjection, it is possible to treat the triad (named set) (A_1, h, A_2) as a fiber bundle with the base A_1 and the bundle space A_2 (cf., for example, Steenrod, 1951; Husemöller, 1994; Burgin, 2011). Even when h is an arbitrary mapping, it is possible to handle the triad (named set) (A_1, h, A_2) as a fiber bundle in the sense of Goldblatt (1984).

In this setting, Mark Burgin used projectivity of abstract prearithmetics to construct the first class of non-Diophantine arithmetics defining innovative operations with whole and natural numbers (Burgin, 1977; 1997) and an outstanding American physicist Paul Benioff used weak projectivity of abstract prearithmetics building his machinery of scaling in gauge theory and geometry (Benioff, 2011; 2015; 2016).

Interestingly, a similar technique is also used for defining operations with hypernumbers when it is impossible to define these operations straightforwardly (Burgin, 2010a; 2011; 2012c; 2015). The main idea is in defining operations in the base A_1 of the bundle by using some section of the fiber bundle (A_1, h, A_2) to go to the bundle space A_2, performing the necessary operation there and then returning back with the result to the base A_1.

These applications of weak projectivity show efficacy of this construction.

Remark 2.2.3. Weak projectivity and projectivity of abstract prearithmetics and arithmetics is also associated with important concepts in communication theory and computer science. Namely, in the process of communication, information is usually coded before transmission and then decoded after reception (cf., for example, Miller, 1966; Shannon, 1993; Burgin, 2010b).

Coding and decoding functions are also used in theoretical computer science for enumeration of Turing machines and partial recursive functions, defining universal Turing machines (cf., for example, Kozen, 1997; Morita, 2017) and constructing reducibility of one automaton or algorithm, such as a Turing machine, to another automaton or algorithm (cf., for example, Burgin, 2010c). Codes of Turing machines or partial recursive functions are often called Gödel numbers.

Coding is also used in the process of software development for writing a program. After the program is ready, computers perform operational decoding of this program in the process of software functioning.

The coding–decoding schema in computer science also finds important application in computational complexity, algorithmic complexity and axiomatic theory of algorithms where this schema determines reducibility of algorithms, programs, problems, automata and processes (Balcazar *et al.*, 1988; Burgin, 2007a; 2010c; 2010d; Hopcroft *et al.*, 2007; Li and Vitanyi, 1997). In turn, reducibility is used for building universal Turing machines, for measuring power of algorithms, programs and automata and for defining hard and complete problems.

Coding and decoding are frequently utilized in mathematics and logic. For instance, natural numbers are coded by numerals and rational numbers are coded by fractions. Numbers are also coded by arithmetical expressions. For instance, the equality $2 + 3 = 5$ is a decoding of the expression $2 + 3$. In logic, propositions are coded by logical formulas. The famous Gödel's incompleteness theorems were proved by coding expression of the formal arithmetic by natural numbers (Gödel, 1931/1932). Now this and similar procedures are called the Gödel numbering. The noted American logician Alonzo Church (1903–1995) encoded data in the lambda calculus by natural numbers (Church, 1932).

In the case of weak projectivity and projectivity of abstract prearithmetics and arithmetics, the projector g plays the role of the coding function and the coprojector h plays the role of the decoding function. Indeed, the projector g codes elements from the first abstract prearithmetic (arithmetic) $\boldsymbol{A}_1$ by elements from the second abstract prearithmetic (arithmetic) $\boldsymbol{A}_2$. Then operation is performed with the elements (codes) in the second abstract prearithmetic (arithmetic) $\boldsymbol{A}_2$ and the result is decoded by the coprojector h in the role of the decoding function coming back to the prearithmetic (arithmetic) $\boldsymbol{A}_1$.

Remark 2.2.4. It is possible to consider partial weak projectivity when projector and coprojector connect only addition or only multiplication (Burgin, 2019). For instance, if an abstract prearithmetic $\boldsymbol{A}_1 = (A_1; +_1, \circ_1, \leq_1)$ is *weakly projective in addition* with respect to an abstract prearithmetic $\boldsymbol{A}_2 = (A_2; +_2, \circ_2, \leq_2)$, then we have

$$a +_1 b = h(g(a) +_2 g(b)).$$

Basically, the concepts of weak projectivity and other kinds of projectivity considered in the next sections allow defining operations in one set using operations in another set and, in particular, it is possible to build new prearithmetics using existing prearithmetics. This construction is described below.

Let us consider an arbitrary set B, an abstract prearithmetic $\boldsymbol{A} = (A; +, \circ, \leq)$ and two functions g: $B \to A$ and h: $A \to B$. Definition 2.2.1 implies the following result.

Proposition 2.2.1. *It is possible to define on B the unique structure of an abstract prearithmetic $\boldsymbol{B}^{g,h}$ that is weakly projective with respect to the abstract prearithmetic $\boldsymbol{A}$ with the projector g and the coprojector h.*

Proof: To prove this statement, we define the relation $\leq_B$ and two operations $+_B$ and $\circ_B$ in the set B by the following rules:

$$a +_B b = h(g(a) + g(b)),$$
$$a \circ_B b = h(g(a) \circ g(b)),$$
$$a \leq_B b \text{ if and only if } g(a) \leq g(b) \text{ [Condition 1] and}$$
$$(g(c) = g(a) \text{ implies } c = a) \text{ and } (g(d) = g(b) \text{ implies } d = b) \text{ [Condition2]},$$

for any elements a and b from B.

Note that only such elements b can be in the relation $\leq_B$ which constitute the whole inverse image of the element $g(b)$. In other words, if two elements from B are mapped into the element from A, then neither of them can be in relation $\leq_B$ with any other element from B.

Now, we need to check that the relation $\leq_B$ is a partial order, i.e., it is reflexive, symmetric and transitive. Taking arbitrary elements a and b from B, we derive the following properties.

Reflexivity. $a \leq_B a$ because $g(a) \leq g(a)$.
Anti-symmetry. Let us have $a \leq_B b$ and $b \leq_B a$. Then $a \leq_B b$ implies $g(a) \leq g(b)$ and $b \leq_B a$ and implies $g(b) \leq g(c)$. As the relation $\leq$ is a partial order, it is antisymmetric (cf. Appendix) and we have $g(b) = g(a)$. Then by definition, we have $b = a$.
Transitivity. Given $a \leq_B b$ and $b \leq_B c$, by definition, we have $g(a) \leq g(b)$ and $g(b) \leq g(c)$. As the relation $\leq$ is a partial order, it is transitive and we have $g(a) \leq g(c)$. Consequently, we have $a \leq_B c$.

Proposition is proved. □

Remark 2.2.5. This approach of building an arithmetic from an arbitrary set using some arithmetic and a projectivity relation was implicitly utilized when people employed operations with tokens, such as rocks, sticks, bones, clay, stones, wood carvings, and knotted ropes, to represent arithmetical operations with arbitrary objects.

We see that the prearithmetic $\boldsymbol{B}^{g,h}$ constructed in Proposition 2.2.1 is the prearithmetic $\boldsymbol{A}_{g,h}$ defined above. This relation implies a possibility to deduce some of its properties from properties of the prearithmetic $\boldsymbol{A}$.

Proposition 2.2.2. *If the mapping g: $B \to A$ is an injection and $\leq$ is a linear (total) order in A, then $\leq_B$ is a linear (total) order in $\boldsymbol{B}^{g,h}$.*

Indeed, if g is an injection, then the third condition, which defines the prearithmetic $\boldsymbol{B}^{g,h}$, is equivalent to the condition

$$a \leq_B b \text{ if and only if } g(a) \leq g(b).$$

Thus, if for any two elements $x, y \in A$, either $x \leq y$ or $x \leq y$ is true, then for any two elements $a, b \in B$, either $a \leq_B b$ or $a \leq_B b$ is true.

The construction of the prearithmetic $\boldsymbol{B}^{g,h}$ in Proposition 2.2.1 implies its relative uniqueness. Indeed, let us consider three mappings g: $B \to A$, f: $A \to B$ and h: $A \to B$.

Proposition 2.2.3. *If mappings f and h coincide on the abstract prearithmetic $PA(g(B))$ generated by image of B in $\boldsymbol{A}$, i.e., $f|_{PA(g(B))} = h|_{PA(g(B))}$, then $\boldsymbol{B}^{g,h} = \boldsymbol{B}^{g,f}$.*

Proof is left as an exercise. □

When an abstract prearithmetic $\boldsymbol{A}_1$ is weakly projective with respect to an abstract prearithmetic $\boldsymbol{A}_2$, it is often possible to deduce some properties of the prearithmetic $\boldsymbol{A}_1$ from properties of the prearithmetic $\boldsymbol{A}_2$. In particular, weak projectivity in the class of prearithmetics implies definite relations between properties of operations in prearithmetics.

Proposition 2.2.4. *If an abstract prearithmetic $\boldsymbol{A}_1 = (A_1; +_1, \circ_1, \leq_1)$ is weakly projective with respect to an abstract prearithmetic $\boldsymbol{A}_2 = (A_2; +_2, \circ_2, \leq_2)$ and the operation $+_2$ is commutative in the prearithmetic $\boldsymbol{A}_2$, then the operation $+_1$ is commutative in the prearithmetic $\boldsymbol{A}_1$.*

Indeed, for any elements a and b from A_1, we have

$$a +_1 b = h(g(a) +_2 g(b)) = h(g(b) +_2 g(a)) = b +_1 a.$$

This shows that inverse weak projectivity preserves commutativity of addition.

A similar property is valid for multiplication in abstract prearithmetics.

Proposition 2.2.5. *If an abstract prearithmetic* $\boldsymbol{A}_1 = (A_1; +_1, \circ_1, \leq_1)$ *is weakly projective with respect to an abstract prearithmetic* $\boldsymbol{A}_2 = (A_2; +_2, \circ_2, \leq_2)$ *and the operation* $\circ_2$ *is commutative in the prearithmetic* $\boldsymbol{A}_2$*, then the operation* $\circ_1$ *is commutative in the prearithmetic* $\boldsymbol{A}_1$*.*

Indeed, for any elements a and b from A_1, we have

$$a \circ_1 b = h(g(a) \circ_2 g(b)) = h(g(b) \circ_2 g(a)) = b \circ_1 a.$$

This shows that inverse weak projectivity preserves commutativity of multiplication.

To preserve associativity of addition in inverse weak projectivity, we need stronger conditions.

Proposition 2.2.6. *If an abstract prearithmetic* $\boldsymbol{A}_1 = (A_1; +_1, \circ_1, \leq_1)$ *is weakly projective with respect to an abstract prearithmetic* $\boldsymbol{A}_2 = (A_2; +_2, \circ_2, \leq_2)$ *with the projector* g *and the coprojector* h*, for which* $gh = 1_{A_2}$*, and the operation* $+_2$ *is associative in the prearithmetic* $\boldsymbol{A}_2$*, then the operation* $+_1$ *is associative in the prearithmetic* $\boldsymbol{A}_1$*.*

Proof: By Definition 2.2.1, for any elements a, b and c from A_1, we have

$$\begin{aligned}(a +_1 b) +_1 c &= h(g(h(g(a) +_2 g(b))) +_2 g(c)),\\ a +_1 (b +_1 c) &= h(g(a) +_2 g(h(g(b) +_2 g(c)))).\end{aligned}$$

If $gh = 1_{A_2}$, then

$$\begin{aligned}(a +_1 b) +_1 c &= h(g(h(g(a) +_2 g(b))) +_2 g(c))\\ &= h((g(a) +_2 g(b)) +_2 g(c)) = h(g(a) +_2 (g(b)) +_2 g(c)))\\ &= h(g(a) +_2 g(h(g(b) +_2 g(c)))) = a +_1 (b +_1 c).\end{aligned}$$

Proposition is proved. □

A similar property is valid for multiplication in abstract prearithmetics.

Proposition 2.2.7. *If an abstract prearithmetic* $\boldsymbol{A}_1 = (A_1; +_1, \circ_1, \leq_1)$ *is weakly projective with respect to an abstract prearithmetic* $\boldsymbol{A}_2 = (A_2; +_2, \circ_2, \leq_2)$ *with the projector* g *and the coprojector* h, *for which* $gh = 1_{A_2}$, *and the operation* $\circ_2$ *is associative in the prearithmetic* $\boldsymbol{A}_2$, *then the operation* $\circ_1$ *is associative in the prearithmetic* $\boldsymbol{A}_1$.

Proof: By Definition 2.2.1, for any elements a, b and c from A_1, we have

$$(a \circ_1 b) \circ_1 c = h(g(h(g(a) \circ_2 g(b))) \circ_2 g(c)),$$
$$a \circ_1 (b \circ_1 c) = h(g(a) \circ_2 g(h(g(b) \circ_2 g(c)))).$$

If $gh = 1_{A_2}$, then

$$\begin{aligned}(a \circ_1 b) \circ_1 c &= h(g(h(g(a) \circ_2 g(b))) \circ_2 g(c)) = h((g(a) \circ_2 g(b)) \circ_2 g(c)) \\ &= h(g(a) \circ_2 (g(b)) \circ_2 g(c))) = h(g(a) \circ_2 g(h(g(b) \circ_2 g(c))))) \\ &= a \circ_1 (b \circ_1 c).\end{aligned}$$

Proposition is proved. □

In the Diophantine arithmetic, multiplication is distributive over addition, i.e., the following identities hold:

$$x \cdot (y + z) = x \cdot y + x \cdot z,$$
$$(y + z) \cdot x = y \cdot x + z \cdot x.$$

Let us remind that in abstract prearithmetics, multiplication is not always commutative and it is necessary to discern three kinds of distributivity. Namely, *distributivity from the left*

$$x \cdot (y + z) = x \cdot y + x \cdot z$$

and *distributivity from the right*

$$(y + z) \cdot x = y \cdot x + z \cdot x$$

In addition, multiplication is *distributive* over addition when both identities hold.

Let us find conditions that allow preserving distributivity under inverse weak projectivity.

Proposition 2.2.8. *If an abstract prearithmetic* $\boldsymbol{A}_1 = (A_1; +_1, \circ_1, \leq_1)$ *is weakly projective with respect to an abstract prearithmetic* $\boldsymbol{A}_2 = (A_2; +_2, \circ_2, \leq_2)$ *with the projector* g *and the coprojector* h, *for which*

$gh = 1_{A_2}$, and multiplication $\circ_2$ is distributive (from the left or from the right) over addition $+_2$ in the prearithmetic $\boldsymbol{A}_2$, then multiplication $\circ_1$ is distributive (from the left or from the right, correspondingly) over addition $+_1$ in the prearithmetic $\boldsymbol{A}_1$.

Proof: Let us check distributivity from the left taking arbitrary elements a, b and c from $\boldsymbol{A}_1$. By Definition 2.2.1, we have

$$\begin{aligned} a \circ_1 (b +_1 c) &= h(g(a) \circ_2 g(h(g(b) +_2 g(c))))) \\ &= h(g(a) \circ_2 (g(b) +_2 g(c))) = h(g(a) \circ_2 g(b) +_2 g(a) \circ_2 g(c)) \\ &= h(g(h(g(a) \circ_2 g(b))) +_2 g(h(g(a) \circ_2 g(c)))) \\ &= h(g(a \circ_1 b) +_2 g(a \circ_1 c)) = (a \circ_1 b) +_1 (a \circ_1 c) \end{aligned}$$

because $gh = 1_{A_2}$ and multiplication $\circ_2$ is distributive from the left over addition $+_2$ in the prearithmetic $\boldsymbol{A}_2$. It means that multiplication $\circ_1$ is distributive from the left over addition $+_2$ in the prearithmetic $\boldsymbol{A}_1$.

Distributivity from the right and complete distributivity are proved in the same way.

Proposition is proved. □

If we consider properties of weak projectivity in the class of abstract prearithmetics, we find that it is a transitive relation.

Proposition 2.2.9. *If an abstract prearithmetic $\boldsymbol{A}_1 = (A_1; +_1, \circ_1, \leq_1)$ is weakly projective with respect to an abstract prearithmetic $\boldsymbol{A}_2 = (A_2; +_2, \circ_2, \leq_2)$ and the abstract prearithmetic $\boldsymbol{A}_2 = (A_2; +_2, \circ_2, \leq_2)$ is weakly projective with respect to an abstract prearithmetic $\boldsymbol{A}_3 = (A_3; +_3, \circ_3, \leq_3)$, then the prearithmetic $\boldsymbol{A}_1$ is weakly projective with respect to the prearithmetic $\boldsymbol{A}_3$.*

Proof: Let us assume that an abstract prearithmetic $\boldsymbol{A}_1 = (A_1;\ +_1,\ \circ_1,\ \leq_1)$ is weakly projective with respect to an abstract prearithmetic $\boldsymbol{A}_2 = (A_2;\ +_2,\ \circ_2,\ \leq_2)$ with the projector g: $A_1 \to A_2$ and the coprojector h: $A_2 \to A_1$ for the pair $(\boldsymbol{A}_1, \boldsymbol{A}_2)$ and the abstract prearithmetic $\boldsymbol{A}_2 = (A_2;\ +_2,\ \circ_2,\ \leq_2)$ is weakly projective with respect to an abstract prearithmetic $\boldsymbol{A}_3 = (A_3;\ +_3,\ \circ_3,\ \leq_3)$ with the projector k: $A_2 \to A_3$ and the coprojector $l : A_3 \to A_2$ for the pair $(\boldsymbol{A}_2, \boldsymbol{A}_3)$. Then we can define mappings $q = kg : A_1 \to A_3$ and $p = hl : A_3 \to A_1$. Let us consider relations between operations and orders:

$$\begin{aligned} a +_1 b &= h(g(a) +_2 g(b)) = h(l(k(g(a)) +_3 k(g(b)))) \\ &= hl(kg(a) +_3 kg(b)) = p(q(a) +_3 q(b)), a \circ_1 b = h(g(a) \circ_2 g(b)) \\ &= h(l(k(g(a)) \circ_3 k(g(b)))) = hl(kg(a) \circ_3 kg(b)) \\ &= p(q(a) \circ_3 q(b)), a \leq_1 b \text{ only if } g(a) \leq_2 g(b) \end{aligned}$$

and

$$g(a) \leq_2 g(b) \quad \text{only if } kg(a) \leq_3 kg(b).$$

Consequently,

$$a \leq_1 b \quad \text{only if } q(a) \leq_3 q(b)$$

for any elements a and b from A_1 because $q = kg$.

Thus, the prearithmetic $\boldsymbol{A}_1$ is weakly projective with respect to the prearithmetic $\boldsymbol{A}_3$ with the projector q and the coprojector p.

Proposition is proved. □

Proposition 2.2.9 allows proving the following result.

Theorem 2.2.1. *Abstract prearithmetics with weak projectivity relations form the category* **APAWP** *where objects are abstract prearithmetics and morphisms are weak projectivity relations.*

Proof is left as an exercise. □

Lemma 2.2.1. *If an abstract prearithmetic* $\boldsymbol{A}_1 = (A_1; +_1, \circ_1, \leq_1)$ *is weakly projective with respect to an abstract prearithmetic* $\boldsymbol{A}_2 = (A_2; +_2, \circ_2, \leq_2)$, *then the projector* g *preserves the order* $\leq_1$.

Indeed, in this case, $a \leq_1 b$ only if $g(a) \leq_2 g(b)$. It means that $a \leq_1 b$ implies $g(a) \leq_2 g(b)$ for any elements a and b from A_1.

Lemma 2.2.1 means that the projector always is a homomorphism of partially ordered sets.

Remark 2.2.6. Abstract prearithmetics are algebraic systems with two operations and one relation. However, in a general case, weak projectivity of prearithmetics is not a homomorphic relation between prearithmetics while projectors are not homomorphisms (cf. Appendix). Thus, it is interesting to find when a projector is a homomorphism of arithmetical structures.

Proposition 2.2.10. *If an abstract prearithmetic* $\boldsymbol{A}_1 = (A_1; +_1, \circ_1, \leq_1)$ *is weakly projective with respect to an abstract prearithmetic* $\boldsymbol{A}_2 = (A_2; +_2, \circ_2, \leq_2)$ *with the projector* g *and the coprojector* h *and the composition* gh *is the identity mapping* $\mathbf{1}_{A_2}$ *of* $\boldsymbol{A}_2$*, then the projector* g *is a homomorphism.*

Proof: By Definition 2.2.1, we have

$$a +_1 b = h(g(a) +_2 g(b)).$$

If $gh = 1_{A_2}$, then

$$g(a +_1 b) = g(h(g(a) +_2 g(b))) = (gh)(g(a) +_2 g(b)) = g(a) +_2 g(b).$$

By the same token, we have

$$a \circ_1 b = h(g(a) \circ_2 g(b)).$$

Then

$$g(a \circ_1 b) = g(h(g(a) \circ_2 g(b))) = (gh)(g(a) \circ_2 g(b)) = g(a) \circ_2 g(b).$$

These equalities mean that g is a homomorphism of $\boldsymbol{A}_1$ as a universal algebra with two operations. Besides, by Lemma 2.2.1, the projector g preserves the order $\leq_1$.

Proposition is proved. □

Corollary 2.2.1. *If an abstract prearithmetic* $\boldsymbol{A}_1 = (A_1; +_1, \circ_1, \leq_1)$ *is weakly projective with respect to an abstract prearithmetic* $\boldsymbol{A}_2 = (A_2; +_2, \circ_2, \leq_2)$ *with the projector* g *and the coprojector* h *and* $h = g^{-1}$*, then the projector* g *is an isomorphism, i.e., prearithmetics* $\boldsymbol{A}_1$ *and* $\boldsymbol{A}_2$ *are isomorphic as algebraic systems.*

Indeed, by Proposition 2.2.10, the projector g is a homomorphism, and if a mapping has the inverse mapping, it is a bijection.

Properties of isomorphisms (cf., for example, Kurosh, 1967) give us the following result.

Corollary 2.2.2. *If an abstract prearithmetic* $\boldsymbol{A}_1 = (A_1; +_1, \circ_1, \leq_1)$ *is weakly projective with respect to an abstract prearithmetic* $\boldsymbol{A}_2 = (A_2; +_2, \circ_2, \leq_2)$ *with the projector* g *and the coprojector* h *and* $h = g^{-1}$*, then identities in prearithmetics* $\boldsymbol{A}_1$ *and* $\boldsymbol{A}_2$ *are the same.*

For instance, if in the prearithmetic $\boldsymbol{A}_1$, addition is associative, then addition in the prearithmetic $\boldsymbol{A}_2$ is also associative, or if in the prearithmetic $\boldsymbol{A}_1$, multiplication is distributive over addition, then multiplication in the prearithmetic $\boldsymbol{A}_2$ is also distributive over addition.

Now let us explicate relations between order relations in abstract prearithmetics connected by weak projectivity.

Proposition 2.2.11. *If an abstract prearithmetic $\boldsymbol{A}_1 = (A_1; +_1, \circ_1, \leq_1)$ is weakly projective with respect to an abstract prearithmetic $\boldsymbol{A}_2 = (A_2; +_2, \circ_2, \leq_2)$, the coprojector h preserves the order $\leq_2$, i.e., $x \leq_2 y$ implies $h(x) \leq_1 h(y)$, and the operation $+_2$ preserves the order $\leq_2$, then the operation $+_1$ preserves the order $\leq_1$.*

Proof: Let us assume that for any elements u, x, y and z from A_2, the inequalities $u \leq x$ and $z \leq y$ imply the inequality $u +_2 z \leq x +_2 y$ and consider elements a, b, c and d from the prearithmetic A_1 such that $a \leq b$ and $d \leq c$. Because $a \leq b$ and $d \leq c$ only if $g(a) \leq g(b)$ and $g(d) \leq g(c)$, we have $g(a) \leq g(b)$ and $g(d) \leq g(c)$. By the initial conditions, $g(a) +_2 g(d) \leq g(b) +_2 g(c)$. As the coprojector h preserves the order $\leq_2$, we have

$$a +_1 d \leq h(g(a) +_2 g(d)) \leq h(g(b) +_2 g(c)) = b +_1 c. \qquad \square$$

It means that the inequalities $a \leq b$ and $d \leq c$ imply the inequality $a +_1 d \leq b +_1 c$.

Proposition is proved as a, b, c and d are arbitrary elements from A_1.

Lemma 2.1.6 and Proposition 2.2.11 give us the following result.

Corollary 2.2.3. *If an abstract prearithmetic $\boldsymbol{A}_1 = (A_1; +_1, \circ_1, \leq_1)$ is weakly projective with respect to an abstract prearithmetic $\boldsymbol{A}_2 = (A_2; +_2, \circ_2, \leq_2)$, the coprojector h preserves the order $\leq_2$, i.e., $x \leq_2 y$ implies $h(x) \leq_1 h(y)$, and the operation $+_2$ is monotone, then the operation $+_1$ is monotone.*

Corollary 2.2.4. *If an abstract prearithmetic $\boldsymbol{A}_1 = (A_1; +_1, \circ_1, \leq_1)$ is weakly projective with respect to an abstract prearithmetic $\boldsymbol{A}_2 = (A_2; +_2, \circ_2, \leq_2)$, which is a partially ordered semigroup with respect to addition, and the coprojector h preserves the order $\leq_2$, then $\boldsymbol{A}_1$ is a partially ordered semigroup with respect to addition.*

Proposition 2.2.12. *If an abstract prearithmetic $\boldsymbol{A}_1 = (A_1; +_1, \circ_1, \leq_1)$ is weakly projective with respect to an abstract prearithmetic $\boldsymbol{A}_2 =$*

$(A_2; +_2, \circ_2, \leq_2)$, the coprojector h preserves the order $\leq_2$, i.e., $x \leq_2 y$ implies $h(x) \leq_1 h(y)$, and the operation $\circ_2$ preserves the order $\leq_2$, then the operation $\circ_1$ preserves the order $\leq_1$.

Proof is similar to the proof of Proposition 2.2.11. □

Lemma 2.1.7 and Proposition 2.1.12 give us the following result.

Corollary 2.2.5. *If an abstract prearithmetic $\boldsymbol{A}_1 = (A_1; +_1, \circ_1, \leq_1)$ is weakly projective with respect to an abstract prearithmetic $\boldsymbol{A}_2 = (A_2; +_2, \circ_2, \leq_2)$, the coprojector h preserves the order $\leq_2$, and the operation $\circ_2$ is monotone, then the operation $\circ_1$ is monotone.*

Corollary 2.2.6. *If an abstract prearithmetic $\boldsymbol{A}_1 = (A_1; +_1, \circ_1, \leq_1)$ is weakly projective with respect to an abstract prearithmetic $\boldsymbol{A}_2 = (A_2; +_2, \circ_2, \leq_2)$, which is a partially ordered semigroup with respect to addition, and the coprojector h preserves the order $\leq_2$, then $\boldsymbol{A}_1$ is a partially ordered semigroup with respect to addition.*

Weak projectivity preserves chains in abstract prearithmetics. Namely, we have the following result.

Proposition 2.2.13. *If an abstract prearithmetic $\boldsymbol{A}_1 = (A_1; +_1, \circ_1, \leq_1)$ is weakly projective with respect to an abstract prearithmetic $\boldsymbol{A}_2 = (A_2; +_2, \circ_2, \leq_2)$ with the projector g and the coprojector h and $a_1 \leq_1 a_2 \leq_1 \cdots \leq_1 a_n$ is a chain in $\boldsymbol{A}_1$, then $g(a_1) \leq_2 g(a_2) \leq_2 \cdots \leq_2 g(a_n)$ is a chain in $\boldsymbol{A}_2$.*

Indeed, by Definition 2.2.1, the inequality $a_j \leq a_{j+1}$ implies the inequality $g(a_j) \leq g(a_{j+1})$ for all $j = 1, 2, 3, \ldots, n - 1$.

Weak projectivity also preserves some properties of chains in abstract prearithmetics.

Proposition 2.2.14. *If an abstract prearithmetic $\boldsymbol{A}_1 = (A_1; +_1, \circ_1, \leq_1)$ is weakly projective with respect to an abstract prearithmetic $\boldsymbol{A}_2 = (A_2; +_2, \circ_2, \leq_2)$ with the projector g and the coprojector h, for any element x from A_2, its coimage $g^{-1}(x)$ is finite and any maximal chain in $\boldsymbol{A}_2$ with distinct elements, which has the beginning and the end, is finite, then any maximal chain in $\boldsymbol{A}_1$ with distinct elements, which has the beginning and the end, is finite.*

Proof: Let us assume that an abstract prearithmetic $\boldsymbol{A}_1 = (A_1; +_1, \circ_1, \leq_1)$ is weakly projective with the projector g and the coprojector h with respect to an abstract prearithmetic $\boldsymbol{A}_2 = (A_2; +_2, \circ_2, \leq_2)$ where any maximal chain in $\boldsymbol{A}_2$ with distinct elements, which has the beginning and the end, is finite, and for any element x from A_2, its coimage $g^{-1}(x)$ is finite.

Let us consider a maximal chain in $\boldsymbol{A}_1$ with distinct elements, which has the beginning a and the end b:

$$a \leq \cdots \leq a_j \leq a_{j+1} \leq \cdots \leq a_{n-1} \leq a_n \leq \cdots \leq b. \tag{2.35}$$

Then by Proposition 2.1.13, there is the chain in $\boldsymbol{A}_2$ with distinct elements, which has the beginning $g(a)$ and the end $g(b)$:

$$g(a \leq \cdots \leq g(a_j) \leq g(a_{j+1}) \leq \cdots \leq g(a_{n-1}) \leq g(a_n) \leq \cdots \leq g(b). \tag{2.36}$$

By the Hausdorff theorem, which is proved using Axiom of Choice, it is possible to include any chain in a partially ordered set into a maximal chain (cf. Kurosh, 1963). By the initial conditions, any such a maximal chain in $\boldsymbol{A}_2$ with distinct elements is finite. This implies that there is only a finite number of distinct elements $g(a_j)$ in the chain (2.36). Because each of these elements $g(a_j)$ has a finite number of elements a_n, which are mapped into $g(a_j)$ by the mapping g, and the whole chain (2.35) is mapped onto the chain (2.36), the chain (2.35) is finite.

Proposition is proved. □

Under definite conditions, weak projectivity preserves finite connectedness.

Proposition 2.2.15. *If an abstract prearithmetic $\boldsymbol{A}_1 = (A_1; +_1, \circ_1, \leq_1)$ is weakly projective with respect to an abstract prearithmetic $\boldsymbol{A}_2 = (A_2; +_2, \circ_2, \leq_2)$ with the projector g and the coprojector h, for any element x from A_2, its coimage $g^{-1}(x)$ is finite and the partially ordered set A_2 is finitely connected, then the partially ordered set A_1 is finitely connected.*

Proof: Let us assume that an abstract prearithmetic $\boldsymbol{A}_1 = (A_1; +_1, \circ_1, \leq_1)$ is weakly projective with the projector g and the coprojector h with respect to an abstract prearithmetic $\boldsymbol{A}_2 = (A_2; +_2, \circ_2, \leq_2)$ where the partially ordered set A_2 is finitely connected, and for any element x from A_2, its coimage $g^{-1}(x)$ is finite. Then by Lemma 2.1.22, any maximal chain in $\boldsymbol{A}_2$, which has the beginning and the end, is finite. By Proposition 2.1.35, any maximal chain in $\boldsymbol{A}_1$, which has the beginning and the end, is finite. Then by Lemma 2.1.21, the partially ordered set A_1 is finitely connected.

Proposition is proved. □

Corollary 2.2.7. *If an abstract prearithmetic* $\boldsymbol{A}_1 = (A_1; +_1, \circ_1, \leq_1)$ *is weakly projective with respect to an abstract prearithmetic* $\boldsymbol{A}_2 = (A_2; +_2, \circ_2, \leq_2)$*, the projector* g *is an injection and the relation* $\leq_2$ *is discrete, then the relation* $\leq_1$ *is discrete.*

Indeed, if the projector g is an injection, then for any element x from A_2, its coimage $g^{-1}(x)$ has, at most, one element and the statement follows from Proposition 2.1.36.

Remark 2.2.7. In contrast to finite connectedness, discreteness of order relation is not preserved by inverse weak projectivity, i.e., analogues of Proposition 2.1.36 and Corollary 2.2.7 are not true in a general case, as the following example demonstrates.

At first, we do not demand that all inverse images $h^{-1}(a)$ are finite.

Example 2.2.2. Let us consider the set N of all natural numbers, which we denote with subscript A, e.g., 3_A or 10_A, and the set Rl[0, 1] of all rational numbers from the interval [0, 1] and define $A_2 = N \cup \text{Rl}[0,\ 1]$. The partial order $\leq_2$ in A_2 is the union of orders in N and in Rl[0, 1] with the additional relation between elements from N and Rl[0, 1], which is defined for any n_A from N and any r from Rl[0, 1] by the following formula:

$$n_A \leq_2 r.$$

Addition and multiplication of elements from N are defined as they are defined for natural numbers. Multiplication of elements from Rl[0, 1] is defined as it is defined for rational numbers. Addition of elements from Rl[0, 1] is defined by the following formula:

$$a +_2 b = \min(a + b, 1).$$

The sum of arbitrary element n_A from N and element r from Rl[0, 1] is defined by the following formula:

$$n_A +_2 r = 1.$$

The product of arbitrary element n_A from N and element r from Rl[0, 1] is defined by the following formula:

$$n_A \circ_2 r = 1.$$

This makes the system $\boldsymbol{A}_2 = (A_2; +_2, \circ_2, \leq_2)$ an abstract prearithmetic.

To build the abstract prearithmetic $\boldsymbol{A}_1$, we take the union $A_1 = \{3_A\} \cup \text{Rl}(0,\ 1]$. In addition, we define all operations and relations in Rl(0, 1] as they are defined in the abstract prearithmetic $\boldsymbol{A}_2$ for the elements from Rl[0, 1]. It is possible to do this because Rl(0, 1] is a subset of Rl[0, 1]. Additionally, we assume that 3_A is less than any element from Rl(0, 1], $3_A +_1 3_A = 3_A$, $3_A \circ_1 3_A = 3_A$, $3_A +_1 r = 3_A \circ_1 r = 1$ for an arbitrary element r from Rl(0, 1].

This makes the system $\boldsymbol{A}_1 = (A_1; +_1, \circ_1, \leq_1)$ an abstract prearithmetic.

To construct a weak projectivity of $\boldsymbol{A}_1$ into $\boldsymbol{A}_2$, we define the projector g and the coprojector h in the following way. The mapping g: $A_1 \to A_2$ is the natural inclusion of A_1 into A_2. The mapping h: $A_2 \to A_1$ maps the set N onto the set $\{3_A\}$, maps 0 from the set Rl[0, 1] onto 1 and is identical on the set Rl(0, 1].

Now let us check conditions of the weak projectivity.

$$3_A +_1 3_A = 3_A = h(6_A) = h(3_A +_2 3_A) = h(g(3_A) +_2 g(3_A)),$$

$$3_A \circ_1 3_A = 3_A = h(3_A) = h(3_A \circ_2 3_A) = h(g(3_A) \circ_2 g(3_A)),$$

$$3_A \leq_1 3_A = 3_A \text{ implies } g(3_A) \leq_2 g(3_A),$$

$$r +_1 q = \min(r + q, 1) = r +_2 q,\ r \circ_1 q = r \cdot q = r \circ_2 q,$$

$$3_A +_1 r = 3_A +_2 r = h(g(3_A) +_2 g(r)) = 1,$$

$$3_A \circ_1 r = 3_A \circ_2 r = h(g(3_A) \circ_2 g(r)) = 1$$

for any elements q and r from Rl(0, 1].

Thus, the abstract prearithmetic $\boldsymbol{A}_1 = (A_1; +_1, \circ_1, \leq_1)$ is weakly projective with respect to the abstract prearithmetic $\boldsymbol{A}_2 = (A_2; +_2, \circ_2, \leq_2)$ with the projector g and the coprojector h while for any element x from A_2, its coimage $g^{-1}(x)$ is finite. However, in the prearithmetic $\boldsymbol{A}_2$, the order is discrete and in particular, any element x but 1 has the successor Sx, while in the prearithmetic $\boldsymbol{A}_1$, the order is not discrete because the element 3_A does not have the successor although $3_A \leq_1 1$.

Example 2.2.3. Let us consider the set Z of all integer numbers, which we denote with subscript A, e.g., 3_A, -4_A or 10_A, and the set Rl[0, 1] of all rational numbers from the interval [0, 1] and define $A_2 = Z \cup$ Rl[0, 1]. The partial order $\leq_2$ in A_2 is the union of orders in Z and in Rt[0, 1] with the additional relation between elements from Z and Rl[0, 1], which makes all elements from Z less than any element from Rl[0, 1] and is defined for any n_A from Z and any r from Rl[0, 1] by the following formula:

$$n_A \leq_2 r.$$

Addition and multiplication of elements from Z are defined as they are defined for integer numbers. Multiplication of elements from Rl[0, 1] is defined as it is defined for rational numbers. Addition of elements from Rl[0, 1] is defined by the following formula:

$$a +_2 b = \min(a + b, 1).$$

The sum of arbitrary element n_A from Z and element r from Rl[0, 1] is defined by the following formula:

$$n_A +_2 r = 1.$$

The product of arbitrary element n_A from N and element r from Rl[0, 1] is defined by the following formula:

$$n_A \circ_2 r = 1.$$

This makes the system $\boldsymbol{A}_2 = (A_2; +_2, \circ_2, \leq_2)$ an abstract prearithmetic.

To build the abstract prearithmetic $\boldsymbol{A}_1$, we take the set Z^- of all non-positive integer numbers and the union $A_1 = \{Z^-\} \cup \text{Rl}(0, 1]$. In addition, we define all operations and relations in Rl(0, 1] as they are defined in the abstract prearithmetic $\boldsymbol{A}_2$ for the elements from Rl[0, 1]. It is possible to do this because Rl(0, 1] is a subset of Rl[0, 1]. Additionally, we define

$$0_A +_1 (-n_A) = -n_A,$$

$$0_A \circ_1 (-n_A) = 0_A,$$

$$(-m_A) +_1 (-n_A) = -(m + n)_A,$$

$$(-m_A) \circ_1 (-n_A) = -(m \cdot n)_A,$$

$$0_A +_1 (-n_A) = -n_A,$$

$$(-n_A) +_1 r = 1,$$

$$(-n_A) \circ_1 r = 1$$

for an arbitrary element r from Rl(0, 1] and arbitrary elements $-m_A$ and $-n_A$ from Z^-.

This makes the system $\boldsymbol{A}_1 = (A_1; +_1, \circ_1, \leq_1)$ an abstract prearithmetic.

To construct a weak projectivity of $\boldsymbol{A}_1$ into $\boldsymbol{A}_2$, we define the projector g and the coprojector h in the following way. The mapping g: $A_1 \to A_2$ is the natural inclusion of A_1 into A_2. The mapping h: $A_2 \to A_1$ maps the set Z^- identically into itself, maps any positive integer number n onto $-n$, maps 0 from the set Rl[0, 1] onto 1 and is identical on the set Rl(0, 1]. Note that $hg = 1_{A_1}$.

Now let us check conditions of the weak projectivity.

$$0_A +_1 (-n_A) = -n_A = h(g(-n_A)) = h(g(0_A) +_2 g(-n_A)),$$
$$0_A \circ_1 (-n_A) = 0_A = h(g(0_A)) = h(g(0_A) \circ_2 g(-n_A)),$$
$$(-m_A) +_2 (-n_A) = -(m+n)_A = h(g(-(m+n)_A))$$
$$= h(g(-m_A) +_2 g(-n_A)),$$
$$(-m_A) \circ_2 (-n_A) = -(m \cdot n)_A,$$
$$0_A +_1 r = 0_A +_2 r = h(g(0_A) +_2 g(r)) = 1,$$
$$0_A \circ_1 r = 0_A \circ_2 r = h(g(0_A) \circ_2 g(r)) = 1,$$
$$(-n_A) +_1 r = (-n_A) +_2 r = h(g(-n_A) \circ_2 g(r)) = 11,$$
$$(-n_A) \circ_1 r = (-n_A) \circ_2 r = h(g(-n_A) \circ_2 g(r)) = 1,$$
$$r +_1 q = \min(r+q, 1) = r +_2 q, r \circ_1 q = r \cdot q = r \circ_2 q$$

for any elements q and r from Rl(0, 1] and arbitrary elements $-m_A$ and $-n_A$ from Z^-.

Thus, the abstract prearithmetic $\boldsymbol{A}_1 = (A_1; +_1, \circ_1, \leq_1)$ is weakly projective with respect to the abstract prearithmetic $\boldsymbol{A}_2 = (A_2; +_2, \circ_2, \leq_2)$ with the projector g and the coprojector h while for any element x from A_2, its coimage $g^{-1}(x)$ is finite. However, in the prearithmetic $\boldsymbol{A}_2$, the order is discrete and in particular, any element x but 1 has the successor Sx, while in the prearithmetic $\boldsymbol{A}_1$, the order is not discrete because the element 0_A does not have the successor although $0_A \leq_1 1$.

Note that for any element a from A_1, its coimage $h^{-1}(a)$ is also finite.

However, if the order relation in $\boldsymbol{A}_2$ is discrete and satisfies additional conditions, then the order relation in $\boldsymbol{A}_1$ is discrete.

Proposition 2.2.16. *If an abstract prearithmetic $\boldsymbol{A}_1 = (A_1; +_1, \circ_1, \leq_1)$ is weakly projective with respect to an abstract prearithmetic $\boldsymbol{A}_2 = (A_2; +_2, \circ_2, \leq_2)$ with the projector g and the coprojector h, for any element x from A_2, its coimage $g^{-1}(x)$ is finite, the relation $\leq_2$ is discrete and any maximal chain in $\boldsymbol{A}_2$ with distinct elements, which has the beginning and the end, is finite, then the relation $\leq_1$ is discrete.*

Proof is left as an exercise. □

If we have projectivity from an abstract prearithmetic $\boldsymbol{A}_1$ to an abstract prearithmetic $\boldsymbol{A}_2$, the order relation is preserved from $\boldsymbol{A}_2$ to $\boldsymbol{A}_1$ (cf. Lemma 2.2.1). In the case, when the order relation is preserved in both direction, i.e., from $\boldsymbol{A}_2$ to $\boldsymbol{A}_1$ and from $\boldsymbol{A}_1$ to $\boldsymbol{A}_2$, we have additional restrictions on the projector.

Proposition 2.2.17. *If an abstract prearithmetic $\boldsymbol{A}_1 = (A_1; +_1, \circ_1, \leq_1)$ is weakly projective with respect to an abstract prearithmetic $\boldsymbol{A}_2 = (A_2; +_2, \circ_2, \leq_2)$ with the projector g and the coprojector h and for any elements a and b from $\boldsymbol{A}_1, g(a) \leq_2 g(b)$ implies $a \leq_1 b$, then g is an injection.*

Proof: Let us take an abstract prearithmetic $\boldsymbol{A}_1 = (A_1; +_1, \circ_1, \leq_1)$, which is weakly projective with respect to an abstract prearithmetic $\boldsymbol{A}_2 = (A_2; +_2, \circ_2, \leq_2)$ with the projector g and the coprojector h, and assume that for any elements a and b from $\boldsymbol{A}_1$, the relation $g(a) \leq_2 g(b)$ implies the relation $a \leq_1 b$. In this case, if $g(a) = g(b)$ for some elements a and b from $\boldsymbol{A}_1$, we have $g(a) \leq_2 g(b)$ and $g(b) \leq_2 g(a)$ because $\leq_2$ is a partial order (cf. Appendix). By the initial conditions, these inequalities imply $a \leq_1 b$ and $b \leq_1 a$. Consequently, $a = b$. As a and b are arbitrary elements from $\boldsymbol{A}_1$, g is an injection.

Proposition is proved. □

Weak projectivity allows introduction of new operations in abstract prearithmetics.

If an abstract prearithmetic $\boldsymbol{A}_1 = (A_1; +_1, \circ_1, \leq_1)$ is weakly projective with respect to an abstract prearithmetic $\boldsymbol{A}_2 = (A_2; +_2, \circ_2, \leq_2)$ with the projector g and the coprojector h, for any element x from A_2, we introduce actions of the element x on an element a from A_1.

Definition 2.2.2. (a) The *additive action from the left* of the element x on an element a is defined as

$$x^{+}a = h(x +_2 g(a)).$$

(b) The *additive action from the right* of the element x on an element a is defined as

$$a^{+}x = h(g(a) +_2 x).$$

(c) The *multiplicative action from the left* of the element x on an element a is defined as

$$xa = h(x \circ_2 g(a)).$$

(d) The *multiplicative action from the right* of the element x on an element a is defined as

$$ax = h(g(a) \circ_2 x).$$

In such a way, the abstract prearithmetic $\boldsymbol{A}_2$ determines a system of operations in abstract prearithmetic $\boldsymbol{A}_1$.

Note that any element from $\boldsymbol{A}_1$ additively and multiplicatively acts on all other elements from $\boldsymbol{A}_1$. In geometry, calculus and functional analysis, such an additive action is called a *shift* or *shift operator* (cf., for example, Burgin, 2008; 2017). Multiplicative action is called *uniform scaling* or *isotropic scaling* and is often used in geometry and physics (cf., for example, Feynman, 1948; Feynman and Hibbs, 1965; Benioff, 2012; 2012a; 2013; 2015; 2016; 2016a; Burgin, 2008b; 2017).

Let us study properties of actions.

Proposition 2.2.18. *If an abstract prearithmetic* $\boldsymbol{A}_1 = (A_1; +_1, \circ_1, \leq_1)$ *is weakly projective with respect to an abstract prearithmetic* $\boldsymbol{A}_2 = (A_2; +_2, \circ_2, \leq_2)$ *with the projector* g *and the coprojector* h, *which preserves addition, and multiplication* $\circ_2$ *is distributive over addition* $+_2$ *in the prearithmetic* $\boldsymbol{A}_2$, *then multiplicative action from the left (from the right) is distributive over addition* $+_1$ *in the prearithmetic* $\boldsymbol{A}_1$.

Proof: By Definitions 2.2.1 and 2.2.2, for any elements a and b from A_1 and x from A_2, we have

$$\begin{aligned} x(a +_1 b) &= h(x \circ_2 (g(a) +_2 g(b))) = h(x \circ_2 g(a) +_2 x \circ_2 g(b)) \\ &= h(x \circ_2 g(a)) +_2 h(x \circ_2 g(b)) = xa +_1 xb. \end{aligned}$$

Distributivity from the right is proved in a similar way.

Proposition is proved. □

Proposition 2.2.19. *If an abstract prearithmetic* $\boldsymbol{A}_1 = (A_1; +_1, \circ_1, \leq_1)$ *is weakly projective with respect to an abstract prearithmetic* $\boldsymbol{A}_2 = (A_2; +_2, \circ_2, \leq_2)$ *with the projector* g *and the coprojector* h, *which preserves addition, and multiplication* $\circ_2$ *is distributive over addition* $+_2$ *in the prearithmetic* $\boldsymbol{A}_2$, *then multiplicative action from the left (from the right) is distributive over addition* $+_2$ *in the prearithmetic* $\boldsymbol{A}_2$.

Proof: By Definitions 2.2.1 and 2.2.2, for any elements x and y from A_2 and a from A_1, we have

$$(x +_2 y)a = h((x +_2 y) \circ_2 g(a)) = h(x \circ_2 g(a) +_2 y \circ_2 g(a))$$
$$= h(x \circ_2 g(a)) +_2 h(y \circ_2 g(a)) = xa +_1 ya.$$

Distributivity from the right is proved in a similar way.

Proposition is proved. □

There are also specific counterparts for associativity with respect to actions.

Proposition 2.2.20. *If an abstract prearithmetic $\boldsymbol{A}_1 = (A_1; +_1, \circ_1, \leq_1)$ is weakly projective with respect to an abstract prearithmetic $\boldsymbol{A}_2 = (A_2; +_2, \circ_2, \leq_2)$ with the projector g and the coprojector h, for which $gh = 1_{A_2}$, and the operation $\circ_2$ is associative in the prearithmetic $\boldsymbol{A}_2$, then we have $x(ya) = (x \circ_2 y)a$ and $(ax)y = a(x \circ_2 y)$ for any elements x and y from A_2 and a from A_1.*

Proof: By Definitions 2.2.1 and 2.2.2, we have

$$x(ya) = x(h(y \circ_2 g(a))) = h(x \circ_2 g(h(y \circ_2 g(a))))$$
$$= h(x \circ_2 (y \circ_2 g(a))) = h((x \circ_2 y) \circ_2 g(a)) = (x \circ_2 y)a$$

The second identity is proved in a similar way.

Proposition is proved. □

Proposition 2.2.21. *If an abstract prearithmetic $\boldsymbol{A}_1 = (A_1; +_1, \circ_1, \leq_1)$ is weakly projective with respect to an abstract prearithmetic $\boldsymbol{A}_2 = (A_2; +_2, \circ_2, \leq_2)$ with the projector g and the coprojector h, for which $gh = 1_{A_2}$, and the operation $\circ_2$ is associative in the prearithmetic $\boldsymbol{A}_2$, then $x(a \circ_2 b) = (x \circ_2 g(a))b$ and $(a \circ_1 b)x = a(g(b) \circ_2 x)$ for any elements a and b from A_1 and x from A_2.*

Proof: By Definitions 2.2.1 and 2.2.2, we have

$$x(a \circ_1 b) = h(x \circ_2 g(a \circ_1 b)) = h(x \circ_2 g(h(g(a) \circ_2 g(b))))$$
$$= h(x \circ_2 (g(a) \circ_2 g(b))) = h((x \circ_2 g(a)) \circ_2 g(b))) = (x \circ_2 g(a))b.$$

The second identity is proved in a similar way.

Proposition is proved. □

Under definite conditions, actions preserve order.

Proposition 2.2.22. *If an abstract prearithmetic $\boldsymbol{A}_1 = (A_1; +_1, \circ_1, \leq_1)$ is weakly projective with respect to an abstract prearithmetic $\boldsymbol{A}_2 = (A_2; +_2, \circ_2, \leq_2)$, the coprojector h preserves the order $\leq_2$, i.e., $x \leq_2 y$ implies $h(x) \leq_1 h(y)$, and the operation $+_2$ preserves the order $\leq_2$, then additive action from the left (from the right) preserves the order $\leq_1$, i.e., additive action from the left (from the right) is a monotone operation in $\boldsymbol{A}_1$.*

Proof: By Definitions 2.2.1 and 2.2.2, for any elements a and b from A_1 and x from A_2, if $a \leq_1 b$, then by Lemma 2.2.1, we have

$$g(a)) \leq_2 g(b).$$

Consequently, as the operation $\circ_2$ preserves the order $\leq_2$

$$x +_2 g(a)) \leq_2 x +_2 g(b).$$

Thus, as the coprojector h preserves the order $\leq_2$

$$x^+ a = h(x +_2 g(a)) \leq_1 h(x +_2 g(b)) = x^+ b.$$

It means that additive action from the left preserves the order $\leq_1$.
The proof for additive action from the right is similar.
Proposition is proved. □

In some cases, action is a monotone operation.

Proposition 2.2.23. *If an abstract prearithmetic $\boldsymbol{A}_1 = (A_1; +_1, \circ_1, \leq_1)$ is weakly projective with respect to an abstract prearithmetic $\boldsymbol{A}_2 = (A_2; +_2, \circ_2, \leq_2)$, the coprojector h preserves the order $\leq_2$, i.e., $x \leq_2 y$ implies $h(x) \leq_1 h(y)$, and the operation $\circ_2$ preserves the order $\leq_2$, then multiplicative action from the left (from the right) preserves the order $\leq_1$, i.e., multiplicative action from the left (from the right) is a monotone operation in $\boldsymbol{A}_1$.*

Proof: By Definitions 2.2.1 and 2.2.2, for any elements a and b from A_1 and x from A_2, if $a \leq_1 b$, then by Lemma 2.2.1, we have

$$g(a)) \leq_2 g(b).$$

Consequently, as the operation $\circ_2$ preserves the order $\leq_2$

$$x \circ_2 g(a)) \leq_2 x \circ_2 g(b).$$

Thus, as the coprojector h preserves the order $\leq_2$

$$xa = h(x \circ_2 g(a)) \leq_1 h(x \circ_2 g(b)) = xb.$$

It means that multiplicative action from the left preserves the order $\leq_1$. The proof for multiplicative action from the right is similar.
Proposition is proved. □

Proposition 2.2.24. *If an abstract prearithmetic* $\boldsymbol{A}_1 = (A_1; +_1, \circ_1, \leq_1)$ *is weakly projective with respect to an abstract prearithmetic* $\boldsymbol{A}_2 = (A_2; +_2, \circ_2, \leq_2)$ *with the projector* g *and the coprojector* h*, then* $g(b)^+a = b +_1 a$ *for any elements* a *and* b *from* A_1*.*

Indeed, we have

$$g(b)^+a = h(g(b) +_2 g(a)) = b +_1 a.$$

Corollary 2.2.8. *If an abstract prearithmetic* $\boldsymbol{A}_1 = (A_1; +_1, \circ_1, \leq_1)$ *with the additive zero* 0_1 *is weakly projective with respect to an abstract prearithmetic* $\boldsymbol{A}_2 = (A_2; +_2, \circ_2, \leq_2)$ *with the projector* g *and the coprojector* h*, then* $g(0_1)^+a = a$ *for any elements* a *and* b *from* A_1*.*

Informally, Proposition 2.2.24 means that additive actions in $\boldsymbol{A}_1$ of images of elements from $\boldsymbol{A}_1$ coincide with additive actions in $\boldsymbol{A}_1$ of the same elements.

Proposition 2.2.25. *If an abstract prearithmetic* $\boldsymbol{A}_1 = (A_1; +_1, \circ_1, \leq_1)$ *is weakly projective with respect to an abstract prearithmetic* $\boldsymbol{A}_2 = (A_2; +_2, \circ_2, \leq_2)$ *with the projector* g *and the coprojector* h*, then* $g(b)a = b \circ_1 a$ *for any elements* a *and* b *from* A_1*.*

Indeed, we have

$$g(b)^+a = h(g(b) \circ_2 g(a)) = b \circ_1 a.$$

Corollary 2.2.9. *If an abstract prearithmetic* $\boldsymbol{A}_1 = (A_1; +_1, \circ_1, \leq_1)$ *with the multiplicative zero* 0_1^{m} *is weakly projective with respect to an abstract prearithmetic* $\boldsymbol{A}_2 = (A_2; +_2, \circ_2, \leq_2)$ *with the projector* g *and the coprojector* h*, then* $g(0_1^m)^+a = 0_1^m$ *for any elements* a *and* b *from* A_1*.*

Corollary 2.2.10. *If an abstract prearithmetic* $\boldsymbol{A}_1 = (A_1; +_1, \circ_1, \leq_1)$ *with the multiplicative one* 1_1 *is weakly projective with respect to an abstract prearithmetic* $\boldsymbol{A}_2 = (A_2; +_2, \circ_2, \leq_2)$ *with the projector* g *and the coprojector* h*, then* $g(1_1)a = a$ *for any elements* a *and* b *from* A_1*.*

Informally, Proposition 2.2.25 means that multiplicative actions in $\boldsymbol{A}_1$ of images of elements from $\boldsymbol{A}_1$ coincide with multiplicative actions in $\boldsymbol{A}_1$ of the same elements.

Let us assume that abstract prearithmetics $\boldsymbol{A}_1$ and $\boldsymbol{A}_2$ have the successor operation S.

Lemma 2.2.2. *If an abstract prearithmetic* $\boldsymbol{A}_1 = (A_1; +_1, \circ_1, \leq_1)$ *is weakly projective with respect to an abstract prearithmetic* $\boldsymbol{A}_2 = (A_2; +_2, \circ_2, \leq_2)$ *with the projector* g *and the coprojector* h*, then*

$$g(a) <_2 Sg(a) \leq_2 g(\mathrm{S}a).$$

Proof: By Definition 2.1.1, we have $g(a) <_2 \mathrm{S}g(a)$. As there are no elements between an element and its successor, we have $\mathrm{S}g(a) \leq_2 g(\mathrm{S}a)$. Transitivity of partial orders concludes the proof.

Lemma is proved. □

Corollary 2.2.11. *If an abstract prearithmetic* $\boldsymbol{A}_1 = (A_1; +_1, \circ_1, \leq_1)$ *with the multiplicative one* 1_1 *is weakly projective with respect to an abstract prearithmetic* $\boldsymbol{A}_2 = (A_2; +_2, \circ_2, \leq_2)$ *with the projector* g *and the coprojector* h*, then for any natural number* n*, we have*

$$\mathrm{S}^n g(a) \leq_2 g(\mathrm{S}^n a).$$

Proof is left as an exercise. □

Additional properties of the projector make it possible to preserve many properties by inverse weak projectivity.

Proposition 2.2.26. *If an abstract prearithmetic* $\boldsymbol{A}_1 = (A_1; +_1, \circ_1, \leq_1)$ *is weakly projective with respect to an abstract prearithmetic* $\boldsymbol{A}_2 = (A_2; +_2, \circ_2, \leq_2)$ *with the coprojector* h *and the projector* g*, which is a homomorphism with respect to addition* $+_1$*, i.e.,* $g(a +_1 b) = g(a) +_2 g(b)$ *for arbitrary elements* a *and* b *from* $\boldsymbol{A}_1$*, and the operation* $+_2$ *is associative* *in the prearithmetic* $\boldsymbol{A}_2$*, then the operation* $+_1$ *is associative in the prearithmetic* $\boldsymbol{A}_1$*.*

Proof: Assuming that g is a homomorphism with respect to addition and the operation $+_2$ is associative in the prearithmetic $\boldsymbol{A}_2$, let us take arbitrary elements a and b from $\boldsymbol{A}_1$. Then by Definition 2.2.1, we have

$$g(a +_1 b) = g(h(g(a) +_2 g(b))) = g(a) +_2 g(b). \tag{2.37}$$

Equality (2.37) implies the following equalities for arbitrary elements a, b and c from $\boldsymbol{A}_1$:

$$(a +_1 b) +_1 c = h(g(h(g(a) +_2 g(b))) +_2 g(c)) = h((g(a) +_2 g(b))) +_2 g(c)),$$

$$a +_1 (b +_1 c) = h(g(a) +_2 g(h(g(b) +_2 g(c)))) = h(g(a) +_2 (g(b) +_2 g(c)))).$$

As the operation $+_2$ is associative in the prearithmetic $\boldsymbol{A}_2$, we have

$$\begin{aligned}(a +_1 b) +_1 c &= h((g(a) +_2 g(b)) +_2 g(c)) \\ &= h(g(a) +_2 (g(b)) +_2 g(c))) = a +_1 (b +_1 c).\end{aligned}$$

Proposition is proved. □

Proposition 2.2.27. *If an abstract prearithmetic* $\boldsymbol{A}_1 = (A_1; +_1, \circ_1, \leq_1)$ *is weakly projective with respect to an abstract prearithmetic* $\boldsymbol{A}_2 = (A_2; +_2, \circ_2, \leq_2)$ *with the coprojector* h *and the projector* g*, which is a homomorphism with respect to multiplication* $\circ_1$*, i.e.,* $g(a \circ_1 b) = g(a) \circ_2 g(b)$ *for arbitrary elements* a *and* b *from* $\boldsymbol{A}_1$*, and the operation* $\circ_2$ *is associative in the prearithmetic* $\boldsymbol{A}_2$*, then the operation* $\circ_1$ *is associative in the prearithmetic* $\boldsymbol{A}_1$*.*

Proof: Assuming that g is a homomorphism with respect to multiplication and the operation $\circ_2$ is associative in the prearithmetic $\boldsymbol{A}_2$, let us take arbitrary elements a and b from $\boldsymbol{A}_1$. Then by Definition 2.2.1, we have

$$g(a \circ_1 b) = g(h(g(a) \circ_2 g(b))) = g(a) \circ_2 g(b). \tag{2.38}$$

Equality (2.38) implies the following equalities for arbitrary elements a, b and c from $\boldsymbol{A}_1$

$$(a \circ_1 b) \circ_1 c = h(g(h(g(a) \circ_2 g(b))) \circ_2 g(c)) = h((g(a) \circ_2 g(b))) \circ_2 g(c)),$$

$$a \circ_1 (b \circ_1 c) = h(g(a) \circ_2 g(h(g(b) \circ_2 g(c)))) = h(g(a) \circ_2 (g(b) \circ_2 g(c))).$$

As the operation $\circ_2$ is associative in the prearithmetic $\boldsymbol{A}_2$, we have

$$(a \circ_1 b) \circ_1 c = h((g(a) \circ_2 g(b)) \circ_2 g(c))$$
$$= h(g(a) \circ_2 (g(b)) \circ_2 g(c))) = a \circ_1 (b \circ_1 c).$$

Proposition is proved. □

Propositions 2.2.26 and 2.2.27 imply the following result.

Proposition 2.2.28. *If an abstract prearithmetic $\boldsymbol{A}_1 = (A_1; +_1, \circ_1, \leq_1)$ is weakly projective with respect to an abstract prearithmetic $\boldsymbol{A}_2 = (A_2; +_2, \circ_2, \leq_2)$ with the coprojector h and the projector g, which is a homomorphism, while addition $+_2$ is associative and multiplication $\circ_2$ is distributive (from the left or from the right) over addition $+_2$ and associative in the prearithmetic $\boldsymbol{A}_2$, then addition $+_1$ is associative and multiplication $\circ_1$ is distributive (from the left or from the right) over addition $+_1$ and associative in the prearithmetic $\boldsymbol{A}_1$.*

Proof: Proposition 2.2.26 implies associativity of addition $+_1$. Proposition 2.2.27 implies associativity of multiplication $\circ_1$. Thus, we need to check only distributivity.

Let us check distributivity from the left taking arbitrary elements a, b and c from $\boldsymbol{A}_1$. By Definition 2.2.1, we have

$$a \circ_1 (b +_1 c) = h(g(a) \circ_2 g(h(g(b) +_2 g(c))))) = h(g(a) \circ_2 (g(b) +_2 g(c)))$$

because it is proved in Proposition 2.2.26 as

$$g(h(g(a) +_2 g(b))) = g(a) +_2 g(b).$$

As multiplication $\circ_2$ is distributive (from the left or from the right) over addition $+_2$, we have

$$h(g(a) \circ_2 (g(b) +_2 g(c))) = h(g(a) \circ_2 g(b) +_2 g(a) \circ_2 g(c)).$$

At the same time, we have

$$(a \circ_1 b) +_1 (a \circ_1 c) = h(g(a \circ_1 b) +_2 g(a \circ_1 c)) = h(g(h(g(a) \circ_2 g(b)))$$
$$+_2 g(h(g(a) \circ_2 g(c)))) = h(g(a) \circ_2 g(b) +_2 g(a) \circ_2 g(c))$$

because it is proved in Proposition 2.2.27 as

$$g(h(g(a) \circ_2 g(b))) = g(a) \circ_2 g(b).$$

Consequently,

$$(a \circ_1 b) +_1 (a \circ_1 c) = h(g(a) \circ_2 g(b) +_2 g(a) \circ_2 g(c)) = a \circ_1 (b +_1 c).$$

It means that multiplication $\circ_1$ is distributive from the left over addition $+_2$ in the prearithmetic $\boldsymbol{A}_1$.

Distributivity from the right and complete distributivity are proved in the same way.

Proposition is proved. □

Although not all abstract prearithmetics have the additive factoring property or strong additive factoring property studied in Section 2.1, weak projectivity provides for preservation of the additive factoring property.

Let us assume that an abstract prearithmetic $\boldsymbol{A}_1 = (A_1; +_1, \circ_1, \leq_1)$ is weakly projective with respect to an abstract prearithmetic $\boldsymbol{A}_2 = (A_2; +_2, \circ_2, \leq_2)$ with the projector g, the coprojector h and a additive zero 0_2 whereas $gh = 1_{A_2}$.

Theorem 2.2.2. *If the abstract prearithmetic $\boldsymbol{A}_2$ has the additive factoring property, then the abstract prearithmetic $\boldsymbol{A}_1$ also has the additive factoring property.*

Proof: Let us suppose that the abstract prearithmetic $\boldsymbol{A}_2 = (A_2; +_2, \circ_2, \leq_2)$ has the additive factoring property. Taking an element a from $\boldsymbol{A}_1$, we see that if $a = d +_1 b$, then by Proposition 2.2.10,

$$g(a) = g(d +_1 b) = g(d) +_2 g(b).$$

At the same time, if elements are additively composite, then it is possible to continue additive decomposition (additive factoring) of the element a. When all elements in the obtained additive decomposition are additively prime (cf. Section 2.1), factoring stops. However, if there is no prime additive decomposition (prime additive factorization) of a, then it has additive decompositions of arbitrary length, in which all additive factors are not equal to 0_2. By Proposition 2.2.10, any additive decomposition of the element a induces an additive decomposition of its image $g(a)$.

However, if we have an additive decomposition of the element x from the abstract prearithmetic $\boldsymbol{A}_2$, it cannot be continued infinitely because $\boldsymbol{A}_2$ has the additive factoring property. This contradiction shows that any additive decomposition of the element a from the abstract prearithmetic $\boldsymbol{A}_1$ cannot be continued infinitely. Thus, if we continue additive decomposition of a, we come to a prime additive decomposition of a. As a is an arbitrary element a from $\boldsymbol{A}_1$, it means that $\boldsymbol{A}_1$ has the additive factoring property.

Theorem is proved. □

The strong additive factoring property necessitates additional properties of weakly projective prearithmetics. Let us assume that an abstract prearithmetic $\boldsymbol{A}_1 = (A_1; +_1, \circ_1, \leq_1)$ is weakly projective with respect to an abstract prearithmetic $\boldsymbol{A}_2 = (A_2; +_2, \circ_2, \leq_2)$ with the projector g, the coprojector h and an additive zero 0_2 whereas $gh = 1_{A_2}$ and $g^{-1}(0_2) = \{0_1\}$.

Theorem 2.2.3. *If the abstract prearithmetic $\boldsymbol{A}_2$ has the strong additive factoring property, then for each non-zero element a from the abstract prearithmetic $\boldsymbol{A}_1$, there is a natural number n such that any additive factorization of a, in which all elements are not equal to 0_1, has less than n elements.*

Proof: Let us suppose that the abstract prearithmetic $\boldsymbol{A}_2 = (A_2; +_2, \circ_2, \leq_2)$ has the strong additive factoring property. Taking an element a from $\boldsymbol{A}_1$, we see that if $a = d +_1 b$, then by Proposition 2.1.15,

$$g(a) = g(d +_1 b) = g(d) +_2 g(b).$$

Thus, any additive factorization (2.39) of a induces an additive factorization (2.40) of $g(a)$ with the same number of elements:

$$a = \sum_{i=1}^{m} a_i, \tag{2.39}$$

$$g(a) = \sum_{i=1}^{m} g(a_i). \tag{2.40}$$

Besides, by the initial conditions, if the additive factorization (2.39) does not have zeros, then the additive factorization (2.40) also does not have zeros. As the abstract prearithmetic $\boldsymbol{A}_2$ has the strong additive factoring property, it is possible to extend the additive factorization (2.40) to the unique additive prime factorization with n elements. As m in the additive factorization (2.40) is always not larger than n, the same is true for the additive factorization (2.39).

Theorem is proved. □

It is possible to obtain similar properties for multiplicative factorization.

Let us assume that an abstract prearithmetic $\boldsymbol{A}_1 = (A_1; +_1, \circ_1, \leq_1)$ is weakly projective with respect to an abstract prearithmetic $\boldsymbol{A}_2 = (A_2; +_2, \circ_2, \leq_2)$ with the projector g, the coprojector h and a multiplicative one 1_2 whereas $gh = 1_{A_2}$.

Theorem 2.2.4. *If the abstract prearithmetic* $\boldsymbol{A}_2$ *has the multiplicative factoring property, then the abstract prearithmetic* $\boldsymbol{A}_1$ *also has the multiplicative factoring property.*

Proof: Let us suppose that the abstract prearithmetic $\boldsymbol{A}_2 = (A_2; +_2, \circ_2, \leq_2)$ has the multiplicative factoring property. Taking an element a from $\boldsymbol{A}_1$, we see that if $a = d +_1 b$, then by Proposition 2.2.10, we have

$$g(a) = g(d +_1 b) = g(d) +_2 g(b).$$

At the same time, if elements are multiplicatively composite, then it is possible to continue multiplicative decomposition (multiplicative factoring) of the element a. Consequently, if there is no prime multiplicative decomposition (prime multiplicative factorization) of a, then it has multiplicative decompositions of arbitrary length, in which all multiplicative factors are not equal to 0_2 or to 1_2. By Proposition 2.2.10, any multiplicative decomposition of the element a induces an multiplicative decomposition of its image $g(a)$.

However, if we have an multiplicative decomposition of the element x from the abstract prearithmetic $\boldsymbol{A}_2$, it cannot be continued infinitely because $\boldsymbol{A}_2$ has the additive factoring property. This contradiction shows that any multiplicative decomposition of the element a from the abstract prearithmetic $\boldsymbol{A}_1$ cannot be continued infinitely. Thus, if we continue additive decomposition of a, we come to a prime multiplicative decomposition of a. As a is an arbitrary element a from $\boldsymbol{A}_1$, it means that $\boldsymbol{A}_1$ has the multiplicative factoring property.

Theorem is proved. □

Let us assume that an abstract prearithmetic $\boldsymbol{A}_1 = (A_1; +_1, \circ_1, \leq_1)$ is weakly projective with respect to an abstract prearithmetic $\boldsymbol{A}_2 = (A_2; +_2, \circ_2, \leq_2)$ with the projector g, the coprojector h and an multiplicative one 1_2 whereas $gh = 1_{A_2}$ and $g^{-1}(1_2) = \{1_1\}$.

Theorem 2.2.5. *If the abstract prearithmetic* $\boldsymbol{A}_2$ *has the strong multiplicative factoring property, then for each element* a *from the abstract prearithmetic* $\boldsymbol{A}_1$ *but zero and multiplicative one, there is a natural number*

n such that any multiplicative factorization of a, in which all elements are not equal to 0_1 or to 1_1, has less than n elements.

Proof is similar to the proof of Theorem 2.2.3. □

Theorems 2.2.2–2.2.5 show that additive and multiplicative factoring properties are preserved by inverse weak projectivity. Thus, it would be interesting to find whether inverse weak projectivity preserves strong additive and multiplicative factoring properties.

2.3. Projectivity of Abstract Prearithmetics

It's important to find what really suits who you are,
because style isn't only what you wear, it's what you project.
Carolina Herrera

Projectivity is a stronger relation between abstract prearithmetics in comparison with weak projectivity.

Definition 2.3.1. An abstract prearithmetic $\boldsymbol{A}_1 = (A_1; +_1, \circ_1, \leq_1)$ is called *projective* with respect to an abstract prearithmetic $\boldsymbol{A}_2 = (A_2; +_2, \circ_2, \leq_2)$ if it is weakly projective with respect to the prearithmetic $\boldsymbol{A}_2$ and $hg = \mathbf{1}_{A_1}$, where $\mathbf{1}_{A_1}$ is the identity mapping of A_1, g: $A_1 \to A_2$ is the *projector* and h: $A_2 \to A_1$ is the *coprojector* for the pair $(\boldsymbol{A}_1, \boldsymbol{A}_2)$, and for any elements a and b from A_1, we have

$$a \leq_1 b \text{ if and only if } g(a) \leq_2 g(b).$$

In this case, we will say that there is *projectivity* between the prearithmetic $\boldsymbol{A}_1$ and the prearithmetic $\boldsymbol{A}_2$ and there is *inverse projectivity* between the prearithmetic $\boldsymbol{A}_2$ and the prearithmetic $\boldsymbol{A}_1$.

When an abstract prearithmetic $\boldsymbol{B}$ is projective with respect to an abstract prearithmetic $\boldsymbol{A}$ with the projector g and the coprojector h, we denote $\boldsymbol{B}$ by $\boldsymbol{A}_{g;h}$ to show this relation between abstract prearithmetics $\boldsymbol{A}$ and $\boldsymbol{B}$.

Definitions imply the following result.

Lemma 2.3.1. *If an abstract prearithmetic $\boldsymbol{A}_1 = (A_1; +_1, \circ_1, \leq_1)$ is projective with respect to an abstract prearithmetic $\boldsymbol{A}_2 = (A_2; +_2, \circ_2, \leq_2)$, then the prearithmetic $\boldsymbol{A}_1$ is weakly projective with respect to a prearithmetic $\boldsymbol{A}_2$, i.e., $\boldsymbol{A}_{g;h}$ is also $\boldsymbol{A}_{g,h}$ for any abstract prearithmetic $\boldsymbol{A}$.*

Proof is left as an exercise. □

This result allows proving many properties of projectivity in the class of abstract prearithmetics. In particular, it is possible to build new prearithmetics using existing prearithmetics.

Let us consider an arbitrary set B, an abstract prearithmetic $\boldsymbol{A} = (A; +, \circ, \leq)$ and two functions g: $B \to A$ and h: $A \to B$ such that $hg = \mathbf{1}_{A_1}$. Definition 2.3.1 implies the following result.

Proposition 2.3.1. *It is possible to define on B the structure of an abstract prearithmetic $\boldsymbol{B}^{g;h}$ that is projective with respect to the abstract prearithmetic $\boldsymbol{A}$ as a unique prearithmetic with the projector g and the coprojector h.*

Proof is similar to the proof of Proposition 2.2.1. □

Note that $\boldsymbol{B}^{g;h}$ is also $\boldsymbol{A}_{g;h}$.

Proposition 2.2.3 and Lemma 2.3.1 imply uniqueness of the construction in Proposition 2.3.1.

Corollary 2.3.1. *If mappings f and h coincide on the image of B, i.e., $f|_{g(B)} = h|_{g(B)}$, then $\boldsymbol{B}^{g;h} = \boldsymbol{B}^{g;f}$.*

Let us study what properties of abstract prearithmetics are preserved by projectivity and inverse projectivity.

Proposition 2.2.4 and Lemma 2.3.1 imply the following result.

Proposition 2.3.2. *If an abstract prearithmetic $\boldsymbol{A}_1 = (A_1; +_1, \circ_1, \leq_1)$ is projective with respect to an abstract prearithmetic $\boldsymbol{A}_2 = (A_2; +_2, \circ_2, \leq_2)$ and the operation $+_2$ is commutative in the prearithmetic $\boldsymbol{A}_2$, then the operation $+_1$ is commutative in the prearithmetic $\boldsymbol{A}_1$.*

Proof is left as an exercise. □

Proposition 2.2.5 and Lemma 2.3.1 imply the following result.

Proposition 2.3.3. *If an abstract prearithmetic $\boldsymbol{A}_1 = (A_1; +_1, \circ_1, \leq_1)$ is projective with respect to an abstract prearithmetic $\boldsymbol{A}_2 = (A_2; +_2, \circ_2, \leq_2)$ and the operation $\circ_2$ is commutative in the prearithmetic $\boldsymbol{A}_2$, then the operation $\circ_1$ is commutative in the prearithmetic $\boldsymbol{A}_1$.*

Proof is left as an exercise. □

Proposition 2.2.13 and Lemma 2.3.1 imply the following result.

Proposition 2.3.4. *If an abstract prearithmetic $\boldsymbol{A}_1 = (A_1; +_1, \circ_1, \leq_1)$ is projective with respect to an abstract prearithmetic $\boldsymbol{A}_2 = (A_2; +_2, \circ_2, \leq_2)$,* *the operation $+_2$ preserves the order $\leq_2$ and either the coprojector h preserves order $\leq_2$ or the projector g is a homomorphism with respect to addition $+_1$, then the operation $+_1$ preserves the order $\leq_1$.*

Proof: Let us assume that an abstract prearithmetic $\boldsymbol{A}_1 = (A_1; +_1, \circ_1, \leq_1)$ is projective with respect to an abstract prearithmetic $\boldsymbol{A}_2 = (A_2; +_2, \circ_2, \leq_2)$, the coprojector h preserves order $\leq_2$ and the operation $+_2$ preserves the order $\leq_2$, while $c \leq_1 a$ and $d \leq_1 b$. Then by the definition of projectivity, we have $g(c) \leq_2 g(a)$ and $g(d) \leq_2 g(b)$. As the operation $+_2$ preserves the order $\leq_2$, we have $g(c) +_2 g(d) \leq_2 g(a) +_2 g(b)$. As the coprojector h preserves order $\leq_2$, this implies $c +_1 d \leq_1 a +_1 b$ because

$$c +_1 d = h(g(c) +_2 g(d)) \leq_1 a +_1 b = h(g(a) +_2 g(b)).$$

In the case when g is a homomorphism, we have

$$g(c +_1 d) = g(c) +_2 g(d) \leq_1 g(a) +_2 g(b) = g(a +_1 b).$$

By the definition of projectivity, this implies

$$c +_1 d \leq_1 a +_1 b.$$

Proposition is proved. □

Proposition 2.2.14 and Lemma 2.3.1 imply the following result.

Proposition 2.3.5. *If an abstract prearithmetic $\boldsymbol{A}_1 = (A_1; +_1, \circ_1, \leq_1)$ is projective with respect to an abstract prearithmetic $\boldsymbol{A}_2 = (A_2; +_2, \circ_2, \leq_2)$, the coprojector h preserves the order $\leq_2$, i.e., $x \leq_2 y$ implies $h(x) \leq_1 h(y)$, or the projector g is a homomorphism or the projector g is a homomorphism with respect to multiplication $\circ_1$, and the operation $\circ_2$ preserves the order $\leq_2$, then the operation $\circ_1$ preserves the order $\leq_1$.*

Proof is similar to the proof of Proposition 2.3.4. □

Proposition 2.2.15 and Lemma 2.3.1 imply the following result.

Proposition 2.3.6. *If an abstract prearithmetic $\boldsymbol{A}_1 = (A_1; +_1, \circ_1, \leq_1)$ is projective with respect to an abstract prearithmetic $\boldsymbol{A}_2 = (A_2; +_2, \circ_2, \leq_2)$ with the projector g and the coprojector h, and the partially ordered set A_2 is finitely connected, then the partially ordered set A_1 is finitely connected.*

Proof is left as an exercise. □

Proposition 2.2.8 and Lemma 2.3.1 imply the following result.

Proposition 2.3.7. *If an abstract prearithmetic* $\boldsymbol{A}_1 = (A_1; +_1, \circ_1, \leq_1)$ *is projective with respect to an abstract prearithmetic* $\boldsymbol{A}_2 = (A_2; +_2, \circ_2, \leq_2)$ *with the projector* g *and the coprojector* h*, which preserves addition, and multiplication* $\circ_2$ *is distributive over addition* $+_2$ *in the prearithmetic* $\boldsymbol{A}_2$*, then multiplicative action from the left (from the right) is distributive over addition* $+_2$ *in the prearithmetic* $\boldsymbol{A}_2$*.*

Proof is left as an exercise. □

Proposition 2.1.20 and Lemma 2.3.1 imply the following result.

Proposition 2.3.8. *If an abstract prearithmetic* $\boldsymbol{A}_1 = (A_1; +_1, \circ_1, \leq_1)$ *is projective with respect to an abstract prearithmetic* $\boldsymbol{A}_2 = (A_2; +_2, \circ_2, \leq_2)$ *with the projector* g *and the coprojector* h*, for which* $gh = 1_{A_2}$*, and the operation* $\circ_2$ *is associative in the prearithmetic* $\boldsymbol{A}_2$*, then* $x(ya) = (x \circ_2 y)a$ *and* $(ax)y = a(x \circ_2 y)$ *for any elements* x *and* y *from* A_2 *and* a *from* A_1*.*

Proof is left as an exercise. □

Proposition 2.2.22 and Lemma 2.3.1 imply the following result.

Proposition 2.3.9. *If an abstract prearithmetic* $\boldsymbol{A}_1 = (A_1; +_1, \circ_1, \leq_1)$ *is projective with respect to an abstract prearithmetic* $\boldsymbol{A}_2 = (A_2; +_2, \circ_2, \leq_2)$*, the coprojector* h *preserves the order* $\leq_2$*, i.e.,* $x \leq_2 y$ *implies* $h(x) \leq_1 h(y)$*, and the operation* $+_2$ *preserves the order* $\leq_2$*, then additive action from the left (from the right) preserves the order* $\leq_1$*, i.e., additive action from the left (from the right) is a monotone operation in* $\boldsymbol{A}_1$*.*

Proof is left as an exercise. □

Proposition 2.2.23 and Lemma 2.3.1 imply the following result.

Proposition 2.3.10. *If an abstract prearithmetic* $\boldsymbol{A}_1 = (A_1; +_1, \circ_1, \leq_1)$ *is projective with respect to an abstract prearithmetic* $\boldsymbol{A}_2 = (A_2; +_2, \circ_2, \leq_2)$*, the coprojector* h *preserves the order* $\leq_2$*, i.e.,* $x \leq_2 y$ *implies* $h(x) \leq_1 h(y)$*, and the operation* $\circ_2$ *preserves the order* $\leq_2$*, then multiplicative action from the left (from the right) preserves the order* $\leq_1$*, i.e., multiplicative action from the left (from the right) is a monotone operation in* $\boldsymbol{A}_1$*.*

Proof is left as an exercise. □

Let us study additional properties of projectors and coprojectors.

Lemma 2.3.2. *If an abstract prearithmetic* $\boldsymbol{A}_1 = (A_1; +_1, \circ_1, \leq_1)$ *is weakly projective with respect to an abstract prearithmetic* $\boldsymbol{A}_2 = (A_2; +_2, \circ_2, \leq_2)$ *with the projector* g *and the coprojector* h, *then the coprojector* h *preserves the order* $\leq_2$ *on the image* $g(A_1)$ *of the set* A_1.

Indeed, in this case by definition, $a \leq_1 b$ if $g(a) \leq_2 g(b)$.

Properties of the category of sets (cf., for example, Herrlich and Strecker, 1973) give us the following results.

Lemma 2.3.3. *If an abstract prearithmetic* $\boldsymbol{A}_1 = (A_1; +_1, \circ_1, \leq_1)$ *is projective with respect to a prearithmetic* $\boldsymbol{A}_2 = (A_2; +_2, \circ_2, \leq_1)$, *then the projector* g *for the pair* $(\boldsymbol{A}_1, \boldsymbol{A}_2)$ *is an injection and the coprojector* h *for the pair* $(\boldsymbol{A}_1, \boldsymbol{A}_2)$ *is a projection (surjection).*

Proof is left as an exercise. □

Lemma 2.3.4. *If an abstract prearithmetic* $\boldsymbol{A}_1 = (A_1; +_1, \circ_1, \leq_1)$ *is projective with respect to a prearithmetic* $\boldsymbol{A}_2 = (A_2; +_2, \circ_2, \leq_1)$, *then the coprojector* h *for the pair* $(\boldsymbol{A}_1, \boldsymbol{A}_2)$ *is a bijection on the image* $g(\boldsymbol{A}_1)$.

Proof: As by Lemma 2.3.3, the coprojector h for the pair $(\boldsymbol{A}_1, \boldsymbol{A}_2)$ is a projection (surjection), we need only to show that the mapping h is an injection (cf. Bourbaki, 1960). Let us suppose that this is not true. Then for some unequal elements $g(a)$ and $g(b)$, we have $h(g(a)) = h(g(b)) = c$. However, by the definition of projectivity, $h(g(a)) = a$ and $h(g(b)) = b$. If $a \neq b$, this is impossible and the Principle of Excluded Middle concludes the proof. □

Projectivity implies definite relations between orders in prearithmetics.

Proposition 2.3.11. *If an abstract prearithmetic* $\boldsymbol{A}_1 = (A_1; +_1, \circ_1, \leq_1)$ *is projective with respect to an abstract prearithmetic* $\boldsymbol{A}_2 = (A_2; +_2, \circ_2, \leq_2)$ *and* $\leq_2$ *is a linear (total) order in the prearithmetic* $\boldsymbol{A}_2$, *then* $\leq_1$ *is a linear (total) order in the prearithmetic* $\boldsymbol{A}_1$.

Proof: Let us take elements a and b from A_1. Then either $g(a) \leq_2 g(b)$ or $g(b) \leq_2 g(a)$ because $\leq_2$ is a linear (total) order. Definition 2.3.1 implies that in this case, we have either $a \leq_1 b$ or $b \leq_1 a$. It means that $\leq_1$ is a linear (total) order in the prearithmetic $\boldsymbol{A}_1$.

Proposition is proved. □

Remark 2.3.1. In contrast to linear order, well-ordering in a prearithmetic does not imply well-ordering in another prearithmetic, which is projective with respect to the first one.

Example 2.3.1. The conventional arithmetic $\boldsymbol{Q}$ of all rational numbers is a well-ordered abstract prearithmetic. However, its subarithmetic $\boldsymbol{Q}^{++}$ of all positive rational numbers is not well ordered although it is projective with respect to $\boldsymbol{Q}$.

Projectivity is a transitive relation.

Proposition 2.3.12. *If an abstract prearithmetic* $\boldsymbol{A}_1 = (A_1; +_1, \circ_1, \leq_1)$ *is projective with respect to an abstract prearithmetic* $\boldsymbol{A}_2 = (A_2; +_2, \circ_2, \leq_2)$ *and the abstract prearithmetic* $\boldsymbol{A}_2$ *is projective with respect to an abstract prearithmetic* $\boldsymbol{A}_3 = (A_3; +_3, \circ_3, \leq_3)$, *then the prearithmetic* $\boldsymbol{A}_1$ *is projective with respect to the prearithmetic* $\boldsymbol{A}_3$.

Proof: Let us assume that an abstract prearithmetic $\boldsymbol{A}_1 = (A_1; +_1, \circ_1, \leq_1)$ is projective with respect to an abstract prearithmetic $\boldsymbol{A}_2 = (A_2; +_2, \circ_2, \leq_2)$ with the projector g: $A_1 \to A_2$ and the coprojector h: $A_2 \to A_1$ for the pair ($\boldsymbol{A}_1$, $\boldsymbol{A}_2$) and the abstract prearithmetic $\boldsymbol{A}_2 = (A_2; +_2, \circ_2, \leq_2)$ is projective with respect to an abstract prearithmetic $\boldsymbol{A}_3 = (A_3; +_3, \circ_3, \leq_3)$ with the projector k: $A_2 \to A_3$ and the coprojector l: $A_3 \to A_2$ for the pair ($\boldsymbol{A}_2$, $\boldsymbol{A}_3$). By Lemma 2.1.3 and Proposition 2.2.9, the prearithmetic $\boldsymbol{A}_1$ is weakly projective with respect to the prearithmetic $\boldsymbol{A}_3$ and we need only to check two additional conditions $pq = 1_{A_1}$ and

$$a \leq_1 b \text{ if and only if } q(a) \leq_2 q(b),$$

where $q = kg : A_1 \to A_3$ and $p = hl : A_3 \to A_1$ for functions g: $A_1 \to A_2$, h: $A_2 \to A_1$, k: $A_2 \to A_3$ and $l : A_3 \to A_2$ considered in Proposition 2.2.9.

Indeed, by the initial conditions, $lk = 1_{A_2}$ and we have

$$pq = (hl)(kg) = h(lk)g = h(1_{A_2})g = hg = 1_{A_1}.$$

Besides,

$$g(a) \leq_2 g(b) \text{ if and only if } kg(a) \leq_3 kg(b).$$

Consequently,

$$a \leq_1 b \text{ if and only if } q(a) \leq_3 q(b)$$

for any elements a and b from A_1 because $q = kg$.

Thus, the prearithmetic $\boldsymbol{A}_1$ is projective with respect to the prearithmetic $\boldsymbol{A}_3$ with the projector q and the coprojector p.

Proposition is proved. □

Proposition 2.3.12 allows proving the following result using the definition of a category (cf. Appendix).

Theorem 2.3.1. *Abstract prearithmetics with projectivity relations form the category* **APAP** *where objects are abstract prearithmetics and morphisms are projectivity relations.*

Proof is left as an exercise. □

Lemma 2.3.1 shows that the category **APAP** is a wide subcategory of the category **APAWP** (cf. Appendix and Section 2.2, Theorem 2.2.1).

Let us study inner properties of abstract prearithmetics. In the conventional Diophantine arithmetic, any number has the successor. This is also true for non-Diophantine arithmetics of natural, whole and integer numbers. However, this is not true in many prearithmetics and in particular, in conventional arithmetics of real and complex numbers. That is why we find in which cases this property is preserved by projective relations.

Theorem 2.3.2. *If an abstract prearithmetic* $\boldsymbol{A}_1 = (A_1; +_1, \circ_1, \leq_1)$ *is projective with respect to an abstract prearithmetic* $\boldsymbol{A}_2 = (A_2; +_2, \circ_2, \leq_2)$, *the inverse to* h *relation* h^{-1} *preserves the order* $\leq_1$ *and any element* x *in* $\boldsymbol{A}_2$, *which is not maximal, has a successor* $\mathrm{S}x$, *then any element* a *in* $\boldsymbol{A}_1$, *which is not maximal, has a successor* $\mathrm{S}a$.

Proof: Let us assume that an abstract prearithmetic $\boldsymbol{A}_1 = (A_1; +_1, \circ_1, \leq_1)$ is projective with respect to an abstract prearithmetic $\boldsymbol{A}_2 = (A_2; +_2, \circ_2, \leq_2)$ with the projector g and coprojector h. and take some element a from A_1. By Lemma 2.3.3, the coprojector h for the pair ($\boldsymbol{A}_1$, $\boldsymbol{A}_2$) is a projection (surjection). Consequently, there is an element x from A_2 such that $h(x) = a$. By Definition 2.3.1, $x = g(a)$.

By the initial conditions, element x has a successor $\mathrm{S}x$. As h is a total function, there is an element b from A_1 such that $h(\mathrm{S}x) = b$. Let us show that $\mathrm{S}a = b$.

The inequality $x \leq_2 \mathrm{S}b$ implies the inequality $a \leq_1 b$ because the coprojector h preserves the order $\leq_2$.

Now let us assume that there is an element c from A_1 such that $a \leq_1 c \leq_1 b$. Denote the element $g(c)$ by y. Then by Lemma 2.3.2, $a \leq_1 c$ implies $g(a) = x \leq_2 g(c) = y$ and $c \leq_1 b$ implies $y \leq_1 \mathrm{S}x$ because the inverse to h relation h^{-1} preserves the order $\leq_1$.

By the definition of a successor, it means that either $y = x$ or $y = \mathrm{S}x$. Consequently, either $c = a$ or $c = b$. This purports that $b = \mathrm{S}a$.

Theorem is proved. □

Theorem 2.3.3. *If an abstract prearithmetic $\boldsymbol{A}_1 = (A_1; +_1, \circ_1, \leq_1)$ is projective with respect to an abstract prearithmetic $\boldsymbol{A}_2 = (A_2; +_2, \circ_2, \leq_2)$, the inverse to h relation h^{-1} preserves the order $\leq_1$ and any element x in $\boldsymbol{A}_2$, which is not minimal, has a predecessor* $\mathrm{P}x$, *then any element a in $\boldsymbol{A}_1$, which is not minimal, has a predecessor* $\mathrm{P}a$.

The proof is similar to the proof of Theorem 2.3.2.

In some cases, inverse projectivity preserves Archimedean properties.

Proposition 2.3.13. *If an abstract prearithmetic $\boldsymbol{A}_1 = (A_1; +_1, \circ_1, \leq_1)$ is projective with respect to an abstract prearithmetic $\boldsymbol{A}_2 = (A_2; +_2, \circ_2, \leq_2)$, the relations $\leq_2$ and $\leq_1$ are discrete and the prearithmetic $\boldsymbol{A}_2$ is Successively Archimedean, then the prearithmetic $\boldsymbol{A}_1$ is Successively Archimedean.*

Proof: Let us consider an abstract prearithmetic $\boldsymbol{A}_1 = (A_1; +_1, \circ_1, \leq_1)$, which has the discrete relation $\leq_1$ and is projective with respect to a Successively Archimedean abstract prearithmetic $\boldsymbol{A}_2 = (A_2; +_2, \circ_2, \leq_2)$ with the discrete relation $\leq_1$. Then taking two elements a and b from A_1, which satisfy the inequality $a \leq_1 b$, we have $g(a) \leq_2 g(b)$ because by Lemma 2.2.1, the projector g preserves the order $\leq_1$.

As the prearithmetic $\boldsymbol{A}_2$ is Successively Archimedean, there is a natural number n such that $g(b) \leq_2 \mathrm{S}^n g(a)$.

Because there are no elements between an element and its successor, we obtain the following inequality:

$$\mathrm{S}g(a) \leq_2 g(\mathrm{S}a).$$

By Corollary 2.2.5,

$$\mathrm{S}^n g(a) \leq_2 g(\mathrm{S}^n a)$$

and by transitivity of the partial order $\leq_2$,

$$g(b) \leq_2 \mathrm{S}^n g(a) \leq_2 g(\mathrm{S}^n a). \tag{2.41}$$

By the definition of projectivity, relation (2.41) implies

$$b \leq_1 \mathrm{S}^n a.$$

Proposition is proved. □

Now let us look at exact Archimedean properties.

Proposition 2.3.14. *If an abstract prearithmetic $\boldsymbol{A}_1 = (A_1; +_1, \circ_1, \leq_1)$ is projective with respect to an abstract prearithmetic $\boldsymbol{A}_2 = (A_2; +_2, \circ_2, \leq_2)$,* *the relations $\leq_2$ and $\leq_1$ are discrete and linear while the prearithmetic $\boldsymbol{A}_2$ is Exactly Successively Archimedean, then the prearithmetic $\boldsymbol{A}_1$ is Exactly Successively Archimedean.*

Proof: Let us consider an abstract prearithmetic $\boldsymbol{A}_1 = (A_1; +_1, \circ_1, \leq_1)$, which has the discrete relation $\leq_1$ and is projective with respect to an Exactly Successively Archimedean abstract prearithmetic $\boldsymbol{A}_2 = (A_2; +_2, \circ_2, \leq_2)$ with the discrete relation $\leq_1$. Then taking two elements a and b from A_1, which satisfy the inequality $a \leq_1 b$, we have by Lemma 2.2.1, $g(a) \leq_2 g(b)$.

As the prearithmetic $\boldsymbol{A}_2$ is Exactly Successively Archimedean, there is a natural number n such that $g(b) = \mathrm{S}^n g(a)$. This equality implies the inequality $g(b) \leq_2 \mathrm{S}^n g(a)$.

Properties of the successor give us the following inequality:

$$\mathrm{S}g(a) \leq_2 g(\mathrm{S}a).$$

As any order is transitive (cf. Appendix), we have

$$g(b) \leq_2 \mathrm{S}^n g(a) \leq_2 g(\mathrm{S}^n a). \tag{2.42}$$

By the definition of projectivity, relation (2.42) implies

$$b \leq_1 \mathrm{S}^n a.$$

Now let us consider the chain of successors

$$a \leq_1 \mathrm{S}a \leq_1 \mathrm{S}^2 a \leq_1 \mathrm{S}^3 a \leq_1 \cdots .$$

Because $b \leq_1 \mathrm{S}^n a$, there is the minimal number j such that $b \leq_1 \mathrm{S}^j a$. Because the order $\leq_1$ is linear, we have

$$\mathrm{S}^{j-1} a \leq_1 b \leq_1 \mathrm{S}^j a.$$

However, as $\mathrm{S}^j a = \mathrm{S}(\mathrm{S}^{j-1} a)$, the element b is either equal to $\mathrm{S}^{j-1} a$ or to $\mathrm{S}^j a$. In both cases, we come to the conclusion that b is equal to some iterative successor of a, which means that the prearithmetic $\boldsymbol{A}_1$ is Exactly Successively Archimedean.

Proposition is proved. □

Numbers in the Diophantine arithmetic (of whole numbers) have the following property:

$$a \leq a + a.$$

Let us explore this property for abstract prearithmetics. We prove that this property can be inherited under projectivity.

Proposition 2.3.15. *If an abstract prearithmetic* $\boldsymbol{A}_1 = (A_1; +_1, \circ_1, \leq_1)$ *is projective with respect to an abstract prearithmetic* $\boldsymbol{A}_2 = (A_2; +_2, \circ_2, \leq_2)$ *with the projector* g *and the coprojector* h*, which preserves the order, and* $x \leq_2 x +_2 x$ *for any element* x *in the prearithmetic* $\boldsymbol{A}_2$*, then* $a \leq_1 a +_1 a$ *for any element* a *in the prearithmetic* $\boldsymbol{A}_1$*.*

Proof: As $x \leq_2 x +_2 x$ for any element x in the prearithmetic $\boldsymbol{A}_2$, we have

$$g(a) \leq_2 g(a) +_2 g(a).$$

As the coprojector h preserves the order, we have

$$h(g(a)) \leq_1 h(g(a) +_2 g(a)).$$

As $hg = 1_{A_1}$, we have

$$a = h(g(a)) \leq_1 h(g(a) +_2 g(a)) = a +_1 a.$$

Proposition is proved. □

Numbers in the Diophantine arithmetic $\boldsymbol{W}$ of whole numbers and in the Diophantine arithmetic $\boldsymbol{N}$ of natural numbers have even the stronger property (2.43) then the one studied in Proposition 2.3.15.

$$a \leq a + b. \tag{2.43}$$

Let us explore this property for abstract prearithmetics.

Lemma 2.3.4. *In an abstract prearithmetic* $\boldsymbol{A} = (A; +, \circ, \leq)$*, which has the additive zero* 0 *and in which addition preserves the strict order, the inequality* $a < a + b$ *is true for any elements* $b > 0$ *and* a *from* $\boldsymbol{A}$*.*

Indeed, $a = a + 0 < a + b$ because $0 < b$ and addition preserves the strict order.

Corollary 2.3.2. *In an abstract prearithmetic* $\boldsymbol{A} = (A; +, \circ, \leq)$*, which has the additive zero* 0 *and in which addition preserves the strict order, the inequality* $a < a + a$ *is true for any element* $a > 0$ *from* $\boldsymbol{A}$*.*

Corollary 2.3.3. *In an abstract prearithmetic* $\boldsymbol{A} = (A; +, \circ, \leq)$, *which has the additive zero* 0 *as its least element and in which addition preserves the strict order, the inequality* $a < a + a$ *is true for any element* $a \neq 0$ *from* $\boldsymbol{A}$.

Corollary 2.3.4. *In an abstract prearithmetic* $\boldsymbol{A} = (A; +, \circ, \leq)$, *which has the additive zero* 0 *and in which addition preserves the strict order, the inequality* $m < n$ *implies*

$$\sum_{i=1}^{m} a_i < \sum_{i=1}^{n} a_i,$$

where all a_i *are equal to an element* $a > 0$.

Corollary 2.3.5. *In an abstract prearithmetic* $\boldsymbol{A} = (A; +, \circ, \leq)$, *which has the additive zero* 0 *as its least element and in which addition preserves the strict order, the inequality* $m < n$ *implies*

$$\sum_{i=1}^{m} a_i < \sum_{i=1}^{n} a_i,$$

where all a_i are equal to any element $a \neq 0$ from $\boldsymbol{A}$.

Corollary 2.3.6. *In an abstract prearithmetic* $\boldsymbol{A} = (A; +, \circ, \leq)$, *which has the additive zero* 0 *as its least element and in which addition preserves the strict order, the inequality* $a < a + b$ *is true for any elements* a *and* $b \neq 0$ *from* $\boldsymbol{A}$.

Proposition 2.3.16. *If an abstract prearithmetic* $\boldsymbol{A}_1 = (A_1; +_1, \circ_1, \leq_1)$ *is projective with respect to an abstract prearithmetic* $\boldsymbol{A}_2 = (A_2; +_2, \circ_2, \leq_2)$ *with the projector* g *and the coprojector* h, *which preserves the order, and* $x \leq_2 x +_2 z$ *for any elements* x *and* z *from the prearithmetic* $\boldsymbol{A}_2$, *then* $a \leq_1 a +_1 b$ *for any elements* a *and* b *from the prearithmetic* $\boldsymbol{A}_1$.

Proof: As $x \leq_2 x +_2 z$ for any elements x and z from the prearithmetic $\boldsymbol{A}_2$, we have

$$g(a) \leq_2 g(a) +_2 g(b).$$

As the coprojector h preserves the order, we have

$$h(g(a)) \leq_1 h(g(a) +_2 g(b)).$$

As $hg = 1_{A_1}$, we have

$$a = h(g(a)) \leq_1 h(g(a) +_2 g(b)) = a +_1 b.$$

Proposition is proved. □

Corollary 2.3.7. *If an abstract prearithmetic* $\boldsymbol{A}_1 = (A_1; +_1, \circ_1, \leq_1)$ *is projective with respect to an abstract prearithmetic* $\boldsymbol{A}_2 = (A_2; +_2, \circ_2, \leq_2)$ *with the projector* g *and the coprojector* h*, which preserves the order, addition* $+_2$ *is commutative and* $x \leq_2 x +_2 z$ *for any elements* x *and* z *in the prearithmetic* $\boldsymbol{A}_2$*, then* $\max(a, b) \leq_1 a +_1 b$ *for any elements* a *and* b *in the prearithmetic* $\boldsymbol{A}_1$.

It is also possible to consider other types of inequalities in abstract prearithmetics.

Remark 2.3.2. The identity $a \leq a + b$ is not true in all abstract prearithmetics. For instance, taking the modular arithmetic $\boldsymbol{Z}_7$ (cf. Example 2.1.3), we have the following inequalities:

$$5 + 6 = 4 < 5 < 6.$$

That is why it is interesting to study the identity

$$a + b \leq a,$$

which can be true in some abstract prearithmetics as the following example demonstrates.

Example 2.3.2. This identity is true in the arithmetic $\boldsymbol{Z}^{--}$ of all negative integer numbers, e.g., $(-5) + (-5) = -10$. Note that in this arithmetic multiplication of two negative numbers gives a negative number, e.g., $(-3) \circ (-5) = -15$.

Thus, it is also interesting to explore the inequality $a + a \leq a$.

Proposition 2.3.17. *If an abstract prearithmetic* $\boldsymbol{A}_1 = (A_1; +_1, \circ_1, \leq_1)$ *is projective with respect to an abstract prearithmetic* $\boldsymbol{A}_2 = (A_2; +_2, \circ_2, \leq_2)$ *with the projector* g *and the coprojector* h*, which preserves the order, and* $x +_2 x \leq_2 x$ *for any element* x *in the prearithmetic* $\boldsymbol{A}_2$*, then* $a +_1 a \leq_1 a$ *for any element* a *in the prearithmetic* $\boldsymbol{A}_1$.

Proof: As $x +_2 x \leq_2 x$ for any element x in the prearithmetic $\boldsymbol{A}_2$, we have

$$g(a) +_2 g(a) \leq_2 g(a).$$

As the coprojector h preserves the order, we have

$$h(g(a) +_2 g(a)) \leq_1 h(g(a)).$$

As $hg = \mathbf{1}_{A_1}$, we have

$$a +_1 a = h(g(a) +_2 g(a)) \leq_1 h(g(a)) = a.$$

Proposition is proved. □

Now let us explore the stronger property (2.44) then the one studied in Proposition 2.3.17.

$$a + b \leq a, \tag{2.44}$$

which is also true in the arithmetic $\boldsymbol{Z}^{--}$.

Lemma 2.3.5. *In an abstract prearithmetic $\boldsymbol{A} = (A; +, \circ, \leq)$, which has the additive zero 0 and in which addition preserves the strict order, then $a > a + b$ for any element $b < 0$ from $\boldsymbol{A}$.*

Indeed, $a = a + 0 > a + b$ because $0 > b$ and addition preserves the strict order.

Corollary 2.3.8. *In an abstract prearithmetic $\boldsymbol{A} = (A; +, \circ, \leq)$, which has the additive zero 0 and in which addition preserves the strict order, then $a > a + a$ for any element $a < 0$ from $\boldsymbol{A}$.*

Corollary 2.3.9. *In an abstract prearithmetic $\boldsymbol{A} = (A; +, \circ, \leq)$, which has the additive zero 0 as its largest element and in which addition preserves the strict order, then $a > a + a$ for any element $a \neq 0$ from $\boldsymbol{A}$.*

Corollary 2.3.10. *In an abstract prearithmetic $\boldsymbol{A} = (A; +, \circ, \leq)$, which has the additive zero 0 and in which addition preserves the strict order, then $m < n$ implies*

$$\sum_{i=1}^{m} a_i > \sum_{i=1}^{n} a_i,$$

where all a_i are equal to an element $a < 0$.

Corollary 2.3.11. *In an abstract prearithmetic $\boldsymbol{A} = (A; +, \circ, \leq)$, which has the additive zero* 0 *as its largest element and in which addition preserves the strict order, then $m < n$ implies*

$$\sum_{i=1}^{m} a_i < \sum_{i=1}^{n} a_i,$$

where all a_i are equal to any element $a \neq 0$ from $\boldsymbol{A}$.

Corollary 2.3.12. *In an abstract prearithmetic $\boldsymbol{A} = (A; +, \circ, \leq)$, which has the additive zero* 0 *as its largest element and in which addition preserves the strict order, then $a < a + b$ for any elements a and $b \neq 0$ from $\boldsymbol{A}$.*

Proposition 2.3.18. *If an abstract prearithmetic $\boldsymbol{A}_1 = (A_1; +_1, \circ_1, \leq_1)$ is projective with respect to an abstract prearithmetic $\boldsymbol{A}_2 = (A_2; +_2, \circ_2, \leq_2)$ with the projector g and the coprojector h, which preserves the order, and $x +_2 z \leq_2 x$ for any elements x and z in the prearithmetic $\boldsymbol{A}_2$, then $a +_1 b \leq_1 a$ for any elements a and b in the prearithmetic $\boldsymbol{A}_1$.*

Proof: As $x +_2 z \leq_2 x$ for any element x in the prearithmetic $\boldsymbol{A}_2$, we have

$$g(a) +_2 g(b) \leq_2 g(a).$$

As the coprojector h preserves the order, we have

$$h(g(a) +_2 g(b)) \leq_1 h(g(a)).$$

As $hg = \mathbf{1}_{A_1}$, we have

$$a +_1 b = h(g(a) +_2 g(b)) \leq_1 h(g(a)) = a.$$

Proposition is proved. □

Corollary 2.3.13. *If an abstract prearithmetic $\boldsymbol{A}_1 = (A_1; +_1, \circ_1, \leq_1)$ is projective with respect to an abstract prearithmetic $\boldsymbol{A}_2 = (A_2; +_2, \circ_2, \leq_2)$ with the projector g and the coprojector h, which preserves the order, addition $+_2$ is commutative and $x +_2 z \leq_2 x$ for any elements x and z in the prearithmetic $\boldsymbol{A}_2$, then $a +_1 b \leq_1 \min\, (a, b)$ for any elements a and b in the prearithmetic $\boldsymbol{A}_1$.*

Elements 0 and 1 have very special properties in the conventional Diophantine arithmetic. Let us explore what is going on with these properties in projectivity of prearithmetics.

Inverse projectivity preserves additive zeros.

Proposition 2.3.19. *If an abstract prearithmetic* $\boldsymbol{A}_1 = (A_1; +_1, \circ_1, \leq_1)$ *is projective with respect to an abstract prearithmetic* $\boldsymbol{A}_2 = (A_2; +_2, \circ_2, \leq_2)$ *with the projector* g, *the coprojector* h *and an additive zero* 0_2, *which belongs to the image* $g(A_1)$ *of* A_1, *then the prearithmetic* $\boldsymbol{A}_1$ *has an additive zero* 0_1 *and if* $g(a) = 0_2$, *then* $a = 0_1$.

Proof: Let us assume that $g(a) = 0_2$. Then by definitions, we have

$$a +_1 b = h(g(a) +_2 g(b)) = h(0_2 +_2 g(b)) = h(g(b)) = b$$

and

$$b +_1 a = h(g(b) +_2 g(a)) = h(g(b) +_2 0_2) = h(g(b)) = b$$

for an arbitrary element b from $\boldsymbol{A}_1$. Consequently, a is an additive zero in $\boldsymbol{A}_1$.

Proposition is proved. □

Inverse projectivity preserves multiplicative zeros.

Proposition 2.3.20. *If an abstract prearithmetic* $\boldsymbol{A}_1 = (A_1; +_1, \circ_1, \leq_1)$ *is projective with respect to an abstract prearithmetic* $\boldsymbol{A}_2 = (A_2; +_2, \circ_2, \leq_2)$ *with the projector* g, *the coprojector* h *and a multiplicative zero* 0_2^{m}, *which belongs to the image* $g(A_1)$ *of* A_1, *then the prearithmetic* $\boldsymbol{A}_1$ *has a multiplicative zero* 0_1^{m} *and if* $g(a) = 0_2^{\mathrm{m}}$, *then* $a = 0_1^{\mathrm{m}}$.

Proof: Let us assume that $g(a) = 0_2^{m}$. Then by definitions, we have

$$a \circ_1 b = h(g(a) \circ_2 g(b)) = h(0_2^{\mathrm{m}} \circ_2 g(b)) = h(0_2^{\mathrm{m}}) = a$$

and

$$b \circ_1 a = h(g(b) \circ_2 g(a)) = h(g(b) \circ_2 0_2^{\mathrm{m}}) = h(0_2^{\mathrm{m}}) = a$$

for an arbitrary element b from $\boldsymbol{A}_1$. Consequently, a is a multiplicative zero, i.e., $a = 0_1^{\mathrm{m}}$ in $\boldsymbol{A}_1$.

Proposition is proved. □

Inverse projectivity preserves multiplicative ones.

Proposition 2.3.21. *If an abstract prearithmetic* $\boldsymbol{A}_1 = (A_1; +_1, \circ_1, \leq_1)$ *is projective with respect to an abstract prearithmetic* $\boldsymbol{A}_2 = (A_2; +_2, \circ_2, \leq_2)$ *with the projector* g, *the coprojector* h *and a multiplicative one* 1_2, *which belongs to the image* $g(A_1)$ *of* A_1, *then the prearithmetic* $\boldsymbol{A}_1$ *has a multiplicative one* 1_1 *and if* $g(a) = 1_2$, *then* $a = 1_1$.

Proof: Let us assume that $g(a) = 1_2$. Then by definitions, we have

$$a \circ_1 b = h(g(a) \circ_2 g(b)) = h(1_2 \circ_2 g(b)) = h(g(b)) = b$$

and

$$b \circ_1 a = h(g(b) \circ_2 g(a)) = h(g(b) \circ_2 1_2) = h(g(b)) = b$$

for an arbitrary element b from $\boldsymbol{A}_1$. Consequently, a is a multiplicative one in $\boldsymbol{A}_1$.

Proposition is proved. □

Proposition 2.3.22. *If an abstract prearithmetic $\boldsymbol{A}_1 = (A_1; +_1, \circ_1, \leq_1)$ is projective with respect to an abstract prearithmetic $\boldsymbol{A}_2 = (A_2; +_2, \circ_2, \leq_2)$ with the projector g, the coprojector h and a multiplicative one 1_2, then $1_2 a = a 1_2 = a$ for any element a from A_1.*

Indeed, $1_2 a = h(1_2 \circ_2 g(a)) = h(g(a)) = a$ and $a 1_2 = h(g(a) \circ_2 1_2) = h(g(a)) = a$.

Proposition 2.3.23. *If an abstract prearithmetic $\boldsymbol{A}_1 = (A_1; +_1, \circ_1, \leq_1)$ is projective with respect to an abstract prearithmetic $\boldsymbol{A}_2 = (A_2; +_2, \circ_2, \leq_2)$ with the projector g, the coprojector h and $\boldsymbol{A}_2$ has a multiplicative zero 0_2^{m}, then $0_2^{\mathrm{m}} a = a 0_2^{\mathrm{m}} = 0_1^{\mathrm{m}}$ for any element a from A_1.*

Indeed, by Proposition 2.3.20, the prearithmetic $\boldsymbol{A}_1$ has a multiplicative zero 0_1^{m} such that $g(0_1^{\mathrm{m}}) = 0_2^{\mathrm{m}}$. Then $0_2^{\mathrm{m}} a = h(0_2^{\mathrm{m}} \circ_2 g(a)) = h(0_2^{\mathrm{m}}) = 0_1^{\mathrm{m}}$ and $a 0_2^{\mathrm{m}} = h(g(a) \circ_2 0_2^{\mathrm{m}}) = h(0_2^{\mathrm{m}}) = 0_1^{\mathrm{m}}$.

Proposition 2.3.24. *If an abstract prearithmetic $\boldsymbol{A}_1 = (A_1; +_1, \circ_1, \leq_1)$ is projective with respect to an abstract prearithmetic $\boldsymbol{A}_2 = (A_2; +_2, \circ_2, \leq_2)$ with the projector g, the coprojector h and an additive zero 0_2^{a}, then $0_2^{\mathrm{a}+} a = a^{+} 0_2^{\mathrm{a}} = a$ for any element a from A_1.*

Indeed, $0_2^{\mathrm{a}+} a = h(0_2^{\mathrm{a}} +_2 g(a)) = h(g(a)) = a$ and $a^{+} 0_2^{\mathrm{a}} = h(g(a) +_2 0_2^{\mathrm{a}}) = h(g(a)) = a$.

Numbers in the Diophantine arithmetic $\boldsymbol{N}$ of natural numbers have the following important property:

$$a \leq a \circ b.$$

Let us explore this property for abstract prearithmetics.

Lemma 2.3.6. *In an abstract prearithmetic $\boldsymbol{A} = (A; +, \circ, \leq)$, which has the multiplicative 1 and in which multiplication preserves the strict order, then $a < a \circ b$ for any element $b > 0$ from $\boldsymbol{A}$.*

Indeed, $a = a \circ 0 < a \circ b$ because $0 < b$ and multiplication preserves the strict order.

Corollary 2.3.14. *In an abstract prearithmetic* $\boldsymbol{A} = (A; +, \circ, \leq)$, *which has the multiplicative* 1 *and in which multiplication preserves the strict order, then* $a < a \circ a$ *for any element* $a > 1$ *from* $\boldsymbol{A}$.

Corollary 2.3.15. *In an abstract prearithmetic* $\boldsymbol{A} = (A; +, \circ, \leq)$, *which has the multiplicative* 1 *as its least element and in which multiplication preserves the strict order, then* $a < a \circ a$ *for any element* $a \neq 0$ *from* $\boldsymbol{A}$.

Corollary 2.3.16. *In an abstract prearithmetic* $\boldsymbol{A} = (A; +, \circ, \leq)$, *which has the multiplicative* 1 *and in which multiplication preserves the strict order, then* $m < n$ *implies*

$$\prod_{i=1}^{m} a_i < \prod_{i=1}^{n} a_i,$$

where all a_i *are equal to an element* $a > 0$.

Corollary 2.3.17. *In an abstract prearithmetic* $\boldsymbol{A} = (A; +, \circ, \leq)$, *which has the multiplicative* 1 *as its least element and in which multiplication preserves the strict order, then* $m < n$ *implies*

$$\prod_{i=1}^{m} a_i < \prod_{i=1}^{n} a_i,$$

where all a_i *are equal to any element* $a \neq 0$ *from* $\boldsymbol{A}$.

Corollary 2.3.18. *In an abstract prearithmetic* $\boldsymbol{A} = (A; +, \circ, \leq)$, *which has the multiplicative* 1 *as its least element and in which multiplication preserves the strict order, then* $a < a \circ b$ *for any elements* a *and* $b \neq 0$ *from* $\boldsymbol{A}$.

Proposition 2.3.25. *If an abstract prearithmetic* $\boldsymbol{A}_1 = (A_1; +_1, \circ_1, \leq_1)$ *is projective with respect to an abstract prearithmetic* $\boldsymbol{A}_2 = (A_2; +_2, \circ_2, \leq_2)$ *with the projector* g *and coprojector* h, *which preserves the order* $\leq_2$, *and* $x \leq_2 x \circ_2 z$ *for any elements* x *and* z *from the prearithmetic* $\boldsymbol{A}_2$, *then* $a \leq_1 a \circ_1 b$ *for any elements* a *and* b *from the prearithmetic* $\boldsymbol{A}_1$.

Proof: As $x \leq_2 x \circ_2 z$ for any elements x and z from the prearithmetic $\boldsymbol{A}_2$, we have

$$g(a) \leq_2 g(a) \circ_2 g(b).$$

As the coprojector h preserves the order, we have

$$h(g(a)) \leq_1 h(g(a) \circ_2 g(b)).$$

As $hg = \mathbf{1}_{A_1}$, we have

$$a = h(g(a)) \leq_1 h(g(a) \circ_2 g(b)) = a \circ_1 b.$$

Proposition is proved. □

Corollary 2.3.19. *If an abstract prearithmetic* $\boldsymbol{A}_1 = (A_1; +_1, \circ_1, \leq_1)$ *is projective with respect to an abstract prearithmetic* $\boldsymbol{A}_2 = (A_2; +_2, \circ_2, \leq_2)$ *with the projector* g *and the coprojector* h*, which preserves the order, multiplication* $\circ_2$ *is commutative and* $x \leq_2 x \circ_2 z$ *for any elements* x *and* z *in the prearithmetic* $\boldsymbol{A}_2$*, then* $max(a, b) \leq_1 a \circ_1 b$ *for any elements* a *and* b *in the prearithmetic* $\boldsymbol{A}_1$*.*

Corollary 2.3.20. *If an abstract prearithmetic* $\boldsymbol{A}_1 = (A_1; +_1, \circ_1, \leq_1)$ *is projective with respect to an abstract prearithmetic* $\boldsymbol{A}_2 = (A_2; +_2, \circ_2, \leq_2)$ *with the projector* g *and the coprojector* h*, which preserves the order, and* $x \leq_2 x \circ_2 x$ *for any element* x *in the prearithmetic* $\boldsymbol{A}_2$*, then* $a \leq_1 a \circ_1 a$ *for any element* a *in the prearithmetic* $\boldsymbol{A}_1$*.*

We remind that a monomorphism of abstract prearithmetics preserves operations and the order relation.

Proposition 2.3.26. *If there is a monomorphism* g *of an abstract prearithmetic* $\boldsymbol{A}_1 = (A_1; +_1, \circ_1, \leq_1)$ *with the total order* $\leq_1$ *into an abstract prearithmetic* $\boldsymbol{A}_2 = (A_2; +_2, \circ_2, \leq_2)$*, then* $\boldsymbol{A}_1$ *is projective with respect to* $\boldsymbol{A}_2$*.*

Proof: To show that $\boldsymbol{A}_1$ is projective with respect to $\boldsymbol{A}_2$, we need to build a mapping h: $A_2 \rightarrow A_1$, which satisfies properties from Definition 2.3.1. For an arbitrary element x from $\boldsymbol{A}_2$, we define h in the following way choosing an arbitrary element a from $\boldsymbol{A}_1$:

$$h(x) = \begin{cases} b & \text{if } g(b) = x, \\ a & \text{if there are no elements } c \text{ in } \boldsymbol{A}_1 \text{ such that } g(c) = x. \end{cases}$$

Let us show that h satisfies conditions from Definition 2.3.1. Indeed, we have

$$a +_1 b = h(g(a) +_2 g(b))$$

because g is a monomorphism and in particular,

$$g(a +_1 b) = g(a) +_2 g(b)$$

and

$$a \circ_1 b = h(g(a) \circ_2 g(b))$$

because g is a monomorphism and in particular,

$$g(a \circ_1 b) = g(a) \circ_2 g(b).$$

Besides, if $a \leq_1 b$, then $g(a) \leq_2 g(b)$, and if $g(a) \leq_2 g(b)$, then $a \leq_1 b$ because the order $\leq_1$ is total.

Proposition is proved. □

Let us consider an abstract prearithmetic $\boldsymbol{A} = (A; +, \circ, \leq)$ and study such properties of its elements as subtractability and divisibility.

Definition 2.3.2.

(a) An element a from $\boldsymbol{A}$ is *subtractable from the right* by an element b from $\boldsymbol{A}$ if $a = d + b$ for some element d from $\boldsymbol{A}$. We denote this by $a \lceil b$.
(b) An element a from $\boldsymbol{A}$ is *subtractable from the left* by an element b from $\boldsymbol{A}$ if $a = b + d$ for some element d from $\boldsymbol{A}$. We denote this by $b \rceil a$.
(c) An element a from $\boldsymbol{A}$ is *subtractable* by an element b if it is subtractable by b from the right and from the left with the same difference, i.e., $b = d + a = a + d$. We denote this by $a\, (\, b$.

For instance, in the conventional Diophantine arithmetic $\boldsymbol{N}$, any number is subtractable from the right and from the left by any smaller number because $n = 1 + (n - 1) = (n - 1) + 1$ for any natural number $n > 1$. However, this is not true for many abstract prearithmetics and non-Diophantine arithmetics.

Example 2.3.3. Let us consider the set N of all natural numbers with the standard order $\leq$ and introduce the following operations:

$$a \oplus b = a \cdot b,$$

$$a \otimes b = a^b.$$

Then the system $\boldsymbol{A} = (N; \oplus, \otimes, \leq)$ is an abstract prearithmetic. Taking number 5_A from this prearithmetic, we see that there is no number n_A in $\boldsymbol{A}$ such that $3_A \oplus b = 5_A$. It means that 5_A is not subtractable by 3_A. Moreover, we can see that in this prearithmetic, subtractability means divisibility.

This example shows that subtractability is an additive counterpart of divisibility.

Definitions 2.3.2 and 2.1.10 imply the following result.

Lemma 2.3.7. *For any elements a and b from an abstract prearithmetic $\boldsymbol{A}$, we have the following conditions:*

(a) $a \lceil b$ *if subtraction from the left* $a \leftharpoondown b$ *is defined.*
(b) $b \rceil a$ *if subtraction from the right* $a \rightharpoondown b$ *is defined.*
(c) $a \,(\, b$ *if full subtraction* $a - b$ *is defined.*

Proof is left as an exercise. □

Lemma 2.3.8. *If addition + is commutative in an abstract prearithmetic $\boldsymbol{A} = (A;\, +,\, \circ,\, \leq)$, then for any elements a and b from $\boldsymbol{A}$, a is subtractable by b if and only if it is subtractable by b from the right or from the left.*

Proof is left as an exercise. □

Proposition 2.3.27. *If addition + is associative in an abstract prearithmetic $\boldsymbol{A} = (A;\, +,\, \circ,\, \leq)$, then for any elements a, b and c from $\boldsymbol{A}$, we have the following conditions:*

(1) $a \lceil b$ *and* $b \lceil c$ *imply* $a \lceil c$.
(2) $b \rceil a$ *and* $c \rceil b$ *imply* $c \rceil a$.
(3) $a \,(\, b$ *and* $b \,(\, c$ *imply* $a \,(\, c$.

Proof: (1) If $a \lceil b$, then $a = d + b$ for some element d from $\boldsymbol{A}$. If $b \rceil c$, then $b = e + c$ for some element e from $\boldsymbol{A}$. Consequently,

$$a = d + b = d + e + c = (d + e) + c.$$

It means that $a \lceil c$.

Statements (2) and (3) are proved in a similar way.

Proposition is proved. □

Proposition 2.3.28. *If addition + is associative in an abstract prearithmetic $\boldsymbol{A} = (A; +, \circ, \leq)$, then for any elements a, b and c from $\boldsymbol{A}$, we have the following conditions:*

(1) $a \lceil b$ *implies* $a + c \lceil b + c$.
(2) $b \rceil a$ *implies* $c + b \rceil c + a$.
(3) $a \,(\, b$ *implies* $a + c \,(\, b + c$ *when addition + is also commutative.*

Proof: (1) If $a \lceil b$, then $a = d + b$ for some element d from $\boldsymbol{A}$. Consequently,

$$a + c = d + b + c = d + (b + c).$$

It means that $a + c \lceil b + c$.

(2) If $b \rceil a$, then $a = b + d$ for some element d from $\boldsymbol{A}$. Consequently,

$$c + a = c + b + d = (c + b) + d.$$

It means that $c + b \rceil c + a$.

By Lemma 2.3.7, the statement (3) follows from statements (1) and (2) when addition + is commutative.

Proposition is proved. □

Proposition 2.3.29. *If addition + is associative in an abstract prearithmetic $\boldsymbol{A} = (A; +, \circ, \leq)$, then for any elements a, b and c from $\boldsymbol{A}$, we have the following conditions:*

(1) $a \lceil b$ *implies* $c + a \lceil b$.
(2) $b \rceil a$ *implies* $b \rceil a + c$.
(3) $a \,(\, b$ *implies* $a + c \,(\, b$ *when addition + is also commutative.*

Proof: (1) If $a \lceil b$, then $a = d + b$ for some element d from $\boldsymbol{A}$. Consequently,

$$c + a = c + d + b = (c + d) + b$$

It means that $c + a \lceil b + c$.

(2) If $b \rceil a$, then $a = b + d$ for some element d from $\boldsymbol{A}$. Consequently,

$$a + c = b + d + c = (c + b) + d + c$$

It means that $a + c \rceil b$.

By Lemma 2.3.7, the statement (3) follows from statements (1) and (2) when addition + is commutative.

Proposition is proved. □

Proposition 2.3.30. *In an Exactly Additively Archimedean abstract prearithmetic $\boldsymbol{A} = (A; +, \circ, \leq)$, for any elements a, b and c from $\boldsymbol{A}$, if $a < b$, then b is subtractable from the left by a.*

Proof is left as an exercise. □

Additive zero impacts subtractability.

Lemma 2.3.9. *If an abstract prearithmetic* $\boldsymbol{A} = (A; +, \circ, \leq)$ *has the additive zero* 0, *then in it, any element is subtractable by itself and by* 0.

Indeed, we have $a = a + 0 = 0 + a$ for any element a from $\boldsymbol{A}$.

This result and Example 2.1.8 show importance of the concepts of prime and composite elements considered in Section 2.1, which serve as counterparts of the well-known concepts of prime and composite numbers in the Diophantine arithmetic.

Note that in the abstract prearithmetic $\boldsymbol{A}$ from Example 2.1.8, an element is additively prime if and only if it is prime in the classical sense in the Diophantine arithmetic $\boldsymbol{W}$ by the traditional definition of prime numbers. At the same time, only numbers 0 and 1 are additively prime in the conventional Diophantine arithmetic $\boldsymbol{W}$. That is why this concept was not studied before. At the same time, Example 2.1.8 shows that an abstract prearithmetic can have infinitely many additively prime elements because there are infinitely many prime numbers in the conventional Diophantine arithmetic $\boldsymbol{W}$ and in the abstract prearithmetic $\boldsymbol{A}$ from Example 2.1.8, an element is additively prime if and only if it is prime in $\boldsymbol{W}$.

Let us find relations between prime elements in abstract prearithmetic, which are weakly projective with respect to one another.

Theorem 2.3.4. *If an abstract prearithmetic* $\boldsymbol{A}_1 = (A_1; +_1, \circ_1, \leq_1)$ *is weakly projective with respect to an abstract prearithmetic* $\boldsymbol{A}_2 = (A_2; +_2, \circ_2, \leq_2)$ *with the projector* g, *the coprojector* h *and a additive zero* 0_2 *whereas* $gh = 1_{A_2}$, *then any element of the inverse image* $g^{-1}(z)$ *of an arbitrary additively prime element* z *from* $\boldsymbol{A}_2$ *is additively prime in* $\boldsymbol{A}_1$.

Proof: Let us assume that an abstract prearithmetic $\boldsymbol{A}_1 = (A_1; +_1, \circ_1, \leq_1)$ is projective with respect to an abstract prearithmetic $\boldsymbol{A}_2 = (A_2; +_2, \circ_2, \leq_2)$ with the projector g, the coprojector h and a multiplicative one 1_2, while z is an additively prime element from $\boldsymbol{A}_2$. Taking an element a from $\boldsymbol{A}_1$, we see that if $a = d +_1 b$, then by Proposition 2.2.10,

$$z = g(a) = g(d +_1 b) = g(d) +_2 g(b).$$

As z is a prime element from $\boldsymbol{A}_2$, either $g(d) = 0_2$ or $g(b) = 0_2$. For convenience, let us assume that $g(d) = 0_2$. Then by Proposition 2.3.20, $d = 0_1$. It means that a is subtractable only by 0_1 and by itself and thus, it is additively prime.

Theorem is proved. □

Corollary 2.3.21. *If an abstract prearithmetic* $\boldsymbol{A}_1 = (A_1; +_1, \circ_1, \leq_1)$ *is weakly projective with respect to an abstract prearithmetic* $\boldsymbol{A}_2 = (A_2; +_2, \circ_2, \leq_2)$ *with the projector* g, *the coprojector* h *and* 0_2 *is an additive zero in* $\boldsymbol{A}_2$ *whereas* $gh = 1_{A_2}$, *then the image* $g(a)$ *of an arbitrary additively composite element* a *from* $\boldsymbol{A}_1$ *is additively composite in* $\boldsymbol{A}_2$.

An important property of numbers is the cancellation law. For instance, if $n + 5 = m + 5$, then $n = m$ and if $n + 5 > m + 5$, then $n > m$ for any whole numbers n and m. Here we consider nine forms of the cancellation law and study them for abstract prearithmetics.

Definition 2.3.3.

(a) $\boldsymbol{A} = (A; +, \circ, \leq)$ is an abstract *prearithmetic with ordered additive cancellation from the right* if $a + c \leq b + c$ implies $a \leq b$.

(b) $\boldsymbol{A} = (A; +, \circ, \leq)$ is an abstract *prearithmetic with additive cancellation from the right* if $a + c = b + c$ implies $a = b$.

(c) $\boldsymbol{A} = (A; +, \circ, \leq)$ is an abstract *prearithmetic with strict additive cancellation from the right* if $a + c < b + c$ implies $a < b$.

(d) $\boldsymbol{A} = (A; +, \circ, \leq)$ is an abstract *prearithmetic with ordered additive cancellation from the left* if $c + a \leq c + b$ implies $a \leq b$.

(e) $\boldsymbol{A} = (A; +, \circ, \leq)$ is an abstract *prearithmetic with additive cancellation from the left* if $c + a = c + b$ implies $a = b$.

(f) $\boldsymbol{A} = (A; +, \circ, \leq)$ is an abstract *prearithmetic with strict additive cancellation from the left* if $c + a < c + b$ implies $a < b$.

(g) $\boldsymbol{A} = (A; +, \circ, \leq)$ is an abstract *prearithmetic with ordered additive cancellation* if it is an abstract prearithmetic with ordered additive cancellation from the left and from the right.

(h) $\boldsymbol{A} = (A; +, \circ, \leq)$ is an abstract *prearithmetic with additive cancellation* if it is an abstract prearithmetic with additive cancellation from the left and from the right.

(j) $\boldsymbol{A} = (A; +, \circ, \leq)$ is an abstract *prearithmetic with strict additive cancellation* if it is an abstract prearithmetic with strict additive cancellation from the left and from the right.

Let us consider some examples.

Example 2.3.4. The arithmetic $\boldsymbol{N}$ of all natural numbers is an abstract prearithmetic with additive cancellation, with ordered additive cancellation and with strict additive cancellation.

Example 2.3.5. However, the Diophantine arithmetic $\boldsymbol{W}$ of all whole numbers is an abstract prearithmetic with ordered additive cancellation but does not have additive cancellation or strict additive cancellation because any number multiplied by 0 is equal to 0.

Lemma 2.3.10. *If an abstract prearithmetic $\boldsymbol{A} = (A; +, \circ, \leq)$ has the additive zero $0_{\boldsymbol{A}}$, then for any element a from $\boldsymbol{A}$, the difference $a - 0_{\boldsymbol{A}}$ is defined.*

Indeed, defining $a - 0_{\boldsymbol{A}} = a$, we have $0_{\boldsymbol{A}} + (a - 0_{\boldsymbol{A}}) = 0_{\boldsymbol{A}} + a = a$ and $(a - 0_{\boldsymbol{A}}) + 0_{\boldsymbol{A}} = a + 0_{\boldsymbol{A}} = a$. By Definition 2.1.12, this means that a is the difference $a - 0_{\boldsymbol{A}}$.

Lemma 2.3.11. *If an abstract prearithmetic $\boldsymbol{A} = (A; +, \circ, \leq)$ with additive cancellation has the additive zero 0_A, then for any element a from $\boldsymbol{A}, a - a$ is defined and equal to $0_{\boldsymbol{A}}$.*

Indeed, by Definition 2.1.5, $a + 0_{\boldsymbol{A}} = a$. Consequently, by Definition 2.1.10, $a - a = 0_{\boldsymbol{A}}$ because if $a + c = a$, then $c = 0_{\boldsymbol{A}}$.

Definition of subtraction allows deducing many other properties of subtraction, which are well known for integer and real numbers (cf., for example, Feferman, 1974), but are also valid in a much more general context of abstract prearithmetics.

Proposition 2.3.31. (a) *An abstract prearithmetic $\boldsymbol{A} = (A; +, \circ, \leq)$ with subtraction from the right satisfies additive cancellation from the right, i.e., $a + c = b + c$ implies $a = b$ for any elements a, b and c from $\boldsymbol{A}$.*

(b) *An abstract prearithmetic $\boldsymbol{A} = (A; +, \circ, \leq)$ with subtraction from the left satisfies additive cancellation from the left, i.e., $c + a = c + b$ implies $a = b$ for any elements a, b and c from $\boldsymbol{A}$.*

Proof: (a) By definition we have

$$a = (a + c) \rightharpoondown c = (b + c) \rightharpoondown c = b,$$

i.e., $a = b$.

The proof for (b) is similar.

Proposition is proved. □

Corollary 2.3.22. *An abstract prearithmetic $\boldsymbol{A} = (A; +, \circ, \leq)$ with subtraction satisfies additive cancellation, i.e., $a + c = b + c$ or $c + a = c + b$ implies $a = b$ for any elements a, b and c from $\boldsymbol{A}$.*

Proposition 2.3.32. *In an Exactly Additively Archimedean abstract prearithmetic* $\boldsymbol{A} = (A; +, \circ, \leq)$ *with additive cancellation from the right (from the left), for any elements* a, b *and* c *from* $\boldsymbol{A}$, *if* $a < b$, *then the difference from the left* $b \leftharpoondown a$ *(the difference from the right* $b \rightharpoondown a$*) is defined.*

Proof is left as an exercise. □

Corollary 2.3.23. *If an abstract prearithmetic* $\boldsymbol{A} = (A; +, \circ, \leq)$ *with additive cancellation has the additive zero* $0_{\boldsymbol{A}}$, *then for any element* a *from* $\boldsymbol{A}, a - 0_{\boldsymbol{A}}$ *is defined and equal to* a.

Let us find conditions for ordered additive cancellation.

Lemma 2.3.12. $\boldsymbol{A} = (A; +, \circ, \leq)$ *is an abstract prearithmetic with ordered additive cancellation if in* $\boldsymbol{A}$, *the order is linear (total) and addition preserves the strict order.*

Indeed, let us assume that $a + c \leq b + c$. If it is not true $a \leq b$, then $b < a$ because the order is total. However, as addition preserves the strict order, it would be $b + c < a + c$. As this contradicts our assumption, we conclude that $a \leq b$.

Lemma 2.3.13. $\boldsymbol{A} = (A; +, \circ, \leq)$ *is an abstract prearithmetic with ordered additive cancellation (from the right or from the left) if it is with additive cancellation (from the right or from the left) and strict additive cancellation (from the right or from the left).*

Proof is left as an exercise. □

Lemma 2.3.14. $\boldsymbol{A} = (A; +, \circ, \leq)$ *is an abstract prearithmetic with ordered additive cancellation (from the right or from the left), then it is an abstract prearithmetic with additive cancellation (from the right or from the left).*

Indeed, $a + c = b + c$ implies $a + c \leq b + c$ and $b + c \leq a + c$. As $\boldsymbol{A}$ is an abstract prearithmetic with ordered additive cancellation, this implies $a \leq b$ and $b \leq a$. Consequently, $a = b$, which means that $\boldsymbol{A}$ is an abstract prearithmetic with additive cancellation.

Cancellation property allows strengthening of results in Lemma 2.3.7.

Lemma 2.3.15. *For any elements* a *and* b *from an abstract prearithmetic* $\boldsymbol{A}$ *with additive cancellation, we have the following lemma:*

(a) $a \lceil b$ *if and only if subtraction from the left* $a \leftharpoondown b$ *is defined.*

(b) $b \rceil a$ *if and only if subtraction from the right* $a \rightharpoondown b$ *is defined.*
(c) $a \,(\, b$ *if and only if full subtraction* $a - b$ *is defined.*

Proof is left as an exercise. □

Lemma 2.3.16. (a) *If subtraction from the left* $\leftharpoondown$ *is defined in an abstract prearithmetic* $\boldsymbol{A}$, *then* $\boldsymbol{A}$ *is with additive cancellation from the left.*

(b) *If subtraction from the right* $\rightharpoondown$ *is defined in an abstract prearithmetic* $\boldsymbol{A}$, *then* $\boldsymbol{A}$ *is with additive cancellation from the right.*

(c) *If full subtraction a – is defined in an abstract prearithmetic* $\boldsymbol{A}$, *then* $\boldsymbol{A}$ *is with additive cancellation.*

Proof is left as an exercise. □

In some cases, additive and multiplicative zeros coincide.

Proposition 2.3.33. *If an abstract prearithmetic* $\boldsymbol{A} = (A; +, \circ, \leq)$ *with additive cancellation has an additive zero* $0_{\boldsymbol{A}}$ *and multiplication is distributive over addition, then* $0_{\boldsymbol{A}}$ *is also a multiplicative zero.*

Proof: Let us take an abstract prearithmetic $\boldsymbol{A} = (A; +, \circ, \leq)$ that satisfies all initial conditions. Taking arbitrary element x from $\boldsymbol{A}$, we have

$$0_{\boldsymbol{A}} \circ x = 0_{\boldsymbol{A}} \circ x + 0_{\boldsymbol{A}} = (0_{\boldsymbol{A}} + 0_{\boldsymbol{A}}) \circ x = 0_{\boldsymbol{A}} \circ x + 0_{\boldsymbol{A}} \circ x. \tag{2.45}$$

As we have additive cancellation in $\boldsymbol{A}$, it is possible to cancel $0_{\boldsymbol{A}} \circ x$ in (2.45). This gives us $0_{\boldsymbol{A}} \circ x = 0_{\boldsymbol{A}}$.

The identity $x \circ 0_{\boldsymbol{A}} = 0_{\boldsymbol{A}}$ is proved in a similar way. Consequently, $0_{\boldsymbol{A}}$ is also a multiplicative zero.

Proposition is proved. □

Corollary 2.3.24. *In a semiring with additive cancellation, additive and multiplicative zeros coincide.*

An important property of numbers is divisibility. Here we study it for abstract prearithmetics.

Definition 2.3.4. (a) An element a from $\boldsymbol{A}$ is *divisible* from the right (from the left) by an element b from $\boldsymbol{A}$ if $a = d \circ b (a = b \circ d)$ for some element d from $\boldsymbol{A}$. We denote this by $a \lfloor$ b (by $b \rfloor a$).

(b) An element a from $\boldsymbol{A}$ is *divisible* by an element b if it is divisible by b from the right and from the left. We denote this by $a|b$.

Remark 2.3.3. If a is divisible by b, it is sometimes denoted by $b|a$.

Note that uniqueness of the element d in Definition 2.3.5 is not demanded. As a result, 0 is divisible by any whole number in the Diophantine arithmetic $\boldsymbol{W}$. At the same time, divisibility of all other numbers in $\boldsymbol{W}$ remains the same as it is conventionally defined.

Lemma 2.3.17. *If multiplication $\circ$ is commutative in an abstract prearithmetic $\boldsymbol{A} = (A; +, \circ, \leq)$ with the multiplicative one 1, then in it, any element a is divisible by b if and only if it is divisible by b from the right or from the left.*

Proof is left as an exercise. □

Proposition 2.3.34. *If multiplication $\circ$ in an abstract prearithmetic $\boldsymbol{A} = (A; +, \circ, \leq)$ is associative, then for any elements a, b and c from $\boldsymbol{A}$, we have the following conditions*:

(1) $a\lfloor b$ *and* $b\lfloor c$ *imply* $a\lfloor c$.
(2) $b\rfloor a$ *and* $c\rfloor b$ *imply* $c\rfloor a$.
(3) $a|b$ $b|c$ *imply* $a|c$.

Proof: (1) If $a\lfloor b$, then $a = d \circ b$ for some element d from $\boldsymbol{A}$. If $b\lfloor c$, then $b = e \circ c$ for some element e from $\boldsymbol{A}$. Consequently,

$$a = d \circ b = d \circ e \circ c = (d \circ e) \circ c.$$

It means that $a\lfloor c$.

Statements (2) and (3) are proved in a similar way.

Proposition is proved. □

Corollary 2.3.25 (Landau, 1999: Theorem 2). *If a, b and c are integer numbers, then $a|b$ and $b|c$ imply $a|c$.*

Proposition 2.3.35. *If multiplication $\circ$ in an abstract prearithmetic $\boldsymbol{A} = (A; +, \circ, \leq)$ is associative, then for any elements a, b and c from $\boldsymbol{A}$, we have the following conditions*:

(1) $a\lfloor b$ *implies* $a \circ c\lfloor b \circ c$.
(2) $b\rfloor a$ *implies* $c \circ b\rfloor c \circ a$.
(3) $a|b$ *implies* $a \circ c|b \circ c$ *when multiplication* $\circ$ *is also commutative.*

Proof: (1) If $a\lfloor b$, then $a = d \circ b$ for some element d from $\boldsymbol{A}$. Consequently,

$$a \circ c = d \circ b \circ c = d \circ (b \circ c).$$

It means that $a \circ c\lfloor b \circ c$.

(2) If $b\rfloor a$, then $a = b \circ d$ for some element d from $\boldsymbol{A}$. Consequently,

$$c \circ a = c \circ b \circ d = (c \circ b) \circ d.$$

It means that $c \circ b\rfloor c \circ a$.

By Lemma 2.3.15, the statement (3) follows from statements (1) and (2) when multiplication $\circ$ is commutative.

Proposition is proved. □

Corollary 2.3.26 (Landau, 1999: Theorem 3b). *If a, b and c are integer numbers, then $a|b$ implies $ac|bc$.*

Proposition 2.3.36. *If multiplication $\circ$ in an abstract prearithmetic $\boldsymbol{A} = (A; +, \circ, \leq)$ is associative, then for any elements a, b and c from $\boldsymbol{A}$, we have the following conditions:*

(1) *$a\lfloor b$ implies $c \circ a\lfloor b$.*
(2) *$b\rfloor a$ implies $b\rfloor a \circ c$.*
(3) *$a|b$ implies $a \circ c|b$ when multiplication $\circ$ is also commutative.*

Proof: (1) If $a\lfloor b$, then $a = d \circ b$ for some element d from $\boldsymbol{A}$. Consequently,

$$c \circ a = c \circ d \circ b = (c \circ d) \circ b$$

It means that $c \circ a\lfloor b$.

(2) If $b\rfloor a$, then $a = b \circ d$ for some element d from $\boldsymbol{A}$. Consequently,

$$a \circ c = b \circ d \circ c = b \circ (d \circ c).$$

It means that $b\rfloor a \circ c$.

By Lemma 2.3.17, the statement (3) follows from statements (1) and (2) when multiplication $\circ$ is commutative.

Proposition is proved. □

Corollary 2.3.27 (Landau, 1999: Theorem 4). *If a, b and c are integer numbers, then $a|b$ implies $ac|b$.*

Proposition 2.3.37. *For any elements a, b and c from an abstract prearithmetic $\boldsymbol{A} = (A; +, \circ, \leq)$, we have the following conditions:*

(1) *$a\lfloor c$ and $b\lfloor c$ imply $a + b\lfloor c$ when multiplication $\circ$ is distributive from the right over addition $+$.*
(2) *$c\rfloor a$ and $b\rfloor c$ imply $c\rfloor a + b$ when multiplication $\circ$ is distributive from the left over addition $+$.*
(3) *$a|c$ and $b|c$ imply $a + b|c$ when multiplication $\circ$ is also commutative.*

Proof: (1) If $a\lfloor c$, then $a = d \circ c$ for some element d from $\boldsymbol{A}$. If $b\lfloor c$, then $b = e \circ c$ for some element e from $\boldsymbol{A}$. Consequently, by distributivity from the right, we have

$$a + b = d \circ c + e \circ c = (e + d) \circ c.$$

It means that $a + b\lfloor c$.

(2) If $c\rfloor a$, then $a = c \circ d$ for some element d from $\boldsymbol{A}$. If $c\rfloor b$, then $b = c \circ e$ for some element e from $\boldsymbol{A}$. Consequently, by distributivity from the left, we have

$$a + b = c \circ d + c \circ e = c \circ (e + d).$$

It means that $c\rfloor a + b$.

The statement (3) follows from statements (1) and (2) when multiplication $\circ$ is commutative.

Proposition is proved. □

Corollary 2.3.28 (Landau, 1999: Theorem 5). *If a, b and c are integer numbers, then $a|c$ and $b|c$ imply $(a + b)|c$.*

Corollary 2.3.29. *For any elements $a_i (i = 1, 2, 3, \ldots, n)$ from an abstract prearithmetic $\boldsymbol{A} = (A; +, \circ, \leq)$, we have the following conditions:*

(1) *$a_i\lfloor c$ for all $i = 1, 2, 3, \ldots, n$ imply $\Sigma_{i=1}^{n} a_i\lfloor c$ when multiplication $\circ$ is distributive from the right over addition $+$.*
(2) *$c\rfloor a_i$ for all $i = 1, 2, 3, \ldots, n$ imply $c\rfloor \Sigma_{i=1}^{n} a_i$ when multiplication $\circ$ is distributive from the left over addition $+$.*
(3) *$a_i|c$ for all $i = 1, 2, 3, \ldots, n$ imply $\Sigma_{i=1}^{n} a_i|c$ when multiplication $\circ$ is also commutative.*

Proposition 2.3.38. *For any elements a, b, k, h and c from an abstract prearithmetic $\boldsymbol{A} = (A; +, \circ, \leq)$, we have the following conditions:*

(1) *$a\lfloor c$ and $b\lfloor c$ imply $(a \circ k + b \circ h)\lfloor c$ when multiplication $\circ$ is associative and distributive from the right over addition $+$.*
(2) *$c\rfloor a$ and $b\rfloor c$ imply $c\rfloor(a \circ k + b \circ h)$ when multiplication $\circ$ is associative and distributive from the left over addition $+$.*
(3) *$a|c$ and $b|c$ imply $(a \circ k + b \circ h)|c$ when multiplication $\circ$ is also commutative.*

Proof: (1) By Proposition 2.3.36, for any elements a, b, k, h and c from $\boldsymbol{A}$, we have

$$(a \circ k) \lfloor c \text{ and } (b \circ h) \lfloor c.$$

Thus, By Proposition 2.3.37,

$$(a \circ k + b \circ h) \lfloor c.$$

Statements (2) and (3) are proved in the same way based on Propositions 2.3.37 and 2.3.36.

Proposition is proved. □

Corollary 2.3.30 (Landau, 1999: Theorem 6). *If a, b, k, h and c are integer numbers, then $a|c$ and $b|c$ imply $(a \circ k + b \circ h)|c$.*

Proposition 2.3.39. *For any elements a, b and c from an abstract prearithmetic $\boldsymbol{A} = (A; +, \circ, \leq)$, we have the following conditions:*

(1) *$a \lfloor c$ and $b \lfloor c$ imply $a \rightharpoondown b \lfloor c$ when the right difference $a \rightharpoondown b$ of a and b exists and multiplication $\circ$ is distributive from the right over the right difference.*
(2) *$c \rfloor a$ and $b \rfloor c$ imply $c \rfloor a \leftharpoondown b$ when the left difference $a \leftharpoondown b$ of a and b exists and multiplication $\circ$ is distributive from the left over the left difference.*
(3) *$a|c$ and $b|c$ imply $a - b|c$ when the difference $a - b$ of a and b exists and multiplication $\circ$ is commutative and distributive over the difference.*

Proof: (1) If $a \lfloor c$, then $a = d \circ c$ for some element d from $\boldsymbol{A}$. If $b \lfloor c$, then $b = e \circ c$ for some element e from $\boldsymbol{A}$. Consequently, by distributivity from the right, we have

$$a \rightharpoondown b = d \circ c \rightharpoondown e \circ c = (e \rightharpoondown d) \circ c.$$

It means that $a \rightharpoondown b \lfloor c$.

(2) If $c \rfloor a$, then $a = c \circ d$ for some element d from $\boldsymbol{A}$. If $c \rfloor b$, then $b = c \circ e$ for some element e from $\boldsymbol{A}$. Consequently, by distributivity from the left, we have

$$a \leftharpoondown b = c \circ d \leftharpoondown c \circ e = c \circ (e \leftharpoondown d).$$

It means that $c \rfloor a \leftharpoondown b$.

The statement (3) follows from statements (1) and (2) when multiplication $\circ$ is commutative and distributive over the difference.

Proposition is proved. □

Corollary 2.3.31 (Landau, 1999: Theorem 5). *If a, b and c are integer numbers, then $a|c$ and $b|c$ imply $(a-b)|c$.*

Lemma 2.3.18. *If an abstract prearithmetic $\boldsymbol{A} = (A; +, \circ, \leq)$ has the multiplicative zero 0^{m}, then 0^{m} is divisible by any element from $\boldsymbol{A}$.*

Indeed, we have $0^{\mathrm{m}} = a \circ 0^{\mathrm{m}} = 0^{\mathrm{m}} \circ a$ for any element a from $\boldsymbol{A}$.

Proposition 2.3.40. *In an Exactly Multiplicatively Archimedean abstract prearithmetic $\boldsymbol{A} = (A; +, \circ, \leq)$, for any elements a, b and c from $\boldsymbol{A}$, if $a < b$, then b is divisible from the left by a.*

Proof is left as an exercise. □

Lemma 2.3.19. *If an abstract prearithmetic $\boldsymbol{A} = (A; +, \circ, \leq)$ with the multiplicative one 1_A, then in it, any element is divisible by itself and by 1_A.*

Indeed, we have $a = a \circ 1_A = 1_A \circ a$ for any element a from $\boldsymbol{A}$.

This result brings us to the concept of a multiplicatively prime element studied in Section 2.1. It is a counterpart of the concept of a prime number in the Diophantine arithmetic $\boldsymbol{N}$, which is extensively investigated in number theory (cf., for example, Dickson, 1932; Friedberg, 1968; Davenport, 1992; Landau, 1999; Crandall and Pomerance, 2001). Although there are infinitely many multiplicatively prime numbers in the conventional Diophantine arithmetic $\boldsymbol{N}$, for an arbitrary abstract prearithmetic, the situation can be essentially different (cf. Section 2.1).

Multiplication in an abstract prearithmetic $\boldsymbol{A} = (A; +, \circ, \leq)$ is *distributive from the left with respect to difference* if for any elements a, b, c and d from $\boldsymbol{A}$, the equality $c \circ a + d = c \circ b$ implies the equality $d = c \circ e$ where the element e is the difference of b and a.

For instance, for integer numbers, it means that if $k = n - m$, $m = uw$ and $n = uv$, then

$$k = n - m = uv - uw = u(v - w).$$

Let us consider an abstract prearithmetic $\boldsymbol{A} = (A; +, \circ, \leq)$, which has multiplicative 1, is totally ordered, additively Archimedean and exactly binary for addition Archimedean and in which addition preserves the order and multiplication is associative and distributive from the left with respect to difference and addition.

Theorem 2.3.5. *If m is the smallest common multiple of elements a and b from $\boldsymbol{A}$, then any common multiple u of elements a and b is divisible by m.*

Proof: As m is the smallest common multiple of elements a and b, we have $m < u$. Then by Theorem 2.1.5, $u = n[m]$ or $u = n[m] + r$ and $r < m$.

In the first case, $u = m \circ n[1]$ because multiplication is distributive from the left with respect to addition and $m \circ 1 = m$. It means that the statement of Theorem 2.3.5 is true.

As it was demonstrated, $n[m]$ is divisible by m and thus, by Proposition 2.3.34, it is divisible by a and b. As u is also divisible by a and b, the element r is divisible by a and b. It means that in the second case, r is a common multiple of elements a and b. However, this contradicts to the condition that m is the smallest common multiple of elements a and b. Consequently, only the first case is possible.

Theorem is proved. □

Corollary 2.3.32 (Landau, 1999: Theorem 9). *If m is the smallest common multiple of natural numbers k and h, then any common multiple n of numbers k and h is divisible by m.*

Let us look how primality is preserved by projectivity relations.

Theorem 2.3.6. *If an abstract prearithmetic $\boldsymbol{A}_1 = (A_1; +_1, \circ_1, \leq_1)$ is weakly projective with respect to an abstract prearithmetic $\boldsymbol{A}_2 = (A_2; +_2, \circ_2, \leq_2)$ with the projector g, the coprojector h and 1_2 is a multiplicative one in $\boldsymbol{A}_2$ whereas $gh = 1_{A_2}$, then any element of the inverse image $g^{-1}(z)$ of an arbitrary prime element z from $\boldsymbol{A}_2$ is prime in $\boldsymbol{A}_1$.*

Proof: Let us assume that an abstract prearithmetic $\boldsymbol{A}_1 = (A_1; +_1, \circ_1, \leq_1)$ is projective with respect to an abstract prearithmetic $\boldsymbol{A}_2 = (A_2; +_2, \circ_2, \leq_2)$ with the projector g, the coprojector h and a multiplicative one 1_2, while z is a prime element from $\boldsymbol{A}_2$. Taking an element a from $\boldsymbol{A}_1$, we see that if $a = d \circ_1 b$, then by Proposition 2.2.10,

$$z = g(a) = g(d \circ_1 b) = g(d) \circ_2 g(b).$$

As z is a prime element from $\boldsymbol{A}_2$, either $g(d) = 1_2$ or $g(b) = 1_2$. For convenience, let us assume that $g(d) = 1_2$. Then by Proposition 2.3.21, $d = 1_1$. It means that a is divisible only by 1_1 and by itself and thus, it is (multiplicatively) prime.

Theorem is proved. □

Corollary 2.3.33. *If an abstract prearithmetic* $\boldsymbol{A}_1 = (A_1; +_1, \circ_1, \leq_1)$ *is weakly projective with respect to an abstract prearithmetic* $\boldsymbol{A}_2 = (A_2; +_2, \circ_2, \leq_2)$ *with the projector* g*, the coprojector* h *and* 1_2 *is a multiplicative one in* $\boldsymbol{A}_2$ *whereas* $gh = 1_{A_2}$*, then the image* $g(a)$ *of an arbitrary composite element* a *from* $\boldsymbol{A}_1$ *is composite in* $\boldsymbol{A}_2$*.*

Although not all abstract prearithmetics have the multiplicative factoring property or strong multiplicative factoring property described in Section 2.1, projectivity provides for preservation of the multiplicative factoring property. Let us assume that an abstract prearithmetic $\boldsymbol{A}_1 = (A_1; +_1, \circ_1, \leq_1)$ is weakly projective with respect to an abstract prearithmetic $\boldsymbol{A}_2 = (A_2; +_2, \circ_2, \leq_2)$ with the projector g, the coprojector h and a multiplicative one 1_2 whereas $gh = 1_{A_2}$.

Theorem 2.3.7. *If the abstract prearithmetic* $\boldsymbol{A}_2$ *has the multiplicative factoring property, then the abstract prearithmetic* $\boldsymbol{A}_1$ *also has the multiplicative factoring property.*

Proof: Let us suppose that the abstract prearithmetic $\boldsymbol{A}_2 = (A_2; +_2, \circ_2, \leq_2)$ has the multiplicative factoring property. Taking an element a from $\boldsymbol{A}_1$, we see that if $a = d +_1 b$, then by Proposition 2.1.15,

$$g(a) = g(d +_1 b) = g(d) +_2 g(b).$$

At the same time, if elements are multiplicatively composite, then it is possible to continue multiplicative decomposition (factoring) of the element a. Consequently, if there is no prime multiplicative decomposition (prime factorization) of a, then it has multiplicative decompositions of arbitrary length, in which all factors are not equal to 1_2. By Proposition 2.1.15, any multiplicative decomposition of the element a induces a multiplicative decomposition of its image $g(a)$.

At the same time, if we have a multiplicative decomposition of the element x from the abstract prearithmetic $\boldsymbol{A}_2$, it cannot be continued infinitely because $\boldsymbol{A}_2$ has the multiplicative factoring property. This contradiction shows that any multiplicative decomposition of the element a from the abstract prearithmetic $\boldsymbol{A}_1$ cannot be continued infinitely. Thus, it leads to a prime multiplicative decomposition of a. As a is an arbitrary element a from $\boldsymbol{A}_1$, it means that $\boldsymbol{A}_1$ has the multiplicative factoring property.

Theorem is proved. □

It is possible to ask the question whether it is possible that an element from an abstract prearithmetic has a prime factorization and at the same time, additive factorizations of an arbitrary big length. We know that this is impossible in the Diophantine arithmetic. However, in the class of abstract prearithmetics, this can happen as the following example demonstrates.

Example 2.3.6. Let us consider the set $F = \{0, 1, 1/3, 1/2, (1/2)^2, (1/2)^3, \ldots, (1/2)^n, \ldots\}$ and the set P_2 of all polynomials of the form $p(1/2) = a_0 + a_1(1/2) + a_2(1/2)^2 + a_3(1/2)^3 + \ldots + a_n(1/2)^n$ where all a_n are natural numbers. We see that the sum and product of these polynomials have the same form. Now, we can define the set A_2 of numbers that can be expressed as polynomials from P. Naturally, some of these polynomials define the same number. For instance, $1 = 2(1/2)$ or $(1/2) \circ (1/2)^2 = (1/2) \cdot (1/2)^2 = (1/2)^3$.

We also take the set P_3 of all expressions of the form $q(1/3) = b_0 + b_1(1/3)$ where b_0 and b_1 are natural numbers. We see that the sum of these polynomials has the same form while the product is defined as

$$(b_0 + b_1(1/3)) \circ (d_0 + d_1(1/3)) = b_0 \cdot d_0 + (b_1 \cdot d_0 + b_0 \cdot d_1)(1/3)$$

In addition, we define

$$1/3 \circ 1/3 = 1/3 \circ (1/2)^n = (1/2)^n \circ 1/3 = 1/3 + (1/2)^n = (1/2)^n + 1/3 = 0.$$

Taking the set P of the sums of polynomials from P_2 and expressions from P_3, we can define the set A of numbers that can be expressed as elements from P.

This gives us the abstract prearithmetic $\boldsymbol{A} = (A; +, \circ, \leq)$, in which order is the same as in the arithmetic $\boldsymbol{R}$ of all real numbers, while addition and multiplication are defined above. In it, any polynomial from P_2 is additively composite. Indeed, we have

$$\begin{aligned} &a_0 + a_1(1/2) + a_2(1/2)^2 + a_3(1/2)^3 + \cdots + a_n(1/2)^n \\ &\quad = (a_0(1/2) + a_1(1/2)^2 + a_2(1/2)^3 + \cdots + a_{n-1}(1/2)^n) \\ &\quad\quad + (a_0(1/2) + a_1(1/2)^2 + a_2(1/2)^3 + \cdots + a_{n-1}(1/2)^n). \end{aligned}$$

At the same time, $1/3$ is an additively prime element. This allows building two additive factorizations for the element 1. Namely, we have the prime additive factorization

$$1 = 1/3 + 1/3 + 1/3$$

and additive factorizations of arbitrary length

$$1 = (1/2)^n + (1/2)^n + (1/2)^n + \cdots + (1/2)^n$$

It is also possible to show that a similar situation can occur for multiplicative decompositions (factorizations) of elements from abstract prearithmetics.

The strong additive factoring property necessitates multiplication properties of weakly projective prearithmetics. Let us assume that an abstract prearithmetic $\boldsymbol{A}_1 = (A_1; +_1, \circ_1, \leq_1)$ is weakly projective with respect to an abstract prearithmetic $\boldsymbol{A}_2 = (A_2; +_2, \circ_2, \leq_2)$ with the projector g, the coprojector h and a multiplicative one 1_2 whereas $gh = 1_{A_2}$ and $g^{-1}(1_2) = \{1_1\}$.

Theorem 2.3.8. *If the abstract prearithmetic $\boldsymbol{A}_2$ has the strong multiplicative factoring property, then for each element a from the abstract prearithmetic $\boldsymbol{A}_1$, there is a natural number n such that any multiplicative factorization of a, in which all elements are not equal to 0_1, has less than n elements.*

Proof: Let us suppose that the abstract prearithmetic $\boldsymbol{A}_2 = (A_2; +_2, \circ_2, \leq_2)$ has the strong multiplicative factoring property. Taking an element a from $\boldsymbol{A}_1$, we see that if $a = d \circ_1 b$, then by Proposition 2.1.15,

$$g(a) = g(d \circ_1 b) = g(d) \circ_2 g(b).$$

Thus, any additive factorization (2.46) of a induces a multiplicative factorization (2.47) of $g(a)$ with the same number of elements.

$$a = \prod_{i=1}^{m} a_i, \tag{2.46}$$

$$g(a) = \prod_{i=1}^{m} g(a_i). \tag{2.47}$$

Besides, by the initial conditions, if (2.47) does not have ones, then (2.46) does not have ones. As the abstract prearithmetic $\boldsymbol{A}_2$ has the strong multiplicative factoring property, it is possible to extend the multiplicative factorization (2.47) to the unique multiplicative prime factorization with n elements. As m in (2.47) is always not larger than n, the same is true for (2.46).

Theorem is proved. □

There are intrinsic relations between the Additively Archimedean Property, monotonicity of addition, Exactly Binary Archimedean Property for addition (cf. Section 2.1.2) and the cancellation property.

Proposition 2.3.41 (Clifford, 1954). *If an abstract prearithmetic $\boldsymbol{A} = (A; +, \circ, \leq)$ with the linear order $\leq$ and associative monotone addition has the Additively Archimedean Property and exactly right Binary Archimedean property for addition but does not have the additive cancellation property from the left, then we have the following conditions:*

(i) *$\boldsymbol{A}$ has the maximal element u.*
(ii) *The equality $a + b = a + c$ and the inequality $a + b \neq u$ imply $b = c$.*
(iii) *For any element a from $\boldsymbol{A}$, there is a natural number n such that*

$$u = \sum_{i=1}^{n} a_i,$$

where all a_i are equal to a.

Proof: If an abstract prearithmetic $\boldsymbol{A} = (A; +, \circ, \leq)$ does not have the additive cancellation property from the left, then there are elements a, b and c from $\boldsymbol{A}$ such that $a + b = a + c$ but $b < c$ because the order in $\boldsymbol{A}$ is linear. As the prearithmetic $\boldsymbol{A}$ has the exactly right Binary Archimedean property for addition, there is an element d from $\boldsymbol{A}$ such that $b + d = c$.

Denoting $a + b$ by u, we have

$$u = a + b = a + c = a + b + d = u + d.$$

This implies that for any natural number n, we have

$$u = u + \sum_{i=1}^{n} d_i,$$

where all d_i are equal to d.

Let us suppose there is an element w from $\boldsymbol{A}$ such that $u < w$. As the prearithmetic $\boldsymbol{A}$ has the Additively Archimedean property, there is a natural number m such that

$$w \leq \sum_{i=1}^{m} d_i,$$

where all d_i are equal to d. As addition is monotone in $\boldsymbol{A}$, we have

$$u = u + \sum_{i=1}^{n} d_i \geq u + w \geq w > u.$$

This contradicts the definition of a linear order and proves statements (i) and (ii). As the prearithmetic $\boldsymbol{A}$ has the Additively Archimedean property, for any element a from $\boldsymbol{A}$, there is a natural number n such that

$$w \leq \sum_{i=1}^{n} a_i,$$

where all a_i are equal to a. However, as u is the maximal element, we have

$$u = \sum_{i=1}^{n} a_i.$$

Proposition is proved. □

Let us study multiplicative cancellation properties of abstract prearithmetics.

Definition 2.3.5.

(a) $\boldsymbol{A} = (A; +, \circ, \leq)$ is an abstract prearithmetic with *ordered multiplicative cancellation from the right (from the left)* if $a \circ c \leq b \circ c$ ($c \circ a \leq c \circ b$) implies $a \leq b$.
(b) $\boldsymbol{A} = (A; +, \circ, \leq)$ is an abstract prearithmetic with *multiplicative cancellation from the right (from the left)* if $a \circ c = b \circ c$ ($c \circ a = c \circ b$) implies $a = b$.
(c) $\boldsymbol{A} = (A; +, \circ, \leq)$ is an abstract prearithmetic with *strict multiplicative cancellation from the right (from the left)* if $a \circ c < b \circ c$ ($c \circ a < c \circ b$) implies $a < b$.
(d) $\boldsymbol{A} = (A; +, \circ, \leq)$ is an abstract prearithmetic with *ordered non-zero multiplicative cancellation from the right (from the left)* if $a \circ c \leq b \circ c$ and $c \neq 0$ ($c \circ a \leq c \circ b$ and $c \neq 0$) imply $a \leq b$.
(e) $\boldsymbol{A} = (A; +, \circ, \leq)$ is an abstract prearithmetic with *non-zero multiplicative cancellation from the right (from the left)* if it has the multiplicative zero 0 and conditions $a \circ c = b \circ c$ ($c \circ a = c \circ b$ and $c \neq 0$) and $c \neq 0$ imply $a = b$.
(f) $\boldsymbol{A} = (A; +, \circ, \leq)$ is an abstract prearithmetic with *strict non-zero multiplicative cancellation from the right (from the left)* if the multiplicative zero 0 and conditions $a \circ c < b \circ c$ ($c \circ a < c \circ b$ and $c \neq 0$) and $c \neq 0$ imply $a < b$.

(g) $\boldsymbol{A} = (A; +, \circ, \leq)$ is an abstract *prearithmetic with ordered (non-zero) multiplicative cancellation* if it is an abstract prearithmetic with ordered (non-zero) multiplicative cancellation from the left and from the right.

(h) $\boldsymbol{A} = (A; +, \circ, \leq)$ is an abstract *prearithmetic with (non-zero) multiplicative cancellation* if it is an abstract prearithmetic with (non-zero) multiplicative cancellation from the left and from the right.

(j) $\boldsymbol{A} = (A; +, \circ, \leq)$ is an abstract *prearithmetic with strict (non-zero) multiplicative cancellation* if it is an abstract prearithmetic with strict (non-zero) multiplicative cancellation from the left and from the right.

Example 2.3.7. The Diophantine arithmetic $\boldsymbol{N}$ is an abstract prearithmetic with strict multiplicative cancellation, with multiplicative cancellation and with ordered multiplicative cancellation both from the right and from the left.

Example 2.3.8. The Diophantine arithmetic $\boldsymbol{W}$ is an abstract prearithmetic with strict non-zero multiplicative cancellation, with non-zero multiplicative cancellation and with ordered non-zero multiplicative cancellation both from the right and from the left.

Lemma 2.3.20. *(a) If multiplication* $\circ$ *in an abstract prearithmetic* $\boldsymbol{A} = (A; +, \circ, \leq)$ *is commutative, then the property of strict multiplicative cancellation from the left (multiplicative cancellation from the left or ordered multiplicative cancellation from the left) coincides with strict multiplicative cancellation from the right (multiplicative cancellation from the right or ordered multiplicative cancellation from the right).*

(b) If multiplication $\circ$ *in an abstract prearithmetic* $\boldsymbol{A} = (A; +, \circ, \leq)$ *is commutative, then the property of strict non-zero multiplicative cancellation from the left (non-zero multiplicative cancellation from the left or ordered non-zero multiplicative cancellation from the left) coincides with strict non-zero multiplicative cancellation from the right (non-zero multiplicative cancellation from the right or ordered non-zero multiplicative cancellation from the right).*

Proof is left as an exercise. □

Lemma 2.3.21. $\boldsymbol{A} = (A; +, \circ, \leq)$ *is an abstract prearithmetic with ordered multiplicative cancellation if the order is linear (total) and multiplication preserves the strict order in* $\boldsymbol{A}$.

Indeed, let us assume that $a \circ c \leq b \circ c$. If it is not true $a \leq b$, then $b < a$ because the order is total. However, as multiplication preserves the strict order, it would be $b \circ c < a \circ c$. As this contradicts our assumption, we conclude that $a \leq b$.

There are definite relations between different cancellation laws in abstract prearithmetics.

Lemma 2.3.22. *$\boldsymbol{A} = (A; +, \circ, \leq)$ is an abstract prearithmetic with ordered multiplicative cancellation if it is with multiplicative cancellation and strict multiplicative cancellation.*

Proof is left as an exercise. □

Lemma 2.3.23. *$\boldsymbol{A} = (A; +, \circ, \leq)$ is an abstract prearithmetic with ordered multiplicative cancellation, then it is an abstract prearithmetic with multiplicative cancellation.*

Indeed, $a \circ c = b \circ c$ implies $a \circ c \leq b + c$ and $b \circ c \leq a \circ c$. As $\boldsymbol{A}$ is an abstract prearithmetic with ordered multiplicative cancellation, this implies $a \leq b$ and $b \leq a$. Consequently, $a = b$, which means that $\boldsymbol{A}$ is an abstract prearithmetic with multiplicative cancellation.

However, the converse of Lemma 2.3.23 is not true as the following example demonstrates.

Example 2.3.9. Let us take the Diophantine arithmetic $\boldsymbol{Z}$ of integer numbers. It is an abstract prearithmetic with multiplicative cancellation. At the same time, the inequality $(-3) \circ (-5) \leq (-3) \circ (-7)$ does not imply $-5 \leq -7$.

Corollary 2.3.34. *An abstract prearithmetic $\boldsymbol{A} = (A; +, \circ, \leq)$ with division satisfies multiplicative cancellation, i.e., $a \circ c = b \circ c$ or $c \circ a = c \circ b$ implies $a = b$ for any elements a, b and c from $\boldsymbol{A}$. such that $c \neq 0^{\mathrm{m}}_{\boldsymbol{A}}$.*

Proposition 2.3.42. *If multiplication $\circ$ in an abstract prearithmetic $\boldsymbol{A} = (A; +, \circ, \leq)$ is associative, then for any elements a, b and c from $\boldsymbol{A}$, we have the following conditions:*

(1) *$a \circ c \lfloor b \circ c$ implies $a \lfloor b$ when $\boldsymbol{A}$ is with non-zero multiplicative cancellation from the right.*
(3) *$c \circ b \rfloor c \circ a$ implies $b \rfloor a$ when $\boldsymbol{A}$ is with non-zero multiplicative cancellation from the left*
(3) *$a \circ c | b \circ c$ implies $a | b$ when $\boldsymbol{A}$ is with non-zero multiplicative cancellation from the left and multiplication $\circ$ is also commutative.*

Proof is left as an exercise. □

Corollary 2.3.35 (Landau, 1999: Theorem 3a). *If a, b and c are integer numbers, then $ac | bc$ implies $a | b$.*

Proposition 2.3.43. *If an abstract prearithmetic $\boldsymbol{A} = (A; +, \circ, \leq)$ has a multiplicative zero 0^{m}, contains an opposite element $-a$ for each element a, and multiplication is distributive over addition, then $\boldsymbol{A}$ has the property of non-zero multiplicative cancellation from the right or from the left if and only if the product of any two non-zero elements is not equal to zero.*

Proof: *Sufficiency.* Let us take an abstract prearithmetic $\boldsymbol{A} = (A; +, \circ, \leq)$ with non-zero multiplicative cancellation from the right that satisfies all initial conditions and suppose that it has two elements $a \neq 0^{\mathrm{m}}$ and $b \neq 0^{\mathrm{m}}$ such that $a \circ b = 0^{\mathrm{m}}$. Then we have

$$a \circ b = 0^{\mathrm{m}} = 0^{\mathrm{m}} \circ b.$$

As $\boldsymbol{A}$ is with non-zero multiplicative cancellation from the right, we can cancel b and have $a = 0^{\mathrm{m}}$. This contradicts our assumption that $a \neq 0^{\mathrm{m}}$ and by the Principle of Excluded Middle, proves that the product of any two non-zero elements is non-zero.

The case when $\boldsymbol{A}$ is with non-zero multiplicative cancellation from the left is proved in a similar way.

Necessity. Let us take an abstract prearithmetic $\boldsymbol{A} = (A; +, \circ, \leq)$ that satisfies all initial conditions and the product of any two non-zero elements is non-zero in $\boldsymbol{A}$. We remind that if x is an element from $\boldsymbol{A}$, then an element y is called opposite to x and denoted by $-x$ when $x + y = 0^{\mathrm{m}}$.

Let us assume that $a \circ c = b \circ c$ and $c \neq 0^{\mathrm{m}}$ for some elements from $\boldsymbol{A}$. Then $b \circ c$ has the opposite element $-(b \circ c)$ and we have

$$a \circ c - b \circ c = 0^{\mathrm{m}} = (a - b) \circ c.$$

As $c \neq 0$, we have $a - b = 0^{\mathrm{m}}$ and consequently, $a = b$. As a, b and $c \neq 0^{\mathrm{m}}$ are arbitrary elements from $\boldsymbol{A}$, the abstract prearithmetic $\boldsymbol{A} = (A; +, \circ, \leq)$ has the property of non-zero multiplicative cancellation from the right.

The proof for non-zero multiplicative cancellation from the left is similar.

Proposition is proved. □

As a corollary, we obtain a well-known result from the theory of rings.

Corollary 2.3.36 (Rotman, 1996: Theorem 3.5). *A commutative ring is a domain if and only if the product of any two non-zero elements is not equal to zero.*

The proof of also gives us the following result.

Corollary 2.3.37. *In an abstract prearithmetic* $\boldsymbol{A} = (A; +, \circ, \leq)$ *with a multiplicative zero* 0^{m} *and non-zero multiplicative cancellation from the right or from the left, the product of any two non-zero elements is not equal to zero.*

Corollary 2.3.38. *If an abstract prearithmetic* $\boldsymbol{A} = (A; +, \circ, \leq)$ *has a multiplicative zero* 0^{m}*, contains an opposite element* $-a$ *for each element* a*, and multiplication is distributive over addition, then* $\boldsymbol{A}$ *has the property of non-zero multiplicative cancellation if and only if the product of any two non-zero elements is not equal to zero.*

When $a \circ b = 0^{\mathrm{m}}$, then a is called a *left divisor* of zero and b is called a *right divisor* of zero (cf., for example, Landau, 1999). Thus, Proposition 2.3.43 states that non-zero multiplicative cancellation from the right or from the left is equivalent to absence of divisors of zero in an abstract prearithmetic $\boldsymbol{A}$ if $\boldsymbol{A}$ has a multiplicative zero 0^{m}, opposite elements and satisfies the distributive law.

Lemma 2.3.24. *If an abstract prearithmetic* $\boldsymbol{A} = (A; +, \circ, \leq)$ *with multiplicative cancellation has the multiplicative one* 1_A*, then for any element* a *from* $\boldsymbol{A}$*,* $a \div a$ *is defined and equal to* $1_{\boldsymbol{A}}$*.*

Indeed, by Definition 2.1.5,$a \circ 1_{\boldsymbol{A}} = a$. As by Definition 2.1.10, $a \div a = 1_A$ because if $a \circ c = a$, then $c = 1_{\boldsymbol{A}}$.

Proposition 2.3.44. *If an abstract prearithmetic* $\boldsymbol{A} = (A; +, \circ, \leq)$ *with multiplicative cancellation has the multiplicative one* $1_{\boldsymbol{A}}$*, then for any element* a *from* $\boldsymbol{A}$*,* $a \div 1_{\boldsymbol{A}}$ *is defined and equal to* a*.*

Proof is left as an exercise. □

These results form an introduction into number theory in abstract prearithmetics.

Let us find algebraic properties of projectivity of abstract prearithmetics.

Proposition 2.3.45. *If an abstract prearithmetic* $\boldsymbol{A}_1 = (A_1; +_1, \circ_1, \leq_1)$ *is projective with respect to an abstract prearithmetic* $\boldsymbol{A}_2 = (A_2; +_2, \circ_2, \leq_2)$ *with the projector* g *and the coprojector* h*, then the coprojector* h *preserves addition and multiplication in the image* $g\ (\boldsymbol{A}_1)$ *of the prearithmetic* $\boldsymbol{A}_1$*, i.e.,* h *is a homomorphism of* $g(\boldsymbol{A}_1)$*.*

Proof: Let us take two elements x and y from the image $g(\boldsymbol{A}_1)$ of the prearithmetic $\boldsymbol{A}_1$. By definition, there are elements a and b in the prearithmetic $\boldsymbol{A}_1$ such that $x = g(a)$ and $y = g(b)$. Then by Definition 2.3.1, we have

$$h(x +_2 y) = h(g(a) +_2 g(b)) = a +_1 b = h(g(a)) +_1 h(g(b)) = h(x) +_1 h(y)$$

and

$$h(x \circ_2 y) = h(g(a) \circ_2 g(b)) = a \circ_1 b = h(g(a)) \circ_1 h(g(b)) = h(x) \circ_1 h(y)$$

because $x = g(a)$ and $y = g(b)$ and the composition hg is equal to $\mathbf{1}_{A_1}$. These equalities mean that h is a homomorphism of $g(\boldsymbol{A}_1)$ because x and y are arbitrary elements from $g(\boldsymbol{A}_1)$. In addition, by Lemma 2.3.2, the coprojector h preserves the order $\leq_2$.

Proposition is proved. □

Corollary 2.3.39. *If an abstract prearithmetic* $\boldsymbol{A}_1 = (A_1; +_1, \circ_1, \leq_1)$ *is projective with respect to an abstract prearithmetic* $\boldsymbol{A}_2 = (A_2; +_2, \circ_2, \leq_2)$ *with the surjective projector* g *and the coprojector* h, *then the coprojector* h *preserves addition and multiplication in the prearithmetic* $\boldsymbol{A}_2$, *i.e.,* h *is a homomorphism of* $\boldsymbol{A}_2$.

It is possible to derive additional relations between properties of abstract prearithmetics connected by the projectivity relation using properties of homomorphisms because homomorphisms preserve algebraic structures of abstract prearithmetics.

2.4. Exact Projectivity of Abstract Prearithmetics

Every project has challenges, and every project has its rewards.

Stephen Schwartz

Here we explore relations between abstract prearithmetic, which are stronger than projectivity.

Definition 2.4.1. An abstract prearithmetic $\boldsymbol{A}_1 = (A_1; +_1, \circ_1, \leq_1)$ is called *exactly projective* with respect to an abstract prearithmetic $\boldsymbol{A}_2 = (A_2; +_2, \circ_2, \leq_2)$ if $\boldsymbol{A}_1$ is projective with respect to the prearithmetic $\boldsymbol{A}_2$ and $gh = 1_{A_2}$ where g: $A_1 \to A_2$ is the *projector* and h: $A_2 \to A_1$ is the *coprojector* for the pair $(\boldsymbol{A}_1, \boldsymbol{A}_2)$.

In this case, we will say that there is *exact projectivity* between the prearithmetic $\boldsymbol{A}_1$ and the prearithmetic $\boldsymbol{A}_2$ and there is *exact inverse projectivity* between the prearithmetic $\boldsymbol{A}_2$ and the prearithmetic $\boldsymbol{A}_1$.

Example 2.4.1. The abstract prearithmetic $3\boldsymbol{W}$ of all whole numbers divisible by 3 is exactly projective with respect to the Diophantine arithmetic $\boldsymbol{W}$ with the projector $g(3n) = n$ and the coprojector $g(n) = 3n$.

Definitions imply the following results.

Lemma 2.4.1. *If an abstract prearithmetic $\boldsymbol{A}_1 = (A_1; +_1, \circ_1, \leq_1)$ is exactly projective with respect to an abstract prearithmetic $\boldsymbol{A}_2 = (A_2; +_2, \circ_2, \leq_2)$, then the prearithmetic $\boldsymbol{A}_1$ is projective with respect to a prearithmetic $\boldsymbol{A}_2$.*

Proof is left as an exercise. □

Corollary 2.4.1. *If an abstract prearithmetic $\boldsymbol{A}_1 = (A_1; +_1, \circ_1, \leq_1)$ is exactly projective with respect to a prearithmetic $\boldsymbol{A}_2 = (A_2; +_2, \circ_2, \leq_2)$, then the prearithmetic $\boldsymbol{A}_1$ is weakly projective with respect to a prearithmetic $\boldsymbol{A}_2$.*

Corollaries 2.2.5 and 2.4.1 imply the following result.

Corollary 2.4.2. *If an abstract prearithmetic $\boldsymbol{A}_1 = (A_1; +_1, \circ_1, \leq_1)$ is exactly projective with respect to an abstract prearithmetic $\boldsymbol{A}_2 = (A_2; +_2, \circ_2, \leq_2)$ and the relation $\leq_2$ is discrete, then the relation $\leq_1$ is discrete.*

Corollary 2.4.3. *If an abstract prearithmetic $\boldsymbol{A}_1 = (A_1; +_1, \circ_1, \leq_1)$ is exactly projective with respect to an abstract prearithmetic $\boldsymbol{A}_2 = (A_2; +_2, \circ_2, \leq_2)$ and any element x in $\boldsymbol{A}_2$, which is not minimal, has a predecessor Px, then any element a in $\boldsymbol{A}_1$, which is not minimal, has a predecessor Pa.*

Corollary 2.4.4. *If an abstract prearithmetic $\boldsymbol{A}_1 = (A_1; +_1, \circ_1, \leq_1)$ is exactly projective with respect to an abstract prearithmetic $\boldsymbol{A}_2 = (A_2; +_2, \circ_2, \leq_2)$ and any element x in $\boldsymbol{A}_2$, which is not maximal, has a successor Sx, then any element a in $\boldsymbol{A}_1$, which is not maximal, has a successor* $\mathrm{S}a$.

Corollary 2.4.5. *If an abstract prearithmetic $\boldsymbol{A}_1 = (A_1; +_1, \circ_1, \leq_1)$ is exactly projective with respect to an abstract prearithmetic $\boldsymbol{A}_2 = (A_2; +_2, \circ_2, \leq_2)$ and any element x in $\boldsymbol{A}_2$, which is not minimal, has the unique predecessor Px, then any element a in $\boldsymbol{A}_1$, which is not minimal, has the unique predecessor Pa.*

Corollary 2.4.6. *If an abstract prearithmetic $\boldsymbol{A}_1 = (A_1; +_1, \circ_1, \leq_1)$ is exactly projective with respect to an abstract prearithmetic $\boldsymbol{A}_2 = (A_2; +_2, \circ_2, \leq_2)$ and any element x in $\boldsymbol{A}_2$, which is not maximal, has the unique* *successor Sx, then any element a in $\boldsymbol{A}_1$, which is not maximal, has the unique successor* Sa.

Proposition 2.3.10 and Lemma 2.4.1 imply the following result.

Corollary 2.4.7. *If an abstract prearithmetic $\boldsymbol{A}_1 = (A_1; +_1, \circ_1, \leq_1)$ is exactly projective with respect to an abstract prearithmetic $\boldsymbol{A}_2 = (A_2; +_2, \circ_2, \leq_2)$ and $\leq_2$ is a linear (total) order in the prearithmetic $\boldsymbol{A}_2$, then $\leq_1$ is a linear (total) order in the prearithmetic $\boldsymbol{A}_1$.*

Properties of the category of sets (cf., for example, Herrlich and Strecker, 1973) give us the following result.

Lemma 2.4.2. *If an abstract prearithmetic $\boldsymbol{A}_1 = (A_1; +_1, \circ_1, \leq_1)$ is exactly projective with respect to an abstract prearithmetic $\boldsymbol{A}_2 = (A_2; +_2, \circ_2, \leq_2)$, then the coprojector for the pair $(\boldsymbol{A}_1, \boldsymbol{A}_2)$ is an injection and the projector for the pair $(\boldsymbol{A}_1, \boldsymbol{A}_2)$ is a projection.*

Proof is left as an exercise. □

Lemmas 2.3.3 and 2.4.2 give us the following result.

Proposition 2.4.1. *If an abstract prearithmetic $\boldsymbol{A}_1 = (A_1; +_1, \circ_1, \leq_1)$ is exactly projective with respect to an abstract prearithmetic $\boldsymbol{A}_2 = (A_2; +_2, \circ_2, \leq_2)$, then both the projector g and coprojector h for the pair $(\boldsymbol{A}_1, \boldsymbol{A}_2)$ are bijections and both sets A_1 and A_2 are equipollent (cf. Appendix).*

Proof is left as an exercise. □

Corollary 2.4.8. *If an abstract prearithmetic $\boldsymbol{A}_1 = (A_1; +_1, \circ_1, \leq_1)$ is exactly projective with respect to an abstract prearithmetic $\boldsymbol{A}_2 = (A_2; +_2, \circ_2, \leq_2)$ with the projector g and the coprojector h, then there are inverse mappings $g^{-1} = h$: $A_2 \to A_1$ and $h^{-1} = g$: $A_1 \to A_2$.*

Corollary 2.4.9. *If an abstract prearithmetic $\boldsymbol{A}_1 = (A_1; +_1, \circ_1, \leq_1)$ is exactly projective with respect to an abstract prearithmetic $\boldsymbol{A}_2 = (A_2; +_2, \circ_2, \leq_2)$ with the projector g and the coprojector h, then the prearithmetic $\boldsymbol{A}_1$ is infinite if and only if the prearithmetic $\boldsymbol{A}_2$ is infinite.*

Example 2.4.2. It is interesting that mathematicians started implicitly using a special case of exact projectivity of numerical prearithmetics and arithmetics long ago with the advent of percent, which were used in ancient Rome. Percents are used to express ratios and fractions, which are usually small, by 100 times larger numbers. This conversion makes addition and subtraction easier and the ratios more graspable.

Time and again the term *percentage* is used instead of the term *percent.* Although they are often used interchangeably, *percentage* is sometimes regarded as the more general term and "percent" as the more specific term (Knapp, 2009).

To convert from a fraction or a proportion to a percentage you multiply by 100 and add the % sign. To convert from a percentage to a proportion you delete the % sign and divide by 100.

Transformation of fractions into percent and back is a kind of exact projectivity of the arithmetic $\boldsymbol{F}$ of fractions representing rational numbers and the arithmetic $\boldsymbol{P}$ of rational numbers representing percents with the projector g: $\boldsymbol{F} \to \boldsymbol{P}$ defined as $g(a) = 100a$ and coprojector h: $\boldsymbol{P} \to \boldsymbol{F}$ defined as $h(n) = n/100$. We see that $h = g^{-1}$.

This exact projectivity does not change addition in both $\boldsymbol{F}$ and $\boldsymbol{P}$. Numbers are added according to the rules of the conventional arithmetic $\boldsymbol{Q}$ of rational numbers. However, multiplication in $\boldsymbol{P}$ is defined differently by the formula

$$n \otimes m = g(h(n) \cdot h(m)) = 100(n/100 \cdot m/100) = n/100 \cdot m.$$

This is exactly the formula of taking n percent of the number m, e.g., of taking 30% percent of \$1000 what gives us

$$30 \otimes 1000 = (30/100) \cdot 1000 = 300.$$

Percents (percentages) are used in many areas such as in finance or business.

Buchanan wrote: "The percentage is the most useful statistic ever invented$\cdots$" (Buchanan, 1974).

Proposition 2.2.10 implies the following result.

Proposition 2.4.2. *If an abstract prearithmetic* $\boldsymbol{A}_1 = (A_1; +_1, \circ_1, \leq_1)$ *is exactly projective with respect to an abstract prearithmetic* $\boldsymbol{A}_2 = (A_2; +_2, \circ_2, \leq_2)$ *with the projector* g *and the coprojector* h, *then the projector* g *preserves addition and multiplication in the prearithmetic* $\boldsymbol{A}_1$, *i.e.,* g *is a homomorphism.*

Proof is left as an exercise. □

Corollary 2.3.33 implies the following result.

Proposition 2.4.3. *If an abstract prearithmetic* $\boldsymbol{A}_1 = (A_1; +_1, \circ_1, \leq_1)$ *is exactly projective with respect to an abstract prearithmetic* $\boldsymbol{A}_2 = (A_2; +_2, \circ_2, \leq_2)$ *with the projector* g *and the coprojector* h, *then the coprojector* h

preserves addition and multiplication in the prearithmetic $\boldsymbol{A}_2$, *i.e.*, h *is a homomorphism.*

Proof: Let us take two elements x and y from the prearithmetic $\boldsymbol{A}_2$. By Proposition 2.4.1, g is a bijection. Consequently, there are elements a and b in the prearithmetic $\boldsymbol{A}_1$ such that $x = g(a)$ and $y = g(b)$. Then by definitions, we have

$$h(x +_2 y) = h(g(a) +_2 g(b)) = a +_1 b = h(g(a)) +_1 h(g(b)) = h(x) +_1 h(y)$$

and

$$h(x \circ_2 y) = h(g(a) \circ_2 g(b)) = a \circ_1 b = h(g(a)) \circ_1 h(g(b)) = h(x) \circ_1 h(y)$$

because and the composition hg is equal to $\mathbf{1}_{A_1}$.

Proposition is proved. □

Propositions 2.4.1–2.4.3 give us the following result.

Theorem 2.4.1. *An abstract prearithmetic* $\boldsymbol{A}_1 = (A_1; +_1, \circ_1, \leq_1)$ *is exactly projective with respect to an abstract prearithmetic* $\boldsymbol{A}_2 = (A_2; +_2, \circ_2, \leq_2)$ *with the projector* g *and the coprojector* h *if and only if the prearithmetics* $\boldsymbol{A}_1$ *and* $\boldsymbol{A}_2$ *are isomorphic as algebraic systems with two operations and one relation.*

Proof is left as an exercise. □

Informally, Theorem 2.4.1 means that both abstract prearithmetics $\boldsymbol{A}_1$ and $\boldsymbol{A}_2$ have the same individual algebraic structures (cf., for example, Kurosh, 1967).

That is why when an abstract prearithmetic $\boldsymbol{B}$ is exactly projective with respect to an abstract prearithmetic $\boldsymbol{A}$ with the projector g and the coprojector h, we denote $\boldsymbol{B}$ by $\boldsymbol{A}_g$ to show this relation between abstract prearithmetics $\boldsymbol{A}$ and $\boldsymbol{B}$. It is not necessary to use the coprojector h in this notation because $h = g^{-1}$.

Corollary 2.4.10. *If an abstract prearithmetic* $\boldsymbol{A}_1 = (A_1; +_1, \circ_1, \leq_1)$ *is exactly projective with respect to an abstract prearithmetic* $\boldsymbol{A}_2 = (A_2; +_2, \circ_2, \leq_2)$ *with the projector* g *and the coprojector* h, *then the prearithmetics* $\boldsymbol{A}_1$ *and* $\boldsymbol{A}_2$ *satisfy the same algebraic laws.*

Corollary 2.4.11. *If an abstract prearithmetic* $\boldsymbol{A}_1 = (A_1; +_1, \circ_1, \leq_1)$ *is exactly projective with respect to an abstract prearithmetic* $\boldsymbol{A}_2 = (A_2; +_2, \circ_2, \leq_2)$ *with the projector* g *and the coprojector* h, *then the prearithmetic* $\boldsymbol{A}_1$ *is a ring if and only if the prearithmetic* $\boldsymbol{A}_2$ *is a ring.*

Corollary 2.4.12. *If an abstract prearithmetic* $\boldsymbol{A}_1 = (A_1; +_1, \circ_1, \leq_1)$ *is exactly projective with respect to an abstract prearithmetic* $\boldsymbol{A}_2 = (A_2; +_2, \circ_2, \leq_2)$ *with the projector* g *and the coprojector* h, *then the prearithmetic* $\boldsymbol{A}_1$ *is a field if and only if the prearithmetic* $\boldsymbol{A}_2$ *is a field.*

Corollary 2.4.13. *If an abstract prearithmetic* $\boldsymbol{A}_1 = (A_1; +_1, \circ_1, \leq_1)$ *is exactly projective with respect to an abstract prearithmetic* $\boldsymbol{A}_2 = (A_2; +_2, \circ_2, \leq_2)$ *and* $\leq_2$ *is a well-ordering in the prearithmetic* $\boldsymbol{A}_2$, *then* $\leq_1$ *is a well-ordering in the prearithmetic* $\boldsymbol{A}_1$.

Remark 2.4.1. In algebra, it is assumed that there are no differences between isomorphic algebraic systems, i.e., in the setting of algebra, they are the same. With arithmetic, situation is different. We can see this from the following examples.

Example 2.4.3. Let us take a natural number k and the function $g(x) = x + k$ as the projector and the function $h(x) = x - k$ as the coprojector of the projectivity of an abstract prearithmetic $\boldsymbol{A} = (Z; \oplus, \circ, \leq)$ with respect to the Diophantine arithmetic $\boldsymbol{Z} = (Z; +, \cdot, \leq)$ of integers, which is also a prearithmetic. As $h = g^{-1}$, these prearithmetics are exactly projective with respect to one another.

Let us take $k = 3$ and consider operations in both prearithmetics. Then in the Diophantine arithmetic $\boldsymbol{Z}$, we have

$$2 + 2 = 4.$$

At the same time, in the prearithmetic $\boldsymbol{A}$, we have

$$2 \oplus 2 = ((2 + 3) + (2 + 3)) - 3 = 10 - 3 = 4 + 3 = 7.$$

We see that addition of the same numbers gives different results in these prearithmetics.

In addition, we have

$$2 \circ 2 = ((2 + 3) \cdot (2 + 3)) - 3 = 5 \cdot 5 - 3 = 25 - 3 = 22.$$

While in $\boldsymbol{Z}$, we have

$$2 \cdot 2 = 4.$$

We see that multiplication of the same numbers gives different results in prearithmetics $\boldsymbol{Z}$ and $\boldsymbol{A}$.

Example 2.4.4. Let us take a natural number k and the function $g(x) = x - k$ as the projector and the function $h(x) = x + k$ as the coprojector of the projectivity of an abstract prearithmetic $\boldsymbol{B} = (Z; \oplus, \circ, \leq)$ with respect to the Diophantine arithmetic $\boldsymbol{Z} = (Z; +, \cdot, \leq)$ of integers, which is also a prearithmetic. As $h = g^{-1}$, these prearithmetics are exactly projective with respect to one another.

Let us take $k = 3$ and consider operations in both prearithmetics. Then in the Diophantine arithmetic $\boldsymbol{Z}$, we have

$$2 + 2 = 4.$$

At the same time, in the prearithmetic $\boldsymbol{B}$, we have

$$2 \oplus 2 = ((2 - 3) + (2 - 3)) + 3 = 4 - 3 = 1.$$

We see that addition of the same numbers gives different results in these prearithmetics.

In addition, we have

$$5 \circ 5 = ((5 - 3) \cdot (5 - 3)) + 3 = 2 \cdot 2 + 3 = 4 + 3 = 7.$$

While in $\boldsymbol{Z}$, we have

$$5 \cdot 5 = 25.$$

We see that multiplication of the same numbers gives different results in prearithmetics $\boldsymbol{Z}$ and $\boldsymbol{B}$.

It is possible to ask a question: "Are not these situations artificial, while in real life, we do not have anything like this?"

It is interesting that we encounter similar situations in real life. For instance, imagine that a person C lives in a country A and has bank accounts in two countries A and B. When C makes a bank transfer from her account in the country A to her account in the country B, the first bank charges her \$10. As a result, when in the second bank, C adds \$100 from the first bank and \$100 from the second bank, she will not have \$200 but only \$190. This corresponds to the operation

$$100 + 100 = 200 - 10 = 190.$$

This is exactly how addition is performed in the prearithmetic $\boldsymbol{B}$ from Example 2.4.3 when $k = 10$.

Here is one more example from real life. Some banks give bonuses when their clients add substantial amounts to their accounts. Let us consider the case when a bank gives the bonus of \$100 for adding \$10,000 to the account. So, if a client A has \$10,000 in her account and adds \$10,000, then her account will not be equal to \$10,000 + \$10,000 = \$20,000 but she will have

$$\$10,000 + \$10,000 + \$100 = \$20,100$$

in her account. This is exactly how addition is performed in the prearithmetic $\boldsymbol{A}$ from Example 2.4.2 when $k = 100$.

In addition to financial computing, rules of non-Diophantine arithmetics and prearithmetics are now used as metaphors in business to reflect important characteristics of economic, business, and social processes. For instance, successful mergers and acquisitions are now and then reflected by the expression $1 + 1 = 3$ (cf., for example, Beechler, 2013; Jude, 2014; Kress, 2015) because this expression describes the fact that after a merger or acquisition, the sum of the two combined entities, 3, is bigger than the two parts considered individually, 1 and 1. In a similar way, the Cambridge Business English Dictionary (2011) defines that when two companies or organizations stick together, they achieve more and are more successful than if they work separately, or in other words, the merger results in $2 + 2 = 5$. In these situations, *synergy* emerges when the teamwork of two systems gives a result greater than the sum of their separate parts. That is why these expressions form the so-called synergy arithmetic (Burgin and Meissner, 2017). Naturally, unsuccessful marketing or ineffective mergers and acquisitions are described by the expressions $1 + 1 = 1$ or even by $1 + 1 = 3/4$ (Ries, 2014).

In contrast to weak projectivity and projectivity, exact projectivity is a symmetric relation.

Theorem 2.4.2. *If an abstract prearithmetic* $\boldsymbol{A}_1 = (A_1; +_1, \circ_1, \leq_1)$ *is exactly projective with respect to an abstract prearithmetic* $\boldsymbol{A}_2 = (A_2; +_2, \circ_2, \leq_1)$ *with the projector* g *and coprojector* h*, then the prearithmetic* $\boldsymbol{A}_2$ *is exactly projective with respect to a prearithmetic* $\boldsymbol{A}_1$ *with the coprojector* g *and projector* h*.*

Proof: Let us assume that an abstract prearithmetic $\boldsymbol{A}_1 = (A_1; +_1, \circ_1, \leq_1)$ is exactly projective with respect to an abstract prearithmetic $\boldsymbol{A}_2 = (A_2; +_2, \circ_2, \leq_2)$ with the projector g and coprojector h, and take arbitrary elements x and y from A_2. As the projector g is a bijection, there are elements a and b from A_1 such that $x = g(a)$ and $y = g(b)$. Then we have

$$a +_1 b = h(g(a) +_2 g(b)).$$

Consequently,

$$h^{-1}(a +_1 b) = g(a) +_2 g(b) = x +_2 y$$

and

$$x +_2 y = h^{-1}(g^{-1}(x) +_1 g^{-1}(y)) = g(h(x) +_1 h(y)).$$

In a similar way, we have

$$a \circ_1 b = h(g(a) \circ_2 g(b)).$$

Consequently,

$$h^{-1}(a \circ_1 b) = g(a) \circ_2 g(b) = x \circ_2 y$$

and

$$x \circ_2 y = h^{-1}(g^{-1}(x) \circ_1 g^{-1}(y)) = g(h(x) \circ_1 h(y)).$$

In addition, for any elements a and b from A_1, we have

$$a \leq_1 b \text{ if and only if } g(a) \leq_2 g(b).$$

Consequently, for any elements x and y from A_2, we have

$$x \leq_2 y \text{ if and only if } h(x) \leq_1 h(y) = g^{-1}(x) \leq_1 g^{-1}(y) = a \leq_1 b.$$

It means that the abstract prearithmetic $\boldsymbol{A}_2$ is exactly projective with respect to the abstract prearithmetic $\boldsymbol{A}_1$ with the coprojector g and projector h.

Theorem is proved. □

Exact projectivity is also a transitive relation.

Proposition 2.4.4. *If an abstract prearithmetic $\boldsymbol{A}_1 = (A_1; +_1, \circ_1, \leq_1)$ is exactly projective with respect to an abstract prearithmetic $\boldsymbol{A}_2 = (A_2; +_2, \circ_2, \leq_2)$ and the abstract prearithmetic $\boldsymbol{A}_2$ is exactly projective with respect to an abstract prearithmetic $\boldsymbol{A}_3 = (A_3; +_3, \circ_3, \leq_3)$, then the abstract prearithmetic $\boldsymbol{A}_1$ is exactly projective with respect to the abstract prearithmetic $\boldsymbol{A}_3$.*

Proof: Let us assume that an abstract prearithmetic $\boldsymbol{A}_1 = (A_1; +_1, \circ_1, \leq_1)$ is exactly projective with respect to an abstract prearithmetic $\boldsymbol{A}_2 = (A_2; +_2, \circ_2, \leq_2)$ with the projector g: $A_1 \to A_2$ and the coprojector h: $A_2 \to A_1$ for the pair $(\boldsymbol{A}_1, \boldsymbol{A}_2)$ and the abstract prearithmetic $\boldsymbol{A}_2 = (A_2; +_2, \circ_2, \leq_2)$ is exactly projective with respect to an abstract prearithmetic $\boldsymbol{A}_3 = (A_3; +_3, \circ_3, \leq_3)$ with the projector k: $A_2 \to A_3$ and the coprojector $l : A_3 \to A_2$ for the pair $(\boldsymbol{A}_2, \boldsymbol{A}_3)$. By Lemma 2.4.1 and Proposition 2.1.10, the prearithmetic $\boldsymbol{A}_1$ is projective with respect to the prearithmetic $\boldsymbol{A}_3$ and we need only to check the additional condition $qp = 1_{A_3}$ where $q = kg : A_1 \to A_3$ and $p = hl : A_3 \to A_1$ for functions g: $A_1 \to A_2$, h: $A_2 \to A_1$, k: $A_2 \to A_3$ and $l : A_3 \to A_2$ considered in Proposition 2.1.6.

Indeed, by the initial conditions,

$$qp = (kg)(hl) = k(gh)l = k(1_{A_2})l = kl = 1_{A_3}.$$

Thus, the prearithmetic $\boldsymbol{A}_1$ is exactly projective with respect to the prearithmetic $\boldsymbol{A}_3$ with the projector q and the coprojector p.

Proposition is proved. □

Proposition 2.4.4 allows proving the following result.

Theorem 2.4.3. *Abstract prearithmetics with exact projectivity relations form the category* **APEP** *where objects are abstract prearithmetics and morphisms are exact projectivity relations.*

Proof is left as an exercise. □

Lemma 2.4.1 shows that the category **APEP** is a wide subcategory of the category **APAP** while all morphisms in it are isomorphisms (cf. Appendix and Section 2.3, Theorem 2.3.1).

Theorem 2.4.1 allows proving the following result.

Proposition 2.4.5. *The category* **APEP** *is isomorphic to the category in which objects are abstract prearithmetics and morphisms are their isomorphisms.*

Proof is left as an exercise. □

One-to-one mappings make possible defining the structure of abstract prearithmetics in arbitrary sets.

Let us consider an arbitrary set B, a prearithmetic $\boldsymbol{A} = (A; +, \circ, \leq)$ and two functions g: $B \to A$ and h: $A \to B$ such that $hg = 1_{A_1}$ and $gh = 1_{A_2}$. Definition 2.4.1 implies the following result.

Proposition 2.4.6. *It is possible to define on B the structure of a prearithmetic $\boldsymbol{B}^g$ that is exactly projective with respect to $\boldsymbol{A}$ as a unique prearithmetic with the projector g and the coprojector h, i.e., $\boldsymbol{B}^g$ is also $\boldsymbol{A}_g$.*

Proof is similar to the proof of Proposition 2.2.1. □

Weak projectivity becomes exact projectivity when the projector g is inverse to the coprojector h.

Theorem 2.4.4. *If an abstract prearithmetic $\boldsymbol{A}_1 = (A_1; +_1, \circ_1, \leq_1)$ with a linear order $\leq_1$ is weakly projective with respect to an abstract prearithmetic $\boldsymbol{A}_2 = (A_2; +_2, \circ_2, \leq_2)$ with the projector g and the coprojector h and $h = g^{-1}$, then the prearithmetic $\boldsymbol{A}_1$ is exactly projective with respect to the prearithmetic $\boldsymbol{A}_2$.*

Proof is left as an exercise. □

Corollary 2.4.14. *If an abstract prearithmetic $\boldsymbol{A}_1 = (A_1; +_1, \circ_1, \leq_1)$ is weakly projective with respect to an abstract prearithmetic $\boldsymbol{A}_2 = (A_2; +_2, \circ_2, \leq_2)$ with the projector g and the coprojector h and $h = g^{-1}$, then the prearithmetic $\boldsymbol{A}_1$ is isomorphic to the prearithmetic $\boldsymbol{A}_2$ as a universal algebra with two operations.*

Corollary 2.4.15. *If an abstract prearithmetic $\boldsymbol{A}_1 = (A_1; +_1, \circ_1, \leq_1)$ is exactly projective with respect to an abstract prearithmetic $\boldsymbol{A}_2 = (A_2; +_2, \circ_2, \leq_2)$ and $\leq_1$ is a well-ordering in the prearithmetic $\boldsymbol{A}_2$, then $\leq_2$ is a well-ordering in the prearithmetic $\boldsymbol{A}_1$.*

Corollary 2.4.16. *If an abstract prearithmetic $\boldsymbol{A}_1 = (A_1; +_1, \circ_1, \leq_1)$ is exactly projective with respect to an abstract prearithmetic $\boldsymbol{A}_2 = (A_2; +_2, \circ_2, \leq_2)$ and the relation $\leq_1$ is discrete, then the relation $\leq_2$ is discrete.*

Corollary 2.4.17. *If an abstract prearithmetic $\boldsymbol{A}_1 = (A_1; +_1, \circ_1, \leq_1)$ is exactly projective with respect to an abstract prearithmetic $\boldsymbol{A}_2 = (A_2; +_2, \circ_2, \leq_2)$ and any element x in $\boldsymbol{A}_1$, which is not minimal, has a predecessor Px, then any element a in $\boldsymbol{A}_2$, which is not minimal, has a predecessor Pa.*

Corollary 2.4.18. *If an abstract prearithmetic $\boldsymbol{A}_1 = (A_1; +_1, \circ_1, \leq_1)$ is exactly projective with respect to an abstract prearithmetic $\boldsymbol{A}_2 = (A_2; +_2, \circ_2, \leq_2)$ and any element x in $\boldsymbol{A}_1$, which is not maximal, has a successor Sx, then any element a in $\boldsymbol{A}_2$, which is not maximal, has a successor* Sa.

Corollary 2.4.19. *If an abstract prearithmetic* $\boldsymbol{A}_1 = (A_1; +_1, \circ_1, \leq_1)$ *is exactly projective with respect to an abstract prearithmetic* $\boldsymbol{A}_2 = (A_2; +_2, \circ_2, \leq_2)$ *and any element* x *in* $\boldsymbol{A}_1$*, which is not minimal, has the unique predecessor* Px*, then any element* a *in* $\boldsymbol{A}_2$*, which is not minimal, has the unique predecessor* Pa*.*

Corollary 2.4.20. *If an abstract prearithmetic* $\boldsymbol{A}_1 = (A_1; +_1, \circ_1, \leq_1)$ *is exactly projective with respect to an abstract prearithmetic* $\boldsymbol{A}_2 = (A_2; +_2, \circ_2, \leq_2)$ *and any element* x *in* $\boldsymbol{A}_1$*, which is not maximal, has the unique successor* Sx*, then any element* a *in* $\boldsymbol{A}_2$*, which is not maximal, has the unique successor* Sa*.*

Corollary 2.4.21. *If an abstract prearithmetic* $\boldsymbol{A}_1 = (A_1; +_1, \circ_1, \leq_1)$ *is exactly projective with respect to an abstract prearithmetic* $\boldsymbol{A}_2 = (A_2; +_2, \circ_2, \leq_2)$ *and* $\leq_2$ *is a linear (total) order in the prearithmetic* $\boldsymbol{A}_1$*, then* $\leq_1$ *is a linear (total) order in the prearithmetic* $\boldsymbol{A}_2$*.*

Some abstract prearithmetics do not have other exactly projective abstract prearithmetics.

Theorem 2.4.5. (*a*) *If an abstract prearithmetic* $\boldsymbol{A} = (A; \oplus, \otimes, \leq)$ *is exactly projective with respect to the Diophantine arithmetic* $\boldsymbol{W}$*, then the prearithmetic* $\boldsymbol{A}$ *coincides with* $\boldsymbol{W}$ *up to renaming its elements.*

(*b*) *If an abstract prearithmetic* $\boldsymbol{A} = (A; \oplus, \otimes, \leq)$ *is exactly projective with respect to the Diophantine arithmetic* $\boldsymbol{N}$*, then the prearithmetic* $\boldsymbol{A}$ *coincides with* $\boldsymbol{N}$ *up to renaming its elements.*

Proof: (a) If an abstract prearithmetic $\boldsymbol{A} = (A; \oplus, \otimes, \leq_A)$ is exactly projective with respect to the Diophantine arithmetic $\boldsymbol{W}$, then there is a one-to-one mapping g: $A \to W$. Let us denote by n_A the element a from $\boldsymbol{A}$ such that $g(a) = n$ for all n from $\boldsymbol{N}$. As g is a one-to-one mapping, all elements from $\boldsymbol{A}$ are uniquely enumerated in such a way. As by Proposition 2.4.2, g is a homomorphism, we have

$$n_A \oplus m_A = (n + m)_A,$$

$$n_A \otimes m_A = (n \cdot m)_A,$$

$$m_A \leq_A n_A \text{ if and only if } m \leq n.$$

Consequently, 0_A is the zero in $\boldsymbol{A}$, 1_A is the one in $\boldsymbol{A}$ and any element n_A in $\boldsymbol{A}$ is a sum of ones. Changing the name n_A by the name n for all whole numbers n, we see that with these names $\boldsymbol{A}$ coincides with $\boldsymbol{W}$. Thus, the statement (a) is proved.

The proof for the statement (b) is similar.

Theorem is proved. □

Corollary 2.4.22. *(a) If an abstract prearithmetic* $\boldsymbol{A} = (W; \oplus, \otimes, \leq)$ *with the natural order on* W *is exactly projective with respect to the Diophantine arithmetic* $\boldsymbol{W}$*, then the prearithmetic* $\boldsymbol{A}$ *coincides with* $\boldsymbol{W}$*.*

(b) If an abstract prearithmetic $\boldsymbol{A} = (N; \oplus, \otimes, \leq)$ *with the natural order on* N *is exactly projective with respect to the Diophantine arithmetic* $\boldsymbol{N}$*, then the prearithmetic* $\boldsymbol{A}$ *coincides with* $\boldsymbol{N}$*.*

Indeed, because the projector $g: W \to W$ preserves order, 0 from $\boldsymbol{A}$ has to be mapped into 0 from $\boldsymbol{W}$. Otherwise, g is not a one-to-one mapping contrary to the definition of exact projectivity. By induction, we can show that g maps any number n from $\boldsymbol{A}$ into the same number n from $\boldsymbol{W}$. Thus, $\boldsymbol{A}$ coincides with $\boldsymbol{W}$.

The same argument holds for the Diophantine arithmetic $\boldsymbol{N}$.

As exact projectivity preserves algebraic properties of mathematical structures, we have the following results.

Proposition 2.4.7. *(a) An abstract prearithmetic* $\boldsymbol{A} = (A; +, \circ, \leq)$ *that is exactly projective with respect to an additive (multiplicative) semigroup is an additive (multiplicative) semigroup.*

(b) An abstract prearithmetic $\boldsymbol{A} = (A; +, \circ, \leq)$ *that is exactly projective with respect to a commutative additive (multiplicative) semigroup is a commutative additive (multiplicative) semigroup.*

Proof is left as an exercise. □

Corollary 2.4.23. *(a) If an additive (multiplicative) semigroup is exactly projective with respect to an abstract prearithmetic* $\boldsymbol{A} = (A; +, \circ, \leq)$*, then* $\boldsymbol{A}$ *is an additive (multiplicative) semigroup.*

(b) If a commutative additive (multiplicative) semigroup is exactly projective with respect to an abstract prearithmetic $\boldsymbol{A} = (A; +, \circ, \leq)$*, then* $\boldsymbol{A}$ *is a commutative additive (multiplicative) semigroup.*

Proposition 2.4.8. *(a) An abstract prearithmetic* $\boldsymbol{A} = (A; +, \circ, \leq)$ *that is exactly projective with respect to an additive (multiplicative) group is an additive (multiplicative) group.*

(b) An abstract prearithmetic $\boldsymbol{A} = (A; +, \circ, \leq)$ *that is exactly projective with respect to a commutative additive (multiplicative) group is a commutative additive (multiplicative) group.*

Proof is left as an exercise. □

Corollary 2.4.24. *(a) If an additive (multiplicative) group is exactly projective with respect to an abstract prearithmetic* $\boldsymbol{A} = (A; +, \circ, \leq)$, *then* $\boldsymbol{A}$ *is an additive (multiplicative) group.*

(b) If a commutative additive (multiplicative) group is exactly projective with respect to an abstract prearithmetic $\boldsymbol{A} = (A; +, \circ, \leq)$, *then* $\boldsymbol{A}$ *is a commutative additive (multiplicative) group.*

Proposition 2.4.9. *(a) An abstract prearithmetic* $\boldsymbol{A} = (A; +, \circ, \leq)$ *that is exactly projective with respect to a ring is a ring.*

(b) An abstract prearithmetic $\boldsymbol{A} = (A; +, \circ, \leq)$ *that is exactly projective with respect to a commutative ring is a commutative ring.*

(c) An abstract prearithmetic $\boldsymbol{A} = (A; +, \circ, \leq)$ *that is exactly projective with respect to an associative ring is an associative ring.*

Proof is left as an exercise. □

Corollary 2.4.25. *(a) If a ring is exactly projective with respect to an abstract prearithmetic* $\boldsymbol{A} = (A; +, \circ, \leq)$, *then* $\boldsymbol{A}$ *is a ring.*

(b) If a commutative ring is exactly projective with respect to an abstract prearithmetic $\boldsymbol{A} = (A; +, \circ, \leq)$, *then* $\boldsymbol{A}$ *is a commutative ring.*

(c) If an associative ring is exactly projective with respect to an abstract prearithmetic $\boldsymbol{A} = (A; +, \circ, \leq)$, *then* $\boldsymbol{A}$ *is an associative ring.*

Proposition 2.4.10. *(a) An abstract prearithmetic* $\boldsymbol{A} = (A; +, \circ, \leq)$ *that is exactly projective with respect to an ordered additive (multiplicative) group is an ordered additive (multiplicative) group.*

(b) An abstract prearithmetic $\boldsymbol{A} = (A; +, \circ, \leq)$ *that is exactly projective with respect to an ordered commutative additive (multiplicative) group is an ordered commutative additive (multiplicative) group.*

Proof is left as an exercise. □

Corollary 2.4.26. *(a) If an ordered additive (multiplicative) group is exactly projective with respect to an abstract prearithmetic* $\boldsymbol{A} = (A; +, \circ, \leq)$, *then* $\boldsymbol{A}$ *is an ordered additive (multiplicative) group.*

(b) If an ordered commutative additive (multiplicative) group is exactly projective with respect to an abstract prearithmetic $\boldsymbol{A} = (A; +, \circ, \leq)$, *then* $\boldsymbol{A}$ *is an ordered commutative additive (multiplicative) group.*

Proposition 2.4.11. *(a) An abstract prearithmetic* $\boldsymbol{A} = (A; +, \circ, \leq)$ *that is exactly projective with respect to an ordered ring is an ordered ring.*

(b) An abstract prearithmetic $\boldsymbol{A} = (A; +, \circ, \leq)$ *that is exactly projective with respect to an ordered commutative ring is an ordered commutative ring.*

(c) *An abstract prearithmetic* $\boldsymbol{A} = (A; +, \circ, \leq)$ *that is exactly projective with respect to an ordered associative ring is an ordered associative ring.*

Proof is left as an exercise. □

Corollary 2.4.27. (a) *If an ordered ring is exactly projective with respect to an abstract prearithmetic* $\boldsymbol{A} = (A; +, \circ, \leq)$, *then* $\boldsymbol{A}$ *is an ordered ring.*

(b) *If an ad ordered commutative ring is exactly projective with respect to an abstract prearithmetic* $\boldsymbol{A} = (A; +, \circ, \leq)$, *then* $\boldsymbol{A}$ *is an ordered commutative ring.*

(c) *If an ordered associative ring is exactly projective with respect to an abstract prearithmetic* $\boldsymbol{A} = (A; +, \circ, \leq)$, *then* $\boldsymbol{A}$ *is an ordered associative ring.*

Proposition 2.4.12. *An abstract prearithmetic* $\boldsymbol{A} = (A; +, \circ, \leq)$ *that is exactly projective with respect to a field is a field.*

Proof is left as an exercise. □

Corollary 2.4.28. *If a field is exactly projective with respect to an abstract prearithmetic* $\boldsymbol{A} = (A; +, \circ, \leq)$, *then* $\boldsymbol{A}$ *is a field.*

Proposition 2.4.13. *An abstract prearithmetic* $\boldsymbol{A} = (A; +, \circ, \leq)$ *that is exactly projective with respect to an ordered field is an ordered field.*

Proof is left as an exercise. □

Corollary 2.4.29. *If an ordered field is exactly projective with respect to an abstract prearithmetic* $\boldsymbol{A} = (A; +, \circ, \leq)$, *then* $\boldsymbol{A}$ *is an ordered field.*

Proposition 2.4.14. *An abstract prearithmetic* $\boldsymbol{A} = (A; +, \circ, \leq)$ *that is exactly projective with respect to a lattice is a lattice.*

Proof is left as an exercise. □

Corollary 2.4.30. *If a lattice is exactly projective with respect to an abstract prearithmetic* $\boldsymbol{A} = (A; +, \circ, \leq)$, *then* $\boldsymbol{A}$ *is a lattice.*

Proposition 2.4.15. *If an abstract prearithmetic* $\boldsymbol{A}_1 = (A_1; +_1, \circ_1, \leq_1)$ *is exactly projective with respect to an abstract prearithmetic* $\boldsymbol{A}_2 = (A_2; +_2, \circ_2, \leq_2)$ *and an abstract prearithmetic* $\boldsymbol{A}_3 = (A_3;\ +_3,\ \circ_3,\ \leq_3)$ *is exactly projective with respect to the abstract prearithmetic* $\boldsymbol{A}_2$, *then the abstract prearithmetic* $\boldsymbol{A}_1$ *is exactly projective with respect to the abstract prearithmetic* $\boldsymbol{A}_3$.

Proof: Let us assume that an abstract prearithmetic $\boldsymbol{A}_1 = (A_1; +_1, \circ_1, \leq_1)$ is exactly projective with respect to an abstract prearithmetic $\boldsymbol{A}_2 = (A_2; +_2, \circ_2, \leq_2)$ with the projector g: $A_1 \to A_2$ and the coprojector h: $A_2 \to A_1$ for the pair $(\boldsymbol{A}_1, \boldsymbol{A}_2)$ and an abstract prearithmetic $\boldsymbol{A}_3 = (A_3; +_3, \circ_3, \leq_3)$ is exactly projective with respect to the abstract prearithmetic $\boldsymbol{A}_2$ with the projector k: $A_3 \to A_2$ and the coprojector $l : A_2 \to A_3$ for the pair $(\boldsymbol{A}_3, \boldsymbol{A}_2)$. By Proposition 2.4.1, mappings k and l have inverse mappings k^{-1} and l^{-1}. This allows us to define the mappings $k^{-1}g$: $A_1 \to A_3$ and $hl^{-1} : A_3 \to A_1$. Then abstract prearithmetic $\boldsymbol{A}_1 = (A_1; +_1, \circ_1, \leq_1)$ is exactly projective with respect to an abstract prearithmetic $\boldsymbol{A}_3 = (A_3; +_3, \circ_3, \leq_3)$ with the projector $k^{-1}g$: $A_1 \to A_3$ and the coprojector $hl^{-1} : A_3 \to A_1$. Indeed, for any elements a and b from $\boldsymbol{A}_1$, we have the following relations:

$$\begin{aligned} a +_1 b &= h(g(a) +_2 g(b)) = h(l^{-1}(k^{-1}(g(a)) +_3 k^{-1}(g(b)))), \\ &= (hl^{-1})((k^{-1}g)(a) +_3 (k^{-1}g)(b)), \\ a \circ_1 b &= h(g(a) \circ_2 g(b)) = h(l^{-1}(k^{-1}(g(a)) \circ_3 k^{-1}(g(b)))) \\ &= (hl^{-1})((k^{-1}g)(a) \circ_3 (k^{-1}g)(b)) \end{aligned}$$

because

$$a +_1 b = h(g(a) +_2 g(b)),$$
$$a \circ_1 b = h(g(a) \circ_2 g(b))$$

and

$$x +_3 y = l(k(x) +_2 k(y)) = x \circ_3 b = l(k(x) \circ_2 g(y))$$

implies

$$l^{-1}(x +_3 y) = k(x) +_2 k(y),$$
$$l^{-1}(x \circ_3 b) = k(x) \circ_2 g(y).$$

As by Proposition 2.4.1, we have equalities $h = g^{-1}$ and $l = k^{-1}$, then

$$hl^{-1} = g^{-1}k = (k^{-1}g)^{-1}$$

Thus, by Theorem 2.4.4, the abstract prearithmetic $\boldsymbol{A}_1$ is exactly projective with respect to the abstract prearithmetic $\boldsymbol{A}_3$.

Proposition is proved. □

2.5. Subprearithmetics

Nothing is thoroughly approved but mediocrity.
The majority has established this, and
it fixes its fangs on whatever gets beyond it either way.

Blaise Pascal

An important structure of the theory of non-Diophantine arithmetics is a subprearithmetic.

Definition 2.5.1.

(a) An abstract prearithmetic $\boldsymbol{A}_1 = (A_1;\ +_1,\ \circ_1,\ \leq_1)$ is a *subprearithmetic* of an abstract prearithmetic $\boldsymbol{A}_2 = (A_2;\ +_2,\ \circ_2,\ \leq_2)$ if $A_1 \subseteq A_2$, the operation $+_1$ is the restriction of the operation $+_2$ onto A_1, the operation $\circ_1$ is the restriction of the operation $\circ_2$ onto A_1, and the relation $\leq_1$ is the restriction of the relation $\leq_2$ onto A_1.

(b) If a subprearithmetic $\boldsymbol{A}_1$ of $\boldsymbol{A}_2$ is an arithmetic, then it is a *subarithmetic* of $\boldsymbol{A}_2$.

Example 2.5.1. The Diophantine arithmetic $\boldsymbol{N}$ of all natural numbers is a subprearithmetic of the Diophantine arithmetic $\boldsymbol{W}$ of all whole numbers.

Example 2.5.2. The Diophantine arithmetic $\boldsymbol{W}$ of all whole numbers is a subprearithmetic of the conventional arithmetic $\boldsymbol{Z}$ of all integer numbers.

Example 2.5.3. The conventional Diophantine arithmetic $\boldsymbol{Z}$ of all integer numbers is a subprearithmetic of the conventional arithmetic $\boldsymbol{Q}$ of all rational numbers.

Example 2.5.4. The conventional Diophantine arithmetic $\boldsymbol{Q}$ of all rational numbers is a subprearithmetic of the conventional arithmetic $\boldsymbol{R}$ of all real numbers.

Example 2.5.5. The conventional Diophantine arithmetic $\boldsymbol{R}$ of all real numbers is a subprearithmetic of the conventional arithmetic $\boldsymbol{C}$ of all complex numbers.

Example 2.5.6. The arithmetic $3\boldsymbol{W}$ of all whole numbers divisible by 3 with the standard arithmetical operations and order is a subprearithmetic of the Diophantine arithmetic $\boldsymbol{W}$.

Remark 2.5.1. It is possible (cf., for example, Yershov and Palyutin, 1979) to treat the conventional (Diophantine) arithmetic as the universal algebra $\boldsymbol{NA} = (N; +, \cdot, \underline{1}, \underline{0})$ where there are two binary operations of addition and multiplication and two null-ary operations $\underline{1}$ and $\underline{0}$ of taking the number 1 and taking the number 0. Operations + and · are commutative and associative while multiplication is distributive over addition. Such a universal algebra is called a semiring.

The universal algebra $\boldsymbol{NA}$ does not have subalgebras. At the same time, Diophantine arithmetic $\boldsymbol{N}$ has subprearithmetics. For instance, it has the arithmetic $\boldsymbol{EN} = (E; +, \cdot, \leq)$ of all even numbers as its subprearithmetic. In it, E is the set of even numbers, + is the conventional addition of numbers, · is the conventional multiplication and $\leq$ is the conventional order, e.g., $4 \leq 8$. This shows that subprearithmetics are not always subalgebras of the prearithmetics to which they belong.

2.5.1. *Operations with and properties of subprearithmetics*

The pleasure we obtain from music comes from counting,
but counting unconsciously.
Music is nothing but unconscious arithmetic.
Gottfried Wilhelm Leibniz

It is possible to perform set-theoretical operations with subprearithmetics. Subprearithmetics are closed with respect to some of these operations.

Proposition 2.5.1. *The set-theoretical intersection of subprearithmetics of an abstract prearithmetic* $\boldsymbol{A}$ *is a subprearithmetic of* $\boldsymbol{A}$.

Proof is left as an exercise. □

It means that the set-theoretical intersection of subprearithmetics is the algebraic intersection of these subprearithmetics. Proposition 2.5.1 also shows how it is possible to build subprearithmetics.

Properties of sets (Abian, 1965) imply the following result.

Proposition 2.5.2. *Intersection of subprearithmetics is a commutative and associative operation.*

Proof is left as an exercise. □

Lemma 2.5.1. *If* $\boldsymbol{A} = (A;\ +, \circ, \leq)$ *is an abstract prearithmetic and* $C \subseteq A$*, then there is the least subprearithmetic* $\mathrm{sa}_C \boldsymbol{A} = (B;\ +, \circ, \leq)$ *with* $C \subseteq B \subseteq A$.

Proof is left as an exercise. □

Thus, the subprearithmetic $\mathrm{sa}_C\ \boldsymbol{A}$ is defined by three conditions:

(1) Any element from C belongs to B.
(2) If elements a and b belong to B, then $a + b$ belongs to B.
(3) If elements a and b belong to B, then $a \circ b$ belongs to B.
(4) There are no other elements in B.

We say that the prearithmetic $\mathrm{sa}_C\ \boldsymbol{A} = (B;\ +,\ \circ,\ \leq)$ is *generated* by the set C in $\boldsymbol{A}$ or that $\mathrm{sa}_C\ \boldsymbol{A} = (B;\ +,\ \circ, \leq)$ is the *arithmetical completion* of C. By construction, $\mathrm{sa}_C\ \boldsymbol{A} = (B;\ +, \circ,\ \leq)$ consists of all sums and products of elements from X.

For instance, if we take the set $\{1\}$ that has only one element 1, then its arithmetical completion in the Diophantine arithmetic $\boldsymbol{R}$ of all real numbers is the Diophantine arithmetic $\boldsymbol{N}$ of all natural numbers because in $\boldsymbol{R}$, any natural number is the sum of some quantity of the number 1. The arithmetical completion in $\boldsymbol{R}$ of the set $\{0, 1\}$ is the Diophantine arithmetic $\boldsymbol{W}$ of all whole numbers.

Lemma 2.5.1 allows defining unions of subprearithmetics. Namely, if $\boldsymbol{B} = (B;\ +,\ \circ, \leq)$ and $\boldsymbol{C} = (C;\ +, \circ, \leq)$ are subprearithmetics of an abstract prearithmetic $\boldsymbol{A}$, then $\boldsymbol{B} \cup \boldsymbol{C} = \mathrm{sa}_{B \cup C} \boldsymbol{A} = (B \cup\ C;\ +,\ \circ,\ \leq)$.

Proposition 2.5.3. *The union of subprearithmetics is a commutative and associative operation.*

Proof is left as an exercise. □

Subprearithmetics inherit many properties of the abstract prearithmetics, which include them. This depends on the logical expression of the properties. Namely, we discern properties represented by logical formulas that contain only universal quantifiers. For instance, the formula

$$\forall x\ \forall y\ (xy = yx)$$

contains only universal quantifiers while the formula

$$\exists x\ \forall y\ (y \leq x)$$

also contains the existential quantifier $\exists$.

In this setting, we have the following results.

Proposition 2.5.4. *All properties of abstract prearithmetics are preserved in their subprearithmetics if the logical formula representing this property contains only universal quantifiers.*

Indeed, if some statement is true for all elements from an abstract prearithmetic $\boldsymbol{A}_1$, then it is also true for all elements from its arbitrary subprearithmetic.

This statement allows us to obtain the following results.

Proposition 2.5.5. *The following order properties are inherited by subprearithmetics:*

(a) *the linear (total) order;*
(b) *well-ordering;*
(c) *the successively Archimedean property when any element a in the subprearithmetic has the successor* $\mathrm{S}a$;
(d) *finite connectedness of the order relation;*
(e) *preservation of order by addition;*
(f) *preservation of order by multiplication;*
(g) *monotonicity of addition;*
(h) *monotonicity of multiplication;*
(i) *ordered additive cancellation property;*
(j) *ordered additive cancellation property from the right (from the left);*
(k) *ordered multiplicative cancellation property;*
(l) *ordered multiplicative cancellation property from the right (from the left);*
(m) *strict additive cancellation property;*
(n) *strict additive cancellation property from the right (from the left);*
(o) *strict multiplicative cancellation property;*
(p) *strict multiplicative cancellation property from the right (from the left).*

Proof: Using Proposition 2.5.4, we check inheritance.

(a) Let us assume that order $\leq_2$ in an abstract prearithmetic $\boldsymbol{A}_2 = (A_2; +_2, \circ_2, \leq_2)$ is total and an abstract prearithmetic $\boldsymbol{A}_1 = (A_1; +_1, \circ_1, \leq_1)$ is a subprearithmetic of an abstract prearithmetic $\boldsymbol{A}_2$. Because order $\leq_1$ is the restriction of the order $\leq_2$, relation $\leq_1$ is defined for any two elements from $\boldsymbol{A}_2$, i.e., order $\leq_1$ is also total.
(b) Let us assume that order $\leq_2$ in an abstract prearithmetic $\boldsymbol{A}_2 = (A_2; +_2, \circ_2, \leq_2)$ is a well-ordering and an abstract prearithmetic $\boldsymbol{A}_1 = (A_1; +_1, \circ_1, \leq_1)$ is a subprearithmetic of an abstract prearithmetic $\boldsymbol{A}_2$. Well-ordering means that any subset of A_2 has a minimal element (cf. Appendix).

Because $A_1 \subseteq A_2$ and order $\leq_1$ is the restriction of the order $\leq_2$, any subset X of A_1 is also a subset of A_2 and thus, has a minimal element.

(c) Let an abstract prearithmetic $\boldsymbol{A}_1 = (A_1; +_1, \circ_1, \leq_1)$ be a subprearithmetic of an abstract prearithmetic $\boldsymbol{A}_2$ with the successively Archimedean property and any element a in the subprearithmetic $\boldsymbol{A}_1$ has the successor $\mathrm{S}_1 a$. As $a \leq_1 \mathrm{S}_1 a$, we have $a \leq_2 \mathrm{S}_1 a$ and $a \leq_2 \mathrm{S}a \leq_2 \mathrm{S}_1 a$. Iterating the operations S and S_1, we obtain the inequality $\mathrm{S}^n a \leq_2 \mathrm{S}_1^n a$ for arbitrary natural number n.

If a, $b \in A_1$ and $a < b$, then there is a natural number n such that $b \leq_2 \mathrm{S}^n a$ as the prearithmetic $\boldsymbol{A}_2$ has the successively Archimedean property. Consequently, $b \leq_2 \mathrm{S}_1^n a$ and as a and b are arbitrary elements from A_1, the prearithmetic $\boldsymbol{A}_1$ has the successively Archimedean property.

(d) Let an abstract prearithmetic $\boldsymbol{A}_1 = (A_1; +_1, \circ_1, \leq_1)$ is a subprearithmetic of an abstract prearithmetic $\boldsymbol{A}_2$.

We remind set X with a partial order $\leq$ is called finitely connected if for any elements a and b from X, the inequality $a < b$ implies that any chain $a < a_1 < a_2 < \cdots < a_n < \cdots < b$ between elements a and b is finite (cf. Section 2.1). As any chain in $\boldsymbol{A}_1$ is also chain in $\boldsymbol{A}_2$, then $\boldsymbol{A}_1$ cannot have infinite chains between two elements when there are no infinite chains between two elements in $\boldsymbol{A}_2$.

(e) Indeed, because the order and addition in a subprearithmetic $\boldsymbol{A}_1$ of an abstract prearithmetic $\boldsymbol{A}_2$ are restrictions of the order and addition in the prearithmetic $\boldsymbol{A}_2$, which contains $\boldsymbol{A}_1$, if addition preserves order in $\boldsymbol{A}_2$, then it preserves order in $\boldsymbol{A}_1$.

(f) Indeed, because the order and multiplication in a subprearithmetic $\boldsymbol{A}_1$ of an abstract prearithmetic $\boldsymbol{A}_2$ are restrictions of the order and multiplication in the prearithmetic $\boldsymbol{A}_2$, which contains $\boldsymbol{A}_1$, if multiplication preserves order in $\boldsymbol{A}_2$, then it preserves order in $\boldsymbol{A}_1$.

Lemma 2.1.6 and (e) imply (g).

Lemma 2.1.7 and (f) imply (h).

(i)–(p) If an abstract prearithmetic $\boldsymbol{A}_1 = (A_1; +_1, \circ_1, \leq_1)$ is a subprearithmetic of an abstract prearithmetic $\boldsymbol{A}_2 = (A_2; +_2, \circ_2, \leq_2)$, then it is possible to cancel in $\boldsymbol{A}_2$, therefore so more in $\boldsymbol{A}_1$ because $A_1 \subseteq A_2$.

Proposition is proved. □

However, subprearithmetics do not always inherit all properties of the abstract prearithmetics that include them. In particular, they do not, as a rule, inherit properties represented by logical formulas that contain existential quantifiers.

Proposition 2.5.6. *The following order properties are not always inherited by subprearithmetics:*

(a) *discreteness of the order relation;*
(b) *having a successor for all elements;*
(c) *having a predecessor for all elements;*
(d) *having the least element;*
(e) *having the largest element;*
(f) *having maximal elements;*
(g) *having minimal elements;*
(h) *the exact successively Archimedean property even if any element a in the subprearithmetic has the successor* $\mathrm{S}a$.

Proof: To show that these statements are true, it is sufficient to find such a prearithmetic, which has one of these properties but some subprearithmetic of which does not have the corresponding property.

(a) & (b) Let us consider the set $A = \{0, 1, 1/2–(1/2)^{n+1}, 1/2 + (1/2)^{n+1}; n = 1, 2, 3, \cdots\}$ with the natural order of numbers and define operations in such a way that any sum and any product of its elements are equal to 0. It gives us an abstract prearithmetic $\boldsymbol{A}$. Then the order in A and, thus, in the constructed prearithmetic $\boldsymbol{A}$ is discrete because any element a in it but 1 has the successor $\mathrm{S}a$ and any element a in it but 0 has the predecessor $\mathrm{P}a$.

At the same time, in the subprearithmetic $\boldsymbol{D}$ of $\boldsymbol{A}$ generated by the set $D = \{0, 1, 1/2 + (1/2)^{n+1}; n = 1, 2, 3, \cdots\}$ the order is not discrete because the element 0 does not have the successor S0 although $0 < 1$.

(c) In the subprearithmetic $\boldsymbol{C}$ of $\boldsymbol{A}$ generated by the set $C = \{0, 1, 1/2–(1/2)^{n+1}; n = 1, 2, 3, \cdots\}$ the element 1 does not have the predecessor P1 although $0 < 1$.

(d) The prearithmetic $\boldsymbol{A}$ has the least element 0 while the subprearithmetic $\boldsymbol{H}$ of $\boldsymbol{A}$ generated by the set $C = \{1/2 + (1/2)^{n+1}; n = 1, 2, 3, \cdots\}$ does not have the least element.

(e) The prearithmetic $\boldsymbol{A}$ has the largest element 1 while the subprearithmetic $\boldsymbol{H}$ of $\boldsymbol{A}$ generated by the set $C = \{1/2–(1/2)^{n+1}; n = 1, 2, 3, \cdots\}$ does not have the largest element.

(f) The prearithmetic $\boldsymbol{A}$ has the maximal element 1 while the subprearithmetic $\boldsymbol{H}$ of $\boldsymbol{A}$ generated by the set $C = \{1/2 - (1/2)^{n+1}; n = 1, 2, 3, \cdots\}$ does not have maximal elements.

(g) The prearithmetic $\boldsymbol{A}$ has the minimal element 0 while the subprearithmetic $\boldsymbol{H}$ of $\boldsymbol{A}$ generated by the set $C = \{1/2 + (1/2)^{n+1}; n = 1, 2, 3, \cdots\}$ does not have minimal elements.

(h) Let us consider the subprearithmetic $\boldsymbol{A}$ of the Diophantine arithmetic $\boldsymbol{N}$ that is generated by numbers 2 and 7. Elements in this subprearithmetic $\boldsymbol{A}$ are defined recursively by applying addition and multiplication to numbers 2, 7 and results of the previous applications. The order in $\boldsymbol{A}$ is the restriction of the order in $\boldsymbol{N}$.

Then it is easy to see that although each element a in $\boldsymbol{A}$ has the successor $\mathrm{S}_A a$ in $\boldsymbol{A}$ and $2 < 7$, there is no natural number n such that $\mathrm{S}_A 2 = 7$. It means that $\boldsymbol{A}$ does not have the exact successively Archimedean property.

Proposition is proved. □

Now let us look at properties of addition and figure out which of them are inherited by subprearithmetics.

Proposition 2.5.7. *The following properties of addition are inherited by subprearithmetics*:

(a) *commutativity of addition*;
(b) *associativity of addition*;
(c) *identities* $a \leq a + a$ *and* $a \leq a + b$ *or* $a + a \leq a$ *and* $a + b \leq a$;
(d) *preservation of order by addition*;
(e) *monotonicity of addition*;
(f) *additive cancellation property*;
(g) *additive cancellation property from the right (from the left)*;
(h) *ordered additive cancellation property*;
(i) *ordered additive cancellation property from the right (from the left)*;
(j) *strict additive cancellation property*;
(k) *strict additive cancellation property from the right (from the left)*;
(l) *the additively Archimedean property when any element* a *in the subprearithmetic has the successor* $\mathrm{S}a$.

Proof is similar to the proof of Proposition 2.5.5 and left as an exercise. □

There are definite relations between order and addition in abstract prearithmetics.

Definition 2.5.2.

(a) A successor $\mathrm{S}x$ of an element x is called *arithmetical* if $\mathrm{S}x = x + 1$ and it is denoted by $\mathrm{AS}x$.
(b) A predecessor $\mathrm{P}x$ of an element x is called *arithmetical* if $\mathrm{P}x = x - 1$, i.e., $x = \mathrm{P}x + 1$, and it is denoted by $\mathrm{AP}x$.

As we know, in the Diophantine arithmetics $\boldsymbol{N}$ and $\boldsymbol{W}$, all successors and predecessors are arithmetical. However, for arbitrary abstract prearithmetics, this is not always true as the following result demonstrates.

Proposition 2.5.8. *There is an abstract prearithmetic, in which successors exist for all elements but there are no arithmetical successors.*

Proof: Let us take the set $N_A = \{n_A; n = 1, 2, 3, \ldots\}$ and define the order $\leq$ as

$$n_A \leq_A m_A \text{ if and only if } n \leq m$$

the function g as

$$g(1_A) = 3,$$
$$g(\mathrm{S}n_A) = 2g(n_A) + 1 \text{ for all } n > 1$$

and the function h as

$$h(m) = n_A \text{ if } g(n_A) \leq m < 2g(n_A) + 1 \text{ for all } n > 0.$$

This gives us the prearithmetic $\boldsymbol{A} = (N_A; +, \circ, \leq_A)$, in which for all n from $\boldsymbol{A}$, we have

$$n_A \leq n_A + 1_A = h(g(n_A) + g(1_A)) = h(g(n_A) + 3) \leq h(2g(n_A)) = n_A.$$

Thus, $n_A + 1_A = n_A$ for all n from $\boldsymbol{A}$. Consequently, all successors in $\boldsymbol{A}$ are not arithmetical.

Proposition is proved. □

Proposition 2.5.9. *For any natural number k, there is an abstract prearithmetic, in which successors exist for all elements and exactly k of them are arithmetical.*

Proof: Let us take the set $N_A = \{n_A; n = 1, 2, 3, \ldots\}$ and define the order $\leq$ as

$$n_A \leq_A m_A \text{ if and only if } n \leq m$$

the function g as

$$g(1_A) = 3,$$
$$g(\mathrm{S}n_A) = 2g(n_A) + 1 \text{ for all } n > 1$$

and the function h as

$$h(m) = n_A \text{ if } g(n_A) \leq m < 2g(n_A) + 1 \text{ for all } n > 0.$$

This gives us the prearithmetic $\boldsymbol{A} = (N_A; +, \circ, \leq_A)$, in which for all n from $\boldsymbol{A}$, we have

$$n_A \leq n_A + 1_A = h(g(n_A) + g(1_A)) = h(g(n_A) + 3) \leq h(2g(n_A)) = n_A.$$

Thus, $n_A + 1_A = n_A$ for all n from $\boldsymbol{A}$. Consequently, all successors in $\boldsymbol{A}$ are not arithmetical.

Proposition is proved. □

However, subprearithmetics do not always inherit all properties of addition in the abstract prearithmetics that include them.

Proposition 2.5.10. *The following properties of addition are not always inherited by subprearithmetics:*

(a) *the additively Archimedean property;*
(b) *the exact additively Archimedean property;*
(c) *having an arithmetical successor* $\mathrm{AS}x = x + 1$ *for all elements;*
(d) *having an arithmetical predecessor* $\mathrm{AP}x = x - 1$ *for all elements;*
(e) *having an additive zero;*
(f) *totality of subtraction;*
(g) *totality of subtraction from the right (from the left).*

Proof is similar to the proof of Proposition 2.5.6 and left as an exercise. □

Now, let us look at properties of multiplication and figure out which of them are inherited by subprearithmetics.

Proposition 2.5.11. *The following properties of multiplication are inherited by subprearithmetics:*

(a) *commutativity of multiplication;*
(b) *associativity of multiplication;*
(c) *distributivity of multiplication with respect to addition;*
(d) *identities* $a \leq a \circ a$ *and* $a \leq a \circ b$ *or* $a \circ a \leq a$ *and* $a \circ b \leq a$;
(e) *preservation of order by multiplication;*
(f) *monotonicity of multiplication;*
(g) *multiplicative cancellation property;*
(h) *multiplicative cancellation property from the right (from the left);*
(i) *ordered multiplicative cancellation property;*
(j) *ordered multiplicative cancellation property from the right (from the left);*
(k) *strict multiplicative cancellation property;*
(l) *strict multiplicative cancellation property from the right (from the left).*

Proof is similar to the proof of Proposition 2.5.5 and left as an exercise. □

However, subprearithmetics do not always inherit all properties of multiplication in the abstract prearithmetics that include them.

Proposition 2.5.12. *The following properties of multiplication are not always inherited by subprearithmetics:*

(a) *the multiplicatively Archimedean property;*
(b) *the exact multiplicatively Archimedean property;*
(c) *the Binary Archimedean Property;*
(d) *having a multiplicative zero* 0;
(e) *having a multiplicative one* 1;
(f) *totality of division;*
(g) *totality of division from the right (from the left).*

Proof is similar to the proof of Proposition 2.5.6 and left as an exercise. □

Theorem 2.5.1. *If an abstract prearithmetic* $\boldsymbol{A}_1 = (A_1; +_1, \circ_1, \leq_1)$ *is a subprearithmetic of an abstract prearithmetic* $\boldsymbol{A}_2 = (A_2; +_2, \circ_2, \leq_2)$, *then the prearithmetic* $\boldsymbol{A}_1$ *is projective with respect to the prearithmetic* $\boldsymbol{A}_2$.

Proof: Let us assume that an abstract prearithmetic $\boldsymbol{A}_1 = (A_1; +_1, \circ_1, \leq_1)$ is a subprearithmetic of an abstract prearithmetic $\boldsymbol{A}_2 = (A_2; +_2, \circ_2, \leq_2)$. As A_1 is a subset of A_2, there is a natural injection $l : A_1 \to A_2$. Then we can take an element a from A_1 and define a mapping $p : A_2 \to A_1$ by the following formula:

$$p(x) = \begin{cases} x & \text{if } x \in A_1, \\ a & \text{if } x \notin A_1. \end{cases}$$

Then we can easily check that the prearithmetic $\boldsymbol{A}_1$ is projective with respect to the prearithmetic $\boldsymbol{A}_2$ with the projector l and coprojector p. Indeed, for any elements a and b from A_1, we have

$$a +_1 b = a +_2 b = p(l(a) +_2 l(b)),$$

$$a \circ_1 b = a \circ_2 b = p(l(a) \ \circ_2 l(b)),$$

$$a \leq_1 b \text{ if and only if } l(a) \leq_2 \ l(b)$$

because $a = l(a) = p(l(a))$ and $b = l(b) = p(l(b))$.

In addition, $pl = \mathbf{1}_{A_1}$.

Theorem is proved. □

As projectivity is a special case of weak projectivity (cf. Section 2.3), we have the following result.

Corollary 2.5.1. *If an abstract prearithmetic $\boldsymbol{A}_1 = (A_1; +_1, \circ_1, \leq_1)$ is a subprearithmetic of an abstract prearithmetic $\boldsymbol{A}_2 = (A_2; +_2, \circ_2, \leq_2)$, then the abstract prearithmetic $\boldsymbol{A}_1$ is weakly projective with respect to the abstract prearithmetic $\boldsymbol{A}_2$.*

The property "to be a subprearithmetic" is transitive. Namely, properties of sets (cf., for example, Jech, 2002) and universal algebras (Cohn, 1965) give us the following result.

Proposition 2.5.13. *If an abstract prearithmetic $\boldsymbol{A}_1 = (A_1; +_1, \circ_1, \leq_1)$ is a subprearithmetic of an abstract prearithmetic $\boldsymbol{A}_2 = (A_2; +_2, \circ_2, \leq_2)$ and the abstract prearithmetic $\boldsymbol{A}_2$ is a subprearithmetic of an abstract prearithmetic $\boldsymbol{A}_3 = (A_3; +_3, \circ_3, \leq_3)$, then the abstract prearithmetic $\boldsymbol{A}_1$ is a subprearithmetic of the abstract prearithmetic $\boldsymbol{A}_3$.*

Proof is left as an exercise. □

This shows that the property "to be a subprearithmetic" is transitive.

Example 2.5.7. The arithmetic $2\boldsymbol{N}$ of all even numbers with the standard arithmetical operations and order is a subprearithmetic of the Diophantine arithmetic $\boldsymbol{N}$, which is a subprearithmetic of the Diophantine arithmetic $\boldsymbol{W}$ of all whole numbers, and thus, $2\boldsymbol{N}$ is a subprearithmetic of the Diophantine arithmetic $\boldsymbol{W}$.

Subprearithmetics inherit weak projectivity.

Proposition 2.5.14. *If an abstract prearithmetic* $\boldsymbol{A}_1 = (A_1; +_1, \circ_1, \leq_1)$ *is a subprearithmetic of an abstract prearithmetic* $\boldsymbol{A}_2 = (A_2; +_2, \circ_2, \leq_2)$ *and the abstract prearithmetic* $\boldsymbol{A}_2$ *is weakly projective with respect to an abstract prearithmetic* $\boldsymbol{A}_3 = (A_3; +_3, \circ_3, \leq_3)$, *then the abstract prearithmetic* $\boldsymbol{A}_1$ *is weakly projective with respect to the abstract prearithmetic* $\boldsymbol{A}_3$.

Proof: Let us suppose that an abstract prearithmetic $\boldsymbol{A}_2 = (A_2;\ +_2,\ \circ_2,\ \leq_2)$ is weakly projective with respect to an abstract prearithmetic $\boldsymbol{A}_3 = (A_3; +_3, \circ_3, \leq_3)$ with the projector g and the coprojector h while $\boldsymbol{A}_1 = (A_1; +_1, \circ_1, \leq_1)$ is a subprearithmetic of $\boldsymbol{A}_2$. As the set A_1 is a subset of the set A_2, it is possible to take an element d from A_1 and the restriction f of the mapping g on the set A_1 defining the function $t : A_3 \to A_1$ as follows:

$$t(x) = \begin{cases} h(x) & \text{if } h(x) \in A_1, \\ d & \text{if } h(x) \notin A_1. \end{cases}$$

Then the prearithmetic $\boldsymbol{A}_1$ is weakly projective with respect to the prearithmetic $\boldsymbol{A}_3$ with the projector f and the coprojector t. Indeed, if a, $b \in A_1$, then $h(g(a) +_2 g(b)) \in A_1$ and $h(g(a) \circ_2 g(b)) \in A_1$ because $\boldsymbol{A}_1$ is a subprearithmetic of $\boldsymbol{A}_2$. Consequently, we have

$$a +_1 b = h(g(a) +_2 g(b)) = t(f(a) +_2 f(b)),$$

$$a \circ_1 b = h(g(a) \circ_2 g(b)) = t(f(a) \circ_2 f(b)).$$

It means that the prearithmetic $\boldsymbol{A}_1$ is weakly projective with respect to the prearithmetic $\boldsymbol{A}_3$ with the projector f and the coprojector t.

Proposition is proved. □

Proposition 2.5.15. *If an abstract prearithmetic* $\boldsymbol{A}_1 = (A_1; +_1, \circ_1, \leq_1)$ *is weakly projective with respect to an abstract prearithmetic* $\boldsymbol{A}_2 = (A_2; +_2, \circ_2, \leq_2)$ *and the abstract prearithmetic* $\boldsymbol{A}_2$ *is a subprearithmetic of an abstract prearithmetic* $\boldsymbol{A}_3 = (A_3; +_3, \circ_3, \leq_3)$, *then the abstract prearithmetic* $\boldsymbol{A}_1$ *is weakly projective with respect to the abstract prearithmetic* $\boldsymbol{A}_3$.

Proof: Let us suppose that an abstract prearithmetic $\boldsymbol{A}_1 = (A_1; +_1, \circ_1, \leq_1)$ is weakly projective with respect to an abstract prearithmetic $\boldsymbol{A}_2 = (A_2; +_2, \circ_2, \leq_2)$ with the projector g and the coprojector h and the abstract prearithmetic $\boldsymbol{A}_2$ is a subprearithmetic of an abstract prearithmetic $\boldsymbol{A}_3 = (A_3; +_3, \circ_3, \leq_3)$. Then by Corollary 2.5.1, the abstract prearithmetic $\boldsymbol{A}_2$ is weakly projective with respect to the abstract prearithmetic $\boldsymbol{A}_3$. Therefore, by Proposition 2.2.9, the prearithmetic $\boldsymbol{A}_1$ is weakly projective with respect to the prearithmetic $\boldsymbol{A}_3$.

Proposition is proved. □

Proposition 2.5.16. *If an abstract prearithmetic $\boldsymbol{A}_1 = (A_1; +_1, \circ_1, \leq_1)$ is a subprearithmetic of an abstract prearithmetic $\boldsymbol{A}_2 = (A_2; +_2, \circ_2, \leq_2)$ and the abstract prearithmetic $\boldsymbol{A}_2$ is projective with respect to an abstract prearithmetic $\boldsymbol{A}_3 = (A_3; +_3, \circ_3, \leq_3)$, then the prearithmetic $\boldsymbol{A}_1$ is projective with respect to the prearithmetic $\boldsymbol{A}_3$.*

Proof: Let us suppose that an abstract prearithmetic $\boldsymbol{A}_2$ is projective with respect to an abstract prearithmetic $\boldsymbol{A}_3$ with the projector g and the coprojector h while $\boldsymbol{A}_1$ is a subprearithmetic of $\boldsymbol{A}_2$. By Theorem 2.5.1, the prearithmetic $\boldsymbol{A}_1$ is projective with respect to the prearithmetic $\boldsymbol{A}_2$ with the projector f and the coprojector t. Then by Proposition 2.3.12, prearithmetic $\boldsymbol{A}_1$ is projective with respect to the prearithmetic $\boldsymbol{A}_3$.

Proposition is proved. □

Proposition 2.5.17. *If an abstract prearithmetic $\boldsymbol{A}_1 = (A_1; +_1, \circ_1, \leq_1)$ is projective with respect to an abstract prearithmetic $\boldsymbol{A}_2 = (A_2; +_2, \circ_2, \leq_2)$ and the abstract prearithmetic $\boldsymbol{A}_2$ is a subprearithmetic of an abstract prearithmetic $\boldsymbol{A}_3 = (A_3; +_3, \circ_3, \leq_3)$, then the prearithmetic $\boldsymbol{A}_1$ is projective with respect to the prearithmetic $\boldsymbol{A}_3$.*

Proof is similar to the proof of Propositions 2.5.15 being based on Theorem 2.5.1 instead of Corollary 2.5.1. □

Remark 2.5.2. For exact projectivity, the statements of Propositions 2.5.14–2.5.17 are not always true.

Theorem 2.5.2. *If an abstract prearithmetic $\boldsymbol{A}_2 = (A_2; +_2, \circ_2, \leq_2)$ is projective with respect to its subprearithmetic $\boldsymbol{A}_1 = (A_1; +_1, \circ_1, \leq_1)$, then the sets A_1 and A_2 are equipotent.*

Proof: Let us assume that an abstract prearithmetic $\boldsymbol{A}_1 = (A_1; +_1, \circ_1, \leq_1)$ is a subprearithmetic of an abstract prearithmetic $\boldsymbol{A}_2 = (A_2; +_2, \circ_2, \leq_2)$ and $\boldsymbol{A}_2$ is projective with respect to its subprearithmetic $\boldsymbol{A}_1$. The first condition implies (cf. Definition 2.5.1) that there is an inclusion of A_1 into A_2. The second condition implies (cf. Definition 2.4.1 and Lemma 2.1.9) that there is an inclusion of A_2 into A_1. Thus, by the Cantor–Bernstein–Schröder theorem (cf., for example, Kuratowski and Mostowski, 1967; Burgin, 2011), sets A_1 and A_2 are equipotent.

Theorem is proved. □

This result brings us to the following problem.

Problem 2.5.1. Find if an abstract prearithmetic $\boldsymbol{A}_2$, which is projective with respect to its subprearithmetic $\boldsymbol{A}_1$, is exactly projective with respect to $\boldsymbol{A}_1$.

In some cases, projectivity of abstract prearithmetics implies their exact projectivity.

Proposition 2.5.18. *If an abstract prearithmetic* $\boldsymbol{A}_1 = (A_1; +_1, \circ_1, \leq_1)$ *is projective with respect to a prearithmetic* $\boldsymbol{A}_2 = (A_2; +_2, \circ_2, \leq_1)$ *and the projector* g *is a homomorphism, then the abstract prearithmetic* $\boldsymbol{A}_1$ *is exactly projective with respect to a subprearithmetic of the prearithmetic* $\boldsymbol{A}_2$.

Proof: Let us assume that an abstract prearithmetic $\boldsymbol{A}_1 = (A_1; +_1, \circ_1, \leq_1)$ is projective with respect to a prearithmetic $\boldsymbol{A}_2 = (A_2; +_2, \circ_2, \leq_1)$ with the projector g and the coprojector h. Then by Lemma 2.3.3, the projector g is an injection.

Let us denote the image $g(A_1)$ of A_1 by B. Then B generates the subprearithmetic $\mathrm{sa}_B \boldsymbol{A}_2 = (C; +_2, \circ_2, \leq)$ in $\boldsymbol{A}_2$. We show that $C = B$ because B is closed with respect to addition and multiplication in $\boldsymbol{A}_2$.

Indeed, as the projector g is a homomorphism, for arbitrary elements a and b from $\boldsymbol{A}_1$, we have

$$g(a) +_2 g(b) = g(a +_1 b)$$

and

$$g(a) \circ_2 g(b) = g(a \circ_1 b).$$

Thus, the abstract prearithmetic $\boldsymbol{A}_1 = (A_1; +_1, \circ_1, \leq_1)$ is exactly projective with respect to the subprearithmetic $\mathrm{sa}_B\ \boldsymbol{A}_2 = (C; +_2, \circ_2, \leq)$ of $\boldsymbol{A}_2$ with the projector g and the coprojector g^{-1}.

Proposition is proved. □

Propositions 2.5.18 and 2.4.2 imply the following result.

Corollary 2.5.2. *If an abstract prearithmetic* $\boldsymbol{A}_1 = (A_1; +_1, \circ_1, \leq_1)$ *is projective with respect to a prearithmetic* $\boldsymbol{A}_2 = (A_2; +_2, \circ_2, \leq_1)$ *and the projector* g *is a homomorphism, then the abstract prearithmetic* $\boldsymbol{A}_1$ *is isomorphic to a subprearithmetic of the prearithmetic* $\boldsymbol{A}_2$.

Subsets of a carrier of an abstract prearithmetic generate subprearithmetics of this prearithmetic.

Lemma 2.5.2. *If* $\boldsymbol{A} = (A;\ +, \circ, \leq)$ *is an abstract prearithmetic and* C *is a subset of* A*, then there is the least subprearithmetic* $\mathrm{sa}_C\boldsymbol{A} = (B;\ +, \circ, \leq)$ *such that* $C \subseteq B \subseteq A$.

Indeed, we take as B all sums and products of the elements from C and define for them the same operations and relations, which exist in $\boldsymbol{A}$.

Thus, the subprearithmetic $\mathrm{sa}_C\ \boldsymbol{A}$ is defined by three conditions:

(1) Any element from C belongs to B.
(2) If elements a and b belong to B, then $a + b$ and $a \circ b$ belong to B.
(3) There are no other elements in B.

We say that the prearithmetic $\mathrm{sa}_C\boldsymbol{A} = (B;\ +, \circ, \leq)$ is generated by C in $\boldsymbol{A}$.

Let us study properties of subprearithmetics of conventional arithmetics.

Proposition 2.5.19. *If a subprearithmetic* $\boldsymbol{A} = (A; +, \circ, \leq)$ *of the Diophantine arithmetic* $\boldsymbol{N}$ *contains* 1*, then* $\boldsymbol{A}$ *coincides with* $\boldsymbol{N}$.

Indeed, as it is possible to obtain any natural number by adding 1 to itself the necessary number of times, $\boldsymbol{A} = \boldsymbol{N}$.

Proposition 2.5.20. *If a subprearithmetic* $\boldsymbol{B} = (B; +, \circ, \leq)$ *of the Diophantine arithmetic* $\boldsymbol{W}$ *contains* 1*, then* $\boldsymbol{B}$ *coincides either with* $\boldsymbol{N}$ *or with* $\boldsymbol{W}$.

Proof is left as an exercise. □

Proposition 2.5.21. *If a subprearithmetic* $\boldsymbol{D} = (D; +, \circ, \leq)$ *of the Diophantine arithmetic* $\boldsymbol{W}$ *contains* 1 *and* 0*, then* $\boldsymbol{D}$ *coincides with* $\boldsymbol{W}$.

Proof is left as an exercise. □

Remark 2.5.3. There are many proper subprearithmetics in arithmetics $\boldsymbol{N}$ and $\boldsymbol{W}$. An example of such a proper subprearithmetic is the set $2N$ of all even numbers with the conventional operations and order.

Proposition 2.5.22. *If a subprearithmetic* $\boldsymbol{H} = (H; +, \circ, \leq)$ *of the arithmetic* $\boldsymbol{Z}$ *of integer numbers contains* 1 *and some negative number, then* $\boldsymbol{H}$ *coincides with* $\boldsymbol{Z}$.

Proof is left as an exercise. □

Proposition 2.5.23. *If a subprearithmetic* $\boldsymbol{G} = (G; +, \circ, \leq)$ *of the arithmetic* $\boldsymbol{Z}$ *of integer numbers contains* -1 *and some positive number, then* $\boldsymbol{G}$ *coincides with* $\boldsymbol{Z}$.

Proof is left as an exercise. □

In the theory of universal algebras, it is proved that if $f\colon A \to B$ is a homomorphism of universal algebras, then the image of a subalgebra of A is a subalgebra of B (Cohn, 1965). This gives us the following result.

Proposition 2.5.24. *If an abstract prearithmetic* $\boldsymbol{A}_1 = (A_1; +_1, \circ_1, \leq_1)$ *is exactly projective with respect to an abstract prearithmetic* $\boldsymbol{A}_2 = (A_2; +_2, \circ_2, \leq_2)$ *with the projector* g *and the coprojector* h, $\boldsymbol{H} = (H; +, \circ, \leq)$ *is a subprearithmetic of* $\boldsymbol{A}_1$ *and* $\boldsymbol{G} = (G; +, \circ, \leq)$ *is a subprearithmetic of* $\boldsymbol{A}_2$*, then the image* $h\boldsymbol{G}$ *is a subprearithmetic of* $\boldsymbol{A}_1$ *and the image* $g\boldsymbol{H}$ *is a subprearithmetic of* $\boldsymbol{A}_2$.

Proof: By Proposition 2.4.3, the mapping h is a homomorphism of the abstract prearithmetic $\boldsymbol{A}_2$ and thus, by properties of homomorphisms, $h\boldsymbol{G}$ is a subprearithmetic of $\boldsymbol{A}_1$. By Proposition 2.4.2, the mapping g is a homomorphism of the abstract prearithmetic $\boldsymbol{A}_1$ and thus, by properties of homomorphisms, $g\boldsymbol{H}$ is a subprearithmetic of $\boldsymbol{A}_2$.

Proposition is proved. □

Corollary 2.5.3. *If an abstract prearithmetic* $\boldsymbol{A}_1 = (A_1; +_1, \circ_1, \leq_1)$ *is exactly projective with respect to an abstract prearithmetic* $\boldsymbol{A}_2 = (A_2; +_2, \circ_2, \leq_2)$*, then there is a one-to-one correspondence between their subprearithmetics.*

It would be interesting to find when subprearithmetics are preserved in weak projectivity and projectivity.

2.5.2. *Embedding abstract prearithmetics*

Flight by machines heavier than air is unpractical and insignificant, if not utterly impossible.

Simon Newcomb

An important relation between prearithmetics is embedding.

Definition 2.5.3. An abstract prearithmetic $\boldsymbol{A}_1 = (A_1; +_1, \circ_1, \leq_1)$ is *embedded* in an abstract prearithmetic $\boldsymbol{A}_2 = (A_2; +_2, \circ_2, \leq_2)$ if $\boldsymbol{A}_1$ is a subprearithmetic $\boldsymbol{A}_2$.

Example 2.5.8. The conventional arithmetic $2\boldsymbol{N}$ of all even numbers is embedded in Diophantine arithmetic $\boldsymbol{N}$ of all natural numbers.

In some cases, it is possible to embed prearithmetics to prearithmetics that have better properties.

Proposition 2.5.25. *Any abstract prearithmetic* $\boldsymbol{A} = (A;\ +, \circ, \leq)$ *without additive and multiplicative zeros can be embedded into an abstract prearithmetic* $\boldsymbol{B} = (B;\ +_2, \circ_2, \leq_2)$ *with additive and multiplicative zeros.*

Proof: Let us take an abstract prearithmetic $\boldsymbol{A} = (A;\ +, \circ, \leq)$ without additive and multiplicative zeros and add the element 0 to the set A. We obtain the set $B = A \cup \{0\}$ and define the following operations and relations in it:

$$0 +_2 a = a +_2 0 = a,$$

$$a \circ_2 0 = 0 \circ_2 a = 0,$$

$$0 +_2 0 = 0 \circ_2 0 = 0,$$

$$a \circ_2 b = a \circ b,$$

$$a +_2 b = a + b,$$

$$0 \leq_2 a,$$

$$b \leq_2 a \text{ if and only if } b \leq a$$

for any elements a and b from $\boldsymbol{A}$. As a result, we obtain an abstract prearithmetic $\boldsymbol{B} = (B; +_2, \circ_2, \leq_2)$, in which 0 is the additive and multiplicative zero.

Proposition is proved. □

Corollary 2.5.4. *Any abstract prearithmetic* $\boldsymbol{A} = (A; +, \circ, \leq)$ *without an additive zero can be embedded into an abstract prearithmetic* $\boldsymbol{D} = (D; +_1, \circ_1, \leq_1)$ *with an additive zero.*

Corollary 2.5.5. *Any abstract prearithmetic* $\boldsymbol{A} = (A; +, \circ, \leq)$ *without a multiplicative zero can be embedded into an abstract prearithmetic* $\boldsymbol{D} = (D; +_1, \circ_1, \leq_1)$ *with a multiplicative zero.*

Note that if an abstract prearithmetic $\boldsymbol{A} = (A; +, \circ, \leq)$ has an additive or multiplicative zero 0_A, then application of Proposition 2.5.25 introduces a new zero 0 making the old zero 0_A a non-zero element.

Proposition 2.5.26. *Any abstract prearithmetic $\boldsymbol{A} = (A; +, \circ, \leq)$ without the multiplicative one can be embedded into an abstract prearithmetic $\boldsymbol{B} = (B; +_2, \circ_2, \leq_2)$ with the multiplicative one.*

Proof: Let us take an abstract prearithmetic $\boldsymbol{A} = (A; \dot{+}, \circ \leq)$ without the multiplicative one, the Diophantine arithmetic $\boldsymbol{N} = (N; +, \circ, \leq)$ and define the abstract prearithmetic $\boldsymbol{B} = (B; +_2, \circ_2, \leq_2)$ in the following way:

$$B = A \cup N,$$

$$0 +_2 a = a +_2 0 = a,$$

$$a \circ_2 1 = 1 \circ_2 a = a,$$

$$n +_2 a = a \circ_2 n = 1 \text{ if } a \in \boldsymbol{A} \text{ and } n \in \boldsymbol{N},$$

$$a \circ_2 b = a \circ b \text{ if } a, b \in \boldsymbol{A},$$

$$a +_2 b = a \dot{+} b \text{ if } a, b \in \boldsymbol{A},$$

$$m \circ_2 n = m \cdot n \text{ if } m, n \in \boldsymbol{N},$$

$$m +_2 n = m + n \text{ if } m, n \in \boldsymbol{N},$$

$$n \leq_2 a \text{ if } a \in \boldsymbol{A} \text{ and } n \in \boldsymbol{N},$$

$$b \leq_2 a \text{ if and only if } a, b \in \boldsymbol{A} \text{ and } b \leq a,$$

$$m \leq_2 n \text{ if and only if } m \leq n \text{ if } m, n \in \boldsymbol{N} \text{ and } m \leq n.$$

As a result, we obtain an abstract prearithmetic $\boldsymbol{B} = (B; +_2, \circ_2, \leq_2)$, in which 1 is the multiplicative one.

Proposition is proved. □

Note that if an abstract prearithmetic $\boldsymbol{A} = (A; +, \circ, \leq)$ has a multiplicative one 1_A, then application of Proposition 2.5.26 introduces a new multiplicative one 1 eliminating corresponding properties of the old multiplicative one 1_A.

Proposition 2.5.9 implies the following result.

Lemma 2.5.3. *If an abstract prearithmetic $\boldsymbol{A}_1 = (A_1; +_1, \circ_1, \leq_1)$ is embedded in an abstract prearithmetic $\boldsymbol{A}_2 = (A_2; +_2, \circ_2, \leq_2)$ and the abstract prearithmetic $\boldsymbol{A}_2 = (A_2; +_2, \circ_2, \leq_2)$ is embedded in an abstract* *prearithmetic* $\boldsymbol{A}_3 = (A_3; +_3, \circ_3, \leq_3)$, *then the prearithmetic* $\boldsymbol{A}_1$ *is embedded in the prearithmetic* $\boldsymbol{A}_3$.

Proof is left as an exercise. □

Let us consider a weaker relation between prearithmetics than embedding.

Definition 2.5.4. An abstract prearithmetic $\boldsymbol{A}_1 = (A_1; +_1, \circ_1, \leq_1)$ is an *inserted* into an abstract prearithmetic $\boldsymbol{A}_2 = (A_2; +_2, \circ_2, \leq_2)$ if $\boldsymbol{A}_1$ is isomorphic to a subprearithmetic $\boldsymbol{A}_2$.

Example 2.5.9. The Diophantine arithmetic $\boldsymbol{N}$ of all natural numbers is inserted into the arithmetic $\boldsymbol{Fr}$ of all fractions in which numerator and denominator are natural numbers by the mapping $f(n) = n/1$ but it is not embedded into $\boldsymbol{Fr}$.

Lemma 2.5.4. *If an abstract prearithmetic* $\boldsymbol{A}_1 = (A_1; +_1, \circ_1, \leq_1)$ *is embedded into an abstract prearithmetic* $\boldsymbol{A}_2 = (A_2; +_2, \circ_2, \leq_2)$, *then* $\boldsymbol{A}_1$ *is inserted into* $\boldsymbol{A}_2$.

Proof is left as an exercise. □

As the composition of isomorphisms is an isomorphism (Kurosh, 1963), we have the following result.

Lemma 2.5.5. *If an abstract prearithmetic* $\boldsymbol{A}_1 = (A_1; +_1, \circ_1, \leq_1)$ *is inserted into an abstract prearithmetic* $\boldsymbol{A}_2 = (A_2; +_2, \circ_2, \leq_2)$ *and the abstract prearithmetic* $\boldsymbol{A}_2 = (A_2; +_2, \circ_2, \leq_2)$ *is inserted into an abstract prearithmetic* $\boldsymbol{A}_3 = (A_3; +_3, \circ_3, \leq_3)$, *then the prearithmetic* $\boldsymbol{A}_1$ *is inserted into the prearithmetic* $\boldsymbol{A}_3$.

Proof is left as an exercise. □

Remark 2.5.4. In algebra, mathematicians usually do not make a distinction between insertion and embedding as well as between isomorphic algebras. However, as it is demonstrated in Section 2.4, these distinctions are essential.

Theorem 2.5.3. *Any abstract prearithmetic* $\boldsymbol{A} = (A;\ +,\ \circ, \leq)$ *with commutative and associative addition, commutative and associative multiplication, which is also distributive with respect to addition, and additive cancellation can be inserted into an abstract prearithmetic* $\boldsymbol{B} = (B; +_2,\ \circ_2, \leq_2)$ *with subtraction, commutative and associative addition and multiplication, and distributive multiplication with respect to addition.*

Proof: Let us consider an abstract prearithmetic $\boldsymbol{A} = (A;\ +, \circ, \leq)$ with commutative and associative addition, distributive multiplication with respect to addition and additive cancellation. Taking the set A^2 of all pairs of elements from A, we define the congruence relation $\sim$ on A^2 by the following rule:

$$(c, d) \sim (a, b) \text{ if } c + b = a + d$$

for any elements a, b, c and d from $\boldsymbol{A}$.

Let us check that $\sim$ is an equivalence relation (cf. Appendix).

Reflexivity. Naturally, $(a, b) \sim (a, b)$ because $a + b = a + b$.

Symmetry. In the same way, $(c, d) \sim (a, b)$ implies $(a, b) \sim (c, d)$ because $c + b = a + d$ implies $a + d = c + b$ as addition is commutative.

Transitivity. If $(c, d) \sim (a, b)$ and $(a, b) \sim (e, f)$, then we have

$$c + b = a + d$$

and

$$a + f = e + b.$$

Consequently,

$$c + b + a + f = a + d + e + b.$$

Commutativity and associativity of addition imply

$$(a + b) + (c + f) = (a + b) + (e + d).$$

Applying additive cancellation, we obtain

$$c + f = e + d.$$

It means that

$$(c, d) \sim (e, f),$$

i.e., $\sim$ is a transitive relation.

Let as also define operations and order in the set A^2.

Addition of pairs is defined as

$$(a, b) + (c, d) = (a + c, b + d).$$

Multiplication is defined as

$$(a, b) \circ (c, d) = (a \circ c + b \circ d,\ a \circ d + b \circ c).$$

In A^2, the relation $\sim$ is invariant with respect to both operations, that is, if pairs of elements are equivalent, then their sum and product are also equivalent. Indeed, if $(c, d) \sim (e, f)$, then

$$(a, b) + (c, d) = (a + c, b + d),$$
$$(a, b) + (e, f) = (a + e, b + f)$$

and

$$a + c + b + f = a + b + c + f = a + b + e + d = a + e + b + d$$

because

$$c + f = e + d.$$

It means that

$$(a, b) + (c, d) \sim (a, b) + (e, f).$$

In a similar way, we have

$$(a, b) \circ (c, d) = (a \circ c + b \circ d,\ a \circ d + b \circ c),$$
$$(a, b) \circ (e, f) = (a \circ e + b \circ f,\ a \circ f + b \circ e).$$

These pairs are equivalent. Indeed, distributivity of multiplication with respect to addition gives us

$$\begin{aligned} &a \circ c + b \circ d + a \circ f + b \circ e \\ &\quad = a \circ c + a \circ f + b \circ d + b \circ e \\ &\quad = a \circ (c + f) + b \circ (d + e) = a \circ (e + d) + b \circ (f + c) \\ &\quad = a \circ e + a \circ d + b \circ f + b \circ c = a \circ e + b \circ f + a \circ d + b \circ c \end{aligned}$$

because

$$c + f = f + c = d + e = e + d$$

It means that

$$(a, b) \circ (c, d) \sim (a, b) \circ (e, f).$$

The congruence $\sim$ allows taking the set B of equivalences classes and defining operations addition $+_2$ and multiplication $\circ_2$ using representatives of these classes (cf., for example, Artin, 1991; Kurosh, 1963). Namely, denoting the equivalence class of an element (a, b) by $\langle(a, b)\rangle$, we define operations in the set B as

$$\langle(a, b)\rangle +_2 \langle(c, d)\rangle = \langle(a, b)\ + (c, d)\rangle$$

and

$$\langle(a, b)\rangle\ \circ_2\ \langle(c, d)\rangle = \langle(a, b) \circ (c, d)\rangle$$

We also define the order

$$(c, d)\ \leq_2\ (a, b) \text{ if } (c, d) \sim (a, b).$$

This gives us an abstract prearithmetic $\boldsymbol{B}_1 = (B; +_2, \circ_2, \leq_2)$, in which the equivalence class where all elements have the form (c, c), is the additive zero 0_B. Indeed,

$$(c, c) + (a, b) = (c + a,\ c + b)$$

and

$$(c + a,\ c + b) \sim (a, b)$$

because

$$c + a + b = a + c + b,$$

i.e., elements $(c, c) + (a, b)$ and (a, b) belong to the same equivalence class. By the same token, elements $(a, b) + (c, c)$ and (a, b) belong to the same equivalence class. It means that addition of the element (c, c) does not change the equivalence class.

In this case, the additive zero 0_B is also the multiplicative zero in $\boldsymbol{B}_1$. Indeed,

$$(c, c) \circ (a, b) = (c \circ a + c \circ b, c \circ a + c \circ b)$$

and

$$(c \circ a + c \circ b, c \circ a + c \circ b) \sim (c, c) = 0_B$$

because

$$c \circ a + c \circ b = c \circ a + c \circ b.$$

If the abstract prearithmetic $\boldsymbol{A} = (A; +, \circ, \leq)$ does not have the additive and multiplicative zero, by Proposition 2.5.1, it is possible to imbed $\boldsymbol{A}$ into an abstract prearithmetic with the additive and multiplicative zero before building the abstract prearithmetic $\boldsymbol{B}$. That is why it is possible to assume from the beginning that the abstract prearithmetic $\boldsymbol{A}$ has the additive and multiplicative zero 0.

This makes it possible to assign the equivalence class of the pair $(c, 0)$ to each element c from $\boldsymbol{A}$. This mapping $f(c) = (c, 0)$ is an injection. Indeed, if two elements a and b from $\boldsymbol{A}$ are mapped into the same equivalence class, then

$$(a, 0) \sim (b, 0).$$

This implies

$$a = a + 0 = b + 0 = b,$$

i.e., elements a and b coincide.

In addition, the mapping f is a homomorphism. Indeed,

$$f(a + b) = (a + b, 0) = (a, 0) + (b, 0) = f(a) + f(b)$$

and

$$\begin{aligned} f(a) \circ f(b) &= (a, 0) \circ (b, 0) \\ &= (a \circ b + 0 \circ 0, a \circ 0 + b \circ 0) = (a \circ b, 0) = f(a \circ b). \end{aligned}$$

We also extend the relation $\leq_2$ by defining

$$(c, 0) \leq_2 (a, 0) \text{ if } c \leq a.$$

This gives us the abstract prearithmetic $\boldsymbol{B} = (B; +_2, \circ_2, \leq_2)$ where the mapping f is an embedding of $\boldsymbol{A}$ into $\boldsymbol{B}$.

We can also define subtraction in $\boldsymbol{B}$ by the following rules:

$$(a, b) - (c, d) = (a + d,\ b + c)$$

and

$$\langle(a, b)\rangle -_2 \langle(c, d)\rangle = \langle(a, b) - -(c, d)\rangle.$$

This definition is correct because if $(c, d) \sim (e, f)$, then

$$(a, b) - (c, d) = (a + d,\ b + c),$$

$$(a, b) - (e, f) = (a + f, b + e)$$

and

$$a + d + b + e = a + b + d + e = a + b + f + c = a + f + b + c$$

because

$$f + c = d + e.$$

Thus,

$$(a, b) - (c, d) \sim (a, b) - (e, f),$$

i.e., the relation $\sim$ is invariant with respect to the operation — and we need to show that — is subtraction in $\boldsymbol{B}$.

To do this, let us take the sum

$$(c, d) + ((a, b) - (c, d)).$$

By definition, we have

$$\begin{aligned}(c, d) + ((a, b) - (c, d)) &= (c, d) + ((a + d, c + b)) \\ &= (c + a + d, d + c + b) = (a + c + d, b + c + d) \\ &= (a, b) + (c + d, c + d) = (a, b)\end{aligned}$$

and

$$\begin{aligned}\langle(c, d)\rangle +_2 (\langle(a, b)\rangle -_2 \langle(c, d)\rangle) &= \langle(c, d)\rangle +_2 (\langle(a, b) - (c, d)\rangle) \\ &= \langle(c, d) + ((a, b) - (c, d))\rangle \\ &= \langle(a, b)\rangle +_2 \langle(c + d, c + d)\rangle = \langle(a, b)\rangle\end{aligned}$$

because $\langle(c + d, c + d)\rangle$ is an additive zero in $\boldsymbol{B}$.

In a similar way, we check that

$$(\langle(a, b)\rangle -_2 \langle(c, d)\rangle) +_2 \langle(c, d)\rangle = \langle(a, b)\rangle.$$

Consequently, the operation $-_2$ is subtraction in $\boldsymbol{B}$, which is, thus, an abstract prearithmetic with subtraction and $\boldsymbol{A}$ is embedded in $\boldsymbol{B}$.

Let us check that the abstract prearithmetic $\boldsymbol{B}$ has commutative and associative addition and multiplication, and distributive multiplication with respect to addition.

Commutativity of addition

$$\begin{aligned}
&\langle(a,b)\rangle +_2 \langle(c,d)\rangle \\
&\quad = \langle(a,b)+(c,d)\rangle = \langle(a+c,b+d)\rangle = \langle(c+a,d+b)\rangle \\
&\quad = \langle(c,d)+(a,b)\rangle = \langle(c,d)\rangle +_2 \langle(a,b)\rangle.
\end{aligned}$$

Commutativity of multiplication

$$\begin{aligned}
&\langle(a,b)\rangle \circ_2 \langle(c,d)\rangle \\
&\quad = \langle(a,b)\circ(c,d)\rangle = \langle(a\circ c+b\circ d, a\circ d+b\circ c)\rangle \\
&\quad = \langle(c\circ a+d\circ b, c\circ b+d\circ a)\rangle = \langle(c,d)\circ(a,b)\rangle \\
&\quad = \langle(c,d)\rangle \circ_2 \langle(a,b)\rangle.
\end{aligned}$$

Associativity of addition and multiplication is verified in a similar way. Let us consider distributivity of multiplication with respect to addition.

$$\begin{aligned}
&\langle(e,f)\rangle \circ_2 (\langle(a,b)\rangle +_2 \langle(c,d)\rangle) \\
&\quad = \langle(e,f)\rangle \circ_2 \langle(a,b)+(c,d)\rangle = \langle(e,f)\rangle \circ_2 \langle(a+c,b+d)\rangle \\
&\quad = \langle(e\circ(a+c)+f\circ(b+d),\ e\circ(b+d)+f\circ(a+c)\rangle \\
&\quad = \langle(e\circ a+e\circ c+f\circ b+f\circ d, e\circ b+e\circ d+f\circ a+f\circ c\rangle \\
&\quad = \langle(e\circ a+f\circ b+e\circ c+f\circ d, e\circ b+f\circ a+e\circ d+f\circ c\rangle \\
&\quad = \langle(e\circ a+f\circ b, e\circ b+f\circ a)\rangle +_2 \langle(e\circ c+f\circ d, e\circ d+f\circ c)\rangle \\
&\quad = \langle(e,f)\rangle \circ_2 \langle(a,b)\rangle +_2 \langle(e,f)\rangle \circ_2 \langle(c,d)\rangle.
\end{aligned}$$

Thus,

$$\begin{aligned}
&\langle(e,f)\rangle \circ_2 (\langle(a,b)\rangle +_2 \langle(c,d)\rangle) \\
&\quad = \langle(e,f)\rangle \circ_2 \langle(a,b)\rangle +_2 \langle(e,f)\rangle \circ_2 \langle(c,d)\rangle,
\end{aligned}$$

i.e., multiplication is distributive in $\boldsymbol{B}$ with respect to addition.

Note that by Lemma 2.3.15, the prearithmetic $\boldsymbol{B}$ also satisfies the cancellation law.

Theorem is proved. □

As a corollary, we obtain a well-known result (cf., for example, Kurosh, 1963).

Corollary 2.5.6. *Any abelian semigroup with cancellation can be inserted into an abelian group.*

Taking into account the definition of a ring (cf. Appendix), we come to the following result.

Corollary 2.5.7. *Any abstract prearithmetic* $\boldsymbol{A} = (A; +, \circ, \leq)$ *with commutative and associative addition and multiplication, distributive multiplication with respect to addition and additive cancellation can be inserted into a ring.*

Taking into account the definition of a semiring (cf. Appendix), we come to the following result.

Corollary 2.5.8. *Any semiring with additive cancellation can be inserted into a ring.*

It is also possible to insert an abstract prearithmetic into an abstract prearithmetic with division.

Theorem 2.5.4. *Any abstract prearithmetic* $\boldsymbol{A} = (A; +, \circ, \leq)$ *with commutative and associative addition, commutative and associative multiplication, which is also distributive with respect to addition, and multiplicative cancellation can be inserted into an abstract prearithmetic* $\boldsymbol{B} = (B; +_2, \circ_2, \leq_2)$ *with division, commutative and associative addition and multiplication, and distributive multiplication with respect to addition.*

Proof: Let us consider an abstract prearithmetic $\boldsymbol{A} = (A; +, \circ, \leq)$ with commutative and associative addition and multiplication, distributive multiplication with respect to addition and multiplicative cancellation. Taking the set Fr A of the form a/b where a and b are elements from A, we define the congruence relation $\sim$ on Fr A by the following rule:

$$c/d \sim a/b \text{ if } c \circ b = a \circ d$$

for any elements a, b, c and d from $\boldsymbol{A}$ where d and b are not equal to 0

Let us check that $\sim$ is an equivalence relation (cf. Appendix).

Reflexivity. Naturally, $(a, b) \sim (a, b)$ because $a \circ b = a \circ b$.

Symmetry. In the same way, $c/d \sim a/b$ implies $a/b \sim c/d$ because $c \circ b = a \circ d$ implies $a \circ d = c \circ b$ as multiplication is commutative.

Transitivity. If $c/d \sim a/b$ and $a/b \sim e/f$, then we have

$$c \circ b = a \circ d$$

and

$$a \circ f = e \circ b.$$

Consequently,

$$c \circ b \circ a \circ f = a \circ d \circ e \circ b.$$

Commutativity and associativity of multiplication imply

$$(a \circ b) \circ (c \circ f) = (a \circ b) \circ (e \circ d).$$

Applying multiplicative cancellation, we obtain

$$c \circ f = e \circ d.$$

It means that

$$c/d \sim e/f,$$

i.e., $\sim$ is a transitive relation.

Let as also define operations and order in the set Fr A.

Multiplication is defined as

$$a/b \circ c/d = a \circ c/b \circ d.$$

Addition is defined as

$$a/b + c/d = (a \circ d + b \circ c)/b \circ d.$$

In Fr A, the relation $\sim$ is invariant with respect to both operations, that is, if pairs of elements are equivalent, then their sums are equivalent and products are also equivalent. Indeed, if $c/d \sim e/f$, then

$$a/b + c/d = (a \circ d + b \circ c)/b \circ d,$$
$$a/b + e/f = (a \circ f + b \circ e)/b \circ f$$

and distributivity of multiplication with respect to addition gives us

$$\begin{aligned}
&(a \circ d + b \circ c) \circ (b \circ f) \\
&\quad = a \circ d \circ b \circ f + b \circ c \circ b \circ f = a \circ b \circ d \circ f + b \circ b \circ c \circ f \\
&\quad = a \circ b \circ f \circ d + b \circ b \circ e \circ d = a \circ f \circ b \circ d + b \circ e \circ b \circ d = (a \circ f + b \circ e) \circ (b \circ d)
\end{aligned}$$

because

$$c \circ f = e \circ d.$$

It means that

$$a/b + c/d \sim a/b + e/f.$$

In a similar way, we have equivalence of products

$$a/b \circ c/d = a \circ c/b \circ d,$$

$$a/b \circ e/f = a \circ e/b \circ f$$

and

$$a \circ c \circ b \circ f = a \circ b \circ c \circ f = a \circ b \circ e \circ d = a \circ e \circ b \circ d$$

because

$$c \circ f = e \circ d.$$

It means that

$$a/b \circ c/d \sim a/b \circ e/f.$$

The congruence $\sim$ allows taking the set B of equivalences classes and defining operations addition $+_2$ and multiplication $\circ_2$ using representatives of these classes (cf., for example, Artin, 1991; Kurosh, 1963). Namely, denoting the equivalence class of an element a/b by $\langle a/b \rangle$, we define operations in the set B as

$$\langle a/b \rangle +_2 \langle c/d \rangle = \langle a/b + c/d \rangle$$

and

$$\langle a/b \rangle \ \circ_2 \ \langle c/d \rangle = \langle a/b \circ c/d \rangle.$$

We also define the order

$$c/d \leq_2 a/b \text{ if } c/d \sim a/b.$$

This gives us an abstract prearithmetic $\boldsymbol{B}_1 = (B; +_2, \circ_2, \leq_2)$, in which the equivalence class where all elements have the form c/c, is the multiplicative one 1_B. Indeed,

$$c/c \circ a/b = c \circ a/c \circ b$$

and

$$c \circ a/c \circ b \sim a/b$$

because

$$c \circ a \circ b = a \circ c \circ b,$$

i.e., elements $c/c \circ a/b$ and a/b belong to the same equivalence class. By the same token, elements $a/b \circ c/c$ and a/b belong to the same equivalence class. It means that multiplication by the element c/c does not change the equivalence class.

If the abstract prearithmetic $\boldsymbol{A} = (A; +, \circ, \leq)$ does not have the multiplicative one 1, by Proposition 2.5.2, it is possible to imbed $\boldsymbol{A}$ into an abstract prearithmetic with the multiplicative one 1_A before building the abstract prearithmetic $\boldsymbol{B}$. That is why it is possible to assume from the beginning that the abstract prearithmetic $\boldsymbol{A}$ has the multiplicative one 1_A.

This makes it possible to assign the equivalence class of the pair $(c, 1_A)$ to each element c from $\boldsymbol{A}$. This mapping $f(c) = c/1_A$ is an injection. Indeed, if two elements a and b from $\boldsymbol{A}$ are mapped into the same equivalence class, then

$$a/1_A \sim b/1_A.$$

This implies

$$a = a \circ 1_A = b \circ 1_A = b,$$

i.e., elements a and b coincide.

In addition, the mapping f is a homomorphism. Indeed,

$$f(a + b) = (a + b)/1_A = a/1_A + b/1_A = f(a) + f(b)$$

and

$$f(a) \circ f(b) = a/1_A \circ b/1_A = (a \circ b)/1_A = f(a \circ b).$$

We also extend the relation $\leq_2$ by defining

$$c/1_A \leq_2 a/1_A \text{ if } c \leq a.$$

This gives us the abstract prearithmetic $\boldsymbol{B} = (B; +_2, \circ_2, \leq_2)$ where the mapping f is an embedding of $\boldsymbol{A}$ into $\boldsymbol{B}$.

We can also define division $\div$ in $\boldsymbol{B}$ by the following rules:

$$a/b \div c/d = a \circ d/c \circ b$$

and

$$\langle a/b \rangle \div_2 \langle c/d \rangle = \langle a/b \div c/d \rangle.$$

This definition is correct because if $c/d \sim e/f$, then

$$a/b \div c/d = a \circ d/c \circ b,$$
$$a/b \div e/f = a \circ f/e \circ b$$

and

$$a \circ d \circ e \circ b = a \circ b \circ d \circ e = a \circ b \circ c \circ f = a \circ f \circ c \circ b$$

because

$$c \circ f = d \circ e.$$

It means that

$$a/b \ \circ \ c/d \sim a/b \ \circ \ e/f$$

because

$$a \circ d \circ e \circ b = a \circ f \circ c \circ b,$$

i.e., the relation $\sim$ is invariant with respect to the operation $\div$ and we need to show that $\div$ is division in $\boldsymbol{B}$.

To do this, let us take the product

$$c/d \ \circ \ (a/b \ \div \ c/d).$$

By definition, we have

$$\begin{aligned} c/d \ \circ \ (a/b \div c/d) &= c/d \ \circ \ (a \circ d/c \circ b) \\ &= c \circ a \circ d/d \circ c \circ b = c \circ d \circ a/c \circ d \circ b \\ &= (c \circ d/c \circ d) \ \circ \ (a/b) \end{aligned}$$

and

$$\begin{aligned} &\langle c/d \rangle \ \circ_2 \ (\langle a/b \rangle \div_2 \langle c/d \rangle) \\ &\quad = \langle c/d \rangle \ \circ_2 \ \langle a/b \div c/d \rangle = \langle c/d \ \circ \ (a/b \div c/d) \rangle \\ &\quad = \langle (c \circ d/c \circ d) \ \circ \ (a/b) \rangle = \langle c \circ d/c \circ d \rangle \ \circ_2 \ \langle a/b \rangle = \langle a/b \rangle \end{aligned}$$

because $\langle c \circ d/c \circ d \rangle$ is the multiplicative one 1_B in $\boldsymbol{B}$.

In a similar way, we check that

$$(\langle a/b \rangle \div_2 \langle c/d \rangle) \ \circ_2 \ \langle c/d \rangle = \langle a/b \rangle.$$

Consequently, the operation $\div_2$ is division in $\boldsymbol{B}$, which is, thus, an abstract prearithmetic with division and $\boldsymbol{A}$ is embedded in $\boldsymbol{B}$.

The abstract prearithmetic $\boldsymbol{B}$ has additional properties. One of them is the *law of fraction reduction*:

For any elements a, b and c from $\boldsymbol{A}$, we have

$$\langle a \circ c/b \circ c\rangle = \langle a/b\rangle.$$

Indeed, as $\langle c/c\rangle$ is the multiplicative one 1_B in $\boldsymbol{B}$, we have

$$\langle a \circ c/b \circ c\rangle = \langle a/b \;\circ\; c/c\rangle = \langle a/b\rangle \;\circ_2\; \langle c/c\rangle = \langle a/b\rangle \;\circ_2\; 1_B = \langle a/b\rangle$$

as $\langle c/c\rangle$ is the multiplicative one 1_B in $\boldsymbol{B}$.

In a similar way, we obtain

$$\langle c \circ a/c \circ b\rangle = \langle a/b\rangle$$

for any elements a, b and c from $\boldsymbol{A}$.

Let us check that the abstract prearithmetic $\boldsymbol{B}$ has commutative and associative addition and multiplication, and distributive multiplication with respect to addition.

Commutativity of addition:

$$\begin{aligned}&\langle a/b\rangle +_2 \langle c/d\rangle = \langle a/b + c/d\rangle = \langle (a \circ d + b \circ c)/b \circ d\rangle\\&= \langle (c \circ b + d \circ a)/d \circ b\rangle = \langle c/d + a/b\rangle = \langle c/d\rangle +_2 \langle a/b\rangle\end{aligned}$$

Commutativity of multiplication

$$\begin{aligned}\langle a/b\rangle \;\circ_2\; \langle c/d\rangle &= \langle a/b \;\circ\; c/d\rangle = \langle a \circ c/b \circ d\rangle\\&= \langle c \circ a/b \circ d\rangle = \langle c/d \;\circ\; a/b\rangle = \langle c/d\rangle \;\circ_2\; \langle a/b\rangle.\end{aligned}$$

Associativity of addition and multiplication is verified in a similar way.

Let us consider distributivity of multiplication with respect to addition.

$$\begin{aligned}&\langle e/f\rangle \;\circ_2\; (\langle a/b\rangle +_2 \langle c/d\rangle) = \langle (e, f)\rangle \;\circ_2\; \langle a/b + c/d\rangle\\&\quad= \langle e/f\rangle \;\circ_2\; \langle (a \circ d + b \circ c)/b \circ d\rangle = \langle e/f \circ_2 (a \circ d + b \circ c)/b \circ d\rangle\\&\quad= \langle e \;\circ\; (a \circ d + b \circ c)/f \circ b \circ d\rangle = \langle (e \circ a \circ d + e \circ b \circ c)/f \circ b \circ d\rangle\\&\quad= \langle e \circ a \circ d/f \circ b \circ d + e \circ b \circ c/f \circ b \circ d\rangle\\&\quad= \langle e \circ a \circ d/f \circ b \circ d\rangle +_2 \langle e \circ b \circ c/f \circ b \circ d\rangle.\end{aligned}$$

By the law of fraction reduction and commutativity of multiplication, we have

$$\begin{aligned}&\langle e \circ a \circ d/f \circ b \circ d\rangle +_2 \langle e \circ b \circ c/f \circ b \circ d\rangle\\&\quad = \langle e \circ a \circ d/f \circ b \circ d\rangle +_2 \langle b \circ e \circ c/b \circ f \circ d\rangle\\&\quad = \langle e \circ a/f \circ b\rangle +_2 \langle e \circ c/f \circ d\rangle = \langle e/f \circ a/b\rangle +_2 \langle e/f \circ c/d\rangle\\&\quad = \langle e/f\rangle \circ_2 \langle a/b\rangle +_2 \langle e/f\rangle \circ_2 \langle c/d\rangle.\end{aligned}$$

Thus,

$$\langle e/f\rangle \circ_2 (\langle a/b\rangle +_2 \langle c/d\rangle) = \langle e/f\rangle \circ_2 \langle a/b\rangle +_2 \langle e/f\rangle \circ_2 \langle c/d\rangle,$$

i.e., multiplication is distributive in $\boldsymbol{B}$ with respect to addition.

Theorem is proved. □

Theorem 2.5.4 shows how the Diophantine arithmetic $\boldsymbol{N}$ is inserted in the arithmetic $\boldsymbol{Q}^{++}$ of all positive rational numbers, which are represented by fractions. However, many prearithmetics and arithmetics, for example, the Diophantine arithmetic $\boldsymbol{W}$, have the multiplicative zero. A prearithmetic with the multiplicative zero does not have the cancellation law but only the non-zero cancellation law (cf. Section 2.3). That is why we also consider this situation.

Theorem 2.5.5. *Any abstract prearithmetic* $\boldsymbol{A} = (A; +, \circ, \leq)$ *with commutative and associative addition, commutative and associative multiplication, which is also distributive with respect to addition, and non-zero multiplicative cancellation can be inserted into an abstract prearithmetic* $\boldsymbol{B} = (B; +_2, \circ_2, \leq_2)$ *with non-zero division, commutative and associative addition and multiplication, and distributive multiplication with respect to addition.*

Proof is similar to the proof is similar to the proof of Theorem 2.5.4 only instead of the set of all fractions, we take the set $\mathrm{Fr}_1\ A$ of fractions that have the form a/b where a and b are elements from A while b is not equal to 0. □

Theorem 2.5.4 allows proving the following result.

Theorem 2.5.6. *Any abstract prearithmetic* $\boldsymbol{A} = (A; +, \circ, \leq)$ *with subtraction, commutative and associative addition, commutative and associative multiplication, which is also distributive with respect to addition, and multiplicative cancellation can be inserted into an abstract prearithmetic* $\boldsymbol{B} = (B; +_2, \circ_2, \leq_2)$ *with division, subtraction, commutative and associative* *addition and multiplication, and distributive multiplication with respect to addition.*

Proof: Let us consider an abstract prearithmetic $\boldsymbol{A} = (A; +, \circ, \leq)$ with subtraction, commutative and associative addition, commutative and associative multiplication, which is also distributive with respect to addition and multiplicative cancellation. Note that by Corollary 2.1.30, the prearithmetic $\boldsymbol{A}$ also has additive cancellation.

By Theorems 2.5.4, it is possible to insert the prearithmetic $\boldsymbol{A}$ into a prearithmetic $\boldsymbol{B}$, which preserves all properties of addition and multiplication in $\boldsymbol{A}$ and has division. To prove Theorem 2.5.6, we need only to show that the prearithmetic $\boldsymbol{B}$ also has subtraction.

At first, we define subtraction of fractions by the following formula:

$$a/b - c/d = (a \circ d - b \circ c)/b \circ d.$$

In Fr A, the relation $\sim$ is invariant with respect to this operation, that is, if pairs of elements are equivalent, then their differences are also equivalent. Indeed, if $c/d \sim e/f$, then

$$a/b - c/d = (a \circ d - b \circ c)/b \circ d,$$

$$a/b - e/f = (a \circ f - b \circ e)/b \circ f.$$

By Corollary 2.1.34, multiplication is distributive with respect to subtraction. This gives us the following equalities:

$$\begin{aligned}(a \circ d - b \circ c) \circ (b \circ f) &= a \circ d \circ b \circ f - b \circ c \circ b \circ f = a \circ b \circ d \circ f - b \circ b \circ c \circ f \\ &= a \circ b \circ f \circ d - b \circ b \circ e \circ d = a \circ f \circ b \circ d - b \circ e \circ b \circ d \\ &= (a \circ f - b \circ e) \circ (b \circ d)\end{aligned}$$

because

$$c \circ f = e \circ d$$

It means that

$$a/b - c/d \sim a/b - e/f.$$

Consequently, it is possible to define operation $-_2$ in the prearithmetic $\boldsymbol{B}$ by the following formula:

$$\langle a/b \rangle -_2 \langle c/d \rangle = \langle a/b - c/d \rangle.$$

Now, we need to check that this operation is subtraction according to Definition 2.1.11. Indeed, we have in the prearithmetic $\boldsymbol{B}$:

$$\begin{aligned}
c/d + (a/b - c/d) &= c/d + ((a \circ d - b \circ c)/b \circ d) \\
&= (c \circ b \circ d + (a \circ d - b \circ c) \circ d)/b \circ d) = (c \circ b \circ d \\
&\quad + (a \circ d \circ d - b \circ c \circ d))/b \circ d) \\
&= (b \circ c \circ d + (a \circ d \circ d - b \circ c \circ d))/b \circ d) = a \circ d \circ d/b \circ d = a/b
\end{aligned}$$

because multiplication is commutative, $-_2$ is subtraction in the prearithmetic $\boldsymbol{B}$ and $\boldsymbol{B}$ has the property of fraction reduction (cf. proof of Theorem 2.5.4).

Consequently,

$$\begin{aligned}
\langle c/d\rangle +_2 (\langle a/b\rangle -_2 \langle c/d\rangle) &= \langle c/d\rangle +_2 \langle a/b - c/d\rangle \\
&= \langle c/d + (a/b - c/d)\rangle = \langle a/b\rangle.
\end{aligned}$$

The equalities

$$\begin{aligned}
(\langle c/d\rangle +_2 \langle a/b\rangle) -_2 \langle c/d\rangle &= \langle a/b\rangle, \\
(\langle a/b\rangle -_2 \langle c/d\rangle) +_2 \langle c/d\rangle &= \langle a/b\rangle, \\
(\langle a/b\rangle +_2 \langle c/d\rangle) -_2 \langle c/d\rangle &= \langle a/b\rangle
\end{aligned}$$

are checked in the same way. Thus, $-_2$ is subtraction in the prearithmetic $\boldsymbol{B}$.

Theorem is proved. □

Theorem 2.5.5 allows proving the following result.

Theorem 2.5.7. *Any abstract prearithmetic* $\boldsymbol{A} = (A; +, \circ, \leq)$ *with subtraction, commutative and associative addition, commutative and associative multiplication, which is also distributive with respect to addition, and non-zero multiplicative cancellation can be inserted into an abstract prearithmetic* $\boldsymbol{B} = (B; +_2, \circ_2, \leq_2)$ *with non-zero division, subtraction, commutative and associative addition and multiplication, and distributive multiplication with respect to addition.*

Proof: Let us consider an abstract prearithmetic $\boldsymbol{A} = (A; +, \circ, \leq)$ with subtraction, commutative and associative addition, commutative and associative multiplication, which is also distributive with respect to addition, and non-zero multiplicative cancellation. Note that by Corollary 2.1.30, the prearithmetic $\boldsymbol{A}$ also has additive cancellation.

By Theorems 2.5.5, it is possible to insert the prearithmetic $\boldsymbol{A}$ into a prearithmetic $\boldsymbol{B}$, which preserves all properties of addition and multiplication in $\boldsymbol{A}$ and has non-zero division. To prove Theorem 2.5.7, we need only to show that $\boldsymbol{B}$ also has subtraction. This is done in the same way as in the proof of Theorems 2.5.6.

Theorem is proved. □

As a corollary, we obtain a well-known result (cf., for example, Kurosh, 1963; Rotman, 1996).

Corollary 2.5.9. *Any domain can be inserted into a field.*

Theorems 2.5.3 and 2.5.4 allow proving the following result.

Theorem 2.5.8. *Any abstract prearithmetic $\boldsymbol{A} = (A; +, \circ, \leq)$ with additive and multiplicative cancellation, commutative and associative addition, commutative and associative multiplication, which is also distributive with respect to addition, can be inserted into an abstract prearithmetic $\boldsymbol{B} = (B; +_2, \circ_2, \leq_2)$ with division, subtraction, commutative and associative addition and multiplication, and distributive multiplication with respect to addition.*

Proof: Let us consider an abstract prearithmetic $\boldsymbol{A} = (A; +, \circ, \leq)$ with additive and multiplicative cancellation, commutative and associative addition, commutative and associative multiplication, which is also distributive with respect to addition. Note that by Corollary 2.1.30, the prearithmetic $\boldsymbol{A}$ also has additive cancellation.

By Theorem 2.5.3, it is possible to insert the prearithmetic $\boldsymbol{A}$ into a prearithmetic $\boldsymbol{B}$, which preserves all properties of addition and multiplication in $\boldsymbol{A}$ and has subtraction. Then by Theorem 2.5.4, it is possible to insert the prearithmetic $\boldsymbol{A}$ into a prearithmetic $\boldsymbol{C}$, which preserves all properties of addition and multiplication in $\boldsymbol{A}$ and has division. By Theorem 2.5.6, $\boldsymbol{C}$ also has subtraction.

Theorem is proved. □

We also have the following result.

Theorem 2.5.9. *Any abstract prearithmetic $\boldsymbol{A} = (A; +, \circ, \leq)$ with additive and non-zero multiplicative cancellation, subtraction, commutative and associative addition, commutative and associative multiplication, which is also distributive with respect to addition, can be inserted into an abstract prearithmetic $\boldsymbol{B} = (B; +_2, \circ_2, \leq_2)$ with non-zero division, subtraction, commutative and associative addition and multiplication, and distributive multiplication with respect to addition.*

Proof is similar to the proof of Theorems 2.5.8 only instead of Theorems 2.5.4, we use Theorems 2.5.5. □

2.5.3. *Partial subprearithmetics*

Divide each difficulty into as many parts as is feasible and necessary to resolve it.

Rene Descartes

Developing theory of prearithmetics and arithmetics, it is also possible to consider partial subprearithmetics.

Definition 2.5.5.

(a) An abstract prearithmetic $\boldsymbol{A}_1 = (A_1; +_1, \circ_1, \leq_1)$ is an *additive subprearithmetic* of an abstract prearithmetic $\boldsymbol{A}_2 = (A_2; +_2, \circ_2, \leq_2)$ if $A_1 \subseteq A_2$ and the operation $+_1$ is the restriction of the operation $+_2$ onto A_1.
(b) An abstract prearithmetic $\boldsymbol{A}_1 = (A_1; +_1, \circ_1, \leq_1)$ is a *multiplicative subprearithmetic* of an abstract prearithmetic $\boldsymbol{A}_2 = (A_2; +_2, \circ_2, \leq_2)$ if $A_1 \subseteq A_2$ and the operation $\circ_1$ is the restriction of the operation $\circ_2$ onto A_1.

Example 2.5.10. Let us use a real number k and the function $f(x) = kx$ to define operations in the abstract prearithmetic $\boldsymbol{A} = (R; \dot{+}, \circ, \leq)$ as projective with respect to the Diophantine arithmetic $\boldsymbol{R}$ with the projector f and coprojector f^{-1}. Namely, for any whole numbers n and m, we have

$$n \dot{+} m = k^{-1}(kn + km) = n + m$$

and

$$n \circ m = k^{-1}(kn \cdot km) = k \cdot n \cdot m.$$

We see that the abstract prearithmetic $\boldsymbol{A}$ is an additive subprearithmetic of the Diophantine arithmetic $\boldsymbol{R}$ although it not a subprearithmetic of $\boldsymbol{R}$.

Example 2.5.11. Let us consider an abstract prearithmetic $\boldsymbol{A} = (N; \oplus, \cdot, \leq)$, in which the carrier N is the set of all natural numbers, where multiplication $\cdot$ and order $\leq$ are the same as in while addition $\oplus$ is defined by the following formula:

$$n \oplus m = n^2 + m^2.$$

We see that the prearithmetic $\boldsymbol{A}$ is a multiplicative subprearithmetic of the Diophantine arithmetic $\boldsymbol{N}$ although it is not a subprearithmetic of $\boldsymbol{N}$.

Lemma 2.5.6. *Any subprearithmetic of an abstract prearithmetic* $\boldsymbol{A}$ *is an additive subprearithmetic of* $\boldsymbol{A}$.

Proof is left as an exercise. □

Proposition 2.5.27. *Intersection of additive subprearithmetics of an abstract prearithmetic* $\boldsymbol{A}$ *is an additive subprearithmetic of* $\boldsymbol{A}$.

Proof is left as an exercise. □

This result shows how it is possible to build additive subprearithmetics.

Lemma 2.5.7. *If* $\boldsymbol{A} = (A; +, \circ, \leq)$ *is an abstract prearithmetic and* $C \subseteq A$, *then there is the least additive subprearithmetic* $\mathrm{asa}_C \boldsymbol{A} = (B; +, \circ, \leq)$ *with* $C \subseteq B \subseteq A$.

Proof is left as an exercise. □

Thus, the additive subprearithmetic $\mathrm{asa}_C \boldsymbol{A}$ is defined by three conditions:

(1) Any element from C belongs to B.
(2) If elements a and b belong to B, then $a + b$ belongs to B.
(3) There are no other elements in B.

We say that the additive subprearithmetic $\mathrm{asa}_C \boldsymbol{A} = (B; +, \circ, \leq)$ is *generated* by C in $\boldsymbol{A}$.

Note that multiplication can be defined in $\mathrm{asa}_C \boldsymbol{A}$ independently of multiplication in $\boldsymbol{A}$. However, order in $\mathrm{asa}_C \boldsymbol{A}$ is the same as order in $\boldsymbol{A}$.

Theorem 2.1.7 implies the following result. Let us assume that an abstract prearithmetic $\boldsymbol{A}_1 = (A_1; +_1, \circ_1, \leq_1)$ is weakly projective with respect to an abstract prearithmetic $\boldsymbol{A}_2 = (A_2; +_2, \circ_2, \leq_2)$ with the projector g, the coprojector h and an additive zero 0_2 whereas $gh = \mathbf{1}_{A_2}$.

Corollary 2.5.10. *If the prearithmetic* $\boldsymbol{A}_1$ *does not have additively prime numbers, then the prearithmetic* $\mathrm{asa}_{g(A1)}\, \boldsymbol{A}_2$ *also does not have additively prime numbers.*

Theorem 2.1.9 implies the following result. Let us assume that an abstract prearithmetic $\boldsymbol{A}_1 = (A_1; +_1, \circ_1, \leq_1)$ is weakly projective with respect to an abstract prearithmetic $\boldsymbol{A}_2 = (A_2; +_2, \circ_2, \leq_2)$ with the projector g, the coprojector h and a multiplicative one 1_2 whereas $gh = \mathbf{1}_{A_2}$.

Corollary 2.5.11. *If the prearithmetic* $\boldsymbol{A}_1$ *does not have multiplicatively prime numbers, then the prearithmetic* $\mathrm{msa}_{g(A_1)}\ \boldsymbol{A}_2$ *also does not have multiplicatively prime numbers.*

Like subprearithmetics, an additive subprearithmetic inherits some properties of the prearithmetic that includes it.

Proposition 2.5.28. *The following properties are inherited by additive subprearithmetics:*

(a) *the linear (total) order;*
(b) *the successively Archimedean property when successors are defined in the subprearithmetic and the prearithmetic that contains it;*
(c) *finite connectedness of the order relation;*
(d) *preservation of order by addition;*
(e) *preservation of order by multiplication;*
(f) *the additively Archimedean property;*
(g) *commutativity of addition;*
(h) *associativity of addition;*
(i) *identities* $a \leq a + a$ and $a \leq a + b$ or $a + a \leq a$ and $a + b \leq a$;
(j) *preservation of order by addition.*

Proof is similar to the proof of Proposition 2.5.4 and left as an exercise. □

Proposition 2.5.29. *The following properties are not always inherited by additive subprearithmetics:*

(a) *well-ordering;*
(b) *discreteness of the order relation;*
(c) *having a successor for all elements;*
(d) *having a predecessor for all elements;*
(e) *having an arithmetical successor* $\mathrm{AS}x = x + 1$ *for all elements;*
(f) *having an arithmetical predecessor* $\mathrm{AP}x = x - 1$ *for all elements;*
(g) *having an additive zero.*

Proof is similar to the proof of Proposition 2.5.5 and left as an exercise. □

When we study projectivity and weak projectivity of one prearithmetic with respect to another, inheritance of addition properties actually depends on properties of additive subprearithmetics generated by the image of the first prearithmetic and not so much on properties of the second prearithmetic. For instance, we have the following results.

Proposition 2.5.30. *If an abstract prearithmetic* $\boldsymbol{A}_1 = (A_1; +_1, \circ_1, \leq_1)$ *is weakly projective with respect to an abstract prearithmetic* $\boldsymbol{A}_2 = (A_2; +_2, \circ_2, \leq_2)$ *and the operation* $+_2$ *is commutative in the additive subprearithmetic of* $\boldsymbol{A}_2$ *generated by the image* $g(A_1)$ *in* $\boldsymbol{A}_2$, *then the operation* $+_1$ *is commutative in the prearithmetic* $\boldsymbol{A}_1$.

Proof is similar to the proof of Proposition 2.2.4. □

Proposition 2.5.31. *If an abstract prearithmetic* $\boldsymbol{A}_1 = (A_1; +_1, \circ_1, \leq_1)$ *is weakly projective with respect to an abstract prearithmetic* $\boldsymbol{A}_2 = (A_2; +_2, \circ_2, \leq_2)$, *the coprojector* h *and the operation* $+_2$ *preserve the order* $\leq_2$ *in the additive subprearithmetic of* $\boldsymbol{A}_2$ *generated by the image* $g(A_1)$ *in* $\boldsymbol{A}_2$, *then the operation* $+_1$ *preserves the order* $\leq_1$.

Proof is similar to the proof of Proposition 2.2.11. □

Now let us study properties of multiplicative subprearithmetics.

Lemma 2.5.8. *Any subprearithmetic of an abstract prearithmetic* $\boldsymbol{A}$ *is a multiplicative subprearithmetics of* $\boldsymbol{A}$.

Proof is left as an exercise. □

Proposition 2.5.32. *Intersection of multiplicative subprearithmetics of an abstract prearithmetic* $\boldsymbol{A}$ *is a multiplicative subprearithmetic of* $\boldsymbol{A}$.

Proof is left as an exercise. □

This result shows how it is possible to build multiplicative subprearithmetics.

Lemma 2.5.9. *If* $\boldsymbol{A} = (A; +, \circ, \leq)$ *is an abstract prearithmetic and* $C \subseteq A$, *then there is the least multiplicative subprearithmetic* $\mathrm{msa}_C \boldsymbol{A} = (B; +, \circ, \leq)$ *with* $C \subseteq B \subseteq A$.

Proof is left as an exercise. □

Thus, the multiplicative subprearithmetic $\mathrm{msa}_C \boldsymbol{A}$ is defined by three conditions:

(1) Any element from C belongs to B.
(2) If elements a and b belong to B, then $a \circ b$ belongs to B.
(3) There are no other elements in B.

We say that the multiplicative prearithmetic msa$_C$ $\boldsymbol{A} = (B; +, \circ, \leq)$ is *generated* by C in $\boldsymbol{A}$.

Note that addition can be defined in msa$_C$ $\boldsymbol{A}$ independently of addition in $\boldsymbol{A}$. However, order in msa$_C$ $\boldsymbol{A}$ is the same as order in $\boldsymbol{A}$.

Example 2.5.12. Let us define operations in the abstract prearithmetic $\boldsymbol{A} = (R; \dot{+}, \circ, \leq)$ so that the sum of any number of numbers from R is equal to 0, while multiplication is the same as in the Diophantine arithmetic $\boldsymbol{R}$. Then it is possible to see the abstract prearithmetic $\boldsymbol{A}$ it is a multiplicative subprearithmetic of the Diophantine arithmetic $\boldsymbol{R}$ although it is not a subprearithmetic of $\boldsymbol{R}$.

Like subprearithmetics, a multiplicative subprearithmetic inherits some properties of the prearithmetic that includes it.

Proposition 2.5.33. *The following properties are inherited by multiplicative subprearithmetics:*

(a) *the linear (total) order;*
(b) *the successively Archimedean property when successors are defined in the subprearithmetic and the prearithmetic that contains it;*
(c) *finite connectedness of the order relation;*
(e) *the multiplicatively Archimedean property;*
(f) *commutativity of multiplication;*
(g) *associativity of multiplication;*
(h) *distributivity of multiplication with respect to addition;*
(i) *identities* $a \leq a \circ a$ and $a \leq a \circ b$ or $a \circ a \leq a$ and $a \circ b \leq a$;
(j) *preservation of order by multiplication.*

Proof is similar to the proof of Proposition 2.5.4 and left as an exercise. □

Proposition 2.5.34. *The following properties are not always inherited by multiplicative subprearithmetics:*

(a) *well-ordering;*
(b) *discreteness of the order relation;*
(c) *having a successor for all elements;*
(d) *having a predecessor for all elements;*
(e) *having a multiplicative zero* 0;
(f) *having a multiplicative one* 1.

Proof is similar to the proof of Proposition 2.5.5 and is left as an exercise. □

When we study projectivity and weak projectivity of one prearithmetic with respect to another, inheritance of multiplication properties actually depends on properties of multiplicative subprearithmetics generated by the image of the first prearithmetic and not so much on properties of the second prearithmetic. For instance, we have the following results.

Proposition 2.5.35. *If an abstract prearithmetic* $\boldsymbol{A}_1 = (A_1; +_1, \circ_1, \leq_1)$ *is weakly projective with respect to an abstract prearithmetic* $\boldsymbol{A}_2 = (A_2; +_2, \circ_2, \leq_2)$ *and the operation* $\circ_2$ *is commutative in the multiplicative subprearithmetic of* $\boldsymbol{A}_2$ *generated by the image* $g(A_1)$ *in* $\boldsymbol{A}_2$*, then the operation* $\circ_1$ *is commutative in the prearithmetic* $\boldsymbol{A}_1$*.*

Proof is similar to the proof of Proposition 2.2.5. □

Proposition 2.5.36. *If an abstract prearithmetic* $\boldsymbol{A}_1 = (A_1; +_1, \circ_1, \leq_1)$ *is weakly projective with respect to an abstract prearithmetic* $\boldsymbol{A}_2 = (A_2; +_2, \circ_2, \leq_2)$*, the coprojector* h *and the operation* $\circ_2$ *preserve the order* $\leq_2$ *in the multiplicative subprearithmetic of* $\boldsymbol{A}_2$ *generated by the image* $g(A_1)$ *in* $\boldsymbol{A}_2$*, then the operation* $\circ_1$ *preserves the order* $\leq_1$*.*

Proof is similar to the proof of Proposition 2.2.12. □

Let us find some properties of additive and multiplicative subprearithmetics of the Diophantine arithmetic $\boldsymbol{N}$ and other conventional arithmetics.

Proposition 2.5.37. *If an additive subprearithmetic* $\boldsymbol{A} = (A; +, \circ, \leq)$ *of the Diophantine arithmetic* $\boldsymbol{N}$ *contains* 1*, then* $\boldsymbol{A}$ *coincides with* $\boldsymbol{N}$*.*

Indeed, as it is possible to obtain any natural number by adding 1 to itself the necessary number of times, we have $\boldsymbol{A} = \boldsymbol{N}$.

Remark 2.5.5. A similar statement for multiplicative subprearithmetics is not true as the following example demonstrates.

Example 2.5.13. The set $\{1\}$ with the conventional multiplication and addition $1 + 1 = 1$ is a multiplicative subprearithmetic of $\boldsymbol{N}$.

Proposition 2.5.38. *If an additive subprearithmetic* $\boldsymbol{B} = (B; +, \circ, \leq)$ *of the Diophantine arithmetic* $\boldsymbol{W}$ *contains* 1*, then* $\boldsymbol{B}$ *coincides either with* $\boldsymbol{N}$ *or with* $\boldsymbol{W}$*.*

Proof is left as an exercise. □

Remark 2.5.6. A similar statement for multiplicative subprearithmetics is not true as Example 2.5.13 demonstrates.

Proposition 2.5.39. *If an additive subprearithmetic* $\boldsymbol{D} = (D; +, \circ, \leq)$ *of the Diophantine arithmetic* $\boldsymbol{W}$ *contains* 1 *and* 0, *then* $\boldsymbol{D}$ *coincides with* $\boldsymbol{W}$.

Proof is left as an exercise. □

Remark 2.5.7. A similar statement for multiplicative subprearithmetics is not true as the following example demonstrates.

Example 2.5.14. The set $\{0, 1\}$ with the conventional multiplication and order $0 \leq 1$ is a multiplicative subprearithmetic of $\boldsymbol{Z}$ when addition is defined by the following formulas:

$$0 + 0 = 0,$$

$$0 + 1 = 1 + 0 = 1,$$

$$1 + 1 = 1.$$

Remark 2.5.8. There are many proper additive subprearithmetics in arithmetics $\boldsymbol{N}$ and $\boldsymbol{W}$. An example of such a proper subprearithmetic is the set $2N$ of all even numbers with the conventional operations and order.

Proposition 2.5.40. *If an additive subprearithmetic* $\boldsymbol{H} = (H; +, \circ, \leq)$ *of the arithmetic* $\boldsymbol{Z}$ *of integer numbers contains* 1 *and some negative number, then* $\boldsymbol{H}$ *coincides with* $\boldsymbol{Z}$.

Proof is left as an exercise. □

Remark 2.5.9. A similar statement for multiplicative subprearithmetics is not true as the following examples demonstrates.

Example 2.5.15. The set $\{1, -1\}$ with the conventional multiplication and order $-1 \leq 1$ is a multiplicative subprearithmetic of $\boldsymbol{Z}$ although addition in it is different being defined by the following formulas:

$$(-1) + (-1) = 1,$$

$$(-1) + 1 = 1 + (-1) = 1,$$

$$1 + 1 = 1.$$

Proposition 2.5.41. *If an additive subprearithmetic* $\boldsymbol{G} = (G; +, \circ, \leq)$ *of the arithmetic* $\boldsymbol{Z}$ *of integer numbers contains* -1 *and some positive number, then* $\boldsymbol{G}$ *coincides with* $\boldsymbol{Z}$.

Proof is left as an exercise. □

Remark 2.5.10. A similar statement for multiplicative subprearithmetics is not true as the Example 2.5.16 demonstrates.

Example 2.5.16. Taking the set $A = \{0, 1, -1\}$, it is possible to define addition in it by the following formulas:

$$0 + 0 = 0,$$

$$0 + 1 = 1 + 0 = 1,$$

$$1 + 1 = 1,$$

$$(-1) + (-1) = 1,$$

$$(-1) + 1 = 1 + (-1) = 0,$$

$$0 + (-1) = (-1) + 0 = -1.$$

At the same time, we can use the conventional multiplication and order $-1 \leq 0 \leq 1$ in this set. In such a way, we obtain an abstract prearithmetic $\boldsymbol{A} = (A; +, \cdot, \leq_A)$. It is a proper multiplicative subprearithmetic of the Diophantine arithmetic $\boldsymbol{Z}$ but not a subprearithmetic and not an additive subprearithmetic of $\boldsymbol{Z}$.

These results show essential difference between additive and multiplicative subprearithmetics.

2.6. Skew Projectivity of Abstract Prearithmetics

In art or architecture, your project is only done when you say it's done.
If you want to rip it apart at the eleventh hour
and start all over again, you never finish.

Maya Lin

Skew projectivity of abstract prearithmetics is a special kind of weak projectivity. By definition, weak projectivity is defined by two mapping of abstract prearithmetics. In skew projectivity, one of these mappings is the identity mapping. In this section, two kinds of skew projectivity between abstract prearithmetics — direct and inverse skew projectivity — are introduced and studied. These types of skew projectivity are interesting as mathematical constructions while their properties are important because some useful mathematical structures, such as the logarithmic scale, relations between percents and their numerical representations or rounding in computations, arise from specific cases of skew projectivity.

2.6.1. *Direct skew projectivity*

When adults first become conscious of something new, they usually either attack or try to escape from it... Attack includes such mild forms as ridicule, and escape includes merely putting out of mind.

W. I. B. Beveridge

Informally, direct skew projectivity means that operations with elements of a prearithmetic, e.g., with numbers, are actually performed with their images in another arithmetic, which has the same elements.

Definition 2.6.1. An abstract prearithmetic $\boldsymbol{A}_1 = (A; +_1, \circ_1, \leq_1)$ is called *directly skew projective* with respect to a prearithmetic $\boldsymbol{A}_2 = (A; +_2, \circ_2, \leq_2)$ if it is weakly projective with respect to $\boldsymbol{A}_2$ and the coprojector h is the identity mapping 1_A of A.

It means that in direct skew projectivity of $\boldsymbol{A}_1$ with respect to $\boldsymbol{A}_2$, we have

$$a +_1 b = g(a) +_2 g(b),$$

$$a \circ_1 \ b = g(a) \circ_2 \ g(b).$$

Example 2.6.1. Mathematicians started implicitly using a special case of direct skew projectivity of abstract prearithmetics and arithmetics several centuries ago when, as it is described in Sections 1.4 and 2.2, logarithms were introduced and employed for reducing wide-ranging quantities to much smaller numbers and reducing multiplication to a much simpler addition (Napier, 1614; Burton, 1997). Logarithmic scales have been utilized in physics, chemistry, and other sciences (Pierce, 1977). Logarithm tables were extensively used in navigation and engineering.

In particular, logarithms were used in slide rules and logarithm tables for several centuries. Now the logarithmic number system (LNS) is employed for number representation in application-specific computer systems (Parhami, 2010). In LNS, a value x is represented by its sign and the logarithm of its absolute value in some base.

In mathematics, logarithmic derivatives play important role in the theory of meromorphic functions (Nevanlinna, 1939) and in the area of differential equations (cf., for example, Bali, 2005; Bird, 1993; Krantz, 2003).

Besides, scientists found that in some cultures, people map symbolic and non-symbolic numbers onto a logarithmic scale operating with them in diverse settings (Burr and Ross, 2008; Dehaene, 2003; Dehaene *et al.*, 1998; 2004; 2008; Crollen *et al.*, 2013; Piazza and Dehaene, 2004). Moreover, researchers conjectured that the concept of a linear number line appears to be a cultural invention that fails to develop in the absence of formal education.

In addition, psychologists discovered that when young children attempt to locate the positions of numerals on a number line, the positions are often logarithmically rather than linearly distributed (Rips, 2013). This finding has been taken as evidence that the children represent numbers on a mental number line that is logarithmically calibrated.

Example 2.6.2. It is interesting that people also started implicitly using a special case of direct skew projectivity of numerical prearithmetics and arithmetics long ago with the advent of percent. It was essentially earlier than logarithms were invented. Percents are used to express ratios, which are usually small, by 100 times larger numbers. This conversion makes addition and subtraction easier and the ratios more graspable.

Although some think that a percentage is a pure number, this is not true because a percentage is the number of percents. For instance, 5 apples is the number of apples and in the same way, 5 percent or 5% is the number of percents.

Transformation of fractions into percent is a direct skew projectivity because to get percent the fraction is multiplied by 100, i.e., the projector is the function $g(a) = 100a$.

These examples show that direct skew projectivity has important applications and it is useful to know its properties.

It is unreasonable to extend direct skew projectivity to projectivity as the following example demonstrates.

Lemma 2.6.1. *If an abstract prearithmetic* $\boldsymbol{A}_1 = (A; +_1, \circ_1, \leq_1)$ *is directly skew projective and projective with respect to a prearithmetic* $\boldsymbol{A}_2 = (A; +_2, \circ_2, \leq_2)$, *then these prearithmetics coincide.*

Proof is left as an exercise. □

Because direct skew projectivity is a special case of weak projectivity, it inherits many properties of weak projectivity.

Proposition 2.6.1. *If an abstract prearithmetic $\boldsymbol{A}_1 = (A; +_1, \circ_1, \leq_1)$ is directly skew projective with respect to an abstract prearithmetic $\boldsymbol{A}_2 = (A; +_2, \circ_2, \leq_2)$ and the operation $+_2$ is commutative in the prearithmetic $\boldsymbol{A}_2$, then the operation $+_1$ is commutative in the prearithmetic $\boldsymbol{A}_1$.*

Indeed, we have

$$a +_1 b = g(a) +_2 g(b) = g(b) +_2 g(c) = b +_1 a.$$

Multiplication in abstract prearithmetics has the same property.

Proposition 2.6.2. *If an abstract prearithmetic $\boldsymbol{A}_1 = (A; +_1, \circ_1, \leq_1)$ is directly skew projective with respect to an abstract prearithmetic $\boldsymbol{A}_2 = (A; +_2, \circ_2, \leq_2)$ and the operation $\circ_2$ is commutative in the prearithmetic $\boldsymbol{A}_2$, then the operation $\circ_1$ is commutative in the prearithmetic $\boldsymbol{A}_1$.*

Proof is similar to the proof of Proposition 2.6.1. □

However, associativity, as a rule, is not preserved in direct skew projectivity.

Example 2.6.3. Let us take the Diophantine arithmetic $\boldsymbol{N} = (N; +, \cdot, \leq)$ of all natural numbers and build the abstract prearithmetic $\boldsymbol{A} = (N; +_1, \circ_1, \leq_1)$ is directly skew projective with respect to $\boldsymbol{N}$ using the mapping g: $N \to N$ as $g(x) = 2x$. Then we have

$$a +_1 (b +_1 c) = g(a) +_2 g(g(b) +_2 g(c)) = 2a + 2(2b +_2 2c) = 2a + 4b +_2 4c$$

while

$$(a +_1 b) +_1 c = g(g(a) +_2 g(b)) +_2 g(c) = 2(2a + 2b) +_2 2c = 4a + 4b +_2 2c.$$

Thus, in a general case,

$$a +_1 (b +_1 c) \neq (a +_1 b) +_1 c.$$

We see that although in $\boldsymbol{N}$, addition is associative and $\boldsymbol{A}$ is directly skew projective with respect to $\boldsymbol{N}$, in $\boldsymbol{A}$, addition is not associative.

The same is true for multiplication in $\boldsymbol{A}$.

However, when direct skew projectivity satisfies additional conditions, it preserves associativity.

Let us consider condition (Z)

$$g(0_2) = 0_2,$$

where 0_2 is an additive zero in $\boldsymbol{A}_2$ and g: $A \to A$.

Proposition 2.6.3. *If an abstract prearithmetic $\boldsymbol{A}_1 = (A; +_1, \circ_1, \leq_1)$ is directly skew projective with respect to an abstract prearithmetic $\boldsymbol{A}_2 = (A; +_2, \circ_2, \leq_2)$ with an additive zero 0_2 and associative addition while the projector g is a homomorphism with respect to the addition $+_2$ and satisfies condition (Z), then addition $+_1$ in $\boldsymbol{A}_1$ is associative if and only if the projector g is idempotent, i.e., $g^2 = g$.*

Proof: *Necessity.* Let us assume that addition $+_1$ in $\boldsymbol{A}_1$ is associative and the abstract prearithmetic $\boldsymbol{A}_2$ has an additive zero 0_2 and associative addition while the projector g is a homomorphism with respect to the addition $+_2$ and satisfies condition (Z). Then for an arbitrary element a from A, we have

$$\begin{aligned} a +_1 (0_2 +_1 0_2) &= g(a) +_2 g(g(0_2) +_2 g(0_2)) = g(a) +_2 g(0_2 +_2 0_2) \\ &= g(a) +_2 g(0_2) = g(a) +_2 0_2 = g(a). \end{aligned}$$

At the same time,

$$(a +_1 c) +_1 c = g(g(a) +_2 g(c)) +_2 g(c) = g(g(a) +_2 0_2) +_2 0_2 = g(g(a)).$$

If addition $+_1$ in $\boldsymbol{A}_1$ is associative, then

$$g(g(a)) = g(a).$$

This means $g^2 = g$ because a is an arbitrary element from A.

Sufficiency. Let us assume that $g^2 = g$ and g is a homomorphism with respect to the addition $+_2$. Then for arbitrary elements a, b and c from A, we have

$$\begin{aligned} a +_1 (b +_1 c) &= g(a) +_2 g(g(b) +_2 g(c)) = g(a) +_2 g(g(b)) +_2 g(g(c)) \\ &= g(a) +_2 (g(b) +_2 g(c)) = (g(a) +_2 g(b)) +_2 g(c) \\ &= (g(g(a) +_2 g(g(b)) +_2 g(c) \\ &= g(g(a) +_2 g(b)) +_2 g(c) = (a +_1 b) +_1 c. \end{aligned}$$

It means that addition $+_1$ in $\boldsymbol{A}_1$ is associative.

Proposition is proved. □

As any homomorphism is an additive homomorphism, we have the following result.

Corollary 2.6.1. *If an abstract prearithmetic* $\boldsymbol{A}_1 = (A; +_1, \circ_1, \leq_1)$ *is directly skew projective with respect to an abstract prearithmetic* $\boldsymbol{A}_2 = (A; +_2, \circ_2, \leq_2)$ *with an additive zero* 0_2 *and associative addition, while the projector g is a homomorphism of the prearithmetic* $\boldsymbol{A}_2$ *and satisfies condition (Z), then addition* $+_1$ *in* $\boldsymbol{A}_1$ *is associative if and only if the projector g is idempotent, i.e.,* $g^2 = g$.

Let us consider condition (M)

$$g(1_2) = 1_2,$$

where 1_2 is a multiplicative one in $\boldsymbol{A}_2$ and $g \colon A \to A$.

Proposition 2.6.4. *If an abstract prearithmetic* $\boldsymbol{A}_1 = (A; +_1, \circ_1, \leq_1)$ *is directly skew projective with respect to an abstract prearithmetic* $\boldsymbol{A}_2 = (A; +_2, \circ_2, \leq_2)$ *with a multiplicative one* 1_2 *and associative multiplication, while the projector g is a homomorphism with respect to the multiplication* $\circ_2$ *and satisfies condition (M), then multiplication* $\circ_1$ *in* $\boldsymbol{A}_1$ *is associative if and only if the projector g is idempotent, i.e.,* $g^2 = g$.

Proof: *Necessity.* Let us assume that multiplication $\circ_1$ in $\boldsymbol{A}_1$ is associative and the abstract prearithmetic $\boldsymbol{A}_2$ has a multiplicative one 1_2 and associative multiplication, while the projector g is a homomorphism with respect to the multiplication $\circ_1$ and satisfies condition (M). Then for an arbitrary element a from A, we have

$$\begin{aligned} a \circ_1 (1_2 \circ_1 1_2) &= g(a) \circ_2 g(g(1_2) \circ_2 \ g(1_2)) = g(a) \circ_2 g(1_2 \circ_2 1_2) \\ &= g(a) \circ_2 g(1_2) = g(a) \circ_2 1_2 = g(a). \end{aligned}$$

At the same time,

$$(a \circ_1 c) \circ_1 c = g(g(a) \circ_2 g(c)) \circ_2 g(c) = g(g(a) \circ_2 1_2) \circ_2 1_2 = g(g(a)).$$

If multiplication $\circ_1$ in $\boldsymbol{A}_1$ is associative, then

$$g(g(a)) = g(a).$$

This means $g^2 = g$ because a is an arbitrary element from A.

Sufficiency. Let us assume that $g^2 = g$ and g is a homomorphism with respect to the multiplication $\circ_2$. Then for arbitrary elements a, b and c from

A, we have

$$\begin{aligned} a \circ_1 (b \circ_1 c) &= g(a) \circ_2 g(g(b) \circ_2 g(c)) = g(a) \circ_2 \; g(g(b)) \circ_2 g(g(c)) \\ &= g(a) \circ_2 (g(b) \circ_2 g(c)) = (g(a) \circ_2 g(b)) \circ_2 g(c) \\ &= (g(g(a) \circ_2 \; g(g(b)) \circ_2 g(c) = g(g(a) \circ_2 g(b)) \circ_2 g(c) \\ &= (a \circ_1 b) \circ_1 c. \end{aligned}$$

It means that multiplication $\circ_1$ in $\boldsymbol{A}_1$ is associative.

Proposition is proved. □

As any homomorphism is a multiplicative, we have the following result.

Corollary 2.6.2. *If an abstract prearithmetic $\boldsymbol{A}_1 = (A; +_1, \circ_1, \leq_1)$ is directly skew projective with respect to an abstract prearithmetic $\boldsymbol{A}_2 = (A; +_2, \circ_2, \leq_2)$ with a multiplicative one 1_2 and associative multiplication while the projector g is a homomorphism of the prearithmetic $\boldsymbol{A}_2$ and satisfies condition (M), then multiplication $\circ_1$ in $\boldsymbol{A}_1$ is associative if and only if the projector g is idempotent, i.e., $g^2 = g$.*

Under definite conditions, skew projectivity preserves order. Namely, Proposition 2.2.11 implies the following result.

Proposition 2.6.5. *If an abstract prearithmetic $\boldsymbol{A}_1 = (A; +_1, \circ_1, \leq_1)$ is directly skew projective with respect to an abstract prearithmetic $\boldsymbol{A}_2 = (A; +_2, \circ_2, \leq_2)$, the projector g preserves the order $\leq_2$, i.e., $x \leq_2 y$ implies $g(x) \leq_1 g(y)$, and the operation $+_2$ preserves the order $\leq_2$, then the operation $+_1$ preserves the order $\leq_1$.*

Proof is left as an exercise. □

Lemma 2.1.6 implies the following result.

Corollary 2.6.3. *If an abstract prearithmetic $\boldsymbol{A}_1 = (A; +_1, \circ_1, \leq_1)$ is directly skew projective with respect to an abstract prearithmetic $\boldsymbol{A}_2 = (A; +_2, \circ_2, \leq_2)$, the projector g preserves the order $\leq_2$, i.e., $x \leq_2 y$ implies $g(x) \leq_1 g(y)$, and the operation $+_2$ is monotone, then the operation $+_1$ is monotone.*

Proposition 2.2.12 implies the following result.

Proposition 2.6.6. *If an abstract prearithmetic $\boldsymbol{A}_1 = (A; +_1, \circ_1, \leq_1)$ is directly skew projective with respect to an abstract prearithmetic $\boldsymbol{A}_2 = (A; +_2, \circ_2, \leq_2)$, the projector g preserves the order $\leq_2$, and the operation $\circ_2$ preserves the order $\leq_2$, then the operation $\circ_1$ preserves the order $\leq_1$.*

Proof is left as an exercise. □

Lemma 2.1.7 implies the following result.

Corollary 2.6.4. *If an abstract prearithmetic* $\boldsymbol{A}_1 = (A; +_1, \circ_1, \leq_1)$ *is directly skew projective with respect to an abstract prearithmetic* $\boldsymbol{A}_2 = (A; +_2, \circ_2, \leq_2)$, *the projector* g *preserves the order* $\leq_2$, *and the operation* $\circ_2$ *is monotone, then the operation* $\circ_1$ *is monotone.*

Now let us study the cancellation property.

Proposition 2.6.7. *If an abstract prearithmetic* $\boldsymbol{A}_1 = (A; +_1, \circ_1, \leq_1)$ *is directly skew projective with respect to an abstract prearithmetic* $\boldsymbol{A}_2 = (A; +_2, \circ_2, \leq_2)$ *with additive cancellation, then* $\boldsymbol{A}_1$ *is an abstract prearithmetic with additive cancellation if and only if the projector* g *is an injection.*

Proof: *Necessity.* Let us assume that an abstract prearithmetic $\boldsymbol{A}_1 = (A; +_1, \circ_1, \leq_1)$ is directly skew projective with respect to an abstract prearithmetic $\boldsymbol{A}_2 = (A; +_2, \circ_2, \leq_2)$ and the projector g is not an injection. Then there are elements a and b from $\boldsymbol{A}_1$ such that $g(a) = g(b)$ while $a \neq b$. Then for any element c from $\boldsymbol{A}_1$, we have

$$a +_1 c = g(a) +_2 g(c) = g(b) +_2 g(c) = b +_1 c$$

although $a \neq b$. It means that $\boldsymbol{A}_1$ is not a prearithmetic with additive cancellation.

Sufficiency. Let us assume that an abstract prearithmetic $\boldsymbol{A}_1 = (A; +_1, \circ_1, \leq_1)$ is directly skew projective with respect to an abstract prearithmetic $\boldsymbol{A}_2 = (A; +_2, \circ_2, \leq_2)$ with additive cancellation and the projector g is an injection. If

$$a +_1 c = g(a) +_2 g(c) = g(b) +_2 g(c) = b +_1 c,$$

then $g(a) = g(b)$ because $\boldsymbol{A}_2$ is an abstract prearithmetic with cancellation. Consequently, $a = b$ as g is an injection. It means that $\boldsymbol{A}_2$ is an abstract prearithmetic with additive cancellation.

Proposition is proved. □

The same is true for multiplication.

Proposition 2.6.8. *If an abstract prearithmetic* $\boldsymbol{A}_1 = (A; +_1, \circ_1, \leq_1)$ *is directly skew projective with respect to an abstract prearithmetic* $\boldsymbol{A}_2 = (A; +_2, \circ_2, \leq_2)$ *with multiplicative cancellation, then* $\boldsymbol{A}_1$ *is an abstract* *prearithmetic with multiplicative cancellation if and only if the projector g is an injection.*

Proof is similar to the proof of Proposition 2.6.7. □

These results show that preservation of the cancellation law for addition or multiplication demands injectivity of the projector.

Direct skew projectivity is a transitive relation.

Proposition 2.6.9. *If an abstract prearithmetic $\boldsymbol{A}_1 = (A_1; +_1, \circ_1, \leq_1)$ is directly skew projective with respect to an abstract prearithmetic $\boldsymbol{A}_2 = (A_2; +_2, \circ_2, \leq_2)$ and the abstract prearithmetic $\boldsymbol{A}_2 = (A_2; +_2, \circ_2, \leq_2)$ is directly skew projective with respect to an abstract prearithmetic $\boldsymbol{A}_3 = (A_3; +_3, \circ_3, \leq_3)$, then the prearithmetic $\boldsymbol{A}_1$ is directly skew projective with respect to the prearithmetic $\boldsymbol{A}_3$.*

Proof is similar to the proof of Proposition 2.4.4. □

Proposition 2.6.6 allows proving the following result.

Theorem 2.6.1. *Abstract prearithmetics with direct skew projectivity relations form a category where objects are abstract prearithmetics and morphisms are direct skew projectivity relations.*

Proof is left as an exercise. □

In a general case, direct skew projectivity does not preserve distributivity of multiplication with respect to addition. However, we can define generalized distributivity, which is intrinsically related to skew projectivity. Namely, having skew projectivity, it is possible to consider skew distributivity. Let us assume that an abstract prearithmetic $\boldsymbol{A}_1 = (A; +_1, \circ_1, \leq_1)$ is skew projective with respect to an abstract prearithmetic $\boldsymbol{A}_2 = (A; +_2, \circ_2, \leq_2)$ with the projector $g(x)$.

Definition 2.6.2. The multiplication $\circ_1$ is skew distributive if the following identity is valid:

$$x \circ_1 (y +_1 z) = g((x \circ_1 y)) +_2 g((x \circ_1 z)).$$

Proposition 2.6.10. *If an abstract prearithmetic $\boldsymbol{A}_1 = (A; +_1, \circ_1, \leq_1)$ is directly skew projective with respect to an abstract prearithmetic $\boldsymbol{A}_2 = (A; +_2, \circ_2, \leq_2)$, in which multiplication is distributive with respect to* *addition, and the projector g is idempotent and is a homomorphism of the prearithmetic $\boldsymbol{A}_2$, then the multiplication $\circ_1$ in $\boldsymbol{A}_1$ is skew distributive.*

Proof: Let us assume that g is a homomorphism. Then for arbitrary elements a, b and c from A, we have

$$\begin{aligned} a \circ_1 (b +_1 c) &= g(a) \circ_2 g(g(b) +_2 g(c)) = g(a) \circ_2 (g(g(b)) +_2 g(g(c))) \\ &= g(a) \circ_2 (g(b) +_2 g(c)) = (g(a) \circ_2 g(b)) +_2 (g(a) \circ_2 \ g(c)) \\ &= (g(g(a) \circ_2 g(g(b))) +_2 (g(g(a)) \circ_2 g(g(c))) \\ &= (g(g(a) \circ_2 g(b))) +_2 (g(g(a)) \circ_2 g(c))) \\ &= g((a \circ_1 b) +_2 g((a \circ_1 c)). \end{aligned}$$

It means that multiplication $\circ_1$ in $\boldsymbol{A}_1$ is skew distributive.

Proposition is proved. □

Proposition 2.2.14 implies the following result.

Proposition 2.6.11. *If an abstract prearithmetic $\boldsymbol{A}_1 = (A; +_1, \circ_1, \leq_1)$ is directly skew projective with respect to an abstract prearithmetic $\boldsymbol{A}_2 = (A; +_2, \circ_2, \leq_2)$, for any element x from A_2, its coimage $g^{-1}(x)$ is finite and any maximal chain in $\boldsymbol{A}_2$ with distinct elements, which has the beginning and the end, is finite, then any maximal chain in $\boldsymbol{A}_1$ with distinct elements, which has the beginning and the end, is finite.*

Proof is left as an exercise. □

Proposition 2.2.15 implies the following result.

Proposition 2.6.12. *If an abstract prearithmetic $\boldsymbol{A}_1 = (A; +_1, \circ_1, \leq_1)$ is directly skew projective with respect to an abstract prearithmetic $\boldsymbol{A}_2 = (A; +_2, \circ_2, \leq_2)$, for any element x from A_2, its coimage $g^{-1}(x)$ is finite and the partially ordered set A_2 is finitely connected, then the partially ordered set A_1 is finitely connected.*

Proof is left as an exercise. □

Let us consider prearithmetics with zeros and ones.

Proposition 2.6.13. (a) *If an abstract prearithmetic $\boldsymbol{A}_1 = (A; +_1, \circ_1, \leq_1)$ is directly skew projective with respect to an abstract prearithmetic $\boldsymbol{A}_2 = (A; +_2, \circ_2, \leq_2)$ and $g(a)$ is an additive zero in $\boldsymbol{A}_2$, then a is an additive zero in $\boldsymbol{A}_1$.*

(b) *If an abstract prearithmetic* $\boldsymbol{A}_1 = (A; +_1, \circ_1, \leq_1)$ *is directly skew projective with respect to an abstract prearithmetic* $\boldsymbol{A}_2 = (A; +_2, \circ_2, \leq_2)$ *and* $g(a)$ *is a multiplicative zero in* $\boldsymbol{A}_2$*, then* a *is a multiplicative zero in* $\boldsymbol{A}_1$*.*

(c) *If an abstract prearithmetic* $\boldsymbol{A}_1 = (A; +_1, \circ_1, \leq_1)$ *is directly skew projective with respect to an abstract prearithmetic* $\boldsymbol{A}_2 = (A; +_2, \circ_2, \leq_2)$ *and* $g(d)$ *is a multiplicative one in* $\boldsymbol{A}_2$*, then* d *is a multiplicative one in* $\boldsymbol{A}_1$*.*

Proof is left as an exercise. □

Proposition 2.6.14. *If an abstract prearithmetic* $\boldsymbol{A}_1 = (A; +_1, \circ_1, \leq_1)$ *is directly skew projective with respect to an abstract prearithmetic* $\boldsymbol{A}_2 = (A; +_2, \circ_2, \leq_2)$ *and* a *is an additively composite number in* $\boldsymbol{A}_1$*, then* a *is also an additively composite number in* $\boldsymbol{A}_2$*.*

Proof: Let us assume that a is an additively composite number in $\boldsymbol{A}_1$. It means that for some numbers $c \neq 0_{A_1}$ and $b \neq 0_{A_1}$, we have $a = c +_1 b$. Then by definition, we have $a = c +_1 b = g(c) +_2 g(b)$. Besides, $g(c) \neq 0_{A_2}$ and $g(b) \neq 0_{A_2}$ because if, for example, $g(c) = 0_{A_2}$, then by Proposition 2.6.13(a), we have $c = 0_{A_1}$, which is not true.

Proposition is proved. □

Corollary 2.6.5. *If an abstract prearithmetic* $\boldsymbol{A}_1 = (A; +_1, \circ_1, \leq_1)$ *is directly skew projective with respect to an abstract prearithmetic* $\boldsymbol{A}_2 = (A; +_2, \circ_2, \leq_2)$*, then any additively prime number in* $\boldsymbol{A}_2$ *is also additively prime in* $\boldsymbol{A}_1$*.*

Proposition 2.6.15. *If an abstract prearithmetic* $\boldsymbol{A}_1 = (A; +_1, \circ_1, \leq_1)$ *is directly skew projective with respect to an abstract prearithmetic* $\boldsymbol{A}_2 = (A; +_2, \circ_2, \leq_2)$ *and* a *is a multiplicatively composite number in* $\boldsymbol{A}_1$*, then* a *is also a multiplicatively composite number in* $\boldsymbol{A}_2$*.*

Proof: Let us assume that a is an additively composite number in $\boldsymbol{A}_1$. It means that for some numbers $c \neq 1_{A_1}$ and $b \neq 1_{A_1}$, we have $a = c \circ_1 b$. Then by definition, we have $a = c \circ_1 \ b = g(c) \circ_2 g(b)$. Besides, $g(c) \neq 1_{A_2}$ and $g(b) \neq 1_{A_2}$ because if, for example, $g(c) = 1_{A_2}$, then by Proposition 2.6.13(c), we have $c = 1_{A_1}$, which is not true.

Proposition is proved. □

Corollary 2.6.6. *If an abstract prearithmetic* $\boldsymbol{A}_1 = (A; +_1, \circ_1, \leq_1)$ *is directly skew projective with respect to an abstract prearithmetic* $\boldsymbol{A}_2 = (A; +_2, \circ_2, \leq_2)$*, then any multiplicatively prime number in* $\boldsymbol{A}_2$ *is also multiplicatively prime in* $\boldsymbol{A}_1$*.*

Let us study subtractability of elements in abstract prearithmetics.

Proposition 2.6.16. *If an abstract prearithmetic* $\boldsymbol{A}_1 = (A; +_1, \circ_1, \leq_1)$ *is directly skew projective with respect to an abstract prearithmetic* $\boldsymbol{A}_2 = (A; +_2, \circ_2, \leq_2)$ *with the projector* g *and a number* a *is subtractable from the right (from the left) in* $\boldsymbol{A}_1$ *by a number* c*, then the number* a *is subtractable from the right (from the left) in* $\boldsymbol{A}_2$ *by the number* $g(c)$.

Proof: Let us assume that a number a is subtractable from the right (from the left) in $\boldsymbol{A}_1$ by a number c. It means (cf. Section 2.3) that for some number d, we have $a = d +_1 c$. Then by definition, we have $a = d +_1 c = g(d) +_2 g(c)$. It means that the number a is subtractable from the right in $\boldsymbol{A}_2$ by the number $g(c)$.

Subtractability from the left is proved in a similar way.

Proposition is proved. □

Corollary 2.6.7. *If an abstract prearithmetic* $\boldsymbol{A}_1 = (A; +_1, \circ_1, \leq_1)$ *is directly skew projective with respect to an abstract prearithmetic* $\boldsymbol{A}_2 = (A; +_2, \circ_2, \leq_2)$ *with the projector* g *and a number* a *is subtractable in* $\boldsymbol{A}_1$ *by a number* c*, then the number* a *is subtractable in* $\boldsymbol{A}_2$ *by the number* $g(c)$.

Corollary 2.6.8. *If an abstract prearithmetic* $\boldsymbol{A}_1 = (A; +_1, \circ_1, \leq_1)$ *is directly skew projective with respect to an abstract prearithmetic* $\boldsymbol{A}_2 = (A; +_2, \circ_2, \leq_2)$ *with the projector* g *and a number* a *is additively even in* $\boldsymbol{A}_1$ *with respect to a number* b*, then the number* a *is additively even in* $\boldsymbol{A}_2$ *with respect to the number* $g(b)$.

Corollary 2.6.9. *If an abstract prearithmetic* $\boldsymbol{A}_1 = (A; +_1, \circ_1, \leq_1)$ *is directly skew projective with respect to an abstract prearithmetic* $\boldsymbol{A}_2 = (A; +_2, \circ_2, \leq_2)$ *with the projector* g *and a number* a *is additively odd in* $\boldsymbol{A}_2$ *with respect to a number* $g(b)$*, then the number* a *is additively odd in* $\boldsymbol{A}_1$ *with respect to the number* b.

Let us study divisibility of elements in abstract prearithmetics.

Proposition 2.6.17. *If an abstract prearithmetic* $\boldsymbol{A}_1 = (A; +_1, \circ_1, \leq_1)$ *is directly skew projective with respect to an abstract prearithmetic* $\boldsymbol{A}_2 = (A; +_2, \circ_2, \leq_2)$ *with the projector* g *and a number* a *is divisible from the right (from the left) in* $\boldsymbol{A}_1$ *by a number* c*, then the number* a *is divisible from the right (from the left) in* $\boldsymbol{A}_2$ *by the number* $g(c)$.

Proof is similar to the proof of Proposition 2.6.16. □

Corollary 2.6.10. *If an abstract prearithmetic $\boldsymbol{A}_1 = (A; +_1, \circ_1, \leq_1)$ is directly skew projective with respect to an abstract prearithmetic $\boldsymbol{A}_2 = (A; +_2, \circ_2, \leq_2)$ with the projector g and a number a is divisible in $\boldsymbol{A}_1$ by a number c, then the number a is divisible in $\boldsymbol{A}_2$ by the number $g(c)$.*

Corollary 2.6.11. *If an abstract prearithmetic $\boldsymbol{A}_1 = (A; +_1, \circ_1, \leq_1)$ is directly skew projective with respect to an abstract prearithmetic $\boldsymbol{A}_2 = (A; +_2, \circ_2, \leq_2)$ with the projector g while a number a is multiplicatively even in $\boldsymbol{A}_1$ with respect to a number b, then the number a is multiplicatively even in $\boldsymbol{A}_2$ with respect to the number $g(b)$.*

Corollary 2.6.12. *If an abstract prearithmetic $\boldsymbol{A}_1 = (A; +_1, \circ_1, \leq_1)$ is directly skew projective with respect to an abstract prearithmetic $\boldsymbol{A}_2 = (A; +_2, \circ_2, \leq_2)$ with the projector g while a number a is multiplicatively odd in $\boldsymbol{A}_2$ with respect to a number $g(b)$, then the number a is multiplicatively odd in $\boldsymbol{A}_1$ with respect to the number b.*

Proposition 2.6.18. *If an abstract prearithmetic $\boldsymbol{A}_1 = (A; +_1, \circ_1, \leq_1)$ is directly skew projective with respect to an abstract prearithmetic $\boldsymbol{A}_2 = (A; +_2, \circ_2, \leq_2)$ with additive cancellation from the right (from the left) and the projector g is an injection, then the prearithmetic $\boldsymbol{A}_1$ is with additive cancellation from the right (from the left).*

Proof: Let us assume that the prearithmetic $\boldsymbol{A}_2$ is with additive cancellation from the left (cf. Section 2.3) and $c +_1 a = c +_1 b$. Then by definition, we have $g(c) +_2 g(a) = g(c) +_2 g(b)$. As the prearithmetic $\boldsymbol{A}_2$ is with additive cancellation from the left, $g(a) = g(b)$. As g is an injection, $a = b$. As a and b are arbitrary elements from A, the prearithmetic $\boldsymbol{A}_1$ is with additive cancellation from the left.

Additive cancellation from the right is proved in a similar way.

Proposition is proved. □

Corollary 2.6.13. *If an abstract prearithmetic $\boldsymbol{A}_1 = (A; +_1, \circ_1, \leq_1)$ is directly skew projective with respect to an abstract prearithmetic $\boldsymbol{A}_2 = (A; +_2, \circ_2, \leq_2)$ with additive cancellation and the projector g is an injection, then the prearithmetic $\boldsymbol{A}_1$ is with additive cancellation.*

Proposition 2.6.19. *If an abstract prearithmetic $\boldsymbol{A}_1 = (A; +_1, \circ_1, \leq_1)$ is directly skew projective with respect to an abstract prearithmetic $\boldsymbol{A}_2 = (A; +_2, \circ_2, \leq_2)$ with multiplicative cancellation from the right (from the left) and the projector g is an injection, then the prearithmetic $\boldsymbol{A}_1$ is also with multiplicative cancellation from the right (from the left).*

Proof: Let us assume that the prearithmetic $\boldsymbol{A}_2$ is with multiplicative cancellation from the left (cf. Section 2.3) and $c \circ_1 a = c \circ_1 b$. Then by definition, we have $g(c) \circ_2 g(a) = g(c) \circ_2 g(b)$. As the prearithmetic $\boldsymbol{A}_2$ is with multiplicative cancellation from the left, $g(a) = g(b)$. As g is an injection, $a = b$. As a and b are arbitrary elements from A, the prearithmetic $\boldsymbol{A}_1$ is with multiplicative cancellation from the left.

Multiplicative cancellation from the right is proved in a similar way.

Proposition is proved. □

Corollary 2.6.14. *If an abstract prearithmetic $\boldsymbol{A}_1 = (A; +_1, \circ_1, \leq_1)$ is directly skew projective with respect to an abstract prearithmetic $\boldsymbol{A}_2 = (A; +_2, \circ_2, \leq_2)$ with multiplicative cancellation and the projector g is an injection, then the prearithmetic $\boldsymbol{A}_1$ is also with multiplicative cancellation.*

Proposition 2.6.20. *If an abstract prearithmetic $\boldsymbol{A}_1 = (A; +_1, \circ_1, \leq_1)$ is directly skew projective with respect to an abstract prearithmetic $\boldsymbol{A}_2 = (A; +_2, \circ_2, \leq_2)$ with ordered additive cancellation from the right (from the left), then the prearithmetic $\boldsymbol{A}_1$ is with ordered additive cancellation from the right (from the left).*

Proof: Let us assume that the prearithmetic $\boldsymbol{A}_2$ is with ordered additive cancellation from the left (cf. Section 2.3) and $c +_1 a \leq_1 c +_1 b$. Then by definition, we have $g(c) +_2 g(a) \leq_2 g(c) +_2 g(b)$. As the prearithmetic $\boldsymbol{A}_2$ is with ordered additive cancellation from the left, $g(a) \leq_2 g(b)$. By the definition of weak projectivity, this implies $a \leq_1 b$. As a and b are arbitrary elements from A, the prearithmetic $\boldsymbol{A}_1$ is with ordered additive cancellation from the left.

Ordered additive cancellation from the right is proved in a similar way.

Proposition is proved. □

Corollary 2.6.15. *If an abstract prearithmetic $\boldsymbol{A}_1 = (A; +_1, \circ_1, \leq_1)$ is directly skew projective with respect to an abstract prearithmetic $\boldsymbol{A}_2 = (A; +_2, \circ_2, \leq_2)$ with ordered additive cancellation, then the prearithmetic $\boldsymbol{A}_1$ is with ordered additive cancellation.*

Proposition 2.6.21. *If an abstract prearithmetic $\boldsymbol{A}_1 = (A; +_1, \circ_1, \leq_1)$ is directly skew projective with respect to an abstract prearithmetic $\boldsymbol{A}_2 = (A; +_2, \circ_2, \leq_2)$ with strict additive cancellation from the right (from the left) and the projector g is a strictly increasing function, then the prearithmetic $\boldsymbol{A}_1$ has strict additive cancellation from the right (from the left).*

Proof: Let us assume that the prearithmetic $\boldsymbol{A}_2$ is with strict additive cancellation from the left (cf. Section 2.3) and $c +_1 a <_1 c +_1 b$. Then we have $g(c) +_2 g(a) <_2 g(c) +_2 g(b)$ because g is a strictly increasing function. As the prearithmetic $\boldsymbol{A}_2$ is with strict additive cancellation from the left, $g(a) <_2 g(b)$. As g is a strictly increasing function, we have $a <_1 b$. As a and b are arbitrary elements from A, the prearithmetic $\boldsymbol{A}_1$ is with strict additive cancellation from the left.

Ordered additive cancellation from the right is proved in a similar way.

Proposition is proved. □

Corollary 2.6.16. *If an abstract prearithmetic $\boldsymbol{A}_1 = (A; +_1, \circ_1, \leq_1)$ is directly skew projective with respect to an abstract prearithmetic $\boldsymbol{A}_2 = (A; +_2, \circ_2, \leq_2)$ with strict additive cancellation and g is a strictly increasing function, then the prearithmetic $\boldsymbol{A}_1$ is with strict additive cancellation.*

Proposition 2.6.22. *If an abstract prearithmetic $\boldsymbol{A}_1 = (A; +_1, \circ_1, \leq_1)$ is directly skew projective with respect to an abstract prearithmetic $\boldsymbol{A}_2 = (A; +_2, \circ_2, \leq_2)$ with ordered multiplicative cancellation from the right (from the left), then the prearithmetic $\boldsymbol{A}_1$ is with ordered multiplicative cancellation from the right (from the left).*

Proof: Let us assume that the prearithmetic $\boldsymbol{A}_2$ is with ordered multiplicative cancellation from the left (cf. Section 2.3) and $c \circ_1 a \leq_1 c \circ_1 b$. Then by definition, we have $g(c) \circ_2 g(a) \leq_2 g(c) \circ_2 g(b)$. As the prearithmetic $\boldsymbol{A}_2$ is with ordered multiplicative cancellation from the left, $g(a) \leq_2 g(b)$. By the definition of weak projectivity, this implies $a \leq_1 b$. As a and b are arbitrary elements from A, the prearithmetic $\boldsymbol{A}_1$ is with ordered multiplicative cancellation from the left.

Ordered multiplicative cancellation from the right is proved in a similar way.

Proposition is proved. □

Corollary 2.6.17. *If an abstract prearithmetic $\boldsymbol{A}_1 = (A; +_1, \circ_1, \leq_1)$ is directly skew projective with respect to an abstract prearithmetic $\boldsymbol{A}_2 = (A; +_2, \circ_2, \leq_2)$ with ordered multiplicative cancellation, then the prearithmetic $\boldsymbol{A}_1$ is with ordered multiplicative cancellation.*

Proposition 2.6.23. *If an abstract prearithmetic $\boldsymbol{A}_1 = (A; +_1, \circ_1, \leq_1)$ is directly skew projective with respect to an abstract prearithmetic $\boldsymbol{A}_2 = (A; +_2, \circ_2, \leq_2)$ with strict multiplicative cancellation from the right (from the left) and the projector g is a strictly increasing function, then the* *prearithmetic* $\boldsymbol{A}_1$ *has strict multiplicative cancellation from the right (from the left).*

Proof: Let us assume that the prearithmetic $\boldsymbol{A}_2$ is with strict multiplicative cancellation from the left (cf. Section 2.3) and $c \circ_1 a <_1 c \circ_1 b$. Then we have $g(c) \circ_2 g(a) <_2 g(c) \circ_2 g(b)$ because g is a strictly increasing function. As the prearithmetic $\boldsymbol{A}_2$ is with strict multiplicative cancellation from the left, $g(a) <_2 g(b)$. As g is a strictly increasing function, we have $a <_1 b$. As a and b are arbitrary elements from A, the prearithmetic $\boldsymbol{A}_1$ is with strict multiplicative cancellation from the left.

Ordered multiplicative cancellation from the right is proved in a similar way.

Proposition is proved. □

Corollary 2.6.18. *If an abstract prearithmetic* $\boldsymbol{A}_1 = (A; +_1, \circ_1, \leq_1)$ *is directly skew projective with respect to an abstract prearithmetic* $\boldsymbol{A}_2 = (A; +_2, \circ_2, \leq_2)$ *with strict multiplicative cancellation and the projector* g *is a strictly increasing function, then the prearithmetic* $\boldsymbol{A}_1$ *has strict multiplicative cancellation.*

2.6.2. *Inverse skew projectivity*

Advances are made by answering questions.
Discoveries are made by questioning answers.

Bernhard Haisch

Informally, inverse skew projectivity means that operations with elements of a prearithmetic, e.g., with numbers, are actually performed in another arithmetic, which has the same elements and the result is then mapped back by some function.

Definition 2.6.3. An abstract prearithmetic $\boldsymbol{A}_1 = (A; +_1, \circ_1, \leq_1)$ is called *inversely skew projective* with respect to a prearithmetic $\boldsymbol{A}_2 = (A; +_2, \circ_2, \leq_2)$ if it is weakly projective with respect to $\boldsymbol{A}_2$ and the projector g is the identity mapping 1_A of A.

It means that in inverse skew projectivity of $\boldsymbol{A}_1$ with respect to $\boldsymbol{A}_2$, we have

$$a +_1 b = h(a +_2 b),$$

$$a \circ_1 b = h(a \circ_2 b).$$

Example 2.6.4. Inverse skew projectivity is implicitly used in computer arithmetics for formalizing and correct utilization of rounding (Kulisch, 1982). Given an arithmetic $\boldsymbol{A} = (A; +, \circ, \leq)$, for example, the Diophantine arithmetic $\boldsymbol{W}$ of all whole numbers or the Diophantine arithmetic $\boldsymbol{Q}$ of all rational numbers, *rounding* in $\boldsymbol{A}$ is defined as a mapping r from A into a definite subset D of A, which satisfies the following properties:

(1) for any element x from D, we have $r(x) = x$, i.e., r is identical on D;
(2) for any elements x and y from A, we have $x \leq y \Rightarrow r(x) \leq r(y)$, i.e., r preserves order;
(3) for any element x from A, we have $r(-x) = -r(x)$.

The mapping r allows defining new operations in A by the following formulas:

$$a \oplus b = r(a + b), \tag{2.48}$$

$$a \otimes b = r(a \circ_2 b). \tag{2.49}$$

With respect to these operations, we obtain another arithmetic $\boldsymbol{D} = (A; \oplus, \otimes, \leq)$ with the same order as in the arithmetic $\boldsymbol{A}$ and we see that $\boldsymbol{D}$ is inversely skew projective with respect to $\boldsymbol{A}$. In this new arithmetic $\boldsymbol{D}$, addition and multiplication are performed with rounding as it is done in the computer arithmetic.

The new arithmetic $\boldsymbol{D}$ is the computer arithmetic, which turns out to be a key property for an automatic error control in numerical analysis (Kulisch, 1982). To work with the computer arithmetic, popular programming languages such as PASCAL, FORTRAN and APL were relevantly extended.

It is unreasonable to extend inverse skew projectivity to projectivity as the following example demonstrates.

Lemma 2.6.2. *If an abstract prearithmetic* $\boldsymbol{A}_1 = (A; +_1, \circ_1, \leq_1)$ *is inversely skew projective and projective with respect to a prearithmetic* $\boldsymbol{A}_2 = (A; +_2, \circ_2, \leq_2)$, *then these prearithmetics coincide.*

Proof is left as an exercise. □

Let us study properties of inverse skew projectivity. Because inverse skew projectivity is a special case of weak projectivity, it inherits many properties of weak projectivity.

Proposition 2.6.24. *If an abstract prearithmetic* $\boldsymbol{A}_1 = (A; +_1, \circ_1, \leq_1)$ *is inversely skew projective with respect to an abstract prearithmetic* $\boldsymbol{A}_2 = (A; +_2, \circ_2, \leq_2)$ *and the operation* $+_2$ *is commutative in the prearithmetic* $\boldsymbol{A}_2$, *then the operation* $+_1$ *is commutative in the prearithmetic* $\boldsymbol{A}_1$.

Indeed, we have

$$a +_1 b = h(a +_2 b) = h(b +_2 a) = b +_1 a.$$

Proposition 2.6.25. *If an abstract prearithmetic* $\boldsymbol{A}_1 = (A; +_1, \circ_1, \leq_1)$ *is inversely skew projective with respect to an abstract prearithmetic* $\boldsymbol{A}_2 = (A; +_2, \circ_2, \leq_2)$ *and the operation* $\circ_2$ *is commutative in the prearithmetic* $\boldsymbol{A}_2$, *then the operation* $\circ_1$ *is commutative in the prearithmetic* $\boldsymbol{A}_1$.

Proof is similar to the proof of Proposition 2.6.24. □

However, associativity of addition, as a rule, is not preserved in inverse skew projectivity.

Example 2.6.5. Let us take the Diophantine arithmetic $\boldsymbol{W} = (W; +, \cdot, \leq)$ of all whole numbers and the arithmetic $\boldsymbol{R}^+ = (R^+; +, \cdot, \leq)$ of all positive real numbers. The mapping h: $R^+ \to W$ is rounding of real numbers. This mapping defines by formulas (2.48) and (2.49) inverse skew projectivity of the new arithmetic $\boldsymbol{D} = (R^+; \oplus, \otimes, \leq)$ with respect to $\boldsymbol{R}^+$. Then we have

$$2.2 \oplus (2.2 \oplus 2.4) = h(2.2 \oplus h(2.2 \oplus 2.4)) = h(2.2 \oplus 5)) = 7$$

while

$$(2.2 \oplus 2.2) \oplus 2.4 = h(h(2.2 \oplus 2.2) \oplus 2.4)) = h(4 \oplus 2.4)) = 6.$$

Thus, in a general case,

$$a \oplus (b \oplus c) \neq (a \oplus b) \oplus c.$$

We see that although addition is associative in $\boldsymbol{R}^+$ and $\boldsymbol{D}$ is inversely skew projective with respect to $\boldsymbol{R}^+$, addition is not associative in $\boldsymbol{D}$.

The same is true for multiplication in $\boldsymbol{D}$.

Proposition 2.2.14 implies the following result.

Proposition 2.6.26. *If an abstract prearithmetic* $\boldsymbol{A}_1 = (A_1; +_1, \circ_1, \leq_1)$ *is inversely skew projective with respect to an abstract prearithmetic* $\boldsymbol{A}_2 = (A_2; +_2, \circ_2, \leq_2)$ *and any maximal chain in* $\boldsymbol{A}_2$ *with distinct elements, which* *has the beginning and the end, is finite, then any maximal chain in* $\boldsymbol{A}_1$ *with distinct elements, which has the beginning and the end, is also finite.*

Proof is left as an exercise. □

Proposition 2.2.15 implies the following result because the projector g is a one-to-one mapping.

Proposition 2.6.27. *If an abstract prearithmetic* $\boldsymbol{A}_1 = (A_1; +_1, \circ_1, \leq_1)$ *is inversely skew projective with respect to an abstract prearithmetic* $\boldsymbol{A}_2 = (A_2; +_2, \circ_2, \leq_2)$ *and the partially ordered set* A_2 *is finitely connected, then the partially ordered set* A_1 *is finitely connected.*

Proof is left as an exercise. □

Proposition 2.2.11 implies the following result.

Proposition 2.6.28. *If an abstract prearithmetic* $\boldsymbol{A}_1 = (A; +_1, \circ_1, \leq_1)$ *is inversely skew projective with respect to an abstract prearithmetic* $\boldsymbol{A}_2 = (A; +_2, \circ_2, \leq_2)$ *and the operation* $+_2$ *preserves the order* $\leq_2$*, then the operation* $+_1$ *preserves the order* $\leq_1$*.*

Proof is left as an exercise. □

Lemma 2.1.6 implies the following result.

Corollary 2.6.19. *If an abstract prearithmetic* $\boldsymbol{A}_1 = (A; +_1, \circ_1, \leq_1)$ *is inversely skew projective with respect to an abstract prearithmetic* $\boldsymbol{A}_2 = (A; +_2, \circ_2, \leq_2)$ *and the operation* $+_2$ *is monotone, then the operation* $+_1$ *is monotone.*

Proposition 2.2.12 implies the following result.

Proposition 2.6.29. *If an abstract prearithmetic* $\boldsymbol{A}_1 = (A; +_1, \circ_1, \leq_1)$ *is inversely skew projective with respect to an abstract prearithmetic* $\boldsymbol{A}_2 = (A; +_2, \circ_2, \leq_2)$ *and the operation* $\circ_2$ *preserves the order* $\leq_2$*, then the operation* $\circ_1$ *preserves the order* $\leq_1$*.*

Proof is left as an exercise. □

Lemma 2.1.7 implies the following result.

Corollary 2.6.20. *If an abstract prearithmetic* $\boldsymbol{A}_1 = (A; +_1, \circ_1, \leq_1)$ *is inversely skew projective with respect to an abstract prearithmetic* $\boldsymbol{A}_2 = (A; +_2, \circ_2, \leq_2)$ *and the operation* $\circ_2$ *is monotone, then the operation* $\circ_1$ *is monotone.*

In some cases, direct and inverse skew projectivity are dual being diametrically connected.

Proposition 2.6.30. *An abstract prearithmetic* $\boldsymbol{A}_1 = (A; +_1, \circ_1, \leq_1)$ *is inversely skew projective with respect to an abstract prearithmetic* $\boldsymbol{A}_2 = (A; +_2, \circ_2, \leq_2)$ *with the coprojector* h*, which is a homomorphism with respect to operations* $+_2$ *and* $\circ_2$*, if and only if the prearithmetic* $\boldsymbol{A}_1$ *is directly skew projective with respect to the prearithmetic* $\boldsymbol{A}_2$ *with the projector* h.

Indeed, we have

$$a +_1 b = h(a +_2 b) = h(a) +_2 h(b).$$

It means that for homomorphisms, direct skew projectivity coincides with inverse skew projectivity only the coprojector becomes the projector.

This result on the one hand, explains in what sense the concept of inverse skew projectivity is dual to the concept of direct skew projectivity, while on the other hand, it allows obtaining many properties of inverse skew projectivity from properties of direct skew projectivity.

Inverse skew projectivity is a transitive relation.

Proposition 2.6.31. *If an abstract prearithmetic* $\boldsymbol{A}_1 = (A_1; +_1, \circ_1, \leq_1)$ *is inversely skew projective with respect to an abstract prearithmetic* $\boldsymbol{A}_2 = (A_2; +_2, \circ_2, \leq_2)$ *and the abstract prearithmetic* $\boldsymbol{A}_2 = (A_2; +_2, \circ_2, \leq_2)$ *is inversely skew projective with respect to an abstract prearithmetic* $\boldsymbol{A}_3 = (A_3; +_3, \circ_3, \leq_3)$*, then the prearithmetic* $\boldsymbol{A}_1$ *is inversely skew projective with respect to the prearithmetic* $\boldsymbol{A}_3$.

Proof is similar to the proof of Proposition 2.4.4. □

Proposition 2.6.6 allows proving the following result.

Theorem 2.6.2. *Abstract prearithmetics with inverse skew projectivity relations form a category where objects are abstract prearithmetics and morphisms are inverse skew projectivity relations.*

Proof is left as an exercise. □

Similar to direct skew projectivity, inverse skew projectivity does not preserve distributivity of multiplication with respect to addition in a general case. However, we can get generalized distributivity. Namely, having skew projectivity, it is possible to consider skew distributivity. Let us assume that an abstract prearithmetic $\boldsymbol{A}_1 = (A; +_1, \circ_1, \leq_1)$ is inversely skew projective with respect to an abstract prearithmetic $\boldsymbol{A}_2 = (A; +_2, \circ_2, \leq_2)$ with the coprojector h.

Definition 2.6.4. The multiplication $\circ_1$ is skew h-distributive if the following identity is valid:

$$x \circ_1 (y +_1 z) = (x \circ_2 h(y)) +_1 (x \circ_2 h(z)).$$

Proposition 2.6.32. *If an abstract prearithmetic $\boldsymbol{A}_1 = (A; +_1, \circ_1, \leq_1)$ is inversely skew projective with respect to an abstract prearithmetic $\boldsymbol{A}_2 = (A; +_2, \circ_2, \leq_2)$, in which multiplication $\circ_2$ is distributive with respect to addition $+_2$, and the coprojector h is a homomorphism of the prearithmetic $\boldsymbol{A}_2$ with respect to addition $+_2$, then the multiplication $\circ_1$ in $\boldsymbol{A}_1$ is skew h-distributive.*

Proof: Let us assume that g is a homomorphism. Then for arbitrary elements a, b and c from A, we have

$$\begin{aligned} a \circ_1 (b +_1 c) &= h(a \circ_2 h(b +_2 c)) = h(a \circ_2 (h(b) +_2 h(c))) \\ &= h(a \circ_2 h(b) +_2 a \circ_2 h(c)) = a \circ_2 h(b) +_1 a \circ_2 \; h(c). \end{aligned}$$

It means that multiplication $\circ_1$ in $\boldsymbol{A}_1$ is skew h-distributive.

Proposition is proved. □

Proposition 2.6.33. *If an abstract prearithmetic $\boldsymbol{A}_1 = (A; +_1, \circ_1, \leq_1)$ is inversely skew projective with respect to an abstract prearithmetic $\boldsymbol{A}_2 = (A; +_2, \circ_2, \leq_2)$ with the one-to-one coprojector h, then prearithmetic $\boldsymbol{A}_2$ is inversely skew projective with respect to the prearithmetic $\boldsymbol{A}_1$ with the coprojector h^{-1}.*

Proof is left as an exercise. □

2.7. Numerical Prearithmetics

Mathematics is concerned only with
the enumeration and comparison of relations.
Carl Friedrich Gauss

In this and next sections, we consider abstract prearithmetics $\boldsymbol{A} = (A; \oplus, \otimes, \leq_A)$, in which the carrier A is a subset of the set W of whole numbers or of the set N of natural numbers. In both cases, $\oplus$ and $\otimes$ denote binary operations of addition and multiplication in $\boldsymbol{A}$ and $\leq_A$ is a partial order in the set A.

To build numerical prearithmetics, we use the following basic Diophantine arithmetics, in which the conventional addition is denoted by $+$, the conventional multiplication is denoted by $\cdot$ and the conventional order is denoted by $\leq$:

- the Diophantine arithmetic $\boldsymbol{N}$ of all natural numbers;
- the Diophantine arithmetic $\boldsymbol{W}$ of all whole numbers;
- the Diophantine arithmetic $\boldsymbol{I}$ of all integer numbers;
- the Diophantine arithmetic $\boldsymbol{Q}$ of all rational numbers;
- the Diophantine arithmetic $\boldsymbol{R}$ of all real numbers;
- the Diophantine arithmetic $\boldsymbol{C}$ of all complex numbers.

The first two standard arithmetics allow us to define whole-number prearithmetics and natural-number prearithmetics.

Definition 2.7.1. An abstract prearithmetic $\boldsymbol{A} = (A; \oplus, \otimes, \leq_A)$ is called a *whole-number prearithmetic*, or W-*prearithmetic*, if A is a subset of the set W of all whole numbers and the prearithmetic $\boldsymbol{A}$ is weakly projective with respect to the (conventional) Diophantine arithmetic $\boldsymbol{W} = (W; +, \cdot, \leq)$ of whole numbers.

It means that there are two functions: the projector g: $A \to W$ and the coprojector h: $W \to A$ such that for any numbers a and b from A, we have

$$a \oplus b = h(g(a) + g(b)),$$
$$a \otimes b = h(g(a) \cdot g(b)).$$

Here a and b are whole numbers, $+$ is conventional addition and $\cdot$ is conventional multiplication of real numbers, while $\oplus$ is addition and $\otimes$ is multiplication of numbers in prearithmetic defined by functions f and g.

Elements of A are called numbers of $\boldsymbol{A}$ and are denoted with subscript A, i.e., number 2 in $\boldsymbol{A}$ is denoted by 2_A and number 5 in $\boldsymbol{A}$ is denoted by 5_A.

In essence, weak projectivity defined in Section 2.2 allows building new prearithmetics on whole numbers using the Diophantine arithmetic $\boldsymbol{W}$ and deducing many properties of these new prearithmetics from properties of the arithmetic $\boldsymbol{W}$.

Example 2.7.1. Any modular arithmetic (cf. Example 2.1.3) is a W-prearithmetic.

Example 2.7.2. The arithmetic $5\boldsymbol{W}$ of all whole numbers divisible by 5 is a W-prearithmetic.

Lemma 2.7.1. *Any whole-number prearithmetic is either finite or countable.*

Indeed, according to set theory (cf., for example, Fraenkel and Bar-Hillel, 1958), a subset of a countable set is either finite or countable.

Remark 2.7.1. Not all abstract prearithmetics are countable. For instance, the arithmetics $\boldsymbol{R}$ of all real numbers or the arithmetics $\boldsymbol{C}$ of all complex numbers are uncountable. Thus, there are abstract prearithmetics that are not W-prearithmetics.

It is possible to derive many properties of W-prearithmetics from properties of abstract prearithmetics and the Diophantine arithmetic $\boldsymbol{W}$.

Let us consider an arbitrary set A and a mapping g: $A \to W$.

Proposition 2.7.1. *It is possible to define on A the structure of a W-prearithmetic $\boldsymbol{A}$.*

Proof is similar to the proof of Proposition 2.2.1 and uses an arbitrary mapping h: $W \to A$. □

Note that if we want to preserve the initial order of numbers, the mapping g has to be an injection.

Remark 2.7.2. Proposition 2.7.1 shows that it is possible to assume that in a W-prearithmetic $\boldsymbol{A}$, its carrier A can be any set for which a mapping g: $A \to W$ exists. This mapping determines a naming of the elements from A by whole numbers (Burgin, 2011). Taking this into account, we can denote the element from A that is mapped onto the whole number n by n_A. Note that if g is not a bijection, it is possible $n_A = m_A$ although $n \neq m$.

Thus, we will distinguish W-prearithmetics and whole-number arithmetics assuming that the carrier of the latter consists of whole numbers. This makes whole-number arithmetics a proper subclass in the class of all W-prearithmetics.

Propositions 2.2.4 and 2.2.5 and properties of the Diophantine arithmetic $\boldsymbol{W}$ imply the following result.

Proposition 2.7.2. *For any W-prearithmetic $\boldsymbol{A} = (A; \oplus, \otimes, \leq_A)$, addition $\oplus$ and multiplication $\otimes$ are commutative.*

Indeed, by Propositions 2.1.4 and 2.1.5, operations $\oplus$ and $\otimes$ are commutative because addition and multiplication in the Diophantine arithmetic $\boldsymbol{W}$ are commutative.

Proposition 2.7.3. *For any W-prearithmetic $\boldsymbol{A} = (A; \oplus, \otimes, \leq_A)$ with an injective projector into W, the order $\leq$ in $\boldsymbol{A}$ is linear.*

Indeed, if the projector g is an injection, then by Proposition 2.1.2, the order $\leq$ in $\boldsymbol{A}$ is linear because the order $\leq$ in $\boldsymbol{W}$ is linear.

Corollary 2.7.1. *In an abstract prearithmetic $\boldsymbol{A} = (A; +, \circ, \leq_A)$, which is weakly projective with respect to a W-prearithmetic $\boldsymbol{D} = (D; \oplus, \otimes, \leq_D)$ with the injective projector g, the relation $\leq$ is a well-ordering.*

Taking a whole-number prearithmetic $\boldsymbol{A} = (A; \oplus, \otimes, \leq)$ in addition to addition and multiplication, we can define other operations in $\boldsymbol{A}$. In particular, we have a system of pairs of n-ary operations for each natural number $n = 3, 4, 5, \ldots$ (cf. Section 2.1):

1. $\Sigma^{\oplus n}(a_{A,1}, a_{A,2}, \ldots, a_{A,n-1}, a_{A,n}) = h(g(a_{A,1}) + g(a_{A,2}) + \cdots + g(a_{A,n}))$;
2. $\Pi^{\otimes n}(a_{A,1}, a_{A,2}, \ldots, a_{A,n-1}, a_{A,n}) = h(g(a_{A,1}) \cdot g(a_{A,2}) \cdot \ldots \cdot g(a_{A,n}))$.

Note that for $n = 2$, these operations coincide with addition and multiplication, i.e., $\Sigma^{\oplus n} = \oplus$ and $\Pi^{\otimes n} = \otimes$.

Proposition 2.7.4. *For any W-prearithmetic $\boldsymbol{A} = (A; \oplus, \otimes, \leq_A)$, operations $\Sigma^{\oplus n}$ and $\Pi^{\otimes n}$ are commutative for all $n = 2, 3, 4, 5, \ldots$.*

Indeed,

$$\begin{aligned}\Sigma^{\oplus n}(a_{A,1}, a_{A,2}, \ldots, a_{A,n-1}, a_{A,n}) &= h(g(a_{A,1}) + g(a_{A,2}) + \cdots + g(a_{A,n}))\\ &= h(g(a_{A,j_1}) + g(a_{A,j_2}) + \cdots + g(a_{A,j_n}))\\ &= \Sigma^{\oplus n}(a_{A,j_1}, a_{A,j_2}, \ldots, a_{A,j_n}),\end{aligned}$$

where $j_1, j_2, j_3, \ldots, j_n$ is an arbitrary permutation of the sequence $1, 2, 3, \ldots, n$.

In a similar way, we have

$$\begin{aligned}&\Pi^{\otimes n}(a_{A,1}, a_{A,2}, \ldots, a_{A,n-1}, a_{A,n}) = h(g(a_{A,1}) \cdot g(a_{A,2}) \cdot \ldots \cdot g(a_{A,n}))\\ &\quad = h(g(a_{A,j_1}) \cdot g(a_{A,j_2}) \cdot \ldots \cdot g(a_{A,j_n})) = \Pi^{\otimes n}(a_{A,j_1}, a_{A,j_2}, \ldots, a_{A,j_n}),\end{aligned}$$

where $j_1, j_2, j_3, \ldots, j_n$ is an arbitrary permutation of the sequence $1, 2, 3, \ldots, n$.

Propositions 2.2.6, 2.2.7 and properties of the conventional arithmetic $\boldsymbol{W}$ imply the following result.

Proposition 2.7.5. *For any W-prearithmetic* $\boldsymbol{A} = (A; \oplus, \otimes, \leq_A)$ *with the projector* g *and the coprojector* h, *for which* $gh = \mathbf{1}_W$, *addition* $\oplus$ *and multiplication* $\otimes$ *are associative.*

Proof is left as an exercise. □

Proposition 2.2.8 and properties of the conventional arithmetic $\boldsymbol{W}$ imply the following result.

Proposition 2.7.6. *For any W-prearithmetic* $\boldsymbol{A} = (A; \oplus, \otimes, \leq_A)$ *with the projector* g *and the coprojector* h, *for which* $gh = \mathbf{1}_W$, *multiplication* $\otimes$ *is distributive over addition* $\oplus$.

Proof is left as an exercise. □

Propositions 2.2.11, 2.2.12 and properties of the conventional arithmetic $\boldsymbol{W}$ imply the following result.

Proposition 2.7.7. *For any W-prearithmetic* $\boldsymbol{A} = (A; \oplus, \otimes, \leq_A)$ *with the coprojector* h, *which preserves the order* $\leq$ *in* $\boldsymbol{W}$, *addition* $\oplus$ *and multiplication* $\otimes$ *preserve the order* $\leq_A$.

Proof is left as an exercise. □

Corollary 2.7.2. *For any W-prearithmetic* $\boldsymbol{A} = (A; \oplus, \otimes, \leq_A)$ *with the coprojector* h, *which preserves the order* $\leq$ *in* $\boldsymbol{W}$, *addition* $\oplus$ *is monotone.*

Corollary 2.7.3. *For any W-prearithmetic* $\boldsymbol{A} = (A; \oplus, \otimes, \leq_A)$ *with the coprojector* h, *which preserves the order* $\leq$ *in* $\boldsymbol{W}$, *multiplication* $\otimes$ *is monotone.*

The property of abstract prearithmetics "to be a W-prearithmetic" is hereditary as the following result demonstrates.

Proposition 2.7.8. *A subprearithmetic* $\boldsymbol{K}$ *of a W-prearithmetic* $\boldsymbol{A} = (A; \oplus, \otimes, \leq)$ *is a W-prearithmetic.*

Indeed, by Proposition 2.5.12, the prearithmetic $\boldsymbol{K}$ is weakly projective with respect to the Diophantine arithmetic $\boldsymbol{W}$ and thus, it is a W-prearithmetic.

Corollary 2.7.4. *A subprearithmetic* $\boldsymbol{D}$ *of a subprearithmetic* $\boldsymbol{K}$ *of a W-prearithmetic* $\boldsymbol{A} = (A; \oplus, \otimes, \leq)$ *is a W-prearithmetic.*

Indeed, by Proposition 2.5.11, the prearithmetic $\boldsymbol{D}$ is a subprearithmetic of a W-prearithmetic $\boldsymbol{A}$ and thus by Proposition 2.7.8, $\boldsymbol{D}$ is a W-prearithmetic.

Proposition 2.7.9. *An abstract prearithmetic* $\boldsymbol{A} = (A; +, \circ, \leq_A)$, *which is weakly projective with respect to a W-prearithmetic* $\boldsymbol{D} = (D; \oplus, \otimes, \leq_D)$ *is a W-prearithmetic.*

Indeed, by Proposition 2.2.9, the prearithmetic $\boldsymbol{A}$ is weakly projective with respect to the Diophantine arithmetic $\boldsymbol{W}$ and thus, it is a W-prearithmetic.

Proposition 2.7.10. *If an abstract prearithmetic* $\boldsymbol{A}_1 = (A_1; +_1, \circ_1, \leq_1)$ *is weakly projective with respect to an abstract prearithmetic* $\boldsymbol{A}_2 = (A_2; +_2, \circ_2, \leq_2)$ *and the abstract prearithmetic* $\boldsymbol{A}_2 = (A_2; +_2, \circ_2, \leq_2)$ *is weakly projective with respect to a W-prearithmetic* $\boldsymbol{A}_3 = (A_3; +_3, \circ_3, \leq_3)$, *then the prearithmetic* $\boldsymbol{A}_1$ *is a W-prearithmetic.*

Indeed, by Proposition 2.7.9, the prearithmetic $\boldsymbol{A}_2$ is a W-prearithmetic and applying once more Proposition 2.7.9, we come to the conclusion that $\boldsymbol{A}_1$ is also a W-prearithmetic.

Proposition 2.7.10 implies the following result.

Theorem 2.7.1. *W-prearithmetics form the category* ***WPAWP*** *where objects are W-prearithmetics and morphisms are weak projectivity relations.*

Proof is left as an exercise. □

The construction from Definition 2.2.2 allows defining actions of whole numbers on elements from a W-prearithmetic $\boldsymbol{A} = (A; \oplus, \otimes, \leq_A)$ with the projector g and the coprojector h.

Definition 2.7.2. (a) The *additive action from the left* of a whole number n on an element a from $\boldsymbol{A}$ is defined as

$$n^{+}a = h(n + g(a)).$$

(b) The *additive action from the right* of a whole number n on an element a from $\boldsymbol{A}$ is defined as

$$a^{+}n = h(g(a) + n).$$

(c) The *multiplicative action from the left* of a whole number n on an element a from $\boldsymbol{A}$ is defined as

$$na = h(n \cdot g(a)).$$

(d) The *multiplicative action from the right* of a whole number n on an element a from $\boldsymbol{A}$ is defined as

$$an = h(g(a) \cdot n).$$

Let us study how whole numbers act on elements of W-prearithmetics using obtained in Section 2.2 properties of actions in general abstract prearithmetics. As addition of whole numbers is commutative, we have the following result.

Proposition 2.7.11. *For any W-prearithmetic* $\boldsymbol{A} = (A; \oplus, \otimes, \leq_A)$, *additive action from the left coincides with additive action from the right.*

Proof is left as an exercise. □

Proposition 2.7.11 allows skipping indication of the side of additive action in W-prearithmetics.

As multiplication of whole numbers is commutative, we have the following result.

Proposition 2.7.12. *For any W-prearithmetic* $\boldsymbol{A} = (A; \oplus, \otimes, \leq_A)$, *multiplicative action from the left coincides with multiplicative action from the right.*

Proof is left as an exercise. □

Proposition 2.7.12 allows skipping indication of the side of multiplicative action in W-prearithmetics.

Proposition 2.2.18 implies the following result.

Proposition 2.7.13. *For any W-prearithmetic* $\boldsymbol{A} = (A; \oplus, \otimes, \leq_A)$ *with the coprojector* h, *which preserves addition, multiplicative action is distributive over addition* $\oplus$.

Proof is left as an exercise. □

Proposition 2.2.19 implies the following result.

Proposition 2.7.14. *For any W-prearithmetic* $\boldsymbol{A} = (A; \oplus, \otimes, \leq_A)$ *with the coprojector* h, *which preserves addition, multiplicative action is distributive over addition* $+$ *in the arithmetic* $\boldsymbol{W}$.

Proof is left as an exercise. □

Proposition 2.2.22 implies the following result.

Proposition 2.7.15. *For any W-prearithmetic* $\boldsymbol{A} = (A; \oplus, \otimes, \leq_A)$ *with the coprojector* h, *which preserves the order of whole numbers, additive action is a monotone operation.*

Proof is left as an exercise. □

Proposition 2.2.23 implies the following result.

Proposition 2.7.16. *For any W-prearithmetic* $\boldsymbol{A} = (A; \oplus, \otimes, \leq_A)$ *with the coprojector* h*, which preserves the order of whole numbers, multiplicative action is a monotone operation.*

Proof is left as an exercise. □

There are W-prearithmetics, which are very different from the Diophantine arithmetic $\boldsymbol{W}$.

Proposition 2.7.17. *For any natural number* n*, there is a W-prearithmetic* $\boldsymbol{A} = (A; \oplus, \otimes, \leq_A)$ *with* n *elements.*

Proof: If $n = 1$, then the subprearithmetic $\mathbf{0}$ of $\boldsymbol{W}$ contains only one element 0 and is a W-prearithmetic. For any $n > 1$, we use modular arithmetics. Indeed, for any natural number n, there is the modular arithmetic $\boldsymbol{Z}_n$, which contains exactly n elements. It is necessary only to show that $\boldsymbol{Z}_n$ is a W-prearithmetic.

Denoting any number m from $\boldsymbol{Z}_n$ by m_{zn} to discern it from the number m in W, we define the projector g as $g(m_{zn}) = m$. For a number r from W, we define the coprojector h as $h(r) = m$ such that $r = m \pmod n$ and $m < n$.

Then taking two numbers m_{zn} and r_{zn} from $\boldsymbol{Z}_n$, we have

$$m_{zn} \oplus r_{zn} = \min\{t; m + r = t(\operatorname{mod} n)\} = h(m + r) = h(g(m) + g(r))$$

and

$$m_{zn} \otimes r_{zn} = \min\{t; m \cdot r = t(\operatorname{mod} n)\} = h(m \cdot r) = h(g(m) \cdot g(r)).$$

Besides, the order $\leq$ in $\boldsymbol{Z}_n$ is the same as the order $\leq$ in $\boldsymbol{W}$.

Thus, the modular arithmetic $\boldsymbol{Z}_n$ is a W-prearithmetic with exactly n elements.

Proposition is proved. □

Let us study properties of subprearithmetics of W-prearithmetics.

Proposition 2.7.18. *Any subprearithmetic* $\boldsymbol{K}$ *of the Diophantine arithmetic* $\boldsymbol{W}$ *either contains only one element or is infinite.*

Indeed, if $\boldsymbol{K}$ contains only 0, then it is a subprearithmetic $\boldsymbol{W}$ with only one element. If $\boldsymbol{K}$ contains a non-zero element n, then $\boldsymbol{K}$ contains all its powers and thus, is infinite.

However, properties of arbitrary W-prearithmetics can be essentially different from the Diophantine arithmetic $\boldsymbol{W}$ as the following result demonstrates.

Proposition 2.7.19. *For any natural number n, there is a W-prearithmetic $\boldsymbol{A} = (A; \oplus, \otimes, \leq_A)$, which contains a proper subprearithmetic with n elements.*

Indeed, as it is proved in Proposition 2.7.6, the modular arithmetic $\boldsymbol{Z}_n$ is a W-prearithmetic with n elements and the modular arithmetic $\boldsymbol{Z}_{2n}$ is a W-prearithmetic with $2n$ elements. In addition, $\boldsymbol{Z}_n$ is a subprearithmetic of the modular arithmetic $\boldsymbol{Z}_{2n}$.

We can prove even a stronger result.

Theorem 2.7.2. *For any natural number n, there is a W-prearithmetic $\boldsymbol{A} = (A; \oplus, \otimes, \leq_A)$ such that for any natural number n, $\boldsymbol{A}$ contains a proper subprearithmetic with n elements.*

Proof: Let us consider the set W of all whole numbers and build a W-prearithmetic $\boldsymbol{A} = (A; \oplus, \otimes, \leq_A)$ with necessary properties. To do this, we construct a special sequence of numbers $\{c_n; n = 3, 4, 5, \ldots\}$ using the following recursion:

$$c_3 = 3,$$
$$c_{n+1} = 2c_n + 1 \quad \text{for all } n = 3, 4, 5, \ldots .$$

In this sequence, we have $c_4 = 7$, $c_5 = 15$, $c_6 = 31$, and $c_7 = 63$.

Note that $2n < n^2 < 10^{c_n}$ for all $n = 3, 4, 5, \ldots$. We show this by induction.

For $n = 3$, we have

$$2 \cdot 3 = 6 < 3^2 = 9 < 10^3 = 1000.$$

Let us assume that $2n < n^2 < 10^{c_n}$ and prove a similar inequality for $n+1$. Indeed,

$$2(n+1) = 2n + 2 < (n+1)^2 = n^2 + 2n + 1 < 2 \cdot 10^{c_n} < 10^{c_{n+1}}.$$

By the principle of the mathematical induction, this is true for all $n = 3, 4, 5, \ldots$.

Now we can define the set A by the following formula:

$$A = \bigcup_{n=1}^{\infty} W_n.$$

In this formula, we have the following sets:

$$W_1 = \{0\};$$
$$W_2 = \{10, 11\};$$
$$W_n = \{10^{c_n}, 10^{c_n} + 1, 10^{c_n} + 2, \ldots, 10^{c_n} + n - 1\} \quad \text{for all } n = 3, 4, 5, \ldots.$$

Note that each set W_n contains the same number of element as the modular arithmetic $\boldsymbol{Z}_n$.

To build a W-prearithmetic $\boldsymbol{A}$ on the set A, we need to define the projector g: $A \to W$ and the coprojector h: $W \to A$.

$$g(m) = m \quad \text{for all } m \text{ from the set } A,$$
$$h(k) = 0 \quad \text{for all numbers } k \text{ less than } 10.$$

For larger than 9 numbers m, we define h as the sequential composition of three functions $u(x)$, $l(x)$ and $p(x)$.

When m belongs to the interval $10 \leq m < 1000$, it has the form $m = q \cdot 10 + k$ where $k < 10$. Then we define

$$p(m) = k,$$
$$l(k) = \min\{t; k = t \pmod 2\},$$
$$u(t) = 10 + t$$

and

$$h(m) = u(l(p(m))).$$

When m belongs to the interval $10^{c_n} \leq m < 10^{c_{n+1}}$, it has the form $m = q \cdot 10^{c_n} + k$ where $k < 10^{c_n}$. Then we define

$$p(m) = k,$$
$$l(k) = \min\{t; k = t \pmod n\},$$
$$u(t) = 10^{c_n} + t$$

and

$$h(m) = u(l(p(m))).$$

This determines the function h for all whole numbers because intervals, where it is defined, cover W.

For instance,

$$h(5433) = u(l(p(5433))) = u(l(p(5{\cdot}10^3{+}433))) = u(l(433)) = u(1) = 10^3{+}1$$

because $433 = 1 \pmod 3$.

For all numbers in A, we keep the relation $\leq$ from the set W. Therefore, to have a W-prearithmetic, we need to define operations of addition $\oplus$ and multiplication $\otimes$.

Taking two numbers m and r from A, we have

$$m \oplus r = h(g(m) + g(r))$$

and

$$m \otimes r = h(g(m) \cdot g(r)).$$

This gives us a W-prearithmetic $\boldsymbol{A} = (A; \oplus, \otimes, \leq)$.

Let us take the set W_n and show that it is closed with respect to operations of addition $\oplus$ and multiplication $\otimes$ being a subprearithmetic of the W-prearithmetic $\boldsymbol{A}$, which is isomorphic to the modular arithmetic $\boldsymbol{Z}_n$.

When $n = 1$, we have

$$0 \oplus 0 = h(g(0) + g(0)) = h(0 + 0) = h(0) = 0$$

and

$$0 \otimes 0 = h(g(0) \cdot g(0)) = h(0 \cdot 0) = h(0) = 0.$$

It means that we have a subprearithmetic $\boldsymbol{W}_1 = (W_1; \oplus, \otimes, \leq)$ of the W-prearithmetic $\boldsymbol{A}$, which consists of a single element.

When $n = 2$, we have the following equalities for addition:

$$p(g(10) + g(10)) = p(10 + 10) = p(20) = 0,$$

$$p(g(10) + g(11)) = p(10 + 11) = p(20 + 1) = 1,$$

$$p(g(11) + g(11)) = p(11 + 11) = p(20 + 2) = 2.$$

Then

$$l(0) = l(2) = 0,$$
$$l(1) = 1$$

and

$$u(0) = 10,$$
$$u(1) = 11.$$

Consequently,

$$\begin{aligned}
10 \oplus 10 &= h(g(10) + g(10)) = h(10 + 10) = h(20) \\
&= u(l(p(20))) = u(l(0)) = u(0) = 10, \\
10 \oplus 11 &= h(g(10) + g(11)) = h(10 + 11) = h(21) \\
&= u(l(p(21))) = u(l(1)) = u(1) = 11, \\
11 \oplus 11 &= h(g(11) + g(11)) = h(11 + 11) = h(21) \\
&= u(l(p(22))) = u(l(2)) = u(0) = 10.
\end{aligned}$$

When $n = 2$, we have the following equalities for multiplication:

$$\begin{aligned}
p(g(10) \cdot g(10)) &= p(10 \cdot 10) = p(100) = 0, \\
p(g(10) \cdot g(11)) &= p(10 \cdot 11) = p(110) = 0, \\
p(g(11) \cdot g(11)) &= p(11 \cdot 11) = p(121) = 1.
\end{aligned}$$

Then

$$l(0) = 0,$$
$$l(1) = 1$$

and

$$u(0) = 10,$$
$$u(1) = 11.$$

Consequently,

$$\begin{aligned}
10 \otimes 10 &= h(g(10) \cdot g(10)) = h(10 \cdot 10) = h(100) \\
&= u(l(p(20))) = u(l(0)) = u(0) = 10,
\end{aligned}$$

$$\begin{aligned}10 \otimes 11 &= h(g(10) \cdot g(11)) = h(10 \cdot 11) = h(110)\\ &= u(l(p(110))) = u(l(0)) = u(0) = 10,\\ 11 \otimes 11 &= h(g(11) \cdot g(11)) = h(11 \cdot 11) = h(121)\\ &= u(l(p(121))) = u(l(1)) = u(1) = 11.\end{aligned}$$

It means that we have a subprearithmetic $\boldsymbol{W}_2 = (W_2; \oplus, \otimes, \leq)$ of the W-prearithmetic $\boldsymbol{A}$, which consists of two elements and is isomorphic to the modular arithmetic $\boldsymbol{Z}_2$.

Now let us consider the case when $n > 2$. If numbers m and r belong to W_n, then $m = 10^{c_n} + k_m$ and $r = 10^{c_n} + k_r$ where $0 \leq k_m < n$ and $0 \leq k_r < n$. In the arithmetic $\boldsymbol{W}$, we have

$$g(m) + g(r) = m + r = 10^{c_n} + k_m + 10^{c_n} + k_r = 2 \cdot 10^{c_n} + k_m + k_r.$$

Then numbers k_m and k_r belong to the modular arithmetic $\boldsymbol{Z}_n$ and

$$p(g(m) + g(r)) = p(2 \cdot 10^{c_n} + k_m + k_r) = k_m + k_r$$

because as it was proved, $k_m + k_r < 2n < 10^{c_n}$. Then

$$l(k_m + k_r) = \min\{t; k_m + k_r = t \pmod n\} = a.$$

It means that the function $l(x)$ defines addition in W_n in the same way as it is defined in the modular arithmetic $\boldsymbol{Z}_n$ (cf. Example 2.1.3). Finally, we have

$$u(l(k_m + k_r)) = u(a) = 10^{c_n} + a,$$

that is

$$m \oplus r = h(g(m) + g(r)) = 10^{c_n} + a.$$

Thus, the set W_n is closed with respect to addition of its elements.

Let us once more take numbers m and r from the set W_n, then $m = 10^{c_n} + k_m$ and $r = 10^{c_n} + k_r$ where $0 \leq k_m < n$ and $0 \leq k_r < n$. In the arithmetic $\boldsymbol{W}$, we have

$$g(m)\cdot g(r) = m\cdot r = (10^{c_n}+k_m)\cdot(10^{c_n}+k_r) = 10^{2c_n}+10^{c_n}(k_m+k_r)+k_m\cdot k_r.$$

Then numbers k_m and k_r belong to the modular arithmetic $\boldsymbol{Z}_n$ and

$$p(g(m)\cdot g(r)) = p(10^{2c_n} + 10^{c_n}(k_m + k_r) + k_m \cdot k_r) = k_m \cdot k_r$$

because as it was proved, $k_m \cdot k_r < n^2 < 10^{c_n}$. Then

$$l(k_m \cdot k_r) = \min\{t; k_m \cdot k_r = t(\operatorname{mod} n)\} = b.$$

It means that the function $l(x)$ defines multiplication in W_n in the same way as it is defined in the modular arithmetic $\boldsymbol{Z}_n$ (cf. Example 2.1.3). Finally we have

$$u(l(k_m \cdot k_r)) = u(b) = 10^{c_n} + b,$$

that is

$$m \otimes r = h(g(m)\cdot g(r)) = 10^{c_n} + b.$$

Thus, the set W_n is closed with respect to multiplication of its elements. Consequently, the W-prearithmetic $\boldsymbol{W}_n = (W_n; \oplus, \otimes, \leq)$ is a subprearithmetic of the constructed W-prearithmetic $\boldsymbol{A} = (A; \oplus, \otimes, \leq)$. As each W-prearithmetic $\boldsymbol{W}_n = (W_n; \oplus, \otimes, \leq)$ contains n elements, the W-prearithmetic $\boldsymbol{A} = (A; \oplus, \otimes, \leq)$ contains a proper subprearithmetic with n elements for any natural number n.

In addition, the W-prearithmetic $\boldsymbol{W}_n = (W_n; \oplus, \otimes, \leq)$ is isomorphic to the modular arithmetic $\boldsymbol{Z}_n$ for all $n = 1, 2, 3, \ldots$.

Theorem is proved. □

Proposition 2.2.26 and properties of the conventional arithmetic $\boldsymbol{W}$ imply the following result.

Proposition 2.7.20. *For any W-prearithmetic $\boldsymbol{A} = (A; \oplus, \otimes, \leq_A)$ with the coprojector h and the projector g, which is an additive homomorphism, addition $\oplus$ is associative.*

Proof is left as an exercise. □

Proposition 2.2.27 and properties of the conventional arithmetic $\boldsymbol{W}$ imply the following result.

Proposition 2.7.21. *For any W-prearithmetic* $\boldsymbol{A} = (A; \oplus, \otimes, \leq_A)$ *with the coprojector* h *and the projector* g, *which is a multiplicative homomorphism, multiplication* $\otimes$ *is associative.*

Proof is left as an exercise. □

Proposition 2.2.28 and properties of the conventional arithmetic $\boldsymbol{W}$ imply the following result.

Proposition 2.7.22. *For any W-prearithmetic* $\boldsymbol{A} = (A; \oplus, \otimes, \leq_A)$ *with the coprojector* h *and the projector* g, *which is a homomorphism, addition* $\oplus$ *is associative and multiplication* $\otimes$ *is associative and distributive over addition* $\oplus$.

Proof is left as an exercise. □

Let us consider W-prearithmetics, properties of which are similar to the properties of the Diophantine arithmetic $\boldsymbol{W}$.

Definition 2.7.3. A whole-number prearithmetic (W-prearithmetic) $\boldsymbol{A} = (A; \oplus, \otimes, \leq_A)$ is:

(a) *broad* if $A \approx W$, i.e., there is a bijection $b : A \to W$;
(b) *total* if its projector g is a surjection, i.e., its image is the whole W;
(c) *complete* if $A = W$.

Example 2.7.3. The arithmetic $2\boldsymbol{W}$, which contains zero and all even numbers with the standard arithmetical operations and order, is a broad whole-number prearithmetic because there is a bijection $b : 2W \to W$. It is not total when the projector g is the natural inclusion of $2W$ into W but it is total when the projector g is defined as $g(2n) = n$.

We see that totality of W-prearithmetics is not an absolute property but depends on the projector.

Example 2.7.4. The arithmetic $3\boldsymbol{W}$ of all whole numbers divisible by 3 and zero with the standard arithmetical operations and order is a broad whole-number prearithmetic because there is a bijection $b : 3W \to W$. It is not total when the projector g is the natural inclusion of $3W$ into W but it is total when the projector g is defined as $g(3n) = n$.

Definition 2.7.3 directly implies the following result.

Lemma 2.7.2. *Any complete W-prearithmetic* $\boldsymbol{A} = (A; \oplus, \otimes, \leq_A)$ *is broad.*

Proof is left as an exercise. □

Proposition 2.7.23. *Any total W-prearithmetic* $\boldsymbol{A} = (A; \oplus, \otimes, \leq_A)$ *is broad.*

Proof: Let us assume that a W-prearithmetic $\boldsymbol{A} = (A; \oplus, \otimes, \leq_A)$ is total. By Definition 2.7.3, there is a surjection g: $A \to W$. Then by properties of sets, there is an injection f: $W \to A$.

At the same time, by Definition 2.7.1, A is a subset of the set W. Consequently, by the Cantor–Bernstein–Schröder theorem (cf., for example, Burgin, 2011), sets A and W are equipollent, i.e., there is a bijection b : $A \to W$. It means that the W-prearithmetic $\boldsymbol{A} = (A; \oplus, \otimes, \leq_A)$ is broad.

Proposition is proved. □

However, not every total W-prearithmetic is complete and not every complete W-prearithmetic is total.

Example 2.7.5. The W-prearithmetic $\boldsymbol{A} = (W; \oplus, \otimes, \leq_A)$ with the projector $g(n) = n + 1$ and coprojector $h(n) = n$ is complete but not total.

Example 2.7.6. The W-prearithmetic $\boldsymbol{A} = (N; \oplus, \otimes, \leq_A)$ with the projector $g(n) = n - 1$ and coprojector $h(n) = n$ is total but not complete.

Note that a total complete W-prearithmetic can be essentially different from the Diophantine arithmetic $\boldsymbol{W}$.

Example 2.7.7. Let us consider a W-prearithmetic $\boldsymbol{A} = (W; \oplus, \otimes, \leq_A)$ with the projector g and coprojector h, which are defined as $h(n) = n$ and

$$g(n) = \begin{cases} m & \text{if } n = 2m, \\ m & \text{if } n = 2m + 1. \end{cases}$$

Operations in $\boldsymbol{A}$ are defined as

$$a \oplus b = h(g(a) + g(b)),$$
$$a \otimes b = h(g(a) \cdot g(b)).$$

We see that the W-prearithmetic $\boldsymbol{A}$ is complete and total as any whole number is a half of some even number. At the same time, operations in $\boldsymbol{A}$

are essentially different from operations in the arithmetic $\boldsymbol{W}$. Indeed,

$$1 \oplus 1 = h(g(1) + g(1)) = h(0 + 0) = h(0) = 0,$$

$$2 \oplus 2 = h(g(2) + g(2)) = h(1 + 1) = h(2) = 2,$$

$$3 \oplus 5 = h(g(3) + g(5)) = h(1 + 2) = h(3) = 3.$$

Example 2.7.8. Let us consider a W-prearithmetic $\boldsymbol{A} = (W; \oplus, \otimes, \leq_A)$ with the projector g and coprojector h, which are defined as $h(n) = n + 1$ and

$$g(n) = \begin{cases} m + 1 & \text{if } n = 2m, \\ m & \text{if } n = 2m + 1. \end{cases}$$

Operations in $\boldsymbol{A}$ are defined as

$$a \oplus b = h(g(a) + g(b)),$$

$$a \otimes b = h(g(a) \cdot g(b)).$$

We see that the W-prearithmetic $\boldsymbol{A}$ is complete and total. At the same time, operations in $\boldsymbol{A}$ are essentially different from operations in the arithmetic $\boldsymbol{W}$. Indeed,

$$1 \oplus 0 = h(g(1) + g(0)) = h(0 + 1) = 1 + 1 = 2,$$

$$2 \oplus 2 = h(g(2) + g(2)) = h(2 + 2) = 4 + 1 = 5,$$

$$0 \oplus 5 = h(g(0) + g(5)) = h(1 + 3) = 4 + 1 = 5,$$

$$0 \oplus 2 = h(g(0) + g(2)) = h(1 + 2) = 3 + 1 = 4.$$

Definitions and properties of sets (Kuratowski and Mostowski, 1967) imply the following results.

Proposition 2.7.24. *If a subprearithmetic $\boldsymbol{K}$ of a W-prearithmetic $\boldsymbol{A} = (A; \oplus, \otimes, \leq_A)$ is broad, then $\boldsymbol{A}$ is also broad.*

Proof is left as an exercise. □

Proposition 2.7.25. *If a subprearithmetic $\boldsymbol{K}$ of a W-prearithmetic $\boldsymbol{A} = (A; \oplus, \otimes, \leq_A)$ is complete, then $\boldsymbol{A}$ is also complete.*

Proof is left as an exercise. □

Proposition 2.7.26. *If a W-prearithmetic $\boldsymbol{A}$ contains numbers n and m such that $g(n) = 0$ and $g(m) = 1$ while its projector g is a homomorphism, then $\boldsymbol{A}$ is total.*

Indeed, the image of $\boldsymbol{A}$ contains 0 and any number n larger than 0 is a sum of 1s and thus, n belongs to the image of $\boldsymbol{A}$ because g is a homomorphism. Consequently, the projector g is a surjection and $\boldsymbol{A}$ is total.

Remark 2.7.3. In a general case, a subprearithmetic $\boldsymbol{K}$ of the Diophantine arithmetic $\boldsymbol{W}$ is not always broad and thus, not always total because the subprearithmetic $\mathbf{0}$ of $\boldsymbol{W}$, which contains only 0, is not broad and thus, is not total.

However, because each infinite W-prearithmetic is either finite or countable (cf. Lemma 2.7.1), we have the following result.

Proposition 2.7.27. *Any infinite W-prearithmetic is broad.*

Proof is left as an exercise. □

Corollary 2.7.5. *Any subprearithmetic $\boldsymbol{K}$ of the Diophantine arithmetic $\boldsymbol{W}$ either contains only one element or is broad.*

Complete W-prearithmetics can have operations that are essentially different from operations in $\boldsymbol{W}$. For instance, we have the following result.

Proposition 2.7.28. *There is a complete W-prearithmetic $\boldsymbol{A} = (W; \oplus, \otimes, \leq_A)$, in which $n_A \oplus n_A = n_A$ for any number n_A.*

Proof: Let us take the set $W_A = \{n_A; n = 0, 1, 2, 3, \ldots\}$ and define the order $\leq$ as

$$n_A \leq_A m_A \text{ if and only if } n \leq m,$$

the function g as

$$g(0_A) = 0,$$
$$g(1_A) = 3,$$
$$g(\mathrm{S}n_A) = 2g(n_A) + 1 \quad \text{for all } n > 1$$

and the function h as

$$h(0) = 0_A,$$
$$h(m) = n_A \text{ if } g(n_A) \leq m < 2g(n_A) + 1 \quad \text{for all } n > 0.$$

This gives us the complete W-prearithmetic $\boldsymbol{A} = (W_A; \oplus, \otimes, \leq_A)$, in which

$$
\begin{aligned}
0_A \oplus 0_A &= 0_A, \\
1_A \oplus 1_A &= h(g(1_A) + g(1_A)) = h(3+3) = h(6) = 1_A, \\
n_A \oplus n_A &= h(g(n_A) + g(n_A)) \\
&= h(2g(n_A)) = h(2g(n_A)) = n_A \quad \text{for all } n > 1.
\end{aligned}
$$

Proposition is proved. □

Corollary 2.7.6. *There is a broad W-prearithmetic* $\boldsymbol{A} = (W; \oplus, \otimes, \leq_A)$, *in which* $n_A \oplus n_A = n_A$ *for any number* n_A.

It is possible to prove a more general result.

Proposition 2.7.29. *There is a complete W-prearithmetic* $\boldsymbol{A} = (W; \oplus, \otimes, \leq_A)$, *in which* $n_A \oplus m_A = \max(n_A, m_A)$ *for any numbers* n_A *and* m_A.

Proof: The W-prearithmetic $\boldsymbol{A} = (W; \oplus, \otimes, \leq_A)$, which is constructed in Proposition 2.7.28, has this property. Indeed, by construction, the coprojector h preserves the order $\leq$ in W. Consequently, by Corollary 2.7.2, the addition $\oplus$ is monotone in $\boldsymbol{A}$. As a result, assuming for certainty that $n_A \leq_A m_A$, we have

$$n_A = n_A \oplus n_A \leq_A n_A \oplus m_A \leq_A m_A \oplus m_A = m_A.$$

If $n_A = m_A$, then

$$n_A \oplus m_A = m_A \oplus m_A = m_A = \max(n_A, m_A).$$

If $n_A <_A m_A$, then

$$m_A = 0_A \oplus m_A \leq_A n_A \oplus m_A \leq_A m_A.$$

Consequently,

$$n_A \oplus m_A = m_A = \max(n_A, m_A).$$

Proposition is proved. □

Informally, it means that there are complete W-prearithmetics in which addition coincides with the operation of taking the maximum.

Corollary 2.7.7. *There is a broad W-prearithmetic $\boldsymbol{A} = (W; \oplus, \otimes, \leq_A)$, in which $n_A \oplus m_A = \max(n_A, m_A)$ for any numbers n_A and m_A.*

Remark 2.7.4. It is possible to deduce Proposition 2.7.28 from Proposition 2.7.29. However, we give independent proofs to provide constructions of different W-prearithmetics used in these proofs.

Similar results are valid for multiplication in W-prearithmetics.

Proposition 2.7.30. *There is a complete W-prearithmetic $\boldsymbol{A} = (W; \oplus, \otimes, \leq_A)$, in which $n_A \otimes n_A = n_A$ for any number n_A.*

Proof: Let us take the set $W_A = \{n_A;\ n = 0, 1, 2, 3, \ldots\}$ and define the following relations and functions

$$n_A \leq_A m_A \text{ if and only if } n \leq m,$$

$$g(0_A) = 0,$$

$$g(1_A) = 3,$$

$$g(\mathrm{S}n_A) = g(n_A)^2 + 1 \quad \text{for } n = 1, 2, 3, \ldots,$$

$$h(0) = h(1) = h(2) = 0_A,$$

$$h(m) = n_A \text{ if } g(n_A) \leq m < g(n_A)^2 + 1 \quad \text{for } n = 1, 2, 3 \ldots.$$

Functions g and h define the complete W-prearithmetic $\boldsymbol{A} = (W; \oplus, \otimes, \leq_A)$, in which

$$n_A \otimes n_A = h(g(n_A) \cdot g(n_A)) = h(g(n_A)^2) = h(g(n_A)) = n_A.$$

Proposition is proved. □

Corollary 2.7.8. *There is a broad W-prearithmetic $\boldsymbol{A} = (W; \oplus, \otimes, \leq_A)$, in which $n_A \otimes n_A = n_A$ for any number n_A.*

It is possible to prove a stronger result.

Proposition 2.7.31. *There is a complete W-prearithmetic $\boldsymbol{A} = (W; \oplus, \otimes, \leq_A)$, in which $n_A \otimes m_A = \max(n_A, m_A)$ for any number $n_A > 0_A$.*

Proof: The W-prearithmetic $\boldsymbol{A} = (W; \oplus, \otimes, \leq_A)$, which is constructed in Proposition 2.7.30, has this property. Indeed, by construction, the coprojector h preserves the order $\leq$ in W. Consequently, by Corollary 2.7.3, the multiplication $\otimes$ is monotone in $\boldsymbol{A}$. As a result, assuming for certainty that $n_A \leq_A m_A$, we have

$$n_A = n_A \otimes n_A \leq_A n_A \otimes m_A \leq_A m_A \otimes m_A = m_A.$$

If $n_A = m_A$, then

$$n_A \otimes m_A = m_A \otimes m_A = m_A = \max(n_A, m_A).$$

If $0_A \leq_A n_A <_A m_A$, then

$$m_A = 1_A \otimes m_A \leq_A n_A \otimes m_A \leq_A m_A.$$

Consequently,

$$n_A \otimes m_A = m_A = \max(n_A, m_A).$$

Proposition is proved. □

Informally, it means that there are complete W-prearithmetics in which multiplication coincides with the operation of taking the maximum.

Corollary 2.7.9. *There is a broad W-prearithmetic $\boldsymbol{A} = (W; \oplus, \otimes, \leq_A)$, in which $n_A \otimes m_A = \max(n_A, m_A)$ for any number $n_A > 0_A$.*

Remark 2.7.5. It is possible to deduce Proposition 2.7.30 from Proposition 2.7.31. However, we give independent proofs to provide constructions of different W-prearithmetics used in these proofs.

Proposition 2.7.32. *A W-prearithmetic $\boldsymbol{A} = (A; \oplus, \otimes, \leq_A)$ with the projector g and the coprojector h such that $gh = \mathbf{1}_W$ is total.*

Indeed, if $gh = \mathbf{1}_W$, then g is a surjection (Bucur and Deleanu, 1968) and by Definition 2.7.3, the W-prearithmetic $\boldsymbol{A} = (A; \oplus, \otimes, \leq_A)$ is total.

As the arithmetic $\boldsymbol{W}$ has the strong additive factoring property (cf. Section 2.2), Theorem 2.2.3 implies the following result.

Proposition 2.7.33. *Any W-prearithmetic $\boldsymbol{A} = (A; \oplus, \otimes, \leq_A)$ has the strong additive factoring property.*

Proof is left as an exercise. □

As the arithmetic $\boldsymbol{W}$ has the strong multiplicative factoring property (cf. Section 2.2), Theorem 2.2.5 implies the following result.

Proposition 2.7.34. *Any W-prearithmetic* $\boldsymbol{A} = (A; \oplus, \otimes, \leq_A)$ *has the strong multiplicative factoring property.*

Proof is left as an exercise. □

Let us define and study abstract prearithmetics, in which operations and order are defined by operations and order in the Diophantine arithmetic $\boldsymbol{N}$.

Definition 2.7.4. An abstract prearithmetic $\boldsymbol{A} = (A; \oplus, \otimes, \leq_A)$ is called a *natural-number prearithmetic*, or N-*prearithmetic*, if A is a subset of the set N of all natural numbers and the prearithmetic $\boldsymbol{A}$ is weakly projective with respect to the conventional arithmetic $\boldsymbol{N} = (N; +, \cdot, \leq)$ of natural numbers.

Example 2.7.9. Any modular arithmetic (cf. Example 2.1.3) is an N-prearithmetic.

Example 2.7.10. The arithmetic $7\boldsymbol{N}$ of all natural numbers divisible by 7 is an N-prearithmetic.

Let us consider an arbitrary set A and a mapping g: $A \to N$.

Proposition 2.7.35. *It is possible to define on* A *the structure of an N-prearithmetic* $\boldsymbol{A}$.

Proof is similar to the proof of Proposition 2.2.1 and uses an arbitrary mapping h: $W \to A$. □

Remark 2.7.6. Proposition 2.7.35 shows that it is possible to assume that in an N-prearithmetic $\boldsymbol{A}$, its carrier A can be any set for which a mapping g: $A \to N$ exists. This mapping determines a naming of the elements from A by natural numbers (Burgin, 2011). Taking this into account, we can denote the element from A that is mapped onto the natural number n by n_A. Note that if g is not a bijection, it is possible $n_A = m_A$ although $n \neq m$.

Thus, we will distinguish N-prearithmetics and natural-number arithmetics assuming that the carrier of the latter consists of natural numbers. This makes natural-number arithmetics a proper subclass in the class of all N-prearithmetics.

Proposition 2.7.36. *Any N-prearithmetic is a W-prearithmetic.*

Proof: The conventional Diophantine arithmetic $\boldsymbol{N}$ of all natural numbers is a subprearithmetic of the conventional Diophantine arithmetic $\boldsymbol{W}$ of all whole numbers (cf. Section 2.1). By Proposition 2.7.2, $\boldsymbol{N}$ is a W-prearithmetic. If $\boldsymbol{A}$ is an N-prearithmetic, then $\boldsymbol{A}$ is weakly projective with respect to the conventional arithmetic $\boldsymbol{N}$. Thus, by Proposition 2.7.3, $\boldsymbol{A}$ is a W-prearithmetic.

Proposition is proved. □

This result allows deducing many properties of N-prearithmetics from already obtained properties of W-prearithmetics. For instance, Propositions 2.7.1 and 2.7.35 imply the following result.

Corollary 2.7.10. *Any N-prearithmetic is a subprearithmetic of a W-prearithmetic.*

Proposition 2.7.37. *For any N-prearithmetic $\boldsymbol{A} = (A; \oplus, \otimes, \leq)$, operations $\oplus$ and $\otimes$ are commutative.*

Proof is similar to the proof of Proposition 2.7.1.

Propositions 2.7.2 and 2.7.35 imply the following result.

Proposition 2.7.38. *For any N-prearithmetic $\boldsymbol{A} = (A; \oplus, \otimes, \leq_A)$ with an injective projector into W, the order $\leq_A$ is linear.*

Indeed, if the projector g is an injection, then by Proposition 2.1.2, the order $\leq$ in $\boldsymbol{A}$ is linear because the order $\leq$ in $\boldsymbol{W}$ is linear.

Corollary 2.7.11. *In an abstract prearithmetic $\boldsymbol{A} = (A; +, \circ, \leq_A)$, which is weakly projective with respect to an N-prearithmetic $\boldsymbol{D} = (D; \oplus, \otimes, \leq_D)$ with the injective projector g, the relation $\leq$ is a well-ordering.*

Propositions 2.7.3 and 2.7.35 imply the following result.

Proposition 2.7.39. *For any N-prearithmetic $\boldsymbol{A} = (A; \oplus, \otimes, \leq_A)$, operations $\Sigma^{\oplus n}$ and $\Pi^{\otimes n}$ are commutative for all $n = 2, 3, 4, 5, \ldots$.*

Proof is left as an exercise. □

Propositions 2.7.4 and 2.7.36 imply the following result.

Proposition 2.7.40. *For any N-prearithmetic $\boldsymbol{A} = (A; \oplus, \otimes, \leq_A)$ with the projector g and the coprojector h, for which $gh = \mathbf{1}_W$, addition $\oplus$ and multiplication $\otimes$ are associative.*

Proof is left as an exercise. □

Proposition 2.2.8 and properties of the Diophantine arithmetic $\boldsymbol{N}$ imply the following result.

Proposition 2.7.41. *For any N-prearithmetic* $\boldsymbol{A} = (A; \oplus, \otimes, \leq_A)$ *with the projector* g *and the coprojector* h, *for which* $gh = \mathbf{1}_W$, *multiplication* $\otimes$ *is distributive over addition* $\oplus$.

Proof is left as an exercise. □

Propositions 2.2.11, 2.2.12 and properties of the Diophantine arithmetic $\boldsymbol{N}$ imply the following result.

Proposition 2.7.42. *For any N-prearithmetic* $\boldsymbol{A} = (A; \oplus, \otimes, \leq_A)$ *with the coprojector* h, *which preserves the order* $\leq$ *in* $\boldsymbol{N}$, *addition* $\oplus$ *and multiplication* $\otimes$ *preserve the order* $\leq_A$.

Proof is left as an exercise. □

The property of abstract prearithmetics "to be an N-prearithmetic" is hereditary as the following result demonstrates.

Proposition 2.7.43. *A subprearithmetic* $\boldsymbol{K}$ *of an N-prearithmetic* $\boldsymbol{A} = (A; \oplus, \otimes, \leq)$ *is an N-prearithmetic.*

Indeed, by Proposition 2.5.12, the prearithmetic $\boldsymbol{K}$ is weakly projective with respect to the prearithmetic $\boldsymbol{N}$ and thus, it is an N-prearithmetic.

Corollary 2.7.12. *A subprearithmetic* $\boldsymbol{K}$ *of an N-prearithmetic* $\boldsymbol{A} = (A; \oplus, \otimes, \leq)$ *is a W-prearithmetic.*

Corollary 2.7.13. *A subprearithmetic* $\boldsymbol{D}$ *of a subprearithmetic* $\boldsymbol{K}$ *of an N-prearithmetic* $\boldsymbol{A} = (A; \oplus, \otimes, \leq)$ *is an N-prearithmetic.*

Indeed, by Proposition 2.5.11, the prearithmetic $\boldsymbol{D}$ is a subprearithmetic $\boldsymbol{K}$ of an N-prearithmetic $\boldsymbol{A}$ and thus by Proposition 2.7.7, $\boldsymbol{D}$ is an N-prearithmetic.

Corollary 2.7.14. *A subprearithmetic* $\boldsymbol{D}$ *of a subprearithmetic* $\boldsymbol{K}$ *of an N-prearithmetic* $\boldsymbol{A} = (A; \oplus, \otimes, \leq)$ *is a W-prearithmetic.*

Proposition 2.7.44. *An abstract prearithmetic* $\boldsymbol{A} = (A; +, \circ, \leq_A)$, *which is weakly projective with respect to an N-prearithmetic* $\boldsymbol{D} = (D; \oplus, \otimes, \leq_D)$ *is an N-prearithmetic.*

Indeed, by Proposition 2.2.9, the prearithmetic $\boldsymbol{A}$ is weakly projective with respect to the prearithmetic $\boldsymbol{N}$ and thus, it is an N-prearithmetic.

Corollary 2.7.15. *An abstract prearithmetic* $\boldsymbol{A} = (A; +, \circ, \leq_A)$, *which is weakly projective with respect to an N-prearithmetic* $\boldsymbol{D} = (D; \oplus, \otimes, \leq_D)$ *is a W-prearithmetic.*

Proposition 2.7.45. *If an abstract prearithmetic* $\boldsymbol{A}_1 = (A_1; +_1, \circ_1, \leq_1)$ *is weakly projective with respect to an abstract prearithmetic* $\boldsymbol{A}_2 = (A_2; +_2, \circ_2, \leq_2)$ *and the abstract prearithmetic* $\boldsymbol{A}_2 = (A_2; +_2, \circ_2, \leq_2)$ *is weakly projective with respect to an N-prearithmetic* $\boldsymbol{A}_3 = (A_3; +_3, \circ_3, \leq_3)$, *then the prearithmetic* $\boldsymbol{A}_1$ *is an N-prearithmetic.*

Indeed, by Proposition 2.7.8, the prearithmetic $\boldsymbol{A}_2$ is an N-prearithmetic and applying once more Proposition 2.7.8, we come to the conclusion that $\boldsymbol{A}_1$ also is an N-prearithmetic.

Proposition 2.7.44 implies the following result.

Theorem 2.7.3. *N-prearithmetics form the category* ***NPAWP*** *where objects are N-prearithmetics and morphisms are weak projectivity relations.*

Proof is left as an exercise. □

Corollary 2.7.16. *The category* ***NPAWP*** *is a full subcategory of the category* ***WPAWP***.

Construction from Definition 2.2.2 allows defining actions of natural numbers on elements from an N-prearithmetic $\boldsymbol{A} = (A; \oplus, \otimes, \leq_A)$ with the projector g and the coprojector h. The rules are exactly the same as in Definition 2.7.2. As a result, properties of actions are also the same. In particular, we have the following results.

Proposition 2.7.46. *For any N-prearithmetic* $\boldsymbol{A} = (A; \oplus, \otimes, \leq_A)$, *additive action from the left coincides with additive action from the right.*

Proof is left as an exercise. □

Proposition 2.7.47. *For any N-prearithmetic* $\boldsymbol{A} = (A; \oplus, \otimes, \leq_A)$, *multiplicative action from the left coincides with multiplicative action from the right.*

Proof is left as an exercise. □

Proposition 2.2.18 implies the following result.

Proposition 2.7.48. *For any N-prearithmetic* $\boldsymbol{A} = (A; \oplus, \otimes, \leq_A)$ *with the coprojector* h, *which preserves addition, multiplicative action is distributive over addition* $\oplus$.

Proof is left as an exercise. □

Let us consider N-prearithmetics, properties of which are more similar to the properties of the Diophantine arithmetic $\boldsymbol{N}$.

Definition 2.7.5. An N-prearithmetic $\boldsymbol{A} = (A; \oplus, \otimes, \leq_A)$ is:

(a) *broad* if $A \approx N$., i.e., there is a bijection $b : A \to N$;
(b) *total* if its projector g is a surjection, i.e., its image is the whole set N;
(c) *complete* if $A = N$.

Example 2.7.11. The arithmetic $2\boldsymbol{N}$ of all even numbers with the standard arithmetical operations and order is a broad N-prearithmetic because there is a bijection $b : 2N \to N$. It is not total when the projector g is the natural inclusion of $2N$ into N but it is total when the projector g is defined as $g(2n) = n$.

Example 2.7.12. The arithmetic $3\boldsymbol{N}$ of all natural numbers $3N$ divisible by 3 with the standard arithmetical operations and order is a broad N-prearithmetic because there is a bijection $b : 3N \to N$. It is not total when the projector is the natural inclusion of $3N$ into N but it is total when the projector g is defined as $g(3n) = n$.

We see that totality of N-prearithmetics is not an absolute property but depends on the projector.

Definition 2.7.5 directly implies the following results.

Lemma 2.7.3. *Any complete N-prearithmetic* $\boldsymbol{A} = (A; \oplus, \otimes, \leq_A)$ *is broad.*

Proof is left as an exercise. □

Lemma 2.7.4. *Any total N-prearithmetic* $\boldsymbol{A} = (A; \oplus, \otimes, \leq_A)$ *is broad.*

Proof is similar to the proof of Proposition 2.7.23. □

However, not every total N-prearithmetic is complete and not every complete N-prearithmetic is total.

Example 2.7.13. The N-prearithmetic $\boldsymbol{A} = (N; \oplus, \otimes, \leq_A)$ with the projector $g(n) = n + 1$ and coprojector $h(n) = n$ is complete but not total.

Example 2.7.14. The W-prearithmetic $\boldsymbol{A} = (A = \{3, 4, 5, \ldots\}; \oplus, \otimes, \leq_A)$ with the projector $g(n) = n - 3$ and coprojector $h(n) = n$ is total but not complete.

Some properties of prearithmetics are induced by properties of their subprearithmetics.

Proposition 2.7.49. *If a subprearithmetic* $\boldsymbol{C}$ *of an N-prearithmetic* $\boldsymbol{A} = (A; \oplus, \otimes, \leq_A)$ *is total, then* $\boldsymbol{A}$ *is also total.*

Proof is left as an exercise. □

Corollary 2.7.17. *If a subprearithmetic* $\boldsymbol{C}$ *of an N-prearithmetic* $\boldsymbol{A} = (A; \oplus, \otimes, \leq_A)$ *is broad, then* $\boldsymbol{A}$ *is also broad.*

Proposition 2.7.50. *If a subprearithmetic* $\boldsymbol{C}$ *of an N-prearithmetic* $\boldsymbol{A} = (A; \oplus, \otimes, \leq_A)$ *is complete, then* $\boldsymbol{A}$ *is also complete.*

Proof is left as an exercise. □

Proposition 2.7.51. *Any subprearithmetic* $\boldsymbol{D}$ *of the arithmetic* $\boldsymbol{N}$ *is complete if it contains number* 1.

Indeed, any natural number n is a sum of 1s and thus, n belongs to $\boldsymbol{D}$. Consequently, $\boldsymbol{D}$ is complete.

However, because each infinite N-prearithmetic is either finite or countable (cf. Lemma 2.7.1), we have the following result.

Proposition 2.7.52. *Any infinite N-prearithmetic is broad.*

Proof is left as an exercise. □

Proposition 2.7.53. *The arithmetic* $\boldsymbol{N}$ *has only infinite subprearithmetics.*

Proof is left as an exercise. □

The situation with arbitrary N-prearithmetic is essentially different.

Theorem 2.7.4. *For any natural number* n*, there is an N-prearithmetic* $\boldsymbol{B} = (B; \oplus, \otimes, \leq_B)$ *such that for any natural number* n, $\boldsymbol{B}$ *contains a proper subprearithmetic with* n *elements.*

Proof: Let us consider the set N of all natural numbers and build an N-prearithmetic $\boldsymbol{B} = (B; \oplus, \otimes, \leq_B)$ with necessary properties. At first, we construct a special sequence of numbers $\{c_n;\ n = 3, 4, 5, \ldots\}$ using the following recursion:

$$c_3 = 3,$$
$$c_{n+1} = 3c_n + 1 \quad \text{for all } n = 3, 4, 5, \ldots .$$

In this sequence, we have $c_4 = 10$, $c_5 = 31$, $c_6 = 94$, and $c_7 = 283$.

Note that $2n < n^2 < 10^{c_n}$ for all $n = 3, 4, 5, \ldots$. We show this by induction.

For $n = 3$, we have

$$2 \cdot 3 = 6 < 3^2 = 9 < 10^3 = 1000.$$

Let us assume that $2n < n^2 < 10^{c_n}$ and prove a similar inequality for $n + 1$. Indeed,

$$2(n+1) = 2n + 2 < (n+1)^2 = n^2 + 2n + 1 < 2 \cdot 10^{c_n} < 10^{c_{n+1}}.$$

Thus, by the principle of the mathematical induction, this is true for all $n = 3, 4, 5, \ldots$.

Now we can define the set B by the following formula:

$$B = \bigcup_{n=1}^{\infty} N_n.$$

In this formula, we have the following sets:

$$N_1 = \{1\},$$
$$N_2 = \{10, 11\},$$
$$N_n = \{10^{c_n}, 10^{c_n} + 1, 10^{c_n} + 2, \ldots, 10^{c_n} + n - 1\} \quad \text{for all } n = 3, 4, 5, \ldots .$$

Note that each set N_n contains the same number of element as the modular arithmetic $\boldsymbol{Z}_n$.

To build an N-prearithmetic $\boldsymbol{B}$ on the set B, we need to define the projector g: $B \to N$ and the coprojector h: $N \to B$.

$$g(m) = m \quad \text{for all } m \text{ from the set } B,$$
$$h(1) = 1 \quad \text{for all numbers less than } 10.$$

For larger numbers m, we define h as the sequential composition of three functions $u(x)$, $l(x)$ and $p(x)$.

When m belongs to the interval $10 \leq m < 1000$, it has the form $m = q \cdot 10 + k$ where $k < 10$. Then we define

$$p(m) = k,$$
$$l(k) = \min\{t; k = t(\text{mod}\, 2)\},$$
$$u(t) = 10 + t$$

and

$$h(m) = u(l(p(m))).$$

When m belongs to the interval $10^{c_n} \leq m < 10^{c_n+1}$, it has the form $m = q \cdot 10^{c_n} + k$ where $k < 10^{c_n}$. Then we define

$$p(m) = k,$$
$$l(k) = \min\{t;\; k = t(\text{mod}\, n)\},$$
$$u(t) = 10^{c_n} + t$$

and

$$h(m) = u(l(p(m))).$$

This determines the function h for all whole numbers because intervals, where it is defined, cover N.

For instance,

$$h(5433) = u(l(p(5433))) = u(l(p(5 \cdot 10^3 + 433))) = u(l(433)) = u(1) = 10^3 + 1$$

because $433 = 1\ (\text{mod}\, 3)$.

For all numbers in B, we keep the relation $\leq$ from the set N. Therefore, to have an N-prearithmetic, we need to define operations of addition $\oplus$ and multiplication $\otimes$.

Taking two numbers m and r from A, we have

$$m \oplus r = h(g(m) + g(r))$$

and

$$m \otimes r = h(g(m) \cdot g(r)).$$

This gives us an N-prearithmetic $\boldsymbol{B} = (B; \oplus, \otimes, \leq_B)$.

Let us take the set N_n and show that it is closed with respect to operations of addition $\oplus$ and multiplication $\otimes$ being a subprearithmetic of the N-prearithmetic $\boldsymbol{B}$, which is isomorphic to the modular arithmetic $\boldsymbol{Z}_n$.

When $n = 1$, we have

$$1 \oplus 1 = h(g(1) + g(1)) = h(1 + 1) = h(2) = 1$$

and

$$1 \otimes 1 = h(g(1) \cdot g(1)) = h(1 \cdot 1) = h(1) = 1.$$

It means that we have a subprearithmetic $\boldsymbol{N}_1 = (N_1; \oplus, \otimes, \leq_1)$ of the N-prearithmetic $\boldsymbol{B}$, which consists of a single element.

When $n = 2$, we have the following equalities for addition:

$$\begin{aligned} p(g(10) + g(10)) &= p(10 + 10) = p(20) = 0, \\ p(g(10) + g(11)) &= p(10 + 11) = p(20 + 1) = 1, \\ p(g(11) + g(11)) &= p(11 + 11) = p(20 + 2) = 2. \end{aligned}$$

Then

$$\begin{aligned} &l(0) = l(2) = 0, \\ &l(1) = 1. \end{aligned}$$

and

$$\begin{aligned} &u(0) = 10, \\ &u(1) = 11. \end{aligned}$$

Consequently,

$$\begin{aligned} 10 \oplus 10 &= h(g(10) + g(10)) = h(10 + 10) = h(20) \\ &= u(l(p(20))) = u(l(0)) = u(0) = 10, \\ 10 \oplus 11 &= h(g(10) + g(11)) = h(10 + 11) = h(21) \\ &= u(l(p(21))) = u(l(1)) = u(1) = 11, \\ 11 \oplus 11 &= h(g(11) + g(11)) = h(11 + 11) = h(21) \\ &= u(l(p(22))) = u(l(2)) = u(0) = 10. \end{aligned}$$

When $n = 2$, we have the following equalities for multiplication:

$$p(g(10) \cdot g(10)) = p(10 \cdot 10) = p(100) = 0,$$
$$p(g(10) \cdot g(11)) = p(10 \cdot 11) = p(110) = 0,$$
$$p(g(11) \cdot g(11)) = p(11 \cdot 11) = p(121) = 1.$$

Then

$$l(0) = 0,$$
$$l(1) = 1.$$

and

$$u(0) = 10,$$
$$u(1) = 11.$$

Consequently,

$$\begin{aligned} 10 \otimes 10 &= h(g(10) \cdot g(10)) = h(10 \cdot 10) = h(100) \\ &= u(l(p(20))) = u(l(0)) = u(0) = 10, \\ 10 \otimes 11 &= h(g(10) \cdot g(11)) = h(10 \cdot 11) = h(110) \\ &= u(l(p(110))) = u(l(0)) = u(0) = 10, \\ 11 \otimes 11 &= h(g(11) \cdot g(11)) = h(11 \cdot 11) = h(121) \\ &= u(l(p(121))) = u(l(1)) = u(1) = 11. \end{aligned}$$

It means that we have a subprearithmetic $\boldsymbol{N}_2 = (N_2; \oplus, \otimes, \leq)$ of the N-prearithmetic $\boldsymbol{B}$, which consists of two elements and is isomorphic to the modular arithmetic $\boldsymbol{Z}_2$.

Now let us consider the case when $n > 2$. If numbers m and r belong to N_n, then $m = 10^{c_n} + k_m$ and $r = 10^{c_n} + k_r$ where $0 \leq k_m < n$ and $0 \leq k_r < n$. In the arithmetic $\boldsymbol{N}$, we have

$$g(m) + g(r) = m + r = 10^{c_n} + k_m + 10^{c_n} + k_r = 2 \cdot 10^{c_n} + k_m + k_r.$$

Then numbers k_m and k_r belong to the modular arithmetic $\boldsymbol{Z}_n$ and

$$p(g(m) + g(r)) = p(2 \cdot 10^{c_n} + k_m + k_r) = k_m + k_r$$

because as it was proved, $k_m + k_r < 2n < 10^{c_n}$. Then

$$l(k_m + k_r) = \min\{t; k_m + k_r = t(\bmod n)\} = a.$$

It means that the function $l(x)$ defines addition in N_n in the same way as it is defined in the modular arithmetic $\boldsymbol{Z}_n$ (cf. Example 2.1.3). Finally, we have

$$u(l(k_m + k_r)) = u(a) = 10^{c_n} + a,$$

that is

$$m \oplus r = h(g(m) + g(r)) = 10^{c_n} + a.$$

Thus, the set N_n is closed with respect to addition of its elements.

Let us once more take numbers m and r from the set N_n, then $m = 10^{c_n} + k_m$ and $r = 10^{c_n} + k_r$ where $0 \leq k_m < n$ and $0 \leq k_r < n$. In the arithmetic $\boldsymbol{N}$, we have

$$\begin{aligned} g(m) \cdot g(r) \\ &= m \cdot r = (10^{c_n} + k_m) \cdot (10^{c_n} + k_r) = 10^{2c_n} + 10^{c_n}(k_m + k_r) + k_m \cdot k_r. \end{aligned}$$

Then numbers k_m and k_r belong to the modular arithmetic $\boldsymbol{Z}_n$ and

$$p(g(m) \cdot g(r)) = p(10^{2c_n} + 10^{c_n}(k_m + k_r) + k_m \cdot k_r) = k_m \cdot k_r$$

because as it was proved, $k_m \cdot k_r < n^2 < 10^{c_n}$. Then

$$l(k_m \cdot k_r) = \min\{t; k_m \cdot k_r = t(\bmod n)\} = b.$$

It means that the function $l(x)$ defines multiplication in N_n in the same way as it is defined in the modular arithmetic $\boldsymbol{Z}_n$ (cf. Example 2.1.3). Finally we have

$$u(l(k_m \cdot k_r)) = u(b) = 10^{c_n} + b,$$

that is

$$m \otimes r = h(g(m) \cdot g(r)) = 10^{c_n} + b.$$

Thus, the set N_n is closed with respect to multiplication of its elements. Consequently, the N-prearithmetic $\boldsymbol{N}_n = (N_n; \oplus, \otimes, \leq_n)$ is a subprearithmetic of the constructed N-prearithmetic $\boldsymbol{B} = (B; \oplus, \otimes, \leq_B)$. As each N-prearithmetic $\boldsymbol{N}_n = (N_n; \oplus, \otimes, \leq_n)$ contains n elements, the N-prearithmetic $\boldsymbol{B} = (B; \oplus, \otimes, \leq)$ contains a proper subprearithmetic with n elements for any natural number n.

In addition, the N-prearithmetic $\boldsymbol{N}_n = (N_n; \oplus, \otimes, \leq_n)$ is isomorphic to the modular arithmetic $\boldsymbol{Z}_n$ for all $n = 1, 2, 3, \ldots$.

Theorem is proved. □

It is possible to build new prearithmetics on whole numbers using the conventional Diophantine arithmetic $\boldsymbol{R} = (R; +, \cdot, \leq)$ of all real numbers.

Definition 2.7.6. An abstract prearithmetic $\boldsymbol{A} = (A; \oplus, \otimes, \leq_A)$ is called a *real-whole-number prearithmetic*, or RW- *prearithmetic*, if A is a subset of the set W of all whole numbers and the prearithmetic $\boldsymbol{A}$ is weakly projective with respect to the conventional arithmetic $\boldsymbol{R} = (R; +, \cdot, \leq)$ of all real numbers.

It means that there are two functions: a projector g: $A \to R$ and a coprojector h: $R \to A$ such that for any numbers a and b from A, we have

$$a \oplus b = h(g(a) + g(b)),$$

$$a \otimes b = h(g(a) \cdot g(b)),$$

$$a \leq b \text{ in } \boldsymbol{A} \text{ only if } g(a) \leq_2 g(b) \text{ in } \boldsymbol{R}.$$

Here a and b are whole numbers, $+$ is addition and $\cdot$ is multiplication of real numbers, while $\oplus$ is addition and $\otimes$ is multiplication of numbers in prearithmetic defined by functions f and g.

Example 2.7.15. RW-prearithmetics allow us to formalize some transformations used by mathematicians as tricks. For instance, to show that it is possible to have

$$2 + 2 = 5 \tag{2.50}$$

mathematicians use the following reasoning. By conventional rules for rounding decimal numbers, the number 2.4 is rounded and becomes equal to 2. At the same time, in the conventional arithmetic of real numbers, we have

$$2.4 + 2.4 = 4.8 \tag{2.51}$$

while the number 4.8 is rounded to 5. Performing rounding to each number in the equality (2.51), we come to the equality (2.50).

This looks artificial but we can build the WR-prearithmetic $\boldsymbol{D}$, which validates this reasoning and where equality (2.50) is true.

Let us consider functions g: $W \to R$ and h: $R \to W$ where $g(n) = n + 0.4$, e.g., $g(5) = 5 + 0.4 = 5.4$, for the whole number $n \in W$ and for the real number $a \in R$, we define $h(a) = [a]$ when $0 \leq a$ and $h(a) = 0$ when $a < 0$, e.g., $h(2.5) = 2$. We take the function g as the projector and the function h as the coprojector defining the RW-prearithmetic $\boldsymbol{D} = (W; \oplus, \otimes, \leq)$, in which

$$a \oplus b = h(g(a) + g(b)),$$

$$a \otimes b = h(g(a) \cdot g(b))$$

and the order $\leq$ of numbers is the same as in the arithmetic $\boldsymbol{W}$.

Then we have

$$2 \oplus 2 = [2.4 + 2.4] = [4.8] = 5.$$

Thus, two plus two is equal to five in the RW-prearithmetic $\boldsymbol{D} = (W; \oplus, \otimes, \leq)$.

Proposition 2.7.54. *Any W-prearithmetic is also an RW-prearithmetic.*

Proof: Let us consider a W-prearithmetic $\boldsymbol{A} = (A; \oplus, \otimes, \leq)$. In it, $A \subseteq W$ and there are a projector g: $A \to W$ and a coprojector h: $W \to A$, which determine operations and order in $\boldsymbol{A}$. To prove our statement, we need to define a projector f: $A \to R$ and a coprojector $q : R \to A$, which determine the same operations and the same order in $\boldsymbol{A}$. As $W \subseteq R$, we can take $f = g$ and for any real number r, define

$$q(r) = \begin{cases} q(r) & \text{if } r \in W, \\ 0 & \text{otherwise.} \end{cases}$$

Then for any a and b from A, we have

$$a \oplus b = h(g(a) + g(b)) = q(f(a) + f(b))$$

and

$$a \otimes b = h(g(a) \cdot g(b)) = q(f(a) \cdot f(b)).$$

Proposition is proved. □

Many properties of RW-prearithmetics are the same as properties of W-prearithmetics studied above because the Diophantine arithmetics $\boldsymbol{W}$ and $\boldsymbol{R}$ have many similar properties. In particular, we have the following results.

Proposition 2.7.55. *For any RW-prearithmetic* $\boldsymbol{A} = (A; \oplus, \otimes, \leq_A)$ *with an injective projector into* W*, the order* $\leq$ *in* $\boldsymbol{A}$ *is linear.*

Indeed, if the projector g is an injection, then by Proposition 2.1.2, the order $\leq$ in $\boldsymbol{A}$ is linear because the order $\leq$ in the Diophantine arithmetic $\boldsymbol{R}$ is linear.

Corollary 2.7.18. *In an abstract prearithmetic $\boldsymbol{A} = (A; +, \circ, \leq_A)$, which is weakly projective with respect to an RW-prearithmetic $\boldsymbol{D} = (D; \oplus, \otimes, \leq_D)$ with the injective projector g, the relation $\leq$ is a well-ordering.*

Proposition 2.7.56. *For any RW-prearithmetic $\boldsymbol{A} = (A; \oplus, \otimes, \leq_A)$, operations $\Sigma^{\oplus n}$ and $\Pi^{\otimes n}$ are commutative for all $n = 2, 3, 4, 5, \ldots$.*

Proof is left as an exercise. □

Proposition 2.7.57. *For any RW-prearithmetic $\boldsymbol{A} = (A; \oplus, \otimes, \leq_A)$ with the projector g and the coprojector h, for which $gh = \mathbf{1}_W$, addition $\oplus$ and multiplication $\otimes$ are associative.*

Proof is left as an exercise. □

Proposition 2.7.58. *For any RW-prearithmetic $\boldsymbol{A} = (A; \oplus, \otimes, \leq_A)$ with the projector g and the coprojector h, for which $gh = \mathbf{1}_W$, multiplication $\otimes$ is distributive with respect to addition $\oplus$.*

Proof is left as an exercise. □

Propositions 2.2.11, 2.2.12 and properties of the conventional Diophantine arithmetic $\boldsymbol{R}$ imply the following result.

Proposition 2.7.59. *For any RW-prearithmetic $\boldsymbol{A} = (A; \oplus, \otimes, \leq_A)$ with the coprojector h, which preserves the order $\leq$ in $\boldsymbol{R}$, addition $\oplus$ and multiplication $\otimes$ preserve the order $\leq_A$.*

Proof is left as an exercise. □

Proposition 2.7.60. *A subprearithmetic $\boldsymbol{K}$ of an RW-prearithmetic $\boldsymbol{A} = (A; \oplus, \otimes, \leq)$ is an RW-prearithmetic.*

Indeed, by Proposition 2.5.12, the prearithmetic $\boldsymbol{K}$ is weakly projective with respect to the Diophantine arithmetic $\boldsymbol{R}$ and thus, it is an RW-prearithmetic.

Corollary 2.7.19. *A subprearithmetic $\boldsymbol{D}$ of a subprearithmetic $\boldsymbol{K}$ of an RW-prearithmetic $\boldsymbol{A} = (A; \oplus, \otimes, \leq)$ is an RW-prearithmetic.*

Indeed, by Proposition 2.5.11, the prearithmetic $\boldsymbol{D}$ is a subprearithmetic of an RW-prearithmetic $\boldsymbol{A}$ and thus by Proposition 2.7.57, $\boldsymbol{D}$ is an RW-prearithmetic.

Proposition 2.7.61. *An abstract prearithmetic* $\boldsymbol{A} = (A; +, \circ, \leq_A)$, *which is weakly projective with respect to an RW-prearithmetic* $\boldsymbol{D} = (D; \oplus, \otimes, \leq_D)$ *is an RW-prearithmetic.*

Indeed, by Proposition 2.2.9, the prearithmetic $\boldsymbol{A}$ is weakly projective with respect to the prearithmetic $\boldsymbol{R}$ and thus, it is an RW-prearithmetic.

Proposition 2.7.62. *If an abstract prearithmetic* $\boldsymbol{A}_1 = (A_1; +_1, \circ_1, \leq_1)$ *is weakly projective with respect to an abstract prearithmetic* $\boldsymbol{A}_2 = (A_2; +_2, \circ_2, \leq_2)$ *and the abstract prearithmetic* $\boldsymbol{A}_2 = (A_2; +_2, \circ_2, \leq_2)$ *is weakly projective with respect to an RW-prearithmetic* $\boldsymbol{A}_3 = (A_3; +_3, \circ_3, \leq_3)$, *then the prearithmetic* $\boldsymbol{A}_1$ *is an RW-prearithmetic.*

Proof is left as an exercise. □

Now let us consider prearithmetics related to arithmetics of natural numbers.

Definition 2.7.7. An abstract prearithmetic $\boldsymbol{A} = (A; \oplus, \otimes, \leq_A)$ is called a *real-natural-number prearithmetic*, or RN-*prearithmetic*, if A is a subset of the set N of natural numbers and the prearithmetic $\boldsymbol{A}$ is weakly projective with respect to the conventional arithmetic $\boldsymbol{R} = (R; +, \cdot, \leq)$ of all real numbers R.

Proposition 2.7.63. *Any RN-prearithmetic is also an RW-prearithmetic.*

Proof is left as an exercise. □

Corollary 2.7.20. *Any RN-prearithmetic is a subprearithmetic of an RW-prearithmetic.*

Proposition 2.7.64. *Any N-prearithmetic is also an RN-prearithmetic.*

Proof is left as an exercise. □

Many properties of RW-prearithmetics are the same as properties of W-prearithmetics studied above because the Diophantine arithmetics $\boldsymbol{N}$ and $\boldsymbol{R}$ have many similar properties. We do not formulate these because a qualified reader can do this without difficulties. At the same time, some properties are essentially different. For instance, we have the following results.

We remind that a number n is *even* in a whole-number prearithmetic $\boldsymbol{A}$ if $n = 2 \otimes m$ for some number m from $\boldsymbol{A}$.

Proposition 2.7.65. *There is an RN-prearithmetic in which all numbers but one are even.*

Proof: Let us consider the whole-number prearithmetic $\boldsymbol{N}_{\text{exp}} = (N; \oplus, \otimes, \leq)$ with the functional projector $f(x)$, where $f(0) = 1$ and $f(n) = 2^n$ when $n > 0$, and the coprojector $\lfloor \log_2 n \rfloor$.

Then multiplication in $\boldsymbol{N}_{\text{exp}}$ is defined in the following way.

If p and q are numbers from $\boldsymbol{N}_{\text{exp}}$, then $p \otimes q = \log_2(2^p \cdot 2^q) = \log_2(2^{p+q}) = p + q$. If n is a number from $\boldsymbol{N}_{\text{exp}}$ and $n > 2$, then $n = 2 \otimes m = 2 + m$ where $m = n - 2$. Thus, by definition, n is evenly divisible by 2 and consequently, it is an even number in $\boldsymbol{N}_{\text{exp}}$.

Proposition is proved. □

We remind that a number n is *multiplicatively prime* in a prearithmetic $\boldsymbol{A} = (A; \oplus, \otimes, \leq)$ if there are no numbers p, $q \neq 1$ in $\boldsymbol{A}$ such that $n = p \otimes q$.

Corollary 2.7.21. *There is an RN-prearithmetic that has only one multiplicatively prime number.*

Indeed, the prearithmetic $\boldsymbol{N}_{\text{exp}}$ has only one prime number 2 because all other numbers are divisible by 2.

Corollary 2.7.22. *There is an RW-prearithmetic in which all numbers but two are even.*

Corollary 2.7.23. *There is an RW-prearithmetic that has only one multiplicatively prime number.*

In some RN-prearithmetics and RW-prearithmetics, divisibility is essentially different from divisibility in Diophantine arithmetics.

Proposition 2.7.66. *There is a whole-number prearithmetic in which, any number is divisible by any smaller number.*

Indeed, in the prearithmetic $\boldsymbol{N}_{\text{exp}}$, any number is divisible by any smaller number because if $m < n$, then there is a natural number k such that $n = k + m$. Consequently, $n = k + m = k \otimes m$.

It is possible to extend the statement of Corollary 2.7.21 from number 2 to any natural number n.

Theorem 2.7.5. *For any $n > 0$, there is a whole-number prearithmetic that has exactly n multiplicatively prime numbers.*

Proof: Let us take the nth prime number p_n from the Diophantine arithmetic $\boldsymbol{N}$ and build two functions $g(k)$ and $h(k)$ by the following rules:

$$g(k) = \begin{cases} k & \text{if } k < p_n + 1, \\ 2^k & \text{if } k > p_n, \end{cases}$$

$$h(k) = \begin{cases} k & \text{if } \lfloor \log_2 k \rfloor < p_n + 1, \\ \lfloor \log_2 k \rfloor & \text{if } \lfloor \log_2 k \rfloor > p_n. \end{cases}$$

Let us consider the prearithmetic $\boldsymbol{N}_{n,\exp} = (N; \oplus, \otimes, \leq)$ with the functional projector $g(x)$ and the coprojector is $h(x)$.

Then multiplication in $\boldsymbol{N}_{n,\exp}$ is defined in the following way.

If p and q are numbers from $\boldsymbol{N}_{n,\exp}$ and $p \cdot q < p_n + 1$, then $p < p_n + 1$ and $q < p_n + 1$. Thus,

$$p \otimes q = h(g(p) \cdot g(q)) = p \cdot q.$$

It means that all composite in $\boldsymbol{N}$ numbers smaller than $p_n + 1$ are composite in $\boldsymbol{N}_{n,\exp}$ and all prime in $\boldsymbol{N}$ numbers smaller than $p_n + 1$ are prime in $\boldsymbol{N}_{n,\exp}$. Consequently, we have exactly n prime numbers in $\boldsymbol{N}_{n,\exp}$, which are smaller than $p_n + 1$.

Let us show that all numbers larger than p_n are composite in $\boldsymbol{N}_{n,\exp}$.

If r is a number from $\boldsymbol{N}_{n,\exp}$ and $r > p_n$, then as $\lfloor \log_2(2^{r+1}) \rfloor = \log_2(2^{r+1}) = r + 1 > p_n$, we have

$$\begin{aligned} 2 \otimes r &= h(g(2) \cdot g(r)) = \lfloor \log_2(2 \cdot 2^r) \rfloor = \lfloor \log_2(2^{r+1}) \rfloor \\ &= \log_2(2^{r+1}) = r + 1. \end{aligned}$$

Thus, by definition, $r + 1$ is an even number in $\boldsymbol{N}_{n,\exp}$ when $r > p_n$. It means that all numbers larger than $p_n + 1$ are even in $\boldsymbol{N}_{n,\exp}$.

Now let us consider the number $m = p_n + 1$. It is an even number in the Diophantine arithmetic $\boldsymbol{N}$, i.e., $m = 2q$ for some natural number q. Besides, $\lfloor \log_2(m) \rfloor < p_n$. Thus, we have

$$2 \otimes q = h(g(2) \cdot g(q)) = 2 \cdot q.$$

It means that m is also a composite number in $\boldsymbol{N}_{n,\exp}$.

Theorem is proved. □

2.8. Weak Numerical Arithmetics and Projective Non-Diophantine Arithmetics

The human senses and understanding, weak as they are, are not to be deprived of their authority, but to be supplied with helps.

Francis Bacon

The Diophantine arithmetic of natural or of whole numbers is a numerical prearithmetic that satisfies extra conditions in comparison with the numerical prearithmetics studied in the previous section. Non-Diophantine arithmetics share many but not all properties of the Diophantine arithmetic. So, in this section at first, we consider numerical prearithmetics, which are called weak numerical arithmetics being even closer to the Diophantine arithmetic, i.e., having more common properties, than numerical prearithmetics studied above. Then we introduce and study projective arithmetics, which are even closer to the Diophantine arithmetic. Finally, when a weak arithmetic contains all natural numbers with the natural order, it is called a non-Diophantine arithmetic of natural numbers. Similarly, when a weak arithmetic contains all whole numbers with the natural order, it is called a non-Diophantine arithmetic of whole numbers.

Definition 2.8.1. A whole-number prearithmetic $\boldsymbol{A} = (A; \oplus, \otimes, \leq_A)$ is called a *weak whole-number arithmetic*, or a *weak* W-*arithmetic*, if it is projective with respect to the conventional (Diophantine) arithmetic $\boldsymbol{W} = (W; +, \cdot, \leq)$ of the whole numbers.

It means that there are two functions: a projector g: $A \to W$ and a coprojector h: $W \to A$ such that $hg = \mathbf{1}_A$ and for any numbers a and b from A, we have

$$a \oplus b = h(g(a) + g(b)),$$

$$a \otimes b = h(g(a) \cdot g(b)),$$

$$a \leq b \text{ in } \boldsymbol{A} \text{ if and only if } g(a) \leq_2 g(b) \text{ in } \boldsymbol{R}.$$

Lemma 2.3.1 implies the following result.

Lemma 2.8.1. *Any weak W-arithmetic is a W-prearithmetic.*

Proof is left as an exercise. □

Lemma 2.8.1 allows getting many properties of weak W-arithmetics from properties of W-prearithmetics. For instance, Proposition 2.7.1 and Lemma 2.8.1 imply the following result.

Proposition 2.8.1. *Both operations, addition $\oplus$ and multiplication $\otimes$, are commutative in any weak W-arithmetic $\boldsymbol{A} = (A; \oplus, \otimes, \leq_A)$.*

Proof is left as an exercise. □

Remark 2.8.1. In some weak W-arithmetics, addition $\oplus$ or/and multiplication $\otimes$ is not associative.

Proposition 2.7.3 and Lemma 2.8.1 imply the following result.

Proposition 2.8.2. *For any weak W-arithmetic $\boldsymbol{A} = (A; \oplus, \otimes, \leq_A)$, operations $\Sigma^{\oplus n}$ and $\Pi^{\otimes n}$ are commutative for all $n = 2, 3, 4, 5, \ldots$.*

Proof is left as an exercise. □

Let us study relation between the order and operations in weak W-arithmetics.

Proposition 2.8.3. *If, in a weak W-arithmetic $\boldsymbol{A} = (A; \oplus, \otimes, \leq_A)$, the coprojector h preserves the order $\leq$, then the operation $\oplus$ preserves the order $\leq_A$.*

Proof: Let us consider a weak W-arithmetic $\boldsymbol{A} = (A; \oplus, \otimes, \leq_A)$, for which the coprojector h preserves the order $\leq$ and take four elements a, b, c and d from A, which satisfy the inequalities $a \leq_A b$ and $c \leq_A d$. Then by Definition 2.2.1, $g(a) \leq g(b)$ and $g(c) \leq g(d)$. Addition of whole numbers preserves the order $\leq$. Thus, $g(a) + g(c) \leq g(b) + g(d)$. As the coprojector h preserves the order $\leq$, we have

$$a \oplus c \leq h(g(a) + g(c)) \leq h(g(b) + g(d)) = b \oplus d.$$

Proposition is proved. □

Corollary 2.8.1. *For any W-prearithmetic $\boldsymbol{A} = (A; \oplus, \otimes, \leq_A)$ with the coprojector h, which preserves the order $\leq$ in the Diophantine arithmetic $\boldsymbol{W}$, addition $\oplus$ is monotone.*

Proposition 2.8.4. *If, in a weak W-arithmetic $\boldsymbol{A} = (A; \oplus, \otimes, \leq_A)$, the coprojector h preserves the order $\leq$, then the operation $\otimes$ preserves the order $\leq_A$.*

Proof: Let us consider a weak W-arithmetic $\boldsymbol{A} = (A; \oplus, \otimes, \leq_A)$, for which the coprojector h preserves the order $\leq$ and take four elements a, b, c and d from A, which satisfy the inequalities $a \leq_A b$ and $c \leq_A d$. Then by Definition 2.2.1, $g(a) \leq g(b)$ and $g(c) \leq g(d)$. Multiplication of whole numbers preserves the order $\leq$. Thus, $g(a) \cdot g(c) \leq g(b) \cdot g(d)$. As the coprojector h preserves the order $\leq$, we have

$$a \otimes c \leq h(g(a) \cdot g(c)) \leq h(g(b) \cdot g(d)) = b \otimes d.$$

Proposition is proved. □

Corollary 2.8.2. *For any W-prearithmetic $\boldsymbol{A} = (A; \oplus, \otimes, \leq_A)$ with the coprojector h, which preserves the order $\leq$ in the Diophantine arithmetic $\boldsymbol{W}$, multiplication $\otimes$ is monotone.*

Proposition 2.7.10 and Lemma 2.8.1 imply the following result.

Proposition 2.8.5. *For any weak W-arithmetic $\boldsymbol{A} = (A; \oplus, \otimes, \leq_A)$, additive action from the left coincides with additive action from the right.*

Proof is left as an exercise. □

Proposition 2.8.5 allows skipping indication of the side of additive action.

Proposition 2.7.11 and Lemma 2.8.1 imply the following result.

Proposition 2.8.6. *For any weak W-arithmetic $\boldsymbol{A} = (A; \oplus, \otimes, \leq_A)$, multiplicative action from the left coincides with multiplicative action from the right.*

Proof is left as an exercise. □

Proposition 2.8.6 allows skipping indication of the side of multiplicative action.

Proposition 2.7.12 and Lemma 2.8.1 imply the following result.

Proposition 2.8.7. *For any W-prearithmetic $\boldsymbol{A} = (A; \oplus, \otimes, \leq_A)$ with the coprojector h, which preserves addition, multiplicative action is distributive over addition $\oplus$.*

Proof is left as an exercise. □

As the ordered set W of all whole numbers is finitely connected, Proposition 2.3.6 implies the following result.

Proposition 2.8.8. *In a weak W-arithmetic $\boldsymbol{A} = (A; \oplus, \otimes, \leq_A)$, the partially ordered set A is finitely connected.*

Proof is left as an exercise. □

Because the natural order in the Diophantine arithmetic $\boldsymbol{W}$ is linear, Proposition 2.3.11 implies the following result.

Proposition 2.8.9. *For any weak W-arithmetic $\boldsymbol{A} = (A; \oplus, \otimes, \leq_A)$, the order $\leq_A$ is linear.*

Proof is left as an exercise. □

Proposition 2.8.8 and Lemma 2.1.22 imply the following result.

Proposition 2.8.10. *For any weak W-arithmetic $\boldsymbol{A} = (A; \oplus, \otimes, \leq_A)$, the order $\leq_A$ is discrete.*

Indeed, by Lemma 2.1.22, if a partially ordered set X is finitely connected, then the order in X is discrete and by Proposition 2.8.8, the order $\leq_A$ is finitely connected.

This result implies that any element x of a weak W-arithmetic $\boldsymbol{A} = (A; \oplus, \otimes, \leq_A)$, which is not maximal, has a successor Sx, and any element y of $\boldsymbol{A}$, which is not minimal, has a predecessor Py.

Proposition 2.3.13 and properties of the Diophantine arithmetic $\boldsymbol{W}$ imply the following result.

Proposition 2.8.11. *Any weak W-arithmetic $\boldsymbol{A} = (A; \oplus, \otimes, \leq_A)$ is Successively Archimedean.*

Proof is left as an exercise. □

Proposition 2.8.11 and Corollary 2.1.13 imply the following result.

Corollary 2.8.3. *Any weak W-arithmetic $\boldsymbol{A} = (A; \oplus, \otimes, \leq_A)$ is constrained.*

Propositions 2.3.14, 2.8.5 and properties of the Diophantine arithmetic $\boldsymbol{W}$ imply the following result.

Proposition 2.8.12. *Any weak W-arithmetic $\boldsymbol{A} = (A; \oplus, \otimes, \leq_A)$ is Exactly Successively Archimedean.*

Proof is left as an exercise. □

The property of abstract prearithmetics "to be a weak W-arithmetic" is hereditary as the following result demonstrates.

Proposition 2.8.13. *A subprearithmetic $\boldsymbol{K}$ of a weak W-arithmetic $\boldsymbol{A} = (A; \oplus, \otimes, \leq)$ is a weak W-arithmetic.*

Indeed, by Proposition 2.5.14, the prearithmetic $\boldsymbol{K}$ is projective with respect to the Diophantine arithmetic $\boldsymbol{W}$ and thus, it is a weak W-arithmetic.

Corollary 2.8.4. *A subprearithmetic $\boldsymbol{D}$ of a subprearithmetic $\boldsymbol{K}$ of a weak W-arithmetic $\boldsymbol{A} = (A; \oplus, \otimes, \leq_A)$ is a weak W-arithmetic.*

Indeed, by Proposition 2.5.11, the prearithmetic $\boldsymbol{D}$ is a subprearithmetic of a weak W-arithmetic $\boldsymbol{A}$ and thus by Proposition 2.8.12, $\boldsymbol{D}$ is a W-prearithmetic.

Proposition 2.8.13 shows that subprearithmetics of weak W-arithmetics are their *subarithmetics* and allows getting many properties of subarithmetics from properties of subprearithmetics and weak W-arithmetics. For instance, the property of abstract prearithmetics "to be a weak whole-number arithmetic" is hereditary as the following result demonstrates.

Corollary 2.8.5. *A weak W-arithmetic $\boldsymbol{A}_1$ is a subarithmetic of a weak W-arithmetic $\boldsymbol{A}_2$ if and only if $\boldsymbol{A}_1$ is a subprearithmetic of $\boldsymbol{A}_2$.*

Corollary 2.8.6. *A subarithmetic $\boldsymbol{D}$ of a subarithmetic $\boldsymbol{K}$ of a weak W-arithmetic $\boldsymbol{A} = (A; \oplus, \otimes, \leq_A)$ is a weak W-arithmetics.*

Corollary 2.8.7. *A subprearithmetic $\boldsymbol{D}$ of a subprearithmetic $\boldsymbol{K}$ of a weak W-arithmetic $\boldsymbol{A} = (A; \oplus, \otimes, \leq)$ is a weak W-arithmetic.*

Proposition 2.8.14. *An abstract prearithmetic $\boldsymbol{A} = (A; +, \circ, \leq_A)$, which is projective with respect to a weak W-arithmetic $\boldsymbol{D} = (D; \oplus, \otimes, \leq_D)$, is a weak W-arithmetic.*

Indeed, by Proposition 2.3.12, the prearithmetic $\boldsymbol{A}$ is projective with respect to the prearithmetic $\boldsymbol{W}$ and thus, it is a weak W-arithmetic.

Proposition 2.8.15. *If an abstract prearithmetic $\boldsymbol{A}_1 = (A_1; +_1, \circ_1, \leq_1)$ is projective with respect to an abstract prearithmetic $\boldsymbol{A}_2 = (A_2; +_2, \circ_2, \leq_2)$ and the abstract prearithmetic $\boldsymbol{A}_2$ is projective with respect to a weak W-arithmetic $\boldsymbol{A}_3 = (A_3; +_3, \circ_3, \leq_3)$, then the abstract prearithmetic $\boldsymbol{A}_1$ is a weak W-arithmetic.*

Indeed, by Proposition 2.8.14, the prearithmetic $\boldsymbol{A}_2$ is a weak W-arithmetic and applying once more Proposition 2.8.14, we come to the conclusion that $\boldsymbol{A}_1$, also is a weak W-arithmetic.

Proposition 2.3.16 and properties of the Diophantine arithmetic $\boldsymbol{W}$ imply the following result.

Proposition 2.8.16. *In a weak W-arithmetic* $\boldsymbol{A} = (A; \oplus, \otimes, \leq_A)$ *with the coprojector h that preserves the order, the inequality* $a \leq_A a \oplus b$ *is true for any elements a and b from* $\boldsymbol{A}$.

Proof is left as an exercise. □

Corollary 2.8.8. *In a weak W-arithmetic* $\boldsymbol{A} = (A; \oplus, \otimes, \leq_A)$ *with the coprojector h that preserves the order, the inequality* $a \leq_A a \oplus a$ *is true for any element a from* $\boldsymbol{A}$.

Corollary 2.8.9. *In a weak W-arithmetic* $\boldsymbol{A} = (A; \oplus, \otimes, \leq_A)$ *with the coprojector h that preserves the order, the inequality* $\max(a, b) \leq_A a \oplus b$ *is true for any elements a and b from* $\boldsymbol{A}$.

In a general case, a weak W-arithmetic $\boldsymbol{A}$ does not contain an additive zero, which is denoted by 0_A. Even if $\boldsymbol{A}$ contains zero 0_A, it is possible that 0_A does not all have properties of zero in the conventional arithmetic $\boldsymbol{W}$. However, additional conditions allow getting these properties.

Proposition 2.8.17. *If a weak W-arithmetic* $\boldsymbol{A} = (A; \oplus, \otimes, \leq_A)$ *with the projector g contains* 0_A *and* $g(0_A) = 0$, *then for any* $n_A \in \boldsymbol{A}$, *the inequality* $0_A \leq n_A$ *and equalities* $0_A \oplus n_A = n_A$ *and* $0_A \otimes n_A = n_A$ *are true.*

Proof: By definition, we have

$$0_A \oplus n_A = h(g(0_A) + g(n_A)) = h(0 + g(n_A)) = h(g(n_A)) = n_A,$$

$$0_A \otimes n_A = h(g(0_A) \cdot g(n_A)) = h(0 \cdot g(n_A)) = h(0)) = h(g(0_A)) = 0_A$$

because the prearithmetic $\boldsymbol{A}$ is projective with respect to the arithmetic $\boldsymbol{W}$, i.e., $h(g(n_A)) = n_A$ for any $n_A \in \boldsymbol{A}$, and $g(0_A) = 0$.

Besides, by the properties of the arithmetic $\boldsymbol{W}$, we have $g(0_A) = 0 \leq g(n_A)$ for any $n_A \in \boldsymbol{A}$. Consequently, $0_A \leq n_A$ for any $n_A \in \boldsymbol{A}$.

Proposition is proved. □

In a general case, a weak W-arithmetic $\boldsymbol{A}$ does not contain number one, which is denoted by 1_A. Even if $\boldsymbol{A}$ contains 1_A, it is possible that 1_A does not have all properties of the number 1 in the conventional arithmetic $\boldsymbol{W}$. However, additional conditions allow getting some of these properties.

Proposition 2.8.18. *If a weak W-arithmetic* $\boldsymbol{A} = (A; \oplus, \otimes, \leq_A)$ *with the projector g contains* 1_A *and* $g(1_A) = 1$, *then for any number* $n_A \in \boldsymbol{A}$, *the equality* $1_A \otimes n_A = n_A$ *is true, i.e.,* 1_A *is a multiplicative one in* $\boldsymbol{A}$.

Proof: By definition, we have

$$1_A \otimes n_A = h(g(1_A) \cdot g(n_A)) = h(1 \cdot g(n_A)) = h(g(n_A)) = n_A$$

because the prearithmetic $\boldsymbol{A}$ is projective with respect to the arithmetic $\boldsymbol{W}$, i.e., $h(g(n_A)) = n_A$ for any $n_A \in \boldsymbol{A}$, and $g(1_A) = 1$.

Proposition is proved. □

Special types of projectors g imply additional properties of weak W-arithmetics.

Theorem 2.8.1. *If 1_A belongs to a weak W-arithmetic $\boldsymbol{A} = (A; \oplus, \otimes, \leq_A)$ with the projector g and the coprojector h, then there is a number k such that for any number n_A from $\boldsymbol{A}$, $g(n_A) = k \cdot n$ if and only if the projector g preserves addition, the coprojector h is an injection on the subarithmetic* $\text{asa}_{g(A)}\boldsymbol{A}$ *generated by the image $g(A)$ of the set A and for any numbers $m_A, n_A \in \boldsymbol{A}$, the equality $m_A \oplus n_A = (m+n)_A$ is true.*

Proof: *Necessity.* Let us assume that 1_A belongs to a weak W-arithmetic $\boldsymbol{A} = (A; \oplus, \otimes, \leq)$ with the projector g and the coprojector h and there is a number k such that for any number $n_A \in \boldsymbol{A}$, $g(n_A) = k \cdot n$. Then we have

$$\begin{aligned} m_A \oplus n_A &= h(g(m_A) + g(n_A)) = h(k \cdot m + k \cdot n) = h(k \cdot (m+n)) \\ &= h(g((m+n)_A)) = (m+n)_A \end{aligned}$$

because $g((m+n)_A = k \cdot (m+n)$ and $hg = \mathbf{1}_A$.

It means that $m_A \oplus n_A = (m+n)_A$ for any numbers n_A and m_A from $\boldsymbol{A}$.

In addition,

$$g(m_A) + g(n_A) = k \cdot m + k \cdot n = k \cdot (m+n) = g((m+n)_A) = g(m_A \oplus n_A).$$

It means that the projector g preserves addition

Sufficiency. Let us assume that 1_A belongs to a weak W-arithmetic $\boldsymbol{A} = (A; \oplus, \otimes, \leq_A)$ the projector g of which preserves addition, the coprojector h of which is injective and for any numbers m_A and n_A from $\boldsymbol{A}$, the equality $m_A \oplus n_A = (m+n)_A$ is true. Let us show that we can take $g(1_A) = k$ as the necessary number k.

At first, let us suppose that the weak W-arithmetic $\boldsymbol{A}$ does not have number 0_A. Then the weak W-arithmetic $\boldsymbol{A}$ contains all natural numbers, i.e., $A = N$. We prove this by induction.

By the initial conditions, 1_A belongs to the weak W-arithmetic $\boldsymbol{A}$.

Suppose the number n_A belongs to the weak W-arithmetic $\boldsymbol{A}$. Then we have

$$n_A \oplus 1_A = (n+1)_A.$$

Thus, as n_A belongs to $\boldsymbol{A}$, then $(n+1)_A$ also belongs to $\boldsymbol{A}$. By the principle of mathematical induction, the weak W-arithmetic $\boldsymbol{A}$ contains all natural numbers.

In a similar way, we prove that all numbers in $\boldsymbol{A}$ are sums of some number of the element 1_A, i.e., if $n_A \in \boldsymbol{A}$, then $n_A = 1_A \oplus 1_A \oplus \cdots \oplus 1_A$.

Now let us prove by induction that $g(n_A) = k \cdot n$ for any number $n_A \in \boldsymbol{A}$.

If $n_A = 1_A$, then $g(1_A) = k \cdot 1_A$. Let us assume that $g(n_A) = k \cdot n$. Then we have

$$g((n+1)_A) = g(n_A \oplus 1_A) = g(n_A) + g(1_A) = k \cdot n + k = k \cdot (n+1).$$

Thus, by the principle of induction the equality $g(n_A) = k \cdot n$ is true for all numbers $n_A \in \boldsymbol{A}$.

Theorem is proved. □

Properties of the coprojector h imply useful properties of weak W-arithmetics.

Proposition 2.8.19. *In a weak W-arithmetic $\boldsymbol{A} = (A; \oplus, \otimes, \leq_A)$, if the coprojector h preserves the order $\leq$, then $a \leq b \oplus a$ and $a \leq a \oplus b$ for any elements a and b from $\boldsymbol{A}$.*

Proof: Let us consider a weak W-arithmetic $\boldsymbol{A} = (A; \oplus, \otimes, \leq_A)$, for which the coprojector h preserves the order $\leq$ and take some elements a and b from A. Then by the properties of the Diophantine arithmetic $\boldsymbol{W}$ we have $g(a) \leq g(b) + g(a)$. As the coprojector h preserves the order $\leq$ and by Definition 2.3.1, hg is the identity mapping, we have

$$a \leq h(g(a)) \leq h(g(b) + g(a)) = b \oplus a.$$

Validity of the inequality $a \leq a \oplus b$ follows from Proposition 2.8.1.

Proposition is proved. □

Corollary 2.8.10. *In a weak W-arithmetic $\boldsymbol{A} = (A; \oplus, \otimes, \leq_A)$, if the coprojector h preserves the order $\leq$, then $a \leq a \oplus a$ for any element a from $\boldsymbol{A}$.*

Proposition 2.8.20. *In a weak W-arithmetic $\boldsymbol{A} = (A; \oplus, \otimes, \leq_A)$, if the coprojector h preserves the order $\leq$, then $a \leq b \otimes a$ and $a \leq a \otimes b$ for any non-zero elements a and b from $\boldsymbol{A}$.*

Proof: Let us consider a weak W-arithmetic $\boldsymbol{A} = (A; \oplus, \otimes, \leq_A)$, for which the coprojector h preserves the order $\leq$ and take some non-zero elements a and b from A. Then by Proposition 2.3.20, either 0 does not belong to the image $g(A)$ and then $g(b) \neq 0$ and $g(a) \neq 0$, or 0 belongs to the image $g(A)$ but because it is the image of the zero from $\boldsymbol{A}$, we also have $g(b) \neq 0$ and $g(a) \neq 0$.

Consequently, by the properties of the Diophantine arithmetic $\boldsymbol{W}$, we have $g(a) \leq g(b) \cdot g(a)$. As the coprojector h preserves the order $\leq$ and by Definition 2.3.1, hg is the identity mapping, we have

$$a \leq h(g(a)) \leq h(g(b) \cdot g(a)) = b \otimes a.$$

Validity of the inequality $a \leq a \otimes b$ follows from Proposition 2.8.1.

Proposition is proved. □

Corollary 2.8.11. *In a weak W-arithmetic $\boldsymbol{A} = (A; \oplus, \otimes, \leq_A)$, if the coprojector h preserves the order $\leq$, then $a \leq a \otimes a$ for any element a from $\boldsymbol{A}$.*

Indeed, for a non-zero element a from $\boldsymbol{A}$, this follows from Proposition 2.8.12, while

$$0_A \leq 0_A = 0_A \otimes 0_A.$$

In addition to binary relations $\leq$ and $<$, there are other kinds of order relations in a weak W-arithmetic $\boldsymbol{A} = (A; +, \circ, \leq_A)$:

(1) When $a, b \in \boldsymbol{A}$, then the formula $a \ll b$ means that a is *much less* than b and in this case, b is *much larger* than a, i.e., $b \gg a$. It is defined as

$$a \ll b \quad \text{if and only if } b \dot{+} a = b.$$

(2) The formula $a_{A,1}, a_{A,2}, \ldots, a_{A,n-1} \ll_n b_A$ means that the group $a_{A,1}$, $a_{A,2}, \ldots, a_{A,n-1}$ is *much less* than b_A and in this case, b_A is *much larger* than the group $a_{A,1}, a_{A,2}, \ldots, a_{A,n-1}$, i.e., $m_A \gg n_A$. It is defined as

$$a_{A,1}, a_{A,2}, \ldots, a_{A,n-1} \ll_n b_A$$
$$\text{if and only if } \Sigma^{\oplus n}(a_{A,1}, a_{A,2}, \ldots, a_{A,n-1}, b_A) = b_A.$$

(3) The formula $a \lll b$ means that a is *much much less* than b and in this case, b is *much much larger* than a, i.e., $b \ggg a$. It is defined as

$$a \lll b \quad \text{if and only if } b \circ a = b.$$

(4) The formula $a_{A,1}, a_{A,2}, \ldots, a_{A,n-1} \lll_n b_A$ means that the group $a_{A,1}$, $a_{A,2}, \ldots, a_{A,n-1}$ is *much much less* than b_A and in this case, b_A is *much much larger* than the group $a_{A,1}, a_{A,2}, \ldots, a_{A,n-1}$, i.e., $b_A \ggg_n a_{A,1}$, $a_{A,2}, \ldots, a_{A,n-1}$. It is defined as

$$a_{A,1}, a_{A,2}, \ldots, a_{A,n-1} \lll_n b_A \text{ if and only if } \Pi^{\otimes n}(a_{A,1}, a_{A,2}, \ldots, a_{A,n-1}, b_A) = b_A.$$

Let us consider a W-prearithmetic $\boldsymbol{A} = (A; \oplus, \otimes, \leq_A)$.

Proposition 2.8.21. *If* $0 < n_A$ *and* $n_A \ll m_A$, *then* m_A *is an additively composite number for any number* n_A *from* $\boldsymbol{A}$.

Indeed, the relation $n_A \ll m_A$ means $m_A \oplus n_A = m_A$.

Corollary 2.8.12. *If* $0_A < n_A$ *and* $n_A \ll n_A$, *then* n_A *is an additively composite number for any number* n_A *from* $\boldsymbol{A}$.

Proposition 2.8.22. *If* $1_A < n_A$ *and* $n_A \lll m_A$, *then* m_A *is a multiplicatively composite number.*

Indeed, the relation $n_A \lll m_A$ means $m_A \otimes n_A = m_A$.

Corollary 2.8.13. *If* $0 < n_A$ *and* $n_A \ll n_A$, *then* n_A *is an additively composite number for any number* n_A *from* $\boldsymbol{A}$.

As we know, in the Diophantine arithmetic $\boldsymbol{W}$, relation $m \ll n$ is true only when m is 0 and relation $m \lll n$ is true only when m is 1. However, this is not true for weak W-arithmetics in a general case.

Theorem 2.8.2. *There are weak W-arithmetics in which the relation* $n \ll n$ *is valid for any number* n.

Proof: Let us consider the set W of all whole numbers and define the following functions:

$$g(n) = \begin{cases} 0 & \text{when } n = 0, \\ 2^{n^2} & \text{when } n > 0 \end{cases}$$

and

$$h(x) = \begin{cases} 0 & \text{when } x = 0, \\ \lfloor (\log_2 x)^{1/2} \rfloor & \text{when } x > 0. \end{cases}$$

As $hg = \mathbf{1}_W$, by Proposition 2.1.20, we can build an abstract prearithmetic $\boldsymbol{A} = (W; \oplus, \otimes, \leq)$ that is projective with respect to the Diophantine arithmetic $\boldsymbol{W}$ with the projector g and the coprojector h. Namely, the following operations are defined for whole numbers m and n larger than 0:

$$m \oplus n = h(g(m) + g(n)) = \lfloor (\log_2(2^{m^2} + 2^{n^2}))^{1/2} \rfloor,$$

$$\begin{aligned} m \otimes n &= h(g(m) \cdot g(n)) = \lfloor (\log_2(2^{m^2} \cdot 2^{n^2}))^{1/2} \rfloor \\ &= \lfloor (\log_2(2^{m^2+n^2}))^{1/2} \rfloor = \lfloor ((m^2 + n^2)^{1/2} \rfloor. \end{aligned}$$

Besides,

$$0 \oplus n = h(g(0) + g(n)) = \lfloor (\log_2(0 + 2^{n^2}))^{1/2} \rfloor = \lfloor (\log_2(2^{n^2}))^{1/2} \rfloor = n.$$

These definitions give us the following relations:

$$\begin{aligned} n \oplus n &= h(g(n) + g(n)) = \lfloor (\log_2(2^{n^2} + 2^{n^2}))^{1/2} \rfloor \\ &= \lfloor (\log_2(2 \cdot 2^{n^2}))^{1/2} \rfloor = \lfloor (\log_2(2^{n^{2+1}}))^{1/2} \rfloor \\ &< \lfloor (\log_2(2^{(n+1)^2}))^{1/2} \rfloor = ((n+1)^2)^{1/2} = n + 1. \end{aligned}$$

Thus, $n \oplus n = n$ and $n \ll n$.

Theorem is proved. □

Corollary 2.8.14. *There are weak W-arithmetics in which $n \oplus n = n$ for any number n.*

Proposition 2.8.23. *There is a weak W-arithmetic $\boldsymbol{A} = (W; \oplus, \otimes, \leq_A)$, in which $n_A \oplus m_A = \max(n_A, m_A)$ for any numbers n_A and m_A.*

Proof: The weak W-arithmetic $\boldsymbol{A} = (W; \oplus, \otimes, \leq_A)$, which is constructed in Theorem 2.8.2, has this property. Indeed, by construction, the coprojector h preserves the order $\leq$ in W because functions $\log_2 x$ and $x^{1/2}$ are monotone. Consequently, by Proposition 2.8.3, the addition $\oplus$ preserves the order $\leq_A$ in $\boldsymbol{A}$. As a result, using Corollary 2.8.9 and assuming for certainty that $n_A \leq_A m_A$, we have

$$n_A = n_A \oplus n_A \leq_A n_A \oplus m_A \leq_A m_A \oplus m_A = m_A.$$

If $n_A = m_A$, then

$$n_A \oplus m_A = m_A \oplus m_A = m_A = \max(n_A, m_A).$$

If $n_A <_A m_A$, then

$$m_A = 0_A \oplus m_A \leq_A n_A \oplus m_A \leq_A m_A.$$

Consequently,

$$n_A \oplus m_A = m_A = \max(n_A, m_A).$$

Proposition is proved. □

Proposition 2.8.23 shows that there are weak W-arithmetics, in which addition coincides with the operation of taking maximum.

Theorem 2.8.3. *There are weak W-arithmetics in which the relation $n \lll n$ is valid for any number n.*

Proof: The prearithmetic $\boldsymbol{A} = (W; \oplus, \otimes, \leq_A)$ constructed in the proof of Proposition 2.7.29 is a weak W-arithmetic in which $n_A \otimes n_A = n_A$ for any number n_A. By definition, this equality means that $n_A \lll n_A$ for any number n_A.

Theorem is proved. □

Corollary 2.8.15. *There are weak W-arithmetics in which $n \otimes n = n$ for any number n.*

Proposition 2.8.24. *There is a weak W-arithmetic $\boldsymbol{A} = (W; \oplus, \otimes, \leq_A)$, in which $n_A \otimes m_A = \max(n_A, m_A)$ for any numbers n_A and m_A.*

Proof is similar to the proof of Proposition 2.8.26. □

Weak W-arithmetics are closed with respect to inverse projectivity.

Proposition 2.8.25. *An abstract prearithmetic $\boldsymbol{A} = (A; +, \circ, \leq_A)$, which is projective with respect to a weak W-arithmetic $\boldsymbol{D} = (D; \oplus, \otimes, \leq_D)$ is a weak W-arithmetic.*

Indeed, by Proposition 2.3.12, the abstract prearithmetic $\boldsymbol{A}$ is projective with respect to the prearithmetic $\boldsymbol{W}$ and thus, it is a weak W-arithmetic.

Proposition 2.8.26. *If an abstract prearithmetic* $\boldsymbol{A}_1 = (A_1; +_1, \circ_1, \leq_1)$ *is projective with respect to an abstract prearithmetic* $\boldsymbol{A}_2 = (A_2; +_2, \circ_2, \leq_2)$ *and the abstract prearithmetic* $\boldsymbol{A}_2 = (A_2; +_2, \circ_2, \leq_2)$ *is projective with respect to a weak W-arithmetic* $\boldsymbol{A}_3 = (A_3; +_3, \circ_3, \leq_3)$, *then the prearithmetic* $\boldsymbol{A}_1$ *is a weak W-arithmetic.*

Indeed, by Proposition 2.8.25, the prearithmetic $\boldsymbol{A}_2$ is a weak W-arithmetic and applying once more Proposition 2.8.25, we come to the conclusion that $\boldsymbol{A}_1$ also is a weak W-arithmetic.

Proposition 2.8.26 implies the following result.

Theorem 2.8.4. *Weak W-arithmetics form the category* ***WAAP*** *where objects are weak W-arithmetics and morphisms are projectivity relations.*

Proof is left as an exercise. □

Corollary 2.8.16. *The category* ***WAAP*** *is a subcategory of the category* ***WPAWP***.

From weak W-arithmetics, we come to W-arithmetics.

Definition 2.8.2. A weak whole-number arithmetic $\boldsymbol{A} = (A; \oplus, \otimes, \leq_A)$ is called a *whole-number arithmetic*, as well as a *projective* W-*arithmetic* or a *projective non-Diophantine arithmetic* of whole numbers if it is complete, i.e., $A = W$.

Example 2.8.1. The W-prearithmetic $\mathbf{2}\,\boldsymbol{W} = (2W; \oplus, \otimes, \leq_A)$ of all even numbers is a weak W-arithmetic but not a projective non-Diophantine arithmetic of whole numbers.

Lemma 2.8.2. *Any projective W-arithmetic is a weak W-arithmetic.*

Proof is left as an exercise. □

Projective W-arithmetics that contain all whole numbers N form a class **PRNDAW** of projective non-Diophantine arithmetics of whole numbers.

Note that in general, projective W-arithmetics do not have all properties of W-non-Diophantine arithmetics of whole numbers. For instance, in a general case, a projective W-arithmetic $\boldsymbol{A}$ does not contain number zero, which is denoted by 0_A, as Example 2.8.1 demonstrates, while any W-non-Diophantine arithmetic of whole numbers contains all natural numbers. Indeed, $\boldsymbol{N}$ is a projective W-arithmetic and does not contain number zero.

Example 2.8.2. Let us consider a W-prearithmetic $\boldsymbol{A} = (N; \oplus, \otimes, \leq_A)$ with the projector $g(n) = n - 1$ and coprojector $h(n) = n + 1$. Then for arbitrary natural numbers n and m, we have

$$n \oplus m = h(g(n) + g(m)) = ((n-1) + (m-1)) + 1 = n + m - 1.$$

Besides, g is a projection of N onto W and $hg = 1_N$. It shows that $\boldsymbol{A}$ is a projective W-arithmetic but not a projective non-Diophantine arithmetic of whole numbers. However, $\boldsymbol{A}$ is a projective non-Diophantine arithmetic of natural numbers.

Definition 2.3.1 of projectivity implies the following result.

Proposition 2.8.27. *The relation $\leq_A$ in a projective non-Diophantine arithmetic $\boldsymbol{A} = (W; \oplus, \otimes, \leq_A)$ is a discrete well-ordering of the set W of all whole numbers coinciding with the natural ordering in W.*

Proof is left as an exercise. □

This result implies that any element x of a non-Diophantine arithmetic $\boldsymbol{A} = (A; \oplus, \otimes, \leq_A)$ of whole numbers has the successor Sx, and any non-zero element y of $\boldsymbol{A}$ has the predecessor Py.

Proposition 2.8.28. *Any projective non-Diophantine arithmetic $\boldsymbol{A} = (W; \oplus, \otimes, \leq_A)$ of whole numbers is Successively Archimedean and Exactly Successively Archimedean.*

Proof is left as an exercise. □

Remark 2.8.2. However, there are projective non-Diophantine arithmetics of whole numbers that are not Additively Archimedean because the weak W-arithmetic constructed in Theorem 2.8.2 is a projective non-Diophantine arithmetic of whole numbers, in which $n \oplus 1 = n$ for any number n.

There are also projective non-Diophantine arithmetics of whole numbers that are not Multiplicatively Archimedean because this property does not hold for number 1 (cf. Section 2.1.2).

It is possible to use real numbers to build non-Diophantine arithmetics of whole numbers.

Definition 2.8.3. An abstract prearithmetic $\boldsymbol{A} = (A; \oplus, \otimes, \leq_A)$ is called a *weak real-whole-number arithmetic*, or *weak* RW-*arithmetic*, if A is a subset of the set W of all whole numbers and the prearithmetic $\boldsymbol{A}$ is projective with respect to the conventional arithmetic $\boldsymbol{R} = (R; +, \cdot, \leq)$ of all real numbers.

It means that there are two functions: a projector g: $A \to R$ and a coprojector h: $R \to A$, such that $hg = \mathbf{1}_A$ and for any numbers a and b from A, we have

$$a \oplus b = h(g(a) + g(b)),$$
$$a \otimes b = h(g(a) \cdot g(b)),$$
$$a \leq b \text{ in } \boldsymbol{A} \text{ if and only if } g(a) \leq_2 g(b) \text{ in } \boldsymbol{R}.$$

Lemma 2.8.3. *Any weak RW-arithmetic is also an RW-prearithmetic.*

Proof is left as an exercise. □

This result allows deducing many properties of weak RW-arithmetics from properties of RW-prearithmetic.

Proposition 2.7.59 and Lemma 2.8.3 imply the following result.

Proposition 2.8.29. *For any weak RW-arithmetic* $\boldsymbol{A} = (A; \oplus, \otimes, \leq_A)$ *with the coprojector* h, *which preserves the order* $\leq$ *in* $\boldsymbol{R}$, *addition* $\oplus$ *and multiplication* $\otimes$ *preserve the order* $\leq_A$.

Proof is left as an exercise. □

Theorem 2.5.1 implies the following result.

Proposition 2.8.30. *A subprearithmetic* $\boldsymbol{K}$ *of a weak RW-arithmetic* $\boldsymbol{A} = (A; \oplus, \otimes, \leq)$ *is a weak RW-arithmetic.*

Indeed, by Theorem 2.5.1, the prearithmetic $\boldsymbol{K}$ is projective with respect to the Diophantine arithmetic $\boldsymbol{R}$ and thus, it is Theorem 2.5.1.

Corollary 2.8.17. *A subprearithmetic* $\boldsymbol{D}$ *of a subprearithmetic* $\boldsymbol{K}$ *of a weak RW-arithmetic* $\boldsymbol{A} = (A; \oplus, \otimes, \leq)$ *is a weak RW-arithmetic.*

Proposition 2.8.31. *An abstract prearithmetic* $\boldsymbol{A} = (A; +, \circ, \leq_A)$, *which is projective with respect to a weak RW-arithmetic* $\boldsymbol{D} = (D; \oplus, \otimes, \leq_D)$ *is an RW-prearithmetic.*

Indeed, by Proposition 2.3.12, the prearithmetic $\boldsymbol{A}$ is projective with respect to the prearithmetic $\boldsymbol{R}$ and thus, it is a weak RW-arithmetic.

Proposition 2.8.32. *If an abstract prearithmetic* $\boldsymbol{A}_1 = (A_1; +_1, \circ_1, \leq_1)$ *is projective with respect to an abstract prearithmetic* $\boldsymbol{A}_2 = (A_2; +_2, \circ_2, \leq_2)$ *and the abstract prearithmetic* $\boldsymbol{A}_2 = (A_2; +_2, \circ_2, \leq_2)$ *is projective with respect to a weak RW-arithmetic* $\boldsymbol{A}_3 = (A_3; +_3, \circ_3, \leq_3)$, *then the prearithmetic* $\boldsymbol{A}_1$ *is a weak RW-arithmetic*

Proof is left as an exercise. □

Definition 2.8.4. A natural-number prearithmetic $\boldsymbol{A} = (A; \oplus, \otimes, \leq_A)$ is called a *weak natural-number arithmetic*, or a *weak* N-*arithmetic*, if it is projective with respect to the Diophantine arithmetic $\boldsymbol{N} = (N; +, \cdot, \leq)$ of natural numbers.

Weak N-arithmetics and weak W-arithmetics are closely related.

Proposition 2.8.33. *Any weak N-arithmetic is a subprearithmetic of a weak W-arithmetic.*

Proof is left as an exercise. □

Proposition 2.5.13 allows proving even a stronger result.

Lemma 2.8.4. *Any weak N-arithmetic is a weak W-arithmetic.*

Proof is left as an exercise. □

Lemma 2.8.4 allows getting many properties of weak W-arithmetics from properties of W-prearithmetics. For instance, Proposition 2.8.1 and Lemma 2.8.4 imply the following result.

Proposition 2.8.34. *Both operations, addition $\oplus$ and multiplication $\otimes$, are commutative in any weak N-arithmetic $\boldsymbol{A} = (A; \oplus, \otimes, \leq_A)$.*

Proof is left as an exercise. □

Remark 2.8.3. In some weak N-arithmetics, addition $\oplus$ or/and multiplication $\otimes$ is not associative.

Proposition 2.8.2 and Lemma 2.8.4 imply the following result.

Proposition 2.8.35. *For any weak N-arithmetic $\boldsymbol{A} = (A; \oplus, \otimes, \leq_A)$, operations $\Sigma^{\oplus n}$ and $\Pi^{\otimes n}$ are commutative for all $n = 2, 3, 4, 5, \ldots$.*

Proof is left as an exercise. □

Let us study relation between the order and operations in weak N-arithmetics.

Proposition 2.8.36. *If, in a weak N-arithmetic $\boldsymbol{A} = (A; \oplus, \otimes, \leq_A)$, the coprojector h preserves the order $\leq$, then the operation $\oplus$ preserves the order $\leq_A$.*

Proof: Let us consider a weak N-arithmetic $\boldsymbol{A} = (A; \oplus, \otimes, \leq_A)$, for which the coprojector h preserves the order $\leq$ and take four elements a, b, c and d from A, which satisfy the inequalities $a \leq_A b$ and $c \leq_A d$. Then by Definition 2.2.1, $g(a) \leq g(b)$ and $g(c) \leq g(d)$. Addition of whole numbers preserves the order $\leq$. Thus, $g(a) + g(c) \leq g(b) + g(d)$. As the coprojector h preserves the order $\leq$, we have

$$a \oplus c \leq h(g(a) + g(c)) \leq h(g(b) + g(d)) = b \oplus d.$$

Proposition is proved. □

Corollary 2.8.18. *For any N-prearithmetic $\boldsymbol{A} = (A; \oplus, \otimes, \leq_A)$ with the coprojector h, which preserves the order $\leq$ in the Diophantine arithmetic $\boldsymbol{N}$, addition $\oplus$ is monotone.*

Proposition 2.8.37. *If, in a weak N-arithmetic $\boldsymbol{A} = (A; \oplus, \otimes, \leq_A)$, the coprojector h preserves the order $\leq$, then the operation $\otimes$ preserves the order $\leq_A$.*

Proof: Let us consider a weak N-arithmetic $\boldsymbol{A} = (A; \oplus, \otimes, \leq_A)$, for which the coprojector h preserves the order $\leq$ and take four elements a, b, c and d from A, which satisfy the inequalities $a \leq_A b$ and $c \leq_A d$. Then by Definition 2.2.1, $g(a) \leq g(b)$ and $g(c) \leq g(d)$. Multiplication of whole numbers preserves the order $\leq$. Thus, $g(a) \cdot g(c) \leq g(b) \cdot g(d)$. As the coprojector h preserves the order $\leq$, we have

$$a \otimes c \leq h(g(a) \cdot g(c)) \leq h(g(b) \cdot g(d)) = b \otimes d.$$

Proposition is proved. □

Corollary 2.8.19. *For any weak N-arithmetic $\boldsymbol{A} = (A; \oplus, \otimes, \leq_A)$ with the coprojector h, which preserves the order $\leq$ in the Diophantine arithmetic $\boldsymbol{N}$, multiplication $\otimes$ is monotone.*

Proposition 2.8.5 and Lemma 2.8.4 imply the following result.

Proposition 2.8.38. *For any weak N-arithmetic $\boldsymbol{A} = (A; \oplus, \otimes, \leq_A)$, additive action from the left coincides with additive action from the right.*

Proof is left as an exercise. □

Proposition 2.8.38 allows skipping indication of the side of additive action.

Proposition 2.8.6 and Lemma 2.8.4 imply the following result.

Proposition 2.8.39. *For any weak N-arithmetic $\boldsymbol{A} = (A; \oplus, \otimes, \leq_A)$, multiplicative action from the left coincides with multiplicative action from the right.*

Proof is left as an exercise. □

Proposition 2.8.39 allows skipping indication of the side of multiplicative action.

Proposition 2.8.7 and Lemma 2.8.4 imply the following result.

Proposition 2.8.40. *For any N-prearithmetic* $\boldsymbol{A} = (A; \oplus, \otimes, \leq_A)$ *with the coprojector h, which preserves addition, multiplicative action is distributive over addition* $\oplus$.

Proof is left as an exercise. □

As the ordered set N of all natural numbers is finitely connected, Proposition 2.3.6 implies the following result.

Proposition 2.8.41. *In a weak N-arithmetic* $\boldsymbol{A} = (A; \oplus, \otimes, \leq_A)$, *the partially ordered set* A *is finitely connected.*

Proof is left as an exercise. □

Because the natural order in the Diophantine arithmetic $\boldsymbol{N}$ is linear, Proposition 2.3.11 implies the following result.

Proposition 2.8.42. *For any weak N-arithmetic* $\boldsymbol{A} = (A; \oplus, \otimes, \leq_A)$, *the order* $\leq_A$ *is linear.*

Proof is left as an exercise. □

Proposition 2.8.7 and Lemma 2.1.22 imply the following result.

Proposition 2.8.43. *For any weak N-arithmetic* $\boldsymbol{A} = (A; \oplus, \otimes, \leq_A)$, *the order* $\leq_A$ *is discrete.*

Indeed, by Lemma 2.1.22, if a partially ordered set X is finitely connected, then the order in X is discrete and by Proposition 2.8.7, the order $\leq_A$ is finitely connected.

This result implies that any element x of a weak N-arithmetic $\boldsymbol{A} = (A; \oplus, \otimes, \leq_A)$, which is not maximal, has a successor $\mathrm{S}x$, and any element y of $\boldsymbol{A}$, which is not minimal, has a predecessor $\mathrm{P}y$.

Proposition 2.3.13 and properties of the Diophantine arithmetic $\boldsymbol{N}$ imply the following result.

Proposition 2.8.44. *Any weak N-arithmetic* $\boldsymbol{A} = (A; \oplus, \otimes, \leq_A)$ *is Successively Archimedean.*

Proof is left as an exercise. □

In a general case, a weak N-arithmetic $\boldsymbol{A}$ does not contain number one, which is denoted by 1_A. Even if $\boldsymbol{A}$ contains 1_A, it is possible that 1_A does not have all properties of 1 in the conventional arithmetic $\boldsymbol{N}$. However, additional conditions allow getting some of these properties.

Proposition 2.8.45. *If a weak N-arithmetic $\boldsymbol{A} = (A; \oplus, \otimes, \leq_A)$ with the projector g contains the number 1_A and $g(1_A) = 1$, then for any number $n_A \in \boldsymbol{A}$, the equality $1_A \otimes n_A = n_A$ and inequality $1_A \leq_A n_A$ are true.*

Proof: By definition, we have

$$1_A \otimes n_A = h(g(1_A) \cdot g(n_A)) = h(1 \cdot g(n_A)) = h(g(1_A)) = n_A$$

because the prearithmetic $\boldsymbol{A}$ is projective with respect to the arithmetic $\boldsymbol{N}$, i.e., $h(g(n_A)) = n_A$ for any $n_A \in \boldsymbol{A}$, and $g(1_A) = 1$.

Besides, by the properties of the arithmetic $\boldsymbol{N}$, $g(1_A) = 1 \leq g(n_A)$ for any $n_A \in \boldsymbol{A}$. Consequently, by the definition of projectivity, $1_A \leq_A n_A$ for any $n_A \in \boldsymbol{A}$.

Proposition is proved. □

Corollary 2.8.5 implies the following result.

Corollary 2.8.20. *A weak arithmetic $\boldsymbol{A}_1$ is a subarithmetic of a weak arithmetic $\boldsymbol{A}_2$ if and only if $\boldsymbol{A}_1$ is a subprearithmetic of $\boldsymbol{A}_2$.*

Corollary 2.8.6 implies the following result.

Corollary 2.8.21. *A subarithmetic $\boldsymbol{D}$ of a subarithmetic $\boldsymbol{K}$ of a weak arithmetic $\boldsymbol{A} = (A; \oplus, \otimes, \leq)$ is a weak arithmetics.*

Corollary 2.8.7 implies the following result.

Corollary 2.8.22. *A subprearithmetic $\boldsymbol{D}$ of a subprearithmetic $\boldsymbol{K}$ of a weak arithmetic $\boldsymbol{A} = (A; \oplus, \otimes, \leq)$ is a weak arithmetic.*

Proposition 2.8.14 implies the following result.

Proposition 2.8.46. *An abstract prearithmetic $\boldsymbol{A} = (A; +, \circ, \leq_A)$, which is projective with respect to a weak arithmetic $\boldsymbol{D} = (D; \oplus, \otimes, \leq_D)$, is a weak arithmetic.*

Proof is left as an exercise. □

Proposition 2.8.15 implies the following result.

Proposition 2.8.47. *If an abstract prearithmetic* $\boldsymbol{A}_1 = (A_1; +_1, \circ_1, \leq_1)$ *is projective with respect to an abstract prearithmetic* $\boldsymbol{A}_2 = (A_2; +_2, \circ_2, \leq_2)$ *and the abstract prearithmetic* $\boldsymbol{A}_2$ *is projective with respect to a weak arithmetic* $\boldsymbol{A}_3 = (A_3; +_3, \circ_3, \leq_3)$, *then the abstract prearithmetic* $\boldsymbol{A}_1$ *is a weak arithmetic.*

Proof is left as an exercise. □

Now we come to non-Diophantine arithmetics.

Definition 2.8.5. A weak N-arithmetic $\boldsymbol{A} = (A; \oplus, \otimes, \leq_A)$ is called a *natural-number arithmetic*, or simply, a *projective non-Diophantine arithmetic* if it is complete.

It means that a natural-number arithmetic is a complete weak natural-number arithmetic. Thus, we have the following result.

Lemma 2.8.5. *Any projective non-Diophantine arithmetic is a weak N-arithmetic.*

Proof is left as an exercise. □

This result allows deducing many properties of projective arithmetics from already obtained properties of weak N-arithmetics. For instance, Propositions 2.8.8 and 2.8.9 imply the following results.

Proposition 2.8.48. *For any projective non-Diophantine arithmetic* $\boldsymbol{A} = (A; \oplus, \otimes, \leq_A)$, *the order* $\leq_A$ *is linear.*

Proof is left as an exercise. □

Proposition 2.8.49. *For any projective non-Diophantine arithmetic* $\boldsymbol{A} = (A; \oplus, \otimes, \leq_A)$, *the order* $\leq_A$ *is discrete.*

Proof is left as an exercise. □

Often the carrier A of a projective non-Diophantine arithmetic $\boldsymbol{A}$ is a set of natural numbers. However, in general, the carrier A of a functional IC-arithmetic $\boldsymbol{A}$ can be any set for which a one-to-one mapping (bijection) $f_A : A \to C$ exists. This bijection determines a one-to-one naming of the elements from A by complex numbers (Burgin, 2011). Taking this into account, we can denote the element from A that is mapped onto the complex number r by r_A.

Projective arithmetics that contain all natural numbers N form a class **PRNDAW** of projective non-Diophantine arithmetics of natural numbers.

Projective arithmetics do not have all properties of projective non-Diophantine arithmetics of natural numbers. For instance, in a general case, a projective arithmetic $\boldsymbol{A}$ does not contain number one, which is denoted by 1_A, while any non-Diophantine arithmetic of natural numbers contains all natural numbers. Indeed, the arithmetic $\boldsymbol{2N}$ of all even numbers is projective and does not contain number one.

Definition 2.3.1 of projectivity implies the following result.

Proposition 2.8.50. *The relation $\leq_A$ in a non-Diophantine arithmetic $\boldsymbol{A} = (N; \oplus, \otimes, \leq_A)$ is a discrete well-ordering of the set N of all whole numbers coinciding with the natural ordering in N.*

This result implies that any element x of a non-Diophantine arithmetic $\boldsymbol{A} = (A; \oplus, \otimes, \leq_A)$ of natural numbers has the successor $\mathrm{S}x$, and any non-zero element y of $\boldsymbol{A}$ has the predecessor $\mathrm{P}y$.

Proposition 2.8.51. *Any non-Diophantine arithmetic $\boldsymbol{A} = (A; \oplus, \otimes, \leq_A)$ of natural numbers is Successively Archimedean and Exactly Successively Archimedean.*

Remark 2.8.4. However, there are non-Diophantine arithmetics of natural numbers that are not Additively Archimedean because the weak W-arithmetic constructed in Theorem 2.8.2 contains a non-Diophantine arithmetic of natural numbers, in which $n \oplus 1 = n$ for any number n.

There are also non-Diophantine arithmetics of natural numbers that are not Multiplicatively Archimedean because this property does not hold for number 1 (cf. Section 2.1.2).

Remark 2.8.5. Theorem 2.4.5 implies that it is impossible to build non-Diophantine arithmetics of whole numbers using isomorphisms of $\boldsymbol{W}$ because any isomorphism $f\colon \boldsymbol{W} \to \boldsymbol{W}$ is the identity mapping of $\boldsymbol{W}$. Indeed, for any whole number n, we have

$$f(0) + f(n) = f(0 + n) = f(n).$$

It implies that $f(0) = 0$.

In the same way, for any whole number n, we have

$$f(1) \cdot f(n) = f(1 \cdot n) = f(n).$$

It implies that $f(1) = 1$.

Finally, if $n > 1$, then

$$f(n) = f(1 + 1 + \cdots + 1) = f(1) + f(1) + \cdots + f(1) = 1 + 1 + \cdots + 1 = n.$$

It means that f is the identity mapping of $\boldsymbol{W}$.

By the same token, any isomorphism $f\colon \boldsymbol{N} \to \boldsymbol{N}$ is the identity mapping of $\boldsymbol{N}$ and thus, it is also impossible to build non-Diophantine arithmetics of natural numbers using isomorphisms of $\boldsymbol{N}$.

As we will see in Chapter 3, the situation is very different for real and complex numbers where it is possible to build non-Diophantine arithmetics using isomorphisms. Isomorphism-generated non-Diophantine arithmetics are very useful in fractal theory (Aerts *et al.*, 2016; 2016a; 2018), psychology (Czachor, 2017a), and physics (Czachor, 2016; 2017; 2017b; Czachor and Posiewnik, 2016).

2.9. Direct Perspective Arithmetics

And all arithmetic and calculation have to do with number?
Yes.
And they both appear to lead the mind towards truth?
Yes, in a very remarkable manner.
Plato *The Republic*

We start our study of direct perspective arithmetics with a description and exploration of their general structure in the form of a prearithmetic. Informally a *direct perspective prearithmetic* is a subset of all natural numbers with two basic operations $\dot{+}$ (addition) and $\circ$ (multiplication), which are defined for all its elements (numbers), and a natural order between its elements (numbers). In what follows, an arbitrary direct perspective prearithmetic is denoted by $\boldsymbol{A} = (A; \dot{+}, \circ, \leq)$ where A is the set of the elements or numbers from $\boldsymbol{A}$. Operations in $\boldsymbol{A}$ are induced by the corresponding operations in the Diophantine arithmetic $\boldsymbol{N}$ or $\boldsymbol{W}$ using weak projectivity and the generating relation or function. This relation, or function, serves as the parameter of the direct perspective prearithmetic $\boldsymbol{A}$. An important feature of this approach is that direct perspective prearithmetics and arithmetics are not described by axioms or by any other conditions (Burgin, 1977; 1997). They were constructed by an explicit procedure using the conventional natural, whole and real numbers. This procedure is explained below.

In this setting, a *direct perspective arithmetic*, which is a kind of non-Diophantine arithmetics, is a direct perspective prearithmetic that satisfies additional conditions making many (although not all) of its properties similar to the properties of the Diophantine arithmetic. For instance, addition and multiplication in direct perspective arithmetics are commutative and each of these arithmetics contains all natural numbers. At the same time, there are many direct perspective arithmetics in which there are no prime numbers or in which each number is a square.

Perspective prearithmetics form a special class **P** of projective prearithmetics studied in the previous section because they are constructed using specific projectors and coprojectors (Burgin, 1977; 1997). In turn, perspective prearithmetics and arithmetics have two types: direct perspective prearithmetics (arithmetics) and dual perspective prearithmetics (arithmetics). Direct perspective arithmetics, which we study in this section, constitute the first discovered class of non-Diophantine arithmetics (Burgin, 1977). In this setting, our main interest is in functional perspective prearithmetics and arithmetics.

Direct perspective arithmetics are built as specializations, i.e., special kinds, of whole-number and natural-number functional direct perspective prearithmetics. The main condition that distinguishes direct whole-number (natural-number) perspective arithmetics from other direct whole-number (natural-number) perspective prearithmetics is inclusion of all whole (correspondingly, all natural) numbers.

2.9.1. *Direct perspective prearithmetics of whole numbers*

Small minds have always lashed out at what they don't understand.

Dan Brown

To construct a general direct perspective prearithmetic, let us consider an interval $[q, r)$ of real numbers with $q \in W$, $r \in R^+ \cup \{\infty\}$ and a binary relation $Q \subseteq [q, r) \times R^+$ that has the domain $\mathrm{Dom}(Q) = [q, r)$, the same definability domain $\mathrm{DDom}(Q) = [q, r)$, the range $\mathrm{Rg}(Q)$ is equal to an infinite interval $[a, \infty)$ in R^+ and the codomain $\mathrm{Codom}(Q) = R^+$. For $q = 0$ and $r = \infty$, we have $Q \subseteq R^+ \times R^+$. By Q^{-1} we denote the relation, which is inverse (or dual) to Q, i.e., $Q^{-1} = \{(x, y); (y, x) \in Q\}$. If $x \in R^+$, then $Q^{-1}(x) = \{y; (y, x) \in Q\}$ is the coimage of the element x and $Q(x) = \{z; (x, z) \in Q\}$ is the image of the element x.

Relation Q allows us to define two functions, which are taking values in $\boldsymbol{W}$

$$Q_{\mathrm{u_T}}(x) = \inf\{n; n \in W \text{ and } \forall z \in Q(x)(n \geq z)\},$$
$$Q_{\mathrm{u^T}}(x) = \sup\{m; m \in W \text{ and } \forall z \in Q^{-1}(x)(m \leq z)\}.$$

Note that if $Q(x)$ is a whole number, then $Q_{\mathrm{u_T}}(x) = Q(x)$, and if $Q^{-1}(x)$ is a whole number, then $Q_{\mathrm{u^T}}(x) = Q^{-1}(x)$.

We see that $Q_{\mathrm{u_T}}$ maps the set $[q, r)$ onto whole numbers from R^+ and $Q_{\mathrm{u^T}}$ maps R^+ onto whole numbers from $[q, r)$. This allows us, taking restrictions of both functions on the set $W_{q,r} = [q, r) \cap W$, to obtain two additional functions $Q_{\mathrm{T}} : W_{q,r} \to R^+$ and $Q^{\mathrm{T}} : R^+ \to W_{q,r}$.

Definition 2.9.1.

(a) A prearithmetic $\boldsymbol{A} = (W_{q,r}; \dot{+}, \circ, \leq)$ is called a *direct perspective* W-*prearithmetic* if it is weakly projective with respect to the conventional (Diophantine) arithmetic $\boldsymbol{W} = (W; +, \cdot, \leq)$ with the *projector* $Q_{\mathrm{T}}(x)$ and the *coprojector* $\mathrm{Q}^{\mathrm{T}}(x)$.

(b) The relation Q is called the *generator* of the *projector* $Q_{\mathrm{T}}(x)$ and of the *coprojector* $\mathrm{Q}^{\mathrm{T}}(x)$ as well as the *relational parameter* of the prearithmetic $\boldsymbol{A}$.

The direct perspective W-prearithmetic with the generator Q is denoted by $\boldsymbol{W}_{wQ}$.

Elements from the set $W_{q,r}$ are called numbers of the prearithmetic $\boldsymbol{A}$ and are denoted with the subscript A, i.e., 2 in $\boldsymbol{A}$ is denoted by 2_A and 5 in $\boldsymbol{A}$ is denoted by 5_A. Numbers of $\boldsymbol{A}$ are ordered by the same order relation $\leq$ by which they are ordered in $\boldsymbol{W}$. This implies the following result.

Lemma 2.9.1. *The unary operation* S *of taking the next number is defined in all direct perspective W-prearithmetics with more than one element coinciding with the operation* S *in the Diophantine arithmetic* $\boldsymbol{W}$.

Proof is left as an exercise. □

Direct perspective W-prearithmetics also inherit some Archimedean properties (cf. Section 2.1) from the Diophantine arithmetic $\boldsymbol{W}$.

Lemma 2.9.2. *Any direct perspective W-prearithmetic* $\boldsymbol{A} = (A; \dot{+}, \circ, \leq)$ *is successively Archimedean and exactly successively Archimedean.*

Proof is left as an exercise. □

As elements of a direct perspective W-prearithmetic $\boldsymbol{A} = (A; \dot{+}, \circ, \leq)$ form an interval of whole numbers, we have the following result.

Lemma 2.9.3. *If a direct perspective W-prearithmetic* $\boldsymbol{A} = (A; \dot{+}, \circ, \leq)$ *with more than one element has zero* 0_A*, then* $S0_A = 1_A$.

Proof is left as an exercise. □

As $\boldsymbol{A}$ is weakly projective with respect to Diophantine arithmetic $\boldsymbol{W}$, for any two numbers n_A and m_A from $W_{q,r}$, the basic binary operations in $\boldsymbol{A}$ are defined by the following formulas:

$$n_A \dot{+} m_A = Q^{\mathrm{T}}(Q_{\mathrm{T}}(n_A) + Q_{\mathrm{T}}(m_A)),$$

$$n_A \circ m_A = Q^{\mathrm{T}}(Q_{\mathrm{T}}(n_A) \cdot Q_{\mathrm{T}}(m_A)).$$

Example 2.9.1. Let us take the set $[0, m)$ and define the relation

$$Q_m = \{(x, z); 0 \leq x < m, z = mn + x, = 1, 2, 3, \ldots\}.$$

It is possible to see a part of the graph of this relation in Figure 2.5.

By definition, if $mn \leq x < m(n+1)$, then the projector is defined as $Q_{m^{\mathrm{T}}}(n) =]x[$ and the coprojector is defined as $Q_{m^{\mathrm{T}}}(n) = [x - mn]$. Then the direct projective prearithmetic $\boldsymbol{A} = (W_m; \dot{+}, \circ, \leq)$ defined by these functions is the modular arithmetic $\boldsymbol{Z}_n$ considered in Section 2.1 (cf. Example 2.1.3). Note that operations in this direct projective prearithmetic $\boldsymbol{A}$ are not monotone.

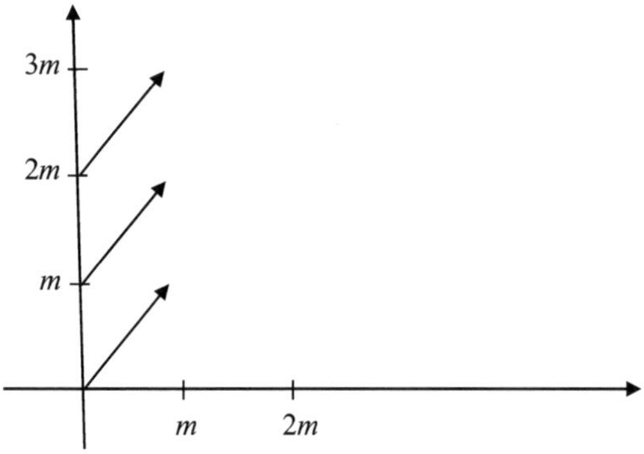

Figure 2.5. A part of the graph of the relation Q_m.

In a general case, as Q is a binary relation between the interval $[q, r)$ and R^+, i.e., $Q \subseteq [q, r) \times R^+$, it is possible to correspond a relation $Q_+ \subseteq R^+ \times R^+$ which consists of the same elements as Q when the first element belongs to $[q, r)$. Other elements Q_+ can be different but the definability domain of Q_+ has to be equal to R^+. Then it is possible to build the direct perspective W-prearithmetic $\boldsymbol{W}_{wQ+}$.

Lemma 2.9.4. *The direct perspective W-prearithmetic $\boldsymbol{W}_{wQ}$ is a subprearithmetic of the direct perspective W-prearithmetic $\boldsymbol{W}_{wQ+}$.*

Proof is left as an exercise. □

This result and results from Section 2.5 show that in many cases it is possible to study the direct perspective W-prearithmetic $\boldsymbol{W}_{wQ+}$ and derive properties of the direct perspective W-prearithmetic $\boldsymbol{W}_{wQ}$ from the properties of properties of $\boldsymbol{W}_{wQ+}$.

Proposition 2.7.1 and properties of the conventional Diophantine arithmetic $\boldsymbol{W}$ imply the following result.

Proposition 2.9.1. *For any direct perspective W-prearithmetic $\boldsymbol{A} = (A; \dot{+}, \circ, \leq_A)$, addition $\dot{+}$ and multiplication $\circ$ are commutative.*

Proof is left as an exercise. □

We can also find other useful properties of operations in direct perspective W-prearithmetics.

Proposition 2.9.2. *In a direct perspective W-prearithmetic $\boldsymbol{A} = (A; \dot{+}, \circ, \leq_A)$ with the relational parameter Q, addition $\dot{+}$ is monotone if functions Q_{T} and Q^{T} are increasing.*

Proof: Let us take elements n_A, l_A and m_A from A such that $l_A \leq_A m_A$. Then we have

$$Q_{\mathrm{T}}(l_A) \leq Q_{\mathrm{T}}(m_A)).$$

Consequently,

$$Q_{\mathrm{T}}(n_A) + Q_{\mathrm{T}}(l_A) \leq_A Q_{\mathrm{T}}(n_A) + Q_{\mathrm{T}}(m_A).$$

and

$$n_A \dot{+} l_A = Q^{\mathrm{T}}(Q_{\mathrm{T}}(n_A) + Q_{\mathrm{T}}(l_A)) \leq_A Q^{\mathrm{T}}(Q_{\mathrm{T}}(n_A) + Q_{\mathrm{T}}(m_A)) = n_A \dot{+} m_A.$$

Proposition is proved. □

Proposition 2.9.2 and Lemma 2.1.6 imply the following result.

Corollary 2.9.1. *For any direct perspective W-prearithmetic $\boldsymbol{A} = (A; \dot{+}, \circ, \leq_A)$, addition $\dot{+}$ preserves the order $\leq_A$ if functions Q_{T} and Q^{T} are increasing.*

Proposition 2.9.3. *In a direct perspective W-prearithmetic $\boldsymbol{A} = (A; \dot{+}, \circ, \leq_A)$ with the relational parameter Q, multiplication $\circ$ is monotone if functions Q_{T} and Q^{T} are increasing.*

Proof is similar to the proof of Proposition 2.9.2. □

Proposition 2.9.3 and Lemma 2.1.7 imply the following result.

Corollary 2.9.2. *For any direct perspective W-prearithmetic $\boldsymbol{A} = (A; \dot{+}, \circ, \leq_A)$, multiplication $\circ$ preserves the order $\leq_A$ if functions Q_{T} and Q^{T} are increasing.*

Different types of relations Q determine specific classes of direct perspective W-prearithmetics. Let us consider one of them, which is the most important in the context of non-Diophantine arithmetics.

Definition 2.9.2.

(a) A direct perspective W-prearithmetic $\boldsymbol{A} = (W_{q,r}; \dot{+}, \circ, \leq)$ with the *projector* $Q_{\mathrm{T}}(x)$ and the *coprojector* $Q^{\mathrm{T}}(x)$ is called *functional* if the relation Q is a function, say f.
(b) The function f called the *generator* of the *projector* $f_{\mathrm{T}}(x)$ and of the *coprojector* $f^{\mathrm{T}}(x)$ as well as the *functional parameter* of the prearithmetic $\boldsymbol{A}$.

The functional direct perspective W-prearithmetic with the generator f is denoted by $\boldsymbol{W}_f$.

Note that as the range $\mathrm{Rg}(f) = [a, \infty) \subseteq R^+$, the function f is an arbitrary projection (surjection) f: $[q, r) \to [a, \infty)$. When $a = q = 0$ and $r = \infty$, we have f: $[0, \infty) = R^+ \to R^+$. By f^{-1}, we denote the relation, which is inverse to the function f. Note that the relation f^{-1} is a function only in some cases, for example, when f is a strictly increasing mapping.

In this setting, we define functions $f_{o\mathrm{T}} : [q, r) \to W$ and $f_o^{\mathrm{T}} : R^+ \to [q, r)$ by the following equalities, in which $f^{-1}(x) = \{z; f(z) = x\}$ is the coimage of the element $x \in R^+$:

$$f_{o\mathrm{T}}(x) = \inf\{n; n \in W \text{ and } n \geq f(x)\} =]f(x)[= \lceil f(x) \rceil,$$

$$f_o^{\mathrm{T}}(x) = \sup\{m; m \in W \text{ and } \forall z \in f^{-1}(x)(m \leq z)\}.$$

The function f_o^{T} maps R into $[q,\ r)$ because f is a projection, which implies that for each element $x \in R^+$, there is an element $z \in [q, r)$ such that $f(z) = x$.

Note that when f is a one-to-one mapping, it has the inverse function f^{-1} and in this case

$$f_o^{\mathrm{T}}(x) = [f^{-1}(x)] = \lfloor f^{-1}(x) \rfloor.$$

Taking functions $f_{o\mathrm{T}}$ and f_o^{T}, we define the function f^{T}: $R^+ \to W_{q,r}$ where $W_{q,r} = [q, r) \cap W$ as the restriction of f_o^{T} onto $W_{q,r}$ and the function $f_{\mathrm{T}} : W_{q,r} \to R^+$ as the restriction of $f_{o\mathrm{T}}$ onto $W_{q,r}$. Besides, we define order $\leq$ in $\boldsymbol{A}$ as the standard order in the interval $[q, r)$.

As before, elements from the set $W_{q,r}$ are called numbers of the functional direct perspective W-prearithmetic $\boldsymbol{A}$ and are denoted using the subscript A, i.e., 2 in $\boldsymbol{A}$ is denoted by 2_A and 5 in $\boldsymbol{A}$ is denoted by 5_A. Numbers of $\boldsymbol{A}$ are ordered by the same order relation $\leq$ as they are ordered in $\boldsymbol{W}$.

In this section, we are mostly interested in functional direct perspective W-prearithmetics, functional direct perspective prearithmetics and functional direct perspective arithmetics, which form a big class of non-Diophantine arithmetics. That is why for clarity, we explicitly describe the whole process of their construction.

As $\boldsymbol{A}$ is weakly projective with respect to Diophantine arithmetic $\boldsymbol{W}$, for any two numbers n_A and m_A from $W_{q,r}$, operations in $\boldsymbol{A}$ are defined as follows:

$$n_A \dot{+} m_A = f^{\mathrm{T}}(f_{\mathrm{T}}(n_A) + f_{\mathrm{T}}(m_A)),$$

$$n_A \circ m_A = f^{\mathrm{T}}(f_{\mathrm{T}}(n_A) \cdot f_{\mathrm{T}}(m_A)).$$

Example 2.9.2. Let us take a natural number k and the function $f(x) = kx$ as the functional parameter of the functional direct perspective W-prearithmetic $\boldsymbol{A} = (W; \dot{+}, \circ, \leq)$. Then for any whole numbers n and m, we have

$$n_A \dot{+} m_A = k^{-1}(kn + km) = (n + m)_A$$

and

$$n_A \circ m_A = k^{-1}(kn \cdot km) = (nmk)_A.$$

For instance, in $\boldsymbol{A}$ with $k = 3$, we have

$$2_A \dot{+} 2_A = 4_A,$$

i.e., two plus two is equal to four, but

$$2_A \circ 2_A = 12_A,$$

i.e., two times two is equal to twelve.

We see that in the constructed prearithmetic, addition coincides with addition in the Diophantine arithmetic but multiplication is essentially different.

Example 2.9.3. Let us take a natural number k and the function $f(0) = 0$ and $f(x) = x + k$ for $x > 0$ as the functional parameter of the functional direct perspective W-prearithmetic $\boldsymbol{A} = (W; \dot{+}, \circ, \leq)$. Then for any whole numbers n and m, we have

$$n_A \dot{+} m_A = (n + k + m + k) - k = (n + m + k)_A$$

and

$$n_A \circ m_A = ((n + k) \cdot (m + k)) - k = (nm + nk + mk + k^2 - k)_A.$$

For instance, taking $k = 1$, we have in the W-prearithmetic $\boldsymbol{A}$

$$2_A \dot{+} 2_A = 5_A,$$

i.e., two plus two is equal to five,

$$1_A \dot{+} 1_A = 3_A,$$

i.e., one plus one is equal to three, and

$$2_A \circ 2_A = (4 + 2 + 2 + 1 - 1)_A = 8_A,$$

i.e., two times two is equal to eight.

We see that in the constructed prearithmetic, both addition and multiplication are essentially different from addition and multiplication in the Diophantine arithmetic.

It is possible to treat the function f: $[q, r) \to [a, \infty)$ as a restriction of a partial function $f_+ : R^+ \times R^+$ defined on the whole set R^+ and build the functional direct perspective W-prearithmetic $\boldsymbol{W}_{wf_+}$.

Lemma 2.9.5. *The functional direct perspective W-prearithmetic $\boldsymbol{W}_{wf}$ is a subprearithmetic of the functional direct perspective W-prearithmetic $\boldsymbol{W}_{wf_+}$.*

Proof is left as an exercise. □

This result and results from Section 2.5 show that in many cases it is possible to study the functional direct perspective W-prearithmetic $\boldsymbol{W}_{wf_+}$ and derive properties of the functional direct perspective W-prearithmetic $\boldsymbol{W}_{wf}$ from the properties of properties of $\boldsymbol{W}_{wf_+}$.

2.9.2. *Properties of functional parameters*

When a thing is new, people say: '*It is not true.*'
Later, when its truth becomes obvious, they say: '*It is not important.*'
Finally, when its importance cannot be denied, they say: '*Anyway, it is not new.*'

William James

At first, we explore functional parameters of functional direct perspective prearithmetics.

If f: $X \to Y$ is a function, then any function g: $Y \to X$ for which $gf = 1_X$ is called *inverse* to f.

Properties of the category of sets (cf., for example, Bucur and Deleanu, 1968) imply the following result.

Lemma 2.9.6. *Any function f: $X \to Y$, which has inverse, is an injection.*

Proof is left as an exercise. □

It is useful to know when a function has the inverse.

Lemma 2.9.7. *Any increasing injection f: $W \to R$ is a strictly increasing function and it is possible to define inverse function $f^{-1} : R \to W$.*

Proof: Let us consider a non-decreasing injection f: $W \to R$ and assume $f(n) \leq f(m)$ for the least whole number n that is larger than m. As the function f is increasing, this implies $f(n) = f(m)$. Then either there is a whole number k such that $n > k > m$ or $n = Sm$.

In the first case, we have $f(k) > f(n) = f(m)$ and as $n > k$, the function f is not increasing. This contradicts the initial conditions.

In the second case, we have $f(\mathrm{S}m) = f(m)$ and consequently, the function f is not an injection. This also contradicts the initial conditions.

Lemma is proved. □

Let us consider the set $W_{q,r} = W \cap [q, r)$ of all whole numbers in the interval $[q, r)$.

Lemma 2.9.8. *Any increasing injection $f\colon W_{q,r} \to R^+$ is a strictly increasing function and it is possible to define inverse function $f^{-1}: R^+ \to W_{q,r}$.*

Proof is left as an exercise. □

Note that any increasing injection $f\colon R^+ \to R^+$ is a bijection between R^+ and $f(R^+)$.

Let us assume that $f\colon W \to R$ or $f\colon W_{q,r} \to R^+$.

Lemma 2.9.9. *If $f(n)$ is a whole number, then $f_{\mathrm{T}}(n) = f(n)$.*

Proof is left as an exercise. □

Corollary 2.9.3. *If for any whole number n (from the set $W_{q,r}$), its image $f(n)$ is also a whole number, then (the restriction of) the function f_{T} coincides with f (on the image $f(W_{q,r})$ of the set $W_{q,r}$).*

Lemma 2.9.10. *If a surjection $f\colon [0, r) \to R^+$ is increasing, then $f_{\mathrm{T}}(0_A) = 0$.*

Indeed, as the function f is a surjection, we have $f(x_A) = 0$ for some number x. Then for all numbers $z_A < x_A$, we have $f(z_A) = 0$ because the function f is increasing. Thus, $f_{\mathrm{T}}(0_A) = 0$.

Lemma 2.9.11. *If a surjection $f\colon [0, r) \to R^+$ is strictly increasing, then $f^{\mathrm{T}}(0) = 0_A$.*

Indeed, as the function f is a strictly increasing surjection, then it is a bijection and has the inverse function f^{-1}. The function f^{-1} is also a strictly increasing bijection. Then

$$\begin{aligned} f^{\mathrm{T}}(0) &= \sup\{n; n \in W \text{ and } n \leq f^{-1}(0)\} \\ &= \sup\{n; n \in W \text{ and } n \leq 0_A\} = 0_A. \end{aligned}$$

Let us consider a direct perspective W-prearithmetic $\boldsymbol{A} = (W_{q,r}; \dot{+}, \circ, \leq)$. Note that it is possible that $W_{q,r} = W$.

Lemma 2.9.12. *If the function f is strictly increasing on whole numbers, i.e., for any n and m from $W_{q,r}$ (from W), the inequality $n < m$ implies* *the inequality* $f(n) < f(m)$, *then for any element* n_A *from* $\boldsymbol{A}$, *we have* $f_{\mathrm{T}}(n_A) \geq n$.

Proof is left as an exercise. □

In some cases, it is possible to derive that the function f_{T} is strictly increasing from information that the function f is strictly increasing. For instance, Lemma 2.9.6 implies the following result.

Lemma 2.9.13. *If the function* f *is (strictly) increasing and maps whole numbers into whole numbers, then the function* f_{T} *is also (strictly) increasing.*

Indeed, in this case, $f_{\mathrm{T}}(n_A) = f(n_A)$ for any whole number n.

However, without the second condition, the statement of Lemma 2.9.8 is not true as the following example demonstrates.

Example 2.9.4. As in Example 2.9.2, let us consider a real number k and the function $f(x) = kx$ as the functional parameter of the functional direct perspective W-prearithmetic $\boldsymbol{A} = (W; \dot{+}, \circ, \leq)$. Then taking $k = 0.1$, we have

$$f_{\mathrm{T}}(1) = f_{\mathrm{T}}(2) = f_{\mathrm{T}}(3) = 1.$$

This shows that although the functional parameter $f(x) = 0.1x$ is strictly increasing, the function f_{T} is not strictly increasing.

Still it is possible to prove a weaker result.

Lemma 2.9.14. *If the function* f *is increasing, then the function* f_{T} *is increasing.*

Proof: Let us take numbers a and b such that $a \leq b$. Then

$$f_{\mathrm{T}}(a) = \inf\{n; n \in W \text{ and } n \geq f(a)\}$$

and

$$f_{\mathrm{T}}(b) = \inf\{n; n \in W \text{ and } n \geq f(b)\}.$$

As f is an increasing function, $f(a) \leq f(b)$ and by the definition of infimum, we have

$$\inf\{n; n \in W \text{ and } n \geq f(a)\} \leq \inf\{n; n \in W \text{ and } n \geq f(b)\},$$

i.e., $f_{\mathrm{T}}(a) \leq f_{\mathrm{T}}(b)$.

Lemma is proved. □

Corollary 2.9.4. *If the function f is strictly increasing, then the function f_{T} is increasing.*

Similar results are valid for the function f^{T}.

Lemma 2.9.15. *If the function f is increasing, then the function f^{T} is increasing.*

Proof: Let us take real numbers x and z such that $x \le z$. Then for any numbers $b \in f^{-1}(z)$ and $a \in f^{-1}(x)$, we have $a \le b$ because otherwise the function f would not be increasing (cf. Appendix). Indeed, if $a > b$, then for an increasing function f, we have

$$x = f(a) > f(b) = z.$$

Thus,

$$\begin{aligned}&\sup\{n; n \in W \text{ and } \forall a \in f^{-1}(x)(m \le a)\}\\ &\quad \le \sup\{n; n \in W \text{ and } \forall b \in f^{-1}(x)(m \le b)\},\end{aligned}$$

i.e., $f^{\mathrm{T}}(x) \le f^{\mathrm{T}}(z)$.

Lemma is proved. □

Corollary 2.9.5. *If the function f is strictly increasing, then the function f^{T} is increasing.*

Strictly increasing functions influence even more properties of direct perspective W-prearithmetics.

Lemma 2.9.16. *If the function f is strictly increasing, then for any element n_A from $\boldsymbol{A}$, we have $f^{\mathrm{T}}(f_{\mathrm{T}}(n_A)) \ge n_A$.*

Proof: By definition, we have $f(n_A) \le f_{\mathrm{T}}(n_A)$. As the function f is strictly increasing, it has the inverse function f^{-1}, which is also strictly increasing. Consequently, $n_A = f^{-1}(f(n_A)) \le f^{-1}(f_{\mathrm{T}}(n_A))$. As n_A is a whole number and f^{-1} is a function,

$$\begin{aligned}f^{\mathrm{T}}(f(n_A)) &= \sup\{m; m \in W \text{ and } \forall z \in f^{-1}(f(n_A))(m \le z)\}\\ &= \sup\{m; m \in W \text{ and } m \le f^{-1}(f(n_A)) = n_A\} = n_A.\end{aligned}$$

By Lemma 2.9.15, the function f^{T} is increasing. Thus, we have

$$n_A = f^{\mathrm{T}}(f(n_A)) \le f^{\mathrm{T}}(f_{\mathrm{T}}(n_A)).$$

Lemma is proved. □

Proposition 2.9.4. *For any functional direct perspective W-prearithmetic* $\boldsymbol{A} = (A; \dot{+}, \circ, \leq)$ *with the increasing functional parameter* f, *if for a whole number* n_A *(from the set* $W_{q,r}$*), the image* $f(n_A)$ *is also a whole number, then* $f^{\mathrm{T}}(f_{\mathrm{T}}(n_A)) = n_A$.

Proof: By Lemma 2.9.8, we have

$$\begin{aligned} f^{\mathrm{T}}(f_{\mathrm{T}}(n_A)) &= f^{\mathrm{T}}(f(n_A)) = \sup\{k_A; k \in W \text{ and } k \leq f^{-1}(f_{\mathrm{T}}(n_A))\} \\ &= \sup\{k_A; k_A \in W \text{ and } k_A \leq f^{-1}(f(n_A))\} \\ &= \sup\{k_A; k_A \in W \text{ and } k_A \leq n_A\} = n_A. \end{aligned}$$

Proposition is proved. □

Corollary 2.9.6. *If a function* f *is increasing and for any whole number* n_A *(from the set* $W_{q,r}$*), the image* $f(n_A)$ *is also a whole number, then* $f^{\mathrm{T}} f_{\mathrm{T}} = 1_W (f^{\mathrm{T}} f_{\mathrm{T}} n_A = 1_{Wq,r})$ *is the identity function.*

Corollary 2.9.7. *If a function* f *is increasing and for any whole number* n_A *(from the set* $W_{q,r}$*), the image* $f(n_A)$ *is also a whole number, then the functional direct perspective W-prearithmetic* $\boldsymbol{A} = (A; \dot{+}, \circ, \leq)$ *with the functional parameter* f *is projective with respect to Diophantine arithmetic* $\boldsymbol{W}$.

Corollary 2.9.8. *If the function* f *is strictly increasing and maps whole numbers into whole numbers, then* $f^{\mathrm{T}}(f_{\mathrm{T}}(n_A)) = n_A$ *for any element* n_A *from* $\boldsymbol{A}$.

However, if we change the places of functions f_{T} and f^{T}, we can obtain only a weaker result.

Proposition 2.9.5. *For any functional direct perspective W-prearithmetic* $\boldsymbol{A} = (A; \dot{+}, \circ, \leq)$ *with the increasing functional parameter* f, *if for a whole number* n_A *(from the set* $W_{q,r}$*), the image* $f(n_A)$ *is also a whole number, then* $f_{\mathrm{T}}(f^{\mathrm{T}}(n)) \leq n$.

Proof: By construction, we have $f^{\mathrm{T}}(n) \leq z$ for any number z from $f^{-1}(n)$. As $f^{\mathrm{T}}(n)$ is a whole number, by Lemma 2.9.8, we have $f_{\mathrm{T}}(f^{\mathrm{T}}(n)) = f(f^{\mathrm{T}}(n))$. As the function f is a increasing, for any number z from $f^{-1}(n)$, we have

$$f_{\mathrm{T}}(f^{\mathrm{T}}(n)) = f(f^{\mathrm{T}}(n)) \leq f(z) = n.$$

Proposition is proved. □

2.9.3. *Properties of addition*

Synergy is what happens when one plus one equals ten or a hundred or even a thousand!

Stephen Covey

Now let us study properties of addition in functional direct perspective W-prearithmetics.

Lemmas 2.9.14 and 2.9.15 allow us to prove the following results.

Proposition 2.9.6. *In a functional direct perspective W-prearithmetic* $\boldsymbol{A} = (A; \dot{+}, \circ, \leq)$ *with the increasing functional parameter* f, *addition* $\dot{+}$ *is monotone.*

Proof: Let us take elements n_A, l_A and m_A from A such that $l_A \leq m_A$. Then by Lemma 2.9.13, we have

$$f_{\mathrm{T}}(l_A) \leq f_{\mathrm{T}}(m_A)).$$

Consequently,

$$f_{\mathrm{T}}(n_A) + f_{\mathrm{T}}(l_A) \leq f_{\mathrm{T}}(n_A) + f_{\mathrm{T}}(m_A)$$

because addition is monotone in the Diophantine arithmetic $\boldsymbol{W}$ and by Lemma 2.9.14, we have

$$n_A \dot{+} l_A = f^{\mathrm{T}}(f_{\mathrm{T}}(n_A) + f_{\mathrm{T}}(l_A)) \leq f^{\mathrm{T}}(f_{\mathrm{T}}(n_A) + f_{\mathrm{T}}(m_A)) = n_A \dot{+} m_A.$$

Proposition is proved. □

Proposition 2.9.6 and Lemma 2.1.12 imply the following result.

Corollary 2.9.9. *For any functional direct perspective W-prearithmetic* $\boldsymbol{A} = (A; \dot{+}, \circ, \leq)$ *with the increasing functional parameter* f, *addition* $\dot{+}$ *preserves the order* $\leq$.

Proposition 2.7.1 and properties of the conventional arithmetic $\boldsymbol{W}$ imply the following result.

Proposition 2.9.7. *For any functional direct perspective W-prearithmetic* $\boldsymbol{A} = (A; \dot{+}, \circ, \leq_A)$, *addition* $\dot{+}$ *is commutative.*

Proof is left as an exercise. □

Remark 2.9.1. In a general case of functional direct perspective W-prearithmetics, addition $\dot{+}$ can be non-associative as the following example demonstrates.

Example 2.9.5. Let us take the functional direct perspective arithmetic $\boldsymbol{A}_{x^2} = (N; \dot{+}, \circ, \leq)$ with the functional parameter $f(x) = x^2$, and look how n-ary operations are performed comparing them with operations in the Diophantine arithmetic $\boldsymbol{N}$. We have

$$3_A \dot{+} (4_A \dot{+} 5_A) = 3_A \dot{+} f^{\mathrm{T}}(16 + 25) = 3_A \dot{+} f^{\mathrm{T}}(41)$$
$$= 3_A \dot{+} 6_A = f^{\mathrm{T}}(9 + 36) = f^{\mathrm{T}}(45) = 6_A$$

while

$$(3_A \dot{+} 4_A) \dot{+} 5_A = f^{\mathrm{T}}(9 + 16) \dot{+} 5_A = 3_A \dot{+} f^{\mathrm{T}}(25) \dot{+} 5_A$$
$$= 5_A \dot{+} 5_A = f^{\mathrm{T}}(25 + 25) = f^{\mathrm{T}}(50) = 7_A.$$

Thus,

$$3_A \dot{+} (4_A \dot{+} 5_A) \neq (3_A \dot{+} 4_A) \dot{+} 5_A$$

and consequently, addition in $\boldsymbol{A}$ is not associative.

Now let us study properties of zero 0_A in functional direct perspective W-prearithmetics. Note that in a general case, 0_A does not necessarily belong to a functional direct perspective W-prearithmetic $\boldsymbol{A}$ and even when it belongs, it may be neither additive zero nor multiplicative zero although it is always the *order zero*, i.e., the least element in $\boldsymbol{A}$.

Let us consider a functional direct perspective W-prearithmetic $\boldsymbol{A} = (W_{0,r}; \dot{+}, \circ, \leq)$ with the functional parameter f and zero 0_A.

Theorem 2.9.1. *If for any number n_A, its image $f(n_A)$ is also a whole number and $f_{\mathrm{T}}(0_A) = 0$, then for any $n_A \in \boldsymbol{A}$, the equality $n_A \dot{+} 0_A = n_A (0_A \dot{+} n_A = n_A)$ is true if and only if $f(x)$ is a strictly increasing function on the set $W_{0,r}$.*

Proof: *Necessity.* Let us assume that $f(x)$ is not a strictly increasing function on the set $W_{0,r}$, i.e., for some numbers m_A and n_A from $\boldsymbol{A}$ with $m_A < n_A$, we have $f(n_A) \leq f(m_A)$. Then by Lemma 2.9.9, the inequality $f(n_A) \leq f(m_A)$ implies the following sequence of equalities:

$$n_A \dot{+} 0_A = f^{\mathrm{T}}(f_{\mathrm{T}}(n_A) + f_{\mathrm{T}}(0_A)) = f^{\mathrm{T}}(f_{\mathrm{T}}(n_A) + 0) = f^{\mathrm{T}}(f_{\mathrm{T}}(n_A))$$
$$= f^{\mathrm{T}}(f(n_A)) = \sup\{m; m \in W \text{ and } \forall z \in f^{-1}(f(n_A))(m \leq z)\}$$
$$\leq \sup\{m; m \in W \text{ and } \forall z \in f^{-1}(f(m_A))(m \leq z)\} \leq m_A$$

because m_A is a whole number which is mapped onto $f_{\mathrm{T}}(m_A)$, i.e., $m_A \in f^{-1}(f(m_A))$.

It means that $n_A \dot{+} 0_A \neq n_A$. Thus, by the Principle of Excluded Middle, $f(x)$ is a strictly increasing function on the set $W_{0,r}$.

The equality $0_A \dot{+} n_A = n_A$ is treated in a similar way.

Sufficiency. Let us assume that $f(x)$ is a strictly increasing function on the set $W_{0,r}$. Then by Proposition 2.9.4, we have the following sequence of equalities:

$$n_A \dot{+} 0_A = f^{\mathrm{T}}(f_{\mathrm{T}}(n_A) + f_{\mathrm{T}}(0_A)) = f^{\mathrm{T}}(f_{\mathrm{T}}(n_A) + 0) = f^{\mathrm{T}}(f_{\mathrm{T}}(n_A)) = n_A.$$

The equality $0_A \dot{+} n_A = n_A$ is derived in a similar way.

Theorem is proved. □

This theorem gives necessary and sufficient conditions for 0_A to be a neutral element (additive zero) with respect to addition in a functional direct perspective W-prearithmetics prearithmetic $\boldsymbol{A}$.

The proof of Theorem 2.9.1 implies the following results.

Corollary 2.9.10. *For any number n_A, if its image $f(n_A)$ is also a whole number and $f(x)$ is a strictly increasing function, then the equalities $0_A \dot{+} n_A = n_A$ and $n_A \dot{+} 0_A = n_A$ are true in the W-prearithmetic $\boldsymbol{A}$.*

Corollary 2.9.11. *If the image $f(n_A)$ is a whole number for any number n_A, but $f(x)$ is not a strictly increasing function on the set $W_{q,r}$, then there is a number m_A such that $0_A \dot{+} m_A < m_A$ and $m_A \dot{+} 0_A < m_A$ in the W-prearithmetic $\boldsymbol{A}$.*

Corollary 2.9.12. *If for any number n_A, its image $f(n_A)$ is also a whole number and $f(x)$ is an increasing function, then for any $n_A \in \boldsymbol{A}$ the equalities $n_A \dot{+} 0_A = n_A$ and $0_A \dot{+} n_A = n_A$ are true if and only if $f_{\mathrm{T}}(x)$ is a strictly increasing function on the set $W_{q,r}$.*

Proposition 2.9.4 and Theorem 2.9.1 imply the following result.

Corollary 2.9.13. *If for any number n_A, if its image $f(n_A)$ is also a whole number, and $f(x)$ is a strictly increasing function, then for any $n_A \in \boldsymbol{A}$ the equalities $0_A \dot{+} n_A = n_A$ and $n_A \dot{+} 0_A = n_A$ are true, i.e., 0_A is an additive zero in the functional direct perspective W-prearithmetic $\boldsymbol{A}$.*

Remark 2.9.2. However, even if $f_{\mathrm{T}}(0_A) = 0$ and $f(x)$ is a strictly increasing function, it is not necessary, in a general case, that 0_A is an additive zero, i.e., the equality $0_A \dot{+} a_A = a_A$ is not always true as the following example demonstrates.

Example 2.9.6. Let us take $f(x) = \sqrt{x}$ and build a functional direct perspective W-prearithmetic $\boldsymbol{A} = (A; \dot{+}, \circ, \leq)$ with the functional parameter f. Then $f^{-1}(x) = x^2$ and we have

$$\begin{aligned} 3_A \dot{+} 0_A &= f^{\mathrm{T}}(f_{\mathrm{T}}(3_A) + f_{\mathrm{T}}(0_A)) \\ &= f^{\mathrm{T}}(\sqrt{3} + 0) = f^{\mathrm{T}}(]\sqrt{3}[+0) = f^{\mathrm{T}}(2) = 4_A \neq 3_A \end{aligned}$$

although $f_{\mathrm{T}}(0_A) = 0$ and $f(x)$ is a strictly increasing function. It means that 0_A is not an additive zero in the W-prearithmetic $\boldsymbol{A}$. At the same time, 0_A is a multiplicative zero in $\boldsymbol{A}$.

Proposition 2.9.8. *For any number n_A from a functional direct perspective W-prearithmetic $\boldsymbol{A} = (A; \dot{+}, \circ, \leq)$ with the strictly increasing functional parameter f and zero 0_A, if $f(n_A)$ is a whole number, then $n_A \leq n_A \dot{+} m_A$ and $n_A \leq m_A \dot{+} n_A$ for any number m_A from $\boldsymbol{A}$.*

Proof: At first, we show that $n_A = 0_A \dot{+} n_A = n_A \dot{+} 0_A$. Indeed, by Lemma 2.9.10, $f_{\mathrm{T}}(0_A)) = 0$. Besides, by Proposition 2.9.4, $f^{\mathrm{T}}(f_{\mathrm{T}}(n_A)) = n_A$ and we have

$$n_A \dot{+} 0_A = f^{\mathrm{T}}(f_{\mathrm{T}}(n_A) + f_{\mathrm{T}}(0_A)) = f^{\mathrm{T}}(f_{\mathrm{T}}(n_A) + 0) = f^{\mathrm{T}}(f_{\mathrm{T}}(n_A)) = n_A.$$

Since $0_A \leq m_A$ for any element m_A from $\boldsymbol{A}$ and by Proposition 2.9.6, addition in $\boldsymbol{A}$ is monotone, we obtain

$$n_A = n_A \dot{+} 0_A \leq n_A \dot{+} m_A.$$

In the same way, we have

$$n_A = 0_A \dot{+} n_A \leq m_A \dot{+} n_A.$$

Proposition is proved. □

Corollary 2.9.14. *If for any number n_A from a functional direct perspective W-prearithmetic $\boldsymbol{A} = (A; \dot{+}, \circ, \leq)$ with the strictly increasing functional parameter f, its image $f(n_A)$ is also a whole number, then for any $n_A \in \boldsymbol{A}$, the equalities $0_A \dot{+} n_A = n_A$ and $n_A \dot{+} 0_A = n_A$ are true, i.e., 0_A is an additive zero in the W-prearithmetic $\boldsymbol{A}$.*

It is possible to obtain a similar result even without the condition that the considered prearithmetic $\boldsymbol{A}$ has the additive zero. Note that even when 0_A belongs to $\boldsymbol{A}$, it is not necessarily the additive zero in the prearithmetic $\boldsymbol{A}$ as it is demonstrated in Example 2.9.6.

Proposition 2.9.9. *If the functional parameter f is strictly increasing, then for any numbers m_A and n_A from a functional direct perspective W-prearithmetic $\boldsymbol{A} = (W; \dot{+}, \circ, \leq)$ with the functional parameter f, we have $m_A \leq m_A \dot{+} n_A$ and $n_A \leq m_A \dot{+} n_A$.*

Proof: By properties of whole numbers, we have

$$f_{\mathrm{T}}(m_A) \leq f_{\mathrm{T}}(m_A) + f_{\mathrm{T}}(n_A).$$

and because f is strictly increasing, by Lemma 2.9.16, we have

$$f^{\mathrm{T}}(f_{\mathrm{T}}(m_A)) \geq m_A.$$

As the functional parameter f is strictly increasing, by Lemma 2.9.15, the function f^{T} is increasing. Consequently,

$$m_A \dot{+} n_A = f^{\mathrm{T}}(f_{\mathrm{T}}(n_A) + f_{\mathrm{T}}(m_A)) \geq f^{\mathrm{T}}(f_{\mathrm{T}}(m_A)) \geq m_A.$$

The inequality $n_A \leq m_A \dot{+} n_A$ is proved in a similar way.

Proposition is proved. □

Note that a function f can be strictly increasing while the function f_{T} is increasing but not strictly increasing. For instance, let us consider the function $f(x) = \log_2 x + 1$. Then we have

$$f_{\mathrm{T}}(3_A) = 3 \quad \text{and} \quad f_{\mathrm{T}}(4_A) = 3.$$

It means that the function f_{T} is not strictly increasing.

Studying relations in W-prearithmetics, we see that in addition to binary relations $\leq$ and $<$, there are other kinds of order relations in a functional direct perspective W-prearithmetic $\boldsymbol{A}$. Let us consider relations determined by properties of addition.

Definition 2.9.3. The expression $n_A \ll m_A$ where n_A and m_A belong to $\boldsymbol{A}$ means that n_A is *much less* than m_A and in this case, m_A is *much larger* than n_A, i.e., $m_A \gg n_A$. It is defined as

$$n_A \ll m_A \text{ if and only if } m_A \dot{+} n_A = m_A.$$

For instance, in the Diophantine arithmetic $\boldsymbol{W}$, zero "0" is much less than any element of $\boldsymbol{W}$. Even more, in any functional direct perspective W-prearithmetic $\boldsymbol{A}$, its additive zero 0_A is much less than any element of $\boldsymbol{A}$.

This correlates with our intuition and with other fields of mathematics. For instance, there are infinitely small numbers in non-standard analysis but any positive infinitely small number is still larger than zero (Robinson, 1966).

Note that in the Diophantine arithmetics, the relations $\ll$ is very rare. Namely, in the Diophantine arithmetic $\boldsymbol{N}$, the relation $\ll$ is empty. In the Diophantine arithmetic $\boldsymbol{W}$, the relation $n \ll m$ is true only when n is equal to 0. However, as we will see, there are functional direct perspective prearithmetics and non-Diophantine arithmetics, in which $\ll$ is a total (linear) order, i.e., for any two numbers n_A and m_A, we have either $n_A \ll m_A$ or $m_A \ll n_A$ (cf. Appendix).

Let us find properties of the introduced relation $\ll$.

Proposition 2.9.10. *For any number m_A from a functional direct perspective W-prearithmetic $\boldsymbol{A}$ with the strictly increasing functional parameter f, if $0_A < n_A$, then the inequality $n_A \ll m_A$ implies the inequality $n_A \leq m_A$.*

Proof: Let us assume that $0_A < n_A \ll m_A$. Then

$$m_A \dot{+} n_A = m_A$$

and because f is strictly increasing, by Proposition 2.9.9, we have

$$n_A \leq m_A \dot{+} n_A = m_A.$$

Thus, $n_A \leq m_A$.

Proposition is proved. □

Corollary 2.9.14 implies the following result.

Proposition 2.9.11. *If for any number n_A from a functional direct perspective W-prearithmetic $\boldsymbol{A} = (A; \dot{+}, \circ, \leq)$ with the strictly increasing functional parameter f, its image $f(n_A)$ is also a whole number, then for any $n_A \in \boldsymbol{A}$, we have $0_A \ll n_A$.*

Proof is left as an exercise. □

Proposition 2.9.12. *For any numbers n_A and m_A from a functional direct perspective W-prearithmetic $\boldsymbol{A}$, if $0 < n_A$ and $n_A \ll m_A$, then m_A is an additively composite number.*

Indeed, the relation $n_A \ll m_A$ means $m_A \dot{+} n_A = m_A$.

Corollary 2.9.15. *If $0 < n_A$ and $n_A \ll n_A$, then n_A is an additively composite number for any number n_A from a functional direct perspective W-prearithmetic $\boldsymbol{A}$.*

We remind that an element a in a universal algebra with a binary operation $\circ$ is called *idempotent* if $a \circ a = a$ (Kurosh, 1963). Thus we have the following statements.

Corollary 2.9.16. *An element n_A of a functional direct perspective W-prearithmetic $\boldsymbol{A}$ is an additive idempotent if and only if $n_A \ll n_A$.*

Relation $\ll$ is determined by properties of the function f_{T}.

Lemma 2.9.17. *For any numbers a and b from a functional direct perspective W-prearithmetic $\boldsymbol{A} = (A; \dot{+}, \circ, \leq)$ with the increasing functional parameter f, which maps whole numbers into whole numbers, the inequality $a \ll b$ is true if and only if $f_{\mathrm{T}}(a) < f_{\mathrm{T}}(\mathrm{S}b) - f_{\mathrm{T}}(b)$ or $f_{\mathrm{T}}(a) + f_{\mathrm{T}}(b) < f_{\mathrm{T}}(\mathrm{S}b)$.*

Proof: *Necessity.* Let us assume that $a \ll b$ for numbers a and b from $\boldsymbol{A}$. It means that

$$\begin{aligned} b \dot{+} a = f^{\mathrm{T}}(f_{\mathrm{T}}(b) + f_{\mathrm{T}}(a)) = \sup\{m; m \in W \text{ and } \forall z \in f^{-1}(f_{\mathrm{T}}(b) \\ + f_{\mathrm{T}}(a))(m \leq z)\} = b. \end{aligned}$$

Thus, by the definition of supremum, any number z from $f^{-1}(f_{\mathrm{T}}(b) + f_{\mathrm{T}}(a))$ is less than $\mathrm{S}b$ because $\mathrm{S}b > b$. By definition, $f_{\mathrm{T}}(b) \geq f(b)$ for any element b from $\boldsymbol{A}$.

Thus, we have

$$f_{\mathrm{T}}(b) + f_{\mathrm{T}}(a) = f(f^{-1}(f_{\mathrm{T}}(b) + f_{\mathrm{T}}(a))) < f(\mathrm{S}b) \leq f_{\mathrm{T}}(\mathrm{S}b).$$

This implies the inequality

$$f_{\mathrm{T}}(b) + f_{\mathrm{T}}(a) < f_{\mathrm{T}}(\mathrm{S}b)$$

or

$$f_{\mathrm{T}}(a) < f_{\mathrm{T}}(\mathrm{S}b) - f_{\mathrm{T}}(b).$$

Sufficiency. Let us assume that $f_{\mathrm{T}}(a) + f_{\mathrm{T}}(b) < f_{\mathrm{T}}(\mathrm{S}b)$. Then

$$\begin{aligned} b \dot{+} a &= f^{\mathrm{T}}(f_{\mathrm{T}}(b) + f_{\mathrm{T}}(a)) \\ &= \sup\{m; m \in W \text{ and } \forall z \in f^{-1}(f_{\mathrm{T}}(b) + f_{\mathrm{T}}(a))(m \leq z)\} \\ &< f^{\mathrm{T}}(f_{\mathrm{T}}(\mathrm{S}b) = \mathrm{S}b. \end{aligned}$$

As by Proposition 2.9.4, Lemma 2.9.9 and Theorem 2.9.1, the inequality $b \dot{+} a \geq b$ is true, we have $b \dot{+} a = b$, i.e., $a \ll b$.

As $f_T(a) + f_T(b) < f_T(Sb)$ is equivalent to $f_T(a) < f_T(Sb) - f_T(b)$, the lemma is proved. □

This result means that the inequality $a \ll b$ is true if and only if when the value of $f_T(a)$ is less than the value of the direct discrete derivative of $f_T(b)$ (cf., for example, Boole, 1860; 1880; Spiegel, 1971).

Corollary 2.9.17. *For any number a from a functional direct perspective W-prearithmetic **A**, we have a* $\ll$ Sa *if and only if* $f_T(\text{Sa}) + f_T(a) < f_T(\text{SSa})$.

There are useful relations between relations, i.e., relations of the second order (Burgin, 2012a), in abstract prearithmetics in general and in W-prearithmetics, in particular.

Definition 2.9.4.

(a) A binary relation P on a set X is called *compatible from the right* with a binary relation Q on X if $P \circ Q \subseteq P$, i.e., $(a, b) \in P$ and $(b, c) \in Q$ imply $(a, c) \in P$.
(b) A binary relation P on a set X is called *compatible from the left* with a binary relation Q on X if $Q \circ P \subseteq P$, i.e., $(a, b) \in Q$ and $(b, c) \in P$ imply $(a, c) \in P$.
(c) A binary relation P on a set X is called *compatible* with a binary relation Q on X if P is compatible both from the right and from the left with Q.

Proposition 2.9.13. *A binary relation P is compatible from the left (from the right) with itself if and only if it is transitive.*

Proof is left as an exercise. □

Corollary 2.9.18. *A binary relation P is compatible with itself if and only if it is transitive.*

These results show that compatibility of relations is a natural generalization of transitivity of relations.

Proposition 2.9.14. *If a binary relation P is compatible from the left (from the right) with a binary relation Q and* $T \subseteq Q$, *then P is compatible from the left (from the right) with T.*

Proof is left as an exercise. □

Corollary 2.9.19. *If a binary relation P is compatible with a binary relation Q and* $T \subseteq Q$, *then P is compatible with T.*

Proposition 2.9.15. *If a binary relation P in a set X is compatible from the left (from the right) with a binary relation Q in X and $Y \subseteq X$, then the restriction P_Y of P on Y is compatible from the left (from the right) with the restriction Q_Y of Q on Y.*

Proof is left as an exercise. □

Corollary 2.9.20. *If a binary relation P in a set X is compatible with a binary relation Q in X and $Y \subseteq X$, then the restriction P_Y of P on Y is compatible with the restriction Q_Y of Q on Y.*

Let us study compatibility of order relations in functional direct perspective W-prearithmetics.

Theorem 2.9.2. *In a functional direct perspective W-prearithmetic $\boldsymbol{A} = (A; \dot{+}, \circ, \leq)$ with the increasing functional parameter f, which maps whole numbers into whole numbers, relation $\ll$ is compatible from the left with the order $\leq$.*

Proof: Let us assume that k_A, n_A and m_A are numbers from $\boldsymbol{A}$ such that $k_A \leq n_A$ and $n_A \ll m_A$. By Lemma 2.9.17, it means that $f_{\mathrm{T}}(n_A) + f_{\mathrm{T}}(m_A) < f_{\mathrm{T}}(\mathrm{S}m_A)$. As $k_A \leq n_A$, Lemma 2.9.12 implies $f_{\mathrm{T}}(k_A) \leq f_{\mathrm{T}}(n_A)$. Thus, $f_{\mathrm{T}}(k_A) + f_{\mathrm{T}}(m_A) < f_{\mathrm{T}}(\mathrm{S}m_A)$. By Lemma 2.9.17, it implies $k_A \ll m_A$.

Theorem is proved. □

Corollary 2.9.21. *In a functional direct perspective W-prearithmetic $\boldsymbol{A} = (A; \dot{+}, \circ, \leq)$ with the increasing functional parameter f, which maps whole numbers into whole numbers, the order $\leq$ is compatible from the right with the relation $\ll$.*

Indeed, $k_A \leq n_A$ and $n_A \ll m_A$ imply $k_A \ll m_A$, while by Proposition 2.9.13, inequality $k_A \ll m_A$ implies inequality $k_A \leq m_A$. Thus, $k_A \leq n_A$ and $n_A \ll m_A$ imply $k_A \leq m_A$.

Remark 2.9.3. At the same time, relation $\ll$ is not always compatible from the right with the order $\leq$.

Example 2.9.7. Let us consider a strictly increasing continuous function $f(x)$ that satisfies the following conditions:

$$f(0) = 0,$$

$$f(1) = 10,$$

$$f(2) = 16,$$

$$f(3) = 32,$$

$$f(n) = 32 + n \text{ when } n > 3.$$

We can build the functional direct perspective W-prearithmetic $\boldsymbol{A} = (A; \dot{+}, \circ, \leq)$, taking f as its functional parameter, i.e., $\boldsymbol{A} = \boldsymbol{W}_f$. Then we have

$$f_{\mathrm{T}}(0) = 0,$$

$$f_{\mathrm{T}}(1) = 10,$$

$$f_{\mathrm{T}}(2) = 16,$$

$$f_{\mathrm{T}}(3) = 32,$$

$$f_{\mathrm{T}}(n) = 32 + n \text{ when } n > 3.$$

The definition of operations in the functional direct perspective W-prearithmetic $\boldsymbol{A} = \boldsymbol{W}_f$ implies the following equalities and inequalities:

$$1_A \dot{+} 2_A = f^{\mathrm{T}}(f_{\mathrm{T}}(1_A) + f_{\mathrm{T}}(2_A)) = f^{\mathrm{T}}(10+16) < f^{\mathrm{T}}(f_{\mathrm{T}}(3_A) = f^{\mathrm{T}}(32) = 3_A.$$

Consequently, by Lemma 2.9.17,

$$1_A \dot{+} 2_A = 2_A \text{ and } 1_A \ll 2_A.$$

At the same time, we have

$$1_{\mathrm{A}} \dot{+} 5_{\mathrm{A}} = f^{\mathrm{T}}(f_{\mathrm{T}}(1_{\mathrm{A}}) + f_{\mathrm{T}}(5_{\mathrm{A}})) = f^{\mathrm{T}}(10 + 37) = f^{\mathrm{T}}(47) = 15_A$$

because $f\ (15) = 47$.

Consequently, it is not true that $1_A \ll 5_A$ although $1_A < 2_A < 5_A$. It means that the relation $\ll$ is not compatible from the right with relations $<$ and $\leq$.

However, in many cases, the relation $\ll$ is compatible with itself from the right and from the left as the following result demonstrates.

Proposition 2.9.16. *In a functional direct perspective W-prearithmetic* $\boldsymbol{A} = (A; \dot{+}, \circ, \leq)$ *with the increasing functional parameter* f, *the relation* $\ll$ *is transitive.*

Indeed, if $k_A \ll n_A$ and $n_A \ll m_A$, then by Proposition 2.9.10, $k_A \leq n_A$ and by Theorem 2.9.2, we obtain $k_A \ll m_A$.

We know that in the Diophantine arithmetic $\boldsymbol{N}$, the relation $\ll$ is empty, i.e., it cannot be true for any pair of natural numbers. In the Diophantine arithmetic $\boldsymbol{W}$, the relation $\ll$ is true only for 0 and any other number. However, in the general case, it is possible that $\ll$ is a total (linear) order in a functional direct perspective W-prearithmetic. Namely, we have the following result.

Proposition 2.9.17. *There is a functional direct perspective W-prearithmetic $\boldsymbol{A} = (W; \dot{+}, \circ, \leq)$, in which $\ll$ is a total (linear) order on W.*

Proof: Let us take $f(n) = 2^{2^n}$ as the functional parameter for a functional direct perspective W-prearithmetic $\boldsymbol{A}$. By Lemma 2.9.9, we have $f_{\mathrm{T}}(n_A) = f(n) = 2^{2^n}$.

Thus, for any $n = 0, 1, 2, 3, \ldots$, we obtain the sequence of equalities and inequalities

$$\begin{aligned} f_{\mathrm{T}}(n_A + 1) + f_{\mathrm{T}}(n_A) &= 2^{2^{n+1}} + 2^{2^n} < 2^{2^{n+1}} + 2^{2^{n+1}} \\ &< 2^{2^{n+1}} \cdot 2^{2^{n+1}} = 2^{2^{n+1}+2^{n+1}} = 2^{2 \cdot 2^{n+1}} = 2^{2^{n+2}} \\ &= f_{\mathrm{T}}(n_A + 2). \end{aligned}$$

Consequently, by Lemma 2.9.17, we have the following relations $0_A \ll 1_A \ll 2_A \ll 3_A \ll \cdots \ll n_A \ll (n+1)_A \ll \cdots$ in the W-prearithmetic $\boldsymbol{A}$ with the projector $f_{\mathrm{T}}(n)$. As by Proposition 2.9.16, the relation $\ll$ is transitive in $\boldsymbol{A}$, this implies that the relation $\ll$ is a total (linear) order in W.

Proposition is proved. □

Note that in this prearithmetic, 0_A is not an additive zero.

Proposition 2.9.18. *For any numbers n_A, m_A and k_A from a functional direct perspective W-prearithmetic $\boldsymbol{A} = (A; \dot{+}, \circ, \leq)$ with the strictly increasing functional parameter f, which maps whole numbers into whole numbers, the inequality $n_A \dot{+} k_A \ll m_A$ implies the inequalities $n_A \ll m_A$ and $k_A \ll m_A$.*

Proof: Let us assume that n_A, m_A and k_A are numbers from $\boldsymbol{A}$ and $n_A + k_A \ll m_A$. By Proposition 2.9.9, $n_A \leq n_A \dot{+} k_A$ and $k_A \leq n_A \dot{+} k_A$. As by Theorem 2.9.2, relation $\ll$ is compatible from the left with the order $\leq$, we have the inequalities $n_A \ll m_A$ and $k_A \ll m_A$.

Proposition is proved. □

Proposition 2.9.19. *For any numbers* n_A, m_A *and* k_A *from an associative functional direct perspective W-prearithmetic* $\boldsymbol{A} = (A; \dot{+}, \circ, \leq)$, *the inequalities* $n_A \ll m_A$ *and* $k_A \ll m_A$ *imply the inequality* $n_A \dot{+} k_A \ll m_A$.

Proof: Let us assume that n_A, m_A and k_A are numbers from $\boldsymbol{A}$, for which the inequalities $n_A \ll m_A$ and $k_A \ll m_A$ are valid. Then we have

$$(n_A \dot{+} k_A) \dot{+} m_A = n_A \dot{+} (k_A \dot{+} m_A) = n_A \dot{+} m_A = m_A.$$

It means that $n_A \dot{+} k_A \ll m_A$.

Proposition is proved. □

Remark 2.9.4. The associativity condition is essential as the following example demonstrates.

Example 2.9.8. Let us take the functional direct perspective W-prearithmetic $\boldsymbol{A} = (A; \dot{+}, \circ, \leq)$ from Example 2.9.7. Then definitions imply the following equalities and inequalities:

$$\begin{aligned} 1_A \dot{+} 2_A &= f^{\mathrm{T}}(f_{\mathrm{T}}(1_A) + f_{\mathrm{T}}(2_A)) \\ &= f^{\mathrm{T}}(10 + 16) < f^{\mathrm{T}}(f_{\mathrm{T}}(3_A) = f^{\mathrm{T}}(32) = 3_A. \end{aligned}$$

Consequently, by Lemma 2.9.17,

$$1_A \dot{+} 2_A = 2_A \quad \text{and} \quad 1_A \ll 2_A.$$

Besides,

$$2_A \dot{+} 2_A = f^{\mathrm{T}}(f_{\mathrm{T}}(2_A) + f_{\mathrm{T}}(2_A)) = f^{\mathrm{T}}(16 + 16) = f^{\mathrm{T}}(32) = 3_A$$

and

$$\begin{aligned} 1_A \dot{+} 1_A &= f^{\mathrm{T}}(f_{\mathrm{T}}(1_A) + f_{\mathrm{T}}(1_A)) \\ &= f^{\mathrm{T}}(10 + 10) < f^{\mathrm{T}}(f_{\mathrm{T}}(3_A) = f^{\mathrm{T}}(32) = 3_A. \end{aligned}$$

Thus,

$$2_A \leq 1_A \dot{+} 1_A < 3_A.$$

Consequently,

$$1_A \dot{+} 1_A = 2_A.$$

This implies

$$1_A \dot{+} (1_A \dot{+} 2_A) = 1_A \dot{+} 2_A = 2_A.$$

At the same time, we have

$$(1_A \dot{+} 1_A) + 2_A = 2_A \dot{+} 2_A = 3_A.$$

This means that the functional direct perspective W-prearithmetic $\boldsymbol{A}$ is not associative.

In addition, $1_A \ll 2_A$ but it is not true that $1_A \dot{+} 1_A = 2_A \ll 2_A$.

Theorem 2.9.3. *Relation $\ll$ in a functional direct perspective W-prearithmetic $\boldsymbol{A} = (A; \dot{+}, \circ, \leq)$ with the increasing functional parameter f, is transitive and disjunctively asymmetric, i.e., only one relation $x \ll y$ or $y \ll x$ can be valid for all different elements x and y from M.*

Indeed, by Proposition 2.9.16, $\ll$ is transitive and if $n_A \ll m_A$ and $m_A \ll n_A$, then

$$m_A = m_A \dot{+} n_A = n_A.$$

Note that relation $\ll$ is not always asymmetric because in some cases, it is possible that $n_A \ll n_A$.

Proposition 2.9.20. *In a functional direct perspective W-prearithmetic $\boldsymbol{A} = (A; \dot{+}, \circ, \leq)$ with the strictly increasing functional parameter f, the relation $n_A \ll m_A$ implies the relation $n_A \leq m_A$.*

Proof: Let us assume that $n_A \ll m_A$ but $n_A > m_A$. Then by Proposition 2.9.9, we have

$$m_A \dot{+} n_A \geq n_A > m_A$$

while $n_A \ll m_A$ implies $m_A \dot{+} n_A = m_A$. We come to a contradiction.

As numbers of $\boldsymbol{A}$ are ordered as they are ordered in $\boldsymbol{W}$, the order $\leq$ is linear. Thus, we have either $n_A \leq m_A$ or $n_A > m_A$. As the latter inequality contradicts the condition $n_A \ll m_A$, by the Principle of Excluded Middle, $n_A \leq m_A$.

Proposition is proved. □

Theorem 2.9.4. *For any $n > 0$, there is a functional direct perspective arithmetic $\boldsymbol{A}, = (N; \oplus, \otimes, \leq)$ in which $m_A \ll q_A$ if and only if $q > n$ and $m < q$ or $q = n$ and $m \leq q$.*

Proof: *Sufficiency.* Let us define the function $f(k)$ as follows:

$$f(k) = \begin{cases} k & \text{if } k < n+1, \\ 2^k & \text{if } k > n. \end{cases}$$

Then its inverse has the following form:

$$f^{-1}(k) = \begin{cases} k & \text{if } \log_2 k < n+1, \\ \log_2 k & \text{if } \log_2 k > n. \end{cases}$$

Then for any $q > m > n$, we have

$$f_{\mathrm{T}}(m) + f_{\mathrm{T}}(q) = 2^m + 2^q < 2 \cdot 2^q = 2^{q+1} = f_{\mathrm{T}}(q+1).$$

Thus, by Lemma 2.9.17, $m_A \ll q_A$.

If $m < n+ 1$ and $q > n$, we have

$$f_{\mathrm{T}}(m) + f_{\mathrm{T}}(q) = m + 2^q < 2 \cdot 2^q = 2^{q+1} = f_{\mathrm{T}}(q+1).$$

Thus, by Lemma 2.9.17, $m_A \ll q_A$. In both cases, we have $q > n$ and $m < q$.

Besides,

$$f_{\mathrm{T}}(n) + f_{\mathrm{T}}(n) = n + n = 2n < 2^{n+1} = f_{\mathrm{T}}(q+1).$$

Thus, by Lemma 2.9.17, $n_A \ll n_A$. As by Theorem 2.9.2, the relation $\ll$ is compatible from the left with the order $\leq$, $m_A \ll n_A$ for any $m \leq n = q$.

Necessity. At first, let us consider the situation when $q \leq n$. Then for any $m < q$, we have

$$q_A \dot{+} m_A = f^{\mathrm{T}}(f_{\mathrm{T}}(q_A) + f_{\mathrm{T}}(m_A)) = f^{\mathrm{T}}(q+m) > f^{\mathrm{T}}(q) = q_A.$$

It means that the inequality $m_A \ll q_A$ is not true.

Besides, the inequality $r_A \ll r_A$ is not true when $r \neq n$. Indeed, if $r < n$ and $r + r < n$, then

$$r_A \dot{+} r_A = f^{\mathrm{T}}(f_{\mathrm{T}}(r_A) + f_{\mathrm{T}}(r_A)) = f^{\mathrm{T}}(r+r) > f^{\mathrm{T}}(r) = r_A.$$

Thus, the inequality $r_A \ll r_A$ is not true. If $r < n$ but $r + r \geq n$, then

$$r_A \dot{+} r_A = f^{\mathrm{T}}(f_{\mathrm{T}}(r_A) + f_{\mathrm{T}}(r_A)) = f^{\mathrm{T}}(r+r) \geq f^{\mathrm{T}}(n) > f^{\mathrm{T}}(r) = r_A.$$

Thus, the inequality $r_A \ll r_A$ is not true. If $r > n$, then

$$\begin{aligned} r_A \dot{+} r_A &= f^{\mathrm{T}}(f_{\mathrm{T}}(r_A) + f_{\mathrm{T}}(r_A)) = f^{\mathrm{T}}(2^r + 2^r) = f^{\mathrm{T}}(2 \cdot 2^r) = f^{\mathrm{T}}(2^{r+1}) \\ &= r + 1. \end{aligned}$$

Thus, the inequality $r_A \ll r_A$ is not true.

This implies that to have $m_A \ll q_A$, we need $q \geq n$, while Theorem 2.9.2 implies $m < q$ when $q > n$ and $m \leq q$ when $q = n$.

Theorem is proved. □

Relation $\ll$ is intrinsically connected to the Archimedean property for addition in direct perspective W-prearithmetics in general and in non-Diophantine arithmetics, in particular, as the following result demonstrates.

Let us consider a functional direct perspective W-prearithmetic $\boldsymbol{A} = (A; \dot{+}, \circ, \leq)$ with the increasing functional parameter f.

Theorem 2.9.5. *The following conditions are equivalent:*

(a) *The W-prearithmetic $\boldsymbol{A}$ with a strictly increasing functional parameter f is additively Archimedean.*
(b) *There are no non-zero elements n_A and m_A in $\boldsymbol{A}$ such that $n_A \ll m_A$.*
(c) *There is no element $m_A > 0$ in $\boldsymbol{A}$ such that $\mathrm{S}0_A \ll m_A$.*
(d) *Relation $n_A \ll m_A$ always implies $n_A = 0$.*

Proof: (b) ⇒ (c): If the statement (c) is not true, i.e., there is an element $m_A > 0$ in $\boldsymbol{A}$ such that $\mathrm{S}0_A \ll m_A$, then taking $n_A = \mathrm{S}0_A$, we see that the statement (b) is not true. Thus, from the Principle of Excluded Middle it follows that (b) implies (c).

(c) ⇒ (b): Let us assume that there are non-zero elements n_A and m_A in $\boldsymbol{A}$ such that $n_A \ll m_A$. Then by Theorem 2.9.2, we have $\mathrm{S}0_A \ll m_A$ because $\mathrm{S}0_A \leq n_A$. From the Principle of Excluded Middle, it follows that (c) implies (b).

(d) ⇔ (b): As $0_A \ll m_A$ for any element m_A in $\boldsymbol{A}$, condition (d) is equivalent to condition (b).

(a) ⇒ (b): When there are non-zero elements n_A and m_A in $\boldsymbol{A}$ such that $n_A \ll m_A$, then whatever number of elements n_A we add to itself, it will always be less than $\mathrm{S}m_A$ because by Proposition 2.9.10, the inequality $n_A \ll m_A$ implies the inequality $n_A \leq m_A$ and $n_A \dot{+} n_A \leq n_A \dot{+} m_A = m_A$.

It means that the W-prearithmetic $\boldsymbol{A}$ is not Additively Archimedean. From the Principle of Excluded Middle it follows that (a) implies (b).

(b) ⇒ (a): Let us assume that for any element $m_A > 0$ in $\boldsymbol{A}$, the inequality $\mathrm{S}0_A \ll m_A$ is not true. It means that for any element $m_A > 0$ in $\boldsymbol{A}$, we have $\mathrm{S}b \leq \mathrm{S}0_A \dot{+} m_A$ and $m_A \leq \mathrm{S}0_A \dot{+} \mathrm{P}m_A$. If the W-prearithmetic $\boldsymbol{A}$ is not Additively Archimedean, then there is the least element $m_A > 0$ in $\boldsymbol{A}$ such that $\sum_{i=1}^{n} a_i < m_A$ for some element $n_A > 0$ in $\boldsymbol{A}$ and any $n = 1, 2, 3, \ldots$ with all a_i are equal to n_A.

As m_A is the least element in $\boldsymbol{A}$ with this property $\mathrm{P}m_A \leq \sum_{i=1}^{n} a_i$ for some n. Then

$$m_A \leq \mathrm{S}0_A \dot{+} \mathrm{P}m_A \leq n_A \dot{+} \mathrm{P}b \leq n_A \dot{+} \sum_{i=1}^{n} a_i \leq \sum_{i=1}^{n+1} a_i,$$

where all a_i are equal to n_A. This contradicts our assumption about the element m_A and by the Principle of Excluded Middle shows that (b) implies (a).

As implication is a transitive binary relation (cf., for example, Kleene, 2002), (c) implies (a) and (a) implies (c).

Theorem is proved. □

Corollary 2.9.22. *A functional direct perspective (non-Diophantine) W-arithmetic* $\boldsymbol{A}$ *has the Archimedean property if and only if the relation* $a \ll b$ *in* $\boldsymbol{A}$ *always implies* $a = 0$.

Archimedean properties also allow obtaining an axiomatics for the Diophantine arithmetic $\boldsymbol{W}$ in the class of direct perspective W-prearithmetics that is independent from the Peano axiomatics explored in Section 2.11.

Theorem 2.9.6. *A functional direct perspective (non-Diophantine) W-prearithmetic* $\boldsymbol{A} = (W; \dot{+}, \circ, \leq)$ *is isomorphic to the Diophantine arithmetic* $\boldsymbol{W}$ *if and only if it satisfies the following conditions:*

(1) $\boldsymbol{A}$ *is additively Archimedean.*
(2) $\boldsymbol{A}$ *has the left Exactly Binary Archimedean Property for addition.*
(3) $\boldsymbol{A}$ *has additive cancellation.*
(4) *Multiplication* $\circ$ *in* $\boldsymbol{A}$ *is defined by the formula*

$$m_A \circ n_A = m[n_A]$$

and in particular,

$$0_A \circ n_A = 0_A.$$

Proof: *Sufficiency.* Let us consider a functional direct perspective W-prearithmetic $\boldsymbol{A} = (W; \dot{+}, \circ, \leq)$ that satisfies conditions (1)–(3).

At first, we show that 0_A is the additive zero in $\boldsymbol{A}$. To do this, let us assume the opposite, namely, that for some number n_A, we have

$n_A \dot{+} 0_A \neq n_A$. Then as all whole numbers are well ordered, we can assume that n_A is the least number with this property.

If $n_A \dot{+} 0_A = m_A < n_A$, then $n_A \dot{+} 0_A = m_A = m_A \dot{+} 0_A$. As $\boldsymbol{A}$ has additive cancellation, canceling 0_A, we obtain $n_A = m_A$. As this is not true, the first situation when $n_A \dot{+} 0_A = m_A < n_A$ is impossible.

If $n_A \dot{+} 0_A = m_A > n_A$, then there is a number k_A less than n_A such that $0_A + k_A = n_A$ because $\boldsymbol{A}$ has the left Exactly Binary Archimedean Property for addition (cf. Section 2.1). It means that $k_A \dot{+} 0_A \neq k_A$ because k_A is less than n_A. However, we assumed that n_A is least number with this property. Consequently, the second situation when $n_A \dot{+} 0_A = m_A > n_A$ is impossible.

As the only possible situation is $n_A \dot{+} 0_A = n_A$ for all n, we come to the conclusion that 0_A is the additive zero in $\boldsymbol{A}$.

Now let us prove that $f_{\mathrm{T}}(x)$ is not a strictly increasing function on the set W. To do this, let us assume the opposite, namely, that $f_{\mathrm{T}}(x)$ is not a strictly increasing function on the set W, i.e., for some numbers k_A and n_A from $\boldsymbol{A}$ with $k_A < n_A$, we have $f_{\mathrm{T}}(n_A) \leq f_{\mathrm{T}}(k_A)$. Then by Lemma 2.9.7, the inequality $f_{\mathrm{T}}(n_A) \leq f_{\mathrm{T}}(k_A)$ implies the following sequence of equalities:

$$\begin{aligned} n_A \dot{+} 0_A &= f^{\mathrm{T}}(f_{\mathrm{T}}(n_A) + f_{\mathrm{T}}(0_A)) = f^{\mathrm{T}}(f_{\mathrm{T}}(n_A) + 0) = f^{\mathrm{T}}(f_{\mathrm{T}}(n_A)) \\ &= f^{\mathrm{T}}(f(n_A)) = \sup\{m; m \in W \text{and} \forall\, z \in f^{-1}(f_{\mathrm{T}}(n_A) \\ &\quad + f_{\mathrm{T}}(0_A))(m \leq z)\} \leq \sup\{m; m \in W \text{and} \forall z \in f^{-1}(f(k_A) \\ &\quad + f_{\mathrm{T}}(0_A))(m \leq z)\} = k_A \dot{+} 0_A = k_A \end{aligned}$$

because inequality $f_{\mathrm{T}}(n_A) \leq f_{\mathrm{T}}(m_A)$ implies inequality $f_{\mathrm{T}}(n_A) + f_{\mathrm{T}}(0_A) \leq f_{\mathrm{T}}(m_A) + f_{\mathrm{T}}(0_A)$ and supremum is a monotone function.

It means that $n_A \dot{+} 0_A \neq n_A$. However, it is proved that $n_A \dot{+} 0_A = n_A$. Thus, by the Principle of Excluded Middle, $f_{\mathrm{T}}(x)$ is a strictly increasing function on the set W.

The next step in our proof is demonstration that addition $\dot{+}$ is monotone in $\boldsymbol{A}$. Let us take elements n_A, l_A and k_A from A such that $l_A \leq k_A$. As we already proved that $f_{\mathrm{T}}(x)$ is a strictly increasing function on the set W, we obtain the inequality

$$f_{\mathrm{T}}(l_A) \leq f_{\mathrm{T}}(k_A)).$$

Consequently,

$$f_{\mathrm{T}}(n_A) + f_{\mathrm{T}}(l_A) \leq f_{\mathrm{T}}(n_A) + f_{\mathrm{T}}(k_A)$$

because addition is monotone in the Diophantine arithmetic $\boldsymbol{W}$. Consequently, by definition, we have

$$\begin{aligned} n_A \dot{+} l_A = f^{\mathrm{T}}(f_{\mathrm{T}}(n_A) + f_{\mathrm{T}}(l_A)) &= \sup\{m; m \in W \text{ and } \forall z \in f^{-1}(f_{\mathrm{T}}(n_A) \\ &+ f_{\mathrm{T}}(l_A))(m \leq z)\} \leq \sup\{m; m \in W \text{ and } \forall z \in f^{-1}(f(n_A) \\ &+ f_{\mathrm{T}}(k_A))(m \leq z)\} = f^{\mathrm{T}}(f_{\mathrm{T}}(n_A) + f_{\mathrm{T}}(k_A)) = n_A \dot{+} k_A. \end{aligned}$$

Now we prove that $n_A \dot{+} 1_A = \mathrm{S}n_A$ for all whole numbers n. If this is not true, we have two possibilities: $n_A \dot{+} 1_A > \mathrm{S}n_A$ or $n_A \dot{+} 1_A < \mathrm{S}n_A$.

If $n_A \dot{+} 1_A > \mathrm{S}n_A$, then as the functional direct perspective (non-Diophantine) W-prearithmetic $\boldsymbol{A}$ has the left Exactly Binary Archimedean Property for addition, there is an element k_A from $\boldsymbol{A}$ such that $n_A \dot{+} k_A = \mathrm{S}n_A$. This element k_A cannot be equal to 0_A because $n_A \dot{+} 0_A = n_A$ and it cannot be equal to 1_A because $n_A \dot{+} 1_A > \mathrm{S}n_A$. Thus, $k_A > 1_A$ and because we have proved that addition $\dot{+}$ is monotone in $\boldsymbol{A}$, we have $n_A \dot{+} k_A > \mathrm{S}n_A$ for any whole number $k > 0$. This contradicts the assumption that $\boldsymbol{A}$ has the left Exactly Binary Archimedean Property for addition and by the Principle of Excluded Middle, shows that the first inequality is impossible as it leads to a contradiction.

If we have the second possibility, i.e., $n_A \dot{+} 1_A < \mathrm{S}n_A$, then $n_A \dot{+} 1_A = n_A$ because monotonicity of addition $\dot{+}$ implies

$$n_A = n_A \dot{+} 0_A \leq n_A \dot{+} 1_A.$$

By definition, it means that $1_A \ll n_A$. In particular, this implies that independently of how many times we add 1_A to n_A, we will always obtain n_A. As all whole numbers are well ordered, we can assume that n_A is the least number with this property. Then for all $m < n$, we have $m_A \dot{+} 1_A = \mathrm{S}m_A$. Indeed, if $m_A \dot{+} 1_A < \mathrm{S}m_A$, then with the same reasoning as before we come to the inequality $1_A \ll n_A$, which contradicts our assumption that n_A is the least number with this property.

Now let us suppose that $m_A \dot{+} 1_A > \mathrm{S}m_A$. Then as the functional direct perspective (non-Diophantine) W-prearithmetic $\boldsymbol{A}$ has the left Exactly Binary Archimedean Property for addition, there is an element k_A from $\boldsymbol{A}$ such that $m_A \dot{+} k_A = \mathrm{S}m_A$. This element k_A cannot be equal to 0_A because $m_A \dot{+} 0_A = m_A$ and it cannot be equal to 1_A because $m_A \dot{+} 1_A > \mathrm{S}m_A$.

Thus, $k_A > 1_A$ and because we have proved that addition $\dot{+}$ is monotone in $\boldsymbol{A}$, we have $m_A \dot{+} k_A > \mathrm{S}m_A$ for any whole number $k > 0$. This contradicts the assumption that $\boldsymbol{A}$ has the left Exactly Binary Archimedean Property for addition (cf. Section 2.1) and by the Principle of Excluded Middle, for all $m < n$, we have $m_A \dot{+} 1_A = \mathrm{S}m_A$. Thus, we have

$$2_A = 1_A \dot{+} 1_A = 2[1_A],$$

$$3_A = 2_A \dot{+} 1_A = 2[1_A] \dot{+} 1_A = 3[1_A].$$

Continuing this process, we come to the equality

$$n_A = n[1_A].$$

Then inequality $1_A \ll n_A$ implies that for all $t > n$, we have

$$n_A = t[1_A] < \mathrm{S}n_A.$$

The same inequality is true when the whole number t is less than or equal to n. At the same time, the functional direct perspective (non-Diophantine) W-prearithmetic $\boldsymbol{A}$ is additively Archimedean, which means (cf. Section 2.1) that there is a natural number q such that $\mathrm{S}n_A \leq q[1_A]$.

Consequently, we come to the conclusion that our assumption was not true and that $n_A \dot{+} 1_A = \mathrm{S}n_A$ for all whole numbers n. Note that it is possible to use Theorem 2.9.5 to prove this.

Besides, as we have already demonstrated, for any number n_A from $\boldsymbol{A}$, we have

$$n_A = n[1_A],$$

i.e., each number in $\boldsymbol{A}$ is the sum of the corresponding number of the number 1_A.

Now we prove the following identity for addition in $\boldsymbol{A}$:

$$n_A \dot{+} \mathrm{S}m_A = \mathrm{S}(n_A \dot{+} m_A). \tag{2.52}$$

If this is not true, we have two possibilities: $n_A \dot{+} \mathrm{S}m_A > \mathrm{S}(n_A \dot{+} m_A)$ or $n_A \dot{+} \mathrm{S}m_A < S(n_A \dot{+} m_A)$.

If $n_A \dot{+} \mathrm{S}m_A > \mathrm{S}(n_A \dot{+} m_A)$, then as the functional direct perspective (non-Diophantine) W-prearithmetic $\boldsymbol{A}$ has the left Exactly Binary Archimedean Property for addition, there is an element k_A from $\boldsymbol{A}$ such that $n_A \dot{+} k_A = \mathrm{S}(n_A \dot{+} m_A)$. This element k_A cannot be equal to 0_A because $n_A \dot{+} 0_A = n_A$ and it cannot be equal to 1_A because $n_A \dot{+} 1_A = \mathrm{S}n_A$. Thus, $k_A > 1_A$ and because we have proved that addition $\dot{+}$ is monotone in $\boldsymbol{A}$, we have $n_A \dot{+} k_A > \mathrm{S}(n_A \dot{+} m_A)$ for any whole number $k > 0$. This contradicts the assumption that $\boldsymbol{A}$ has the left Exactly Binary Archimedean Property for addition (cf. Section 2.1) and by the Principle of Excluded Middle, shows that the first inequality is impossible as it leads to a contradiction.

If we have the second possibility, i.e., $n_A \dot{+} \mathrm{S}m_A < \mathrm{S}(n_A \dot{+} m_A)$, then monotonicity of addition $\dot{+}$ implies

$$n_A = n_A \dot{+} 0_A \leq n_A \dot{+} \mathrm{S}m_A.$$

Consequently,

$$n_A \dot{+} \mathrm{S}m_A = n_A \dot{+} k_A \tag{2.53}$$

for some $0 < k < \mathrm{S}m_A$. However, the equality (2.53) implies $\mathrm{S}m_A = k_A$ because $\boldsymbol{A}$ has additive cancellation. Thus, the second inequality is impossible and the Principle of Excluded Middle shows that for all whole numbers m and n, the equality (2.52) is valid.

Addition in the Diophantine arithmetic $\boldsymbol{W}$ is defined by formula (2.52) (cf., for example, Rasiowa and Sikorski, 1963; Kleene, 2002). Consequently, there is a natural isomorphism $f\colon \boldsymbol{A} \to \boldsymbol{W}$ between the functional direct perspective (non-Diophantine) W-prearithmetic $\boldsymbol{A} = (W; \dot{+}, \circ, \leq)$ and the Diophantine arithmetic $\boldsymbol{W}$ with respect to addition. This isomorphism is defined by the formula $f(n_A) = n$.

To show that f is an isomorphism with respect to addition, we use induction. Indeed, for any whole number n, we have

$$f(n_A \dot{+} 0_A) = f(n_A) = n = n + 0$$

and

$$f(n_A \dot{+} 1_A) = f(\mathrm{S}n_A) = \mathrm{S}n = n + 1.$$

Now let us suppose that for some whole number m, we have

$$f(n_A \dot{+} m_A) = n + m.$$

Then because we proved that $t_A \dot{+} 1_A = \mathrm{S}t_A$ when t is an arbitrary whole number, for the next number $\mathrm{S}m_A$, we obtain

$$\begin{aligned} f(n_A \dot{+} \mathrm{S}m_A) &= f(\mathrm{S}(n_A \dot{+} m_A)) = f((n_A \dot{+} m_A) \dot{+} 1_A) = f(n_A \dot{+} m_A) + 1 \\ &= (n + m) + 1 = n + (m + 1) = n + \mathrm{S}m. \end{aligned}$$

The axiom of induction (cf. Section 2.11) implies that for any whole numbers n and m, we have

$$f(n_A \dot{+} m_A) = n + m.$$

Now let us prove that f is a natural isomorphism between the functional direct perspective (non-Diophantine) W-prearithmetic $\boldsymbol{A} = (W; \dot{+}, \circ, \leq)$ and the Diophantine arithmetic $\boldsymbol{W}$ with respect to multiplication. Indeed, by condition (4),

$$f(0_A \circ n_A) = f(0_A) = 0 = 0 \cdot n$$

and because f is an isomorphism with respect to addition, we obtain

$$f(m_A \circ n_A) = f(m[n_A]) = f\,\underbrace{(n_A \dot{+} \cdots \dot{+} n_A)}_{m} = \underbrace{n + \cdots + n}_{m} = m \cdot n.$$

It means that f is an isomorphism between $\boldsymbol{A}$ and $\boldsymbol{W}$ with respect to multiplication and thus, an isomorphism between $\boldsymbol{A}$ and $\boldsymbol{W}$ as universal algebras with two operations and an order relation.

Necessity. Let us assume that there is an isomorphism between a functional direct perspective W-prearithmetic $\boldsymbol{A} = (W; \dot{+}, \circ, \leq)$ and the Diophantine arithmetic $\boldsymbol{W}$. We know that $\boldsymbol{W}$ is additively Archimedean, Exactly Binary Archimedean for addition and has additive cancellation. Isomorphism preserves all these properties. Thus, $\boldsymbol{A}$ is additively Archimedean, Exactly Binary Archimedean for addition and has additive cancellation.

Besides, multiplication in $\boldsymbol{W}$ is expressed by means of addition, i.e., for any whole number, we have

$$m \cdot n = \underbrace{n + \cdots + n}_{m}.$$

Since isomorphism preserves this relation, the W-prearithmetic $\boldsymbol{A}$ satisfies condition (4).

Theorem is proved. □

Corollary 2.9.23. *A functional direct perspective (non-Diophantine) W-arithmetic* $\boldsymbol{A} = (W; \dot{+}, \circ, \leq)$ *is isomorphic to the Diophantine arithmetic* $\boldsymbol{W}$ *if and only if it satisfies the following conditions*:

(1) $\boldsymbol{A}$ *is additively Archimedean.*
(2) $\boldsymbol{A}$ *has the left Exactly Binary Archimedean Property for addition.*

(3) ***A*** *has additive cancellation.*
(4) *Multiplication* $\circ$ *in* ***A*** *is defined by the formula*

$$m_A \circ n_A = m[n_A].$$

Corollary 2.9.24. *A functional direct perspective prearithmetic* ***A*** *is isomorphic to the Diophantine arithmetic* ***N*** *if and only if it satisfies the following conditions:*

(1) ***A*** *is additively Archimedean.*
(2) ***A*** *has the left Exactly Binary Archimedean Property for addition.*
(3) ***A*** *has additive cancellation.*
(4) *Multiplication* $\circ$ *in* ***A*** *is defined by the formula*

$$m_A \circ n_A = m[n_A].$$

Corollary 2.9.25. *A functional direct perspective (non-Diophantine) arithmetic* ***A*** *is isomorphic to the Diophantine arithmetic* ***N*** *if and only if it satisfies the following conditions*:

(1) ***A*** *is additively Archimedean.*
(2) ***A*** *has the left Exactly Binary Archimedean Property for addition.*
(3) ***A*** *has additive cancellation.*
(4) *Multiplication* $\circ$ *in* ***A*** *is defined by the formula*

$$m_A \circ n_A = m[n_A].$$

2.9.4. *Properties of multiplication*

Multiplication is a shortcut for addition.

Anonymous

Now let us study multiplication $\circ$ in functional direct perspective W-prearithmetics. Many properties of multiplication are similar to properties of addition.

Proposition 2.7.1 and properties of the conventional arithmetic ***W*** imply the following result.

Proposition 2.9.21. *For any functional direct perspective W-prearithmetic* $\boldsymbol{A} = (A; \dot{+}, \circ, \leq_A)$, *multiplication* $\circ$ *is commutative.*

Proof is left as an exercise. □

Remark 2.9.5. In a general case of functional direct perspective W-prearithmetics, multiplication $\circ$ can be non-associative as Example 2.9.5 demonstrates.

Proposition 2.9.22. *In a functional direct perspective W-prearithmetic* $\boldsymbol{A} = (A; \dot{+}, \circ, \leq)$ *with the increasing functional parameter* f, *multiplication* $\circ$ *is monotone.*

Proof is similar to the proof of Proposition 2.9.6. □

Proposition 2.9.22 and Lemma 2.1.7 imply the following result.

Corollary 2.9.26. *For any functional direct perspective W-prearithmetic* $\boldsymbol{A} = (A; \dot{+}, \circ, \leq)$ *with the increasing functional parameter* f, *multiplication* $\circ$ *preserves the order* $\leq$.

Let us consider a functional direct perspective W-prearithmetic $\boldsymbol{A} = (W_{0,r};\ \dot{+}\ , \circ, \leq)$ with the functional parameter f.

Theorem 2.9.7. *If for any number* n_A, *its image* $f(n_A)$ *is also a whole number and* $f_{\mathrm{T}}(1_A) = 1$, *then for any larger than* 1_A *number* n_A *from* $\boldsymbol{A}$, *the equality* $n_A \circ 1_A = n_A (1_A \circ n_A = n_A)$ *is true if and only if* $f(x)$ *is a strictly increasing function.*

Proof: *Necessity.* Let us assume that $f(x)$ is not a strictly increasing function, i.e., for some numbers m_A and n_A from $\boldsymbol{A}$ with $m_A < n_A$, we have $f(n_A) \leq f(m_A)$. Then by Lemma 2.9.9, the inequality $f(n_A) \leq f(m_A)$ implies the following sequence of equalities:

$$\begin{aligned} n_A \circ 1_A &= f^{\mathrm{T}}(f_{\mathrm{T}}(n_A) \cdot f_{\mathrm{T}}(0_A)) = f^{\mathrm{T}}(f_{\mathrm{T}}(n_A) \cdot 1) = f^{\mathrm{T}}(f_{\mathrm{T}}(n_A)) \\ &= f^{\mathrm{T}}(f(n_A)) = \sup\{m; m \in W \text{ and } \forall z \in f^{-1}(f(n_A))(m \leq z)\} \\ &\leq \sup\{m; m \in W \text{ and } \forall\, z \in f^{-1}(f(m_A))(m \leq z)\} \leq m_A \end{aligned}$$

because m_A is a whole number which is mapped onto $f_{\mathrm{T}}(m_A)$, i.e., $m_A \in f^{-1}(f(m_A))$.

It means that $n_A \circ 1_A \neq n_A$. Thus, by the Principle of Excluded Middle, $f(x)$ is a strictly increasing function.

The equality $1_A \circ n_A = n_A$ is treated in a similar way.

Sufficiency. Let us assume that $f_{\mathrm{T}}(x)$ is a strictly increasing function. Then by Proposition 2.9.4, we have the following sequence of equalities:

$$n_A \circ 1_A = f^{\mathrm{T}}(f_{\mathrm{T}}(n_A) \cdot f_{\mathrm{T}}(1_A)) = f^{\mathrm{T}}(f_{\mathrm{T}}(n_A) \cdot 1) = f^{\mathrm{T}}(f_{\mathrm{T}}(n_A)) = n_A.$$

The equality $1_A \circ n_A = n_A$ is derived in a similar way.

Theorem is proved. □

This theorem gives necessary and sufficient conditions for 1_A to be a neutral element (additive zero) with respect to multiplication in a functional direct perspective W-prearithmetics prearithmetic $\boldsymbol{A}$.

The proof of Theorem 2.9.7 gives us the following results.

Corollary 2.9.27. *If the image $f(n_A)$ of a number n_A is also a whole number, then the equalities $1_A \circ n_A = n_A$ and $n_A \circ 1_A = n_A$ are true in the functional direct perspective W-prearithmetic $\boldsymbol{A}$ with the strictly increasing functional parameter f.*

Corollary 2.9.28. *If the image $f(n_A)$ is a whole number for any number n_A, but $f(x)$ is not a strictly increasing function on whole numbers, then there is a number m_A such that $n_A \circ 1_A < n_A$ and $1_A \circ n_A < n_A$ in the functional direct perspective W-prearithmetic $\boldsymbol{A}$ with the strictly increasing functional parameter f.*

Corollary 2.9.29. *If for any number n_A, its image $f(n_A)$ is also a whole number and $f(x)$ is an increasing function, then for any n_A from the functional direct perspective W-prearithmetic $\boldsymbol{A}$ with the functional parameter f the equalities $1_A \circ n_A = n_A$ and $n_A \circ 1_A = n_A$ are true if and only if $f_{\mathrm{T}}(x)$ is a strictly increasing function.*

Proposition 2.9.4 and Theorem 2.9.1 imply the following result.

Corollary 2.9.30. *If for any number n_A, if its image $f(n_A)$ is also a whole number, and $f(x)$ is a strictly increasing function, then for any $n_A \in \boldsymbol{A}$ the equalities $1_A \circ n_A = n_A$ and $n_A \circ 1_A = n_A$ are true, i.e., 1_A is a multiplicative one in the functional direct perspective W-prearithmetic $\boldsymbol{A}$.*

Although 0_A is not always an additive zero (cf. Remark 2.9.2), 0_A is always a multiplicative zero in a functional direct perspective W-prearithmetic.

Proposition 2.9.23. *For any number m_A from a functional direct perspective W-prearithmetic $\boldsymbol{A}$ with the strictly increasing functional parameter f, the equality $0_A \circ m_A = 0_A$ is true, i.e., 0_A is a multiplicative zero in $\boldsymbol{A}$.*

Indeed, by Lemma 2.9.10, $f_{\mathrm{T}}(0_A) = 0$ and by Lemma 2.9.11, $f^{\mathrm{T}}(0) = 0_A$. Thus, the initial conditions imply the following sequence of equalities:

$$0_A \circ m_A = f^{\mathrm{T}}(f_{\mathrm{T}}(0_A) \cdot f_{\mathrm{T}}(m_A)) = f^{\mathrm{T}}(0 \cdot f_{\mathrm{T}}(m_A)) = f^{\mathrm{T}}(0) = 0_A.$$

It means that 0_A is a multiplicative zero.

Proposition 2.9.24. *For any non-zero number n_A from a functional direct perspective W-prearithmetic $\boldsymbol{A} = (A; \dot{+}, \circ, \leq)$ with the strictly increasing functional parameter f, if $f(n_A)$ is a whole number, then $n_A \leq n_A \circ m_A$ and $n_A \leq m_A \circ n_A$ for any non-zero number m_A from $\boldsymbol{A}$.*

Proof: At first, we show that $n_A \leq n_A \circ 1_A$. Indeed, as by Lemma 2.9.12, $f_{\mathrm{T}}(1_A) \geq 1$ and by Proposition 2.9.4, $f^{\mathrm{T}}(f_{\mathrm{T}}(n_A)) = n_A$, we have

$$n_A \circ 1_A = f^{\mathrm{T}}(f_{\mathrm{T}}(n_A) \cdot f_{\mathrm{T}}(1_A)) \geq f^{\mathrm{T}}(f_{\mathrm{T}}(n_A) \cdot 1) = f^{\mathrm{T}}(f_{\mathrm{T}}(n_A)) = n_A.$$

Since $1_A \leq m_A$ for any non-zero element m_A from $\boldsymbol{A}$ and by Proposition 2.9.7, multiplication in $\boldsymbol{A}$ is monotone, we obtain

$$n_A \leq n_A \circ 1_A \leq n_A \circ m_A.$$

In the same way, we prove

$$n_A \leq 1_A \circ n_A \leq n_A \circ m_A.$$

Proposition is proved. □

Corollary 2.9.31. *If for any number n_A from a functional direct perspective W-prearithmetic $\boldsymbol{A} = (A; \dot{+}, \circ, \leq)$ with the strictly increasing functional parameter f, its image $f(n_A)$ is also a whole number, then for any $n_A \in \boldsymbol{A}$, the inequalities $n_A \leq 1_A \circ n_A$ and $n_A \leq 1_A \circ n_A$ are true.*

It is possible to obtain a similar result even without the condition that the considered prearithmetic $\boldsymbol{A}$ has the multiplicative one 1_A. Note that even 1_A belongs to $\boldsymbol{A}$, it is not necessarily the multiplicative one in the prearithmetic $\boldsymbol{A}$.

Proposition 2.9.25. *If the function f_{T} is strictly increasing, then for any non-zero numbers m_A and n_A from a functional direct perspective W-prearithmetic $\boldsymbol{A} = (A; \dot{+}, \circ, \leq)$ with the strictly increasing functional parameter f, we have $m_A \leq m_A \circ n_A$ and $n_A \leq m_A \circ n_A$.*

Proof: By properties of whole numbers, we have

$$f_{\mathrm{T}}(m_A) \leq f_{\mathrm{T}}(m_A) \cdot f_{\mathrm{T}}(n_A)$$

and because f is strictly increasing, by Lemma 2.9.16, we have

$$f^{\mathrm{T}}(f_{\mathrm{T}}(m_A)) \geq m_A.$$

As the functional parameter f is strictly increasing, by Lemma 2.9.15, the function f^{T} is increasing. Consequently,

$$m_A \circ n_A = f^{\mathrm{T}}(f_{\mathrm{T}}(n_A) \cdot f_{\mathrm{T}}(m_A)) \geq f^{\mathrm{T}}(f_{\mathrm{T}}(m_A)) \geq m_A.$$

The inequality $n_A \leq m_A \circ n_A$ is proved in a similar way.

Proposition is proved. □

Corollary 2.9.32. *In functional direct perspective W-prearithmetics with the strictly increasing functional parameter f, there are no divisors of zero.*

Indeed, if $m_A \neq 0$ and $n_A \neq 0$, then $m_A > 0$ and by Proposition 2.9.25, $m_A \circ n_A \geq m_A > 0$.

Let us consider additional relations in W-prearithmetics determined by properties of multiplication.

Definition 2.9.5. The expression $n_A \lll m_A$ where n_A and m_A belong to $\boldsymbol{A}$ means that n_A is *much much less* than m_A and in this case, m_A is *much much larger* than n_A, i.e., $m_A \ggg n_A$. It is defined as

$$n_A \lll m_A \text{ if and only if } m_A \circ n_A = m_A.$$

For instance, in the Diophantine arithmetic $\boldsymbol{W}$, one 1_A is much much less than any element of $\boldsymbol{W}$.

Note that in the Diophantine arithmetics, these relations are very rare. Namely, in the Diophantine arithmetic $\boldsymbol{N}$, the relation $n \lll m$ is true only when n is equal to 1. In the Diophantine arithmetic $\boldsymbol{W}$, the relation $n \lll m$ is true only when n is equal to 1 or when m is equal to 0. However, as we will see, there are functional direct perspective prearithmetics and non-Diophantine arithmetics, in which $\lll$ is a total (linear) order.

Let us study properties of $\lll$. Theorem 2.9.7 implies the following result.

Proposition 2.9.26. *If for any number n_A from a functional direct perspective W-prearithmetic $\boldsymbol{A} = (A; \dot{+}, \circ, \leq)$ with the strictly increasing functional parameter f, its image $f(n_A)$ is also a whole number and $f(1_A) = 1$, then for any $n_A \in \boldsymbol{A}$, we have $1_A \lll n_A$.*

Proof is left as an exercise. □

Proposition 2.9.27. *If $1_A < n_A$ and $n_A \lll m_A$, then m_A is a multiplicatively composite number in a functional direct perspective W-prearithmetic $\boldsymbol{A}$.*

Indeed, the relation $n_A \lll m_A$ means $m_A \circ n_A = m_A$.

Corollary 2.9.33. *If* $1_A < n_A$ *and* $n_A \lll n_A$, *then the number* n_A *from* $\boldsymbol{A}$ *is multiplicatively composite.*

Corollary 2.9.34. *An element* n_A *of a functional direct perspective W-prearithmetic* $\boldsymbol{A}$ *is a multiplicative idempotent if and only if* $n_A \lll n_A$.

Proposition 2.9.28. *For any number* m_A *from a functional direct perspective W-prearithmetic* $\boldsymbol{A}$ *with the strictly increasing functional parameter* f, *if* $0_A < n_A$, *then inequality* $n_A \lll m_A$ *implies inequality* $n_A \leq m_A$.

Proof: Let us assume that $0_A < n_A \lll m_A$. Then

$$m_A \circ n_A = m_A$$

and because f is strictly increasing, by Proposition 2.9.25, we have

$$n_A \leq m_A \circ n_A.$$

Thus, $n_A \leq m_A$.

Proposition is proved. □

Relation $\lll$ is also determined by properties of the function f_T.

Lemma 2.9.18. *For any numbers* $n_A \neq 0_A$ *and* $m_A \neq 0_A$ *from a functional direct perspective W-prearithmetic* $\boldsymbol{A} = (A; \dot{+}, \circ, \leq)$ *with the functional parameter* f, *which maps whole numbers into whole numbers, the inequality* $n_A \lll m_A$ *is true if and only if* $f_{\mathrm{T}}(n_A) < f_{\mathrm{T}}(\mathrm{S}m_A)/f_{\mathrm{T}}(m_A)$ *or* $f_{\mathrm{T}}(n_A) \cdot f_{\mathrm{T}}(m_A) < f_{\mathrm{T}}(\mathrm{S}m_A)$.

Proof: *Necessity.* Let us assume that the inequality $n_A \lll m_A$ is true for numbers n_A and m_A from $\boldsymbol{A}$. It means that

$$\begin{aligned} m_A \circ n_A &= f^{\mathrm{T}}(f_{\mathrm{T}}(m_A) \cdot f_{\mathrm{T}}(n_A)) \\ &= \sup\{m; m \in W \text{ and } \forall z \in f^{-1}(f_{\mathrm{T}}(m_A) \cdot f_{\mathrm{T}}(n_A))(m \leq z)\} \\ &= m_A. \end{aligned}$$

Thus, any number z from $f^{-1}(f_{\mathrm{T}}(m_A) \cdot f_{\mathrm{T}}(n_A))$ is less than $\mathrm{S}m_A$ because $\mathrm{S}m_A > m_A$. This implies the inequality

$$f_{\mathrm{T}}(m_A) \cdot f_{\mathrm{T}}(n_A) < f_{\mathrm{T}}(\mathrm{S}m_A)$$

or

$$f_{\mathrm{T}}(n_A) < f_{\mathrm{T}}(\mathrm{S}m_A)/f_{\mathrm{T}}(m_A).$$

Sufficiency. Let us assume that $f_T(m_A) \cdot f_T(n_A) < f_T(Sm_A)$. Then

$$\begin{aligned} m_A \circ n_A &= f^T(f_T(m_A) \cdot f_T(n_A)) \\ &= \sup\{m; m \in W \text{ and } \forall z \in f^{-1}(f_T(m_A) \cdot f_T(n_A))(m \leq z)\} \\ &< Sm_A. \end{aligned}$$

As by Proposition 2.9.5, Lemma 2.9.9 and Theorem 2.9.1, $m_A \circ n_A \geq m_A$ when $n_A \neq 0_A$, we have $m_A \circ n_A = m_A$, i.e., $n_A \lll m_A$.

As $f_T(n_A) \cdot f_T(m_A) < f_T(Sm_A)$ is equivalent to $f_T(n_A) < f_T(Sm_A)/f_T(m_A)$, the lemma is proved. □

This result means that the inequality $n_A \lll m_A$ is true if and only if when the value of $f_T(n_A)$ is less than the value of the direct discrete logarithmic derivative of $f_T(m_A)$ plus 1. Indeed, logarithmic derivative is defined as

$$[\ln f]' = f'/f$$

and the direct discrete logarithmic derivative of $f_T(b)$ is equal to

$$[f_T(Sm_A) - f_T(m_A)]/f_T(m_A).$$

Thus,

$$f_T(Sm_A)/f_T(m_A) - 1 = [f_T(Sm_A) - f_T(m_A)]/f_T(m_A)$$

and

$$f_T(Sm_A)/f_T(m_A) = [f_T(Sm_A) - f_T(m_A)]/f_T(m_A) + 1.$$

Logarithmic differentiation or differentiation by taking the derivative of the logarithm of a function is used to transform products into sums and divisions into subtractions (Bali, 2005; Bird, 1993; Krantz, 2003).

Corollary 2.9.35. *For any number m_A from a direct perspective W-prearithmetic $\boldsymbol{A}$, we have $m_A \lll Sm_A$ if and only if $f_T(Sm_A) \cdot f_T(m_A) < f_T(SSm_A)$.*

Theorem 2.9.8. *In a functional direct perspective W-prearithmetic $\boldsymbol{A} = (A; \dot{+}, \circ, \leq)$ with the increasing functional parameter f, which maps whole numbers into whole numbers, relation $\lll$ is compatible from the left with the order $\leq$.*

Proof: Let us assume that k_A, n_A and m_A are numbers from $\boldsymbol{A}$ such that $k_A \leq n_A$ and $n_A \lll m_A$. By Lemma 2.9.18, it means that $f_T(n_A) \cdot$

$f_{\mathrm{T}}(m_A) < f_{\mathrm{T}}(\mathrm{S}m_A)$. As $k_A \leq n_A$, by Lemma 2.9.9, $f_{\mathrm{T}}(k_A) \leq f_{\mathrm{T}}(n_A)$. Thus, $f_{\mathrm{T}}(k_A) \cdot f_{\mathrm{T}}(m_A) < f_{\mathrm{T}}(\mathrm{S}m_A)$. By Lemma 2.9.18, it means $k_A \lll m_A$.

Theorem is proved. □

Corollary 2.9.36. *In a functional direct perspective W-prearithmetic* $\boldsymbol{A} = (A; \dot{+}, \circ, \leq)$ *with the increasing functional parameter* f*, which maps whole numbers into whole numbers, the order* $\leq$ *is compatible from the right with the relation.*

Remark 2.9.6. At the same time, relation $\lll$ is not always compatible from the right with the order $\leq$.

Example 2.9.9. Let us take the following function:

$$f_{\mathrm{T}}(n) = \begin{cases} 2^{2^n} & \text{if } n = 1, 2, 3, 4, \\ 2^{n+12} & \text{if } n > 4. \end{cases}$$

as the functional parameter for a projective arithmetic $\boldsymbol{A}$. Then we have

$$\begin{aligned} 2_A \circ 3_A &= f^{\mathrm{T}}(f_{\mathrm{T}}(2_A) \cdot f_{\mathrm{T}}(3_A)) = f^{\mathrm{T}}(2^{2^2} \cdot 2^{2^3}) = f^{\mathrm{T}}(2^4 \cdot 2^8) \\ &= f^{\mathrm{T}}(2^{4+8}) = f^{\mathrm{T}}(2^{12}) < f^{\mathrm{T}}(2^{16}) = 4_\mu. \end{aligned}$$

Consequently, $2_A \circ 3_A = 3_A$ and $2_A \lll 3_A$.

At the same time, we have

$$\begin{aligned} 2_A \circ 5_A &= f^{\mathrm{T}}(f_{\mathrm{T}}(2_A) \cdot f_{\mathrm{T}}(3_A)) = f^{\mathrm{T}}(2^{2^2} \cdot 2^{17}) \\ &= f^{\mathrm{T}}(2^4 \cdot 2^{17}) = f^{\mathrm{T}}(2^{21}) = 9_A. \end{aligned}$$

Consequently, it is not true that $2_A \lll 5_A$ although $2_A < 3_A < 5_A$. It means that the relation $\lll$ is not compatible from the left with relations $<$ and $\leq$.

Proposition 2.9.29. *In a functional direct perspective W-prearithmetic* $\boldsymbol{A} = (A; \dot{+}, \circ, \leq)$ *with the increasing functional parameter* f*, the relation* $\lll$ *is transitive.*

Indeed, if $k_A \lll n_A$ and $n_A \lll m_A$, then by Proposition 2.9.21, $k_A \leq n_A$ and by Theorem 2.9.8, we obtain $k_A \lll m_A$.

We know that in the Diophantine arithmetics $\boldsymbol{N}$ and $\boldsymbol{W}$, the relation $\lll$ is true only for 1 with respect to any other number. However, in the general case, it is possible that $\lll$ is a total (linear) order in a functional direct perspective prearithmetic. Namely, we have the following result.

Proposition 2.9.30. *There is a functional direct perspective W-prearithmetic* $\boldsymbol{A} = (W; \dot{+}, \circ, \leq)$, *in which* $\lll$ *is a total (linear) order and in particular,* $n_A \lll \mathrm{S}n_A$ *for all* $n = 1, 2, 3, \ldots$.

Proof: Let us take the function f, such that $f(0) = 0$ and $f(n) = 2^{2^{2^n}}$ for $n = 1, 2, 3, \ldots$, as the functional parameter for a functional direct perspective W-prearithmetic $\boldsymbol{A}$. By Lemma 2.9.9, $f_{\mathrm{T}}(n) = f(n) = 2^{2^{2^n}}$.

Thus, for any n, we have

$$f_{\mathrm{T}}(\mathrm{S}n_A) \cdot f_{\mathrm{T}}(n_A) = 2^{2^{2^{n+1}}} \cdot 2^{2^{2^n}} = 2^{(2^{2^{n+1}} + 2^{2^n})} < 2^{(2^{2^{n+1}} \cdot 2^{n+1})}$$
$$= 2^{2^{(2^{n+1} + 2^{n+1})}} = 2^{2^{(2 \cdot 2^{n+1})}} = 2^{2^{2^{n+2}}} = f_{\mathrm{T}}(\mathrm{SS}n_A).$$

In particular for 0, we have

$$f_{\mathrm{T}}(1_A) \cdot f_{\mathrm{T}}(0_A) = 2^{2^{2^1}} \cdot 2^{2^1} = 16 \cdot 4 = 64 < 2^{16} = 2^{2^{2^2}} = f_{\mathrm{T}}(1_A).$$

This implies the inequalities $f_{\mathrm{T}}(\mathrm{S}n) \cdot f_{\mathrm{T}}(n) < f_{\mathrm{T}}(\mathrm{SS}n)$ and $f_{\mathrm{T}}(n) < f_{\mathrm{T}}(\mathrm{SS}n)/f_{\mathrm{T}}(\mathrm{S}n)$. Consequently, by Lemma 2.9.18, we have the following sequence of relations $0_A \lll 1_A \lll 2_A \lll 3_A \lll \cdots \lll n_A \lll \mathrm{S}n_A \lll \cdots$ in this W-prearithmetic $\boldsymbol{A}$ with the projector $f_{\mathrm{T}}(n)$. As by Proposition 2.9.29, the relation is $\lll$ transitive in $\boldsymbol{A}$, this implies that the relation $\lll$ is a total (linear) order in W.

Proposition is proved. □

Note that in this prearithmetic, the number 0_A is not a multiplicative zero.

Let us study relations between relations $\lll$ and $\ll$.

Proposition 2.9.31. *In a functional direct perspective W-prearithmetic* $\boldsymbol{A} = (A; \dot{+}, \circ, \leq)$ *with the strictly increasing on whole numbers functional parameter* f, *which maps whole numbers into whole numbers, relation* $\lll$ *implies relation* $\ll$ *for all elements larger than* 1_A.

Proof: Let us take elements n_A and m_A from a functional direct perspective W-prearithmetic $\boldsymbol{A} = (A; \dot{+}, \circ, \leq)$ with the strictly increasing on whole numbers functional parameter f, which maps whole numbers into whole numbers, and assume $n_A, m_A > 1_A$ and $n_A \lll m_A$. By Lemma 2.9.18, the relation $n_A \lll m_A$ implies $f_{\mathrm{T}}(n_A) \cdot f_{\mathrm{T}}(m_A) < f_{\mathrm{T}}(\mathrm{S}m_A)$. By Proposition 2.9.28, the relation $n_A \lll m_A$ implies the inequality $n_A \leq m_A$ because $0_A < n_A$.

As $1 < n$ and f is strictly increasing on whole numbers, we have $2 \leq f_{\mathrm{T}}(n_A) \leq f_{\mathrm{T}}(m_A)$. Then by properties of the Diophantine arithmetic $\boldsymbol{W}$, the product of numbers $f_{\mathrm{T}}(n_A)$ and $f_{\mathrm{T}}(m_A)$ is larger than their sum. Consequently, we have

$$f_{\mathrm{T}}(n_A) + f_{\mathrm{T}}(m_A) < f_{\mathrm{T}}(n_A) \cdot f_{\mathrm{T}}(m_A) < f_{\mathrm{T}}(\mathrm{S}m_A).$$

By Lemma 2.9.17, the inequality $f_{\mathrm{T}}(n_A) + f_{\mathrm{T}}(m_A) < f_{\mathrm{T}}(\mathrm{S}m_A)$ implies $n_A \ll m_A$.

Proposition is proved. □

It means that relation $\lll$ is stronger than relation $\ll$.

Theorem 2.9.9. *In a functional direct perspective W-prearithmetic* $\boldsymbol{A} = (A; \dot{+}, \circ, \leq)$ *with the increasing functional parameter* f*, which maps whole numbers into whole numbers, the relation* $\lll$ *is compatible from the left with the relation* $\ll$.

Indeed, let us assume that k_A, n_A and m_A are numbers from $\boldsymbol{A}$ such that $k_A \ll n_A$ and $n_A \lll m_A$. Then by Proposition 2.9.10, $k_A \leq n_A$ and by Theorem 2.9.8, we obtain $k_A \lll m_A$.

Corollary 2.9.37. *In a functional direct perspective W-prearithmetic* $\boldsymbol{A} = (A; \dot{+}, \circ, \leq)$ *with the increasing functional parameter* f*, which maps whole numbers into whole numbers, the relation* $\ll$ *is compatible from the right with the relation* $\lll$.

Theorem 2.9.10. *Relation* $\lll$ *in a functional direct perspective W-prearithmetic* $\boldsymbol{A} = (A; \dot{+}, \circ, \leq_A)$ *is transitive and disjunctively asymmetric, i.e., only one relation* $x \lll y$ *or* $y \lll x$ *is valid for different elements* x *and* y *from* M.

Indeed, by Proposition 2.9.29, $\lll$ is transitive and if $n_A \lll m_A$ and $m_A \lll n_A$, then

$$m_A = m_A \circ n_A = n_A.$$

In the Diophantine arithmetic $\boldsymbol{W}$, there are as many even numbers as there are odd numbers. This is not true for functional direct perspective W-prearithmetics in a general case.

Proposition 2.9.32. *There is a functional direct perspective W-prearithmetic* $\boldsymbol{A} = (A; \dot{+}, \circ, \leq_A)$*, in which all numbers but one are even.*

Proof: Let $\boldsymbol{A}_{\text{exp}}$ be the functional direct perspective W-prearithmetic with the functional parameter $f(x) = 2^{2^x}$ when $x > 0$ and $f(0_A) = 0$. In it, for any $n > 2$, we have

$$f_{\text{T}}(n_A) \cdot f_{\text{T}}(2_A) = 2^{2^n} \cdot 2^{2^2} = 2^{2^n+4} < 2^{2^{n+1}} = f_{\text{T}}((n+1)_A).$$

Consequently, by Lemma 2.9.18, we have the following relations $2_A \lll n_A$ and $2_A \circ n_A = n_A$ in this W-prearithmetic $\boldsymbol{A}$ with the projector $f_{\text{T}}(n)$. In addition, we have

$$f_{\text{T}}(2_A) = 2^{2^2} = 2^4 < f_{\text{T}}(1_A) \cdot f_{\text{T}}(2_A) = 2^2 \cdot 2^{2^2} = 2^{2+4} = 2^6 < 2^8$$
$$= 2^{2^3} = f_{\text{T}}(3_A).$$

This implies $2_A \circ 1_A = 2_A$. Besides, $2_A \circ 0_A = 0_A$. Consequently, all numbers but 1_A are even in the W-prearithmetic $\boldsymbol{A}_{\text{exp}}$.

Proposition is proved. □

Proposition 2.9.33. *The W-prearithmetic $\boldsymbol{A}_{\text{exp}}$ has only one multiplicatively prime number 2_A.*

Proof: By Proposition 2.9.32, all numbers larger than 1_A are even in $\boldsymbol{A}_{\text{exp}}$. Thus, because all even numbers in $\boldsymbol{A}_{\text{exp}}$ but 2_A are composite, we have only to show that 2_A is a multiplicatively prime number. Definitions give us the following sequence of equalities:

$$f_{\text{T}}(2_A) \cdot f_{\text{T}}(2_A) = 2^{2^2} \cdot 2^{2^2} = 2^{4+4} = 2^8 < 2^{2^3} = f_{\text{T}}((3_A).$$

Consequently, $2_A \circ 2_A = 3_A$. As by Proposition 2.9.25, can be divisible only by 2_A and 1_A, the number 2_A is multiplicatively prime.

Proposition is proved. □

Corollary 2.9.38. *There are functional direct perspective W-prearithmetics that have only one multiplicatively prime number.*

Corollary 2.9.39. *In the W-prearithmetic $\boldsymbol{A}_{\text{exp}}$, all numbers but two are multiplicatively composite.*

Corollary 2.9.40. *There are functional direct perspective W-prearithmetics in which all numbers but two are multiplicatively composite.*

It is possible to generalize Proposition 2.9.32 for any natural number larger than 1.

Proposition 2.9.34. *For any natural number $n > 1$, there is a functional direct perspective W-prearithmetic $\boldsymbol{A} = (A; \dot{+}, \circ, \leq_A)$, in which all numbers larger than n are divisible by n.*

The proof is similar to the proof of Proposition 2.9.32.

We see that many properties of multiplication in functional direct perspective W-prearithmetics are similar to properties of addition in these W-prearithmetics.

2.9.5. *Additional operations*

In order to penetrate into the inner and further recesses of nature, it is necessary that both notions and axioms be derived from things by a more sure and guarded way; and that a method of intellectual operation be introduced altogether better and more certain.

Francis Bacon

It is possible to define other operations in direct perspective W-prearithmetics. For instance, as it is done in Section 2.7, taking a functional direct perspective W-prearithmetic $\boldsymbol{A} = (A; \dot{+}, \circ, \leq)$ in addition to addition and multiplication, we can define other operations in $\boldsymbol{A}$. In particular, we have a system of pairs of n-ary operations (n-ary addition and n-ary multiplication), which are defined for each natural number $n = 2, 3, 4, 5, \ldots$:

(1) $\sum^{\wedge^n}(a_{A,1}, a_{A,2}, \ldots, a_{A,n-1}, a_{A,n}) = f^{\mathrm{T}}(f_{\mathrm{T}}(a_{A,1}) + f_{\mathrm{T}}(a_{A,2}) + \cdots + f_{\mathrm{T}}(a_{A,n}))$;

(2) $\Pi^{\wedge^n}(a_{A,1}, a_{A,2}, \ldots, a_{A,n-1}, a_{A,n}) = f^{\mathrm{T}}(f_{\mathrm{T}}(a_{A,1}) \cdot f_{\mathrm{T}}(a_{A,2}) \cdot \ldots \cdot f_{\mathrm{T}}(a_{A,n}))$.

Note that for $n = 2$, these operations coincide with addition and multiplication, i.e., $\sum^{\wedge^2} = +$ and $\Pi^{\wedge^2} = \circ$.

Let us look at some examples.

Example 2.9.10. Let us take the functional direct projective W-prearithmetic $\boldsymbol{A}_{x^2} = (N; \dot{+}, \circ, \leq)$ with the functional parameter $f(x) = x^2$, and look how n-ary operations are performed comparing them with corresponding operations in the Diophantine arithmetic $\boldsymbol{N}$. Using ternary addition, we have

$$\Sigma^{\wedge^3}(2_A, 2_A, 2_A) = f^{\mathrm{T}}(4+4+4) = f^{\mathrm{T}}(12)$$
$$= \sup\{m; m \in W \text{ and } m \leq \sqrt{12}\}$$
$$= \sup\{m; m \in W \text{ and } m \leq 3.4\} = 3_A.$$

Iterated addition gives us a different result

$$(2_A \dot{+} 2_A) \dot{+} 2_A = f^{\mathrm{T}}(4+4) \dot{+} 2_A$$
$$= f^{\mathrm{T}}(8) \dot{+} 2_A = 2_A \dot{+} 2_A = f^{\mathrm{T}}(4+4) = f^{\mathrm{T}}(8) = 2_A$$

because

$$2_A \dot{+} 2_A = f^{\mathrm{T}}(8) = \sup\{m; m \in W \text{ and } m \leq \sqrt{8}\}$$
$$= \sup\{m; m \in W \text{ and } m \leq 2.9\} = 2_A.$$

At the same time, in the Diophantine arithmetic $\boldsymbol{N}$, we have a different result

$$2 + 2 + 2 = 6.$$

Ternary addition of other numbers gives us

$$\Sigma^{\wedge^3}(2_A, 2_A, 3_A) = f^{\mathrm{T}}(4+4+9) = f^{\mathrm{T}}(17) = 4_A.$$

Iterated addition gives us a different result

$$(2_A \dot{+} 2_A) \dot{+} 3_A = f^{\mathrm{T}}(4+4) \dot{+} 3_A$$
$$= f^{\mathrm{T}}(8) \dot{+} 3_A = 2_A \dot{+} 3_A = f^{\mathrm{T}}(4+9) = f^{\mathrm{T}}(13) = 3_A.$$

At the same time, in the Diophantine arithmetic $\boldsymbol{N}$, we have a different result

$$2 + 2 + 3 = 7.$$

Besides,

$$\Sigma^{\wedge^6}(1_A, 1_A, 1_A, 1_A, 1_A, 3_A) = f^{\mathrm{T}}(1+1+1+1+1+9) = f^{\mathrm{T}}(14) = 3_A$$

while in the Diophantine arithmetic $\boldsymbol{N}$

$$1 + 1 + 1 + 1 + 1 + 3 = 8.$$

We see that although $8 > 7$, we have $3_A < 4_A$ in the W-prearithmetic $\boldsymbol{A}_{x^2}$.

At the same time, multiplication in $\boldsymbol{A}$ remains the same as in $\boldsymbol{N}$:

$$\Pi^{\wedge^3}(2_A, 5_A, 8_A) = f^{\mathrm{T}}(4 \cdot 25 \cdot 64) = f^{\mathrm{T}}(6400) = 80_A$$

and

$$2 \cdot 5 \cdot 8 = 80.$$

This is a general case for the operation $\Pi^{\wedge^n}$ in the arithmetic $\boldsymbol{A}_{x^2}$ because

$$m_1^2 \cdot m_1^2 \cdot \ldots \cdot m_n^2 = (m_1 \cdot m_1 \cdot \ldots \cdot m_n)^2.$$

However, in other direct projective arithmetics, n-ary multiplication can be also different from multiplication in the Diophantine arithmetic $\boldsymbol{N}$.

At the same time, n-ary addition and multiplication can preserve some properties of addition and multiplication in the Diophantine arithmetic $\boldsymbol{N}$.

Proposition 2.9.35. *Operations $\Sigma^{\wedge^n}$ and $\Pi^{\wedge^n}$ are commutative for all $n = 2, 3, 4, 5, \ldots$ in a functional direct perspective W-prearithmetic $\boldsymbol{A}$.*

Proof is left as an exercise. □

At the same time, it is possible that n-ary operations are not associative and/or distributive.

It is possible to extend Proposition 2.9.9 to the operation $\Sigma^{\wedge^n}$.

Proposition 2.9.36. *If the functional parameter f is strictly increasing, for any numbers $a_{A,1}, a_{A,2}, \ldots, a_{A,n-1}, b_A$ from a functional direct perspective W-prearithmetic $\boldsymbol{A} = (W; \dot{+}, \circ, \leq)$ with the functional parameter f, and $k < n$, we have*

$$\Sigma^{\wedge^k}(a_{A,1}, a_{A,2}, \ldots, a_{A,k-1}, b_A) \leq \Sigma^{\wedge^n}(a_{A,1}, a_{A,2}, \ldots, a_{A,n-1}, b_A).$$

Proof: By properties of whole numbers, we have

$$f_{\mathrm{T}}(a_{A,1}) + f_{\mathrm{T}}(a_{A,2}) + \cdots + f_{\mathrm{T}}(a_{A,k-1}) + f_{\mathrm{T}}(b_A)$$
$$\leq f_{\mathrm{T}}(a_{A,1}) + f_{\mathrm{T}}(a_{A,2}) + \cdots + f_{\mathrm{T}}(a_{A,n-1}) + f_{\mathrm{T}}(b_A).$$

As the functional parameter f is strictly increasing, by Lemma 2.9.15, the function f^{T} is increasing. Consequently,

$$\begin{aligned}\Sigma^{\wedge^k}(a_{A,1}, a_{A,2}, \ldots, a_{A,k-1}, b_A) = f^{\mathrm{T}}(&f_{\mathrm{T}}(a_{A,1}) + f_{\mathrm{T}}(a_{A,2}) + \cdots \\ &+ f_{\mathrm{T}}(a_{A,k-1}) + f_{\mathrm{T}}(b_A)) \leq f^{\mathrm{T}}(f_{\mathrm{T}}(a_{A,1}) \\ &+ f_{\mathrm{T}}(a_{A,2}) + \cdots + f_{\mathrm{T}}(a_{A,n-1}) + f_{\mathrm{T}}(b_A)) \\ &= \Sigma^{\wedge^n}(a_{A,1}, a_{A,2}, \ldots, a_{A,n-1}, b_A).\end{aligned}$$

Proposition is proved. □

Corollary 2.9.41. *If the functional parameter f is strictly increasing, for any $1 \leq i < n$ and any non-zero numbers $a_{A,1}, a_{A,2}, \ldots, a_{A,n-1}, b_A$ from a functional direct perspective W-prearithmetic $\boldsymbol{A} = (W; \dot{+}, \circ, \leq)$ with the functional parameter f, we have*

$$a_{A,i} \dot{+} b_A \leq \Sigma^{\wedge^n}(a_{A,1}, a_{A,2}, \ldots, a_{A,n-1}, b_A).$$

It is also possible to extend Proposition 2.9.25 to the operation $\Pi^{\wedge^n}$.

Proposition 2.9.37. *If the functional parameter f is strictly increasing, for any non-zero numbers $a_{A,1}, a_{A,2}, \ldots, a_{A,n-1}, b_A$ from a functional direct perspective W-prearithmetic $\boldsymbol{A} = (W; \dot{+}, \circ, \leq)$ with the functional parameter f, and any k and n such that $1 < k < n$, we have*

$$\Pi^{\wedge^k}(a_{A,1}, a_{A,2}, \ldots, a_{A,k-1}, b_A) \leq \Pi^{\wedge^n}(a_{A,1}, a_{A,2}, \ldots, a_{A,n-1}, b_A).$$

Proof: By properties of whole numbers, we have

$$\begin{aligned}&f_{\mathrm{T}}(a_{A,1}) \cdot f_{\mathrm{T}}(a_{A,2}) \cdot \ldots \cdot f_{\mathrm{T}}(a_{A,k-1}) \cdot f_{\mathrm{T}}(b_A) \\ &\leq f_{\mathrm{T}}(a_{A,1}) \cdot f_{\mathrm{T}}(a_{A,2}) \cdot \ldots \cdot f_{\mathrm{T}}(a_{A,n-1}) \cdot f_{\mathrm{T}}(b_A).\end{aligned}$$

As the functional parameter f is strictly increasing, by Lemma 2.9.15, the function f^{T} is increasing. Consequently,

$$\begin{aligned}&\Pi^{\wedge^k}(a_{A,1}, a_{A,2}, \ldots, a_{A,k-1}, b_A) \\ &= f^{\mathrm{T}}(f_{\mathrm{T}}(a_{A,1}) \cdot f_{\mathrm{T}}(a_{A,2}) \cdot \ldots \cdot f_{\mathrm{T}}(a_{A,k-1}) \cdot f_{\mathrm{T}}(b_A)) \\ &\leq f^{\mathrm{T}}(f_{\mathrm{T}}(a_{A,1}) \cdot f_{\mathrm{T}}(a_{A,2}) \cdot \ldots \cdot f_{\mathrm{T}}(a_{A,n-1}) \cdot f_{\mathrm{T}}(b_A)) \\ &= \Pi^{\wedge^n}(a_{A,1}, a_{A,2}, \ldots, a_{A,n-1}, b_A).\end{aligned}$$

Proposition is proved. □

Corollary 2.9.42. *If the functional parameter f is strictly increasing, for any $1 \leq i < n$ and any non-zero numbers $a_{A,1}, a_{A,2}, \ldots, a_{A,n-1}, b_A$ from* *a functional direct perspective W-prearithmetic* $\boldsymbol{A} = (W; \dot{+}, \circ, \leq)$ *with the functional parameter* f, *we have*

$$a_{A,i} \circ b_A \leq \Pi^{\wedge^n}(a_{A,1}, a_{A,2}, \ldots, a_{A,n-1}, b_A).$$

It is possible to define relations $\ll$ and $\lll$ for n-ary operations.

(1) The expression $n_{A,1}, n_{A,2}, \ldots, n_{A,n-1} \ll_n m_A$ where $n_{A,1}, n_{A,2}, \ldots, n_{A,n-1}$ and m_A belong to $\boldsymbol{A}$ means that the group $n_{A,1}, n_{A,2}, \ldots, n_{A,n-1}$ is *much less* than m_A and in this case, m_A is *much larger* than the group $n_{A,1}, n_{A,2}, \ldots, n_{A,n-1}$, i.e., $m_{A\,n} \gg n_{A,1}, n_{A,2}, \ldots, n_{A,n-1}$. It is defined as $n_{A,1}, n_{A,2}, \ldots, n_{A,n-1} \ll_n m_A$ if and only if $\Sigma^{\wedge^n}(n_{A,1}, n_{A,2}, \ldots, n_{A,n-1}, m_A) = m_A$.

(2) The expression $n_{A,1}, n_{A,2}, \ldots, n_{A,n-1} \lll_n m_A$ where $n_{A,1}, n_{A,2}, \ldots, n_{A,n-1}$ and m_A are elements from $\boldsymbol{A}$ means that the group $n_{A,1}, n_{A,2}, \ldots, n_{A,n-1}$ is *much much less* than m_A and in this case, m_A is *much much larger* than the group $n_{A,1}, n_{A,2}, \ldots, n_{A,n-1}$, i.e., $m_A \ggg_n n_{A,1}, n_{A,2}, \ldots, n_{A,n-1}$. It is defined as

$$n_{A,1}, n_{A,2}, \ldots, n_{A,n-1} \lll_n m_A \text{ if and only if}$$
$$\Pi^{\wedge^n}(n_{A,1}, n_{A,2}, \ldots, n_{A,n-1}, m_A) = m_A.$$

Let us study properties of these relations.

Proposition 2.9.36 and Corollary 2.9.42 imply the following result.

Proposition 2.9.38. *If the functional parameter* f *is strictly increasing, for any* $1 \leq i < n$ *and any non-zero numbers* $a_{A,1}, a_{A,2}, \ldots, a_{A,n-1}, b_A$ *from a functional direct perspective W-prearithmetic* $\boldsymbol{A} = (W; \dot{+}, \circ, \leq)$ *with the functional parameter* f, *then the inequality* $a_{A,1}, a_{A,2}, \ldots, a_{A,n-1}, \ll_n b_A$ *implies the inequality* $a_{A,i} \ll b_A$.

Proof is left as an exercise. □

Proposition 2.9.37 and Corrollary 2.9.42 imply the following result.

Proposition 2.9.39. *If the functional parameter* f *is strictly increasing, for any* $1 \leq i < n$ *and any non-zero numbers* $a_{A,1}, a_{A,2}, \ldots, a_{A,n-1}, b_A$ *from a functional direct perspective W-prearithmetic* $\boldsymbol{A} = (W; \dot{+}, \circ, \leq)$ *with the functional parameter* f, *then the inequality* $a_{A,1}, a_{A,2}, \ldots, a_{A,n-1} \lll_n b_A$ *implies the inequality* $a_{A,i} \lll b_A$.

Proof is left as an exercise. □

However, the system of relations $a_{A,1} \lll b_A, a_{A,2} \lll b_A, \ldots, a_{A,n-1} \lll b_A$ does not always imply the inequality $a_{A,1}, a_{A,2}, \ldots, a_{A,n-1} \lll_n b_A$ as the following example demonstrates.

Example 2.9.11. Let us take a strictly increasing continuous function $f(x)$ that satisfies the following conditions as functional parameter for the functional direct perspective W-prearithmetic $\boldsymbol{A} = (A; \dot{+}, \circ, \leq)$, i.e., $\boldsymbol{A} = \boldsymbol{W}_f$.

$$f(0) = 0,$$

$$f(1) = 10,$$

$$f(2) = 16,$$

$$f(n) = 32 + n \text{ when } n > 2.$$

Then we have

$$\begin{aligned} f_{\mathrm{T}}(0) &= 0, \\ f_{\mathrm{T}}(1) &= 10, \\ f_{\mathrm{T}}(2) &= 16, \\ f_{\mathrm{T}}(3) &= 35, \\ f_{\mathrm{T}}(n) &= 32 + n \text{ when } n > 3. \end{aligned}$$

Then definitions imply the following equalities and inequalities:

$$\begin{aligned} 1_A \dot{+} 2_A &= f^{\mathrm{T}}(f_{\mathrm{T}}(1_A) + f_{\mathrm{T}}(2_A)) \\ &= f^{\mathrm{T}}(10 + 16) < f^{\mathrm{T}}(f_{\mathrm{T}}(3_A) = f^{\mathrm{T}}(32) = 3_A. \end{aligned}$$

Consequently, by Lemma 2.9.17,

$$1_A \dot{+} 2_A = 2_A \quad \text{and} \quad 1_A \ll 2_A.$$

At the same time, we have

$$\begin{aligned} \Sigma^{\wedge^3}(1_A, 1_A, 2_A) &= f^{\mathrm{T}}(f_{\mathrm{T}}(1_A) + f_{\mathrm{T}}(1_A) + f_{\mathrm{T}}(2_A)) \\ &= f^{\mathrm{T}}(10 + 10 + 16) = f^{\mathrm{T}}(36) = 3_A. \end{aligned}$$

Consequently, it is not true that $1_A \dot{+} 1_A \lll_3 2_A$.

Lemma 2.9.19. *For any numbers $n_{A,1}, n_{A,2}, \ldots, n_{A,n-1}, m_A$ from a functional direct perspective W-prearithmetic $\boldsymbol{A} = (A; \dot{+}, \circ, \leq)$ with an increasing functional parameter f, which maps whole numbers into whole* *numbers, the inequality* $n_{A,1}, n_{A,2}, \ldots, n_{A,n-1} \ll_n m_A$ *is true if and only if* $f_T(n_{A,1}) + f_T(n_{A,2}) + \cdots + f_T(n_{A,n-1}) < f_T(\mathrm{S}m_A) - f_T(m_A)$.

Proof is similar to the proof of Lemma 2.9.17. □

Theorem 2.9.11. *For any* $n \in N$ *and any* a_A *and* b_A *from a functional direct perspective W-prearithmetic* $\boldsymbol{A} = (A; \dot{+}, \circ, \leq)$ *with the increasing functional parameter* f*, which maps whole numbers into whole numbers, for any elements* a_A *and* b_A *from* A *such that* $a_A \leq b_A$*, from the inequality* $n_{A,1}, n_{A,2}, \ldots, n_{A,n-1} \ll_n a_A$ *follows the inequality* $n_{A,1}, n_{A,2}, \ldots, n_{A,n-1} \ll_n b_A$ *if* $f_T(\mathrm{S}a_A) - f_T(a_A) \leq f_T(\mathrm{S}b_A) - f_T(b_A)$.

Proof: Let us assume that $n_{A,1}$, $n_{A,2}, \ldots, n_{A,n-1} \ll_n a_A$. By Lemma 2.9.19,

$$f_T(n_{A,1}) + f_T(n_{A,2}) + \cdots + f_T(n_{A,n-1}) < f_T(\mathrm{S}a_A) - f_T(a_A).$$

As $f_T(\mathrm{S}a) - f_T(a) \leq f_T(\mathrm{S}b) - f_T(b)$, we have

$$f_T(n_{A,1}) + f_T(n_{A,2}) + \cdots + f_T(n_{A,n-1}) < f_T(\mathrm{S}a_A) - f_T(a_A).$$

Thus, by Lemma 2.9.19, $n_{A,1}, n_{A,2}, \ldots, n_{A,n-1} \ll_n b_A$.
Theorem is proved. □

Corollary 2.9.43. *For any* $n \in N$ *and any* a_A *and* b_A *from a functional direct perspective W-prearithmetic* $\boldsymbol{A} = (A; \dot{+}, \circ, \leq)$ *with the increasing functional parameter* f*, which maps whole numbers into whole numbers, for any elements* a_A *and* b_A *from* A *such that* $a_A \leq b_A$*, from the inequality* $\underbrace{1_A, 1_A, \ldots, 1_A}_{n-1} \ll_n a_A$ *follows the inequality* $\underbrace{1_A, 1_A, \ldots, 1_A}_{n-1} \ll_n b_A$ *if* $f_T(\mathrm{S}a_A) - f_T(a_A) \leq f_T(\mathrm{S}b_A) - f_T(b_A)$.

Lemma 2.9.20. *For any numbers* $n_{A,1}, n_{A,2}, \ldots, n_{A,n-1}, m_A$ *from a functional direct perspective W-prearithmetic* $\boldsymbol{A} = (A; \dot{+}, \circ, \leq)$ *with an increasing functional parameter* f*, which maps whole numbers into whole numbers, the inequality* $n_{A,1}, n_{A,2}, \ldots, n_{A,n-1} \lll_n m_A$ *is true if and only if* $f_T(n_{A,1}) \cdot f_T(n_{A,2}) \cdot \ldots \cdot f_T(n_{A,n-1}) < f_T(\mathrm{S}m_A)/f_T(m_A)$.

Proof is similar to the proof of Lemma 2.9.18. □

Theorem 2.9.12. *For any* $n \in N$ *and any* a_A and b_A *from a functional direct perspective W-prearithmetic* $\boldsymbol{A} = (A; \dot{+}, \circ, \leq)$ *with an increasing functional parameter* f*, which maps whole numbers into whole numbers, from* $n_{A,1}, n_{A,2}, \ldots, n_{A,n-1} \lll_n a_A$ *follows* $n_{A,1}, n_{A,2}, \ldots, n_{A,n-1} \lll_n b_A$

if for any elements a_A and b_A from M, the inequality $a_A \leq b_A$ implies $f_T(\mathrm{S}a_A)/f_T(a_A) \leq f_T(\mathrm{S}b_A)/f_T(b_A)$.

Proof: Let us assume that $n_{A,1}, n_{A,2}, \ldots, n_{A,n-1} \lll_n a_A$. By Lemma 2.9.17,

$$f_T(n_{A,1}) \cdot f_T(n_{A,2}) \cdot \ldots \cdot f_T(n_{A,n-1}) < f_T(\mathrm{S}a_A)/f_T(a_A).$$

As $f_T(\mathrm{S}a)/f_T(a) \leq f_T(\mathrm{S}b)/f_T(b)$, we have

$$f_T(n_{A,1}) \cdot f_T(n_{A,2}) \cdot \ldots \cdot f_T(n_{A,n-1}) < f_T(Sb_A)/f_T(b_A).$$

Thus, by Lemma 2.9.20, $n_{A,1}, n_{A,2}, \ldots, n_{A,n-1} \lll_n b_A$.

Theorem is proved. □

Corollary 2.9.44. *For any $n \in N$ and any a_A and b_A from a functional direct perspective W-prearithmetic $\boldsymbol{A} = (A; \dot{+}, \circ, \leq)$ with an increasing functional parameter f, which maps whole numbers into whole numbers, for any elements a_A, d_A and b_A from A, from the inequality $\underbrace{d_A, d_A, \ldots, d_A}_{n-1} \lll_n a_A$ follows $\underbrace{d_A, d_A, \ldots, d_A}_{n-1} \lll_n b_A$ if inequality $a_A \leq b_A$ implies inequality $f_T(\mathrm{S}a_A)/f_T(a_A) \leq f_T(\mathrm{S}b_A)/f_T(b_A)$.*

2.9.6. *Subprearithmetics*

When the world is at its best, when we are at our best, when life feels fullest,
one and one equals three.

Bruce Springsteen

Definition 2.9.6. A subprearithmetic $\boldsymbol{B} = (K; \dot{+}, \circ, \leq)$ of a direct perspective W-prearithmetic $\boldsymbol{A} = (W; \dot{+}, \circ, \leq)$ is:

(a) *connected* if $K = \{m, m+1, m+2, \ldots, n\}$ with $m \geq 0$ or $K = \{m, m+1, m+2, \ldots\}$,
(b) *completely connected* if $K = \{0, 1, 2, \ldots, n\}$ or $K = W$.

This definition directly implies the following result.

Lemma 2.9.21. *Any completely connected subprearithmetic is connected.*

Proof is left as an exercise. □

Example 2.9.12. The subprearithmetic of the Diophantine arithmetic $\boldsymbol{W}$ containing only 0 is completely connected.

Example 2.9.13. All numbers from the subprearithmetic of the Diophantine arithmetic $\boldsymbol{W}$ that are larger than some number n form connected but not completely connected subprearithmetic of $\boldsymbol{W}$.

Example 2.9.14. Let us take the interval $[0, m)$ with $m > 1$ and define the function f: $[0, m) \to R^+$ by the following formula:

$$f(x) = \begin{cases} x & \text{for } x \in [0, m-1], \\ x - m + m^2 & \text{for } x \geq m+1, \\ (m^2+1)(x-m) + m(m+1-x) & \text{for } x \in [m, m+1]. \end{cases}$$

Taking functions $f_{\mathrm{T}}(x)$ and $f^{\mathrm{T}}(x)$, we can define the direct perspective W-prearithmetic $\boldsymbol{B}_m = (W; \dot{+}, \circ, \leq)$. According to Definition 2.9.1, operations in $\boldsymbol{B}_m$ are defined in the following way:

$$k \dot{+} h = \begin{cases} k + h & \text{if } k + h \in [0, m], \\ m & \text{if } k + h \in [m, m+1), \\ k + h - m + m^2 & \text{if } k + h \geq m + 1, \end{cases}$$

$$k \circ h = \begin{cases} k \circ h & \text{if } k \cdot h \in [0, m], \\ m & \text{if } k \cdot h \in [m, m+1), \\ k \cdot h - m + m^2 & \text{if } k \cdot h \geq m + 1. \end{cases}$$

The subset $W_m = \{0, 1, 2, \ldots, m\}$ of the set W forms a completely connected subprearithmetic $\boldsymbol{A}_m$ of the functional direct perspective W-prearithmetic $\boldsymbol{B}_m$.

Example 2.9.15. Let us take the interval [0, m) with $m > 10$ and define the function f: $[0, m) \to R^+$ by the following formula:

$$f(x) = \begin{cases} x & \text{for } x \in [0, m-1], \\ m - 1 + \mathrm{tg}(x - m + 1)\pi & \text{for } x > m - 1 \text{ and } x \in [0, m). \end{cases}$$

Then $f_{\mathrm{T}}(x) = \lceil f(x) \rceil$ for all $x \in [0, m)$ while $f^{\mathrm{T}}(x) = \lfloor x \rfloor$ for $x \in [0, m-1]$ and $f^{\mathrm{T}}(x) = m - 1$ for $x \in [0, m)$ and $x > m - 1$.

Taking the set W_m for $m = 12$, i.e., the set $W_{12} = \{0, 1, 2, \ldots, m - 1 = 12\}$, we can define the functional direct perspective W-prearithmetic $\boldsymbol{A}_{12} = (W_{12}; \dot{+}, \circ, \leq)$. The subset $K = \{0, 2, 4, 6, 8, 10, 12\}$ of the set W_{12} forms a subprearithmetic of $\boldsymbol{A}_m$ but it is not connected.

Proposition 2.9.40. *The Diophantine arithmetic* $\boldsymbol{W}$ *has only two completely connected subprearithmetics.*

Indeed, the subprearithmetic $\boldsymbol{C}$ of $\boldsymbol{W}$, which has only one element 0, is connected. When the subprearithmetic $\boldsymbol{C}$ of $\boldsymbol{W}$ contains more than one element and is connected, it contains the number 1 and in this case coincides with $\boldsymbol{W}$ because all elements in $\boldsymbol{W}$ are sums of number 1.

The same is true for any modular arithmetic $\boldsymbol{Z}_m$.

Proposition 2.9.41. *Any modular arithmetic* $\boldsymbol{Z}_m$ *has only two completely connected subprearithmetics.*

Proof is similar to the proof of Proposition 2.9.40. □

However, in a general case, the situation is very different because we have the following result.

Proposition 2.9.42. *There is a functional direct perspective W-prearithmetic* $\boldsymbol{A} = (A; \dot{+}, \circ, \leq)$ *with a strictly increasing functional parameter* f, *in which for any set* B *of whole numbers, there is a subprearithmetic* $\boldsymbol{B} = (B; \dot{+}, \circ, \leq)$.

Proof: Let us take the functional direct perspective W-prearithmetic $\boldsymbol{A} = (W; , \dot{+}, \circ, \leq)$ with the functional parameter $f(x) = 2^{2^{2^x}}$. Then if $n_A \leq m_A$, we have

$$\begin{aligned} n_A \circ m_A &= f^{\mathrm{T}}(f_{\mathrm{T}}(n_A) \cdot f_{\mathrm{T}}(m_A)) = f^{\mathrm{T}}(2^{2^{2^n}} \cdot 2^{2^{2^m}}) \\ &= f^{\mathrm{T}}(2^{2^{2^n}+2^{2^m}}) \leq f^{\mathrm{T}}(2^{2 \cdot 2^{2^m}}) < f^{\mathrm{T}}(2^{2^{2^{m+1}}}) = \mathrm{S}m_A. \end{aligned}$$

Consequently, $n_A \circ m_A = m_A$ and $n_A \lll m_A$ for any number $n_A \leq m_A$.

In a similar way, we have

$$\begin{aligned} n_A \dot{+} m_A &= f^{\mathrm{T}}(f_{\mathrm{T}}(n_A) + f_{\mathrm{T}}(m_A)) = f^{\mathrm{T}}(2^{2^{2^n}} + 2^{2^{2^m}}) < f^{\mathrm{T}}(2^{2^{2^n}} + 2^{2^{2^m}}) \\ &\leq f^{\mathrm{T}}(2^{2 \cdot 2^{2^m}}) < f^{\mathrm{T}}(2^{2^{2^{m+1}}}) = \mathrm{S}m_A. \end{aligned}$$

Consequently, $n_A \dot{+} m_A = m_A$ and $n_A \ll m_A$.

It means that any set B of whole numbers is closed with respect to operations $\dot{+}$ and $\circ$, i.e., with these operations and the natural order, B is the carrier of a subprearithmetic of the prearithmetic $\boldsymbol{A}$.

Proposition is proved. □

Corollary 2.9.45. *There is a functional direct perspective W-prearithmetic* $\boldsymbol{A} = (A; \dot{+}, \circ, \leq)$ *with the strictly increasing functional parameter*

f, in which for any set $B = \{0, 1, 2, \ldots, n\}$ of whole numbers, there is a connected subprearithmetic $\boldsymbol{B} = (B; \dot{+}, \circ, \leq)$.

The property of abstract prearithmetics "to be a W-prearithmetic" is hereditary as the following result demonstrates.

Proposition 2.9.43. *A subprearithmetic* $\boldsymbol{K}$ *of a (functional) direct perspective W-prearithmetic* $\boldsymbol{A} = (W_{q,r}; \dot{+}, \circ, \leq)$ *is a (functional) direct perspective W-prearithmetic if and only if* $\boldsymbol{K}$ *is connected.*

Proof employs the construction from Example 2.9.8 and is left as an exercise. □

Proposition 2.9.44. *A subprearithmetic* $\boldsymbol{K}$ *of a (functional) direct perspective W-prearithmetic* $\boldsymbol{A} = (W_{q,r}; \dot{+}, \circ, \leq)$ *is a (functional) direct perspective W-prearithmetic with zero if and only if* $\boldsymbol{K}$ *is completely connected.*

Proof employs the construction from Example 2.9.8 and is left as an exercise. □

2.9.7. *Direct perspective prearithmetics of natural numbers*

Nothing is thoroughly approved but mediocrity.
The majority has established this, and
it fixes its fangs on whatever gets beyond it either way.
Blaise Pascal

Now we can introduce and study non-Diophantine prearithmetics and arithmetics of natural numbers.

Let us consider an interval $[q, r)$ of real numbers with $q \in N, r \in [1, \infty)$ and a relation $Q \subseteq [q, r) \times R^{++}$ that has the domain $\mathrm{Dom}(Q) = [q, r)$, the same definability domain $\mathrm{DDom}(Q) = [q, r)$, the range $\mathrm{Rg}(Q) = [1, \infty)$ and the same codomain $\mathrm{Codom}(Q) = [1, \infty)$. When $q = 0$ and $r = \infty$, we have $Q \subseteq [1, \infty) \times [1, \infty)$. By Q^{-1} we denote the relation, which is inverse to Q, i.e., $Q^{-1} = \{(x, y); (y, x) \in Q\}$. If $x \in [1, \infty)$, then $Q^{-1}(x) = \{y; (y, x) \in Q\}$ is the coimage of the element x and $Q(x) = \{z; (x, z) \in Q\}$ is the image of the element x. Then we can define two functions

$$Q_{\circ \mathrm{T}}(x) = \inf\{n; n \in N \text{ and } \forall z \in Q(x)(n \geq z)\},$$

$$Q_{\circ}^{\mathrm{T}}(x) = \sup\{m; m \in N \text{ and } \forall z \in Q^{-1}(x)(m \leq z)\}.$$

We see that $Q_{\circ\mathrm{T}}$ maps N_r where $N_{q,r} = [q, r) \cap N$ into the infinite interval $[q, \infty)$ and $Q_{\circ}^{\mathrm{T}}$ maps $[1, \infty)$ into N_r. This allows us taking restrictions of both functions on the set $N_{q,r}$. It gives us two functions $Q_{\mathrm{T}} : N_{q,r} \to [q, \infty)$ and $Q^{\mathrm{T}} : [q, \infty) \to N_{q,r}$.

Definition 2.9.7.

(a) A prearithmetic $\boldsymbol{A} = (N_{q,r}; \dot{+}, \circ, \leq)$ is called a *direct perspective prearithmetic* if it is weakly projective with respect to the conventional Diophantine arithmetic $\boldsymbol{N} = (N; +, \cdot, \leq)$ with the *projector* $Q_{\mathrm{T}}(x)$ and the *coprojector* $Q^{\mathrm{T}}(x)$.

(b) The relation Q is called the *generator* of the *projector* $Q_{\mathrm{T}}(x)$ and of the *coprojector* $Q^{\mathrm{T}}(x)$ as well as the *relational parameter* of the prearithmetic $\boldsymbol{A}$.

The direct perspective prearithmetic with the generator Q is denoted by $\boldsymbol{N}_{wQ}$.

Elements of N_r are called numbers of the prearithmetic $\boldsymbol{A}$ and are ordered by the same order relation $\leq$ by which they are ordered in the Diophantine arithmetic $\boldsymbol{N}$. This implies the following result.

Lemma 2.9.22. *The unary operation* S *of taking the next number is defined in all direct perspective prearithmetics with more than one element coinciding with the operation* S *in the Diophantine arithmetic* $\boldsymbol{N}$.

Proof is left as an exercise. □

Direct perspective prearithmetics also inherit some Archimedean properties (cf. Section 2.1) from the Diophantine arithmetic $\boldsymbol{N}$.

Lemma 2.9.23. *Any direct perspective prearithmetic* $\boldsymbol{A} = (A; \dot{+}, \circ, \leq)$ *is successively Archimedean and exactly successively Archimedean.*

Proof is left as an exercise. □

As elements of a direct perspective prearithmetic $\boldsymbol{A} = (A; \dot{+}, \circ, \leq)$ form an interval of whole numbers, we have the following result.

Lemma 2.9.24. *If a direct perspective prearithmetic* $\boldsymbol{A} = (A; \dot{+}, \circ, \leq)$ *with more than one element has one* 1_A, *then* $S1_A = 2_A$.

Proof is left as an exercise. □

As $\boldsymbol{A}$ is weakly projective with respect to Diophantine arithmetic $\boldsymbol{N}$, for any two numbers n_A and m_A from $N_{q,r}$, operations in $\boldsymbol{A}$ are defined as follows:

$$n_A \dot{+} m_A = Q^{\mathrm{T}}(Q_{\mathrm{T}}(n_A) + Q_{\mathrm{T}}(m_A)),$$
$$n_A \circ m_A = Q^{\mathrm{T}}(Q_{\mathrm{T}}(n_A) \cdot Q_{\mathrm{T}}(m_A)).$$

There are natural relations between direct perspective W-prearithmetics and direct perspective prearithmetics.

Lemma 2.9.25. *If $Q \subseteq [q, r) \times [1, \infty)$ and $1 \leq q$, then any direct perspective W-prearithmetic with the relational parameter Q is a direct perspective prearithmetic.*

Proof is left as an exercise. □

Taking $Q \subseteq [q, r) \times [1, \infty)$ and $1 \leq q$, we obtain the following result.

Proposition 2.9.45. *Any direct perspective prearithmetic is a direct perspective W-prearithmetic.*

Proof is left as an exercise. □

Corollary 2.9.46. *Any direct perspective prearithmetic is a subprearithmetic of a direct perspective W-prearithmetic.*

These results allow deducing many properties of direct perspective prearithmetics from properties of direct perspective W-prearithmetics. For instance, Propositions 2.9.2 and 2.9.33 imply the following result.

Proposition 2.9.46. *In a direct perspective prearithmetic with the relational parameter Q, addition $\dot{+}$ is monotone if functions $Q_{\circ\mathrm{T}}$ and $Q_{\circ}^{\mathrm{T}}$ are increasing.*

Proof is left as an exercise. □

Proposition 2.9.44 and Lemma 2.1.6 imply the following result.

Corollary 2.9.47. *For any direct perspective prearithmetic $\boldsymbol{A} = (A; \dot{+}, \circ, \leq)$ with the relational parameter Q, addition preserves the order $\leq$ if functions $Q_{\circ\mathrm{T}}$ and $Q_{\circ}^{\mathrm{T}}$ are increasing.*

Multiplication $\circ$ in functional direct perspective prearithmetics has similar properties.

Proposition 2.9.47. *In a direct perspective prearithmetic* $\boldsymbol{A} = (A; \dot{+}, \circ, \leq)$ *with the relational parameter* Q*, multiplication* $\circ$ *is monotone if functions* $Q_{\circ_{\mathrm{T}}}$ *and* $Q_{\circ}^{\mathrm{T}}$ *are increasing*

Proof is left as an exercise. □

Proposition 2.9.47 and Lemma 2.1.6 imply the following result.

Corollary 2.9.48. *For any direct perspective prearithmetic* $\boldsymbol{A} = (A; \dot{+}, \circ, \leq)$ *with the relational parameter* Q*, multiplication* $\circ$ *preserves the order* $\leq$ *if functions* $Q_{\circ_{\mathrm{T}}}$ *and* $Q_{\circ}^{\mathrm{T}}$ *are increasing.*

An important property of Diophantine arithmetic $\boldsymbol{N}$ is that it satisfies three forms of the Archimedean axiom, i.e., it is Successively Archimedean (SAAR), Additively Archimedean (AAAR) and Binary Archimedean (BAAR). At the same time, many direct perspective arithmetics and prearithmetics do not have these properties.

Example 2.9.16. Let us take the following function:

$$f(x) = \begin{cases} x & \text{for } x \leq 10, \\ 2^x & \text{for } x > 10 \end{cases}$$

as a functional parameter for a direct perspective prearithmetic $\boldsymbol{A} = (N; \dot{+}, \circ, \leq)$. Then we have

$$f^{-1}(x) = \begin{cases} x & \text{for } x \leq 10, \\ \log_2 x & \text{for } x > 10 \end{cases}$$

and

$$\begin{aligned} 10_A \dot{+} 10_A &= f^{\mathrm{T}}(f_{\mathrm{T}}(10_A) + f_{\mathrm{T}}(10_A)) = f^{\mathrm{T}}(10 + 10) \\ &= f^{\mathrm{T}}(20) = [\log_2(20)] = 4_A. \end{aligned}$$

Thus, adding any number of elements less than 11_A, we will never get 11_A or a number larger than 11_A because if numbers n_A and m_A are less than or equal to 10_A, then their sum $n_A \dot{+} m_A$ is less than or equal to 4_A. It means that the direct perspective prearithmetic $\boldsymbol{A}$ is not additively Archimedean.

In addition, we have

$$10_A \circ 10_A = f^{\mathrm{T}}(f_{\mathrm{T}}(10_A) \cdot f_{\mathrm{T}}(10_A)) = f^{\mathrm{T}}(10 \cdot 10) = f^{\mathrm{T}}(100)$$
$$= [\log_2(100)] = 6_A.$$

Thus, multiplying any number of elements less than 11_A, we will never get 11_A or a number larger than 11_A because if numbers n_A and m_A are less than or equal to 10_A, then their product $n_A \circ m_A$ is less than or equal to 6_A. It means that the direct perspective prearithmetic $\boldsymbol{A}$ is not Multiplicatively Archimedean and Binary Archimedean.

In this context, functional direct perspective prearithmetics are the most important for Non-Diophantine arithmetics of natural numbers.

Definition 2.9.8.

(a) A direct perspective prearithmetic $\boldsymbol{A} = (N_{q,r}; \dot{+}, \circ, \leq)$ with the *projector* $Q_{\mathrm{T}}(x)$ and the *coprojector* $Q^{\mathrm{T}}(x)$ is called *functional* if Q is a function f.

(b) The function $f(x)$ is called the *generator* of the *projector* $f_{\mathrm{T}}(x)$ and of the *coprojector* $f^{\mathrm{T}}(x)$ as well as the *functional parameter* of the prearithmetic $\boldsymbol{A}$.

Operations in $\boldsymbol{A}$ are defined as follows:

$$n_A \dot{+} m_A = f^{\mathrm{T}}(f_{\mathrm{T}}(n_A) + f_{\mathrm{T}}(m_A)),$$
$$n_A \circ m_A = f^{\mathrm{T}}(f_{\mathrm{T}}(n_A) \cdot f_{\mathrm{T}}(m_A)).$$

The direct perspective prearithmetic with the generator f is denoted by $\boldsymbol{N}_f$.

Elements of $N_{q,r}$ are called numbers of the arithmetic $\boldsymbol{A}$ and are denoted using the subscript A, i.e., 2 in $\boldsymbol{A}$ is denoted by 2_A and 5 in $\boldsymbol{A}$ is denoted by 5_A. Numbers of $\boldsymbol{A}$ are ordered by the same order relation $\leq$ as they are ordered in $\boldsymbol{N}$.

Lemma 2.9.26. *If a function f: $[1, r) \to [1, \infty)$ is a strictly increasing surjection, then $f_{\mathrm{T}}(1_A) = 1$.*

Indeed, as the function f is a strictly increasing surjection, we have $f(x_A) = 1$ for some number x. Then for all numbers $z_A < x_A$, we have $f(z_A) < 1$. Because 1 is the least element in $N_{1,r}$ and N, we have $x_A = 1_A$, i.e., $f(1_A) = 1$. Consequently, by Lemma 2.9.5, $f_{\mathrm{T}}(1_A) = 1$.

Lemma 2.9.27. *If* $f\colon [q, r) \to [1, \infty)$ *and* $1 \le q$, *then any functional direct perspective W-prearithmetic with the functional parameter* f *is a functional direct perspective prearithmetic.*

Proof is left as an exercise. □

Taking $f\colon [q, r) \to [1, \infty)$ and $1 \le q$, we obtain the following result.

Proposition 2.9.48. *Any functional direct perspective prearithmetic is a functional direct perspective W-prearithmetic.*

Proof is left as an exercise. □

Corollary 2.9.49. *Any functional direct perspective prearithmetic is a subprearithmetic of a functional direct perspective W-prearithmetic.*

These results allow deducing many properties of functional direct perspective prearithmetics from properties of functional direct perspective W-prearithmetics. For instance, Propositions 2.9.4 and 2.9.37 imply the following result.

Proposition 2.9.49. *In a functional direct perspective prearithmetic* $\boldsymbol{A} = (A; \dot{+}, \circ, \le)$ *with an increasing functional parameter* f, *addition* $\dot{+}$ *is monotone.*

Proof is left as an exercise. □

Proposition 2.9.49 and Lemma 2.1.6 imply the following result.

Corollary 2.9.50. *For any functional direct perspective prearithmetic* $\boldsymbol{A} = (A; \dot{+}, \circ, \le)$, *addition* $\dot{+}$ *preserves the order* $\le$.

Multiplication $\circ$ in functional direct perspective prearithmetics has similar properties.

Proposition 2.9.50. *In a functional direct perspective prearithmetic* $\boldsymbol{A} = (A; \dot{+}, \circ, \le)$ *with an increasing functional parameter* f, *multiplication* $\circ$ *is monotone.*

Proof is left as an exercise. □

Proposition 2.9.50 and Lemma 2.1.6 imply the following result.

Corollary 2.9.51. *For any functional direct perspective prearithmetic* $\boldsymbol{A} = (A; \dot{+}, \circ, \le)$, *multiplication* $\circ$ *preserves the order* $\le$.

Remark 2.9.7. However, in a general case, even with a strictly increasing functional parameter f, addition $\dot{+}$ is not monotone with respect to the order $\ll$ in functional direct perspective prearithmetics as the following example demonstrates.

Example 2.9.17. Let us take a functional direct projective arithmetic $\boldsymbol{A}_{x^2} = (N; \dot{+}, \circ, \leq)$ with the functional parameter $f(x) = 5x$, which is strictly increasing. However, we have in $\boldsymbol{A}$: $0_A \ll 1_A$ because $1_A \dot{+} 0_A = 1_A$, but it is not true $0_A \dot{+} 1_A \ll 1_A \dot{+} 1_A$ because $0_A \dot{+} 1_A = 1_A$, $1_A \dot{+} 1_A = 2_A$ and $2_A \dot{+} 1_A \neq 2_A$.

Proposition 2.7.1 and properties of the Diophantine arithmetic $\boldsymbol{N}$ imply the following result.

Proposition 2.9.51. *For any functional direct perspective prearithmetic $\boldsymbol{A} = (A; \dot{+}, \circ, \leq_A)$, addition $\dot{+}$ and multiplication $\circ$ are commutative.*

Proof is left as an exercise. □

Proposition 2.9.15 and Theorem 2.9.2 imply the following result.

Theorem 2.9.13. *In a functional direct perspective prearithmetic $\boldsymbol{A} = (A; \dot{+}, \circ, \leq)$ with the increasing functional parameter f, which maps whole numbers into whole numbers, relation $\ll_{\mathrm{w}}$ is compatible from the left with the order $\leq$.*

Proof is left as an exercise. □

Proposition 2.9.15 and Theorem 2.9.8 imply the following result.

Theorem 2.9.14. *In a functional direct perspective W-prearithmetic $\boldsymbol{A} = (A; \dot{+}, \circ, \leq)$ with the increasing functional parameter f, which maps whole numbers into whole numbers, relation $\lll$ is compatible from the left with the order $\leq$.*

Proof is left as an exercise. □

2.9.8. *Non-Diophantine arithmetics of whole numbers*

Great spirits have always found violent opposition from mediocre minds.
The latter cannot understand it when a [person]
does not thoughtlessly submit to hereditary prejudices
but honestly and courageously uses their intelligence.

Albert Einstein

Adding relevant conditions, we come from functional direct perspective W-prearithmetic to the class of functional direct perspective arithmetics (non-Diophantine arithmetics) of whole numbers.

Definition 2.9.9. A functional direct perspective W-prearithmetic $\boldsymbol{A} = (W; \dot{+}, \circ, \leq)$ with the projector $f_{\mathrm{T}}(x)$ and the coprojector $f^{\mathrm{T}}(x)$ is called a *direct perspective* W-*arithmetic* or a *direct perspective arithmetic of whole numbers* if the following conditions are satisfied:

(1) $f_{\mathrm{T}}(0_A) = 0$;
(2) $f_{\mathrm{T}}(x)$ is a strictly increasing function;
(3) for any elements n and m from W, the inequality $n \leq m$ implies the inequality

$$f_{\mathrm{T}}(\mathrm{S}n_A) - f_{\mathrm{T}}(n_A) \leq f_{\mathrm{T}}(\mathrm{S}m_A) - -f_{\mathrm{T}}(m_A).$$

Here, Sa is the successor of the element a (cf. Definition 2.1.2).

The direct perspective W-arithmetic with the generator f is denoted by $\boldsymbol{W}_f$.

Direct perspective W-arithmetics form a class **DRNDAW** of non-Diophantine arithmetics of whole numbers.

The third condition in Definition 2.9.9 means that the direct discrete derivative (cf., for example, Boole, 1860; 1880; Spiegel, 1971) of the function $f_{\mathrm{T}}(x)$ is an increasing function.

Thus, the basic operations in $\boldsymbol{A}$ are defined for any two its numbers a_A and b_A as follows:

$$a_A \dot{+} b_A = f^{\mathrm{T}}(f_{\mathrm{T}}(a) + f_{\mathrm{T}}(b)),$$
$$a_A \circ b_A = f^{\mathrm{T}}(f_{\mathrm{T}}(a) \cdot f_{\mathrm{T}}(b)).$$

Besides, a natural order relation is defined on A:

$$a_A \leq b_A \quad \text{if and only if } a \leq b.$$

Example 2.9.18. Let us take a simple function, for example, x^2, as the functional parameter $f(x)$ for a direct projective arithmetic $\boldsymbol{A} = (N; \dot{+}, \circ, \leq)$, and look how operations are performed:

$$\begin{aligned} 2_A \dot{+} 2_A &= f^{\mathrm{T}}(f_{\mathrm{T}}(2_A) + f_{\mathrm{T}}(2_A)) = f^{\mathrm{T}}(4+4) = f^{\mathrm{T}}(8) \\ &= \sup\{m; m \in N \text{ and } m \leq \sqrt{8}\} = 2_A \end{aligned}$$

because $f^{\mathrm{T}}(3) = 9 > 8$ or $3 > \sqrt{8}$.

$$\begin{aligned} 2_A \dot{+} 3_A &= f^{\mathrm{T}}(f_{\mathrm{T}}(2_A) + f_{\mathrm{T}}(3_A)) = f^{\mathrm{T}}(4+9) = f^{\mathrm{T}}(13) \\ &= \sup\{m; m \in N \text{ and } m \leq \sqrt{13}\} = 3_A. \end{aligned}$$

because $f^{\mathrm{T}}(4) = 16 > 13$ or $4 > \sqrt{13}$.

$$\begin{aligned} 10_A \dotplus 11_A &= f^{\mathrm{T}}(f_{\mathrm{T}}(10_A) + f_{\mathrm{T}}(11_A)) = f^{\mathrm{T}}(100 + 121) = f^{\mathrm{T}}(221) \\ &= \sup\{m; m \in N \ \& \ m \leq \sqrt{221}\} = 14_A \end{aligned}$$

because $f^{\mathrm{T}}(15) = 225 > 221$ and $f^{\mathrm{T}}(14) = 196 < 221$.

In a similar way, we find that $2_A \dotplus 11_A = 11_A$, $3_A \dotplus 11_A = 11_A$, $4_A \dotplus 11_A = 11_A$, but $5_A \dotplus 11_A = 12_A$, $6_A \dotplus 11_A = 12_A$, $7_A \dotplus 11_A = 13_A$, $8_A \dotplus 11_A = 15_A$, and $11_A \dotplus 11_A = 15_A$.

Moreover,

$$\begin{aligned} 2_A \circ 2_A &= f^{\mathrm{T}}(f_{\mathrm{T}}(2_A) \cdot f_{\mathrm{T}}(2_A)) = f^{\mathrm{T}}(4 \cdot 4) = f^{\mathrm{T}}(16) = 4_A, \\ 2_A \circ 3_A &= f^{\mathrm{T}}(f_{\mathrm{T}}(2_A) \cdot f_{\mathrm{T}}(3_A)) = f^{\mathrm{T}}(4 \cdot 9) = f^{\mathrm{T}}(36) = 6_A. \end{aligned}$$

In general, we have $n^2 \cdot m^2 = (n \cdot m)^2$. Consequently, in this arithmetic $\boldsymbol{A}$, multiplication of numbers is the same as in the Diophantine arithmetic, i.e., $n_A \circ m_A = (n \cdot m)_A$, while addition is essentially different. As we have seen, two times two is still four, while two plus two is only two.

Example 2.9.19. Let us take a simple function, such as $10x$ as the functional parameter for a direct projective arithmetic $\boldsymbol{A} = (N; \dotplus, \circ, \leq)$.

$$\begin{aligned} 2_A \dotplus 2_A &= f^{\mathrm{T}}(f_{\mathrm{T}}(2_A) + f_{\mathrm{T}}(2_A)) = f^{\mathrm{T}}(20 + 20) = f^{\mathrm{T}}(40) = 4_A, \\ 2_A \dotplus 3_A &= f^{\mathrm{T}}(f_{\mathrm{T}}(2_A) + f_{\mathrm{T}}(3_A)) = f^{\mathrm{T}}(20 + 30) = f^{\mathrm{T}}(50) = 5_A. \end{aligned}$$

This is a general case for multiplication of numbers in $\boldsymbol{A}$ as

$$\begin{aligned} n_A \dotplus m_A &= f^{\mathrm{T}}(f_{\mathrm{T}}(n_A) + f_{\mathrm{T}}(m_A)) = f^{\mathrm{T}}(10n + 10m) \\ &= f^{\mathrm{T}}(10(n + m)) = (n + m)_A. \end{aligned}$$

At the same time, we have

$$\begin{aligned} 2_A \circ 2_A &= f^{\mathrm{T}}(f_{\mathrm{T}}(2_A) \cdot f_{\mathrm{T}}(2_A)) = f^{\mathrm{T}}(20 \cdot 20) = f^{\mathrm{T}}(400) = 40_A, \\ 2_A \circ 3_A &= f^{\mathrm{T}}(f_{\mathrm{T}}(2_A) \cdot f_{\mathrm{T}}(3_A)) = f^{\mathrm{T}}(20 \cdot 30) = f^{\mathrm{T}}(600) = 60_A. \end{aligned}$$

This is a general case for addition of numbers in $\boldsymbol{A}$ as

$$\begin{aligned} n_A \circ m_A &= f^{\mathrm{T}}(f_{\mathrm{T}}(n_A) \cdot f_{\mathrm{T}}(m_A)) = f^{\mathrm{T}}(10n \cdot 10m) \\ &= f^{\mathrm{T}}(10(n \cdot m)) = (n \cdot m)_A. \end{aligned}$$

In this arithmetic $\boldsymbol{A}$, addition of numbers is the same as in the Diophantine arithmetic, while multiplication is essentially different. As we have seen, two plus two is still four, while two times two is equal to forty.

Remark 2.9.8. It is possible to define the functional parameter (function f) only for natural numbers, but in many cases, its definition for real numbers makes the analytic expression for f simpler.

Lemma 2.9.28. *If the function f_{T} is strictly increasing, then the function f is also strictly increasing on whole numbers.*

Proof: Let us assume that the function f is not strictly increasing. Then there are two whole numbers n_A and m_A such that $n_A < m_A$ but $f(n_A) \geq f(m_A)$. This implies

$$\begin{aligned} f_{\mathrm{T}}(n_A) &= \inf\{n; n \in W \text{ and } n \geq f(n_A) \geq f(m_A)\} \geq f_{\mathrm{T}}(m_A) \\ &= \inf\{n; n \in W \text{ and } n \geq f(m_A)\}. \end{aligned}$$

It means that the function f_{T} is not strictly increasing. This contradicts the assumption of the lemma and by the Principle of Excluded Middle concludes the proof.

Lemma is proved. □

Corollary 2.9.52. *The functional parameter f of a direct perspective W-arithmetic is a strictly increasing on whole numbers function.*

Similar result is valid for the function f^{T}.

Corollary 2.9.53. *In a direct perspective W-arithmetic with the functional parameter f, the function f^{T} is increasing.*

Proposition 2.9.52. *Any direct perspective W-arithmetic is also a functional direct perspective W-prearithmetic.*

Proof is left as an exercise. □

This result allows obtaining many properties of direct perspective W-arithmetics from already proved properties of functional direct perspective W-prearithmetics.

Theorem 2.9.1 implies the following result.

Theorem 2.9.15. *If for any number n_A, its image $f(n_A)$ is also a whole number, then 0_A is an additive zero in a direct perspective W-arithmetic $\boldsymbol{A} = (W; \dot{+}, \circ, \leq)$, i.e., for any $n_A \in \boldsymbol{A}$ the equalities $n_A \dot{+} 0_A = n_A$ and $0_A \dot{+} n_A = n_A$ are true.*

Proof: By Lemma 2.9.9, $f_{\mathrm{T}}(n_A) = f(n)$ and by Definition 2.9.9, $f_{\mathrm{T}}(x)$ is a strictly increasing function. Then by Proposition 2.9.4, $f^{\mathrm{T}}(f_{\mathrm{T}}(n_A)) = n_A$ and we have the following sequence of equalities:

$$n_A \dot{+} 0_A = f^{\mathrm{T}}(f_{\mathrm{T}}(n_A) + f_{\mathrm{T}}(0_A)) = f^{\mathrm{T}}(f_{\mathrm{T}}(n_A) + 0) = f^{\mathrm{T}}(f_{\mathrm{T}}(n_A)) = n_A.$$

The equality $0_A \dot{+} n_A = n_A$ is derived in a similar way.

Theorem is proved. □

Theorem 2.9.2 implies the following result.

Theorem 2.9.16. *For any direct perspective W-arithmetic $\boldsymbol{A} = (W; \dot{+}, \circ, \leq)$ with the functional parameter f, which maps whole numbers into whole numbers, the relation $\ll$ is compatible with the order $\leq$.*

Proof: (a) At first, we treat compatibility from the left. Let us assume that n_A and m_A are numbers from $\boldsymbol{A}$ and $n_A \ll m_A$. It means that $m_A \dot{+} n_A = m_A$. If $k_A \leq n_A$, then by Lemma 2.9.17,

$$f_{\mathrm{T}}(n_A) + f_{\mathrm{T}}(m_A) < f_{\mathrm{T}}(\mathrm{S}m_A).$$

As $k_A \leq n_A$, Definition 2.9.9 implies $f_{\mathrm{T}}(k_A) \leq f_{\mathrm{T}}(n_A)$. Thus,

$$f_{\mathrm{T}}(k_A) + f_{\mathrm{T}}(m_A) < f_{\mathrm{T}}(\mathrm{S}m_A).$$

By Lemma 2.9.17, it means $k_A \ll m_A$. As n_A and m_A are arbitrary numbers from $\boldsymbol{A}$, relation $\ll$ is compatible from the left with the order $\leq$.

(b) Now we prove compatibility from the right. Let us assume that n_A and m_A are numbers from $\boldsymbol{A}$ and $n_A \ll m_A$. It means that $m_A \dot{+} n_A = m_A$ and by Lemma 2.9.17,

$$f_{\mathrm{T}}(n_A) < f_{\mathrm{T}}(\mathrm{S}m_A) - f_{\mathrm{T}}(m_A).$$

If $m_A \leq k_A$, then by definition 2.9.9,

$$f_{\mathrm{T}}(\mathrm{S}m_A) - f_{\mathrm{T}}(m_A) \leq f_{\mathrm{T}}(\mathrm{S}k_A) - f_{\mathrm{T}}(k_A).$$

Thus,

$$f_{\mathrm{T}}(n_A) < f_{\mathrm{T}}(Sk_A) - f_{\mathrm{T}}(k_A).$$

Then by Lemma 2.9.17, $n_A \ll k_A$.
Theorem is proved. □

Proposition 2.9.28 implies the following result.

Proposition 2.9.53. *For any number m_A from a direct perspective W-arithmetic $\boldsymbol{A}$, if $0_A < n_A$, then inequality $n_A \lll m_A$ implies inequality $n_A \leq m_A$.*

Proof is left as an exercise. □

Theorem 2.9.12 implies the following result.

Theorem 2.9.17. *For any $n \in N$ and any numbers a_A and b_A from a direct perspective W-arithmetic $\boldsymbol{A} = (W; \dot{+}, \circ, \leq)$, from $a_A < b_A$ and $n_{A,1}$, $n_{A,2}, \ldots, n_{A,n-1} \ll_n a_A$ follows $n_{A,1}, n_{A,2}, \ldots, n_{A,n-1} \ll {}_n b_A$.*

Proof is left as an exercise. □

Corollary 2.9.54. *The relation $\ll_n$ is compatible from the left with the order $\leq$.*

We have similar results for the relation $\lll$. In particular, Theorem 2.9.8 implies the following consequence.

Theorem 2.9.18. *In a direct perspective W-arithmetic $\boldsymbol{A} = (A; \dot{+}, \circ, \leq)$, the relation $\lll$ is compatible from the left with the order $\leq$.*

Proof is left as an exercise. □

Proposition 2.9.54. *In a direct perspective W-arithmetic $\boldsymbol{A} = (W; \dot{+}, \circ, \leq)$, the element 0_A is an multiplicative zero.*

Indeed, by Proposition 2.9.4 and definition 2.9.9, we have the following sequence of equalities:

$$n_A \circ 0_A = f^{\mathrm{T}}(f_{\mathrm{T}}(n_A) \cdot f_{\mathrm{T}}(0_A)) = f^{\mathrm{T}}(f_{\mathrm{T}}(n_A) \cdot 0) = f^{\mathrm{T}}(0) = 0_A.$$

The equality $0_A \circ n_A = 0_A$ is derived in a similar way.
Propositions 2.9.6–2.9.6 imply the following result.

Theorem 2.9.19. *In a direct perspective W-arithmetic $\boldsymbol{A} = (W; \dot{+}, \circ, \leq)$, addition $\dot{+}$ and multiplication $\circ$ are commutative and monotone.*

Proof is left as an exercise. □

Remark 2.9.9. In some direct perspective W-arithmetics, addition $\dot{+}$ or/and multiplication $\circ$ can be not associative.

Example 2.9.20. Let us take a projective W-arithmetic $\boldsymbol{A}_{x^2} = (N\dot{+}; \circ, \leq)$ with the functional parameter $f(x) = x^2$, and define $(1_A\dot{+}2_A)\dot{+}3_A$ and $1_A\dot{+}(2_A\dot{+}3_A)$. We have

$$\begin{aligned}(1_A\dot{+}2_A)\dot{+}3_A &= \sup\{r; r \in W \text{ and } r \\ &\leq (\sup\{m; m \in W \text{and}\, m \leq (1\dot{+}4)^{1/2}\} + 9)^{1/2}\} \\ &= \sup\{r; r \in W \text{ and } r \leq (2+9)^{1/2}\} = 3\end{aligned}$$

and

$$\begin{aligned}1_A\dot{+}(2_A\dot{+}3_A) = \sup\{r; r \in W \text{ and } r \leq (1 + \sup\{m; m \in W \text{ and } m \\ \leq (4\dot{+}9)^{1/2}\})^{1/2}\} = \sup\{r; r \in W \text{ and } r \\ \leq (1+3)^{1/2}\} = 2.\end{aligned}$$

Thus, $(1_A\dot{+}2_A)\dot{+}3_A \neq 1_A\dot{+}(2_A\dot{+}3_A)$, which implies that addition $\dot{+}$ is not associative.

Proposition 2.9.55. *In a direct perspective W-arithmetic* $\boldsymbol{A} = (W; \dot{+}, \circ, \leq)$ *with the functional parameter* f, *which maps whole numbers into whole numbers, relation* $\ll$ *is transitive.*

Proof: Let us assume that k_A, n_A and m_A are numbers from $\boldsymbol{A}$, $k_A \ll n_A$ and $n_A \ll m_A$. It means that $n_A \dot{+} m_A = m_A$, and if $k_A \dot{+} n_A = n_A$, then by Lemma 2.9.17, we have

$$f_{\mathrm{T}}(n_A) < f_{\mathrm{T}}(\mathrm{S}m_A) - f_{\mathrm{T}}(m_A)$$

and

$$f_{\mathrm{T}}(k_A) < f_{\mathrm{T}}(\mathrm{S}n_A) - f_{\mathrm{T}}(n_A).$$

By Proposition 2.9.17, the inequality $n_A \ll m_A$ implies inequality $n_A \leq m_A$. Thus, by Definition 2.9.9, we obtain

$$f_{\mathrm{T}}(k_A) < f_{\mathrm{T}}(\mathrm{S}n_A) - f_{\mathrm{T}}(n_A) < f_{\mathrm{T}}(\mathrm{S}m_A) - f_{\mathrm{T}}(m_A).$$

Consequently, by Lemma 2.9.17, we have $k_A \ll m_A$.

Proposition is proved. □

Proposition 2.9.56. *There is a direct perspective W-arithmetic (non-Diophantine arithmetic of whole numbers)* $\boldsymbol{A} = (W; \dot{+}, \circ, \leq)$, *in which* $\ll$ *is a total relation on* W.

Proof: Let us consider the function

$$f(n) = \begin{cases} 0 & \text{when } n = 0, \\ 2^{2^n} & \text{when } n > 0. \end{cases}$$

It satisfies conditions from Definition 2.9.9. Indeed, $f_{\mathrm{T}}(0_A) = f(0) = 0$. The function $f_{\mathrm{T}}(x)$ is a strictly increasing because by Lemma 2.9.9, $f_{\mathrm{T}}(n_A) = f(n) = 2^{2^n}$. In addition, for any elements n and m from W, if $n \leq m$, then

$$\begin{aligned} &f_{\mathrm{T}}(\mathrm{S}n_A) - f_{\mathrm{T}}(n_A) \\ &\quad = 2^{2^{n+1}} - 2^{2^n} = 2^{2 \cdot 2^n} - 2^{2^n} = (2^{2^n})^2 - 2^{2^n} = 2^{2^n} \cdot (2^{2^n} - 1) \\ &\quad \leq f_{\mathrm{T}}(\mathrm{S}m_A) - f_{\mathrm{T}}(m_A) = 2^{2^{m+1}} - 2^{2^m} = 2^{2 \cdot 2^m} - 2^{2^m} \\ &\quad = (2^{2^m})^2 - 2^{2^m} = 2^{2^m} \cdot (2^{2^m} - 1). \end{aligned}$$

Thus, we can take $f(n)$ as the functional parameter for a direct perspective W-arithmetic $\boldsymbol{A} = (W; \dot{+}, \circ, \leq)$. By Lemma 2.9.9, $f_{\mathrm{T}}(n_A) = f(n) = 2^{2^n}$.

Consequently, for any $n = 1, 2, 3, \ldots$, we have

$$\begin{aligned} f_{\mathrm{T}}((n+1)_A) + f_{\mathrm{T}}(n) &= 2^{2^{n+1}} + 2^{2^n} < 2^{2^{n+1}} + 2^{2^{n+1}} \\ &\leq 2^{2^{n+1}} \cdot 2^{2^{n+1}} = 2^{2^{n+1} + 2^{n+1}} = 2^{2 \cdot 2^{n+1}} = 2^{2^{n+2}} \\ &= f_{\mathrm{T}}((n+2)_A). \end{aligned}$$

Besides,

$$f_{\mathrm{T}}(1_A) + f_{\mathrm{T}}(0_A) = 2^{2^1} + 0 = 4 < 2^{2^2} = 16 = f_{\mathrm{T}}(2_A).$$

By Lemma 2.9.17, this gives the following sequence of relations $0_A \ll 1_A \ll 2_A \ll 3_A \ll \cdots \ll n_A \ll (n+1)_A \ll \cdots$ in the W-prearithmetic $\boldsymbol{A}$ with the projector $f_{\mathrm{T}}(n)$. As by Proposition 2.9.55, the relation $\ll$ is transitive in $\boldsymbol{A}$, this implies that the relation $\ll$ is a total relation in the direct perspective W-arithmetic (non-Diophantine arithmetic of whole numbers) $\boldsymbol{A}$. □

Proposition is proved.

Proposition 2.9.36 implies the following result.

Proposition 2.9.57. *In a direct perspective W-arithmetic* $\boldsymbol{A} = (W; \dot{+}, \circ, \leq)$ *with the functional parameter* f, *relation* $\lll$ *is transitive.*

Proof is left as an exercise. □

We know that in the Diophantine arithmetics $\boldsymbol{N}$ and $\boldsymbol{W}$, the relation $\lll$ is true only for 1 with respect to any other number. However, in the general case, it is possible that $\lll$ is a total relation in a functional direct perspective prearithmetic. Namely, we have the following result.

Proposition 2.9.58. *There is a direct perspective W-arithmetic $\boldsymbol{A} = (W; \dot{+}, \circ, \leq)$, in which $\lll$ is a total relation and in particular, $n_A \lll \mathrm{S}n_A$ for all $n = 1, 2, 3, \ldots$.*

Proof: Let us consider the function

$$f(n) = \begin{cases} 0 & \text{when } n = 0, \\ 2^{2^{2^n}} & \text{when } n > 0. \end{cases}$$

It satisfies conditions from Definition 2.9.9. Indeed, $f_{\mathrm{T}}(0_A) = f(0) = 0$. The function $f_{\mathrm{T}}(x)$ is a strictly increasing because by Lemma 2.9.9, $f_{\mathrm{T}}(n_A) = f(n) = 2^{2^{2^n}}$. In addition, for any elements n and m from W, if $n \leq m$, then

$$\begin{aligned} f_{\mathrm{T}}(\mathrm{S}n_A) - f_{\mathrm{T}}(n_A) &= 2^{2^{2^{n+1}}} - 2^{2^{2^n}} = 2^{2^{2 \cdot 2^n}} - 2^{2^{2^n}} = 2^{(2^{2^n})^2} - 2^{2^{2^n}} \\ &= 2^{(2^{2^n}) \cdot (2^{2^n})} - 2^{2^{2^n}} = 2^{(2^{2^n})^{2^{2^n}}} - 2^{2^{2^n}} \\ &= -2^{2^{2^n}} \cdot ((2^{2^{2^n}})^{2^{2^n}-1} - 1) \leq f_{\mathrm{T}}(\mathrm{S}m_A) - f_{\mathrm{T}}(m_A) \\ &= 2^{2^{2^{m+1}}} - 2^{2^{2^m}} = 2^{2^{2 \cdot 2^m}} - 2^{2^{2^m}} = 2^{(2^{2^m})^2} - 2^{2^{2^m}} \\ &= 2^{(2^{2^m}) \cdot (2^{2^m})} - 2^{2^{2^m}} = 2^{(2^{2^m})^{2^{2^m}}} - 2^{2^{2^m}} \\ &= 2^{2^{2^m}} \cdot ((2^{2^{2^m}})^{2^{2^m}-1} - 1). \end{aligned}$$

Thus, we can take $f(n)$ as the functional parameter for a direct perspective W-arithmetic $\boldsymbol{A} = (W; \dot{+}, \circ, \leq)$. By Lemma 2.9.9, $f_{\mathrm{T}}(n_A) = f(n) = 2^{2^n}$.

Consequently, for any $n = 1, 2, 3, \ldots$, we have

$$\begin{aligned} f_{\mathrm{T}}((n+1)_A) \cdot f_{\mathrm{T}}(n_A) &= 2^{2^{2^{n+1}}} - 2^{2^{2^n}} = 2^{(2^{2^{n+1}} + 2^{2^{2n}})} < 2^{(2^{2^{n+1}} \cdot 2^{2^{n+1}})} \\ &= 2^{2^{(2^{n+1} + 2^{n+1})}} = 2^{2^{(2 \cdot 2^{n+1})}} = 2^{2^{2n+2}} = f_{\mathrm{T}}((n+2)_A). \end{aligned}$$

Besides for 0, we have

$$f_{\mathrm{T}}(1_A) \cdot f_{\mathrm{T}}(0_A) = 2^{2^{2^1}} \cdot 2^{2^1} = 16\ 4 = 64 < 2^{16} = 2^{2^{2^2}} = f_{\mathrm{T}}(2_A).$$

By Lemma 2.9.18, this gives us the following sequence of relations $0_A \lll 1_A \lll 2_A \lll 3_A \lll \cdots \lll n_A \lll (n+1)_A \lll \cdots$ in the W-arithmetic $\boldsymbol{A}$ with the projector $f_{\mathrm{T}}(n_A)$. As by Proposition 2.9.57, the relation $\lll$ is transitive in $\boldsymbol{A}$, this implies that the relation $\lll$ is a total relation in the direct perspective W-arithmetic (non-Diophantine arithmetic of whole numbers) $\boldsymbol{A}$.

Proposition is proved. □

Theorem 2.9.20. *In a direct perspective W-arithmetic $\boldsymbol{A} = (W; \dot{+}, \circ, \leq)$ with the functional parameter f, which maps whole numbers into whole numbers, relation is compatible with the order $\leq$ if for any elements m and n from W, the inequality $m \leq n$ implies the inequality $f_{\mathrm{T}}(Sm_A)/f_{\mathrm{T}}(m_A) \leq f_{\mathrm{T}}(Sn_A)/f_{\mathrm{T}}(n_A)$.*

Proof: (a) At first, we treat compatibility from the left. Let us assume that n_A and m_A are numbers from $\boldsymbol{A}$ and $n_A \lll m_A$. It means that $m_A \circ n_A = m_A$. If $k_A \leq n_A$, then by Lemma 2.9.18,

$$f_{\mathrm{T}}(n_A) \cdot f_{\mathrm{T}}(m_A) < f_{\mathrm{T}}(\mathrm{S}m_A).$$

As $k_A \leq n_A$, Definition 2.9.9 implies $f_{\mathrm{T}}(k_A) \leq f_{\mathrm{T}}(n_A)$. Thus,

$$f_{\mathrm{T}}(k_A) \cdot f_{\mathrm{T}}(m_A) < f_{\mathrm{T}}(\mathrm{S}m_A).$$

By Lemma 2.9.18, it means $k_A \lll m_A$. As n_A and m_A are arbitrary numbers from $\boldsymbol{A}$, relation $\lll$ is compatible from the left with the order $\leq$.

(b) Now we prove compatibility from the right. Let us assume that n_A and m_A are numbers from $\boldsymbol{A}$ and $n_A \lll m_A$. It means that $m_A \circ n_A = m_A$ and by Lemma 2.9.15,

$$f_{\mathrm{T}}(n_A) < f_{\mathrm{T}}(\mathrm{S}m_A)/f_{\mathrm{T}}(m_A).$$

If $m_A \leq k_A$, then by the conditions of the theorem,

$$f_{\mathrm{T}}(Sm_A)/f_{\mathrm{T}}(m_A) \leq f_{\mathrm{T}}(Sk_A)/f_{\mathrm{T}}(k_A).$$

Thus,

$$f_{\mathrm{T}}(n_A) < f_{\mathrm{T}}(\mathrm{S}k_A)/f_{\mathrm{T}}(k_A).$$

Then by Lemma 2.9.18, $n_A \lll k_A$.

Theorem is proved. □

Theorem 2.9.12 implies the following result.

Theorem 2.9.21. *For any $n \in N$ and any a_A and b_A from a direct perspective W-arithmetic $\boldsymbol{A} = (A; \dot{+}, \circ, \leq)$ with an increasing functional parameter f, from $n_{A,1}, n_{A,2}, \ldots, n_{A,n-1} \lll_n a_A$ follows $n_{A,1}, n_{A,2}, \ldots, n_{A,n-1} \lll_n b_A$ if for any elements a and b from M, the inequality $a \leq b$ implies $f_{\mathrm{T}}(Sa)/f_{\mathrm{T}}(a) \leq f_{\mathrm{T}}(Sb)/f_{\mathrm{T}}(b)$.*

Proof is left as an exercise. □

We see that many properties of direct perspective W-arithmetics (non-Diophantine arithmetics of whole numbers) can be essentially different from properties of the Diophantine arithmetic $\boldsymbol{W}$ although some properties, such as commutativity of addition and subtraction, are still the same.

2.9.9. *Non-Diophantine arithmetics of natural numbers*

In questions of science, the authority of a thousand is not worth the humble reasoning of a single individual.

Galileo Galilei

From non-Diophantine arithmetics of whole numbers, we come to a class of non-Diophantine arithmetics of natural numbers, which can be considered as proper non-Diophantine arithmetics because according to the classical understanding arithmetic is the Diophantine arithmetics of natural numbers

Definition 2.9.10. A functional direct perspective prearithmetic $\boldsymbol{A} = (N; \dot{+}, \circ, \leq)$ with the projector $f_{\mathrm{T}}(x)$ and the coprojector $f^{\mathrm{T}}(x)$ is called a *direct perspective arithmetic* if the following conditions are satisfied:

(1) $f_{\mathrm{T}}(1_A) = 1$;
(2) $f_{\mathrm{T}}(x)$ is a strictly increasing function;
(3) for any elements a and b from N, the inequality $a \leq b$ implies the inequality

$$f_{\mathrm{T}}(Sa) - f_{\mathrm{T}}(a) \leq f_{\mathrm{T}}(Sb) - f_{\mathrm{T}}(b).$$

Here, Sa is the successor of a.

The direct perspective arithmetic with the generator f is denoted by $\boldsymbol{N}_f$.

Direct perspective arithmetics form a class **DRNDAN** of non-Diophantine arithmetics of natural numbers.

The third condition in Definition 2.9.10 means that the direct discrete derivative (cf., for example, Boole, 1860; 1880; Spiegel, 1971) of the function $f_{\mathrm{T}}(x)$ is an increasing function.

Operations in $\boldsymbol{A}$ are defined as follows:

$$n_A \dot{+} m_A = f^{\mathrm{T}}(f_{\mathrm{T}}(n_A) + f_{\mathrm{T}}(m_A)),$$
$$n_A \circ m_A = f^{\mathrm{T}}(f_{\mathrm{T}}(n_A) \cdot f_{\mathrm{T}}(m_A)).$$

Besides, a natural order relation is defined on A:

$$n_A \leq m_A \text{ if and only if } n \leq m.$$

Lemma 2.9.29. *Any direct perspective arithmetic is a subprearithmetic of a direct perspective W-arithmetic.*

Proof is left as an exercise. □

These results allow deducing many properties of direct perspective arithmetics from properties of direct perspective W-arithmetics. For instance, Proposition 2.9.4 and Lemma 2.9.29 imply the following result.

Theorem 2.9.22. *If for any number n_A, its image $f(n_A)$ is also a whole number, then the element 1_A is an multiplicative one in a direct perspective arithmetic $\boldsymbol{A} = (N; \dot{+}, \circ, \leq)$, i.e., for any $n_A \in \boldsymbol{A}$, the equalities $n_A \circ 1_A = n_A$ and $1_A \circ n_A = n_A$ are true.*

Proof: By Lemma 2.9.9, $f_{\mathrm{T}}(n) = f(n)$ and by Definition 2.9.10, $f_{\mathrm{T}}(x)$ is a strictly increasing function. Then by Proposition 2.9.4, $f^{\mathrm{T}}(f_{\mathrm{T}}(n_A)) = n_A$ and we have the following sequence of equalities:

$$n_A \circ 1_A = f^{\mathrm{T}}(f_{\mathrm{T}}(n_A) \cdot f_{\mathrm{T}}(1_A)) = f^{\mathrm{T}}(f_{\mathrm{T}}(n_A) + 1) = f^{\mathrm{T}}(f_{\mathrm{T}}(n_A)) = n_A.$$

The equality $1_A \circ n_A = n_A$ is derived in a similar way.

Theorem is proved. □

Theorem 2.9.11 implies the following result.

Theorem 2.9.23. *For any $n \in N$ and any a_A and $b_A \in \boldsymbol{A}$, from 1_A, $1_A, \ldots, 1_A \ll_n a_A$ and $a_A \leq b_A$ follows 1_A, $1_A, \ldots, 1_A \ll_n b_A$.*

Proof is left as an exercise. □

Propositions 2.9.4 and 2.9.34 imply the following result.

Theorem 2.9.24. *Operations addition $\dot{+}$ and multiplication $\circ$ are commutative in any direct perspective arithmetic.*

Proof is left as an exercise. □

Remark 2.9.10. As Example 2.9.5 demonstrates, in some direct perspective arithmetics, addition or/and multiplication $\circ$ is not associative.

Proposition 2.9.55 implies the following result.

Proposition 2.9.59. *There is a direct perspective arithmetic (non-Diophantine arithmetic)* $\boldsymbol{A} = (N; \dot{+}, \circ, \leq)$, *in which* $\ll$ *is a total relation in* $\boldsymbol{A}$.

Proof is left as an exercise. □

Proposition 2.9.60. *In a direct perspective arithmetic* $\boldsymbol{A} = (A\dot{+}; \circ, \leq)$ *with an increasing functional parameter* f, *addition and multiplication* $\circ$ *are monotone.*

Proof is similar to the proof of Proposition 2.9.4. □

Proposition 2.9.60 and Lemma 2.1.6 imply the following result.

Corollary 2.9.55. *For any direct perspective arithmetic* $\boldsymbol{A} = (A; \dot{+}, \circ, \leq)$, *addition* $\dot{+}$ *preserves the order* $\leq$.

Proposition 2.9.55 imply the following result.

Proposition 2.9.61. *In a direct perspective arithmetic* $\boldsymbol{A} = (N; \dot{+}, \circ, \leq)$, *relation* $\ll$ *is transitive.*

Proof is left as an exercise. □

Proposition 2.9.57 implies the following result.

Proposition 2.9.62. *In a direct perspective arithmetic* $\boldsymbol{A} = (N; \dot{+}, \circ, \leq)$ *with the functional parameter* f, *relation* $\lll$ *is transitive.*

Proof is left as an exercise. □

Theorem 2.9.14 implies the following result.

Proposition 2.9.63. *In a direct perspective arithmetic* $\boldsymbol{A} = (N; \dot{+}, \circ, \leq)$, *relation* $\ll$ *is compatible with the order* $\leq$.

Proof is left as an exercise. □

Non-Diophantine arithmetics can be essentially different from the Diophantine arithmetic $\boldsymbol{N}$. For instance, $\boldsymbol{N}$ has infinitely many prime numbers, while for non-Diophantine arithmetics, we have the following result.

Theorem 2.9.25. *For any $n > 0$, there is a direct perspective arithmetic (non-Diophantine arithmetic) $\boldsymbol{A} = (N; \oplus, \otimes, \leq_A)$ that has exactly n multiplicatively prime numbers.*

Proof: Let us take the nth prime number p_n from the Diophantine arithmetic $\boldsymbol{N}$ and define the function $f(k)$ as follows:

$$f(k) = \begin{cases} k & \text{if } k < p_n + 2, \\ 3^k & \text{if } k > p_n + 1. \end{cases}$$

Then its inverse has the following form:

$$f^{-1}(k) = \begin{cases} k & \text{if } \log_3 k < p_n + 2, \\ \log_3 k & \text{if } \log_3 k > p_n + 1. \end{cases}$$

Then for any $k > p_n + 1$, we have

$$f_{\mathrm{T}}(2) \cdot f_{\mathrm{T}}(k) = 2 \cdot 3^k < 3 \cdot 3^k = 3^{k+1} = f_{\mathrm{T}}(k+1).$$

Thus, by Lemma 2.9.18, we have $2_A \lll k_A$ for any $k > p_n + 1$. Consequently, k_A is divisible by 2_A and thus is a composite number.

At the same time, all composite in $\boldsymbol{N}$ numbers smaller than $p_n + 1$ are composite in $\boldsymbol{A}$ and all prime in $\boldsymbol{N}$ numbers smaller than $p_n + 1$ are prime in $\boldsymbol{A}$. At the same time, if the product of two numbers in $\boldsymbol{A}$ is less than $p_n + 1$, this product coincides with the product of the same numbers in $\boldsymbol{N}$. Consequently, we have exactly n prime numbers in $\boldsymbol{A}$, and they are smaller than $p_n + 1$.

Theorem is proved. □

Corollary 2.9.56. *For any $n > 0$, there is a direct perspective arithmetic $\boldsymbol{A} = (N; \oplus, \otimes, \leq_A)$ in which all numbers larger than n are even.*

The difference between direct perspective W-arithmetics and direct perspective arithmetics is that the former contain 0 and the latter does not contain 0. Consequently, Theorem 2.9.25 implies the following results.

Corollary 2.9.57. *For any $n > 0$, there is a direct perspective W-arithmetic $\boldsymbol{D} = (W; \oplus, \otimes, \leq_A)$ in which all numbers larger than n are even.*

Corollary 2.9.58. *For any $n > 0$, there is a direct perspective W-arithmetic $\boldsymbol{B} = (W; \oplus, \otimes, \leq_A)$ that has exactly n multiplicatively prime numbers.*

Changing p_n to $2n$ in the definition of the function $f(k)$ from the proof of Theorem 2.9.25, we obtain the following result.

Proposition 2.9.64. *For any $n > 0$, there is a direct perspective arithmetic that has exactly n odd numbers.*

Proof is left as an exercise. □

Corollary 2.9.59. *For any $n > 0$, there is a direct perspective W-arithmetic that has exactly n odd numbers.*

It is possible to show that a variety of properties of the Diophantine arithmetic $\boldsymbol{N}$ can be false in non-Diophantine arithmetics. To achieve this goal, we build a special non-Diophantine arithmetics $\boldsymbol{W}_{\text{eexp}k}$ and $\boldsymbol{N}_{\text{eexp}k}$, in which many properties of conventional natural and whole numbers are invalid.

Let us take a natural number $k > 1$ and the function $f_k(n) = k^{2^{2^n}}$ for $n = 0, 1, 2, 3, \ldots$, as the functional parameter for the functional direct perspective W-prearithmetic $\boldsymbol{W}_{\text{eexp}k} = (W; \dot{+}, \circ, \leq)$. The inverse function $f_k^{-1}(q)$ is $\log_2 \log_2 \log_k (q)$.

Note that $\boldsymbol{W}_{\text{eexp}k}$ contains all whole numbers with the natural order $\leq$ between them. If n is a whole number, then its counterpart in $\boldsymbol{W}_{\text{eexp}k}$ is denoted by n_W.

Now let us find properties of this arithmetic as well as properties of the the functional direct perspective prearithmetic $\boldsymbol{N}_{\text{eexp}k} = (N; \dot{+}, \circ, \leq)$ of all natural numbers, which is a subprearithmetic of the prearithmetic $\boldsymbol{W}_{\text{eexp}k}$.

By Lemma 2.9.9, $f_{kT}(n_W) = f_k(n) = k^{2^{2^n}}$. Thus, for any $n > 0$, we have

$$\begin{aligned} f_{k\mathrm{T}}(n_W) \cdot f_{k\mathrm{T}}(n_W) &= k^{2^{2^n}} \cdot k^{2^{2^n}} = k^{(2^{2^n}+2^{2^n})} k^{(2^{2^{2n}}+2^{2^{2n}})} \\ &= k^{2 \cdot 2^n} < k^{(2^{2^n})^2} = k^{(2^{2 \cdot 2^n})} = k^{2^{2n+1}} = f_{k\mathrm{T}}(\mathrm{S}n_W) \end{aligned}$$

because $2 \cdot 2^{2^n} < (2^{2^n})^2$.

It means that $f_{k\mathrm{T}}(n_W) \cdot f_{k\mathrm{T}}(n_W) < f_{k\mathrm{T}}(\mathrm{S}n_W)$ for all $n = 1, 2, 3, 4, 5, \ldots$ Then by Lemma 2.9.18, we have the following inequality $n_W \lll n_W$, i.e., $n_W \circ n_W = n_W$ for all $n = 1, 2, 3, 4, 5, \ldots$. By the definition of subprearithmetics, these relations are also true in the the prearithmetic $\boldsymbol{N}_{\text{eexp}\,k}$ of the W-prearithmetic $\boldsymbol{W}_{\text{eexp}\,k}$. Note that for all $n = 1, 2, 3, 4, 5, \ldots, n_W + 1_W = n_W \neq \mathrm{S}n_W$ in the functional direct perspective W-prearithmetic $\boldsymbol{W}_{\text{eexp}\,k}$.

Besides,

$$f_{k\mathrm{T}}(1_W) \cdot f_{k\mathrm{T}}(0_W) = k^{2^{2^1}} \cdot k^{2^1} = k^4 \cdot k^2 = k^6 < k^{16} = 2^{2^{2^2}} = f_{k\mathrm{T}}(2_W).$$

Then Lemma 2.9.18 implies $0_W \lll 1_W$. As by Theorem 2.9.8, relation $\lll$ is compatible from the left with the order $\leq$, relation $Sn_W \lll Sn_W$ implies relation $n_W \lll Sn_W$. As by Proposition 2.9.29, relation $\lll$ is transitive, we obtain the following sequence of relations:

$$0_W \lll 1_W \lll 2_W \lll 3_W \lll \cdot \lll n_W \lll Sn_W \lll \cdots \tag{9.1}$$

in the W-prearithmetic $\boldsymbol{W}_{\mathrm{eexp}\,k}$.

In addition, we have

$$f_{k\mathrm{T}}(0_W) \cdot f_{k\mathrm{T}}(0_W) = k^{2^1} \cdot k^{2^1} = k^{2+2} = k^4 = k^{2^{2^1}} = f_{k\mathrm{T}}(1_W).$$

By definition, it means that $0_W \circ 0_W = 1_W$. Indeed,

$$0_W \circ 0_W = f_k^{\mathrm{T}}(f_{k\mathrm{T}}(0_W) \cdot f_{k\mathrm{T}}(0_W)) = f_k^{\mathrm{T}}(f_{k\mathrm{T}}(1_W)) = 1_W. \tag{9.2}$$

This shows that 0_W is not a multiplicative zero in $\boldsymbol{W}_{\mathrm{eexp}\,k}$.

By Proposition 2.9.31, relation implies relation $\ll$ for all elements larger than 1_A because the functional parameter f_k is strictly increasing on whole numbers and maps whole numbers into whole numbers. Consequently, sequence (9.1) entails sequence (9.3).

$$2_W \ll 3_W \ll \cdots \ll n_W \ll Sn_W \ll \cdots \tag{9.3}$$

In addition, we have

$$f_{k\mathrm{T}}(0_W) + f_{k\mathrm{T}}(0_W) = k^{2^1} + k^{2^1} = 2k^2 < k^4 = k^{2^{2^1}} = f_{k\mathrm{T}}(1_W)$$

and

$$\begin{aligned} f_{k\mathrm{T}}(1_W) + f_{k\mathrm{T}}(1_W) &= k^{2^2} + k^{2^2} = 2k^{2^2} = 2k^4 < k^{16} \\ &= k^{24} = k^{2^{2^2}} = f_{k\mathrm{T}}(2_W). \end{aligned}$$

It means by Lemma 2.9.17 that $0_W \ll 0_W$ and $1_W \ll 1_W$. Consequently, we have the following inequality $n_W \ll n_W$, i.e., $n_W \dot{+} n_W = n_W$

for all $n = 1, 2, 3, 4, 5, \ldots$. Together with sequence (9.3), this gives us sequence (9.4).

$$0_W \ll 1_W \ll 2_W \ll 3_W \ll \cdots \ll n_W \ll Sn_W \ll \cdots. \tag{9.4}$$

By the definition of subprearithmetics, relations $n_N \ll n_N$ are also true in the subprearithmetic $\boldsymbol{N}_{\text{eexp}k}$ of the W-prearithmetic $\boldsymbol{W}_{\text{eexp}k}$.

Note that equalities (9.2) show that 0_W is not a multiplicative zero in the functional direct perspective W-prearithmetic $\boldsymbol{W}_{\text{eexp}k}$ because $f_k(0) = k^{2^{2^0}} = k^2 \neq 0$. At the same time, 0_A is an additive zero in $\boldsymbol{W}_{\text{eexp}k}$.

To make 0_W a multiplicative zero, we slightly change the W-prearithmetic $\boldsymbol{W}_{\text{eexp}k}$ building the functional direct perspective W-arithmetic $\boldsymbol{W}^0_{\text{eexp}k}$.

To do this, we use the following function:

$$f^0_k(n) = \begin{cases} 0 & \text{for } n = 0, \\ k^{2^{2^n}} & \text{for } n = 1, 2, 3, \ldots. \end{cases}$$

Taking the function f^0_k as the functional parameter, we build the functional direct perspective W-arithmetic $\boldsymbol{W}^0_{\text{eexp}k} = (W; \dot{+}, \circ, \leq)$. Note that $\boldsymbol{W}^0_{\text{eexp}k}$ contains all whole numbers with the natural order $\leq$ between them. If n is a whole number, then its counterpart in $\boldsymbol{W}^0_{\text{eexp}k}$ is denoted by n_W.

All operations and relations between numbers $1_W, 2_W, 3_W, \ldots, n_W, Sn_W, \ldots$ in the W-arithmetic $\boldsymbol{W}^0_{\text{eexp}k}$ are the same as in the W-prearithmetic $\boldsymbol{W}_{\text{eexp}k}$ and we need only to check operations with 0_W. For any whole number n_W, we have

$$0_W \circ n_W = f^0_{k\mathrm{T}}(f^0_{k\mathrm{T}}(0_W) \cdot f^0_{k\mathrm{T}}(n_W)) = f^0_{k\mathrm{T}}(0 \cdot f^0_{k\mathrm{T}}(n_W)) = f^0_{k\mathrm{T}}(0) = 0_W.$$

These equalities mean that 0_W is a multiplicative zero in the functional direct perspective W-arithmetic $\boldsymbol{W}^0_{\text{eexp}k}$. Besides, $0_W \ll n_W$ and $n_W \lll 0_W$ for any n_W from the W-arithmetic $\boldsymbol{W}^0_{\text{eexp}k}$.

Note that $\boldsymbol{N}_{\text{eexp}k}$ is a subprearithmetic both in $\boldsymbol{W}_{\text{eexp}k}$ and $\boldsymbol{W}^0_{\text{eexp}k}$.

Let us check conditions from the definition of functional direct perspective W-arithmetics for $\boldsymbol{W}^0_{\text{eexp}k}$.

The first condition$f_{\mathrm{T}}(0_A) = 0$ is satisfied because $f_{\mathrm{T}}(0_A) = f(0) = 0$.

The second condition, which states that $f_{\mathrm{T}}(x)$ is a strictly increasing function, is satisfied because $f_{\mathrm{T}}(x) = f(x)$ and $f(x)$ is a strictly increasing function.

The third condition $f_{\mathrm{T}}(\mathrm{S}n_A) - f_{\mathrm{T}}(n_A) \leq f_{\mathrm{T}}(\mathrm{S}m_A) - f_{\mathrm{T}}(m_A)$ for any $n \leq m$ is satisfied because if $0 < n \leq m$, then $k^{2^{2^n}} \leq k^{2^{2^m}}$, $((k^{2^{2^n}})^{2^{2^n}-1} - 1) \leq ((k^{2^{2^m}})^{2^{2^m}-1} - 1)$ and we have the following sequence of inequalities:

$$\begin{aligned} f_{\mathrm{T}}(\mathrm{S}n_A) - f_{\mathrm{T}}(n_A) &= k^{2^{2^{n+1}}} - k^{2^{2^n}} = k^{2^{2\cdot(2^n)}} - k^{2^{2^n}} = k^{2^{2^n}\cdot 2^{2^n}} - k^{2^{2^n}} \\ &= (k^{2^{2^n}})^{2^{2^n}} - k^{2^{2^n}} = ((k^{2^{2^n}})^{2^{2^n}-1} - 1)\cdot k^{2^{2^n}} \\ &\leq f_{\mathrm{T}}(\mathrm{S}m_A) - f_{\mathrm{T}}(m_A) = k^{2^{2^{m+1}}} - k^{2^{2^m}} \\ &= k^{2^{2\cdot(2^m)}} - k^{2^{2^m}} = k^{2^{2^m}\cdot 2^{2^m}} - k^{2^{2^m}} \\ &= (k^{2^{2^m}})^{2^{2^m}} - k^{2^{2^m}} = ((k^{2^{2^m}})^{2^{2^{m}-1}} - 1)\cdot k^{2^{2^m}}. \end{aligned}$$

When $n = 0$, the third condition is evident.

As a result, we see that $\boldsymbol{W}^0_{\mathrm{eexp}k}$ is a functional direct perspective W-arithmetic.

Note that $\boldsymbol{W}_{\mathrm{eexp}k}$ does not satisfy the first condition and thus, it is only a functional direct perspective W-prearithmetic.

As $\boldsymbol{N}_{\mathrm{eexp}k}$ is also only a prearithmetic, we need to modify it. We build functional direct perspective arithmetic $\boldsymbol{N}^1_{\mathrm{eexp}k}$ denoting its numbers n by n_N and using the following function as its generator (functional parameter):

$$f_k^1(n) = \begin{cases} 1 & \text{for } n = 1, \\ k^{2^{2^n}} & \text{for } n = 2, 3, \ldots. \end{cases}$$

Taking $\boldsymbol{N}^1_{\mathrm{eexp}k}$, we can check all conditions from the definition of functional direct perspective arithmetics in the same way as we did this for $\boldsymbol{W}^0_{\mathrm{eexp}k}$. As a result, we see that $\boldsymbol{N}^0_{\mathrm{eexp}k}$ is a functional direct perspective arithmetic. In particular, $\boldsymbol{N}^1_{\mathrm{eexp}k}$ contains all numbers 1_N, 2_N, $3_N, \ldots, n_N, \ldots$

At the same time, in $\boldsymbol{N}^1_{\mathrm{eexp}k}$, the following inequality $n_N \lll n_N$, i.e., $n_N \circ n_N = n_N$, is true for all $n = 1, 2, 3, 4, 5, \ldots$ because we can check that $1_N \circ 1_N = 1_N$, while all other numbers are multiplied in the same way as in the functional direct perspective W-arithmetic $\boldsymbol{W}^0_{\mathrm{eexp}k}$. By the same token, we have $n_N \lll n_N$, i.e., $n_N \circ n_N = n_N$, for all $n = 1, 2, 3, 4, 5, \ldots$.

We use constructed arithmetics to prove the following simple but important results demonstrating unusual properties of non-Diophantine arithmetics.

Theorem 2.9.26. *There are infinitely many functional direct perspective arithmetics (non-Diophantine arithmetics of natural numbers), in which any number is a square.*

Indeed, in each functional direct perspective arithmetic $\boldsymbol{N}^1_{\text{eexp}k}$, we have the equality $n_W \circ n_W = n_W$ for all its numbers. In other words, each number is the square of itself. As this is true for any $k > 1$, we have infinitely many functional direct perspective arithmetics $\boldsymbol{N}^1_{\text{eexp}k}$ because k can be equal to 2, 3, 4, 5 and so on.

Corollary 2.9.60. *There are infinitely many functional direct perspective W-arithmetics arithmetics (non-Diophantine arithmetics of whole numbers), in which any number is a square.*

Indeed, in each functional direct perspective W-arithmetic $\boldsymbol{W}^0_{\text{eexp}k}$, we have the equality $n_W \circ n_W = n_W$ for all its numbers n larger than 0 because all the numbers belong to its subarithmetic $\boldsymbol{N}_{\text{eexp}k}$. Besides, in Corollary 2.9.59, it is proved that $0_W \circ 0_W = 0_W$. Consequently, each number is the square of itself in $\boldsymbol{W}^0_{\text{eexp}k}$. As this is true for any $k > 1$, we have infinitely many non-Diophantine arithmetics $\boldsymbol{W}^0_{\text{eexp}k}$ because k can be equal to 2, 3, 4, 5 and so on.

Theorem 2.9.27. *For a given natural number q, there are infinitely many functional direct perspective arithmetics (non-Diophantine arithmetics of natural numbers) $\boldsymbol{A}$, in which $n_A = (n_A)^q$ for all its numbers, i.e., any number is the nth power.*

Indeed, in any functional direct perspective arithmetic $\boldsymbol{N}^1_{\text{eexp}k}$, we have the equality $n_N \circ n_N = n_N$, which by induction implies the equality $n_N = (n_N)^q$ for all its numbers n_N. As there are infinitely many such arithmetics, we have the necessary result.

Theorem 2.9.28. *There are infinitely many functional direct perspective arithmetics (non-Diophantine arithmetics of natural numbers) with only one multiplicatively prime number.*

Proof: To prove the theorem, we at first show that the inequality $m \leq n$ implies the inequality $f_{k\text{T}}(Sm_N)/f_{k\text{T}}(m_N) \leq f_{k\text{T}}(Sn_N)/f_{k\text{T}}(n_N)$ in the functional direct perspective arithmetic $\boldsymbol{N}^1_{\text{eexp}k}$. Indeed, for $m < n$, we have

$$f_{k\text{T}}(Sm_N)/f_{k\text{T}}(m_N) = (k^{2^{2^{m+1}}})/(k^{2^{2^m}}) = k^{2^{2^{m+1}}-2^{2^m}}$$

$$f_{k\text{T}}(Sn_N)/f_{k\text{T}}(n_N) = (k^{2^{2^{n+1}}})/(k^{2^{2^n}}) = k^{2^{2^{n+1}}-2^{2^n}}$$

Using properties of the Diophantine arithmetic $\boldsymbol{N}$, we obtain

$$\begin{aligned}&(2^{2^{n+1}} - 2^{2^n}) - (2^{2^{m+1}} - 2^{2^m})\\ &\quad = (2^{2^{n+1}} - 2^{2^{m+1}}) - (2^{2^n} - 2^{2^m})\\ &\quad = 2^{2^{m+1}}(2^{2^{n+1}-2^{m+1}} - 1) - 2^{2^m}(2^{2^n-2^m} - 1) > 0\end{aligned}$$

because $2^{2^{m+1}} > 2^{2^m}$ and $2^{2^{n+1}-2^{m+1}} > 2^{2^n-2^m} > 1$. Consequently,

$$k^{2^{2^{m+1}}} - 2^{2^m} < k^{2^{2^{n+1}}} - 2^{2^n}$$

and

$$f_{k\mathrm{T}}(Sm_N)/f_{k\mathrm{T}}(m_N) \le f_{k\mathrm{T}}(Sn_N)/f_{k\mathrm{T}}(n_N).$$

By construction, the functional parameter f_k is strictly increasing on natural numbers and maps natural numbers into natural numbers. Consequently, by Theorem 2.9.16, relation $\lll$ is compatible with the order $\le$. It means that $m_N \lll n_N$ if $m \le n$. Thus, $m_N \circ n_N = n_N$ when $m \le n$, i.e., any number n_N with $n > 1$ is divisible by m_N with $m \le n$. As by definition, 1 is not a prime number, any functional direct perspective arithmetic $\boldsymbol{N}^1_{\mathrm{eexp}k}$ has only one multiplicatively prime number 2_N because 2_N is divisible only by 1_N and by itself while all numbers larger than 2_N are divisible by 2_N.

As before, there are infinitely many such non-Diophantine arithmetics $\boldsymbol{N}^1_{\mathrm{eexp}k}$ because k can be equal to 2, 3, 4, 5 and so on.

Theorem is proved. □

Theorem 2.9.29. *There are infinitely many functional direct perspective W-arithmetics (non-Diophantine arithmetics of whole numbers) with only one odd number.*

Indeed, in each functional direct perspective W-arithmetic $\boldsymbol{W}^0_{\mathrm{eexp}k}$ ($k = 2, 3, 4, 5, \ldots$), only 1_W is an odd number because all numbers larger than 1_N are divisible by 2_N, and there are infinitely many such arithmetics.

Corollary 2.9.61. *There are infinitely many functional direct perspective arithmetics (non-Diophantine arithmetics of natural numbers) with only one odd number.*

Indeed, in each functional direct perspective arithmetic $\boldsymbol{N}^1_{\mathrm{eexp}k}$ ($k = 2, 3, 4, 5, \ldots$), only 1_W is an odd number.

Proposition 2.9.65. *There are infinitely many functional direct perspective W-prearithmetics in which all numbers but one are multiplicatively composite.*

Indeed, in each functional direct perspective W-prearithmetic $\boldsymbol{W}_{\text{eexp}k}$ ($k = 2, 3, 4, 5, \ldots$), only 0_W is not a multiplicatively composite number. Note that 1_W is a multiplicatively composite number because it is divisible by 0_W. As there are infinitely many such arithmetics, we have the necessary result.

Theorem 2.9.30. *There are infinitely many functional direct perspective W-prearithmetics (non-Diophantine arithmetics of whole numbers) in which all numbers but one are multiplicatively composite.*

Indeed, in each functional direct perspective W-arithmetic $\boldsymbol{W}^0_{\text{eexp}k}$ ($k = 2, 3, 4, 5, \ldots$), only 1_W and 2_W are not a composite numbers because 1_W is divisible only by itself while 2_W is divisible only by 1_W and by itself. Note that 0_W is a multiplicatively composite number because it is divisible by any number n_W. As there are infinitely many such arithmetics, we have the necessary result.

Corollary 2.9.62. *There are infinitely many functional direct perspective arithmetics (non-Diophantine arithmetics of natural numbers) in which all numbers but two are multiplicatively composite.*

Indeed, in each functional direct perspective arithmetic $\boldsymbol{N}^1_{\text{eexp}k}$ ($k = 2, 3, 4, 5, \ldots$), only 1_N and 2_N are not a composite numbers because 1_N is divisible only by itself while 2_N is divisible only by 1_N and by itself.

Theorem 2.9.31. *There are infinitely many functional direct perspective W-arithmetics (non-Diophantine arithmetics of whole numbers) in which for all $n = 1, 2, 3, 4, 5, \ldots$, all numbers larger than n are divisible by n.*

Proof: Let us consider the non-Diophantine W-arithmetic $\boldsymbol{W}^0_{\text{eexp}k}$ where k can be any number larger than 1. It is proved that $n_W \lll n_W$ in the non-Diophantine arithmetic $\boldsymbol{W}^0_{\text{eexp}k}$ for all $n = 1, 2, 3, 4, 5, \ldots$. By construction the functional parameter f_k is strictly increasing on natural numbers and maps whole numbers into whole numbers. Consequently, by Theorem 2.9.16, relation $\lll$ is compatible with the order $\leq$. It means that $m_W \lll n_W$ if $m \leq n$. Thus, $m_W \circ n_W = n_W$ when $m \leq n$, i.e., any number n_W with $n > 1$ is divisible by m_W with $m \leq n$. Consequently, any number larger than n is divisible by n in the non-Diophantine arithmetic $\boldsymbol{W}^0_{\text{eexp}k}$. As there are infinitely many such arithmetics, we have the necessary result.

Theorem is proved. □

Corollary 2.9.63. *There are infinitely many functional direct perspective arithmetics (non-Diophantine arithmetics of natural numbers), in which for all $n = 1, 2, 3, 4, 5, \ldots$, all numbers larger than n are divisible by n.*

Theorem 2.9.32. *There are infinitely many functional direct perspective W-arithmetics (non-Diophantine arithmetics of whole numbers), in which $\ll$ is a total (linear) order.*

Indeed, as it is already proved, $\ll$ is a total (linear) order in the non-Diophantine W-arithmetic $\boldsymbol{W}^0_{\text{eexp}k}$ for any $k = 2, 3, 4, 5, \ldots$.

Corollary 2.9.64. *There are infinitely many functional direct perspective arithmetics (non-Diophantine arithmetics of natural numbers), in which $\ll$ is a total (linear) order.*

Theorem 2.9.33. *There are infinitely many functional direct perspective W-arithmetics (non-Diophantine arithmetics of whole numbers), in which is a total (linear) order.*

Indeed, as it is already proved, $\lll$ is a total (linear) order in the non-Diophantine W-arithmetic $\boldsymbol{W}_{\text{eexp}k}$ for any $k = 2, 3, 4, 5, \ldots$.

Corollary 2.9.65. *There are infinitely many functional direct perspective arithmetics (non-Diophantine arithmetics of natural numbers), in which is a total (linear) order.*

There is one more essential difference between the Diophantine arithmetic $\boldsymbol{N}$ and non-Diophantine arithmetics of natural numbers. We know that according to Fermat's Last Theorem, also called Fermat's conjecture, which was proved by the English mathematician Andrew Wiles with the help of Richard Taylor, the equation $x^n = y^n + z^n$ cannot have positive integer solutions for any natural number n greater than 2 (Taylor and Wiles, 1995; Wiles, 1995). In general, this is not true non-Diophantine arithmetics as the following result demonstrates.

Theorem 2.9.34. *There are infinitely many functional direct perspective arithmetics (non-Diophantine arithmetics of natural numbers), in which for any natural number n, the equation $x^n = y^n + z^n$ has infinitely many solutions.*

Proof: Let us consider the non-Diophantine arithmetic $\boldsymbol{N}^1_{\text{eexp}k}$ where k can be any natural number larger than 1. It was proved that in it, $m_N \circ m_N = m_N$ for all $m = 1, 2, 3, 4, 5, \ldots$. Consequently, $m_N^n = \underbrace{(\ldots(m_N \circ m_N) \circ m_N) \ldots m_N) \circ m_N}_{n} = m_N$ for all $m, n = 1, 2, 3, 4, 5, \ldots$.

In addition, in the non-Diophantine arithmetic $\boldsymbol{N}^1_{\text{eexp}k}$, we have $m_N \dot{+} m_N = m_N$ for all $m = 1, 2, 3, 4, 5, \ldots$. Thus, we obtain

$$m_N^n \dot{+} m_N^n = m_N \dot{+} m_N = m_N = m_N^n$$

This gives us infinitely many solutions of the equation $x^n = y^n + z^n$ in any non-Diophantine arithmetic $\boldsymbol{N}^1_{\text{eexp}k}$. Besides, we already know that there are infinitely many such non-Diophantine arithmetics.

Theorem is proved. □

Corollary 2.9.66. *There are infinitely many functional direct perspective W-arithmetics (non-Diophantine arithmetics of whole numbers), in which for any natural number n, the equation $x^n = y^n + z^n$ has infinitely many solutions.*

We see that many properties of direct perspective arithmetics (non-Diophantine arithmetics of natural numbers) can be essentially different from properties of the Diophantine arithmetic $\boldsymbol{N}$ and many properties of direct perspective W-arithmetics (non-Diophantine arithmetics of whole numbers) can be essentially different from properties of the Diophantine arithmetic $\boldsymbol{W}$.

2.10. Dual Perspective Arithmetics

The pressure for conformity is enormous.
I have experienced it in editors rejection of submitted papers,
based on venomous criticism of anonymous referees.
Julian Schwinger

Dual perspective arithmetics form the second discovered or constructed class of non-Diophantine arithmetics of whole and natural numbers. To define these arithmetics, we at first build dual perspective prearithmetics by the following procedure.

Let us consider an interval $[q, r)$ of real numbers with $r \in R^+ \cup \infty$ and a relation $Q \subseteq [q, r) \times R^+$ that has the domain $\text{Dom}(Q) = [q, r)$, the same definability domain $\text{DDom}(Q) = [q, r)$, the range $\text{Rg}(Q) = R^+$ and the codomain Codom$(Q) = R^+$. When $r = \infty$, we have $Q \subseteq R^+ \times R^+$. When $r = \infty$, we have $Q \subseteq R^+ \times R^+$. By Q^{-1} we denote the relation, which is inverse to Q, i.e., $Q^{-1} = \{(x, y); (y, x) \in Q\}$. If $x \in R^+$, then $Q^{-1}(x) = \{y; (y, x) \in Q\}$ is the coimage of the element x and $Q(x) = \{z; (x, z) \in Q\}$ is the image of the element x.

Relation Q allows us to define two functions, which are taking values in $\boldsymbol{W}$

$$Q^{\mathrm{T}}_{\mathrm{uo}}(x) = \sup\{m; m \in W \text{ and } \forall z \in Q(x)(m \leq z)\},$$

$$Q^{\mathrm{o}}_{\mathrm{uT}}(x) = \inf\{n; n \in W \text{ and } \forall z \in Q^{-1}(x)(n \geq z)\}.$$

Note that if $Q(x)$ is a whole number, then $Q^{\mathrm{T}}_{\mathrm{uo}}(x) = Q(x)$, and if $Q^{-1}(x)$ is a whole number, then $Q^{\mathrm{o}}_{\mathrm{uT}}(x) = Q^{-1}(x)$.

We see that $Q^{\mathrm{o}}_{\mathrm{uT}}$ maps the set $[q, r)$ onto whole numbers from R^+ and $Q^{\mathrm{T}}_{\mathrm{uo}}$ maps R^+ onto whole numbers from $[q, r)$. This allows us, taking restrictions of both functions on the set $W_{q,r} = [q, r) \cap W$, to obtain two additional functions $Q^{\mathrm{o}}_{\mathrm{T}} : W_{q,r} \to R^+$ and $Q^{\mathrm{T}}_{\mathrm{o}}$: $R^+ \to W_{q,r}$.

Definition 2.10.1.

(a) A prearithmetic $\boldsymbol{A} = (A; \dot{+}, \circ, \leq)$ is called a *dual perspective* W-*prearithmetic* if it is weakly projective with respect to the conventional (Diophantine) arithmetic $\boldsymbol{W} = (W; +, \cdot, \leq)$ with the *projector* $Q^{\mathrm{T}}_{\mathrm{o}}(x)$ and the *coprojector* $Q^{\mathrm{o}}_{\mathrm{T}}(x)$.

(b) The relation $Q(x)$ is called the *relational parameter* of the prearithmetic $\boldsymbol{A}$, as well as the *generator* of the *projector* $Q^{\mathrm{T}}_{\mathrm{o}}(x)$, of the *coprojector* $Q^{\mathrm{o}}_{\mathrm{T}}(x)$ and the *relational parameter* of the prearithmetic $\boldsymbol{A}$.

The dual perspective W-prearithmetic with the generator Q is denoted by $\boldsymbol{W}^{Q}_{w}$.

Elements of A are called numbers of the dual perspective W-prearithmetics $\boldsymbol{A}$ and are denoted using the subscript A, i.e., 2 in $\boldsymbol{A}$ is denoted by 2_A and 5 in $\boldsymbol{A}$ is denoted by 5_A. Numbers in $\boldsymbol{A}$ are ordered by the same order relation $\leq$ by which they are ordered in $\boldsymbol{N}$. This implies the following result.

Lemma 2.10.1. *The unary operation* S *of taking the next number is defined in all dual perspective W-prearithmetics with more than one element coinciding with the operation* S *in the Diophantine arithmetic* $\boldsymbol{W}$.

Proof is left as an exercise. □

Dual perspective W-prearithmetics also inherit some Archimedean properties (cf. Section 2.1) from the Diophantine arithmetic $\boldsymbol{W}$.

Lemma 2.10.2. *Any dual perspective W-prearithmetic* $\boldsymbol{A} = (A; \dot{+}, \circ, \leq)$ *is successively Archimedean and exactly successively Archimedean.*

Proof is left as an exercise. □

As elements of a dual perspective W-prearithmetic $\boldsymbol{A} = (A; \dot{+}, \circ, \leq)$ form an interval of whole numbers, we have the following result.

Lemma 2.10.3. *If a dual perspective W-prearithmetic* $\boldsymbol{A} = (A; \dot{+}, \circ, \leq)$ *with more than one element has zero* 0_A, *then* $\mathrm{S}0_A = 1_A$.

Proof is left as an exercise. □

As $\boldsymbol{A}$ is weakly projective with respect to Diophantine arithmetic $\boldsymbol{W}$, for any two numbers n_A and m_A from $W_{q,r}$, operations in $\boldsymbol{A}$ are defined as follows:

$$n_A \dot{+} m_A = Q^{\mathrm{o}}_{\mathrm{T}}(Q^{\mathrm{T}}_{\mathrm{o}}(n_A) + Q^{\mathrm{T}}_{\mathrm{o}}(m_A)),$$
$$n_A \circ m_A = Q^{\mathrm{o}}_{\mathrm{T}}(Q^{\mathrm{T}}_{\mathrm{o}}(n_A) \cdot Q^{\mathrm{T}}_{\mathrm{o}}(m_A)).$$

To understand the difference between direct and dual perspective W-prearithmetics, let us build direct and dual perspective W-prearithmetics with the same relational parameter.

Example 2.10.1. Let us take number 0.3 and the relation $Q(x) = \{(x, x+0.3)$ for all $x = 0, 1, 2, 3, \ldots\}$ as the relational parameter of the dual perspective W-prearithmetic $\boldsymbol{B} = (B; \dot{+}, \circ, \leq)$. Then $Q^{-1}(x) = \{(x, x-0.3)$ for all $x = 0, 1, 2, 3, \ldots\}$.

Taking whole numbers $n > 0$ and $k > 0$, we have

$$Q^{\mathrm{T}}_{\mathrm{uo}}(n) = \sup\{m; m \in W \,\&\, m \leq n + 0.3)\} = n,$$
$$Q^{\mathrm{T}}_{\mathrm{uo}}(k) = \sup\{m; m \in W \,\&\, m \leq k + 0.3)\} = k,$$
$$Q^{\mathrm{T}}_{\mathrm{uo}}(n) + Q^{\mathrm{T}}_{\mathrm{uo}}(k) = n + k.$$

Thus,

$$n_A \dot{+} k_A = Q^{\mathrm{o}}_{\mathrm{uT}}(n + k) = \inf\{m; m \in W \text{ and } m \geq n + k)\} = (n + k)_A.$$

In particular, we have

$$2_A \dotplus 2_A = 4_A,$$

$$3_A \dotplus 3_A = 6_A.$$

Similar calculations give us

$$n_A \circ k_A = Q^{\circ}_{\mathrm{uT}}(n \cdot k) = \inf\{m; m \in W \text{ and } m \geq n \cdot k)\} = (n \cdot k)_A.$$

In particular, we have

$$2_A \circ 2_A = 4_A,$$

$$3_A \circ 3_A = 9_A.$$

We can see that addition and multiplication in the dual perspective W-prearithmetic $\boldsymbol{B}$ is the same as in the Diophantine arithmetic $\boldsymbol{W}$.

Now let us build the direct perspective W-prearithmetic $\boldsymbol{A} = (A; \dotplus, \circ, \leq)$ with the same relational parameter Q.

Taking whole numbers $n > 0$ and $k > 0$, we have

$$Q_{\mathrm{uT}}(n) = \inf\{m; m \in W \& \ m \geq n + 0.3)\} = n + 1,$$

$$Q_{\mathrm{uT}}(k) = \inf\{m; m \in W \& \ m \geq k + 0.3)\} = k + 1,$$

$$Q_{\mathrm{uT}}(n) + Q_{\mathrm{uT}}(k) = n + k + 2.$$

Thus,

$$n_A \dotplus k_A = Q^{\mathrm{T}}_{\mathrm{u}}(n+k) = \sup\{m; m \in W \text{ and } m \leq n+k+2)\} = (n+k+2)_A.$$

In particular, we have

$$2_A \dotplus 2_A = 6_A,$$

$$3_A \dotplus 3_A = 8_A.$$

Similar calculations give us

$$n_A \circ k_A = Q^{\mathrm{T}}_{\mathrm{u}}((n + 1) \cdot (k + 1))$$
$$= \sup\{m; m \in W \text{ and } m \leq n \cdot k + n + k + 1\} = (n \cdot k + n + k + 1)_A.$$

In particular, we have

$$2_A \circ 2_A = 9_A,$$

$$3_A \circ 3_A = 16_A.$$

We see that addition and multiplication of the same numbers gives different results in perspective W-prearithmetics $\boldsymbol{A}$ and $\boldsymbol{B}$ although they have the same functional parameter.

In a general case, as Q is a binary relation between the interval $[q, r)$ and R^+, i.e., $Q \subseteq [q, r) \times R^+$, it is possible to correspond a relation $Q_+ \subseteq R^+ \times R^+$ which consists of the same elements as Q when the first element belongs to $[q, r)$. Other elements Q_+ can be different but the definability domain of Q_+ has to be equal to R^+. Then it is possible to build the dual perspective W-prearithmetic $\boldsymbol{W}_w^{Q+}$.

Lemma 2.10.4. *The dual perspective W-prearithmetic* $\boldsymbol{W}_w^Q$ *is a subprearithmetic of the dual perspective W-prearithmetic* $\boldsymbol{W}_w^{Q+}$.

Proof is left as an exercise. □

This result and results from Section 2.5, show that in many cases, it is possible to study the dual perspective W-prearithmetic $\boldsymbol{W}_w^{Q+}$ and derive properties of the dual perspective W-prearithmetic $\boldsymbol{W}_w^Q$ from the properties of properties of $\boldsymbol{W}_w^{Q+}$.

Let us study properties of dual perspective W-prearithmetics. Although dual perspective W-arithmetics are different from direct perspective W-arithmetics, some of their properties are similar. For instance, Proposition 2.7.1 and properties of the conventional Diophantine arithmetic $\boldsymbol{W}$ imply the following result.

Proposition 2.10.1. *For any dual perspective W-prearithmetic* $\boldsymbol{A} = (A; \dotplus, \circ, \leq_A)$, *addition* $\dotplus$ *and multiplication* $\circ$ *are commutative.*

Proof is left as an exercise. □

We can also find other useful properties of operations in dual perspective W-prearithmetics.

Proposition 2.10.2. *In a dual perspective W-prearithmetic* $\boldsymbol{A} = (A; \dotplus, \circ, \leq_A)$ *with the relational parameter* Q, *addition* $\dotplus$ *is monotone if functions* $Q_{\mathrm{T}}^{\mathrm{o}}$ *and* $Q_{\mathrm{o}}^{\mathrm{T}}$ *are increasing.*

Proof: Let us take elements n_A, l_A and m_A from A such that $l_A \leq_A m_A$. Then we have

$$Q_{\mathrm{o}}^{\mathrm{T}}(l_A) \leq Q_{\mathrm{o}}^{\mathrm{T}}(m_A)).$$

Consequently,

$$Q_{\mathrm{o}}^{\mathrm{T}}(n_A) + Q_{\mathrm{o}}^{\mathrm{T}}(l_A) \leq_A Q_{\mathrm{o}}^{\mathrm{T}}(n_A) + Q_{\mathrm{o}}^{\mathrm{T}}(m_A)$$

and

$$n_A \dot{+} l_A = Q_{\mathrm{T}}^{\mathrm{o}}(Q_{\mathrm{o}}^{\mathrm{T}}(n_A) + Q_{\mathrm{o}}^{\mathrm{T}}(l_A)) \leq_A Q_{\mathrm{T}}^{\mathrm{o}}(Q_{\mathrm{o}}^{\mathrm{T}}(n_A) + Q_{\mathrm{o}}^{\mathrm{T}}(m_A)) = n_A \dot{+} m_A.$$

Proposition is proved. □

Proposition 2.10.2 and Lemma 2.1.6 imply the following result.

Corollary 2.10.1. *For any dual perspective W-prearithmetic* $\boldsymbol{A} = (A; \dot{+}, \circ, \leq_A)$, *addition* $\dot{+}$ *preserves the order* $\leq_A$ *if functions* $Q_{\mathrm{T}}^{\mathrm{o}}$ *and* $Q_{\mathrm{o}}^{\mathrm{T}}$ *are increasing.*

Proposition 2.10.3. *In a dual perspective W-prearithmetic* $\boldsymbol{A} = (A; \dot{+}, \circ, \leq_A)$ *with the relational parameter* Q, *multiplication* $\circ$ *is monotone if functions* $Q_{\mathrm{T}}^{\mathrm{o}}$ *and* $Q_{\mathrm{o}}^{\mathrm{T}}$ *are increasing.*

Proof is similar to the proof of Proposition 2.10.2. □

Proposition 2.10.3 and Lemma 2.1.7 imply the following result.

Corollary 2.10.2. *For any dual perspective W-prearithmetic* $\boldsymbol{A} = (A; \dot{+}, \circ, \leq_A)$, *multiplication* $\circ$ *preserves the order* $\leq_A$ *if functions* $Q_{\mathrm{T}}^{\mathrm{o}}$ *and* $Q_{\mathrm{o}}^{\mathrm{T}}$ *are increasing.*

Different types of relation Q determine specific classes of dual perspective W-prearithmetics. Let us consider one of them.

Definition 2.10.2.

(a) A dual perspective W-prearithmetic $\boldsymbol{A} = (W_{q,r}; \dot{+}, \circ, \leq)$ with the *projector* $Q_{\mathrm{o}}^{\mathrm{T}}(x)$ and the *coprojector* $Q_{\mathrm{T}}^{\mathrm{o}}(x)$ is called *functional* if Q is a function, say f.
(b) The function $f(x)$ is called the *generator* of the *projector* $f_{\mathrm{o}}^{\mathrm{T}}(x)$ and of the *coprojector* $f_{\mathrm{T}}^{\mathrm{o}}(x)$ as well as the *functional parameter* of the prearithmetic $\boldsymbol{A}$.

The functional dual perspective W-prearithmetic with the generator f is denoted by $\boldsymbol{W}_w^f$.

Thus, we build functional dual perspective W-prearithmetics in the same way as it is done in Section 2.9 for functional direct perspective W-prearithmetics. Namely, we consider an interval $[q, r)$ of real numbers with $r \in R^+ \cup \infty$ and an arbitrary projection (surjection) f: $[q, r) \to R^+$.

When $q = 0$ and $r = \infty$, we have $f\colon [0, \infty) = R^+ \to R^+$. By f^{-1}, we denote the relation, which is inverse to the function f. Note that the relationf^{-1} is a function only in some cases, for example, when f is a strictly increasing mapping.

In this setting, we define functions $f_{\mathrm{u_T}}^{\mathrm{o}} : [q, r) \to W$ and $f_{\mathrm{uo}}^{\mathrm{T}} : R^+ \to [q, r)$ by the following expressions, in which $f^{-1}(x) = \{z; f(z) = x\}$ is the coimage of the element $x \in R^+$:

$$f_{\mathrm{uo}}^{\mathrm{T}}(x) = \sup\{m; m \in W \text{ and } m \leq f(x)\} = [f(x)] = \rfloor f(x) \lfloor,$$

$$f_{\mathrm{u_T}}^{\mathrm{o}}(x) = \inf\{n; n \in W \text{ and } \forall z \in f^{-1}(x)(n \geq z)\}.$$

The function $f_{\mathrm{uo}}^{\mathrm{T}}$ maps R into $[q, r)$ because f is a projection, which implies that for each element $x \in R^+$, there is an element $z \in [q, r)$ such that $f(z) = x$.

Taking functions $f_{\mathrm{u_T}}^{\mathrm{o}}$ and $f_{\mathrm{uo}}^{\mathrm{T}}$, we define the function $f_{\mathrm{o}}^{\mathrm{T}} : R^+ \to W_{q,r}$ where $W_{q,r} = [q, r) \cap W$ as the restriction of $f_{\mathrm{uo}}^{\mathrm{T}}$ onto $W_{q,r}$ and the function $f_{\mathrm{T}}^{\mathrm{o}} : W_{q,r} \to R^+$ as the restriction of $f_{\mathrm{u_T}}^{\mathrm{o}}$ onto $W_{q,r}$. Besides, we define order $\leq$ in $\boldsymbol{A}$ as the standard order in the interval $[q, r)$. Functions $f_{\mathrm{o}}^{\mathrm{T}}$ and $f_{\mathrm{T}}^{\mathrm{o}}$ are used to define functional dual perspective arithmetics.

Namely, for any two numbers n_A and m_A from $W_{q,r}$, operations in $\boldsymbol{A}$ are defined as follows:

$$n_A \dotplus m_A = f_{\mathrm{T}}^{\mathrm{o}}(f_{\mathrm{o}}^{\mathrm{T}}(n_A) + f_{\mathrm{o}}^{\mathrm{T}}(m_A)),$$

$$n_A \circ m_A = f_{\mathrm{T}}^{\mathrm{o}}(f_{\mathrm{o}}^{\mathrm{T}}(n_A) \cdot f_{\mathrm{o}}^{\mathrm{T}}(m_A)).$$

Besides, we define order $\leq$ in $\boldsymbol{A}$ as the standard order in the interval $W_{q,r} = [q, r)$.

Let us consider some examples.

Example 2.10.2. Let us take the function $f(x) = kx$ where k is a natural number as the functional parameter of the dual perspective W-prearithmetic $\boldsymbol{A} = (A; \dotplus, \circ, \leq)$. Then for any whole numbers $n = rk$ and $m = qk$, we have

$$n_A \dotplus m_A = k^{-1}(kn + km) = (n + m)_A$$

and

$$n_A \circ m_A = k^{-1}(kr \cdot kq) = (krq)_A.$$

For instance, in $\boldsymbol{A}$ with $k = 3$, we have

$$3_A \dot{+} 3_A = \frac{1}{3}(3 \cdot 3 + 3 \cdot 3) = \frac{1}{3} \cdot 18 = 6_A$$

and

$$3_A \circ 3_A = \frac{1}{3}(3 \cdot 3 \cdot 3 \cdot 3) = \frac{1}{3} \cdot 81 = 27_A.$$

This shows that addition in the dual perspective W-prearithmetic $\boldsymbol{A}$ is the same as in the Diophantine arithmetic $\boldsymbol{W}$, while multiplication is essentially different.

Example 2.10.3. Let us take the function $f(x) = \log_2 x$ and its inverse function $f^{-1}(x) = 2^x$. This allows us to build the dual perspective W-prearithmetic $\boldsymbol{A} = (N; \dot{+}, \circ, \leq)$ and perform summation and multiplication in it, finding some sums and products:

$$1_A \dot{+} 1_A = 2^{(\log_2 1 + \log_2 1)} = 2^{(0+0)} = 2^0 = 1_A,$$
$$1_A \dot{+} 4_A = 2^{(\log_2 1 + \log_2 4)} = 2^{(0+2)} = 2^2 = 4_A,$$
$$2_A \dot{+} 2_A = 2^{(\log_2 2 + \log_2 2)} = 2^{(1+1)} = 2^2 = 4_A,$$
$$2_A \dot{+} 4_A = 2^{(\log_2 2 + \log_2 4)} = 2^{(1+2)} = 2^3 = 8_A,$$

while

$$1_{A}\circ 1_A = 2^{(\log_2 1 \cdot \log_2 1)} = 2^{(0 \cdot 0)} = 2^0 = 1_A,$$
$$2_{A}\circ 2_A = 2^{(\log_2 2 \cdot \log_2 2)} = 2^{(1 \cdot 1)} = 2^1 = 2_A,$$
$$2_{A}\circ 4_A = 2^{(\log_2 2 \cdot \log_2 4)} = 2^{(1 \cdot 2)} = 2^2 = 4_A.$$

We can see that addition and multiplication in the dual perspective W-prearithmetic $\boldsymbol{A}$ are essentially different from addition and multiplication in the Diophantine arithmetic $\boldsymbol{W}$. In particular we see that $1_A \ll 1_A, 1_A \ll 4_A, 2_A \lll 2_A$ and $2_A \lll 4_A$.

Note that in the Diophantine arithmetic $\boldsymbol{W}$, the relation $n \ll m$ is only when n is equal to 0 while $n \lll m$ is true only when n is equal to 1 or when m is equal to 0.

It is possible to treat the function f: $[q, r) \rightarrow [a, \infty)$ as a restriction of a partial function $f_+ : R^+ \times R^+$ defined on the whole set R^+ and build the functional dual perspective W-prearithmetic $\boldsymbol{W}_w^{f+}$.

Lemma 2.10.5. *The functional dual perspective W-prearithmetic $\boldsymbol{W}_w^f$ is a subprearithmetic of the functional dual perspective W-prearithmetic $\boldsymbol{W}_w^{f^+}$.*

Proof is left as an exercise. □

This result and results from Section 2.5, show that in many cases, it is possible to study the functional dual perspective W-prearithmetic $\boldsymbol{W}_w^{f+}$ and derive properties of the functional dual perspective W-prearithmetic $\boldsymbol{W}_w^{f}$ from the properties of $\boldsymbol{W}_w^{f+}$.

Let us consider a dual perspective W-prearithmetic $\boldsymbol{A} = (A; \dot{+}, \circ, \leq)$ with the functional parameter f.

Lemma 2.10.6. *If $f(n)$ is a whole number, then $f_{\mathrm{o}}^{\mathrm{T}}(n) = f(n)$.*

Proof is left as an exercise. □

Corollary 2.10.3. *If for any whole number n, its image $f(n)$ is also a whole number, then (the restriction of) the function $f_{\mathrm{o}}^{\mathrm{T}}$ coincides with f.*

Lemma 2.10.7. *If the function f is increasing, then $f_{\mathrm{o}}^{\mathrm{T}}({}_A) = 0$.*

Indeed, as the function f is a surjection, we have $f(x_A) = 0$ for some number x. Then for all numbers $z_A < x_A$, we have $f(z_A) = 0$ because the function f is increasing. Thus, $f_{\mathrm{o}}^{\mathrm{T}}(0_A) = 0$.

Lemma 2.10.8. *If the function f: $R^+ \to R^+$ is strictly increasing, then $f_{\mathrm{T}}^{\mathrm{o}}(0) = 0_A$.*

Indeed, as the function f is a strictly increasing surjection, then it is a bijection and has the inverse function f^{-1}. The function f^{-1} is also a strictly increasing bijection. Then

$$f_{\mathrm{T}}^{\mathrm{o}}(0) = \inf\{n; n \in W \text{ and } n \geq f^{-1}(0)\} = \inf\{n; n \in W \text{ and } n \geq 0_A\} = 0_A.$$

Lemma 2.10.9. *If the function f is strictly increasing on whole numbers, i.e., for any n and m from $W_{q,r}$, the inequality $n < m$ implies the inequality $f(n) < f(m)$, then for any element n_A from $\boldsymbol{A}$, we have $f_{\mathrm{o}}^{\mathrm{T}}(n_A) \geq n$.*

Indeed, as the function f is a strictly increasing on whole numbers, for any number n from$W_{q,r}$, we have $f(n) \geq n$. Consequently,

$$\begin{aligned} f_{\mathrm{o}}^{\mathrm{T}}(n_A) &= \sup\{m; m \in W \text{ and } m \leq f(n)\} \\ &\geq \sup\{m; m \in W \text{ and } m \leq n\} = n, \end{aligned}$$

i.e., $f_{\mathrm{o}}^{\mathrm{T}}(n_A) \geq n$.

In some cases, it is possible to derive that the function $f_{\mathrm{T}}^{\mathrm{o}}$ is strictly increasing from information that the function f is strictly increasing.

Lemma 2.10.10. *If the function f is strictly increasing and f^{-1} maps whole numbers into whole numbers, then the function f°_{T} is also strictly increasing and $f^{\circ}_{\mathrm{T}} = f$.*

Proof: As the function f is a strictly increasing surjection, then it is a bijection and has the inverse function f^{-1}. The function f^{-1} is also a strictly increasing bijection. As f^{-1} maps whole numbers into whole numbers, i.e., $f^{-1}(n)$ is a whole number whenever n is a whole number, we have

$$f^{\circ}_{\mathrm{T}}(n) = \inf\{m; m \in W \text{ and } m \geq f^{-1}(n)\} = f^{-1}(n).$$

If $n < m$ for numbers n and m from $W_{q,r}$, then we have $f^{-1}(n) < f^{-1}(m)$. Consequently,

$$f^{\circ}_{\mathrm{T}}(n) = f^{-1}(n) < f^{-1}(m) = f^{\circ}_{\mathrm{T}}(m).$$

It means that the function f°_{T} is also strictly increasing.

Lemma is proved. □

Lemma 2.10.11. *If the function f is strictly increasing and f^{-1} maps whole numbers into whole numbers, then for any number n from $W_{q,r}$, we have $f^{\circ}_{\mathrm{T}}(n) \geq n$.*

Proof: As the function f is a strictly increasing surjection, then it is a bijection and has the inverse function f^{-1}. The function f^{-1} is also a strictly increasing bijection. As the function f is a strictly increasing on whole numbers, for any number n from $W_{q,r}$, we have $f^{-1}(n) \geq n$. Consequently,

$$\begin{aligned} f^{\circ}_{\mathrm{T}}(n) &= \inf\{m; m \in W \text{ and } m \geq f^{-1}(n)\} \\ &\geq \inf\{m; m \in W \text{ and } m \geq n\} = n \end{aligned}$$

that is, $f^{\mathrm{T}}_{\circ}(n_A) \geq n$.

Lemma is proved. □

However, without the second condition, the statements of Lemmas 2.10.10 and 2.10.11 are not true as the following example demonstrates.

Example 2.10.4. Let us take, as we did in Example 2.10.2, a natural number k and the function $f(x) = kx$ as the functional parameter of the dual perspective W-prearithmetic $\boldsymbol{A} = (A; \dotplus, \circ, \leq)$. Then the inverse function is $f^{-1}(x) = k^{-1}x$ and taking $k = 10$, we have

$$f^{\circ}_{\mathrm{T}}(1) = \inf\{n; n \in W \text{ and } n \geq f^{-1}(1)\} = \inf\{n; n \in W \text{ and } n \geq 0.1\} = 1,$$

$$f_{\mathrm{T}}^{\mathrm{o}}(2) = \inf\{n; n \in W \text{ and } n \geq f^{-1}(2)\} = \inf\{n; n \in W \text{ and } n \geq 0.2\} = 1,$$

$$f_{\mathrm{T}}^{\mathrm{o}}(3) = \inf\{n; n \in W \text{ and } n \geq f^{-1}(3)\} = \inf\{n; n \in W \text{ and } n \geq 0.3\} = 1.$$

This shows that although the function $f(x) = 10x$ is strictly increasing, the function $f_{\mathrm{T}}^{\mathrm{o}}$ is not strictly increasing. Besides, $f_{\mathrm{T}}^{\mathrm{o}}(3) = 1 < 3$.

Still it is possible to prove a weaker result without additional conditions on the function (relation) f^{-1}.

Lemma 2.10.12. *If the function f is increasing, then the function $f_{\mathrm{T}}^{\mathrm{o}}$ is increasing.*

Proof: Let us consider numbers m and n such that $m < n$. Then by definition, we have

$$f_{\mathrm{T}}^{\mathrm{o}}(m) = \inf\{k; k \in W \text{ and } \forall z \in f^{-1}(m)(k \leq z)\}$$

and

$$f_{\mathrm{T}}^{\mathrm{o}}(n) = \inf\{k; k \in W \text{ and } \forall z \in f^{-1}(n)(k \leq z)\}.$$

As f is an increasing function, we have the following condition:

For any number z from $f^{-1}(n)$, there is any number x from $f^{-1}(m)$ such that $x \leq z$. (*)

Indeed, if z belongs to $f^{-1}(n)$ and x belongs to $f^{-1}(m)$ while $z < x$, then $f(z) = n \leq f(x) = m$ because f is an increasing function. However, this inequality contradicts the initial condition that $m < n$. By the Principle of Excluded Middle, condition (*) is true.

In turn, condition (*) implies

$$f_{\mathrm{T}}^{\mathrm{o}}(m) = \inf\{k; k \in W \text{ and } \forall z \in f^{-1}(m)(k \leq z)\}$$
$$\leq f_{\mathrm{T}}^{\mathrm{o}}(n) = \inf\{k; k \in W \text{ and } \forall z \in f^{-1}(n)(k \leq z)\}.$$

It means that the function $f_{\mathrm{T}}^{\mathrm{o}}$ is increasing.

Lemma is proved. □

Corollary 2.10.4. *If the function f is strictly increasing, then the function $f_{\mathrm{T}}^{\mathrm{o}}$ is increasing.*

Similar results are valid for the function $f_{\mathrm{o}}^{\mathrm{T}}$.

Lemma 2.10.13. *If the function f is increasing, then the function $f_{\mathrm{o}}^{\mathrm{T}}$ is increasing.*

Proof: Let us take numbers x and z such that $x \leq z$. Then

$$f_{\mathrm{o}}^{\mathrm{T}}(x) = \sup\{n; n \in W \text{ and } n \leq f(x)\}$$

and

$$f_{\mathrm{o}}^{\mathrm{T}}(y) = \sup\{n; n \in W \text{ and } n \leq f(y)\}.$$

As f is an increasing function, we have $f(x) \leq f(y)$ and by the definition of infimum, we have

$$\sup\{n; n \in W \text{ and } n \leq f(x) \leq f(y)\} \leq \sup\{n; n \in W \text{ and } n \leq f(y)\},$$

i.e., $f_{\mathrm{o}}^{\mathrm{T}}(a) \leq f_{\mathrm{o}}^{\mathrm{T}}(b)$.

Lemma is proved. □

Corollary 2.10.5. *If the function f is strictly increasing, then the function $f_{\mathrm{o}}^{\mathrm{T}}$ is increasing.*

Note that even when the function f is strictly increasing, then the function $f_{\mathrm{o}}^{\mathrm{T}}$ is not always strictly increasing.

Example 2.10.5. Let us take a natural number k and the function $f(x) = kx$ as the functional parameter of the dual perspective W-prearithmetic $\boldsymbol{A} = (A; \dot{+}, \circ, \leq)$. When $k = 0.1$, we have

$$f_{\mathrm{o}}^{\mathrm{T}}(1) = \sup\{n; n \in W \text{ and } n \leq f(1)\} = \sup\{n; n \in W \text{ and } n \leq 0.1\} = 0,$$

$$f_{\mathrm{o}}^{\mathrm{T}}(2) = \sup\{n; n \in W \text{ and } n \leq f(2)\} = \sup\{n; n \in W \text{ and } n \leq 0.2\} = 0,$$

$$f_{\mathrm{o}}^{\mathrm{T}}(3) = \sup\{n; n \in W \text{ and } n \leq f(3)\} = \sup\{n; n \in W \text{ and } n \leq 0.3\} = 0.$$

This shows that although the function $f(x) = 10x$ is strictly increasing, the function $f_{\mathrm{o}}^{\mathrm{T}}$ is not strictly increasing.

Lemma 2.10.14. *If f is a strictly increasing function and for a whole number n (from the set $W_{q,r}$), the image $f(n)$ is also a whole number, then $f_{\mathrm{T}}^{\mathrm{o}}(f_{\mathrm{o}}^{\mathrm{T}}(n_A)) = n_A$.*

Proof: As the function f is a strictly increasing surjection, then it is a bijection and has the inverse function f^{-1}. Then by Lemma 2.10.6, $f_{\mathrm{o}}^{\mathrm{T}}(n) = f(n)$ and we have

$$f_{\mathrm{T}}^{\mathrm{o}}(f_{\mathrm{o}}^{\mathrm{T}}(n_A)) = f_{\mathrm{T}}^{\mathrm{o}}(f(n_A)) = \inf\{k_A; k \in W \text{ and } k \geq f^{-1}(f(n_A))\}$$
$$= \inf\{k_A; k_A \in W \text{ and } k_A \leq n_A\} = n_A.$$

Lemma is proved. □

Corollary 2.10.6. *If for any whole number n (from the set $W_{q,r}$), the image $f(n_A)$ is also a whole number, then $f^{\circ}_{\mathrm{T}} f^{\mathrm{T}}_{\circ} = 1_W (f^{\circ}_{\mathrm{T}} f^{\mathrm{T}}_{\circ} = 1_{Wq,r})$ is the identity function.*

However, without the second condition, the statements from Lemma 2.10.14 are not true as the Example 2.10.5 demonstrates because for the function f considered in that example, $f^{\circ}_{\mathrm{T}}(f^{\mathrm{T}}_{\circ}(3_A)) = 0$.

Proposition 2.7.1 and properties of the conventional arithmetic $\boldsymbol{W}$ imply the following result.

Proposition 2.10.4. *For any dual perspective W-prearithmetic $\boldsymbol{A} = (A; \dot{+}, \circ, \leq_A)$, addition $\dot{+}$ and multiplication $\circ$ are commutative.*

Proof is left as an exercise. □

Remark 2.10.1. For some dual perspective W-prearithmetics, addition $\dot{+}$ or/and multiplication $\circ$ can be non-associative as the following example demonstrates.

Example 2.10.6. Let us take the function

$$f(x) = \begin{cases} 2x & \text{for } 0 \leq x \leq 3, \\ 10x - 24 & \text{for } 3 \leq x \leq 4, \\ 4x & \text{for } x \geq 4. \end{cases}$$

Then its inverse function is

$$f^{-1}(x) = \begin{cases} 1/2\, x & \text{for } 0 \leq x \leq 6, \\ 0.1x + 2.4 & \text{for } 6 \leq x \leq 16, \\ 1/4\, x & \text{for } x \geq 16. \end{cases}$$

In this case, $f^{\circ}_{\mathrm{T}}(a) = f(a)$ for all whole numbers a and if $c = 2d$, then $f^{\mathrm{T}}_{\circ}(c) = f^{-1}(c)$. This allows us to build the functional dual perspective W-prearithmetic $\boldsymbol{A} = (A; \dot{+}, \circ, \leq)$ with the functional parameter f and perform summation and multiplication in it finding some sums and products:

$$1_A \dot{+} 1_A = f^{\circ}_{\mathrm{T}}(2 \cdot 1 + 2 \cdot 1) = 1/2\,(2 + 2) = 1/2 \cdot 4 = 2_A,$$

$$1_A \dot{+} 2_A = f^{\circ}_{\mathrm{T}}(2 \cdot 1 + 2 \cdot 2) = 1/2\,(2 + 4) = 1/2 \cdot 6 = 3_A,$$

$$\begin{aligned} 1_A \dot{+} 3_A &= f^{\circ}_{\mathrm{T}}(2 \cdot 1 + 2 \cdot 3) = f^{\circ}_{\mathrm{T}}(2 + 6) = f^{\circ}_{\mathrm{T}}(8) \\ &= \inf\{n; n \in W \text{ and } n \geq 0.1 \cdot 8 + 2.4\} \\ &= \inf\{n; n \in W \text{ and } n \geq 3.2\} = 4_A, \end{aligned}$$

$$\begin{aligned}
2_A \dot{+} 2_A &= f_{\mathrm{T}}^{\mathrm{o}}(2 \cdot 2 + 2 \cdot 2) = f_{\mathrm{T}}^{\mathrm{o}}(4+4) = f_{\mathrm{T}}^{\mathrm{o}}(8) = 4_A,\\
2_A \dot{+} 3_A &= f_{\mathrm{T}}^{\mathrm{o}}(2 \cdot 2 + 2 \cdot 3)\\
&= \inf\{n; n \in W \text{ and } n \geq 0.1 \cdot 10 + 2.4\}\\
&= \inf\{n; n \in W \text{ and } n \geq 3.4\} = 4_A,\\
1_A \dot{+} 4_A &= f_{\mathrm{T}}^{\mathrm{o}}(2 \cdot 1 + 4 \cdot 4) = \inf\{n; n \in W \text{ and } n \geq {}^1\!/\!_4\,(2+16)\\
&= \inf\{n; n \in W \text{ and } n \geq 4.25\} = 5_A,\\
1_A \dot{+} 5_A &= f_{\mathrm{T}}^{\mathrm{o}}(2 \cdot 1 + 4 \cdot 5) = \inf\{n; n \in W \text{ and } n \geq {}^1\!/\!_4\,(2+20)\\
&= \inf\{n; n \in W \text{ and } n \geq 5.25\} = 6_A,\\
2_A \dot{+} 4_A &= f_{\mathrm{T}}^{\mathrm{o}}(2 \cdot 2 + 4 \cdot 4) = \inf\{n; n \in W \text{ and } n \geq {}^1\!/\!_4\,(4+16)\\
&= \inf\{n; n \in W \text{ and } n \geq 5\} = 5_A
\end{aligned}$$

and

$$\begin{aligned}
1_A \circ 1_A &= f_{\mathrm{T}}^{\mathrm{o}}((2 \cdot 1) \cdot (2 \cdot 1)) = {}^1\!/\!_2\,(2 \cdot 2) = {}^1\!/\!_2 \cdot 4 = 2_A,\\
1_A \circ 2_A &= f_{\mathrm{T}}^{\mathrm{o}}((2 \cdot 1) \cdot (2 \cdot 2)) = f_{\mathrm{T}}^{\mathrm{o}}(2 \cdot 4) = f_{\mathrm{T}}^{\mathrm{o}}(8) = 4_A,\\
1_A \circ 4_A &= f_{\mathrm{T}}^{\mathrm{o}}((2 \cdot 1) \cdot (4 \cdot 4)) = f_{\mathrm{T}}^{\mathrm{o}}(2 \cdot 16) = f_{\mathrm{T}}^{\mathrm{o}}(32) = {}^1\!/\!_4 \cdot 32 = 8_A,\\
2_A \circ 2_A &= f_{\mathrm{T}}^{\mathrm{o}}((2 \cdot 2) \cdot (2 \cdot 2)) = f_{\mathrm{T}}^{\mathrm{o}}(4 \cdot 4) = {}^1\!/\!_4 \cdot 16 = 4_A.
\end{aligned}$$

This shows that in the dual perspective W-prearithmetic $\boldsymbol{A}$, addition and multiplication are essentially different from their counterparts in the Diophantine arithmetic $\boldsymbol{W}$. In particular, 1_A is not a multiplicative one in $\boldsymbol{A}$ because $1_A \circ 4_A = 8_A$.

Besides, we have

$$(1_A \circ 1_A) \circ 2_A = 2_A \circ 2_A = 4_A$$

while

$$1_A \circ (1_A \circ 2_A) = 1_A \circ 4_A = 8_A$$

and

$$(1_A \dot{+} 1_A) \dot{+} 4_A = 2_A \dot{+} 4_A = 5_A$$

while

$$1_A \dot{+} (1_A \dot{+} 4_A) = 1_A \dot{+} 5_A = 6_A.$$

We see that both addition and multiplication are not associative in the functional dual perspective W-prearithmetic $\boldsymbol{A}$.

At the same time, there are functional dual perspective W-prearithmetics where addition and/or multiplication is associative.

Example 2.10.7. Let us take the function $f(x) = x + 1$ and its inverse function $f^{-1}(x) = x - 1$ when $x > 0$. In this case, $f_{\mathrm{T}}(a) = f(a)$ for all whole numbers a and $f^{\mathrm{T}}(c) = f^{-1}(c)$. This allows us to build the dual perspective W-prearithmetic $\boldsymbol{A} = (A; \dotplus, \circ, \leq)$ and perform summation and multiplication in it finding some sums and products:

$$1_A \dotplus 0_A = ((1 + 1) + (0 + 1)) - 1 = (2 + 1) - 1 = 3 - 1 = 2_A,$$

$$2_A \dotplus 2_A = ((2 + 1) + (2 + 1)) - 1 = (3 + 3) - 1 = 6 - 1 = 5_A$$

and

$$1_A \circ 0_A = ((1 + 1) \cdot (0 + 1)) - 1 = (2 \cdot 1) - 1 = 2 - 1 = 1_A,$$

$$1_A \circ 1_A = ((1 + 1) \cdot (1 + 1)) - 1 = (2 \cdot 2) - 1 = 4 - 1 = 3_A,$$

$$2_A \circ 2_A = ((2 + 1) \cdot (2 + 1)) - 1 = (3 \cdot 3) - 1 = 9 - 1 = 8_A,$$

$$2_A \circ 1_A = ((2 + 1) \cdot (1 + 1)) - 1 = (3 \cdot 2) - 1 = 6 - 1 = 5_A,$$

$$2_A \circ 3_A = ((2 + 1) \cdot (3 + 1)) - 1 = (3 \cdot 4) - 1 = 12 - 1 = 5_A.$$

This shows that addition and multiplication in the functional dual perspective W-prearithmetic $\boldsymbol{A}$ are different from addition and multiplication in the Diophantine arithmetic $\boldsymbol{W}$. In particular, 0_A is not an additive zero and 1_A is not a multiplicative one in $\boldsymbol{A}$.

In a general case, we have

$$m_A + n_A = ((m + 1) + (n + 1)) - 1 = (m + n + 1)_A,$$

$$m_A \circ n_A = ((m + 1) \circ (n + 1)) - 1 = (mn + m + n)_A.$$

Consequently,

$$(m_A + n_A) + q_A = (m + n + 1)_A + q_A = (m + n + q + 2)_A,$$

$$m_A + (n_A + q_A) = m_A + (n + q + 1)_A = (m + n + q + 2)_A$$

and

$$\begin{aligned}(m_A \circ n_A) \circ q_A &= (mn + m + n)_A \circ q_A \\ &= (mnq + mq + nq + mn + m + n + q)_A, \\ m_A \circ (n_A \circ q_A) &= m_A \circ (nq + n + q)_A \\ &= (mnq + mq + nq + mn + m + n + q)_A.\end{aligned}$$

We see that in this functional dual perspective W-prearithmetic, both addition and multiplication are associative.

Proposition 2.10.5. *For any numbers m_A and n_A from a functional dual perspective W-prearithmetic $\boldsymbol{A}$ with the strictly increasing functional parameter f, if $f(m_A)$ is a whole number, $f_{\text{o}}^{\text{T}}(n_A) \neq 0$ and the function f_{T}^{o} is strictly increasing, then $\text{S}m_A \leq m_A \dotplus n_A$.*

Proof: By properties of whole numbers,

$$f_{\text{o}}^{\text{T}}(m_A) < f_{\text{o}}^{\text{T}}(m_A) + f_{\text{o}}^{\text{T}}(n_A)$$

and because f_{T}^{o} is strictly increasing, by Lemma 2.10.14, we have

$$m_A \dotplus n_A = f_{\text{T}}^{\text{o}}(f_{\text{o}}^{\text{T}}(n_A) + f_{\text{o}}^{\text{T}}(m_A)) > f_{\text{T}}^{\text{o}}(f_{\text{o}}^{\text{T}}(m_A)) = m_A.$$

Consequently,

$$\text{S}m_A \leq m_A \dotplus n_A.$$

Proposition is proved. □

As $\text{S}x > x$ for any element in an arbitrary prearithmetic (cf. Section 2.1), we have the following result.

Corollary 2.10.7. *If the functional parameter f maps whole numbers into whole numbers, the function f_{T}^{o} is strictly increasing, and $f_{\text{o}}^{\text{T}}(n_A) = 0_A$ implies $n_A = 0_A$, then $m_A \dotplus n_A > m_A$ for any non-zero numbers m_A and n_A from an arbitrary dual arithmetic $\boldsymbol{A}$ with the functional parameter f.*

Corollary 2.10.8. *If the function f_{o}^{T} is strictly increasing, $m_A \ll n_A$ and $f(n_A)$ is a whole number, then $m_A = 0$.*

Remark 2.10.2. In the dual perspective W-prearithmetic $\boldsymbol{A} = (A; \dotplus, \circ, \leq)$ from Example 2.10.3, we have relations $1_A \ll 1_A$ and $1_A \ll 4_A$, which violate the conclusions from Corollaries 2.10.7 and 2.10.8. However, this does not invalidate these corollaries because the function $f_{\mathrm{o}}^{\mathrm{T}}$ is not strictly increasing. Indeed,

$$f_{\mathrm{uo}}^{\mathrm{T}}(4) = \sup\{m; m \in W \text{ and } m \leq \log_2 4\} = \log_2 4 = 2$$

and

$$f_{\mathrm{uo}}^{\mathrm{T}}(5) = \sup\{m; m \in W \text{ and } m \leq \log_2 5\} = 2.$$

At the same time, Proposition 2.10.5 shows that in many interesting cases of functional dual perspective W-prearithmetics, the relation $\ll$ is trivial. Thus, it is reasonable to consider the relation $\ll_{\mathrm{w}}$, which is weaker than the relation $\ll$. Let us consider a functional dual perspective W-prearithmetic $\boldsymbol{A} = (A; \dot{+}, \circ, \leq)$.

(1) The formula $n_A \lll_{\mathrm{w}} m_A$ where n_A and m_A belong to $\boldsymbol{A}$ means that n_A is *roughly much less* than m_A and in this case, m_A is *roughly much larger* than n_A, i.e., $m_{A\,\mathrm{w}} \ggg n_A$. It is defined as

$$n_A \ll_{\mathrm{w}} m_A \quad \text{if and only if} \quad m_A \dot{+} n_A \leq \mathrm{S}m_A.$$

(2) The formula $a_{A,1}, a_{A,2}, \ldots, a_{A,n-1} \ll_{\mathrm{w}n} b_A$ where $a_{A,1}, a_{A,2}, \ldots, a_{A,n-1}$ and b_A belong to $\boldsymbol{A}$ means that the group $a_{A,1}, a_{A,2}, \ldots, a_{A,n-1}$ is *roughly much less* than b_A and in this case, b_A is *roughly much larger* than the group $a_{A,1}, a_{A,2}, \ldots, a_{A,n-1}$, i.e., $b_{A\,n\mathrm{w}} \gg a_{A,1}, a_{A,2}, \ldots, a_{A,n-1}$. It is defined as

$$a_{A,1}, a_{A,2}, \ldots, a_{A,n-1} \lll_{\mathrm{w}n} b_A$$

$$\text{if and only if } \sum^{\wedge n}(a_{A,1}, a_{A,2}, \ldots, a_{A,n-1}, b_A) \leq Sb_A.$$

For instance, in any functional dual perspective W-prearithmetic $\boldsymbol{A}$, its additive zero 0_A is much less than any element of $\boldsymbol{A}$. In the Diophantine arithmetic $\boldsymbol{W}$, numbers zero 0 and one 1 are roughly much less than any element of $\boldsymbol{W}$.

Lemma 2.10.15. *The relation $\ll$ is stronger than the relation $\ll_{\mathrm{w}}$, i.e., $n_A \ll m_A$ implies $n_A \ll_{\mathrm{w}} m_A$ for any numbers m_A and n_A from a functional dual perspective W-prearithmetic $\boldsymbol{A}$.*

Indeed, if $n_A \ll m_A$, then $m_A \dot{+} n_A \leq m_A$. As $m_A < \mathrm{S}m_A$, we have $m_A \dot{+} n_A \leq \mathrm{S}m_A$. Then by definition, it means $n_A \ll_{\mathrm{w}} m_A$.

Lemma 2.10.16. *For any numbers m_A and n_A from a functional dual perspective W-prearithmetic $\boldsymbol{A}$ with the strictly increasing functional parameter f, if $f(m_A)$ is a whole number, $f_{\text{o}}^{\text{T}}(n_A) \neq 0$, the function f_{T}^{o} is strictly increasing, $0 < n_A$ and $n_A \ll_{\text{w}} m_A$, then $m_A \dot{+} n_A = \text{S}m_A$ for any numbers n_A and m_A from $\boldsymbol{A}$.*

Indeed, the inequality $n_A \ll_{\text{w}} m_A$ implies $m_A \dot{+} n_A \leq \text{S}m_A$, while by Proposition 2.10.5, $\text{S}m_A \leq m_A \dot{+} n_A$.

Proposition 2.10.6. *For any numbers m_A and n_A from a functional dual perspective W-prearithmetic $\boldsymbol{A}$ with the strictly increasing functional parameter f, if $f(m_A)$ is a whole number, $f_{\text{o}}^{\text{T}}(n_A) \neq 0$, the function f_{T}^{o} is strictly increasing, $0 < n_A$ and $n_A \ll_{\text{w}} m_A$, then $\text{S}m_A$ is an additively composite number for any number m_A from $\boldsymbol{A}$.*

Indeed, by Lemma 2.10.16, the relation $n_A \ll_{\text{w}} m_A$ means $m_A \dot{+} n_A = \text{S}m_A$, i.e., $\text{S}m_A$ is an additively composite number.

Corollary 2.10.9. *If $0 < n_A$ and $n_A \ll_{\text{w}} n_A$, then $\text{S}n_A$ is an additively composite number for any number n_A from $\boldsymbol{A}$.*

As in the case of the relation $\ll_{\text{w}}$ (cf. Section 2.9), relation $\ll$ is determined by relations between values of the function $f_{\text{o}}^{\text{T}}(x)$. Namely, we have the following result.

Lemma 2.10.17. *For any numbers n_A and m_A from a functional dual perspective W-prearithmetic $\boldsymbol{A} = (A; \dot{+}, \circ, \leq)$ with the strictly increasing functional parameter f, which maps whole numbers into whole numbers, the inequality $n_A \ll_{\text{w}} m_A$ is true if and only if $f_{\text{o}}^{\text{T}}(n_A) \leq f_{\text{o}}^{\text{T}}(\text{S}m_A) - f_{\text{o}}^{\text{T}}(m_A)$ or $f_{\text{o}}^{\text{T}}(n_A) + f_{\text{o}}^{\text{T}}(m_A) \leq f_{\text{o}}^{\text{T}}(\text{S}m_A)$.*

Proof: *Necessity.* Let us assume that $n_A \ll_{\text{w}} m_A$ for numbers n_A and m_A from $\boldsymbol{A}$. It means that

$$\begin{aligned} m_A \dot{+} n_A &= f_{\text{T}}^{\text{o}}(f_{\text{o}}^{\text{T}}(m_A) + f_{\text{o}}^{\text{T}}(n_A)) \\ &= \inf\{m; m \in W \text{ and } \forall z \in f^{-1}(f_{\text{o}}^{\text{T}}(m_A) + f_{\text{o}}^{\text{T}}(n_A))(z \leq m)\} \\ &\leq \text{S}m_A. \end{aligned}$$

Thus, by the definition of infimum, any number z from $f^{-1}(f_{\text{o}}^{\text{T}}(m_A) + f_{\text{o}}^{\text{T}}(n_A))$ is less than or equal to $\text{S}m_A$.

By definition, $f_{\text{o}}^{\text{T}}(m_A) = f(m_A)$ for any whole number m_A from $\boldsymbol{A}$.

Thus, we have

$$f_{\mathrm{o}}^{\mathrm{T}}(m_A) + f_{\mathrm{o}}^{\mathrm{T}}(n_A) = f(f^{-1}(f_{\mathrm{o}}^{\mathrm{T}}(m_A) + f_{\mathrm{o}}^{\mathrm{T}}(n_A))) \leq f(\mathrm{S}m_A) = f_{\mathrm{o}}^{\mathrm{T}}(\mathrm{S}m_A).$$

This implies the inequality

$$f_{\mathrm{o}}^{\mathrm{T}}(m_A) + f_{\mathrm{o}}^{\mathrm{T}}(n_A) \leq f_{\mathrm{o}}^{\mathrm{T}}(\mathrm{S}m_A)$$

or

$$f_{\mathrm{T}}(n_A) \leq f_{\mathrm{o}}^{\mathrm{T}}(\mathrm{S}m_A) - f_{\mathrm{o}}^{\mathrm{T}}(m_A).$$

Sufficiency. Let us assume that $f_{\mathrm{o}}^{\mathrm{T}}(n_A) + f_{\mathrm{o}}^{\mathrm{T}}(m_A) \leq f_{\mathrm{o}}^{\mathrm{T}}(\mathrm{S}m_A)$. Then

$$\begin{aligned} m_A \dotplus n_A &= f_{\mathrm{T}}^{\mathrm{o}}(f_{\mathrm{o}}^{\mathrm{T}}(m_A) + f_{\mathrm{o}}^{\mathrm{T}}(n_A)) \\ &= \inf\{m; m \in W \text{ and for } \forall z \in f^{-1}(f_{\mathrm{o}}^{\mathrm{T}}(b) + f_{\mathrm{o}}^{\mathrm{T}}(n_A))(z \leq m)\} \\ &\leq f_{\mathrm{T}}^{\mathrm{o}}(f_{\mathrm{o}}^{\mathrm{T}}(\mathrm{S}m_A) = \mathrm{S}m_A. \end{aligned}$$

As by Proposition 2.10.5, the inequality $m_A \dotplus n_A \geq \mathrm{S}m_A$ is true, we have $m_A \dotplus n_A = \mathrm{S}m_A$, i.e., $n_A \ll_{\mathrm{w}} m_A$.

As $f_{\mathrm{o}}^{\mathrm{T}}(n_A) \leq f_{\mathrm{o}}^{\mathrm{T}}(\mathrm{S}m_A) - f_{\mathrm{o}}^{\mathrm{T}}(m_A)$ is equivalent to $f_{\mathrm{o}}^{\mathrm{T}}(n_A) + f_{\mathrm{o}}^{\mathrm{T}}(m_A) \leq f_{\mathrm{o}}^{\mathrm{T}}(\mathrm{S}m_A)$, the lemma is proved. □

This result means that the inequality $n_A \ll_{\mathrm{w}} m_A$ is true if and only if when the value of $f_{\mathrm{o}}^{\mathrm{T}}(n_A)$ is equal to or less than the value of the direct discrete derivative of $f_{\mathrm{o}}^{\mathrm{T}}(m_A)$ (cf., for example, Boole, 1860; 1880; Spiegel, 1971).

Corollary 2.10.10. *For any number a from a functional dual perspective W-prearithmetic **A**, we have* $n_A \ll_{\mathrm{w}} Sn_A$ *if and only if* $f_{\mathrm{o}}^{\mathrm{T}}(Sn_A) + f_{\mathrm{o}}^{\mathrm{T}}(n_A) < f_{\mathrm{o}}^{\mathrm{T}}(SSn_A)$.

Proposition 2.10.7. *If* m_A *and* n_A *are larger than* 1_A *numbers from a functional dual perspective W-prearithmetic **A** with the strictly increasing functional parameter f that maps whole numbers into whole numbers and the function* $f_{\mathrm{T}}^{\mathrm{o}}$ *is strictly increasing, then the inequality* $n_A \ll_{\mathrm{w}} m_A$ *implies the inequality* $n_A \leq m_A$.

Proof: Let us assume that $1_A < n_A \ll_{\mathrm{w}} m_A$ and $1_A < m_A$. Then the relation $n_A \ll_{\mathrm{w}} m_A$ means

$$m_A \dotplus n_A \leq \mathrm{S}m_A$$

and because f is strictly increasing, by Proposition 2.10.5, we have

$$\mathrm{S}n_A \leq m_A \dotplus n_A.$$

Consequently,

$$\mathrm{S}n_A \leq m_A \dotplus n_A \leq \mathrm{S}m_A.$$

Thus, $n_A \leq m_A$.

Proposition is proved. □

Proposition 2.10.7 shows that relation $\ll_{\mathrm{w}}$ is stronger than relation $\leq$.

Now let us study multiplication in functional dual perspective W-prearithmetics.

Proposition 2.10.8. *For any numbers m_A and n_A from a functional dual perspective W-prearithmetic **A** with the strictly increasing functional parameter f, if $f(m_A)$ is a whole number, $f_{\mathrm{o}}^{\mathrm{T}}(n_A) > 1$ and the function $f_{\mathrm{T}}^{\mathrm{o}}$ is strictly increasing, then $\mathrm{S}m_A \leq m_A \circ n_A$.*

Proof: By properties of whole numbers,

$$f_{\mathrm{o}}^{\mathrm{T}}(m_A) < f_{\mathrm{o}}^{\mathrm{T}}(m_A) \cdot f_{\mathrm{o}}^{\mathrm{T}}(n_A)$$

and because $f_{\mathrm{T}}^{\mathrm{o}}$ is strictly increasing, by Lemma 2.10.14, we have

$$m_A \circ n_A = f_{\mathrm{T}}^{\mathrm{o}}(f_{\mathrm{o}}^{\mathrm{T}}(n_A) \cdot f_{\mathrm{o}}^{\mathrm{T}}(m_A)) > f_{\mathrm{T}}^{\mathrm{o}}(f_{\mathrm{o}}^{\mathrm{T}}(m_A)) = m_A.$$

Consequently,

$$\mathrm{S}m_A \leq m_A \circ n_A.$$

Proposition is proved. □

As $\mathrm{S}x > x$ for any element in an arbitrary prearithmetic (cf. Section 2.1), we have the following result.

Corollary 2.10.11. *If the functional parameter f maps whole numbers into whole numbers, the function $f_{\mathrm{T}}^{\mathrm{o}}$ is strictly increasing, and $f_{\mathrm{o}}^{\mathrm{T}}(n_A) = 0_A$ implies $n_A = 0_A$, then $m_A \circ n_A > m_A$ for any non-zero numbers m_A and n_A from an arbitrary functional dual arithmetic **A** with the functional parameter f.*

Corollary 2.10.12. *If the function $f_{\mathrm{o}}^{\mathrm{T}}$ is strictly increasing, and $m_A \lll n_A$ and $f(n_A)$ is a whole number, then $m_A = 1$.*

At the same time, Proposition 2.10.8 shows that in many interesting cases of functional dual perspective W-prearithmetics, the relation $\lll$ is trivial. Thus, it is reasonable to consider the relation $\lll_{\mathrm{w}}$, which is weaker than the relation $\lll$. Let us consider a functional dual perspective W-prearithmetic $\boldsymbol{A} = (A; \dotplus, \circ, \leq)$ and introduce relations $\lll_{\mathrm{w}}$ and $\lll_{\mathrm{w}n}$.

(1) The formula $n_A \lll_{\mathrm{w}} m_A$ where n_A and m_A belong to $\boldsymbol{A}$ means that n_A is *roughly much much less* than m_A and in this case, m_A is *roughly much much larger* than n_A, i.e., $m_A \ {}_{\mathrm{w}}\!\ggg n_A$. It is defined as

$$n_A \lll_{\mathrm{w}} m_A \text{ if and only if } m_A \circ n_A \leq \mathrm{S}m_A.$$

(2) The formula $a_{A,1}, a_{A,2}, \ldots, a_{A,n-1} \lll_{\mathrm{w}n} b_A$ where $a_{A,1}, a_{A,2}, \ldots, a_{A,n-1}$ and b_A belong to $\boldsymbol{A}$ means that the group $a_{A,1}, a_{A,2}, \ldots, a_{A,n-1}$ is *roughly much much less* than b_A and in this case, b_A is *roughly much much larger* than the group $a_{A,1}, a_{A,2}, \ldots, a_{A,n-1}$, i.e., $b_A \ {}_{n\mathrm{w}}\!\ggg a_{A,1}, a_{A,2}, \ldots, a_{A,n-1}$. It is defined as

$$a_{A,1}, a_{A,2}, \ldots, a_{A,n-1} \lll_{\mathrm{w}n} b_A$$
$$\text{if and only if } \Pi^{\wedge n}(a_{A,1}, a_{A,2}, \ldots, a_{A,n-1}, b_A) \leq \mathrm{S}b_A.$$

For instance, in any functional dual perspective W-prearithmetic $\boldsymbol{A}$, its multiplicative one 1_A is roughly much much less than any element of $\boldsymbol{A}$.

Let us find some properties of the relation $\lll_{\mathrm{w}}$.

Proposition 2.10.9. *If $1 < n_A$ and $n_A \lll_{\mathrm{w}} m_A$, then $\mathrm{S}m_A$ is a multiplicatively composite number.*

Indeed, the relation $n_A \lll m_A$ means $m_A \circ n_A = \mathrm{S}m_A$.

Corollary 2.10.13. *If $0 < n_A$ and $n_A \lll_{\mathrm{w}} n_A$, then $\mathrm{S}n_A$ is an multiplicatively composite number for any number n_A from $\boldsymbol{A}$.*

In many functional dual perspective W-prearithmetics, relation $\lll_{\mathrm{w}}$ is determined by properties of the function f_{T}.

Lemma 2.10.18. *For any numbers $n_A \neq 0_A$ and $m_A \neq 0_A$ from a functional dual perspective W-prearithmetic $\boldsymbol{A} = (A; \dotplus, \circ, \leq)$ with the functional parameter f, which maps whole numbers into whole numbers, the inequality $n_A \lll_{\mathrm{w}} m_A$ is true if and only if $f_{\mathrm{T}}(n_A) \leq f_{\mathrm{T}}(\mathrm{S}m_A)/f_{\mathrm{T}}(m_A)$ or $f_{\mathrm{T}}(n_A) \cdot f_{\mathrm{T}}(m_A) \leq f_{\mathrm{T}}(\mathrm{S}m_A)$.*

Proof: *Necessity.* Let us assume that $n_A \lll_w b$ for numbers n_A and m_A from $\boldsymbol{A}$. It means that

$$\begin{aligned} m_A \circ n_A &= f^{\mathrm{o}}_{\mathrm{T}}(f^{\mathrm{T}}_{\mathrm{o}}(m_A) \cdot f^{\mathrm{T}}_{\mathrm{o}}(n_A)) \\ &= \inf\{m; m \in W \text{ and } \forall z \in f^{-1}(f^{\mathrm{T}}_{\mathrm{o}}(m_A) \cdot f^{\mathrm{T}}_{\mathrm{o}}(n_A))(z \leq m)\} \\ &\leq \mathrm{S}m_A. \end{aligned}$$

Thus, by the definition of infimum, any number z from $f^{-1}(f^{\mathrm{T}}_{\mathrm{o}}(m_A) \cdot f^{\mathrm{T}}_{\mathrm{o}}(n_A))$ is less than or equal to $\mathrm{S}m_A$.

By definition, $f^{\mathrm{T}}_{\mathrm{o}}(m_A) = f(m_A)$ for any whole number m_A from $\boldsymbol{A}$. Thus, we have

$$\begin{aligned} f^{\mathrm{T}}_{\mathrm{o}}(m_A) \cdot f^{\mathrm{T}}_{\mathrm{o}}(n_A) &= f(f^{-1}(f^{\mathrm{T}}_{\mathrm{o}}(m_A) \cdot f^{\mathrm{T}}_{\mathrm{o}}(n_A))) \\ &\leq f(\mathrm{S}m_A) = f^{\mathrm{T}}_{\mathrm{o}}(\mathrm{S}m_A). \end{aligned}$$

This implies the inequality

$$f^{\mathrm{T}}_{\mathrm{o}}(m_A) \cdot f^{\mathrm{T}}_{\mathrm{o}}(n_A) \leq f^{\mathrm{T}}_{\mathrm{o}}(\mathrm{S}m_A)$$

or

$$f_{\mathrm{T}}(n_A) \leq f^{\mathrm{T}}_{\mathrm{o}}(\mathrm{S}m_A)/f^{\mathrm{T}}_{\mathrm{o}}(m_A).$$

Sufficiency. Let us assume that $f^{\mathrm{T}}_{\mathrm{o}}(m_A) \cdot f^{\mathrm{T}}_{\mathrm{o}}(n_A) \leq f^{\mathrm{T}}_{\mathrm{o}}(\mathrm{S}m_A)$. Then

$$\begin{aligned} m_A \circ n_A &= f^{\mathrm{o}}_{\mathrm{T}}(f^{\mathrm{T}}_{\mathrm{o}}(m_A) \cdot f^{\mathrm{T}}_{\mathrm{o}}(n_A)) \\ &= \inf\{m; m \in W \text{ and } \forall z \in f^{-1}(f^{\mathrm{T}}_{\mathrm{o}}(b) \cdot f^{\mathrm{T}}_{\mathrm{o}}(n_A))(z \leq m)\} \\ &\leq f^{\mathrm{o}}_{\mathrm{T}}(f^{\mathrm{T}}_{\mathrm{o}}(\mathrm{S}m_A) = \mathrm{S}m_A. \end{aligned}$$

As by Proposition 2.10.5, the inequality $m_A \circ n_A \geq \mathrm{S}m_A$ is true when $m_A \neq 0_A$ and $n_A \neq 0_A$, we have $m_A \circ n_A = \mathrm{S}m_A$, i.e., $n_A \lll_w m_A$.

As $f^{\mathrm{T}}_{\mathrm{o}}(n_A) \leq f^{\mathrm{T}}_{\mathrm{o}}(\mathrm{S}m_A)/f^{\mathrm{T}}_{\mathrm{o}}(m_A)$ is equivalent to $f^{\mathrm{T}}_{\mathrm{o}}(n_A) \cdot f^{\mathrm{T}}_{\mathrm{o}}(m_A) \leq f^{\mathrm{T}}_{\mathrm{o}}(\mathrm{S}m_A)$, the lemma is proved. □

This result means that the inequality $n_A \lll_w m_A$ is true if and only if when the value of $f_{\mathrm{T}}(n_A)$ is less than the value of the direct discrete logarithmic derivative of $f_{\mathrm{T}}(m_A)$ plus 1. Indeed, logarithmic derivative is defined as

$$[\ln f]' = f'/f$$

and the direct discrete logarithmic derivative of $f_{\mathrm{T}}(b)$ is equal to

$$[f_{\mathrm{T}}(\mathrm{S}m_A) - f_{\mathrm{T}}(m_A)]/f_{\mathrm{T}}(m_A).$$

Thus,

$$f_{\mathrm{T}}(\mathrm{S}m_A)/f_{\mathrm{T}}(m_A) - 1 = [f_{\mathrm{T}}(\mathrm{S}m_A) - f_{\mathrm{T}}(m_A)]/f_{\mathrm{T}}(m_A)$$

and

$$f_{\mathrm{T}}(\mathrm{S}m_A)/f_{\mathrm{T}}(m_A) = [f_{\mathrm{T}}(\mathrm{S}m_A) - f_{\mathrm{T}}(m_A)]/f_{\mathrm{T}}(m_A) + 1.$$

Logarithmic differentiation or differentiation by taking the derivative of the logarithm of a function is used to transform products into sums and divisions into subtractions (Bali, 2005; Bird, 1993; Krantz, 2003).

Logarithmic differentiation or differentiation by taking the derivative of the logarithm of a function is used to transform products into sums and divisions into subtractions (Bali, 2005; Bird, 1993; Krantz, 2003).

Corollary 2.10.14. *For any number a from a dual perspective W-prearithmetic $\boldsymbol{A}$, we have $a \lll Sa$ if and only if $f_{\mathrm{T}}(Sa) \cdot f_{\mathrm{T}}(a) < f_{\mathrm{T}}(SSa)$.*

Let us study relation between different types of inequalities.

Proposition 2.10.10. *If m_A and n_A are larger than 1_A numbers from a functional dual perspective W-prearithmetic $\boldsymbol{A}$ with the strictly increasing functional parameter f that maps whole numbers into whole numbers and the function f_{T}° is strictly increasing, then the inequality $n_A \lll_{\mathrm{w}} m_A$ implies the inequality $n_A \leq m_A$.*

Proof: Let us assume that $1_A < n_A \lll_{\mathrm{w}} m_A$ and $1_A < m_A$. Then by definition,

$$m_A \circ n_A \leq \mathrm{S}m_A$$

and because f is strictly increasing, by Proposition 2.10.8, we have

$$\mathrm{S}n_A \leq m_A \circ n_A.$$

Consequently,

$$\mathrm{S}n_A \leq m_A \circ n_A \leq \mathrm{S}m_A.$$

Thus, $n_A \leq m_A$.

Proposition is proved. □

Proposition 2.10.10 means that relation $\lll_{\mathrm{w}}$ is stronger than relation $\leq$.

Similar relations exist between inequalities $\lll_{\mathrm{w}}$ and $\ll_{\mathrm{w}}$.

Proposition 2.10.11. *In a functional dual perspective W-prearithmetic* $\boldsymbol{A} = (A; \dot{+}, \circ, \leq)$ *with the strictly increasing on whole numbers functional parameter* f, *which maps whole numbers into whole numbers, relation* $\lll_{\mathrm{w}}$ *implies relation* $\ll_{\mathrm{w}}$ *for all elements larger than* 1_A.

Proof: Let us take elements n_A and m_A from a functional dual perspective W-prearithmetic $\boldsymbol{A} = (A; \dot{+}, \circ, \leq)$ with the strictly increasing on whole numbers functional parameter f, which maps whole numbers into whole numbers, and assume n_A, $m_A > 1_A$ and $n_A \lll_{\mathrm{w}} m_A$. By Lemma 2.10.18, the relation $n_A \lll_{\mathrm{w}} m_A$ implies $f_{\mathrm{T}}(n_A) \cdot f_{\mathrm{T}}(m_A) \leq f_{\mathrm{T}}(\mathrm{S}m_A)$. By Proposition 2.10.10, the relation $n_A \lll m_A$ implies the inequality $n_A \leq m_A$ because $0_A < n_A$.

As $1 < n$ and f is strictly increasing on whole numbers, we have $2 \leq f_{\mathrm{T}}(n_A) \leq f_{\mathrm{T}}(m_A)$. Then by properties of the Diophantine arithmetic $\boldsymbol{W}$, the product of numbers $f_{\mathrm{T}}(n_A)$ and $f_{\mathrm{T}}(m_A)$ is larger than their sum. Consequently, we have

$$f_{\mathrm{T}}(n_A) + f_{\mathrm{T}}(m_A) \leq f_{\mathrm{T}}(n_A) \cdot f_{\mathrm{T}}(m_A) \leq f_{\mathrm{T}}(\mathrm{S}m_A).$$

By Lemma 2.10.17, the inequality $f_{\mathrm{T}}(n_A) + f_{\mathrm{T}}(m_A) \leq f_{\mathrm{T}}(\mathrm{S}m_A)$ implies $n_A \ll_{\mathrm{w}} m_A$.

Proposition is proved. □

Proposition 2.10.11 means that relation $\lll_{\mathrm{w}}$ is stronger than relation $\ll_{\mathrm{w}}$.

Let us consider a functional dual perspective W-prearithmetic $\boldsymbol{A} = (W; \dot{+}, \circ, \leq)$ with the functional parameter $f{:}[0, r) \to R^+$ such that $f(0) = 0$.

Theorem 2.10.1. *If for any number* n_A *in* $\boldsymbol{A}$, *its image* $f(n_A)$ *is also a whole number,* $f^{\circ}_{\mathrm{T}}(x)$ *is an increasing function and* $f^{\mathrm{T}}_{\circ}(0_A) = 0$, *then for any* $n_A \in \boldsymbol{A}$, *the equality* $0_A \dot{+} n_A = n_A$ $(n_A \dot{+} 0_A = n_A)$ *is true if and only if* $f^{\mathrm{T}}_{\circ}(x)$ *is a strictly increasing function.*

Proof: *Necessity.* Let us assume that $n_A + 0_A = n_A$ for any number n_A from $\boldsymbol{A}$ but for some numbers m_A and n_A from $\boldsymbol{A}$ with $m_A < n_A$, we have

$f_\circ^{\mathrm{T}}(n_A) \leq f_\circ^{\mathrm{T}}(m_A)$. Then because $f_\circ^{\mathrm{T}}(m_A)$ is a whole number, we have

$$\begin{aligned} n_A \dotplus 0_A &= f_{\mathrm{T}}^{\circ}(f_\circ^{\mathrm{T}}(n_A) + f_\circ^{\mathrm{T}}(0_A)) = f_{\mathrm{T}}^{\circ}(f_\circ^{\mathrm{T}}(n_A) + 0) \\ &= f_{\mathrm{T}}^{\circ}(f_\circ^{\mathrm{T}}(n_A)) \leq f_{\mathrm{T}}^{\circ}(f_\circ^{\mathrm{T}}(m_A)) = m_A < n_A. \end{aligned}$$

It means that $n_A \dotplus 0_A \neq n_A$. Thus, by the Principle of Excluded Middle, $f_\circ^{\mathrm{T}}(x)$ is a strictly increasing function.

The case $n_A \dotplus 0_A = n_A$ is considered in a similar way.

Sufficiency. Let us assume that $f_\circ^{\mathrm{T}}(x)$ is a strictly increasing function. Then by Proposition 2.10.4, we have the following sequence of equalities

$$n_A \dotplus 0_A = f_{\mathrm{T}}^{\circ}(f_\circ^{\mathrm{T}}(n_A) + f_\circ^{\mathrm{T}}(0_A)) = f_{\mathrm{T}}^{\circ}(f_\circ^{\mathrm{T}}(n_A) + 0) = f_{\mathrm{T}}^{\circ}(f_\circ^{\mathrm{T}}(n_A)) = n_A.$$

Theorem is proved. □

In contrast to the Diophantine arithmetic $\boldsymbol{W}$, some functional dual perspective W-prearithmetics have divisors of zero. To show this, let us consider a functional dual perspective W-prearithmetic $\boldsymbol{A} = (W; \dotplus, \circ, \leq)$ with the functional parameter $f(x) = 0.1x$. Then we have

$$f_\circ^{\mathrm{T}}(2_A) = \sup\{m; m \in W \text{ and } m \leq 0.1 \cdot 2 = 0.2)\} = 0$$

and

$$f_\circ^{\mathrm{T}}(3_A) = \sup\{m; m \in W \text{ and } m \leq 0.1 \cdot 3 = 0.3)\} = 0.$$

Consequently

$$2_A \circ 3_A = f_{\mathrm{T}}^{\circ}(f_\circ^{\mathrm{T}}(2_A) \cdot f_\circ^{\mathrm{T}}(0_A)) = f_{\mathrm{T}}^{\circ}(0) = 0_A.$$

It means that numbers 2_A and 3_A are divisors of zero in the W-prearithmetic $\boldsymbol{A}$.

In the Diophantine arithmetic $\boldsymbol{W}$, there are as many even numbers as there are odd numbers. This is not true for functional dual perspective W-prearithmetics in a general case.

Proposition 2.10.12. *There is a functional dual perspective W-prearithmetic $\boldsymbol{A} = (A; \dotplus, \circ, \leq_A)$, in which all numbers but one are even.*

Proof: Let $\boldsymbol{A}$ be an arithmetic with the functional parameter $f(x) = 2^{2^x}$ when $x > 0$ and $f(0_A) = 0$. Its inverse function is $f^{-1}(x) = \log_2(\log_2(x))$.

In $\boldsymbol{A}$, for any $n > 2$, we have

$$f_{\text{o}}^{\text{T}}(n_A) \cdot f_{\text{o}}^{\text{T}}(2_A) = f(n_A) \cdot f(2_A) = 2^{2^n} \cdot 2^{2^2} = 2^{2^n+2^2} = 2^{2^n+4} < 2^{2^{n+1}}.$$

Consequently, by definition, we have

$$\begin{aligned} 2_A \circ n_A &= f_{\text{T}}^{\text{o}}(f_{\text{o}}^{\text{T}}(n_A) \cdot f_{\text{o}}^{\text{T}}(2_A)) \\ &= \inf\{m; m \in W \text{ and } \forall z \in \log_2(\log_2(2^{2^n+4})(m \geq z)\} \\ &\leq \inf\{m; m \in W \text{ and } \forall z \in \log_2(\log_2(2^{2^{n+1}})(m \geq z)\} \\ &= \inf\{m; m \in W \text{ and } m \geq n+1\} = (n+1)_A. \end{aligned}$$

It means that $2_A \circ n_A = Sn_A$ and $2_A \lll_{\text{w}} n_A$. Consequently, all numbers $4_A, 5_A, \ldots$ are even in the constructed W-prearithmetic $\boldsymbol{A}$.

Besides,

$$f_{\text{o}}^{\text{T}}(2_A) \cdot f_{\text{o}}^{\text{T}}(2_A) = 2^{2^2} \cdot 2^{2^2} = 2^4 \cdot 2^4 = 16 \cdot 16 = 256 = 2^8 = f_{\text{o}}^{\text{T}}(3_A).$$

As $f_{\text{T}}^{\text{o}}(2^8) = 3_A$, we have $2_A \circ 2_A = 3_A$. Consequently, all numbers but 1_A are even in the constructed W-prearithmetic $\boldsymbol{A}$.

Proposition is proved. □

Because all even numbers but 2 are composite, we have the following result.

Corollary 2.10.15. *The constructed W-prearithmetic* $\boldsymbol{A}$ *has only one multiplicatively prime number* 2_A.

Corollary 2.10.16. *There are functional dual perspective W-prearithmetics that have only one multiplicatively prime number.*

Corollary 2.10.17. *In the constructed W-prearithmetic* $\boldsymbol{A}$, *all numbers but two are composite.*

Corollary 2.10.18. *There are functional dual perspective W-prearithmetics in which all numbers but two are composite.*

Let us study compatibility of order relations in functional dual perspective W-prearithmetics.

Theorem 2.10.2. *In a functional dual perspective W-prearithmetic* $\boldsymbol{A} = (A; \dot{+}, \circ, \leq)$ *with the increasing functional parameter* f, *which maps whole numbers into whole numbers, relation* $\ll_{\text{w}}$ *is compatible from the left with the order* $\leq$.

Proof: Let us assume that k_A, n_A and m_A are numbers from $\boldsymbol{A}$ such that $k_A \leq n_A$ and $n_A \ll_{\mathrm{w}} m_A$. By Lemma 2.10.17, it means that $f_{\mathrm{o}}^{\mathrm{T}}(n_A) + f_{\mathrm{o}}^{\mathrm{T}}(m_A) \leq f_{\mathrm{o}}^{\mathrm{T}}(\mathrm{S}m_A)$. As $k_A \leq n_A$, Lemma 2.10.13 implies the inequality $f_{\mathrm{o}}^{\mathrm{T}}(k_A) \leq f_{\mathrm{o}}^{\mathrm{T}}(n_A)$. As addition is monotone in $\boldsymbol{N}$, we have $f_{\mathrm{o}}^{\mathrm{T}}(k_A) + f_{\mathrm{o}}^{\mathrm{T}}(m_A) \leq f_{\mathrm{o}}^{\mathrm{T}}(\mathrm{S}m_A)$. By Lemma 2.10.17, it implies $k_A \ll_{\mathrm{w}} m_A$.

Theorem is proved. □

Corollary 2.10.19. *In a functional dual perspective W-prearithmetic $\boldsymbol{A} = (A; \dot{+}, \circ, \leq)$ with the increasing functional parameter f, which maps whole numbers into whole numbers, the order $\leq$ is compatible from the right with the relation $\ll_{\mathrm{w}}$.*

Theorem 2.10.3. *In a functional dual perspective W-prearithmetic $\boldsymbol{A} = (A; \dot{+}, \circ, \leq)$ with the increasing functional parameter f, which maps whole numbers into whole numbers, relation $\lll_{\mathrm{w}}$ is compatible from the left with the order $\leq$.*

Proof: Let us assume that k_A, n_A and m_A are numbers from $\boldsymbol{A}$ such that $k_A \leq n_A$ and $n_A \lll_{\mathrm{w}} m_A$. By Lemma 2.10.18, it means that $f_{\mathrm{o}}^{\mathrm{T}}(n_A) \cdot f_{\mathrm{o}}^{\mathrm{T}}(m_A) \leq f_{\mathrm{o}}^{\mathrm{T}}(\mathrm{S}m_A)$. The inequality $k_A \leq n_A$ implies the inequality, $f_{\mathrm{o}}^{\mathrm{T}}(k_A) \leq f_{\mathrm{o}}^{\mathrm{T}}(n_A)$ because by Lemma 2.10.10 the function $f_{\mathrm{o}}^{\mathrm{T}}$ is increasing. Thus, $f_{\mathrm{o}}^{\mathrm{T}}(k_A) \cdot f_{\mathrm{o}}^{\mathrm{T}}(m_A) \leq f_{\mathrm{o}}^{\mathrm{T}}(\mathrm{S}m_A)$. By Lemma 2.10.18, it means $k_A \lll_{\mathrm{w}} m_A$.

Theorem is proved. □

Corollary 2.10.20. *In a functional dual perspective W-prearithmetic $\boldsymbol{A} = (A; \dot{+}, \circ, \leq)$ with the increasing functional parameter f, which maps whole numbers into whole numbers, the order $\leq$ is compatible from the right with the relation $\lll_{\mathrm{w}}$.*

Now we come to dual perspective prearithmetics of natural numbers building them in the same way as dual perspective W-prearithmetics. Namely, we consider an interval $[q, r)$ of real numbers with $r \in [1, \infty]$ and a relation $Q \subseteq [q, r) \times [1, \infty)$ that has the domain $\mathrm{Dom}(Q) = [q, r)$, the same definability domain $\mathrm{DDom}(Q) = [q, r)$, the range $\mathrm{Rg}(Q) = [1, \infty)$ and the same codomain $\mathrm{Codom}(Q) = [1, \infty)$, we define dual perspective prearithmetics using functions $Q_{\mathrm{o}}^{\mathrm{T}}(x)$ and $Q_{\mathrm{T}}^{\mathrm{o}}(x)$.

Definition 2.10.3.

(a) A prearithmetic $\boldsymbol{A} = (A; \dot{+}, \circ, \leq)$ is called a *dual perspective prearithmetic* if it is weakly projective with respect to the conventional

(Diophantine) arithmetic $\boldsymbol{N} = (N; +, \cdot, \leq)$ with the *projector* $Q_{\mathrm{o}}^{\mathrm{T}}(x)$ and the *coprojector* $Q_{\mathrm{T}}^{\mathrm{o}}(x)$.

(b) The relation $Q(x)$ is called the *relational parameter* of the prearithmetic $\boldsymbol{A}$, as well as the *generator* of the *projector* $Q_{\mathrm{o}}^{\mathrm{T}}(x)$ and of the *coprojector* $Q_{\mathrm{T}}^{\mathrm{o}}(x)$.

The dual perspective prearithmetic with the generator Q is denoted by $\boldsymbol{N}_w^Q$.

Numbers of $\boldsymbol{A}$ are ordered by the same order relation $\leq$ by which they are ordered in $\boldsymbol{N}$. This implies the following result.

Lemma 2.10.19. *The unary operation S of taking the next number is defined in all dual perspective prearithmetics with more than one element coinciding with the operation S in the Diophantine arithmetic $\boldsymbol{N}$.*

Proof is left as an exercise. □

Dual perspective prearithmetics also inherit some Archimedean properties (cf. Section 2.1) from the Diophantine arithmetic $\boldsymbol{N}$.

Lemma 2.10.20. *Any dual perspective prearithmetic $\boldsymbol{A} = (A; \dot{+}, \circ, \leq)$ is successively Archimedean and exactly successively Archimedean.*

Proof is left as an exercise. □

As elements of a dual perspective prearithmetic $\boldsymbol{A} = (A; \dot{+}, \circ, \leq)$ form an interval of whole numbers, we have the following result.

Lemma 2.10.21. *If a dual perspective prearithmetic $\boldsymbol{A} = (A; \dot{+}, \circ, \leq)$ with more than one element has one 1_A, then $\mathrm{S}1_A = 2_A$.*

Proof is left as an exercise. □

Operations in $\boldsymbol{A}$ are defined by the following formulas:

$$n_A \dot{+} m_A = Q_{\mathrm{T}}^{\mathrm{o}}(Q_{\mathrm{o}}^{\mathrm{T}}(n_A) + Q_{\mathrm{o}}^{\mathrm{T}}(m_A)),$$

$$n_A \circ m_A = Q_{\mathrm{T}}^{\mathrm{o}}(Q_{\mathrm{o}}^{\mathrm{T}}(n_A) \cdot Q_{\mathrm{o}}^{\mathrm{T}}(m_A)).$$

There are natural relations between dual perspective W-prearithmetics and dual perspective prearithmetics.

Lemma 2.10.22. *If $Q \subseteq [q, r) \times [1, \infty)$ and $1 \leq q$, then any dual perspective W-prearithmetic with the relational parameter Q is a dual perspective prearithmetic.*

Proof is left as an exercise. □

Taking $Q \subseteq [q, r) \times [1, \infty)$ and $1 \leq q$, we obtain the following result.

Proposition 2.10.13. *Any dual perspective prearithmetic is a dual perspective W-prearithmetic.*

Proof is left as an exercise. □

Corollary 2.10.21. *Any dual perspective prearithmetic is a subprearithmetic of a dual perspective W-prearithmetic.*

These results allow deducing many properties of dual perspective prearithmetics from properties of dual perspective W-prearithmetics. For instance, Propositions 2.10.2 and 2.10.9 imply the following result.

Proposition 2.10.14. *In a dual perspective prearithmetic with the relational parameter Q, addition $\dot{+}$ is monotone if functions $Q_{\circ\mathrm{T}}$ and $Q_{\circ}^{\mathrm{T}}$ are increasing.*

Proof is left as an exercise. □

Proposition 2.10.14 and Lemma 2.1.6 imply the following result.

Corollary 2.10.22. *For any dual perspective prearithmetic $\boldsymbol{A} = (A; \dot{+}, \circ, \leq)$ with the relational parameter Q, addition $\dot{+}$ preserves the order $\leq$ if functions $Q_{\circ\mathrm{T}}$ and $Q_{\circ}^{\mathrm{T}}$ are increasing.*

Multiplication $\circ$ in functional dual perspective prearithmetics has similar properties.

Proposition 2.10.15. *In a dual perspective prearithmetic $\boldsymbol{A} = (A; \dot{+}, \circ, \leq)$ with the relational parameter Q, multiplication $\circ$ is monotone if functions $Q_{\circ\mathrm{T}}$ and $Q_{\circ}^{\mathrm{T}}$ are increasing.*

Proof is left as an exercise. □

Let us study functional dual perspective prearithmetics of natural numbers.

Definition 2.10.4.

(a) A dual perspective prearithmetic $\boldsymbol{A} = (N_{q,r}; \dot{+}, \circ, \leq)$ with the *projector* $Q_{\circ}^{\mathrm{T}}(x)$ and the *coprojector* $Q_{\mathrm{T}}^{\circ}(x)$ is called *functional* if Q is a function f.

(b) The function $f(x)$ is called the *generator* of the *projector* $f_{\circ}^{\mathrm{T}}(x)$, of the *coprojector* $f_{\mathrm{T}}^{\circ}(x)$ and the *functional parameter* of the prearithmetic $\boldsymbol{A}$.

The functional dual perspective prearithmetic with the generator f is denoted by $\boldsymbol{N}_{w}^{f}$.

For convenience, we describe the construction in more detail. We take an interval $[q, r)$ of real numbers with $r \in [1, \infty]$ and an arbitrary projection (surjection) $f: [q, r) \to [1, \infty)$. When $q = 1$ and $r = \infty$, we have $f: [1, \infty) \to [1, \infty)$. By f^{-1}, we denote the relation, which is inverse to the function f. Note that the relation f^{-1} is a function only in some cases, for example, when f is a strictly increasing mapping.

In this setting, we define functions $f^{\mathrm{o}}_{\mathrm{uT}} : [q, r) \to N$ and $f^{\mathrm{T}}_{\mathrm{uo}} : [1, \infty) \to [q, r)$ by the following expressions, in which $f^{-1}(x) = \{z; f(z) = x\}$ is the coimage of the element $x \in R^+$:

$$f^{\mathrm{T}}_{\mathrm{uo}}(x) = \sup\{m; m \in N \text{ and } m \leq f(x)\} = [f(x)] = \rfloor f(x) \lfloor,$$

$$f^{\mathrm{o}}_{\mathrm{uT}}(x) = \inf\{n; n \in N \text{ and } \forall z \in f^{-1}(x)(n \geq z)\}.$$

The function $f^{\mathrm{T}}_{\mathrm{uo}}$ maps R into $[q, r)$ because f is a projection, which implies that for each element $x \in R^+$, there is an element $z \in [q, r)$ such that $f(z) = x$.

Taking functions $f^{\mathrm{o}}_{\mathrm{uT}}$ and $f^{\mathrm{T}}_{\mathrm{uo}}$, we define the function $f^{\mathrm{T}}_{\mathrm{o}} : [1, \infty) \to N_{q,r}$ where $N_{q,r} = [q, r) \cap W$ as the restriction of $f^{\mathrm{T}}_{\mathrm{uo}}$ onto $N_{q,r}$ and the function $f^{\mathrm{o}}_{\mathrm{T}} : N_{q,r} \to [1, \infty)$ as the restriction of $f^{\mathrm{o}}_{\mathrm{uT}}$ onto $N_{q,r}$.

Operations in $\boldsymbol{A}$ are defined as follows:

$$n_A \dotplus m_A = f^{\mathrm{o}}_{\mathrm{T}}(f^{\mathrm{T}}_{\mathrm{o}}(n_A) + f^{\mathrm{T}}_{\mathrm{o}}(m_A)),$$

$$n_A \circ m_A = f^{\mathrm{o}}_{\mathrm{T}}(f^{\mathrm{T}}_{\mathrm{o}}(n_A) \cdot f^{\mathrm{T}}_{\mathrm{o}}(m_A)).$$

Besides, we define order $\leq$ in $\boldsymbol{A}$ as the standard order in the interval $N_{q,r} = [q, r)$.

There are natural relations between functional dual perspective W-prearithmetics and functional dual perspective prearithmetics.

Lemma 2.10.23. *If $f: [q, r) \to [1, \infty)$ and $1 \leq q$, then any functional dual perspective W-prearithmetic with the relational parameter Q is a functional dual perspective prearithmetic.*

Proof is left as an exercise. □

Taking $Q \subseteq [q, r) \times [1, \infty)$ and $1 \leq q$, we obtain the following result.

Proposition 2.10.16. *Any functional dual perspective prearithmetic is a functional dual perspective W-prearithmetic.*

Proof is left as an exercise. □

Corollary 2.10.23. *Any functional dual perspective prearithmetic is a subprearithmetic of a functional dual perspective W-prearithmetic.*

These results allow deducing many properties of dual perspective prearithmetics from properties of dual perspective W-prearithmetics. For instance, Propositions 2.10.2 and 2.10.16 imply the following result.

Proposition 2.10.17. *In a dual perspective prearithmetic with the relational parameter Q, addition $\dot{+}$ is monotone if functions $Q_{\circ \mathrm{T}}$ and $Q_{\circ}^{\mathrm{T}}$ are increasing.*

Proof is left as an exercise. □

Propositions 2.10.3 and 2.10.16 imply the following result.

Proposition 2.10.18. *In a dual perspective prearithmetic with the relational parameter Q, multiplication $\circ$ is monotone if functions $Q_{\circ \mathrm{T}}$ and $Q_{\circ}^{\mathrm{T}}$ are increasing.*

Proof is left as an exercise. □

Proposition 2.10.5 allows proving the following result

Proposition 2.10.19. *A functional dual perspective prearithmetic $\boldsymbol{A}$ with the strictly increasing function f_{T}° and functional parameter f, which maps natural numbers into natural numbers, is additively Archimedean.*

Proof: Let us consider the functional dual perspective arithmetic $\boldsymbol{A} = (A; \dot{+}, \circ, \leq)$ that satisfies conditions from Proposition 2.10.19 and take two numbers $m_A < q_A$ from it. Then by Proposition 2.10.5, we have $\mathrm{S}m_A \leq m_A \dot{+} m_A$ and $\mathrm{S}m_A \leq k_A \dot{+} m_A$ for any k. Consequently, by induction, we obtain

$$\mathrm{S}^n m_A \leq \underbrace{(\cdots(m_A \dot{+} m_A) \dot{+} m_A) \cdots m_A) \dot{+} m_A}_{n} = n[m_A]$$

for all $m, n = 1, 2, 3, 4, 5, \ldots$.

As by Lemma 2.10.20, the dual perspective prearithmetic $\boldsymbol{A}$ is successively Archimedean, for some natural number n, we have

$$q_A \leq \mathrm{S}^n m_A.$$

Consequently, we have

$$q_A \leq n[m_A].$$

Proposition is proved. □

This result shows essential difference between dual and direct perspective prearithmetics because according to Theorems 2.9.5 and 2.9.32, there are infinitely many functional direct perspective arithmetics, which satisfy all conditions from Proposition 2.10.11 but are not additively Archimedean. In essence, there are more additively Archimedean dual perspective prearithmetics than additively Archimedean direct perspective prearithmetics, i.e., infinitely many functional parameters, which define additively Archimedean dual perspective prearithmetics, define direct perspective prearithmetics that are not additively Archimedean.

We can also prove a multiplicative counterpart of Proposition 2.10.19.

Proposition 2.10.20. *A functional dual perspective prearithmetic* $\boldsymbol{A}$ *with the strictly increasing function* f_{T}° *and functional parameter* f, *which maps natural numbers into natural numbers, is multiplicatively Archimedean if it does not have the multiplicative one* 1_A.

Proof is similar to the proof of Proposition 2.10.19. □

Proposition 2.10.15 and Lemma 2.1.6 imply the following result.

Corollary 2.10.24. *For any functional dual perspective prearithmetic* $\boldsymbol{A} = (A; \dot{+}, \circ, \leq)$ *with the functional parameter* f, *multiplication* $\circ$ *preserves the order* $\leq$ *if functions* $f_{\circ\mathrm{T}}$ *and* $f_{\circ}^{\mathrm{T}}$ *are increasing.*

As before arithmetics form a special class of prearithmetics.

Definition 2.10.5. A functional dual perspective W-prearithmetic $\boldsymbol{A} = (W; \dot{+}, \circ, \leq)$ with the projector $f^{\mathrm{T}}(x)$ and the coprojector $f_{\mathrm{T}}^{\circ}(x)$ is called a *dual perspective W-arithmetic* if the following conditions are satisfied:

(1) $f_{\circ}^{\mathrm{T}}(0_A) = 0$;
(2) $f_{\circ}^{\mathrm{T}}(x)$ is a strictly increasing function;
(3) for any elements a and b from W, the inequality $a \leq b$ implies the inequality

$$f_{\circ}^{\mathrm{T}}(Sa) - f_{\circ}^{\mathrm{T}}(a) \leq f_{\circ}^{\mathrm{T}}(Sb) - f_{\circ}^{\mathrm{T}}(b).$$

The functional dual perspective arithmetic with the generator f is denoted by $\boldsymbol{N}^f$.

Functional dual perspective arithmetics form a class **DLNDA** of non-Diophantine arithmetics of natural numbers.

Proposition 2.10.21. *Any dual perspective W-arithmetic is also a functional dual perspective W-prearithmetic.*

Proof is left as an exercise. □

This result allows obtaining many properties of dual perspective W-arithmetics from already proved properties of functional dual perspective W-prearithmetics.

Theorem 2.10.1 implies the following result.

Proposition 2.10.22. *In a dual perspective W-arithmetic* $\boldsymbol{A}$, *the element* 0_A *is the additive zero, i.e., for any* $n_A \in \boldsymbol{A}$, *the equality* $0_A \dot{+} n_A = n_A \dot{+} 0_A = n_A$ *is valid.*

Proof is left as an exercise. □

Theorem 2.10.2 implies the following result.

Proposition 2.10.23. *In a dual perspective W-arithmetic* $\boldsymbol{A}$ *with the functional parameter* f, *which maps whole numbers into whole numbers, relation* $\ll_{\mathrm{w}}$ *is compatible from the left with the order* $\leq$.

Proof is left as an exercise. □

Corollary 2.10.25. *In a dual perspective W-arithmetic* $\boldsymbol{A}$ *with the functional parameter* f, *which maps whole numbers into whole numbers, the order* $\leq$ *is compatible from the right with the relation* $\ll_{\mathrm{w}}$.

Theorem 2.10.3 implies the following result.

Proposition 2.10.24. *In a dual perspective W-arithmetic* $\boldsymbol{A}$ *with the functional parameter* f, *which maps whole numbers into whole numbers, relation* $\lll_{\mathrm{w}}$ *is compatible from the left with the order* $\leq$.

Proof is left as an exercise. □

Corollary 2.10.26. *In a dual perspective W-arithmetic* $\boldsymbol{A}$ *with the functional parameter* f, *which maps whole numbers into whole numbers, the order* $\leq$ *is compatible from the right with the relation* $\lll_{\mathrm{w}}$.

Dual perspective arithmetics of natural numbers are similar to dual perspective arithmetics of whole numbers.

Definition 2.10.6. A functional dual perspective prearithmetic $\boldsymbol{A} = (N; \dot{+}, \circ, \leq)$ with the projector $f_{\mathrm{T}}^{\mathrm{o}}(x)$ and the coprojector $f_{\mathrm{o}}^{\mathrm{T}}(x)$ is called a *dual perspective arithmetic* if the following conditions are satisfied:

(1) $f_{\mathrm{o}}^{\mathrm{T}}(1_A) = 1$;
(2) $f_{\mathrm{o}}^{\mathrm{T}}(x)$ is a strictly increasing function;

(3) for any elements a and b from N, the inequality $a \leq b$ implies the inequality

$$f_{\circ}^{\mathrm{T}}(Sa) - f_{\circ}^{\mathrm{T}}(a) \leq f_{\circ}^{\mathrm{T}}(Sb) - f_{\circ}^{\mathrm{T}}(b).$$

This arithmetic of whole numbers is denoted by ${}_f\boldsymbol{N}$.

Dual perspective arithmetics form a class **DLNDAN** of non-Diophantine arithmetics of natural numbers.

Proposition 2.10.25. *Any dual perspective arithmetic is also a functional dual perspective prearithmetic.*

Proof is left as an exercise. □

This result allows obtaining many properties of dual perspective arithmetics from already proved properties of functional dual perspective prearithmetics.

Lemma 2.10.20 and Proposition 2.10.25 imply the following result.

Proposition 2.10.26. *Any dual perspective arithmetic* $\boldsymbol{A} = (A; \dot{+}, \circ, \leq)$ *is successively Archimedean and exactly successively Archimedean.*

Proof is left as an exercise. □

Let us take a natural number $k > 1$ and the function $f_k(n) = k^{2^{2n}}$ for $n = 1, 2, 3, \ldots$, and $f_k(n) = 0$ as the functional parameter for the functional dual perspective W-arithmetic $\boldsymbol{W}^{\mathrm{d}}_{\mathrm{eexp}k} = (W; \dot{+}, \circ, \leq)$. We also have the inverse function $f_k^{-1}(m) = \log_2 \log_2 \log_k m$ for $m > 0$ and $f_k^{-1}(0) = 0$.

Note that $\boldsymbol{W}^{\mathrm{d}}_{\mathrm{eexp}k}$ contains all whole numbers with the natural order $\leq$ between them. If n is a whole number, then its counterpart in $\boldsymbol{W}^{\mathrm{d}}_{\mathrm{eexp}k}$ is denoted by n_W.

Now let us find properties of this arithmetic as well as properties of the functional dual perspective arithmetic $\boldsymbol{N}^{\mathrm{d}}_{\mathrm{eexp}k} = (N; \dot{+}, \circ, \leq)$ of all natural numbers, which is a subarithmetic of the arithmetic $\boldsymbol{W}^{d}_{\mathrm{eexp}k}$.

By Lemma 2.10.3, $f_{k\circ}^{\mathrm{T}}(n_W) = f_k(n) = k^{2^{2n}}$. Besides, by Definition 2.10.2,

$$f_{\mathrm{uT}}^{\circ}(x) = \inf\{m; m \in W \text{ and } m \geq \log_2 \log_2 \log_k x\}.$$

Then for any $n > 0$, we have

$$\begin{aligned} n_W \circ n_W &= f_{\mathrm{uT}}^{\circ}(f_{k\mathrm{T}}(n_W) \cdot f_{k\mathrm{T}}(n_W)) \\ &= f_{\mathrm{uT}}^{\circ}(k^{2^{2^n}} \cdot k^{2^{2^n}}) = f_{\mathrm{uT}}^{\circ}(k^{(2^{2^n} + 2^{2^n})}) \end{aligned}$$

$$
\begin{aligned}
&= f^{\circ}_{\mathrm{u_T}}(k^{2 \cdot 2^{2^n}}) = \inf\{m; m \in W \text{ and } m \geq \log_2 \log_2 \log_k k^{2 \cdot 2^{2^n}}\} \\
&= \inf\{m; m \in W \text{ and } m \geq \log_2 \log_2 (2 \cdot 2^{2^n})\} \\
&= \inf\{m; m \in W \text{ and } m \geq \log_2 (\log_2 2 + \log_2 2^{2^n})\} \\
&= \inf\{m; m \in W \text{ and } m \geq \log_2 (1 + 2^n)\} = (n+1)_W,
\end{aligned}
$$

i.e., $n_W \circ n_W = \mathrm{S}n_W$. It means that $n_W \lll_{\mathrm{w}} n_W$.

By Proposition 2.10.11, relation $\lll_{\mathrm{w}}$ implies relation $\ll_{\mathrm{w}}$ for all elements larger than 1_A because the functional parameter f_k is strictly increasing on whole numbers and maps whole numbers into whole numbers. Consequently, we have $m_W \ll_{\mathrm{w}} n_W$ for all $m_W \leq n_W$. In particular, we obtain $n_W \dotplus n_W = \mathrm{S}n_W$.

Let us show that the functional dual perspective W-arithmetics for $\boldsymbol{W}^{\mathrm{d}}_{\mathrm{eexp}k}$ satisfies conditions from Definition 2.10.5.

The first condition $f_{\mathrm{T}}(0_A) = 0$ is satisfied because $f_{\mathrm{T}}(0_A) = f(0) = 0$.

The second condition, which states that $f^{\mathrm{T}}_{\mathrm{o}}(x)$ is a strictly increasing function, is satisfied because $f^{\mathrm{T}}_{\mathrm{o}}(x) = f(x)$ and $f(x)$ is a strictly increasing function.

The third condition $f^{\mathrm{T}}_{\mathrm{o}}(\mathrm{S}n_A) - f^{\mathrm{T}}_{\mathrm{o}}(n_A) \leq f^{\mathrm{T}}_{\mathrm{o}}(\mathrm{S}m_A) - f^{\mathrm{T}}_{\mathrm{o}}(m_A)$ for any $n \leq m$ is satisfied because if $0 < n \leq m$, then $k^{2^{2^n}} \leq k^{2^{2^m}}$, $((k^{2^{2^n}})^{2^{2^n-1}} - 1) \leq ((k^{2^{2m}})^{2^{2^m-1}} - 1)$ and we have the following sequence of inequalities:

$$
\begin{aligned}
f^{\mathrm{T}}_{\mathrm{o}}(\mathrm{S}n_A) - f^{\mathrm{T}}_{\mathrm{o}}(n_A) &= k^{2^{2^{n+1}}} - k^{2^{2^n}} = k^{2^{2 \cdot (2^n)}} - k^{2^{2^n}} = k^{2^{2^n} \cdot 2^{2^n}} - k^{2^{2^n}} \\
&= (k^{2^{2^n}})^{2^{2^n}} - k^{2^{2^n}} = ((k^{2^{2^n}})^{2^{2^n-1}} - 1) \cdot k^{2^{2^n}} \\
&\leq f^{\mathrm{T}}_{\mathrm{o}}(\mathrm{S}m_A) - f^{\mathrm{T}}_{\mathrm{o}}(m_A) = k^{2^{2^{m+1}}} - k^{2^{2^m}} \\
&= k^{2^{2 \cdot (2^m)}} - k^{2^{2^m}} = k^{2^{2^m} \cdot 2^{2^m}} - k^{2^{2^m}} \\
&= (k^{2^{2^m}})^{2^{2^m}} - k^{2^{2^m}} = ((k^{2^{2^m}})^{2^{2^m-1}} - 1) \cdot k^{2^{2^m}}.
\end{aligned}
$$

When $n = 0$, the third condition is evident.

Thus, we proved that $\boldsymbol{W}^{\mathrm{d}}_{\mathrm{eexp}k}$ is a functional dual perspective W-arithmetic.

In addition to the functional dual perspective W-arithmetic $\boldsymbol{W}^{\mathrm{d}}_{\mathrm{eexp}k}$, we build the functional dual perspective arithmetic $\boldsymbol{N}^{\mathrm{d}}_{\mathrm{eexp}k}$ using the following function as its generator (functional parameter):

$$
f^{1}_{k}(n) = \begin{cases} 1 & \text{for } n = 1, \\ k^{2^{2^n}} & \text{for } n = 2, 3, \ldots. \end{cases}
$$

Taking $\boldsymbol{N}^{\mathrm{d}}_{\mathrm{eexp}k}$, we can check all conditions from the definition of functional dual perspective arithmetics in the same way as we did this for $\boldsymbol{W}^{\mathrm{d}}_{\mathrm{eexp}k}$. As a result, we see that $\boldsymbol{N}^{\mathrm{d}}_{\mathrm{eexp}k}$ is a functional dual perspective arithmetic, in which its numbers $1, 2, 3, \ldots, n, \ldots$ are denoted by 1_N, 2_N, 3_N, $\ldots$, n_N, $\ldots$.

Properties of the functional dual perspective W-arithmetic $\boldsymbol{W}^{\mathrm{d}}_{\mathrm{eexp}k}$ and the functional dual perspective arithmetic $\boldsymbol{N}^{\mathrm{d}}_{\mathrm{eexp}k}$ give us the following results.

Lemma 2.10.24. *In the functional dual perspective W-arithmetic* $\boldsymbol{W}^{\mathrm{d}}_{\mathrm{eexp}k}$, *any number larger than* 1_W *is a square.*

Proof is left as an exercise. □

Lemma 2.10.25. *In the functional dual perspective W-arithmetic* $\boldsymbol{N}^{\mathrm{d}}_{\mathrm{eexp}k}$, *any number larger than* 1_N *is a square.*

Proof is left as an exercise. □

Remark 2.10.3. These results look similar to the results from Section 2.9. However, the constructed W-arithmetic $\boldsymbol{W}^{\mathrm{d}}_{\mathrm{eexp}k}$ demonstrates essential differences between functional dual perspective W-arithmetics and functional direct perspective W-arithmetics studied in Section 2.9. Indeed, the functional direct perspective W-arithmetic $\boldsymbol{W}_{\mathrm{eexp}k}$ constructed in Section 2.9 has the same generator $f(n)$ as the functional dual perspective W-arithmetic $\boldsymbol{W}^{\mathrm{d}}_{\mathrm{eexp}k}$. At the same time, the identity $n_W \circ n_W = n_W$ is valid in $\boldsymbol{W}_{\mathrm{eexp}k}$ while it is invalid in $\boldsymbol{W}^{\mathrm{d}}_{\mathrm{eexp}k}$. In a similar way, the identity $n_W \circ n_W = \mathrm{S}n_W$ is valid in $\boldsymbol{W}^{\mathrm{d}}_{\mathrm{eexp}k}$ while it is invalid in $\boldsymbol{W}_{\mathrm{eexp}k}$.

Properties of arithmetics $\boldsymbol{W}^{\mathrm{d}}_{\mathrm{eexp}k}$ and $\boldsymbol{N}^{\mathrm{d}}_{\mathrm{eexp}k}$ allow us to show many unusual properties of non-Diophantine arithmetics.

Theorem 2.10.4. *There are infinitely many functional dual perspective arithmetics, in which almost all numbers are squares.*

Indeed, in each functional dual perspective arithmetic $\boldsymbol{N}^{\mathrm{d}}_{\mathrm{eexp}k}$, we have the equality $n_W \circ n_W = \mathrm{S}n_W$ for all numbers n larger than 0. In other words, in $\boldsymbol{N}^{\mathrm{d}}_{\mathrm{eexp}k}$, each number larger than 1 is the square of its predecessor. As this is true for any $k > 1$, we have infinitely many non-Diophantine arithmetics $\boldsymbol{N}^{\mathrm{d}}_{\mathrm{eexp}k}$ because k can be equal to 2, 3, 4, 5 and so on.

Corollary 2.10.27. *There are infinitely many functional dual perspective W-arithmetics, in almost all numbers are squares.*

Although functional dual perspective arithmetics are different from functional direct perspective arithmetics, some of their properties are similar. For instance, we have the following result.

Theorem 2.10.5. *There are infinitely many functional dual perspective arithmetics (non-Diophantine arithmetics of natural numbers), in which for any natural number n, the equation $x^n = y^n + z^n$ has infinitely many solutions.*

Proof: Let us consider the functional dual perspective arithmetic $\boldsymbol{N}^1_{\text{eexp}k}$ where k can be any natural number larger than 1. It was proved that in $\boldsymbol{N}^1_{\text{eexp}k}$, $m_W \circ m_W = \mathrm{S}m_W$ for all $m = 1, 2, 3, 4, 5, \ldots$. Consequently, by induction, we obtain

$$m_W^n = \underbrace{(\cdots(m_W \circ m_W) \circ m_W)\cdots m_W) \circ m_W}_{n} = \mathrm{S}^n m_W$$

for all $m, n = 1, 2, 3, 4, 5, \ldots$.

In addition, we have demonstrated that in the functional dual perspective arithmetic $\boldsymbol{N}^1_{\text{eexp}k}$, we have $q_W \dot{+} q_W = \mathrm{S}q_W$ for all $q = 1, 2, 3, 4, 5, \ldots$. Thus, we obtain

$$m_W^n \dot{+} m_W^n = \mathrm{S}^n m_W \dot{+} \mathrm{S}^n m_W = \mathrm{S}^{n+1} m_W = \mathrm{S}^n(\mathrm{S}m_W) = (\mathrm{S}m_W)^n.$$

This gives us infinitely many solutions of the equation $x^n = y^n + z^n$, consequently, in any non-functional dual perspective $\boldsymbol{N}^1_{\text{eexp}k}$. Besides, we already know that there are infinitely many such non-Diophantine arithmetics.

Theorem is proved. □

2.11. Axiomatic Theories of Arithmetic

Axioms, when rightly investigated and established,
prepare us not for a limited but abundant practice,
and bring in their train whole troops of effects.

Francis Bacon

Any mathematical theory can be presented in three basic forms: structural, explorative and axiomatic.

- In the *structural form*, mathematical knowledge is oriented on structures, their properties and operations with them.

- In the *explorative form*, mathematics goes from problems to their solutions via operations, algorithms and procedures.
- In the *axiomatic form*, mathematics goes from axioms and definitions to theorems, lemmas and propositions via inferences and proofs.

In Sections 2.1–2.10, we have studied arithmetics and prearithmetics utilizing the structural approach. In this section, we elucidate axiomatic representation and exploration of arithmetics.

When logicians construct and study axiomatic theories of arithmetic, they always consider either the Diophantine arithmetic $\boldsymbol{N}$ of all natural numbers or sometimes the Diophantine arithmetic $\boldsymbol{W}$ of all whole numbers. The reason for this is that all other arithmetics, e.g., arithmetic of integer numbers or arithmetic of real numbers, are constructed using the Diophantine arithmetic $\boldsymbol{N}$ of all natural numbers or the Diophantine arithmetic $\boldsymbol{W}$ of all whole numbers as the base. This gives additional evidence to the fact that just these two arithmetics, $\boldsymbol{N}$ and $\boldsymbol{W}$, impersonate the traditional understanding of the term *arithmetic*.

At the same time, it is necessary to distinguish axiomatic theories of natural and whole numbers as tools for counting from axiomatic theories of their arithmetics as systems of numbers with operations.

2.11.1. *The original axiom systems for arithmetic: Grassmann, Peirce, Dedekind and Peano*

All truth passes through three stages:
First, it is ridiculed;
Second, it is violently opposed; and
Third, it is accepted as self-evident.

Arthur Schopenhauer

While geometry was axiomatized by ancient Greeks, arithmetic has to wait to the 19th century. Only then mathematicians started its axiomatization. The outstanding German mathematician and polymath Hermann Günther Grassmann (1809–1877) was the first to develop an axiomatics for arithmetic (Grassmann, 1861). He was also a linguist, physicist, neohumanist, general scholar, and publisher. In his work, Grassmann introduced mathematical systems with operations, which later were called vector spaces. He studied the concepts of linear maps, scalar products, linear independence and dimension although Peano was the first to give the modern definition of vector spaces and linear maps (Peano, 1888). Grassmann even exceeded the framework of vector spaces as he used multiplication bringing him to another new concept of a linear algebra. Nevertheless, in spite of all his achievement and innovations, the mathematical work of Hermann Grassmann was little noted until he was in his sixties.

According to Hao Wang, traditional recursive definitions of addition and multiplication are also due to Hermann Grassmann (Hao Wang, 1957). Namely, he introduced the following expressions characterizing both operations as derivatives from the unary operation $x + 1$:

$$x + 0 = x,$$
$$x + (y + 1) = (x + y) + 1,$$
$$x \cdot 0 = 0,$$
$$x \cdot (y + 1) = (x \cdot y) + x.$$

Hermann's brother Robert Grassmann (1815–1901) aimed to further develop this approach to arithmetic (Grassmann, 1872, 1891). According to Robert Grassmann, arithmetic, and more generally, mathematics as a whole, did not need syllogistic logic, and, even more strongly, no logical theory at all as its foundation. On the contrary, logic as a discipline had to be a chapter of mathematics.

In this context, Robert Grassmann treated mathematics as a science of forms (Formenlehre). Interpreting forms as structures (Burgin, 2012), we see that Hermann and Robert Grassmann were the first mathematicians to suggest structural account of mathematics. In the 20th century, this approach became very popular under the name structuralism in mathematics (Hellman, 2001; Carter, 2008).

The next step in axiomatizing arithmetic was done by the outstanding American philosopher, logician, mathematician, and versatile scientist Charles Sanders Peirce (1839–1914), who published his axiomatic system for arithmetic (Peirce, 1881). According to Shields, it is not generally known that his paper provided the first abstract formulation of the notions of partial and total linear order introducing recursive definitions for arithmetical operations independently of Grassmann (Shields, 1997). Peirce also originated semiotics as a scientific discipline and pragmatism as a philosophical direction. In his scientific and philosophical works, Peirce was an enthusiastic classificator developing a host of triadic typologies.

One more axiomatization of arithmetic was developed by the prominent German mathematician Richard Julius Wilhelm Dedekind (1831–1916).

However, as in his other works, he was ahead of his time aiming at generalization of existing structures and making important contributions to abstract algebra, axiomatic foundation for natural numbers, algebraic number theory and the definition of the real numbers.

In developing axiomatics for arithmetic, Dedekind starts with logical foundations of his approach. In it, logic is based on three notions: object (*Ding*), set or system (*System*), and function or mapping (*Abbildung*). Being fundamental for human thought — they are not definable in terms of anything even more basic. In particular, Dedekind does not reduce functions to sets as it is done in Zermelo-Fraenkel set theory **ZF** or sets to functions as it is done in von Neumann set theory **VN** (cf. Fraenkel *et al.*, 1973). After this, Dedekind shows how it is possible to reconstruct arithmetic in terms of objects, systems and mappings (Dedekind, 1888).

In his reconstruction, Dedekind actually builds chains of objects calling these chains by the name "simple infinity" or "simply infinite systems" (Gillies, 1982; Belna, 1996; Ferreirós, 2005). The construction is described by four axioms assuming existence of a set S, a function f and its subset N, which is called "simple infinity" and can be equal to S:

(i) f maps N into itself;
(ii) N is the chain in S starting with 1 and generated by f;
(iii) 1 does not belong to the image of N under f;
(iv) f is an injection.

Axiom (ii) is a rudimentary form of mathematical induction, while these axioms described the chain (linearly ordered set) of natural numbers.

Having established his axioms, Dedekind proves that every infinite set contains a simply infinite subset while any two simply infinite systems are isomorphic, i.e., that his axiom system is categorical. These results connect infinite sets in general and the set of natural numbers.

We remind that two sets X and Y are *equipotent* if there is a one-to-one correspondence between X and Y (cf. Appendix) and a set is *infinite*, or more exactly, *Dedekind infinite*, if it is equipotent to its proper part (cf., for example, Fraenkel and Bar-Hillel, 1958; Burgin, 2011).

Let us consider the following important result, which connects arbitrary infinite sets and the set N of all natural numbers.

Theorem 2.11.1 (Dedekind). *If an infinite set exists, then a set equipotent to the set N of all natural numbers exists.*

Proof: Let us take an infinite set L. By definition, there is a proper subset M of the set L such that there is a one-to-one mapping $f\colon L \to M$. As M is a proper subset of L, there is an element a that belongs to L but does not belong to M.

Then the element $f(a)$ belongs to M because f maps L onto M but does not belong to the image $f(M)$ of M. Indeed, suppose $f(a) = b$ belongs to $f(M)$. As a does not belong to M, there is an element c from M such that $f(c) = b$. However, this contradicts the condition that f is a one-to-one mapping meaning that

$$f(a) \in M \backslash f^n(M).$$

Now let us prove by induction that

$$f^n(a) \in f^{n-1}(M) \backslash f^n(M). \tag{2.54}$$

For $n = 1$, this statement is already proved. Assume that this statement is already proved for $n - 1$, i.e.,

$$f^{n-1}(a) \in f^{n-2}(M) \backslash f^{n-1}(M)$$

and

$$f^n(a) = f(f^{n-1}(a)) = d \in f^n(M)$$

As $f^{n-1}(a)$ does not belong to $f^{n-1}(M)$, there is an element e from $f^{n-1}(M)$ such that $f(e) = f^n(a)$. However, this contradicts the condition that f is a one-to-one mapping meaning that

$$f^{n-1}(a) \in f^{n-2}(M) \backslash f^{n-1}(M).$$

So, by the principle (axiom) of mathematical induction, the statement (2.54) is true.

Thus, for all n,

$$f^n(a) \neq f^{n-1}(a)$$

and we can put $a_1 = a$ and $a_n = f^n(a)$ for all $n = 1, 2, 3, \ldots$. This procedure gives us the set $\{a_1, a_2, a_3, \ldots, a_n\}$, which is equipotent to the set N by construction.

Theorem is proved. □

As it is possible to treat any set X, which is equipotent to N, as the set N itself by giving appropriate numerical names to the elements from X, Theorem 2.11.1 implies the following result.

Corollary 2.11.1. *If some infinite set exists, then the set N of all natural numbers also exists.*

In addition, Dedekind shows how several basic, and formerly unproven, arithmetic facts can be proved too using his axiom system (Dedekind, 1888).

However, the most popular formal axiomatics for arithmetic was developed by the Italian mathematician Giuseppe Peano because he directly gave axioms for natural numbers constituting the Diophantine arithmetic in a clear and formalized form (Peano, 1889). This shows that often the attempt to build a more general system results in more hardships for the reader.

Giuseppe Peano (1858–1932) was a very active researcher. He authorized more than 200 papers and books, created his own school of mathematics, became one of the founders of mathematical logic and essentially developed logical notation. As a part of his research, Peano made significant contributions to the contemporary precise and systematic treatment of mathematical induction. The traditional axiomatization of natural numbers, which is presented below, is named the Peano axioms in his honor and consists of five axioms (Peano, 1989).

His system includes a specific natural number, 1, and a successor function, which to each natural number x assigns its successor $\mathrm{S}x$. Note that in the Diophantine arithmetic, we have relations $\mathrm{S}x > x$ and $\mathrm{S}x = x+1$.

Peano axioms

(P1) 1 is a natural number.
(P2) If x is a natural number, then $\mathrm{S}x$ is a natural number.
(P3) If x is a natural number, then $\mathrm{S}x$ is not 1.
(P4) If $\mathrm{S}x = \mathrm{S}y$, then $x = y$.
(P5) If M is a set of natural numbers including 1, and if for every x in M the successor $\mathrm{S}x$ is also in M, then every natural number is in M.

Note that although the statements from the system (P1)–(P5) are called Peano axioms, the statement (P5) is an axiom schema, which represents an infinite number of axioms, and not a single axiom. Moreover, it is proved that it is impossible to define the Diophantine arithmetic $\boldsymbol{N}$ as a formal first-order theory by a finite number of axioms (Ryll-Nardzawski, 1952).

These axioms and Dedekind axioms describe natural or counting numbers but not the arithmetic of these numbers, which demands additional axioms. To get the Diophantine arithmetic, researchers augment Peano axioms with the operations of addition and multiplication and the usual total (linear) ordering in $\boldsymbol{N}$, while its axiomatic representation is called *Peano arithmetic* or *Peano–Dedekind arithmetic*.

Comparing the works of Hermann Grassmann, Charles Peirce, Richard Dedekind, and Giuseppe Peano, we can see that often the attempt to build a more general system results in more hardships for the reader. In turn, this explains why the system of Peano was better accepted and became so popular in the mathematical community.

Observe that at that time and much later, mathematicians and other people knew only one arithmetic — the Diophantine arithmetic of natural numbers, or of whole numbers as some mathematicians added zero to the set of natural numbers (cf., for example, Mac Lane and Birkhoff, 1999).

It is interesting that Dedekind described axioms similar to the axioms of Peano in his letter to Dr. H. Keferstein dated 27 February 1890 (cf. Hao Wang, 1957). That is why now Peano axioms (postulates) are often called Peano–Dedekind axioms. However, Peano axioms and Dedekind axioms from the letter are not equivalent because Peano axioms characterize natural numbers in a unique way up to isomorphism, while non-standard arithmetics also satisfy Dedekind axioms as Dedekind himself writes in his letter (Di Giorgio, 2010).

As we know, natural and whole numbers are ordered. This order is straightforwardly derived from the successor operation S by an inductive (recursive) definition. For example, it is possible to define:

(1) For any natural number x, it is $x < \mathrm{S}x$.
(2) For any natural numbers x, y and z, if $x < y$ and $y < z$, then $x < z$.
(3) For any natural numbers x and y, if $x \leq y$ if either $x < y$ or $x = y$.

In comparison with Peano postulates, David Hilbert (1862–1943) and Paul Bernays (1888–1977) suggested a slightly different axiom system for natural numbers calling it system B (Hilbert and Bernays, 1968). They use a binary relation $<$ and a function that maps x to x', which are defined only by the axioms from B. In their case, it is natural to interpret the symbol $<$

as a strict order on the set of all whole numbers and x' as Sx.

$$x = x,$$
$$x = y \Rightarrow (A(x) \Rightarrow A(y)),$$
$$\neg x < x,$$
$$x < y \;\&\; y < z \Rightarrow x < z,$$
$$x < x',$$
$$A(0)\&\forall x(A(x) \Rightarrow A(x')) \Rightarrow \forall x(A(x).$$

In addition, Hilbert and Bernays introduce arithmetical operations by means of recursive definitions, which are considered in Section 2.11.4.

It is necessary to remark that Frege also suggested an axiom system for arithmetic but it was observed that his system was inconsistent (cf., for example, Boolos, 1987; Schirn, 1995).

2.11.2. *Equivalence, equality and identity*

It isn't that they can't see the solution.
It's that they can't see the problem.
G. K. Chesterton

In addition to his structural axioms (postulates) of natural (counting) numbers, Peano also included axioms for equality for natural numbers. However, axioms for equality are not specific axioms for arithmetic. That is why they and other logical axioms are not always included in the axiomatics of arithmetic. To follow the example of Euclid, it might be reasonable to call axioms for equality and other logical axioms by the name *axiom*, while giving axioms (P1)–(P5) the name *postulate*. We consider Peano axioms for equality later while now focusing on the concept of equality in general.

Equality is used not only for natural numbers but it is also one of the most important relations in mathematics and logic. Importance of the notion of equality has been emphasized by many mathematicians and logicians, who have methodically studied properties of equality trying to achieve more clarity and accuracy in its understanding (cf., for example, Frege, 1892; Church, 1956; Kleene, 1967; Kauffman, 1995; 1999; *Homotopy Type Theory*, 2013). However, in spite of their efforts, the equality enigma persists without consensus on any solution (Parsons, 2000).

Although the notion of equality has existed from ancient times, the mathematical sign = denoting equality was introduced by the Welsh physician and mathematician Robert Recorde (ca. 1512–1558) in 1557 (Recorde, 1557). Much later Gottlob Frege (1848–1925) systematically explained the difficulties posed by the notion of equality (Frege, 1892). Looking rather simple and intuitive, the concept of equality demands more considerations than it is usually done because what seems equal from some point of view can look very different from another. For instance, all letters a printed on a piece of paper are equal as symbols of an alphabet, i.e., from the linguistic perspective. However, the same letters are different as physical objects being situated at different places of the paper.

These considerations bring us to the idea of equality relativity meaning what is equal from one point of view or in one situation can be not equal from another point of view or in another situation. Similar idea already appears in the dialogue *Phaedo* of Plato where Socrates asks:

Is it not true that equal stones and sticks sometimes, without changing in themselves, appear equal to one person and unequal to another? (Plato, 1961)

However, explaining relativity and subjective comprehension of equality of things, Socrates suggests:

We admit, I suppose, that there is such a thing as equality — not the equality of stick to stick and stone to stone, and so on, but something beyond all that and distinct from it — absolute equality. (Plato, 1961)

It is reasonable to suggest that this point of view represents the approach of Plato.

Likewise, two great mathematicians Leonhard Euler (1707–1783) and Gottfried Wilhelm Leibniz (1646–1716) appear to have realized more clearly than their contemporaries that there is more than one relation falling under the general heading of "equality" (cf. Bair *et al.*, 2017).

Later philosophers, logicians and mathematicians started investigating the concept of equality trying to find an adequate definition for this evasive structure. They encountered many difficulties due to the problems caused by the notion of equality.

There are different approaches to the concept of equality. For instance, according to Frege, equality he studied was related to names, or signs, of objects, and not to objects themselves (Frege, 1879). Note that in terms of named set theory, a description, definition or portrayal of an object is a name of this object (Burgin, 2011). Thus, Frege studied equality of names relating them to objects they designate. This implies the natural definition of equality, namely, two names are equal if they are names of the same object.

In his studies of equality, Frege treated equality as a proposition and genuinely assumed that the proposition $a = b$ is true if and only if the object with the name a is identical to the object with the name b. For instance, the proposition $1 + 1 = 2$ is true if and only if the number $1 + 1$ just is the number 2. At the same time, the equality "*Mark Twain is Samuel Clemens*" is true if and only if the person Mark Twain just is the person Samuel Clemens. This assumption means that Frege equated the concepts of equality and identity.

However, Frege noticed that the proposition $a = a$ has a cognitive significance (or meaning) that must be different from the cognitive significance of the proposition $a = b$. Indeed it is possible to learn that "*Mark Twain is Mark Twain*" is true simply by inspecting it, but it is impossible to learn the truth of the proposition "*Mark Twain is Samuel Clemens*" simply by inspecting it. Specifically, it is necessary to have additional information to see whether the two persons are the same. Besides, the first of these equalities "*Mark Twain=Mark Twain*" does not give us information while the second one "*Mark Twain= Samuel Clemens*" informs us what was the real name of the famous writer Mark Twain.

This distinction shows that there are different types of equality and analyzing equality it is necessary to take into account this aspect.

In logics and semiotics, meanings related to terms or names are often called *intensions* while things designated by these terms or names are specified as *extensions*. Consequently, contexts in which extension is all that matters are called *extensional*, while contexts in which extension is not enough are identified as *intensional*. In mathematics, equality is typically extensional. For instance, the equality $1 + 7 = 5 + 3$ is true in the conventional Diophantine arithmetic even though the two terms involved may differ in meaning when the knowledge of small children is involved. This causes intentionality of many arithmetical equalities and demonstrates that mathematical pedagogy differs from mathematics as a science.

Hilbert and Bernays introduced and studied *deductive equality* of formulas in logics. Namely, formulas A and B are *deductively equal* in a logic L if A can be deduced from B and B can be deduced from A in the logic L (Hilbert and Bernays, 1968).

Traditionally logicians studied equality in three forms — as a binary relation in a set (class), as a binary predicate/proposition on pairs of elements from a set (class), and as a characteristic function or indicator function identifying those pair elements of which are equal. Recently, a new homotopic approach to equality was suggested in homotopy type theory by treating equality as the type $\mathrm{IdA}(a, b)$ or $a = b$ representing the proposition of equality between a and b. As the basic structure in homotopy type theory is the path, the type $a = b$ is the type of all paths from the object a to the object b. In a formal theory, objects are propositions and a path from the object a to the object b is interpreted as a proof that a proposition a is equal a proposition b. In homotopy type theory, a type IdA is treated as a space and a proof of the equality $a = b$ is interpreted as a path between a and b (*Homotopy Type Theory*, 2013).

Properties of paths allow getting properties of equalities such as reflexivity, symmetry and transitivity (cf. Appendix). Indeed, for any object a, there exists a path of type $a = a$, corresponding to the reflexive property of equality. The symmetric property of equality follows from the property of paths stating that a path of type $a = b$ can be inverted, forming a path of type $b = a$. The transitive property of equality follows from the property of paths stating that two paths of type $a = b$ and $b = c$ can be concatenated forming a path from a to b.

Note that in a type theory, all objects, e.g., sets or elements of sets, have types. This makes each object a named set or fundamental triad, the name of which is its type (Burgin, 2011). Moreover, *homotopy* is a continuous transformation or function (say, f) of one topological object A into another B. Consequently, homotopy is also a named set (fundamental triad) of the form (A, f, B). Homotopy Type Theory is used as Univalent Foundations of mathematics (*Homotopy Type Theory*, 2013). Thus, Univalent Foundations of mathematics are based on named sets demonstrating once more that named set theory forms the most basic unified foundations of mathematics (Burgin, 2004).

To better understand equality, here we consider three levels of objects' connection going from the strongest to the weakest (Burgin, 2018a):

(1) *Identity* determines when it is the same object represented in different observations. It is denoted by $a \equiv b$.
(2) *Equality* determines when it is possible to consider (treat) two objects as the same. It is denoted by $a = b$.
(3) *Equivalence* determines when it is possible to consider (treat) two objects as interchangeable (in some situations). It is denoted by $a \approx b$.

It is necessary to remark that in literature in general and even in mathematical publications, the terms *equality* and *identity* are used interchangeably (cf., for example, Feferman, 1974; Parsons, 2000). Besides, the terms *equality* and *equivalence* are also often used interchangeably although in some contexts, equality is sharply distinguished from equivalence or isomorphism.

To be able to define these terms in an exact fashion, it is necessary to discern identity, equality and equivalence in logic, as well as in mathematical and scientific contexts.

Equivalence or the equivalence relation on a set is defined axiomatically by three properties — reflexivity, symmetry, and transitivity. Namely, a binary relation Q on X is an equivalence if it satisfies the following axioms:

E1. Q is *reflexive*, i.e. $x\mathrm{Q}x$ for all x from X.
E2. Q is *symmetric*, i.e., $x\mathrm{Q}y$ implies $y\mathrm{Q}x$ for all x and y from X.
E3. Q is *transitive*, i.e., $x\mathrm{Q}y$ and $y\mathrm{Q}z$ imply $x\mathrm{Q}z$ for all $x, y, z \in X$.

There are different kinds of the equivalence relation.

In algebra, the major equivalence relation is called *isomorphism.* Namely, two algebraic systems A and B are *isomorphic* if there is a one-to-one mapping between elements of A and B such that it and its inverse preserve algebraic operations.

In topology, the major equivalence relation is called *homeomorphism.* Namely, two topological spaces A and B are *homeomorphic* if there is a one-to-one mapping between elements of A and B such that it and its inverse are continuous.

In set theory, the major equivalence relation is called *equipotence.* Namely, two sets X and Y are *equipotent* if there is a one-to-one correspondence between X and Y (cf. Appendix)

In geometry, the major equivalence relation is called *isometry.* Namely, two geometric objects A and B are *isometric* if there is a one-to-one mapping between elements of A and B such that it and its inverse preserve distances.

Remark 2.11.1. In algebra, equalities with variables are often called identities. For instance,

$$x + y = y + x$$

is the commutativity identity or law.

Axiomatic systems in mathematics, as a rule, use equalities. This is also true for arithmetic. That is why in addition to his axioms for arithmetic, Peano suggested axioms for equality (Peano, 1989), which are usually presented in the following form:

$$x = x,$$
$$x = y \Leftrightarrow y = x,$$
$$x = y \,\&\, y = z \Rightarrow x = z,$$
$$x = y \Rightarrow (A(x) \Rightarrow A(y)).$$

These axioms represent the concept of *predicative equality* and are consistent and independent (cf., for example, Church, 1956). In the contemporary expositions, the fourth axiom is usually presented by two axioms (cf., for example, Rasiowa and Sikorski, 1963):

$$x_1 = y_1 \,\&\, x_2 = y_2 \,\&\ldots\&x_n = y_n$$
$$\Rightarrow (F(x_1, x_2, \ldots, x_n) \Rightarrow F(y_1, y_2, \ldots, y_n)),$$
$$x_1 = y_1 \,\&\, x_2 = y_2 \,\&\ldots\&\, x_n = y_n$$
$$\Rightarrow (P(x_1, x_2, \ldots, x_n) \Rightarrow P(y_1, y_2, \ldots, y_n)).$$

Here F is an arbitrary function with n arguments and P is an arbitrary predicate with n arguments.

Stolz gave axioms for equality together with axioms for strict order relation $<$ and its dual relation $>$ in the following form (Stolz, 1885):

(1) If $A = B$, then $B = A$.
(2) If $A > B$, then $B < A$ (and conversely).
(3) $A = B$ or $A > B$ or $A < B$ and only one of these relations holds.
(4) If $A = B$ and $B = C$, then $A = C$.
(5) If $A = B$ and $B > C$, then $A > C$.
(6) If $A > B$ and $B > C$, then $A > C$.

Similar to other authors who wrote on equality prior to the works of Peano on arithmetic, Stolz did not include the reflexivity of the equality relation (cf. Appendix) in his system of axioms.

Later logicians build a formal first-order theory of pure equality, the signature of which consists of the equality relation symbol $=$ and which does not include non-logical axioms (Monk 1976). This theory is consistent because any set with the usual equality relation provides its interpretation. Löwenheim proved decidability of the first-order theory of pure equality (Löwenheim, 1915). By adding either the axiom saying that for a fixed natural number m, there are exactly m objects in the considered set, or an axiom schema stating that the set is infinite, the theory of pure equality is made complete.

To describe basic forms of equality, we assume that identity of objects is already, i.e., *a priory*, defined for elements of a chosen set. In each case, a specific set is chosen, in which it is possible to uniquely identify objects, i.e., to find when what seems to be two objects is in essence one and the same object.

The definition of the predicative equality originated from the suggestion of Leibniz, who described equality in the following way (cf., for example, Blok and Pigozzi, 1989):

Given any objects a and b, the equality $a = b$ is true if and only if, given any predicate P, we have $P(a)$ if and only if $P(b)$.

In a similar way, Hermann Grassmann wrote "two things are said to be equal, if in each statement you can substitute the one for the other" (Grassmann, 1861).

However, these definitions contain a vicious circle because in them the equality $a = b$ is defined using the equality $P(a) = P(b)$ of predicate values or the equalities $P(a)$ is (i.e., equal to) true and $P(b)$ is (i.e., equal to) true. Besides, the condition that includes any predicate P is not constructive. We eliminate these shortcomings by making the definition relative with respect to a set of predicates and grounding it on the concept of identity, which is assumed being already defined for the range of considered predicates.

Thus, to define *predicative equality*, let us consider a class $\mathbf{R}$ of objects and a set $\mathbf{P}$ of predicates that take values in a set C, for elements of which identity is defined. Classical predicates take values in the set $\{\text{True}, \text{False}\}$ or $\{1, 0\}$.

Definition 2.11.1. Two objects a and b from $\mathbf{R}$ are $\mathbf{P}$-*equal*, i.e., equal with respect to $\mathbf{P}$, if for any predicate $P(x)$ from $\mathbf{P}$, the values $P(a)$ and $P(b)$ are the same (are identical).

We denote it $a =_{\mathbf{P}} b$.

Note that the scale of the predicates in $\mathbf{P}$ is not always $\{\text{True}, \text{False}\}$ because $\mathbf{P}$ can contain predicates from a multivalued logic. Besides, predicates are special cases of abstract properties (Burgin, 1985; 1986). That is why it is possible to define predicative equality using abstract properties instead of predicates.

Example 2.11.1. If we have a mapping $f\colon X \to Y$, then the equality $f(x) = y$ implies that for any predicate $P(a)$ defined for elements from Y, the value $P(f(x))$ is identical to (the same as) $P(y)$.

Example 2.11.2. In ZF-axiomatic theory of sets, there are two approaches to defining equality of sets in the class **Set** of all sets (Fraenkel and Bar-Hillel, 1958; Kuratowski and Mostowski, 1967):

(*Intensional equality*) $\forall x$ and y ($x = y$ if $\forall z$ ($x \in z$ if and only if $y \in z$)).

(*Extensional equality*) $\forall x$ and y ($x = y$ if $\forall z$ ($z \in x$ if and only if $z \in y$)).

Intensional equality is a predicative equality defined by the system of predicates $\mathbf{P}_{\mathrm{I}} = \{P_z(x)\}$ of the form $x \in z$.

Extensional equality is a predicative equality defined by the system of predicates $\mathbf{P}_{\mathrm{E}} = \{Q_z(x)\}$ of the form $z \in x$.

Note that extensional equality is based on the inner structure of a set while intensional equality is based on the outer structure of a set (Burgin, 2017).

Lemma 2.11.1. *Predicative equality is an equivalence relation in* $\boldsymbol{R}$.

Proof: According to the properties of equivalence relation (cf. Section 3), we have to check that the relation $=_{\mathbf{P}}$ is reflexive, symmetric and transitive.

Reflexivity. Indeed, for any object a and any system $\mathbf{P}$ of predicates and any predicate P from $\mathbf{P}$, we have $P(a) = P(a)$ implying $a =_{\mathbf{P}} a$.

Symmetry. Indeed, for any system $\mathbf{P}$ of predicates and any predicate P from $\mathbf{P}$, we have if $P(a) = P(b)$, then $P(b) = P(a)$ and thus, $a =_{\mathbf{P}} b$ implies $b =_{\mathbf{P}} a$.

Transitivity. Let us assume that for some objects a, b and c from $\mathbf{R}$, we have $a =_{\mathbf{P}} b$ and $b =_{\mathbf{P}} c$. It means that for any predicate P from $\mathbf{P}$, we have $P(a) = P(b)$ and $P(b) = P(c)$. As identity is a transitive relation, we obtain $P(a) = P(b)$. Consequently, $P(a) = P(c)$ and thus, $a =_{\mathbf{P}} c$.

Lemma is proved. □

It is possible to find a more formalized proof of this result in Kleene (1967).

Note that objects from $\mathbf{R}$ can be equal with respect to one set of predicates, say $\mathbf{P}$, and different with respect to another set of predicates, say $\mathbf{Q}$.

Example 2.11.3. Let take the class **Fr** of fractions, the set $\mathbf{P}$, which consists of one predicate (abstract property) *Nb* that assigns to each fraction the rational number represented by this fraction, and the set $\mathbf{Q}$, which consists of one predicate (abstract property) *Gr* that assigns to each fraction its graphical representation. Then fractions $^2/_4$ and $^1/_2$ are equal with respect to $\mathbf{P}$ but are not equal with respect to $\mathbf{Q}$.

To define *semiotic equality*, let us consider a semantic function (mapping) $S : \mathbf{R} \to M$ defined for the class **R** of objects and having the field of meanings M, for elements of which identity is defined.

Definition 2.11.2. Two objects a and b from **R** are *S-equal*, i.e., equal with respect to the semantics S, if $S(a) = S(b)$.

We denote it $a =_S b$.

Example 2.11.4. Let take a class **R**, which consists of texts while the semantic S assigns meaning to the texts. Then, texts are semantically equal if they have the same meaning.

Lemma 2.11.2. *Semiotic equality is an equivalence relation in* ***R***.

Proof is similar to the proof of Lemma 2.11.1. □

A particular case of semiotic equality is *nominalistic equality.* It is defined for objects that play the role of names for other objects. That is, the class **R** consists of names of objects from the set M.

Definition 2.11.3. Two names a and b from **R** are *nominalistically equal* if they are names of the same object from M.

We denote it $a =_M b$.

Example 2.11.5. Logicians often like to consider the following example suggested by Frege (Frege, 1892).

The morning star is equal to the evening star. (1)

For a long time, people saw a bright star in the morning and a bright star in the evening. For instance, ancient Egyptians and Greeks believed that those were actually two separate objects, a morning star and an evening star. Consequently, ancient Greeks called the morning star *Phosphorus*, which meant *the bringer of light*, while they called the evening star *Hesperus*, which meant *the star of the evening.* When astronomers started their observations, they found that the names "the morning star" and "the evening star" both designate the planet Venus. Thus, equality (1) means that "the morning star" and "the evening star" are names of the same physical object.

So, why one object had in some sense opposite names? The cause is that the orbit of the Venus is inside the orbit of the Earth. As a result, the

Venus is always relatively close to the Sun in the sky. Thus, when the Venus is on one side of the Sun, it is following the Sun in the sky and comes into view soon after the Sun sets, while the sky is already dark enough for it to be observable. Thus, the Venus is at its brightest only minutes after the Sun disappears implying that the Venus is the evening "star".

However, being on the other side of the Sun, the Venus go in front of the Sun as it travels across the sky. In particular, the Venus rises in the morning a few hours before the Sun and when the Sun rises, the sky brightens and the Venus disappears gradually in the daytime sky, which implies that the Venus the morning "star".

In this context, the morning star and the evening star are the same in one way and not the same in another (cf., for example, Fitting, 2015). Indeed, the terms "the morning star" and "the evening star" convey different information about the planet Venus and thus, have different meaning.

Lemma 2.11.3. *Nominalistic equality is an equivalence relation in* $\boldsymbol{R}$.

Proof is similar to the proof of Lemma 2.11.1. □

To define *systemic equality*, let us consider a class **R** of objects and a class of systems **K**.

Definition 2.11.4. Two objects a and b from **R** are equal with respect to **K** or **K**-*equal* if for any system A from **K** in which a is its element (component or part), substitution of some of objects a by objects b in A does not change this system A and for any system B from **K** in which b is its element (component or part), substitution of some of objects b by objects a in B does not change this system B.

We denote it $a =_{\boldsymbol{K}} b$.

Example 2.11.6. In the world "alphabet", the first letter a is equal to the second letter a because it is possible to change one for another without changing the word "alphabet". In this case, the class **K** consists of words and two physical symbols (letters) are **K**-equal if any word stays the same after we change one of them by another one.

Lemma 2.11.4. *Systemic equality is an equivalence relation in* $\boldsymbol{R}$.

Proof is similar to the proof of Lemma 2.11.1. □

Note that systemic equality assumes that objects from **R** can be elements (components or parts) of systems **K**. However, if all objects from

R cannot be elements (components or parts) of systems **K**, then all elements from **R** are equal with respect to **K** because by the properties of predicative reasoning, any two objects from **R** that cannot be elements (components or parts) of systems **K** are equal with respect to **K**.

It is also possible to introduce *lexical equality* as a specific kind of systemic equality taking texts as systems and words as objects.

Let us consider a set **W** of words and a set of texts **T**. Note that it is possible to consider any situation or system as a text (in some generalized way) and objects as words from such generalized texts.

Definition 2.11.5. Two words v and u from **W** are equal with respect to **T** or **T**-*equal* if for any text A from **T** in which v is its element, substitution of v by u in A does not change the meaning of text A and for any text B from **K** in which u is its element, substitution of u by v in B does not change the meaning of text B.

We denote it $a =_{\mathbf{T}} b$.

Lemma 2.11.5. *Lexical equality is an equivalence relation in* **W**.

Proof is similar to the proof of Lemma 2.11.1. □

It is possible to demonstrate that all these concepts of equality are reducible to one another, i.e., they are equivalent.

Let us consider a class **R** of objects. Then it is also possible to represent S-equality by **P**-equality.

Theorem 2.11.2. *For any semantics* S: $\mathbf{R} \to M$, *there is a system* **P** *of binary predicates such that any two objects* a *and* b *from* **R** *are* S-*equal if and only if they are* **P**-*equal.*

Proof: Let us consider a semantics S: $\mathbf{R} \to M$. For each element m from M, we build the following predicate on the class **R**:

$$P_m(x) = \begin{cases} 1 & \text{if } S(x) = m, \\ 0 & \text{if } S(x) \neq m \end{cases}$$

and define $\mathbf{P} = \{P_m(x); m \in M\}$. Then for any objects a and b from **R**, $S(a) = S(b)$ if and only if $P_m(a) = P_m(b)$ for all $m \in M$. It means that a and b are S-equal if and only if they are **P**-equal.

Theorem is proved. □

It is also possible to represent **P**-equality by S-equality.

Theorem 2.11.3. *For any system* **P** *of predicates, there is a semantics* S*:* $\mathbf{R} \to M$ *such that any two objects* a *and* b *from* **R** *are* **P***-equal if and only if they are* S*-equal.*

Proof: Let us consider a set **P** of predicates P_i with the scale C_i of its values ($i \in I$). Note that all scales C_i can coincide. For instance, they all can be equal to the set $\{\text{True}, \text{False}\}$.

We define the set M equal to the Cartesian product $\Pi_{i \in I} C_i$ and semantics $S : \mathbf{R} \to M$ by the rule

$$S(a) = (P_i(a); i \in I).$$

Then for any objects a and b from the class **R**, $S(a) = S(b)$ if and only if $P_i(a) = P_i(b)$ for all $i \in I$. It means that a and b are S-equal if and only if they are **P**-equal.

In such a way, we build the necessary logical semantics.

Theorem is proved. □

Additionally it is possible to represent **P**-equality by **K**-equality.

Theorem 2.11.4. *For any set* **P** *of predicates, there is a set* **K** *of systems such that any two objects* a *and* b *from the class* **R** *are* **P***-equal if and only if they are* **K***-equal.*

Proof: Let us consider a set **P** of predicates $P_i(x)$ with the scale C_i of its values ($i \in I$). Note that all scales C_i can coincide. For instance, all of scales can be equal to the set $\{\text{True}, \text{False}\}$. Each predicate $P_i(x)$ is an abstract system and x is an element of this system. Thus, we can take **K** equal to $\{(P_i(x), P_i(a));\ a \in \mathbf{R},\ i \in I\}$ where $P_i(a)$ is the value of for the element a. Then a and b are **K**-equal if and only if they are **P**-equal because both in the case of **P**-equality and the case of **K**-equal, $a = b$ if and only if $P_i(a) = P_i(b)$ for all $i \in I$.

Theorem is proved. □

In addition it is possible to represent S-equality by **K**-equality.

Theorem 2.11.5. *For any set* **K** *of systems, there is a semantics* S*:* $\mathbf{R} \to M$ *such that any two objects* a *and* b *from* **R** *are* S*-equal if and only if they are* **K***-equal.*

Proof: Let us consider a set **K** of systems, a class (set) M equal to the set $2^{\mathbf{R}}$ of all subsets of **R** and build semantics S: $\mathbf{R} \to M$ by the following rule:

$$S(a) = \{b_i;\ a =_{\boldsymbol{K}} b_i\}.$$

Note that any object a from $\mathbf{R}$ belongs to $S(a)$ because by Lemma 2.11.1, $=_{\boldsymbol{K}}$ is an equivalence relation.

Then for any objects a and b from $\mathbf{R}$, $S(a) = S(b)$ if and only if $a =_{\boldsymbol{K}} b$. Consequently, objects a and b are S-equal if and only if they are $\mathbf{K}$-equal.

Theorem is proved. □

Furthermore, it is possible to represent $\mathbf{K}$-equality by $\mathbf{P}$-equality.

Theorem 2.11.6. *For any set* $\mathbf{K}$ *of systems, there is a system* $\mathbf{P}$ *of binary predicates such that any two objects* a *and* b *from* $\mathbf{R}$ *are* $\mathbf{K}$*-equal if and only if they are* $\mathbf{P}$*-equal.*

Proof: Let us consider a set $\mathbf{K}$ of systems. Then by Theorem 2.11.5, there is a semantics S: $\mathbf{R} \to M$ such that any two objects a and b from $\mathbf{R}$ are S-equal if and only if they are $\mathbf{K}$-equal. In addition, by Theorem 2.11.2, there is a system $\mathbf{P}$ of binary predicates such that any two objects a and b from $\mathbf{R}$ are S-equal if and only if they are $\mathbf{P}$-equal. Consequently, any two objects a and b from $\mathbf{R}$ are $\mathbf{K}$-equal if and only if they are $\mathbf{P}$-equal.

Theorem is proved. □

Finally, it is possible to represent S-equality by $\mathbf{K}$-equality.

Theorem 2.11.7. *For any semantics* S: $\mathbf{R} \to M$*, there is a set* $\mathbf{K}$ *of systems such that any two objects* a *and* b *from the class* $\mathbf{R}$ *are* S*-equal if and only if they are* $\mathbf{K}$*-equal.*

Proof: Let us consider a semantics $S : \mathbf{R} \to M$. Then by Theorem 2.11.2, there is a system $\mathbf{P}$ of binary predicates such that any two objects a and b from $\mathbf{R}$ are S-equal if and only if they are $\mathbf{P}$-equal. At the same time, by Theorem 2.11.4, there is a set $\mathbf{K}$ of systems such that any two objects a and b from the class $\mathbf{R}$ are $\mathbf{P}$-equal if and only if they are $\mathbf{K}$-equal. Consequently, any two objects a and b from $\mathbf{R}$ are $\mathbf{K}$-equal if and only if they are S-equal.

Theorem is proved. □

Homotopy type equality was introduced by Voevodsky and studied in homotopy type theory (Homotopy Type Theory, 2013). There are also: *transportational equality*, *substitutional equality* and *judgmental* or *definitional equality*. In addition, there are higher equalities in homotopy type theory (Shulman, 2016).

Philosophers and logicians tried to characterize a unique absolute equality. Even Plato discussed a unique absolute idea, or structure in contemporary understanding (Burgin, 2017), of equality in general (Plato, 1961). However, relativity of equality considered above shows that there is a variety of different types and forms of equality. To organize this variety in a structured system, we additionally introduce *modalities of equality*. There are several types of such modalities.

Existential modalities of equality

(1) A tentative (potential) equality is not completely established.
(2) An existing (situational) equality is completely established but depends on the situation — in one situation it is true, while in another it is false.
(3) An imperative (necessary) equality is always true.

Note that the concept of a potential (tentative) equality depends on the individual (the group of individuals) who regards this equality.

Example 2.11.7. Now, i.e., in 2018, the statement "P = NP" expresses a potential equality for the majority of mathematicians and computer scientists.

Example 2.11.8. The expression "2 + 2 = 4" expresses situational equality because it is true in the conventional Diophantine arithmetic but is not true in some non-Diophantine arithmetics (Burgin, 1997; 2007).

Example 2.11.9. The expression "2 + 2 = 5" is a potential equality for those who do not know non-Diophantine arithmetics and is an actual equality for those who know non-Diophantine arithmetics (Burgin, 1997; 2007).

Example 2.11.10. The expression "3 = 3" expresses imperative equality.

Inclusive modalities of equality

(1) Relative equality is equality that depends on its context.
(2) Absolute in a class (invariant) equality is equality that is invariant in the given class.
(3) Absolute equality is equality that is invariant in general.

Example 2.11.11. The equality $2 + 2 = 4$ is relative in the class of all arithmetics but is absolute in conventional Diophantine arithmetic, while the equality $3 = 3$ is absolute.

Temporal modalities of equality

(1) Static equality is equality between two unchanging objects.
(2) Dynamic equality is equality between two changing objects.
(3) Evolutionary equality is equality between two developing objects.

Example 2.11.12. Equality $2+2=4$ is static in conventional Diophantine arithmetic. Equality between you today and you yesterday is dynamic.

Example 2.11.13. Equality between you as a child and you as an adult is evolutionary.

Modalities of equality are complemented by roles of equality. It is possible to discern three basic roles:

(1) Equality plays the *definitive role* when one side of it (usually, right) is known while the other side (usually, left) is unknown and defined by the first (right) side.
(2) Equality plays the *equivalizing role* when both sides of it are known and they are treated as the same.
(3) Equality plays the *reduction role* when both sides of it are known and equality shows that it is possible to use one side instead of the other one.

Example 2.11.14. When we say "arithmetic is a science of numbers", it is equality in the definitive role.

Example 2.11.15. Equality $2+2=4$ is in the equivalizing role.

Example 2.11.16. Equality $^2/_4 = {}^1/_2$ is plays the reduction role demonstrating how the fraction $^2/_4$ is reduced to the fraction $^1/_2$.

Equality is also classified by the domain where it is applied. People have discussed and studied:

- *Political equality*, which means that all have equal voice in making laws and selecting political leaders;
- *Social equality*, which means that all have equal access to those things necessary to leading a decent or good life;
- *Moral equality*, which means for all equal human value or worth;
- *Legal equality*, which means that all are subject to the same laws;
- *Linguistic equality*, which is the equality of languages (Griffiths, 1990);
- *Logical equality*, which is an operation on two logical values;
- *Mathematical equality*, which is the equality of mathematical objects, such as numbers, or of their names, such as fractions or sequences of decimal digits.

Here we are dealing only with mathematical equalities because in this section, we study axiomatic systems in mathematics.

2.11.3. *Axioms for natural/counting numbers*

Logic is the beginning of wisdom, not the end.

Leonard Nimoy

The system that satisfies postulates (P1)–(P5) is usually called a *Peano system* (cf., for example, Feferman, 1974). When additional postulates defining addition and multiplication are also included, a Peano system is usually called the *Peano arithmetic*. However, there is an essential difference between a Peano system and a Peano arithmetic. Namely, in contrast to the former, the latter also has two binary operations — addition and multiplication.

For instance, Shapiro writes:

"*The subject matter of arithmetic is a single abstract structure, the pattern common to any infinite collection of objects that has a successor relation with a unique initial object and satisfies the (second-order) induction principle*" (Shapiro, 1997).

As we will see later not only the Diophantine arithmetic $\boldsymbol{N}$ satisfies this definition but also all non-Diophantine arithmetics (of natural numbers).

Note that a Peano system as a universal algebra has only one unary operation S and one null-ary operation 1. We remind that a null-ary operation is selection of a specific element in the universal algebra (Kurosh, 1963). Thus, they are neither prearithmetics nor arithmetics (cf. Section 2.1). To make an arithmetic of such a system, it is necessary to define addition and multiplication in it.

Then Peano postulates together with axioms for arithmetical operations of addition and multiplication as well as with conventional logical rules of reasoning constitute a formal basis for the subject of arithmetic, and all formal proofs are explicitly or implicitly based on them.

In Peano's original presentation, the induction axiom (P5) is a second-order axiom. It is now common to replace this second-order principle with a weaker first-order induction schema. As logicians know well there are important differences between the second-order and first-order formulations. That is why there is a distinction between first-order and second-order Peano arithmetics (Shoenfield, 2001; Simpson, 2009).

Let us consider some properties of a system P, which satisfies postulates (P1)–(P5) and elements of which we will call natural numbers.

Lemma 2.11.6 (Feferman, 1974). *For any element x from P, either $x = 1$ or there is an element y in P such that $x = \mathrm{S}y$.*

Proof: To prove this statement, we use mathematical induction, which is formalized by Axiom P5.

The base of induction. For the number 1, the statement is evidently true.

For the number 2, the statement is true because $2 = \mathrm{S}1$.

The general step of induction. Assume that some number n, we have $n = \mathrm{S}m$. Then $\mathrm{S}n = \mathrm{S}(\mathrm{S}m)$. It means that the statement of the lemma is true for $\mathrm{S}n$. Thus by the axiom of induction, it is true for all natural numbers.

Lemma is proved. □

Lemma 2.11.7. *For any element x from P, the statement $Sx = x$ is not true.*

Proof: To prove this statement, we use mathematical induction, which is formalized by Axiom P5.

The base of induction. For the number 1, the equality $1 = \mathrm{S}1$ is not true because by Axiom P3 stating that 1 is not a successor of any natural number.

The general step of induction. By Peano axioms, an arbitrary element x from P has the form S^n1. Let us take a natural number $x = \mathrm{S}^n1$ that is not equal to 1 and assume that $\mathrm{S}x = x$, while for all $m < n$ and all natural numbers $z = S^m1$ the equality $z = \mathrm{S}z$ is not true. As $x \neq 1$, by Lemma 2.11.6, we have $x = \mathrm{S}y$ for some natural number y because by Peano axioms, an arbitrary element x from P has the form S^n1. Consequently, $\mathrm{S}x = \mathrm{S}y$. Then by Axiom P4, $x = y$ and by the first part of our assumption $\mathrm{S}x = x$. However, this contradicts the second part of our assumption because $y = \mathrm{S}^{n-1}1$. This contradiction shows that the equality $\mathrm{S}x = x$ is not true.

Thus, by Axiom P5, the equality $\mathrm{S}x = x$ is not true for all natural numbers from P.

Lemma is proved. □

Proposition 2.11.1. *For any Peano system P, it is possible to introduce a linear order $\leq$ compatible with the operation S, i.e., such that $x \leq \mathrm{S}x$.*

Proof: For any element x from N, let us define $x \leq x$ and $x \leq \mathrm{S}x$. Then we take the transitive closure of this relation denoting it by the same symbol $\leq$. Let us demonstrate that this is an order relation (cf. Appendix).

Reflexivity follows from definition because $x \leq x$ for any element x from N.

Transitivity follows from construction.

Now let us prove that $\leq$ is a total relation. By Peano axioms, an arbitrary element x from P has the form $\mathrm{S}^n 1$. Taking two elements x and y from P, we have $x = \mathrm{S}^n 1$ and $y = \mathrm{S}^m 1$ for some natural numbers m and n. If $m < n$, then $x = \mathrm{S}^{n-m} y$. Thus, by the definition of the relation $\leq$, we have $y \leq x$, i.e., $\leq$ is a total relation.

Let us show that the relation $\leq$ is antisymmetric. Suppose $y \leq x$ and $x \leq y$ for two elements x and y from P. As before, $x = \mathrm{S}^n 1$ and $y = \mathrm{S}^m 1$ for some natural numbers m and n. Then $m \leq n$ and $n \leq m$. Thus, $m = n$ and consequently, $x = y$.

This means that $\leq$ is a linear order compatible with the operation S.

Proposition is proved. □

Proposition 2.11.1 and Lemma 2.11.7 imply the following result.

Corollary 2.11.2. *For any element x from N, the statement $Sx \leq x$ is not true.*

Axioms (P1)–(P5) characterize Peano systems up to isomorphism of them as universal algebras with one unary operation S and one null-ary operation 1. This statement is proved in Feferman (1974). Here we give a much simpler proof of this result.

Theorem 2.11.8. *For any two Peano systems $P_1 = (A_1, \mathrm{S}_1, 1_1)$ and $P_2 = (A_2, \mathrm{S}_2, 1_2)$, there is a one-to-one mapping $f\colon A_1 \to A_2$ such that $f(1_1) = 1_2$ and $f(\mathrm{S}_1 x) = \mathrm{S}_2 f(x)$ for any element x from A_1.*

Proof: By Peano axioms, an arbitrary element x from A_1 has the form $\mathrm{S}_1^n 1_1$. Then we can define $f(1_1) = 1_2$ and $f(x) = f(\mathrm{S}_1^n 1_1) = \mathrm{S}_2^n 1_2$ for all n. Now to prove that f has all necessary properties, we can use mathematical induction, which is formalized by Axiom P5.

The base of induction. For the number 1, we have $f(1_1) = 1_2$ by definition.

The general step of induction. Let us assume that $f(\mathrm{S}_1(\mathrm{S}_1^m 1_1)) = f(\mathrm{S}_1^{m+1} 1_1) = \mathrm{S}_2^{m+1} 1_2 = \mathrm{S}_2^{m+1} f(1_1)$ for all $m < n$ and take $x = \mathrm{S}^n 1_1$. Then we have

$$f(\mathrm{S}_1 x) = f(\mathrm{S}_1(\mathrm{S}_1^{n-1} 1_1)) = f(\mathrm{S}_1^{n+1} 1_1) = f(\mathrm{S}_1(\mathrm{S}_1^n 1_1)) = \mathrm{S}_2 f(\mathrm{S}_1^n 1_1)$$
$$= \mathrm{S}_2(\mathrm{S}_2^n 1_2) = \mathrm{S}_2^{n+1} f(1_1) = \mathrm{S}_2 f(x).$$

Thus, by Axiom P5, $f(\mathrm{S}_1x) = \mathrm{S}_2f(x)$ for any element x from A_1.
Theorem is proved. □

It is natural to take postulates (P1)–(P4) without assuming that elements are conventional natural numbers. This gives us the following system:

(P1g) 1 is a number.
(P2g) If x is a number then Sx is a number.
(P3g) If x is a number, then Sx is not 1.
(P4g) If Sx = Sy, then $x = y$.

It is also possible to consider more general fifth postulate changing it to the axiom of induction (cf., for example, Hilbert and Bernays, 1968; Kleene, 2002).

Let us also consider a predicate (property) $A(x)$ with one variable.

(P5g) If a set P includes 1, and if $A(0)$ and for every x in P, $A(x)$ implies $A(\mathrm{S}x)$, then $A(x)$ is true for all elements in P.

This is an axiom for every predicate (property) $A(x)$. However, a set of axioms, represented by one expression is called an axiom schema. Thus, (P5g) is the induction schema.

Then it is possible to build a Peano system P, which is not isomorphic (as an algebraic system) to the system $\boldsymbol{N}$ of all natural numbers but satisfies postulates (P1g)–(P5g).

Example 2.11.17. Let us consider the set $P = N \cup Z_o$ where Z_o is isomorphic to the set Z of all integer numbers. If n is an element from Z_o, we denote it by n_o, for example, 5_o or -10_o. The order relation $\leq$ is defined in the standard way in P and Z_o, Besides, we assume that any element from Z_o is larger than any element (natural number) from N. The successor Sx is defined in the standard way, for example, S5 = 6 or S$(-10_o) = -9_o$. It is easy to check that the system P satisfies postulates (P1g)–(P5g).

Example 2.11.18. It is possible to find systems isomorphic to P, which consist of rational or real numbers. Indeed, let us take the set $QP = A \cup B$ where $A = \{1, 1^1\!/\!_2, 1^1\!/\!_4, \ldots, 1^1\!/\!_{2^n}, \ldots\}$ and $A = \{\ldots, 2^1\!/\!_{2^n}, \ldots, 2^1\!/\!_4, 2^1\!/\!_2, 3, 3^1\!/\!_2, 3^1\!/\!_4, \ldots, 3^1\!/\!_{2^n}, \ldots\}$. Then we define $\mathrm{S}1 = 1^1\!/\!_2$, $\mathrm{S}1^1\!/\!_2 = 1^1\!/\!_4, \ldots, \mathrm{S}1^1\!/\!_{2^n} = 1^1\!/\!_{2^{n+1}}$, $\mathrm{S}2^1\!/\!_{2^n} = 2^1\!/\!_{2^{n-1}}$, $\mathrm{S}2^1\!/\!_4 = 2^1\!/\!_2$, $\mathrm{S}2^1\!/\!_2 = 3$, $\mathrm{S}3 = 3^1\!/\!_2$, $\mathrm{S}3^1\!/\!_2 = 3^1\!/\!_4, \ldots$, $\mathrm{S}3^1\!/\!_{2^n} = 3^1\!/\!_{2^{n+1}}, \ldots$ We see that as a universal algebra with one unary operation, QP is isomorphic to P and thus, satisfies postulates (P1g)–(P5g).

An interesting peculiarity of Peano postulates is that they do not ascribe any conditions on operations in a system that satisfies postulates (P1)–(P5). Such axioms have been added by different researchers later and we will consider them but now we want to attract attention to the fact that instead of understanding that these postulates allow for different kinds of arithmetical operations, the researchers simply axiomatized operations from the Diophantine arithmetic overlooking other opportunities.

The opinion that there is only one arithmetic was so strong that the great French mathematician Henri Poincaré once pointed out that one does not "prove" $2+2=4$, one "checks" it (cf. Gonthier, 2008). Naturally, when we check this model equality and see that it is sometimes violated, we come to the conclusion that other arithmetics are necessary. Many such situations are described in Chapter 1 (cf. also Helmholtz, 1887; Kline, 1967; 1980; Davis, 1972; Rashevsky, 1973; Blehman *et al.*, 1983; Davis and Hersh, 1986; Burgin, 2001; 2001a).

For millennia, mathematicians have been inside the "box" of the Diophantine arithmetic and instead of going out of this artificial "box", they tried to canonize it as an absolute ideal epitome. Some even proved uniqueness of the Diophantine arithmetic remaining inside it. Naturally, they got the result they expected. For instance, Feferman proves for the Peano system N of all natural numbers, that any of its subsystems, say, P, is isomorphic to N if it satisfies postulates (P1)–(P5) (Feferman, 1974). However, as we can see from Examples 2.11.17 and 2.11.18, there are other than N Peano systems.

Moreover, while Peano defines the concept of the successor $\mathrm{S}n$ of a number n as a function without using other concepts, later his axioms were "improved" by characterizing the successor of a number n in terms of addition as the number $n+1$ (cf., for example, Ross, 1996). Peirce, who also axiomatically described arithmetic, defined the successor of n as $n+1$ (Peirce, 1881). This is true for Diophantine arithmetic but eliminates existence possibilities of non-Diophantine arithmetics.

This situation is very similar to the history of geometry when for a long time mathematicians tried to derive the fifth postulate of the Euclidean geometry from the first four postulates overlooking the possibility to discover (build) non-Euclidean geometries, which were discovered only in the 19th century. It is possible to find the history of this groundbreaking discovery in many books (cf., for example, Carroll, 2009; Burton, 1997; Eves, 1990; Trudeau, 1987; Struik, 1987).

We can see lost and found opportunities with Peano postulates as the following simple but important result demonstrates that all non-Diophantine arithmetics also satisfy Peano postulates (P1)–(P5).

Theorem 2.11.9. *All projective, direct and dual perspective arithmetics of natural numbers satisfy Peano postulates* (P1)–(P5).

Indeed, as any non-Diophantine arithmetic $\boldsymbol{A} = (N; +, \circ, \leq)$ of natural numbers consists of all natural numbers, it has number 1 (postulate P1), each number in it has the successor (postulate P2), 1 is not a successor of any number (postulate P3), any successor uniquely defines its predecessor (postulate P4), and if a subset X of N in $\boldsymbol{A}$ contains 1 and successors of all elements from X, then X contains all natural numbers N (postulate P5).

Note that direct and dual perspective, projective and other non-Diophantine arithmetics contain all natural numbers but have atypical operations with these numbers.

It might be useful to note that other kinds of non-Diophantine arithmetics also satisfy Peano axioms.

An interesting analysis of properties of Peano systems in general and the set N of all natural numbers is performed by Pierce. He considers the following fundamental properties of the set N (Pierce, 2012):

(1) *N admits proofs by induction*: Every subset that contains 1 and that contains Sk whenever it contains k is the whole set.
(2) *N is well-ordered*: Every non-empty subset has a least element.
(3) *N admits proofs by "complete" or "strong" induction*: Every subset that contains 1 and that contains Sk whenever it contains $1, \ldots, k$ is the whole set.
(4) *N admits recursive definitions of functions*: There is a unique function h: $N \to N$ such that $h(1) = a$ and $h(\mathrm{S}k) = f(h(k))$, where f is a given operation on a set that has a given element a.

It is often assumed that these properties are equivalent. But they are not equivalent. Indeed, it is "not even wrong" to say that they are equivalent, since properties (1) and (4) are possible properties of algebras in the signature $\{1, \mathrm{S}\}$, while property (2) is a property of some ordered sets, and property (3) is a property of some ordered algebras. However, properties (1)–(4) become equivalent under additional assumptions (Pierce, 2012). Interestingly, following the traditional approach and writing about the set N without arithmetical operations, i.e., as a Peano system, Pierce nevertheless identifies the next element Sk with $k + 1$. This is true in the Diophantine arithmetic $\boldsymbol{N}$ but it is not true in many non-Diophantine arithmetics as we have seen in the previous sections.

2.11.4. *Axioms for arithmetic and arithmetical operations*

Simply pushing harder within the old boundaries will not do.

Karl Weick

We see that the imperative property of natural numbers is the existence of successors and the initial object. However, some mathematicians and philosophers expressed the opinion that this is not enough. For instance, Russell wrote:

"[I]*t is impossible that the* [*numbers*] *should be, as Dedekind suggests, nothing but the terms of such relations as constitute a progression* [*i.e., a natural number system*]. *If they are to be anything at all, they must be intrinsically something; they must differ from other entities as points from instants, or colours from sounds*". (Russell, 1903)

Indeed, as we have seen, an additional key property of arithmetics is existence of, at least, two operations — addition and multiplication. There are several axiomatics of the Diophantine arithmetic $\boldsymbol{N}$, in which axioms defining these two operations are also included. Often the arithmetic that satisfies these axioms is too called Peano arithmetic. It is natural to call axiomatic systems that include Peano axioms (P1)–(P5) by the name *extended Peano axioms.*

Besides, in many cases, axioms characterize the arithmetic $\boldsymbol{W}$ of whole numbers and not only the arithmetic $\boldsymbol{N}$ of natural numbers. This creates some confusion in this area and to eliminate it, we suggest, as it is done in this book, distinguishing arithmetics of whole numbers and arithmetics of natural numbers. When we write simply *arithmetic*, it means an arithmetic of natural numbers.

Let us consider some of the suggested axiomatics for the Diophantine arithmetic $\boldsymbol{N}$.

Charles Sanders Peirce (1839–1914) also gave an independent of Peano axiomatic description of the Diophantine arithmetic $\boldsymbol{N}$ (Peirce, 1881). Peirce called his statements describing properties of the arithmetic not axioms but definitions. Let us look at them.

(1) The minimum number is called one.
(2) By $x + y$ is meant, in case $x = 1$, the number next greater than y; and in other cases, the number next greater than $x' + y$, where x' is the number next smaller than x.
(3) By $x \times y$ is meant, in case $x = 1$, the number y, and in other cases $y + x'y$, where x' is the number next smaller than x.

Peirce also writes that it may be remarked that the symbols + and × are triple relatives, their two correlates being placed one before and the other after the symbols themselves.

It is possible to convert Peirce's definitions into formal axioms of the Diophantine arithmetic $\boldsymbol{N}$. To do this, we define the inverse to Peano's successor function S the predecessor function P. Namely, $\mathrm{P}x = y$ if and only if $\mathrm{S}y = x$. Then we have the following definitions:

AA1. $\exists x = 1 \& \rceil \exists y(y = \mathrm{P}1)$.
AA2. Operations + and × are ternary relations.
AA3. (A) $1 + y = \mathrm{S}y$;
(B) If $x \neq 1$, then $x + y = \mathrm{S}(\mathrm{P}x + y)$.
AA4. (A) $1 \times y = y$;
(B) If $x \neq 1$, then $x \times y = y + \mathrm{P}x \times y$.

We see that the system of Peirce is incomplete lacking the Axiom of Induction. Consequently, there are arithmetics of ordinal numbers, which satisfy the axioms (have these properties).

Stephen Cole Kleene (1909–1994) gives a logically oriented system of axioms (Kleene, 2002):

$$Sx = Sy \Rightarrow x = y,$$
$$x = y \Rightarrow (x = z \Rightarrow y = z),$$
$$x + 0 = x,$$
$$x \cdot 0 = 0,$$
$$Sx \neq 0,$$
$$x = y \Rightarrow Sx = Sy,$$
$$x + Sy = S(x + y),$$
$$x \cdot Sy = x \cdot y + x,$$
$$A(0) \& \forall x(A(x) \Rightarrow A(Sx)) \Rightarrow \forall x(A(x)).$$

Note that Kleene does not include axioms of equality in his system and includes 0 into arithmetic.

In contrast to Kleene, Yuri Manin gives an algebraically oriented system of axioms (Manin, 1991):

(A) Axioms of equality:

$$x = x,$$
$$x = y \Leftrightarrow y = x,$$
$$x = y \;\&\; y = z \Rightarrow x = z,$$
$$x = y \Rightarrow (A(x, x) \Rightarrow A(x, y)).$$

(B) Axioms of addition:

$$x + 0 = x,$$
$$x + y = y + x,$$
$$x + (y + z) = (x + y) + z,$$
$$x + y = z + y \Rightarrow x = z.$$

(C) Axioms of multiplication:

$$x \cdot 0 = 0,$$
$$x \cdot 1 = x,$$
$$x \cdot y = y \cdot x,$$
$$x \cdot (y \cdot z) = (x \cdot y) \cdot z.$$

(D) Axiom of distributivity:

$$x \cdot (y + z) = x \cdot y + x \cdot z.$$

(E) Axiom of induction:

If A is a predicate, then

$$A(0)\&\forall x(A(x) \Rightarrow A(x + 1)) \Rightarrow \forall x(A(x).$$

One more axiomatics for the Diophantine arithmetic $\boldsymbol{N}$ is given in (Rasiowa and Sikorski, 1963):

(1) Axioms of equality:

$$x = x,$$
$$x = y \Leftrightarrow y = x,$$
$$x = y \;\&\; y = z \Rightarrow x = z,$$
$$(x = y \;\&\; u = w) \Rightarrow x + u = z + w,$$
$$(x = y \;\&\; u = w) \Rightarrow x \cdot u = z \cdot w.$$

(2) Axioms of order and operations:

$$\text{It is not true that} 1 = x + 1,$$
$$x + 1 = y + 1 \Rightarrow x = y,$$
$$x + (y + 1) = (x + y) + 1,$$
$$x \cdot 1 = x,$$
$$x \cdot (y + 1) = (x \cdot y) + x.$$

(3) Axiom of induction.

An important question is to what extent all these axioms characterize the Diophantine arithmetic $\boldsymbol{N}$. We remind that a system of axioms A is called *categorical* if there is only one object (up to the equality) that satisfies these axioms. Logicians found that this characterization is not categorical and there are non-standard models of arithmetic, which contain non-standard natural numbers. For instance, it is possible consistently to add to the Diophantine arithmetic $\boldsymbol{N}$ an element x that is larger than any natural number.

Dedekind already wrote about possibility to build different systems that satisfied his axioms. However, the first rigorous construction of non-standard models of arithmetic is due to the Norwegian mathematician Thoralf Skolem (1887–1963), who proved that the first-order theory of arithmetic has models that are not isomorphic to $\boldsymbol{N}$ (Skolem, 1934).

It is necessary to remark that all non-Diophantine arithmetics of natural numbers, which have been studied, contain only standard natural numbers. However, it might be interesting to construct and study non-standard models of non-Diophantine arithmetics.

As the Diophantine arithmetic $\boldsymbol{N}$ is one of the most basic objects in mathematics, logicians have been especially interested in its properties. After $\boldsymbol{N}$ had been axiomatized, logicians became concerned whether those axioms completely characterize truth in $\boldsymbol{N}$, i.e., whether it was possible to prove all true sentences in this basic and relatively simple mathematical object. One of the great mathematicians of the 20th century David Hilbert (1862–1943) proclaimed his faith in the solvability of every mathematical problem. According to Hilbert's program, all truths in arithmetic and even in more complex mathematical structures have to be deductible from properly chosen axioms using conventional deduction.

Different mathematicians tried to prove this. When their efforts did not bring success, they started exploring simpler similar systems. In this search, Mojzesz Presburger (1904–1943) built a new kind of axiomatic arithmetics with the language L_1, which was later called *Presburger's arithmetic* and denoted by $\boldsymbol{PA}_1$ (Presburger, 1929). Presburger proved the following statements:

(a) There is an algorithm that allows deciding whether an arbitrary closed formula in the language L_1 is true in $\boldsymbol{PA}_1$ or not.
(b) Using the axioms of $\boldsymbol{PA}_1$ and conventional deduction, it is possible to prove each true in $\boldsymbol{PA}_1$ formula of the language L_1.
(c) It is also possible to prove consistency of $\boldsymbol{PA}_1$ using conventional deduction allowed in Hilbert's program.

Based on this and some other results, mathematicians continued to believe in Hilbert's program. Nevertheless, the groundbreaking results of the famous Austrian/American logician Kurt Gödel (1906–1978) astonished all mathematicians and logicians because according to the first Gödel theorem, any consistent finite system of axioms cannot allow deduction of all true sentences in the Diophantine arithmetic $\boldsymbol{N}$ (Gödel, 1931/1932).

An interesting peculiarity of Gödel's result is that it is based on utilization of recursive algorithms, such as conventional logical deduction or Turing machine, for proving theorems in mathematics. However, application of super-recursive algorithms, such as inductive Turing machine, allows proving all true sentences in the Diophantine arithmetic $\boldsymbol{N}$ due to the fact that there is, for example, a hierarchy of inductive Turing machines, which can compute and decide the whole arithmetical hierarchy (Burgin, 2003a; 2005). It means that what is impossible to do with a system of deduction algorithms is possible to achieve using more powerful induction algorithms based on inductive Turing machines.

Arithmetical truth is not only non-provable by recursive algorithms but it is also not representable in the standard arithmetic. Indeed, Tarski's undefinability theorem states that arithmetical truth cannot be defined in formal arithmetic (Tarski, 1936). Now this result is also called Gödel–Tarski theorem because Gödel also discovered it in 1930, while proving his incompleteness theorems published in 1931 (Murawski, 1998). Note that this theorem is essentially based on the formal language used in arithmetic. Building a hierarchy of languages and axioms for arithmetic, which form a logical variety in the sense of (Burgin, 1997d; Burgin and de Vey Mestdagh, 2015), it is possible to define and prove arithmetical truth in this system.

One of the most important conditions for a mathematical system is its consistency. As a result, essential attention has been directed at consistency of the Peano arithmetic, or more exactly, the Diophantine arithmetic $\boldsymbol{N}$, as a basic mathematical system. For instance, this problem was included by Hilbert as the second of the main problems in his famous list (Hilbert, 1902). In his lecture where he formulated ten out of 23 later published problems, Hilbert explained that this problem is crucial for building safe and sound foundations for the whole mathematics.

The first result in this direction was obtained by the German mathematician Wilhelm Friedrich Ackermann (1896–1962), who gave a finitistic proof of the consistency of Peano arithmetic without the axiom scheme of induction, which was a much weaker system than Peano arithmetic (Ackermann, 1924).

Next attempt was by the Hungarian-American mathematician John von Neumann (1903–1957), who proved consistency of a larger fragment of Peano arithmetic with definite restrictions on the induction (von Neumann, 1927).

One more consistency proof was obtained by the French mathematician Jacques Herbrand (1908–1931) for arithmetic with induction for formulas containing no bounded variables and induction for formulas containing bounded variables but containing no function symbols except eventually the successor function (Herbrand, 1931).

However, at the same time, Gödel published his groundbreaking paper where his second incompleteness theorem stated that it was impossible to prove consistency of the Diophantine arithmetic $\boldsymbol{N}$ (Peano arithmetic) using the language of $\boldsymbol{N}$ and traditional deduction (Gödel, 1931/1932). As the Diophantine arithmetic is the most indispensable mathematical structure and traditional deduction was considered as the universal inference mechanism of mathematics, many believed that it was unquestionably impossible to prove consistency of the Diophantine arithmetic of natural numbers and Gödel theorems put absolute boundaries on what is possible to prove in mathematics. However, as the history of mathematics shows, when mathematicians cannot solve some problems, they invent new structures and innovative techniques to overcome this "impossibility" (Burgin, 2012).

The same happened with problem of consistency of the Diophantine arithmetic of natural numbers. The German mathematician and logician Gerhard Karl Erich Gentzen (1909–1945) suggested several proofs of consistency of the Diophantine arithmetic of natural numbers using transfinite induction (Gentzen, 1936; 1938; 1943).

Ackermann also proved the consistency of the Diophantine arithmetic of natural numbers using methods from his paper (Ackermann, 1924) and the transfinite induction (Ackermann, 1940). There were also other proofs the consistency of the Diophantine arithmetic of natural numbers using the transfinite induction (cf., for example, Lorenzen, 1951; Schütte, 1951, 1960; Hlodovskii, 1959; Gauthier, 2000).

It was also found that the possibility of proving consistency of the Diophantine arithmetic depended on the way in which the metamathematical property of consistency is expressed in the language of the considered theory (Murawski, 2001).

To conclude, it is necessary to explain that all these undecidability and incompleteness results related to arithmetic are about names and incompleteness of definite named sets in the sense of (Burgin, 2011). Indeed, proving his famous incompleteness theorem, Gödel used enumeration, which is now called Gödel numbering, of all arithmetical formulas (Gödel, 1931/1932; Rogers, 1958). However, enumeration is a specific naming where numbers are names of formulas. Formulas of a logical language are names of relations and sets (Church, 1956). Thus, Tarski's undefinability theorem asserts that arithmetical truth does not have a name in the set of arithmetical formulas. In the most explicit form, the naming context of undecidability theorems was presented by Huizing, Kuiper and Verhoeff, who proved two interesting generalizations of Rice's theorem, which demonstrate undecidability of many problems in the theory of algorithms and abstract automata (Huizing *et al.*, 2012).

Chapter 3

Non-Diophantine Arithmetics of Real and Complex Numbers

Ideas are real.
Plato

As we know, conventional (Diophantine) arithmetics of integer, rational, real and complex numbers have additional operations in comparison with the conventional Diophantine arithmetics of natural and whole numbers. That is why to build and study non-Diophantine arithmetics of real and complex numbers, we at first extend the concept of an abstract prearithmetic studied in Chapter 2 augmenting its structure by additional operations.

In Section 3.1, we introduce and investigate *partially extended abstract prearithmetics* and *arithmetics*, which in addition to two basic operations of abstract prearithmetics — addition and multiplication — have one more operation *subtraction* as the Diophantine arithmetic of integers has. We introduce and explore properties of and relations between partially extended prearithmetics and arithmetics. Although we consider regular types of relations, such as homomorphisms and isomorphisms, the main emphasis is made on new types of relations, such as weak projectivity, projectivity and exact projectivity. In particular, we develop tools for construction of new partially extended (abstract) prearithmetics and arithmetics using already existing partially extended prearithmetics and arithmetics. We also show what properties of new partially extended prearithmetics and arithmetics are deducible from properties of partially extended prearithmetics and arithmetics used for construction.

However, the main emphasis in this chapter is made on *wholly extended abstract prearithmetics* and *arithmetics*, which in addition to two basic operations of abstract prearithmetics, namely, addition and multiplication, have two more operations *subtraction* $\ominus$ and *division* $\oslash$ as the Diophantine arithmetics of real and complex numbers have. Traditionally, these operations are treated as *basic arithmetical operations*. Therefore, in Section 3.2, we introduce and explore properties of and relations between wholly extended abstract prearithmetics and arithmetics. Although we consider regular types of relations, such as homomorphisms and isomorphisms, the main emphasis is made on new types of relations, such as weak projectivity, projectivity and exact projectivity. In particular, we develop tools of construction of new wholly extended (abstract) prearithmetics and arithmetics using already existing wholly extended prearithmetics and arithmetics. We also show what properties of new wholly extended prearithmetics and arithmetics are deducible from properties of wholly extended prearithmetics and arithmetics used for construction.

Tools developed in Sections 3.1 and 3.2 are used for construction and study of non-Diophantine arithmetics of real (in Section 3.3) and complex (in Section 3.4) numbers. Specifically, in Section 3.3, we introduce and explore properties of and relations between wholly extended prearithmetics and arithmetics of real numbers. In this setting, non-Diophantine arithmetics of real numbers are wholly extended prearithmetics of real numbers that contain all real numbers with the conventional order but have non-conventional arithmetical operations. We show how to build non-Diophantine arithmetics of real numbers using the Diophantine (conventional) arithmetic of real numbers.

In Section 3.4, we introduce and explore properties of and relations between wholly extended prearithmetics and arithmetics of complex numbers. In this setting, non-Diophantine arithmetics of complex numbers are wholly extended prearithmetics of real numbers that contain all complex numbers with the conventional order but have non-conventional arithmetical operations. We show how to build non-Diophantine arithmetics of complex numbers using the Diophantine (conventional) arithmetic of complex numbers. According to their construction, prearithmetics and non-Diophantine arithmetics of complex numbers form three classes: functional, operational and stratified prearithmetics and arithmetics of complex numbers.

In Section 3.5, we use non-Diophantine arithmetics of real and complex numbers to build and study quasilinear spaces and algebras in the context of prearithmetics. Non-Grassmannian linear spaces and algebras are important cases of quasilinear spaces and algebras.

A conventional linear space has the linear structure over a field. In a similar way, a quasilinear space has the linear structure over a prearithmetic or arithmetic. Properties of quasilinear and non-Grassmannian linear spaces and algebras are studied. Many of these properties are similar to the properties of linear spaces. For instance, it is proved that the sequential composition of linear mappings of left quasilinear spaces is also a linear mapping of left quasilinear spaces (Proposition 3.5.23) or that in a left quasilinear space L that has dimension n, any system of quasivectors with more than n elements is linearly dependent (Theorem 3.5.3).

3.1. Partially Extended Abstract Prearithmetics and Their Projectivity

Creativity is a natural extension of our enthusiasm.

Earl Nightingale

A *partially extended abstract prearithmetic* is a set (often a set of numbers) A with a partial order $\leq$ and three binary operations $+$ (addition), $-$ (subtraction) and $\circ$ (multiplication), which are defined for all its elements. It is denoted by $\boldsymbol{A} = (A; +, \circ, -, \leq)$. As before, the set A is called either the *set of elements* or the *set of numbers* or the *carrier* of the prearithmetic $\boldsymbol{A}$. In a usual way, if $x \leq y$ and $x \neq y$, then we denote it by $x < y$. Operation $+$ is called *addition*, operation $-$ is called *subtraction* and operation $\circ$ is called *multiplication* in the prearithmetic $\boldsymbol{A}$.

Note that a partially extended abstract prearithmetic can have more operations and order relations but the three basic operations are always defined. Besides, in a general case, subtraction in partially extended abstract prearithmetics is not necessarily subtraction introduced for abstract prearithmetics in Section 2.1. Indeed, subtraction introduced in Section 2.1 is inverse to addition in general abstract prearithmetics, while subtraction in partially extended abstract prearithmetics is independent from addition and multiplication in a general case.

Example 3.1.1. Naturally, the conventional arithmetic $\boldsymbol{Z}$ of all integer numbers is a partially extended abstract prearithmetic.

Example 3.1.2. Another example of partially extended abstract prearithmetics is a *modular arithmetic*, which sometimes known as *residue arithmetic* or *clock arithmetic* (cf. Section 2.1).

Example 3.1.3. The conventional arithmetics $\boldsymbol{Q}$ of all rational numbers, $\boldsymbol{Z}$ of all integer numbers, $\boldsymbol{R}$ of all real numbers and $\boldsymbol{C}$ of all complex numbers are partially extended abstract prearithmetics.

All these examples show that many conventional arithmetics are partially extended abstract prearithmetics. However, there are also many unusual partially extended abstract prearithmetics.

Example 3.1.4. Let us consider the set N of all natural numbers with the standard order $\leq$ and introduce the following operations:

$$a \oplus b = a \cdot b,$$

$$a \otimes b = a^b,$$

$$a \ominus b = a + b.$$

Then the system $\boldsymbol{A} = (N; \oplus, \otimes, \ominus, \leq)$ is a partially extended abstract prearithmetic with addition $\oplus$, subtraction $\ominus$ and multiplication $\otimes$

Example 3.1.5. Let us consider the set R of all real numbers without 0 and with the standard order $\leq$ and introduce the following operations:

$$a \boxplus b = a - b,$$

$$a ※ b = a \div b,$$

$$a \ominus b = a + b.$$

Then the system $\boldsymbol{B} = (R; \boxplus, ※, \ominus, \leq)$ is a partially extended abstract prearithmetic with addition $\boxplus$, subtraction $\ominus$ and multiplication ※.

Example 3.1.6. Any ring, semiring, module, vector space, linear algebra or field is a partially extended abstract prearithmetic with a trivial order (cf. Appendix). Any ordered ring, linear algebra, semiring or field is also a partially extended abstract prearithmetic (Kurosh, 1963). In particular, algebras (arithmetics) of quaternions, octonions, sedenions and other hypercomplex numbers (Kantor and Solodovnikov, 1989; Dixon, 1994; Schlote, 1996) are partially extended abstract prearithmetics.

There are motivations of three kinds toward extending abstract prearithmetics with subtraction and negative elements, or as Gauss suggested to call them *inverse* or *opposite elements* (Gauss, 1831). One motivation is theoretical, another is pragmatic and the third one is practical.

Theoretical motivation. For any abstract prearithmetic, there are two total operations — addition and multiplication — and a partial operation of subtraction (cf. Section 2.1). Thus, to make arithmetic more operational and powerful as a tool for solving problems, it would be reasonable to make the latter operation also total.

Pragmatic motivation. In many problems, the quantity or quantities to be determined cannot be obtained by a direct arithmetical calculation but instead have to be deduced from given conditions or restrictions on these quantities. Often these conditions have the form of equations. For instance, we have an equation $10 + x = 5$ in conventional numbers and need to find its solution. As we know, to solve this equation, we need negative numbers while to operate with negative numbers, we need subtraction. Similar equations arise in abstract prearithmetics and arithmetics. To solve them, we also need subtraction and negative elements (cf. Section 2.1).

Practical motivation. Numbers are used in many practical and theoretical areas — in physics, finance, business, biology, etc. As we know in all these and many other areas, negative numbers play an important role. In a similar way, application of abstract prearithmetics and arithmetics, also demands negative numbers and subtraction.

Now let us study properties of partially extended abstract prearithmetics. In particular, properties of partially extended abstract prearithmetics allow assembling other algebraic constructions using partially extended abstract prearithmetics as it is done in the case of Diophantine arithmetics $\boldsymbol{R}$ and $\boldsymbol{C}$. Here we consider only such constructions as vectors, matrices and their prearithmetics although it is possible to construct similar structures for arrays of arbitrary dimensions and explore what properties they inherit from the initial partially extended abstract prearithmetic. For instance, taking a partially extended abstract prearithmetic $\boldsymbol{A} = (A; +, \circ, -, \leq)$ and a natural number n, it is possible to build the partially extended abstract prearithmetic of n-dimensional A-vectors $\mathrm{V}^n\boldsymbol{A} = (\mathrm{V}^n A; +, \circ, -, \leq)$, elements of which are vectors in A. Namely, elements of the partially extended n-dimensional A-vector prearithmetic $\mathrm{V}^n\boldsymbol{A} = (\mathrm{V}^n A; +, \circ, -, \leq)$, i.e., A-vectors, have the form $(a_1, a_2, \ldots, a_n)$ where $a_1, a_2, \ldots, a_n$ are elements from the abstract prearithmetic $\boldsymbol{A}$.

In a similar way, taking a partially extended abstract prearithmetic $\boldsymbol{A} = (A; +, \circ, -, \leq)$ and a pair of natural numbers n and m, it is also possible to build the partially extended abstract prearithmetic of $n \times m$-dimensional A-matrices $\mathrm{M}\boldsymbol{A} = (\mathrm{M}A; +, \circ, -, \leq)$, elements of which are matrices in A. Namely, elements of the partially extended $n \times m$-dimensional A-matrix prearithmetic $\mathrm{M}^{n\times m}\boldsymbol{A} = (\mathrm{M}^{n\times m}A; +, \circ, -, \leq)$, i.e., A-matrices, have the form

$$\begin{pmatrix} a_{11} & a_{12} & a_{13} & \dots & a_{1n} \\ a_{21} & a_{22} & a_{23} & \dots & a_{2n} \\ \vdots & \vdots & \vdots & \vdots & \vdots \\ a_{m1} & a_{m2} & a_{m3} & \dots & a_{mn} \end{pmatrix},$$

where all $a_{ij}(i = 1, 2, 3, \dots, m; j = 1, 2, 3, \dots, n)$ are elements from the abstract prearithmetic $\boldsymbol{A}$.

Addition, subtraction and multiplication in these prearithmetics are defined coordinate-wise. For instance, taking two-dimensional Z-vectors (2, 3) and (4, 5) from the prearithmetic $\mathrm{V}^2\boldsymbol{Z}$ of Z-vectors, we define their sum as $(2, 3) + (4, 5) = (6, 8)$, their difference as $(2, 3) - (4, 5) = (-2, -2)$, and their product as $(2, 3) \circ (4, 5) = (8, 15)$.
Order is defined by the following condition:

If $(a_1, a_2, \dots, a_n)$ and $(b_1, b_2, \dots, b_n)$ are vectors from V^nA, then $(a_1, a_2, \dots, a_n) \leq (b_1, b_2, \dots, b_n)$ if and only if $a_j \leq b_j$ for all $j = 1, 2, 3, \dots, n$.

For matrices, addition, subtraction, multiplication and order are defined in a similar way.

Note that the defined multiplication is *scalar multiplication* of vectors and matrices, which is different from vector and matrix multiplication.

Partially extended prearithmetics $\mathrm{V}^n\boldsymbol{A}$ and $\mathrm{M}^{n\times m}\boldsymbol{A}$ preserve many properties of the partially extended abstract prearithmetic $\boldsymbol{A}$. For instance, we have the following results.

Proposition 3.1.1. *If addition is commutative in a partially extended abstract prearithmetic $\boldsymbol{A}$, then addition is commutative in the partially extended vector prearithmetic $V^n\boldsymbol{A}$ and in the partially extended matrix prearithmetic $\mathrm{M}^{n\times m}\boldsymbol{A}$.*

Proof is left as an exercise. □

The same is true for multiplication.

Proposition 3.1.2. *If multiplication is commutative in a partially extended abstract prearithmetic* $\boldsymbol{A}$*, then multiplication is commutative in the partially extended vector prearithmetic* $V^n\boldsymbol{A}$ *and in the partially extended matrix prearithmetic* $\mathrm{M}^{n\times m}\boldsymbol{A}$.

Proof is left as an exercise. □

Proposition 3.1.3. *If addition is associative in a partially extended abstract prearithmetic* $\boldsymbol{A}$*, then addition is associative in the partially extended vector prearithmetic* $V^n\boldsymbol{A}$ *and in the partially extended matrix prearithmetic* $\mathrm{M}^{n\times m}\boldsymbol{A}$.

Proof is left as an exercise. □

The same is true for multiplication.

Proposition 3.1.4. *If multiplication is associative in a partially extended abstract prearithmetic* $\boldsymbol{A}$*, then multiplication is associative in the partially extended vector prearithmetic* $V^n\boldsymbol{A}$ *and in the partially extended matrix prearithmetic* $\mathrm{M}^{n\times m}\boldsymbol{A}$.

Proof is left as an exercise. □

In the Diophantine arithmetic, multiplication is distributive with respect to addition, i.e., the following identities hold:

$$x \cdot (y + z) = x \cdot y + x \cdot z,$$

$$(y + z) \cdot x = y \cdot x + z \cdot x.$$

However, in abstract prearithmetics, multiplication is not always commutative and we need to discern three kinds of distributivity. Namely, *distributivity from the left*

$$x \cdot (y + z) = x \cdot y + x \cdot z$$

and *distributivity from the right*

$$(y + z) \cdot x = y \cdot x + z \cdot x.$$

In addition, multiplication is *distributive* with respect to addition when both identities hold.

Note that in the case of commutative multiplication, all three kinds of distributivity coincide.

Proposition 3.1.5. *If multiplication is distributive (distributive from the left or distributive from the right) with respect to addition in a partially extended abstract prearithmetic* $\boldsymbol{A}$, *then multiplication is distributive (distributive from the left or distributive from the right) with respect to addition in the partially extended vector prearithmetic* $V^n\boldsymbol{A}$ *and in the partially extended matrix prearithmetic* $\mathrm{M}^{n\times m}\boldsymbol{A}$.

Proof is left as an exercise. □

Remark 3.1.1. Having a partially extended abstract prearithmetic $\boldsymbol{A} = (A; +, \circ, -, \leq)$, it is possible to build not only partially extended abstract prearithmetics of A-vectors and A-matrices but also partially extended abstract prearithmetics of multidimensional matrices or arrays in A, i.e., multidimensional A-matrices or A-arrays.

Let us study relations between addition and subtraction in partially extended abstract prearithmetics. In a general case, these operations are completely independent. However, in the Diophantine arithmetics $\boldsymbol{Z}$, $\boldsymbol{R}$ or $\boldsymbol{C}$, subtraction is *inverse* to addition and vice versa. That is (cf. Section 2.1), the following identities are true

$$(x - y) + y = y + (x - y) = x.$$

This property of partially extended abstract prearithmetics is intrinsically related to existence of opposite numbers.

Definition 3.1.1. An element a is *opposite* to an element b if the following equality (cf. Section 2.1) is valid

$$b + a = 0.$$

We denote an element opposite to b by $-b$.

Lemma 3.1.1. *In an abstract prearithmetic* $\boldsymbol{A}$ *with additive cancellation, the opposite element to any element from* $\boldsymbol{A}$ *is unique.*

Proof is left as an exercise. □

Proposition 3.1.6. *In an abstract prearithmetic* $\boldsymbol{A}$ *with additive cancellation and commutative addition, opposite to the opposite of an element is the same element, i.e.,* $-(-a) = a$ *for any* a *from* $\boldsymbol{A}$.

Indeed, by definition,

$$(-a) + -(-a) = -(-a) + (-a) = 0$$

and

$$a + (-a) = 0$$

i.e.,

$$-(-a) + (-a) = a + (-a)$$

By additive cancellation, we have $-(-a) = a$.

In an abstract prearithmetic $\boldsymbol{A}$ with the additive zero 0, we can separate positive and negative elements in the conventional way.

Definition 3.1.2.

(a) An element a from $\boldsymbol{A}$ is *positive* if $a > 0$.
(b) An element a from $\boldsymbol{A}$ is *negative* if $a < 0$.

Lemma 3.1.2. *If an element a from $\boldsymbol{A}$ is not equal to* 0*, then its opposite $-a$ (when it exists) also is not equal to* 0.

Proof: Let us assume that $-a = 0$. Then

$$0 = a + (-a) = a + 0 = a.$$

This contradicts out assumption $a \neq 0$ and, by the Principle of Excluded Middle, proves the lemma. □

Let us assume that an abstract prearithmetic $\boldsymbol{A}$ with monotone addition and total order has the additive zero 0.

Lemma 3.1.3. *If an element a from $\boldsymbol{A}$ is positive (negative), then its opposite $-a$ (when it exists) is negative (positive).*

Proof: Let us take a positive element a and assume that its opposite $-a$ exists and is also positive. Note that by Lemma 3.1.2, the element $-a$ cannot be equal to 0. So, because the order is total, $-a$ is either negative or positive.

As addition is monotone, we have

$$a + (-a) \geq a > 0.$$

This contradicts our assumption that $a+(-a) = 0$ and, by the Principle of Excluded Middle, proves the lemma. □

Let us assume that an abstract prearithmetic $\boldsymbol{A}$ with commutative and associative addition has the additive zero 0.

Proposition 3.1.7. *The abstract prearithmetic $\boldsymbol{A}$ is partially extended and subtraction is inverse to addition if and only if for any element a from $\boldsymbol{A}$, there is an opposite element $-a$.*

Proof: *Necessity.* Let us consider a partially extended abstract prearithmetic $\boldsymbol{A}$ with subtraction inverse to addition and an element a from $\boldsymbol{A}$. Then it is possible to define

$$-a = 0 - a.$$

As subtraction inverse to addition in $\boldsymbol{A}$, we have

$$a + (0 - a) = 0.$$

It means that the element $0 - a$ is opposite to the element a.
Sufficiency. Let us consider an abstract prearithmetic $\boldsymbol{A}$ in which for any element a from $\boldsymbol{A}$, there is an opposite element $-a$. Then it is possible to define

$$b - a = b + (-a).$$

This makes the abstract prearithmetic $\boldsymbol{A}$ partially extended and we need to check that subtraction is inverse to addition, i.e., it is necessary to verify that for any element a and b from $\boldsymbol{A}$, we have

$$(a - b) + b = b + (a - b) = a.$$

Indeed, as addition is commutative and associative, we have

$$(a - b) + b = (a + (-b)) + b = a + ((-b) + b) = a + 0 = a.$$

As addition is commutative, we also have

$$b + (a - b) = a.$$

Proposition is proved. □

It would be interesting to find if commutativity and associativity of addition are necessary for validity of Proposition 3.1.7.

Let us study relations between partially extended abstract prearithmetics. One of the most important of them is weak projectivity.

Let us take two partially extended abstract prearithmetics $\boldsymbol{A}_1 = (A_1; +_1, \circ_1, -_1, \leq_1)$ and $\boldsymbol{A}_2 = (A_2; +_2, \circ_2, -_2, \leq_2)$ and consider two mappings g: $A_1 \to A_2$ and h: $A_2 \to A_1$.

Definition 3.1.3. (a) A partially extended abstract prearithmetic $\boldsymbol{A}_1 = (A_1; +_1, \circ_1, -_1, \leq_1)$ is called *weakly projective* with respect to a partially extended abstract prearithmetic $\boldsymbol{A}_2 = (A_2; +_2, \circ_2, -_2, \leq_2)$ if there are following relations between orders and operations in $\boldsymbol{A}_1$ and in $\boldsymbol{A}_2$:

$$a +_1 b = h(g(a) +_2 g(b)),$$

$$a \circ_1 b = h(g(a) \circ_2 g(b)),$$

$$a -_1 b = h(g(a) -_2 g(b)),$$

$$a \leq_1 b \text{ only if } g(a) \leq_2 g(b).$$

(b) The mapping g is called the *projector* and the mapping h is called the *coprojector* for the pair $(\boldsymbol{A}_1, \boldsymbol{A}_2)$.

In this case, we will say there is *weak projectivity* between the partially extended abstract prearithmetic $\boldsymbol{A}_1$ and the partially extended abstract prearithmetic $\boldsymbol{A}_2$ and there is *weak inverse projectivity* between the partially extended abstract prearithmetic $\boldsymbol{A}_2$ and the partially extended abstract prearithmetic $\boldsymbol{A}_1$.

When a partially extended abstract prearithmetic $\boldsymbol{B}$ is weakly projective with respect to a partially extended abstract prearithmetic $\boldsymbol{A}$ with the projector g and the coprojector h, we denote $\boldsymbol{B}$ by $\boldsymbol{A}_{g,h}$ to show this relation between partially extended abstract prearithmetics $\boldsymbol{A}$ and $\boldsymbol{B}$.

As in the general case of abstract prearithmetics, weak projectivity between partially extended abstract prearithmetics determines a bidirectional named set (cf. Remark 2.2.1) and a fiber bundle (cf. Remark 2.2.2).

Quintessentially, the idea of projectivity allows defining operations in one set using operations in another set. This technique is demonstrated by the following result.

Let us consider an arbitrary set B, a partially extended abstract prearithmetic $\boldsymbol{A} = (A; +, \circ, -, \leq)$ and two functions g: $B \to A$ and h: $A \to B$. Definition 3.1.3 implies the following result.

Proposition 3.1.8. *It is possible to define on B the unique structure of a partially extended abstract prearithmetic $\boldsymbol{B}^{g,h}$ that is weakly projective with respect to the partially extended abstract prearithmetic $\boldsymbol{A}$ with the projector g and the coprojector h.*

Proof: To prove this statement, we define the relation $\leq_B$ and three operations $+_B, -_B$ and $\circ_B$ in the set B by the following rules:

$$a +_B b = h(g(a) + g(b)),$$

$$a -_B b = h(g(a) - g(b)),$$

$$a \circ_B b = h(g(a) \circ g(b)),$$

$a \leq_B b$ if and only if $g(a) \leq g(b)$ and ($g(c) = g(a)$ implies $c = a$) and ($g(d) = g(b)$ implies $d = b$) for any elements a and b from B.

Now, we need to check that the relation $\leq_B$ is a partial order, i.e., it is reflexive, antisymmetric and transitive. Taking arbitrary elements a and b from B, we derive the following properties.

- *Reflexivity.* $a \leq_B a$ because $g(a) \leq g(a)$.
- *Anti-symmetry.* Let us have $a \leq_B b$ and $b \leq_B a$. Then $a \leq_B b$ implies $g(a) \leq g(b)$ and $b \leq_B a$ and implies $g(b) \leq g(c)$. As the relation $\leq$ is a partial order, it is antisymmetric (cf. Appendix) and we have $g(b) = g(a)$. Then by definition, we have $b = a$.
- *Transitivity.* Given $a \leq_B b$ and $b \leq_B c$, by definition, we have $g(a) \leq g(b)$ and $g(b) \leq g(c)$. As the relation $\leq$ is a partial order, it is transitive and we have $g(a) \leq g(c)$. Consequently, we have $a \leq_B c$.

Proposition is proved. □

Note that $\boldsymbol{B}^{g,h}$ is also denoted by $\boldsymbol{A}_{g,h}$.

Proposition 3.1.9. *If the mapping* g: $B \to A$ *is an injection and* $\leq$ *is a linear (total) order in* A, *then* $\leq_B$ *is a linear (total) order in* B.

Indeed, if g is an injection, then the third condition, which defines the prearithmetic $\boldsymbol{B}$, is equivalent to the condition

$$a \leq_B b \text{ if and only if } g(a) \leq g(b).$$

Thus, if for any two elements $x, y \in A$, either $x \leq y$ or $x \leq y$ is true, then for any two elements $a, b \in B$, either $a \leq_B b$ or $a \leq_B b$ is true.

The construction of the prearithmetic $\boldsymbol{B}$ in Proposition 3.1.8 implies its uniqueness. Indeed, let us consider three mappings g: $B \to A$, f: $A \to B$ and h: $A \to B$.

Proposition 3.1.10. *If mappings* f *and* h *coincide on the partially extended abstract prearithmetic* $\mathrm{PA}(g(B))$ *generated by image of* B *in* $\boldsymbol{A}$, *i.e.,* $f|_{\mathrm{PA}(g(B))} = h|_{\mathrm{PA}(g(B))}$, *then* $\boldsymbol{B}_{g,h} = \boldsymbol{B}_{g,f}$.

Proof is left as an exercise. □

When a partially extended abstract prearithmetic $\boldsymbol{A}_1$ is weakly projective with respect to a partially extended abstract prearithmetic $\boldsymbol{A}_2$, it is often possible to deduce some properties of the prearithmetic $\boldsymbol{A}_1$ from properties of the prearithmetic $\boldsymbol{A}_2$. In particular, weak projectivity in the class of partially extended abstract prearithmetics implies definite relations between properties of operations in prearithmetics.

Proposition 2.2.4 implies the following result.

Proposition 3.1.11. *If a partially extended abstract prearithmetic* $\boldsymbol{A}_1 = (A_1; +_1, \circ_1, -_1, \leq_1)$ *is weakly projective with respect to a partially extended abstract prearithmetic* $\boldsymbol{A}_2 = (A_2; +_2, \circ_2, -_2, \leq_2)$ *and the operation* $+_2$ *is commutative in the prearithmetic* $\boldsymbol{A}_2$*, then the operation* $+_1$ *is commutative in the prearithmetic* $\boldsymbol{A}_1$*.*

Proof is left as an exercise. □

This shows that inverse weak projectivity preserves commutativity of addition.

Proposition 2.2.5 implies a similar property for multiplication in abstract prearithmetics.

Proposition 3.1.12. *If a partially extended abstract prearithmetic* $\boldsymbol{A}_1 = (A_1; +_1, \circ_1, -_1, \leq_1)$ *is weakly projective with respect to a partially extended abstract prearithmetic* $\boldsymbol{A}_2 = (A_2; +_2, \circ_2, -_2, \leq_2)$ *and the operation* $\circ_2$ *is commutative in the prearithmetic* $\boldsymbol{A}_2$*, then the operation* $\circ_1$ *is commutative in the prearithmetic* $\boldsymbol{A}_1$*.*

Proof is left as an exercise. □

This shows that inverse weak projectivity preserves commutativity of multiplication.

Let us consider two partially extended abstract prearithmetics $\boldsymbol{A}_1 = (A_1; +_1, \circ_1, -_1, \leq_1)$ and $\boldsymbol{A}_2 = (A_2; +_2, \circ_2, -_2, \leq_2)$, which have additive zeros 0_1 and 0_2, respectively.

Proposition 3.1.13. *If the prearithmetic* $\boldsymbol{A}_1$ *is weakly projective with respect to the prearithmetic* $\boldsymbol{A}_2$ *with the projector* g *and the coprojector* $h, h(0_2) = 0_1$ *and* $x -_2 x = 0_2$ *for any element* x *from the prearithmetic* $\boldsymbol{A}_2$*, then* $a -_1 a = 0_1$ *for any element* a *from the prearithmetic* $\boldsymbol{A}_1$*.*

Indeed,

$$a -_1 a = h(g(a) -_2 g(a)) = h(0_2) = 0_1.$$

Weak projectivity is a transitive relation in the class of partially extended abstract prearithmetics.

Proposition 3.1.14. *If a partially extended abstract prearithmetic* $\boldsymbol{A}_1 = (A_1; +_1, \circ_1, -_1, \leq_1)$ *is weakly projective with respect to a partially extended abstract prearithmetic* $\boldsymbol{A}_2 = (A_2; +_2, \circ_2, -_2, \leq_2)$ *and the partially extended abstract prearithmetic* $\boldsymbol{A}_2$ *is weakly projective with respect to a partially extended abstract prearithmetic* $\boldsymbol{A}_3 = (A_3; +_3, \circ_3, -_3, \leq_3)$, *then the prearithmetic* $\boldsymbol{A}_1$ *is weakly projective with respect to the prearithmetic* $\boldsymbol{A}_3$.

Proof is similar to the proof of Proposition 2.2.9.

Proposition 3.1.14 allows proving the following result.

Theorem 3.1.1. *Partially extended abstract prearithmetics with weak projectivity relations form the category* **PEAPAWP**, *in which objects are partially extended abstract prearithmetics and morphisms are weak projectivity relations.*

Proof is left as an exercise. □

Now let us explicate relations between order relations in partially extended abstract prearithmetics connected by weak projectivity. Proposition 2.2.11 implies the following result.

Proposition 3.1.15. *If a partially extended abstract prearithmetic* $\boldsymbol{A}_1 = (A_1; +_1, \circ_1, -_1, \leq_1)$ *is weakly projective with respect to a partially extended abstract prearithmetic* $\boldsymbol{A}_2 = (A_2; +_2, \circ_2, -_2, \leq_2)$, *the coprojector* h *preserves the order* $\leq_2$, *i.e.,* $x \leq_2 y$ *implies* $h(x) \leq_1 h(y)$, *and the operation* $+_2$ *preserves the order* $\leq_2$, *then the operation* $+_1$ *preserves the order* $\leq_1$.

Proof is left as an exercise. □

Corollary 3.1.1. *If a partially extended abstract prearithmetic* $\boldsymbol{A}_1 = (A_1; +_1, \circ_1, -_1, \leq_1)$ *is weakly projective with respect to a partially extended abstract prearithmetic* $\boldsymbol{A}_2 = (A_2; +_2, \circ_2, -_2, \leq_2)$, *the coprojector* h *preserves the order* $\leq_2$, *i.e.,* $x \leq_2 y$ *implies* $h(x) \leq_1 h(y)$, *and the operation* $+_2$ *is monotone, then the operation* $+_1$ *is monotone.*

Proposition 2.2.12 implies the following result.

Proposition 3.1.16. *If a partially extended abstract prearithmetic* $\boldsymbol{A}_1 = (A_1; +_1, \circ_1, -_1, \leq_1)$ *is weakly projective with respect to a partially extended abstract prearithmetic* $\boldsymbol{A}_2 = (A_2; +_2, \circ_2, \leq_2)$*, the coprojector* h *preserves the order* $\leq_2$*, i.e.,* $x \leq_2 y$ *implies* $h(x) \leq_1 h(y)$*, and the operation* $\circ_2$ *preserves the order* $\leq_2$*, then the operation* $\circ_1$ *preserves the order* $\leq_1$*.*

Proof is left as an exercise. □

Corollary 3.1.2. *If a partially extended abstract prearithmetic* $\boldsymbol{A}_1 = (A_1; +_1, \circ_1, -_1, \leq_1)$ *is weakly projective with respect to an abstract prearithmetic* $\boldsymbol{A}_2 = (A_2; +_2, \circ_2, -_2, \leq_2)$*, the coprojector* h *preserves the order* $\leq_2$*, i.e.,* $x \leq_2 y$ *implies* $h(x) \leq_1 h(y)$*, and the operation* $\circ_2$ *is monotone, then the operation* $\circ_1$ *is monotone.*

Definition 3.1.4.

(a) A binary operation $*$, e.g., addition $+$ or multiplication $\circ$, in an ordered structure is *monotone with respect to the first (second) variable* if for any elements a, b and d from A, the inequality $a \leq b$ implies the inequalities $a * d \leq b * d(d * a \leq d * b)$.

(b) A binary operation $*$ in an ordered structure is *strictly monotone with respect to the first (second) variable* if for any elements a, b and d from A, the inequality $a < b$ implies the inequalities $a*d < b*d(d*a < d*b)$.

(c) A binary operation $*$, e.g., addition $+$ or multiplication $\circ$, in an ordered structure is *antitone with respect to the first (second) variable* if for any elements a, b and d from A, the inequality $a \leq b$ implies the inequalities $a * d \geq b * d(d * a \geq d * b)$.

(d) A binary operation $*$ in an ordered structure is *strictly antitone with respect to the first (second) variable* if for any elements a, b and d from A, the inequality $a < b$ implies the inequalities $a*d > b*d(d*a > d*b)$.

For instance, in Diophantine arithmetics, subtraction is monotone with respect to the first variable and antitone with respect to the second variable. For partially extended abstract prearithmetics, we have the following results.

Proposition 3.1.17. *If a partially extended abstract prearithmetic* $\boldsymbol{A}_1 = (A_1; +_1, \circ_1, -_1, \leq_1)$ *is weakly projective with respect to an abstract prearithmetic* $\boldsymbol{A}_2 = (A_2; +_2, \circ_2, -_2, \leq_2)$*, the coprojector* h *preserves the order* $\leq_2$*, i.e.,* $x \leq_2 y$ *implies* $h(x) \leq_1 h(y)$*, and the operation* $-_2$ *is monotone with respect to the first variable, then the operation* $-_1$ *is monotone with respect to the first variable.*

Proof: Let us assume that for any elements u, y and z from A_2, the inequalities $u \leq y$ imply the inequality $u -_2 z \leq y -_2 z$ and consider elements a, b and d from the prearithmetic A_1 such that $a \leq b$. Because $a \leq b$ only if $g(a) \leq g(b)$, we have $g(a) \leq g(b)$. By the initial conditions, $g(a) -_2 g(d) \leq g(b) -_2 g(d)$. As the coprojector h preserves the order $\leq_2$, we have

$$a -_1 d \leq h(g(a) -_2 g(d)) \leq h(g(b) -_2 g(d)) = b -_1 d.$$

It means that the inequality $a \leq b$ implies the inequality $a -_1 d \leq b -_1 c$.

Proposition is proved. □

Proposition 3.1.18. *If a partially extended abstract prearithmetic $\boldsymbol{A}_1 = (A_1; +_1, \circ_1, -_1, \leq_1)$ is weakly projective with respect to an abstract prearithmetic $\boldsymbol{A}_2 = (A_2; +_2, \circ_2, -_2, \leq_2)$, the coprojector h preserves the order $\leq_2$, i.e., $x \leq_2 y$ implies $h(x) \leq_1 h(y)$, and the operation $-_2$ is antitone with respect to the second variable, then the operation $-_1$ is antitone with respect to the second variable.*

Proof: Let us assume that for any elements u, y and z from A_2, the inequalities $u \leq y$ imply the inequality $z -_2 y \leq z -_2 u$ and consider elements a, b and d from the prearithmetic A_1 such that $a \leq b$. Because $a \leq b$ only if $g(a) \leq g(b)$, we have $g(a) \leq g(b)$. By the initial conditions, $g(d) -_2 g(b) \leq g(d) -_2 g(a)$. As the coprojector h preserves the order $\leq_2$, we have

$$d -_1 b \leq h(g(d) -_2 g(b)) \leq h(g(d) -_2 g(a)) = d -_1 a.$$

It means that the inequality $a \leq b$ implies the inequality $d -_1 b \leq d -_1 a$.

Proposition is proved. □

One more important property of Diophantine arithmetics, e.g., $\boldsymbol{N}$ or $\boldsymbol{R}$, is distributivity. Let us explore this property for partially extended abstract prearithmetics. Proposition 2.2.8 implies the following result.

Proposition 3.1.19. *If a partially extended abstract prearithmetic $\boldsymbol{A}_1 = (A_1; +_1, \circ_1, -_1, \leq_1)$ is weakly projective with respect to a partially extended abstract prearithmetic $\boldsymbol{A}_2 = (A_2; +_2, \circ_2, -_2, \leq_2)$ with the projector g and the coprojector h, for which $gh = \mathbf{1}_{A_2}$, and multiplication $\circ_2$ is distributive (from the left or from the right) over addition $+_2$ in the prearithmetic $\boldsymbol{A}_2$, then multiplication $\circ_1$ is distributive (from the left or from the right, respectively) over addition $+_1$ in the prearithmetic $\boldsymbol{A}_1$.*

Proof is left as an exercise. □

We can prove that a similar property for subtraction instead of addition. Indeed, as we know, in the Diophantine arithmetic $\boldsymbol{Z}$, multiplication is distributive with respect to subtraction, i.e., the following identities hold:

$$x \cdot (y - z) = x \cdot y - x \cdot z,$$

$$(y - z) \cdot x = y \cdot x - z \cdot x.$$

Let us remind that in abstract prearithmetics, multiplication is not always commutative and it is necessary to discern three kinds of distributivity. Namely, *distributivity from the left with respect to subtraction*

$$x \cdot (y - z) = x \cdot y - x \cdot z$$

and *distributivity from the right with respect to subtraction*

$$(y - z) \cdot x = y \cdot x - z \cdot x.$$

In addition, multiplication is *distributive* with respect to subtraction when both identities hold.

Proposition 3.1.20. *If a partially extended abstract prearithmetic* $\boldsymbol{A}_1 = (A_1; +_1, \circ_1, -_1, \leq_1)$ *is weakly projective with respect to a partially extended abstract prearithmetic* $\boldsymbol{A}_2 = (A_2; +_2, \circ_2, -_2, \leq_2)$ *with the projector* g *and the coprojector* h, *for which* $gh = \mathbf{1}_{A_2}$, *and multiplication* $\circ_2$ *is distributive (from the left or from the right) with respect to subtraction* $-_2$ *in the prearithmetic* $\boldsymbol{A}_2$, *then multiplication* $\circ_1$ *is distributive (from the left or from the right, respectively) with respect to subtraction* $-_1$ *in the prearithmetic* $\boldsymbol{A}_1$.

Proof: Let us check distributivity from the left taking arbitrary elements a, b and c from $\boldsymbol{A}_1$. By Definition 3.1.3, we have

$$\begin{aligned} a \circ_1 (b -_1 c) &= h(g(a) \circ_2 g(h(g(b) -_2 g(c))))) \\ &= h(g(a) \circ_2 (g(b) -_2 g(c))) = h(g(a) \circ_2 g(b) -_2 g(a) \circ_2 g(c)) \\ &= h(g(h(g(a) \circ_2 g(b))) -_2 g(h(g(a) \circ_2 g(c)))) \\ &= h(g(a \circ_1 b) -_2 g(a \circ_1 c)) = (a \circ_1 b) -_1 (a \circ_1 c) \end{aligned}$$

because $gh = \mathbf{1}_{A_2}$ and multiplication $\circ_2$ is distributive from the left with respect to subtraction $-_2$ in the prearithmetic $\boldsymbol{A}_2$. It means that multiplication $\circ_1$ is distributive from the left with respect to subtraction $-_1$ in the prearithmetic $\boldsymbol{A}_1$.

Distributivity from the right and complete distributivity are proved in the same way.

Proposition is proved. □

Remark 3.1.2. In some partially extended abstract prearithmetics, multiplication can be distributive from the right or from both sides with respect to addition or/and subtraction. In these cases, inverse weak projectivity preserves corresponding distributivity under the condition $gh = \mathbf{1}_{A_2}$.

Condition $gh = \mathbf{1}_{A_2}$ implies additional properties of the projector g.

Proposition 3.1.21. *If a partially extended abstract prearithmetic* $\boldsymbol{A}_1 = (A_1; +_1, \circ_1, -_1, \leq_1)$ *is weakly projective with respect to a partially extended abstract prearithmetic* $\boldsymbol{A}_2 = (A_2; +_2, \circ_2, -_2, \leq_2)$ *with the projector* g *and the coprojector* h *and the composition* gh *is the identity mapping* $\mathbf{1}_{A_2}$ *of* $\boldsymbol{A}_2$*, then the projector* g *is a homomorphism.*

Proof: By Definition 3.1.3, we have

$$a +_1 b = h(g(a) +_2 g(b)).$$

If $gh = \mathbf{1}_{A_2}$, then

$$g(a +_1 b) = g(h(g(a) +_2 g(b))) = (gh)(g(a) +_2 g(b)) = g(a) +_2 g(b).$$

By the same token, we have

$$a \circ_1 b = h(g(a) \circ_2 g(b)).$$

If $gh = \mathbf{1}_{A_2}$, then

$$g(a \circ_1 b) = g(h(g(a) \circ_2 g(b))) = (gh)(g(a) \circ_2 g(b)) = g(a) \circ_2 g(b).$$

In addition, we obtain

$$a -_1 b = h(g(a) -_2 g(b)).$$

If $gh = \mathbf{1}_{A_2}$, then

$$g(a -_1 b) = g(h(g(a) -_2 g(b))) = (gh)(g(a) -_2 g(b)) = g(a) -_2 g(b).$$

Besides, by Lemma 2.2.1, the projector g preserves the order $\leq_1$.

Proposition is proved. □

Corollary 3.1.3. *If a partially extended abstract prearithmetic* $\boldsymbol{A}_1 = (A_1; +_1, \circ_1, -_1, \leq_1)$ *is weakly projective with respect to a partially extended abstract prearithmetic* $\boldsymbol{A}_2 = (A_2; +_2, \circ_2, -_2, \leq_2)$ *with the projector* g *and* *the coprojector h while $h = g^{-1}$, then the projector g is an isomorphism, i.e., partially extended prearithmetics $\boldsymbol{A}_1$ and $\boldsymbol{A}_2$ are isomorphic as algebraic systems.*

Let us consider one more relation between partially extended abstract prearithmetics as we did for abstract prearithmetics in Section 2.3.

Definition 3.1.5. A partially extended abstract prearithmetic $\boldsymbol{A}_1 = (A_1; +_1, \circ_1, \leq_1)$ is called *projective* with respect to a partially extended abstract prearithmetic $\boldsymbol{A}_2 = (A_2; +_2, \circ_2, \leq_2)$ if it is weakly projective with respect to the prearithmetic $\boldsymbol{A}_2$ and $hg = \mathbf{1}_{A_1}$, where $\mathbf{1}_{A_1}$ is the identity mapping of A_1, g: $A_1 \to A_2$ is the *projector* and h: $A_2 \to A_1$ is the *coprojector* for the pair ($\boldsymbol{A}_1$, $\boldsymbol{A}_2$), and for any elements a and b from A_1, we have

$$a \leq_1 b \text{ if and only if } g(a) \leq_2 g(b).$$

When a partially extended abstract prearithmetic $\boldsymbol{B}$ is projective with respect to a partially extended abstract prearithmetic $\boldsymbol{A}$ with the projector g and the coprojector h, we denote $\boldsymbol{B}$ by $\boldsymbol{A}_{g;h}$ to show this relation between partially extended abstract prearithmetics $\boldsymbol{A}$ and $\boldsymbol{B}$.

Definitions imply the following result.

Lemma 3.1.4. *If a partially extended abstract prearithmetic $\boldsymbol{A}_1 = (A_1; +_1, \circ_1, -_1, \leq_1)$ is projective with respect to a partially extended abstract prearithmetic $\boldsymbol{A}_2 = (A_2; +_2, \circ_2, -_2, \leq_2)$, then the prearithmetic $\boldsymbol{A}_1$ is weakly projective with respect to the prearithmetic $\boldsymbol{A}_2$.*

Proof is left as an exercise. □

This result allows proving many properties of projectivity in the class of partially extended abstract prearithmetics.

Let us consider an arbitrary set B, a partially extended abstract prearithmetic $\boldsymbol{A} = (A; +, \circ, \leq)$ and two functions g: $B \to A$ and h: $A \to B$ such that $hg = \mathbf{1}_{A_1}$. Then as in the case of abstract prearithmetics, it is possible to define on B the structure of a partially extended abstract prearithmetic $\boldsymbol{B}^{g;h}$ that is projective with respect to the partially extended abstract prearithmetic $\boldsymbol{A}$ as a unique prearithmetic with the projector g and the coprojector h.

Note that $\boldsymbol{B}^{g;h}$ is also $\boldsymbol{A}_{g;h}$.

Let us explore other standard properties of subtraction in partially extended abstract prearithmetics.

One of the most important properties of Diophantine arithmetics $\boldsymbol{Z}$, $\boldsymbol{R}$ and $\boldsymbol{C}$ is that subtraction is inverse to addition, i.e., $b + a - a = b$.

However, weak projectivity and inverse weak projectivity do not preserve this property as the following example demonstrates.

Example 3.1.7. Let us take a natural number k and the function $g(x) = x+7$ as the projector and the function $h(x) = x-5$ as the coprojector of the weak projectivity of a partially extended abstract prearithmetic $\boldsymbol{A} = (Z; \oplus, \circ, \ominus, \leq)$ with respect to the Diophantine arithmetic $\boldsymbol{Z} = (Z; +, \cdot, -, \leq)$ of integers, which is also a prearithmetic.

Then in the prearithmetic $\boldsymbol{A}$, we have

$$(2 \oplus 2) \ominus 2 = \{([((2+7)+(2+7)) - 5] + 7) - (2+7)\} - 5 = \{20 - 9\} - 5 = 6.$$

We see that

$$(2 \oplus 2) \ominus 2 = 6 \neq 2.$$

It means that in $\boldsymbol{A}$, subtraction is not inverse to addition, i.e., inverse weak projectivity does not preserve this property.

Note that the Diophantine arithmetic $\boldsymbol{Z} = (Z; +, \cdot, -, \leq)$ is weakly projective with respect to the abstract prearithmetic $\boldsymbol{A} = (Z; \oplus, \circ, \ominus, \leq)$ with the function $g(x) = x + 7$ as the coprojector and the function $h(x) = x - 5$ as the projector. Consequently, weak projectivity also does not preserve the conventional property of subtraction, which is inverse to addition.

However, under some conditions, inverse projectivity preserves this property.

Proposition 3.1.22. *If the partially extended abstract prearithmetic* $\boldsymbol{A}_1$ *is projective with respect to the partially extended abstract prearithmetic* $\boldsymbol{A}_2$ *with the projector* g *and the coprojector* h*, for which* $gh = 1_{A_2}$*, and* $(z +_2 x) -_2 x = z$ *for any elements* x *and* z *from the prearithmetic* $\boldsymbol{A}_2$*, then* $(b +_2 a) -_2 a = b$ *for any elements* a *and* b *from the prearithmetic* $\boldsymbol{A}_1$*.*

Proof: By definition and initial conditions, we have

$$\begin{aligned}(b +_2 a) -_2 a &= h(g(h(g(b) +_2 g(a)) -_2 g(a))) \\ &= h((g(b) +_2 g(a)) -_2 g(a)) = h(g(b)) = b\end{aligned}$$

as $gh = 1_{A_2}$, and $hg = 1_{A_1}$.

Proposition is proved. □

We also know that conventional addition is inverse to conventional subtraction, i.e., $b - a + a = b$.

However, weak projectivity and inverse weak projectivity do not preserve this property as the following example demonstrates.

Example 3.1.8. Let us take the partially extended abstract prearithmetic $\boldsymbol{A} = (Z; \oplus, \circ, \ominus, \leq)$ from Example 3.1.7, which is weakly projective with respect to the Diophantine arithmetic $\boldsymbol{Z} = (Z; +, \cdot, -, \leq)$.

Then in the prearithmetic $\boldsymbol{A}$, we have

$$(8 \ominus 2) \oplus 2 = \{([((8+7)-(2+7))-5]+7)+(2+7)\}-5 = \{8+9\}-5 = 12.$$

We see that

$$(8 \ominus 2) \oplus 2 = 12 \neq 8.$$

It means that in $\boldsymbol{A}$, addition is not inverse to subtraction, i.e., inverse weak projectivity does not preserve this property.

Note that the Diophantine arithmetic $\boldsymbol{Z} = (Z; +, \cdot, -, \leq)$ is weakly projective with respect to the abstract prearithmetic $\boldsymbol{A} = (Z; \oplus, \circ, \ominus, \leq)$ with the function $g(x) = x + 7$ as the coprojector and the function $h(x) = x - 5$ as the projector. Consequently, weak projectivity also does not preserve the property of addition being inverse to subtraction.

However, under some conditions, inverse weak projectivity also preserves this property.

Proposition 3.1.23. *If the partially extended abstract prearithmetic $\boldsymbol{A}_1$ is projective with respect to the partially extended abstract prearithmetic $\boldsymbol{A}_2$ with the projector g and the coprojector h, for which $gh = 1_{A_2}, x +_2 (z -_2 x) = z$ and $(z -_2 x) +_2 x = z$ for any elements x and z from the prearithmetic $\boldsymbol{A}_2$, then $(b -_2 a) +_2 a = b$ and $a +_2 (b -_2 a) = b$ for any elements a and b from the prearithmetic $\boldsymbol{A}_1$.*

Proof: By definition and initial conditions, we have

$$\begin{aligned}(b -_2 a) +_2 a &= h(g(h(g(b) -_2 g(a)) +_2 g(a))) \\ &= h((g(b) -_2 g(a)) +_2 g(a)) = h(g(b)) = b\end{aligned}$$

as $gh = 1_{A_2}$, and $hg = 1_{A_1}$.

The second identity is proved in a similar way.

Proposition is proved. □

Let us consider two partially extended abstract prearithmetics $\boldsymbol{A}_1 = (A_1; +_1, \circ_1, -_1, \leq_1)$ and $\boldsymbol{A}_2 = (A_2; +_2, \circ_2, -_2, \leq_2)$, which has an additive zero 0_2. Then Proposition 2.3.19 implies the following result.

Proposition 3.1.24. *If a partially extended abstract prearithmetic* $\boldsymbol{A}_1$ *is projective with respect to a partially extended abstract prearithmetic* $\boldsymbol{A}_2$ *with the projector* g *and the coprojector* h *and* $g(a) = 0_2$*, then* a *is an additive zero* 0_1 *in the prearithmetic* $\boldsymbol{A}_1$*.*

Indeed, let us take an arbitrary element b from $\boldsymbol{A}_1$. Then we have

$$b +_1 a = h(g(b) +_2 g(a)) = h(g(b) +_2 0_2) = h(g(b) = b$$

and

$$a +_1 b = h(g(a) +_2 g(b)) = h(0_2 +_2 g(b)) = h(g(b) = b.$$

It means that a is an additive zero 0_1 in the prearithmetic $\boldsymbol{A}_1$.

Here is one more property.

Proposition 3.1.25. *If a partially extended abstract prearithmetic* $\boldsymbol{A}_1$ *is projective with respect to a partially extended abstract prearithmetic* $\boldsymbol{A}_2$ *with the projector* g *and the coprojector* h*, for which* $gh = 1_{A_2}$*, and* $(z -_2 x) -_2 y = z -_2 (x +_2 y)$ *for any elements* x, y *and* z *from the prearithmetic* $\boldsymbol{A}_2$*, then* $(b -_1 a) -_1 c = b -_1 (a +_1 c)$ *for any elements* a *and* b *from the prearithmetic* $\boldsymbol{A}_1$*.*

Indeed, by definition and initial conditions, we have

$$\begin{aligned}(b -_1 a) -_1 c &= h(g(h(g(b) -_2 g(a)) -_2 g(c))) = h((g(b) -_2 g(a)) -_2 g(c)) \\ &= h(g(b) -_2 (g(a)) +_2 g(c))) = h(g(b) -_2 g(h(g(a)) \\ &\quad +_2 g(c)))) = b -_1 (a +_1 c).\end{aligned}$$

Projectivity is a transitive relation in the class of partially extended abstract prearithmetics.

Proposition 3.1.26. *If a partially extended abstract prearithmetic* $\boldsymbol{A}_1 = (A_1; +_1, \circ_1, -_1, \leq_1)$ *is projective with respect to a partially extended abstract prearithmetic* $\boldsymbol{A}_2 = (A_2; +_2, \circ_2, -_2, \leq_2)$ *and the partially extended abstract prearithmetic* $\boldsymbol{A}_2 = (A_2; +_2, \circ_2, -_2, \leq_2)$ *is projective with respect to a partially extended abstract prearithmetic* $\boldsymbol{A}_3 = (A_3; +_3, \circ_3, -_3, \leq_3)$*, then the prearithmetic* $\boldsymbol{A}_1$ *is projective with respect to the prearithmetic* $\boldsymbol{A}_3$*.*

Proof is similar to the proof of Proposition 2.3.11. □

Proposition 3.1.26 allows proving the following result.

Theorem 3.1.2. *Partially extended abstract prearithmetics with projectivity relations form the category* **PEAPAP** *where objects are partially extended abstract prearithmetics and morphisms are projectivity relations.*

Proof is left as an exercise. □

Corollary 3.1.4. *The category* **PEAPAP** *is a subcategory of the category* **PEAPAWP**.

By Lemma 3.1.4, projectivity of partially extended abstract prearithmetics is a special case of their weak projectivity. Proposition 3.1.13 implies the following result.

Proposition 3.1.27. *If a partially extended abstract prearithmetic* $\boldsymbol{A}_1 = (A_1; +_1, \circ_1, -_1, \leq_1)$ *is projective with respect to a partially extended abstract prearithmetic* $\boldsymbol{A}_2 = (A_2; +_2, \circ_2, -_2, \leq_2)$ *with the projector* g *and the coprojector* h, *for which* $gh = \mathbf{1}_{A_2}$, *and multiplication* $\circ_2$ *is distributive (from the left or from the right) with respect to addition* $+_2$ *in the prearithmetic* $\boldsymbol{A}_2$, *then multiplication* $\circ_1$ *is distributive (from the left or from the right, respectively) with respect to addition* $+_1$ *in the prearithmetic* $\boldsymbol{A}_1$.

Proof is left as an exercise. □

We can prove that a similar property for subtraction instead of addition. Indeed, as we know, in the Diophantine arithmetic $\boldsymbol{Z}$, multiplication is distributive with respect to subtraction, i.e., the following identities hold:

$$x \cdot (y - z) = x \cdot y - x \cdot z,$$

$$(y - z) \cdot x = y \cdot x - z \cdot x.$$

Let us remind that in abstract prearithmetics, multiplication is not always commutative and it is necessary to discern three kinds of distributivity. Namely, *distributivity from the left with respect to subtraction*

$$x \cdot (y - z) = x \cdot y - x \cdot z$$

and *distributivity from the right with respect to subtraction*

$$(y - z) \cdot x = y \cdot x - z \cdot x.$$

In addition, multiplication is *distributive* with respect to subtraction when both identities hold.

Proposition 3.1.16 implies the following result.

Proposition 3.1.28. *If a partially extended abstract prearithmetic* $\boldsymbol{A}_1 = (A_1; +_1, \circ_1, -_1, \leq_1)$ *is projective with respect to a partially extended abstract prearithmetic* $\boldsymbol{A}_2 = (A_2; +_2, \circ_2, -_2, \leq_2)$ *with the projector* g *and the coprojector* h*, for which* $gh = \mathbf{1}_{A_2}$*, and multiplication* $\circ_2$ *is distributive (from the left or from the right) with respect to subtraction* $-_2$ *in the prearithmetic* $\boldsymbol{A}_2$*, then multiplication* $\circ_1$ *is distributive (from the left or from the right, respectively) with respect to subtraction* $-_1$ *in the prearithmetic* $\boldsymbol{A}_1$*.*

Proof is left as an exercise. □

Proposition 3.1.21 and Lemma 3.1.4 imply the following result.

Proposition 3.1.29. *If a partially extended abstract prearithmetic* $\boldsymbol{A}_1 = (A_1; +_1, \circ_1, -_1, \leq_1)$ *is projective with respect to a partially extended abstract prearithmetic* $\boldsymbol{A}_2 = (A_2; +_2, \circ_2, -_2, \leq_2)$ *with the projector* g *and the coprojector* h *and the composition* gh *is the identity mapping* $\mathbf{1}_{A_2}$ *of* $\boldsymbol{A}_2$*, then the projector* g *is a homomorphism.*

Proof is left as an exercise. □

Corollary 3.1.5. *If a partially extended abstract prearithmetic* $\boldsymbol{A}_1 = (A_1; +_1, \circ_1, -_1, \leq_1)$ *is projective with respect to a partially extended abstract prearithmetic* $\boldsymbol{A}_2 = (A_2; +_2, \circ_2, -_2, \leq_2)$ *with the projector* g *and the coprojector* h *while* $h = g^{-1}$*, then the projector* g *is an isomorphism, i.e., partially extended prearithmetics* $\boldsymbol{A}_1$ *and* $\boldsymbol{A}_2$ *are isomorphic as algebraic systems.*

Proposition 3.1.30. *If a partially extended abstract prearithmetic* $\boldsymbol{A}_1 = (A_1; +_1, \circ_1, -_1, \leq_1)$ *is projective with respect to a partially extended abstract prearithmetic* $\boldsymbol{A}_2 = (A_2; +_2, \circ_2, -_2, \leq_2)$ *with the projector* g *and the coprojector* h*, then the coprojector* h *preserves the order and all three operations in the image* $g(\boldsymbol{A}_1)$ *of the prearithmetic* $\boldsymbol{A}_1$*, i.e.,* h *is a homomorphism of* $g(\boldsymbol{A}_1)$*.*

Proof: Let us take two elements x and y from the image $g(\boldsymbol{A}_1)$ of the partially extended abstract prearithmetic $\boldsymbol{A}_1$. By definition, there are elements a and b in the prearithmetic $\boldsymbol{A}_1$ such that $x = g(a)$ and $y = g(b)$. Then by Definition 3.1.3, we have

$$h(x +_2 y) = h(g(a) +_2 g(b)) = a +_1 b = h(g(a)) +_1 h(g(b)) = h(x) +_1 h(y),$$

$$h(x -_2 y) = h(g(a) -_2 g(b)) = a -_1 b = h(g(a)) -_1 h(g(b)) = h(x) -_1 h(y),$$

$$h(x \circ_2 y) = h(g(a) \circ_2 g(b)) = a \circ_1 b = h(g(a)) \circ_1 h(g(b)) = h(x) \circ_1 h(y)$$

because the composition hg is equal to $\mathbf{1}_{A_1}$. These equalities mean that h is a homomorphism of $g(\boldsymbol{A}_1)$ with respect to all three operations. In addition, by Lemma 2.3.2, the coprojector h preserves the order $\leq_2$.

Proposition is proved. □

A stronger relation between partially extended abstract prearithmetics is exact projectivity.

Definition 3.1.6. A partially extended abstract prearithmetic $\boldsymbol{A}_1 = (A_1; +_1, \circ_1, -_1, \leq_1)$ is called *exactly projective* with respect to a partially extended abstract prearithmetic $\boldsymbol{A}_2 = (A_2; +_2, \circ_2, -_2, \leq_2)$ if it is projective with respect to the prearithmetic $\boldsymbol{A}_2$ and $gh = \mathbf{1}_{A_2}$, where $\mathbf{1}_{A_2}$ is the identity mapping of A_2, g: $A_1 \to A_2$ is the *projector* and h: $A_2 \to A_1$ is the *coprojector* for the pair $(\boldsymbol{A}_1, \boldsymbol{A}_2)$.

When a partially extended abstract prearithmetic $\boldsymbol{B}$ is exactly projective with respect to a partially extended abstract prearithmetic $\boldsymbol{A}$ with the projector g and the coprojector h, we denote $\boldsymbol{B}$ by $\boldsymbol{A}_g$ to show this relation between partially extended abstract prearithmetics $\boldsymbol{A}$ and $\boldsymbol{B}$. It is not necessary to use the coprojector h in this notation because $h = g^{-1}$ (cf. Section 2.4).

Example 3.1.9. The partially extended abstract prearithmetic $7\boldsymbol{Z}$ of all integer numbers divisible by 7 is exactly projective with respect to the Diophantine arithmetic $\boldsymbol{Z}$ with the projector $g(7n) = n$ and the coprojector $g(n) = 7n$.

The definitions imply the following result.

Lemma 3.1.5. *If a partially extended abstract prearithmetic* $\boldsymbol{A}_1 = (A_1; +_1, \circ_1, -_1, \leq_1)$ *is exactly projective with respect to a partially extended abstract prearithmetic* $\boldsymbol{A}_2 = (A_2; +_2, \circ_2, -_2, \leq_2)$, *then the prearithmetic* $\boldsymbol{A}_1$ *is projective with respect to a prearithmetic* $\boldsymbol{A}_2$.

Proof is left as an exercise. □

Corollary 3.1.6. *If a partially extended abstract prearithmetic* $\boldsymbol{A}_1 = (A_1; +_1, \circ_1, -_1, \leq_1)$ *is exactly projective with respect to a partially extended abstract prearithmetic* $\boldsymbol{A}_2 = (A_2; +_2, \circ_2, -_2, \leq_2)$, *then the prearithmetic* $\boldsymbol{A}_1$ *is weakly projective with respect to a prearithmetic* $\boldsymbol{A}_2$.

Corollaries 2.2.5 and 3.1.3 imply the following result.

Corollary 3.1.7. *If a partially extended abstract prearithmetic* $\boldsymbol{A}_1 = (A_1; +_1, \circ_1, -_1, \leq_1)$ *is exactly projective with respect to a partially extended abstract prearithmetic* $\boldsymbol{A}_2 = (A_2; +_2, \circ_2, -_2, \leq_2)$ *and the relation* $\leq_2$ *is discrete, then the relation* $\leq_1$ *is discrete.*

Corollary 3.1.8. *If a partially extended abstract prearithmetic* $\boldsymbol{A}_1 = (A_1; +_1, \circ_1, -_1, \leq_1)$ *is exactly projective with respect to a partially extended abstract prearithmetic* $\boldsymbol{A}_2 = (A_2; +_2, \circ_2, -_2, \leq_2)$ *and any element* x *in* $\boldsymbol{A}_2$*, which is not minimal, has a predecessor* Px*, then any element* a *in* $\boldsymbol{A}_1$*, which is not minimal, has a predecessor* Pa*.*

Corollary 3.1.9. *If a partially extended abstract prearithmetic* $\boldsymbol{A}_1 = (A_1; +_1, \circ_1, -_1, \leq_1)$ *is exactly projective with respect to an abstract prearithmetic* $\boldsymbol{A}_2 = (A_2; +_2, \circ_2, -_2, \leq_2)$ *and any element* x *in* $\boldsymbol{A}_2$*, which is not maximal, has a successor* Sx*, then any element* a *in* $\boldsymbol{A}_1$*, which is not maximal, has a successor* Sa*.*

Corollary 3.1.10. *If a partially extended abstract prearithmetic* $\boldsymbol{A}_1 = (A_1; +_1, \circ_1, -_1, \leq_1)$ *is exactly projective with respect to a partially extended abstract prearithmetic* $\boldsymbol{A}_2 = (A_2; +_2, \circ_2, -_2, \leq_2)$ *and any element* x *in* $\boldsymbol{A}_2$*, which is not minimal, has the unique predecessor* Px*, then any element* a *in* $\boldsymbol{A}_1$*, which is not minimal, has the unique predecessor* Pa*.*

Corollary 3.1.11. *If a partially extended abstract prearithmetic* $\boldsymbol{A}_1 = (A_1; +_1, \circ_1, -_1, \leq_1)$ *is exactly projective with respect to a partially extended abstract prearithmetic* $\boldsymbol{A}_2 = (A_2; +_2, \circ_2, -_2, \leq_2)$ *and any element* x *in* $\boldsymbol{A}_2$*, which is not maximal, has the uniquc successor* Sx*, then any element* a *in* $\boldsymbol{A}_1$*, which is not maximal, has the unique successor* Sa*.*

Proposition 2.3.10 and Lemma 3.1.1 imply the following result.

Corollary 3.1.12. *If a partially extended abstract prearithmetic* $\boldsymbol{A}_1 = (A_1; +_1, \circ_1, -_1, \leq_1)$ *is exactly projective with respect to a partially extended abstract prearithmetic* $\boldsymbol{A}_2 = (A_2; +_2, \circ_2, -_2, \leq_2)$ *and* $\leq_2$ *is a linear (total) order in the prearithmetic* $\boldsymbol{A}_2$*, then* $\leq_1$ *is a linear (total) order in the prearithmetic* $\boldsymbol{A}_1$*.*

Properties of the category of sets (cf., for example, Herrlich and Strecker, 1973) give us the following result.

Lemma 3.1.6. *If a partially extended abstract prearithmetic* $\boldsymbol{A}_1 = (A_1; +_1, \circ_1, -_1, \leq_1)$ *is exactly projective with respect to a partially extended abstract prearithmetic* $\boldsymbol{A}_2 = (A_2; +_2, \circ_2, -_2, \leq_2)$*, then the coprojector for the pair* $(\boldsymbol{A}_1, \boldsymbol{A}_2)$ *is an injection and the projector for the pair* $(\boldsymbol{A}_1, \boldsymbol{A}_2)$ *is a projection.*

Proof is left as an exercise. □

Lemmas 2.3.3 and 3.1.6 give us the following result.

Proposition 3.1.31. *If a partially extended abstract prearithmetic* $\boldsymbol{A}_1 = (A_1; +_1, \circ_1, -_1, \leq_1)$ *is exactly projective with respect to a partially extended abstract prearithmetic* $\boldsymbol{A}_2 = (A_2; +_2, \circ_2, -_2, \leq_2)$*, then both the projector* g *and coprojector* h *for the pair* $(\boldsymbol{A}_1, \boldsymbol{A}_2)$ *are bijections and both sets* A_1 *and* A_2 *are equipollent (cf. Appendix).*

Proof is left as an exercise. □

Corollary 3.1.13. *If a partially extended abstract prearithmetic* $\boldsymbol{A}_1 = (A_1; +_1, \circ_1, -_1, \leq_1)$ *is exactly projective with respect to a partially extended abstract prearithmetic* $\boldsymbol{A}_2 = (A_2; +_2, \circ_2, -_2, \leq_2)$ *with the projector* g *and coprojector* h*, then* $g = h^{-1}$ *and* $h = g^{-1}$*.*

Exact projectivity also implies additional algebraic properties of projectors and coprojectors.

Proposition 3.1.32. *If a partially extended abstract prearithmetic* $\boldsymbol{A}_1 = (A_1; +_1, \circ_1, -_1, \leq_1)$ *is exactly projective with respect to a partially extended abstract prearithmetic* $\boldsymbol{A}_2 = (A_2; +_2, \circ_2, -_2, \leq_2)$ *with the projector* g *and the coprojector* h*, then the projector* g *preserves addition, subtraction and multiplication in the prearithmetic* $\boldsymbol{A}_1$*, i.e.,* g *is a isomorphism.*

Proof: Let us take two elements a and b from the arithmetic A_1. Then, by definition, we obtain the following sequences of equalities:

$$g(a +_1 b) = g(h(g(a) +_2 g(b))) = (gh)(g(a)) +_2 g(b)) = g(a) +_2 g(b), \tag{3.1}$$

$$g(a \circ_1 b) = g(h(g(a)) \circ_2 g(b))) = (gh)(g(a)) \circ_2 g(b)) = g(a) \circ_2 g(b), \tag{3.2}$$

$$g(a -_1 b) = g(h(g(a)) -_2 g(b))) = (gh)(g(a)) -_2 g(b)) = g(a) -_2 g(b) \tag{3.3}$$

because the composition gh is equal to $\mathbf{1}_{A_2}$. These equalities mean that g is a homomorphism of partially extended abstract prearithmetics as universal algebras with three operations.

Proposition is proved. □

Corollary 3.1.14. *If a partially extended abstract prearithmetic* $\boldsymbol{A}_1 = (A_1; +_1, \circ_1, -_1, \leq_1)$ *is weakly projective with respect to a partially extended abstract prearithmetic* $\boldsymbol{A}_2 = (A_2; +_2, \circ_2, -_2, \leq_2)$ *with the projector* g *and the coprojector* h *while* $h = g^{-1}$, *then the projector* g *preserves addition, subtraction and multiplication in the prearithmetic* $\boldsymbol{A}_1$, *i.e.,* g *is a homomorphism.*

The coprojector h has a similar property.

Proposition 3.1.33. *If a partially extended abstract prearithmetic* $\boldsymbol{A}_1 = (A_1; +_1, \circ_1, -_1, \leq_1)$ *is exactly projective with respect to a partially extended abstract prearithmetic* $\boldsymbol{A}_2 = (A_2; +_2, \circ_2, -_2, \leq_2)$ *with the projector* g *and the coprojector* h, *then the coprojector* h *preserves addition, subtraction, division and multiplication in the prearithmetic* $\boldsymbol{A}_2$, *i.e.,* h *is a homomorphism.*

Proof: Let us take two elements $x = g(a)$ and $y = g(b)$ from the prearithmetic $\boldsymbol{A}_2$. By Proposition 3.1.13, g is a bijection. Consequently, there are elements a and b in the prearithmetic $\boldsymbol{A}_1$ such that $x = g(a)$ and $y = g(b)$. Then by definitions, we obtain the following sequences of equalities:

$$h(x +_2 y) = h(g(a) +_2 g(b)) = a +_1 b = h(g(a)) +_1 h(g(b)) = h(x) +_1 h(y), \tag{3.4}$$

$$h(x \circ_2 y) = h(g(a) \circ_2 g(b)) = a \circ_1 b = h(g(a)) \circ_1 h(g(b)) = h(x) \circ_1 h(y), \tag{3.5}$$

$$h(x -_2 y) = h(g(a) -_2 g(b)) = a -_1 b = h(g(a)) -_1 h(g(b)) = h(x) -_1 h(y) \tag{3.6}$$

because the composition hg is equal to $\mathbf{1}_{A_1}$. These equalities mean that h is a homomorphism of partially extended abstract prearithmetics.

Proposition is proved. □

Corollary 3.1.15. *If a partially extended abstract prearithmetic* $\boldsymbol{A}_1 = (A_1; +_1, \circ_1, -_1, \leq_1)$ *is weakly projective with respect to a partially extended* *abstract prearithmetic* $\boldsymbol{A}_2 = (A_2; +_2, \circ_2, -_2, \leq_2)$ *with the projector* g *and the coprojector* h *while* $h = g^{-1}$, *then the coprojector* h *preserves addition, subtraction, division and multiplication in the prearithmetic* $\boldsymbol{A}_1$, *i.e.,* h *is a homomorphism.*

Corollary 3.1.16. *If a partially extended abstract prearithmetic* $\boldsymbol{A}_1 = (A_1; +_1, \circ_1, -_1, \leq_1)$ *is weakly projective with respect to a partially extended abstract prearithmetic* $\boldsymbol{A}_2 = (A_2; +_2, \circ_2, -_2, \leq_2)$ *with the projector* g *and the coprojector* h *while* $h = g^{-1}$, *then the prearithmetics* $\boldsymbol{A}_1$ *and* $\boldsymbol{A}_2$ *are isomorphic as algebraic systems with three operations and one relation.*

In contrast to weak projectivity and projectivity, exact projectivity is a symmetric relation.

Theorem 3.1.3. *If a partially extended abstract prearithmetic* $\boldsymbol{A}_1 = (A_1;+_1, \circ_1, -_1, \leq_1)$ *is exactly projective with respect to a partially extended abstract prearithmetic* $\boldsymbol{A}_2 = (A_2; +_2, \circ_2, -_2, \leq_1)$ *with the projector* g *and coprojector* h, *then the prearithmetic* $\boldsymbol{A}_2$ *is exactly projective with respect to a prearithmetic* $\boldsymbol{A}_1$ *with the coprojector* g *and projector* h.

Proof: Let us assume that a partially extended abstract prearithmetic $\boldsymbol{A}_1 = (A_1; +_1, \circ_1, -_1, \leq_1)$ is exactly projective with respect to a partially extended abstract prearithmetic $\boldsymbol{A}_2 = (A_2; +_2, \circ_2, -_2, \leq_2)$ with the projector g and coprojector h, and take arbitrary elements x and y from A_2. As the projector g is a bijection, there are elements a and b from A_1 such that $x = g(a)$ and $y = g(b)$. Then we have

$$a +_1 b = h(g(a) +_2 g(b)).$$

Consequently,

$$h^{-1}(a +_1 b) = g(a) +_2 g(b) = x +_2 y$$

and

$$x +_2 y = h^{-1}(g^{-1}(x) +_1 g^{-1}(y)) = g(h(x) +_1 h(y)).$$

Additionally, we have

$$a -_1 b = h(g(a) -_2 g(b)).$$

Consequently,

$$h^{-1}(a -_1 b) = g(a) -_2 g(b) = x -_2 y$$

and

$$x -_2 y = h^{-1}(g^{-1}(x) -_1 g^{-1}(y)) = g(h(x) -_1 h(y))$$

In a similar way, we have

$$a \circ_1 b = h(g(a) \circ_2 g(b)).$$

Consequently,

$$h^{-1}(a \circ_1 b) = g(a) \circ_2 g(b) = x \circ_2 y$$

and

$$x \circ_2 y = h^{-1}(g^{-1}(x) \circ_1 g^{-1}(y)) = g(h(x) \circ_1 h(y)).$$

In addition, for any elements a and b from A_1, we have

$$a \leq_1 b \text{ if and only if } g(a) \leq_2 g(b).$$

Consequently, for any elements x and y from A_2, we have

$$x \leq_2 y \text{ if and only if } h(x) \leq_1 h(y) = g^{-1}(x) \leq_1 g^{-1}(y) = a \leq_1 b.$$

It means that the partially extended prearithmetic $\boldsymbol{A}_2$ is exactly projective with respect to the partially extended prearithmetic $\boldsymbol{A}_1$ with the coprojector g and projector h.

Theorem is proved. □

Obtained results show that exact projectivity coincides with isomorphisms between partially extended abstract prearithmetics. Namely, Theorem 3.1.3 and Propositions 3.1.31–3.1.33 give us the following result.

Theorem 3.1.4. *A partially extended abstract prearithmetic* $\boldsymbol{A}_1 = (A_1; +_1, \circ_1, -_1, \leq_1)$ *is exactly projective with respect to a partially extended abstract prearithmetic* $\boldsymbol{A}_2 = (A_2; +_2, \circ_2, -_2, \leq_2)$ *with the projector* g *and the coprojector* h *if and only if the prearithmetics* $\boldsymbol{A}_1$ *and* $\boldsymbol{A}_2$ *are isomorphic as algebraic systems with three operations and one relation.*

Proof is left as an exercise. □

Exact projectivity also is a transitive relation in the class of partially extended abstract prearithmetics.

Proposition 3.1.34. *If a partially extended abstract prearithmetic* $\boldsymbol{A}_1 = (A_1; +_1, \circ_1, -_1, \leq_1)$ *is exactly projective with respect to a partially extended abstract prearithmetic* $\boldsymbol{A}_2 = (A_2; +_2, \circ_2, -_2, \leq_2)$ *and the partially extended* *abstract prearithmetic* $\boldsymbol{A}_2$ *is exactly projective with respect to a partially extended abstract prearithmetic* $\boldsymbol{A}_3 = (A_3; +_3, \circ_3, -_3, \leq_3)$*, then the prearithmetic* $\boldsymbol{A}_1$ *is exactly projective with respect to the prearithmetic* $\boldsymbol{A}_3$.

Proof is similar to the proof of Proposition 2.4.4.

Proposition 3.1.34 allows proving the following result.

Theorem 3.1.5. *Partially extended abstract prearithmetics with exact projectivity relations form a category* **PEAPAEP**, *in which objects are partially extended abstract prearithmetics and morphisms are exact projectivity relations.*

Proof is left as an exercise. □

Corollary 3.1.17. *The category* **PEAPAEP** *is a subcategory of the category* **PEAPAP**.

Theorem 3.1.4 allows proving the following result.

Proposition 3.1.35. *The category* **PEAPAEP** *is isomorphic to the category in which objects are partially extended abstract prearithmetics and morphisms are their isomorphisms.*

Proof is left as an exercise. □

Distributivity of multiplication with respect to addition is an important property of Diophantine arithmetics. Proposition 3.1.15 implies the following result.

Proposition 3.1.36. *If a partially extended abstract prearithmetic* $\boldsymbol{A}_1 = (A_1; +_1, \circ_1, -_1, \leq_1)$ *is exactly projective with respect to a partially extended abstract prearithmetic* $\boldsymbol{A}_2 = (A_2; +_2, \circ_2, -_2, \leq_2)$ *and multiplication* $\circ_2$ *is distributive (from the left or from the right) with respect to addition* $+_2$ *in the prearithmetic* $\boldsymbol{A}_2$*, then multiplication* $\circ_1$ *is distributive (from the left or from the right, respectively) with respect to addition* $+_1$ *in the prearithmetic* $\boldsymbol{A}_1$.

Proof is left as an exercise. □

Proposition 3.1.36 and Theorem 3.1.4 imply the following consequence.

Corollary 3.1.18. *If a partially extended abstract prearithmetic* $\boldsymbol{A}_1 = (A_1; +_1, \circ_1, -_1, \leq_1)$ *is exactly projective with respect to a partially extended abstract prearithmetic* $\boldsymbol{A}_2 = (A_2; +_2, \circ_2, -_2, \leq_2)$ *and multiplication* $\circ_1$ *is distributive (from the left or from the right, respectively) with respect to* *addition $+_1$ in the prearithmetic $\boldsymbol{A}_1$, then multiplication $\circ_2$ is distributive (from the left or from the right) with respect to addition $+_2$ in the prearithmetic $\boldsymbol{A}_2$.*

Distributivity of multiplication with respect to subtraction is an important property of Diophantine arithmetics. Proposition 3.1.20 implies the following result.

Proposition 3.1.37. *If a partially extended abstract prearithmetic $\boldsymbol{A}_1 = (A_1; +_1, \circ_1, -_1, \leq_1)$ is exactly projective with respect to a partially extended abstract prearithmetic $\boldsymbol{A}_2 = (A_2; +_2, \circ_2, -_2, \leq_2)$ and multiplication $\circ_2$ is distributive (from the left or from the right) with respect to subtraction $-_2$ in the prearithmetic $\boldsymbol{A}_2$, then multiplication $\circ_1$ is distributive (from the left or from the right, respectively) with respect to subtraction $-_1$ in the prearithmetic $\boldsymbol{A}_1$.*

Proof is left as an exercise. □

Proposition 3.1.37 and Theorem 3.1.4 imply the following consequence.

Corollary 3.1.19. *If a partially extended abstract prearithmetic $\boldsymbol{A}_1 = (A_1; +_1, \circ_1, -_1, \leq_1)$ is exactly projective with respect to a partially extended abstract prearithmetic $\boldsymbol{A}_2 = (A_2; +_2, \circ_2, -_2, \leq_2)$ and multiplication $\circ_1$ is distributive (from the left or from the right, respectively) with respect to subtraction $-_1$ in the prearithmetic $\boldsymbol{A}_1$, then multiplication $\circ_2$ is distributive (from the left or from the right) with respect to subtraction $-_2$ in the prearithmetic $\boldsymbol{A}_2$.*

Remark 3.1.3. In some partially extended abstract prearithmetics, division can be distributive from the right or from both sides with respect to addition or/and subtraction. In these cases, exact projectivity preserves corresponding distributivity.

Proposition 3.1.38. *If a partially extended abstract prearithmetic $\boldsymbol{A}_1 = (A_1; +_1, \circ_1, -_1, \leq_1)$ is exactly projective with respect to a partially extended abstract prearithmetic $\boldsymbol{A}_2 = (A_2; +_2, \circ_2, -_2, \leq_2)$ and a partially extended abstract prearithmetic $\boldsymbol{A}_3 = (A_3; +_3, \circ_3, -_3, \leq_3)$ is exactly projective with respect to the partially extended abstract prearithmetic $\boldsymbol{A}_2$, then the partially extended abstract prearithmetic $\boldsymbol{A}_1$ is exactly projective with respect to the partially extended abstract prearithmetic $\boldsymbol{A}_3$.*

Proof: Let us assume that an abstract prearithmetic $\boldsymbol{A}_1 = (A_1; +_1, \circ_1, -_1, \leq_1)$ is exactly projective with respect to an abstract prearithmetic $\boldsymbol{A}_2 = (A_2; +_2, \circ_2, -_2, \leq_2)$ with the projector g: $A_1 \to A_2$ and the coprojector h: $A_2 \to A_1$ for the pair $(\boldsymbol{A}_1, \boldsymbol{A}_2)$ and the abstract prearithmetic $\boldsymbol{A}_3 = (A_3; +_3, \circ_3, -_3, \leq_3)$ is exactly projective with respect to the abstract prearithmetic $\boldsymbol{A}_2$ with the projector k: $A_3 \to A_2$ and the coprojector $l : A_2 \to A_3$ for the pair $(\boldsymbol{A}_3, \boldsymbol{A}_2)$. By Proposition 3.1.31, mappings k and l have inverse mappings k^{-1} and l^{-1}. This allows us to define the mappings $k^{-1}g$: $A_1 \to A_3$ and $hl^{-1} : A_3 \to A_1$. Then abstract prearithmetic $\boldsymbol{A}_1 = (A_1; +_1, \circ_1, -_1, \leq_1)$ is exactly projective with respect to an abstract prearithmetic $\boldsymbol{A}_3 = (A_3; +_3, \circ_3, -_3, \leq_3)$ with the projector $k^{-1}g$: $A_1 \to A_3$ and the coprojector $hl^{-1} : A_3 \to A_1$. Indeed, for any elements a and b from $\boldsymbol{A}_1$, we have the following relations:

$$\begin{aligned}
a +_1 b &= h(g(a) +_2 g(b)) = h(l^{-1}(k^{-1}(g(a)) +_3 k^{-1}(g(b)))) \\
&= (hl^{-1})((k^{-1}g)(a) +_3 (k^{-1}g)(b)), \\
a \circ_1 b &= h(g(a) \circ_2 g(b)) = h(l^{-1}(k^{-1}(g(a)) \circ_3 k^{-1}(g(b)))) \\
&= (hl^{-1})((k^{-1}g)(a) \circ_3 (k^{-1}g)(b)), \\
a -_1 b &= h(g(a) -_2 g(b)) = h(l^{-1}(k^{-1}(g(a)) -_3 k^{-1}(g(b)))) \\
&= (hl^{-1})((k^{-1}g)(a) -_3 (k^{-1}g)(b))
\end{aligned}$$

because

$$\begin{aligned}
a +_1 b &= h(g(a) +_2 g(b)), \\
a \circ_1 b &= h(g(a) \circ_2 g(b)), \\
a -_1 b &= h(g(a) -_2 g(b))
\end{aligned}$$

and

$$\begin{aligned}
x +_3 y &= l(k(x) +_2 k(y)), \\
x \circ_3 b &= l(k(x) \circ_2 g(y)), \\
x -_3 y &= l(k(x) -_2 k(y))
\end{aligned}$$

imply

$$\begin{aligned}
l^{-1}(x +_3 y) &= k(x) +_2 k(y), \\
l^{-1}(x \circ_3 b) &= k(x) \circ_2 g(y), \\
l^{-1}(x -_3 y) &= k(x) -_2 k(y).
\end{aligned}$$

As by Proposition 3.1.31, we have equalities $h = g^{-1}$ and $l = k^{-1}$, then

$$hl^{-1} = g^{-1}k = (k^{-1}g)^{-1}.$$

Thus, by Theorem 3.1.4, the abstract prearithmetic $\boldsymbol{A}_1$ is exactly projective with respect to the abstract prearithmetic $\boldsymbol{A}_3$.

Proposition is proved. □

As exact projectivity preserves algebraic properties of mathematical structures because it is an isomorphism, we have the following results.

Proposition 3.1.39.

(*a*) *A partially extended abstract prearithmetic* $\boldsymbol{A} = (A; +, \circ, -, \leq)$ *that is exactly projective with respect to a ring is a ring.*
(*b*) *An abstract prearithmetic* $\boldsymbol{A} = (A; +, \circ, -, \leq)$ *that is exactly projective with respect to a commutative ring is a commutative ring.*
(*c*) *An abstract prearithmetic* $\boldsymbol{A} = (A; +, \circ, -, \leq)$ *that is exactly projective with respect to an associative ring is an associative ring.*

Proof is left as an exercise. □

Proposition 3.1.40.

(*a*) *A partially extended abstract prearithmetic* $\boldsymbol{A} = (A; +, \circ, -, \leq)$ *that is exactly projective with respect to an additive (multiplicative) group is an additive (multiplicative) group.*
(*b*) *A partially extended abstract prearithmetic* $\boldsymbol{A} = (A; +, \circ, -, \leq)$ *that is exactly projective with respect to a commutative additive (multiplicative) group is a commutative additive (multiplicative) group.*

Proof is left as an exercise. □

Proposition 3.1.41. *A partially extended abstract prearithmetic* $\boldsymbol{A} = (A; +, \circ, -, \leq)$ *that is exactly projective with respect to a field is a field.*

Proof is left as an exercise. □

Proposition 3.1.42.

(*a*) *A partially extended abstract prearithmetic* $\boldsymbol{A} = (A; +, \circ, -, \leq)$ *that is exactly projective with respect to an ordered ring is an ordered ring.*
(*b*) *A partially extended abstract prearithmetic* $\boldsymbol{A} = (A; +, \circ, -, \leq)$ *that is exactly projective with respect to an ordered commutative ring is an ordered commutative ring.*

(*c*) *A partially extended abstract prearithmetic* $\boldsymbol{A} = (A; +, \circ, -, \leq)$ *that is exactly projective with respect to an ordered associative ring is an ordered associative ring.*

Proof is left as an exercise. □

Proposition 3.1.43.

(*a*) *A partially extended abstract prearithmetic* $\boldsymbol{A} = (A; +, \circ, -, \leq)$ *that is exactly projective with respect to an ordered additive group is an ordered additive group.*
(*b*) *A partially extended abstract prearithmetic* $\boldsymbol{A} = (A; +, \circ, -, \leq)$ *that is exactly projective with respect to an ordered commutative additive* (*multiplicative*) *group is an ordered commutative additive* (*multiplicative*) *group.*

Proof is left as an exercise. □

Proposition 3.1.44. *A partially extended abstract prearithmetic* $\boldsymbol{A} = (A; +, \circ, -, \leq)$ *that is exactly projective with respect to an ordered field is an ordered field.*

Proof is left as an exercise. □

We see that many properties of partially extended abstract prearithmetics are similar or even the same as properties of abstract prearithmetics in general. However, existence of an additional operation–and its relations to addition and multiplication adds new properties of partially extended abstract prearithmetics.

Definition 3.1.7. A partially extended abstract prearithmetic $\boldsymbol{A}_1 = (A_1; +_1, \circ_1, -_1, \leq_1)$ is a *subprearithmetic* of a partially extended abstract prearithmetic $\boldsymbol{A}_2 = (A_2; +_2, \circ_2, -_2, \leq_2)$ if $A_1 \subseteq A_2$, the operation $+_1$ is the restriction of the operation $+_2$ onto A_1, the operation $-_1$ is the restriction of the operation $-_2$ onto A_1, the operation $\circ_1$ is the restriction of the operation $\circ_2$ onto A_1, and the relation $\leq_1$ is the restriction of the relation $\leq_2$ onto A_1.

Example 3.1.10. The conventional Diophantine arithmetic $\boldsymbol{Z}$ of all integer numbers is a subprearithmetic of the conventional Diophantine arithmetics $\boldsymbol{Q}$ of all rational numbers and $\boldsymbol{R}$ of all real numbers.

Proposition 3.1.45. *If a partially extended abstract prearithmetic* $\boldsymbol{A}_1 = (A_1; +_1, \circ_1, -_1, /_1, \leq_1)$ *is a subprearithmetic of a partially extended abstract* *prearithmetic* $\boldsymbol{A}_2 = (A_2; +_2, \circ_2, -_2, /_2, \leq_2)$, *then the partially extended abstract prearithmetic* $\boldsymbol{A}_1$ *is projective with respect to the partially extended abstract prearithmetic* $\boldsymbol{A}_2$.

Proof is similar to the proof of Theorem 2.5.1. □

As projectivity is a special case of weak projectivity (cf. Section 2.3), we have the following result.

Corollary 3.1.20. *If a partially extended abstract prearithmetic* $\boldsymbol{A}_1 = (A_1; +_1, \circ_1, -_1, /_1, \leq_1)$ *is a subprearithmetic of a partially extended abstract prearithmetic* $\boldsymbol{A}_2 = (A_2; +_2, \circ_2, -_2, /_2, \leq_2)$, *then the partially extended abstract prearithmetic* $\boldsymbol{A}_1$ *is weakly projective with respect to the partially extended abstract prearithmetic* $\boldsymbol{A}_2$.

The property "to be a subprearithmetic" is transitive.

Proposition 3.1.46. *If a partially extended abstract prearithmetic* $\boldsymbol{A}_1 = (A_1; +_1, \circ_1, -_1, \leq_1)$ *is a subprearithmetic of a partially extended abstract prearithmetic* $\boldsymbol{A}_2 = (A_2; +_2, \circ_2, -_2, \leq_2)$ *and the abstract prearithmetic* $\boldsymbol{A}_2$ *is a subprearithmetic of a partially extended abstract prearithmetic* $\boldsymbol{A}_3 = (A_3; +_3, \circ_3, -_3, \leq_3)$ *then the partially extended abstract prearithmetic* $\boldsymbol{A}_1$ *is a subprearithmetic of the partially extended abstract prearithmetic* $\boldsymbol{A}_3$.

Proof is left as an exercise. □

Subprearithmetics inherit weak projectivity.

Proposition 3.1.47. *If a partially extended abstract prearithmetic* $\boldsymbol{A}_1 = (A_1; +_1, \circ_1, -_1, \leq_1)$ *is a subprearithmetic of a partially extended abstract prearithmetic* $\boldsymbol{A}_2 = (A_2; +_2, \circ_2, -_2, \leq_2)$ *and the abstract prearithmetic* $\boldsymbol{A}_2$ *is weakly projective with respect to an abstract prearithmetic* $\boldsymbol{A}_3 = (A_3; +_3, \circ_3, \leq_3)$, *then the abstract prearithmetic* $\boldsymbol{A}_1$ *is weakly projective with respect to the abstract prearithmetic* $\boldsymbol{A}_3$.

Proof is similar to the proof of Proposition 2.5.14. □

Proposition 3.1.48. *If a partially extended abstract prearithmetic* $\boldsymbol{A}_1 = (A_1; +_1, \circ_1, -_1, \leq_1)$ *is weakly projective with respect to a partially extended abstract prearithmetic* $\boldsymbol{A}_2 = (A_2; +_2, \circ_2, -_2, \leq_2)$ *and the abstract prearithmetic* $\boldsymbol{A}_2$ *is a subprearithmetic of an abstract prearithmetic* $\boldsymbol{A}_3 = (A_3; +_3, \circ_3, \leq_3)$, *then the abstract prearithmetic* $\boldsymbol{A}_1$ *is weakly projective with respect to the abstract prearithmetic* $\boldsymbol{A}_3$.

Proof is similar to the proof of Proposition 2.5.15. □

Proposition 3.1.49. *If a partially extended abstract prearithmetic* $\boldsymbol{A}_1 = (A_1; +_1, \circ_1, -_1, \leq_1)$ *is a subprearithmetic of a partially extended abstract prearithmetic* $\boldsymbol{A}_2 = (A_2; +_2, \circ_2, -_2, \leq_2)$ *and the abstract prearithmetic* $\boldsymbol{A}_2$ *is projective with respect to an abstract prearithmetic* $\boldsymbol{A}_3 = (A_3; +_3, \circ_3, \leq_3)$, *then the prearithmetic* $\boldsymbol{A}_1$ *is projective with respect to the prearithmetic* $\boldsymbol{A}_3$.

Proof is similar to the proof of Proposition 2.5.16. □

Proposition 3.1.50. *If a partially extended abstract prearithmetic* $\boldsymbol{A}_1 = (A_1; +_1, \circ_1, -_1, \leq_1)$ *is projective with respect to a partially extended abstract prearithmetic* $\boldsymbol{A}_2 = (A_2; +_2, \circ_2, -_2, \leq_2)$ *and the partially extended abstract prearithmetic* $\boldsymbol{A}_2$ *is a subprearithmetic of an abstract prearithmetic* $\boldsymbol{A}_3 = (A_3; +_3, \circ_3, \leq_3)$, *then the prearithmetic* $\boldsymbol{A}_1$ *is projective with respect to the prearithmetic* $\boldsymbol{A}_3$.

Proof is similar to the proof of Proposition 2.5.17. □

Remark 3.1.4. For exact projectivity, the statements of Propositions 3.1.47–3.1.50 are not always true because a subprearithmetic of an arbitrary abstract prearithmetic $\boldsymbol{A}$ is not always exactly projective with respect to $\boldsymbol{A}$.

3.2. Wholly Extended Abstract Prearithmetics and Their Projectivity

To be great, be whole; exclude nothing, exaggerate nothing that is not you.
Be whole in everything. Put all you are into the smallest thing you do.

Fernando Pessoa

A *wholly extended abstract prearithmetic* is a set (often a set of numbers) A with a partial order $\leq$, three binary operations $+$ (addition), $-$ (subtraction) and $\circ$ (multiplication), which are defined for all its elements, and the partial binary operation/(division). It is denoted by $\boldsymbol{A} = (A; +, \circ, -, /, \leq)$. As before, the set A is called either the *set of elements* or the *set of numbers* or the *carrier* of the prearithmetic $\boldsymbol{A}$. In a usual way, if $x \leq y$ and $x \neq y$, then we denote it by $x < y$. Operation $+$ is called *addition*, operation $-$ is called *subtraction*, operation / is called *division* and operation $\circ$ is called *multiplication* in the prearithmetic $\boldsymbol{A}$. It is naturally to assume that division/is not defined only when the second argument is zero. Thus, when $\boldsymbol{A}$ does not have zeros,/is a total operation.

Note that a wholly extended abstract prearithmetic can have more operations and order relations but the four basic operations are always defined. Besides, in a general case, division in wholly extended abstract prearithmetics is not necessarily division introduced for abstract prearithmetics in Section 2.1. Indeed, division introduced in Section 2.1 is inverse to multiplication while division in wholly extended abstract prearithmetics is independent from addition and multiplication in a general case. In addition, division can be a total operation in some wholly extended abstract prearithmetics.

Example 3.2.1. The conventional arithmetic $\boldsymbol{R}$ of all real numbers is a wholly extended abstract prearithmetic.

Example 3.2.2. The conventional arithmetic $\boldsymbol{C}$ of all complex numbers is a wholly extended abstract prearithmetic.

Example 3.2.3. The conventional arithmetic $\boldsymbol{Q}$ of all rational numbers is a wholly extended abstract prearithmetic.

All these examples show that many conventional arithmetics are partially extended abstract prearithmetics. However, there are also many unusual partially extended abstract prearithmetics.

All these examples show that some conventional arithmetics are wholly extended abstract prearithmetics. However, there are also unusual wholly extended abstract prearithmetics.

Example 3.2.4. Let us consider the set N of all natural numbers with the standard order $\leq$ and introduce the following operations:

$$a \oplus b = a \cdot b,$$

$$a \otimes b = a^b,$$

$$a \ominus b = a + b,$$

$$a \oslash b = a - b.$$

Then the system $\boldsymbol{A} = (N; \oplus, \otimes, \ominus, \oslash, \leq)$ is a wholly extended abstract prearithmetic with addition $\oplus$, subtraction $\ominus$, multiplication $\otimes$ and division $\oslash$.

Example 3.2.5. Any field is a wholly extended abstract prearithmetic with a trivial order (cf. Appendix). Any ordered field is a wholly extended abstract prearithmetic (Kurosh, 1963).

There are three kinds of motivations toward extending abstract prearithmetics and arithmetics with division. One is theoretical, the second is practical and the third is geometric.

Theoretical motivation. For any abstract prearithmetic or arithmetic, there are two total operations — addition and multiplication — and a partial operation of division (cf. Section 2.1). Thus, it would be reasonable to make the latter operation also total.

Practical motivation. In many problems, the quantity or quantities to be determined cannot be obtained by a direct arithmetical calculation but instead have to be deduced from some required conditions. Often these conditions have the form of equations. For instance, we have an equation $7x = 5$. As we know, to solve this equation, we need rational numbers and to have rational numbers, we need division. Similar equations arise in abstract prearithmetics and arithmetics. To solve them, we need division (cf. Section 2.1).

Geometric motivation. Numbers are used for measurement. There are also measures with more general values. In particular, it is possible to use abstract prearithmetics and arithmetics for measurement. At the same time, very often the object that is measured can be divided into smaller objects. For instance, the typical object of measurement is a straight line segment. As we know, any line segment can be subdivided into smaller equal segments. That is why, if we want to use abstract prearithmetics or arithmetics for measurement, we need operation *division* in prearithmetic or arithmetic to represent division of measured objects.

Let us study properties wholly extended abstract prearithmetics. In particular, properties of wholly extended abstract prearithmetics allow assembling algebraic constructions using wholly extended abstract prearithmetics as it is done in the case of Diophantine arithmetics $\boldsymbol{R}$ and $\boldsymbol{C}$. For instance, taking a wholly extended abstract prearithmetic $\boldsymbol{A} = (A; +, \circ, -, /, \leq)$ and a natural number n, it is possible to build the wholly extended abstract prearithmetic of n-dimensional A-vectors $\mathrm{V}^n\boldsymbol{A} = (\mathrm{V}^n A; +, \circ, -, /, \leq)$, elements of which are vectors in A. Namely, elements of the wholly extended n-dimensional A-vector prearithmetic $\mathrm{V}^n\boldsymbol{A} = (\mathrm{V}^n A; +, \circ, -, /, \leq)$, i.e., A-vectors, have the form $(a_1, a_2, \ldots, a_n)$ where $a_1, a_2, \ldots, a_n$ are elements from the abstract prearithmetic $\boldsymbol{A}$.

In a similar way, taking a wholly extended abstract prearithmetic $\boldsymbol{A} = (A; +, \circ, -, /, \leq)$ and a pair of natural numbers n and m, it is also possible to build the wholly extended abstract prearithmetic of $n \times m$-dimensional A-matrices $\mathrm{M}^{n\times m}\boldsymbol{A} = (\mathrm{M}^{n\times m} A; +, \circ, -, /, \leq)$, elements of which are matrices in A. Namely, elements of the wholly extended $n \times m$-dimensional A-matrix prearithmetic $\mathrm{M}^{n\times m}\boldsymbol{A} = (\mathrm{M}^{n\times m} A; +, \circ, -, /, \leq)$, i.e., A-matrices, have the form

$$\begin{pmatrix} a_{11} & a_{12} & a_{13} & \dots & a_{1n} \\ a_{21} & a_{22} & a_{23} & \dots & a_{2n} \\ \vdots & \vdots & \vdots & \vdots & \vdots \\ a_{m1} & a_{m2} & a_{m3} & \dots & a_{mn} \end{pmatrix},$$

where all $a_{ij} (i = 1, 2, 3, \dots, m; j = 1, 2, 3, \dots, n)$ are elements from the abstract prearithmetic $\boldsymbol{A}$.

Addition, subtraction, division and multiplication are defined coordinate-wise. For instance, taking vectors (10, 8) and (2, 4) from the prearithmetic $\mathrm{V}^2\boldsymbol{R}$ of vectors, we define their sum as (10, 8) + (2, 4) = (12, 12), their difference as (10, 8) – (2, 4) = (8, 4), their product as (10, 8) ∘ (2, 4) = (20, 32) and their quotient (10, 8)/(2, 4) = (5, 2).

Order is defined by the following condition:

If $(a_1, a_2, \dots, a_n)$ and $(b_1, b_2, \dots, b_n)$ are vectors from $\mathrm{V}^n A$, then $(a_1, a_2, \dots, a_n) \le (b_1, b_2, \dots, b_n)$ if and only if $a_j \le b_j$ for all $j = 1, 2, 3, \dots, n$.

For matrices, addition, subtraction, multiplication and order are defined in a similar way.

Note that the defined multiplication is *scalar multiplication* of vectors and matrices, which is different from vector and matrix multiplication.

Wholly extended prearithmetics $\mathrm{V}^n\boldsymbol{A}$ and $\mathrm{M}^{n\times m}\boldsymbol{A}$ preserve many properties of the wholly extended abstract prearithmetic $\boldsymbol{A}$. For instance, we have the following results.

Proposition 3.2.1. *If addition is commutative in an abstract prearithmetic $\boldsymbol{A}$, then addition is commutative in the wholly extended vector prearithmetic $V^n\boldsymbol{A}$ and in the wholly extended matrix prearithmetic $M\boldsymbol{A}$.*

Proof is left as an exercise. □

The same is true for multiplication.

Proposition 3.2.2. *If multiplication is commutative in a wholly extended abstract prearithmetic $\boldsymbol{A}$, then multiplication is commutative in the wholly extended vector prearithmetic $V^n\boldsymbol{A}$ and in the wholly extended matrix prearithmetic $M^{n\times m}\boldsymbol{A}$.*

Proof is left as an exercise. □

Proposition 3.2.3. *If addition is associative in a wholly extended abstract prearithmetic* $\boldsymbol{A}$*, then addition is associative in the wholly extended vector prearithmetic* $V^n\boldsymbol{A}$ *and in the wholly extended matrix prearithmetic* $M^{n\times m}\boldsymbol{A}$.

Proof is left as an exercise. □

The same is true for multiplication.

Proposition 3.2.4. *If multiplication is associative in a wholly extended abstract prearithmetic* $\boldsymbol{A}$*, then multiplication is associative in the wholly extended vector prearithmetic* $V^n\boldsymbol{A}$ *and in the wholly extended matrix prearithmetic* $M^{n\times m}\boldsymbol{A}$.

Proof is left as an exercise. □

In the Diophantine arithmetic, multiplication is distributive with respect to addition, i.e., the following identities hold:

$$x \cdot (y + z) = x \cdot y + x \cdot z,$$

$$(y + z) \cdot x = y \cdot x + z \cdot x.$$

However, in abstract prearithmetics, multiplication is not always commutative and we need to discern three kinds of distributivity. Namely, *distributivity from the left*

$$x \cdot (y + z) = x \cdot y + x \cdot z$$

and *distributivity from the right*

$$(y + z) \cdot x = y \cdot x + z \cdot x.$$

In addition, multiplication is *distributive* with respect to addition when both identities hold.

Proposition 3.2.5. *If multiplication is distributive (distributive from the left or distributive from the right) with respect to addition in a wholly extended abstract prearithmetic* $\boldsymbol{A}$*, then multiplication is distributive (distributive from the left or distributive from the right) with respect to addition in the wholly extended vector prearithmetic* $V^n\boldsymbol{A}$ *and in the wholly extended matrix prearithmetic* $M^{n\times m}\boldsymbol{A}$.

Proof is left as an exercise. □

Remark 3.2.1. Having a wholly extended abstract prearithmetic $\boldsymbol{A} = (A; +, \circ, -, /, \leq)$, it is possible to build not only wholly extended abstract prearithmetics of A-vectors and A-matrices but also multidimensional matrices or arrays in A, i.e., multidimensional A-matrices or A-arrays and build their prearithmetics exploring what properties they inherit from the initial wholly extended abstract prearithmetic.

Let us study relations between wholly extended abstract prearithmetics. One of the most important is weak projectivity.

Let us take two wholly extended abstract prearithmetics $\boldsymbol{A}_1 = (A_1; +_1, \circ_1, -_1, \leq_1)$ and $\boldsymbol{A}_2 = (A_2; +_2, \circ_2, -_2, \leq_2)$ and consider two mappings g: $A_1 \to A_2$ and h: $A_2 \to A_1$.

Definition 3.2.1. (a) A wholly extended abstract prearithmetic $\boldsymbol{A}_1 = (A_1; +_1, \circ_1, -_1, /_1, \leq_1)$ is called *weakly projective* with respect to a wholly extended abstract prearithmetic $\boldsymbol{A}_2 = (A_2; +_2, \circ_2, -_2, /_2, \leq_2)$ if there are following relations between orders and operations in $\boldsymbol{A}_1$ and in $\boldsymbol{A}_2$:

$$\begin{aligned} a +_1 b &= h(g(a) +_2 g(b)), \\ a \circ_1 b &= h(g(a) \circ_2 g(b)), \\ a -_1 b &= h(g(a) -_2 g(b)), \\ a/_1 b &= h(g(a)/_2 g(b)), \end{aligned}$$
$$a \leq_1 b \text{ only if } g(a) \leq_2 g(b).$$

(b) The mapping g is called the *projector* and the mapping h is called the *coprojector* for the pair ($\boldsymbol{A}_1$, $\boldsymbol{A}_2$).

In this case, we will say that there is *weak projectivity* between the wholly extended abstract prearithmetic $\boldsymbol{A}_1$ and the wholly extended abstract prearithmetic $\boldsymbol{A}_2$ and there is *weak inverse projectivity* between the wholly extended abstract prearithmetic $\boldsymbol{A}_2$ and the wholly extended abstract prearithmetic $\boldsymbol{A}_1$.

When a wholly extended abstract prearithmetic $\boldsymbol{B}$ is weakly projective with respect to a wholly extended abstract prearithmetic $\boldsymbol{A}$ with the projector g and the coprojector h, we denote $\boldsymbol{B}$ by $\boldsymbol{A}_{g,h}$ to show this relation between wholly extended abstract prearithmetics $\boldsymbol{A}$ and $\boldsymbol{B}$.

As in the case of abstract prearithmetics, weak projectivity between the wholly extended abstract prearithmetics determines a bidirectional named set (cf. Remark 2.2.1) and a fiber bundle (cf. Remark 2.2.2).

In essence, the idea of projectivity is to define operations in one set using operations in another set. This is demonstrated by the following result.

Let us consider an arbitrary set B, a wholly extended abstract prearithmetic $\boldsymbol{A} = (A; +, \circ, -, /, \leq)$ and two functions g: $B \to A$ and h: $A \to B$. Definition 3.2.1 implies the following result.

Proposition 3.2.6. *It is possible to define on B the unique structure of a wholly extended abstract prearithmetic $\boldsymbol{B}$ that is weakly projective with respect to the wholly extended abstract prearithmetic $\boldsymbol{A}$ with the projector g and the coprojector h.*

Proof: To prove this statement, we define the relation $\leq_B$ and four operations $+_B, -_B, /_B$ and $\circ_B$ in the set B by the following rules:

$$a +_B b = h(g(a) + g(b)),$$

$$a -_B b = h(g(a) - g(b)),$$

$$a \circ_B b = h(g(a) \circ g(b)),$$

$$a /_B b = h(g(a)/g(b)),$$

$a \leq_B b$ if and only if $g(a) \leq g(b)$ and ($g(c) = g(a)$ implies $c = a$) and ($g(d) = g(b)$ implies $d = b$) for any elements a and b from B.

Now we need to check that the relation $\leq_B$ is a partial order, i.e., it is reflexive, symmetric and transitive. Taking arbitrary elements a and b from B, we derive the following properties.

- *Reflexivity.* $a \leq_B a$ because $g(a) \leq g(a)$.
- *Anti-symmetry.* Let us have $a \leq_B b$ and $b \leq_B a$. Then $a \leq_B b$ implies $g(a) \leq g(b)$ and $b \leq_B a$ and implies $g(b) \leq g(c)$. As the relation $\leq$ is a partial order, it is symmetric (cf. Appendix) and we have $g(b) = g(a)$. Then by definition, we have $b = a$.
- *Transitivity.* Given $a \leq_B b$ and $b \leq_B c$, by definition, we have $g(a) \leq g(b)$ and $g(b) \leq g(c)$. As the relation $\leq$ is a partial order, it is transitive and we have $g(a) \leq g(c)$. Consequently, we have $a \leq_B c$.

Proposition is proved. □

We denote the wholly extended abstract prearithmetic $\boldsymbol{B}$ constructed in Proposition 3.2.6 by $\boldsymbol{B}^{g,h}$.

Proposition 3.2.7. *If the mapping* g: $B \to A$ *is an injection and* $\leq$ *is a linear (total) order in* A, *then* $\leq_B$ *is a linear (total) order in* B.

Indeed, if g is an injection, then the third condition, which defines the prearithmetic $\boldsymbol{B}$, is equivalent to the condition

$$a \leq_B b \text{ if and only if } g(a) \leq g(b).$$

Thus, if for any two elements $x, y \in A$, either $x \leq y$ or $x \leq y$ is true, then for any two elements $a, b \in B$, either $a \leq_B b$ or $a \leq_B b$ is true.

The construction of the prearithmetic $\boldsymbol{B}$ in Proposition 3.2.6 implies its uniqueness. Indeed, let us consider three mappings g: $B \to A$, f: $A \to B$ and h: $A \to B$.

Proposition 3.2.8. *If mappings* f *and* h *coincide on the wholly extended abstract prearithmetic* $PA(g(B))$ *generated by image of* B *in* $\boldsymbol{A}$, *i.e.,* $f_{|PA(g(B)} = h_{|PA(g(B)}$, *then* $\boldsymbol{B}_{g,h} = \boldsymbol{B}_{g,f}$.

Proof is left as an exercise. □

When a wholly extended abstract prearithmetic $\boldsymbol{A}_1$ is weakly projective with respect to a wholly extended abstract prearithmetic $\boldsymbol{A}_2$, it is often possible to deduce some properties of the prearithmetic $\boldsymbol{A}_1$ from properties of the prearithmetic $\boldsymbol{A}_2$. In particular, weak projectivity in the class of wholly extended abstract prearithmetics implies definite relations between properties of operations in prearithmetics.

Proposition 2.2.4 implies the following result.

Proposition 3.2.9. *If a wholly extended abstract prearithmetic* $\boldsymbol{A}_1 = (A_1; +_1, \circ_1, -_1, /_1, \leq_1)$ *is weakly projective with respect to a wholly extended abstract prearithmetic* $\boldsymbol{A}_2 = (A_2; +_2, \circ_2, -_2, /_2, \leq_2)$ *and the operation* $+_2$ *is commutative in the prearithmetic* $\boldsymbol{A}_2$, *then the operation* $+_1$ *is commutative in the prearithmetic* $\boldsymbol{A}_1$.

Proof is left as an exercise. □

This shows that inverse weak projectivity preserves commutativity of addition.

Proposition 2.2.5 implies a similar property for multiplication in abstract prearithmetics.

Proposition 3.2.10. *If a wholly extended abstract prearithmetic* $\boldsymbol{A}_1 = (A_1; +_1, \circ_1, -_1, /_1, \leq_1)$ *is weakly projective with respect to a wholly extended abstract prearithmetic* $\boldsymbol{A}_2 = (A_2; +_2, \circ_2, -_2, /_2, \leq_2)$ *and the operation* $\circ_2$ *is commutative in the prearithmetic* $\boldsymbol{A}_2$*, then the operation* $\circ_1$ *is commutative in the prearithmetic* $\boldsymbol{A}_1$*.*

Proof is left as an exercise. □

This shows that inverse weak projectivity preserves commutativity of multiplication.

Let us consider two wholly extended abstract prearithmetics $\boldsymbol{A}_1 = (A_1; +_1, \circ_1, -_1, /_1, \leq_1)$ and $\boldsymbol{A}_2 = (A_2; +_2, \circ_2, -_2, /_2, \leq_2)$, which have additive zeros 0_1 and 0_2, respectively.

Proposition 3.2.11. *If the prearithmetic* $\boldsymbol{A}_1$ *is weakly projective with respect to the prearithmetic* $\boldsymbol{A}_2$ *with the projector* g *and the coprojector* $h, h(0_2) = 0_1$ *and* $x -_2 x = 0_2$ *for any element* x *from the prearithmetic* $\boldsymbol{A}_2$*, then* $a -_1 a = 0_1$ *for any element* a *from the prearithmetic* $\boldsymbol{A}_1$*.*

Indeed,

$$a -_1 a = h(g(a) -_2 g(a)) = h(0_2) = 0_1.$$

Distributivity of multiplication with respect to addition is an important property of Diophantine arithmetics. Proposition 3.1.19 implies the following result.

Proposition 3.2.12. *If a wholly extended abstract prearithmetic* $\boldsymbol{A}_1 = (A_1; +_1, \circ_1, -_1, /_1, \leq_1)$ *is weakly projective with respect to a wholly extended abstract prearithmetic* $\boldsymbol{A}_2 = (A_2; +_2, \circ_2, -_2, /_2, \leq_2)$ *with the projector* g *and the coprojector* h*, for which* $gh = \mathbf{1}_{A_2}$*, and multiplication* $\circ_2$ *is distributive (from the left or from the right) with respect to addition* $+_2$ *in the prearithmetic* $\boldsymbol{A}_2$*, then multiplication* $\circ_1$ *is distributive (from the left or from the right, respectively) with respect to addition* $+_1$ *in the prearithmetic* $\boldsymbol{A}_1$*.*

Proof is left as an exercise. □

Distributivity of multiplication with respect to subtraction is an important property of Diophantine arithmetics. Proposition 3.1.20 implies the following result.

Proposition 3.2.13. *If a wholly extended abstract prearithmetic* $\boldsymbol{A}_1 = (A_1; +_1, \circ_1, -_1, /_1, \leq_1)$ *is weakly projective with respect to a wholly extended* *abstract prearithmetic* $\boldsymbol{A}_2 = (A_2; +_2, \circ_2, -_2, /_2, \leq_2)$ *with the projector* g *and the coprojector* h*, for which* $gh = \mathbf{1}_{A_2}$*, and multiplication* $\circ_2$ *is distributive (from the left or from the right) with respect to subtraction* $-_2$ *in the prearithmetic* $\boldsymbol{A}_2$*, then multiplication* $\circ_1$ *is distributive (from the left or from the right, respectively) with respect to subtraction* $-_1$ *in the prearithmetic* $\boldsymbol{A}_1$*.*

Proof is left as an exercise. □

Remark 3.2.2. In some wholly extended abstract prearithmetics, multiplication can be distributive from the right or from both sides with respect to addition or/and subtraction. In these cases, inverse weak projectivity preserves corresponding distributivity under the condition $gh = \mathbf{1}_{A_2}$.

Division is also distributive from the right with respect to addition in Diophantine arithmetics $\boldsymbol{R}$ or $\boldsymbol{C}$. Let us explore this property for wholly extended abstract prearithmetics.

Proposition 3.2.14. *If a wholly extended abstract prearithmetic* $\boldsymbol{A}_1 = (A_1; +_1, \circ_1, -_1, /_1, \leq_1)$ *is weakly projective with respect to a wholly extended abstract prearithmetic* $\boldsymbol{A}_2 = (A_2; +_2, \circ_2, -_2, /_2, \leq_2)$ *with the projector* g *and the coprojector* h*, for which* $gh = \mathbf{1}_{A_2}$*, and division* $\circ_2$ *is distributive from the right with respect to addition* $+_2$ *in the prearithmetic* $\boldsymbol{A}_2$*, then division* $\circ_1$ *is distributive from the right with respect to addition* $+_1$ *in the prearithmetic* $\boldsymbol{A}_1$*.*

Proof: Let us check distributivity from the right taking arbitrary elements a, b and c from $\boldsymbol{A}_1$. By Definition 3.2.1, we have

$$\begin{aligned}(b +_1 c)/_1 a &= h((g(h(g(b) +_2 g(c))))/_2 g(a)) \\ &= h((g(b) +_2 g(c)))/_2 g(a)) = h(g(b)/_2 g(a) +_2 g(c)/_2 g(a)) \\ &= h(g(h(g(a)/_2 g(b))) +_2 g(h(g(a)/_2 g(c)))) \\ &= h(g(a/_1 b) +_2 g(a/_1 c)) = (a/_1 b) +_1 (a/_1 c)\end{aligned}$$

because $gh = \mathbf{1}_{A_2}$ and division $/_2$ is distributive from the right with respect to addition $+_2$ in the prearithmetic $\boldsymbol{A}_2$. It means that division $/_1$ is distributive from the right with respect to addition $+_1$ in the prearithmetic $\boldsymbol{A}_1$.

Proposition is proved. □

Proposition 3.2.15. *If a wholly extended abstract prearithmetic* $\boldsymbol{A}_1 = (A_1; +_1, \circ_1, -_1, /_1, \leq_1)$ *is weakly projective with respect to a wholly extended abstract prearithmetic* $\boldsymbol{A}_2 = (A_2; +_2, \circ_2, -_2, /_2, \leq_2)$ *with the projector* g *and the coprojector* h*, for which* $gh = \mathbf{1}_{A_2}$*, and division* $\circ_2$ *is distributive from the right with respect to subtraction* $-_2$ *in the prearithmetic* $\boldsymbol{A}_2$*, then division* $\circ_1$ *is distributive from the right with respect to subtraction* $-_1$ *in the prearithmetic* $\boldsymbol{A}_1$*.*

Proof: Let us check distributivity from the right taking arbitrary elements a, b and c from $\boldsymbol{A}_1$. By Definition 3.2.1, we have

$$\begin{aligned}(b -_1 c)/_1 a &= h((g(h(g(b) -_2 g(c))))/_2 g(a))\\ &= h((g(b) -_2 g(c)))/_2 g(a)) = h(g(b)/_2 g(a) -_2 g(c)/_2 g(a))\\ &= h(g(h(g(a)/_2 g(b))) -_2 g(h(g(a)/_2 g(c))))\\ &= h(g(a/_1 b) -_2 g(a/_1 c)) = (a/_1 b) -_1 (a/_1 c)\end{aligned}$$

because $gh = \mathbf{1}_{A_2}$ and division $/_2$ is distributive from the right with respect to subtraction $-_2$ in the prearithmetic $\boldsymbol{A}_2$. It means that division $/_1$ is distributive from the right with respect to subtraction $-_1$ in the prearithmetic $\boldsymbol{A}_1$.

Proposition is proved. □

Remark 3.2.3. In some wholly extended abstract prearithmetics, division can be distributive from the right or from both sides with respect to addition or/and subtraction. In these cases, inverse weak projectivity preserves corresponding distributivity under the condition $gh = \mathbf{1}_{A_2}$.

Weak projectivity is a transitive relation in the class of wholly extended abstract prearithmetics.

Proposition 3.2.16. *If a wholly extended abstract prearithmetic* $\boldsymbol{A}_1 = (A_1; +_1, \circ_1, -_1, /_1, \leq_1)$ *is weakly projective with respect to a wholly extended abstract prearithmetic* $\boldsymbol{A}_2 = (A_2; +_2, \circ_2, -_2, /_2, \leq_2)$ *and the wholly extended abstract prearithmetic* $\boldsymbol{A}_2$ *is weakly projective with respect to a wholly extended abstract prearithmetic* $\boldsymbol{A}_3 = (A_3; +_3, \circ_3, -_3, /_3, \leq_3)$*, then the prearithmetic* $\boldsymbol{A}_1$ *is weakly projective with respect to the prearithmetic* $\boldsymbol{A}_3$*.*

Proof is similar to the proof of Proposition 2.2.9.

Proposition 3.2.16 allows proving the following result.

Theorem 3.2.1. *Wholly extended abstract prearithmetics with weak projectivity relations form a category* **WEAPAWP**, *in which objects are wholly extended abstract prearithmetics and morphisms are weak projectivity relations.*

Proof is left as an exercise. □

Proposition 3.2.17. *If a wholly extended abstract prearithmetic* $\boldsymbol{A}_1 = (A_1; +_1, \circ_1, -_1, /_1, \leq_1)$ *is weakly projective with respect to a wholly extended abstract prearithmetic* $\boldsymbol{A}_2 = (A_2; +_2, \circ_2, -_2, /_2, \leq_2)$ *with the projector* g *and the coprojector* h *and* $gh = \mathbf{1}_{A_2}$, *then the projector* g *is a homomorphism.*

Proof: The definition of weak projectivity and the equality $gh = \mathbf{1}_{A_2}$ imply the following systems of equalities:

$$g(a +_1 b) = g(h(g(a) +_2 g(b))) = (gh)(g(a) +_2 g(b)) = g(a) +_2 g(b),$$

$$g(a \circ_1 b) = g(h(g(a) \circ_2 g(b))) = (gh)(g(a) \circ_2 g(b)) = g(a) \circ_2 g(b),$$

$$g(a -_1 b) = g(h(g(a) -_2 g(b))) = (gh)(g(a) -_2 g(b)) = g(a) -_2 g(b),$$

$$g(a/_1 b) = g(h(g(a)/_2 g(b))) = (gh)(g(a)/_2 g(b)) = g(a)/_2 g(b).$$

These equalities mean that g is a homomorphism of $\boldsymbol{A}_1$ as a universal algebra with four operations. Besides, by Lemma 2.2.1, the projector g preserves the order $\leq_1$.

Proposition is proved. □

Corollary 3.2.1. *If an abstract prearithmetic* $\boldsymbol{A}_1 = (A_1; +_1, \circ_1, -_1, /_1, \leq_1)$ *is weakly projective with respect to an abstract prearithmetic* $\boldsymbol{A}_2 = (A_2; +_2, \circ_2, -_2, /_2, \leq_2)$ *with the projector* g *and the coprojector* h *and* $h = g^{-1}$, *then the projector* g *is an isomorphism, i.e., prearithmetics* $\boldsymbol{A}_1$ *and* $\boldsymbol{A}_2$ *are isomorphic as algebraic systems.*

Now let us study projectivity of wholly extended abstract prearithmetics.

Definition 3.2.2. A wholly extended abstract prearithmetic $\boldsymbol{A}_1 = (A_1; +_1, \circ_1, -_1, /_1, \leq_1)$ is called *projective* with respect to a wholly extended abstract prearithmetic $\boldsymbol{A}_2 = (A_2; +_2, \circ_2, -_2, /_2, \leq_2)$ if it is weakly projective with respect to a wholly extended $\boldsymbol{A}_2$ and $hg = \mathbf{1}_{A_1}$, where $\mathbf{1}_{A_1}$ is the identity mapping of A_1, g: $A_1 \to A_2$ is the *projector* and h: $A_2 \to A_1$ is the *coprojector* for the pair $(\boldsymbol{A}_1, \boldsymbol{A}_2)$, and for any elements a and b from A_1, we have

$$a \leq_1 b \text{ if and only if } g(a) \leq_2 g(b).$$

When a wholly extended abstract prearithmetic $\boldsymbol{B}$ is projective with respect to a wholly extended abstract prearithmetic $\boldsymbol{A}$ with the projector g and the coprojector h, we denote $\boldsymbol{B}$ by $\boldsymbol{A}_{g;h}$.

Let us consider an arbitrary set B, a wholly extended abstract prearithmetic $\boldsymbol{A} = (A; +, \circ, -, /, \leq)$ and two functions g: $B \to A$ and h: $A \to B$ such that $hg = \mathbf{1}_{A_1}$. Definition 3.2.2 implies the following result.

Proposition 3.2.18. *It is possible to define on B the structure of a wholly extended abstract prearithmetic $\boldsymbol{B}^{g;h}$ that is projective with respect to the wholly extended abstract prearithmetic $\boldsymbol{A}$ as a unique prearithmetic with the projector g and the coprojector h.*

Proof is similar to the proof of Proposition 2.2.1.

Note that $\boldsymbol{B}^{g;h}$ is also $\boldsymbol{A}_{g;h}$.

Let us consider a wholly extended abstract prearithmetics $\boldsymbol{A}_1 = (A_1; +_1, \circ_1, -_1, /_1, \leq_1)$ and a wholly extended abstract prearithmetics $\boldsymbol{A}_2 = (A_2; +_2, \circ_2, -_2, /_2, \leq_2)$, which has an additive zero 0_2.

Proposition 3.2.19. *If the wholly extended abstract prearithmetic $\boldsymbol{A}_1$ is projective with respect to the wholly extended abstract prearithmetic $\boldsymbol{A}_2$ with the projector g and the coprojector h and $g(a) = 0_2$, then a is an additive zero 0_1 in the wholly extended abstract prearithmetic $\boldsymbol{A}_1$.*

Indeed, let us take an arbitrary element b from $\boldsymbol{A}_1$. Then we have

$$b +_1 a = h(g(b) +_2 g(a)) = h(g(b) +_2 0_2) = h(g(b)) = b.$$

Proposition 3.1.25 implies the following result.

Proposition 3.2.20. *If the wholly extended abstract prearithmetic $\boldsymbol{A}_1$ is projective with respect to the wholly extended abstract prearithmetic $\boldsymbol{A}_2$ with the projector g and the coprojector h, for which $gh = 1_{A_2}$, and $(z -_2 x) -_2 y = z -_2 (x +_2 y)$ for any elements x, y and z from the wholly extended abstract prearithmetic $\boldsymbol{A}_2$, then $(b -_1 a) -_1 c = b -_1 (a +_1 c)$ for any elements a and b from the wholly extended abstract prearithmetic $\boldsymbol{A}_1$.*

Proof is left as an exercise. □

Proposition 3.2.21. *If the wholly extended abstract prearithmetic* $\boldsymbol{A}_1$ *is projective with respect to the wholly extended abstract prearithmetic* $\boldsymbol{A}_2$ *with the projector* g *and the coprojector* h, *for which* $gh = 1_{A_2}$, *and* $(z/_2x)/_2y = z/_2(x \circ_2 y)$ *for any elements* x, y *and* z *from the wholly extended abstract prearithmetic* $\boldsymbol{A}_2$, *then* $(b/_1a)/_1c = b/_1(a \circ_1 c)$ *for any elements* a *and* b *from the wholly extended abstract prearithmetic* $\boldsymbol{A}_1$.

Indeed, by definition and initial conditions, we have

$$\begin{aligned}(b/_1a)/_1c &= h(g(h(g(b)/_2g(a))/_2g(c))) = h((g(b)/_2g(a))/_2g(c)) \\ &= h(g(b)/_2(g(a)) \circ_2 g(c))) = h(g(b)/_2g(h(g(a)) \circ_2 g(c)))) \\ &= b/_1(a \circ_1 c).\end{aligned}$$

The next result shows that projectivity of wholly extended abstract prearithmetics is a transitive relation.

Proposition 3.2.22. *If a wholly extended abstract prearithmetics* $\boldsymbol{A}_1 = (A_1; +_1, \circ_1, -_1, /_1, \leq_1)$ *is projective with respect to a wholly extended abstract prearithmetics* $\boldsymbol{A}_2 = (A_2; +_2, \circ_2, -_2, /_2, \leq_2)$, *and the abstract prearithmetic* $\boldsymbol{A}_2$ *is projective with respect to an abstract prearithmetic* $\boldsymbol{A}_3 = (A_3; +_3, \circ_3, \leq_3)$, *then the prearithmetic* $\boldsymbol{A}_1$ *is projective with respect to the prearithmetic* $\boldsymbol{A}_3$.

Proof is similar to the proof of Proposition 2.3.12. □

Proposition 3.2.22 allows proving the following result using the definition of a category (cf. Appendix).

Theorem 3.2.2. *Wholly extended abstract prearithmetics with projectivity relations form the category* **WEAPAP**, *where objects are wholly extended abstract prearithmetics and morphisms are projectivity relations.*

Proof is left as an exercise. □

Lemma 2.3.1 shows that the category **WEAPAP** is a wide subcategory of the category **WEAPAWP**.

Proposition 3.2.23. *If a wholly extended abstract prearithmetic* $\boldsymbol{A}_1 = (A_1; +_1, \circ_1, -_1, /_1, \leq_1)$ *is projective with respect to a wholly extended abstract prearithmetic* $\boldsymbol{A}_2 = (A_2; +_2, \circ_2, -_2, /_2, \leq_2)$ *with the projector* g *and the coprojector* h, *then the coprojector* h *preserves all four operations* *in the image $g(\boldsymbol{A}_1)$ of the prearithmetic $\boldsymbol{A}_1$, i.e., h is a homomorphism of $g(\boldsymbol{A}_1)$.*

Proof: Let us take two elements x and y from the image $g(\boldsymbol{A}_1)$ of the prearithmetic $\boldsymbol{A}_1$. By definition, there are elements a and b in the prearithmetic $\boldsymbol{A}_1$ such that $x = g(a)$ and $y = g(b)$. Then by Definition 2.3.1, we have

$$h(x +_2 y) = h(g(a) +_2 g(b)) = a +_1 b = h(g(a)) +_1 h(g(b)) = h(x) +_1 h(y)$$

and

$$h(x \circ_2 y) = h(g(a) \circ_2 g(b)) = a \circ_1 b = h(g(a)) \circ_1 h(g(b)) = h(x) \circ_1 h(y)$$

because $x = g(a)$ and $y = g(b)$ and the composition hg is equal to $\mathbf{1}_{A_1}$. These equalities mean that h is a homomorphism of $g(\boldsymbol{A}_1)$ because x and y are arbitrary elements from $g(\boldsymbol{A}_1)$.

Proposition is proved. □

Corollary 3.2.2. *If an abstract prearithmetic $\boldsymbol{A}_1 = (A_1; +_1, \circ_1, \leq_1)$ is projective with respect to an abstract prearithmetic $\boldsymbol{A}_2 = (A_2; +_2, \circ_2, \leq_2)$ with the surjective projector g and the coprojector h, then the coprojector h preserves addition and multiplication in the prearithmetic $\boldsymbol{A}_2$, i.e., h is a homomorphism of $\boldsymbol{A}_2$.*

Now let us study a stronger case of projectivity of wholly extended abstract prearithmetics.

Definition 3.2.3. A wholly extended abstract prearithmetic $\boldsymbol{A}_1 = (A_1; +_1, \circ_1, -_1, /_1, \leq_1)$ is called *exactly projective* with respect to a wholly extended abstract prearithmetic $\boldsymbol{A}_2 = (A_2; +_2, \circ_2, -_2, /_2, \leq_2)$ if it is projective with respect to the prearithmetic $\boldsymbol{A}_2$ and $gh = \mathbf{1}_{A_2}$, where $\mathbf{1}_{A_2}$ is the identity mapping of A_2, g: $A_1 \to A_2$ is the *projector* and h: $A_2 \to A_1$ is the *coprojector* for the pair $(\boldsymbol{A}_1, \boldsymbol{A}_2)$.

Definitions imply the following results.

Lemma 3.2.1. *If a wholly extended abstract prearithmetic $\boldsymbol{A}_1 = (A_1; +_1, \circ_1, -_1, /_1, \leq_1)$ is exactly projective with respect to a wholly extended abstract prearithmetic $\boldsymbol{A}_2 = (A_2; +_2, \circ_2, -_2, /_2, \leq_2)$, then the wholly extended abstract prearithmetic $\boldsymbol{A}_1$ is projective with respect to the wholly extended abstract prearithmetic $\boldsymbol{A}_2$.*

Proof is left as an exercise. □

Properties of the category of sets (cf., for example, Herrlich and Strecker, 1973) give us the following result.

Lemma 3.2.2. *If a wholly extended abstract prearithmetic* $\boldsymbol{A}_1 = (A_1; +_1, \circ_1, -_1, /_1, \leq_1)$ *is exactly projective with respect to a wholly extended abstract prearithmetic* $\boldsymbol{A}_2 = (A_2; +_2, \circ_2, -_2, /_2, \leq_2)$, *then the coprojector for the pair* $(\boldsymbol{A}_1, \boldsymbol{A}_2)$ *is an injection and the projector for the pair* $(\boldsymbol{A}_1, \boldsymbol{A}_2)$ *is a projection.*

Proof is left as an exercise. □

Lemmas 2.3.3 and 3.2.2 give us the following result.

Proposition 3.2.24. *If a wholly extended abstract prearithmetic* $\boldsymbol{A}_1 = (A_1; +_1, \circ_1, -_1, /_1, \leq_1)$ *is exactly projective with respect to a wholly extended abstract prearithmetic* $\boldsymbol{A}_2 = (A_2; +_2, \circ_2, -_2, /_2, \leq_2)$, *then both the projector* g *and coprojector* h *for the pair* $(\boldsymbol{A}_1, \boldsymbol{A}_2)$ *are bijections and both sets* A_1 *and* A_2 *are equipollent (cf. Appendix).*

Proof is left as an exercise. □

Corollary 3.2.3. *If an abstract prearithmetic* $\boldsymbol{A}_1 = (A_1; +_1, \circ_1, \leq_1)$ *is exactly projective with respect to an abstract prearithmetic* $\boldsymbol{A}_2 = (A_2; +_2, \circ_2, \leq_2)$ *with the projector* g *and the coprojector* h, *then there are inverse mappings* $g^{-1} = h\colon A_2 \to A_1$ *and* $h^{-1} = g\colon A_1 \to A_2$.

Proposition 3.2.25. *If a wholly extended abstract prearithmetic* $\boldsymbol{A}_1 = (A_1; +_1, \circ_1, -_1, /_1, \leq_1)$ *is exactly projective with respect to a wholly extended abstract prearithmetic* $\boldsymbol{A}_2 = (A_2; +_2, \circ_2, -_2, /_2, \leq_2)$ *with the projector* g *and the coprojector* h, *then the projector* g *preserves addition, subtraction, division and multiplication in the prearithmetic* $\boldsymbol{A}_1$, *i.e.,* g *is a homomorphism.*

Proof: Let us take two elements a and b from the arithmetic A_1. Then, by definition, we obtain the following sequences of equalities:

$$g(a +_1 b) = g(h(g(a) +_2 g(b))) = (gh)(g(a)) +_2 g(b)) = g(a) +_2 g(b),$$

$$g(a \circ_1 b) = g(h(g(a)) \circ_2 g(b))) = (gh)(g(a)) \circ_2 g(b)) = g(a) \circ_2 g(b),$$

$$g(a -_1 b) = g(h(g(a)) -_2 g(b))) = (gh)(g(a)) -_2 g(b)) = g(a) -_2 g(b),$$

$$g(a/_1 b) = g(h(g(a))/_2 g(b))) = (gh)(g(a))/_2 g(b)) = g(a)/_2\, g(b)$$

and the composition gh is equal to $\mathbf{1}_{A_2}$. These equalities mean that g is a homomorphism of $\boldsymbol{A}_1$ as a universal algebra with four operations. In addition, by Lemma 2.2.1, the projector g preserves the order $\leq_1$.

Proposition is proved. □

Corollary 3.2.4. *If a wholly extended abstract prearithmetic $\boldsymbol{A}_1 = (A_1; +_1, \circ_1, -_1, /_1, \leq_1)$ is weakly projective with respect to a wholly extended abstract prearithmetic $\boldsymbol{A}_2 = (A_2; +_2, \circ_2, -_2, /_2, \leq_2)$ with the projector g and the coprojector h while $h = g^{-1}$, then the projector g preserves addition, subtraction, division and multiplication in the prearithmetic $\boldsymbol{A}_1$, i.e., g is a homomorphism.*

In projectivity, the coprojector has similar properties to properties of the projector.

Proposition 3.2.26. *If a wholly extended abstract prearithmetic $\boldsymbol{A}_1 = (A_1; +_1, \circ_1, -_1, \leq_1)$ is projective with respect to a wholly extended abstract prearithmetic $\boldsymbol{A}_2 = (A_2; +_2, \circ_2, -_2, \leq_2)$ with the projector g and the coprojector h, then the coprojector h preserves all four operations in the image $g(\boldsymbol{A}_1)$ of the prearithmetic $\boldsymbol{A}_1$, i.e., h is a homomorphism of $g(\boldsymbol{A}_1)$.*

Proof: Let us take two elements x and y from the image $g(\boldsymbol{A}_1)$ of the partially extended abstract prearithmetic $\boldsymbol{A}_1$. By definition, there are elements a and b in the prearithmetic $\boldsymbol{A}_1$ such that $x = g(a)$ and $y = g(b)$. Then by Definition 3.1.3, we have

$$h(x +_2 y) = h(g(a) +_2 g(b)) = a +_1 b = h(g(a)) +_1 h(g(b)) = h(x) +_1 h(y), \tag{3.7}$$

$$h(x \circ_2 y) = h(g(a) \circ_2 g(b)) = a \circ_1 b = h(g(a)) \circ_1 h(g(b)) = h(x) \circ_1 h(y), \tag{3.8}$$

$$h(x -_2 y) = h(g(a) -_2 g(b)) = a -_1 b = h(g(a)) -_1 h(g(b)) = h(x) -_1 h(y), \tag{3.9}$$

$$h(x /_2 y) = h(g(a) /_2 g(b)) = a /_1 b = h(g(a)) /_1 h(g(b)) = h(x) /_1 h(y) \tag{3.10}$$

and the composition hg is equal to $\mathbf{1}_{A_1}$. These equalities mean that h is a homomorphism of $g(\boldsymbol{A}_1)$.

Proposition is proved. □

As exact projectivity is a kind of projectivity, we have the following result.

Corollary 3.2.5. *If a wholly extended abstract prearithmetic* $\boldsymbol{A}_1 = (A_1; +_1, \circ_1, -_1, /_1, \leq_1)$ *is exactly projective with respect to a wholly extended abstract prearithmetic* $\boldsymbol{A}_2 = (A_2; +_2, \circ_2, -_2, /_2, \leq_2)$ *with the projector* g *and the coprojector* h*, then the coprojector* h *preserves addition, subtraction, division and multiplication in the prearithmetic* $\boldsymbol{A}_2$*, i.e.,* h *is a homomorphism.*

Corollary 3.2.6. *If a wholly extended abstract prearithmetic* $\boldsymbol{A}_1 = (A_1; +_1, \circ_1, -_1, /_1, \leq_1)$ *is weakly projective with respect to a wholly extended abstract prearithmetic* $\boldsymbol{A}_2 = (A_2; +_2, \circ_2, -_2, /_2, \leq_2)$ *with the projector* g *and the coprojector* h *while* $h = g^{-1}$*, then the projector* g *preserves addition, subtraction, division and multiplication in the prearithmetic* $\boldsymbol{A}_1$*, i.e.,* h *is a homomorphism.*

Corollary 3.2.7. *If a wholly extended abstract prearithmetic* $\boldsymbol{A}_1 = (A_1; +_1, \circ_1, -_1, /_1, \leq_1)$ *is weakly projective with respect to a wholly extended abstract prearithmetic* $\boldsymbol{A}_2 = (A_2; +_2, \circ_2, -_2, /_2, \leq_2)$ *with the projector* g *and the coprojector* h *while* $h = g^{-1}$*, then the prearithmetics* $\boldsymbol{A}_1$ *and* $\boldsymbol{A}_2$ *are isomorphic as algebraic systems with four operations and one relation.*

In other words, we have the following result.

Proposition 3.2.27. *An isomorphism* f *of wholly extended abstract prearithmetics defines their exact projectivity with the projector* f *and coprojector* f^{-1}*.*

Proof is left as an exercise. □

Propositions 3.2.24–3.2.26 show that the converse of Proposition 3.2.27 is also true.

Theorem 3.2.3. *A wholly extended abstract prearithmetic* $\boldsymbol{A}_1 = (A_1; +_1, \circ_1, -_1, /_1, \leq_1)$ *is exactly projective with respect to a wholly extended abstract prearithmetic* $\boldsymbol{A}_2 = (A_2; +_2, \circ_2, -_2, /_2, \leq_2)$ *with the projector* g *and the coprojector* h *if and only if the prearithmetics* $\boldsymbol{A}_1$ *and* $\boldsymbol{A}_2$ *are isomorphic as algebraic systems with four operations and one relation.*

Proof: *Sufficiency.* Assume wholly extended abstract prearithmetic $\boldsymbol{A}_1 = (A_1; +_1, \circ_1, -_1, /_1, \leq_1)$ is exactly projective with respect to a wholly extended abstract prearithmetic $\boldsymbol{A}_2 = (A_2; +_2, \circ_2, -_2, /_2, \leq_2)$ with the projector g and the coprojector h. Then, by Proposition 3.2.24, the projector g is one-to-one mapping and the coprojector h is its inverse, i.e., $h = g^{-1}$. By Proposition 3.2.25, the projector g is a homomorphism. By Proposition 3.2.26, the coprojector h is a homomorphism. Consequently, both h and g are isomorphisms (Kurosh, 1963). Thus, $\boldsymbol{A}_1$ and $\boldsymbol{A}_2$ are isomorphic.

Necessity. Suppose that a wholly extended abstract prearithmetic $\boldsymbol{A}_1 = (A_1; +_1, \circ_1, -_1, /_1, \leq_1)$ is isomorphic to a wholly extended abstract prearithmetic $\boldsymbol{A}_2 = (A_2; +_2, \circ_2, -_2, /_2, \leq_2)$ and g: $A_1 \to A_2$ is the isomorphism between them. Then we have

$$g(a +_1 b) = g(a) +_2 g(b),$$

$$g(a \circ_1 b) = g(a) \circ_2 g(b),$$

$$g(a -_1 b) = g(a) -_2 g(b),$$

$$g(a /_1 b) = g(a) /_2 \, g(b).$$

Consequently,

$$a +_1 b = g^{-1}(a +_1 b) = g^{-1}(g(a) +_2 g(b)),$$

$$a \circ_1 b = g^{-1}(a \circ_1 b) = g^{-1}(g(a) \circ_2 g(b)),$$

$$a -_1 b = g^{-1}(a -_1 b) = g^{-1}(g(a) -_2 g(b)),$$

$$a /_1 b = g^{-1}(a /_1 b) = g^{-1}(g(a) /_2 \, g(b)).$$

Besides,

$$a \leq_1 b \text{ only if } g(a) \leq_2 g(b).$$

It means that the wholly extended abstract prearithmetic $\boldsymbol{A}_1 = (A_1; +_1, \circ_1, -_1, /_1, \leq_1)$ is exactly projective with respect to a wholly extended abstract prearithmetic $\boldsymbol{A}_2 = (A_2; +_2, \circ_2, -_2, /_2, \leq_2)$ with the projector g and the coprojector g^{-1}.

Theorem is proved. □

In algebra, isomorphic algebras are considered completely equivalent because they have the same identities (cf., for example, Artin, 1991; Birkhoff and Bartee, 1967; Kurosh, 1963). In arithmetic, the situation is essentially different as the following example demonstrates.

Example 3.2.6. Let us take a real number k and the function $g(x) = x + k$ as the projector and the function $h(x) = x - k$ as the coprojector of the projectivity of a wholly extended abstract prearithmetic $\boldsymbol{B} = (R; \oplus, \otimes, \ominus, \oslash, \leq)$ with respect to the Diophantine arithmetic $\boldsymbol{R} = (R; +, \cdot, -, /, \leq)$ of real numbers, which is also a wholly extended abstract prearithmetic. As $h = g^{-1}$, these prearithmetics are exactly projective with respect to one another and thus, isomorphic.

Let us take $k = 5$ and consider operations in both prearithmetics. Then in the Diophantine arithmetic $\boldsymbol{Z}$, we have

$$2 + 2 = 4.$$

At the same time, in the prearithmetic $\boldsymbol{B}$, we have

$$2 \oplus 2 = ((2 + 5) + (2 + 5)) - 5 = 14 - 5 = 9.$$

We see that addition of the same numbers gives different results in these prearithmetics.

In addition, we have

$$5 \otimes 5 = ((5 + 5) \cdot (5 + 5)) - 5 = 10 \cdot 10 - 5 = 95.$$

While in $\boldsymbol{Z}$, we have

$$5 \cdot 5 = 25.$$

We see that multiplication of the same numbers gives different results in the wholly extended abstract prearithmetics $\boldsymbol{R}$ and $\boldsymbol{B}$.

This prearithmetic $\boldsymbol{B}$ reflects some situations in real life.

Example 3.2.7. Imagine you pay for car insurance \$1,000 per year. To do this you have three options:

(1) To pay the whole amount.
(2) To pay the half of the whole amount two times.
(3) To pay the quarter of the whole amount four times.

For installment of each payment, the insurance company charges \$5. Consequently, the first option demands payment of \$1,005, the second option demands payment of \$1,010, and the third option demands payment of \$1,020. These amounts exactly correspond to calculations in the constructed wholly extended abstract prearithmetic $\boldsymbol{B}$.

Indeed, when you pay the whole amount, it is addition:

$$0 \oplus 1,000 = ((1,000 + 5) + (1,000 + 5)) - 5 = 1,000 + 5 = 1,005.$$

When you pay two halves of the whole amount, it is addition:

$$(0 \oplus 500) \oplus 500 = ((((0 + 5) + (500 + 5)) - 5) + 5) + (500 + 5)) - 5$$
$$= 1,000 + 10 = 1,010.$$

When you pay four quarters of the whole amount, it is addition:

$$(((0 \oplus 250) \oplus 250) \oplus 250) \oplus 250 =$$
$$((((((((((0 + 5) + (250 + 5)) - 5) + 5) + (250 + 5)) - 5) + 5)$$
$$+(250 + 5)) - 5) + 5) + (250 + 5)) - 5 = 1,000 + 20 = 1,020.$$

As it is explained in Section 3.3, the constructed prearithmetic $\boldsymbol{B}$ is a non-Diophantine arithmetic of real numbers. So, the situation with car insurance is properly modeled by a non-Diophantine arithmetic.

As exact projectivity preserves algebraic properties of mathematical structures, we have the following results.

Proposition 3.2.28. *A wholly extended abstract prearithmetic* $\boldsymbol{A} = (A; +, \circ, -, /, \leq)$ *that is exactly projective with respect to an additive group is an additive group.*

Proof is left as an exercise. □

Proposition 3.2.29. *A wholly extended abstract prearithmetic* $\boldsymbol{A} = (A; +, \circ, -, /, \leq)$ *that is exactly projective with respect to a ring is a ring.*

Proof is left as an exercise. □

Proposition 3.2.30. *A wholly extended abstract prearithmetic* $\boldsymbol{A} = (A; +, \circ, -, /, \leq)$ *that is exactly projective with respect to a field is a field.*

Proof is left as an exercise. □

Proposition 3.2.31. *A wholly extended abstract prearithmetic* $\boldsymbol{A} = (A; +, \circ, -, /, \leq)$ *that is exactly projective with respect to an ordered additive group is an ordered additive group.*

Proof is left as an exercise. □

Proposition 3.2.32. *A wholly extended abstract prearithmetic* $\boldsymbol{A} = (A; +, \circ, -, /, \leq)$ *that is exactly projective with respect to an ordered ring is an ordered ring.*

Proof is left as an exercise. □

Proposition 3.2.33. *A wholly extended abstract prearithmetic* $\boldsymbol{A} = (A; +, \circ, -, /, \leq)$ *that is exactly projective with respect to an ordered field is an ordered field.*

Proof is left as an exercise. □

Proposition 3.2.34. *If a wholly extended abstract prearithmetic* $\boldsymbol{A}_1 = (A_1; +_1, \circ_1, -_1, /_1, \leq_1)$ *is exactly projective with respect to a wholly extended abstract prearithmetic* $\boldsymbol{A}_2 = (A_2; +_2, \circ_2, -_2, /_2, \leq_2)$ *and a wholly extended abstract prearithmetic* $\boldsymbol{A}_3 = (A_3; +_3, \circ_3, -_3, /_3, \leq_3)$ *is exactly projective with respect to the wholly extended abstract prearithmetic* $\boldsymbol{A}_2$*, then the wholly extended abstract prearithmetic* $\boldsymbol{A}_1$ *is exactly projective with respect to the wholly extended abstract prearithmetic* $\boldsymbol{A}_3$.

Proof is similar to the proof of Proposition 3.1.38. □

Definition 3.2.4. A wholly extended abstract prearithmetic $\boldsymbol{A}_1 = (A_1; +_1, \circ_1, -_1, /_1, \leq_1)$ is a *subprearithmetic* of a wholly extended abstract prearithmetic $\boldsymbol{A}_2 = (A_2; +_2, \circ_2, -_2, /_2, \leq_2)$ if $A_1 \subseteq A_2$, the operation $+_1$ is the restriction of the operation $+_2$ onto A_1, the operation $-_1$ is the restriction of the operation $-_2$ onto A_1, the operation $\circ_1$ is the restriction of the operation $\circ_2$ onto A_1, the operation$/_1$ is the restriction of the operation $/_2$ onto A_1 and the relation $\leq_1$ is the restriction of the relation $\leq_2$ onto A_1.

Example 3.2.8. The conventional Diophantine arithmetic $\boldsymbol{Q}$ of all rational numbers is a subprearithmetic of the conventional Diophantine arithmetics $\boldsymbol{C}$ of all complex numbers and $\boldsymbol{R}$ of all real numbers.

Proposition 3.2.35. *If a wholly extended abstract prearithmetic* $\boldsymbol{A}_1 = (A_1; +_1, \circ_1, -_1, /_1, \leq_1)$ *is a subprearithmetic of a wholly extended abstract prearithmetic* $\boldsymbol{A}_2 = (A_2; +_2, \circ_2, -_2, /_2, \leq_2)$*, then the wholly extended abstract prearithmetic* $\boldsymbol{A}_1$ *is projective with respect to the wholly extended abstract prearithmetic* $\boldsymbol{A}_2$.

Proof: Let us assume that a wholly extended abstract prearithmetic $\boldsymbol{A}_1 = (A_1; +_1, \circ_1, -_1, /_1, \leq_1)$ is a subprearithmetic of a wholly extended abstract prearithmetic $\boldsymbol{A}_2 = (A_2; +_2, \circ_2, -_2, /_2, \leq_2)$. As A_1 is a subset of

A_2, there is a natural injection $l : A_1 \to A_2$. Then we can take an element a from A_1 and define a mapping $p : A_2 \to A_1$ by the following formula:

$$p(x) = \begin{cases} x & \text{if } x \in A_1, \\ a & \text{if } x \notin A_1. \end{cases}$$

Then we can easily check that the wholly extended abstract prearithmetic $\boldsymbol{A}_1$ is projective with respect to the wholly extended abstract prearithmetic $\boldsymbol{A}_2$ with the projector l and coprojector p. Indeed, for any elements a and b from A_1, we have

$$\begin{aligned} a +_1 b &= a +_2 b = p(l(a) +_2 l(b)), \\ a \circ_1 b &= a \circ_2 b = p(l(a) \circ_2 l(b)), \\ a -_1 b &= a -_2 b = p(l(a) -_2 l(b)), \\ a /_1 b &= a /_2 b = p(l(a) /_2 l(b)), \\ &a \leq_1 b \text{ if and only if } l(a) \leq_2 l(b) \end{aligned}$$

because $a = l(a) = p(l(a))$ and $b = l(b) = p(l(b))$.

In addition, $pl = \mathbf{1}_{A_1}$.

Proposition is proved. □

As projectivity is a special case of weak projectivity (cf. Section 2.3), we have the following result.

Corollary 3.2.8. *If a wholly extended abstract prearithmetic* $\boldsymbol{A}_1 = (A_1; +_1, \circ_1, -_1, /_1, \leq_1)$ *is a subprearithmetic of a wholly extended abstract prearithmetic* $\boldsymbol{A}_2 = (A_2; +_2, \circ_2, -_2, /_2, \leq_2)$*, then the wholly extended abstract prearithmetic* $\boldsymbol{A}_1$ *is weakly projective with respect to the wholly extended abstract prearithmetic* $\boldsymbol{A}_2$.

The property "to be a subprearithmetic" is transitive.

Proposition 3.2.36. *If a wholly extended abstract prearithmetic* $\boldsymbol{A}_1 = (A_1; +_1, \circ_1, -_1, /_1, \leq_1)$ *is a subprearithmetic of a wholly extended abstract prearithmetic* $\boldsymbol{A}_2 = (A_2; +_2, \circ_2, -_2, /_2, \leq_2)$ *and the abstract prearithmetic* $\boldsymbol{A}_2$ *is a subprearithmetic of a wholly extended abstract prearithmetic* $\boldsymbol{A}_3 = (A_3; +_3, \circ_3, -_3, /_3, \leq_3)$ *then the wholly extended abstract prearithmetic* $\boldsymbol{A}_1$ *is a subprearithmetic of the wholly extended abstract prearithmetic* $\boldsymbol{A}_3$.

Indeed, if $A_1 \subseteq A_2$ and $A_2 \subseteq A_3$, the $A_1 \subseteq A_3$ (Abian, 1965). Besides, if $f\colon X \to Y$ is a mapping, $A \subseteq B \subseteq X$, a mapping g is a restriction of f onto B and a mapping h is a restriction of g onto A, then h is a restriction of f onto A. In a similar way, if $R \subseteq X \times X$ is a binary relation, $A \subseteq B \subseteq X$, a binary relation P is a restriction of R onto $B \times B$ and a binary relation Q is a restriction of P onto $A \times A$, then Q is a restriction of R onto $A \times A$. This shows that all four operations and the relation in $\boldsymbol{A}_1$ are restrictions of the corresponding operations and relation in $\boldsymbol{A}_3$, and thus, the wholly extended abstract prearithmetic $\boldsymbol{A}_1$ is a subprearithmetic of the wholly extended abstract prearithmetic $\boldsymbol{A}_3$.

Subprearithmetics inherit weak projectivity.

Proposition 3.2.37. *If a wholly extended abstract prearithmetic* $\boldsymbol{A}_1 = (A_1; +_1, \circ_1, -_1, /_1, \leq_1)$ *is a subprearithmetic of a wholly extended abstract prearithmetic* $\boldsymbol{A}_2 = (A_2; +_2, \circ_2, -_2, /_2, \leq_2)$ *and the wholly extended abstract prearithmetic* $\boldsymbol{A}_2$ *is weakly projective with respect to a wholly extended abstract prearithmetic* $\boldsymbol{A}_3 = (A_3; +_3, \circ_3, -_3, /_3, \leq_3)$, *then the wholly extended abstract prearithmetic* $\boldsymbol{A}_1$ *is weakly projective with respect to the wholly extended abstract prearithmetic* $\boldsymbol{A}_3$.

Proof: Let us suppose that a wholly extended abstract prearithmetic $\boldsymbol{A}_2 = (A_2; +_2, \circ_2, -_2, /_2, \leq_2)$ is weakly projective with respect to a wholly extended abstract prearithmetic $\boldsymbol{A}_3 = (A_3; +_3, \circ_3, -_3, /_3, \leq_3)$ with the projector g and the coprojector h while a wholly extended abstract prearithmetic $\boldsymbol{A}_1 = (A_1; +_1, \circ_1, -_1, /_1, \leq_1)$ is a subprearithmetic of $\boldsymbol{A}_2$. As the set A_1 is a subset of the set A_2, it is possible to take an element d from A_1 and the restriction f of the mapping g on the set A_1 defining the function $t : A_3 \to A_1$ as follows:

$$t(x) = \begin{cases} h(x) & \text{if } h(x) \in A_1, \\ d & \text{if } h(x) \notin A_1. \end{cases}$$

Then the wholly extended abstract prearithmetic $\boldsymbol{A}_1$ is weakly projective with respect to the wholly extended abstract prearithmetic $\boldsymbol{A}_3$ with the projector f and the coprojector t. Indeed, if $a, b \in A_1$, then $h(g(a) +_2 g(b)) \in A_1$ and $h(g(a) \circ_2 g(b)) \in A_1$ because $\boldsymbol{A}_1$ is a subprearithmetic of $\boldsymbol{A}_2$. Consequently, we have

$$a +_1 b = h(g(a) +_2 g(b)) = t(f(a) +_2 f(b)),$$

$$a \circ_1 b = h(g(a) \circ_2 g(b)) = t(f(a) \circ_2 f(b)),$$

$$a -_1 b = h(g(a) -_2 g(b)) = t(f(a) -_2 f(b)),$$

$$a /_1 b = h(g(a) /_2 g(b)) = t(f(a) /_2 f(b)).$$

It means that the wholly extended abstract prearithmetic $\boldsymbol{A}_1$ is weakly projective with respect to the wholly extended abstract prearithmetic $\boldsymbol{A}_3$ with the projector f and the coprojector t.

Proposition is proved. □

Proposition 3.2.38. *If a wholly extended abstract prearithmetic* $\boldsymbol{A}_1 = (A_1; +_1, \circ_1, -_1, /_1, \leq_1)$ *is weakly projective with respect to a wholly extended abstract prearithmetic* $\boldsymbol{A}_2 = (A_2; +_2, \circ_2, -_2, /_2, \leq_2)$ *and the abstract prearithmetic* $\boldsymbol{A}_2$ *is a subprearithmetic of a wholly extended abstract prearithmetic* $\boldsymbol{A}_3 = (A_3; +_3, \circ_3, -_3, /_3, \leq_3)$*, then the abstract prearithmetic* $\boldsymbol{A}_1$ *is weakly projective with respect to the abstract prearithmetic* $\boldsymbol{A}_3$*.*

Proof: Let us suppose that a wholly extended abstract prearithmetic $\boldsymbol{A}_1 = (A_1; +_1, \circ_1, -_1, /_1, \leq_1)$ is weakly projective with respect to a wholly extended abstract prearithmetic $\boldsymbol{A}_2 = (A_2; +_2, \circ_2, -_2, /_2, \leq_2)$ with the projector g and the coprojector h and the wholly extended abstract prearithmetic $\boldsymbol{A}_2$ is a subprearithmetic of a wholly extended abstract prearithmetic $\boldsymbol{A}_3 = (A_3; +_3, \circ_3, -_3, /_3, \leq_3)$. Then by Corollary 3.2.8, the wholly extended abstract prearithmetic $\boldsymbol{A}_2$ is weakly projective with respect to the wholly extended abstract prearithmetic $\boldsymbol{A}_3$. Therefore, by Proposition 3.2.16, the wholly extended abstract prearithmetic $\boldsymbol{A}_1$ is weakly projective with respect to the wholly extended abstract prearithmetic $\boldsymbol{A}_3$.

Proposition is proved. □

Proposition 3.2.39. *If a wholly extended abstract prearithmetic* $\boldsymbol{A}_1 = (A_1; +_1, \circ_1, -_1, /_1, \leq_1)$ *is a subprearithmetic of a wholly extended abstract prearithmetic* $\boldsymbol{A}_2 = (A_2; +_2, \circ_2, -_2, /_2, \leq_2)$ *and the abstract prearithmetic* $\boldsymbol{A}_2$ *is projective with respect to a wholly extended abstract prearithmetic* $\boldsymbol{A}_3 = (A_3; +_3, \circ_3, -_3, /_3, \leq_3)$*, then the prearithmetic* $\boldsymbol{A}_1$ *is projective with respect to the prearithmetic* $\boldsymbol{A}_3$*.*

Proof: Let us suppose that a wholly extended abstract prearithmetic $\boldsymbol{A}_2 = (A_2; +_2, \circ_2, -_2, /_2, \leq_2)$ is projective with respect to a wholly extended abstract prearithmetic $\boldsymbol{A}_3 = (A_3; +_3, \circ_3, -_3, /_3, \leq_3)$ with the projector g and the coprojector h while a wholly extended abstract prearithmetic $\boldsymbol{A}_1 = (A_1; +_1, \circ_1, -_1, /_1, \leq_1)$ is a subprearithmetic of $\boldsymbol{A}_2$. By Proposition 3.2.35, the prearithmetic $\boldsymbol{A}_1$ is projective with respect to the prearithmetic $\boldsymbol{A}_2$ with the projector f and the coprojector t. Then by Proposition 3.2.22, prearithmetic $\boldsymbol{A}_1$ is projective with respect to the prearithmetic $\boldsymbol{A}_3$.

Proposition is proved. □

Proposition 3.2.40. *If a wholly extended abstract prearithmetic* $\boldsymbol{A}_1 = (A_1; +_1, \circ_1, -_1, /_1, \leq_1)$ *is projective with respect to a wholly extended abstract prearithmetic* $\boldsymbol{A}_2 = (A_2; +_2, \circ_2, -_2, /_2, \leq_2)$ *and the wholly extended abstract prearithmetic* $\boldsymbol{A}_2$ *is a subprearithmetic of a wholly extended abstract prearithmetic* $\boldsymbol{A}_3 = (A_3; +_3, \circ_3, -_3, /_3, \leq_3)$*, then the prearithmetic* $\boldsymbol{A}_1$ *is projective with respect to the prearithmetic* $\boldsymbol{A}_3$.

Proof: Let us suppose that a wholly extended abstract prearithmetic $\boldsymbol{A}_1 = (A_1; +_1, \circ_1, -_1, /_1, \leq_1)$ is projective with respect to a wholly extended abstract prearithmetic $\boldsymbol{A}_2 = (A_2; +_2, \circ_2, -_2, /_2, \leq_2)$ with the projector g and the coprojector h and the wholly extended abstract prearithmetic $\boldsymbol{A}_2$ is a subprearithmetic of a wholly extended abstract prearithmetic $\boldsymbol{A}_3 = (A_3; +_3, \circ_3, -_3, /_3, \leq_3)$. Then by Proposition 3.2.35, the wholly extended abstract prearithmetic $\boldsymbol{A}_2$ is projective with respect to the wholly extended abstract prearithmetic $\boldsymbol{A}_3$. Therefore, by Proposition 3.2.22, the wholly extended abstract prearithmetic $\boldsymbol{A}_1$ is projective with respect to the wholly extended abstract prearithmetic $\boldsymbol{A}_3$.

Proposition is proved. □

Remark 3.2.4. For exact projectivity, the statements of Propositions 3.2.37–3.2.40 are not always true.

Obtained properties of wholly extended abstract prearithmetics allow us to introduce and study prearithmetics and arithmetics of real and complex numbers including non-Diophantine arithmetics of real and complex numbers in the following two sections.

3.3. Non-Diophantine Arithmetics of Real Numbers

He who loves practice without theory
is like the sailor who boards ship
without a rudder and compass
and never knows where he may cast.

Leonardo da Vinci

In this section, we study numerical prearithmetics and arithmetics of real numbers in general and Diophantine arithmetics of real numbers, in particular. Numerical prearithmetics and arithmetics of real numbers have various applications in a diversity of areas including physics, economics, psychology, decision making, optimization, dynamical systems, differential equations, chaos theory, marketing, finance, fractal geometry, image analysis, electrical engineering, optimal design of computer systems and media, optimal organization of data processing, dynamic programming, computer science, discrete mathematics, and mathematical logic.

3.3.1. *Real-number prearithmetics*

All that passes for knowledge can be arranged in a hierarchy of degrees of certainty, with arithmetic and the facts of perception at the top.

Bertrand Russell

At first, we consider abstract wholly extended prearithmetics and arithmetics $\boldsymbol{A} = (A; \oplus, \otimes, \ominus, \oslash, \leq_A)$, in which the carrier A is a subset of the set R of all real numbers and which are weakly projective with respect to the (conventional) Diophantine arithmetic $\boldsymbol{R}$. In $\boldsymbol{A}$, symbols $\oplus, \otimes, \ominus$ and $\oslash$ denote binary operations of addition, multiplication, subtraction and division in $\boldsymbol{A}$ and $\leq_A$ is a partial order in the set A. In this setting, the arithmetic $\boldsymbol{R}$ of all real numbers allows defining real-number prearithmetics on arbitrary sets of real numbers.

Definition 3.3.1. A wholly extended abstract prearithmetic $\boldsymbol{A} = (A; \oplus, \otimes, \ominus, \oslash, \leq_A)$ is called a *real-number prearithmetic*, or R-*prearithmetic*, if A is a subset of the set R of all real numbers and the prearithmetic $\boldsymbol{A}$ is weakly projective with respect to the (conventional) Diophantine arithmetic $\boldsymbol{R} = (R; +, \cdot, -, /, \leq)$ of real numbers.

It means that taking the projector g: $A \to R$ and coprojector h: $R \to A$ of $\boldsymbol{A}$ with respect to $\boldsymbol{R}$, operations in the prearithmetic $\boldsymbol{A}$ are expressed through the conventional operations with real numbers in the following way:

$$a \oplus b = h(g(a) + g(b)),$$

$$a \otimes b = h(g(a) \cdot g(b)),$$

$$a \ominus b = h(g(a) - g(b)),$$

$$a \oslash b = h(g(a) \div g(b)).$$

This R-prearithmetic is denoted by $\boldsymbol{R}_{g,h}$.

As we already saw in previous chapters, people in general and mathematicians in particular already utilized some special cases of non-Diophantine arithmetics of real numbers. Here we give two more examples of such utilization.

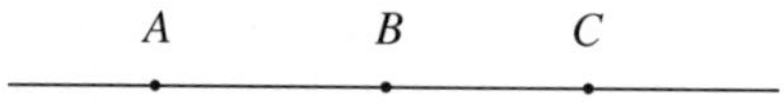

Figure 3.1. A one-dimensional Euclidean space.

Example 3.3.1. The second oldest branch of mathematics is geometry, which extensively uses numbers. In ancient times, Pythagoras tried reducing geometry to arithmetic, which as we already know is the oldest branch of mathematics (cf. Section 1.3). However, he was not able to achieve this because at that time mathematicians knew only natural numbers and positive rational numbers in the form of fractions. Much later, René Descartes (1596–1650) created analytic geometry as an ample reduction of geometry to arithmetic as well as to its further development — algebra (Descartes, 1637).

The basic concept of geometry is distance. Taking points A, B and C in the one-dimensional Euclidean space (Figure 3.1), which is represented by the real line, we know that the distance from A to C is the sum of the distances from A to B and from B to C, i.e.,

$$d(A, C) = d(A, B) + d(B, C).$$

It shows that the length of the longest interval is the sum of the lengths of two other intervals.

The same is true for higher dimensional Euclidean spaces only the length between two points is the sum of distances between projections of these points while this sum is taken in a specific non-Diophantine arithmetic. Note that, on a straight line, points coincide with their projections. So, the sum of distances between the points coincides with the sum of distances between the projections of these points.

For simplicity, let us show this for the distances in a plane, which is, as we know, a two-dimensional Euclidean space. To do this, consider the conventional Diophantine arithmetic $\boldsymbol{R} = (R; +, \cdot, -, /, \leq)$ of all real numbers and the mapping f: $R \to R$ defined as $f(x) = x^2$. This function allows us to build the R-prearithmetic $\boldsymbol{A} = (R^+; \oplus, \otimes, \ominus, \oslash, \leq)$, which is also a non-Diophantine arithmetic of non-negative real numbers, using the function f in the role of the projector and its partially inverse function $f^{-1} : R^+ \to R^+$ defined as $f(x) = x^{\frac{1}{2}}$ in the role of the coprojector as it is done in Sections 2.2 and 2.3. The R-prearithmetic $\boldsymbol{A}$ is exactly projective with respect to the Diophantine arithmetic $\boldsymbol{R}^+$ of non-negative real numbers. In this setting, addition in $\boldsymbol{A}$ is defined by the following formula:

$$a \oplus b = \sqrt{a^2 + b^2} = (a^2 + b^2)^{\frac{1}{2}}.$$

This is exactly the expression of the Pythagorean theorem. Besides, it gives us the formula for the distance between points in the Euclidean plane, i.e.,

$$\begin{aligned} d(A, B) &= \sqrt{d(pr_x\, A, pr_x\, B)^2 + d(pr_y\, A, pr_y\, B)^2} \\ &= d(pr_x\, A, pr_x\, B) \oplus d(pr_y\, A, pr_y\, B). \end{aligned}$$

It means that calculations of distances in the Euclidean plane have been performed in this R-prearithmetic, i.e., distance between two points in the Euclidean plane is the non-Diophantine sum of distances between projections of these points. Besides, the length of the hypotenuse (the longest side) in a right triangle is the non-Diophantine sum of the lengths of the legs (two other sides) of this triangle.

This shows that transition from a one-dimensional Euclidean space to a two-dimensional Euclidean space is performed by utilization of a non-Diophantine arithmetic.

Note that multiplication in the arithmetic $\boldsymbol{A}$ is the same as in the Diophantine arithmetic $\boldsymbol{R}^+$ of non-negative real numbers.

In an n-dimensional Euclidean space E_n, mathematicians use the same non-Diophantine arithmetic $\boldsymbol{A}$ and n-ary addition $\sum$ introduced in Section 2.1, which coincides with the binary addition when $n = 2$, i.e.,

$$d(A, B) = \sqrt{d(pr_1\, A, pr_1 B)^2 + d(pr_2\, A, pr_2\, B)^2 + \cdots + d(pr_n\, A, pr_n\, B)^2}.$$

This shows that transition from a one-dimensional Euclidean space to an n-dimensional Euclidean space is performed for all n by utilization of a non-Diophantine arithmetic.

Example 3.3.2. In a similar way, the well-known operation of taking the absolute value is presented here and considered in the previous example R-prearithmetic $\boldsymbol{A}$ by non-Diophantine multiplication by the number 1. In addition, multiplication of absolute values of two numbers is also non-Diophantine multiplication of these numbers in the considered arithmetic $\boldsymbol{A}$. Indeed, we have the following formulas:

$$a \otimes b = \sqrt{a^2 \cdot b^2} = |a \cdot b|$$

and

$$a \otimes 1 = \sqrt{a^2 \cdot 1^2} = |a|.$$

Note that the considered R-prearithmetic $\boldsymbol{A} = (R; \oplus, \otimes, \ominus, \oslash, \leq)$ is weakly projective with respect to the Diophantine arithmetic $\boldsymbol{R}$ of real numbers but these arithmetics are not isomorphic as algebraic systems and are not exactly projective with respect to one another (cf. Section 2.4) because the inverse function f^{-1} is not defined for negative numbers.

At the same time, the R-prearithmetic $\boldsymbol{A} = (R^+; \oplus, \otimes, \ominus, \oslash, \leq)$ is exactly projective with respect to the Diophantine arithmetic $\boldsymbol{R}^+$ of all non-negative real numbers.

Let us consider an arbitrary set B of real numbers and a mapping $g: B \to R$. In general, the set B is not an arithmetic and g is not necessarily the natural inclusion. For instance, we can take the set Pr of all prime numbers as B and define $g(p_n) = n$ where p_n is the nth prime number, e.g., $p_3 = 5$ and $g(5) = 3$. The mapping g allows obtaining the following result.

Proposition 3.3.1. *It is possible to define on B the structure of an R-prearithmetic $\boldsymbol{B} = (B; \oplus, \otimes, \ominus, \oslash, \leq)$.*

Proof is similar to the proof of Proposition 2.2.1 and uses an arbitrary mapping $h: R \to B$. □

Note that if we want to preserve the initial order of numbers, the mapping g has to be an injection.

Propositions 2.2.4, 2.2.5 and properties of the Diophantine arithmetic $\boldsymbol{R}$ imply the following result.

Proposition 3.3.2. *For any R-prearithmetic $\boldsymbol{A} = (A; \oplus, \otimes, \ominus, \oslash, \leq_A)$, addition $\oplus$ and multiplication $\otimes$ are commutative.*

Indeed, by Propositions 2.1.4 and 2.1.5, operations $\oplus$ and $\otimes$ are commutative because addition and multiplication in the Diophantine arithmetic R are commutative.

Proposition 3.3.3. *For any R-prearithmetic $\boldsymbol{A} = (A; \oplus, \otimes, \ominus, \oslash, \leq_A)$ with an injective projector into R, the order $\leq_A$ in $\boldsymbol{A}$ is linear.*

Indeed, if the projector g is an injection, then by Proposition 2.1.2, the order $\leq_A$ in $\boldsymbol{A}$ is linear because the order $\leq$ in $\boldsymbol{R}$ is linear.

Corollary 3.3.1. *In a wholly extended abstract prearithmetic* $\boldsymbol{A} = (A; +, \circ, -, \div, \leq_A)$, *which is weakly projective with respect to an R-prearithmetic* $\boldsymbol{D} = (D; \oplus, \otimes, \ominus, \oslash, \leq_D)$ *with the injective projector* g, *the relation* $\leq$ *is a well ordering.*

Proposition 2.5.12 implies the following result.

Proposition 3.3.4. *A subprearithmetic* $\boldsymbol{D} = (D; \oplus, \otimes, \ominus, \oslash, \leq_D)$ *of an R-prearithmetic* $\boldsymbol{A} = (A; \oplus, \otimes, \ominus, \oslash, \leq_A)$ *is an R-prearithmetic.*

Corollary 3.3.2. *The Diophantine arithmetic* $\boldsymbol{Q}$ *is an R-prearithmetic.*

Taking an R-prearithmetic $\boldsymbol{A} = (A; \oplus, \otimes, \ominus, \oslash, \leq_A)$ in addition to addition $\oplus$, multiplication $\otimes$, subtraction $\ominus$ and division $\oslash$, we can define other operations in $\boldsymbol{A}$. In particular, we have a system of pairs of n-ary operations for each natural number $n = 2, 3, 4, 5, \ldots$ (cf. Section 2.1):

(1) $\Sigma^{\oplus n}(a_{A,1}, a_{A,2}, \ldots, a_{A,n-1}, a_{A,n}) = h(g(a_{A,1}) + g(a_{A,2}) + \cdots + g(a_{A,n}))$,
(2) $\Pi^{\otimes n}(a_{A,1}, a_{A,2}, \ldots, a_{A,n-1}, a_{A,n}) = h(g(a_{A,1}) \cdot g(a_{A,2}) \cdot \ldots \cdot \quad g(a_{A,n}))$.

For $n = 2$, these operations coincide with addition and multiplication, respectively, i.e., $\Sigma^{\oplus 2} = \oplus$ and $\Pi^{\otimes 2} = \otimes$.

When the number n is not specified, $\Sigma^{\oplus}$ and $\Pi^{\otimes}$ are integral operations in the sense of (Burgin and Karasik, 1976; Burgin, 1982), that is, they can be applied to any finite number of elements from the R-prearithmetic $\boldsymbol{A}$. When $\Sigma^{\oplus}(\Pi^{\otimes})$ is applied to n elements, it coincides with $\Sigma^{\oplus n}(\Pi^{\otimes n})$.

Remark 3.3.1. In the Diophantine arithmetic $\boldsymbol{R}$ of real numbers, operation $\Sigma^{\oplus n}$ coincides with iteration of addition, i.e.,

$$\Sigma^{\oplus n}(a_1, a_2, \ldots, a_{n-1}, a_n) = a_1 + a_2 + \cdots + a_{n-1} + a_n.$$

It is not true for R-prearithmetics and non-Diophantine arithmetics of real and complex numbers in a general case as the following example demonstrates.

Example 3.3.3. Let us take the R-arithmetic $\boldsymbol{A} = (R; \oplus, \otimes, \ominus, \oslash, \leq_A)$ in which operation $\oplus$ is defined by the following rule:

$$x \oplus y = 2x + 2y.$$

Then, we have

$$(1 \oplus 2) \oplus 3 = (2 + 4) \oplus 3 = 2 \cdot 6 + 6 = 18,$$

while

$$1 \oplus (2 \oplus 3) = 2 + 2 \cdot (2 \oplus 3) = 2 + 2 \cdot (4 + 6) = 2 + 20 = 22.$$

Thus, in a general case,

$$(a \oplus b) \oplus c \neq a \oplus (b \oplus c).$$

Proposition 3.3.5. *For any R-prearithmetic $\boldsymbol{A} = (A; \oplus, \otimes, \ominus, \oslash, \leq_A)$, operations $\Sigma^{\oplus n}$ and $\Pi^{\otimes n}$ are commutative for all $n = 2, 3, 4, 5, \ldots$.*

Indeed,

$$\begin{aligned}
&\Sigma^{\oplus n}(a_{A,1}, a_{A,2}, \ldots, a_{A,n-1}, a_{A,n})\\
&\quad = h(g(a_{A,1}) + g(a_{A,2}) + \cdots + g(a_{A,n}))\\
&\quad = h(g(a_{A,j1}) + g(a_{A,j2}) + \cdots + g(a_{A,jn}))\\
&\quad = \Sigma^{\oplus n}(a_{A,j1}, a_{A,j2}, \ldots, a_{A,jn}),
\end{aligned}$$

where $j_1, j_2, j_3, \ldots, j_n$ is an arbitrary permutation of the sequence $1, 2, 3, \ldots, n$.

In a similar way, we have

$$\begin{aligned}
\Pi^{\otimes n}(a_{A,1}, a_{A,2}, \ldots, a_{A,n-1}, a_{A,n}) &= h(g(a_{A,1}) \cdot g(a_{A,2}) \cdot \ldots \cdot g(a_{A,n}))\\
&= h(g(a_{A,j1}) \cdot g(a_{A,j2}) \cdot \ldots \cdot g(a_{A,jn}))\\
&= \Pi^{\otimes n}(a_{A,j1}, a_{A,j2}, \ldots, a_{A,jn}),
\end{aligned}$$

where $j_1, j_2, j_3, \ldots, j_n$ is an arbitrary permutation of the sequence $1, 2, 3, \ldots, n$.

Propositions 2.2.6, 2.2.7 and properties of the conventional arithmetic $\boldsymbol{R}$ imply the following result.

Proposition 3.3.6. *For any R-prearithmetic $\boldsymbol{A} = (A; \oplus, \otimes, \ominus, \oslash, \leq_A)$ with the projector g and the coprojector h, for which $gh = \mathbf{1}_R$, addition $\oplus$ and multiplication $\otimes$ are associative.*

Proof is left as an exercise. □

Proposition 2.2.8 and properties of the conventional arithmetic $\boldsymbol{R}$ imply the following result.

Proposition 3.3.7. *For any R-prearithmetic* $\boldsymbol{A} = (A; \oplus, \otimes, \ominus, \oslash, \leq_A)$ *with the projector* g *and the coprojector* h*, for which* $gh = \mathbf{1}_R$*, multiplication* $\otimes$ *is distributive over addition* $\oplus$*.*

Proof is left as an exercise. □

Propositions 2.2.11, 2.2.12 and properties of the conventional arithmetic $\boldsymbol{R}$ imply the following result.

Proposition 3.3.8. *For any R-prearithmetic* $\boldsymbol{A} = (A; \oplus, \otimes, \ominus, \oslash, \leq_A)$ *with the coprojector* h*, which preserves the order* $\leq$ *in* $\boldsymbol{R}$*, addition* $\oplus$ *and multiplication* $\otimes$ *preserve the order* $\leq_A$*.*

Proof is left as an exercise. □

Corollary 3.3.3. *For any R-prearithmetic* $\boldsymbol{A} = (A; \oplus, \otimes, \ominus, \oslash, \leq_A)$ *with the coprojector* h*, which preserves the order* $\leq$ *in* $\boldsymbol{R}$*, addition* $\oplus$ *is monotone.*

Corollary 3.3.4. *For any R-prearithmetic* $\boldsymbol{A} = (A; \oplus, \otimes, \ominus, \oslash, \leq_A)$ *with the coprojector* h*, which preserves the order* $\leq$ *in* $\boldsymbol{R}$*, multiplication* $\otimes$ *is monotone.*

The property of wholly extended abstract prearithmetics "to be a subprearithmetic" is hereditary as the following result demonstrates.

Proposition 3.3.9. *A subprearithmetic* $\boldsymbol{K}$ *of an R-prearithmetic* $\boldsymbol{A} = (A; \oplus, \otimes, \ominus, \oslash, \leq)$ *is an R-prearithmetic.*

Indeed, by Proposition 2.5.12, the prearithmetic $\boldsymbol{K}$ is weakly projective with respect to the Diophantine arithmetic $\boldsymbol{R}$ and thus, it is an R-prearithmetic.

Corollary 3.3.5. *A subprearithmetic* $\boldsymbol{D}$ *of a subprearithmetic* $\boldsymbol{K}$ *of an R-prearithmetic* $\boldsymbol{A} = (A; \oplus, \otimes, \ominus, \oslash, \leq)$ *is an R-prearithmetic.*

Indeed, by Proposition 2.5.11, the prearithmetic $\boldsymbol{D}$ is a subprearithmetic of the wholly extended R-prearithmetic $\boldsymbol{A}$, and thus, by Proposition 3.3.8, $\boldsymbol{D}$ is an R-prearithmetic.

Proposition 3.3.10. *A wholly extended abstract prearithmetic* $\boldsymbol{A} = (A; +, \circ, -, \div, \leq_A)$*, which is weakly projective with respect to an R-prearithmetic* $\boldsymbol{D} = (D; \oplus, \otimes, \ominus, \oslash, \leq_D)$*, is an R-prearithmetic.*

Indeed, by Proposition 2.2.9, the prearithmetic $\boldsymbol{A}$ is weakly projective with respect to the Diophantine arithmetic $\boldsymbol{R}$ and thus, it is an R-prearithmetic.

Proposition 3.3.11. *If a wholly extended abstract prearithmetic* $\boldsymbol{A}_1 = (A_1; +_1, \circ_1, -_1, \div_1, \leq_1)$ *is weakly projective with respect to a wholly extended abstract prearithmetic* $\boldsymbol{A}_2 = (A_2; +_2, \circ_2, -_2, \div_2, \leq_2)$ *and the abstract prearithmetic* $\boldsymbol{A}_2 = (A_2; +_2, \circ_2, -_2, \div_2, \leq_2)$ *is weakly projective with respect to an R-prearithmetic* $\boldsymbol{A}_3 = (A_3; +_3, \circ_3, -_3, \div_3, \leq_3)$, *then the prearithmetic* $\boldsymbol{A}_1$ *is an R-prearithmetic.*

Indeed, by Proposition 3.3.10, the prearithmetic $\boldsymbol{A}_2$ is an R-prearithmetic and applying once more Proposition 3.3.10, we come to the conclusion that $\boldsymbol{A}_1$ also is an R-prearithmetic.

Proposition 3.3.11 implies the following result.

Theorem 3.3.1. *R-prearithmetics form a category where objects are R-prearithmetics and morphisms are weak projectivity relations.*

Proof is left as an exercise. □

The construction from Definition 2.2.2 allows defining actions of real numbers on elements from a wholly extended R-prearithmetic $\boldsymbol{A} = (A; \oplus, \otimes, \ominus, \oslash, \leq_A)$ with the projector g and the coprojector h.

Definition 3.3.2.

(a) The *additive action from the left* of a real number n on an element a from $\boldsymbol{A}$ is defined as

$$n^+a = h(n + g(a)).$$

(b) The *additive action from the right* of a real number n on an element a from $\boldsymbol{A}$ is defined as

$$a^+n = h(g(a) + n).$$

(c) The *multiplicative action from the left* of a real number n on an element a from $\boldsymbol{A}$ is defined as

$$na = h(n \cdot g(a)).$$

(d) The *multiplicative action from the right* of a real number n on an element a from $\boldsymbol{A}$ is defined as

$$an = h(g(a) \cdot n).$$

Let us study how real numbers act on elements of R-prearithmetics using the properties of actions obtained in Section 2.2 in general wholly extended abstract prearithmetics. As addition of real numbers is commutative, we have the following result.

Proposition 3.3.12. *For any R-prearithmetic* $\boldsymbol{A} = (A; \oplus, \otimes, \ominus, \oslash, \leq_A)$, *additive action from the left coincides with additive action from the right.*

Proof is left as an exercise. □

Proposition 3.3.12 allows skipping indication of the side of additive action in R-prearithmetics.

As multiplication of real numbers is commutative, we have the following result.

Proposition 3.3.13. *For any R-prearithmetic* $\boldsymbol{A} = (A; \oplus, \otimes, \ominus, \oslash, \leq_A)$, *multiplicative action from the left coincides with multiplicative action from the right.*

Proof is left as an exercise. □

Proposition 3.3.13 allows skipping indication of the side of multiplicative action in R-prearithmetics.

Proposition 2.2.18 implies the following result.

Proposition 3.3.14. *For any R-prearithmetic* $\boldsymbol{A} = (A; \oplus, \otimes, \ominus, \oslash, \leq_A)$ *with the coprojector* h, *which preserves addition, multiplicative action is distributive over addition* $\oplus$.

Proof is left as an exercise. □

Proposition 2.2.19 implies the following result.

Proposition 3.3.15. *For any R-prearithmetic* $\boldsymbol{A} = (A; \oplus, \otimes, \ominus, \oslash, \leq_A)$ *with the coprojector* h, *which preserves addition, multiplicative action is distributive over addition* $+$ *in the arithmetic* $\boldsymbol{R}$.

Proof is left as an exercise. □

Proposition 2.2.22 implies the following result.

Proposition 3.3.16. *For any R-prearithmetic* $\boldsymbol{A} = (A; \oplus, \otimes, \ominus, \oslash, \leq_A)$ *with the coprojector* h*, which preserves the order of real numbers, additive action is a monotone operation.*

Proof is left as an exercise. □

Proposition 2.2.23 implies the following result.

Proposition 3.3.17. *For any R-prearithmetic* $\boldsymbol{A} = (A; \oplus, \otimes, \ominus, \oslash, \leq_A)$ *with the coprojector* h: $R \to A$*, which preserves the order of real numbers, multiplicative action is a monotone operation.*

Proof is left as an exercise. □

Let us study identities in R-prearithmetics. Proposition 3.1.25 implies the following result.

Proposition 3.3.18. *For any R-prearithmetic* $\boldsymbol{A} = (A; \oplus, \otimes, \ominus, \oslash, \leq_A)$ *with the projector* g: $A \to R$ *and the coprojector* h: $R \to A$*, for which* $gh = 1_R$*, we have* $(b \ominus a) \ominus c = b \ominus (a \oplus c)$ *for any elements* a, b *and* c *from the prearithmetic* $\boldsymbol{A}_1$.

Proof is left as an exercise. □

Let us consider R-prearithmetics, properties of which are similar to the properties of the Diophantine arithmetic $\boldsymbol{R}$.

Definition 3.3.3. A real-number prearithmetic (R-prearithmetic) $\boldsymbol{A} = (A; \oplus, \otimes, \ominus, \oslash, \leq_A)$ is:

(a) *broad* if $A \approx R$, i.e., there is a bijection $b : A \to R$;
(b) *total* if its projector g is a surjection, i.e., its image is the whole R;
(c) *complete* if $A = R$.

Example 3.3.4. Let us take the interval $I = (-\pi/2, \pi/2)$ and define $g(x) = \operatorname{tg} x$ and $h(x) = \operatorname{arctg} x$. We remind that is the angle between $-\pi/2$ and $\pi/2$ tangent of which is equal to x. We can define the real-number prearithmetic (R-prearithmetic) $\boldsymbol{A} = (A; \oplus, \otimes, \ominus, \oslash, \leq_A)$, which is projective with respect to the conventional arithmetic $\boldsymbol{N} = (N; +, \cdot, \leq)$ of natural numbers with the projector g and the coprojector h. Then the R-prearithmetic $\boldsymbol{A}$ is total.

Note that totality and even completeness do not mean similarity to the arithmetic $\boldsymbol{R}$, i.e., a total or complete R-prearithmetic can be essentially different from the arithmetic $\boldsymbol{R}$.

Example 3.3.5. Let us consider an R-prearithmetic $\boldsymbol{A} = (A; \oplus, \otimes, \ominus, \oslash, \leq_A)$ with the projector g and coprojector h, which are defined as $h(n) = n$ and

$$g(n) = \begin{cases} n+1 & \text{if } n \text{ is an integer number,} \\ n & \text{otherwise.} \end{cases}$$

Operations in $\boldsymbol{A}$ are defined as

$$a \oplus b = h(g(a) + g(b)),$$

$$a \ominus b = h(g(a) - g(b)),$$

$$a \otimes b = h(g(a) \cdot g(b)),$$

$$a \oslash b = h(g(a) \div g(b)).$$

We see that the R-prearithmetic $\boldsymbol{A}$ is total. At the same time, operations in $\boldsymbol{A}$ are essentially different from operations in the arithmetic $\boldsymbol{W}$. Indeed,

$$1 \oplus 1.5 = h(g(1) + g(1.5)) = h(2 + 1.5) = 3.5,$$

$$2 \oplus 2.1 = h(g(2) + g(2.1)) = h(3 + 2.1) = 5.1,$$

while

$$2 \oplus 2 = h(g(2) + g(2)) = h(3 + 3) = 6,$$

$$2.001 \oplus 2.001 = h(g(2) + g(2)) = h(2.001 + 2.001) = 4.002.$$

This shows that sums of very close numbers can be essentially different. Even larger differences can be between products of very close numbers.

Let us study relations between introduced properties of R-prearithmetics.

Proposition 3.3.19. *Any total R-prearithmetic* $\boldsymbol{A} = (A; \oplus, \otimes, \ominus, \oslash, \leq_A)$ *is broad.*

Proof: Let us assume that an R-prearithmetic $\boldsymbol{A} = (A; \oplus, \otimes, \ominus, \oslash, \leq_A)$ is total. By Definition 3.3.3, there is a surjection $g: A \to R$. Then by properties of sets, there is an injection $g: R \to A$.

At the same time, by Definition 3.3.1, A is a subset of the set R. Consequently, by the Cantor–Bernstein–Schröder theorem (cf., for example, Burgin, 2011), sets A and R are equipollent, i.e., there is a bijection

$b: A \to R$. It means that the R-prearithmetic $\boldsymbol{A} = (A; \oplus, \otimes, \ominus, \oslash, \leq_A)$ is broad.

Proposition is proved. □

Constructions of set theory and properties of sets (Kuratowski and Mostowski, 1967) imply the following results.

Proposition 3.3.20. *If a subprearithmetic* $\boldsymbol{K}$ *of a R-prearithmetic* $\boldsymbol{A} = (A; \oplus, \otimes, \ominus, \oslash, \leq_A)$ *is broad, then* $\boldsymbol{A}$ *is also broad.*

Proof is left as an exercise. □

Proposition 3.3.21. *An R-prearithmetic* $\boldsymbol{A} = (A; \oplus, \otimes, \ominus, \oslash, \leq_A)$ *with the projector* g *and the coprojector* h *such that* $gh = \mathbf{1}_R$ *is complete.*

Indeed, if $gh = \mathbf{1}_R$, then g is a surjection (Bucur and Deleanu, 1968) and by Definition 3.3.3, the R-prearithmetic $\boldsymbol{A} = (A; \oplus, \otimes, \leq_A)$ is total.

Definition 3.3.4. A real-number prearithmetic $\boldsymbol{A} = (A; \oplus, \otimes, \ominus, \oslash, \leq_A)$ is called a *weak real-number arithmetic*, or a *weak R-arithmetic*, if it is projective with respect to the conventional (Diophantine) arithmetic $\boldsymbol{R} = (R; +, \cdot, -, /, \leq)$ of all real numbers.

It means that there are two functions — a projector g: $A \to W$ and a coprojector h: $W \to A$ — such that $hg = \mathbf{1}_A$ and for any numbers a and b from A, we have

$$a \oplus b = h(g(a) + g(b)),$$

$$a \otimes b = h(g(a) \cdot g(b)),$$

$$a \leq b \text{ in } \boldsymbol{A} \text{ if and only if } g(a) \leq_2 g(b) \text{ in } \boldsymbol{R}.$$

Lemma 2.3.1 implies the following result.

Lemma 3.3.1. *Any weak R-arithmetic is an R-prearithmetic.*

Proof is left as an exercise. □

Lemma 3.3.1 allows getting many properties of weak R-arithmetics from properties of R-prearithmetics. For instance, Proposition 3.3.2 and Lemma 3.3.1 imply the following result.

Proposition 3.3.22. *Both operations, addition* $\oplus$ *and multiplication* $\otimes$, *are commutative in any weak R-arithmetic* $\boldsymbol{A} = (A; \oplus, \otimes, \ominus, \oslash, \leq_A)$.

Proof is left as an exercise. □

Remark 3.3.2. In some weak R-arithmetics, either addition $\oplus$ or multiplication $\otimes$ or both are not associative.

Proposition 3.3.5 and Lemma 3.3.1 imply the following result.

Proposition 3.3.23. *For any weak R-arithmetic* $\boldsymbol{A} = (A; \oplus, \otimes, \leq_A)$, *operations* $\Sigma^{\oplus n}$ *and* $\Pi^{\otimes n}$ *are commutative for all* $n = 2, 3, 4, 5, \ldots$.

Proof is left as an exercise. □

Let us study relation between the order and operations in weak R-arithmetics.

Proposition 3.3.24. *If, in a weak R-arithmetic* $\boldsymbol{A} = (A; \oplus, \otimes, \leq_A)$, *the coprojector* h *preserves the order* $\leq$, *then the operation* $\oplus$ *preserves the order* $\leq_A$.

Proof: Let us consider a weak R-arithmetic $\boldsymbol{A} = (A; \oplus, \otimes, \ominus, \oslash, \leq_A)$, for which the coprojector h preserves the order $\leq$ and take four elements a, b, c and d from A, which satisfy the inequalities $a \leq_A b$ and $c \leq_A d$. Then by Definition 2.2.1, $g(a) \leq g(b)$ and $g(c) \leq g(d)$. Addition of whole numbers preserves the order $\leq$. Thus, $g(a) + g(c) \leq g(b) + g(d)$. As the coprojector h preserves the order $\leq$, we have

$$a \oplus c \leq h(g(a) + g(c)) \leq h(g(b) + g(d)) = b \oplus d.$$

Proposition is proved. □

Corollary 3.3.6. *For any R-prearithmetic* $\boldsymbol{A} = (A; \oplus, \otimes, \ominus, \oslash, \leq_A)$ *with the coprojector* h, *which preserves the order* $\leq$ *in the Diophantine arithmetic* $\boldsymbol{W}$, *addition* $\oplus$ *is monotone.*

Proposition 3.3.25. *If, in a weak R-arithmetic* $\boldsymbol{A} = (A; \oplus, \otimes, \ominus, \oslash, \leq_A)$, *the coprojector* h *preserves the order* $\leq$, *then the operation* $\otimes$ *preserves the order* $\leq_A$.

Proof: Let us consider a weak R-arithmetic $\boldsymbol{A} = (A; \oplus, \otimes, \ominus, \oslash, \leq_A)$, for which the coprojector h preserves the order $\leq$ and take four elements a, b, c and d from A, which satisfy the inequalities $a \leq_A b$ and $c \leq_A d$. Then by Definition 2.2.1, $g(a) \leq g(b)$ and $g(c) \leq g(d)$. Multiplication of whole numbers preserves the order $\leq$. Thus, $g(a) \cdot g(c) \leq g(b) \cdot g(d)$. As the coprojector h preserves the order $\leq$, we have

$$a \otimes c \leq h(g(a) \cdot g(c)) \leq h(g(b) \cdot g(d)) = b \otimes d.$$

Proposition is proved. □

Corollary 3.3.7. *For any R-prearithmetic* $\boldsymbol{A} = (A; \oplus, \otimes, \ominus, \oslash, \leq_A)$ *with the coprojector* h*, which preserves the order* $\leq$ *in the Diophantine arithmetic* $\boldsymbol{W}$*, multiplication* $\otimes$ *is monotone.*

Proposition 3.3.12 and Lemma 3.3.1 imply the following result.

Proposition 3.3.26. *For any weak R-arithmetic* $\boldsymbol{A} = (A; \oplus, \otimes, \ominus, \oslash, \leq_A)$*, additive action from the left coincides with additive action from the right.*

Proof is left as an exercise. □

Proposition 3.3.26 allows skipping indication of the side of additive action.

Proposition 3.3.13 and Lemma 3.3.1 imply the following result.

Proposition 3.3.27. *For any weak R-arithmetic* $\boldsymbol{A} = (A; \oplus, \otimes, \ominus, \oslash, \leq_A)$*, multiplicative action from the left coincides with multiplicative action from the right.*

Proof is left as an exercise. □

Proposition 3.3.27 allows skipping indication of the side of multiplicative action.

Proposition 3.3.12 and Lemma 3.3.1 imply the following result.

Proposition 3.3.28. *For any weak R-arithmetic* $\boldsymbol{A} = (A; \oplus, \otimes, \ominus, \oslash, \leq_A)$ *with the coprojector* h*, which preserves addition, multiplicative action is distributive over addition* $\oplus$*.*

Proof is left as an exercise. □

Because the natural order in the Diophantine arithmetic $\boldsymbol{R}$ is linear, Proposition 2.3.11 implies the following result.

Proposition 3.3.29. *For any weak R-arithmetic* $\boldsymbol{A} = (A; \oplus, \otimes, \ominus, \oslash, \leq_A)$*, the order* $\leq_A$ *is linear.*

Proof is left as an exercise. □

The property of abstract prearithmetics "to be a weak R-arithmetic" is hereditary as the following result demonstrates.

Proposition 3.3.30. *A subprearithmetic* $\boldsymbol{K}$ *of a weak R-arithmetic* $\boldsymbol{A} = (A; \oplus, \otimes, \ominus, \oslash, \leq_A)$ *is a weak R-arithmetic.*

Indeed, by Proposition 2.5.14, the prearithmetic $\boldsymbol{K}$ is projective with respect to the Diophantine arithmetic $\boldsymbol{R}$ and thus, it is a weak R-arithmetic.

Corollary 3.3.8. *A subprearithmetic $\boldsymbol{D}$ of a subprearithmetic $\boldsymbol{K}$ of a weak R-arithmetic $\boldsymbol{A} = (A; \oplus, \otimes, \ominus, \oslash, \leq_A)$ is a weak R-arithmetic.*

Indeed, by Proposition 2.5.11, the prearithmetic $\boldsymbol{D}$ is a subprearithmetic of a weak R-arithmetic $\boldsymbol{A}$ and thus by Proposition 3.3.30, $\boldsymbol{D}$ is a weak R-arithmetic.

Proposition 3.3.31. *A wholly extended abstract prearithmetic $\boldsymbol{A} = (A; \oplus, \otimes, \ominus, \oslash, \leq_A)$ that is projective with respect to a weak R-arithmetic $\boldsymbol{D} = (D; \oplus, \otimes, \ominus, \oslash, \leq_D)$ is a weak R-arithmetic.*

Indeed, by Proposition 2.3.12, the prearithmetic $\boldsymbol{A}$ is projective with respect to the prearithmetic $\boldsymbol{R}$ and thus, it is a weak R-arithmetic.

Proposition 3.3.32. *If a wholly extended abstract prearithmetic $\boldsymbol{A}_1 = (A_1; \oplus_1, \otimes_1, \ominus_1, \oslash_1, \leq_1)$ is projective with respect to a wholly extended abstract prearithmetic $\boldsymbol{A}_2 = (A_2; \oplus_2, \otimes_2, \ominus_2, \oslash_2, \leq_2)$ and the prearithmetic $\boldsymbol{A}_2$ is projective with respect to a weak R-arithmetic $\boldsymbol{A}_3 = (A_3; \oplus_3, \otimes_3, \ominus_3, \oslash_3, \leq_3)$, then the prearithmetic $\boldsymbol{A}_1$ is a weak R-arithmetic.*

Indeed, by Proposition 3.3.31, the prearithmetic $\boldsymbol{A}_2$ is a weak R-arithmetic and applying once more Proposition 3.3.31, we come to the conclusion that $\boldsymbol{A}_1$ is also a weak R-arithmetic.

Let us explore properties of special elements in weak R-arithmetics.

In a general case, a weak R-arithmetic $\boldsymbol{A}$ contains neither additive zero nor a multiplicative zero. However, additional conditions allow demonstrating that a weak R-arithmetic $\boldsymbol{A}$ has zero, which is denoted by 0_A.

Proposition 3.3.33. *If a weak R-arithmetic $\boldsymbol{A} = (A; \oplus, \otimes, \ominus, \oslash, \leq_A)$ with the projector g contains an element d such that $g(d) = 0$, then for any $n_A \in \boldsymbol{A}$, the equalities $d \oplus n_A = n_A$ and $d \otimes n_A = d$ are true, i.e., d is the zero 0_A in $\boldsymbol{A}$.*

Proof: By definition, we have

$$d \oplus n_A = h(g(d) + g(n_A)) = h(0 + g(n_A)) = h(g(n_A)) = n_A$$

because the weak R-arithmetic $\boldsymbol{A}$ is projective with respect to the arithmetic $\boldsymbol{R}$, i.e., $h(g(n_A)) = n_A$ for any $n_A \in \boldsymbol{A}$, and $g(d) = 0$.

In a similar way, we show that $d \otimes n_A = d$ for any $n_A \in \boldsymbol{A}$. Namely, we have

$$d \otimes n_A = h(g(d) \cdot g(n_A)) = h(0 \cdot g(n_A)) = h(0) = h(g(d)) = d.$$

Proposition is proved. □

Corollary 3.3.9. *An R-prearithmetic $\boldsymbol{A} = (A; \oplus, \otimes, \ominus, \oslash, \leq_A)$ with the projector g and the coprojector h such that $gh = \mathbf{1}_R$ has zero 0_A.*

In a general case, a weak R-arithmetic $\boldsymbol{A}$ does not contain number one, which is denoted by 1_A. Even if $\boldsymbol{A}$ contains 1_A, it is possible that 1_A does not have all properties of the number 1 in the conventional arithmetic $\boldsymbol{W}$. However, additional conditions allow getting some of these properties.

Proposition 3.3.34. *If a weak R-arithmetic $\boldsymbol{A} = (A; \oplus, \otimes, \ominus, \oslash, \leq_A)$ with the projector g contains an element b such that $g(b) = 1$, then for any number $n_A \in \boldsymbol{A}$, the equality $1_A \otimes n_A = n_A$ is true, i.e., 1_A is a multiplicative one in $\boldsymbol{A}$.*

Proof: By definition, we have

$$1_A \otimes n_A = h(g(1_A) \cdot g(n_A)) = h(1 \cdot g(n_A)) = h(g(n_A)) = n_A$$

because the prearithmetic $\boldsymbol{A}$ is projective with respect to the arithmetic $\boldsymbol{R}$, i.e., $h(g(n_A)) = n_A$ for any $n_A \in \boldsymbol{A}$, and $g(1_A) = 1$.

Proposition is proved. □

Corollary 3.3.10. *An R-prearithmetic $\boldsymbol{A} = (A; \oplus, \otimes, \ominus, \oslash, \leq_A)$ with the projector g and the coprojector h such that $gh = \mathbf{1}_R$ has multiplicative one 1_A.*

Definition 3.3.5. An element a from a weak R-arithmetic $\boldsymbol{A} = (A; \oplus, \otimes, \ominus, \oslash, \leq_A)$ is called *positive* (*negative*) if the projector g maps it into a positive (negative) real number.

Properties of the coprojector h imply useful properties of weak R-arithmetics.

Proposition 3.3.35. *In a weak R-arithmetic $\boldsymbol{A} = (A; \oplus, \otimes, \ominus, \oslash, \leq_A)$, if the coprojector h preserves the order $\leq$, then $a \leq a \oplus b$ for any positive* *elements a and b from $\boldsymbol{A}$ and $c \geq c \oplus d$ for any negative elements c and d from $\boldsymbol{A}$.*

Proof: Let us consider a weak R-arithmetic $\boldsymbol{A} = (A; \oplus, \otimes, \ominus, \oslash, \leq_A)$ for which the coprojector h preserves the order $\leq$ and take some positive elements a and b from A. Then by the properties of the Diophantine arithmetic $\boldsymbol{R}$, we have $g(a) \leq g(a) + g(b)$ because $g(a)$ and $g(b)$ are larger than 0. As the coprojector h preserves the order $\leq$ and by Definition 2.3.1, hg is the identity mapping, we have

$$a = h(g(a)) \leq h(g(a) + g(b)) = a \oplus b.$$

For negative elements c and d from $\boldsymbol{A}$, $g(a)$ and $g(b)$ are less than 0. Consequently, $g(c) \geq g(c) + g(d)$ and thus,

$$c = h(g(c)) \geq h(g(c) + g(d)) = c \oplus d.$$

Proposition is proved. □

Corollary 3.3.11. *In a weak R-arithmetic $\boldsymbol{A} = (A; \oplus, \otimes, \ominus, \oslash, \leq_A)$, if the coprojector h preserves the order $\leq$, then $a \leq a \oplus a$ for any positive element a from $\boldsymbol{A}$.*

Proposition 3.3.36. *In a weak R-arithmetic $\boldsymbol{A} = (A; \oplus, \otimes, \ominus, \oslash, \leq_A)$, if the coprojector h preserves the order $\leq$, then $a \leq b \otimes a$ for any positive elements a and b from $\boldsymbol{A}$.*

Proof: Let us consider a weak R-arithmetic $\boldsymbol{A} = (A; \oplus, \otimes, \ominus, \oslash, \leq_A)$, for which the coprojector h preserves the order $\leq$ and take some positive elements a and b from A. Consequently, by the properties of the Diophantine arithmetic $\boldsymbol{R}$, we have $g(a) \leq g(b) \cdot g(a)$ because $g(a)$ and $g(b)$ are larger than 0. As the coprojector h preserves the order $\leq$ and hg is the identity mapping, we have

$$a = h(g(a)) \leq h(g(b) \cdot g(a)) = b \otimes a.$$

Proposition is proved. □

Corollary 3.3.12. *In a weak R-arithmetic $\boldsymbol{A} = (A; \oplus, \otimes, \ominus, \oslash, \leq_A)$, if the coprojector h preserves the order $\leq$, then $a \leq a \otimes a$ for any positive element a from $\boldsymbol{A}$.*

In addition to binary relations $\leq$ and $<$, there are other kinds of order relations in a weak R-arithmetic $\boldsymbol{A} = (A; +, \circ, \leq_A)$:

(1) When $a, b \in \boldsymbol{A}$, then the formula $a \ll b$ means that a is *much less* than b and in this case, b is *much greater* than a, i.e., $b \gg a$. It is defined as

$$a \ll b \text{ if and only if } b \dotplus a = b.$$

(2) The formula $a_{A,1}, a_{A,2}, \ldots, a_{A,n-1} \ll_n b_A$ means that the group $a_{A,1}, a_{A,2}, \ldots, a_{A,n-1}$ is *much less* than b_A and in this case, b_A is *much greater* than the group $a_{A,1}, a_{A,2}, \ldots, a_{A,n-1}$, i.e., $m_A \gg n_A$. It is defined as

$$a_{A,1}, a_{A,2}, \ldots, a_{A,n-1} \ll_n b_A \text{ if and only if}$$

$$\Sigma^{\oplus n}(a_{A,1}, a_{A,2}, \ldots, a_{A,n-1}, b_A) = b_A.$$

(3) The formula $a \lll b$ means that a is *much much less* than b and in this case, b is *much much greater* than a, i.e., $b \ggg a$. It is defined as

$$a \lll b \text{ if and only if } b \circ a = b$$

(4) The formula $a_{A,1}, a_{A,2}, \ldots, a_{A,n-1} \lll_n b_A$ means that the group $a_{A,1}, a_{A,2}, \ldots, a_{A,n-1}$ is *much much less* than b_A and in this case, b_A is *much much greater* than the group $a_{A,1}, a_{A,2}, \ldots, a_{A,n-1}$, i.e., $b_A \ggg_n a_{A,1}, a_{A,2}, \ldots, a_{A,n-1}$. It is defined as

$$a_{A,1}, a_{A,2}, \ldots, a_{A,n-1} \lll_n b_A \text{ if and only if}$$

$$\Pi^{\otimes n}(a_{A,1}, a_{A,2}, \ldots, a_{A,n-1}, b_A) = b_A.$$

As we know, in the Diophantine arithmetic $\boldsymbol{R}$, relation $m \ll n$ is true only when m is 0 and relation $m \lll n$ is true only when m is 1. However, this is not true for weak R-arithmetics in a general case.

Theorem 3.3.2. *There are weak R-arithmetics in which relation $n \ll n$ is valid for any natural number n.*

Proof: Let us consider the set W of all whole numbers and define the following functions:

$$g(r) = \begin{cases} 0 & \textit{when } r = 0, \\ 2^{r^2} & \textit{when } r > 0 \end{cases}$$

and

$$h(x) = \begin{cases} 0 & when\ x \leq 0, \\ [(\log_2 x)^{\frac{1}{2}}] & when\ x > 0. \end{cases}$$

As $hg = \mathbf{1}_W$, by Proposition 3.2.16, we can build an abstract prearithmetic $\boldsymbol{A} = (W; \oplus, \otimes, \ominus, \oslash, \leq)$ that is projective with respect to the Diophantine arithmetic $\boldsymbol{R}$ with the projector g and the coprojector h. Namely, the following operations are defined for natural numbers m and n

$$\begin{aligned} m \oplus n &= h(g(m) + g(n)) = \lfloor (\log_2(2^{m^2} + 2^{n^2}))^{\frac{1}{2}} \rfloor, \\ m \otimes n &= h(g(m) \cdot g(n)) = \lfloor (\log_2(2^{m^2} \cdot 2^{n^2}))^{\frac{1}{2}} \rfloor \\ &= \lfloor (\log_2(2^{m^2+n^2}))^{\frac{1}{2}} \rfloor = \lfloor ((m^2 + n^2)^{\frac{1}{2}} \rfloor. \end{aligned}$$

Besides,

$$0 \oplus n = h(g(0) + g(n)) = \lfloor (\log_2(0 + 2^{n^2}))^{\frac{1}{2}} \rfloor = \lfloor (\log_2(2^{n^2}))^{\frac{1}{2}} \rfloor = n.$$

These definitions give us the following relations:

$$\begin{aligned} n \oplus n &= h(g(n) + g(n)) = \lfloor (\log_2(2^{n^2} + 2^{n^2}))^{\frac{1}{2}} \rfloor \\ &= \lfloor (\log_2(2 \cdot 2^{n^2}))^{\frac{1}{2}} \rfloor = \lfloor (\log_2(2^{n^2+1}))^{\frac{1}{2}} \rfloor \\ &< \lfloor (\log_2(2^{(n+1)^2}))^{\frac{1}{2}} \rfloor = ((n+1)^2)^{\frac{1}{2}} = n + 1. \end{aligned}$$

Thus, $n \oplus n = n$ and $n \ll n$.

Theorem is proved. □

Corollary 3.3.13. *There are weak R-arithmetics in which $n \oplus n = n$ for any natural number n.*

Proposition 3.3.37. *There is a weak R-arithmetic $\boldsymbol{A} = (W; \oplus, \otimes, \leq_A)$, in which $n_A \oplus m_A = \max(n_A, m_A)$ for any natural numbers n and m.*

Proof: The weak R-arithmetic $\boldsymbol{A} = (W; \oplus, \otimes, \leq_A)$, which is constructed in Theorem 3.3.2, has this property. Indeed, by construction, the coprojector h preserves the order $\leq$ in W because functions $\log_2 x$ and $x^{\frac{1}{2}}$ are monotone. Consequently, by Proposition 3.3.3, the addition $\oplus$ preserves the order $\leq_A$ in $\boldsymbol{A}$. As a result, using Corollary 3.3.13 and assuming for certainty that $n_A \leq_A m_A$, we have

$$n_A = n_A \oplus n_A \leq_A n_A \oplus m_A \leq_A m_A \oplus m_A = m_A.$$

If $n_A = m_A$, then

$$n_A \oplus m_A = m_A \oplus m_A = m_A = \max(n_A, m_A).$$

If $n_A <_A m_A$, then

$$m_A = 0_A \oplus m_A \leq_A n_A \oplus m_A \leq_A m_A.$$

Consequently,

$$n_A \oplus m_A = m_A = \max(n_A, m_A).$$

Proposition is proved. □

Proposition 3.3.37 shows that there are weak R-arithmetics in which addition coincides with the operation of taking the maximum of two real numbers. These weak R-arithmetics are efficiently used in idempotent analysis (Maslov, 1987; Maslov and Samborskii, 1992; Maslov and Kolokoltsov, 1997) and tropical analysis (Litvinov, 2007; Speyer and Sturmfels, 2009).

Theorem 3.3.3. *There are weak R-arithmetics in which the relation $n \lll n$ is valid for any number n.*

Proof: The prearithmetic $\boldsymbol{A} = (W; \oplus, \otimes, \leq_A)$ constructed in the proof of Proposition 2.7.29 is a weak R-arithmetic in which $n_A \otimes n_A = n_A$ for any number n_A. By definition, this equality means that $n_A \lll n_A$ for any number n_A.

Theorem is proved. □

Corollary 3.3.14. *There are weak R-arithmetics in which $n \otimes n = n$ for any number n.*

Proposition 3.3.38. *There is a weak R-arithmetic $\boldsymbol{A} = (W; \oplus, \otimes, \leq_A)$, in which $n_A \otimes m_A = \max(n_A, m_A)$ for any numbers n_A and m_A.*

Proof is similar to the proof of Proposition 3.3.37. □

In Section 2.5, we introduced the concept of a subprearithmetic. In the case of weak arithmetics and arithmetics, it gives us the concept of a subarithmetic.

Definition 3.3.6. A weak R-arithmetic $\boldsymbol{A}_1 = (A; +_1, \circ_1, \leq_1)$ is a *subarithmetic* of a weak R-arithmetic $\boldsymbol{A}_2 = (W; +_2, \circ_2, \leq_2)$ if $A_1 \subseteq A_2$, operation $+_1$ is the restriction of the operation $+_2$ onto $g(W)$, the operation $\circ_1$ is the restriction of the operation $\circ_2$ onto $g(W)$, and the relation $\leq_1$ is the restriction of the relation $\leq_2$ onto $g(W)$.

The following lemma shows relations between subprearithmetics and subarithmetics.

Lemma 3.3.2. *An R-prearithmetic* $\boldsymbol{A}_1$ *is a subprearithmetic of a weak R-arithmetic* $\boldsymbol{A}_2$ *if and only if* $\boldsymbol{A}_1$ *is a subarithmetic of* $\boldsymbol{A}_2$.

Proof: *Sufficiency.* Indeed, any weak R-arithmetic is an R-prearithmetic and thus, any subarithmetic is a subprearithmetic.

Necessity. Let us assume that an R-prearithmetic $\boldsymbol{A}_1$ is a subprearithmetic of a weak R-arithmetic $\boldsymbol{A}_2$. Then by Theorem 2.5.1, $\boldsymbol{A}_1$ is projective with respect to $\boldsymbol{A}_2$. By Definition 3.3.4, $\boldsymbol{A}_2$ is projective with respect to the arithmetic $\boldsymbol{R} = (R; +, \cdot, -, /, \leq)$. By Proposition 2.3.12, $\boldsymbol{A}_1$ is projective with respect to the arithmetic $\boldsymbol{R}$. Thus, by Definition 3.3.4, $\boldsymbol{A}_1$ is a weak R-arithmetic and by Definition 3.3.6, $\boldsymbol{A}_1$ is a subarithmetic of $\boldsymbol{A}_2$.

Lemma is proved. □

This result allows getting many properties of subarithmetics from properties of subprearithmetics and weak R-arithmetics. For instance, the property of abstract prearithmetics "to be a weak R-arithmetic" is hereditary as the following result demonstrates.

Corollary 3.3.15. *A subarithmetic* $\boldsymbol{K}$ *of a weak R-arithmetic* $\boldsymbol{A} = (A; \oplus, \otimes, \ominus, \oslash, \leq)$ *is a weak R-arithmetic.*

Corollary 3.3.16. *A subarithmetic* $\boldsymbol{D}$ *of a subarithmetic* $\boldsymbol{K}$ *of a weak R-arithmetic* $\boldsymbol{A} = (A; \oplus, \otimes, \ominus, \oslash, \leq)$ *is a weak R-arithmetic.*

Corollary 3.3.17. *A subprearithmetic* $\boldsymbol{D}$ *of a subprearithmetic* $\boldsymbol{K}$ *of a weak R-arithmetic* $\boldsymbol{A} = (A; \oplus, \otimes, \ominus, \oslash, \leq)$ *is a weak R-arithmetic.*

Proposition 3.3.39. *A wholly extended abstract prearithmetic* $\boldsymbol{A} = (A; +, \circ, -, \div, \leq_A)$, *which is projective with respect to a weak R-arithmetic* $\boldsymbol{D} = (D; \oplus, \otimes, \ominus, \oslash, \leq_D)$, *is a weak R-arithmetic.*

Indeed, by Proposition 2.3.12, the abstract prearithmetic $\boldsymbol{A}$ is projective with respect to the prearithmetic $\boldsymbol{W}$ and thus, it is a weak R-arithmetic.

Proposition 3.3.40. *If a wholly extended abstract prearithmetic* $\boldsymbol{A}_1 = (A_1; +_1, \circ_1, -_1, \div_1, \leq_1)$ *is projective with respect to an abstract prearithmetic* $\boldsymbol{A}_2 = (A_2; +_2, \circ_2, -_2, \div_2, \leq_2)$ *and the wholly extended abstract prearithmetic* $\boldsymbol{A}_2 = (A_2; +_2, \circ_2, \leq_2)$ *is projective with respect to a weak R-arithmetic* $\boldsymbol{A}_3 = (A_3; +_3, \circ_3, -_3, \div_3, \leq_3)$*, then the prearithmetic* $\boldsymbol{A}_1$ *is a weak R-arithmetic.*

Indeed, by Proposition 2.8.3, the prearithmetic $\boldsymbol{A}_2$ is a weak R-arithmetic and applying once more Proposition 3.3.3, we come to the conclusion that $\boldsymbol{A}_1$ is also a weak R-arithmetic.

Proposition 3.3.40 implies the following result.

Theorem 3.3.4. *Weak R-arithmetics form a category* **WAAP** *where objects are weak R-arithmetics and morphisms are projectivity relations.*

Proof is left as an exercise. □

3.3.2. *Projective non-Diophantine arithmetics of real numbers*

Arithmetic is the first of the sciences and the mother of safety.

Charles Victor Cherbuliez

Definition 3.3.7. A weak R-arithmetic $\boldsymbol{A} = (A; \oplus, \otimes, \ominus, \oslash, \leq_A)$ is called a *real-number arithmetic*, or simply, a *projective non-Diophantine arithmetic* of real numbers if it is complete.

Example 3.3.6. Let us consider an R-prearithmetic $\boldsymbol{A} = ([-\pi/2, \pi/2]; \oplus, \otimes, \ominus, \oslash, \leq_A)$ with the projector $g(x) = tg\, x$ and coprojector $h(x) = arctg\, x$. Then g is a projection of $[-\pi/2, \pi/2]$ onto R and $hg = 1_N$. It shows that $\boldsymbol{A}$ is a projective R-arithmetic but not a projective non-Diophantine arithmetic of real numbers.

As projective non-Diophantine arithmetics of real numbers are special cases of weak R-arithmetics, they inherit many properties from weak R-arithmetics and R-prearithmetics. In particular, we have the following results.

Proposition 3.3.41. *For any projective non-Diophantine arithmetic* $\boldsymbol{A} = (A; \oplus, \otimes, \ominus, \oslash, \leq_A)$:

(*a*) *Addition* $\oplus$ *and multiplication* $\otimes$ *are commutative;*
(*b*) *Additive action from the left coincides with additive action from the right;*

(*c*) *Multiplicative action from the left coincides with multiplicative action from the right;*
(*d*) *Order* $\leq_A$ *is linear.*

Proof is left as an exercise. □

Proposition 3.3.42. *For any projective non-Diophantine arithmetic* $\boldsymbol{A} = (A; \oplus, \otimes, \ominus, \oslash, \leq_A)$ *with the coprojector* h, *which preserves the order* $\leq$ *in* $\boldsymbol{R}$, *addition* $\oplus$ *and multiplication* $\otimes$ *preserve the order* $\leq_A$.

Proof is left as an exercise. □

Definition 3.3.8. A wholly extended abstract prearithmetic $\boldsymbol{A} = (R; \oplus, \otimes, \ominus, \oslash, \leq_A)$ is called: (a) An IR-*arithmetic*, or a *functional non-Diophantine arithmetic* of real numbers, if A is exactly projective with respect to the (conventional) Diophantine arithmetic $\boldsymbol{R} = (R; +, \cdot, -, \div, \leq)$ of real numbers.

(b) A *positive* IR-*arithmetic*, if A is exactly projective with respect to the (conventional) Diophantine arithmetic $\boldsymbol{R}^+ = (R^+; +, \cdot, -, \div, \leq)$ of non-negative real numbers, in which subtraction is a partial operation.

IR-arithmetics form a distinct class of non-Diophantine arithmetics of real numbers.

Positive IR-arithmetics form a distinct class of non-Diophantine arithmetics of non-negative real numbers.

IR-arithmetics have many properties similar to properties of the Diophantine arithmetic $\boldsymbol{R}$ of real numbers. For instance, Theorem 3.2.3 implies the following results.

Theorem 3.3.5.

(*a*) *Any IR-arithmetic* $\boldsymbol{A} = (R; \oplus_A, \otimes_A, \ominus_A, \oslash_A, \leq_A)$ *is isomorphic to the Diophantine arithmetic* $\boldsymbol{R}$.
(*b*) *Any positive IR-arithmetic* $\boldsymbol{A} = (R; \oplus_A, \otimes_A, \ominus_A, \oslash_A, \leq_A)$ *is isomorphic to the Diophantine arithmetic* $\boldsymbol{R}^+$.

Proof is left as an exercise. □

This result allows describing many properties of IR-arithmetics.

Corollary 3.3.18. *Any IR-arithmetic* $\boldsymbol{A}$ *has the following properties:*

(*a*) *Addition is commutative;*
(*b*) *Multiplication is commutative;*
(*c*) *Addition is associative;*

(*d*) *Multiplication is associative*;
(*e*) *Multiplication is distributive with respect to addition*;
(*f*) *Multiplication is distributive with respect to subtraction*;
(*g*) ***A*** *has the additive and multiplicative zero* 0_A;
(*h*) ***A*** *has the multiplicative one* 1_A;
(*i*) ***A*** *has continuum of elements.*

Corollary 3.3.19. *Any positive IR-arithmetic* ***A*** *has the following properties*:

(*a*) *Addition is commutative*;
(*b*) *Multiplication is commutative*;
(*c*) *Addition is associative*;
(*d*) *Multiplication is associative*;
(*e*) *Multiplication is distributive with respect to addition*;
(*f*) ***A*** *has the additive and multiplicative zero* 0_A, *which is the least element in* ***A***;
(*g*) ***A*** *has the multiplicative one* 1_A;
(*h*) ***A*** *has continuum of elements.*

Corollary 3.3.20. *Any IR-arithmetic* ***A*** *is an ordered field.*

Definition 3.3.9. The isomorphism f between an IR-arithmetic ***A*** and ***R*** is called the *functional parameter* of ***A***.

It is often denoted as $f_A : A \to R$.

Often the carrier A of an IR-arithmetic ***A*** is a set of real numbers. However, in general, the carrier A of an IR-arithmetic ***A*** can be any set for which a one-to-one mapping (bijection) $f_A : A \to R$ exists. This bijection determines a one-to-one naming of the elements from A by real numbers (Burgin, 2011). Taking this into account, we can denote the element from A that is mapped onto the real number r by r_A.

Example 3.3.7. Let us consider the interval $(-\frac{1}{2}\pi, \frac{1}{2}\pi)$ and its mapping $f(x) = \operatorname{tg} x$. This mapping determines that the IR-arithmetic ***A*** is with the interval $(-\frac{1}{2}\pi, \frac{1}{2}\pi)$ as its carrier (domain) and the functional parameter tg x.

An important operation in non-Diophantine arithmetics is exponent or power. For whole powers, it is constructed using multiplication in the same way as it is done in Diophantine arithmetics. Namely, in an IR-arithmetic (non-Diophantine arithmetic of real numbers) $\boldsymbol{A} = (A; \oplus_A, \otimes_A, \ominus_A, \oslash_A, \leq_A)$, we have

$$x^{n_A} = \underbrace{x \otimes_A x \otimes_A x \otimes_A \cdots \otimes_A x \otimes_A x}_{n}.$$

Note that exponents (powers) are efficiently used in the calculus in mathematics and in a variety of its applications, in particular, in physics.

Lemma 3.3.3. *For an arbitrary natural number n, the power $x \mapsto x^{n_A}$ defines a monomorphism with respect to multiplication of the non-Diophantine arithmetic $\boldsymbol{A}$ into itself, i.e., the following identity is true:*

$$x^{n_A} \otimes_A z^{n_A} = (x \otimes_A z)^{n_A}.$$

In the same way as in the Diophantine arithmetic $\boldsymbol{R}$, we define roots in the non-Diophantine arithmetic $\boldsymbol{A} = (A; \oplus_A, \otimes_A, \ominus_A, \oslash_A, \leq_A)$.

Definition 3.3.10. A number z is a root of power n of element $x \in A$ if $z^{n_A} = x$. It is denoted by $\sqrt[n_A]{x}$.

As in the Diophantine arithmetic $\boldsymbol{R}$, we have the following result.

Proposition 3.3.43. *The power is inverse operation to the root, i.e., we have the following identities*:

$$\left(\sqrt[n_A]{x}\right)^{n_A} = x,$$
$$\sqrt[n_A]{x^{n_A}} = x.$$

In addition to conventional operations, non-Diophantine arithmetics of real numbers have unusual operations.

3.3.3. *Functional powers in non-Diophantine arithmetics of real numbers*

The effect of a single piece of apparatus given to one man is also additive only, but when a group of men are cooperating, as distinct from merely operating, their work raises with some higher power of the number than the first power.

Willis R. Whitney

Let us take two sets A and B and a mapping h: $A \to B$ which can be described by a convergent power series. When A and B are equipped with basic arithmetical operations, we construct *power* or *exponent* of elements from one arithmetic with respect to another arithmetic. To do this, let us consider two bijections $f_A : A \to R, f_B : B \to R$, and $f = f_B^{-1} \circ f_A$. For the mapping f: $A \to B$, we use the following notation $f(x) = x^{1_{AB}}$ and call $x^{1_{AB}}$ the first B-valued functional power of the element $x \in A$ with respect to functions f_A and f_B. The notation reflects utilized sets (or spaces as often they have additional structures) A and B as well as the degree of the exponent.

Remark 3.3.3. It is possible to build functional powers in a more general case. Indeed, taking two mappings g: $A \to C$ and h: $C \to B$, we can define the sequential composition of these functions $f = h \circ g$ and $x^{1_{gh}}$ as the first B-valued functional power of the element $x \in A$ with respect to functions g and h.

This approach allows defining functional powers not only for non-Diophantine arithmetics of real numbers as it is done above but actually for any abstract prearithmetic.

However, here we consider only powers $x^{1_{AB}}$ for arbitrary bijections f_A and f_B and study their properties.

Lemma 3.3.4. *For any $x \in A$, we have*

$$f_A(x) = f_B(x^{1_{AB}}).$$

Indeed,

$$f_B(x^{1_{AB}}) = f_B(f_B^{-1} \circ f_A(x)) = f_B(f_B^{-1}(f_A(x)) = f_A(x)$$

because f_B is a bijection.

Let us consider three bijections $f_A : A \to R, f_B : B \to R$ and $f_C : C \to R$. They define functional powers $x^{1_{AB}}, y^{1_{BC}}$ and $x^{1_{AC}}$ for elements $x \in A$ and $y \in B$.

Proposition 3.3.44. *The following identities are true for the functional powers:*

$$(x^{1_{AB}})^{1_{BC}} = x^{1_{AC}},$$

$$(x^{1_{AB}})^{1_{BA}} = x^{1_{AA}} = x.$$

Proof: As f_A, f_B and f_C are bijections, we have

$$(x^{1_{AB}})^{1_{BC}} = (f_C^{-1} \circ f_B)(f_B^{-1} \circ f_A(x)) = f_C^{-1} \circ f_B \circ f_B^{-1} \circ f_A(x)$$
$$= f_C^{-1} \circ f_A(x) = x^{1_{AC}}.$$

In the same way, we obtain

$$(x^{1_{AB}})^{1_{BA}} = (f_A^{-1} \circ f_B)(f_B^{-1} \circ f_A(x)) = f_A^{-1} \circ f_B \circ f_B^{-1} \circ f_A(x)$$
$$= f_A^{-1} \circ f_A(x) = x.$$

Proposition is proved. □

Now let us assume that A and B are carriers while $f_A : A \to R$ and $f_B : B \to R$ are functional parameters of IR-arithmetics $\boldsymbol{A} = (A; \oplus_A, \otimes_A, \ominus_A, \oslash_A, \leq_A)$ and $\boldsymbol{B} = (B; \oplus_B, \otimes_B, \ominus_B, \oslash_B, \leq_B)$, respectively.

Proposition 3.3.45. *The first functional power $x \mapsto x^{1_{AB}}$ defines an isomorphism between the non-Diophantine arithmetics $\boldsymbol{A}$ and $\boldsymbol{B}$, i.e., the following identities are true:*

$$(x \oplus_A z)^{1_{AB}} = x^{1_{AB}} \oplus_B z^{1_{AB}},$$
$$(x \otimes_A z)^{1_{AB}} = x^{1_{AB}} \otimes_B z^{1_{AB}},$$
$$(x \ominus_{Az})^{1_{AB}} = x^{1_{AB}} \ominus_B z^{1_{AB}},$$
$$(x \oslash_A z)^{1_{AB}} = x^{1_{AB}} \oslash_B z^{1_{AB}}.$$

Proof: The inverse of an isomorphism is an isomorphism and the composition of two isomorphisms is an isomorphism (cf., for example, Kurosh, 1963). Consequently, the mapping $f = f_B^{-1} \circ f_A$, which defines the first functional power, is an isomorphism.

Proposition is proved. □

Corollary 3.3.21. *The following equalities are true:*

$$0_A^{1_{AB}} = 0_B,$$
$$1_A^{1_{AB}} = 1_B.$$

To define functional power n of an element x from A, we can use two constructions. In one of them, we take nth power

$$\underbrace{x \otimes_A x \otimes_A x \otimes_A \cdots \otimes_A x \otimes_A x}_{n} = x^{n_A}$$

of x in $\boldsymbol{A}$ and apply to it the first functional power obtaining

$$(x^{n_A})^{1_{AB}} = x^{n_{AB}}.$$

Another way to define functional power n of an element x from A is at first to take the first functional power $x^{1_{AB}}$ of the element x and then take nth power of $x^{1_{AB}}$ in $\boldsymbol{B}$ obtaining

$$\underbrace{x^{1_{AB}} \otimes_B x^{1_{AB}} \otimes_B x^{1_{AB}} \otimes_B \cdots \otimes_B x^{1_{AB}} \otimes_B x^{1_{AB}}}_{n} = {}^{n_{AB}}x.$$

As by Proposition 3.3.2, the first functional power is an isomorphism, we have the following result.

Lemma 3.3.5. *For an arbitrary natural number n and any $x \in A$, we have*

$$x^{n_{AB}} = {}^{n_{AB}}x.$$

Proof is left as an exercise. □

Lemma 3.3.6. *For an arbitrary natural number n, the functional power $x \mapsto x^{n_A}$ defines a monomorphism with respect to multiplication of the non-Diophantine arithmetic $\boldsymbol{A}$ into the non-Diophantine arithmetic $\boldsymbol{B}$, i.e., the following identity is true:*

$$x^{n_{AB}} \otimes_B z^{n_{AB}} = (x \otimes_A z)^{n_{AB}}.$$

Proof is left as an exercise. □

It is also possible to define negative powers:

$$x^{(-1)_{AB}} = (1_A \oslash_A x)^{1_{AB}},$$

$$x^{(-n)_{XB}} = \underbrace{(1_A \oslash_A x) \otimes_A \cdots \otimes_A (1_A \oslash_A x)}_{n}.$$

Remark 3.3.4. It is also possible to define functional roots and fractional exponents (powers).

Let consider properties of functional power.

Proposition 3.3.46. *Functional power satisfies the same identities as the power (exponent) in the Diophantine arithmetic* $\boldsymbol{R}$*, in particular, for arbitrary natural numbers* m *and* n*, we have*

$$\begin{aligned} 0_A^{1_{AB}} &= 0_B, \\ 1_A^{1_{AB}} &= 1_B, \\ x^{n_{AB}} \otimes_A x^{m_{AB}} &= x^{(n+m)_{AB}}, \\ (x^{n_{AB}})^{m_{BC}} &= (x^{m_{AB}})^{n_{BC}} = x^{(n \cdot m)_{AC}}, \\ x^{(-n)_{AB}} &= (1_B \oslash_B x^{n_{AB}}). \end{aligned}$$

Proof is left as an exercise. □

Having defined and studied prearithmetics and arithmetics of real numbers, now we can develop a theory of prearithmetics and arithmetics of complex numbers in general and non-Diophantine arithmetics of complex numbers in particular.

3.4. Non-Diophantine Arithmetics of Complex Numbers

For every complex problem, there is a solution
that is simple, neat, and wrong.
H. L. Mencken

We build arithmetics and prearithmetics of complex numbers using three approaches: functional, operational and stratified. In the *functional approach*, a prearithmetic or a non-Diophantine arithmetic of complex numbers is constructed using projectivity with respect to the Diophantine arithmetic $\boldsymbol{C}$ of complex numbers. In the *operational approach*, a prearithmetic or a non-Diophantine arithmetic of complex numbers is composed as an extension of a non-Diophantine arithmetic or a prearithmetic of real numbers. In the *stratified approach*, prearithmetics or a non-Diophantine arithmetic of complex numbers is composed as an extension of a pair of prearithmetics or non-Diophantine arithmetics of real numbers.

3.4.1. *Functional arithmetics and prearithmetics of complex numbers*

He who is ignorant of the art of arithmetic is but half a man.

Charles XII

At first, we build functional arithmetics and prearithmetics of complex numbers taking abstract wholly extended prearithmetics and arithmetics $\boldsymbol{A} = (A; \oplus, \otimes, \ominus, \oslash, \leq_A)$, in which the carrier A is a subset of the set C of all complex numbers and which are weakly projective with respect to the (conventional) Diophantine arithmetic $\boldsymbol{C}$. Here $\oplus$, $\otimes$, $\ominus$, and $\oslash$ are binary operations of addition, multiplication, subtraction and division in $\boldsymbol{A}$ and $\leq_A$ is a partial order in the set A. In this setting, the arithmetic $\boldsymbol{C}$ of all complex numbers allows defining complex-number prearithmetics on arbitrary sets of complex numbers.

Definition 3.4.1. A wholly extended abstract prearithmetic $\boldsymbol{A} = (A; \oplus, \otimes, \ominus, \oslash, \leq_A)$ is called a *complex-number prearithmetic*, or *functional* C-*prearithmetic*, if A is a subset of the set C of all complex numbers and the prearithmetic $\boldsymbol{A}$ is weakly projective with respect to the (conventional) Diophantine arithmetic $\boldsymbol{C} = (C; +, \cdot, -, \div, \leq)$ of complex numbers.

It means that operations in $\boldsymbol{A}$ are expressed through the conventional operations in the following way:

$$a \oplus b = h(g(a) + g(b)), \tag{3.11}$$

$$a \otimes b = h(g(a) \cdot g(b)), \tag{3.12}$$

$$a \ominus b = h(g(a) - g(b)), \tag{3.13}$$

$$a \oslash b = h(g(a) \div g(b)). \tag{3.14}$$

This functional C-prearithmetic is denoted by $\boldsymbol{C}_{g,h}$.

Let us consider an arbitrary set B and a mapping g: $B \to C$.

Proposition 3.4.1. *It is possible to define on B the structure of a functional C-prearithmetic $\boldsymbol{B} = (B; \oplus, \otimes, \ominus, \oslash, \leq_A)$.*

Proof is similar to the proof of Proposition 2.2.1 and uses an arbitrary mapping h: $C \to B$. □

Note that if we want to preserve the initial order of numbers, the mapping g has to be an injection.

Propositions 2.2.4, 2.2.5 and properties of the Diophantine arithmetic $\boldsymbol{C}$ imply the following result.

Proposition 3.4.2. *For any functional C-prearithmetic* $\boldsymbol{A} = (A; \oplus, \otimes, \ominus, \oslash, \leq_A)$*, addition* $\oplus$ *and multiplication* $\otimes$ *are commutative.*

Indeed, by Propositions 2.1.4 and 2.1.5, operations $\oplus$ and $\otimes$ are commutative because addition and multiplication in the Diophantine arithmetic $\boldsymbol{C}$ are commutative.

If a functional C-prearithmetic $\boldsymbol{A}$ is projective with respect to the Diophantine arithmetic $\boldsymbol{C}$, then it has more common properties with $\boldsymbol{C}$.

Proposition 3.4.3. *If a functional C-prearithmetic* $\boldsymbol{A} = (A; \oplus, \otimes, \ominus, \oslash, \leq_A)$ *is projective with respect to the Diophantine arithmetic* $\boldsymbol{C}$ *with the projector* g *and the coprojector* h*, and there is an element* e *such that* $g(e) = i$*, then any element* a *in* $\boldsymbol{A}$ *has a representation* $a = c \oplus (e \otimes d)$ *where* $c, d \in h(\boldsymbol{R})$ *and formulas for addition, subtraction and multiplication in* $\boldsymbol{A}$ *are similar to formulas of these operations in* $\boldsymbol{C}$.

Proof: Let us take an element a from the prearithmetic $\boldsymbol{A}$. Then for some real numbers x and y, we have

$$g(a) = x + iy.$$

Here, as usually, $i = \sqrt{-1}$.

By Proposition 3.2.22, h is a homomorphism of $g(\boldsymbol{A})$. Consequently, we have

$$\begin{aligned} a &= h(g(a)) = h(x + iy) = h(x) + h(iy) = h(x) \oplus (h(i) \otimes h(y)) \\ &= h(x) \oplus (e \otimes h(y)) \end{aligned}$$

because the composition hg is equal to $\mathbf{1}_A$. In this representation of the element a, elements $c = h(x)$ and $d = h(y)$ belong to the inverse image $h(\boldsymbol{R})$ of the real number subarithmetic $\boldsymbol{R}$ of the Diophantine arithmetic $\boldsymbol{C}$ and play the role of real numbers in $\boldsymbol{A}$, while e plays the role of the imaginary number i. We see that the representation $a = h(x) \oplus (e \otimes h(y))$ of elements from $\boldsymbol{A}$ is similar to the vector representation of complex numbers.

We remind that when complex numbers are represented in the form $x + iy$, the following formulas describe operations with them:

$$(x + iy) + (u + iz) = (x + u) + i(y + z),$$

$$(x + iy) - (u + iz) = (x - u) + i(y - z),$$

$$(x + iy) \cdot (u + iz)) = (xu - yz) + i(xz + yu).$$

Now let us look how this representation $a = h(x) \oplus (e \otimes h(y))$ behaves in operations of addition, subtraction and multiplication of two arbitrary elements $a = h(x) \oplus (e \otimes h(y)) = c \oplus (e \otimes d)$ and $q = h(u) \oplus (e \otimes h(z)) = r \oplus (e \otimes p)$ from $\boldsymbol{A}$ taking into account that h is a homomorphism and $g(q) = u + iz$ for some real numbers u and z.

For addition, we have

$$\begin{aligned} a \oplus q &= h(g(a) + g(q)) = h((x + iy) + (u + iz)) = h((x + u) + i(y + z)) \\ &= h(x + u) \oplus h(i(y + z)) = [h(x) \oplus h(u)] \oplus [h(i) \otimes h(y + z)] \\ &= [h(x) \oplus h(u)] \oplus [h(g(e)) \otimes (h(y) \oplus h(z))] \\ &= [h(x) \oplus h(u)] \oplus [e \otimes (h(y) \oplus h(z))] = [c \oplus r] \oplus [e \otimes (d \oplus p)]. \end{aligned}$$

At the same time,

$$\begin{aligned} a \oplus q &= [h(x) \oplus (e \otimes h(y))] \oplus [h(u) \oplus (e \otimes h(z))] \\ &= [c \oplus (e \otimes d)] \oplus [r \oplus (e \otimes p)]. \end{aligned}$$

Thus, we obtain

$$[c \oplus (c \otimes d)] \oplus [r \oplus (e \otimes p)] = [c \oplus r] \oplus [e \otimes (d \oplus p)].$$

This shows that addition in $\boldsymbol{A}$ is performed componentwise similar to addition in $\boldsymbol{C}$.

For subtraction, we have

$$\begin{aligned} a \ominus q &= h(g(a) - g(q)) = h((x + iy) - (u + iz)) = h((x - u) + i(y - z)) \\ &= h(x - u) \oplus h(i(y - z)) \\ &= [h(x) \ominus h(u)] \oplus [h(i) \otimes h(y - z)] \\ &= [h(x) \ominus h(u)] \oplus [h(g(e)) \otimes (h(y) \ominus h(z))] \\ &= [h(x) \ominus h(u)] \oplus [e \otimes (h(y) \ominus h(z))] = [c \ominus r] \oplus [e \otimes (d \ominus p)]. \end{aligned}$$

At the same time,

$$\begin{aligned} a \ominus q &= [h(x) \oplus (e \otimes h(y))] \ominus [h(u) \oplus (e \otimes h(z))] \\ &= [c \oplus (e \otimes d)] \ominus [r \oplus (e \otimes p)]. \end{aligned}$$

Thus, we obtain

$$[c \oplus (e \otimes d)] \ominus [r \oplus (e \otimes p)] = [c \ominus r] \oplus [e \otimes (d \ominus p)].$$

This shows that subtraction in $\boldsymbol{A}$ is performed componentwise similar to addition in $\boldsymbol{C}$.

For multiplication, we have

$$\begin{aligned} a \otimes q &= h(g(a) \cdot g(q)) = h((x + iy) \cdot (u + iz)) = h((xu - yz) + i(xz + yu)) \\ &= h(xu - yz) \oplus h(i(xz + yu)) = [h(xu) \ominus h(yz)] \\ &\quad \oplus [h(i) \otimes h(xz + yu))] \\ &= [(h(x) \otimes h(u)) \ominus (h(y) \otimes h(z))] \oplus [e \otimes (h(xz) \oplus h(yu)))] \\ &= [(h(x) \otimes h(u)) \ominus (h(y) \otimes h(z))] \oplus [e \otimes ((h(x) \otimes h(z)) \\ &\quad \oplus (h(y) \otimes h(u)))] \\ &= [(c \otimes r) \ominus (d \otimes p)] \oplus [e \otimes ((c \otimes p) \oplus (d \otimes r))]. \end{aligned}$$

At the same time,

$$\begin{aligned} a \otimes q &= [h(x) \oplus (e \otimes h(y))] \otimes [h(u) \oplus (e \otimes h(z))] \\ &= [c \oplus (e \otimes d)] \otimes [r \oplus (e \otimes p)]. \end{aligned}$$

Thus, we obtain

$$[c \oplus (e \otimes d)] \otimes [r \oplus (e \otimes p)] = [(c \otimes r) \ominus (d \otimes p)] \oplus [e \otimes ((c \otimes p) \oplus (d \otimes r))].$$

This shows that multiplication in A is performed componentwise similar to addition in $\boldsymbol{C}$.

Proposition is proved.

Additional conditions on mappings g and h allow obtaining additional similarities between functional C-prearithmetics and the Diophantine arithmetic $\boldsymbol{C}$.

Proposition 3.4.4. *If a functional C-prearithmetic* $\boldsymbol{A} = (A; \oplus, \otimes, \ominus, \oslash, \leq_A)$ *is projective with respect to the Diophantine arithmetic* $\boldsymbol{C}$ *with* *the projector g and the coprojector h, $g(h(-1)) = -1$ and there are elements e and b such that $g(e) = i$ and $g(b) = 1$, then $e \otimes e = h(-1), b \otimes a = a$ and $h(-1) \otimes a = -a$ for any element a in $\boldsymbol{A}$, i.e., e plays the role of the imaginary number i and $h(1)$ plays the role of the multiplicative one 1_A in $\boldsymbol{A}$.*

Proof: Let us take an element a from the prearithmetic $\boldsymbol{A}$. Then for some real numbers x and y, we have

$$e \otimes e = h(g(e) \cdot g(e)) = h(i \cdot i) = h(-1),$$

$$b \otimes a = h(g(b) \cdot g(a))) = h(1 \cdot g(a)) = h(g(a)) = a$$

because by Lemma 2.3.3, the mapping g is an injection, $hg = \mathbf{1}_A$ and thus,

$$g(h(1)) = g(h(g(b))) = g(b) = 1.$$

As by Lemma 2.3.4, the mapping h is an injection on the image $g(\boldsymbol{A})$ of $\boldsymbol{A}$, $h(1) = b$ and $h(i) = e$. □

Taking a functional C-prearithmetic $\boldsymbol{A} = (A; \oplus, \otimes, \leq)$ in addition to addition and multiplication, we can define similar operations in $\boldsymbol{A}$ with more arguments. In particular, we have a system of pairs of n-ary operations for each natural number $n = 3, 4, 5, \ldots$ (cf. Section 2.1):

(3) $\Sigma^{\oplus n}(a_{A,1}, a_{A,2}, \ldots, a_{A,n-1}, a_{A,n}) = h(g(a_{A,1}) + g(a_{A,2}) + \cdots + g(a_{A,n}))$,
(4) $\Pi^{\otimes n}(a_{A,1}, a_{A,2}, \ldots, a_{A,n-1}, a_{A,n}) = h(g(a_{A,1}) \cdot g(a_{A,2}) \cdot \ldots \cdot g(a_{A,n}))$.

Note that for $n = 2$, these operations coincide with addition and multiplication, i.e., $\Sigma^{\oplus 2} = \oplus$ and $\Pi^{\otimes 2} = \otimes$.

Proposition 3.4.5. *For any C-prearithmetic $\boldsymbol{A} = (A; \oplus, \otimes, \ominus, \oslash, \ominus, \oslash, \leq_A)$, operations $\Sigma^{\oplus n}$ and $\Pi^{\otimes n}$ are commutative for all $n = 2, 3, 4, 5, \ldots$.*

Indeed,

$$\Sigma^{\oplus n}(a_{A,1}, a_{A,2}, \ldots, a_{A,n-1}, a_{A,n}) = h(g(a_{A,1}) + g(a_{A,2}) + \cdots + g(a_{A,n}))$$
$$= h(g(a_{A,j1}) + g(a_{A,j2}) + \cdots + g(a_{A,jn}))$$
$$= \Sigma^{\oplus n}(a_{A,j1}, a_{A,j2}, \ldots, a_{A,jn}),$$

where $j_1, j_2, j_3, \ldots, j_n$ is an arbitrary permutation of the sequence $1, 2, 3, \ldots, n$.

In a similar way, we have

$$\Pi^{\otimes n}(a_{A,1}, a_{A,2}, \ldots, a_{A,n-1}, a_{A,n}) = h(g(a_{A,1}) \cdot g(a_{A,2}) \cdot \ldots \cdot g(a_{A,n}))$$
$$= h(g(a_{A,j1}) \cdot g(a_{A,j2}) \cdot \ldots \cdot g(a_{A,jn})) = \Pi^{\otimes n}(a_{A,j1}, a_{A,j2}, \ldots, a_{A,jn}),$$

where $j_1, j_2, j_3, \ldots, j_n$ is an arbitrary permutation of the sequence $1, 2, 3, \ldots, n$.

Propositions 2.2.6, 2.2.7 and properties of the conventional arithmetic $\boldsymbol{C}$ imply the following result.

Proposition 3.4.6. *For any functional C-prearithmetic* $\boldsymbol{A} = (A; \oplus, \otimes, \ominus, \oslash, \leq_A)$ *with the projector* g *and the coprojector* h, *for which* $gh = \mathbf{1}_C$, *addition* $\oplus$ *and multiplication* $\otimes$ *are associative.*

Proof is left as an exercise. □

Proposition 2.2.8 and properties of the conventional arithmetic $\boldsymbol{C}$ imply the following result.

Proposition 3.4.7. *For any functional C-prearithmetic* $\boldsymbol{A} = (A; \oplus, \otimes, \ominus, \oslash, \leq_A)$ *with the projector* g *and the coprojector* h, *for which* $gh = \mathbf{1}_C$, *multiplication* $\otimes$ *is distributive over addition* $\oplus$.

Proof is left as an exercise. □

Corollary 3.4.1. *If a functional C-prearithmetic* $\boldsymbol{A} = (A; \oplus, \otimes, \ominus, \oslash, \leq_A)$ *is exactly projective with respect (isomorphic) to the conventional arithmetic* $\boldsymbol{C}$, *then addition* $\oplus$ *and multiplication* $\otimes$ *are commutative and associative while multiplication* $\otimes$ *is distributive over addition* $\oplus$.

Propositions 2.2.11, 2.2.12 and properties of the conventional arithmetic $\boldsymbol{C}$ imply the following result.

Proposition 3.4.8. *For any functional C-prearithmetic* $\boldsymbol{A} = (A; \oplus, \otimes, \ominus, \oslash, \leq_A)$ *with the coprojector* h, *which preserves the order* $\leq$ *in* $\boldsymbol{C}$, *addition* $\oplus$ *and multiplication* $\otimes$ *preserve the order* $\leq_A$.

Proof is left as an exercise. □

Corollary 3.4.2. *For any functional C-prearithmetic* $\boldsymbol{A} = (A; \oplus, \otimes, \ominus, \oslash, \leq_A)$ *with the coprojector* h, *which preserves the order* $\leq$ *in* $\boldsymbol{C}$, *addition* $\oplus$ *is monotone.*

Corollary 3.4.3. *For any functional C-prearithmetic* $\boldsymbol{A} = (A; \oplus, \otimes, \ominus, \oslash, \leq_A)$ *with the coprojector* h, *which preserves the order* $\leq$ *in* $\boldsymbol{C}$, *multiplication* $\otimes$ *is monotone.*

The property of abstract prearithmetics "to be a functional C-prearithmetic" is hereditary as the following result demonstrates.

Proposition 3.4.9. *A subprearithmetic $\boldsymbol{K}$ of a functional C-prearithmetic $\boldsymbol{A} = (A; \oplus, \otimes, \ominus, \oslash, \leq)$ is a functional C-prearithmetic.*

Indeed, by Proposition 2.5.12, the prearithmetic $\boldsymbol{K}$ is weakly projective with respect to the Diophantine arithmetic $\boldsymbol{C}$ and thus, it is a functional C-prearithmetic.

Corollary 3.4.4. *A subprearithmetic $\boldsymbol{D}$ of a subprearithmetic $\boldsymbol{K}$ of a functional C-prearithmetic $\boldsymbol{A} = (A; \oplus, \otimes, \ominus, \oslash, \leq)$ is a functional C-prearithmetic.*

Indeed, by Proposition 2.5.11, the prearithmetic $\boldsymbol{D}$ is a subprearithmetic of a functional C-prearithmetic $\boldsymbol{A}$ and thus by Proposition 3.4.9, $\boldsymbol{D}$ is a functional C-prearithmetic.

Proposition 3.4.10. *A wholly extended abstract prearithmetic $\boldsymbol{A} = (A; +, \circ, -, \div, \leq_A)$, which is weakly projective with respect to a functional C-prearithmetic $\boldsymbol{D} = (D; \oplus, \otimes, \ominus, \oslash, \leq_D)$, is a functional C-prearithmetic.*

Indeed, by Proposition 2.2.9, the prearithmetic $\boldsymbol{A}$ is weakly projective with respect to the Diophantine arithmetic $\boldsymbol{C}$ and thus, it is a functional C-prearithmetic.

Proposition 3.4.11. *If a wholly extended abstract prearithmetic $\boldsymbol{A}_1 = (A_1; +_1, \circ_1, -_1, \div_1, \leq_1)$ is weakly projective with respect to a wholly extended abstract prearithmetic $\boldsymbol{A}_2 = (A_2; +_2, \circ_2, -_1, \div_2, \leq_2)$ and the wholly extended abstract prearithmetic $\boldsymbol{A}_2 = (A_2; +_2, \circ_2, -_2, \div_2, \leq_2)$ is weakly projective with respect to a functional C-prearithmetic $\boldsymbol{A}_3 = (A_3; +_3, \circ_3, -_3, \div_3, \leq_3)$, then the prearithmetic $\boldsymbol{A}_1$ is a functional C-prearithmetic.*

Indeed, by Proposition 3.4.10, the prearithmetic $\boldsymbol{A}_2$ is a functional C-prearithmetic and applying once more Proposition 3.4.10, we come to the conclusion that $\boldsymbol{A}_1$ is also a functional C-prearithmetic.

Proposition 3.3.11 implies the following result.

Theorem 3.4.1. *C-prearithmetics form a category where objects are functional C-prearithmetics and morphisms are weak projectivity relations.*

Proof is left as an exercise. □

The construction from Definition 2.2.2 allows defining actions of complex numbers on elements from a functional C-prearithmetic $\boldsymbol{A} = (A; \oplus, \otimes, \ominus, \oslash, \leq_A)$ with the projector g and the coprojector h.

Definition 3.4.2.

(a) The *additive action from the left* of a complex number n on an element a from $\boldsymbol{A}$ is defined as

$$n^{+}a = h(n + g(a)).$$

(b) The *additive action from the right* of a complex number n on an element a from $\boldsymbol{A}$ is defined as

$$a^{+}n = h(g(a) + n).$$

(c) The *multiplicative action from the left* of a complex number n on an element a from $\boldsymbol{A}$ is defined as

$$na = h(n \cdot g(a)).$$

(d) The *multiplicative action from the right* of a complex number n on an element a from $\boldsymbol{A}$ is defined as

$$an = h(g(a) \cdot n).$$

Let us study how complex numbers act on elements of C-prearithmetics using the properties of actions obtained in Section 2.2 in general abstract prearithmetics. As addition of complex numbers is commutative, we have the following result.

Proposition 3.4.12. *For any functional C-prearithmetic* $\boldsymbol{A} = (A; \oplus, \otimes, \ominus, \oslash, \leq_A)$*, additive action from the left coincides with additive action from the right.*

Proof is left as an exercise. □

Proposition 3.4.12 allows omitting indication of the side of additive action in C-prearithmetics.

As multiplication of complex numbers is commutative, we have the following result.

Proposition 3.4.13. *For any functional C-prearithmetic* $\boldsymbol{A} = (A; \oplus, \otimes, \ominus, \oslash, \leq_A)$*, multiplicative action from the left coincides with multiplicative action from the right.*

Proof is left as an exercise. □

Proposition 3.4.13 allows omitting indication of the side of multiplicative action in C-prearithmetics.

Proposition 2.2.18 implies the following result.

Proposition 3.4.14. *For any functional C-prearithmetic* $\boldsymbol{A} = (A; \oplus, \otimes, \ominus, \oslash, \leq_A)$ *with the coprojector* h*, which preserves addition, multiplicative action is distributive over addition* $\oplus$*.*

Proof is left as an exercise. □

Proposition 2.2.19 implies the following result.

Proposition 3.4.15. *For any functional C-prearithmetic* $\boldsymbol{A} = (A; \oplus, \otimes, \ominus, \oslash, \leq_A)$ *with the coprojector* h*, which preserves addition, multiplicative action is distributive over addition* $+$ *in the arithmetic* $\boldsymbol{C}$*.*

Proof is left as an exercise. □

Proposition 2.2.22 implies the following result.

Proposition 3.4.16. *For any functional C-prearithmetic* $\boldsymbol{A} = (A; \oplus, \otimes, \ominus, \oslash, \leq_A)$ *with the coprojector* h*, which preserves the order of complex numbers, additive action is a monotone operation.*

Proof is left as an exercise. □

Proposition 2.2.23 implies the following result.

Proposition 3.4.17. *For any functional C-prearithmetic* $\boldsymbol{A} = (A; \oplus, \otimes, \ominus, \oslash, \leq_A)$ *with the coprojector* h*, which preserves the order of complex numbers, multiplicative action is a monotone operation.*

Proof is left as an exercise. □

Proposition 3.1.21 implies the following result.

Proposition 3.4.18. *In a functional C-prearithmetic* $\boldsymbol{A} = (A; \oplus, \otimes, \ominus, \oslash, \leq_A)$*, we have* $(b \ominus a) \ominus c = b \ominus (a \oplus c)$ *for any elements* a, b *and* c *from the prearithmetic* $\boldsymbol{A}_1$*.*

Proof is left as an exercise. □

In applications, we as a rule use non-Diophantine arithmetics of complex numbers. One class of these arithmetics is formed by functional IC-arithmetics.

Definition 3.4.3. A wholly extended abstract prearithmetic $\boldsymbol{A} = (A; \oplus, \otimes, \ominus, \oslash, \leq_A)$ is called a *functional* IC-*arithmetic*, if A is exactly projective with respect to the Diophantine arithmetic $\boldsymbol{C} = (C; +, \cdot, -, \div, \leq)$ of complex numbers.

Functional IC-arithmetics form a distinct class of non-Diophantine arithmetics of complex numbers.

Functional IC-arithmetics have many properties similar to properties of the Diophantine arithmetic $\boldsymbol{C}$ of complex numbers. For instance, Theorem 3.2.3 implies the following results.

Theorem 3.4.2. *(a) Any functional IC-arithmetic* $\boldsymbol{A} = (A; \oplus_A, \otimes_A, \ominus_A, \oslash_A, \leq_A)$ *is isomorphic to the Diophantine arithmetic* $\boldsymbol{C}$.

Proof is left as an exercise. □

This result allows describing many properties of IR-arithmetics.

Corollary 3.4.5. *Any functional IC-arithmetic* $\boldsymbol{A}$ *has the following properties*:

(a) *addition is commutative*;
(b) *multiplication is commutative*;
(c) *addition is associative*;
(d) *multiplication is associative*;
(e) *multiplication is distributive with respect to addition*;
(f) *multiplication is distributive with respect to subtraction*;
(g) $\boldsymbol{A}$ *has the additive and multiplicative zero* 0_A;
(h) $\boldsymbol{A}$ *has the multiplicative one* 1_A;
(i) $\boldsymbol{A}$ *has continuum of elements.*

Definition 3.4.4. The isomorphism f between a functional IC-arithmetic $\boldsymbol{A}$ and $\boldsymbol{C}$ is called the *functional parameter* of $\boldsymbol{A}$.

It is often denoted as $f_A : A \to C$.

Often the carrier A of a functional IC-arithmetic $\boldsymbol{A}$ is a set of complex numbers. However, in general, the carrier A of a functional IC-arithmetic $\boldsymbol{A}$ can be any set for which a one-to-one mapping (bijection) $f_A : A \to C$ exists. This bijection determines a one-to-one naming of the elements from A by complex numbers (Burgin, 2011). Taking this into account, we can denote the element from A that is mapped onto the complex number r by r_A.

3.4.2. *Operational arithmetics and prearithmetics of complex numbers*

I have no satisfaction in formulas unless I feel their arithmetical magnitude.

William Thomson Kelvin

Now we construct and study operational prearithmetics and arithmetics of complex numbers. Their construction is based on the following idea.

The Diophantine arithmetic $\boldsymbol{C}$ of complex numbers is traditionally constructed as an extension of the Diophantine arithmetic $\boldsymbol{R}$ of real numbers. The most popular way of building this extension is by representing complex numbers in the form $a + bi$ where a and b are real numbers and $i = \sqrt{-1}$ and defining operations with complex numbers using operations with real numbers.

An equivalent approach is representing complex numbers by pairs of real numbers and properly defining arithmetical operations. It is possible to build operational C-prearithmetics by a similar technique taking the set of all pairs of elements from an R-prearithmetic and defining four operations and a partial order in this set. There are different ways to define the necessary operations and a partial order but here we do this according to the schema used for the construction of the Diophantine arithmetic $\boldsymbol{C}$ of complex numbers.

At first, we apply this schema to wholly extended abstract prearithmetics.

Let us consider a wholly extended abstract prearithmetic $\boldsymbol{A} = (A; \oplus_A, \otimes_A, \ominus_A, \oslash_A, \leq_A)$ and the set $B = \{(a, b); a, b \in A\}$. Then we can define operations and a partial order as

$$\begin{aligned}(a, b) \oplus (c, d) &= (a \oplus_A c, b \oplus_A d),\\(a, b) \ominus (c, d) &= (a \ominus_A c, b \ominus_A d),\\(a, b) \otimes (c, d) &= ((a \otimes_A c) \ominus_A (b \otimes_A d), (a \otimes_A d) \oplus_A (b \otimes_A c)),\end{aligned}$$

$$(a, b) \leq_A (c, d) \text{ if } a \leq_A c \text{ and } b \leq_A d.$$

Division is defined later after we study properties of the so defined whole abstract prearithmetics.

Using constructed operations and a partial order, we obtain the wholly extended abstract prearithmetic $\boldsymbol{A}_{\text{ext}} = (A^2; \oplus, \otimes, \ominus, \oslash, \leq)$ on the set A^2 of all pairs of elements from $\boldsymbol{A}$. It is called the *complexification* of $\boldsymbol{A}$. If $\boldsymbol{A}$ is an R-prearithmetic, then $\boldsymbol{A}_{\text{ext}}$ is called an operational C-prearithmetic over $\boldsymbol{A}$. If $\boldsymbol{A}$ is a weak R-arithmetic, then $\boldsymbol{A}_{\text{ext}}$ is called an operational weak C-arithmetic over $\boldsymbol{A}$.

To explicate relations between the Diophantine arithmetic of complex numbers $\boldsymbol{C}$ and the operational C-prearithmetic $\boldsymbol{A}_{\text{ext}}$, we use the following notation. In a pair (complex number) $\boldsymbol{a} = (a, b)$, we regard its first component a as the real part of the element (complex number) $\boldsymbol{a}$ from $\boldsymbol{A}_{\text{ext}}$ denoting it by $\mathscr{R}\boldsymbol{a}$, and treat its second component b as the imaginary part of the element (complex number) $\boldsymbol{a}$ denoting it by $\mathscr{I}\boldsymbol{a}$. This gives another representation of the complex number $\boldsymbol{a}$:

$$\boldsymbol{a} = (a, b) = \mathscr{R}\boldsymbol{a} \oplus_A \boldsymbol{i}\mathscr{I}\boldsymbol{a}.$$

Here the element $\boldsymbol{i} = (0_A, 1_A)$ plays the role of the imaginary unit i in $\boldsymbol{E}_{A,D}$.

When $\boldsymbol{A} = (A; \oplus_A, \otimes_A, \ominus_A, \oslash_A, \leq_A)$ is an R-prearithmetic, i.e., $\boldsymbol{A}$ is weakly projective with respect to the Diophantine arithmetic $\boldsymbol{R}$ with the projector g and coprojector h, we can express operations in the operational C-prearithmetic $\boldsymbol{A}_{\text{ext}}$ through the conventional operations with real numbers in the following way:

$$\begin{aligned}
(a, b) \oplus (c, d) &= (a \oplus_A c, b \oplus_A d) = (h(g(a) + g(c)), h(g(b) + g(d))),\\
(a, b) \ominus (c, d) &= (a \ominus_A c, b \ominus_A d) = (h(g(a) - g(c)), h(g(b) - g(d))),\\
(a, b) \otimes (c, d) &= ((a \otimes_A c) \ominus_A (b \otimes_A d), (a \otimes_A d) \oplus_A (b \otimes_A c)),\\
&= (h(g(h(g(a) \cdot g(c))) - g(h(g(b) \cdot g(d)))),\\
&\qquad h(g(h(g(a) \cdot g(d))) + g(h(g(b) \cdot g(c))))).
\end{aligned}$$

In addition, we define the order $\leq$ in $\boldsymbol{A}_{\text{ext}}$ by the following rule:

$$(a, b) \leq (c, d) \text{ if and only if } a \leq_A c \text{ and } b \leq_A d.$$

Using the component notation for elements $\boldsymbol{a}$ and $\boldsymbol{c}$ from $\boldsymbol{A}_{\text{ext}}$, we have

$$\begin{aligned}
&\boldsymbol{a} \oplus \boldsymbol{c} = (\mathscr{R}\boldsymbol{a} \oplus_A \mathscr{R}\boldsymbol{c}, \mathscr{I}\boldsymbol{a} \oplus_A \mathscr{I}\boldsymbol{c}),\\
&\boldsymbol{a} \ominus \boldsymbol{c} = (\mathscr{R}\boldsymbol{a}_A\mathscr{R}\boldsymbol{c}, \mathscr{I}\boldsymbol{a}_A\mathscr{I}\boldsymbol{c}),\\
&\boldsymbol{a} \otimes \boldsymbol{c} = ((\mathscr{R}\boldsymbol{a} \otimes_A \mathscr{R}\boldsymbol{c}) \ominus_A (\mathscr{I}\boldsymbol{a} \otimes_A \mathscr{I}\boldsymbol{c}), (\mathscr{R}\boldsymbol{a} \otimes_A \mathscr{I}\boldsymbol{c}) \oplus_A (\mathscr{I}\boldsymbol{a} \otimes_A \mathscr{R}\boldsymbol{c})).
\end{aligned}$$

Consequently,

$$\mathscr{R}(\boldsymbol{a} \oplus \boldsymbol{c}) = \mathscr{R}\boldsymbol{a} \oplus_A \mathscr{R}\boldsymbol{c},$$

$$\mathscr{I}(\boldsymbol{a} \oplus \boldsymbol{c}) = \mathscr{I}\boldsymbol{a} \oplus_A \mathscr{I}\boldsymbol{c},$$

$$\mathscr{R}(\boldsymbol{a} \ominus \boldsymbol{c}) = \mathscr{R}\boldsymbol{a} \ominus_A \mathscr{R}\boldsymbol{c},$$

$$\mathscr{I}(\boldsymbol{a} \ominus \boldsymbol{c}) = \mathscr{I}\boldsymbol{a} \ominus_A \mathscr{I}\boldsymbol{c},$$

$$\mathscr{R}(\boldsymbol{a} \otimes \boldsymbol{c}) = (\mathscr{R}\boldsymbol{a} \otimes_A \mathscr{R}\boldsymbol{c}) \ominus_A (\mathscr{I}\boldsymbol{a} \otimes_A \mathscr{I}\boldsymbol{c}),$$

$$\mathscr{I}(\boldsymbol{a} \otimes \boldsymbol{c}) = (\mathscr{R}\boldsymbol{a} \otimes_A \mathscr{I}\boldsymbol{c}) \oplus_A (\mathscr{I}\boldsymbol{a} \otimes_A \mathscr{R}\boldsymbol{c}),$$

$$\boldsymbol{a} \leq \boldsymbol{c} \text{ if and only if } \mathscr{R}\boldsymbol{a} \leq_A \mathscr{R}\boldsymbol{c} \text{ and } \mathscr{I}\boldsymbol{a} \leq_A \mathscr{I}\boldsymbol{c}.$$

Proposition 3.4.19. *A wholly extended abstract prearithmetic* $\boldsymbol{A} = (A; \oplus_A, \otimes_A, \ominus_A, \oslash_A, \leq_A)$ *with zero* 0 *is isomorphic (is exactly projective with respect) to a subprearithmetic of the wholly extended abstract prearithmetic* $\boldsymbol{A}_{\text{ext}}$.

Proof: Let us take the set $Q = \{(a, 0); a \in A\}$. This set is closed with respect to operations in $\boldsymbol{A}_{\text{ext}}$. Indeed,

$$(a, 0) \oplus (b, 0) = (a \oplus_A b, 0 \oplus_A 0) = (a \oplus_A b, 0),$$

$$(a, 0) \ominus (b, 0) = (a \oplus_A b, 0 \ominus_A 0) = (a \ominus_A b, 0),$$

$$\begin{aligned}(a, 0) \otimes (b, 0) &= ((a \otimes_A c) \ominus_A (0 \otimes_A 0), (a \otimes_A 0) \oplus_A (0 \otimes_A c)) \\ &= ((a \otimes_A c) \ominus_A 0), 0 \oplus_A 0) = (a \otimes_A c, 0).\end{aligned}$$

In the same way, we prove that Q is closed with respect to division using formula (3.15), which is derived later.

Thus, Q is the carrier of the wholly extended abstract prearithmetic $\boldsymbol{Q} = (Q; \oplus, \otimes, \ominus, \oslash, \leq)$, which is a subprearithmetic of the wholly extended abstract prearithmetic $\boldsymbol{A}_{\text{ext}}$.

Let us define the mapping $q : A \to Q$ by the formula $q(a) = (a, 0)$. By construction, it is a bijection. In addition, by the definition of $\boldsymbol{A}_{\text{ext}}$, we have

$$q(a \oplus_A b) = (a \oplus_A b, 0) = (a \oplus_A b, 0 \oplus_A 0) = (a,0) \oplus (b,0) = q(a) \oplus q(b),$$

$$q(a \ominus_A b) = (a \ominus_A b, 0) = (a \oplus_A b, 0 \ominus_A 0) = (a,0) \ominus (b,0) = q(a) \ominus q(b),$$

$$\begin{aligned} q(a \otimes_A b) &= (a \otimes_A b, 0) = (a \oplus_A b, 0 \oplus_A 0) = q(a) \otimes q(b) = (a,0) \otimes (b,0) \\ &= ((a \otimes_A c) \ominus_A (0 \otimes_A 0), (a \otimes_A 0) \oplus_A (0 \otimes_A c)) \\ &= ((a \otimes_A c) \ominus_A 0), 0 \oplus_A 0) = (a \otimes_A c, 0) = q(a \otimes_A c). \end{aligned}$$

It means that q is a homomorphism with respect to addition, subtraction and multiplication.

In the same way, we prove that q is a homomorphism with respect to division using formula (3.15), which is derived later. As a result, we see that q defines an isomorphism between $\boldsymbol{A}$ and $\boldsymbol{Q}$.

Proposition is proved. □

The wholly extended abstract prearithmetic $\boldsymbol{A}_{\text{ext}}$ inherits many properties of the wholly extended abstract prearithmetic $\boldsymbol{A}$. In particular, the wholly extended abstract prearithmetic $\boldsymbol{A}_{\text{ext}}$ inherits zeros and ones.

Proposition 3.4.20. *If a wholly extended abstract prearithmetic* $\boldsymbol{A} = (A; \oplus_A, \otimes_A, \ominus_A, \oslash_A, \leq_A)$ *has zero* 0_A*, then the wholly extended abstract prearithmetic* $\boldsymbol{A}_{\text{ext}}$ *also has zero* $\mathbf{0} = (0_A, 0_A)$.

Proof: Let us take the element $(0_A, 0_A)$ from $\boldsymbol{A}_{\text{ext}}$ and show that it is zero in $\boldsymbol{A}_{\text{ext}}$. Indeed,

$$(a,b) \oplus (0_A, 0_A) = (a \oplus_A 0_A, b \oplus_A 0_A) = (a,b).$$

It means that $\mathbf{0} = (0_A, 0_A)$ is an additive zero in $\boldsymbol{A}_{\text{ext}}$. In addition,

$$\begin{aligned} (a,b) \otimes (0_A, 0_A) &= ((a \otimes_A 0_A) \ominus_A (b \otimes_A 0_A), (a \otimes_A 0_A) \oplus_A (b \otimes_A 0_A)) \\ &= (0_A, 0_A). \end{aligned}$$

It means that $\mathbf{0} = (0_A, 0_A)$ is a multiplicative zero and thus zero in $\boldsymbol{A}_{\text{ext}}$.

Proposition is proved. □

Propositions 3.3.15 and 3.4.17 imply the following results.

Corollary 3.4.6. *If a weak R-arithmetic* $\boldsymbol{A} = (A; \oplus, \otimes, \ominus, \oslash, \leq_A)$ *with the projector* g *contains an element* d *such that* $g(d) = 0$*, then the operational C-prearithmetic* $\boldsymbol{A}_{\text{ext}}$ *has zero* $\mathbf{0}$.

Corollary 3.4.7. *If $gh = \mathbf{1}_R$, for an R-prearithmetic $\boldsymbol{A} = (A; \oplus, \otimes, \ominus, \oslash, \leq_A)$ with the projector g and the coprojector h, then the operational C-prearithmetic $\boldsymbol{A}_{\text{ext}}$ has zero $\mathbf{0}$.*

Similar statements are true for the multiplicative one.

Proposition 3.4.21. *If a wholly extended abstract prearithmetic $\boldsymbol{A} = (A; \oplus_A, \otimes_A, \ominus_A, \oslash_A, \leq_A)$ with zero 0_A has the multiplicative one 1_A, then the wholly extended abstract prearithmetic $\boldsymbol{A}_{\text{ext}}$ also has the multiplicative one $\mathbf{1}$.*

Proof: Let us take the element $(1_A, 0_A)$ from $\boldsymbol{A}_{\text{ext}}$ and show that it is zero in $\boldsymbol{A}_{\text{ext}}$. Indeed,

$$(a, b) \otimes (1_A, 0_A) = ((a \otimes_A 1_A) \ominus_A (b \otimes_A 0_A), (a \otimes_A 0_A) \oplus_A (b \otimes_A 1_A))$$
$$= (a \ominus_A 0_A, 0_A \oplus_A b) = (a, b).$$

It means that $\mathbf{1} = (1_A, 0_A)$ is the multiplicative one in $\boldsymbol{A}_{\text{ext}}$.
Proposition is proved. □

Propositions 3.3.21 and 3.4.18 imply the following results.

Corollary 3.4.8. *If a weak R-arithmetic $\boldsymbol{A} = (A; \oplus, \otimes, \ominus, \oslash, \leq_A)$ with the projector g contains an element d such that $g(d) = 1$, then the operational weak C-arithmetic $\boldsymbol{A}_{\text{ext}}$ has the multiplicative one $\mathbf{1}$.*

Corollary 3.4.9. *If $gh = \mathbf{1}_R$, for an R-prearithmetic $\boldsymbol{A} = (A; \oplus, \otimes, \ominus, \oslash, \leq_A)$ with the projector g and the coprojector h, then the operational C-prearithmetic $\boldsymbol{A}_{\text{ext}}$ has the multiplicative one $\mathbf{1}$.*

The prearithmetic $\boldsymbol{A}_{\text{ext}}$ also inherits opposite elements.

Proposition 3.4.22. *If elements a and b from a wholly extended abstract prearithmetic $\boldsymbol{A} = (A; \oplus_A, \otimes_A, \ominus_A, \oslash_A, \leq_A)$ with zero 0_A have opposite elements $-a$ and $-b$ (cf. Section 2.1), then the element $\boldsymbol{a} = (a, b)$ from the wholly extended abstract prearithmetic $\boldsymbol{A}_{\text{ext}}$ also has the opposite element $-\boldsymbol{a} = (-a, -b)$.*

Indeed, we have

$$(a, b) \oplus (-a, -b) = (a \oplus_A -a, b \oplus_A -b) = (0_A, 0_A).$$

It means that $(-a, -b)$ is the opposite element to (a, b), i.e., $(-a, -b) = -(a, b)$.

Corollary 3.4.10. *If any element a from a wholly extended abstract prearithmetic* $\boldsymbol{A} = (A; \oplus_A, \otimes_A, \ominus_A, \oslash_A, \leq_A)$ *has the opposite element* $-a$, *then any element* (a, b) *from the wholly extended abstract prearithmetic* $\boldsymbol{A}_{\text{ext}}$ *also has the opposite element.*

Corollary 3.4.11. *If a wholly extended abstract prearithmetic* $\boldsymbol{A} = (A; \oplus_A, \otimes_A, \ominus_A, \oslash_A, \leq_A)$ *is a quasigroup with respect to addition, then the wholly extended abstract prearithmetic* $\boldsymbol{A}_{\text{ext}}$ *is also a quasigroup with respect to addition.*

Commutativity of operations in $\boldsymbol{A}$ implies commutativity of operations in $\boldsymbol{A}_{\text{ext}}$.

Proposition 3.4.23. *If addition in a wholly extended abstract prearithmetic* $\boldsymbol{A} = (A; \oplus_A, \otimes_A, \ominus_A, \oslash_A, \leq_A)$ *is commutative, then addition in the wholly extended abstract prearithmetic* $\boldsymbol{A}_{\text{ext}}$ *is also commutative.*

Indeed,

$$(a, b) \oplus (c, d) = (a \oplus_A c, b \oplus_A d) = (c \oplus_A a, d \oplus_A b) = (c, d) \oplus (a, b).$$

Corollary 3.4.12. *If* $\boldsymbol{A} = (A; \oplus_A, \otimes_A, \ominus_A, \oslash_A, \leq_A)$ *is R-prearithmetic, then addition in the operational C-prearithmetic* $\boldsymbol{A}_{\text{ext}}$ *is commutative.*

Indeed, by Proposition 3.3.2, addition in $\boldsymbol{A}$ is commutative. Then $\boldsymbol{A}_{\text{ext}}$ is an operational C-prearithmetic and by Proposition 3.4.23, addition in the operational C-prearithmetic $\boldsymbol{A}_{\text{ext}}$ is commutative.

Corollary 3.4.13. *If a wholly extended abstract prearithmetic* $\boldsymbol{A} = (A; \oplus_A, \otimes, \ominus_A, \oslash, \leq_A)$ *is a commutative quasigroup with respect to addition, then the wholly extended abstract prearithmetic* $\boldsymbol{A}_{\text{ext}}$ *is also a commutative quasigroup with respect to addition.*

Associativity of operations in $\boldsymbol{A}$ implies associativity of operations in $\boldsymbol{A}_{\text{ext}}$.

Proposition 3.4.24. *If addition in a wholly extended abstract prearithmetic* $\boldsymbol{A} = (A; \oplus_A, \otimes_A, \ominus_A, \oslash_A, \leq_A)$ *is associative, then addition in the wholly extended abstract prearithmetic* $\boldsymbol{A}_{\text{ext}}$ *is also associative.*

Indeed, we have

$$\begin{aligned}((a,b)\oplus(c,d))\oplus(e,l) &= (a\oplus_A c, b\oplus_A d)\oplus(e,l)\\ &= ((a\oplus_A c)\oplus_A e, (b\oplus_A d)\oplus_A l)\\ &= (a\oplus_A (c\oplus_A e), b\oplus_A (d\oplus_A l))\\ &= (a,b)\oplus(c\oplus_A e, d\oplus_A l) = (a,b)\oplus((c,d)\oplus(e,l)).\end{aligned}$$

Corollary 3.4.14. *If a wholly extended abstract prearithmetic* $\boldsymbol{A} = (A; \oplus_A, \otimes_A, \ominus_A, \oslash_A, \leq_A)$ *is a semigroup with respect to addition, then the wholly extended abstract prearithmetic* $\boldsymbol{A}_{\text{ext}}$ *is also a semigroup with respect to addition.*

Corollary 3.4.15. *If a wholly extended abstract prearithmetic* $\boldsymbol{A} = (A; \oplus_A, \otimes_A, \ominus_A, \oslash_A, \leq_A)$ *is a commutative semigroup with respect to addition, then the wholly extended abstract prearithmetic* $\boldsymbol{A}_{\text{ext}}$ *is also a commutative semigroup with respect to addition.*

Corollary 3.4.16. *If a wholly extended abstract prearithmetic* $\boldsymbol{A} = (A; \oplus_A, \otimes_A, \ominus_A, \oslash_A, \leq_A)$ *is a group with respect to addition, then the wholly extended abstract prearithmetic* $\boldsymbol{A}_{ext}$ *is also a group with respect to addition.*

Corollary 3.4.17. *If a wholly extended abstract prearithmetic* $\boldsymbol{A} = (A; \oplus_A, \otimes_A, \ominus_A, \oslash_A, \leq_A)$ *is a commutative group with respect to addition, then the wholly extended abstract prearithmetic* $\boldsymbol{A}_{\text{ext}}$ *is also a commutative group with respect to addition.*

Corollary 3.4.18. *If* $\boldsymbol{A} = (A; \oplus_A, \otimes_A, \ominus_A, \oslash_A, \leq_A)$ *is R-prearithmetic with the projector g and the coprojector h, for which* $gh = \mathbf{1}_R$, *then addition in the operational C-prearithmetic* $\boldsymbol{A}_{\text{ext}}$ *is associative.*

Indeed, by Proposition 3.3.6, addition in $\boldsymbol{A}$ is associative. Then $\boldsymbol{A}_{\text{ext}}$ is an operational C-prearithmetic and by Proposition 3.4.21, addition in the operational C-prearithmetic $\boldsymbol{A}_{\text{ext}}$ is associative.

Proposition 3.4.25. *If multiplication in a wholly extended abstract prearithmetic* $\boldsymbol{A} = (A; \oplus_A, \otimes_A, \ominus_A, \oslash_A, \leq_A)$ *is commutative, then multiplication in the wholly extended abstract prearithmetic* $\boldsymbol{A}_{\text{ext}}$ *is also commutative.*

Indeed,

$$\begin{aligned}(a,b)\otimes(c,d) &= ((a\otimes_A c)\ominus_A(b\otimes_A d),(a\otimes_A d)\oplus_A(b\otimes_A c))\\ &= ((c\otimes_A a)\ominus_A(d\otimes_A b),(d\otimes_A a)\oplus_A(c\otimes_A b))\\ &= (c,d)\otimes(a,b).\end{aligned}$$

Corollary 3.4.19. *If $\boldsymbol{A} = (A;\oplus_A,\otimes_A,\ominus_A,\oslash_A,\leq_A)$ is R-prearithmetic, then multiplication in the operational C-prearithmetic $\boldsymbol{A}_{\text{ext}}$ is commutative.*

Indeed, by Proposition 3.3.2, multiplication in $\boldsymbol{A}$ is commutative. Then $\boldsymbol{A}_{\text{ext}}$ is an operational C-prearithmetic and by Proposition 3.4.25, multiplication in the operational C-prearithmetic $\boldsymbol{A}_{\text{ext}}$ is commutative.

Proposition 3.4.26. *If multiplication in a wholly extended abstract prearithmetic $\boldsymbol{A} = (A;\oplus_A,\otimes_A,\ominus_A,\oslash_A,\leq_A)$ is associative, then multiplication in the wholly extended abstract prearithmetic $\boldsymbol{A}_{\text{ext}}$ is also associative.*

Proof is left as an exercise. □

Corollary 3.4.20. *If $\boldsymbol{A} = (A;\oplus_A,\otimes_A,\ominus_A,\oslash_A,\leq_A)$ is R-prearithmetic with the projector g and the coprojector h, for which $gh = \mathbf{1}_R$, then multiplication in the operational C-prearithmetic $\boldsymbol{A}_{\text{ext}}$ is associative.*

Indeed, by Proposition 3.3.5, multiplication in $\boldsymbol{A}$ is associative. Then $\boldsymbol{A}_{\text{ext}}$ is a operational C-prearithmetic and by Proposition 3.4.26, multiplication in the C-prearithmetic $\boldsymbol{A}_{\text{ext}}$ is associative.

Proposition 3.4.27. *If multiplication in a wholly extended abstract prearithmetic $\boldsymbol{A} = (A;\oplus_A,\otimes_A,\ominus_A,\oslash_A,\leq_A)$ is distributive with respect to addition, then multiplication in the wholly extended abstract prearithmetic $\boldsymbol{A}_{\text{ext}}$ is also distributive with respect to addition.*

Proof: For any elements a, c, d and b from a weak R-arithmetic $\boldsymbol{A}$, we have

$$\begin{aligned}(a,b)\otimes((c,d))\oplus(e,l)) &= (a,b)\otimes(c\oplus_A e, d\oplus_A l)\\ &= ((a\otimes_A(c\oplus_A e))\ominus_A(b\otimes_A(d\oplus_A l)),\\ &\quad (a\otimes_A(d\oplus_A l))\oplus(b\otimes_A(c\oplus_A e)))\end{aligned}$$

$$= ((a \otimes_A c) \oplus_A (a \otimes_A e)) \ominus_A ((b \otimes_A d) \oplus_A (b \otimes_A l)),$$
$$((a \otimes_A d) \oplus_A (a \otimes_A l)) \oplus_A ((b \otimes_A c) \oplus_A (b \otimes_A e)))$$
$$= ((a, b) \otimes (c, d)) \oplus ((a, b) \otimes (e, l)).$$

Proposition is proved. □

Corollary 3.4.21. *If $\boldsymbol{A} = (A; \oplus_A, \otimes_A, \ominus_A, \oslash_A, \leq_A)$ is R-prearithmetic with the projector g and the coprojector h, for which $gh = \mathbf{1}_R$, then multiplication in the operational C-prearithmetic $\boldsymbol{A}_{\text{ext}}$ is associative.*

Indeed, by Proposition 3.3.6, multiplication in $\boldsymbol{A}$ is distributive with respect to addition. Then $\boldsymbol{A}_{\text{ext}}$ is an operational C-prearithmetic and by Proposition 3.4.27, multiplication in the operational C-prearithmetic $\boldsymbol{A}_{\text{ext}}$ is distributive with respect to addition.

Corollary 3.4.22. *If a wholly extended abstract prearithmetic $\boldsymbol{A} = (A; \oplus_A, \otimes_A, \ominus_A, \oslash_A, \leq_A)$ is a ring with respect to addition, then the wholly extended abstract prearithmetic $\boldsymbol{A}_{\text{ext}}$ is also a ring with respect to addition.*

Corollary 3.4.23. *If a wholly extended abstract prearithmetic $\boldsymbol{A} = (A; \oplus_A, \otimes_A, \ominus_A, \oslash_A, \leq_A)$ is a commutative ring with respect to addition, then the wholly extended abstract prearithmetic $\boldsymbol{A}_{\text{ext}}$ is also a commutative ring with respect to addition.*

Let us study relation between addition and subtraction in wholly extended abstract prearithmetics $\boldsymbol{A}$ and $\boldsymbol{A}_{\text{ext}}$.

Proposition 3.4.28. *If for elements a, b, c and d from a wholly extended abstract prearithmetic $\boldsymbol{A} = (A; \oplus_A, \otimes_A, \ominus_A, \oslash_A, \leq_A)$, equalities $a \ominus_A c = a \oplus_A -c$ and $b \ominus_A d = b \oplus_A -d$ are true in $\boldsymbol{A}$, then $(a, b) \ominus_A (c, d) = (a, b) \oplus -(c, d)$ in the wholly extended abstract prearithmetic $\boldsymbol{A}_{\text{ext}}$.*

Indeed, we have

$$(a, b) \oplus -(c, d) = (a, b) \oplus (-c, -d) = (a \oplus_A -c, b \oplus_A -d)$$
$$= (a \ominus_A c, b \ominus_A d) = (a, b) \ominus (c, d).$$

Corollary 3.4.24. *If for all elements a and d from a wholly extended abstract prearithmetic $\boldsymbol{A} = (A; \oplus_A, \otimes_A, \ominus_A, \oslash_A, \leq_A)$ with zero 0_A and multiplicative one 1_A, identities $a \ominus_A d = a \oplus_A -d$ are true in $\boldsymbol{A}$, then*

$(a,b)\ominus(c,d) = (a,b)\oplus -(c,d)$ in the wholly extended abstract prearithmetic $\boldsymbol{A}_{\text{ext}}$.

Corollary 3.4.25. *If for all elements a and d from a wholly extended abstract prearithmetic $\boldsymbol{A} = (A; \oplus_A, \otimes_A, \ominus_A, \oslash_A, \leq_A)$ with zero 0_A and multiplicative one 1_A, identities $a \ominus_A d = a \oplus_A -d$ are true in $\boldsymbol{A}$, then $(0,0) \ominus (c,d) = -(c,d)$ in the wholly extended abstract prearithmetic $\boldsymbol{A}_{\text{ext}}$.*

Indeed, by Propositions 3.4.19 and 3.4.20, we have

$$(0_A, 0_A) \ominus (c,d) = (0_A \ominus_A c, 0_A \ominus_A d) = (0_A \oplus_A -c, 0_A \oplus_A -d)$$

$$= (-a, -b) = -(c,d).$$

Proposition 3.4.29. *If for all elements a and d from a wholly extended abstract prearithmetic $\boldsymbol{A} = (A; \oplus_A, \otimes_A, \ominus_A, \oslash_A, \leq_A)$ with zero 0_A and multiplicative one 1_A, identities $a \ominus_A d = a \oplus_A -d$ are true in $\boldsymbol{A}$, then $(0_A, 1_A)\otimes(0_A, 1_A) = (-1_A, 0_A) = -(1_A, 0_A)$ in the wholly extended abstract prearithmetic $\boldsymbol{A}_{\text{ext}}$.*

Indeed, by Propositions 3.4.27 and 3.4.28, we have

$$(0_A, 1_A) \otimes (0_A, 1_A) = ((0_A \otimes_A 0_A) \ominus_A (1_A \otimes_A 1_A), (0_A \otimes_A 1_A)$$

$$\oplus_A (1_A \otimes_A 0_A)),$$

$$(0_A \ominus_A 1_A, 0_A \oplus_A 0_A) = (-1_A, 0_A) = -(1_A, 0_A).$$

In other words, the square of the element $(0_A, 1_A)$ is equal to negative one because $(1_A, 0_A)$ is one in the wholly extended abstract prearithmetic $\boldsymbol{A}_{\text{ext}}$. It means that the element $\boldsymbol{i} = (0_A, 1_A)$ plays the role of the imaginary number i.

Now we study inverse elements in complexifications of abstract prearithmetics.

Let us consider an associative distributive wholly extended abstract prearithmetic $\boldsymbol{A} = (A; \oplus_A, \otimes_A, \ominus_A, \oslash_A, \leq_A)$ with zero 0_A and multiplicative one 1_A.

Proposition 3.4.30. *If for the sum $(a \otimes_A a) \oplus_A (b \otimes_A b)$ of squares of elements a and b from $\boldsymbol{A}$, there is the inverse element $((a \otimes_A a) \oplus_A$*

$(b \otimes_A b))^{-1}$, *then the element* $\boldsymbol{a} = (a, b)$ *from the wholly extended abstract prearithmetic* $\boldsymbol{A}_{\text{ext}}$ *also has the inverse element* $\boldsymbol{a}^{-1} = (a, b)^{-1}$.

Proof: Let us take the element $(a \otimes_A ((a \otimes_A a) \oplus_A (b \otimes_A b))^{-1}, -b \otimes_A ((a \otimes_A a) \oplus_A (b \otimes_A b))^{-1})$ from $\boldsymbol{A}_{\text{ext}}$ and show that it is inverse to the element (a, b) from $\boldsymbol{A}_{\text{ext}}$. Indeed, by Lemmas 2.1.16 and 2.1.17 and Propositions 3.4.27 and 3.4.28, we have the following sequence of equalities:

$$
\begin{aligned}
&(a, b) \otimes (a \otimes_A ((a \otimes_A a) \oplus_A (b \otimes_A b))^{-1}, -b \otimes_A ((a \otimes_A a) \oplus_A (b \otimes_A b))^{-1})\\
&\quad = (a \otimes_A (a \otimes_A ((a \otimes_A a) \oplus_A (b \otimes_A b))^{-1}) \ominus_A b \otimes_A\\
&\qquad \times (-b \otimes_A ((a \otimes_A a) \oplus_A (b \otimes_A b))^{-1}), b \otimes_A (a \otimes_A ((a \otimes_A a) \oplus_A\\
&\qquad \times (b \otimes_A b))^{-1}) \oplus_A a \otimes_A (-b \otimes_A ((a \otimes_A a) \oplus_A (b \otimes_A b))^{-1}))\\
&\quad = ((a \otimes_A a) \otimes_A ((a \otimes_A a) \oplus_A (b \otimes_A b))^{-1}) \ominus_A\\
&\qquad \times (-b \otimes_A b) \otimes_A ((a \otimes_A a) \oplus_A (b \otimes_A b))^{-1}), (b \otimes_A a) \otimes_A ((a \otimes_A a) \oplus_A\\
&\qquad \times (b \otimes_A b))^{-1}) \oplus_A (-b \otimes_A a) \otimes_A ((a \otimes_A a) \oplus_A (b \otimes_A b))^{-1}))\\
&\quad = (((a \otimes_A a) \ominus_A (-b \otimes_A b)) \otimes_A ((a \otimes_A a) \oplus_A (b \otimes_A b))^{-1},\\
&\qquad \times ((b \otimes_A a) \oplus_A (-b \otimes_A a)) \otimes_A ((a \otimes_A a) \oplus_A (b \otimes_A b))^{-1})\\
&\quad = (((a \otimes_A a) \oplus_A (b \otimes_A b)) \otimes_A ((a \otimes_A a) \oplus_A (b \otimes_A b))^{-1},\\
&\qquad \times ((b \otimes_A a) \oplus_A (-b \otimes_A a)) \otimes_A ((a \otimes_A a) \oplus_A (b \otimes_A b))^{-1}) = (1_A, 0_A)
\end{aligned}
$$

because

$$((a \otimes_A a) \oplus_A (b \otimes_A b)) \otimes_A ((a \otimes_A a) \oplus_A (b \otimes_A b))^{-1} = 1_A$$

and

$$
\begin{aligned}
&((b \otimes_A a) \oplus_A (-b \otimes_A a)) \otimes_A ((a \otimes_A a) \oplus_A (b \otimes_A b))^{-1}\\
&\quad = ((b \otimes_A a) \ominus_A (b \otimes_A a)) \otimes_A ((a \otimes_A a) \oplus_A (b \otimes_A b))^{-1}\\
&\quad = 0_A \otimes_A ((a \otimes_A a) \oplus_A (b \otimes_A b))^{-1} = 0_A.
\end{aligned}
$$

It means that the element $(a \otimes_A ((a \otimes_A a) \oplus_A (b \otimes_A b))^{-1}, -b \otimes_A ((a \otimes_A a) \oplus_A (b \otimes_A b))^{-1})$ from $\boldsymbol{A}_{\text{ext}}$ is inverse to the element (a, b) from $\boldsymbol{A}_{\text{ext}}$.

Proposition is proved. □

Corollary 3.4.26. *If any element* a *from a wholly extended abstract prearithmetic* $\boldsymbol{A} = (A; \oplus_A, \otimes_A, \ominus_A, \oslash_A, \leq_A)$ *has the inverse element* a^{-1}, *then any element* (a, b) *from the wholly extended abstract prearithmetic* $\boldsymbol{A}_{\text{ext}}$ *also has the inverse element.*

Using inverse elements, we can define division $\oslash$ in the wholly extended abstract prearithmetic $\boldsymbol{A}_{\text{ext}}$by the following formula:

$$(a, b) \oslash_A (c, d) = (a, b) \otimes (c, d)^{-1}. \tag{3.15}$$

The properties of four operations defined for the operational weak C-arithmetic $\boldsymbol{A}_{\text{ext}}$ and results describing representations of the Diophantine arithmetic $\boldsymbol{C}$ from (Feferman, 1974) allow obtaining the following result.

Theorem 3.4.3. *The wholly extended abstract prearithmetic* $\boldsymbol{R}_{\text{ext}}$ *coincides with the Diophantine arithmetic* $\boldsymbol{C}$ *of complex numbers.*

Proof is left as an exercise. □

Let us study relations between operational and functional C-prearithmetics.

Theorem 3.4.4. *An operational C-prearithmetic* $\boldsymbol{A}_{\text{ext}}$ *is weakly projective with respect to the (conventional) Diophantine arithmetic* $\boldsymbol{C} = (C; +, \cdot, -, \div, \leq)$ *of complex numbers if* $gh = \mathbf{1}_R$ *for the projector* g *and the coprojector* h *of the R-prearithmetic* $\boldsymbol{A} = (R; \oplus, \otimes, \ominus, \oslash, \leq)$, *i.e.,* $\boldsymbol{A}_{\text{ext}}$ *is a functional C-prearithmetic.*

Proof: If $\boldsymbol{A}_{\text{ext}}$ is an operational C-prearithmetic over $\boldsymbol{A}$, then by definition, the wholly extended abstract prearithmetic $\boldsymbol{A} = (A; \oplus_A, \otimes_A, \ominus_A, \oslash_A, \leq_A)$ is an R-prearithmetic, i.e., it is weakly projective with respect to the (conventional) Diophantine arithmetic $\boldsymbol{R} = (R; +, \cdot, -, \div, \leq)$ of real numbers. It means that if g: $A \to R$ is the projector and coprojector h: $R \to A$ is the coprojector of $\boldsymbol{A}$ with respect to $\boldsymbol{R}$, operations in $\boldsymbol{A}$ are expressed through the conventional operations with real numbers in the following way:

$$a \oplus_A b = h(g(a) + g(b)),$$

$$a \otimes_A b = h(g(a) \cdot g(b)),$$

$$a \ominus_A b = h(g(a) - g(b)),$$

$$a \oslash_A b = h(g(a) \div g(b)).$$

We can also define weak projectivity of the prearithmetic $\boldsymbol{A}_{\text{ext}} = (A \times A; \oplus_A, \otimes_A, \ominus_A, \oslash_A, \leq_A)$ with respect to the (conventional) Diophantine arithmetic $\boldsymbol{C}$, taking the projector $g^2 : A \to C$ and coprojector $h^2 : C \to A$, which are defined as the mappings $g^2 : A \times A \to C$ and $h^2 : C \to A \times A$ where $g^2(a,b) = g(a) + ig(b)$ and $h^2(a+ib) = (h(a), h(b))$. We see that $h^2(g^2(a,b)) = (h(g(a)), h(g(b)))$.

Let us look how the mappings g^2 and h^2 allow representing addition $\oplus$ by means of addition $+$ in the Diophantine arithmetic $\boldsymbol{C}$.

$$\begin{aligned}(a,b) \oplus (c,d) &= (a \oplus_A c, b \oplus_A d) = (h(g(a) + g(c)), h(g(b) + g(d)))\\ &= h^2(g(a) + g(c) + i(g(b) + g(d))) = h^2(g(a) + g(c) + ig(b) + ig(d))\\ &= h^2(g(a) + ig(b) + g(c) + ig(d)) = h^2(g^2(a,b) + g^2(c,d))\end{aligned}$$

It means that addition $\oplus$ in the weak C-arithmetic $\boldsymbol{A}_{\text{ext}}$ is weakly projective with respect to the Diophantine arithmetic $\boldsymbol{C}$ with the projector g^2 and coprojector h^2, i.e., addition $\oplus$ satisfies identity (3.11).

Let us look how the mappings g^2 and h^2 allow representing subtraction by means of subtraction $-$ in the Diophantine arithmetic $\boldsymbol{C}$.

$$\begin{aligned}(a,b) \ominus (c,d) &= (a \ominus_A c, b \ominus_A d) = (h(g(a) - g(c)), h(g(b) - g(d)))\\ &= h^2(g(a) - g(c) + i(g(b) - g(d))) = h^2(g(a) - g(c) + ig(b) - ig(d))\\ &= h^2((g(a) + ig(b)) - (g(c) + ig(d))) = h^2(g^2(a,b) - g^2(c,d)).\end{aligned}$$

It means that subtraction $\ominus$ in the operational C-prearithmetic $\boldsymbol{A}_{\text{ext}}$ is weakly projective with respect to the Diophantine arithmetic $\boldsymbol{C}$ with the projector g^2 and coprojector h^2, i.e., subtraction satisfies identity (3.12).

Now let us look how the mappings g^2 and h^2 allow representing multiplication $\otimes$ by means of multiplication $\cdot$ in the Diophantine arithmetic $\boldsymbol{C}$ taking into account that $gh = \mathbf{1}_R$.

$$\begin{aligned}(a,b) \otimes (c,d) &= ((a \otimes_A c) \ominus_A (b \otimes_A d), (a \otimes_A d) \oplus_A (b \otimes_A c))\\ &= (h(g(a) \cdot g(c)) \ominus_A h(g(b) \cdot g(d)),\\ &\qquad h(g(a) \cdot g(d)) \oplus_A h(g(b) \cdot g(c)))\\ &= (h(g(h(g(a) \cdot g(c))) - g(h(g(b) \cdot g(d)))),\\ &\qquad h(g(h(g(a) \cdot g(d))) + g(h(g(b) \cdot g(c)))))\end{aligned}$$

$$= h^2((g(h(g(a) \cdot g(c))) - g(h(g(b) \cdot g(d)))$$
$$+ i(g(h(g(a) \cdot g(d))) + g(h(g(b) \cdot g(c)))))$$
$$= h^2((g(a) \cdot g(c)) - (g(b) \cdot g(d))$$
$$+ i((g(a) \cdot g(d)) + (g(b) \cdot g(c)))$$
$$= h^2((g(a) + ig(b)) \cdot (g(c)) + ig(d)))$$
$$= h^2(g^2(a, b) \cdot g^2(c, d)).$$

It means that multiplication $\otimes$ in the operational C-prearithmetic $\boldsymbol{A}_{\text{ext}}$ is weakly projective with respect to the Diophantine arithmetic $\boldsymbol{C}$ with the projector g^2 and coprojector h^2, i.e., multiplication $\otimes$ satisfies identity (3.13).

In a similar way, we can check identity (3.14) for division $\oslash$ taking into account that division $\oslash$ is expressed (cf. Proposition 3.4.30) by the formula (3.15) and $gh = \mathbf{1}_R$.

As operations in the operational C-prearithmetic $\boldsymbol{A}_{\text{ext}}$ satisfy identities (3.11)–(3.14), it is weakly projective with respect to the Diophantine arithmetic $\boldsymbol{C}$.

Theorem is proved. □

In such a way, Theorem 3.4.4 establishes relations between operational and functional C-prearithmetics.

Corollary 3.4.27. *If for the projector g and the coprojector h of the R-prearithmetic $\boldsymbol{A} = (R; \oplus, \otimes, \ominus, \oslash, \leq)$, the equality $gh = \mathbf{1}_R$ is true, then the operational C-prearithmetic $\boldsymbol{A}_{\text{ext}}$ is a functional C-prearithmetic with the projector $g^2 : A \to C$ and coprojector $h^2 : C \to A$.*

Exact projectivity provides even closer relations between operational and functional C-prearithmetics.

Theorem 3.4.5. *If an R-prearithmetic $\boldsymbol{A} = (R; \oplus, \otimes, \ominus, \oslash, \leq_A)$ is exactly projective with respect to the Diophantine arithmetic $\boldsymbol{R}$ of real numbers, then the C-arithmetic $\boldsymbol{A}_{\text{ext}}$ is exactly projective with respect to the Diophantine arithmetic $\boldsymbol{C}$ of complex numbers.*

Proof: Let us assume that an R-prearithmetic $\boldsymbol{A} = (R; \oplus, \otimes, \ominus, \oslash, \leq_A)$ is (exactly) projective with respect to the Diophantine arithmetic $\boldsymbol{R}$ of real numbers with the projector g and the coprojector h. It means that there are mappings g: $A \to R$ and h: $R \to A$ such that hg and gh are equal to the identity mapping of R, i.e., $hg = gh = \mathbf{1}_R$. Then we can define two mappings $g^2 : A \times A \to C$ and $h^2 : C \to A \times A$ where $g^2(a, b) = g(a) + ig(b)$ and $h^2(a + ib) = (h(a), h(b))$.

By definition, we have

$$h^2(g^2(a, b)) = h^2(g(a) + ig(b)) = (h(g(a)), h(g(b))) = (a, b),$$

$$g^2(h^2(a + ib)) = g^2((h(a), h(b))) = g(h(a)) + ig(h(b)) = a + ib.$$

It means that $h^2g^2 = \mathbf{1}_A$ and $g^2h^2 = \mathbf{1}_C$. Consequently, by Theorem 3.4.4, the C-arithmetic $\boldsymbol{A}_{\text{ext}}$ is exactly projective with respect to the Diophantine arithmetic $\boldsymbol{C}$ of complex numbers.

Theorem is proved. □

Theorems 3.4.5 and 3.2.2 imply the following result.

Corollary 3.4.28. *If an R-prearithmetic $\boldsymbol{A} = (R; \oplus, \otimes, \ominus, \oslash, \leq_A)$ is isomorphic to the Diophantine arithmetic $\boldsymbol{R}$ of real numbers, then the C-arithmetic $\boldsymbol{A}_{\text{ext}}$ is isomorphic to the Diophantine arithmetic $\boldsymbol{C}$ of complex numbers.*

Now we come to non-Diophantine arithmetics of complex numbers.

Definition 3.4.5. A weak C-arithmetic $\boldsymbol{A} = (R; \oplus, \otimes, \ominus, \oslash, \leq_A)$ is called a *complex-number arithmetic*, or simply, a *projective non-Diophantine arithmetic* of complex numbers if it is projective with respect to the Diophantine arithmetic $\boldsymbol{C}$ of complex numbers.

Proposition 3.4.3 allows proving the following result, which shows additional relations between operational and functional C-prearithmetics. Let us treat $\boldsymbol{R}$ as the subarithmetic of all real numbers in the Diophantine arithmetic $\boldsymbol{C}$.

Theorem 3.4.6. *If a projective non-Diophantine arithmetic $\boldsymbol{B} = (B; \oplus, \otimes, \ominus, \oslash, \leq_A)$ is projective with respect to the Diophantine arithmetic $\boldsymbol{C}$ with the projector g and the coprojector h, satisfies the following conditions: $g(h(\boldsymbol{R})) \subseteq \boldsymbol{R}$ and there is an element e such that $g(e) = i$, then as a partially extended abstract prearithmetic, $\boldsymbol{B}$ is isomorphic to the operational C-prearithmetic $\boldsymbol{A}_{\text{ext}} = (A^2; \oplus_A, \otimes_A, \ominus_A, \oslash_A, \leq_A)$ where $A = h(\boldsymbol{R})$.*

Proof: Let us assume that a functional C-prearithmetic $\boldsymbol{B} = (B; \oplus, \otimes, \ominus, \oslash, \leq_A)$ is projective with respect to the Diophantine arithmetic $\boldsymbol{C}$ with the projector g and the coprojector h, and there are elements b, e in $\boldsymbol{B}$ such that $g(b) = 1, g(e) = i$, and $g(h(\boldsymbol{R})) \subseteq \boldsymbol{R}$.

At first, we show that the image $h(\boldsymbol{R})$ of the real number subarithmetic $\boldsymbol{R}$ of the Diophantine arithmetic $\boldsymbol{C}$ is a subprearithmetic of $\boldsymbol{B}$. To do this, we take two elements a and b from the image $h(\boldsymbol{R})$ and consider operations with them. As $a = h(x)$ and $c = h(y)$ for some elements x and y from $\boldsymbol{R}$, we have

$$a \oplus c = h(g(a) + g(c)) = h(g(h(x)) + g(h(y))) = h(u + v),$$

$$a \ominus c = h(g(a) - g(c)) = h(g(h(x)) - g(h(y))) = h(u - v),$$

$$a \otimes c = h(g(a) \cdot g(c)) = h(g(h(x)) \cdot g(h(y))) = h(u \cdot v),$$

$$a \oslash c = h(g(a) \div g(c)) = h(g(h(x)) \div g(h(y))) = h(u \div v)$$

because $g(h(x)) = u$ and $g(h(y)) = v$ are elements from $\boldsymbol{R}$, and we see that $a \oplus c, a \ominus c, a \otimes c$ and $a \oslash c$ belong to $h(\boldsymbol{R})$, i.e., $h(\boldsymbol{R})$ is a subprearithmetic of $\boldsymbol{B}$.

Taking the restriction k of the mapping h onto $\boldsymbol{R}$, we see that operations in $h(\boldsymbol{R})$ are defined by the following formulas:

$$a \oplus b = k(g(a) + g(b)),$$

$$a \otimes b = k(g(a) \cdot g(b)),$$

$$a \ominus b = k(g(a) - g(b)),$$

$$a \oslash b = k(g(a) \div g(b)).$$

It means that the wholly extended abstract prearithmetic $\boldsymbol{A} = (A; \oplus_A, \otimes_A, \ominus_A, \oslash_A, \leq_A)$, where $A = h(\boldsymbol{R})$ is projective with respect to the Diophantine arithmetic $\boldsymbol{R}$ with the projector g and the coprojector k, is an operational R-prearithmetic.

Theorem is proved. □

Definition 3.4.6. An operational prearithmetic $\boldsymbol{B} = (C; \oplus, \otimes, \ominus, \oslash, \leq_A)$ is called an *operational* IC-*arithmetic* if it is the complexification of an IR-arithmetic $\boldsymbol{A} = (R; +, \cdot, -, \div, \leq)$.

Operational IC-arithmetics form a distinct class of non-Diophantine arithmetics of complex numbers.

Operational IC-arithmetics have many properties similar to properties of the Diophantine arithmetic $\boldsymbol{R}$ of complex numbers. For instance, Theorem 3.4.5 implies the following results.

Theorem 3.4.7. *Any operational IC-arithmetic* $\boldsymbol{A} = (R; \oplus_A, \otimes_A, \ominus_A, \oslash_A, \leq_A)$ *is isomorphic to the Diophantine arithmetic* $\boldsymbol{C}$.

Proof is left as an exercise. □

This result allows describing many properties of IC-arithmetics.

Theorem 3.4.8. *Any operational IC-arithmetic* $\boldsymbol{B}$ *has the following properties*:

(*a*) *addition is commutative*;
(*b*) *multiplication is commutative*;
(*c*) *addition is associative*;
(*d*) *multiplication is associative*;
(*e*) *multiplication is distributive with respect to addition*;
(*f*) *multiplication is distributive with respect to subtraction*;
(*g*) $\boldsymbol{B}$ *has the additive and multiplicative zero* 0_A;
(*h*) $\boldsymbol{B}$ *has the multiplicative one* 1_A;
(*i*) $\boldsymbol{B}$ *has continuum of elements.*

Corollary 3.4.29. *Any operational IC-arithmetic* $\boldsymbol{B}$ *is a partially ordered field.*

Definition 3.4.7. The isomorphism f between an operational IC-arithmetic $\boldsymbol{B}$ and $\boldsymbol{C}$ is called the *functional parameter* of $\boldsymbol{B}$.

It is often denoted as $f_B : B \to C$.

Often the carrier B of an operational IC-arithmetic $\boldsymbol{B}$ is a set of real numbers. However, in general, the carrier B of an operational IC-arithmetic $\boldsymbol{B}$ can be any set for which a one-to-one mapping (bijection) $f_B : B \to C$ exists. This bijection determines a one-to-one naming of the elements from B by complex numbers (Burgin, 2011). Taking this into account, we can denote the element from B that is mapped onto the complex number q by q_B.

3.4.3. *Stratified arithmetics and prearithmetics of complex numbers*

It seems to me, that if statesmen had a little more arithmetic, or were accustomed to calculation, wars would be much less frequent.

Benjamin Franklin

Now we construct and study stratified prearithmetics and arithmetics of complex numbers.

As in the operational approach, we represent complex numbers by pairs of real numbers, for which we properly define arithmetical operations. Only in this case, we take the set of all pairs, in which the first element is from one wholly extended abstract prearithmetic $\boldsymbol{A}$ and the second element is from another wholly extended abstract prearithmetic $\boldsymbol{D}$. Then we define four operations and a partial order in this set. There are different ways to define the necessary operations and a partial order but here we follow the schema of the Diophantine arithmetic $\boldsymbol{C}$ of complex numbers.

Note that wholly extended abstract prearithmetics $\boldsymbol{A}$ and $\boldsymbol{D}$ can coincide but in a general case, they are different.

Let us consider two wholly extended abstract prearithmetics $\boldsymbol{A} = (A; \oplus_A, \otimes_A, \ominus_A, \oslash_A, \leq_A)$ and $\boldsymbol{D} = (D; \oplus_D, \otimes_D, \ominus_D, \oslash_D, \leq_D)$, mappings $g\colon A \to D$ and $h\colon D \to A$ and the set $B = \{(a, d); a \in A \text{ and } d \in D\}$ of pairs (a, d). Then we build the stratified prearithmetic $\boldsymbol{E}_{A,D} = (B; \oplus, \otimes, \ominus, \oslash, \leq)$ of complex numbers defining operations and a partial order by the following formulas:

$$(a, b) \oplus (c, d) = (a \oplus_A c, b \oplus_D d),$$

$$(a, b) \ominus (c, d) = (a \ominus_A c, b \ominus_D d),$$

$$(a, b) \otimes (c, d) = ((a \otimes_A c) \ominus_A (h(b) \otimes_A h(d)), (g(a) \otimes_D d) \oplus_D (b \otimes_D g(c))),$$

$$(a, b) \leq (c, d) \text{ if and only if } a \leq_A c \text{ and } b \leq_D d.$$

Division is defined later after we study properties of the defined C-prearithmetic $\boldsymbol{E}_{A,D}$.

To explicate relations between the Diophantine arithmetic of complex numbers $\boldsymbol{C}$ and the stratified prearithmetic $\boldsymbol{E}_{A,D}$, we use notation used for operational prearithmetics of complex numbers. Namely, in a pair (complex number) $\boldsymbol{a} = (a, d)$, we regard its first component a as the real part of the element (complex number) $\boldsymbol{a}$ from $\boldsymbol{E}_{A,D}$ denoting it by $\mathscr{R}\boldsymbol{a}$, and treat its second component d as the imaginary part of the element (complex number) $\boldsymbol{a}$ denoting it by $\mathscr{I}\boldsymbol{a}$. This gives another representation of the complex number $\boldsymbol{a}$:

$$\boldsymbol{a} = (a, d) = \mathscr{R}\boldsymbol{a} \oplus \boldsymbol{i}\mathscr{I}\boldsymbol{a}.$$

Here the element $\boldsymbol{i} = (0_A, 1_D)$ plays the role of the imaginary unit i in $\boldsymbol{E}_{A,D}$.

This makes it possible to treat the prearithmetic $\boldsymbol{E}_{A,D}$ as $\boldsymbol{A} \oplus \boldsymbol{iD}$ and the set B as $A \oplus \boldsymbol{i}D$.

Using the component notation for elements $\boldsymbol{a}$ and $\boldsymbol{c}$ from $\boldsymbol{E}_{A,D}$, we have the following definition of operations and order:

$$\boldsymbol{a} \oplus \boldsymbol{c} = (\mathscr{R}\boldsymbol{a} \oplus_A \mathscr{R}\boldsymbol{c}, \mathscr{I}(\boldsymbol{a} \oplus \boldsymbol{c}) = \mathscr{I}\boldsymbol{a} \oplus_D \mathscr{I}\boldsymbol{c}),$$

$$\boldsymbol{a} \ominus \boldsymbol{c} = (\mathscr{R}\boldsymbol{a} \ominus_A \mathscr{R}\boldsymbol{c}, \mathscr{I}(\boldsymbol{a} \ominus_D \mathscr{I}\boldsymbol{c}) = \mathscr{I}\boldsymbol{a} \ominus_D \mathscr{I}\boldsymbol{c}),$$

$$\boldsymbol{a} \otimes \boldsymbol{c} = ((\mathscr{R}\boldsymbol{a} \otimes_A \mathscr{R}\boldsymbol{c}) \ominus_A (h(\mathscr{I}\boldsymbol{a}) \otimes_A h(\mathscr{I}\boldsymbol{c})), (g(\mathscr{R}\boldsymbol{a}) \otimes_D \mathscr{I}\boldsymbol{c})$$
$$\oplus_D(\mathscr{I}\boldsymbol{a} \otimes_D g(\mathscr{R}\boldsymbol{c}))).$$

Consequently,

$$\mathscr{R}(\boldsymbol{a} \oplus \boldsymbol{c}) = \mathscr{R}\boldsymbol{a} \oplus_A \mathscr{R}\boldsymbol{c},$$

$$\mathscr{I}(\boldsymbol{a} \oplus \boldsymbol{c}) = \mathscr{I}\boldsymbol{a} \oplus_D \mathscr{I}\boldsymbol{c},$$

$$\mathscr{R}(\boldsymbol{a} \ominus \boldsymbol{c}) = \mathscr{R}\boldsymbol{a} \ominus_A \mathscr{R}\boldsymbol{c},$$

$$\mathscr{I}(\boldsymbol{a} \ominus \boldsymbol{c}) = \mathscr{I}\boldsymbol{a} \ominus_D \mathscr{I}\boldsymbol{c},$$

$$\mathscr{R}(\boldsymbol{a} \otimes \boldsymbol{c}) = (\mathscr{R}\boldsymbol{a} \otimes_A \mathscr{R}\boldsymbol{c}) \ominus_A (h(\mathscr{I}\boldsymbol{a}) \otimes_A h(\mathscr{I}\boldsymbol{c})),$$

$$\mathscr{I}(\boldsymbol{a} \otimes \boldsymbol{c}) = (g(\mathscr{R}\boldsymbol{a}) \otimes_D \mathscr{I}\boldsymbol{c}) \oplus_D (\mathscr{I}\boldsymbol{a} \otimes_D g(\mathscr{R}\boldsymbol{c})).$$

In addition, we define the order $\leq$ in $\boldsymbol{E}_{A,D}$ by the following rule:

$$\boldsymbol{a} \leq \boldsymbol{c} \text{ if and only if } \mathscr{R}\boldsymbol{a} \leq_A \mathscr{R}\boldsymbol{c} \text{ and } \mathscr{I}\boldsymbol{a} \leq_D \mathscr{I}\boldsymbol{c}.$$

Thus, using constructed operations and a partial order, we obtain the wholly extended abstract prearithmetic $\boldsymbol{E}_{A,D} = (B; \oplus, \otimes, \ominus, \oslash, \leq)$ on the set B of all pairs of elements from $\boldsymbol{A}$ and $\boldsymbol{D}$.

Definition 3.4.8.

(a) If $\boldsymbol{A}$ and $\boldsymbol{D}$ are R-prearithmetics, then $\boldsymbol{E}_{A,D}$ is called a stratified C-prearithmetic over the pair $(\boldsymbol{A}, \boldsymbol{D})$.
(b) If $\boldsymbol{A}$ and $\boldsymbol{D}$ are weak R-arithmetics, then $\boldsymbol{E}_{A,D}$ is called a weak stratified C-arithmetic over the pair $(\boldsymbol{A}, \boldsymbol{D})$.

Naturally, if $\boldsymbol{A} = \boldsymbol{D}$ and g and h are identical mappings of $\boldsymbol{A}$, then $\boldsymbol{E}_{A,D} = \boldsymbol{A}_{\text{ext}}$.

Taking the projector q and coprojector r of $\boldsymbol{A}$ with respect to $\boldsymbol{R}$ and the projector f and coprojector p of $\boldsymbol{D}$ with respect to $\boldsymbol{R}$, we can express operations in the stratified C-prearithmetic $\boldsymbol{E}_{A,D}$ through the conventional operations with real numbers in the following way:

$$(a, b) \oplus (c, d) = (q(r(a) + r(c)), p(f(b) + f(d))),$$

$$(a, b) \ominus (c, d) = (q(r(a) - r(c)), p(f(b) - f(d))),$$

$$(a, b) \otimes (c, d) = (q(r(q(r(a) \cdot r(c))) - r(h(p(f(b) \cdot f(d)))),$$
$$p(f(p(f(g(a)) \cdot f(d)))) + p(f(g(b) \cdot f(g(c))))).$$

The wholly extended abstract prearithmetic $\boldsymbol{E}_{A,D}$ inherits many properties of the R-prearithmetics $\boldsymbol{A}$ and $\boldsymbol{D}$. In particular, the wholly extended abstract prearithmetic $\boldsymbol{E}_{A,D}$ inherits zeros and ones.

Proposition 3.4.31. *If a wholly extended abstract prearithmetic $\boldsymbol{A} = (A; \oplus_A, \otimes_A, \ominus_A, \oslash_A, \leq_A)$ has zero 0_A, a wholly extended abstract prearithmetic $\boldsymbol{D} = (D; \oplus_D, \otimes_D, \ominus_D, \oslash_D, \leq_D)$ has zero $0_D, g(0_A) = 0_D$ and $h(0_D) = 0_A$, then the wholly extended abstract prearithmetic $\boldsymbol{E}_{A,D}$ also has zero.*

Proof: Let us take the element $(0_A, 0_D)$ from $\boldsymbol{E}_{A,D}$ and show that it is zero in it. Indeed,

$$(a, b) \oplus (0_A, 0_D) = (a \oplus_A 0_A, b \oplus_D 0_D) = (a, b).$$

It means that $(0_A, 0_D)$ is an additive zero in $\boldsymbol{E}_{A,D}$. In addition,

$$(a, b) \otimes (0_A, 0_D) = ((a \otimes_A 0_A) \ominus_A (h(b) \otimes_A h(0_D)), ((g(a) \otimes_D 0_D) \oplus_D$$
$$\times (b \otimes_D (g(0_A))) = (0_A, 0_D)$$

because $g(0_A) = 0_D, h(0_D) = 0_A, a \otimes_A 0_A = 0_A, h(b) \otimes_A 0_D = 0_A, b \otimes_D 0_D = 0_D, g(a) \otimes_D 0_D = 0_D, 0_A \ominus_A 0_A = 0_A$, and $0_D \oplus_D 0_D = 0_D$.
It means that $(0_A, 0_D)$ is a multiplicative zero and thus zero in $\boldsymbol{E}_{A,D}$.

Proposition is proved. □

Propositions 3.3.15 and 3.4.31 imply the following results.

Corollary 3.4.30. *If an R-prearithmetic $\boldsymbol{A} = (A; \oplus_A, \otimes_A, \ominus_A, \oslash_A, \leq_A)$ with the projector p contains an element d such that $p(d) = 0$, an R-prearithmetic $\boldsymbol{D} = (D; \oplus_D, \otimes_D, \ominus_D, \oslash_D, \leq_D)$ with the projector q contains an element c such that $q(c) = 0, g(d) = c$ and $h(c) = d$, then the stratified C-prearithmetic $\boldsymbol{E}_{A,D}$ also has zero.*

Corollary 3.4.31. *If $qr = \mathbf{1}_R$ for an R-prearithmetic $\boldsymbol{A} = (A; \oplus_A, \otimes_A, \ominus_A, \oslash_A, \leq_A)$ with the projector q and the coprojector $r, fp = \mathbf{1}_R$ for an R-prearithmetic $\boldsymbol{D} = (D; \oplus_D, \otimes_D, \ominus_D, \oslash_D, \leq_D)$ with the projector f and the coprojector $p, g(0_A) = 0_D$ and $h(0_D) = 0_A$, then the stratified C-prearithmetic $\boldsymbol{E}_{A,D}$ has zero 0.*

Corollary 3.4.32. *If a weak R-arithmetics $\boldsymbol{A} = (A; \oplus_A, \otimes_A, \ominus_A, \oslash_A, \leq_A)$ and $\boldsymbol{D} = (D; \oplus_D, \otimes_D, \ominus_D, \oslash_D, \leq_D)$ are exactly projective with respect to the Diophantine arithmetic $\boldsymbol{R}$ of real numbers and g and h are isomorphisms, then the weak stratified C-arithmetic $\boldsymbol{E}_{A,D}$ has zero.*

Similar statements are true for the multiplicative one.

Proposition 3.4.32. *If a wholly extended abstract prearithmetic$\boldsymbol{A} = (A; \oplus_A, \otimes_A, \ominus_A, \oslash_A, \leq_A)$ with zero 0_A has the multiplicative one 1_A, a wholly extended abstract prearithmetic $\boldsymbol{D} = (D; \oplus_D, \otimes_D, \ominus_D, \oslash_D, \leq_D)$ with zero 0_D has the multiplicative one $1_D, g(1_A) = 1_D$ and $h(0_D) = 0_A$, then the weak stratified C-arithmetic $\boldsymbol{E}_{A,D}$ also has the multiplicative one.*

Proof: Let us take the element $(1_A, 0_D)$ from $\boldsymbol{E}_{A,D}$ and show that it is zero in $\boldsymbol{E}_{A,D}$. Indeed,

$$(a, b) \otimes (1_A, 0_D) = ((a \otimes_A 1_A) \ominus_A (h(b) \otimes_A h(0_D)), ((g(a) \otimes_D 0_D)$$

$$\oplus_D (b \otimes_D (g(1_A))) = (a \ominus_A 0_A, 0_D \oplus_D b) = (a, b).$$

It means that $(1_A, 0_D)$ is the multiplicative one in $\boldsymbol{E}_{A,D}$.
Proposition is proved. □

Propositions 3.3.16 and 3.4.32 imply the following results.

Corollary 3.4.33. *If an R-prearithmetic $\boldsymbol{A} = (A; \oplus, \otimes, \ominus, \oslash, \leq_A)$ with the projector p contains elements d and a such that $p(a) = 1$ and $p(d) = 0$, an R-prearithmetic $\boldsymbol{D} = (D; \oplus, \otimes, \ominus, \oslash, \leq_A)$ with the projector q contains elements b and c such that $q(b) = 1$ and $q(c) = 0, g(a) = b, h(b) = a, g(d) = c$, and $h(c) = d$, then the stratified C-prearithmetic $\boldsymbol{E}_{A,D}$ has the multiplicative one.*

Corollary 3.4.34. *If $qr = \mathbf{1}_R$ for an R-prearithmetic $\boldsymbol{A} = (A; \oplus, \otimes, \ominus, \oslash, \leq_A)$ with the projector q and the coprojector $r, fp = \mathbf{1}_R$ for an R-prearithmetic $\boldsymbol{D} = (D; \oplus, \otimes, \ominus, \oslash, \leq_A)$ with the projector f and the coprojector $p, g(1_A) = 1_D$ and $h(0_D) = 0_A$, then the stratified C-prearithmetic $\boldsymbol{E}_{A,D}$ has the multiplicative one.*

Corollary 3.4.35. *If a R-prearithmetics* $\boldsymbol{A} = (A; \oplus_A, \otimes_A, \ominus_A, \oslash_A, \leq_A)$ *and* $\boldsymbol{D} = (D; \oplus_D, \otimes_D, \ominus_D, \oslash_D, \leq_D)$ *are exactly projective with respect to the Diophantine arithmetic* $\boldsymbol{R}$ *of real numbers and* g *and* h *are isomorphisms, then the stratified C-prearithmetic* $\boldsymbol{E}_{A,D}$ *has the multiplicative one.*

Let us consider an R-prearithmetic $\boldsymbol{A} = (A; \oplus_A, \otimes_A, \ominus_A, \oslash_A, \leq_A)$ with the projector $g_A : A \to R$ and coprojector $h_A : R \to A$, an R-prearithmetic $\boldsymbol{D} = (D; \oplus_D, \otimes_D, \ominus_D, \oslash_D, \leq_D)$ with the projector $g_D : A \to R$ and coprojector $h_D : R \to A$, mappings g: $A \to D, h$: $D \to A$ and the stratified prearithmetic $\boldsymbol{E}_{A,D} = (B; \oplus, \otimes, \ominus, \oslash, \leq)$ of complex numbers.

Proposition 3.4.33. *A stratified C-prearithmetic* $\boldsymbol{E}_{A,D}$ *is weakly projective in addition and subtraction with respect to the (conventional) Diophantine arithmetic* $\boldsymbol{C} = (C; +, \cdot, -, /, \leq)$ *of complex numbers.*

Proof: By definition, operations in $\boldsymbol{A}$ are expressed by means of the conventional operations with real numbers in the following way. Let us take arbitrary elements a and b from A. Then we have

$$a \oplus_A b = h_A(g_A(a) + g_A(b)),$$

$$a \otimes_A b = h_A(g_A(a) \cdot g_A(b)),$$

$$a \ominus_A b = h_A(g_A(a) - g_A(b)),$$

$$a \oslash_A b = h_A(g_A(a) \div g_A(b)).$$

Likewise, operations in $\boldsymbol{D}$ are expressed by means of the conventional operations with real numbers in the following way. Let us take arbitrary elements c and d from D. Then we have

$$c \oplus_D d = h_D(g_D(c) + g_D(d)),$$

$$c \otimes_D d = h_D(g_D(c) \cdot g_D(d)),$$

$$c \ominus_D d = h_D(g_D(c) - g_D(d)),$$

$$c \oslash_D d = h_D(g_D(c) \div g_D(d)).$$

This allows defining weak projectivity of the prearithmetic $\boldsymbol{E}_{A,D} = (A \times D; \oplus, \otimes, \ominus, \oslash, \leq)$ with respect to the (conventional) Diophantine arithmetic $\boldsymbol{R}$, taking the projector $f^2 : A \times D \to C$ and coprojector $k^2 : C \to A \times D$, which are defined as $f^2(a, d) = g_A(a) + ig_D(d)$ and $k^2(a + id) = (h_A(a), h_D(d))$ for arbitrary elements a from A and d from D. We see that $k^2(f^2(a, d)) = (h_A(g_A(a)), h_D(g_D(d)))$.

Let us look how the mappings f^2 and k^2 allow representing addition $\oplus$ by means of addition $+$ in the Diophantine arithmetic $\boldsymbol{C}$.

$$\begin{aligned}(a,b)\oplus(c,d) &= (a\oplus_A c, b\oplus_D d) = (h_A(g_A(a)+g_A(c)), h_D(g_D(b)+g_D(d)))\\ &= k^2(g_A(a)+g_A(c)+i(g_D(b)+g_D(d)))\\ &= k^2(g_A(a)+g_A(c)+ig_D(b)+ig_D(d))\\ &= k^2(g_A(a)+ig_A(b)+g_D(c)+ig_D(d))\\ &= k^2(f^2(a,b)+f^2(c,d)).\end{aligned}$$

It means that addition $\oplus$ in the prearithmetic $\boldsymbol{E}_{A,D}$ is weakly projective with respect to the Diophantine arithmetic $\boldsymbol{C}$ with the projector f^2 and coprojector k^2, i.e., addition $\oplus$ satisfies identity (3.11).

Let us look how the mappings f^2 and k^2 allow representing subtraction $\ominus$ by means of subtraction – in the Diophantine arithmetic $\boldsymbol{C}$.

$$\begin{aligned}(a,b)\ominus(c,d) &= (a\ominus_A c, b\ominus_D d) = (h_A(g_A(a)-g_A(c)), h_D(g_D(b)-g_D(d)))\\ &= k^2(g_A(a)-g_A(c)+i(g_D(b)-g_D(d)))\\ &= k^2(g_A(a)-g_A(c)+ig_D(b)-ig_D(d))\\ &= k^2(g_A(a)+ig_A(b)-g_D(c)-ig_D(d))\\ &= k^2(f^2(a,b)-f^2(c,d)).\end{aligned}$$

It means that subtraction $\ominus$ in the prearithmetic $\boldsymbol{E}_{A,D}$ is weakly projective with respect to the Diophantine arithmetic $\boldsymbol{C}$ with the projector f^2 and coprojector k^2, i.e., subtraction $\ominus$ satisfies identity (3.12).

Proposition is proved. □

Let us consider two wholly extended abstract prearithmetics $\boldsymbol{A} = (R; \oplus_A, \otimes_A, \ominus_A, \oslash_A, \leq_A)$ and $\boldsymbol{D} = (D; \oplus_D, \otimes_D, \ominus_D, \oslash_D, \leq_D)$, mappings g: $A \to D$ and h: $D \to A$ and the set $B = \{(a,d); a \in A \text{ and } d \in D\}$ of pairs (a,b).

Definition 3.4.9. A stratified C-prearithmetic $\boldsymbol{B} = (B; \oplus, \otimes, \ominus, \oslash, \leq_A)$ over the pair $(\boldsymbol{A}, \boldsymbol{D})$ is called a *stratified* IC-*arithmetic*, if both $\boldsymbol{A}$ and $\boldsymbol{D}$ are IR-arithmetics.

Stratified IC-arithmetics form a distinct class of non-Diophantine arithmetics of complex numbers.

Theorem 3.4.9. *Any stratified IC-arithmetic* $\boldsymbol{B} = (B; \oplus_A, \otimes_A, \ominus_A, \oslash_A, \leq_A)$ *is isomorphic to the Diophantine arithmetic* $\boldsymbol{C}$.

Proof: By Theorem 3.3.2, both $\boldsymbol{A}$ and $\boldsymbol{D}$ are isomorphic to the Diophantine arithmetic $\boldsymbol{R}$. It means that there are isomorphisms f_A: $\boldsymbol{A} \to \boldsymbol{R}$ and f_D: $\boldsymbol{D} \to \boldsymbol{R}$. As $B = \{(a, d); a \in A \text{ and } d \in D\}$, we can define the mapping $f_B : B \to R$ as

$$f_B(a, d) = f_A(a) + if_D(d).$$

As $\boldsymbol{A}$ and $\boldsymbol{D}$ are isomorphic to the Diophantine arithmetic $\boldsymbol{R}$ and complex numbers form a two-dimensional vector space over $\boldsymbol{R}$, f_B is one-to-one mapping. Let us show that it is an isomorphism.

$$\begin{aligned}
&f_B((a, d) \oplus_B (c, e)) = f_B(a \oplus_A c, d \oplus_D e)\\
&\quad = f_A((a \oplus_A c) + if_D(d \oplus_D e)) = (f_A(a) + f_A(c)) + (if_D(d) + if_D(e))\\
&\quad = (f_A(a) + if_D(d)) + (f_A(c) + if_D(e)) = f_B(a, d) + f_B(c, e).
\end{aligned}$$

Thus,

$$f_B((a, d) \oplus_B (c, e)) = f_B(a, d) + f_B(c, e).$$

In a similar way, we have

$$\begin{aligned}
&f_B((a, d) \ominus_B (c, e)) = f_B(a \ominus_A c, d \ominus_D e)\\
&\quad = f_A((a \ominus_A c) + if_D(d \ominus_D e)) = (f_A(a) - f_A(c)) + (if_D(d) - if_D(e))\\
&\quad = (f_A(a) + if_D(d)) - (f_A(c) + if_D(e)) = f_B(a, d) - f_B(c, e).
\end{aligned}$$

Thus,

$$f_B((a, d) \ominus_B (c, e)) = f_B(a, d) - f_B(c, e).$$

Equalities (3.16) and (3.17) are proved in a similar way.

$$f_B((a, d) \otimes_B (c, e)) = f_B(a, d) \cdot f_B(c, e), \tag{3.16}$$

$$f_B((a, d) \oslash_B (c, e)) = f_B(a, d)/f_B(c, e). \tag{3.17}$$

Theorem is proved. □

Stratified IC-arithmetics have many properties similar to properties of the Diophantine arithmetic $\boldsymbol{R}$ of complex numbers.

This result allows describing many properties of IC-arithmetics.

Theorem 3.4.10. *Any stratified IC-arithmetic* $\boldsymbol{B}$ *has the following properties*:

(*a*) *addition is commutative*;
(*b*) *multiplication is commutative*;
(*c*) *addition is associative*;
(*d*) *multiplication is associative*;
(*e*) *multiplication is distributive with respect to addition*;
(*f*) *multiplication is distributive with respect to subtraction*;
(*g*) $\boldsymbol{B}$ *has the additive and multiplicative zero* 0_A;
(*h*) $\boldsymbol{B}$ *has the multiplicative one* 1_A;
(*i*) $\boldsymbol{B}$ *has continuum of elements.*

Corollary 3.4.36. *Any stratified IC-arithmetic* $\boldsymbol{B}$ *is a partially ordered field.*

Definition 3.4.10. The isomorphism f between a stratified IC-arithmetic $\boldsymbol{B}$ and $\boldsymbol{C}$ is called the *functional parameter* of $\boldsymbol{B}$.

It is often denoted as $f_B : B \to C$.

Often the carrier B of a stratified IC-arithmetic $\boldsymbol{B}$ is a set of real numbers. However, in general, the carrier B of a stratified IC-arithmetic $\boldsymbol{B}$ can be any set for which a one-to-one mapping (bijection) $f_B : B \to C$ exists. This bijection determines a one-to-one naming of the elements from B by complex numbers (Burgin, 2011). Taking this into account, we can denote the element from B that is mapped onto the complex number q by q_B.

All IC-arithmetics are also called *functional non-Diophantine arithmetics* of complex numbers. In Chapter 4, we use functional *non-Diophantine* arithmetics of real and complex numbers for building a non-Newtonian calculus.

3.5. Non-Grassmannian Linear Spaces in the Framework of Non-Diophantine Arithmetics

The whole problem with the world is that fools and fanatics are always so certain of themselves, and wiser people are so full of doubts.

Bertrand Russell

The concept of a linear or vector space is rooted in solutions of systems of linear equations. The first methods of solving such equations are attributed to Babylonian and Egyptian mathematicians working almost four millennia ago. Some special cases of linear spaces were present in the analytic geometry introduced by the outstanding French mathematicians René Descartes and Pierre de Fermat (1607–1665) in the 17th century, in the theory of matrices and in the geometrical representation of complex numbers as vectors in the complex plane elaborated by the Parisian bookkeeper Jean-Robert Argand (1768–1822) and the great German mathematician Carl Friedrich Gauss (1777–1855) and published in (Argand, 1806) and (Gauss, 1831).

However, the outstanding German mathematician and polymath Hermann Günther Grassmann (1809–1877) was the first to introduce systems with operations, which later were called linear (or vector) spaces. In these spaces, he studied the concepts of linear maps, scalar products, linear independence and dimension (Grassmann, 1844). He even exceeded the framework of vector spaces when he used multiplication bringing forth linear algebras. Approximately at the same time, the extremely prolific English mathematician Arthur Cayley (1821–1895) wrote about "analytical geometry of n dimensions" also bringing in a specific case of n-dimensional vector spaces (Cayley, 1843). Giuseppe Peano was the first to give the modern definition of vector spaces and linear maps (Peano, 1888).

The term "vector" was introduced by the renowned Irish mathematician William Rowan Hamilton (1805–1865) when he called the *vector part*, as opposed to the *scalar part*, of a quaternion (Hamilton, 1844). However, the term "radius vector" ("rayon vecteur") appeared essentially earlier in the article "Rayon vecteur" by the French astronomer Joseph Jérôme Lefrançois de Lalande (1732–1807) published in 1776 in the renowned Encyclopédie edited by Denis Diderot (1713–1784). Besides, the idea of a vector as a directed straight line segment was used by several mathematicians: August Ferdinand Möbius (1790–1868) employed it in his barycentric calculus (Möbius, 1827), Giusto Bellavitis (1803–1880) utilized it in his the calculus of equipollence (Bellavitis, 1835), and Hermann Günther Grassmann brought it into play in his the calculus of extension, or "Ausdehnungslehre" (Grassmann, 1844).

In contemporary mathematics, linear spaces are mathematical structures built over fields (cf. Appendix). Here, we extend this concept building linear and quasilinear spaces over abstract and numerical prearithmetics and arithmetics.

Let us consider an abstract prearithmetic $\boldsymbol{A} = (A; +, \circ, \leq)$ with addition $+$ and multiplication $\circ$.

Definition 3.5.1. A *left quasilinear space*, which is also called a *left quasivector space* or a *left quasilinear vector space*, $\boldsymbol{L} = (L, +, \mapsto)$ over an abstract prearithmetic $\boldsymbol{A}$ is a set L with two operations:

- *Addition* $+$: $\boldsymbol{L} \times \boldsymbol{L} \to \boldsymbol{L}$ denoted by $\boldsymbol{x} + \boldsymbol{y}$ where $\boldsymbol{x}$ and $\boldsymbol{y}$ belong to $\boldsymbol{L}$.
- *Scalar action* $\mapsto$: $\boldsymbol{A} \times \boldsymbol{L} \to \boldsymbol{L}$ of elements from $\boldsymbol{A}$ on elements from $\boldsymbol{L}$, which is denoted by $a\boldsymbol{x}$ or by $a \mapsto \boldsymbol{x}$ where $a \in \boldsymbol{A}$ and $\boldsymbol{x} \in \boldsymbol{L}$.

Elements from the space $\boldsymbol{L}$ are called *quasivectors*. That is why it is also possible to call $\boldsymbol{L}$ a *left quasivector space*.

In the same way, we can define *right quasilinear spaces* over abstract prearithmetics, in which scalar action is performed from the right. However, here we consider only left quasilinear spaces because right quasilinear spaces have similar properties and are studied by analogous techniques.

Note that, in a general case, there are no additional conditions on operations in quasilinear spaces.

Example 3.5.1. Any abstract prearithmetic is a left quasilinear space over itself and over any of its subprearithmetics. Consequently (cf. Section 2.1), any ring, lattice, Boolean algebra, linear algebra, field, Ω-group, Ω-ring, Ω-algebra (Burgin, 1970; 1972; Burgin and Baranovich, 1975), topological ring, topological field, ordered ring, normed ring (Naimark, 1959), and normed field is a left quasilinear space over itself.

Example 3.5.2. Any vector space over a field, normed vector space over a field, superspace (Misner *et al.*, 1973; Gates *et al.*, 1983), hypernormed vector space over a field (Burgin, 2017d), module over a ring, linear algebra over a field, ordered linear algebra, superalgebra (Deligne and Morgan, 1999; Varadarajan, 2004a), hyperspace (Burgin, 2011a) or normed linear algebra (Naimark, 1972) is a left quasilinear space.

Example 3.5.3. The abstract prearithmetic of n-dimensional A-vectors $\mathrm{V}^n\boldsymbol{A} = (V^nA; +, \circ, \leq)$, elements of which are vectors in an abstract prearithmetic $\boldsymbol{A} = (A; +, \circ, \leq)$ (cf. Section 2.1) is a left quasilinear space over $\boldsymbol{A}$.

Indeed, it is possible to define actions of elements from $\boldsymbol{A}$ on vectors from the prearithmetic $\mathrm{V}^n\boldsymbol{A}$ by the conventional rule

$$a \mapsto (a_1, a_2, \ldots, a_n) = (a \circ a_1, a \circ a_2, \ldots, a \circ a_n).$$

Example 3.5.4. The abstract prearithmetic of $n \times m$-dimensional A-matrices $\mathrm{M}^{n\times m}\ \boldsymbol{A} = (M^{n\times m}A; +, \circ, \leq)$, elements of which are matrices in an abstract prearithmetic $\boldsymbol{A} = (A; +, \circ, \leq)$ (cf. Section 2.1), is a left quasilinear space over $\boldsymbol{A}$.

Indeed, it is possible to define actions of elements from $\boldsymbol{A}$ on matrices from the matrix prearithmetic $\mathrm{M}^{n\times m}\ \boldsymbol{A}$ by the conventional rule

$$a \rightarrowtail \begin{pmatrix} a_{11} & a_{12} & a_{13} & \cdots & a_{1n} \\ a_{21} & a_{22} & a_{23} & \cdots & a_{2n} \\ \vdots & \vdots & \vdots & \vdots & \vdots \\ a_{m1} & a_{m2} & a_{m3} & \cdots & a_{mn} \end{pmatrix} = \begin{pmatrix} a \circ a_{11} & a \circ a_{12} & a \circ a_{13} & \cdots & a \circ a_{1n} \\ a \circ a_{21} & a \circ a_{22} & a \circ a_{23} & \cdots & a \circ a_{2n} \\ \vdots & \vdots & \vdots & \vdots & \vdots \\ a \circ a_{m1} & a \circ a_{m2} & a \circ a_{m3} & \cdots & a \circ a_{mn} \end{pmatrix}$$

Example 3.5.5. Taking a left quasilinear space $\boldsymbol{L}$ over an abstract prearithmetic $\boldsymbol{A} = (A; +, \circ, \leq)$ and a natural number n, it is possible to build the left quasilinear space of n-dimensional L-vectors $\mathrm{V}^n\boldsymbol{L}$. Namely, elements of n-dimensional L-vector quasilinear space $\mathrm{V}^n\boldsymbol{L}$ have the form $(a_1, a_2, \ldots, a_n)$ where $a_1, a_2, \ldots, a_n$ are elements from the left quasilinear space $\boldsymbol{L}$. Operations in $\mathrm{V}^n\boldsymbol{L}$ are defined coordinate-wise. Namely, we have

$$(a_1, a_2, \ldots, a_n) + (b_1, b_2, \ldots, b_n) = (a_1 + b_1, a_2 + b_2, \ldots, a_n + b_n)$$

and

$$a \mapsto (a_1, a_2, \ldots, a_n) = (a \mapsto a_1, a \mapsto a_2, \ldots, a \mapsto a_n).$$

Example 3.5.6. Taking a left quasilinear space $\boldsymbol{L}$ over an abstract prearithmetic $\boldsymbol{A} = (A; +, \circ, \leq)$ and a pair of natural numbers n and m, it is possible to build the left quasilinear space of $n \times m$-dimensional L-matrices $\mathrm{M}^{n\times m}\ \boldsymbol{L}$. Operations in $\mathrm{M}^{n\times m}\ \boldsymbol{L}$ are defined coordinate-wise in the same way as it is done for n-dimensional L-vector quasilinear space $\mathrm{V}^n\boldsymbol{L}$ in the previous example.

Example 3.5.7. A *left semimodule* over a semiring R or simply *left R-semimodule* consists of an additively written commutative monoid in which R acts (Golan, 1999).

A *right semimodule* over a semiring R or *right R-semimodule* is defined in a similar way.

Any left semimodule is a left quasilinear space and any right semimodule is a right quasilinear space.

Definition 3.5.2. A left quasilinear space $\boldsymbol{L}$ over the abstract prearithmetic $\boldsymbol{A}$ is a *quasilinear space* or *quasivector space* over $\boldsymbol{A}$ if it is also a right quasilinear space over $\boldsymbol{A}$, in which the left action of elements from $\boldsymbol{A}$ coincides with their right action.

In Section 2.2, we considered weakly projective abstract arithmetics and defined actions of one of them on the elements from another one (cf. Definition 2.2.2). This gives us the following result.

Proposition 3.5.1. *If an abstract prearithmetic* $\boldsymbol{A}_1 = (A_1; +_1, \circ_1, \leq_1)$ *is weakly projective with respect to an abstract prearithmetic* $\boldsymbol{A}_2 = (A_2; +_2, \circ_2, \leq_2)$, *then* $\boldsymbol{A}_1$ *is a left and right quasilinear space over* $\boldsymbol{A}_2$.

Proof is left as an exercise. □

Proposition 3.5.2. *A left (right) quasilinear space* $\boldsymbol{L}$ *over an abstract prearithmetic* $\boldsymbol{A}$ *is a left (right) quasilinear space over any subprearithmetic of* $\boldsymbol{A}$.

Proof is left as an exercise. □

Relation of weak projectivity introduced in (Burgin, 1997) and studied in Section 2.2 establishes relations between left quasilinear spaces.

Proposition 3.5.3. *If an abstract prearithmetic* $\boldsymbol{A}_1 = (A_1; +_1, \circ_1, \leq_1)$ *is weakly projective with respect to an abstract prearithmetic* $\boldsymbol{A}_2 = (A_2; +_2, \circ_2, \leq_2)$ *with the projector g and the coprojector h, and* $\boldsymbol{L}$ *is left quasilinear space over* $\boldsymbol{A}_1$, *then* $\boldsymbol{L}$ *is left quasilinear space over* $\boldsymbol{A}_2$.

Indeed, we can define scalar action $\mapsto: \boldsymbol{A}_2 \times \boldsymbol{L} \to \boldsymbol{L}$ of elements from $\boldsymbol{A}_2$ on elements from $\boldsymbol{L}$ by the formula $a\boldsymbol{x} = h(a)\boldsymbol{x}$ or by taking the action $a \mapsto \boldsymbol{x}$ equal to the action $h(a) \mapsto \boldsymbol{x}$ where $a \in \boldsymbol{A}_2$ and $\boldsymbol{x} \in \boldsymbol{L}$. This action is defined for all elements from $\boldsymbol{A}_2$ because $h(a)a \in \boldsymbol{A}_2$ and h is a function.

In Section 2.2, it is described (cf. Proposition 2.2.1) how to convert an arbitrary set into an abstract prearithmetic using a given abstract prearithmetic. This allows defining quasilinear spaces over arbitrary sets using Proposition 3.5.3.

There are different types of left quasilinear spaces defined by their properties.

Definition 3.5.3. A left quasilinear space $\boldsymbol{L}$ over an abstract prearithmetic $\boldsymbol{A}$ is *commutative* if it has commutative addition, i.e., for all elements $\boldsymbol{x}$ and $\boldsymbol{y}$ from $\boldsymbol{L}$, we have

$$\boldsymbol{x} + \boldsymbol{y} = \boldsymbol{y} + \boldsymbol{x}.$$

Any vector space is a commutative left and right quasilinear space.

Definition 3.5.4. A left quasilinear space $\boldsymbol{L}$ over an abstract prearithmetic $\boldsymbol{A}$ is *arithmetically distributive* if for any element $\boldsymbol{x}$ from $\boldsymbol{L}$ and all elementsa, b from $\boldsymbol{A}$, we have

$$(a+b)\boldsymbol{x} = a\boldsymbol{x} + b\boldsymbol{x}.$$

Any vector space is an arithmetically distributive left and right quasilinear space.

Definition 3.5.5. A left quasilinear space $\boldsymbol{L}$ over an abstract prearithmetic $\boldsymbol{A}$ is *additively distributive* if for any element a from $\boldsymbol{A}$ and all elements $\boldsymbol{x}$ and $\boldsymbol{y}$ from $\boldsymbol{L}$, we have

$$a(\boldsymbol{x}+\boldsymbol{y}) = a\boldsymbol{x} + a\boldsymbol{y}.$$

Any vector space is an additively distributive left and right quasilinear space.

There are additively distributive left quasilinear spaces, which are not arithmetically distributive as the following example demonstrates.

Example 3.5.8. Let us take the Diophantine arithmetic $\boldsymbol{R}$ of all real numbers and convert it into a left quasilinear space defining action by the following formula:

$$a\boldsymbol{x} = \boldsymbol{x}.$$

Here a is an arbitrary real number and $\boldsymbol{x}$ is any element of $\boldsymbol{R}$ as the left quasilinear space.

Then we see that

$$a(\boldsymbol{x}+\boldsymbol{y}) = \boldsymbol{x}+\boldsymbol{y} = a\boldsymbol{x} + a\boldsymbol{y},$$

while

$$(a+b)\boldsymbol{x} = \boldsymbol{x} \neq \boldsymbol{x}+\boldsymbol{x} = a\boldsymbol{x} + b\boldsymbol{x}$$

when $\boldsymbol{x}$ is not equal to $\mathbf{0}$.

There are also arithmetically distributive left quasilinear spaces, which are not additively distributive as the following example demonstrates.

Example 3.5.9. Let us take the Diophantine arithmetic $\boldsymbol{R}$ of all real numbers and convert it into a left quasilinear space defining action by the following formula:

$$a\boldsymbol{x} = \boldsymbol{a}.$$

Here a is an arbitrary real number, $\boldsymbol{a}$ is the same number treated as an element of $\boldsymbol{R}$ as the left quasilinear space and $\boldsymbol{x}$ is any element of $\boldsymbol{R}$ as the left quasilinear space.

Then we see that

$$a(\boldsymbol{x} + \boldsymbol{y}) = \boldsymbol{a} \neq \boldsymbol{a} + \boldsymbol{a} = a\boldsymbol{x} + a\boldsymbol{y}$$

when a is not equal to 0, while

$$(a + b)\boldsymbol{x} = \boldsymbol{a} + \boldsymbol{b} = a\boldsymbol{x} + b\boldsymbol{x}.$$

Definition 3.5.6. A left quasilinear space $\boldsymbol{L}$ over an abstract prearithmetic $\boldsymbol{A}$ is *associative* if it has associative addition, i.e., for all elements $\boldsymbol{x}$, $\boldsymbol{y}$ and $\boldsymbol{z}$ from $\boldsymbol{L}$, we have

$$(\boldsymbol{x} + \boldsymbol{y}) + \boldsymbol{z} = \boldsymbol{x} + (\boldsymbol{y} + \boldsymbol{z}).$$

Any vector space is an associative left and right quasilinear space.

An important property of quasilinear spaces is covering.

Definition 3.5.7. A left quasilinear space $\boldsymbol{L}$ over an abstract prearithmetic $\boldsymbol{A}$ is *covered by* $\boldsymbol{A}$ if any element $\boldsymbol{x}$ from $\boldsymbol{L}$ has the form $a\boldsymbol{y}$ for some element a from $\boldsymbol{A}$ and some element $\boldsymbol{y}$ from $\boldsymbol{L}$.

Any vector space over a field $\boldsymbol{F}$ is a left quasilinear space over $\boldsymbol{F}$ covered by $\boldsymbol{F}$.

Proposition 3.5.4. *A left quasilinear space* $\boldsymbol{L} = (L, +, \mapsto)$ *over an abstract prearithmetic* $\boldsymbol{A}$ *is covered by* $\boldsymbol{A}$ *if and only if the scalar action* $\mapsto$ *is a projection.*

Proof is left as an exercise. □

In the same way as prearithmetics and arithmetics, some quasilinear spaces have zeros.

Definition 3.5.8. An element $\mathbf{0}$ from $\boldsymbol{L}$ is called an *additive zero* in $\boldsymbol{L}$ if $\boldsymbol{x} + \mathbf{0} = \mathbf{0} + \boldsymbol{x} = \boldsymbol{x}$ for any element $\boldsymbol{x}$ from $\boldsymbol{L}$.

Any vector space over a field and any ring have an additive zero.

Definition 3.5.9. A left quasilinear space $\boldsymbol{L}$ over an abstract prearithmetic $\boldsymbol{A}$ has *additive cancellation* if the equality $\boldsymbol{z} + \boldsymbol{x} = \boldsymbol{z} + \boldsymbol{y}$ implies the equality $\boldsymbol{x} = \boldsymbol{y}$ for any elements $\boldsymbol{x}$, $\boldsymbol{y}$ and $\boldsymbol{z}$ from $\boldsymbol{L}$.

Any vector space over a field has additive cancellation.

Proposition 3.5.5. *If an abstract prearithmetic $\boldsymbol{A}$ has an additive zero 0 and an arithmetically distributive left quasilinear space $\boldsymbol{L}$ over $\boldsymbol{A}$ has an additive zero $\mathbf{0}$ and additive cancellation, then $0\boldsymbol{x} = \mathbf{0}$ for any element $\boldsymbol{x}$ from $\boldsymbol{L}$.*

Proof: Let us assume that an abstract prearithmetic $\boldsymbol{A} = (A; +, \circ, \leq)$ has an additive zero 0 and an arithmetically distributive left quasilinear space $\boldsymbol{L}$ over $\boldsymbol{A}$ has an additive zero $\mathbf{0}$ and additive cancellation. Then for an arbitrary element $\boldsymbol{x}$, we can show that $0\boldsymbol{x}$ is an additive zero $\mathbf{0}$ in $\boldsymbol{L}$. Indeed, arithmetical distributivity gives us the following sequence of equalities

$$a\boldsymbol{x} + 0\boldsymbol{x} = (a + 0)\boldsymbol{y} = a\boldsymbol{x}.$$

By the definition of an additive zero $\mathbf{0}$, we have

$$a\boldsymbol{x} + 0\boldsymbol{x} = a\boldsymbol{x} + \mathbf{0}.$$

As $\boldsymbol{L}$ has additive cancellation, i.e., $\boldsymbol{z} + \boldsymbol{x} = \boldsymbol{z} + \boldsymbol{y}$ implies $\boldsymbol{x} = \boldsymbol{y}$ for any elements $\boldsymbol{x}$, $\boldsymbol{y}$ and $\boldsymbol{z}$ from $\boldsymbol{L}$, we obtain $0\boldsymbol{x} = \mathbf{0}$.

Proposition is proved. □

Corollary 3.5.1. *If an abstract prearithmetic $\boldsymbol{A}$ has an additive zero 0 and an arithmetically distributive left quasilinear space $\boldsymbol{L}$ over $\boldsymbol{A}$ has an additive zero $\mathbf{0}$ and additive cancellation, then $0\boldsymbol{x} = 0\boldsymbol{z}$ for any elements $\boldsymbol{x}$ and $\boldsymbol{z}$ from $\boldsymbol{L}$.*

Proposition 3.5.6. *If an abstract prearithmetic $\boldsymbol{A}$ has an additive zero 0, then an arithmetically distributive covered by $\boldsymbol{A}$ left quasilinear space $\boldsymbol{L}$ over $\boldsymbol{A}$ has an additive zero $\mathbf{0}$ if $0\boldsymbol{x} = 0\boldsymbol{z}$ for any elements $\boldsymbol{x}$ and $\boldsymbol{z}$ from $\boldsymbol{L}$.*

Proof: Let us assume that an abstract prearithmetic $\boldsymbol{A} = (A; +, \circ, \leq)$ has an additive zero 0 and an arithmetically distributive left quasilinear space $\boldsymbol{L}$ over $\boldsymbol{A}$ is covered by $\boldsymbol{A}$. Then for an arbitrary element $\boldsymbol{x}$, we can show that $0\boldsymbol{x}$ is an additive zero $\mathbf{0}$ in $\boldsymbol{L}$. Indeed, arithmetical distributivity gives us the following sequence of equalities:

$$\boldsymbol{x} + 0\boldsymbol{z} = \boldsymbol{x} + 0\boldsymbol{y} = a\boldsymbol{y} + 0\boldsymbol{y} = (a + 0)\boldsymbol{y} = a\boldsymbol{y} = \boldsymbol{x}.$$

As $\boldsymbol{x}$ is arbitrary element in $\boldsymbol{L}$, the element $0\boldsymbol{z}$ is an additive zero $\mathbf{0}$ in $\boldsymbol{L}$.

Proposition is proved. □

Corollary 3.5.2. *If an abstract prearithmetic* $\boldsymbol{A}$ *has an additive zero* 0, *then in an arithmetically distributive covered by* $\boldsymbol{A}$ *left quasilinear space* $\boldsymbol{L}$ *over* $\boldsymbol{A}$ *the additive zero* $\mathbf{0}$ *is unique.*

Proposition 3.5.7. *If an abstract prearithmetic* $\boldsymbol{A}$ *has an additive zero* 0 *and an additively distributive left quasilinear space* $\boldsymbol{L}$ *over* $\boldsymbol{A}$ *has an additive zero* $\mathbf{0}$ *and additive cancellation, then* $a\mathbf{0} = \mathbf{0}$ *for any element* a *from* $\boldsymbol{A}$.

Proof: Let us assume that an abstract prearithmetic $\boldsymbol{A} = (A; +, \circ, \leq)$ has an additive zero 0 and an arithmetically distributive left quasilinear space $\boldsymbol{L}$ over $\boldsymbol{A}$ is covered by $\boldsymbol{A}$. Then taken an arbitrary element $\boldsymbol{x}$ from $\boldsymbol{L}$ and applying additive distributivity, we obtain the following sequence of equalities:

$$a\mathbf{0} + a\boldsymbol{x} = a(\mathbf{0} + \boldsymbol{x}) = a\boldsymbol{x} = \mathbf{0} + a\boldsymbol{x}.$$

As $\boldsymbol{L}$ has additive cancellation, i.e., $\boldsymbol{z} + \boldsymbol{x} = \boldsymbol{z} + \boldsymbol{y}$ implies $\boldsymbol{x} = \boldsymbol{y}$ for any elements $\boldsymbol{x}$, $\boldsymbol{y}$ and $\boldsymbol{z}$ from $\boldsymbol{L}$, we obtain $a\mathbf{0} = \mathbf{0}$.

Proposition is proved. □

Definition 3.5.10. A left quasilinear space $\boldsymbol{L}$ over an abstract prearithmetic $\boldsymbol{A}$ is *scalar associative* if for any elements a and b from $\boldsymbol{A}$ and any element $\boldsymbol{x}$ from $\boldsymbol{L}$, we have

$$a(b\boldsymbol{x}) = (a \cdot b)\boldsymbol{x}.$$

Often existence of a multiplicative one in an abstract prearithmetic$\boldsymbol{A}$ implies existence of a multiplicative one in a left quasilinear space $\boldsymbol{L}$ over $\boldsymbol{A}$.

Proposition 3.5.8. *If an abstract prearithmetic* $\boldsymbol{A}$ *has a multiplicative one* 1 *and a left quasilinear space* $\boldsymbol{L}$ *over* $\boldsymbol{A}$ *is a scalar associative covered by* $\boldsymbol{A}$ *left quasilinear space over* $\boldsymbol{A}$, *then* $1\ \boldsymbol{x} = \boldsymbol{x}$ *for any element* $\boldsymbol{x}$ *from* $\boldsymbol{L}$.

Proof: Let us assume that an abstract prearithmetic $\boldsymbol{A} = (A; +, \circ, \leq)$ has a multiplicative one 1 and $\boldsymbol{L}$ is an associative covered by $\boldsymbol{A}$ left quasilinear space over $\boldsymbol{A}$. Then an arbitrary element $\boldsymbol{x}$ from $\boldsymbol{L}$ is equal to $a\boldsymbol{y}$ for some element a from $\boldsymbol{A}$ and some element$\boldsymbol{y}$ from $\boldsymbol{L}$. As $\boldsymbol{L}$ is scalar associative, we have

$$1\boldsymbol{x} = 1(a\boldsymbol{y}) = (1 \cdot a)\boldsymbol{y} \ = a\boldsymbol{y} \ = \boldsymbol{x}.$$

Proposition is proved. □

Often existence of an additive zero in an abstract prearithmetic $\boldsymbol{A}$ implies existence of an additive zero in a left quasilinear space $\boldsymbol{L}$ over $\boldsymbol{A}$.

Proposition 3.5.9. *If any element of an abstract prearithmetic $\boldsymbol{A}$ with an additive zero 0 has its opposite and $\boldsymbol{L}$ is an arithmetically distributive covered by $\boldsymbol{A}$ left quasilinear space with an additive zero $\mathbf{0}$ over $\boldsymbol{A}$, then any element of $\boldsymbol{L}$ has its opposite.*

Proof: Let any element of an abstract prearithmetic $\boldsymbol{A}$ has its opposite and $\boldsymbol{L}$ is an arithmetically distributive covered by $\boldsymbol{A}$ left quasilinear space with an additive zero $\mathbf{0}$ over $\boldsymbol{A}$. Then an arbitrary element $\boldsymbol{x}$ from $\boldsymbol{L}$ is equal to $a\boldsymbol{y}$ for some element a from $\boldsymbol{A}$ and some element $\boldsymbol{y}$ from $\boldsymbol{L}$. Besides, the element a has its opposite $-a$, that is, $a + (-a) = 0$. Arithmetical distributivity of $\boldsymbol{L}$ gives us

$$\boldsymbol{x} + (-a)\boldsymbol{y} = a\boldsymbol{y} + (a + (-a))\boldsymbol{y} = 0\boldsymbol{y}.$$

By Proposition 3.5.5, $0\boldsymbol{y}$ is an additive zero $\mathbf{0}$. It means that $(-a)\boldsymbol{y}$ is an opposite element to $\boldsymbol{x}$.

Proposition is proved. □

Corollary 3.5.3. *If any element of an abstract prearithmetic $\boldsymbol{A}$ with an additive zero 0 has its opposite and $\boldsymbol{L}$ is an arithmetically distributive left quasilinear space with an additive zero $\mathbf{0}$ over $\boldsymbol{A}$, then for any element $\boldsymbol{x}$ from $\boldsymbol{L}$ and any element a from $\boldsymbol{A}$, the element $-a\boldsymbol{x}$ is opposite to the element $a\boldsymbol{x}$.*

Let us study cancellation laws in left quasilinear spaces.

Proposition 3.5.10. *If any element of an associative left quasilinear space $\boldsymbol{L}$ has its opposite, then $\boldsymbol{L}$ has additive cancellation.*

Indeed, if $\boldsymbol{x} + \boldsymbol{z} = \boldsymbol{y} + \boldsymbol{z}$ and $-\boldsymbol{z}$ is the opposite element to $\boldsymbol{z}$, then we have

$$\boldsymbol{x} = \boldsymbol{x} + \mathbf{0} = \boldsymbol{x} + \boldsymbol{z} + (-\boldsymbol{z}) = \boldsymbol{y} + \boldsymbol{z} + (-\boldsymbol{z}) = \boldsymbol{y} + \mathbf{0} = \boldsymbol{y},$$

that is, $\boldsymbol{x} = \boldsymbol{y}$.

Propositions 3.5.9 and 3.5.10 imply the following result.

Corollary 3.5.4. *If any element of an abstract prearithmetic* $\boldsymbol{A}$ *with an additive zero* 0 *has its opposite and* $\boldsymbol{L}$ *is an arithmetically distributive covered by* $\boldsymbol{A}$ *left quasilinear space with an additive zero* $\mathbf{0}$ *over* $\boldsymbol{A}$*, then any element of* $\boldsymbol{L}$ *has its opposite.*

Similar properties are true for multiplicative cancellation.

Definition 3.5.11. (a) A left quasilinear space $\boldsymbol{L}$ over an abstract prearithmetic $\boldsymbol{A}$ has *left multiplicative cancellation* if the equality $a\boldsymbol{z} = a\boldsymbol{x}$ with $a \neq 0$ implies $\boldsymbol{z} = \boldsymbol{x}$ for all elements $\boldsymbol{x}$ and $\boldsymbol{z}$ from $\boldsymbol{L}$.

(b) A left quasilinear space $\boldsymbol{L}$ over an abstract prearithmetic $\boldsymbol{A}$ has *right multiplicative cancellation* if the equality $a\boldsymbol{x} = b\boldsymbol{x}$ with $\boldsymbol{x} \neq \mathbf{0}$ implies $a = b$ for all elements a and b from $\boldsymbol{A}$.

Any conventional linear (vector) space has right and left multiplicative cancellation.

Proposition 3.5.11. *If an abstract prearithmetic* $\boldsymbol{A}$ *has an additive zero* 0 *and an arithmetically and additively distributive left quasilinear space* $\boldsymbol{L}$ *over* $\boldsymbol{A}$ *has an additive zero* $\mathbf{0}$*, scalar cancellation and additive cancellation, then* $a\boldsymbol{x} = \mathbf{0}$ *if and only if either* $a = 0$ *or* $\boldsymbol{x} = \mathbf{0}$.

Proof: Let us assume that an abstract prearithmetic $\boldsymbol{A} = (A; +, \circ, \leq)$ has an additive zero 0 and an arithmetically and additively distributive left quasilinear space $\boldsymbol{L}$ over $\boldsymbol{A}$ has an additive zero $\mathbf{0}$, scalar cancellation and additive cancellation.

Sufficiency. If $a = 0$, then by Proposition 3.5.5, $a\boldsymbol{x} = \mathbf{0}$. If $\boldsymbol{x} = \mathbf{0}$, then by Proposition 3.5.7, $a\boldsymbol{x} = \mathbf{0}$.

Necessity. Let us assume that $a\boldsymbol{x} = \mathbf{0}$ and $a \neq 0$. Then $a\boldsymbol{x} = \mathbf{0} = a\mathbf{0}$. As $\boldsymbol{L}$ has scalar cancellation, we obtain $\boldsymbol{x} = \mathbf{0}$.

Proposition is proved. □

A subspace is an important structure in the theory of linear and quasilinear spaces.

Definition 3.5.12.

(a) A subset $\boldsymbol{M}$ of a left quasilinear space $\boldsymbol{L}$ over an abstract prearithmetic $\boldsymbol{A}$ is called a *linear (vector)subspace* of $\boldsymbol{L}$ if $\boldsymbol{M}$ is closed with respect to addition and scalar action.

(b) A subset $\boldsymbol{M}$ of a left quasilinear space $\boldsymbol{L}$ over an abstract prearithmetic $\boldsymbol{A}$ is called a *weak linear (vector)subspace* of $\boldsymbol{L}$ if $\boldsymbol{M}$ is closed with respect to addition and scalar action of elements from some subarithmetic $\boldsymbol{B}$ of $\boldsymbol{A}$.

Example 3.5.10. A subspace of a vector space $\boldsymbol{L}$ over a field is a linear (vector) subspace of $\boldsymbol{L}$.

Example 3.5.11. A submodule of a left module $\boldsymbol{M}$ over a ring is a linear (vector) subspace of $\boldsymbol{M}$.

Example 3.5.12. A subsemimodule of a left semimodule $\boldsymbol{K}$ over a semiring is a linear (vector) subspace of $\boldsymbol{K}$.

Lemma 3.5.1. *A (weak) linear subspace* $\boldsymbol{M}$ *of a left quasilinear space* $\boldsymbol{L}$ *over an abstract prearithmetic* $\boldsymbol{A}$ *is a left quasilinear space over* $\boldsymbol{A}$ *(over some subarithmetic* $\boldsymbol{B}$ *of* $\boldsymbol{A}$*).*

Proof is left as an exercise. □

Proposition 3.5.5 implies the following result.

Proposition 3.5.12. *If an abstract prearithmetic* $\boldsymbol{A}$ *has an additive zero* 0 *and* $\boldsymbol{L}$ *is an arithmetically distributive covered by* $\boldsymbol{A}$ *left quasilinear space over* $\boldsymbol{A}$*, then any linear subspace* $\boldsymbol{M}$ *of* $\boldsymbol{L}$ *has an additive zero* $\mathbf{0}$ *and additive cancellation.*

Proof is left as an exercise. □

Proposition 3.5.13. *If a left quasilinear space* $\boldsymbol{L}$ *over an abstract prearithmetic* $\boldsymbol{A}$ *has additive cancellation, then any linear subspace* $\boldsymbol{M}$ *of* $\boldsymbol{L}$ *has additive cancellation.*

Proof is left as an exercise. □

Proposition 3.5.14. *If a left quasilinear space* $\boldsymbol{L}$ *over an abstract prearithmetic* $\boldsymbol{A}$ *is commutative (associative, scalar associative, arithmetically distributive and/or additively distributive), then any linear subspace* $\boldsymbol{M}$ *of* $\boldsymbol{L}$ *is commutative (associative, scalar associative, arithmetically distributive and/or additively distributive, respectively).*

Proof is left as an exercise. □

For simplicity in what follows, let us assume that all quasilinear spaces are associative and scalar associative,

Definition 3.5.13. An expression

$$\sum_{i=1}^{n} a_i \boldsymbol{x}_i = a_1\boldsymbol{x}_1 + a_2\boldsymbol{x}_2 + \cdots + a_n\boldsymbol{x}_n,$$

where $a_i (i = 1, 2, 3, \ldots, n)$ are elements from an abstract prearithmetic $\boldsymbol{A}$ and $\boldsymbol{x}_1, \boldsymbol{x}_2, \ldots, \boldsymbol{x}_n$ are quasivectors from an associative left quasilinear space $\boldsymbol{L}$ over $\boldsymbol{A}$, is called a *linear combination* of quasivectors $\boldsymbol{x}_1, \boldsymbol{x}_2, \ldots, \boldsymbol{x}_n$.

Linear combinations of vectors define operations in quasilinear spaces using induction on n. Namely, we have

$$\sum_{i=1}^{1} a_i \boldsymbol{x}_i = a_1\boldsymbol{x}_1,$$

$$\sum_{i=1}^{2} a_i \boldsymbol{x}_i = a_1\boldsymbol{x}_1 + a_2\boldsymbol{x}_2.$$

If $\sum_{i=1}^{n-1} a_i$ is defined, then

$$\sum_{i=1}^{n} a_i \boldsymbol{x}_i = \left(\sum_{i=1}^{n-1} a_i \boldsymbol{x}_i\right) + a_n\boldsymbol{x}_n.$$

The result of this operation is naturally called the sum of the elements $a_i\boldsymbol{x}_i$.

When addition + is associative, it is possible to remove parentheses and we have

$$\sum_{i=1}^{n} a_i \boldsymbol{x}_i = a_1\boldsymbol{x}_1 + a_2\boldsymbol{x}_2 + \cdots + a_n\boldsymbol{x}_n.$$

As we know, linear spaces over fields and left modules over rings are left quasilinear spaces. In these spaces, any element or any sum of elements is a linear combination. This not true for left quasilinear spaces in the general case as the following example demonstrates.

Example 3.5.13. The Diophantine arithmetic $\boldsymbol{N}$ is a left quasilinear space over an abstract prearithmetic $2\boldsymbol{N}$ of all even natural numbers. Then any linear combination in $\boldsymbol{N}$ contains only even numbers and thus, 1 is not a linear combination in $\boldsymbol{N}$.

Lemma 3.5.2. *Any linear combination in a left quasilinear space* $\boldsymbol{L}$ *defines a unique element from* $\boldsymbol{L}$.

Proof is left as an exercise. □

Lemma 3.5.3. *If a linear subspace $\boldsymbol{M}$ of a left quasilinear space $\boldsymbol{L}$ contains quasivectors $\boldsymbol{x}_1, \boldsymbol{x}_2, \ldots, \boldsymbol{x}_n$, then it contains all linear combinations of $\boldsymbol{x}_1, \boldsymbol{x}_2, \ldots, \boldsymbol{x}_n$.*

Proof is left as an exercise. □

Lemma 3.5.4. *The sum of linear combinations from $\boldsymbol{L}$ is a linear combination from $\boldsymbol{L}$.*

Proof is left as an exercise. □

Let us assume that left quasilinear space $\boldsymbol{L}$ is additively distributive.

Lemma 3.5.5. *The product of an element a from $\boldsymbol{A}$ and a linear combination from $\boldsymbol{L}$ is a linear combination from $\boldsymbol{L}$.*

Proof is left as an exercise. □

These lemmas allow proving the following result.

Proposition 3.5.15. *Given any quasivectors $\boldsymbol{x}_1, \boldsymbol{x}_2, \ldots, \boldsymbol{x}_n$ from a left quasilinear space $\boldsymbol{L}$ over an abstract prearithmetic $\boldsymbol{A}$ that has a multiplicative one 1, the set $L(\boldsymbol{x}_1, \boldsymbol{x}_2, \ldots, \boldsymbol{x}_n)$ of all linear combinations of $\boldsymbol{x}_1, \boldsymbol{x}_2, \ldots, \boldsymbol{x}_n$ is the least linear subspace of $\boldsymbol{L}$, which contains all $\boldsymbol{x}_1, \boldsymbol{x}_2, \ldots, \boldsymbol{x}_n$.*

Proof is left as an exercise. □

Definition 3.5.14. The set $L(\boldsymbol{x}_1, \boldsymbol{x}_2, \ldots, \boldsymbol{x}_n)$ is called the *linear subspace* of $\boldsymbol{L}$ generated by quasivectors $\boldsymbol{x}_1, \boldsymbol{x}_2, \ldots, \boldsymbol{x}_n$ or the *span* of the quasivectors $\boldsymbol{x}_1, \boldsymbol{x}_2, \ldots, \boldsymbol{x}_n$.

We also say that the space $L(\boldsymbol{x}_1, \boldsymbol{x}_2, \ldots, \boldsymbol{x}_n)$ is *spanned* or *generated* by quasivectors $\boldsymbol{x}_1, \boldsymbol{x}_2, \ldots, \boldsymbol{x}_n$.

Proposition 3.5.16. *If in an arithmetically and additively distributive, associative and commutative left quasilinear space $\boldsymbol{L}$ over an additively associative abstract prearithmetic $\boldsymbol{A}$, a quasivector $\boldsymbol{x}$ is a linear combination of quasivectors $\boldsymbol{x}_1, \boldsymbol{x}_2, \ldots, \boldsymbol{x}_n$ and each quasivector $\boldsymbol{x}_i$ is a linear combination of quasivectors $\boldsymbol{y}_1, \boldsymbol{y}_2, \ldots, \boldsymbol{y}_m (i = 1, 2, 3, \ldots, n)$, then the quasivector $\boldsymbol{x}$ is a linear combination of quasivectors $\boldsymbol{y}_1, \boldsymbol{y}_2, \ldots, \boldsymbol{y}_m$.*

Proof: Given $\boldsymbol{x} = a_1\ \boldsymbol{x}_1 + a_2\ \boldsymbol{x}_2 + a_3\ \boldsymbol{x}_3 + \cdots + a_n\ \boldsymbol{x}_n$ and $\boldsymbol{x}_i = b_{i1}\ \boldsymbol{y}_1 + b_{i2}\ \boldsymbol{y}_2 + b_{i3}\ \boldsymbol{y}_3 + \cdots + b_{im}\ \boldsymbol{y}_m (i = 1, 2, 3, \ldots, n)$, properties of $\boldsymbol{L}$ allow obtaining the following sequence of equalities:

$$\begin{aligned}
\boldsymbol{x} &= a_1\boldsymbol{x}_1 + a_2\boldsymbol{x}_2 + a_3\boldsymbol{x}_3 + \cdots + a_n\boldsymbol{x}_n \\
&= a_1(b_{11}\boldsymbol{y}_1 + b_{12}\boldsymbol{y}_2 + b_{13}\boldsymbol{y}_3 + \cdots + b_{1m}\boldsymbol{y}_m) + a_2(b_{21}\boldsymbol{y}_1 + b_{22}\boldsymbol{y}_2 + b_{23}\boldsymbol{y}_3 \\
&\quad + \cdots + b_{2m}\boldsymbol{y}_m) + a_3(b_{31}\boldsymbol{y}_1 + b_{32}\boldsymbol{y}_2 + b_{33}\boldsymbol{y}_3 + \cdots + b_{3m}\boldsymbol{y}_m) + \cdots \\
&\quad + a_n(b_{n1}\boldsymbol{y}_1 + b_{n2}\boldsymbol{y}_2 + b_{n3}\boldsymbol{y}_3 + \cdots + b_{nm}\boldsymbol{y}_m) \\
&= a_1b_{11}\boldsymbol{y}_1 + a_1b_{12}\boldsymbol{y}_2 + a_1b_{13}\boldsymbol{y}_3 + \cdots + a_1b_{1m}\boldsymbol{y}_m + a_2b_{21}\boldsymbol{y}_1 + a_2b_{22}\boldsymbol{y}_2 \\
&\quad + a_2b_{23}\boldsymbol{y}_3 + \cdots + a_2b_{2m}\boldsymbol{y}_m + a_3b_{31}\boldsymbol{y}_1 + a_3b_{32}\boldsymbol{y}_2 + a_3b_{33}\boldsymbol{y}_3 + \cdots \\
&\quad + a_3b_{3m}\boldsymbol{y}_m + \cdots + a_nb_{n1}\boldsymbol{y}_1 + a_nb_{n2}\boldsymbol{y}_2 + a_nb_{n3}\boldsymbol{y}_3 + \cdots + a_nb_{nm}\boldsymbol{y}_m \\
&= (a_1b_{11}\boldsymbol{y}_1 + a_2b_{21}\boldsymbol{y}_1 + a_3b_{31}\boldsymbol{y}_1 + \cdots + a_nb_{n1}\boldsymbol{y}_1) + (a_1b_{12}\boldsymbol{y}_2 + a_2b_{22}\boldsymbol{y}_2 \\
&\quad + a_3b_{32}\boldsymbol{y}_2 + \cdots + a_nb_{n2}\boldsymbol{y}_2) + (a_1b_{13}\boldsymbol{y}_3 + a_2b_{23}\boldsymbol{y}_3 + a_3b_{33}\boldsymbol{y}_3 + \cdots \\
&\quad + a_3b_{3m}\boldsymbol{y}_m) + \cdots + (a_1b_{1m}\mathbf{y}_m + a_2b_{2m}\boldsymbol{y}_m + a_3b_{3m}\boldsymbol{y}_m + \cdots + a_nb_{nm}\boldsymbol{y}_m) \\
&= (a_1b_{11} + a_2b_{21} + a_3b_{31} + \cdots + a_nb_{n1})\boldsymbol{y}_1 + (a_1b_{12} + a_2b_{22} + a_3b_{32} + \cdots \\
&\quad + a_nb_{n2})\boldsymbol{y}_2 + (a_1b_{13} + a_2b_{23} + a_3b_{33} + \cdots + a_3b_{3m})\boldsymbol{y}_3 + \cdots + (a_1b_{1m} \\
&\quad + a_2b_{2m} + a_nb_{3m} + \cdots + a_nb_{nm})\boldsymbol{y}_m \\
&= c_1\boldsymbol{y}_1 + c_2\boldsymbol{y}_2 + c_3\boldsymbol{y}_3 + \cdots + c_m\boldsymbol{y}_m.
\end{aligned}$$

i.e., we have

$$\boldsymbol{x} = c_1\boldsymbol{y}_1 + c_2\boldsymbol{y}_2 + c_3\boldsymbol{y}_3 + \cdots + c_m\boldsymbol{y}_m$$

Proposition is proved. □

Corollary 3.5.5. *If $\boldsymbol{x}$ is a quasivector in an arithmetically and additively distributive, associative and commutative left quasilinear space $\mathbf{L}$ over an additively associative abstract prearithmetic $\boldsymbol{A}$, and $\boldsymbol{x} \in L$ $(\boldsymbol{x}_1, \boldsymbol{x}_2, \ldots, \boldsymbol{x}_n)$, $\boldsymbol{x}_i \in L$ $(\boldsymbol{y}_1, \boldsymbol{y}_2, \ldots, \boldsymbol{y}_m)$ for all $i = 1, 2, 3, \ldots, n$, then $\boldsymbol{x} \in L$ $(\boldsymbol{y}_1, \boldsymbol{y}_2, \ldots, \boldsymbol{y}_m)$.*

Corollary 3.5.6. *If in an arithmetically and additively distributive, associative and commutative left quasilinear space $\mathbf{L}$ over an additively associative abstract prearithmetic $\boldsymbol{A}$, $\boldsymbol{x}_i \in L(\boldsymbol{y}_1, \boldsymbol{y}_2, \ldots, \boldsymbol{y}_m)$ for all $i = 1, 2, 3, \ldots, n$, then $L(\boldsymbol{x}_1, \boldsymbol{x}_2, \ldots, \boldsymbol{x}_n)$ is a linear subspace of $L(\boldsymbol{y}_1, \boldsymbol{y}_2, \ldots, \boldsymbol{y}_m)$.*

Definition 3.5.15.

(a) Quasivectors $\boldsymbol{x}_1, \boldsymbol{x}_2, \ldots, \boldsymbol{x}_n$ from an associative left quasilinear space $\boldsymbol{L}$ and additive zero $\mathbf{0}$ over an abstract prearithmetic $\boldsymbol{A}$ with additive zero 0 are called *linearly independent* in $\boldsymbol{L}$ if the equality $\Sigma_{i=1}^{n} a_i\ \boldsymbol{x}_i = \mathbf{0}$ implies that all elements a_i $(i = 1, 2, 3, \ldots, n)$ from $\boldsymbol{A}$ are equal to 0.
(b) Otherwise, quasivectors $\boldsymbol{x}_1, \boldsymbol{x}_2, \ldots, \boldsymbol{x}_n$ are called *linearly dependent*, i.e., if there is an equality $\Sigma_{i=1}^{n} a_i\ \boldsymbol{x}_i = \mathbf{0}$ where a_i are elements from $\boldsymbol{A}$ and not all of them are equal to 0.

Let us find some properties of linearly dependent and independent systems of quasivectors.

Lemma 3.5.6. *Any system of quasivectors that contains* $\mathbf{0}$ *is linearly dependent.*

Proof is left as an exercise. □

Proposition 3.5.17. *If in a left quasilinear space* $\boldsymbol{L}$ *over an abstract prearithmetic* $\boldsymbol{A}$*, a system* R *of quasivectors contains a linearly dependent subsystem, then* R *is also linearly dependent, i.e. if quasivectors* $\boldsymbol{x}_1, \boldsymbol{x}_2, \ldots, \boldsymbol{x}_n$ *are linearly dependent, then any system of quasivectors* $\boldsymbol{x}_1, \boldsymbol{x}_2, \ldots, \boldsymbol{x}_n, \boldsymbol{y}_1, \boldsymbol{y}_2, \ldots, \boldsymbol{y}_m$ *is also linearly dependent.*

Indeed, if the system of quasivectors $\boldsymbol{x}_1, \boldsymbol{x}_2, \ldots, \boldsymbol{x}_n$ are linearly dependent, then there is an equality $\sum_{i=1}^{n} a_i\ \boldsymbol{x}_i = \mathbf{0}$ where a_i are elements from $\boldsymbol{A}$ and not all of them are equal to 0. Consequently, there is an equality $\sum_{i=1}^{n} a_i\ \boldsymbol{x}_i + \sum_{j=1}^{m} d_i\ \boldsymbol{y}_j = \mathbf{0}$ in which all d_i are equal to 0 and at least, one a_i is not equal to 0. It means that system of quasivectors $\boldsymbol{x}_1, \boldsymbol{x}_2, \ldots, \boldsymbol{x}_n, \boldsymbol{y}_1, \boldsymbol{y}_2, \ldots, \boldsymbol{y}_m$ is also linearly dependent.

Proposition 3.5.18. *In a left quasilinear space* $\boldsymbol{L}$ *over an abstract prearithmetic* $\boldsymbol{A}$*, a subsystem of a linearly independent system of quasivectors is also linearly independent.*

Indeed, by Proposition 3.5.17, any system of quasivectors containing a linearly dependent subsystem is also linearly dependent.

Note that a union of linearly independent systems can be either linearly independent or linearly dependent while the intersection of linearly independent systems is always linearly independent when it is not empty.

Now, we study relations between left quasilinear spaces such as weak projectivity and homomorphism.

Let us consider a left quasilinear space $\boldsymbol{L}_1 = (L_1, +_1, \mapsto_1)$ over an abstract prearithmetic $\boldsymbol{A}_1 = (A_1; \oplus_1, \circ_1, \leq_1)$, a left quasilinear space $\boldsymbol{L}_2 = (L_2, +_2, \mapsto_2)$ over an abstract prearithmetic $\boldsymbol{A}_2 = (A_2; \oplus_2, \circ_2, \leq_2)$ and four mappings $p : L_1 \to L_2, q : L_2 \to L_1, g$: $A_1 \to A_2$ and h: $A_2 \to A_1$.

Definition 3.5.16. (a) The left quasilinear space $\boldsymbol{L}_1 = (L_1, +_1, \mapsto_1)$ is called *weakly projective* with respect to the left quasilinear space $\boldsymbol{L}_2 = (L_2, +_2, \mapsto_2)$ if there are following relations between orders and operations in $\boldsymbol{L}_1$ and in $\boldsymbol{L}_2$:

$$\boldsymbol{a} +_1 \boldsymbol{b} = q(p(\boldsymbol{a}) +_2 p(\boldsymbol{b})),$$

$$d \mapsto_1 \boldsymbol{b} = q(g(d) \mapsto_2 p(\boldsymbol{b})),$$

$$d \oplus_1 c = h(g(d) \oplus_2 g(c)),$$

$$d \circ_1 c = h(g(d) \circ_2 g(c)),$$

$$d \leq_1 c \text{ only if } g(d) \leq_2 g(c).$$

(b) The pair of mappings (p, g) is called the *projector* and the pair of mappings (q, h) is called the *coprojector* for the pair $(\boldsymbol{L}_1, \boldsymbol{L}_2)$.

In this case, we will say that there is *weak projectivity* between the left quasilinear space $\boldsymbol{L}_1$ and the left quasilinear space $\boldsymbol{L}_2$ and there is *weak inverse projectivity* between the left quasilinear space $\boldsymbol{L}_2$ and the left quasilinear space $\boldsymbol{L}_1$.

When a left quasilinear space $\boldsymbol{H}$ is weakly projective with respect to a left quasilinear space $\boldsymbol{L}$ with the projector (p, g) and the coprojector (q, h), we denote $\boldsymbol{H}$ by $\boldsymbol{L}_{p,q;g,h}$ to show this relation between left quasilinear spaces $\boldsymbol{L}$ and $\boldsymbol{H}$.

We see that if a left quasilinear space $\boldsymbol{H}$ is weakly projective with respect to a left quasilinear space $\boldsymbol{L}$ with the projector (p, g) and the coprojector (q, h), then the abstract prearithmetic $\boldsymbol{A}_1$ is weakly projective with respect to the abstract prearithmetic $\boldsymbol{A}_2$ with the projector g and the coprojector h.

Remark 3.5.1. In addition, it is possible to define projectivity and exact projectivity of left quasilinear space. However, here we consider only weak projectivity.

Definitions imply the following result.

Proposition 3.5.19. *If a left quasilinear space* $\boldsymbol{L}_1 = (L_1, +_1, \mapsto_1)$ *over an abstract prearithmetic* $\boldsymbol{A}_1$ *is weakly projective with respect to the left quasilinear space* $\boldsymbol{L}_2 = (L_2, +_2, \mapsto_2)$ *over an abstract prearithmetic* $\boldsymbol{A}_2$*, then the abstract prearithmetic* $\boldsymbol{A}_1$ *is weakly projective with respect to the abstract prearithmetic* $\boldsymbol{A}_2$*.*

Proof is left as an exercise. □

Let us consider a set M, an abstract prearithmetic $\boldsymbol{A}_1 = (A_1; \oplus_1, \circ_1, \leq_1)$, a left quasilinear space $\boldsymbol{L} = (L, +, \mapsto)$ over an abstract prearithmetic $\boldsymbol{A}_2 = (A_2; \oplus_2, \circ_2, \leq_2)$ and four mappings $p : M \to L, q : L \to M, g\colon A_1 \to A_2$ and $h\colon A_2 \to A_1$ such that the abstract prearithmetic $\boldsymbol{A}_1$ is weakly projective with respect to the abstract prearithmetic $\boldsymbol{A}_2$ with the projector g and coprojector h.

Proposition 3.5.20. *It is possible to define on* M *the unique structure of a left quasilinear space* $\boldsymbol{M}^{p,q}$ *over the abstract prearithmetic* $\boldsymbol{A}_1$ *that is weakly projective with respect to the left quasilinear space* $\boldsymbol{L}_2$ *with the projector* g *and the coprojector* h*.*

Proof is similar to the proof of Proposition 2.2.1. □

Weak projectivity in the class of left quasilinear spaces implies definite relations between properties of operations in left quasilinear spaces.

Proposition 3.5.21. *If a left quasilinear space* $\boldsymbol{L}_1 = (L_1, +_1, \mapsto_1)$ *is weakly projective with respect to a commutative left quasilinear space* $\boldsymbol{L}_2 = (L_2, +_2, \mapsto_2)$ *with the projector* p *and the coprojector* q*, then the left quasilinear space* $\boldsymbol{L}_1$ *is also commutative.*

Indeed, for any elements $\boldsymbol{a}$ and $\boldsymbol{b}$ from L_1, we have

$$\boldsymbol{a} +_1 \boldsymbol{b} = p(q(\boldsymbol{a}) +_2 q(\boldsymbol{b})) = p(q(\boldsymbol{b}) +_2 q(\boldsymbol{a})) = \boldsymbol{b} +_1 \boldsymbol{a}.$$

Proposition 3.5.22. *If a left quasilinear space* $\boldsymbol{L}_1 = (L_1, +_1, \mapsto_1)$ *is weakly projective with respect to an associative left quasilinear space* $\boldsymbol{L}_2 = (L_2, +_2, \mapsto_2)$ *with the projector* p *and the coprojector* q*, for which* $pq = \mathbf{1}L_2$*, then the left quasilinear space* $\boldsymbol{L}_1$ *is also associative.*

Proof: By Definition 3.5.16, for any elements $\boldsymbol{a}$,$\boldsymbol{b}$ and $\boldsymbol{c}$ from L_1, we have

$$(\boldsymbol{a} +_1 \boldsymbol{b}) +_1 \boldsymbol{c} = q(p(q(p(\boldsymbol{a}) +_2 p(\boldsymbol{b}))) +_2 p(\boldsymbol{c})),$$

$$\boldsymbol{a} +_1 (\boldsymbol{b} +_1 \boldsymbol{c}) = q(p(\boldsymbol{a}) +_2 p(q(p(\boldsymbol{b}) +_2 p(\boldsymbol{c}))))).$$

If $pq = \mathbf{1}_{L2}$, then

$$(\boldsymbol{a} +_1 \boldsymbol{b}) +_1 \boldsymbol{c} = q(p(q(p(\boldsymbol{a}) +_2 p(\boldsymbol{b}))) +_2 p(\boldsymbol{c})) = q((p(\boldsymbol{a}) +_2 p(\boldsymbol{b})) +_2 p(\boldsymbol{c}))$$
$$= q(p(\boldsymbol{a}) +_2 (p(\boldsymbol{b})) +_2 p(\boldsymbol{c}))) = q(p(\boldsymbol{a}) +_2 p(q(p(\boldsymbol{b})$$
$$+ {}_2p(\boldsymbol{c}))))) = \boldsymbol{a} +_1 (\boldsymbol{b} +_1 \boldsymbol{c}).$$

Proposition is proved.

Mappings of left quasilinear spaces into left quasilinear spaces are also called *operators*. A mapping of a left quasilinear space over an abstract prearithmetic $\boldsymbol{A}$ into $\boldsymbol{A}$ or another abstract prearithmetic is called a *functional*.

As in the case of vector spaces, linear mappings of left quasilinear spaces are specifically important because they are homomorphisms of left quasilinear spaces as universal algebras as well as morphism in the category of left quasilinear spaces. That is why we study some properties of linear mappings of left quasilinear spaces.

Let us consider a left quasilinear space $\boldsymbol{L}_1 = (L_1, +_1, \mapsto_1)$ over an abstract prearithmetic $\boldsymbol{A}_1 = (A_1; \oplus_1, \circ_1, \leq_1)$, a left quasilinear space $\boldsymbol{L}_2 = (L_2, +_2, \mapsto_2)$ over an abstract prearithmetic $\boldsymbol{A}_2 = (A_2; \oplus_2, \circ_2, \leq_2)$, and a mapping g: $A_1 \to A_2$.

Definition 3.5.17. A pair (f, g) which consists of a mapping f: $L_1 \to L_2$ and a mapping g: $A_1 \to A_2$ is called a *linear mapping* of $\boldsymbol{L}_1$ into $\boldsymbol{L}_2$ if the following identities are valid:

$$f(\boldsymbol{a} +_1 \boldsymbol{b}) = f(\boldsymbol{a}) +_2 f(\boldsymbol{b}),$$

$$f(d \mapsto_1 \boldsymbol{b}) = g(d) \mapsto_2 f(\boldsymbol{b})).$$

Example 3.5.14. If X are Y are real vector spaces, then a mapping f: $X \to Y$ is *linear* if $f(c \cdot u + d \cdot v) = c \cdot f(u) + d \cdot f(v)$ for any elements u and v from X and any real numbers c and d (Artin, 1991). It means that the pair $(f, 1_{\boldsymbol{R}})$ is a linear mapping of X into Y treated as quasilinear spaces over $\boldsymbol{R}$.

Proposition 3.5.23. *The sequential composition of linear mappings of left quasilinear spaces is also a linear mapping of left quasilinear spaces.*

Proof is similar to the proof of the analogous property of linear mappings of vector spaces over fields.

Indeed, let us consider a left quasilinear space $\boldsymbol{L}_1 = (L_1, +_1, \mapsto_1)$ over an abstract prearithmetic $\boldsymbol{A}_1 = (A_1; \oplus_1, \circ_1, \leq_1)$, a left quasilinear space $\boldsymbol{L}_2 = (L_2, +_2, \mapsto_2)$ over an abstract prearithmetic $\boldsymbol{A}_2 = (A_2; \oplus_2, \circ_2, \leq_2)$ and a left quasilinear space $\boldsymbol{L}_3 = (L_3, +_3, \mapsto_3)$ over an abstract prearithmetic $\boldsymbol{A}_3 = (A_3; \oplus_3, \circ_3, \leq_3)$ and four mappings f: $L_1 \to L_2, h$: $L_2 \to L_3, g$: $A_1 \to A_2$ and $q : A_2 \to A_3$ such that (f, g) is called a linear mapping of $\boldsymbol{L}_1$ into $\boldsymbol{L}_2$ and (h, q) is a linear mapping of $\boldsymbol{L}_2$ into $\boldsymbol{L}_3$. It means that the following identities are valid:

$$f(\boldsymbol{a} +_1 \boldsymbol{b}) = f(\boldsymbol{a}) +_2 f(\boldsymbol{b}),$$
$$f(d \mapsto_1 \boldsymbol{b}) = g(d) \mapsto_2 f(\boldsymbol{b}))$$

and

$$h(\boldsymbol{a} +_1 \boldsymbol{b}) = h(\boldsymbol{a}) +_2 h(\boldsymbol{b}),$$
$$h(d \mapsto_1 \boldsymbol{b}) = q(d) \mapsto_2 h(\boldsymbol{b})).$$

Then we have

$$\begin{aligned}(h \circ f)(\boldsymbol{a} +_1 \boldsymbol{b}) &= h(f(\boldsymbol{a} +_1 \boldsymbol{b})) = h(f(\boldsymbol{a}) +_2 f(\boldsymbol{b}))\\ &= h(f(\boldsymbol{a})) +_3 h(f(\boldsymbol{b})) = (h \circ f)(\boldsymbol{a})) +_3 (h \circ f)(\boldsymbol{b}))\end{aligned}$$

and

$$\begin{aligned}(h \circ f)(d \mapsto_1 \boldsymbol{b}) &= h(f(d \mapsto_1 \boldsymbol{b})) = h(g(d) \mapsto_2 f(\boldsymbol{b}))\\ &= q(g(d) \mapsto_3 h(f(\boldsymbol{b})) = (q \circ g)(\boldsymbol{a})) \mapsto_3 (h \circ f)(\boldsymbol{b})).\end{aligned}$$

Proposition is proved. □

Corollary 3.5.7. *Left quasilinear spaces form a category with linear mappings as morphisms.*

Corollary 3.5.8. *Left quasilinear spaces over an abstract prearithmetic* $\boldsymbol{A}$ *form a category with linear mappings of the form* $(f, 1_{\boldsymbol{A}})$ *as morphisms.*

Linear mappings of left quasilinear spaces have other good properties.

Proposition 3.5.24.

(*a*) *If* (f, g) *is a linear mapping of a left quasilinear space* $\boldsymbol{L}_1$ *over an abstract prearithmetic* $\boldsymbol{A}_1$, *into a left quasilinear space* $\boldsymbol{L}_2$ *over an abstract prearithmetic* $\boldsymbol{A}_2$ *and the mapping* $g\colon A_1 \to A_2$ *is a homomorphism, then the image of* f *is a weak linear subspace of* $\boldsymbol{L}_2$.

(*b*) *If* (f, g) *is a linear mapping of a left quasilinear space* $\boldsymbol{L}_1$ *over an abstract prearithmetic* $\boldsymbol{A}_1$ *into a left quasilinear space* $\boldsymbol{L}_2$ *over an abstract prearithmetic* $\boldsymbol{A}_2$ *and the mapping* $g\colon A_1 \to A_2$ *is a epimorphism, then the image of* f *is a linear subspace of* $\boldsymbol{L}_2$.

Proof is left as an exercise. □

Corollary 3.5.9. *If* $(f, 1_{\boldsymbol{A}})$ *is a linear mapping of a left quasilinear space* $\boldsymbol{L}_1$ *over an abstract prearithmetic* $\boldsymbol{A}$ *into a left quasilinear space* $\boldsymbol{L}_2$ *over* $\boldsymbol{A}$, *then the image of* f *is a linear subspace of* $\boldsymbol{L}_2$.

Linear mappings of left quasilinear spaces preserve some identities.

Proposition 3.5.25. (*a*) *If* (f, g) *is a linear mapping of a commutative (associative) left quasilinear space* $\boldsymbol{L}_1$ *over an abstract prearithmetic* $\boldsymbol{A}_1$ *into a left quasilinear space* $\boldsymbol{L}_2$ *over an abstract prearithmetic* $\boldsymbol{A}_2$ *and the mapping* $g\colon A_1 \to A_2$ *is a homomorphism, then the image of* f *is a commutative (associative) left quasilinear space over the image of* g.

(*b*) *If* (f, g) *is a linear mapping of a left quasilinear space* $\boldsymbol{L}_1$ *over an abstract prearithmetic* $\boldsymbol{A}_1$ *into a left quasilinear space* $\boldsymbol{L}_2$ *over an abstract prearithmetic* $\boldsymbol{A}_2$ *and the mapping* $g\colon A_1 \to A_2$ *is a epimorphism, then the image of* f *is a commutative (associative) left quasilinear space over* $\boldsymbol{A}_2$.

Proof is left as an exercise. □

Definition 3.5.18.

(a) A left quasilinear space $\boldsymbol{L}$ over a wholly extended arithmetic is called a *non-Grassmannian left linear space.*

(b) A left quasilinear space $\boldsymbol{L}$ over an IR-arithmetic is called a *real non-Grassmannian left linear space.*

(c) A left quasilinear space $\boldsymbol{L}$ over an IC-arithmetic is called a *complex non-Grassmannian left linear space.*

Example 3.5.15. Any vector space over the real field $\boldsymbol{R}$ is a real non-Grassmannian left and right linear space.

Example 3.5.16. Any vector space over the complex field $\boldsymbol{C}$ is a complex non-Grassmannian left and right linear space.

Example 3.5.17. Any vector space over a field is a non-Grassmannian left and right linear space.

In an analogous way, it is possible to define *non-Grassmannian right quasilinear spaces* over abstract prearithmetics, in which scalar action is performed from the right. However, here we consider only non-Grassmannian left quasilinear space because non-Grassmannian right quasilinear spaces have similar properties and are studied by similar techniques.

All properties of general left (right) quasilinear spaces remain true for non-Grassmannian left (right) quasilinear spaces. At the same time, non-Grassmannian left quasilinear spaces have additional properties. Here are some examples.

Proposition 3.5.26. *If the operation* $-$ *is inverse to the operation* $+$ *in a non-Grassmannian left quasilinear space* $\boldsymbol{L}$, *i.e.,* $(\boldsymbol{z} + \boldsymbol{x}) - \boldsymbol{x} = \boldsymbol{z}$ *for any elements* $\boldsymbol{x}$, $\boldsymbol{y}$ *and* $\boldsymbol{z}$ *from* $\boldsymbol{L}$, *then* $\boldsymbol{L}$ *has additive cancellation.*

Proof is left as an exercise. □

Definition 3.5.19. Operations $+$ and $-$ are associative $+$ in a non-Grassmannian left quasilinear space $\boldsymbol{L}$, i.e., $(\boldsymbol{z} + \boldsymbol{x}) - \boldsymbol{y} = \boldsymbol{z} + (\boldsymbol{x} - \boldsymbol{y})$ for any elements $\boldsymbol{x}$, $\boldsymbol{y}$ and $\boldsymbol{z}$ from $\boldsymbol{L}$, then $\boldsymbol{L}$ has additive cancellation.

Proposition 3.5.27. *If operations* $+$ *and* $-$ *are associative* $+$ *the operation* $-$ *is inverse to the operation* $+$ *in a non-Grassmannian left quasilinear space* $\boldsymbol{L}$, *then for any element* $\boldsymbol{x}$ *from* $\boldsymbol{L}$, *the element* $\boldsymbol{x} - \boldsymbol{x}$ *is an additive zero in* $\boldsymbol{L}$.

Proof is left as an exercise. □

Note that projectivity relations and homomorphisms of non-Grassmannian left quasilinear spaces satisfy more conditions than projectivity relations and homomorphisms of left quasilinear spaces.

Let us continue studying linear dependence and independence assuming in what follows that an abstract prearithmetic $\boldsymbol{A} = (A; +, \circ, \leq)$ is additively associative with opposite elements, has a multiplicative one 1, additive 0 and division while $\boldsymbol{L}$ is associative, commutative, arithmetically and additively distributive, covered by $\boldsymbol{A}$ and scalar associative left quasilinear space over $\boldsymbol{A}$ with additive zero $\mathbf{0}$. Note that any field $\boldsymbol{F}$ and in particular, the Diophantine arithmetic $\boldsymbol{R}$ of all real numbers satisfy all these conditions for the abstract prearithmetic $\boldsymbol{A}$.

Proposition 3.5.28. *For any linearly dependent system of quasivectors $\boldsymbol{x}_1, \boldsymbol{x}_2, \ldots, \boldsymbol{x}_n$ from the left quasilinear space $\boldsymbol{L}$, at least, one quasivector $\boldsymbol{x}_i$ is a linear combination of others.*

Proof: Taking a linearly dependent system of quasivectors $\boldsymbol{x}_1, \boldsymbol{x}_2, \ldots, \boldsymbol{x}_n$ from $\boldsymbol{L}$, we have the equality

$$a_1\boldsymbol{x}_1 + a_2\boldsymbol{x}_2 + a_3\boldsymbol{x}_3 + \cdots + a_n\boldsymbol{x}_n = \mathbf{0}. \tag{3.18}$$

In it, at least, one element a_i is not equal to 0. As the space $\boldsymbol{L}$ is commutative, it is possible to presume that a_1 is not equal to 0.

As by the initial conditions, any element a in $\boldsymbol{A}$ has its opposite $-a$, we can take elements $-a_i\,(i = 2, 3, \ldots, n)$. Then by Corollary 3.5.3, each element $(-a_i\boldsymbol{x}_i)$ is opposite to the element $a_i\boldsymbol{x}_i\,(i = 2, 3, \ldots, n)$ because it is assumed that $\boldsymbol{L}$ is an arithmetically distributive left quasilinear space over an abstract prearithmetic $\boldsymbol{A}$ with opposite elements and an additive zero 0. Adding these elements to both sides of equality (3.18), we obtain equality (3.19).

$$a_1\boldsymbol{x}_1 = -a_2\boldsymbol{x}_2 + (-a_3\boldsymbol{x}_3) + \cdots + (-a_n\boldsymbol{x}_n). \tag{3.19}$$

As the prearithmetic $\boldsymbol{A}$ has division, there is an element c in $\boldsymbol{A}$ such that $ca_1 = 1$ (cf. Section 2.1). Multiplying both sides of equality (3.19) by c, we obtain equality (3.20) because by Proposition 3.5.8, $1\,\boldsymbol{x}_1 = \boldsymbol{x}_1$.

$$\boldsymbol{x}_1 = c(-a_2\boldsymbol{x}_2) + c(-a_3\boldsymbol{x}_3) + \cdots + c(-a_n\boldsymbol{x}_n) = (c(-a_2))\boldsymbol{x}_2 + (c(-a_3))\boldsymbol{x}_3 + \cdots + (c(-a_n))\boldsymbol{x}_n. \tag{3.20}$$

Proposition is proved. □

Corollary 3.5.10. *If $\sum_{i-1}^{n} a_i\,\boldsymbol{x}_i = \mathbf{0}$ and $a_i \neq 0$, then the quasivector $\boldsymbol{x}_i$ is a linear combination of $\boldsymbol{x}_1, \boldsymbol{x}_2, \ldots, \boldsymbol{x}_{i-1}, \boldsymbol{x}_{i+1}, \ldots, \boldsymbol{x}_n$.*

This also gives us usual properties of linear spaces over fields.

Corollary 3.5.11. *For any linearly dependent system of vectors $\boldsymbol{x}_1, \boldsymbol{x}_2, \ldots, \boldsymbol{x}_n$ from a linear space $\boldsymbol{L}$, at least one vector $\boldsymbol{x}_i$ is a linear combination of others.*

Corollary 3.5.12. *If $\sum_{i=1}^{n} a_i\,\boldsymbol{x}_i = \mathbf{0}$ and $a_i \neq 0$ in a linear space $\boldsymbol{L}$, then the quasivector $\boldsymbol{x}_i$ is a linear combination of $\boldsymbol{x}_1, \boldsymbol{x}_2, \ldots, \boldsymbol{x}_{i-1}, \boldsymbol{x}_{i+1}, \ldots, \boldsymbol{x}_n$.*

Adding a quasivector to a linearly independent system of quasivectors can make it linearly dependent.

Proposition 3.5.29. *If in a left quasilinear space* $\boldsymbol{L}$ *over an abstract prearithmetic* $\boldsymbol{A}$, *a quasivector* $\boldsymbol{x}$ *is a linear combination of quasivectors* $\boldsymbol{x}_1$, $\boldsymbol{x}_2, \ldots, \boldsymbol{x}_n$, *then the system* $\boldsymbol{x}$, $\boldsymbol{x}_1$, $\boldsymbol{x}_2, \ldots, \boldsymbol{x}_n$ *is linearly dependent.*

Proof: By the initial conditions, we have

$$\boldsymbol{x} = c_1\boldsymbol{x}_1 + c_2\boldsymbol{x}_2 + c_3\boldsymbol{x}_3 + \cdots + c_m\boldsymbol{x}_m. \tag{3.21}$$

As any element a in $\boldsymbol{A}$ has its opposite $-a$, we can take elements $-c_i(i = 2, 3, \ldots, n)$. Then by Corollary 3.5.3, each element $(-c_i\boldsymbol{x}_i)$ is opposite to the element $c_i\boldsymbol{x}_i(i = 2, 3, \ldots, n)$ because it is assumed that $\boldsymbol{L}$ is an arithmetically distributive left quasilinear space over an abstract prearithmetic $\boldsymbol{A}$ with opposite elements and an additive zero 0. Adding these elements to both sides of equality (3.21), we obtain equality (3.22):

$$\boldsymbol{x} + (-c_1\boldsymbol{x}_1) + (-c_2\boldsymbol{x}_2) + (-c_3\boldsymbol{x}_3) + \cdots + (-c_n\boldsymbol{x}_n) = \boldsymbol{0}. \tag{3.22}$$

It means that the system $\boldsymbol{x}$, $\boldsymbol{x}_1$, $\boldsymbol{x}_2, \ldots, \boldsymbol{x}_n$, is linearly dependent.

Proposition is proved. □

Corollary 3.5.13. *A quasivector* $\boldsymbol{x}$ *is a linear combination of linearly independent system* Q *of quasivectors* $\boldsymbol{x}_1$, $\boldsymbol{x}_2, \ldots, \boldsymbol{x}_n$ *if and only if adding* $\boldsymbol{x}$ *to* Q *makes the new system linearly dependent.*

Proposition 3.5.30. *If in a left quasilinear space* $\boldsymbol{L}$ *over an abstract prearithmetic* $\boldsymbol{A}$, *each quasivector* $\boldsymbol{x}_i$ *from a linearly independent system* $\boldsymbol{x}_1$, $\boldsymbol{x}_2, \ldots, \boldsymbol{x}_n$ *is a linear combination of quasivectors* $\boldsymbol{y}_1$, $\boldsymbol{y}_2, \ldots, \boldsymbol{y}_m(i = 1, 2, 3, \ldots, n)$, *then* $n \leq m$.

Proof: Let us assume that in a system of quasivectors $\boldsymbol{x}_1$, $\boldsymbol{x}_2, \ldots, \boldsymbol{x}_n$ from $\boldsymbol{L}$, each quasivector $\boldsymbol{x}_i$ from a system $\boldsymbol{x}_1$, $\boldsymbol{x}_2, \ldots, \boldsymbol{x}_n$ is a linear combination of quasivectors $\boldsymbol{y}_1$, $\boldsymbol{y}_2, \ldots, \boldsymbol{y}_m(i = 1, 2, 3, \ldots, n)$. Then by Proposition 3.5.29, the system $\boldsymbol{x}_1$, $\boldsymbol{y}_1$, $\boldsymbol{y}_2, \ldots, \boldsymbol{y}_m$, is linearly dependent, i.e., we have

$$\boldsymbol{x}_1 + \sum_{j=1}^{m} d_i\boldsymbol{y}_j = \boldsymbol{0}.$$

As $\boldsymbol{x}_1 \neq \boldsymbol{0}$, at least one element d_j is not equal to 0. As the quasilinear space $\boldsymbol{L}$ is commutative, it is possible to presume that d_1 is not equal to 0. Then by Corollary 3.5.3, the quasivector $\boldsymbol{y}_1$ is a linear combination of quasivectors $\boldsymbol{x}_1$, $\boldsymbol{y}_2, \ldots, \boldsymbol{y}_m$. Consequently, by Proposition 3.5.15, each quasivector $\boldsymbol{x}_i$ from a system $\boldsymbol{x}_2, \ldots, \boldsymbol{x}_n$ is a linear combination of quasivectors $\boldsymbol{x}_1$, $\boldsymbol{y}_2, \ldots, \boldsymbol{y}_m$.

Then by Proposition 3.5.29, the system $\boldsymbol{x}_2$, $\boldsymbol{x}_1$, $\boldsymbol{y}_2, \ldots, \boldsymbol{y}_m$, is linearly dependent, i.e., we have

$$\boldsymbol{x}_2 + c_1\boldsymbol{x}_1 + \sum_{j=2}^{m} c_i \boldsymbol{y}_j = \mathbf{0}.$$

As $\boldsymbol{x}_2 \neq \mathbf{0}$, at least one element c_j is not equal to 0. If only $c_1 \neq 0$, then the system $\{\boldsymbol{x}_2, \boldsymbol{x}_1\}$ would be linearly dependent. By Proposition 3.5.18, this is impossible because the system $\{\boldsymbol{x}_1, \boldsymbol{x}_2, \ldots, \boldsymbol{x}_n\}$ is linearly independent.

As the quasilinear space $\boldsymbol{L}$ is commutative, it is possible to presume that c_2 is not equal to 0. Then by Corollary 3.5.3, the quasivector $\boldsymbol{y}_2$ is a linear combination of quasivectors $\boldsymbol{x}_2$, $\boldsymbol{x}_1$, $\boldsymbol{y}_3, \ldots, \boldsymbol{y}_m$. Consequently, by Proposition 3.5.15, each quasivector $\boldsymbol{x}_i$ from a system $\boldsymbol{x}_3, \ldots, \boldsymbol{x}_n$ is a linear combination of quasivectors $\boldsymbol{x}_2$, $\boldsymbol{x}_1$, $\boldsymbol{y}_3, \ldots, \boldsymbol{y}_m$.

We can continue this process changing quasivectors $\boldsymbol{y}_j$ by quasivector $\boldsymbol{x}_i$. If m is less than n, then we would have linear dependence of the quasivectors $\boldsymbol{x}_1$, $\boldsymbol{x}_2, \ldots, \boldsymbol{x}_n$. As by the initial conditions, this system of quasivectors is linearly independent, then by the Principle of Excluded Middle, we come to the conclusion $n \leq m$.

Proposition is proved. □

Corollary 3.5.14. *If for two linearly independent systems of quasivectors* $\boldsymbol{x}_1, \boldsymbol{x}_2, \ldots, \boldsymbol{x}_n$ *and* $\boldsymbol{y}_1, \boldsymbol{y}_2, \ldots, \boldsymbol{y}_m$ *in a left quasilinear space* $\boldsymbol{L}$ *over an abstract prearithmetic* $\boldsymbol{A}$, *we have* $L(\boldsymbol{x}_1, \boldsymbol{x}_2, \ldots, \boldsymbol{x}_n) = L(\boldsymbol{y}_1, \boldsymbol{y}_2, \ldots, \boldsymbol{y}_m)$, *then* $n = m$.

Proposition 3.5.31. *Given two linearly independent systems of quasivectors* $\boldsymbol{x}_1, \boldsymbol{x}_2, \ldots, \boldsymbol{x}_n$ *and* $\boldsymbol{y}_1, \boldsymbol{y}_2, \ldots, \boldsymbol{y}_n$ *in a left quasilinear space* $\boldsymbol{L}$ *over an abstract prearithmetic* $\boldsymbol{A}$, *it is possible to substitute all quasivectors* $\boldsymbol{y}_i$ *by quasivectors* $\boldsymbol{x}_i$ *in such a way that each step of the substitution process gives a linearly independent system.*

Proof is done by induction and left as an exercise. □

Definition 3.5.20. A system B of quasivectors from a left quasilinear space $\boldsymbol{L}$ is called a *weak basis* or *generative system* of $\boldsymbol{L}$ if any element $\boldsymbol{x}$ from $\boldsymbol{L}$ is equal to a sum $\sum_{i=1}^{n} a_i\, \boldsymbol{x}_i$ where n is some natural number, $\boldsymbol{x}_i$ are elements from B and a_i are elements from $\boldsymbol{A}$.

Example 3.5.18. In a two-dimensional vector space over a field, any system of vectors that contains two linearly independent vectors is a weak basis.

Lemma 3.5.7. *Any system of quasivectors that contains a weak basis of* $\boldsymbol{L}$ *is a weak basis of* $\boldsymbol{L}$.

Proof is left as an exercise. □

Proposition 3.5.32. *Any finite maximal linearly independent system of quasivectors* $\boldsymbol{x}_1, \boldsymbol{x}_2, \ldots, \boldsymbol{x}_n$ *from a left quasilinear space* $\boldsymbol{L}$ *is a weak basis of* $\boldsymbol{L}$.

Proof: Let us take a maximal linearly independent system X of quasivectors $\boldsymbol{x}_1, \boldsymbol{x}_2, \ldots, \boldsymbol{x}_n$ from $\boldsymbol{L}$ and an arbitrary quasivector $\boldsymbol{x}$ from $\boldsymbol{L}$. Because X is a maximal linearly independent system in $\boldsymbol{L}$, the system of quasivectors $\boldsymbol{x}, \boldsymbol{x}_1, \boldsymbol{x}_2, \ldots, \boldsymbol{x}_n$ is linearly dependent. It means that there is the following equality:

$$a\boldsymbol{x} + a_1\boldsymbol{x}_1 + a_2\boldsymbol{x}_2 + a_3\boldsymbol{x}_3 + \cdots + a_n\boldsymbol{x}_n = \mathbf{0}. \tag{3.23}$$

In (3.23), the element a is not equal to zero because otherwise the system X of quasivectors $\boldsymbol{x}_1, \boldsymbol{x}_2, \ldots, \boldsymbol{x}_n$ would be linearly dependent. Thus, by Corollary 3.5.3, the quasivector $\boldsymbol{x}$ is equal to a sum $\sum_{i=1}^{n} d_i\boldsymbol{x}_i$.

As $\boldsymbol{x}$ is an arbitrary quasivector from $\boldsymbol{L}$, Proposition 3.5.32 is proved. □

Corollary 3.5.15. *Any system of quasivectors that contains a maximal linearly independent system is a weak basis of* $\boldsymbol{L}$.

Proposition 3.5.33. *If the system* B *of quasivectors* $\boldsymbol{x}_1, \boldsymbol{x}_2, \ldots, \boldsymbol{x}_n$ *from a left quasilinear space* $\boldsymbol{L}$ *is a weak basis of* $\boldsymbol{L}$ *and each quasivector* $\boldsymbol{x}_i$ *is a linear combination of quasivectors* $\boldsymbol{y}_1, \boldsymbol{y}_2, \ldots, \boldsymbol{y}_m (i = 1, 2, 3, \ldots, n)$, *then the system* $\boldsymbol{y}_1, \boldsymbol{y}_2, \ldots, \boldsymbol{y}_m$ *is a weak basis of* $\boldsymbol{L}$.

Indeed, if a quasivector $\boldsymbol{x}$ is a linear combination of quasivectors $\boldsymbol{x}_1, \boldsymbol{x}_2, \ldots, \boldsymbol{x}_n$, then by Proposition 3.5.15, the quasivector $\boldsymbol{x}$ is a linear combination of quasivectors $\boldsymbol{y}_1, \boldsymbol{y}_2, \ldots, \boldsymbol{y}_m$. It means that the system $\boldsymbol{y}_1, \boldsymbol{y}_2, \ldots, \boldsymbol{y}_m$ is a weak basis of $\boldsymbol{L}$.

Corollary 3.5.16. *If the system* B *of quasivectors* $\boldsymbol{x}_1, \boldsymbol{x}_2, \ldots, \boldsymbol{x}_n$ *from a left quasilinear space* $\boldsymbol{L}$ *is a weak basis of* $\boldsymbol{L}$ *and quasivector* $\boldsymbol{x}_1$ *is a linear* *combination of quasivectors* $\boldsymbol{x}_2, \ldots, \boldsymbol{x}_n$*, then the system* $\boldsymbol{x}_2, \ldots, \boldsymbol{x}_n$ *is a weak basis of* $\boldsymbol{L}$*.*

Definition 3.5.21. A system B of vectors from a left quasilinear space $\boldsymbol{L}$ is called a *basis* of $\boldsymbol{L}$ if any element $\boldsymbol{x}$ from $\boldsymbol{L}$ is equal to a unique sum $\sum_{i=1}^{n} a_i\ \boldsymbol{x}_i$ where n is some natural number, all $\boldsymbol{x}_i$ are elements from B and a_i are non-zero elements from an abstract prearithmetic $\boldsymbol{A}$.

Lemma 3.5.8. *Any basis of* $\boldsymbol{L}$ *is a weak basis of* $\boldsymbol{L}$*.*

Proof is left as an exercise. □

Under definite conditions, a weak basis of a left quasilinear space is its basis.

Theorem 3.5.1. *A weak basis* B *of a left quasilinear space* $\boldsymbol{L}$ *is a basis of* $\boldsymbol{L}$ *if and only if it consists of linearly independent quasivectors.*

Proof: *Sufficiency.* At first, we prove that any weak basis B of $\boldsymbol{L}$ that consists of linearly independent quasivectors is a basis of $\boldsymbol{L}$. Let us assume that the system B of quasivectors $\boldsymbol{x}_1, \boldsymbol{x}_2, \ldots, \boldsymbol{x}_n$ is linearly independent and is a weak basis of $\boldsymbol{L}$ and we have equalities (3.24) and (3.25) in which there is, at least one i such that $a_i \neq b_i$. As the space $\boldsymbol{L}$ is commutative, it is possible to presume that a_1 is not equal to b_1:

$$\boldsymbol{x} = \sum_{i=1}^{n} a_i \boldsymbol{x}_i = \sum_{i=1}^{n} b_i \boldsymbol{x}_i, \tag{3.24}$$

$$\sum_{i=1}^{n} a_i \boldsymbol{x}_i = \sum_{i=1}^{n} b_i \boldsymbol{x}_i. \tag{3.25}$$

As by the initial conditions, any element a in $\boldsymbol{A}$ has its opposite $-a$, we can take elements $-a_i\ (i = 1, 2, 3, \ldots, n)$. Then by Corollary 3.5.3, each element $-a_i\boldsymbol{x}_i$ is opposite to the element $-a_i\boldsymbol{x}_i(i = 1, 2, 3, \ldots, n)$. Adding these elements to both sides of equality (3.25), we obtain equality (3.26) as the space $\boldsymbol{L}$ is scalar distributive.

$$\begin{aligned}\sum_{i=1}^{n} b_i \boldsymbol{x}_i + \sum_{i=1}^{n} (-a_i)\boldsymbol{x}_i &= \sum_{i=1}^{n} (b_i + (-a_i))\boldsymbol{x}_i \\ &= \sum_{i=1}^{n} (a_i + (-a_i))\boldsymbol{x}_i = \sum_{i=1}^{n} 0\boldsymbol{x}_i = \mathbf{0}. \end{aligned} \tag{3.26}$$

Since $a_1 \neq b_1$, we have $b_i + (-a_i) \neq 0$. It means that contrary to our assumption, the system B of quasivectors $\boldsymbol{x}_1, \boldsymbol{x}_2, \ldots, \boldsymbol{x}_n$ is linearly dependent. By the Principle of Excluded Middle, the representation of any element $\boldsymbol{x}$ from $\boldsymbol{L}$ as a sum $\sum_{i=1}^{n} a_i\ \boldsymbol{x}_i$ is unique, i.e., B is a basis of $\boldsymbol{L}$.

The sufficiency is proved.

Necessity. Let us assume that a weak basis B of $\boldsymbol{L}$ is a basis of $\boldsymbol{L}$ but its elements are linearly dependent, for some quasivectors $\boldsymbol{x}_1, \boldsymbol{x}_2, \ldots, \boldsymbol{x}_n$ from B, we have $\sum_{i=1}^{n} a_i\ \boldsymbol{x}_i = \mathbf{0}$ where not all a_i are equal to 0. As the space $\boldsymbol{L}$ is commutative, it is possible to presume that a_1 is not equal to 0. Then by Proposition 3.5.19, the quasivector $\boldsymbol{x}_1$ is a linear combination of $\boldsymbol{x}_2, \ldots, \boldsymbol{x}_n$, i.e.,

$$\boldsymbol{x}_1 = \sum_{i=2}^{n} b_i \boldsymbol{x}_i.$$

It means that the quasivector $\boldsymbol{x}_1$ has two representations in the form $\sum_{i=1}^{n} a_i \boldsymbol{x}_i$. In one of them, $a_1 = 1$ and $a_i = 0$ for $i = 2, 3, \ldots, n$. In the second representation, $a_1 = 0$ and $a_i = b_i$ for $i = 2, 3, \ldots, n$. This contradicts the definition of a basis and by the Principle of Excluded Middle, shows that all quasivectors in B are linearly independent.

Theorem is proved. □

Corollary 3.5.17. *Any basis B of $\boldsymbol{L}$ consists of linearly independent quasivectors.*

Theorems 3.5.1 and Proposition 3.5.32 imply the following result.

Corollary 3.5.18. *Any finite maximal linearly independent system of quasivectors $\boldsymbol{x}_1, \boldsymbol{x}_2, \ldots, \boldsymbol{x}_n$ from a left quasilinear space $\boldsymbol{L}$ is a basis of $\boldsymbol{L}$.*

Corollary 3.5.6 implies the following result.

Proposition 3.5.34. *If in a left quasilinear space $\boldsymbol{L}$ over an abstract prearithmetic $\boldsymbol{A}$, the number of elements in linearly independent systems of quasivectors is a bounded, then $\boldsymbol{L}$ has a finite basis.*

Proof is left as an exercise. □

It is possible to characterize a basis of a left quasilinear space as a minimal weak basis.

Theorem 3.5.2. *A weak basis B of $\boldsymbol{L}$ is a basis of $\boldsymbol{L}$ if and only if it is minimal.*

Proof: *Sufficiency.* Let us consider a minimal weak basis B of $\boldsymbol{L}$. If B is a linearly dependent system of quasivectors, then by Corollary 3.5.3, one of the elements from B is a linear combination of other elements from B. Excluding this element, we still obtain a weak basis B of $\boldsymbol{L}$. As B is a minimal weak basis of $\boldsymbol{L}$, this is impossible and we can conclude that B is a linearly independent system of quasivectors. Then by Theorem 3.5.1, B is a basis of $\boldsymbol{L}$.

Necessity. Let us consider a basis B of $\boldsymbol{L}$. It is linearly independent. Thus, the system obtained by excluding even one element from B cannot be a weak basis. Thus, B is a minimal weak basis B of $\boldsymbol{L}$.

Theorem is proved. □

Definition 3.5.22. Two systems of quasivectors from a left quasilinear space $\boldsymbol{L}$ are called *equivalent* if each element of one of them is expressed as a linear combination of quasivectors from another onc.

Lemma 3.5.9. *Any two weak bases of* $\boldsymbol{L}$ *are equivalent.*

Proof is left as an exercise. □

Corollary 3.5.19. *Any two bases of* $\boldsymbol{L}$ *are equivalent.*

Equivalence of bases results in the equality of their cardinalities. Namely, Corollary 3.5.14 implies the following result.

Proposition 3.5.35. *Any two bases of* $\boldsymbol{L}$ *have the same number of elements.*

Proof is left as an exercise. □

Equivalence of systems of quasivectors allows finding weak bases of left quasilinear spaces.

Proposition 3.5.36. *A system of quasivectors from a left quasilinear space* $\boldsymbol{L}$ *equivalent to a weak basis of* $\boldsymbol{L}$ *is weak basis of* $\boldsymbol{L}$.

Proof is left as an exercise. □

Note that a system of quasivectors from a left quasilinear space $\boldsymbol{L}$ equivalent to a basis of $\boldsymbol{L}$ is not always a basis of $\boldsymbol{L}$.

Proposition 3.5.37. *Equivalent systems of quasivectors from a left quasilinear space* $\boldsymbol{L}$ *generate the same linear subspace of* $\boldsymbol{L}$.

Proof is left as an exercise. □

Proposition 3.5.38. *The linear subspace* $L\{\boldsymbol{x}_1, \boldsymbol{x}_2, \ldots, \boldsymbol{x}_n\}$ *generated by quasivectors* $\boldsymbol{x}_1, \boldsymbol{x}_2, \ldots, \boldsymbol{x}_n$ *from a left quasilinear space* $\boldsymbol{L}$ *coincides with the linear subspace generated by the largest linearly independent subsystem of the system*$\boldsymbol{x}_1, \boldsymbol{x}_2, \ldots, \boldsymbol{x}_n$.

Proof is left as an exercise. □

Definition 3.5.23. The number of elements in a basis of a left quasilinear space $\boldsymbol{L}$ is called the *dimension* of $\boldsymbol{L}$.

Proposition 3.5.36 implies that for each left quasilinear space $\boldsymbol{L}$, its dimension is defined in a unique way.

Proposition 3.5.39. (*a*) *The dimension of a subspace $\boldsymbol{H}$ of a left quasilinear space $\mathbf{L}$ is not larger than the dimension of $\mathbf{L}$.*

(*b*) *The dimension of a subspace $\boldsymbol{H}$ of a left quasilinear space $\mathbf{L}$ is equal to the dimension of $\mathbf{L}$ if and only if $\boldsymbol{H} = \mathbf{L}$.*

Proof is left as an exercise. □

Theorem 3.5.3. *In a left quasilinear space $\mathbf{L}$ that has dimension n, any system of quasivectors with more than n elements is linearly dependent.*

Indeed, if a system of quasivectors in $\boldsymbol{L}$ that has dimension n consists of more than n elements, it cannot be linearly independent by Proposition 3.5.35.

Chapter 4

From Non-Diophantine Arithmetic to Non-Newtonian Calculus

New opinions are always suspected, and usually opposed, without any other reason but because they are not already common.

John Locke

The credit for inventing an arithmetic-inspired non-Newtonian calculus is due to Michael Grossman and Robert Katz. They developed differential and integral calculi for functions from one non-Diophantine arithmetic of real numbers into another non-Diophantine arithmetic of real numbers. Their little book *Non-Newtonian Calculus* (Grossman and Katz, 1972) went, unfortunately, basically unnoticed by the mainstream mathematical community.

Some elements of non-Newtonian calculus (the product integral) can be found already in the works of the outstanding Italian mathematician Vito Volterra (1860–1940) in his study of differential equations (Volterra, 1887; 1887a). The concept of the product integral was further developed and utilized by other mathematicians (cf., for example, Rasch, 1934; Birkhoff, 1938). Another root of the non-Newtonian calculus is logarithmic differentiation (cf., for example, Bali, 2005; Bird, 1993; Krantz, 2003). Cepstral signal analysis and nonlinear filtering can be regarded as examples of applications of non-Newtonian integral calculus (Oppenheim, Schafer and Stockham, 1968; Childers, Skinner and Kemarit, 1977). Some types of non-Newtonian calculus, called g-calculus, were discovered by Endre Pap, who developed differential and integral calculi for functions from a non-Diophantine arithmetic of real numbers into itself (Pap, 1993).

More recently, the structures of non-Newtonian calculus were independently rediscovered by one of us in a more general situation where domains and ranges of studied functions could be sufficiently big infinite sets of any nature (Czachor, 2015). These functions, real or complex, map one non-Diophantine arithmetic into another. The original motivation for this came from attempts of formulating a calculus on fractals, but it soon became clear that the obtained structure of the calculus is in fact very general and flexible. The freedom of choice of arithmetic plays a role of a fundamental symmetry in any natural science.

4.1. Non-Newtonian Derivatives and Integrals of Real Functions

The high-minded man must care more for the truth than for what people think.

Aristotle

For any set whose cardinality is continuum, there exists a bijection mapping it onto R. Let A and B be two such sets, and consider two arithmetics $\boldsymbol{A} = (A; \oplus_A, \otimes_A, \ominus_A, \oslash_A, \leq_A)$, and $\boldsymbol{B} = (B; \oplus_B, \otimes_B, \ominus_B, \oslash_B, \leq_B)$ with bijections $f_A : A \to R$ and $f_B : B \to R$, which are playing the roles of functional parameters of $\boldsymbol{A}$ and $\boldsymbol{B}$. As it is proved in Section 2.4, both arithmetics are isomorphic to the Diophantine arithmetic $\boldsymbol{R}$ of real numbers, i.e., $\boldsymbol{A}$ and $\boldsymbol{B}$ are ordered fields. Operations in the arithmetic $\boldsymbol{A}$ are defined by the following formulas:

$$x \oplus_A y = f_A^{-1}(f_A(x) + f_A(y)),$$

$$x \ominus_A y = f_A^{-1}(f_A(x) - f_A(y)),$$

$$x \otimes_A y = f_A^{-1}(f_A(x) f_A(y)),$$

$$x \oslash_A y = f_A^{-1}(f_A(x) \div f_A(y)),$$

$$x \leq_A y \Longleftrightarrow f_A(x) \leq f_A(y).$$

In a similar way, operations in the arithmetic $\boldsymbol{B}$ are defined. Because f_A and f_B are bijections, multiplication and addition so defined are commutative and associative, and multiplication is distributive with respect to addition (cf. Section 2.4).

A mapping (function) g: $A \to B$ defines a unique function $\tilde{g} = f_B \circ g \circ f_A^{-1} : R \to R$:

$$\begin{array}{ccc} A & \xrightarrow{g} & B \\ f_A\downarrow & & \downarrow f_B \\ R & \xrightarrow{\tilde{g}} & R \end{array}$$

To build a non-Newtonian differential calculus, we define a non-Newtonian derivative of g: $A \to B$, by the formula (4.1) assuming that $\mathrm{d}\tilde{g}(f_A(x))/\mathrm{d}f_A(x)$ is the classical (conventional, Newtonian) derivative and the right part of this formula exists,

$$\frac{\mathrm{D}g(x)}{\mathrm{D}x} = f_B^{-1}\left(\frac{\mathrm{d}}{\mathrm{d}f_A(x)}\underbrace{f_B \circ g \circ f_A^{-1}}_{\tilde{g}}[f_A(x)]\right) \tag{4.1}$$

$$= f_B^{-1}\left(\frac{\mathrm{d}}{\mathrm{d}f_A(x)}f_B \circ g(x)\right), \tag{4.2}$$

where $f_A^{-1} : R \to A$ and $f_B^{-1} : R \to B$ are bijections continuous at $0 \in R$.

We will prove that the non-Newtonian derivative is linear with respect to $\oplus_B$, satisfies the Leibniz rule for $\otimes_B$, and satisfies an appropriate chain rule for composition of functions. We will also see that (4.1) can be alternatively written as

$$\frac{\mathrm{D}g(x)}{\mathrm{D}x} = \lim_{h\to 0}(g(x \oplus_A h_A) \ominus_B g(x)) \oslash_B h_B,$$

where the limit is appropriately defined. Note that $0_A = f_A^{-1}(0)$ and $0_B = f_B^{-1}(0)$ are the neutral elements of addition with respect to $\oplus_A$ and $\oplus_B$, respectively.

Let us begin with some examples.

Example 4.1.1. Let $A = B = R$, $f_A(x) = f_B(x) = x^3 = f(x)$. The arithmetic operations read

$$x \oplus y = f^{-1}(f(x) + f(y)) = \sqrt[3]{x^3 + y^3}, \tag{4.3}$$

$$x \ominus y = f^{-1}(f(x) - f(y)) = \sqrt[3]{x^3 - y^3}, \tag{4.4}$$

$$x \otimes y = f^{-1}(f(x)f(y)) = \sqrt[3]{x^3y^3} = xy, \tag{4.5}$$

$$x \oslash y = f^{-1}(f(x)/f(y)) = \sqrt[3]{x^3/y^3} = x/y. \tag{4.6}$$

The neutral elements of addition and multiplication are the standard ones: $0_A = f^{-1}(0) = \sqrt[3]{0} = 0$, $1_A = f^{-1}(1) = \sqrt[3]{1} = 1$. Although the multiplication is unchanged, the link between addition and multiplication is a subtle one, as can be seen in the following example:

$$\begin{aligned} x \oplus \cdots \oplus x &= \sqrt[3]{x^3 + \cdots + x^3} \quad (n \text{ times}) \\ &= \sqrt[3]{n}x = f^{-1}(n)x. \end{aligned}$$

The inverse bijection $f^{-1}(x) = \sqrt[3]{x}$ is continuous but not differentiable (in the Newtonian sense!) at $x = 0$. Still, the derivative

$$\frac{\mathrm{D}g(x)}{\mathrm{D}x} = \lim_{h\to 0} (g(x \oplus h) \ominus g(x)) \oslash h = f^{-1}\left(\frac{\mathrm{d}\tilde{g}[f(x)]}{\mathrm{d}f(x)}\right) \tag{4.7}$$

is well defined and satisfies all the basic rules of differentiation. It is very instructive to perform calculations by means of the limit form of the derivative when $h \to 0$. The limit itself does not require any further comments in such a simple example.

(a) *The Leibniz rule.*

$$\begin{aligned} &\frac{\mathrm{D}g_1(x) \otimes g_2(x)}{\mathrm{D}x} \\ &= \lim_{h\to 0} (g_1(x \oplus h) \otimes g_2(x \oplus h) \ominus g_1(x) \otimes g_2(x)) \oslash h \\ &= \lim_{h\to 0} \sqrt[3]{g_1(x \oplus h)^3 g_2(x \oplus h)^3 - g_1(x)^3 g_2(x)^3}/h \\ &= \lim_{h\to 0} \sqrt[3]{\frac{g_1(x \oplus h)^3 - g_1(x)^3}{h^3} g_2(x \oplus h)^3 + g_1(x)^3 \frac{g_2(x \oplus h)^3 - g_2(x)^3}{h^3}} \\ &= \lim_{h\to 0} \sqrt[3]{\left(\frac{\sqrt[3]{g_1(x \oplus h)^3 - g_1(x)^3}}{h}\right)^3 g_2(x)^3 + g_1(x)^3 \left(\frac{\sqrt[3]{g_2(x \oplus h)^3 - g_2(x)^3}}{h}\right)^3} \\ &= \sqrt[3]{\left(\lim_{h\to 0} \frac{g_1(x \oplus h) \ominus g_1(x)}{h}\right)^3 g_2(x)^3 + g_1(x)^3 \left(\lim_{h\to 0} \frac{g_2(x \oplus h) \ominus g_2(x)}{h}\right)^3} \\ &= \sqrt[3]{\left(\frac{\mathrm{D}g_1(x)}{\mathrm{D}x} g_2(x)\right)^3 + \left(g_1(x) \frac{\mathrm{D}g_2(x)}{\mathrm{D}x}\right)^3} \\ &= \frac{\mathrm{D}g_1(x)}{\mathrm{D}x} \otimes g_2(x) \oplus g_1(x) \otimes \frac{\mathrm{D}g_2(x)}{\mathrm{D}x}. \end{aligned}$$

(b) *Linearity.* It is enough to prove additivity since 1-homogeneity will follow from the Leibniz rule.

$$\begin{aligned}
&\frac{\mathrm{D}g_1(x) \oplus g_2(x)}{\mathrm{D}x} \\
&\quad = \lim_{h\to 0}(g_1(x\oplus h) \oplus g_2(x\oplus h) \ominus (g_1(x) \oplus g_2(x))) \\
&\quad = \lim_{h\to 0} \frac{\sqrt[3]{(g_1(x\oplus h) \oplus g_2(x\oplus h))^3 - (g_1(x) \oplus g_2(x))^3}}{h} \\
&\quad = \lim_{h\to 0} \frac{\sqrt[3]{g_1(x\oplus h)^3 + g_2(x\oplus h)^3 - g_1(x)^3 - g_2(x)^3}}{h} \\
&\quad = \lim_{h\to 0} \sqrt[3]{\frac{g_1(x\oplus h)^3 - g_1(x)^3}{h^3} + \frac{g_2(x\oplus h)^3 - g_2(x)^3}{h^3}} \\
&\quad = \lim_{h\to 0} \sqrt[3]{\left(\frac{\sqrt[3]{g_1(x\oplus h)^3 - g_1(x)^3}}{h}\right)^3 + \left(\frac{\sqrt[3]{g_2(x\oplus h)^3 - g_2(x)^3}}{h}\right)^3} \\
&\quad = \sqrt[3]{\left(\lim_{h\to 0}\frac{g_1(x\oplus h) \ominus g_1(x)}{h}\right)^3 + \left(\lim_{h\to 0}\frac{g_2(x\oplus h) \ominus g_2(x)}{h}\right)^3} \\
&\quad = \sqrt[3]{\left(\frac{\mathrm{D}g_1(x)}{\mathrm{D}x}\right)^3 + \left(\frac{\mathrm{D}g_2(x)}{\mathrm{D}x}\right)^3} = \frac{\mathrm{D}g_1(x)}{\mathrm{D}x} \oplus \frac{\mathrm{D}g_2(x)}{\mathrm{D}x}.
\end{aligned}$$

(c) *The chain rule.* Consider

$$\begin{aligned}
\frac{\mathrm{D}g_1(g_2(x))}{\mathrm{D}x} &= \lim_{h\to 0}(g_1(g_2(x\oplus h)) \ominus g_1(g_2(x))) \oslash h \\
&= \lim_{h\to 0} \sqrt[3]{\frac{(g_1(g_2(x\oplus h)))^3 - (g_1(g_2(x)))^3}{h^3}}.
\end{aligned}$$

Denoting

$$\delta = g_2(x\oplus h) \ominus g_2(x),$$

or equivalently,

$$g_2(x\oplus h) = g_2(x) \oplus \delta,$$

we obtain

$$\begin{aligned}
\frac{\mathrm{D}g_1(g_2(x))}{\mathrm{D}x} &= \lim_{h\to 0} \sqrt[3]{\frac{(g_1(g_2(x)\oplus\delta))^3 - (g_1(g_2(x)))^3}{h^3}} \\
&= \lim_{h\to 0} \sqrt[3]{\frac{(g_1(g_2(x)\oplus\delta))^3 - (g_1(g_2(x)))^3}{\delta^3} \frac{(g_2(x\oplus h)\ominus g_2(x))^3}{h^3}} \\
&= \lim_{\delta\to 0} \frac{\sqrt[3]{(g_1(g_2(x)\oplus\delta))^3 - (g_1(g_2(x)))^3}}{\delta} \\
&\quad \times \lim_{h\to 0} \frac{g_2(x\oplus h)\ominus g_2(x)}{h} \\
&= \lim_{\delta\to 0} \frac{g_1(g_2(x)\oplus\delta)\ominus g_1(g_2(x))}{\delta} \lim_{h\to 0} \frac{g_2(x\oplus h)\ominus g_2(x)}{h} \\
&= \frac{\mathrm{D}g_1(g_2(x))}{\mathrm{D}g_2(x)} \frac{\mathrm{D}g_2(x)}{\mathrm{D}x} = \frac{\mathrm{D}g_1(g_2(x))}{\mathrm{D}g_2(x)} \otimes \frac{\mathrm{D}g_2(x)}{\mathrm{D}x}.
\end{aligned}$$

Example 4.1.2 (A non-Newtonian differential equation). Consider

$$\frac{\mathrm{D}g(x)}{\mathrm{D}x} = g(x), \quad g(0) = 1, \tag{4.8}$$

where the arithmetics are the same as in the previous example. The solution of (4.8) is

$$g(x) = e^{x^3/3} = f^{-1}(e^{f(x)}), \tag{4.9}$$

as one can verify directly from definition (4.7):

$$\begin{aligned}
\frac{\mathrm{D}g(x)}{\mathrm{D}x} &= \lim_{h\to 0} (g(x\oplus h)\ominus g(x)) \oslash h \\
&= \lim_{h\to 0} \sqrt[3]{(g(x\oplus h)^3 - g(x)^3)}/h \\
&= \lim_{h\to 0} \sqrt[3]{(e^{(x\oplus h)^3/3})^3 - (e^{x^3/3})^3}/h \\
&= \lim_{h\to 0} \sqrt[3]{\frac{e^{(\sqrt[3]{x^3+h^3})^3} - e^{x^3}}{h^3}} \\
&= \sqrt[3]{\lim_{\delta\to 0} \frac{e^{x^3+\delta} - e^{x^3}}{\delta}} \\
&= \sqrt[3]{\frac{\mathrm{d}\exp(x^3)}{\mathrm{d}(x^3)}} = e^{x^3/3} = g(x).
\end{aligned}$$

The solution is unique.

Example 4.1.3. One can similarly verify that

$$\begin{aligned} \operatorname{Sin} x &= \sqrt[3]{\sin(x^3)}, \\ \operatorname{Cos} x &= \sqrt[3]{\cos(x^3)} \end{aligned}$$

satisfy the following equalities:

$$\frac{\mathrm{DSin}\ x}{\mathrm{D}x} = \operatorname{Cos} x, \tag{4.10}$$

$$\frac{\mathrm{D}\operatorname{Cos}\ x}{\mathrm{D}x} = \ominus \operatorname{Sin} x = -\operatorname{Sin} x, \tag{4.11}$$

where $\ominus x = 0 \ominus x = \sqrt[3]{0 - x^3} = -x$, and

$$\operatorname{Sin}^2 x \oplus \operatorname{Cos}^2 x = \sqrt[3]{\sin^2(x^3) + \cos^2(x^3)} = 1.$$

$\operatorname{Sin} x$ and $\operatorname{Cos} x$ are essentially the chirp signals known from signal analysis (Figure 4.1).

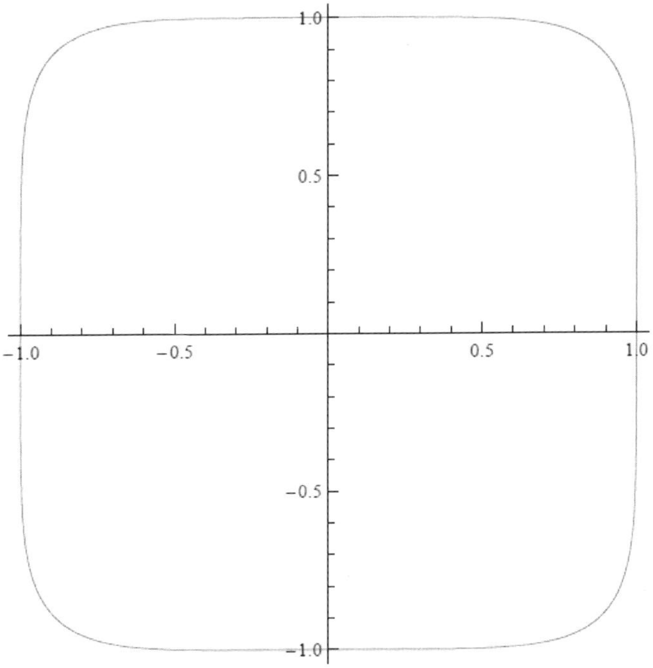

Figure 4.1. The circle $x \mapsto (\operatorname{Cos} x, \operatorname{Sin} x)$, $0 \leq x < (2\pi)^{1/3}$.

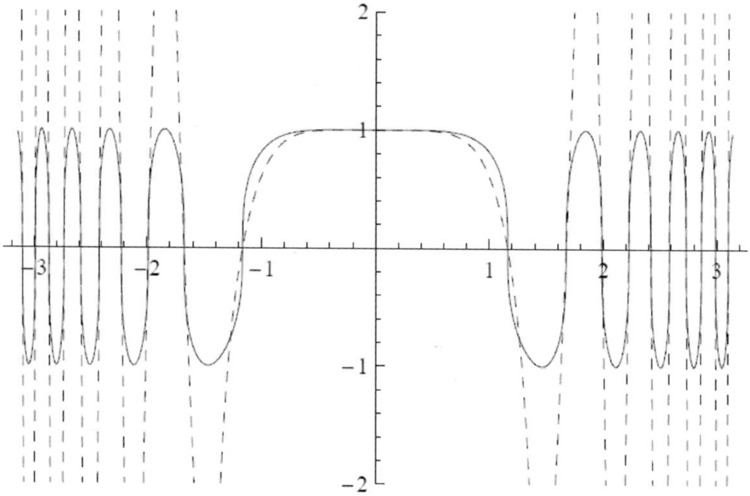

Figure 4.2. The non-Newtonian derivative $\mathrm{DSin}\,x/\mathrm{D}x = \mathrm{Cos}\,x$ (full, Eq. (4.10)), as compared to the Newtonian derivative $\mathrm{dSin}\,x/dx$ (dashed, Eq. (4.12)). The singular behavior of the dashed curve follows from Newtonian non-differentiability of $f^{-1}(x) = \sqrt[3]{x}$ at $x = 0$. In contrast, the non-Newtonian derivative is non-singular since neither f nor f^{-1} is differentiated (in a Newtonian sense).

It is instructive to compare (4.10) with the derivative

$$\frac{\mathrm{d\,Sin}\,x}{\mathrm{d}x} = \frac{x^2 \cos(x^3)}{\sin^{\frac{2}{3}}(x^3)}, \tag{4.12}$$

defined with respect to the "standard" Diophantine arithmetic (Figure 4.2).

Example 4.1.4 (Taylor series). Let us return to (4.8) and look for its solution in a series form

$$g(x) = \sum_{n=0}^{\infty}{}^{\oplus} a_n x^n = \sqrt[3]{\sum_{n=0}^{\infty} (a_n x^n)^3}, \quad g(0) = a_0 = 1.$$

The derivative is linear with respect to $\oplus$, but we have to compute

$$\begin{aligned}\frac{\mathrm{D}x^n}{\mathrm{D}x} &= \frac{\mathrm{D}x}{\mathrm{D}x}x^{n-1} \oplus \cdots \oplus \frac{\mathrm{D}x}{\mathrm{D}x}x^{n-1} \quad (n \text{ times}) \\ &= \sqrt[3]{n}x^{n-1}.\end{aligned}$$

Let us cross-check this result directly from definition:

$$\begin{aligned}
\frac{\mathrm{D}x^n}{\mathrm{D}x} &= \lim_{h\to 0}\frac{(x\oplus h)^n \ominus x^n}{h}\\
&= \lim_{h\to 0}\frac{(\sqrt[3]{x^3+h^3})^n \ominus x^n}{h}\\
&= \lim_{h\to 0}\frac{\sqrt[3]{((\sqrt[3]{x^3+h^3})^n)^3-(x^n)^3}}{h}\\
&= \lim_{h\to 0}\frac{\sqrt[3]{((\sqrt[3]{x^3+h^3})^n)^3-(x^n)^3}}{h}\\
&= \lim_{h\to 0}\sqrt[3]{\frac{(x^3+h^3)^n-(x^n)^3}{h^3}}\\
&= \sqrt[3]{\lim_{\delta\to 0}\frac{(x^3+\delta)^n-(x^3)^n}{\delta}}\\
&= \sqrt[3]{\frac{\mathrm{d}(x^3)^n}{\mathrm{d}(x^3)}} = \sqrt[3]{n(x^3)^{n-1}} = \sqrt[3]{n}x^{n-1}.
\end{aligned}$$

The characteristic expression $n' = f^{-1}(n) = \sqrt[3]{n}$ plays a role of a "non-Diophantine natural number", since $n' \oplus m' = f^{-1}(n+m) = (n+m)'$, and thus, in particular

$$1' \oplus \cdots \oplus 1' = 1 \oplus \cdots \oplus 1 = \sqrt[3]{n} = n' \quad (n \text{ times}).$$

In the 21st century, expression $1 + 1 = 3$ has become a symbol of synergy (cf., for example, Riedell *et al.*, 2002; Brodsky, 2004; Klees, 2006; Marie, 2007; Marks and Mirvis, 2010; Derboven, 2011; Trabacca *et al.*, 2012; Ritchie, 2014; Brown, 2015; Lea, 2016; Burgin and Meissner, 2017). In our case, instead of the synergetic rule "one plus one equals three", here we find the anti-synergetic rule

$$1\oplus 1\oplus 1\oplus 1\oplus 1\oplus 1\oplus 1\oplus 1 = 2,$$

and simultaneously maintaining the standard-looking formula

$$2 = 8' = 1'\oplus 1'\oplus 1'\oplus 1'\oplus 1'\oplus 1'\oplus 1'\oplus 1'.$$

Differentiating the series term by term

$$\frac{\mathrm{D}g(x)}{\mathrm{D}x} = \sum_{n=0}^{\infty}{}^{\oplus} a_n(n)^{1/3}x^{n-1}$$

$$= \sum_{n=0}^{\infty}{}^{\oplus} a_n(n+1)^{1/3}x^{n}$$

$$= \sum_{n=0}^{\infty}{}^{\oplus} a_n x^{n},$$

we get the recurrence relation $a_0 = 1$, $a_n = a_{n+1}\sqrt[3]{n+1}$, with the unique solution

$$a_n = \frac{1}{\sqrt[3]{n!}},$$

$$g(x) = \sum_{n=1}^{\infty}{}^{\oplus} \frac{x^n}{\sqrt[3]{n!}} = \sqrt[3]{\sum_{n=0}^{\infty} \frac{(x^3)^n}{n!}} = e^{x^3/3}.$$

Example 4.1.5. Let $A = R^+$, $B = R$, $f_A(x) = \ln x$, $f_B(x) = x^3$. The arithmetic operations in B remain as in Example 4.1.1. The ones in A read

$$x_1 \oplus_A x_2 = f_A^{-1}(f_A(x_1) + f_A(x_2)) = e^{\ln x_1 + \ln x_2} = x_1 x_2,$$

$$x_1 \ominus_A x_2 = f_A^{-1}(f_A(x_1) - f_A(x_2)) = e^{\ln x_1 - \ln x_2} = x_1/x_2,$$

$$x_1 \otimes_A x_2 = f_A^{-1}(f_A(x_1) f_A(x_2)) = e^{\ln x_1 \ln x_2} = x_1^{\ln x_2} = x_2^{\ln x_1},$$

$$x_1 \oslash_A x_2 = f_A^{-1}(f_A(x_1)/f_A(x_2)) = e^{\ln x_1/\ln x_2} = x_1^{1/\ln x_2}.$$

Neutral elements in A are as follows: $0_A = f_A^{-1}(0) = e^0 = 1$, $1_A = f_A^{-1}(1) = e^1 = e$. A negative of $x \in \boldsymbol{A}$ is given by

$$\ominus_A x = 0_A \ominus_A x = f_A^{-1}[-f_A(x)] = e^{-\ln x} = 1/x \in R^+.$$

So, the numbers negative with respect to the arithmetic $\boldsymbol{A}$ are positive if treated in the usual Diophantine sense. The unique solution $g\colon A \to B$ of the equation

$$\frac{\mathrm{D}g(x)}{\mathrm{D}x} = g(x), \quad g(0_A) = 1_B, \tag{4.13}$$

is

$$g(x) = f_B^{-1}(e^{f_A(x)}) = \sqrt[3]{e^{\ln x}} = \sqrt[3]{x}. \tag{4.14}$$

Indeed, first of all, $g(0_A) = \sqrt[3]{1} = 1 = 1_B$. Recalling that multiplication in B is unchanged, we check directly from definitions:

$$\begin{aligned}
\frac{\mathrm{D}g(x)}{\mathrm{D}x} &= \lim_{h\to 0}(g(x \oplus_A f_A^{-1}(h)) \ominus_B g(x)) \oslash_B f_B^{-1}(h) \\
&= \lim_{h\to 0}(\sqrt[3]{x \oplus_A e^h} \ominus_B \sqrt[3]{x})/\sqrt[3]{h} \\
&= \lim_{h\to 0}\sqrt[3]{x\frac{e^h - 1}{h}} = \sqrt[3]{x} = g(x).
\end{aligned}$$

The exponent satisfies

$$\begin{aligned}
g(x_1 \oplus_A x_2) &= g(x_1 x_2) = \sqrt[3]{x_1 x_2} \\
&= \sqrt[3]{x_1}\sqrt[3]{x_2} = g(x_1)g(x_2) = g(x_1) \otimes_B g(x_2),
\end{aligned}$$

as expected. The results are counterintuitive but consistent.

Example 4.1.6. Let us now replace in the previous example f_B just by the identity $f_B(x) = x$, and solve the same differential equation. We find that the unique solution g: $A \to B$ of

$$\frac{\mathrm{D}g(x)}{\mathrm{D}x} = g(x), \quad g(0_A) = 1_B, \tag{4.15}$$

is

$$g(x) = f_B^{-1}(e^{f_A(x)}) = e^{\ln x} = x. \tag{4.16}$$

Again, from the definition,

$$\begin{aligned}
\frac{\mathrm{D}g(x)}{\mathrm{D}x} &= \lim_{h\to 0}(g(x \oplus_A f_A^{-1}(h)) \ominus_B g(x)) \oslash_B f_B^{-1}(h) \\
&= \lim_{h\to 0}(x \oplus_A f_A^{-1}(h) - x)/h \\
&= \lim_{h\to 0}(xe^h - x)/h = x = g(x).
\end{aligned}$$

Similar to the previous example,

$$\begin{aligned} g(x_1 \oplus_A x_2) &= g(x_1 x_2) = x_1 x_2 \\ &= g(x_1)g(x_2) = g(x_1) \otimes_B g(x_2). \end{aligned}$$

Let us stress that this is not the same as the derivative of a function $g(x) = x$, but where g: $A \to A$ (i.e., with $g(x) \in \boldsymbol{A}$, as opposed to g: $A \to B$, i.e., with $g(x) \in \boldsymbol{B}$). The two cases differ by the choice of arithmetic in the image $g(A)$. For the mapping g: $A \to A$, we have the following derivative:

$$\begin{aligned} \frac{\mathrm{D}g(x)}{\mathrm{D}x} &= \lim_{h\to 0}(g(x \oplus_A f_A^{-1}(h)) \ominus_A g(x)) \oslash_A f_A^{-1}(h) \\ &= \lim_{h\to 0}(x \oplus_A f_A^{-1}(h) \ominus_A x) \oslash_A f_A^{-1}(h) \\ &= \lim_{h\to 0} f_A^{-1}(h) \oslash_A f_A^{-1}(h) = 1_A = e. \end{aligned}$$

Let us cross-check our result

$$\begin{aligned} \frac{\mathrm{D}g(x)}{\mathrm{D}x} &= \lim_{h\to 0}(x \oplus_A f_A^{-1}(h) \ominus_A x) \oslash_A f_A^{-1}(h) \\ &= \lim_{h\to 0} e^{\ln(xe^h/x)/\ln e^h} = e = 1_A. \end{aligned}$$

Example 4.1.7. Let $f_A(x) = x$, $f_B(x) = \ln x$. The non-Newtonian derivative,

$$\frac{\mathrm{D}g(x)}{\mathrm{D}x} = \lim_{h\to 0}(g(x+h) \ominus_B g(x)) \oslash_B h_B \tag{4.17}$$

$$= \lim_{h\to 0} e^{(\ln g(x+h) - \ln g(x))/h} = e^{g'(x)/g(x)}, \tag{4.18}$$

is known as the geometric derivative (Grossman and Katz, 1972).

Here $g'(x) = \mathrm{d}g(x)/\mathrm{d}x$ is the Newtonian derivative. Let us now solve the equation

$$\frac{\mathrm{D}g(x)}{\mathrm{D}x} = g(x), \quad g(0_A) = 1_B = f_B^{-1}(1) = e, \tag{4.19}$$

which is equivalent to

$$e^{g'(x)/g(x)} = g(x). \tag{4.20}$$

Taking logarithm of both sides of (4.20) and noting that $g'(x)/g(x) = (\ln g(x))'$ we find $g(x) = \exp(\exp x)$. The result could be inferred directly from the general formula for non-Newtonian exponential functions as we will discuss later,

$$g(x) = f_B^{-1}(e^{f_A(x)}) = e^{e^x}. \tag{4.21}$$

Example 4.1.8. Let $f_A(x) = f_B(x) = \ln x$. The derivative

$$\frac{\mathrm{D}g(x)}{\mathrm{D}x} = e^{xg'(x)/g(x)} \tag{4.22}$$

is known as the bigeometric derivative (Grossman, 1983). Here values of non-Newtonian and Newtonian exponential functions coincide,

$$g(x) = f_B^{-1}(e^{f_A(x)}) = e^{e^{\ln x}} = e^x, \tag{4.23}$$

but their domains are different. Both geometric and bigeometric differentiations have been extensively studied in the literature, with numerous applications (Aniszewska, 2007; Bashirov, Mısırlı and Ozyapıcı, 2008; Florack and van Assen, 2012; Filip and Piatecki, 2014; Ozyapıcı and Bilgehan, 2016; Yalcina, Celikb and Gokdogana, 2016). The variety of applications, from signal processing to economics, is not that surprising if one realizes that $\ln x$ is a simple representation of a neuronal information channel (see Chapter 7). The two non-Newtonian derivatives represent here a perception of change, and not the change itself.

Theorem 4.1.1. *The non-Newtonian derivative satisfies the Leibniz rule.*

Proof:

$$\begin{aligned}
\frac{\mathrm{D}\left(g_1 \otimes_B g_2\right)(x)}{\mathrm{D}x} &= f_B^{-1}\left(\frac{\mathrm{d}}{\mathrm{d}f_A(x)}\tilde{g}_1(f_A(x))\tilde{g}_2(f_A(x))\right)\\
&= f_B^{-1}\left(\frac{\mathrm{d}\tilde{g}_1(f_A(x))}{\mathrm{d}f_A(x)}\tilde{g}_2(f_A(x)) + \tilde{g}_1(f_A(x))\frac{\mathrm{d}\tilde{g}_2(f_A(x))}{\mathrm{d}f_A(x)}\right)\\
&= f_B^{-1}\left(\frac{\mathrm{d}\tilde{g}_1(f_A(x))}{\mathrm{d}f_A(x)}\tilde{g}_2(f_A(x))\right)\\
&\quad \oplus f_B^{-1}\left(\tilde{g}_1(f_A(x))\frac{\mathrm{d}\tilde{g}_2(f_A(x))}{\mathrm{d}f_A(x)}\right)\\
&= \frac{\mathrm{D}g_1(x)}{\mathrm{D}x} \otimes_B g_2(x) \oplus_B g_1(x) \otimes_B \frac{\mathrm{D}g_2(x)}{\mathrm{D}x}.
\end{aligned}$$

Here, we do not need specifying concrete properties of the bijections.

Theorem is proved. □

Theorem 4.1.2. *The non-Newtonian derivative is a linear mapping, i.e., it is additive and* 1*-homogeneous.*

Proof: The diagrams

$$\begin{array}{ccc} A & \overset{g_1 \oplus_B g_2}{\longrightarrow} & B \\ f_A \downarrow & & \downarrow f_B \\ R & \overset{\tilde{g}_1 + \tilde{g}_2}{\longrightarrow} & R \end{array}, \qquad \begin{array}{ccc} A & \overset{g_1 \otimes_B g_2}{\longrightarrow} & B \\ f_A \downarrow & & \downarrow f_B \\ R & \overset{\tilde{g}_1 \cdot \tilde{g}_2}{\longrightarrow} & R \end{array}$$

are used to show additivity. Namely, we have

$$\begin{aligned} &\frac{\mathrm{D}\,(g_1 \oplus_B g_2)\,(x)}{\mathrm{D}x} \\ &= f_B^{-1}\left(\frac{\mathrm{d}\tilde{g}_1(f_A(x))}{\mathrm{d}f_A(x)} + \frac{\mathrm{d}\tilde{g}_2(f_A(x))}{\mathrm{d}f_A(x)}\right) \\ &= f_B^{-1}\left(f_B \circ f_B^{-1}\left(\frac{\mathrm{d}\tilde{g}_1(f_A(x))}{\mathrm{d}f_A(x)}\right) + f_B \circ f_B^{-1}\left(\frac{\mathrm{d}\tilde{g}_2(f_A(x))}{\mathrm{d}f_A(x)}\right)\right) \\ &= f_B^{-1}\left(\frac{\mathrm{d}\tilde{g}_1(f_A(x))}{\mathrm{d}f_A(x)}\right) \oplus f_B^{-1}\left(\frac{\mathrm{d}\tilde{g}_2(f_A(x))}{\mathrm{d}f_A(x)}\right) \\ &= \frac{\mathrm{D}g_1(x)}{\mathrm{D}x} \oplus_B \frac{\mathrm{D}g_2(x)}{\mathrm{D}x}. \end{aligned}$$

1-homogeneity under multiplication by a constant function follows directly from the Leibniz rule.

Theorem is proved. □

Corollary 4.1.1. *A non-Newtonian derivative is distributive with respect to subtraction* $\ominus_B$ *of functions.*

Corollary 4.1.2. *A non-Newtonian derivative of a constant function is equal to zero* 0_B*.*

Corollary 4.1.3. *For any functions* $g\colon A \to B$ *and* $h\colon A \to B$*, and any constants* $a, b \in \boldsymbol{B}$*, we have*

$$\frac{\mathrm{D}}{\mathrm{D}x}\,(a \otimes_B g \oplus_B b \otimes_B h)\,(x) = \left(a \otimes_B \frac{\mathrm{D}g(x)}{\mathrm{D}x}\right) \oplus_B \left(b \otimes_B \frac{\mathrm{D}h(x)}{\mathrm{D}x}\right).$$

Theorem 4.1.3 (Non-Newtonian chain rule). *Consider two functions* g_1, g_2 *defined by the diagram,*

$$\begin{array}{ccccc} A & \xrightarrow{g_1} & B & \xrightarrow{g_2} & C \\ f_A\downarrow & & f_B\downarrow & & f_C\downarrow \\ R & \xrightarrow{\tilde{g}_1} & R & \xrightarrow{\tilde{g}_2} & R \end{array}$$

and three arithmetics $\boldsymbol{A}$, $\boldsymbol{B}$, $\boldsymbol{C}$, *with functional parameters* f_A, f_B, f_C, *respectively. Then*

$$\frac{\mathrm{D}\left(g_2 \circ g_1\right)(x)}{\mathrm{D}x} = f_C^{-1}\left[f_C\left(\frac{\mathrm{D}g_2\left(g_1(x)\right)}{\mathrm{D}g_1(x)}\right) f_B\left(\frac{\mathrm{D}g_1(x)}{\mathrm{D}x}\right)\right], \tag{4.24}$$

if the derivatives exist.

Proof:

$$\begin{aligned} \frac{\mathrm{D}(g_2 \circ g_1)(x)}{\mathrm{D}x} &= f_C^{-1}\left(\frac{\mathrm{d}}{\mathrm{d}f_A(x)}\tilde{g}_2 \circ \tilde{g}_1(f_A(x))\right) \\ &= f_C^{-1}\left(\frac{\mathrm{d}\tilde{g}_2[\tilde{g}_1(f_A(x))]}{\mathrm{d}\tilde{g}_1(f_A(x))}\,\frac{\mathrm{d}\tilde{g}_1(f_A(x))}{\mathrm{d}f_A(x)}\right) \\ &= f_C^{-1}\left(f_C \circ f_C^{-1}\left(\frac{\mathrm{d}\tilde{g}_2[f_B \circ f_B^{-1} \circ \tilde{g}_1 \circ f_A(x)]}{\mathrm{d}[f_B \circ f_B^{-1} \circ \tilde{g}_1 \circ f_A(x)]}\right)\right. \\ &\qquad \left.\times f_B \circ f_B^{-1}\left(\frac{\mathrm{d}\tilde{g}_1(f_A(x))}{\mathrm{d}f_A(x)}\right)\right) \\ &= f_C^{-1}\left[f_C\left(\frac{\mathrm{D}\,g_2(g_1(x))}{\mathrm{D}g_1(x)}\right) f_B\left(\frac{\mathrm{D}g_1(x)}{\mathrm{D}x}\right)\right]. \end{aligned}$$

Theorem is proved. □

Corollary 4.1.4 (The chain rule for a composition of three functions). *If*

$$Z \xrightarrow{g_0} A \xrightarrow{g_1} B \xrightarrow{g_2} C, \tag{4.25}$$

then

$$\begin{aligned} &\frac{\mathrm{D}\left(g_2 \circ g_1 \circ g_0\right)(x)}{\mathrm{D}x} \\ &= f_C^{-1}\left[f_C\left(\frac{\mathrm{D}g_2[g_1(g_0(x))]}{\mathrm{D}g_1(g_0(x))}\right) f_B\left(\frac{\mathrm{D}g_1(g_0(x))}{\mathrm{D}g_0(x)}\right) f_A\left(\frac{\mathrm{D}g_0(x)}{\mathrm{D}x}\right)\right]. \end{aligned} \tag{4.26}$$

Proof: Denote $g_{01} = g_1 \circ g_0$. The chain rule (4.24) implies

$$\begin{aligned}
\frac{\mathrm{D}(g_2 \circ g_1 \circ g_0)(x)}{\mathrm{D}x} &= \frac{\mathrm{D}(g_2 \circ g_{01})(x)}{\mathrm{D}x} \\
&= f_C^{-1}\left[f_C\left(\frac{\mathrm{D}g_2(g_{01}(x))}{\mathrm{D}g_{01}(x)}\right) f_B\left(\frac{\mathrm{D}g_{01}(x)}{\mathrm{D}x}\right)\right] \\
&= f_C^{-1}\left[f_C\left(\frac{\mathrm{D}g_2[g_1(g_0(x))]}{\mathrm{D}g_1(g_0(x))}\right)\right. \\
&\quad \left.\times f_B\left(f_B^{-1}\left[f_B\left(\frac{\mathrm{D}g_1(g_0(x))}{\mathrm{D}g_0(x)}\right) f_A\left(\frac{\mathrm{D}g_0(x)}{\mathrm{D}x}\right)\right]\right)\right] \\
&= f_C^{-1}\left[f_C\left(\frac{\mathrm{D}g_2[g_1(g_0(x))]}{\mathrm{D}g_1(g_0(x))}\right)\right. \\
&\quad \left.\times f_B\left(\frac{\mathrm{D}g_1(g_0(x))}{\mathrm{D}g_0(x)}\right) f_A\left(\frac{\mathrm{D}g_0(x)}{\mathrm{D}x}\right)\right].
\end{aligned}$$

□

Example 4.1.9 (Derivatives of the functions f_A, f_B, f_A^{-1}, and f_B^{-1}). As an important application of Corollary 4.1.4, let us compute the derivative of $f_A : A \to R$, where the arithmetic of R is Diophantine, i.e., $f_R(x) = x$:

$$\begin{aligned}
\frac{\mathrm{D}f_A(x)}{\mathrm{D}x} &= \lim_{h\to 0}(f_A(x \oplus_A f_A^{-1}(h)) \ominus_R f_A(x)) \oslash_R f_R^{-1}(h) \\
&= \lim_{h\to 0}(f_A(x \oplus_A f_A^{-1}(h)) - f_A(x))/h \\
&= \lim_{h\to 0}(f_A(x) + h - f_A(x))/h = 1.
\end{aligned} \tag{4.27}$$

Analogously,

$$\begin{aligned}
\frac{\mathrm{D}f_B(x)}{\mathrm{D}x} &= \lim_{h\to 0}(f_B(x \oplus_B f_B^{-1}(h)) \ominus_R f_B(x)) \oslash_R f_R^{-1}(h) \\
&= \lim_{h\to 0}(f_B(x \oplus_B f_B^{-1}(h)) - f_B(x))/h \\
&= \lim_{h\to 0}(f_B(x) + h - f_B(x))/h = 1,
\end{aligned} \tag{4.28}$$

$$\begin{aligned}
\frac{\mathrm{D}f_A^{-1}(x)}{\mathrm{D}x} &= \lim_{h\to 0}(f_A^{-1}(x \oplus_R f_R^{-1}(h)) \ominus_A f_A^{-1}(x)) \oslash_A f_A^{-1}(h) \\
&= \lim_{h\to 0}(f_A^{-1}(x + h) \ominus_A f_A^{-1}(x)) \oslash_A f_A^{-1}(h)
\end{aligned}$$

$$= \lim_{h\to 0} f_A^{-1}(f_A(f_A^{-1}(x+h) \ominus_A f_A^{-1}(x))/h)$$
$$= \lim_{h\to 0} f_A^{-1}((x+h-x)/h) = f_A^{-1}(1) = 1_A, \tag{4.29}$$

and

$$\frac{\mathrm{D} f_B^{-1}(x)}{\mathrm{D}x} = 1_B.$$

The formulas show that from the point of view of non-Newtonian differentiation the functional parameters of the arithmetics behave as identity mappings.

Example 4.1.10 (The chain rule versus the form of non-Newtonian derivative). Now consider g: $A \to B$, but written explicitly as the composition

$$A \xrightarrow{f_A} R \xrightarrow{\tilde{g}} R \xrightarrow{f_B^{-1}} B,$$

and then apply the chain rule,

$$\frac{\mathrm{D}g(x)}{\mathrm{D}x} = \frac{\mathrm{D}\left(f_B^{-1} \circ \tilde{g} \circ f_A\right)(x)}{\mathrm{D}x}$$
$$= f_B^{-1}\left[f_B\left(\frac{\mathrm{D}f_B^{-1}\left[\tilde{g}\left(f_A(x)\right]\right.}{\mathrm{D}\tilde{g}(f_A(x))}\right) f_R\left(\frac{\mathrm{D}\tilde{g}(f_A(x))}{\mathrm{D}f_A(x)}\right) f_R\left(\frac{\mathrm{D}f_A(x)}{\mathrm{D}x}\right)\right].$$

The arithmetic $\boldsymbol{R}$ is Diophantine, i.e., $f_R(x) = x$, and thus, the derivative

$$\frac{\mathrm{D}\tilde{g}(f_A(x))}{\mathrm{D}f_A(x)} = \frac{\mathrm{d}\tilde{g}(f_A(x))}{\mathrm{d}f_A(x)}$$

is Newtonian. Moreover,

$$\frac{\mathrm{D}f_B^{-1}[\tilde{g}(f_A(x))]}{\mathrm{D}\tilde{g}(f_A(x))} = 1_B,$$
$$\frac{\mathrm{D}f_A(x)}{\mathrm{D}x} = 1.$$

Finally,

$$\frac{\mathrm{D}g(x)}{\mathrm{D}x} = f_B^{-1}\left(f_B(1_B)\frac{\mathrm{d}\tilde{g}(f_A(x))}{\mathrm{d}f_A(x)}\right) = f_B^{-1}\left(\frac{\mathrm{d}\tilde{g}(f_A(x))}{\mathrm{d}f_A(x)}\right),$$

and we reconstruct our definition of the derivative.

Remark 4.1.1. Now, we can understand why in the formula

$$\frac{\mathrm{D}g(x)}{\mathrm{D}x} = f_B^{-1}\left(\frac{\mathrm{d}\tilde{g}(f_A(x))}{\mathrm{d}f_A(x)}\right)$$

we apparently do not differentiate f_A and f_B^{-1}. In fact, we *do* differentiate these two bijections but in a non-Newtonian way. The Newtonian chain rule for the composition $g = f_B^{-1} \circ \tilde{g} \circ f_A$,

$$\frac{\mathrm{d}g(x)}{\mathrm{d}x} = \frac{\mathrm{d}f_B^{-1}(\tilde{g}(f_A(x)))}{\mathrm{d}\tilde{g}(f_A(x))} \frac{\mathrm{d}\tilde{g}(f_A(x))}{\mathrm{d}f_A(x)} \frac{\mathrm{d}f_A(x)}{\mathrm{d}x},$$

requires differentiability of both f_B^{-1} and f_A. The situation does *not* change in the non-Newtonian calculus, but it turns out that f_B^{-1} and f_A are always non-Newtonian differentiable (with respect to the derivatives and arithmetics they define), while their derivatives are just the appropriate unit elements. This remark is especially important for fractal applications where the functions f_A and f_B are highly non-trivial. Moreover, f_A and f_B are always continuous in topologies they induce in A and B, even if they are discontinuous in metric topologies of A and B. The latter observation explains how to understand the notion of a limit $x \to x$ if $x, x_0 \in \boldsymbol{A}$.

Corollary 4.1.5. *The definition of non-Newtonian derivative is consistent with the limit $x \to x_0$ where $x, x_0 \in \boldsymbol{A}$ understood as follows:*

$$\lim_{x \to x_0} g(x) = f_B^{-1}\left(\lim_{r \to f_A(x_0)} \tilde{g}(r)\right).$$

Note that the limit $r \to f_A(x_0)$ where $r, f_A(x_0) \in R$ at the right-hand side is the one we know from the Newtonian analysis of real functions. Consequently, this formula *defines* the limit $x \to x_0$ at the left side of the equality. We can denote these two limits by the same symbol lim, since they are anyway distinguished by the domains of the symbols associated with the arrow $\to$.

Corollary 4.1.6. *The derivative can be expressed in terms of limits in three equivalent ways*

$$\frac{\mathrm{D}g(x)}{\mathrm{D}x} = \lim_{h \to 0_R} (g(x \oplus_A f_A^{-1}(h)) \ominus_B g(x)) \oslash_B f_B^{-1}(h) \tag{4.30}$$

$$= \lim_{h \to 0_A} (g(x \oplus_A h) \ominus_B g(x)) \oslash_B f(h) \tag{4.31}$$

$$= \lim_{h \to 0} (g(x \oplus_A h_A) \ominus_B g(x)) \oslash_B h_B, \tag{4.32}$$

where $f = f_B^{-1} \circ f_A$.

To build a non-Newtonian integral calculus, we define a *definite non-Newtonian Riemann* (or *Lebesgue*) *integral* of a mapping (function) g: $A \to B$ by

$$\int_a^b g(x)\mathrm{D}x = f_B^{-1}\left(\int_{f_A(a)}^{f_A(b)} \tilde{g}(r)\mathrm{d}r\right)$$
$$= f_B^{-1}\left(\int_{f_A(a)}^{f_A(b)} f_B \circ g \circ f_A^{-1}(r)\mathrm{d}r\right), \tag{4.33}$$

assuming that the definite Riemann (or Lebesgue) integral

$$\int_{f_A(a)}^{f_A(b)} \tilde{g}(r)\mathrm{d}r \tag{4.34}$$

exists. When the upper value b in the integral (4.33) is a fixed element from $\boldsymbol{A} \subseteq \boldsymbol{R}$, the integral (4.33) is a functional or more exactly, B-valued functional. When the upper value b in the integral (4.33) is a variable taking values in $\boldsymbol{A}$, the integral (4.33) is an operator and is called the indefinite non-Newtonian Riemann (Lebesgue) integral. Using the conventional Henstock–Kurzweil integral (Bartle, 2001), it is possible to define the non-Newtonian Henstock–Kurzweil integral.

Remark 4.1.2. It is possible to define integrals in a more general case, namely, for mappings g: $A \to B$, when $\boldsymbol{A} = (A; \oplus_A, \otimes_A, \ominus_A, \oslash_A, \leq_A)$ and $\boldsymbol{B} = (B; \oplus_B, \otimes_B, \ominus_B, \oslash_B, \leq_B)$ are real-number prearithmetics studied in Section 3.3. As a result, we can construct a more general calculus based not only on non-Diophantine arithmetics but even on real-number prearithmetics.

Example 4.1.11. Let us compute some integrals of functions g: $R \to R$, with the arithmetics from Example 4.1.1. Recall that multiplication is here unchanged, so the mapping $x \mapsto x^n$ is the usual one.

$$\int_0^a x^n \mathrm{D}x = f^{-1}\left(\int_{f(0)}^{f(a)} f(f^{-1}(r)^n)\mathrm{d}r\right)$$
$$= \sqrt[3]{\int_0^{a^3} (\sqrt[3]{r^n})^3 \mathrm{d}r} = \sqrt[3]{\int_0^{a^3} r^n \mathrm{d}r} = \sqrt[3]{\frac{a^{3(n+1)}}{(n+1)}} = \frac{a^{n+1}}{\sqrt[3]{n+1}},$$

$$\begin{aligned}\int_0^a \operatorname{Cos} x \,\mathrm{D}x &= f^{-1}\left(\int_{f(0)}^{f(a)} f(\operatorname{Cos}(f^{-1}(r)))\mathrm{d}r\right)\\ &= \sqrt[3]{\int_0^{a^3}\left(\sqrt[3]{\cos(\sqrt[3]{r})^3}\right)^3 \mathrm{d}r}\\ &= \sqrt[3]{\int_0^{a^3}\cos r\mathrm{d}r} = \sqrt[3]{\sin(a^3)} = \operatorname{Sin} a.\end{aligned}$$

Employing Examples 4.1.3 and 4.1.4 we check that the integrals satisfy a non-Newtonian version of the first fundamental law of calculus. We will later see that this is not coincidental.

Theorem 4.1.4. *The definite non-Newtonian Riemann (Lebesgue) integral of* $g\colon A \to B$ *is a linear mapping, i.e., it is additive with respect to* $\oplus_B$ *and* 1*-homogeneous with respect to* $\otimes_B$ *multiplication by a constant* $b \in \boldsymbol{B}$.

Proof: At first, we have

$$\begin{aligned}\int_y^x (g_1 \oplus_B g_2)(x')\mathrm{D}x' &= f_B^{-1}\left(\int_{f_A(y)}^{f_A(x)} (\tilde{g}_1 + \tilde{g}_2)(r)\mathrm{d}r\right)\\ &= f_B^{-1}\left(\int_{f_A(y)}^{f_A(x)} \tilde{g}_1(r)\mathrm{d}r\right) \oplus_B f_B^{-1}\left(\int_{f_A(y)}^{f_A(x)} \tilde{g}_2(r)\mathrm{d}r\right)\\ &= \int_y^x g_1(x')\mathrm{D}x' \oplus_B \int_y^x g_2(x')\mathrm{D}x'.\end{aligned}$$

Then, taking $b \in \boldsymbol{B}$, we obtain

$$\begin{aligned}b \otimes_B \int_y^x g(x')\mathrm{D}x' &= b \otimes_B f_B^{-1}\left(\int_{f_A(y)}^{f_A(x)} \tilde{g}(r)\mathrm{d}r\right)\\ &= f_B^{-1}\left(f_B(B)\int_{f_A(y)}^{f_A(x)} \tilde{g}(r)\mathrm{d}r\right).\end{aligned}$$

Since $\tilde{g}(r) = f_B[g(f_A^{-1}(r))]$, one finds

$$\begin{aligned}f_B(b)\tilde{g}(r) &= f_B(b)f_B(g(f_A^{-1}(r))) = f_B(b \otimes_B g(f_A^{-1}(r)))\\ &= \widetilde{b \otimes_B g}(r),\end{aligned}$$

and thus,

$$b \otimes_B \int_y^x g(x')\mathrm{D}x' = f_B^{-1}\left(\int_{f_A(y)}^{f_A(x)} \widetilde{b \otimes_B g}(r)\mathrm{d}r\right)$$
$$= \int_y^x (b \otimes_B g)(x')\mathrm{D}x'.$$

Theorem is proved. □

Corollary 4.1.7. *A definite non-Newtonian Riemann (Lebesgue) integral is distributive with respect to subtraction* $\ominus_B$ *of functions.*

Corollary 4.1.8. *A definite non-Newtonian Riemann (Lebesgue) integral of the function that is equal to zero* 0_B *is equal to zero* 0_B*.*

Corollary 4.1.9. *For any functions* g: $A \to B$ *and* h: $A \to B$, *and any constants* $a, b \in \boldsymbol{B}$ *we have*

$$\int_r^s (a \otimes_B g \oplus_B b \otimes_B h)(x)\mathrm{D}x = \left(a \otimes_B \int_r^s g(x)\mathrm{D}x\right) \oplus_B \left(b \otimes_B \int_r^s h(x)\mathrm{D}x\right).$$

Theorem 4.1.5. *For any function* g: $A \to B$ *and any elements* $a, b, c \in \boldsymbol{A}$, *we have*

$$\int_a^b g(x)\mathrm{D}x \oplus_B \int_b^c g(x)\mathrm{D}x = \int_a^c g(x)\mathrm{D}x. \tag{4.35}$$

Proof is left as an exercise. □

One of the central results of the classical integral calculus is the fundamental theorem of calculus, which is usually divided into two parts: the first fundamental theorem of calculus and the second fundamental theorem of calculus. Namely, we have the following result, which for the Riemann integral, is proved in numerous books on calculus and taught early in elementary calculus courses (cf., for example, Apostol, 1967). We remind that if $g(x)$ is a real-valued function, its Newtonian antiderivative $G(x)$ is a real-valued function such that $\mathrm{d}G(x)/\mathrm{d}x = g(x)$. If $\mathrm{D}G(x)/\mathrm{D}x = g(x)$, then $G(x)$ is a non-Newtonian antiderivative of g.

Theorem 4.1.6 (The first fundamental theorem of non-Newtonian calculus). *Consider a continuous function* $\tilde{g}$: $R \to R$. *Let* $g = f_B^{-1} \circ \tilde{g} \circ f_A : A \to B$, *where* $f_A : A \to R$ *and* $f_B : B \to R$ *are the functional parameters* *of arithmetics* **A** *and* **B**, *respectively. Then, the non-Newtonian Riemann integral of g is a non-Newtonian antiderivative of g,*

$$\frac{\mathrm{D}}{\mathrm{D}x}\int_y^x g(x')\mathrm{D}x' = g(x). \tag{4.36}$$

Proof: Denoting

$$b(z) = \int_{f_A(y)}^z \tilde{g}(r)\mathrm{d}r$$

we rewrite the integral as

$$\int_y^x g(x')\mathrm{D}x' = f_B^{-1}(b(f_A(x)))$$

and thus

$$\frac{\mathrm{D}}{\mathrm{D}x}\int_y^x g(x')\mathrm{D}x' = f_B^{-1}\left(\frac{\mathrm{d}b(f_A(x))}{\mathrm{d}f_A(x)}\right) = f_B^{-1}(\tilde{g}(f_A(x))) = g(x).$$

Theorem is proved. □

Theorem 4.1.7 (The second fundamental theorem of non-Newtonian calculus). *Let $f_A : A \to R$ and $f_B : B \to R$ be the functional parameters of arithmetics* **A** *and* **B**, *respectively. Consider a Newtonian-differentiable function $\tilde{g}$: $R \to R$ whose Newtonian derivative is Newtonian Riemann integrable on $[f_A(y), f_A(x)]$. Then the non-Newtonian Riemann integral satisfies*

$$\int_y^x \frac{\mathrm{D}g(x')}{\mathrm{D}x'}\mathrm{D}x' = g(x) \ominus_B g(y), \tag{4.37}$$

where $g = f_B^{-1} \circ \tilde{g} \circ f_A : A \to B$.

Proof: Denote $\tilde{g}'(r) = \mathrm{d}\tilde{g}(r)/\mathrm{d}r$. Then

$$\begin{aligned}
\int_y^x \frac{\mathrm{D}g(x')}{\mathrm{D}x'}\mathrm{D}x' &= \int_y^x f_B^{-1} \circ \tilde{g}' \circ f_A(x')\mathrm{D}x' \\
&= f_B^{-1}\left(\int_{f_A(y)}^{f_A(x)} f_B \circ f_B^{-1} \circ \tilde{g}' \circ f_A \circ f_A^{-1}(r)\mathrm{d}r\right) \\
&= f_B^{-1}\left(\int_{f_A(y)}^{f_A(x)} \tilde{g}'(r)\mathrm{d}r\right)
\end{aligned}$$

$$
\begin{aligned}
&= f_B^{-1}(\tilde{g}(f_A(x)) - \tilde{g}(f_A(y))) \\
&= f_B^{-1}(f_B \circ g \circ f_A^{-1}(f_A(x)) - f_B \circ g \circ f_A^{-1}(f_A(y))) \\
&= f_B^{-1}(f_B(g(x)) - f_B(g(y))) \\
&= g(x) \ominus_B g(y).
\end{aligned}
$$

Theorem is proved. □

Remark 4.1.3. Combining this theorem with the Leibniz rule we obtain the formula of integration by parts

$$
\int_y^x \frac{\mathrm{D}g_1(x')}{\mathrm{D}x'} \otimes_B g_2(x')\mathrm{D}x' \oplus_B \int_y^x g_1(x') \otimes_B \frac{\mathrm{D}g_2(x')}{\mathrm{D}x'}\mathrm{D}x'
$$

$$
= g_1(x) \otimes_B g_2(x) \ominus_B g_1(y) \otimes_B g_2(y). \tag{4.38}
$$

Example 4.1.12. As an example where the above rules occur simultaneously, consider the differential equation

$$
\frac{\mathrm{D}g(t)}{\mathrm{D}t} = a(t) \otimes_B g(t), \quad g(0_A) = 1_B, \tag{4.39}
$$

with a, g: $A \to B$. The general chain rule

$$
\frac{\mathrm{D}\,(g_2 \circ g_1)\,(t)}{\mathrm{D}t} = f_C^{-1}\left[f_C\left(\frac{\mathrm{D}g_2(g_1(t))}{\mathrm{D}g_1(t)}\right) f_B\left(\frac{\mathrm{D}g_1(t)}{\mathrm{D}t}\right)\right] \tag{4.40}
$$

when restricted to $B = C$ yields

$$
\frac{\mathrm{D}g_2(g_1(t))}{\mathrm{D}t} = f_B^{-1}\left[f_B\left(\frac{\mathrm{D}g_2(g_1(t))}{\mathrm{D}g_1(t)}\right) f_B\left(\frac{\mathrm{D}g_1(t)}{\mathrm{D}t}\right)\right] \tag{4.41}
$$

$$
= \frac{\mathrm{D}g_2(g_1(t))}{\mathrm{D}g_1(t)} \otimes_B \frac{\mathrm{D}g_1(t)}{\mathrm{D}t} \tag{4.42}
$$

$$
= \frac{\mathrm{D}g_1(t)}{\mathrm{D}t} \otimes_B \frac{\mathrm{D}g_2(g_1(t))}{\mathrm{D}g_1(t)}. \tag{4.43}
$$

Choosing $g_2 : B \to B$, $g_2(x) = \mathrm{Exp}_B(x) = f_B^{-1}[\exp f_B(x)]$ we find

$$
\frac{\mathrm{D}g_2(g_1(t))}{\mathrm{D}g_1(t)} = g_2(g_1(t)) \tag{4.44}
$$

and

$$
\frac{\mathrm{D}g_2(g_1(t))}{\mathrm{D}t} = \frac{\mathrm{D}g_1(t)}{\mathrm{D}t} \otimes_B g_2(g_1(t)). \tag{4.45}
$$

It remains to find g_1 by integrating

$$\frac{\mathrm{D}g_1(t)}{\mathrm{D}t} = a(t), \tag{4.46}$$

so

$$g_1(t) = g_1(0_A) \oplus_B \int_{0_A}^{t} a(t')\mathrm{D}t'. \tag{4.47}$$

Taking $g_1(0_A) = 0_B$, we obtain

$$g(t) = \mathrm{Exp}_B \int_{0_A}^{t} a\left(t'\right)\mathrm{D}t', \tag{4.48}$$

which solves (4.39).

Remark 4.1.4. The first and second fundamental theorems of calculus remain true with some modifications for the Lebesgue integral and even for the Henstock–Kurzweil integral (Bartle, 2001). This allows obtaining the first and second fundamental theorems of the non-Newtonian calculus for the non-Newtonian Lebesgue integral and for the non-Newtonian Henstock–Kurzweil integral.

Remark 4.1.5. There are generalizations to higher dimensions of the first and second fundamental theorems of calculus.

Remark 4.1.6. It is possible to build a non-Newtonian calculus in non-Grassmannian linear spaces described in Section 3.5 and obtain the first and second fundamental theorems for this calculus.

4.2. Relation to Manifolds and Fiber Bundles

The longer you look at an object,
the more abstract it becomes, and,
ironically, the more real.

Lucian Freud

As it is explained in Section 2.2, weak projectivity and projectivity of abstract prearithmetics and arithmetics, and thus of real-number arithmetics, is related to basic topological constructions such as fiber spaces and manifolds.

For example, consider the diagram

$$\begin{array}{ccc} R & \xrightarrow{\tilde{g}_2} & R \\ g_A\uparrow & & \uparrow g_B \\ A & \xrightarrow{g} & B \\ f_A\downarrow & & \downarrow f_B \\ R & \xrightarrow{\tilde{g}_1} & R \end{array}$$

with four bijections f_A, f_B, g_A, and g_B, as well as two functions, $\tilde{g}_1$ and $\tilde{g}_2$, defined by them. It is natural to think of A and B in terms of one-dimensional manifolds whose global charts are defined by the bijections. However, there is a difference.

In differential geometry, one would treat f_A, f_B, g_A, and g_B, as global charts on the manifolds A and B, whereas the derivative of g could be the usual Newtonian derivative of $\tilde{g}_1$, say. Since,

$$\begin{aligned} \tilde{g}_1 &= f_B \circ g_B^{-1} \circ \tilde{g}_2 \circ g_A \circ f_A^{-1} \\ &= \varphi_B^{-1} \circ \tilde{g}_2 \circ \varphi_A, \end{aligned}$$

the derivative of g could be equivalently defined in terms of $\tilde{g}_2$, and the link between them would be given by the usual chain rule.

The arithmetic philosophy is different. "Diophantine vs. non-Diophantine", or "Newtonian vs. non-Newtonian", is treated here in exact analogy to "Euclidean vs. non-Euclidean" in geometry. None of these structures is given *a priori*. The four combinations of bijections define four different pairs of arithmetic. Each combination of arithmetics defines a different derivative of g. In principle, in physics or neuroscience, a choice of arithmetic (and thus, on the corresponding calculus) can be determined by some natural law, e.g., an arithmetic analog of Einstein equations of general relativity. Yet, similarly to the change of charts on a manifold, we can nevertheless still change variables in non-Newtonian derivatives and integrals, even if the arithmetics are fixed. A recipe for the change is given by the non-Newtonian chain rule.

In general, a single global non-Diophantine arithmetic with the smooth parameter may not exist, meaning that different subsets of A behave in qualitatively different ways (cf., for example, Kolmogorov, 1961). However, taking a piecewise-defined function as a functional parameter of a non-Diophantine arithmetic, we can represent these properties of numbers in a consistent rigorous formal way. This type of construction is analogous to a non-trivial manifold, a fiber bundle, or a fiber space.

4.3. Partial Derivatives and Multiple Integrals

Progress is measured by the degree of differentiation within a society.

Herbert Read

One can naturally generalize the elaborated construction to Cartesian products of sets A_a, $a = 1, \dots, n$, B_b, $b = 1, \dots, m$, with bijections $f_{A_a} : A_a \to R$, $f_{B_b} : B_b \to R$. Let $A = A_1 \times \cdots \times A_n$, $B = B_1 \times \cdots \times B_m$. Then, for $x \in A$, $x_a \in A_a$, $y \in B$, $y_b \in B_b$, one defines

$$f_A(x) = f_{A_1}(x_1) \times \cdots \times f_{A_n}(x_n)$$

$$f_B(y) = f_{B_1}(y_1) \times \cdots \times f_{B_m}(y_m),$$

Let g and $\tilde{g}$ be defined by the diagram,

$$\begin{array}{ccc} A & \xrightarrow{g} & B \\ f_A\downarrow & & \downarrow f_B \\ R^n & \xrightarrow{\tilde{g}} & R^m \end{array} \tag{4.49}$$

Definition 4.3.1. Let g and $\tilde{g}$ be defined by (4.49), where $\tilde{g}$ is differentiable in a Newtonian sense with respect to its ath variable. A non-Newtonian partial derivative of g is given by

$$\frac{\mathrm{D}g(x)}{\mathrm{D}x_a} = f_B^{-1}\left(\frac{\partial \tilde{g}(f_A(x))}{\partial f_{A_a}(x_a)}\right).$$

Definition 4.3.2. Let g and $\tilde{g}$ be defined by (4.49), where $\tilde{g}$ is Riemann (Lebesgue) integrable in a Newtonian sense with respect to its ath variable. A non-Newtonian Riemann (Lebesgue) integral of g over x_a reads

$$\int_{y_a}^{x_a} g(x')\mathrm{D}x'_a = f_B^{-1}\left(\int_{f_{A_a}(y_a)}^{f_{A_a}(x_a)} \tilde{g}(f_A(x'))\mathrm{d}f_{A_a}(x'_a)\right).$$

Here (x') is a brief notation for $(x_1, \dots, x_{a-1}, x'_a, x_{a+1}, \dots, x_n)$.

In exact analogy to a one-dimensional case, the integral over x'_a is the inverse of the partial derivative. Assuming that the non-Newtonian Riemann integral exists and the function $\tilde{g}$ defined by the formula (4.49) is continuous with respect to the ath variable, we can perform the standard calculation

$$\begin{aligned}\frac{\mathrm{D}}{\mathrm{D}x_a}\int_{y_a}^{x_a} g(x')\mathrm{D}x'_a &= \frac{\mathrm{D}}{\mathrm{D}x_a} f_B^{-1}\left(\int_{f_{A_a}(y_a)}^{f_{A_a}(x_a)} \tilde{g}(f_A(x'))\mathrm{d}f_{A_a}(x'_a)\right)\\ &= f_B^{-1}\left(\frac{\partial}{\partial f_{A_a}(x_a)}\int_{f_{A_a}(y_a)}^{f_{A_a}(x_a)} \tilde{g}(f_A(x'))\mathrm{d}f_{A_a}(x'_a)\right)\\ &= f_B^{-1}(\tilde{g}(\dots f_{A_{a-1}}(x_{a-1}), f_{A_a}(x_a), f_{A_{a+1}}(x_{a+1})\dots))\\ &= g(x_0,\dots,x_{a-1},x_a,x_{a+1},\dots,x_{n-1}).\end{aligned}$$

This means that the integral is a partial antiderivative of $g(x)$.

Mixed second derivative and double integral read

$$\frac{\mathrm{D}}{\mathrm{D}x_b}\frac{\mathrm{D}g(x)}{\mathrm{D}x_a} = f_B^{-1}\left(\frac{\partial}{\partial f_{A_b}(x_b)}\frac{\partial\tilde{g}(f_A(x))}{\partial f_{A_a}(x_a)}\right),$$

$$\begin{aligned}&\int_{y_b}^{x_b}\mathrm{D}x'_b\int_{y_a}^{x_a} g(x')\mathrm{D}x'_a\\ &\quad = f_B^{-1}\left(\int_{f_{A_b}(y_b)}^{f_{A_b}(x_b)}\mathrm{d}f_{A_b}(x'_b)\int_{f_{A_a}(y_a)}^{f_{A_a}(x_a)}\tilde{g}(f_A(x'))\mathrm{d}f_{A_a}(x'_a)\right).\end{aligned}$$

Here (x') stands for $(x_1,\dots,x_{a-1},x'_a,x_{a+1},\dots,x_{b-1},x'_b,x_{b+1},\dots,x_n)$. The generalization to higher orders is evident.

The formulas will allow us to consider relativistic field theory in space–times consisting of Cartesian products of different fractals, such as $A = A_1 \times R$, where $A_2 = R$ represents time and A_1 is a Koch curve.

4.4. Elementary Functions

Voltaire called the calculus "the Art of numbering and measuring exactly a Thing whose Existence cannot be conceived".

Carl B. Boyer

Let us consider non-Newtonian derivatives and integrals of such functions as monomials, trigonometric functions and exponential function.

4.4.1. *Monomials*

When all is said and done, monotony may after all be the best condition for creation.

Margaret Sackville

A first B-valued power $g(x) = x^{1_{AB}}$, g: $A \to B$, satisfies

$$\frac{\mathrm{D}g(x)}{\mathrm{D}x} = 1_B, \quad g(0_A) = 0_B. \tag{4.50}$$

Let us reverse the problem and define a first B-valued power by (4.50). In order to derive the explicit form of g consider g_1: $A \to B$, defined as

$$g_1(x) = 1_B = f_B^{-1}(1) = f_B^{-1} \circ \tilde{g}_1 \circ f_A(x). \tag{4.51}$$

The last equality in (4.51) implies that $\tilde{g}_1 : R \to R$ is the constant function $\tilde{g}_1(r) = 1$. Integrating (4.50), we obtain

$$\begin{aligned} g(x) = g(x) \ominus_B 0_B = g(x) \ominus_B g(0_A) &= \int_{0_A}^{x} \frac{\mathrm{D}\, g(x')}{\mathrm{D}x} \mathrm{D}x' \\ &= \int_{0_A}^{x} 1_B \mathrm{D}x' = \int_{0_A}^{x} g_1(x') \mathrm{D}x' = f_B^{-1}\left(\int_{f_A(0_A)}^{f_A(x)} \tilde{g}_1(r) \mathrm{d}r \right) \\ &= f_B^{-1}\left(\int_{0}^{f_A(x)} \mathrm{d}r \right) = f_B^{-1}(f_A(x)), \end{aligned}$$

which implies

$$g(x) = f_B^{-1} \circ f_A(x) = x^{1_{AB}},$$

as required. By definition, we have

$$\begin{aligned} x^{n_{AB}} &= x^{1_{AB}} \otimes_B \cdots \otimes_B x^{1_{AB}} \quad (n \text{ times}) \\ &= f_B^{-1}(f_B(x^{1_{AB}})^n) = f_B^{-1}(f_A(x)^n). \end{aligned}$$

The Leibniz rule implies

$$\begin{aligned}\frac{\mathrm{D}x^{n_{AB}}}{\mathrm{D}x} &= x^{(n-1)_{AB}} \oplus_B \cdots \oplus_B x^{(n-1)_{AB}} \quad (n \text{ times})\\ &= f_B^{-1}(n f_B(x^{(n-1)_{AB}})) = f_B^{-1}(f_B(n_B) f_B(x^{(n-1)_{AB}}))\\ &= n_B \otimes_B x^{(n-1)_{AB}},\end{aligned}$$

and thus

$$\int_{0_A}^{x} x_1^{n_{AB}} \mathrm{D}x_1 = x^{(n+1)_{AB}} \oslash_B (n+1)_B.$$

4.4.2. *Exponential function*

In an age of exponential change,
we need the power of diverse thinking, and
we cannot afford to leave any talent untapped.
Cathy Engelbert

The exponential function $\mathrm{Exp} : A \to B$ is defined as the unique solution of

$$\frac{\mathrm{D}\,\mathrm{Exp}(x)}{\mathrm{D}x} = \mathrm{Exp}(x), \quad \mathrm{Exp}(0_A) = 1_B. \tag{4.52}$$

This function will be sometimes denoted by Exp_{AB}. For $A = B$, one can write $\mathrm{Exp}_{AA} = \mathrm{Exp}_A$. Assume

$$\begin{aligned}\mathrm{Exp}(x) &= \sum_{n=0}^{\infty}{}^{\oplus_B} a_n \otimes_B x^{n_{AB}} = f_B^{-1}\left(\sum_{n=0}^{\infty} f_B(a_n) f_B(x^{n_{AB}})\right)\\ &= f_B^{-1}\left(\sum_{n=0}^{\infty} f_B(a_n) f_A(x)^n\right).\end{aligned}$$

Term-by-term differentiating $\mathrm{Exp}(x)$ n times at $x = 0_A$, one gets

$$a_n \otimes_B n_B \otimes_B (n-1)_B \otimes_B \cdots \otimes_B 2_B \otimes_B 1_B = a_n \otimes_B n!_B = 1_B$$

so

$$\mathrm{Exp}(x) = \sum_{n=0}^{\infty}{}^{\oplus_B} x^{n_{AB}} \oslash_B n!_B.$$

Note that

$$\begin{aligned} n!_B &= n_B \otimes_B (n-1)_B \otimes_B \cdots \otimes_B 2_B \otimes_B 1_B \\ &= f_B^{-1}(f_B(n_B) \cdots f_B(2_B) f_B(1_B)) = f_B^{-1}(n!), \end{aligned}$$

while

$$\begin{aligned} 1 = f_B(1_B) &= f_B(a_n \otimes_B n!_B) \\ &= f_B(a_n) f_B(n!_B) = f_B(a_n) n! \end{aligned}$$

implies $f_B(a_n) = 1/n!$. So, finally we have

$$\begin{aligned} \text{Exp}(x) &= f_B^{-1}\left(\sum_{n=0}^{\infty} f_B(a_n)\, f_A(x)^n\right) \\ &= f_B^{-1}\left(\sum_{n=0}^{\infty} f_A(x)^n/n!\right) = f_B^{-1}(e^{f_A(x)}) \end{aligned}$$

This is what we encountered before in various examples.

Alternatively, one could repeatedly integrate (4.52),

$$\begin{aligned} \text{Exp}(x) &= 1_B \oplus_B \int_{0_A}^{x} \text{Exp}(x_1) \text{D}x_1 \\ &= 1_B \oplus_B \int_{0_A}^{x} \left(1_B \oplus_B \int_{0_A}^{x_1} \text{Exp}(x_2) \text{D}x_2\right) \text{D}x_1 \\ &= 1_B \oplus_B \int_{0_A}^{x} 1_B \text{D}x_1 \oplus_B \int_{0_A}^{x} \left(\int_{0_A}^{x_1} \text{Exp}(x_2) \text{D}x_2\right) \text{D}x_1 \\ &= 1_B \oplus_B f_B^{-1}(f_A(x) - f_A(0_A)) \oplus_B \int_{0_A}^{x} x_1^{1_{AB}} \text{D}x_1 \oplus_B \cdots \\ &= 1_B \oplus_B x^{1_{AB}} \oplus_B (x^{2_{AB}} \oslash_B 2_B) \oplus_B \cdots \\ &= \sum_{n=0}^{\infty}{}^{\oplus_B} x^{n_{AB}} \oslash_B n!_B = f_B^{-1}(e^{f_A(x)}). \end{aligned}$$

The inverse function is $\text{Ln} : B \to A$, such that $\text{Ln}(y) = f_A^{-1}(\ln f_B(y))$. One checks that

$$\text{Exp}(x_1 \oplus_A x_2) = \text{Exp}\, x_1 \otimes_B \text{Exp}\, x_2, \tag{4.53}$$

$$\text{Ln}(y_1 \otimes_B y_2) = \text{Ln}\, y_1 \oplus_A \text{Ln}\, y_2. \tag{4.54}$$

4.4.3. *Trigonometric functions*

To a scholar, mathematics is music.
Amit Kalantri

Trigonometric functions can be defined through solutions of the harmonic-oscillator equation

$$\begin{aligned}\frac{\mathrm{D}}{\mathrm{D}x}\frac{\mathrm{D}g(x)}{\mathrm{D}x} &= f_B^{-1}\left(\frac{\mathrm{d}^2\tilde{g}(f_A(x))}{\mathrm{d}f_A(x)^2}\right)\\ &= \ominus_B g(x) = f_B^{-1}(-\tilde{g}(f_A(x))),\end{aligned}$$

which immediately implies

$$\begin{aligned}\operatorname{Cos} x &= f_B^{-1}(\cos f_A(x)) = \frac{\mathrm{D}\operatorname{Sin} x}{\mathrm{D}x},\\ \operatorname{Sin} x &= f_B^{-1}(\sin f_A(x)) = \ominus_B\frac{\mathrm{D}\operatorname{Cos} x}{\mathrm{D}x}.\end{aligned}$$

These functions satisfy a number of standard formulas, such as

$$\operatorname{Cos}^{2_B} x \oplus_B \operatorname{Sin}^{2_B} x = 1_B,$$

defining a circle of unit radius, and

$$\begin{aligned}\operatorname{Sin}(x_1 \oplus_A x_2) &= \operatorname{Cos} x_1 \otimes_B \operatorname{Sin} x_2 \oplus_B \operatorname{Sin} x_1 \otimes_B \operatorname{Cos} x_2,\\ \operatorname{Cos}(x_1 \oplus_A x_2) &= \operatorname{Cos} x_1 \otimes_B \operatorname{Cos} x_2 \ominus_B \operatorname{Sin} x_1 \otimes_B \operatorname{Sin} x_2,\end{aligned}$$

which allow to define rotations as a Lie group with non-Diophantine arithmetic of its parameters.

4.4.4. *Application of trigonometric functions: Rotations in a plane*

Crystals grew inside rock like arithmetic flowers. They lengthened and spread, added plane to plane in an awed and perfect obedience to an absolute geometry that even stones — maybe only the stones — understood.
Annie Dillard

Here a plane is identified with the Cartesian product $A = A_1 \times A_2$ where A_j are equipped with their own arithmetics defined by $f_{Aj}(j = 1, 2)$. The plane can play a role of a set of complex numbers, but whose real and imaginary parts can, in principle, involve different arithmetics.

Let $f_A(x) = f_A(x_1, x_2) = f_{A_1}(x_1) \times f_{A_2}(x_2) \in R^2$, and let

$$\begin{pmatrix} \tilde{L}_{11} & \tilde{L}_{12} \\ \tilde{L}_{21} & \tilde{L}_{22} \end{pmatrix}$$

be a real matrix. Consider the mapping $L : A \to A$, defined by

$$\begin{aligned} L(x) &= \begin{pmatrix} f_{A_1}^{-1}(\tilde{L}_{11} f_{A_1}(x_1) + \tilde{L}_{12} f_{A_2}(x_2)) \\ f_{A_2}^{-1}(\tilde{L}_{21} f_{A_1}(x_1) + \tilde{L}_{22} f_{A_2}(x_2)) \end{pmatrix} \\ &= \begin{pmatrix} f_{A_1}^{-1}(f_{A_1} \circ f_{A_1}^{-1}(\tilde{L}_{11} f_{A_1}(x_1)) + f_{A_1} \circ f_{A_1}^{-1}(\tilde{L}_{12} f_{A_2}(x_2))) \\ f_{A_2}^{-1}(f_{A_2} \circ f_{A_2}^{-1}(\tilde{L}_{21} f_{A_1}(x_1)) + f_{A_2} \circ f_{A_2}^{-1}(\tilde{L}_{22} f_{A_2}(x_2))) \end{pmatrix} \\ &= \begin{pmatrix} f_{A_1}^{-1}(\tilde{L}_{11} f_{A_1}(x_1)) \oplus_{A_1} f_{A_1}^{-1}(\tilde{L}_{12} f_{A_2}(x_2)) \\ f_{A_2}^{-1}(\tilde{L}_{21} f_{A_1}(x_1)) \oplus_{A_2} f_{A_2}^{-1}(\tilde{L}_{22} f_{A_2}(x_2)) \end{pmatrix} \\ &= \begin{pmatrix} L_{11}(x_1) \oplus_{A_1} L_{12}(x_2) \\ L_{21}(x_1) \oplus_{A_2} L_{22}(x_2) \end{pmatrix} = \begin{pmatrix} \tilde{x}_1 \\ \tilde{x}_2 \end{pmatrix} = \tilde{x}, \end{aligned}$$

with $L_{jk} : A_k \to A_j$. Let $\tilde{L}_{11} = \cos x = \tilde{L}_{22}$ and $\tilde{L}_{12} = -\sin x = -\tilde{L}_{21}$. The rotation in A is defined by

$$\begin{aligned} \begin{pmatrix} \tilde{x}_1 \\ \tilde{x}_2 \end{pmatrix} &= \begin{pmatrix} f_{A_1}^{-1}(f_{A_1}(x_1)\cos\alpha - f_{A_2}(x_2)\sin\alpha) \\ f_{A_2}^{-1}(f_{A_1}(x_1)\sin\alpha + f_{A_2}(x_2)\cos\alpha) \end{pmatrix} \\ &= \begin{pmatrix} f_{A_1}^{-1}(f_{A_1}(x_1) f_{A_1} \circ f_{A_1}^{-1}(\cos\alpha) \\ \quad - f_{A_1} \circ f_{A_1}^{-1} \circ f_{A_2}(x_2) f_{A_1} \circ f_{A_1}^{-1}(\sin\alpha)) \\ f_{A_2}^{-1}(f_{A_2} \circ f_{A_2}^{-1} \circ f_{A_1}(x_1) f_{A_2} \circ f_{A_2}^{-1}(\sin\alpha) \\ \quad + f_{A_2}(x_2) f_{A_2} \circ f_{A_2}^{-1}(\cos\alpha)) \end{pmatrix} \\ &= \begin{pmatrix} x_1 \otimes_{A_1} f_{A_1}^{-1}(\cos\alpha) \ominus_{A_1} f_{A_1}^{-1} \circ f_{A_2}(x_2) \otimes_{A_1} f_{A_1}^{-1}(\sin\alpha) \\ f_{A_2}^{-1} \circ f_{A_1}(x_1) \otimes_{A_2} f_{A_2}^{-1}(\sin\alpha) \oplus_{A_2} x_2 \otimes_{A_2}^{-1} f_{A_2}^{-1}(\cos\alpha) \end{pmatrix} \\ &= \begin{pmatrix} x_1 \otimes_{A_1} f_{A_1}^{-1}(\cos\alpha) \ominus_{A_1} x_2^{1_{A_2 A_1}} \otimes_{A_1} f_{A_1}^{-1}(\sin\alpha) \\ x_1^{1_{A_1 A_2}} \otimes_{A_2} f_{A_2}^{-1}(\sin\alpha) \oplus_{A_2} x_2 \otimes_{A_2} f_{A_2}^{-1}(\cos\alpha) \end{pmatrix}. \end{aligned}$$

Let $\boldsymbol{A}_0 = (A_0; \oplus_0, \otimes_0, \ominus_0, \oslash_0, \leq_0)$ be some arithmetic with the functional parameter f_{A_0}. For $\alpha_0 \in \boldsymbol{A}_0$, the mapping

$$\begin{pmatrix} \tilde{x}_1 \\ \tilde{x}_2 \end{pmatrix} = \begin{pmatrix} x_1 \otimes_{A_1} \mathrm{Cos}_{10}\, \alpha_0 \ominus_{A_1} x_2^{1_{A_2 A_1}} \otimes_{A_1} \mathrm{Sin}_{10}\, \alpha_0 \\ x_1^{1_{A_1 A_2}} \otimes_{A_2} \mathrm{Sin}_{20}\, \alpha_0 \oplus_{A_2} x_2 \otimes_{A_2} \mathrm{Cos}_{20}\, \alpha_0 \end{pmatrix}$$

is a general form of rotation in the Cartesian product of the two sets, each of them equipped with its own arithmetic, while the angle of rotation fulfills its own non-Diophantine arithmetic rules. The trigonometric functions are here defined by

$$\mathrm{Cos}_{k0}\,\alpha_0 = f_{A_k}^{-1}(\cos f_{A_0}(\alpha_0)),$$
$$\mathrm{Sin}_{k0}\,\alpha_0 = f_{A_k}^{-1}(\sin f_{A_0}(\alpha_0)).$$

4.5. Complex-Valued Functions

The supreme function of reason is to show man that some things are beyond reason.

Blaise Pascal

Many results for complex-valued functions are essentially the same as those known for real-valued functions. The proofs typically reduce to appropriate repetitions of the proofs we have already encountered in previous sections, but applied separately to real and imaginary parts. The main peculiarity of the non-Newtonian formalism is here the possibility of dealing with different arithmetics of real and imaginary parts of non-Diophantine complex numbers. We will thus concentrate on selected topics, important for further applications to Fourier transforms and quantum mechanics on fractals.

Consider three arithmetics, $\boldsymbol{A} = (A; \oplus_A, \otimes_A, \ominus_A, \oslash_A, \leq_A)$ and $\boldsymbol{B}_j = (B_j; \oplus_{B_j}, \otimes_{B_j}, \ominus_{Bj}, \oslash_{B_j}, \leq_{B_j})$ and functional parameters f_A, f_{B_j} for $j = 1, 2$, respectively.

Let us recall that a complex number is a pair $(b_1, b_2) \in B_1 \times B_2 = B$. A complex-valued function is a mapping g: $A \to B$. The arithmetic operations in B are defined by

$$\begin{pmatrix} x_1 \\ x_2 \end{pmatrix} \oplus \begin{pmatrix} y_1 \\ y_2 \end{pmatrix} = \begin{pmatrix} x_1 \oplus_{B_1} y_1 \\ x_2 \oplus_{B_2} y_2 \end{pmatrix}, \tag{4.55}$$

$$\begin{pmatrix} x_1 \\ x_2 \end{pmatrix} \otimes \begin{pmatrix} y_1 \\ y_2 \end{pmatrix} = \begin{pmatrix} x_1 \otimes_{B_1} y_1 \ominus_{B_1} x_2^{1_{B_2 B_1}} \otimes_{B_1} y_2^{1_{B_2 B_1}} \\ x_2 \otimes_{B_2} y_1^{1_{B_1 B_2}} \oplus_{B_2} x_1^{1_{B_1 B_2}} \otimes_{B_2} y_2 \end{pmatrix} \tag{4.56}$$

(for notational convenience we represent pairs from B either by rows or columns). The pair $b \in B$ can be written in terms of its real and imaginary parts,

$$b = (b_1, b_2) = (\Re b, \Im b).$$

The complex number b is real if

$$b = (b_1, 0_{B_2})$$

and imaginary if

$$b = (0_{B_1}, b_2).$$

Complex conjugations maps $b = (b_1, b_2)$ into

$$b^* = (b_1, \ominus_{B_2} b_2).$$

The neutral elements of addition and multiplication as well as the imaginary unit are given, respectively, by

$$\begin{aligned} 0_B &= (0_{B_1}, 0_{B_2}), \\ 1_B &= (1_{B_1}, 0_{B_2}), \\ i_B &= (0_{B_1}, 1_{B_2}). \end{aligned}$$

The modulus $|b|$ of b is the real complex number defined by

$$\begin{aligned} |b|^{2_B} &= |b| \otimes_B |b| = b \otimes_B b^* \\ &= \begin{pmatrix} x_1 \otimes_{B_1} x_1 \oplus_{B_1} x_2^{1_{B_2 B_1}} \otimes_{B_1} y_2^{1_{B_2 B_1}} \\ 0_{B_2} \end{pmatrix}, \\ 0_{B_1} &\leq_{B_1} \Re|b|. \end{aligned}$$

4.5.1. *Differentiation*

The freedom of thought is a sacred right of every individual man, and diversity will continue to increase with the progress, refinement, and differentiation of the human intellect.

Felix Adler

Let $x \in \boldsymbol{A}$ and $g(x) = (g_1(x), g_2(x)) \in B = B_1 \times B_2$ be defined by the following commutative diagram:

$$\begin{array}{ccc} A & \xrightarrow{g} & B_1 \times B_2 \\ f_A \downarrow & & \downarrow f_{B_1} \times f_{B_2} \\ R & \xrightarrow{\tilde{g}} & R^2 \end{array} \tag{4.57}$$

Explicitly,

$$g(x) = (f_{B_1}^{-1}(\tilde{g}_1(f_A(x))), f_{B_2}^{-1}(\tilde{g}_2(f_A(x)))).$$

In the following paragraphs whenever a Newtonian derivative of a real function occurs, we assume that the function is differentiable. The same concerns integrability of functions occurring in the Newtonian integrals.

Definition 4.5.1. Non-Newtonian derivative of a complex-valued function represented by (4.57) is a pair of non-Newtonian derivatives or real and imaginary parts of g,

$$\begin{aligned}\frac{\mathrm{D}g(x)}{\mathrm{D}x} &= \left(\frac{\mathrm{D}g_1(x)}{\mathrm{D}x}, \frac{\mathrm{D}g_2(x)}{\mathrm{D}x}\right)\\ &= \left(f_{B_1}^{-1}\left(\frac{\mathrm{d}\tilde{g}_1(f_A(x))}{\mathrm{d}f_A(x)}\right), f_{B_2}^{-1}\left(\frac{\mathrm{d}\tilde{g}_2(f_A(x))}{\mathrm{d}f_A(x)}\right)\right).\end{aligned}$$

Theorem 4.5.1 (The Leibniz rule for complex-valued functions).

$$\frac{\mathrm{D}(g\otimes h)(x)}{\mathrm{D}x} = \frac{\mathrm{D}g(x)}{\mathrm{D}x}\otimes h(x)\oplus g(x)\otimes\frac{\mathrm{D}h(x)}{\mathrm{D}x}.$$

Proof: Taking the definition

$$\frac{\mathrm{D}g\otimes h(x)}{\mathrm{D}x} = \left(\frac{\mathrm{D}(g\otimes h)_1(x)}{\mathrm{D}x}, \frac{\mathrm{D}(g\otimes h)_2(x)}{\mathrm{D}x}\right)$$

we apply formulas (4.55) and (4.56). Then the real part satisfies the following equalities:

$$\begin{aligned}\frac{\mathrm{D}(g\otimes h)_1(x)}{\mathrm{D}x} &= \frac{\mathrm{D}}{\mathrm{D}x}f_{B_1}^{-1}(f_{B_1}(g_1(x))f_{B_1}(h_1(x)) - f_{B_2}(g_2(x))f_{B_2}(h_2(x)))\\ &= \frac{\mathrm{D}}{\mathrm{D}x}f_{B_1}^{-1}(\tilde{g}_1(f_A(x))\tilde{h}_1(f_A(x)) - \tilde{g}_2(f_A(x))\tilde{h}_2(f_A(x)))\\ &= f_{B_1}^{-1}\left(\frac{\mathrm{d}}{\mathrm{d}f_A(x)}\tilde{g}_1(f_A(x))\tilde{h}_1(f_A(x))\right.\\ &\quad\left. - \frac{\mathrm{d}}{\mathrm{d}f_A(x)}\tilde{g}_2(f_A(x))\tilde{h}_2(f_A(x))\right)\\ &= f_{B_1}^{-1}\left(f_{B_1}\circ f_{B_1}^{-1}\left(\frac{\mathrm{d}\tilde{g}_1(f_A(x))}{\mathrm{d}f_A(x)}\right)f_{B_1}\circ f_{B_1}^{-1}(\tilde{h}_1(f_A(x)))\right.\\ &\quad - f_{B_1}\circ f_{B_1}^{-1}\circ f_{B_2}\circ f_{B_2}^{-1}\left(\frac{d\tilde{g}_2(f_A(x))}{df_A(x)}\right)\\ &\quad \times f_{B_1}\circ f_{B_1}^{-1}\circ f_{B_2}\circ f_{B_2}^{-1}(\tilde{h}_2(f_A(x)))\end{aligned}$$

$$+ f_{B_1} \circ f_{B_1}^{-1}(\tilde{g}_1(f_A(x))) f_{B_1} \circ f_{B_1}^{-1} \left(\frac{\mathrm{d}\tilde{h}_1(f_A(x))}{\mathrm{d}f_A(x)} \right)$$
$$- f_{B_1} \circ f_{B_1}^{-1} \circ f_{B_2} \circ f_{B_2}^{-1}(\tilde{g}_2(f_A(x)))$$
$$\times f_{B_1} \circ f_{B_1}^{-1} \circ f_{B_2} \circ f_{B_2}^{-1} \left(\frac{\mathrm{d}\tilde{h}_2(f_A(x))}{\mathrm{d}f_A(x)} \right) \Bigg).$$

Recalling the definition of a first-degree monomial, we obtain

$$\frac{\mathrm{D}(g \otimes h)_1(x)}{\mathrm{D}x} = f_{B_1}^{-1} \left(f_{B_1} \left(\frac{\mathrm{D}g_1(x)}{\mathrm{D}x} \right) f_{B_1}(h_1(x)) \right.$$
$$- f_{B_1} \circ f_{B_1}^{-1} \circ f_{B_2} \left(\frac{\mathrm{D}g_2(x)}{\mathrm{D}x} \right) f_{B1} \circ f_{B1}^{-1} \circ f_{B2}(h_2(x))$$
$$+ f_{B_1}(g_1(x)) f_{B_1} \left(\frac{\mathrm{D}h_1(x)}{\mathrm{D}x} \right)$$
$$\left. - f_{B_1} \circ f_{B_1}^{-1} \circ f_{B_2}(g_2(x)) f_{B_1} \circ f_{B_1}^{-1} \circ f_{B_2} \left(\frac{\mathrm{D}h_2(x)}{\mathrm{D}x} \right) \right)$$
$$= \frac{\mathrm{D}g_1(x)}{\mathrm{D}x} \otimes_{B_1} h_1(x) \ominus_{B_1} \left(\frac{\mathrm{D}g_2(x)}{\mathrm{D}x} \right)^{1_{B_2 B_1}} \otimes_{B_1} h_2(x)^{1_{B2} B_1}$$
$$\oplus_{B_1} g_1(x) \otimes_{B_1} \frac{\mathrm{D}h_1(x)}{\mathrm{D}x} \ominus_{B_1} g_2(x)^{1_{B_2 B_1}} \otimes_{B_1} \left(\frac{\mathrm{D}h_2(x)}{\mathrm{D}x} \right)^{1_{B_2 B_1}}$$
$$= \left(\frac{\mathrm{D}g(x)}{\mathrm{D}x} \otimes h(x) \right)_1 \oplus_{B_1} \left(g(x) \otimes \frac{\mathrm{Dh}(x)}{\mathrm{D}x} \right)_1$$
$$= \left(\frac{\mathrm{D}g(x)}{\mathrm{D}x} \otimes h(x) \oplus g(x) \otimes \frac{\mathrm{Dh}(x)}{\mathrm{D}x} \right)_1.$$

In a similar way, we have

$$\frac{\mathrm{D}(\mathrm{g} \otimes h)_2(x)}{\mathrm{D}x} = \frac{\mathrm{D}}{\mathrm{D}x} f_{B_2}^{-1}(\tilde{g}_1(f_A(x))\tilde{h}_2(f_A(x)) + \tilde{g}_2(f_A(x))\tilde{h}_1(f_A(x)))$$
$$= f_{B_2}^{-1} \left(\frac{\mathrm{d}}{\mathrm{d}f_A(x)} \tilde{g}_1(f_A(x))\tilde{h}_2(f_A(x)) \right.$$
$$\left. + \frac{\mathrm{d}}{\mathrm{d}f_A(x)} \tilde{g}_2(f_A(x))\tilde{h}_1(f_A(x)) \right)$$
$$= \left(\frac{\mathrm{D}g(x)}{\mathrm{D}x} \otimes h(x) \oplus g(x) \otimes \frac{\mathrm{D}h(x)}{\mathrm{D}x} \right)_2.$$

Theorem is proved. □

Theorem 4.5.2. *The derivative of a complex-valued function is a linear mapping.*

Proof: A sum of derivatives equals a derivative of a sum,

$$\frac{\mathrm{D}g(x)}{\mathrm{D}x} \oplus \frac{\mathrm{D}h(x)}{\mathrm{D}x} = \left(\frac{\mathrm{D}g_1(x)}{\mathrm{D}x} \oplus_{B_1} \frac{\mathrm{D}h_1(x)}{\mathrm{D}x}, \frac{\mathrm{D}g_2(x)}{\mathrm{D}x} \oplus_{B_2} \frac{\mathrm{D}h_2(x)}{\mathrm{D}x}\right)$$
$$= \left(\frac{\mathrm{D}(g_1 \oplus_{B_1} h_1)(x)}{\mathrm{D}x}, \frac{\mathrm{D}(g_2 \oplus h_2 h_2)(x)}{\mathrm{D}x}\right)$$
$$= \left(\frac{\mathrm{D}(g \oplus h)_1(x)}{\mathrm{D}x}, \frac{\mathrm{D}(g \oplus h)_2(x)}{\mathrm{D}x}\right)$$
$$= \frac{\mathrm{D}(g \oplus h)(x)}{Dx}.$$

The 1-homogeneity under multiplication by a complex number follows from the Leibniz rule, if one sets $g(x) = \text{const}$.

Theorem is proved. □

4.5.2. *Integration*

The Internet is ultimately about innovation and integration.

Louis V. Gerstner, Jr.

Once we have derivative, it is clear how to define integral.

Definition 4.5.2. The integral of a complex-valued function is defined by the following pair of non-Newtonian integrals

$$\int_y^x g(x')\mathrm{D}x' = \left(\int_y^x g_1(x')\mathrm{D}x', \int_y^x g_2(x')\mathrm{D}x'\right)$$
$$= \left(f_{B_1}^{-1}\left(\int_{f_A(y)}^{f_A(x)} \tilde{g}_1(r)\mathrm{d}r\right), f_{B_2}^{-1}\left(\int_{f_A(y)}^{f_A(x)} \tilde{g}_2(r)\mathrm{d}r\right)\right).$$

Theorem 4.5.3 (Fundamental theorems of non-Newtonian calculus for complex-valued functions). *Let $\tilde{g}_1(r)$ and $\tilde{g}_2(r)$ satisfy all the assumptions needed in the fundamental theorems of Newtonian calculus. Then*

$$\frac{\mathrm{D}}{\mathrm{D}x}\int_y^x g(x')\mathrm{D}x' = g(x), \tag{4.58}$$

$$\int_y^x \frac{\mathrm{D}g(x')}{\mathrm{D}x'}\mathrm{D}x' = g(x) \ominus_B g(y). \tag{4.59}$$

Proof is obtained by application of Theorems 4.1.6 and 4.1.7 to real and imaginary parts of both expressions. □

Theorem 4.5.4. *The integral of a complex-valued function is a linear mapping.*

Proof: Linearity under addition follows from Theorem 4.1.2 applied to real and imaginary parts. To prove 1-homogeneity under multiplication by a complex number, we proceed as follows. Denote

$$z = \int_y^x g(x')\mathrm{D}x' = \begin{pmatrix} z_1 \\ z_2 \end{pmatrix} = \begin{pmatrix} f_{B_1}^{-1}\left(\int_{f_A(y)}^{f_A(x)} \tilde{g}_1(r)\mathrm{d}r\right) \\ f_{B_2}^{-1}\left(\int_{f_A(y)}^{f_A(x)} \tilde{g}_2(r)\mathrm{d}r\right) \end{pmatrix} = \begin{pmatrix} f_{B_1}^{-1}(\tilde{z}_1) \\ f_{B_2}^{-1}(\tilde{z}_2) \end{pmatrix},$$

$$\lambda = \begin{pmatrix} \lambda_1 \\ \lambda_2 \end{pmatrix} = \begin{pmatrix} f_{B_1}^{-1}\left(\tilde{\lambda}_1\right) \\ f_{B_2}^{-1}\left(\tilde{\lambda}_2\right) \end{pmatrix}.$$

Then

$$\lambda \otimes \int_y^x g(x')\mathrm{D}x' = \begin{pmatrix} \lambda_1 \\ \lambda_2 \end{pmatrix} \otimes \begin{pmatrix} z_1 \\ z_2 \end{pmatrix} = \begin{pmatrix} f_{B_1}^{-1}\left(\tilde{\lambda}_1\tilde{z}_1 - \tilde{\lambda}_2\tilde{z}_2\right) \\ f_{B_2}^{-1}\left(\tilde{\lambda}_1\tilde{z}_2 + \tilde{\lambda}_2\tilde{z}_1\right) \end{pmatrix}$$

$$= \begin{pmatrix} f_{B_1}^{-1}\left(\tilde{\lambda}_1 \int_{f_A(y)}^{f_A(x)} \tilde{g}_1(r)\mathrm{d}r - \tilde{\lambda}_2 \int_{f_A(y)}^{f_A(x)} \tilde{g}_2(r)\mathrm{d}r\right) \\ f_{B_2}^{-1}\left(\tilde{\lambda}_1 \int_{f_A(y)}^{f_A(x)} \tilde{g}_2(r)\mathrm{d}r + \tilde{\lambda}_2 \int_{f_A(y)}^{f_A(x)} \tilde{g}_1(r)\mathrm{d}r\right) \end{pmatrix}$$

$$= \begin{pmatrix} f_{B_1}^{-1}\left(\int_{f_A(y)}^{f_A(x)} \left(\tilde{\lambda}_1\tilde{g}_1(r) - \tilde{\lambda}_2\tilde{g}_2(r)\right)\mathrm{d}r\right) \\ f_{B_2}^{-1}\left(\int_{f_A(y)}^{f_A(x)} \left(\tilde{\lambda}_1\tilde{g}_2(r) + \tilde{\lambda}_2\tilde{g}_1(r)\right)\mathrm{d}r\right) \end{pmatrix}$$

$$= \begin{pmatrix} f_{B_1}^{-1}\left(\int_{f_A(y)}^{f_A(x)} \tilde{b}_1(r)\mathrm{d}r\right) \\ f_{B_2}^{-1}\left(\int_{f_A(y)}^{f_A(x)} \tilde{b}_2(r)\mathrm{d}r\right) \end{pmatrix} = \int_y^x b(x)\mathrm{D}x,$$

where

$$b(x) = \begin{pmatrix} f_{B_1}^{-1}\left(\tilde{b}_1(f_A(x))\right) \\ f_{B_2}^{-1}\left(\tilde{b}_2(f_A(x))\right) \end{pmatrix} = \begin{pmatrix} f_{B_1}^{-1}\left(\tilde{\lambda}_1\tilde{g}_1(f_A(x)) - \tilde{\lambda}_2\tilde{g}_2(f_A(x))\right) \\ f_{B_2}^{-1}\left(\tilde{\lambda}_1\tilde{g}_2(f_A(x)) - \tilde{\lambda}_2\tilde{g}_1(f_A(x))\right) \end{pmatrix}$$

$$= \begin{pmatrix} \lambda_1 \\ \lambda_2 \end{pmatrix} \otimes \begin{pmatrix} f_{B_1}^{-1}\left(\tilde{g}_1(f_A(x))\right) \\ f_{B_2}^{-1}\left(\tilde{g}_2(f_A(x))\right) \end{pmatrix} = \lambda \otimes g(x).$$

Finally,

$$\lambda \otimes \int_y^x g(x')\mathrm{D}x' = \int_y^x (\lambda \otimes g)(x')\mathrm{D}x'.$$

Theorem is proved. □

4.6. Non-Newtonian Fourier Transforms

We are in a constant state of transformation.

Alejandro Gonzalez Inarritu

Let us develop Fourier analysis of real-valued or complex-valued functions $g\colon A \to B$, with arithmetics $\boldsymbol{A}$ and $\boldsymbol{B}$, represented by either of the diagrams.

$$\begin{array}{ccc} A & \xrightarrow{g} & B \\ f_A\downarrow & & \downarrow f_B \\ R & \xrightarrow{\tilde{g}} & R \end{array} \quad \text{or} \quad \begin{array}{ccc} A & \xrightarrow{g} & B = B_1 \times B_2 \\ f_A\downarrow & & \downarrow f_B = f_{B_1} \times f_{B_2} \\ R & \xrightarrow{\tilde{g}} & C = R^2 \end{array} \tag{4.60}$$

The non-Newtonian formalism becomes especially useful for signal analysis on fractals as it automatically circumvents various impossibility theorems known from the more traditional approaches. In particular, the old and difficult problem of Fourier analysis on arbitrary Cantor sets finds here a solution which is almost trivial.

4.6.1. *Scalar product*

A formally harmonious product needs no decoration;
it should be elevated through pure form.

Ferdinand Porsche

In the space of complex-valued functions, one can define a non-Newtonian scalar product in exact analogy to the Newtonian case. The product is essential for signal analysis and quantum mechanics on fractals.

Consider two functions $g, h\colon A \to B_1 \times B_2$ associated with diagrams of the form (4.60). Let $\tilde{g}(r) = \tilde{g}_1(r) + i\tilde{g}_2(r), \tilde{h}(r) = \tilde{h}_1(r) + i\tilde{h}_2(r)$, and

$$\langle \tilde{g}|\tilde{h}\rangle = \int_{-f_A(T)/2}^{f_A(T)/2} \overline{\tilde{g}(r)}\tilde{h}(r)\mathrm{d}r,$$

where $f_A(T)$ can be finite or infinite. Here "i" denotes the usual Diophantine imaginary unit and the bar over $\tilde{g}(r)$ denotes the usual Diophantine complex conjugation.

Definition 4.6.1. A scalar product of two functions $g, h\colon A \to B_1 \times B_2$, is defined as

$$\langle g|h\rangle = \int_{\ominus_A T \oslash_A 2_A}^{T \oslash_A 2_A} g(x)^* \otimes h(x) \mathrm{D}x. \tag{4.61}$$

The asterisk denotes the non-Diophantine complex conjugation. The same symbol of the scalar product for both $\langle g|h\rangle$ and $\langle \tilde{g}|\tilde{h}\rangle$ will not lead to ambiguities.

Theorem 4.6.1 (Properties of the scalar product).

$$\langle g|h\rangle = \begin{pmatrix} f_{B_1}^{-1}(\Re\langle \tilde{g}|\tilde{h}\rangle) \\ f_{B_2}^{-1}(\Im\langle \tilde{g}|\tilde{h}\rangle) \end{pmatrix}, \tag{4.62}$$

$$\langle g|h\rangle^* = \langle h|g\rangle, \tag{4.63}$$

$$\langle g|h_1 \oplus h_2\rangle = \langle g|h_1\rangle \oplus \langle g|h_2\rangle, \tag{4.64}$$

$$\langle g|\lambda \otimes h\rangle = \lambda \otimes \langle g|h\rangle. \tag{4.65}$$

Proof: We have

$$\begin{aligned}
g(x)^* \otimes h(x) &= \begin{pmatrix} g_1(x) \\ \ominus_{B_2} g_2(x) \end{pmatrix} \otimes \begin{pmatrix} h_1(x) \\ h_2(x) \end{pmatrix} \\
&= \begin{pmatrix} f_{B_1}^{-1}(f_{B_1}(g_1(x)) f_{B_1}(h_1(x)) - f_{B_2}(\ominus_{B_2} g_2(x)) f_{B_2}(h_2(x))) \\ f_{B_2}^{-1}(f_{B_1}(g_1(x)) f_{B_2}(h_2(x)) + f_{B_2}(\ominus_{B_2} g_2(x)) f_{B_1}(h_1(x))) \end{pmatrix} \\
&= \begin{pmatrix} f_{B_1}^{-1}(\tilde{g}_1(f_A(x)) \tilde{h}_1(f_A(x)) + \tilde{g}_2(f_A(x)) \tilde{h}_2(f_A(x))) \\ f_{B_2}^{-1}(\tilde{g}_1(f_A(x)) \tilde{h}_2(f_A(x)) - \tilde{g}_2(f_A(x)) \tilde{h}_1(f_A(x))) \end{pmatrix} \\
&= \begin{pmatrix} f_{B_1}^{-1}(\tilde{j}_1(f_A(x))) \\ f_{B_2}^{-1}(\tilde{j}_2(f_A(x))) \end{pmatrix} = j(x).
\end{aligned}$$

So,

$$\langle g|h\rangle = \int_{\ominus_A T \oslash_A 2_A}^{T \oslash_A 2_A} j(x) \mathrm{D}x = \begin{pmatrix} f_{B_1}^{-1}\left(\int_{f_A(\ominus_A T \oslash_A 2_A)}^{f_A(T \oslash_A 2_A)} \tilde{j}_1(r) \mathrm{d}r\right) \\ f_{B_1}^{-1}\left(\int_{f_A(\ominus_A T \oslash_A 2_A)}^{f_A(T \oslash_A 2_A)} \tilde{j}_2(r) \mathrm{d}r\right) \end{pmatrix}$$

$$= \begin{pmatrix} f_{B_1}^{-1}\left(\int_{-f_A(T)/2}^{f_A(T)/2} \tilde{j}_1(r)\mathrm{d}r\right) \\ f_{B_2}^{-1}\left(\int_{-f_A(T)/2}^{f_A(T)/2} \tilde{j}_2(r)\mathrm{d}r\right) \end{pmatrix} = \begin{pmatrix} f_{B_1}^{-1}\left(\int_{-f_A(T)/2}^{f_A(T)/2} \Re(\overline{\tilde{g}(r)}\tilde{h}(r))\mathrm{d}r\right) \\ f_{B_2}^{-1}\left(\int_{-f_A(T)/2}^{f_A(T)/2} \Im(\overline{\tilde{g}(r)}\tilde{h}(r))\mathrm{d}r\right) \end{pmatrix}$$

$$= \begin{pmatrix} f_{B_1}^{-1}(\Re\langle\tilde{g}|\tilde{h}\rangle) \\ f_{B_2}^{-1}(\Im\langle\tilde{g}|\tilde{h}\rangle) \end{pmatrix}. \tag{4.66}$$

Real and imaginary parts are, of course, defined here in the usual Diophantine sense.

Next, by definition of complex conjugation,

$$\langle g|h\rangle^* = \begin{pmatrix} f_{B_1}^{-1}(\Re\langle\tilde{g}|\tilde{h}\rangle) \\ \ominus_{B_2} f_{B_2}^{-1}(\Im\langle\tilde{g}|\tilde{h}\rangle) \end{pmatrix} = \begin{pmatrix} f_{B_1}^{-1}(\Re\langle\tilde{g}|\tilde{h}\rangle) \\ f_{B_2}^{-1}(-\Im\langle\tilde{g}|\tilde{h}\rangle) \end{pmatrix} = \begin{pmatrix} f_{B_1}^{-1}(\Re\langle\tilde{h}|\tilde{g}\rangle) \\ f_{B_2}^{-1}(\Im\langle\tilde{h}|\tilde{g}\rangle) \end{pmatrix}$$
$$= \langle h|g\rangle. \tag{4.67}$$

The remaining two properties follow from associativity and distributivity of the arithmetic operations, supplemented by linearity of the integral.

Theorem is proved. □

4.6.2. *Sine and cosine transforms of real-valued signals*

Some painters transform the sun into a yellow spot,
others transform a yellow spot into the sun.

Pablo Picasso

Definition 4.6.2. The function $g: A \to B$ defined by (4.60) is $L^p(A, \mathrm{D}x)$ function, or just L^p function, if the Lebesgue integral

$$\int_{-f_A(T)/2}^{f_A(T)/2} |\tilde{g}(r)|^p \mathrm{d}r$$

is finite. If g is both L^p and L^q functions, we say that g is $L^p \cap L^q$ function.

Now consider a function g corresponding to the left diagram (4.60). Let $f_A(T)$ be finite. We can apply the standard representation of Dirac's delta-function in the space of $L^1 \cap L^2$ functions on the segment

$[-f_A(T)/2, f_A(T)/2]$,

$$\delta(r-r') = \frac{1}{f_A(T)} \sum_{n\in Z} e^{i2n\pi(r-r')/f_A(T)} \tag{4.68}$$

$$= \frac{1}{f_A(T)} + \frac{2}{f_A(T)} \sum_{n>0} \times \left(\cos\frac{2n\pi r}{f_A(T)} \cos\frac{2n\pi r'}{f_A(T)} + \sin\frac{2n\pi r}{f_A(T)} \sin\frac{2n\pi r'}{f_A(T)} \right). \tag{4.69}$$

Denoting

$$c_n(r) = \sqrt{\frac{2}{f_A(T)}} \cos\frac{2n\pi r}{f_A(T)}, \quad n > 0, \tag{4.70}$$

$$s_n(r) = \sqrt{\frac{2}{f_A(T)}} \sin\frac{2n\pi r}{f_A(T)}, \quad n > 0, \tag{4.71}$$

$$c_0(r) = \sqrt{\frac{1}{f_A(T)}}, \tag{4.72}$$

$$s_0(r) = 0, \tag{4.73}$$

one finds

$$\delta(r-r') = \sum_{n\geq 0} (c_n(r)c_n(r') + s_n(r)s_n(r')). \tag{4.74}$$

with the standard Newtonian Fourier reconstruction formula

$$\tilde{g}(r) = \int_{-f_A(T)/2}^{f_A(T)/2} \delta(r-r')\tilde{g}(r')\mathrm{d}r' = \sum_{n\geq 0} c_n(r)\langle c_n|\tilde{g}\rangle + \sum_{n\geq 0} s_n(r)\langle s_n|\tilde{g}\rangle, \tag{4.75}$$

which is true up to the Gibbs phenomenon (Hewitt and Hewitt, 1979). Accordingly,

$$g(x) = f_B^{-1}(\tilde{g}(f_A(x))) = f_B^{-1}\left(\sum_{n\geq 0} c_n(f_A(x))\langle c_n|\tilde{g}\rangle + \sum_{n\geq 0} s_n(f_A(x))\langle s_n|\tilde{g}\rangle \right)$$

$$= f_B^{-1}\left(\sum_{n\geq 0} f_B \circ \underbrace{f_B^{-1}(c_n(f_A(x)))}_{C_n(x)} f_B \circ \underbrace{f_B^{-1}(\langle c_n|\tilde{g}\rangle)}_{\langle C_n|g\rangle} + \cdots\right).$$

Here, the dots stand for the part involving $S_n(x)$,

$$C_n(x) = f_B^{-1}(c_n(f_A(x))), \tag{4.76}$$

$$S_n(x) = f_B^{-1}(s_n(f_A(x))), \tag{4.77}$$

and we have used the analog of the scalar product from Section 4.6.1, but here formulated for real-valued functions,

$$\langle g|h\rangle = f_B^{-1}(\langle\tilde{g}|\tilde{h}\rangle) \tag{4.78}$$

Theorem 4.6.2 (A non-Newtonian Fourier reconstruction theorem).

$$g(x) = \sum_{n\geq 0}{}^{\oplus_B} C_n(x) \otimes_B \langle C_n|g\rangle \oplus_B \sum_{n\geq 0}{}^{\oplus_B} S_n(x) \otimes_B \langle S_n|g\rangle. \tag{4.79}$$

Proof is left as an exercise. □

The orthogonality conditions

$$\langle c_n|c_m\rangle = \langle s_n|s_m\rangle = \delta_{nm}, \tag{4.80}$$

$$\langle c_n|s_m\rangle = 0 \tag{4.81}$$

imply

$$\langle C_n|C_m\rangle = f_B^{-1}(\langle c_n|c_m\rangle) = f_B^{-1}(\delta_{nm}) = \delta_{Bnm} = \begin{cases} 1_B & \text{for } n = m, \\ 0_B & \text{for } n \neq m, \end{cases} \tag{4.82}$$

$$\langle S_n|S_m\rangle = \delta_{Bnm}, \tag{4.83}$$

$$\langle S_n|C_m\rangle = 0_B. \tag{4.84}$$

As a consequence, we have the following result.

Theorem 4.6.3 (A non-Newtonian Fourier Parseval formula). *Let g and h be $L^1 \cap L^2$ functions. Then*

$$\langle g|h\rangle = \sum_{n\geq 0}{}^{\oplus_B}(\langle g|C_n\rangle \otimes_B \langle C_n|h\rangle \oplus_B \langle g|S_n\rangle \otimes_B \langle S_n|h\rangle). \tag{4.85}$$

Proof is left as an exercise. □

4.6.3. *Complex-valued Fourier transforms*

Transformation is not a future event. It is a present activity.
Jillian Michaels

Complex-valued Fourier transforms are needed for quantum mechanics on fractal spaces. Consider the trigonometric functions,

$$\mathrm{Cos}_j\, x = f_{B_j}^{-1}(\cos f_A(x)), \quad j = 1, 2, \tag{4.86}$$

$$\mathrm{Sin}_j\, x = f_{B_j}^{-1}(\sin f_A(x)), \tag{4.87}$$

The functions have the following symmetry properties:

$$\mathrm{Cos}_j(\ominus_A x) = f_{B_j}^{-1}(\cos f_A(\ominus_A x)) = f_{B_j}^{-1}(\cos(-f_A(x))) = \mathrm{Cos}_j\, x, \tag{4.88}$$

$$\mathrm{Sin}_j(\ominus_A x) = f_{B_j}^{-1}(\sin f_A(\ominus_A x)) = f_{B_j}^{-1}(\sin(-f_A(x))) = \ominus_{B_j} \mathrm{Sin}_j\, x. \tag{4.89}$$

We will also need the complex exponent

$$\mathrm{Exi}\ x = \begin{pmatrix} \mathrm{Cos}_1\, x \\ \mathrm{Sin}_2\, x \end{pmatrix} = \begin{pmatrix} \mathrm{Exi}_1(x) \\ \mathrm{Exi}_2(x) \end{pmatrix} = \begin{pmatrix} f_{B_1}^{-1}(\widetilde{\mathrm{Exi}}_1(f_A(x))) \\ f_{B_e}^{-1}(\widetilde{\mathrm{Exi}}_2(f_A(x))) \end{pmatrix}, \tag{4.90}$$

satisfying

$$(\mathrm{Exi}\ x)^* = \mathrm{Exi}(\ominus_A x), \tag{4.91}$$

$$\mathrm{Exi}\ 0_A = 1_B, \tag{4.92}$$

where 1_B is the neutral element of complex multiplication in $B = B_1 \times B_2$. Moreover

$$\begin{aligned}
\mathrm{Exi}\ x \otimes \mathrm{Exi}\ y &= \begin{pmatrix} \mathrm{Cos}_1\, x \\ \mathrm{Sin}_2\, x \end{pmatrix} \otimes \begin{pmatrix} \mathrm{Cos}_1\, y \\ \mathrm{Sin}_2\, y \end{pmatrix} \\
&= \begin{pmatrix} f_{B_1}^{-1}(f_{B_1}(\mathrm{Cos}_1\, x) f_{B_1}(\mathrm{Cos}_1\, y) - f_{B_2}(\mathrm{Sin}_2\, x) f_{B_2}(\mathrm{Sin}_2\, y)) \\ f_{B_2}^{-1}(f_{B_1}(\mathrm{Cos}_1\, x) f_{B_2}(\mathrm{Sin}_2\, y) + f_{B_2}(\mathrm{Sin}_2\, x) f_{B_1}(\mathrm{Cos}_1\, y)) \end{pmatrix} \\
&= \begin{pmatrix} f_{B_1}^{-1}(\cos f_A(x) \cos f_A(y) - \sin f_A(x) \sin f_A(y)) \\ f_{B_2}^{-1}(\cos f_A(x) \sin f_A(y) + \sin f_A(x) \cos f_A(y)) \end{pmatrix} \\
&= \begin{pmatrix} f_{B_1}^{-1}(\cos(f_A(x) + f_A(y))) \\ f_{B_2}^{-1}(\sin(f_A(x) + f_A(y))) \end{pmatrix} = \mathrm{Exi}(x \oplus_A y),
\end{aligned} \tag{4.93}$$

implying in particular

$$(\text{Exi } x)^* \otimes \text{Exi } x = \text{Exi}(\ominus_A x \oplus_A x) = 1_B. \tag{4.94}$$

Definition 4.6.3. The complex Fourier transform of a $L^1 \cap L^2$ function g: $A \to B$ is a function $\hat{g}$: $A \to B$ defined by the integral

$$\hat{g}(k) = \begin{pmatrix} \hat{g}_1(k) \\ \hat{g}_2(k) \end{pmatrix} = \int_{\ominus_A T \oslash_A 2_A}^{T \oslash_A 2_A} g(x) \otimes \text{Exi}(\ominus_A k \otimes_A x) \text{D}x. \tag{4.95}$$

Denoting

$$\begin{aligned} h[k](x) &= \begin{pmatrix} h[k]_1(x) \\ h[k]_2(x) \end{pmatrix} = \text{Exi}(k \otimes_A x) = \begin{pmatrix} \text{Cos}_1(k \otimes_A x) \\ \text{Sin}_2(k \otimes_A x) \end{pmatrix} \\ &= \begin{pmatrix} f_{B_1}^{-1}(\widetilde{h[k]}_1(f_A(x))) \\ f_{B_e}^{-1}(\widetilde{h[k]}_2(f_A(x))) \end{pmatrix} = \begin{pmatrix} f_{B_1}^{-1}(\cos(f_A(k) f_A(x))) \\ f_{B_e}^{-1}(\sin(f_A(k) f_A(x))) \end{pmatrix}, \end{aligned}$$

we find

$$\widetilde{h[k]}_1(r) = \cos(f_A(k) r), \tag{4.96}$$

$$\widetilde{h[k]}_2(r) = \sin(f_A(k) r), \tag{4.97}$$

and

$$\widetilde{h[k]}(r) = \widetilde{h[k]}_1(r) + i\widetilde{h[k]}_2(r) = e^{i f_A(k) r}. \tag{4.98}$$

Accordingly, we obtain an equivalent form of the Fourier transform

$$\begin{aligned} \hat{g}(k) = \langle h[k] | g \rangle &= \begin{pmatrix} f_{B_1}^{-1}(\Re \langle \widetilde{h[k]} | \tilde{g} \rangle) \\ f_{B_2}^{-1}(\Im \langle \widetilde{h[k]} | \tilde{g} \rangle) \end{pmatrix} \\ &= \begin{pmatrix} f_{B_1}^{-1}\left(\Re \int_{-f_A(T)/2}^{f_A(T)/2} \overline{\widetilde{h[k]}(r)} \tilde{g}(r) \text{d}r\right) \\ f_{B_2}^{-1}\left(\Im \int_{-f_A(T)/2}^{f_A(T)/2} \overline{\widetilde{h[k]}(r)} \tilde{g}(r) \text{d}r\right) \end{pmatrix} \\ &= \begin{pmatrix} f_{B_1}^{-1}\left(\Re \int_{-f_A(T)/2}^{f_A(T)/2} \tilde{g}(r) e^{-i f_A(k) r} \text{d}r\right) \\ f_{B_2}^{-1}\left(\Im \int_{-f_A(T)/2}^{f_A(T)/2} \tilde{g}(r) e^{-i f_A(k) r} \text{d}r\right) \end{pmatrix}. \end{aligned}$$

This gives us the following result.

Corollary 4.6.1. *The complex Fourier transform can be equivalently expressed as*

$$\hat{g}(k) = \begin{pmatrix} f_{B_1}^{-1}\left(\Re \int_{-f_A(T)/2}^{f_A(T)/2} \tilde{g}(r) e^{-if_A(k)r} \mathrm{d}r\right) \\ f_{B_2}^{-1}\left(\Im \int_{-f_A(T)/2}^{f_A(T)/2} \tilde{g}(r) e^{-if_A(k)r} \mathrm{d}r\right) \end{pmatrix}.$$

Example 4.6.1 (Plane waves). Let us return to

$$h[k](x) = \begin{pmatrix} h[k]_1(x) \\ h[k]_2(x) \end{pmatrix} = \mathrm{Exi}(k \otimes_A x) = \begin{pmatrix} f_{B_1}^{-1}(\cos(f_A(k) f_A(x))) \\ f_{B_e}^{-1}(\sin(f_A(k) f_A(x))) \end{pmatrix}$$

$$= \begin{pmatrix} f_{B_1}^{-1}(\widetilde{h[k]}_1(f_A(x))) \\ f_{B_e}^{-1}(\widetilde{h[k]}_2(f_A(x))) \end{pmatrix}.$$

The derivative

$$\frac{\mathrm{D}h[k](x)}{\mathrm{D}x} = \begin{pmatrix} f_{B_1}^{-1}\left(\frac{\mathrm{d}\cos(f_A(k) f_A(x))}{\mathrm{d}f_A(x)}\right) \\ f_{B_e}^{-1}\left(\frac{\mathrm{d}\sin(f_A(k) f_A(x))}{\mathrm{d}f_A(x)}\right) \end{pmatrix}$$

$$= \begin{pmatrix} f_{B_1}^{-1}(-f_A(k) \sin(f_A(k) f_A(x))) \\ f_{B_2}^{-1}(f_A(k) \cos(f_A(k) f_A(x))) \end{pmatrix}$$

$$= \begin{pmatrix} f_{B_1}^{-1}(-f_{B_1} \circ f_{B_1}^{-1} \circ f_A(k) f_{B_1} \circ f_{B_1}^{-1} \circ \sin(f_A(k) f_A(x))) \\ f_{B_2}^{-1}(f_{B_2} \circ f_{B_2}^{-1} \circ f_A(k) f_{B_2} \circ f_{B_2}^{-1} \circ \cos(f_A(k) f_A(x))) \end{pmatrix}$$

$$= \begin{pmatrix} f_{B_1}^{-1}(-f_{B_1}(k^{1_{AB_1}}) f_{B_1}(\mathrm{Sin}_1(k \otimes_A x))) \\ f_{B_2}^{-1}(f_{B_2}(k^{1_{AB_2}}) f_{B_2}(\mathrm{Cos}_2(k \otimes_A x))) \end{pmatrix}$$

$$= \begin{pmatrix} \ominus_{B_1} k^{1_{AB_1}} \otimes_{B_1} \mathrm{Sin}_1(k \otimes_A x) \\ k^{1_{AB_2}} \otimes_{B_2} \mathrm{Cos}_2(k \otimes_A x) \end{pmatrix}$$

$$= \begin{pmatrix} \ominus_{B_1} (k^{1_{AB_2}} \otimes_{B_2} \mathrm{Sin}_2(k \otimes_A x))^{1_{B_2 B_1}} \\ (k^{1_{AB_1}} \otimes_{B_1} \mathrm{Cos}_1(k \otimes_A x))^{1_{B_1 B_2}} \end{pmatrix}. \tag{4.99}$$

Since

$$i_B \otimes_B \begin{pmatrix} y_1 \\ y_2 \end{pmatrix} = \begin{pmatrix} \ominus_{B_1} {y_2}^{1_{B_2 B_1}} \\ {y_1}^{1_{B_1 B_2}} \end{pmatrix} \tag{4.100}$$

we rewrite

$$\begin{aligned} \frac{\mathrm{D}h[k](x)}{\mathrm{D}x} &= i_B \otimes_B \begin{pmatrix} k^{1_{AB_1}} \otimes_{B_1} \mathrm{Cos}_1(k \otimes_A x) \\ k^{1_{AB_2}} \otimes_{B_2} \mathrm{Sin}_2(k \otimes_A x) \end{pmatrix} \\ &= i_B \otimes_B \begin{pmatrix} k^{1_{AB_1}} \otimes_{B_1} \mathrm{Cos}_1(k \otimes_A x) \\ (k^{1_{AB_1}})^{1_{B_1 B_2}} \otimes_{B_2} \mathrm{Sin}_2(k \otimes_A x) \end{pmatrix} \\ &= i_B \otimes_B \begin{pmatrix} k^{1_{AB_1}} \\ 0_{B_2} \end{pmatrix} \otimes_B \begin{pmatrix} \mathrm{Cos}_1(k \otimes_A x) \\ \mathrm{Sin}_2(k \otimes_A x) \end{pmatrix} \\ &= i_B \otimes_B \begin{pmatrix} k^{1_{AB_1}} \\ 0_{B_2} \end{pmatrix} \otimes_B h[k](x), \end{aligned} \tag{4.101}$$

where

$$\begin{pmatrix} k_1 \\ k_2 \end{pmatrix} = \begin{pmatrix} k^{1_{AB_1}} \\ 0_{B_2} \end{pmatrix} \tag{4.102}$$

is a complex number with vanishing imaginary part (hence a real number in complex representation).

Let us note that (4.101) could be alternatively written as

$$\frac{\mathrm{D}h[k](x)}{\mathrm{D}x} = i_B \otimes_B k \otimes_B h[k](x), \tag{4.103}$$

where

$$k \otimes_B h[k](x) = \begin{pmatrix} k^{1_{AB_1}} \otimes_{B_1} h[k]_1(x) \\ k^{1_{AB_2}} \otimes_{B_2} h[k]_2(x) \end{pmatrix}. \tag{4.104}$$

Since

$$\begin{aligned} \widetilde{h[k]}(r) &= \widetilde{h[k]}_1(r) + i\widetilde{h[k]}_2(r) \\ &= \cos(f_A(k)r) + i\sin(f_A(k)r) = e^{if_A(k)r}, \end{aligned} \tag{4.105}$$

the scalar product of two plane waves is given by

$$\langle h[k]|h[l]\rangle = \begin{pmatrix} f_{B_1}^{-1}\left(\Re\int_{-f_A(T)/2}^{f_A(T)/2}\overline{\widetilde{h[k]}(r)}\widetilde{h[l]}(r)\mathrm{d}r\right) \\ f_{B_2}^{-1}\left(\Im\int_{-f_A(T)/2}^{f_A(T)/2}\overline{\widetilde{h[k]}(r)}\widetilde{h[l]}(r)\mathrm{d}r\right) \end{pmatrix}$$

$$= \begin{pmatrix} f_{B_1}^{-1}\left(\Re\int_{-f_A(T)/2}^{f_A(T)/2} e^{i(f_A(l)-f_A(k))r}\mathrm{d}r\right) \\ f_{B_2}^{-1}\left(\Im\int_{-f_A(T)/2}^{f_A(T)/2} e^{i(f_A(l)-f_A(k))r}\mathrm{d}r\right) \end{pmatrix}$$

$$= \begin{pmatrix} f_{B_1}^{-1}\left(\frac{e^{i(f_A(l)-f_A(k))f_A(T)/2}-e^{-i(f_A(l)-f_A(k))f_A(T)/2}}{i(f_A(l)-f_A(k))}\right) \\ f_{B_2}^{-1}(0) \end{pmatrix}$$

$$= \begin{pmatrix} f_{B_1}^{-1}\left(f_A(T)\frac{\sin(f_A(l)-f_A(k))f_A(T)/2}{(f_A(l)-f_A(k))f_A(T)/2}\right) \\ 0_{B_2} \end{pmatrix}.$$

For $k = l$

$$\langle h[k]|h[k]\rangle = \begin{pmatrix} f_{B_1}^{-1}\circ f_A(T) \\ 0_{B_2} \end{pmatrix} = \begin{pmatrix} T_{AB_1} \\ 0_{B_2} \end{pmatrix}.$$

Additionally, $f_A(l) \neq f_A(k)$ for $k \neq l$ since f_A is a bijection. Imposing $\langle h[k]|h[l]|\rangle = 0_B$ we obtain a quantization condition,

$$(f_A(l) - f_A(k))f_A(T)/2 = n\pi, \quad n \in Z, \tag{4.106}$$

for all $k, l \in A$. Its solution is given by

$$f_A(k)f_A(T) = 2\pi n_k, \quad n_k \in Z. \tag{4.107}$$

Equivalently,

$$\begin{aligned} k = k_n &= f_A^{-1}(2\pi n/f_A(T)) = f_A^{-1}(2\pi n) \oslash_A T \\ &= (2\pi n)_A \oslash_A T = 2_A \otimes_A \pi_A \otimes_A n_A \oslash_A T, \quad n \in Z \end{aligned} \tag{4.108}$$

where $n_A = f_A^{-1}(n), \pi_A = f_A^{-1}(\pi)$, etc. Now, consider the sum

$$F(x) = \sum_{n\in Z}{}^{\oplus_B} h[k_n](x) \otimes_B \langle h[k_n]|g\rangle \oslash_B \begin{pmatrix} T^{1_{AB_1}} \\ 0_{B_2} \end{pmatrix}. \tag{4.109}$$

One can check that

$$F(x) = \begin{pmatrix} f_{B_1}^{-1}\left(\Re \frac{1}{f_A(T)} \sum_n e^{i2\pi n f_A(x)/f_A(T)} \int_{-f_A(T)/2}^{f_A(T)/2} \tilde{A}(r) e^{-i2\pi n r/f_A(T)} \mathrm{d}r\right) \\ f_{B_2}^{-1}\left(\Im \frac{1}{f_A(T)} \sum_n e^{i2\pi n f_A(x)/f_A(T)} \int_{-f_A(T)/2}^{f_A(T)/2} \tilde{A}(r) e^{-i2\pi n r/f_A(T)} \mathrm{d}r\right) \end{pmatrix}.$$

Denoting $\tilde{x} = f_A(x)$, $\tilde{\omega}_n = n/f_A(T)$, $\Delta\tilde{\omega} = \tilde{\omega}_1 = 1/f_A(T)$, and

$$\begin{aligned} \tilde{F}(\tilde{x}) &= \frac{1}{f_A(T)} \sum_n e^{i2\pi n f_A(x)/f_A(T)} \int_{-f_A(T)/2}^{f_A(T)/2} \tilde{A}(r) e^{-i2\pi n r/f_A(T)} \mathrm{d}r \\ &= \Delta\tilde{\omega} \sum_n e^{i2\pi \tilde{\omega}_n \tilde{x}} \int_{-f_A(T)/2}^{f_A(T)/2} \tilde{A}(r) e^{-i2\pi \tilde{\omega}_n r} \mathrm{d}r \end{aligned} \tag{4.110}$$

we obtain the usual Fourier-series reconstruction formula for the function $\tilde{g}(\tilde{x}) = \tilde{F}(\tilde{x})$. Regarding the sum in (4.110) as the Riemann sum we can consider the limit $f_A(T) \to \infty$, arriving at the usual Fourier-transform reconstruction formulas,

$$\begin{aligned} \tilde{g}(s) &= \int_{-\infty}^{\infty} \mathrm{d}\omega e^{i2\pi\omega s} \int_{-\infty}^{\infty} \tilde{g}(r) e^{-i2\pi\omega r} \mathrm{d}r \\ &= \int_{-\infty}^{\infty} \hat{\tilde{g}}(\omega) e^{i2\pi\omega s} \mathrm{d}\omega, \\ \hat{\tilde{g}}(\omega) &= \int_{-\infty}^{\infty} \tilde{g}(r) e^{-i2\pi\omega r} \mathrm{d}r. \end{aligned}$$

We conclude that

$$\begin{aligned} g(x) &= \begin{pmatrix} f_{B_1}^{-1}\left(\Re \int_{-\infty}^{\infty} \hat{\tilde{g}}(\omega) e^{i2\pi\omega f_A(x)} \mathrm{d}\omega\right) \\ f_{B_2}^{-1}\left(\Im \int_{-\infty}^{\infty} \hat{\tilde{g}}(\omega) e^{i2\pi\omega f_A(x)} \mathrm{d}\omega\right) \end{pmatrix} = \begin{pmatrix} f_{B_1}^{-1}\left(\tilde{g}_1\left(f_A(x)\right)\right) \\ f_{B_2}\left(\tilde{g}_2\left(f_A(x)\right)\right) \end{pmatrix} \\ &= \begin{pmatrix} f_{B_1}^{-1}\left(\Re\tilde{g}\left(f_A(x)\right)\right) \\ f_{B_2}^{-1}\left(\Im\tilde{g}\left(f_A(x)\right)\right) \end{pmatrix}, \end{aligned}$$

where

$$\tilde{g}(r) = \tilde{g}_1(r) + i\tilde{g}_2(r) = \int_{-\infty}^{\infty} \hat{\tilde{g}}(\omega) e^{i2\pi\omega r} \mathrm{d}\omega.$$

Chapter 5

Non-Diophantine Arithmetics and Fractals

The greatest advances almost always come
not from new inventions per se, but
from novel combinations of already existing inventions
— making the familiar unfamiliar.
Eric Haseltine

In this chapter, we apply the formalism developed in Chapter 4 to two old problems of fractal analysis: Fourier transform on arbitrary Cantor sets (Jorgensen and Pedersen, 1998; Jorgensen, 2006; Aerts, Czachor and Kuna, 2016), and wave equations on fractal space–times (Golmankhaneh, Golmankhaneh and Baleanu, 2015; Czachor, 2019). The analysis on Cantor sets can be easily generalized to Sierpiński-type sets (Aerts, Czachor and Kuna, 2018). The construction of the Cantor and Sierpiński sets is somewhat different from what can be found in the typical literature of the subject. On the one hand, the sets we consider do not have to possess any kind of self-similarity. On the other hand, however, we need to construct an appropriate bijection $f_X : X \to C$ as required by the non-Newtonian formalism considered in the previous chapter. The bijections always exist since the cardinality of the Cantor, Sierpiński and Koch fractals equals that of the continuum. Depending on the way we define the fractals of interest, the bijections are more or less natural. A bijection can be explicitly constructed also for the so-called F^{α}-calculus (Parvate and Gangal, 2005, 2009, 2011; Parvate, Satin and Gangal, 2011; Satin and Gangal, 2016), but its arithmetic aspects still await a detailed study and thus, will not be discussed here.

5.1. Cantor Sets

What music is to the heart, mathematics is to the mind.

Amit Kalantri

The usual triadic middle-third Cantor set is constructed by the algorithm from Figure 5.1(a). In the first step, one removes the interior of the middle one-third of the segment $C_0 = [0, 1]$. The result is $C_1 = [0, 1/3] \bigcup [2/3, 1]$. In the second step, one performs the same operation on [0,1/3] and [2/3,1], arriving at $C_2 = [0, 1/9] \bigcup [2/9, 3/9] \bigcup [6/9, 7/9] \bigcup [8/9, 1]$, and so on, *ad infinitum*. The sets are embedded in one another: $C_0 \supset C_1 \supset C_2 \supset \cdots$. The Cantor set is the limit $C = \bigcap_{n=0}^{\infty} C_n$. The set is self-similar. The Lebesgue measure μ of C_n satisfies the following condition:

$$\mu(C_{n+1}) = (2/3)\mu(C_n) = (2/3)^{n+1}\mu(C_0).$$

It implies $\mu(C) = \lim_{n\to\infty}(2/3)^n \mu(C_0) = 0$. The set C is sometimes called the *Cantor dust* (Edgar, 2008).

The basic intuition behind self-similarity is that decreasing three times the size of C is equivalent to keeping the first half of C. In order to make the idea more formal assume that we perform the construction with $C_0 = [0, L]$. Let us denote the resulting Cantor set by $C(L)$. If L is the "size" of the dust, then $\mu_d(C(L)) = L^d$ can play the role of its "volume". Self-similarity implies

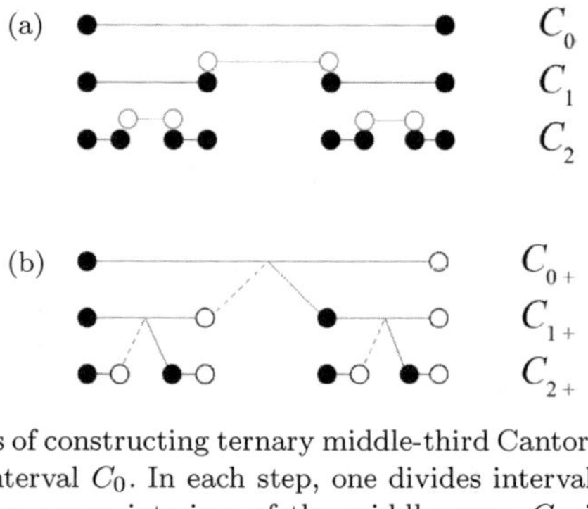

Figure 5.1. Two ways of constructing ternary middle-third Cantor-type fractals. (a) One starts with a closed interval C_0. In each step, one divides intervals into three segments of equal lengths, and removes interiors of the middle ones. C_n is the set obtained in the nth step. After n steps the number of endpoints equals 2^{n+1}. The Cantor set is $C = \bigcap_{n=0}^{\infty} C_n$. (b) One starts with a right-open interval. Each interval is cut in the middle and split, forming two right-open intervals, three times shorter than the split one. C_n is the set obtained in the nth step. After n steps the number of endpoints equals 2^{n+}. The Cantor set is $C_+ = \bigcap_{n=0}^{\infty} C_{n+}$. The difference $C \backslash C_+$ is countable. Both sets are self-similar, with the same similarity dimension $\log_3 2$. The second procedure defines a bijection $f\colon C_+ \to C_0$.

that $\mu_d(C(L/3)) = (1/2)\mu_d(C(L))$, thus $L^d/3^d = L^d/2$. The parameter $d = \log_3 2 \approx 0.63$ is called the similarity dimension of $C(L)$. C is an example of a set whose dimension is not given by a natural number, hence the name fractal coined by Mandelbrot. Volume of a 3-meter long Cantor dust equals 2 "Cantorian meters", i.e., $2m^{0.63}$.

One can perform an analogous construction for $C_{0+} = [0, L)$ (Figure 5.1(b)), but in each step removing a left-closed interval, so that $C_{1+} = [0, L/3) \bigcup [2L/3, L)$, etc. The resulting set C_+ is self-similar with the same similarity dimension $d = \log_3 2$. Alternatively, one can start with $C_{0-} = (0, L]$ and remove right-closed intervals, arriving at C_-. The three sets, C, C_+ and C_-, have the same similarity dimensions. Each of them can be called a ternary Cantor set. All three sets are uncountable. The bijections $g_\pm : C_\pm \to C_0$ where $C_\pm$ is equal either to C_+ or to C_- are simple and natural. They will be discussed in the context of arithmetic on triadic Cantor sets.

Figure 5.1(b) illustrates the essence of the one-to-one map $g_+ : C_+ \to C_0$. Assume $C_{0+} = [0, 1)$. The black endpoints in C_{2+} correspond to numbers x from $[0, 1)$ whose ternary (base 3) representations are: 0.0_3, 0.02_3, 0.2_3, 0.22_3. A ternary digit 1 does not occur. Since practically only two digits are employed, we could replace 2_3 by 1_2, and treat the resulting sequence of 0s and 1s as a binary representation. Therefore, the black endpoints from C_2 could represent the numbers 0.0_2, 0.01_2, 0.1_2, 0.11_2 in binary notation. Now, the subtlety is that the three non-zero numbers could be alternatively written in the binary form as $0.01_2 = 0.00(1)_2$, $0.1_2 = 0.0(1)_2$, $0.11_2 = 0.10(1)_2$. If we now employ the inverse recipe and replace all 1s by 2s, i.e., performing transformations $0.00(1)_2 \to 0.00(2)_3 = 0.01_3$, $0.0(1)_2 \to 0.0(2)_3 = 0.1_3$, $0.10(1)_2 \to 0.20(2)_3 = 0.21_3$, we obtain precisely the *white* endpoints, which do not belong to C_+. They do, however, belong to C. One concludes that the map g_+, based on $0_2 \leftrightarrow 0_3$, $1_2 \leftrightarrow 2_3$, is one-to-one in the case of C_+, but would be two-to-one in the case of C.

This is why it is much more convenient to work with C_+ (or C_-) than with the standard C. The bijection is a natural starting point for non-Newtonian calculus on Cantorian dusts. The situation does not change if one considers a completely random Cantor-type subset of R depicted in Figure 5.2.

Example 5.1 (Arithmetic on the ternary Cantor line). Let us start with the right-open interval $[0, 1) \subset R$, and let the (countable) set $Y_2 \subset [0, 1)$ consist of those numbers that have two different binary

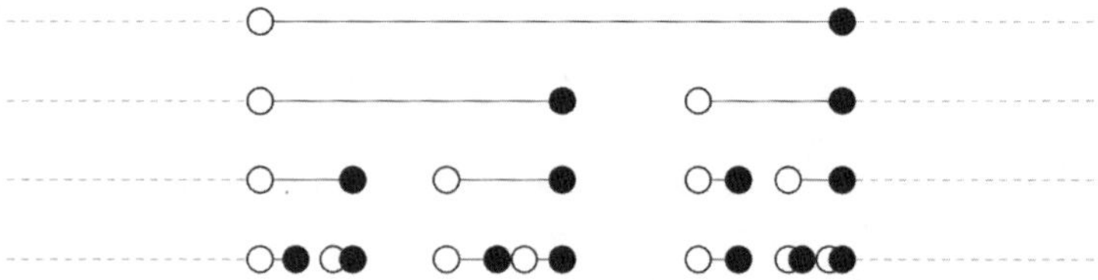

Figure 5.2. Decomposition of a line into a randomly constructed Cantor-type set. The set R is decomposed into a sum of disjoint left-open intervals of arbitrary lengths. Then each of the intervals becomes a C_0 of a certain Cantor type set obtained by splitting. The proportions of the resulting smaller left-open intervals are arbitrary. The set may not be self-similar, and its dimension may not be defined, even locally. In spite of this, there exists a bijection of the resulting fractal onto R.

representations. Denote by $0.t_1t_2\ldots$ a ternary representation of some $x \in [0,1)$. If $y \in Y_1 = [0,1)\backslash Y_2$, then y has a unique binary representation, say $y = 0.b_1b_2\ldots$. One then sets $g_\pm(y) = 0.t_1t_2\ldots$, $t_j = 2b_j$. The index $\pm$ appears for the following reason. Let $y = 0.b_1b_2\ldots = 0.b'_1b'_2\ldots$ be the two representations of the element $y \in Y_2$. There are two options, so we define $g_-(y) = \min\{0.t_1t_2\ldots, 0.t'_1t'_2\ldots\}$ and $g_+(y) = \max\{0.t_1t_2\ldots, 0.t'_1t'_2\ldots\}$, where $t_j = 2b_j$, $t'_j = 2b'_j$. We have therefore constructed two injective maps $g_\pm : [0,1) \to [0,1)$. Ternary Cantor-like sets are defined as the images $C_\pm(0,1) = g_\pm([0,1))$, and $f_\pm : C_\pm(0,1) \to [0,1)$, $f_\pm = g_\pm^{-1}$, is a bijection between $C_\pm(0,1)$ and the interval. For example, $1/2 \in Y_2$ since $1/2 = 0.1_2 = 0.0(1)_2$. We find

$$g_-(1/2) = \min\{0.2_3 = 2/3, 0.0(2)_3 = 1/3\} = 1/3,$$

$$g_+(1/2) = \max\{0.2_3 = 2/3, 0.0(2)_3 = 1/3\} = 2/3.$$

Accordingly, we have $1/3 \in C_-(0,1)$ while $2/3 \notin C_-(0,1)$, and vice versa, $1/3 \notin C_+(0,1)$, $2/3 \in C_+(0,1)$. The standard Cantor set is the sum $\hat{C} = C_-(0,1) \bigcup C_+(0,1)$. All irrational elements of $\hat{C}$ belong to $C_\pm(0,1)$ (an irrational number has a unique binary form), so sets $\hat{C}$ and $C_\pm(0,1)$ differ on a countable set. Note that $0 \in C_\pm(0,1)$, with $f_\pm(0) = 0$. In Czachor (2016) and Aerts *et al.* (2016a, 2016b), the set $C_-(0,1)$ is considered, so let us concentrate on this case. Let $C_-(k,k+1)$, $k \in Z$, be the copy of $C_-(0,1)$ but shifted by k. We construct a fractal $X = \bigcup_{k \in Z} C_-(k,k+1)$, and the bijection f: $X \to R$. Explicitly, if $x \in C_-(0,1)$, then $x + k \in C_-(k,k+1)$, and $f(x+k) = f(x) + k$ by definition. The set X is termed the Cantor line, and f is the Cantor-line function (Czachor, 2016; Aerts, Czachor and Kuna, 2016a, 2016b).

The set $X \cap [k, k+1)$ is self-similar, but X as a whole is not-self similar. Figure 5.3 (a) shows the plot of $g = f^{-1}$. Completely irregular generalizations of the Cantor line will be discussed in Chapter 6.

Let us make a remark that in the literature one typically considers Cantor sets $\hat{C}$ so that the resulting function g: $\hat{C} \to [0, 1)$ is non-invertible on a countable subset. In Semadeni (1982), one employs the map g to define the Haar basis on $\hat{C}$ "up to a countable set of points". In our formalism, we have to work with a bijective mapping g since we need its inverse.

Example 5.2 (Arithmetic on the quaternary Cantor line). Here we construct a Cantor set, which is analogous to the one employed in (Jorgensen and Pedersen, 1998). In a single step of the algorithm, it splits an interval into four identical segments and retains only the first and the third. Similar to the triadic set one needs to remove a countable subset of right or left endpoints of the sub-intervals in order to have a one-to-one map onto $[0, 1)$. Then, we extend the construction in a self-similar way to the whole space R. So, as opposed to the previous paragraph, we will not consider the sum of translated copies, but rather the sum of the rescaled copies.

Consider a number $y \in R^+$. Let Y_1 denote those y that have a unique binary representation $y = (b_m \dots b_1 b_0 . b_{-1} \dots b_{-k} \dots)_2$. We define

$$g_\pm(y) = (2b_m \dots 2b_1 2b_0 . 2b_{-1} \dots 2b_{-k} \dots)_4. \tag{5.1}$$

Here $g_\pm(y)$ is a number whose quaternary (i.e., base-four) representation contains only 0s and 2s. The ternary set had the same property, but in the ternary (base-3) representation.

Now, if $y \in Y_2 = R \backslash Y_1$, we have the uncertainty which of the two binary forms of y to take. Applying to the two forms the recipe (5.1) we get two numbers, say x and x'. Then $g_-(y) = \min\{x, x'\}$ and $g_+(y) = \max\{x, x'\}$. Finally, we extend the maps by anti-symmetry to negative y, i.e., $g_\pm(-y) = -g_\pm(y)$.

The images $X_\pm = g_\pm(R)$ define two quaternary Cantor sets, and $f_\pm = g_\pm^{-1}$ are the required bijections $f_\pm : X_\pm \to R$. Figure 5.3 (b) shows the plot of g_+. Note that

$$\begin{aligned} 1'_+ &= f_+^{-1}(1) = g_+(1.(0)_2) = g_+(0.(1)_2) \\ &= \max\{2.(0)_4 = 2, \quad 0.(2)_4 = 2/3\} = 2. \end{aligned}$$

(a)

(b)

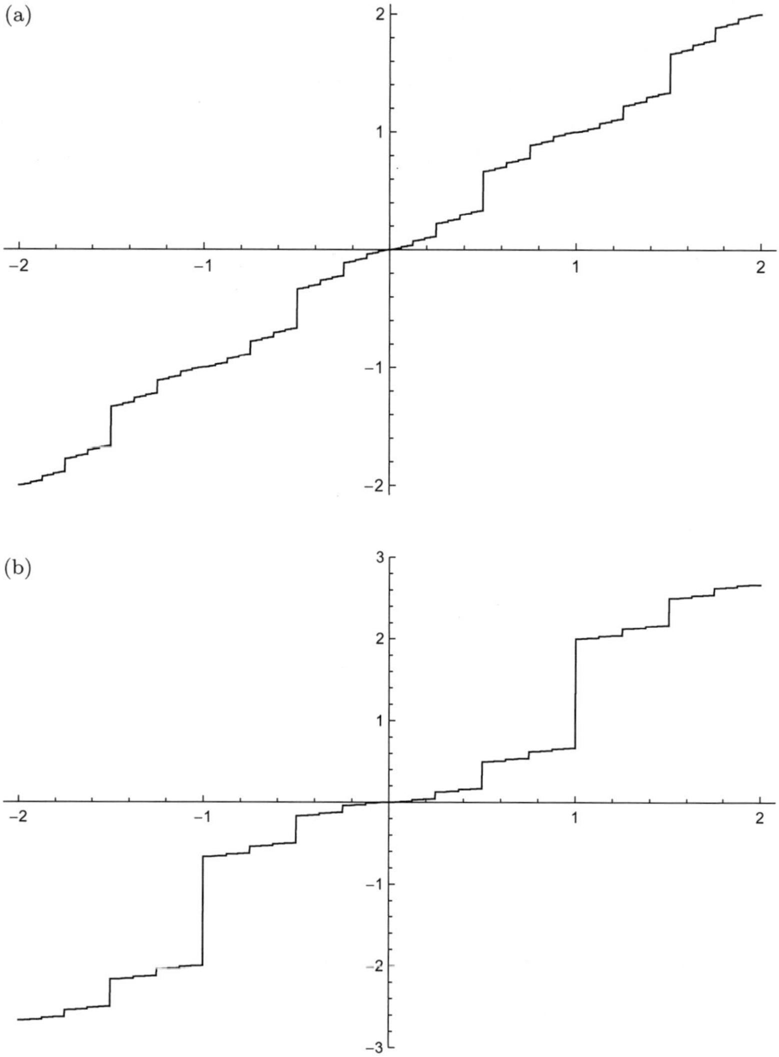

Figure 5.3. f^{-1} for the ternary Cantor line X (a), and f_+^{-1} for the quaternary Cantor set $X_+ = f_+^{-1}(R)$ (b). f and f_+ are used in construction of non-Diophantine arithmetic on both Cantor sets.

Analogously

$$\begin{aligned} 1'_- &= f_-^{-1}(1) = g_-(1.(0)_2) = g_-(0.(1)_2) \\ &= \min\{2.(0)_4 = 2, 0.(2)_4 = 2/3\} = 2/3. \end{aligned}$$

As we can see, both unit elements $1'_\pm$ differ from 1, so it is not clear which of the two bijections, and thus which of the two Cantor sets $X_\pm$, is more "natural". Calculations are simpler with X_+.

A particularly striking application of non-Newtonian calculus is the question of Fourier transform of functions with Cantor-dust domains. In the standard Newtonian framework, the problem of a Fourier analysis on the triadic Cantor sets is regarded as difficult (Jorgensen and Pedersen, 1997), and some results are known only for special types of Cantor sets, such as the quaternary one (C_1 is obtained by dividing C_0 into four equal parts and removing interiors of the second and the fourth segment). The non-Newtonian framework makes the problem trivial. Let us see how it works.

Example 5.3 (Fourier transform on a middle-third Cantor set). Consider the Cantor line defined by a periodic repetition of the ternary middle-third Cantor set, defined in Example 5.1. Now consider the sawtooth function $\tilde{A} : R \to R$ and its ternary Cantor-line analog, $A : X \to X$, $A = f_X^{-1} \circ \tilde{A} \circ f_X$, depicted in Figure 5.4. Let us perform the Fourier transform with $f_X(T) = 1$, i.e., $T = 1_X$. Figure 5.5 shows two finite-sum reconstructions of A,

$$A(x) = \sum_{n \le m}{}^{\oplus_X} (C_n(x) \otimes_X \langle C_n|A\rangle \oplus_X S_n(x) \otimes_X \langle S_n|A\rangle), \tag{5.2}$$

with $m = 5$ and $m = 30$ Fourier terms, respectively. The Gibbs phenomenon is clearly visible.

5.2. Double Cover of the Sierpiński Set

To a scholar, mathematics is music.

Amit Kalantri

Consider $x \in R^+$ and its ternary representation $x = (t_n \dots t_1 t_0 . t_{-1} t_{-2} \dots)_3$. If x has two different ternary representations, we choose the one that ends with infinitely many 2s. Keeping the digits unchanged let us change the base from 3 to 4, i.e.,

$$x = (t_n \dots t_1 t_0 . t_{-1} t_{-2} \dots)_3 \mapsto (t_n \dots t_1 t_0 . t_{-1} t_{-2} \dots)_4 = y.$$

The quaternary representation of y is unique, and it does not involve the digit 3. Next, let us parameterize the quaternary digits in a binary way, but

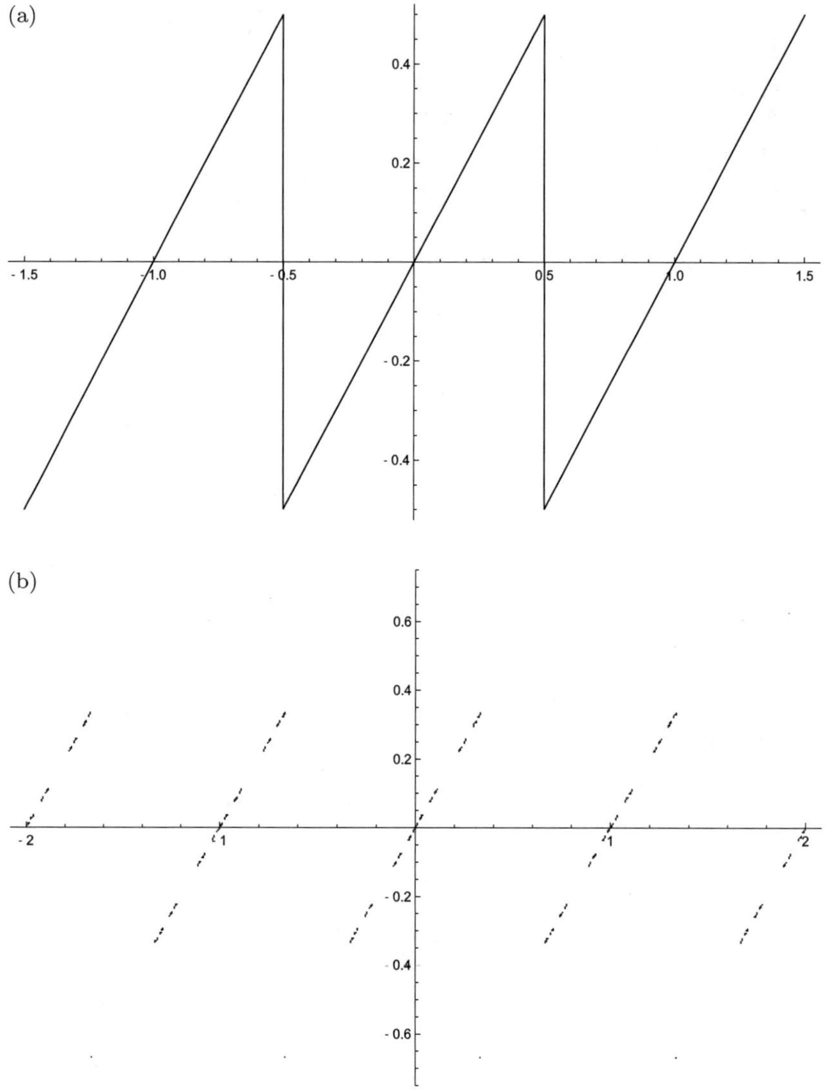

Figure 5.4. The sawtooth function $\tilde{A}$ (a), and its Cantorian analogue $A = f_X^{-1} \circ \tilde{A} \circ f_X$, where f_X is the ternary Cantor-line function (b).

written in a column form:

$$0 = \frac{0}{0}, \quad 1 = \frac{0}{1}, \quad 2 = \frac{1}{0}, \quad 3 = \frac{1}{1}.$$

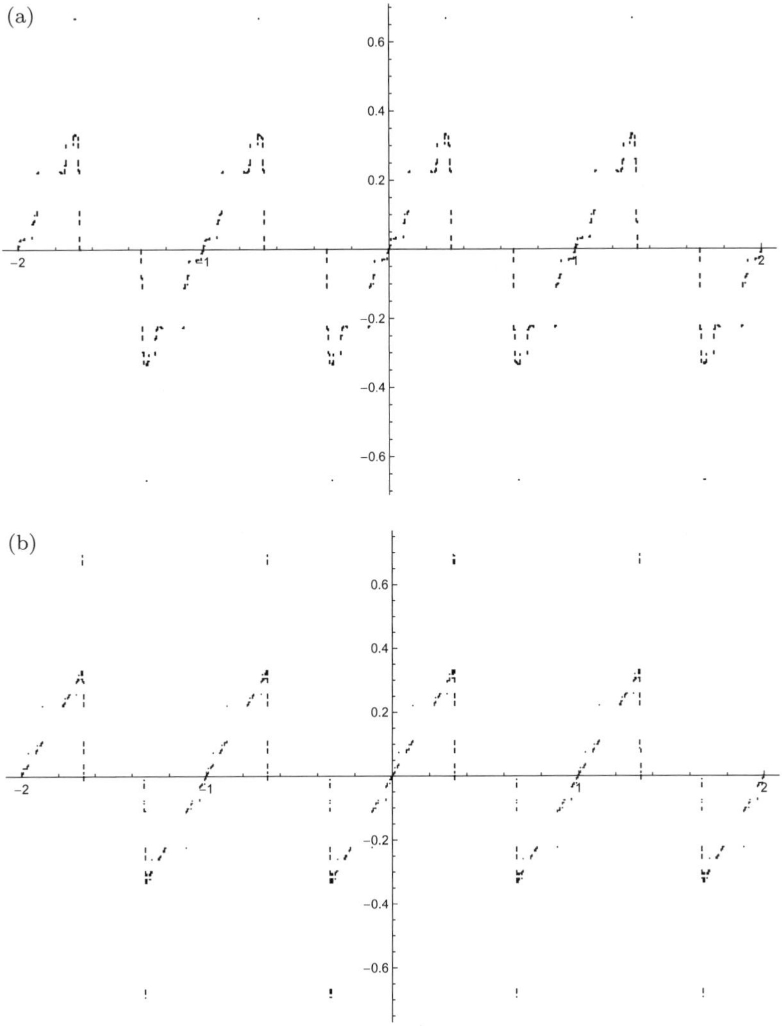

Figure 5.5. Finite-sum Fourier reconstructions of the sawtooth function, with 5 (a) and 30 terms (b). The Gibbs effect is enhanced by jumps of the Cantor-line function.

In effect, the element y has been converted into a pair of binary sequences,

$$(t_n \dots t_0.t_{-1}t_{-2}\dots)_4 \mapsto \begin{pmatrix} a_n \dots a_0.a_{-1}a_{-2}\dots \\ b_n \dots b_0.b_{-1}b_{-2}\dots \end{pmatrix}_2,$$

where $(a_j, b_j) \neq (1, 1)$ for any j. The resulting sequences are in a one-to-one relation with the x we have started with. Each of the two sequences defines a number in binary notation: we have mapped x into a point of the plane $R^+ \times R^+$. The image of R^+ under our algorithm defines a Sierpiński-type set. The algorithm is not invertible. Indeed, take the point $(1, 1)$. Depending on the way we represent it in a binary form, we find

$$\begin{pmatrix} 1.(0) \\ 0.(1) \end{pmatrix}_2 \mapsto (2.(1))_4 \mapsto (2.(1))_3 = 2.5,$$

and

$$\begin{pmatrix} 0.(1) \\ 1.(0) \end{pmatrix}_2 \mapsto (1.(2))_4 \mapsto (1.(2))_3 = 2.$$

The ambiguity comes from the two identifications:

$$(1, 1) = (1.(0)_2,\ 0.(1)_2) \quad \text{and} \quad (1, 1) = (0.(1)_2,\ 1.(0)_2).$$

However, if we write the above two relations as

$$(1, 1) = (1.(0)_2,\ 0.(1)_2) \quad \text{and} \quad (1, 1)_+ = (0.(1)_2,\ 1.(0)_2), \tag{5.3}$$

and treat the two points $(1, 1)_\pm$ as belonging to two different sides of an oriented plane, the ambiguity of the inverse algorithm disappears. The relation

$$(1, 1)_- \leftrightarrow 2.5 \quad \text{and} \quad (1, 1)_+ \leftrightarrow 2$$

is one-to-one.

Therefore, let us use index $+$ (respectively, $-$) for those pairs (a, b) where a is a rational number represented by a binary sequence involving $(1)_2$ (respectively, $(0)_2$), and b is a rational number whose binary representation contains $(0)_2$ (respectively, $(1)_2$). In both cases, we have $(a_j, b_j) \neq (1, 1)$ by construction. The corresponding rational numbers involve, respectively, $(2)_3$ and $(1)_3$.

But what about the other cases, such as a, b irrational, or a irrational but b rational? It turns out that the ambiguity is absent (Aerts *et al.*, 2018). In order to prove it, first of all note that we did not have to consider the cases

$$(1, 1) = (1.(0)_2, 1.(0)_2)$$

and

$$(1, 1) = (0.(1)_2, 0.(1)_2)$$

since the pairs $(a_j, b_j) = (1, 1)$ cannot appear as a result of the algorithm, and two infinite sequences of 0s would imply that $(t_n \dots t_1 t_0.t_{-1}t_{-2}.\dots)_4$ ends with an infinite sequence of 0s while this form is excluded by the algorithm.

The same mechanism eliminates all the remaining ambiguities:

(A) If a, b are both irrational, or a is irrational and b rational-periodic, their binary forms are unique.
(B) If a is irrational (or rational-periodic), but b rational non-periodic, then b cannot end with infinitely many 1s, as it would mean that a ends with infinitely many 0s. So these cases are again unique. Conclusions are unchanged if one interchanges a and b.
(C) The only ambiguity appears if a ends with infinitely many 0s, but b with infinitely many 1s (or the other way around). But this is the case we have started with.

In cases (A) and (B) we identify $(a, b)_+ = (a, b) = (a, b)$. Only the (countable) case (C) requires a two-sided plane $(a, b)_+ \neq (a, b)$. The case (C) occurs for those $x \in R$ whose ternary representation ends with $(2)_3$ or $(1)_3$. Only the latter numbers are mapped into (a, b).

As we can see, what we have constructed is a version of a double cover of the Sierpiński set.

Our algorithm defines an injective map g_+ of R^+ into a two-sided plane, with the above mentioned identifications. Let us extend g_+ to g by $g(|x|) = g_+(|x|)$, $g(-|x|) = -g_+(|x|)$. The image $S = g(R)$ is our definition of the Sierpiński set. Denoting $f = g^{-1}$, f: $S \to R$ we obtain non-Diophantine arithmetic intrinsic to S.

$$x \oplus y = f^{-1}(f(x) + f(y)),$$

$$x \ominus y = f^{-1}(f(x) - f(y)),$$

$$x \otimes y = f^{-1}(f(x)f(y)),$$

$$x \oslash y = f^{-1}(f(x)/f(y)).$$

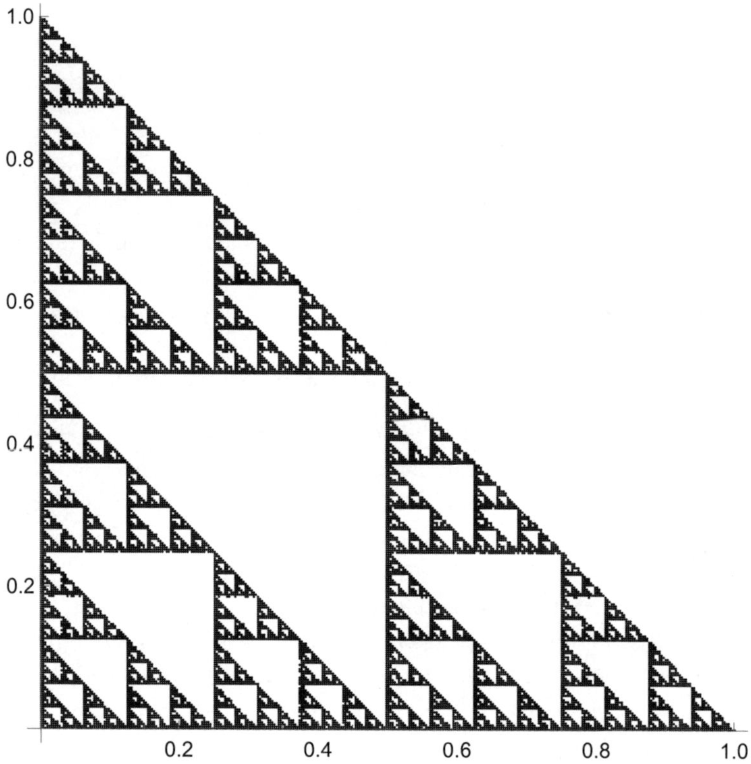

Figure 5.6. The inverse image $f^{-1}([0,1))$ view from the positive side of the oriented plane.

Figure 5.6 shows the set $f^{-1}([0,1))$. The set is self-similar and its similarity dimension is $\log_2 3$.

Example 5.4 (Arithmetic on the Sierpiński set). Let us begin with neutral elements of addition and multiplication in S. By definition, $0' \oplus x = x$, $1' \otimes x = x$, where

$$0' = f^{-1}(0) = (0,0) \in S,$$
$$1' = f^{-1}(1) = (1,0)_+ \in S.$$

A power function $A(x) = x \otimes \cdots \otimes x$ (n times) is denoted by $x^{n'}$, which is consistent with

$$x^{n\prime} \otimes x^{m\prime} = x^{(n+m)\prime} = x^{n' \oplus m'}.$$

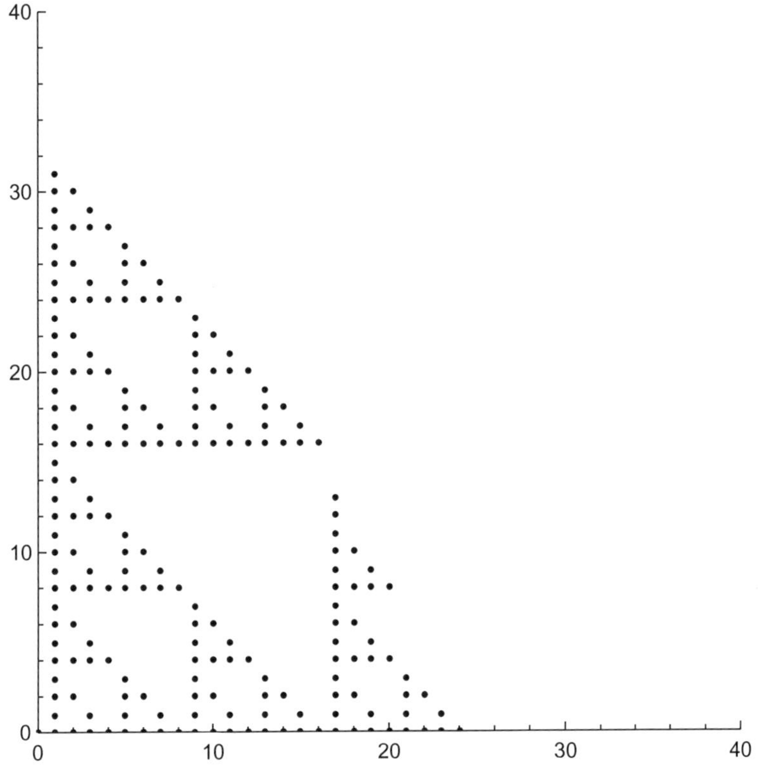

Figure 5.7. The image of the first 200 natural numbers, $f^{-1}(\{1, \ldots, 200\})$. All natural numbers are mapped into the positive side of the oriented plane.

Let $n' = f^{-1}(n)$. All Sierpińskian integers are represented by pairs of integers (Figure 5.7), a representation somewhat similar to complex numbers, but with different rules of addition and multiplication, as illustrated by

$$3' \oplus 4' = (2, 0)_+ \oplus (1, 2)_+ = 7' = (3, 0)_+.$$

Example 5.5 (Fourier transform on the double cover of the Sierpiński set). Consider the Sierpiński set constructed in Section 5.2. Let $X = S$ and $Y = R$. Consider the function $A : S \to R$ (Figure 5.8),

$$A(x) = \begin{cases} 1 & \text{for } x \in f_S^{-1}((0, 1)), \\ -1 & \text{for } x \in f_S^{-1}((-1, 0)), \\ 0 & \text{otherwise.} \end{cases} \tag{5.4}$$

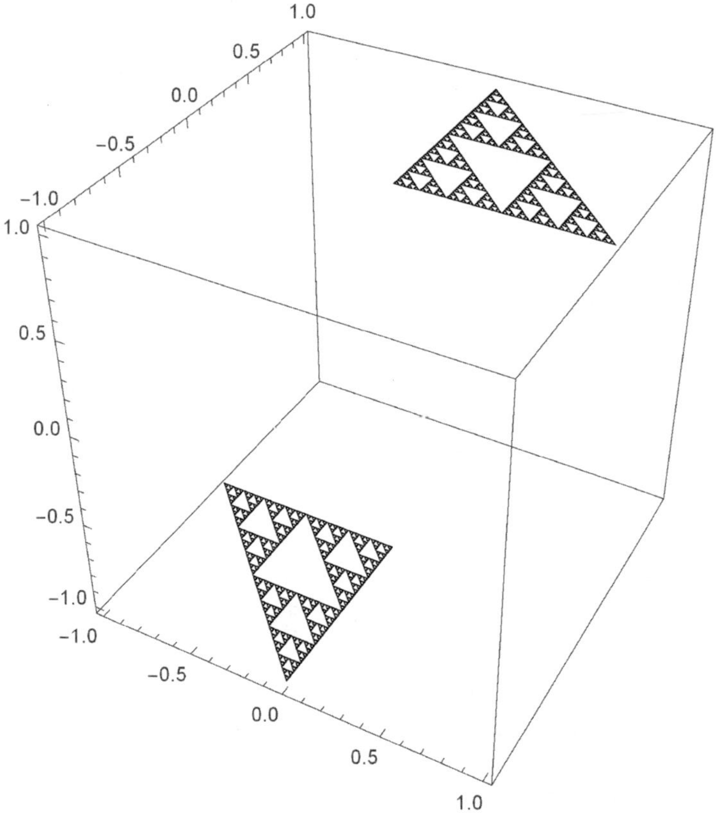

Figure 5.8. Function $A(x)$ defined in (5.4).

Since $f_S^{-1}((-1,0)) = S \cap (-1,0)^2$, $f_S^{-1}((0,1)) = S \cap (0,1)^2$, $S \cap \{(0,0)\} = f_S^{-1}(0)$, we introduce $\tilde{A} : R \to R$,

$$\tilde{A}(r) = \begin{cases} 1 & \text{for } r \in (0,1), \\ -1 & \text{for } r \in (-1,0), \\ 0 & \text{otherwise.} \end{cases} \tag{5.5}$$

Employing $f_X = f_S$, $f_Y = \mathrm{id}_R$, we get

$$\tilde{A} = f_Y \circ A \circ f_X^{-1} = A \circ f_S^{-1}. \tag{5.6}$$

In order to perform Fourier analysis of A, we have to introduce the basis of non-Newtonian sines and cosines. The scalar product of two functions

$\eta_j : S \to R$ and $\tilde{\eta}_j : R \to R$, $j = 1, 2$, is taken here as

$$\langle \eta_1 | \eta_2 \rangle = \int_{\ominus_Y T}^{T} \eta_1(x) \otimes_Y \eta_2(x) \mathrm{D}x \tag{5.7}$$

$$= f_Y^{-1} \left(\int_{-f_S(T)}^{f_S(T)} \tilde{\eta}_1(x) \tilde{\eta}_2(x) \mathrm{d}x \right) \tag{5.8}$$

$$= \int_{-1}^{1} \tilde{\eta}_1(x) \tilde{\eta}_2(x) \mathrm{d}x = \langle \tilde{\eta}_1 | \tilde{\eta}_2 \rangle, \tag{5.9}$$

where $\ominus_S T = 0_S \ominus_S T = f_S^{-1}(-1)$. In our case $T = 1_S = f_S^{-1}(1)$ (the neutral element of multiplication in S). Denoting

$$c_n(y) = \cos n\pi y, \quad n > 0, \tag{5.10}$$

$$s_n(y) = \sin n\pi y, \quad n > 0, \tag{5.11}$$

$$c_0(y) = 1/\sqrt{2}, \tag{5.12}$$

$$s_0(y) = 0, \tag{5.13}$$

$$C_n(x) = c_n(f_S(x)), \tag{5.14}$$

$$S_n(x) = s_n(f_S(x)), \tag{5.15}$$

we apply the resolution of unity,

$$\delta(x - y) = \sum_{n \geq 0} (c_n(x) c_n(y) + s_n(x) s_n(y)), \tag{5.16}$$

and finally obtain

$$A(x) = \sum_{n \geq 0} (C_n(x) \langle C_n | A \rangle + S_n(x) \langle S_n | A \rangle) \tag{5.17}$$

$$= \sum_{n \geq 0} (C_n(x) \langle c_n | \tilde{A} \rangle + S_n(x) \langle s_n | \tilde{A} \rangle) \tag{5.18}$$

$$= \sum_{n > 0} \frac{2(1 - (-1)^n)}{n\pi} S_n(x). \tag{5.19}$$

Figures 5.9 and 5.10 illustrate finite-sum Fourier reconstructions of the function plotted in Figure 5.8. The negative-side of S occurs only as the image of those elements $r \in R$ whose ternary representation ends with infinitely many 1s, whereas all finite-ternary-digit numbers are mapped into the positive side of S. This leads to the practical moral: Plotting functions

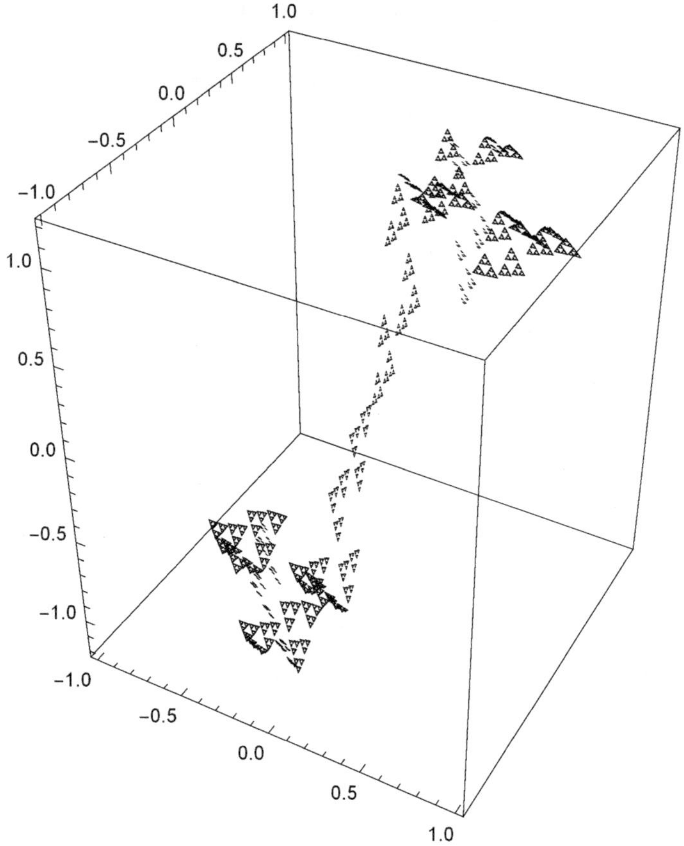

Figure 5.9. Finite-sum reconstruction of $A(x)$ with five Fourier terms in (5.19).

with domains in S, we can concentrate exclusively on the positive side of S, unless one employs a symbolic algorithm that recognizes numbers involving $(1)_3$.

5.3. Koch Curves

Everyone knows what a curve is, until he has studied enough mathematics to become confused through the countless number of possible exceptions.

Felix Klein

For convenience, we represent R^2 by C. Let us begin with the Koch curve $K_{[0,1]} \subset C$, beginning at 0 and ending at 1 (Figure 5.11).

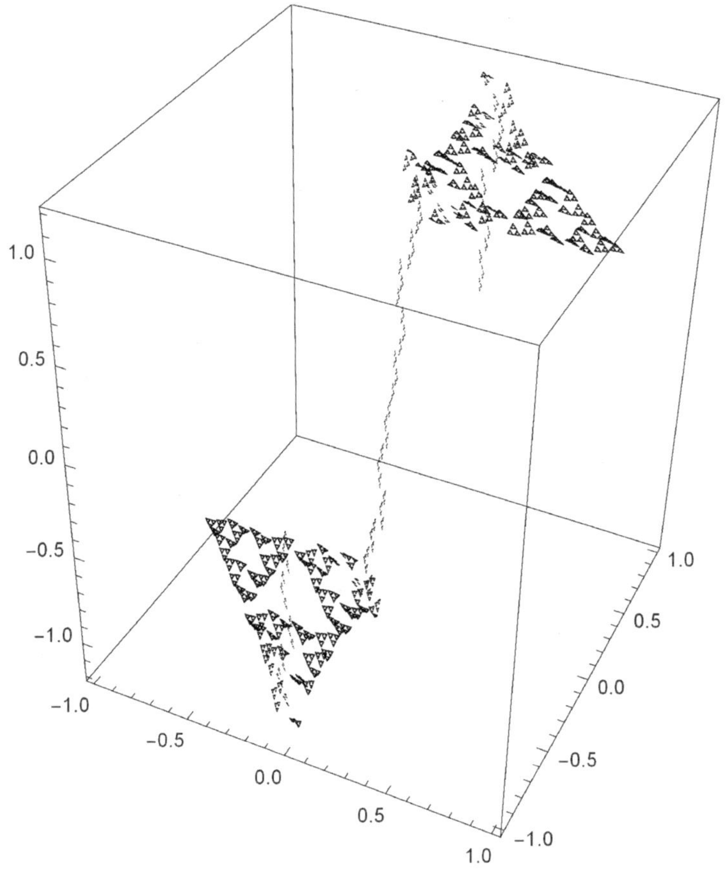

Figure 5.10. Finite-sum reconstruction of $A(x)$ with 50 terms. The Gibbs phenomenon is clearly visible.

A point $z \in K_{[0,1]}$ can be parameterized by a real number in quaternary representation,

$$y = (0.q_1, \ldots, q_j, \ldots)_4 \in [0, 1],$$

where $q_k = 0, 1, 2, 3$. The parameterization is defined by a bijection g: $[0, 1] \to K_{[0,1]}$, $z = g(y)$, constructed as follows. Consider $a = e^{i\alpha}$, $0 \leq \alpha \leq \pi/2$, $L = 1/(2 + 2\cos\alpha)$, and

$$\begin{aligned} \hat{0}(z) &= Lz, \\ \hat{1}(z) &= L(1 + az), \end{aligned} \tag{5.20}$$

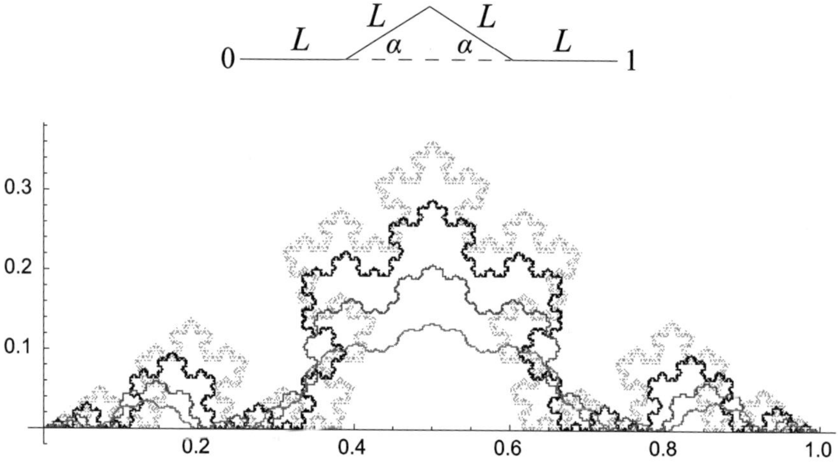

Figure 5.11. Koch curves and their generator parameterized by α and corresponding to (5.20)–(5.22), from top to bottom: $\alpha = \pi/2.5$, $\alpha = \pi/3$, $\alpha = \pi/4$, $\alpha = \pi/6$.

$$\begin{aligned} \hat{2}(z) &= L(1 + a + \bar{a}z), \\ \hat{3}(z) &= L(1 + 2\cos\alpha + z). \end{aligned} \tag{5.21}$$

An n-digit point $z \in K_{[0,1]}$ corresponding to $y = (0.q_1, \ldots, q_n)_4$, $q_n \neq 0$, is given by

$$\hat{q}_1 \circ \cdots \circ \hat{q}_n(0) = g(y) \tag{5.22}$$

(value at 0 of the composition of maps). If $y_n = (0.q_1, \ldots, q_n)_4$, is a Cauchy sequence convergent to $y = \lim_{n\to\infty} y_n$, then $g(y) = \lim_{n\to\infty} g(y_n)$. Curves from Figure 5.11 are the images $g([0,1])$ for various α. g is one-to-one, so it defines the inverse bijection $g^{-1} = f$: $K_{[0,1]} \to [0,1]$.

For $\alpha = \pi/3$, we obtain the standard curve, generated by equilateral triangles. Similarity dimension of a curve generated by (5.20)–(5.22) is given by Figure 5.12

$$D = \log 4/\log(2 + 2\cos\alpha). \tag{5.23}$$

There are many ways of extending the Koch curve from $K_{[0,1]}$ to K_R. For example, let $K_{[k,k+1]}$ be the curve $K_{[0,1]}$ shifted according to $z \mapsto z + k$, $k \in Z$. Then $K_R = \bigcup_{k\in Z} K_{[k,k+1]}$ is a periodic Koch curve, with the bijection f: $K_R \to R$ constructed from appropriately shifted maps g defined above. Non-periodic but self-similar extensions can be obtained by shifts

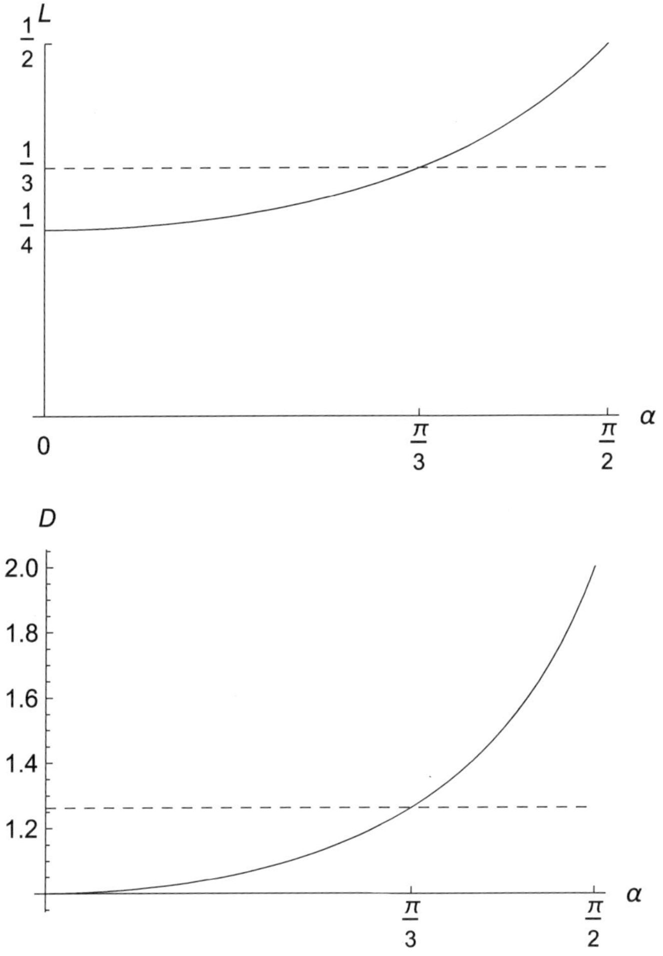

Figure 5.12. Similarity dimension D and the length L of the generator from Figure 5.11 as functions of α. The horizontal lines show the values for the standard $\pi/3$ Koch curve.

and rescalings. From our point of view the only condition we impose on f is the continuity of $g = f^{-1}$ at 0, i.e., $\lim_{y\to 0-} g(y) = \lim_{y\to 0+} g(y) = g(0)$. We take $g(0) = 0$.

Combining the generalized Koch curves we can construct a curve which is in a one-to-one relation with R, with explicitly given bijection f, and whose fractal dimensions vary from segment to segment in a prescribed way. This type of generalization may be useful for applications involving realistic coastlines, whose fractal dimensions coincide with the data described by the Richardson law (Mandelbrot, 1967).

Example 5.6 (Wave equation on a Koch curve (Czachor, 2019)). Assume we discuss a real-valued field whose evolution on the Koch curve $X = K_R$ is described with respect to the usual non-fractal time t. The field is thus represented by $R \times X \mapsto \Phi_t(x) \in R$, with $x \in X$. Since $Y = R$ we take $f_Y = \mathrm{id}_R$. The wave equation is

$$\frac{1}{c^2}\frac{\mathrm{d}^2}{\mathrm{d}t^2}\Phi_t(x) - \frac{\mathrm{D}^2}{\mathrm{D}x^2}\Phi_t(x) = 0, \tag{5.24}$$

where

$$\frac{\mathrm{d}}{\mathrm{d}t}\Phi_t(x) = \lim_{h\to 0}(\Phi_{t+h}(x) - \Phi_t(x))/h, \tag{5.25}$$

$$\frac{\mathrm{D}}{\mathrm{D}x}\Phi_t(x) = \lim_{h\to 0}(\Phi_t(x \oplus_X f_X^{-1}(h)) - \Phi_t(x))/h. \tag{5.26}$$

We search solutions in the form (here $y = ct$)

$$\Phi_t(x) = A(x, y) + B(x, y), \tag{5.27}$$

where

$$\left(\frac{\mathrm{d}}{\mathrm{d}y} - \frac{\mathrm{D}}{\mathrm{D}x}\right) A(x, y) = \left(\frac{\mathrm{d}}{\mathrm{d}y} + \frac{\mathrm{D}}{\mathrm{D}x}\right) B(x, y) \equiv 0, \tag{5.28}$$

suggesting simply

$$A(x, y) = a(f_X(x) + y), \tag{5.29}$$

$$B(x, y) = b(f_X(x) - y), \tag{5.30}$$

for some twice differentiable $a, b : R \to R$.

Indeed, from definitions

$$\begin{aligned}\frac{\mathrm{D}}{\mathrm{D}x}A(x, y) &= \lim_{h\to 0}\frac{A(x \oplus_X f_X^{-1}(h), y) - A(x, y)}{h}\\ &= \lim_{h\to 0}\frac{a(f_X(x) + h + y) - a(f_X(x) + y)}{h}\\ &\equiv \frac{\mathrm{d}}{\mathrm{d}y}a(f_X(x) + y) = \frac{\mathrm{d}}{\mathrm{d}y}A(x, y).\end{aligned} \tag{5.31}$$

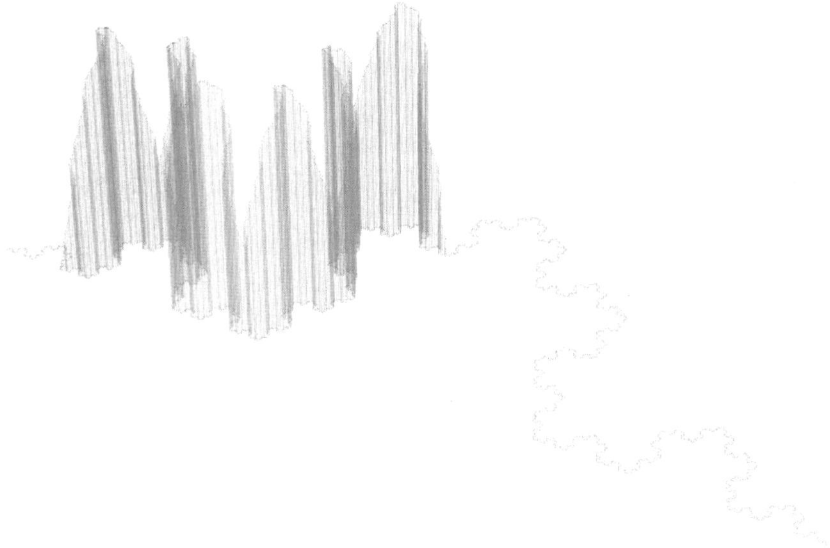

Figure 5.13. "Aurora borealis wave": Six snapshots of $\Phi_t(x)$ propagating to the right along the Koch curve. The inset shows the corresponding function b occurring in (5.30).

One similarly verifies that $\mathrm{d}/\mathrm{d}y$ and $\mathrm{D}/\mathrm{D}x$ commute, and

$$\frac{\mathrm{D}}{\mathrm{D}x}B(x,y) \equiv -\frac{\mathrm{d}}{\mathrm{d}y}B(x,y). \tag{5.32}$$

Figure 5.13 shows the dynamics of $\Phi_t(x)$ with $a = 0$.

The energy of the wave is given by

$$E = \frac{1}{2}\int_{f_X^{-1}(-\infty)}^{f_X^{-1}(\infty)} \left(\frac{1}{c^2}\left|\frac{\mathrm{d}\Phi_t(x)}{\mathrm{d}t}\right|^2 + \left|\frac{\mathrm{D}\Phi_t(x)}{\mathrm{D}x}\right|^2\right)\mathrm{D}x, \tag{5.33}$$

where the integral is defined by (4.2).

Let us explicitly check the time independence of E for the particular case of $\Phi_t(x) = a(f_X(x) + ct)$. Let $a'(x) = \mathrm{d}a(x)/\mathrm{d}x$. Then,

$$E = \int_{-\infty}^{\infty} |a'(f_X \circ f_X^{-1}(x) + ct)|^2 \mathrm{d}x \tag{5.34}$$

$$= \int_{-\infty}^{\infty} |a'(x)|^2 \mathrm{d}x \tag{5.35}$$

is independent of time, as it should be.

5.4. Generalization to Non-self-similar Fractals

A new idea is delicate.
It can be killed by a sneer or a yawn;
it can be stabbed to death by a joke, or
worried to death by a frown on the right person's brow.

Charles Brower

Although Cantor-type sets are homeomorphic to the idealized fully symmetric triadic Cantor set, it is clear that fractal-like sets one encounters in real life are highly non-symmetric and non-regular. Their effective dimensions vary with resolution and are position dependent. The mathematical notion that seems close to natural fractals is associated with the concept of a multifractal. However, in order to apply the idea of fractal arithmetic to a multifractal, one needs a bijection f, and it is by no means evident that such an f always exists.

So, we propose to reverse the problem. Namely, can we describe a class of fractals that, on the one hand, have the irregularities typical of multifractals, but on the other hand are equipped with the necessary bijection f?

5.4.1. *Multi-resolution representation of real numbers*

Unity, not uniformity, must be our aim. We attain unity only through variety.
Differences must be integrated, not annihilated, not absorbed.

Mary Parker Follett

To begin with, let us make the trivial remark that geometry of physical space–time involves objects that have "dimension of length" (x or $x_0 = ct$ are expressed in meters, inches, parsecs, Planck lengths, ...). In pure mathematics, the element $1 \in R$ is just the neutral element of multiplication in the real "line" and, obviously, does not have a "physical unit". We are interested in physical-space fractals constructed by means of a map f satisfying $f(1_X) = 1$, where the 1s are understood as neutral elements of multiplication. We will treat the physical space as an object which is dimensionless, and this can be obtained only for the price of introducing a fundamental unit of length, λ say.

With this observation in mind let us split a one-dimensional physical "position-space line" X into a countable union of disjoint intervals of length ℓ. In order to model it mathematically, we identify $X/\ell = R = \bigcup_{j \in Z}[j, j+1)$.

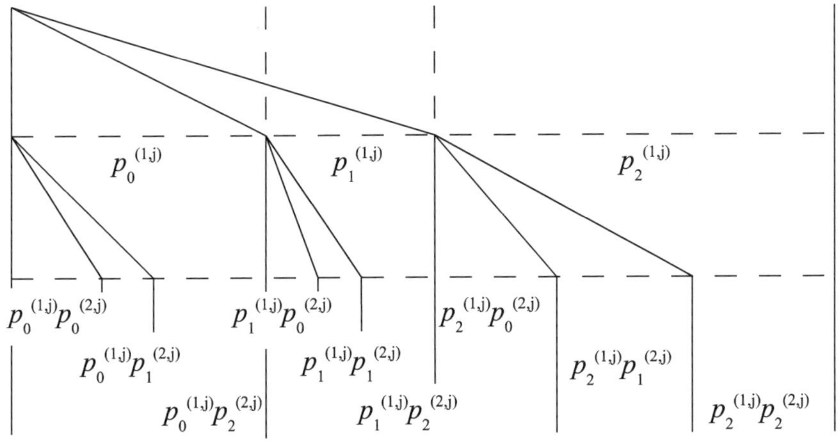

Figure 5.14. A jth interval and its splitting into intervals whose proportions differ from resolution to resolution (indexed by n in (n,j)).

The diagram shown in Figure 5.14 shows a jth interval $[j, j+1)$. The interval is split into three (right-open) segments of length $p_k^{(1,j)}$, $k = 0, 1, 2$. Each of the three segments is yet further split into three right-open intervals whose mutual proportions are determined by $p_k^{(2,j)}$, $k = 0, 1, 2$, and so on. For each $(n,j) \in N \times Z$ we assume $\sum_{k=0}^{2} p_k^{(n,j)} = 1$, $0 \leq p_k^{(n,j)} \leq 1$. Denoting

$$r_0^{(n,j)} = 0, \; r_1^{(n,j)} = p_0^{(n,j)}, \; r_2^{(n,j)} = p_0^{(n,j)} + p_1^{(n,j)}, \text{ and}$$
$$s^j_{a1\ldots.an} = p_{a1}^{(1,j)} \ldots p_{a(n-1)}^{(n-1,j)} r_{an}^{(n,j)},$$

one can associate each node of the diagram with the real number

$$x = x_{j,a_1\ldots a_n} = j + s^j_{a_1} + \cdots + s^j_{a_1\ldots a_n} \in [j, j+1), \tag{5.36}$$

where $a_k = 0, 1, 2$, $j \in Z$. There are two extreme cases of (5.36). On the one hand, if all $p_k^{(n,j)} = 1/3$, then we obtain a representation where the non-integer part $x - \lfloor x \rfloor$ has the standard ternary form. On the other hand, we have the case where the proportions $p_k^{(n,j)}$ are completely unrelated to one another for different choices of (n,j). The intermediate case where $\lim_{n\to\infty} p_k^{(n,j)} = p_k$ exists and is independent of j is, as we will see later, of particular interest.

Theorem 5.1. *There exist numbers that have exactly two different representations* (5.36),

$$x_{j.a_1\ldots a_n 1} = x_{j.a_1\ldots a_n 1(0)} = x_{j.a_1\ldots a_n 0(2)}. \tag{5.37}$$

The set of such numbers is countable. If $x = j \in Z$, *then* $x = x_{j.(0)} = x_{j-1.(2)} \in [j, j+1)$.

Proof: As usual, the symbol "(2)" denotes the infinite sequence of 2s. Let us first show that

$$r_2^{(n,j)} + p_2^{(n,j)} r_2^{(n+1,j)} + p_2^{(n,j)} p_2^{(n+1,j)} r_2^{(n+2,j)} + \cdots = 1. \tag{5.38}$$

Figure 5.14 shows that

$$r_2^{(1,j)} + p_2^{(1,j)} r_2^{(2,j)} + p_2^{(1,j)} p_2^{(2,j)} r_2^{(3,j)} + \cdots = 1$$

by definition. So

$$\begin{aligned}
1 &= r_2^{(1,j)} + p_2^{(1,j)} r_2^{(2,j)} + p_2^{(1,j)} p_2^{(2,j)} r_2^{(3,j)} + \cdots \\
&= 1 - p_2^{(1,j)} + p_2^{(1,j)} r_2^{(2,j)} + p_2^{(1,j)} p_2^{(2,j)} r_2^{(3,j)} + \cdots \\
&= 1 + p_2^{(1,j)}(-1 + r_2^{(2,j)} + p_2^{(2,j)} r_2^{(3,j)} + \cdots) \\
&= 1 + p_2^{(1,j)} p_2^{(2,j)}(-1 + r_2^{(3,j)} + p_2^{(3,j)} r_2^{(4,j)} + \cdots) \\
&= 1 + \prod_{l=1}^{n-1} p_2^{(l,j)}(-1 + r_2^{(n,j)} + p_2^{(n,j)} r_2^{(n,j)} + \cdots),
\end{aligned}$$

which implies (5.38) for any n. Now consider

$$\begin{aligned}
x &= x_{j.a_1\ldots a_n 0(2)} \\
&= j + s_{a_1}^j + \cdots + s_{a_1\ldots a_n}^j + s_{a_1\ldots a_n 0(2)}^j \\
&= j + s_{a_1}^j + \cdots + s_{a_1\ldots a_n}^j \\
&\quad + p_{a_1}^{(1,j)} \cdots p_{a_n}^{(n,j)} p_0^{(n+1,j)} r_2^{(n+2,j)} \\
&\quad + p_{a_1}^{(1,j)} \cdots p_{a_n}^{(n,j)} p_0^{(n+1,j)} p_2^{(n+2,j)} r_2^{(n+3,j)} + \cdots \\
&= j + s_{a_1}^j + \cdots + s_{a_1\ldots a_n}^j \\
&\quad + p_{a_1}^{(1,j)} \cdots p_{a_n}^{(n,j)} p_0^{(n+1,j)} \\
&\quad \times (r_2^{(n+2,j)} + p_2^{(n+2,j)} r_2^{(n+3,j)} + \cdots)
\end{aligned}$$

Employing (5.38) and $r_1^{(n+1,j)} = p_0^{(n+1,j)}$, we get

$$\begin{aligned}
x &= x_{j.a_1\dots a_n 0(2)} \\
&= j + s^j_{a_1} + \cdots + s^j_{a_1\dots a_n} \\
&\quad + p^{(1,j)}_{a_1} \cdots p^{(n,j)}_{a_n} r_1^{(n+1,j)} \\
&= j + s^j_{a_1} + \cdots + s^j_{a_1\dots a_n} + s^j_{a_1\dots a_n 1} \\
&= x_{j.a_1\dots a_n 1}.
\end{aligned}$$

Theorem is proved. □

5.4.2. *Multi-Resolution Cantor Line*

The Greeks ... adopted a geometrical procedure wherever it was possible, and they even treated arithmetic as a branch of geometry by means of the device of representing numbers by lines.

W. W. R. Ball

Here we generalize the construction of the Cantor line given in Example 5.1 (for $p_k^{(n,j)} = 1/3$). Our goal is to have a fractal $C \subset R$ and a bijection $f: C \to R$ which will be used in definition of fractal arithmetic, an essential ingredient of fractal derivatives and integrals. Fractal C so constructed is, to some extent, reminiscent of a multifractal but, as opposed to standard multifractals, is equipped with the natural bijection f. This is why we speak of multi-resolution fractals, distinguishing them from multifractals where arithmetic operations and derivatives are difficult to introduce.

To begin with, consider a real number $y = j + \varepsilon \in [j, j+1)$. The number $\varepsilon \in [0,1)$ has at most two different binary representations,

$$\varepsilon = (0.b_1 \dots b_n \dots)_2 = (0.b'_1 \dots b'_n \dots)_2, \tag{5.39}$$

which defines two sequences of bits. The two sequences allow us to define two numbers of the form (5.36):

$$x = x_{j.2b_1\dots 2b_n\dots} \quad \text{and} \quad x' = x_{j.2b'_1\dots 2b'_n\dots}, \tag{5.40}$$

both belonging to $[j, j+1)$. Note that out of the three possible digits $a_k = 0, 1, 2$, occurring in (5.36), formula (5.40) involves only two of them: $2b_k = 0, 2$, $2b'_k = 0, 2$. The absence of ternary 1 is typical of Cantor-like sets.

The injective map $g: R \to R$ is defined by $g(y) = \min\{x, x'\}$. The image $C = g(R)$ will be termed the multi-resolution Cantor line. The inverse map

$f: C \to R$, $f = g^{-1}$, defines the bijection we need in order to construct arithmetic in C. Let us check that $f(0) = 0$, $f(1) = 1$. Zero occurring in the argument of $f(0)$ corresponds to $x_{0.(0)} = x_{-1.(2)} \in C$. By definition $f(x_{0.(0)}) = 0 + 0.(0)_2 = 0$, $f(x_{-1.(2)}) = -1 + 0.(1)_2 = 0$. 1 in the argument of $f(1)$ corresponds to $x_{1.(0)} = x_{0.(2)} \in C$. Again, by definition $f(x_{1.(0)}) = 1 + 0.(0)_2 = 1$, $f(x_{0.(2)}) = 0 + 0.(1)_2 = 1$.

One similarly shows that $f(j) = f(x_{j.(0)}) = j + 0.(0)_2 = f(x_{j-1.(2)}) = j - 1 + 0.(1)_2 = j \in Z$. Thus, for integer x one finds $f(x) = x = f^{-1}(x)$. In particular, $f(-1) = -1$. This is a peculiarity of this concrete f, and further generalizations are possible.

5.4.3. *Relation to Multifractals*

Mathematics is the queen of the sciences and arithmetic [number theory] is the queen of mathematics. She often condescends to render service to astronomy and other natural sciences, but in all relations, she is entitled to first rank.

Carl Friedrich Gauss

Let us put what we do in the context of multifractals, concentrating only on multifractals of a Cantor type. Assume, first of all, that $p_k^{(n,j)} = p_k$ for any (n, j), but not all p_k are equal. At resolution n one deals with segments of length $p_0^{n-m} p_2^m$, $m = 0, \ldots, n$, and each interval $[j, j+1)$ contains $n!/[m!(n-m)!]$ segments of an mth type. The overall length of all the segments of the mth type is $p_0^{n-m} p_2^m n!/[m!(n-m)!]$ and the sum over all m is $(1 - p_1)^n$. So, if $p_1 > 0$ then $\lim_{n\to\infty}(1 - p_1)^n = 0$. Removing in each step a non-zero proportion p_1 of $[j, j + 1)$ we get in the limit a set of Lebesgue measure zero.

The Hausdorff dimension D is defined by

$$\sum_{m=0}^{n} \frac{n!}{m!(n-m)!} (p_0^{n-m} p_2^m)^D = (p_0^D + p_2^D)^n = 1. \tag{5.41}$$

Hence, $p_0^D + p_2^D = 1$ and D coincides with $\lim_{n\to\infty} D^{(n,j)}$ discussed in the next section.

In order to introduce the multifractal formalism (Hentschel and Procaccia, 1983; Halsey *et al.*, 1986) one additionally assumes that there exists some random process with probabilities P_0, P_2, $P_0 + P_2 = 1$ such that the algorithm of generating the fractal may be regarded as a kind of random walk. One introduces a parameter q and a function $\tau(q)$, and demands that

$$P_0^q p_0^{-\tau(q)} + P_2^q p_2^{-\tau(q)} = 1. \tag{5.42}$$

For $q = 0$ and $\tau(0) = -D$ one finds that $-\tau(0)$ is the Hausdorff dimension. The so-called generalized dimensions are defined by $D(q) = \tau(q)/(q-1)$.

5.4.4. *Dimensions of C*

I am further inclined to think, that when our views are sufficiently extended, to enable us to reason with precision concerning the proportions of elementary atoms, we shall find the arithmetical relation alone will not be sufficient to explain their mutual action, and that we shall be obliged to acquire a geometric conception of their relative arrangement in all three dimensions of solid extension.

William Hyde Wollaston

With each node $x_{j.a1...an}$ from Figure 5.14 one can associate the length $l_{j.a1...an}$ of the interval extending to the right till its nearest sibling,

$$l_{j,a_1...a_n} = p_{a_1}^{(1,j)} \dots p_{a_n}^{(n,j)}, \tag{5.43}$$

satisfying

$$\sum_{a_1=0}^{2} \cdots \sum_{a_n=0}^{2} l_{j.a_1...a_n} = \prod_{k=1}^{n} \sum_{a=0}^{2} p_a^{(k,j)} = 1. \tag{5.44}$$

In Cantor-like sets the indices $a = 1$ would be missing in sums (5.44), but one can find numbers $D^{(k,j)}$ such that

$$\prod_{k=1}^{n} \sum_{a \neq 1} (p_a^{(k,j)})^{D^{(k,j)}} = 1. \tag{5.45}$$

Putting $n = 1$ in (5.45) we get

$$\sum_{a \neq 1} (p_a^{(1,j)})^{D^{(1,j)}} = 1, \tag{5.46}$$

which implies by induction that (5.45) is equivalent to

$$(p_0^{(k,j)})^{D^{(k,j)}} + (p_2^{(k,j)})^{D^{(k,j)}} = 1, \tag{5.47}$$

which has a unique solution $D^{(k,j)}$ for any (k, j) (the proof is standard; cf. the analysis of similarity dimension in Edgar (2008)).

Alternatively, one can consider $D_{(n,j)}$ defined by

$$1 = \sum_{a_1 \neq 1} \cdots \sum_{a_n \neq 1} (l_{j.a_1 \ldots a_n})^{D_{(n,j)}}. \tag{5.48}$$

Equation (5.48) possesses a unique solution $D_{(n,j)}$ which, however, in general differs from $D^{(n,j)}$. The limiting case $\lim_{n\to\infty} D_{(n,j)}$ equals the Hausdorff dimension of the jth interval.

Dimensions $D^{(n,j)}$ and $D_{(n,j)}$ are the two effective similarity dimensions that can be associated with resolution n in the jth segment of C. Note that $D^{(n,j)} = 1 = D_{(n,j)}$ if and only if $p_1^{(n,j)} = 0$, independently of the choice of $p_0^{(n,j)}$ and $p_2^{(n,j)}$. In infinite resolution the dimension $D^{(n,j)}$ is well defined if $\lim_{n\to\infty} p_k^{(n,j)}$ exists. If the limit does not exist then $D^{(n,j)}$ fluctuates at large resolutions.

5.4.5. *Irregularities of C Violate Parity Invariance at Large Resolutions*

Progress is man's ability to complicate simplicity.

Thor Heyerdahl

The readers may have noticed that for $\ominus x \neq -x$, i.e., $-f(x) \neq f(-x)$, one implicitly violates parity invariance, a property that leads to a reasonable estimate of $\ell < 10^{-18}$ m, which is the electroweak range. Plots such as those from Figures 6.2 and 6.5 show that an antisymmetric f implies an unphysical-looking symmetry around $x^1 = 0$ of space–time fractals.

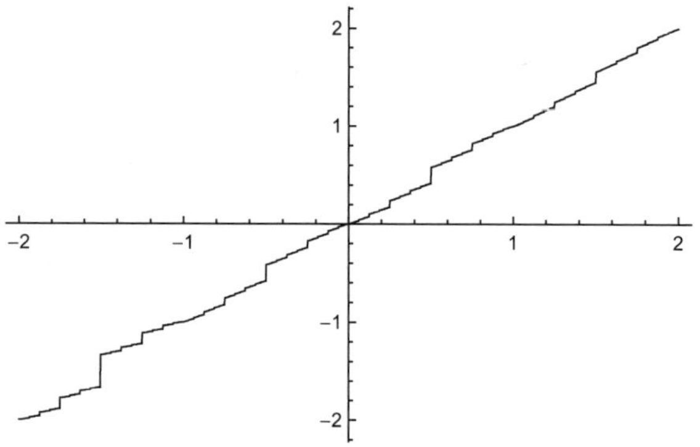

Figure 5.15. Future cone generated by fractal-arithmetic proper-time SO(1, 1) homogeneous spaces. From top to bottom: f_C^{-1}, and the corresponding 1, 2, and 20 hyperbolas. Note the lack of exact reflection symmetry, typical of multi-resolution Cantor sets.

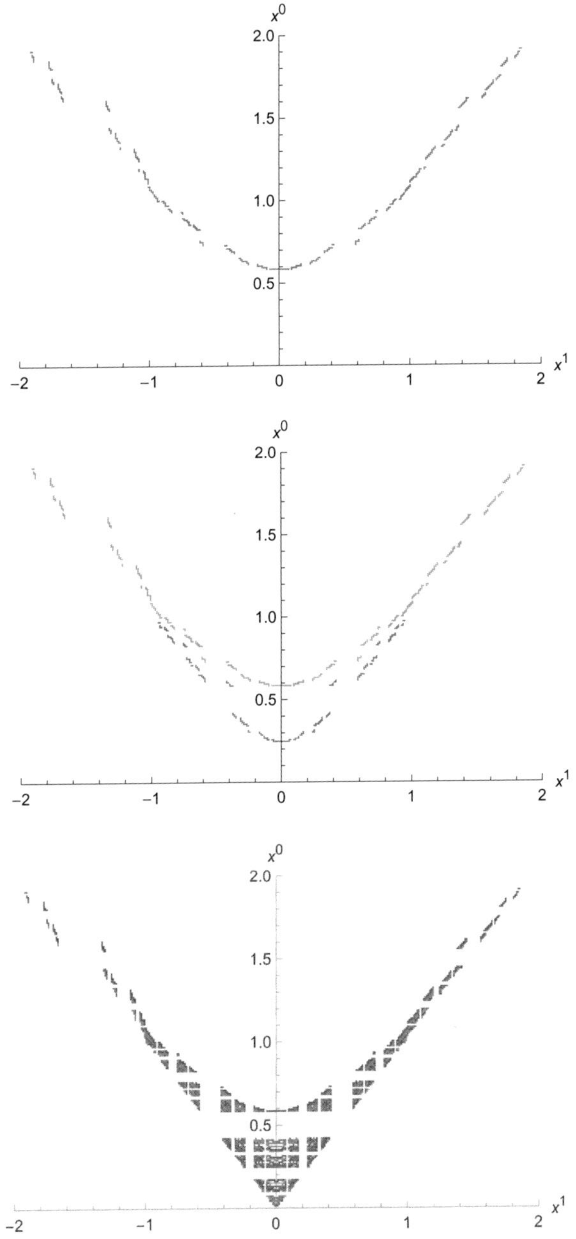

Figure 5.15. *(Continued)*

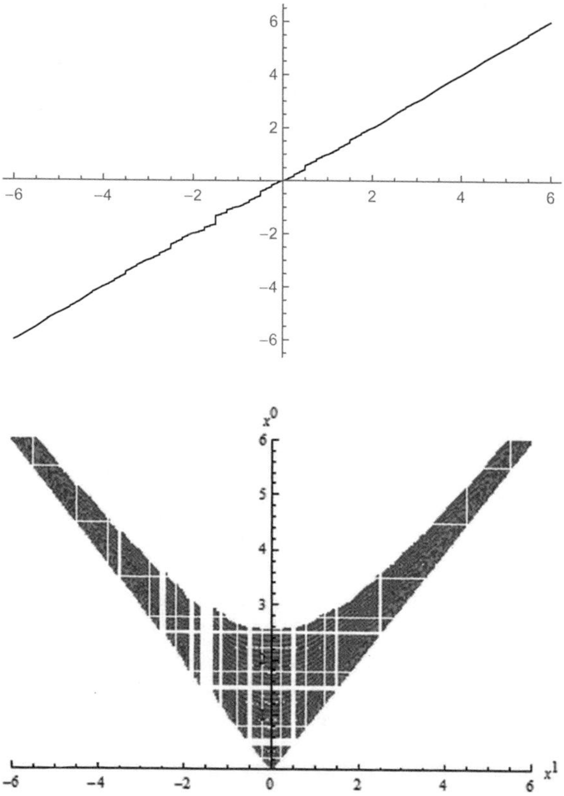

Figure 5.16. The same homogeneous spaces as shown in Figure 5.15 but seen from a larger perspective: 100 proper-time hyperboles. Parity non-invariance is less pronounced at larger length-scales.

According to the Copernican principle no preferred x^1 should be *a priori* assumed. This can be achieved either by translation invariance of space, which is excluded if a fractal structure is present, or by a complete irregularity of f. This is the main reason why the notion of multi-resolution Cantor line is introduced. Figures 5.15 and 5.16 show an example of f constructed by means of a slightly less trivial $p_k^{(n,j)}$. Here, we have chosen

$$p_0^{(n,j)} = p_2^{(n,j)} = \frac{1}{2}\left(1 - \frac{1}{3(|j+1|+1)}\right), \tag{5.49}$$

$$p_1^{(n,j)} = \frac{1}{3(|j+1|+1)}. \tag{5.50}$$

Independence of n makes the $n \to \infty$ limit trivial but the effective dimension is j dependent. Solving

$$(p_0^{(n,j)})^{D^{(n,j)}} + (p_2^{(n,j)})^{D^{(n,j)}} = 1$$

for $D^{(n,j)}$ we find that the minimal dimension $\log_3 2$ is for $j = -1$, and with $j \to \pm\infty$ the dimensions tend to 1. Figure 5.16 shows the same plot as in Figure 5.15, but from a wider perspective, illustrating the effective disappearance of irregularities at distances much larger than λ.

Chapter 6

Non-Diophantine Arithmetics in Physics

Science is the knowledge of the consequences of names;
the dependence of one fact upon another;
the ability to make things happen.
T. Hobbes

Fractal space–times, Rényi entropies, and the problem of dark energy are naturally related to non-Diophantine arithmetics and non-Newtonian calculi. This chapter gives an outline of results and several possible research projects in this area.

6.1. Dimensional vs. Dimensionless Variables

Man's mind stretched to a new idea never goes back to its original dimension.
Oliver Wendell Holmes

Conceptual difficulties and subtleties related to the fundamental length ℓ are well known and have been discussed in the literature for more than a century (Amelino-Camelia, 2002; Magueijo, 2003; Duff, 2004). To analyze this problem, we consider here the use of dimensional quantities, such as length, in non-Diophantine arithmetics in general and in fractal arithmetics in particular. Here the term *dimension* is understood in the context of systems of physical units, and is not related to Hausdorff dimensions or similar topological concepts. First of all, there exist non-Diophantine arithmetics defined by functions from the class that do not lead to any difficulties with dimensional quantities. Namely, we can take functions of the form $f(x) = x^q$. For such a function f used as the functional parameter of a non-Diophantine arithmetic, one defines operations as $x \otimes y = xy$ and $x \oplus y = (x^q + y^q)^{1/q}$. It means that multiplication and division also remain unchanged in the new arithmetic. Rules such as $1\,\mathrm{km} = 1000\,\mathrm{m}$ are also unaffected by the change of the utilized arithmetic. Addition also does not create problems for dimensional quantities. For instance, we have

$$\begin{aligned} 20\,\mathrm{m} \oplus 3\,\mathrm{km} &= 20\,\mathrm{m} \oplus 3000\,\mathrm{m} = [(20\,\mathrm{m})^q + (3000\,\mathrm{m})^q]^{1/q} \\ &= (20^q + 3000^q)^{1/q}\,\mathrm{m} = (20 \oplus 3000)\,\mathrm{m}. \end{aligned} \tag{6.1}$$

Alternatively,

$$\begin{aligned} 20\,\mathrm{m} \oplus 3\,\mathrm{km} &= [(0.02\,\mathrm{km})^q + (3\,\mathrm{km})^q]^{1/q} = (0.02^q + 3^q)^{1/q}\,\mathrm{km} \\ &= (0.02^q + 3^q)^{1/q} 1000\,\mathrm{m} = (20 \oplus 3000)\,\mathrm{m} \end{aligned} \tag{6.2}$$

because

$$(x \oplus y) \otimes z = (x \oplus y)z = (xz) \oplus (yz).$$

Note that formula for non-Diophantine addition is used for defining metrics in two-dimensional spaces.

Let us also mention that the quantum harmonic oscillator formulated in terms of the non-Diophantine arithmetic with of the functional parameter $f(x) = x^q$ has energy levels $E_n = (2n+1)^{1/q}\hbar\omega/2$ (Czachor, 2016).

Example 6.1.1 (Benioff's number scaling). Number-scaling theory (Benioff, 2011, 2015, 2016) can be regarded as a particular case of the above formalism with $f(x) = px$, $p \neq 0$. Indeed, $x \otimes y = (1/p)(pxpy) = pxy$, $x \oplus y = (1/p)(px + py) = x + y$, $x \oslash y = (1/p)(px)/(py) = x/(py)$, but $f(1/p) = 1$. Since $(1/p) \otimes x = (1/p)\ (p(1/p)px) = x$, one infers that $1' = f^{-1}(1) = 1/p$ is the unit element of multiplication in Benioff's non-Diophantine arithmetic.

Benioff's function $f(x) = px$ rescales multiplication, i.e., $x \otimes y = pxy$, but keeps addition unchanged, i.e., $x \oplus y = x + y$. Now the oscillator has energy levels $E_n = p\hbar\omega(n + 1/2)$. The example shows that a change of the utilized arithmetic may have non-trivial consequences for the issue of varying fundamental constants.

The case of a linear function f was extensively studied by Benioff also in a somewhat different context. The departure point was the observation that the set N of natural numbers can be identified with any countable well-ordered set whose first two elements define 0 and 1 of the arithmetic. For example, the set $N_2 = \{0, 2, 4, \ldots\}$ of even natural numbers may be regarded as a representation of N. The *value function* $f(x) = x/2$ maps a natural number $x \in N_2$ into its value, but to make the structure self-consistent one has to redefine the multiplication: $x \otimes y = f^{-1}(f(x)f(y)) = xy/2$. So, in this perspective, numbers by themselves are just elements of some formal axiomatic structure, while their values are determined by appropriate linear value maps, which are simultaneously used to redefine the arithmetic operations.

Non-trivial functions f typically require dimensionless arguments in $f(x)$. Still, physical variables x are dimensional. One has to associate with a physically meaningful variable x a dimensionless number, and we come to the problem described by Benioff (2011, 2015, 2016). Indeed, a dimensionless x is obtained for the price of including a physical unit (of length, say), and units can be chosen arbitrarily. The choice of units effectively introduces a value map, and a change of the scale changes this map. In the non-Newtonian calculus, f can be an arbitrary bijection, so it can be composed with any value mapping with no loss of bijectivity of the composition. Benioff's value mappings are thus intrinsically related to these bijections, but are not necessarily equivalent to them.

A dimensional quantity is a pair (x, ℓ) say, which is denoted by $x\ell$ in the standard notation, although x and ℓ are not objects of the same type: x is dimensionless while ℓ keeps track of the type of physical quantity. The fundamental unit ℓ plays the role of an abstract index, analogous to the names "Alice" and "Bob" in cryptography (cf., for example, Mollin, 2005), or the Penrose spinor/tensor abstract indices (Penrose and Rindler, 1984). The change of the scale by λ is mathematically achieved by the identification $\lambda(x, \ell) = (x, \lambda\ell) = (\lambda x, \ell)$. So, dimensional quantities belong to the quotient space obtained by factoring the Cartesian product of spaces by an appropriate equivalence relation. This is in fact how one defines a tensor product in abstract algebra. We can thus say that the dimensional quantity (x, ℓ) is a tensor product $x \otimes_T \ell$ (the more standard tensor-product symbol $\otimes$ is in this book reserved for non-Diophantine multiplication). But now we deal with three different sets: the dimensionless set $X = \{x\}$, the collection of all the possible fundamental lengths $\Lambda = \{\ell\}$, and the tensor product $X \otimes_T \Lambda$. In principle, in each of these sets, we can define different arithmetic operations, provided they are mutually consistent. So, let $\otimes_X$, $\oplus_X$ be the arithmetical operations in X, $\otimes_\Lambda$, $\oplus_\Lambda$ be the arithmetical operations in Λ, and let the arithmetical operations $\otimes_{X \otimes_T \Lambda}$, $\oplus_{X \otimes_T \Lambda}$ act in the space $X \otimes_T \Lambda$. In order to identify

$$(\lambda \otimes_X x) \otimes_T \ell = x \otimes_T (\lambda \otimes_\Lambda \ell) = \lambda \otimes_{X \otimes_T \Lambda} (x \otimes_T \ell) \tag{6.3}$$

we have to use only elements λs such that $\lambda \otimes_X x$ and $\lambda \otimes_\Lambda \ell$ simultaneously make sense. For example, in the Cantor line C, introduced in the previous chapter, one finds $1/3 \in C$ and $2/3 \notin C$. The change of units $\ell \to \ell/3$ is then meaningful, but $\ell \to 2\ell/3$ is not. Let us note that this notion of meaningfulness is different from the one discussed in Falmagne and Doble (2015), where a possibility of arbitrary changes of units is treated as a *sine qua non* natural principle. The question of dimensional vs. dimensionless variables will be also considered in Chapter 7, in the context of psychophysics.

6.2. Non-Diophantine Relativity

Shallow ideas can be assimilated;
ideas that require people to reorganize
their picture of the world provoke hostility.

James Gleick

Knowing how to perform linear transformations in Cartesian products of sets equipped with different arithmetical operations, we can formulate relativistic physics in non-Diophantine space–time. Let us begin with several examples.

Example 6.2.1 (Lorentz covariance of wave equation defined on the Cartesian product of ordinary time and Koch-type space). In our model, space–time consists of points $(x^0, x^1) = (ct, x) \in R \times X$, with $(x^0, f_X(x^1)) \in R^2$. A Lorentz transformation $x' = \Lambda(x)$, $\Lambda : R \times X \to R \times X$, is defined by

$$\begin{pmatrix} x'^0 \\ x'^1 \end{pmatrix} = \begin{pmatrix} L^0_0 x^0 + L^0_1 f_X(x^1) \\ f_X^{-1}(L^1_0 x^0 + L^1_1 f_X(x^1)) \end{pmatrix}, \tag{6.4}$$

or, equivalently, by

$$\begin{pmatrix} x'^0 \\ f_X(x'^1) \end{pmatrix} = \begin{pmatrix} L^0_0 & L^0_1 \\ L^1_0 & L^1_1 \end{pmatrix} \begin{pmatrix} x^0 \\ f_X(x^1) \end{pmatrix}, \tag{6.5}$$

where $L \in \mathrm{SO}(1, 1)$. Formula (6.4) implements a nonlinear action of the group $\mathrm{SO}(1, 1)$, and reduces to the usual representation if $X = R$ and $f_X(x^1) = x^1$. Transformations described by (6.4) form a group.

In order to prove Lorentz invariance of the wave equation, let us at first note that its solution

$$\begin{aligned}\Phi_t(x) &= a(f_X(x^1) + x^0) + b(f_X(x^1) - x^0)\\ &= \phi(x^0, f_X(x^1)) \end{aligned} \tag{6.6}$$

defines a function ϕ, satisfying (due to triviality of f_Y)

$$\frac{\mathrm{D}\Phi_t(x)}{\mathrm{D}x} = \frac{\partial\phi(x^0, f_X(x^1))}{\partial f_X(x^1)}, \tag{6.7}$$

$$\frac{1}{c}\frac{\mathrm{d}\Phi_t(x)}{\mathrm{d}t} = \frac{\partial\phi(x^0, f_X(x^1))}{\partial x^0}. \tag{6.8}$$

Accordingly, the wave equation takes the standard form

$$\left(\frac{1}{c^2}\frac{\partial^2}{\partial t^2} - \frac{\partial^2}{\partial f_X(x^1)^2}\right)\phi(x^0, f_X(x^1)) = 0. \tag{6.9}$$

It is invariant under (6.5) if ϕ transforms by

$$\phi'(x'^0, f_X(x'^1)) = \phi(x^0, f_X(x^1)), \tag{6.10}$$

which is equivalent to the scalar-field transformation $\Phi'_{t'}(x') = \Phi_t(x)$.

Replacing $R \times X$ by a more general case $X_0 \times X_1$, $f_{Xj} : X_j \to R$, one arrives at a Lorentz invariant wave equation (with both space–time derivatives appropriately defined), and Lorentz transformations

$$\begin{pmatrix} x'^0 \\ x'^1 \end{pmatrix} = \begin{pmatrix} f_{X_0}^{-1}(L^0_0 f_{X_0}(x^0) + L^0_1 f_{X_1}(x^1)) \\ f_{X_1}^{-1}(L^1_0 f_{X_0}(x^0) + L^1_1 f_{X_1}(x^1)) \end{pmatrix}. \tag{6.11}$$

A generalization to space–times constructed by Cartesian products of arbitrary numbers of fractals is now obvious.

Example 6.2.2 (Rotations in a Cartesian product of two Cantor sets). Let $X_1 = X_2 = C$ be the Cantor line obtained in Example 5.1 by the periodic repetition of middle-third Cantor sets $C_-(0,1)$. Consider the plane $C^2 = X_1 \times X_2$, and let $(x_1, x_2) \in C^2$. A rotation by a Diophantine

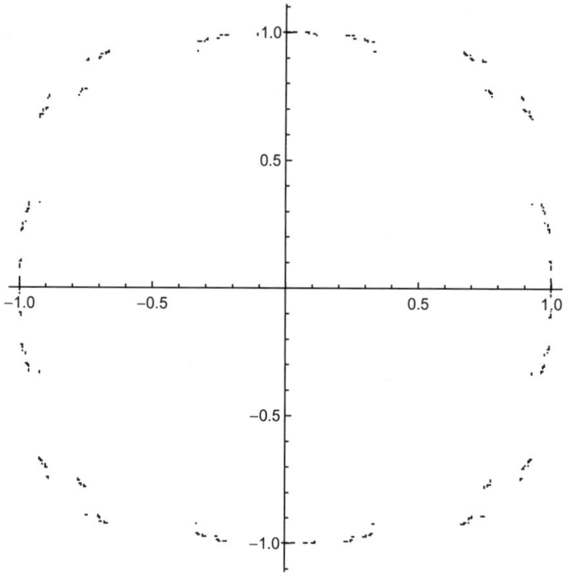

Figure 6.1. A new type of rotational symmetry. Circles in Cantorian plane C^2, for various radii. From top to bottom: 1, 10, and 50 circles.

angle $\alpha \in R$, $f_R(\alpha) = \alpha$, is defined by

$$x_1' = x_1 \otimes \mathrm{Cos}\,\alpha \oplus x_2 \otimes \mathrm{Sin}\,\alpha, \tag{6.12}$$

$$x_2' = \ominus x_1 \otimes \mathrm{Sin}\,\alpha \oplus x_2 \otimes \mathrm{Cos}\,\alpha, \tag{6.13}$$

The utilized non-Diophantine arithmetic is defined in C, and

$$\mathrm{Cos}\,\alpha = f_C^{-1}(\cos\alpha), \tag{6.14}$$

$$\mathrm{Sin}\,\alpha = f_C^{-1}(\sin\alpha). \tag{6.15}$$

Figure 6.1 shows circles of various radii, defined parametrically by

$$R \ni \alpha \mapsto (r \otimes \mathrm{Cos}\,\alpha, r \otimes \mathrm{Sin}\,\alpha) \in C^2, \quad r \in C. \tag{6.16}$$

The bijection f_C is antisymmetric $f_C(\ominus x) = f_C(-x) = -f_C(x)$. The circles are examples of fractal homogeneous spaces, here corresponding to the rotation group in C^2.

Example 6.2.3 (Lorentz transformations in (1 + 1)-dimensional Cantor–Minkowski space). The Cantor–Minkowski space is defined as the same Cartesian product as in the previous example, with the

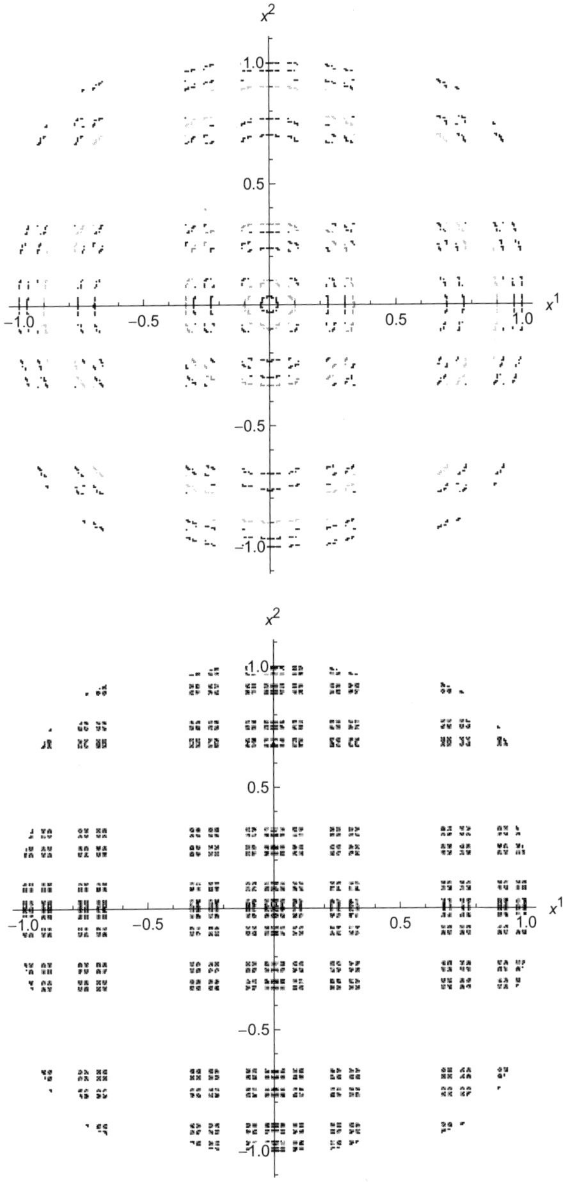

Figure 6.1. (*Continued*)

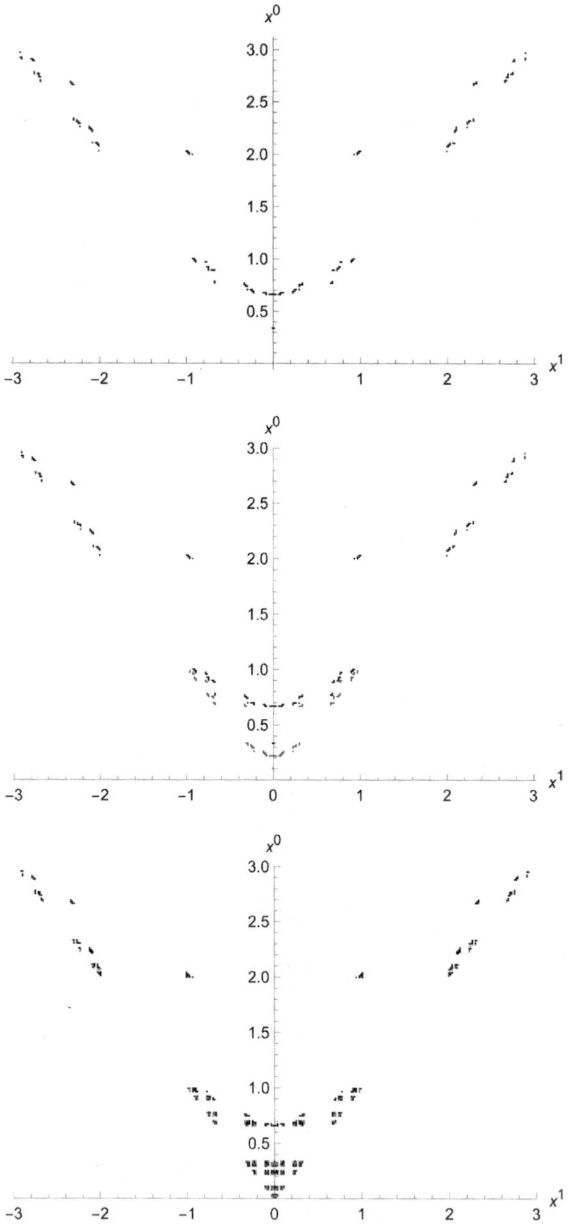

Figure 6.2. Hyperbolic symmetry. Proper-time hyperbolas in (1 + 1)-dimensional Cantorian Minkowski space–time C^2. From top to bottom: 1, 2, and 20 hyperbolas.

same arithmetics, but with a different metric structure. The Lorentz transformations in $1+1$ dimensions are given by hyperbolic rotations

$$x'^0 = x^0 \otimes \operatorname{Cosh}\alpha \ominus x^1 \otimes \operatorname{Sinh}\alpha, \tag{6.17}$$

$$x'^1 = \ominus x^0 \otimes \operatorname{Sinh}\alpha \oplus x^1 \otimes \operatorname{Cosh}\alpha, \tag{6.18}$$

where

$$\operatorname{Cosh}\alpha = f_C^{-1}(\cosh\alpha), \tag{6.19}$$

$$\operatorname{Sinh}\alpha = f_C^{-1}(\sinh\alpha). \tag{6.20}$$

Homogeneous spaces of the $(1 + 1)$-dimensional Lorentz group are hyperbolas,

$$R \ni \alpha \mapsto (r \otimes \operatorname{Cosh}\alpha, r \otimes \operatorname{Sinh}\alpha) \in C^2, \quad r \in C. \tag{6.21}$$

Example 6.2.4 (Plane as a Cartesian product of two different Koch curves). Consider $K_R(\alpha) = \bigcup_{k \in Z} K_{[k,k+1]}(\alpha)$, where $K_{[k,k+1]}(\alpha)$ is the Koch curve on the unit interval $[k, k+1]$ discussed in Example 5.3, and with the value α of the angle that defines the curve (cf. Figure 5.11). Now, let $X = K_R(\alpha_1) \times K_R(\alpha_2)$. There is no unique way of mapping the Cartesian product of two curves into a surface in R^3, so various geometric objects may represent the same Cartesian product. However, whichever representation is selected, one can proceed further by introducing symmetry groups, homogeneous spaces generated by them, and so on. In what follows, let us employ the following embedding. Consider two curves in $R^2 : u \mapsto (x_1(u), y_1(u))$, $v \mapsto (x_2(v), y_2(v))$. The Cartesian product of the curves will be modeled by the vector (Diophantine) sum

$$(x_1(u), 0, y_1(u)) + (0, x_2(v), y_2(v)). \tag{6.22}$$

Figure 6.3 shows the surface representing the Cartesian product $K_R(\pi/5) \times K_R(\pi/3)$.

Example 6.2.5 (Circles in $K_R(\pi/5) \times K_R(\pi/3)$). Let us now consider circles in $X = K_R(\pi/5) \times K_R(\pi/3)$, obtained by means of rotations

$$\begin{aligned} \theta &\mapsto (r^{1_{K_R(\pi/5)}} \otimes_{K_R(\pi/5)} \operatorname{Cos}\theta, r^{1_{K_R(\pi/3)}} \otimes_{K_R(\pi/3)} \operatorname{Sin}\theta) \\ &= (f^{-1}_{K_R(\pi/5)}(r\cos\theta), f^{-1}_{K_R(\pi/3)}(r\sin\theta)), \end{aligned} \tag{6.23}$$

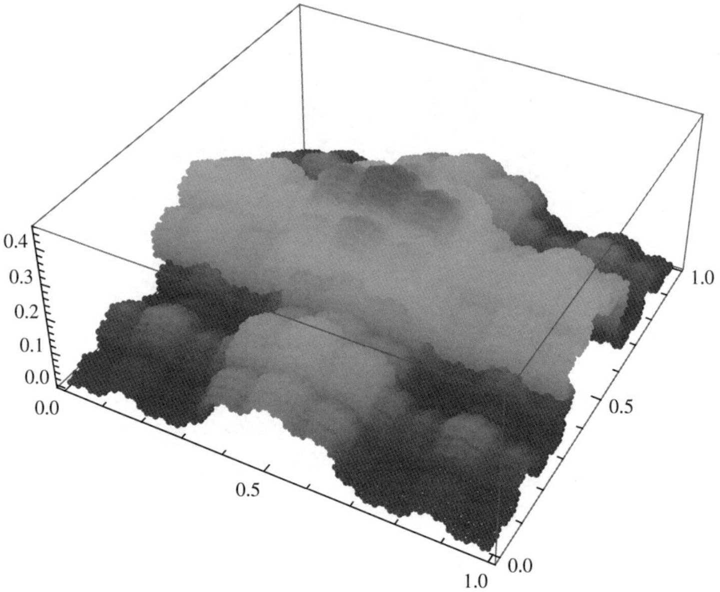

Figure 6.3. The Cartesian product of two Koch curves with $\alpha_1 = \pi/5$ and $\alpha_2 = \pi/3$.

where

$$\operatorname{Cos}\theta = f^{-1}_{K_R(\pi/5)}(\cos\theta), \tag{6.24}$$

$$\operatorname{Sin}\theta = f^{-1}_{K_R(\pi/3)}(\sin\theta). \tag{6.25}$$

We assume for simplicity that $\theta \in [0, 2\pi)$, which is equipped with the operations from the Diophantine arithmetic $\boldsymbol{R}$.

Figure 6.4 shows four circles for different radii. The circles are examples homogeneous spaces of the rotation group in X, with parameters θ satisfying the ordinary Diophantine arithmetic of real numbers.

Example 6.2.6 (Special relativity in Cantor Minkowski space–time). Figure 6.2 shows proper-time hyperbolas defined by

$$g_{ab} \otimes x^a \otimes x^b = x^0 \otimes x^0 \ominus x^1 \otimes x^1 = s^{2'}. \tag{6.26}$$

Note that in $(1+1)$-dimensional Minkowski space, one finds

$$x_0 = x^0 = g_{00} \otimes x^0 = 1_C \otimes x^0, \tag{6.27}$$

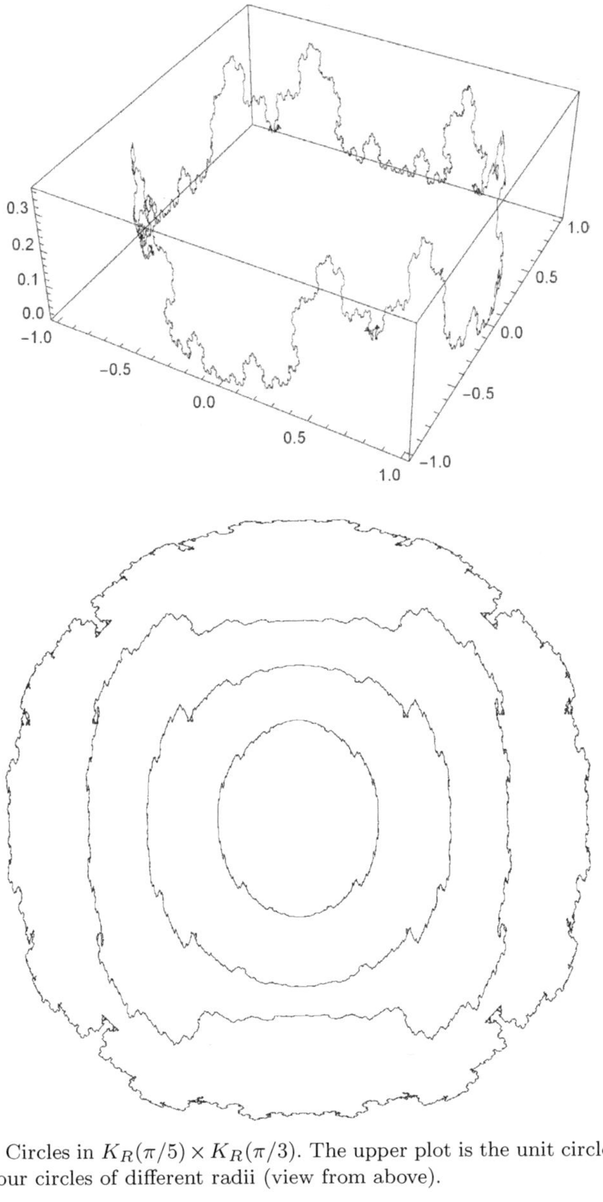

Figure 6.4. Circles in $K_R(\pi/5) \times K_R(\pi/3)$. The upper plot is the unit circle. The lower plot shows four circles of different radii (view from above).

so $g_{00} = 1_C$ is the neutral element of multiplication. Now we obtain

$$x_1 = g_{11} \otimes x^1 = f_C^{-1}(f_C(g_{11}) f_C(x^1)). \tag{6.28}$$

Putting it differently, we find

$$x_1 = \ominus x^1 = 0_C \ominus x^1 = f_C^{-1}(f_C(0) - f_C(x^1)). \tag{6.29}$$

Accordingly, we have

$$f_C(g_{11}) f_C(x^1) = -f_C(x^1), \; g_{11} = f_C^{-1}(-1),$$

while in general,

$$x_1 = f_C^{-1}(-f_C(x^1)) \neq -x^1. \tag{6.30}$$

The Lorentz transformations, defined by (6.17)–(6.18) satisfy

$$g_{ab} \otimes x'^a \otimes x'^b = g_{ab} \otimes x^a \otimes x^b. \tag{6.31}$$

The characteristic Cantor-like structure visible at the lowest plot at Figure 6.4 could be equivalently generated by plotting a bunch of "straight" world half-lines, as shown in Figure 6.5,

$$x^a(s) = u^a \otimes s, \quad 0_C \leq s \tag{6.32}$$

($0_C \leq s$ if and only if $0 \leq f_C^{-1}(s)$). The uppermost plot involves only three world half-lines, two null and one timelike (a world vector $x^a \in X^4$ is spacelike, null, or timelike if, respectively, $\sum_{ab} f_X(g_{ab}) f_X(x^a) f_X(x^b)$ is negative, zero, or positive). The null lines look "ordinary", i.e., comply with the intuitive picture of a straight line. The timelike world line is also "straight" in the sense of formula (6.32), and for inhabitants of Cantorian Minkowski space would appear as "ordinarily straight" as generators of the light cone.

By definition of the derivative, we find

$$\frac{\mathrm{D}x^a(s)}{\mathrm{D}s} = \lim_{h \to 0_C} (u^a \otimes (s \oplus h) \ominus u^a \otimes s) \oslash h = u^a \tag{6.33}$$

and thus the straight line (6.32) is a space–time trajectory in the usual sense, with four-velocity u^a. Such a simple family of world lines is enough to formulate a fractal analog of the twin paradox.

Lorentz transformations (6.17)–(6.18) define coordinate axes as the world lines

$$x^0 = \beta \otimes x^1 \tag{6.34}$$

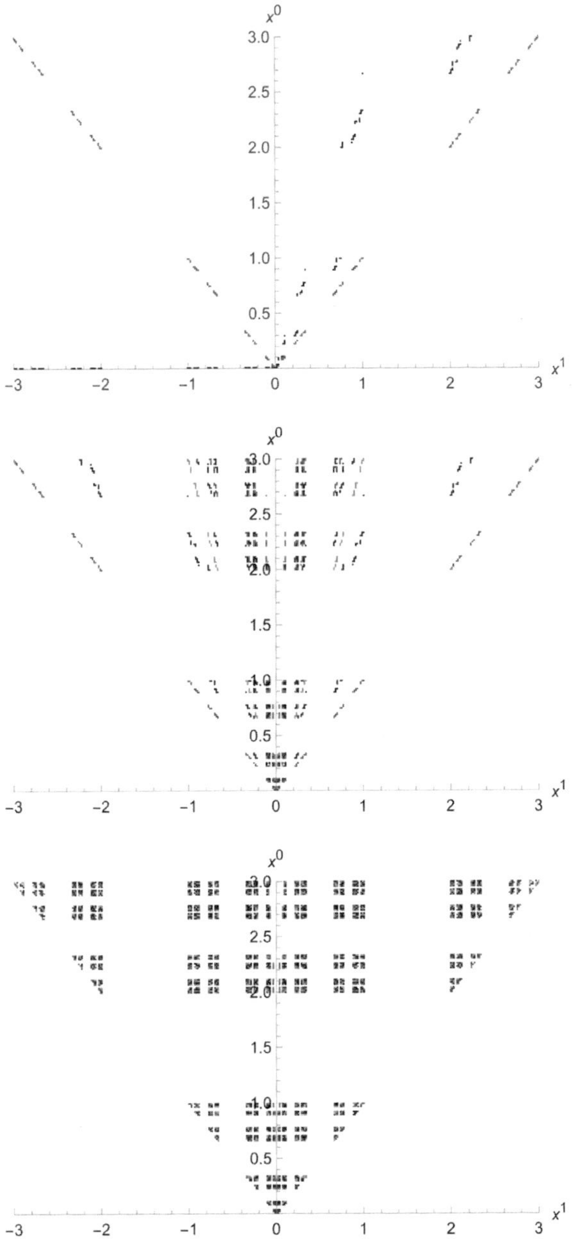

Figure 6.5. World half-lines in (1 + 1)-dimensional Cantorian Minkowski space–time C^2. From top to bottom: Light cone plus 1, 22, and 400 timelike world lines. The same f_C as shown in Figure 6.1.

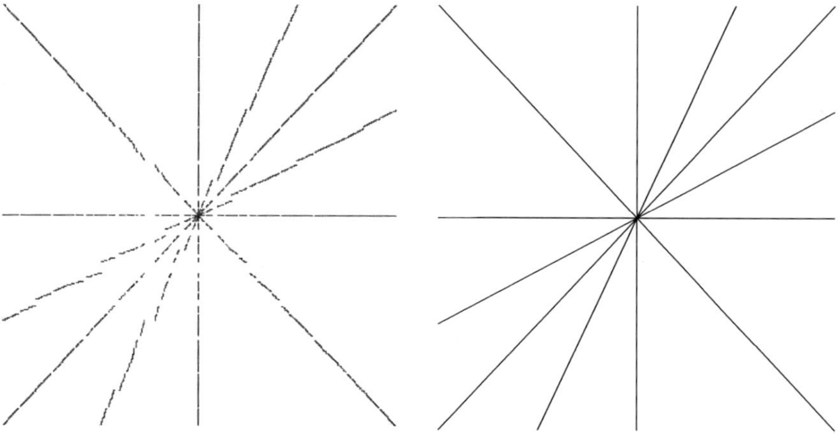

Figure 6.6. Left: fractal light cone and two fractal coordinate systems for the multi-resolution Cantor-line-Minkowski $(1+1)$-dimensional space C^2, defined by means of f_C from Section 5.4. Right: the same lines but mapped by f_C into the usual Minkowski space.

(Lorentz-transformed simultaneity hyperplane of the event $x_a = x'_a = 0_C$) and

$$x^1 = \beta \otimes x^0 \tag{6.35}$$

(Lorentz-transformed time axis), where $\beta = \operatorname{Tanh} \alpha = \operatorname{Sinh} \alpha \oslash \operatorname{Cosh} \alpha$.

The left part of Figure 6.6 shows two coordinate systems in fractal Minkowski space C^2. Coordinate axes correspond to $\beta = 0_C$ (vertical and horizontal axes) and $\beta = f_C^{-1}(1/2)$ (diagonal broken lines), together with the light cone defined by $\beta = 1_C$ and $\beta = \ominus 1_C$. The right part shows the result of applying f_C to C^2. The bijection f_C is taken in the irregular "multi-resolution" form, discussed in detail in Section 5.4.

Example 6.2.7 (Twin paradox in space–time with arbitrary arithmetic). Now let us consider the twin paradox in space–time with arithmetics in space and time defined by an arbitrary bijection $f_X = f$. We assume that both space and time involve the same arithmetic. Consider two world lines. The traveling twin corresponds to

$$x^a(s) = \begin{cases} u^a \otimes s & \text{for } 0_X \leq s < s_1, \\ u^a \otimes s_1 \oplus v^a \otimes (s \ominus s_1) & \text{for } s_1 \leq s \leq s_2. \end{cases} \tag{6.36}$$

The twin "at rest" is described by

$$y^a(s) = (u^a \otimes s_1 \oplus v^a \otimes (s_2 \ominus s_1)) \otimes s \oslash s_2, \tag{6.37}$$

for $0_X \le s \le s_2$. Here x^a, y^a, u^a, and v^a are position and 4-velocity world vectors, respectively, with $u_a \otimes u^a = v_a \otimes v^a = 1_X$. Since $x(0_X) = y(0_X)$ and $x(s_2) = y(s_2)$, the two trajectories can be used to derive the paradox. The Cantorian Minkowski-space length of $s \mapsto x(s)$ is $S_x = s_1 \oplus (s_2 \ominus s_1) = s_2$ whereas the one of $y(s)$ satisfies $S_y^{2\prime} = g_{ab} \otimes y^a(s_2) \otimes y^b(s_2)$. Assume for simplicity that $S_x = s_2 = s_1 \oplus s_1 = 2' \otimes s_1$:

$$\begin{aligned} y^a(s) &= (u^a \oplus v^a) \otimes s_1 \otimes s \oslash s_2, \\ S_y^{2'} &= (u_a \oplus v_a) \otimes (u^a \oplus v^a) \otimes s_1^{2'} \\ &= S_x^{2'} \otimes (1_X \oplus u_a \otimes v^a) \oslash 2'. \end{aligned} \tag{6.38}$$

In order to cross-check (6.38), take the trivial case when $f(x) = x$, $(u^0, u^1) = (1, \beta)/(1-\beta^2)^{1/2}$, $(v^0, v^1) = (1, -\beta)/(1-\beta^2)^{1/2}$. Then

$$(1_X \oplus u_a \otimes v^a) \oslash 2' = 1/(1-\beta^2) = u_0^2 \tag{6.39}$$

i.e., $S_x = S_y(1-\beta^2)^{1/2} = S_y/u_0$, as expected. For a general f let us first note that the normalization

$$1_X = u_a \otimes u^a = u^0 \otimes u^0 \ominus u^1 \otimes u^1 = f^{-1}(f(u^0)^2 - f(u^1)^2) \tag{6.40}$$

together with $f(1_X) = 1$ implies $f(u^0)^2 - f(u^1)^2 = 1$. If $v^0 = u^0$ and $v^1 = \ominus u^1$, then

$$(1_X \oplus u_a \otimes v^a) \oslash 2' = f^{-1}(f(u^0)^2) = u_0^{2'}. \tag{6.41}$$

Equation (6.41) is exactly analogous to (6.40), so finally we get the simple formula for the time delay which is valid in any f-arithmetic Minkowski space,

$$S_x = S_y \oslash u_0. \tag{6.42}$$

Since

$$1_X \oslash u_0 = f^{-1}[(1 - f(\beta)^2)^{1/2}], \quad \beta = u^1 \oslash u^0, \tag{6.43}$$

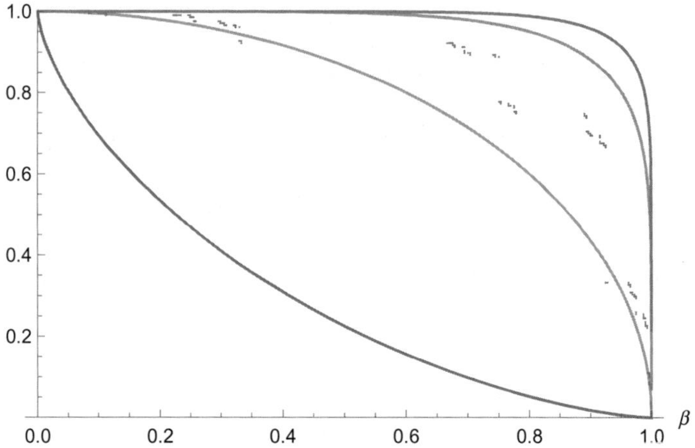

Figure 6.7. The time-delay factor $1_X \oslash u_0$ of Eq. (6.43), for various choices of f. The continuous curves correspond to $f(x) = x^q$, with $q = 1/3$ (leftmost), $q = 1$, $q = 3$, and $q = 5$ (rightmost). The discrete points correspond to $f = f_C$ of the standard triadic Cantor set.

we can alternatively write

$$f(S_x) = f(S_y)(1 - f(\beta)^2)^{1/2}. \tag{6.44}$$

Comparing S_x and S_y in a space–time neighborhood of a given $x = x(0_X) = y(0_X)$, we can *in principle* directly probe the form of f (Figure 6.7).

The problem is which of the two formulas, (6.42) or (6.44), should be employed in comparison of experimental data with the theory? Which of the two velocity parameters, $0_X \leq \beta \leq 1_X$ or $0 \leq f(\beta) \leq 1$, is the one employed in the experiment if we assume that the observers live in the fractal space–time?

6.3. Minkowski Space–Time R^{+4}

Who never walks save where he sees men's tracks makes no discoveries.

J. G. Holland

Let $f(x) = \mu \ln x + \nu$, $\mu > 0$, be the bijection $f\colon R^+ \to R$. Accordingly, $X = R^+$. $f^{-1}(x) = e^{(x-\nu)/\mu}$, and thus $0' = f^{-1}(0) = e^{-\nu/\mu}$, $1' = f^{-1}(1) = e^{(1-\nu)/\mu}$.

6.3.1. *Arithmetic*

The numbers may be said to rule the whole world of quantity, and the four rules of arithmetic may be regarded as the complete equipment of the mathematician.

James C. Maxwell

Let us begin with the explicit form of arithmetic operations. Addition and subtraction explicitly read

$$x \oplus y = f^{-1}(f(x) + f(y)) = xye^{\nu/\mu}, \tag{6.45}$$

$$x \ominus y = f^{-1}(f(x) - f(y)) = e^{-\nu/\mu}x/y. \tag{6.46}$$

The arithmetic operations occurring at the right sides of (6.45) and (6.46) are those from R and not from X (the latter occur at the left sides of these formulas). For example, $x \oplus 0' = xe^{-\nu/\mu}e^{\nu/\mu} = x$.

Note that although $x > 0$ in $f(x) = \mu \ln x + \nu$, one nevertheless has a well-defined negative number $\ominus x = 0' \ominus x = e^{-2\nu/\mu}/x \in X = R^+$, which is positive from the point of view of the arithmetic of R. Let us cross-check the negativity of $\ominus x$:

$$\ominus x \oplus x = (\ominus x)xe^{\nu/\mu} \tag{6.47}$$

$$= (e^{-2\nu/\mu}/x)xe^{\nu/\mu} \tag{6.48}$$

$$= e^{-\nu/\mu} = 0'. \tag{6.49}$$

Now $(X, \oplus) = (R^+, \oplus)$ is a group, as opposed to $(R^+, +)$. In consequence, the Minkowski space R^{+4} is invariant under the non-Diophantine Poincaré group.

The multiplication in X is explicitly given by

$$x \otimes y = f^{-1}(f(x)f(y)) = e^{\mu \ln x \ln y + \nu \ln x + \nu \ln y + \nu^2/\mu - \nu/\mu}, \tag{6.50}$$

$$x \oslash y = f^{-1}(f(x)/f(y)) = e^{(\ln x + \nu/\mu)/(\mu \ln y + \nu) - \nu/\mu}. \tag{6.51}$$

Again the expressions at the right-hand sides of (6.50) and (6.51) involve the arithmetic of R.

6.3.2. *Light cone*

A new scientific truth does not triumph by convincing its opponents and making them see the light, but rather because its opponents eventually die and a new generation grows up that is familiar with it.

M. Planck

The light cone in R^{+4} consists of vectors satisfying

$$G_{ab} \otimes x^a \otimes x^b = f^{-1}(f(G_{ab})f(x^a)f(x^b)) \tag{6.52}$$
$$= f^{-1}(g_{ab}f(x^a)f(x^b)) \tag{6.53}$$
$$= 0' = f^{-1}(0) = e^{-\nu/\mu}. \tag{6.54}$$

This is equivalent to $f(x^0)^2 = f(x^1)^2 + f(x^2)^2 + f(x^3)^2$, i.e.,

$$x^0 = f^{-1}(\pm\sqrt{f(x^1)^2 + f(x^2)^2 + f(x^3)^2}) \tag{6.55}$$
$$= e^{(\pm\sqrt{(\ln x^1+\nu/\mu)^2+(\ln x^2+\nu/\mu)^2+(\ln x^3+\nu/\mu)^2}-\nu/\mu)}. \tag{6.56}$$

Figure 6.8 shows the light cone $G_{ab} \otimes x^a \otimes x^b = 0'$ in $(1+2)$-dimensional Minkowski space R^{+3}.

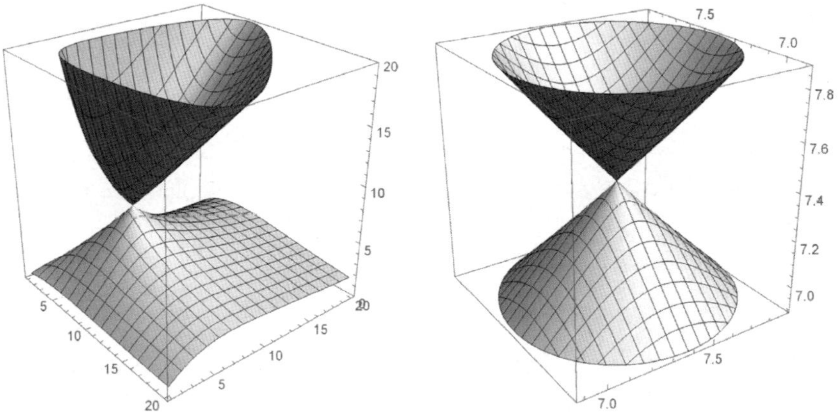

Figure 6.8. Light cone $x_a \otimes x^a = 0'$ in $(1+2)$-dimensional Minkowski space R^{+3}, for $\mu = 10$, $\nu = -20$. Close-up of a neighborhood of $(0', 0', 0')$, where $0' = e^{-\nu/\mu} = e^2 \approx 7.39$ (right).

An arbitrarily located light cone

$$G_{ab} \otimes (x^a \ominus y^a) \otimes (x^b \ominus y^b) = 0' \tag{6.57}$$

corresponds to

$$g_{ab}(f(x^a) - f(y^a))(f(x^b) - f(y^b)) = 0, \tag{6.58}$$

that is

$$x^0 = y^0 e^{\pm\sqrt{\ln^2(x^1/y^1)+\ln^2(x^2/y^2)+\ln^2(x^3/y^3)}}. \tag{6.59}$$

For $y^0 = \cdots = y^3 = 0' = e^{-\nu/\mu}$ we reconstruct (6.56).

Figure 6.9 shows the light cones (6.59) in small neighborhoods of various origins y^a. The plots suggest that a Lorentzian geometry is typical of both the standard (1 + 2)-dimensional space–time R^3 (where it is just globally Minkowskian), and of X^3. Recall that the latter is also globally Minkowskian, but with respect to the non-Newtonian calculus. Interestingly, when we employ in X^3 the mismatched formalism taken from R^3, the formulas are locally Lorentzian. The further away from the "walls" of $X^3 = R^{+3}$ the observation is performed, the more Minkowskian the geometry appears, provided one does not observe objects that are too far away from the observer.

To prove the local Lorentzian structure analytically, let $x^a = y^a + \varepsilon^a$. Here the operation + is taken from the Diophantine arithmetic $\boldsymbol{R}$ since the observer is assumed to perform his analysis in the "wrong" formalism. Then, for $|\varepsilon^a/y^a| \ll 1$,

$$g_{ab}f'(y^a)\epsilon^a f'(y^b)\epsilon^b = \hat{g}_{ab}(y)\epsilon^a\epsilon^b \approx 0$$

$$\text{(there is no summation over repeated indices)}, \tag{6.60}$$

where

$$\hat{g}_{ab} = g_{ab}f'(y^a)f'(y^b) \quad \text{(no sum)} \tag{6.61}$$

$$= \text{diag}((y^0)^{-2}, -(y^1)^{-2}, -(y^2)^{-2}, -(y^3)^{-2}) \tag{6.62}$$

is the Lorentzian metric. The metric (6.62) becomes just a conformally rescaled Minkowskian g_{ab} if $y^0 = y^1 = y^2 = y^3$. The same effect occurs for general hyperboloids $G_{ab} \otimes (x^a \ominus y^a) \otimes (x^b \ominus y^b) \neq 0'$.

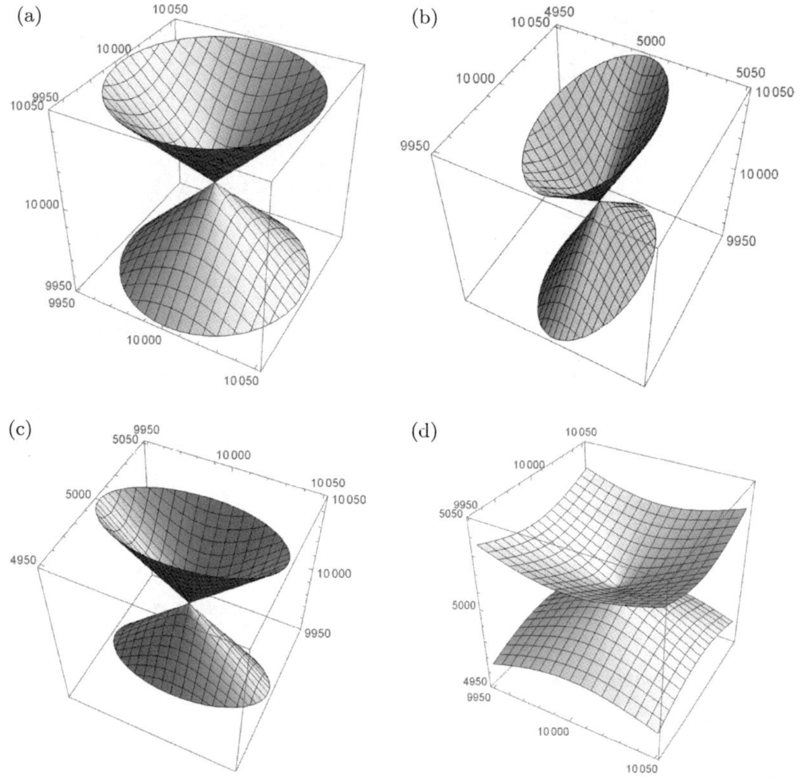

Figure 6.9. Minkowskian correspondence principle in (1 + 2)-dimensional space–time R^{+3}. Light cone $G_{ab} \otimes (x^a \ominus y^a) \otimes (x^b \ominus y^b) = 0'$ for (a) $(y^0, y^1, y^3) = (10000, 10001, 10002)$, (b) $(y^0, y^1, y^3) = (10000, 5000, 10000)$, (c) $(y^0, y^1, y^3) = (10000, 10000, 5000)$, and (d) $(y^0, y^1, y^3) = (5000, 10000, 10001)$. The further away from the boundaries of R^{+3}, the more Minkowskian-looking the light cones are.

6.4. Minkowski Space–Time $(-L/2, L/2)^4$

I do not define time, space, place, and motion, as being well known to all.

Isaac Newton

Let $X = (-L/2, L/2)$ and $f(x) = \tan(\pi x/L)$, $f^{-1}(x) = (L/\pi)\arctan x$. The functional parameter f: $X \to R$ is a bijection between an interval of real numbers and $\boldsymbol{R}$, with $0' = 0$ and $1' = L/4$. Such a non-Newtonian calculus where the set X is an interval of real numbers is studied in great detail by Endre Pap and his collaborators (Pap, 1993, 2008; Agahi *et al.*, 2011; Pap *et al.*, 2014; Štrboja *et al.*, 2018, 2019).

6.4.1. *Light cone*

We've all got both light and dark inside us.
What matters is the part we choose to act on.
That's who we really are.
J. K. Rowling

Figures 6.10 and 6.11 show the light cone $G_{ab} \otimes (x^a \ominus y^a) \otimes (x^b \ominus y^b) = 0'$ corresponding to

$$x^0 = f^{-1}\left(f(y^0) \pm \sqrt{(f(x^1)-f(y^1))^2 + (f(x^2)-f(y^2))^2 + (f(x^3)-f(y^3))^2}\right)$$
$$= \frac{L}{\pi}\arctan\left(\tan\frac{\pi y^0}{L}\right.$$
$$\left.\pm\sqrt{\left(\tan\frac{\pi x^1}{L} - \tan\frac{\pi y^1}{L}\right)^2 + \cdots + \left(\tan\frac{\pi x^3}{L} - \tan\frac{\pi y^3}{L}\right)^2}\right). \quad (6.63)$$

Now let $x^a = y^a + \varepsilon^a$, where $|\varepsilon^a/L| \ll 1$. The effective Lorentzian metric in a neighborhood of y reads

$$\hat{g}_{ab}(y) = \frac{\pi^2}{L^2}\operatorname{diag}\left(\cos^{-4}\frac{\pi y^0}{L}, -\cos^{-4}\frac{\pi y^1}{L}, -\cos^{-4}\frac{\pi y^2}{L}, -\cos^{-4}\frac{\pi y^3}{L}\right). \quad (6.64)$$

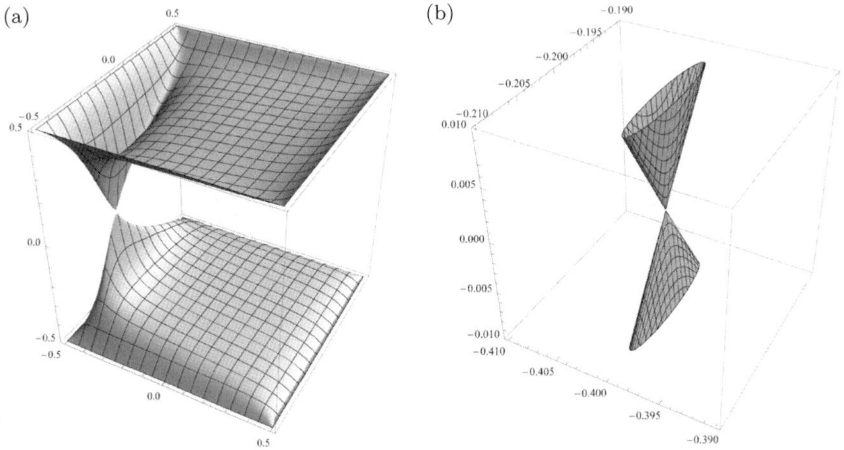

Figure 6.10. Light-cone with the origin at $(y^0, y^1, y^3) = (0, -0.4, -0.2)$ in $(1+2)$-dimensional Minkowski space $(-0.5, 0.5)^3 (L = 1)$. (a) The global picture, and (b) the close-up of the origin of the cone.

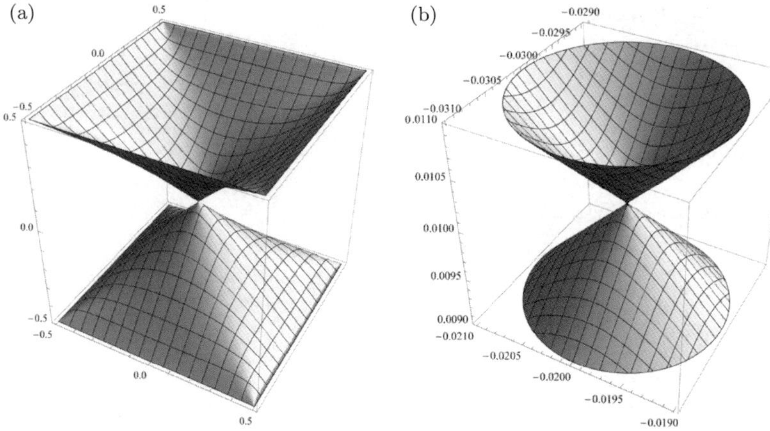

Figure 6.11. The same as in Figure 6.10, but with $(y^0, y^1, y^3) = (0.01, -0.02, -0.03)$.

Another example of a bijection $f\colon X \to R$ is provided by $f(x) = \operatorname{arctanh}(2x/L)$, $f^{-1}(x) = (L/2)\tanh x$.

6.5. Friedman Equation

The discovery of truth is prevented more effectively
not by the false appearance of things present and which mislead into error,
not directly by weakness of the reasoning powers,
but by preconceived opinion, by prejudice.

Arthur Schopenhauer

Taylor expansions of (6.87), (6.93) in a neighborhood of $y^1 = 0$ begin with third-order terms $\sim (y^1/L)^3$. The effect is small. However, when we switch to non-Diophantine generalized Einstein equations, the correction should become visible at large distances.

So, let us consider the Friedman equation for a flat, matter-dominated FRW model with exactly vanishing cosmological constant (Hartle, 2003). In matter-dominated cosmology ($\Omega_{\text{matter}} = 1, \Omega_{\text{radiation}} = 0$), with no dark energy ($\Omega_{\text{vacuum}} = 0$), the scale factor is given by $a(t) = (t/t_0)^{2/3}$. In the non-Newtonian notation the solution reads

$$A(t) = (t \oslash t_1)^{2' \oslash 3'} \tag{6.65}$$

$$= f^{-1}((f(t)/f(t_1))^{2/3}). \tag{6.66}$$

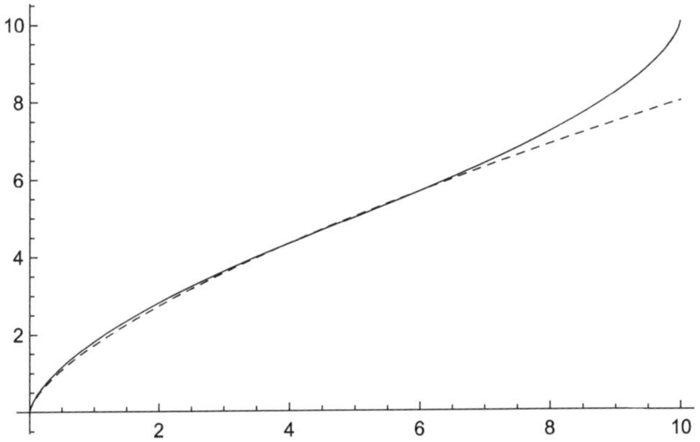

Figure 6.12. The plot of (6.66) (full) as compared with $(t/t_0)^{2/3}$ for $f(x) = \tan(\pi x/L)$. Both models involve no dark energy, i.e., $\Omega_{\text{vacuum}} = 0 = 0'$.

We choose initial conditions at t_0 and t_1, respectively, in order to have an additional fitting parameter. To make (6.66) more explicit, let us return to $X = (-L/2, L/2)$ and $f(x) = \tan(\pi x/L)$, $f^{-1}(x) = (L/\pi)\arctan x$. Figure 6.12 shows (6.66) as compared with the standard $(t/t_0)^{2/3}$, for $L = 20$, $t_1 = 1'$, and t_0 chosen in a way guaranteeing a reasonable fit of the two plots.

The curve bends up in a characteristic way, typical of dark-energy models of the accelerating universe. The effect is of a purely arithmetic origin, with no need of utilization of the physical concept *dark energy*. It can be shown (Czachor, 2020a) that there exists a non-Diophantine arithmetic which exactly reconstructs, with zero cosmological constant, the observable expansion of the universe. As a result, arithmetic becomes as physical as geometry.

6.6. Generalized Entropies and Non-Diophantine Probability

We do not know how to predict what would happen under given conditions and we believe today that it is impossible — that the only thing that can be predicted is the probability of various events.

Richard Feynman

Example 6.6.1 (Non-Diophantine interpretation of Kolmogorov–Nagumo averages). Consider a set of (Diophantine) probabilities p_k, $\sum_k p_k = 1$. A Kolmogorov–Nagumo average of a real random variable a

(Kolmogorov, 1930; Nagumo, 1930; Grabisch *et al.*, 2009), is described by the formula

$$\langle a \rangle_f = f^{-1}\left(\sum_k f(a_k)p_k\right). \tag{6.67}$$

Here f is any strictly monotonic function. The Kolmogorov–Nagumo average is invariant under the affine transformation $f(x) \mapsto Af(x) + B$, with constants A, B. Now, let c be a constant random variable. Demanding

$$\langle a + c \rangle_f = \langle a \rangle_f + c, \tag{6.68}$$

we find that f is either a linear or exponential function (up to $f\ (x) \mapsto Af(x) + B$, of course). Let $f_q(x) = e^{(1-q)x}$, $f_q^{-1}(x) = (1-q)^{-1}\ln x$, $q \in R$. With this particular choice of f

$$\langle a \rangle_{f_q} = \frac{1}{1-q}\ln\left(\sum_k p_k e^{(1-q)a_k}\right), \tag{6.69}$$

one obtains the standard linear average as the limit

$$\lim_{q\to 1}\langle a \rangle_{f_q} = \sum_k p_k a_k. \tag{6.70}$$

The entropy of Shannon (1948)

$$S = -\sum_k p_k \ln p_k = \sum_k p_k \ln(1/p_k) \tag{6.71}$$

is therefore the limiting $q \to 1$ case of

$$\begin{aligned} S_q &= \frac{1}{1-q}\ln\left(\sum_k p_k e^{(1-q)\ln(1/p_k)}\right) = \frac{1}{1-q}\ln\left(\sum_k p_k e^{\ln p_k^{q-1}}\right) \\ &= \frac{1}{1-q}\ln\sum_k p_k^q, \end{aligned} \tag{6.72}$$

which is known as the Rényi entropy (Rényi, 1960). Since their invention, the Rényi entropies found numerous applications in statistical physics, theory of information, fractals, etc. However, what is interesting from our point of view is their link to a non-Diophantine arithmetic. Let us introduce the equality $p'_k = f^{-1}(p_k)$ for all k. Now the Kolmogorov–Nagumo average is expressed as

$$\langle a \rangle_f = f^{-1}\left(\sum_k f(a_k) f(p'_k)\right) = \sum_k{}^{\oplus} a_k \otimes p'_k. \tag{6.73}$$

The normalization

$$\sum_k{}^{\oplus} p'_k = f^{-1}\left(\sum_k f(p'_k)\right) = f^{-1}\left(\sum_k p_k\right) = f^{-1}(1) = 1' \tag{6.74}$$

shows that the primed probabilities are normalized to the non-Diophantine neutral element of multiplication. This type of normalization will be employed also in non-Diophantine quantum mechanics. Note that

$$\langle a \oplus b \rangle_f = \langle a \rangle_f \oplus \langle b \rangle_f \tag{6.75}$$

for any f. From this perspective, the Rényi entropy f_q is not more special than any other f. Still, the Rényi entropy f_q is an interesting special case. For example,

$$p'_k = f_q^{-1}(p_k) = \frac{1}{1-q} \ln p_k = \frac{1}{q-1} \ln(1/p_k). \tag{6.76}$$

Random variable $a_k = \log_b(1/p_k)$ is, according to Shannon, the amount of information obtained by observing an event whose probability is p_k. The choice of b defines units of information. The non-Diophantine probability p'_k is the amount of information encoded in p_k. Now, consider the case where we have n successes in N trials (here n and N are the usual Diophantine natural numbers). The corresponding frequency is $p = n/N$. An observer who analyzes the same sequence of events by means of a non-Diophantine arithmetic would obtain

$$p' = n' \oslash N' = f^{-1}(f(n')/f(N')) = f^{-1}(n/N) = f^{-1}(p), \tag{6.77}$$

arriving at the same rule as the one, which is implicitly present in the Kolmogorov–Nagumo averaging. Probability based on non-Diophantine arithmetic sheds interesting new light on the old problem of completeness of quantum mechanics. Hidden-variable theories based on non-Diophantine arithmetics circumvent (Czachor, 2020b,c) certain limitations of Bell's theorem (Bell, 1964).

Any non-Diophantine arithmetic can be consistently employed in the construction of a probability theory. The Rényi case is the one where non-Diophantine probabilities directly represent Shannon's quantity of information.

Example 6.6.2 (Non-Diophantine Shannon entropy). The amount of non-Diophantine information in an event the non-Diophantine probability of which is p'_k, with the normalization $\sum_k{}^{\oplus_X} p'_k = 1_X$, is naturally defined as

$$S = \sum_k{}^{\oplus_Y} p_k'^{1XY} \otimes_Y \mathrm{Ln}_Y(1_X \oslash_X p'_k) = f_Y^{-1}\left(-\sum_k f_X(p'_k)\ln f_X(p'_k)\right)$$

$$= f_Y^{-1}\left(-\sum_k p_k \ln p_k\right), \tag{6.78}$$

where $\sum_{k=1}^n f_X(p'_k) = \sum_{k=1}^n p_k = 1$.

Generalized arithmetics were explicitly employed in the context of thermodynamics by Kaniadikis (2001, 2002, 2005, 2011, 2012), Kaniadikis and Scarfone (2002), Borges (2004), Nivanen, *et al.*, (2003), Sunehag (2007), Chung and Hassanabadi (2019), Chung and Hounkonnou (2020), Jizba and Korbel (2020). Classes of generalized exponents were introduced in the same context by Naudts (2002).

6.7. Non-Newtonian Quantum Mechanics

Reality is that which, when you stop believing in it, doesn't go away.

Phillip K. Dick

Example 6.7.1 (One-dimensional stationary Schrödinger equation with an arbitrary arithmetic). In this simple example one does not need complex numbers. Let $\eta : X \to Y$ be a solution of

$$E \otimes_Y \eta(x) = \ominus_Y \varepsilon \otimes_Y \Delta\eta(x) \oplus_Y U(x) \otimes_Y \eta(x), \tag{6.79}$$

where Δ is a one-dimensional non-Newtonian Laplacian (i.e., a non-Newtonian second derivative), and $\varepsilon \in Y$ a parameter. Normalization of states is performed in the form

$$1_Y = \langle\eta|\eta\rangle = \int_{f_Y^{-1}(-\infty)}^{f_Y^{-1}(\infty)} \eta(x)^{2_Y} \mathrm{D}x. \tag{6.80}$$

Probability of finding a particle in $[a,b] \subset X$ is equal to

$$P(a,b) = \int_a^b \eta(x)^{2_Y} \mathrm{D}x. \tag{6.81}$$

As usually, $\eta = f_Y^{-1} \circ \tilde{\eta} \circ f_X$, $U = f_Y^{-1} \circ \tilde{U} \circ f_X$. Let $\tilde{\eta}''(f_X(x))$ be the Newtonian second derivative of $\tilde{\eta}$ with respect to $f_X(x)$, so that the Schrödinger equation is equivalent to the standard equation

$$f_Y(E)\tilde{\eta}(f_X(x)) = -f_Y(\varepsilon)\tilde{\eta}''(f_X(x)) + \tilde{U}(f_X(x))\tilde{\eta}(f_X(x)) \tag{6.82}$$

but with redefined parameters. In the simplified notation, we have

$$f_Y(E)\tilde{\eta}(r) = -f_Y(\varepsilon)\tilde{\eta}''(r) + \tilde{U}(r)\tilde{\eta}(r) \tag{6.83}$$

$$1 = \langle\tilde{\eta}|\tilde{\eta}\rangle = \int_{-\infty}^{\infty} \tilde{\eta}(r)^2 \mathrm{d}r. \tag{6.84}$$

Now let us consider our well-known example when $X = Y = R$, $f_X(x) = x^3$, $f_Y(x) = x$. Then $\eta = \tilde{\eta} \circ f_X$, $U = \tilde{U} \circ f_X$, and the Schrödinger equation has the form

$$E\tilde{\eta}(r) = -\varepsilon\tilde{\eta}''(r) + \tilde{U}(r)\tilde{\eta}(r). \tag{6.85}$$

So, apparently the problem is completely equivalent to the standard one. However, the probability is

$$P(a,b) = \int_{f_X(a)}^{f_X(b)} \tilde{\eta}(r)^2 \mathrm{d}r = \int_{a^3}^{b^3} \tilde{\eta}(r)^2 \mathrm{d}r. \tag{6.86}$$

As we can see, in spite of mathematical triviality of the problem, the non-Diophantine arithmetic of X does influence the probability of finding the particle in the interval $[a,b]$. Figure 6.13 shows the probability of finding the particle in $[-a,a]$ as a function of a.

Example 6.7.2. Now let $f_X(x) = x^3 = f_Y(x)$, and consider the same system as in Figure 6.13. Then

$$E^3\tilde{\eta}(r) = -\varepsilon^3\tilde{\eta}''(r) + r^2\tilde{\eta}(r) \tag{6.87}$$

with unchanged normalization

$$1 = \langle\tilde{\eta}|\tilde{\eta}\rangle = \int_{-\infty}^{\infty} \tilde{\eta}(r)^2 \mathrm{d}r. \tag{6.88}$$

Leaving aside the fact that the eigenvalue E is arithmetic dependent (since $E^3 = f_Y(E)$, and not directly E occurs in the equation), the probability reads

$$P(a,b) = \sqrt[3]{\int_{a^3}^{b^3} \tilde{\eta}(r)^2 \mathrm{d}r}. \tag{6.89}$$

Obviously, $P(-\infty,\infty) = 1$, and if $a < b < c$ we get the additivity

$$\begin{aligned} P(a,b) \oplus_Y P(b,c) &= \sqrt[3]{P(a,b)^3 + P(b,c)^3} \\ &= P(a,c). \end{aligned} \tag{6.90}$$

The ground state is $\tilde{\eta}(r) = \pi^{-1/4}\exp(-r^2/2)$, but the probabilities from Figure 6.13 get modified, as shown in Figure 6.14. A change of arithmetic does not influence the mathematical problem itself, but changes its physical interpretation.

Example 6.7.3 (Analogy to quantum mechanical amplitudes). Formula (6.90) strikingly resembles the rule for addition of probability amplitudes in quantum mechanics. Let us make the analogy even stronger. Consider $f_X(x) = x$, $f_Y(x) = x/|x|^{-1/2}$ and $f_Y^{-1}(x) = x^3/|x|$. Now,

$$P(a,b) = \left(\int_a^b \tilde{\eta}(r)^2 \mathrm{d}r\right)^2. \tag{6.91}$$

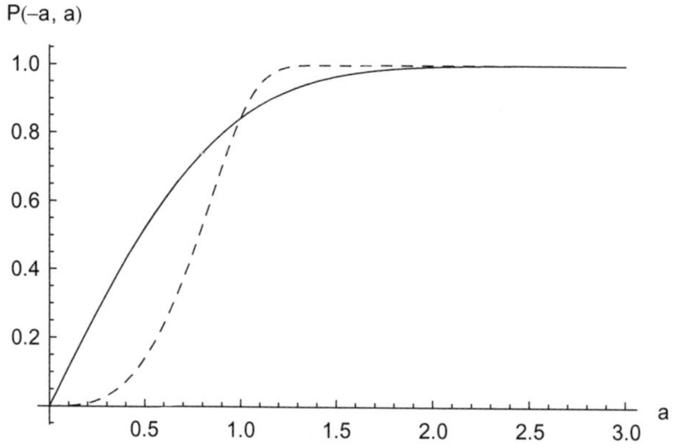

Figure 6.13. Probability of finding a particle in $[-a,a]$ for $0 \le a \le 3$, with $\tilde{\eta}(r) = \pi^{-1/4}\exp(-r^2/2)$ representing the ground state of a quantum harmonic oscillator, in dimensionless units, $\varepsilon = 1$, $\tilde{U}(r) = r^2$, for the Diophantine (full) and non-Diophantine arithmetic in R (dashed) defined by $f_X(x) = x^3$. The Diophantine arithmetic is experimentally distinguishable from a non-Diophantine one.

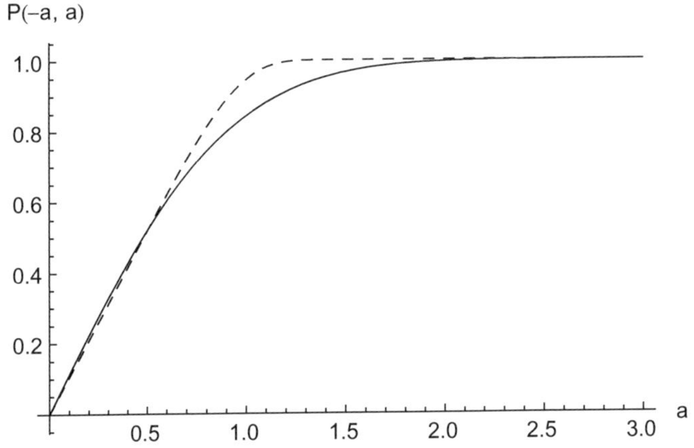

Figure 6.14. The same situation as in Figure 6.13, but now with $f_X(x) = x^3 = f_Y(x)$.

Non-Diophantine addition looks like a superposition principle from quantum mechanics,

$$P(a,b) \oplus_Y P(b,c) = (\sqrt{P(a,b)} + \sqrt{P(b,c)})^2 = P(a,c). \tag{6.92}$$

6.8. A Non-Diophantine Correspondence Principle

It is as fatal as it is cowardly to blink facts because they are not to our taste.

John Tyndall

Example 6.8.1 (Non-Diophantine-to-Diophantine correspondence principle). Non-relativistic physics is obtained from Einsteinian special relativity theory in the limit where c, the velocity of light in vacuum, is regarded as infinitely great as compared to other physically meaningful velocities implying $v/c \to 0$. Classical mechanics is obtained from quantum mechanics in the limit where the Planck constant h is regarded as infinitely small. Recipes based on $v/c \to 0$ and $h \to 0$ are known as correspondence principles. Thermodynamic limits, occurring in very large systems, can be regarded as correspondence principles linking discrete and continuous media. The non-relativistic limit, when represented at the level of Lie algebras, is defined through the Inonu–Wigner contraction (Inonu and Wigner, 1953) of the Lie algebra SO(3,1), which is a form of a correspondence principle. It is natural to ask if one can construct an analogous principle, linking non-Diophantine and Diophantine arithmetics, or non-Newtonian and Newtonian calculi.

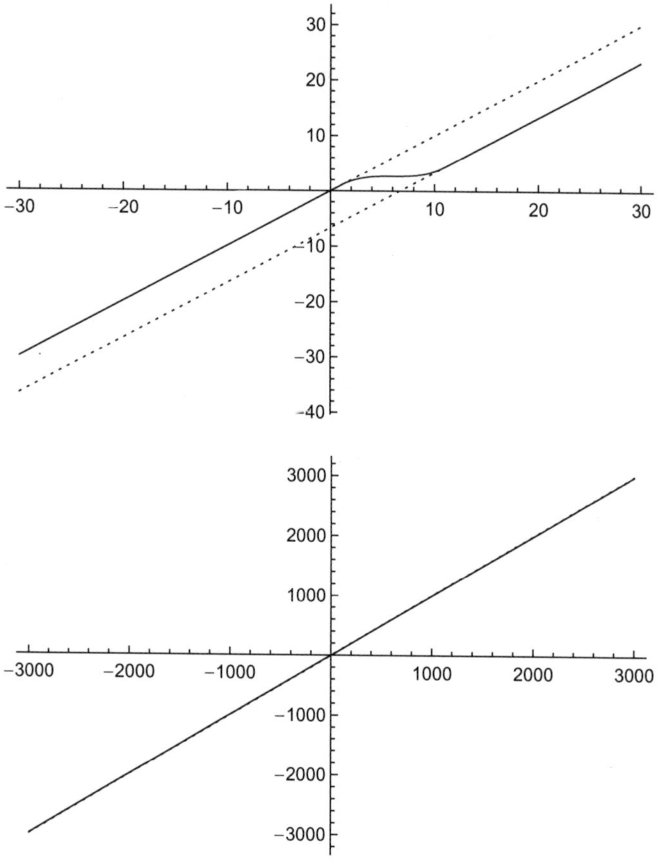

Figure 6.15. Diophantine correspondence principle.
Top graph: The function $f(x)$ that lies between $f_0(x) = x$ (for $x \leq 1$) and $f_1(x) = x - 20/3$ (for $x \geq 11$). In the transition region $f(x) = (-16 + 108x - 18x^2 + x^3)/75$, the function $f\colon R \to R$ is bijective and satisfies conditions $f(0) = 0$ and $f(1) = 1$. Arithmetic defined by the functional parameter f is non-Diophantine, but is well approximated by the standard (Diophantine) functional parameter $f_0(x) = x$ when x is large enough.
Bottom graph: The same function $f(x)$ plotted in a different scale. Here $f(x)$ is practically indistinguishable from the function $f_0(x) = x$ and thus, the resulting arithmetic is effectively Diophantine.

Let us consider the two plots of the same function $f(x)$ presented in Figure 6.15. The right one is practically indistinguishable from $f(x) = x$, which means that numbers that are sufficiently large will satisfy the laws of the Diophantine arithmetic and Newtonian calculus, with $x \oplus y \approx x + y$, $x \otimes y \approx xy$, $\mathrm{D}/\mathrm{D}x \approx \mathrm{d}/\mathrm{d}x$, etc. Functional parameters f that entail such a

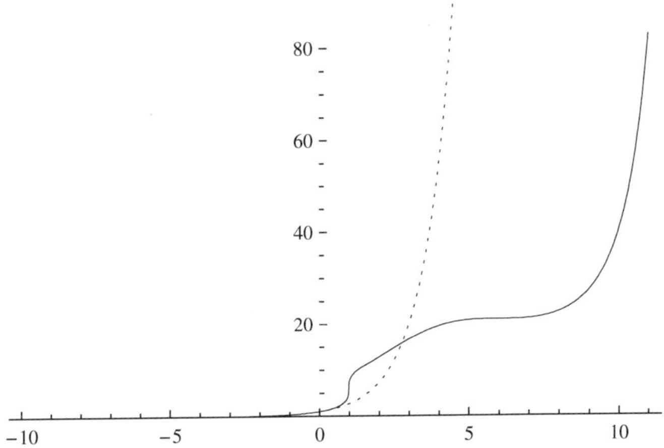

Figure 6.16. Ordinary e^x (dotted) versus Exp x defined in terms of the function f from Figure 6.15. Note the "inflation" and "dark energy" parts of Exp x.

correspondence principle can be characterized by the condition

$$\lim_{\lambda \to \infty} f(\lambda x)/\lambda = f(x). \tag{6.93}$$

The exponential function Exp x defined in the non-Diophantine arithmetic with the functional parameter $f(x)$ from Figure 6.15 is presented in Figure 6.16. As it is possible to see, the resulting exponent Exp x involves two standard exponential regimes separated by a "non-exponential" part. Still, one has to keep in mind that this awkward-looking function does nevertheless satisfy the fundamental property

$$\text{Exp}(x_1 \oplus x_2) = \text{Exp}\, x_1 \otimes \text{Exp}\, x_2. \tag{6.94}$$

So, this is just *the* exponential function in the non-Diophantine arithmetic and calculus with the functional parameter f. The exponent Exp x from Figure 6.16 has striking similarity to the expansion factor occurring in the ΛCDM standard model of cosmology (Frieman *et al.*, 2008). The two standard exponential regimes thus can be termed the inflationary and dark-energy parts of Exp x.

Chapter 7

Non-Diophantine Arithmetic in Psychophysics

Our ability to handle life's challenges is a measure of our strength of character.

Les Brown

Human and animal nervous systems are information channels, creating filters between *us* and the physical reality. A change of a physical stimulus, $x \to x + \Delta(x)$, is detectable by a subject if $\Delta(x)$ exceeds a certain *just-noticeable difference* Δ_{jnd} (Luce and Edwards, 1958). In arithmetic terms, we have the equality $x + \Delta = x$ when $0 < \Delta < \Delta_{\text{jnd}}$. This equality, as we have seen, looks as an equality from a non-Diophantine arithmetic (cf. Chapter 2). Moreover, the Weber–Fechner law states that the difference between physical stimuli $x + kx$ and x is perceived as being independent of x (Fechner, 1860; Baird and Noma, 1978). The law suggests that our nervous system implements a non-Diophantine subtraction $\ominus$, making $(x + kx) \ominus x$ independent of x. Recent psychophysical studies of acoustic stimuli (Friedrich and Heil, 2017; Friedrich, 2019) are explicitly based on a generalized arithmetic. Modern approaches to psychophysics, going in general far beyond the Fechner methods, are experimentally subtle and mathematically precise, and thus are very interesting from our non-Diophantine perspective.

7.1. Weber and Sensitivity Functions

The sensitivity of men to small matters,
and their indifference to great ones,
indicates a strange inversion.

Blaise Pascal

Given a stimulus value x, what is the smallest increment $\Delta(x)$ such that $x + \Delta(x)$ *just noticeably* exceeds x? The question, as such, is ill posed and depends on the meaning we give to the expression "just noticeably". In the modern literature (Falmagne, 1985), one replaces $\Delta(x)$ by a *Weber function* of two variables, $(x, p) \mapsto \Delta_p(x)$, where $0 < p < 1$ is a probability related to some experimental criterion. For example, $x + \Delta_{0.75}(x)$ is experimentally judged as exceeding x on 75% of trials. Let us assume that the *constant error* $\Delta_{0.5}(x) = 0$ for all x. If the constant error is zero, then this element is in essence a neutral element of addition in some arithmetic. If one expects that $\Delta_p(x) < 0$ for $0 < p < 0.5$, then, in most cases, $x + \Delta_p(x)$ has to be smaller than x.

If $x + \Delta_p(x) \in X \subset R^+$ is with probability p judged as exceeding x, it means that there exists a function $p_x \colon X \to (0, 1)$ such that

$$p_x(y) = \text{"probability that } y \text{ is judged as greater than } x\text{"}, \tag{7.1}$$

$$p_x(x + \Delta_p(x)) = p. \tag{7.2}$$

The function p_x is called a *psychometric function.* Psychometric functions are assumed to be strictly increasing. For $x \in R^+$, one may require the following additional conditions:

$$\lim_{y \to 0} p_x(y) = 0, \tag{7.3}$$

$$\lim_{y \to \infty} p_x(y) = 1, \tag{7.4}$$

$$p_x(x + \Delta_{1/2}(x)) = p_x(x) = 1/2. \tag{7.5}$$

Sometimes $p_x(y)$ is interpreted as the probability of detecting a stimulus y over a noisy background x. In such a case, variables x and y can represent different types of physical variables. However, in what follows, we assume that x, $y \in R^+$. For any variable $x \in X$ and any number p with $0 < p < 1$, the psychometric function is invertible and the formula

$$p_x^{-1}(p) = \xi_p(x) \tag{7.6}$$

defines the *sensitivity function* $x \mapsto \xi_p(x)$. Weber functions are in one-to-one correspondence with sensitivity functions,

$$\xi_p(x) = p_x^{-1}(p) = p_x^{-1}[p_x(x + \Delta_p(x))] = x + \Delta_p(x), \tag{7.7}$$

and can be experimentally determined by a number of methods. If one increases the value of some stimulus from x to y, then $y = \xi_p(x)$ is the value it achieves, at the moment the observer notices that x has changed (in $100p\%$ of cases). Let us assume that for any pair of non-equal stimuli x, $y \in X \subset R^+$, which are greater than a certain detection threshold $x_0 > 0$, there exists $p \in (0, 1)$ such that $y = \xi_p(x)$.

A *Weber fraction* is defined as

$$\Delta_p(x)/x = \xi_p(x)/x - 1. \tag{7.8}$$

Example 7.1 (The Weber law). The traditional Weber law states that the Weber fraction $\Delta_p(x)/x = \text{const}$ if $x > 0$ is an intensity of a stimulus. This law is approximately valid in some experimental cases, and its more modern versions have the form

$$\Delta_p(\lambda x) = \lambda \Delta_p(x) \tag{7.9}$$

and

$$p_{\lambda x}(\lambda y) = p_x(y). \tag{7.10}$$

The latter means that the map $(x, y) \mapsto p_x(y)$ is 0-homogeneous. In terms of measurement, equation (7.10) is typically expressed by the statement that the Weber fraction does not vary with the intensity of the stimulus. Real data show that the Weber law fails for relatively small or large values of x or y (Figure 7.1). Instead, what one finds is the *Weber inequality*

$$p_{\lambda x}(\lambda y) \geq p_x(y) \tag{7.11}$$

whenever $y \geq x$ and $\lambda \geq 1$.

Example 7.2 (Weber and sensitivity functions for the Weber law). Assume $p_{\lambda x}(\lambda y) = p_x(y)$. Setting $\lambda = 1/x$ or $\lambda = 1/y$ we get $p_x(y) = p_1(y/x) = F(y/x)$ or $p_x(y) = p_{x/y}(1) = G(x/y)$. Let $p_x(y) = F(y/x)$. Inverting $y \mapsto p_x(y) = p$ we find $xF^{-1}(p) = y$, so

$$p_x^{-1}(p) = xF^{-1}(p) = \xi_p(x) = x + \Delta_p(x), \tag{7.12}$$

$$\Delta_p(x) = \xi_p(x) - x = (F^{-1}(p) - 1)x. \tag{7.13}$$

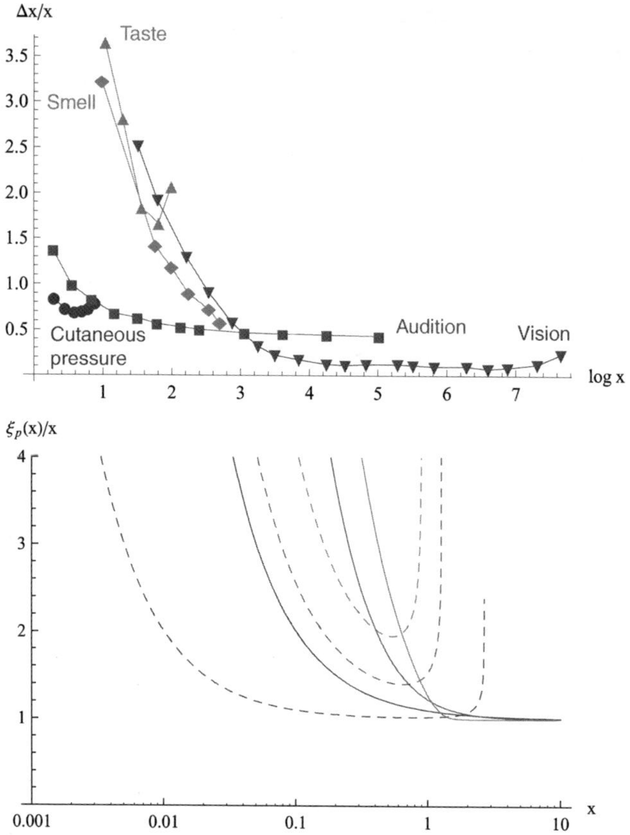

Figure 7.1. Typical Weber-fraction data versus sensation level of the stimulus, normalized to 1 at the origin (adapted from Luce *et al.* (1963)). The traditional Weber law corresponds to the flat parts of the curves. Below, there are the plots of $\xi_p(x)/x$ for different $u(x)$. Full curves: $u(x) = x$, $\omega_p = 0.1$ (leftmost), $u(x) = x^2$, $\omega_p = 0.5$ (middle), $u(x) = x^{10}$, $\omega_p = 10$ (rightmost). Dashed curves: $u(x) = \tanh x$ with $\omega_p = 0.01$ (lowest), $\omega_p = 0.15$ (middle), $\omega_p = 0.29$ (upper). The bending-up occurring for the dashed lines is a consequence of bounding $u(x)$ from above. The plots for $\Delta_p(x)/x$ can be obtained by formula (7.8).

The normalization $\Delta_{1/2}(x) = 0$ implies that $F^{-1}(1/2) = 1$, so $p_x(x) = F(1) = 1/2$. For $p = {}^1\!/\!_2$, the sensitivity function is the identity map $\xi_{1/2}(x) = xF^{-1}({}^1\!/\!_2) = x$. Furthermore, $\lim_{y\to 0} p_x(y) = \lim_{y\to 0} F(y/x) = F(0) = 0$ and $\lim_{y\to\infty} p_x(y) = \lim_{y\to\infty} F(y/x) = 1$ imply that the function $F\colon R_0^+ \to [0, 1)$ satisfies $F(0) = 0$, $\lim_{x\to\infty} F(x) = 1$ and must be strictly increasing. The Weber "constant" is equal to $\Delta_p(x)/x = F^{-1}(p) - 1$.

This brings us to a non-Diophantine arithmetic.

Example 7.3 (Fechnerian arithmetic). Consider $x \in R^+$. Let us find a generalized arithmetic such that $(x + kx) \ominus x$ is independent of x for any $k > 0$. In other words, we have to find f by solving the equation

$$(x + kx) \ominus x = f^{-1}(f(x + kx) - f(x)) = \delta_k \tag{7.14}$$

with x-independent δ_k. Acting with f on both sides of (7.14), we get the following equation:

$$f(k'x) - f(x) = f(\delta_k), \quad k' = 1 + k, \tag{7.15}$$

whose unique solution is given by $f(x) = \mu \ln x + \nu$ where μ, $\nu \in R$ and $\mu \neq 0$. The function $f(x)$ is the bijection that defines a non-Diophantine arithmetic in R^+, which is called the Fechnerian arithmetic. The neutral elements of this arithmetic are $0' = f^{-1}(0) = e^{-\nu/\mu}$ and $1' = f^{-1}(1) = e^{(1-\nu)/\mu}$. We considered this arithmetic in Section 6.3. The function $f(x)$ is an example of the so-called sensory scale.

Example 7.4 (Weber–Fechner exponential function). The exponential function is defined as the solution of

$$\frac{\mathrm{D}\,\mathrm{Exp}\,x}{\mathrm{D}x} = \mathrm{Exp}\,x, \tag{7.16}$$

$$\mathrm{Exp}(0') = 1'. \tag{7.17}$$

where

$$\frac{\mathrm{D}\ \mathrm{Exp}\ x}{\mathrm{D}x} = \lim_{h \to 0'} (\mathrm{Exp}(x \oplus h) \ominus \mathrm{Exp}(x)) \oslash h \tag{7.18}$$

is defined in terms of the Fechnerian arithmetic. The unique solution is $\mathrm{Exp}\ x = f^{-1}(e^{f(x)})$, and thus

$$\mathrm{Exp}\,x = e^{(e^{\mu \ln x + \nu} - \nu)/\mu} = e^{e^{\nu} x^{\mu}/\mu} e^{-\nu/\mu}. \tag{7.19}$$

Let us check the initial condition:

$$\mathrm{Exp}(0') = \mathrm{Exp}(e^{-\nu/\mu}) = e^{(e^{\nu}(e^{-\nu/\mu})^{\mu} - \nu)/\mu} = e^{(1-\nu)/\mu} = 1'. \tag{7.20}$$

It is an instructive exercise to compute the derivative directly from definition:

$$\frac{\mathrm{D}\,\mathrm{Exp}\,x}{\mathrm{D}x} = \lim_{h\to 0'} (\mathrm{Exp}(x \oplus h) \ominus \mathrm{Exp}(x)) \oslash h \tag{7.21}$$

$$= \lim_{h\to 0'} (e^{e^{\nu}(x\oplus h)^{\mu}/\mu} e^{-\nu/\mu} \ominus e^{e^{\nu}x^{\mu}/\mu} e^{-\nu/\mu}) \oslash h \tag{7.22}$$

$$= \lim_{h\to e^{-\nu/\mu}} (e^{e^{\nu}(xhe^{\nu/\mu})^{\mu}/\mu} e^{-\nu/\mu} \ominus e^{e^{\nu}x^{\mu}/\mu} e^{-\nu/\mu}) \oslash h \tag{7.23}$$

$$= \lim_{h\to e^{-\nu/\mu}} \left(e^{-\nu/\mu} \frac{e^{e^{\nu}(xhe^{\nu/\mu})^{\mu}/\mu} e^{-\nu/\mu}}{e^{e^{\nu}x^{\mu}/\mu} e^{-\nu/\mu}} \right) \oslash h \tag{7.24}$$

$$= \lim_{h\to e^{-\nu/\mu}} e^{(\ln e^{-\nu/\mu + x^{\mu}h^{\mu}e^{2\nu}/\mu - e^{\nu}x^{\mu}/\mu} + \nu/\mu)/(\mu\ln h + \nu) - \nu/\mu} \tag{7.25}$$

$$= \lim_{h\to e^{-\nu/\mu}} e^{(-\nu/\mu + x^{\mu}h^{\mu}e^{2\nu}/\mu - e^{\nu}x^{\mu}/\mu + \nu/\mu)/(\mu\ln h + \nu) - \nu/\mu} \tag{7.26}$$

$$= \lim_{h\to e^{-\nu/\mu}} e^{(x^{\mu}h^{\mu}e^{2\nu}/\mu - e^{\nu}x^{\mu}/\mu)/(\mu\ln h + \nu) - \nu/\mu} \tag{7.27}$$

$$= \lim_{c\to\nu} e^{(x^{\mu}(e^{-c/\mu})^{\mu}e^{2\nu}/\mu - e^{\nu}x^{\mu}/\mu)/(\mu\ln e^{-c/\mu} + \nu) - \nu/\mu} \tag{7.28}$$

$$= \lim_{c\to\nu} e^{\frac{e^{\nu-c}-1}{\nu-c} e^{\nu}x^{\mu}/\mu - \nu/\mu} \tag{7.29}$$

$$= e^{e^{\nu}x^{\mu}/\mu - \nu/\mu} = \mathrm{Exp}\,x. \tag{7.30}$$

One further finds that $\mathrm{Exp}(x \oplus y) = \mathrm{Exp}\,x \otimes \mathrm{Exp}\,y$. The inverse function

$$\mathrm{Ln}\,x = f^{-1}(\ln f(x)) = e^{(\ln f(x) - \nu)/\mu} = e^{-\nu/\mu} f(x) = e^{-\nu/\mu}(\mu \ln x + \nu) \tag{7.31}$$

satisfies $\mathrm{Ln}(x \otimes y) = \mathrm{Ln}\,x \oplus \mathrm{Ln}\,y$.

7.2. Fechner Problems and Sensory Scales

We live in an ascending scale when we live happily, one thing leading to another in an endless series.

Robert Louis Stevenson

A *Fechner problem* is to find a *sensory scale* such that two stimuli differing by $\Delta_p(x)$ will be perceived as differing by a number which is independent of x. Let x_0 be a positive real number that plays a role of a detection threshold in psychophysical measurements. A sensory scale u is any solution of the following problem.

Problem 7.1. For a fixed p, $0 < p < 1$, find those invertible functions $u : R^+ \supset X \to Y \subset R$ such that, for all $x > x_0 > 0$,

$$u(x + \Delta_p(x)) - u(x) = u(\xi_p(x)) - u(x) = \omega_p \tag{7.32}$$

is independent of x, where ω_p is a strictly monotone increasing function of p.

Equation (7.32) is an example of Abel's equation (Luce and Edwards, 1958; Kuczma *et al.*, 1990). Rewriting (7.32) as

$$u^{-1}(u(x + \Delta_p(x)) - u(x)) = u^{-1}(u(\xi_p(x)) - u(x)) = u^{-1}(\omega_p), \tag{7.33}$$

it is possible to ask the previous question in a different way.

Problem 7.2. For a fixed p, $0 < p < 1$, find a non-Diophantine arithmetic such that, for all $x > x_0 > 0$,

$$\xi_p(x) \ominus x = \tilde{\omega}_p \tag{7.34}$$

is independent of x, where $\tilde{\omega}_p$ is an invertible function of p.

A solution of Problem 7.2 means that the sensitivity function can be represented as a non-Diophantine sum $\xi_p(x) = x \oplus \tilde{\omega}_p$. Example 7.3 showed that $u(x) = \mu \ln x + \nu$ is a solution of both problems for a constant Weber fraction. For a more general case, we have the following uniqueness theorem (Luce *et al.*, 1963).

Theorem 7.1. *For a given ω_p, let u be an invertible solution of* (7.32). *Then $\tilde{u}$ is another solution of* (7.32) *if and only if*

$$\tilde{u}(x) = u(x) + \nu(u(x)), \tag{7.35}$$

where ν is periodic with period ω_p.

Proof: Let $u(x)$ be a solution of (7.32) and ν be a periodic function with the period ω_p. Then, taking the function $\tilde{u}(x) = u(x) + \nu(u(x))$, we have

$$\begin{aligned}\tilde{u}(x + \Delta_p(x)) - \tilde{u}(x) &= u(x + \Delta_p(x)) - u(x) + \nu(u(x + \Delta_p(x))) - \nu(u(x)) \\ &= \omega_p + \nu(u(x) + \omega_p) - \nu(u(x)) = \omega_p \end{aligned} \tag{7.36}$$

since $\nu(u(x) + \omega_p) = \nu(u(x))$. It means that $\tilde{u}(x)$ is a solution of (7.32).

Now, assume that

$$u(x + \Delta_p(x)) - u(x) = \omega_p, \tag{7.37}$$

$$\tilde{u}(x + \Delta_p(x)) - \tilde{u}(x) = \omega_p, \tag{7.38}$$

and define $w = \tilde{u} - u$. Accordingly,

$$w(x + \Delta_p(x)) - w(x) = \tilde{u}(x + \Delta_p(x)) - \tilde{u}(x) - u(x + \Delta_p(x)) + u(x)$$
$$= \omega_p - \omega_p = 0. \tag{7.39}$$

The inverse function u^{-1} exists by the assumption, so

$$u(x + \Delta_p(x)) = u(x) + \omega_p, \tag{7.40}$$

$$x + \Delta_p(x) = u^{-1}(u(x) + \omega_p). \tag{7.41}$$

Set $\nu = w \circ u^{-1}$. Then, we obtain

$$0 = w(x + \Delta_p(x)) - w(x) = w(u^{-1}(u(x) + \omega_p)) - w(u^{-1}(u(x)))$$
$$= \nu(u(x) + \omega_p) - \nu(u(x)). \tag{7.42}$$

This proves periodicity of the function ν with period ω_p. Finally,

$$\tilde{u}(x) = u(x) + \tilde{u}(x) - u(x) = u(x) + w(x)$$
$$= u(x) + w(u^{-1}(u(x))) = u(x) + \nu(u(x)). \tag{7.43}$$

The proof is complete. □

Let us note that for a non-trivial function ν, the new function $\tilde{u}$ may not be invertible, and thus, it will not lead to an acceptable sensory scale. A non-invertible function $\tilde{u}$ will not define a non-Diophantine arithmetic either. So, following Luce and Galanter, let us consider a *revised Fechner problem.*

Problem 7.3. Find those invertible functions u: $R^+ \supset X \to Y \subset R$, such that, for all $x > x_0 > 0$, and for all p, $0 < p < 1$,

$$u(x + \Delta_p(x)) - u(x) = \omega_p \tag{7.44}$$

is independent of x, where ω_p is a strictly monotone increasing function of p.

For a given ω_p, let u and $\tilde{u}$ be two different solutions of Problem 7.3.

Theorem 7.2. *The difference* $\nu(x) = \tilde{u}(x) - u(x)$ *is a constant function.*

Proof: The function $\nu(x)$ is periodic with the period ω_p for any $0 < p < 1$. Since the function $p \mapsto \omega_p$ is strictly monotone, $\nu(x)$ must be a constant. □

If we relax the condition that ω_p is a given concrete function of p, but demand only that ω_p be a strictly monotone function of p, which is independent of x, then $u(x)$ is given uniquely up to an arbitrary affine transformation $u(x) \mapsto \alpha u(x) + \beta$. The same type of arbitrariness is characteristic for Kolmogorov–Nagumo averages (Kolmogorov, 1930; Nagumo, 1930).

7.3. Fechnerian Psychometric Functions

You have to have confidence in your ability,
and then be tough enough to follow through.

Rosalynn Carter

Almost all models for discrimination of data considered in modern psychophysics involve psychometric functions of the form

$$p_x(y) = F(u(y) - u(x)) \tag{7.45}$$

for some strictly increasing F, $u : R \to R$. The difference $u(y) - u(x)$ measures *subjective dissimilarity* of stimuli x and y (Dzhafarov and Colonius, 2011). In some cases, one encounters the following generalization of (7.45),

$$p_x(y) = F(u(y) - u_1(x)). \tag{7.46}$$

Typically, one demands a continuity of F, u, and u_1, but this notion is topology dependent, of course, a fact of special importance for fractal applications.

Example 7.5 (Strict utility model). Let $n(x)$ be a number of neural channels triggered by stimulus x. Under certain technical conditions (cf. Yellot, 1977), one arrives at

$$\begin{aligned} p_x(y) &= \frac{n(y)}{n(y) + n(x)} = \frac{1}{1 + e^{-(\ln n(y) - \ln n(y))}} \\ &= F(u(y) - u(x)), \end{aligned} \tag{7.47}$$

$$F(s) = \frac{1}{1+e^{-s}}, \tag{7.48}$$

$$u(x) = \ln n(x). \tag{7.49}$$

The function $F(s)$ is known as the logistic distribution. The sensory scale $u(x)$ is essentially the number of digits needed to code the value of $n(x)$.

Psychometric functions (7.45) are naturally related to a non-Diophantine addition,

$$p_x(y) \oplus p_y(z) = F(F^{-1}(p_x(y)) + F^{-1}(p_y(z))) = p_x(z). \tag{7.50}$$

The equation has the same form as the non-Diophantine formulas (6.126)–(6.128) for probabilities, which we encountered in the context of quantum mechanics and Kolmogorov–Nagumo averages (Example 6.9).

If Problem 7.1 has a solution (u, ω_p) for a given sensitivity function $\xi_p(x)$, we say that $\xi_p(x)$ is Fechnerian. The psychometric function $p_x(y)$ is called Fechnerian if $\xi_p(x) = p_x^{-1}(p)$ is Fechnerian. The obvious question (Luce *et al.*, 1963; Falmagne, 1971) is as follows.

Problem 7.4. Is there any link between Fechnerian psychometric functions and functions of the form (7.45)?

Assume, for any x, y, there exists p such that $y = \xi_p(x)$.

Theorem 7.3. (a) *The function $p_x(y)$ has the form* (7.45) *if and only if $p_x(y)$ is Fechnerian.*

(b) *if u and ω_p are solutions* (7.32) *for a Fechnerian function $\xi_p(x)$, then there exists a function F such that $p_x(y)$ has the form* (7.45).

Proof: By assumption, for any x, y, there exists p such that $y = \xi_p(x)$. Since $\xi_p(x)$ is Fechnerian, the problem $u(\xi_p(x)) - u(x) = \omega_p = g(p)$ has a solution. The map $p \mapsto g(p)$ is invertible, so

$$g^{-1}(u(\xi_p(x)) - u(x)) = p = p_x(\xi_p(x)) \tag{7.51}$$

and thus,

$$p_x(y) = g^{-1}(u(y) - u(x)) \tag{7.52}$$

implying $F = g^{-1}$. Now assume $p_x(y) = F(u(y) - u(x))$. For $y = \xi_p(x)$,

$$p = p_x(\xi_p(x)) = F(u(\xi_p(x)) - u(x)), \tag{7.53}$$

so

$$u(\xi_p(x)) - u(x) = F^{-1}(p) = \omega_p, \tag{7.54}$$

which ends the proof. □

7.4. Does the Scale Entail Arithmetic?

It's a really paradoxical thing.
We want to think big, but start small.
And then scale fast.

Eric Ries

Weber and sensitivity functions are definable in experiments. From our perspective, it is especially interesting if experiments can tell us anything about the form of arithmetic implicitly employed by real-life neural systems.

Let us begin with the more general form of $p_x(y)$ given by (7.46). A psychophysicist interprets u and u_1 as representing a rescaling of physical variables by a sensory mechanism. These functions are typically regarded as more fundamental than F, which may be affected by variables of a cognitive type (order in which x and y are presented to the subject, motivation, etc.). For this reason, in search for a fundamental non-Diophantine aspect of psychophysics, one probably should not concentrate on F-dependent expressions such as (7.50).

Now, let us fix the value $p = p_x(y)$, so that $F^{-1}(p) = u(y) - u_1(x)$, and

$$y = u^{-1}(u_1(x) + F^{-1}(p)). \tag{7.55}$$

Since $y = p_x^{-1}(p) = \xi_p(x)$, we get

$$\begin{aligned}\xi_p(x) &= y = u^{-1}(u_1(x) + F^{-1}(p))\\ &= u^{-1}[u_1(x) + u_1(u_1^{-1} \circ F^{-1}(p))] = x \oplus a_p,\end{aligned} \tag{7.56}$$

where $a_p = u_1^{-1} \circ F^{-1}(p)$. Here operation $\oplus$ is an example of addition in a general non-Diophantine pre-arithmetic. In the Fechnerian case, we have $u = u_1$ and

$$\xi_p(x) = u^{-1}[u(x) + u(u^{-1} \circ F^{-1}(p))] = x \oplus a_p, \tag{7.57}$$

where $a_p = u^{-1} \circ F^{-1}(p)$. Note that

$$\xi_{1/2}(x) = x = x \oplus a_{1/2}. \tag{7.58}$$

So, $a_{1/2} = 0'$ is the neutral element of non-Diophantine Fechnerian addition, which generalizes the logarithmic case.

The parameter p can be kept constant in an experiment, say $p = 0.75$, and thus the sensitivity function is effectively determined by two functions, u and u_1, and the parameter $F^{-1}(p)$. One of crucial assumptions behind psychophysical measurements is that u and u_1 are unaffected by nonsensory variables. Apparently, there is little experimental evidence that the assumption is invalid (Falmagne, 1985).

As we can see, Fechnerian sensitivity and psychometric functions are naturally related to a non-Diophantine arithmetic, defined in terms of the sensory scales. The sensory scales are given uniquely up to an affine transformation $u(x) \mapsto \alpha u(x) + \beta$. This *per se* does not yet imply that the arithmetic defined by $u(x)$ is the one employed in neural processing.

7.5. Inverse Fechner Problem

What [software] must not do
is not the inverse of what it must do.
Nancy Leveson

Given a concrete $\xi_p(x)$ it is in general difficult, or even impossible, to find a sensory scale u satisfying

$$u(\xi_p(x)) - u(x) = \omega_p, \quad x > 0, \tag{7.59}$$

with ω_p independent of x. An inverse problem is simple and sometimes useful (cf. Krantz, 1971). Namely, given a ω_p and an invertible function $u(x)$, one can transform (7.59) into

$$\xi_p(x) = u^{-1}(u(x) + \omega_p). \tag{7.60}$$

The formula in (7.60) determines the sensitivity function $\xi_p(x)$ such that u and ω_p solve equation (7.59). The corresponding Weber fractions

$$\Delta_p(x)/x = \xi_p(x)/x - 1 = u^{-1}(u(x) + \omega_p)/x - 1 \tag{7.61}$$

can be compared with experiment.

Example 7.6. It is clear that $\Delta_p(x)/x$ is a constant only for $u(x) = \mu \ln x + \nu$,

$$\Delta_p(x)/x = u^{-1}(u(x) + \omega_p)/x - 1 = \exp(\omega_p/\mu) - 1. \tag{7.62}$$

Example 7.7. Figure 7.1 shows examples of sensitivity functions originating from $u(x) = x^n$ and $u(x) = \tanh x$ for different values of ω_p. The curves corresponding to $u(x) = \tanh x$ bend up for large x in a way characteristic of real data. The bending is caused by the bound $\tanh x < 1$. It is natural to expect that true $u(x)$ should be bounded from above since sufficiently strong signals are indistinguishable for physiological reasons.

Comparison of theory with experiment encounters in psychophysics the same problem we have in physics: arguments of $u(x)$, in general, have to be dimensionless. If the measured value of $\xi_p(x)$ has physical units (decibels, say), then a constant with the dimension of the unit is required. Let us denote this constant by ℓ. Then, a natural convention is

$$\xi_p(x) = u^{-1}(u(x/\ell) + \omega_p)\ell', \tag{7.63}$$

where $\ell' = k\ell$ for some $k > 0$, so that the Abel equation reads

$$u(\xi_p(x)/\ell') - u(x/\ell) = \omega_p. \tag{7.64}$$

Looking at the plots from Figure 7.1, we see that the ratio $\xi_p(x)/x$ based on (7.63) tends asymptotically to k, for $x \gg \ell$, independently of the choice of units, $u(x) \sim (x/\ell)^q$. On the other hand, for $u(x) = \tanh x$, we find the late increase of $\xi_p(x)/x$, similar to the one observed in real data.

Example 7.8. Let us investigate the tanh case more closely. It is often stressed in the literature (Luce *et al.*, 1963) that most of the Weber-fraction data are correctly described by the affine function

$$\Delta_p(x) = \alpha_p x + \beta_p, \tag{7.65}$$

$$\xi_p(x) = x + \Delta_p(x) = (\alpha_p + 1)x + \beta_p, \tag{7.66}$$

$$\Delta_p(x)/x = \alpha_p + \beta_p/x. \tag{7.67}$$

The generalized Weber law (7.67) correctly describes the increase of the fraction for small x. However, the characteristic increase for large x does not occur, since asymptotically the fraction is constant,

$$\Delta_p(x)/x \approx \alpha_p, \quad \text{for } x \gg \ell, \tag{7.68}$$

in agreement with the standard Weber law. So, the first question we have to understand is if $u(x) = \tanh x$ can reconstruct the small values of x

and the plateau parts of (7.67) for any α_p and β_p. The explicit form of the sensitivity function

$$\begin{aligned}\xi_p(x) &= u^{-1}(u(x/\ell) + \omega_p)\ell' = \ell' \tanh^{-1}(\tanh(x/\ell) + \omega_p) \\ &= \frac{\ell'}{2} \ln \frac{1 + \tanh(x/\ell) + \omega_p}{1 - \tanh(x/\ell) - \omega_p}\end{aligned} \tag{7.69}$$

shows that

$$\xi_p : (-\infty, \ell \tanh^{-1}(1 - \omega_p)) \to \left(\frac{\ell'}{2} \ln \frac{\omega_p}{2 - \omega_p}, \infty\right) \tag{7.70}$$

is a bijection. So, although we have considered non-negative values of x and $\xi_p(x)$, the solution can be continued to negative values of both the input and the output. It is difficult to judge if this property can be given some psychophysical interpretation, or is it just a mathematical curiosity.

The asymptotic behavior of (7.69), for small x, is determined by the first two terms of the Maclaurin expansion

$$\xi_p(x) = \ell' \tanh^{-1} \omega_p + \frac{x\ell'}{(1 - \omega_p^2)\ell} + O(x^2). \tag{7.71}$$

Accordingly, the parameters have to satisfy

$$\alpha_p = \frac{k}{(1 - \omega_p^2)} - 1, \tag{7.72}$$

$$\beta_p = \ell' \tanh^{-1} \omega_p = \frac{\ell'}{2} \ln \frac{1 + \omega_p}{1 - \omega_p} \tag{7.73}$$

with $0 \leq \omega_p < 1$ (Figure 7.2).

The inverse relations

$$\omega_p = \sqrt{\frac{1 + \alpha_p - k}{1 + \alpha_p}}, \quad 0 < k < 1 + \alpha_p, \tag{7.74}$$

$$\ell = \frac{\beta_p}{k \tanh^{-1} \sqrt{(1 + \alpha_p - k)/(1 + \alpha_p)}} \tag{7.75}$$

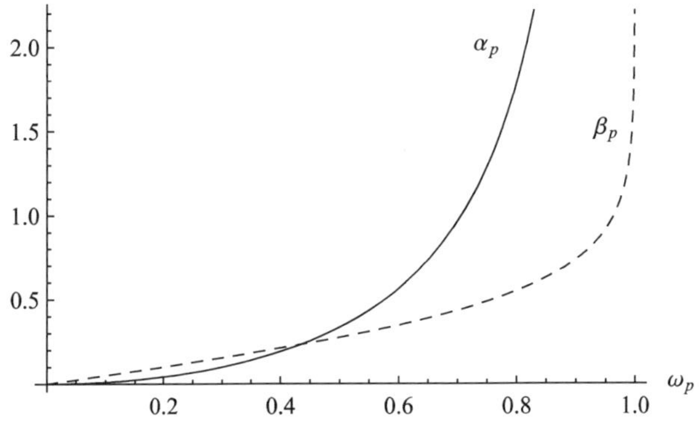

Figure 7.2. Parameters α_p and β_p given by (7.72) and (7.73) as functions of ω_p with $\ell = \ell' = 1$. The crossing point moves from left to right as ℓ increases from 0 to ∞.

define ω_p and the unit ℓ. Finally,

$$\frac{\xi_p(x)}{x} = \frac{\beta_p/x}{\tanh^{-1}\sqrt{\frac{1+\alpha_p-k}{1+\alpha_p}}} \times \tanh^{-1}\left[\tanh\left(\frac{kx}{\beta_p}\tanh^{-1}\sqrt{\frac{1+\alpha_p-k}{1+\alpha_p}}\right)+\sqrt{\frac{1+\alpha_p-k}{1+\alpha_p}}\right] \tag{7.76}$$

should be compared with (7.66). The right asymptote (a vertical line) of (7.76) in Figure 7.3 occurs at the maximal value $x = x_{\max}$,

$$x_{\max} = \ell\tanh^{-1}(1-\omega_p) = \frac{\beta_p}{k}\frac{\tanh^{-1}\left(1-\sqrt{\frac{1+\alpha_p-k}{1+\alpha_p}}\right)}{\tanh^{-1}\sqrt{\frac{1+\alpha_p-k}{1+\alpha_p}}}. \tag{7.77}$$

The smaller the value of ω_p is, the greater the maximal acceptable value of the stimulus $x = x_{\max}$ becomes.

In the context of fitting experimental data, it is relevant to discuss the affine freedom of the scale, $u(x) \mapsto \tilde{u}(x) = Cu(x) + D$. Since $\tilde{u}^{-1}(x) = u^{-1}((x-D)/C)$, we find

$$\begin{aligned}\tilde{\xi}_p(x) &= \tilde{u}^{-1}(\tilde{u}(x/\ell)+\tilde{\omega}_p)\ell' = u^{-1}\left(\frac{Cu(x/\ell)+D+\tilde{\omega}_p-D}{C}\right)\ell' \\ &= u^{-1}(u(x/\ell)+\tilde{\omega}_p/C)\ell'.\end{aligned} \tag{7.78}$$

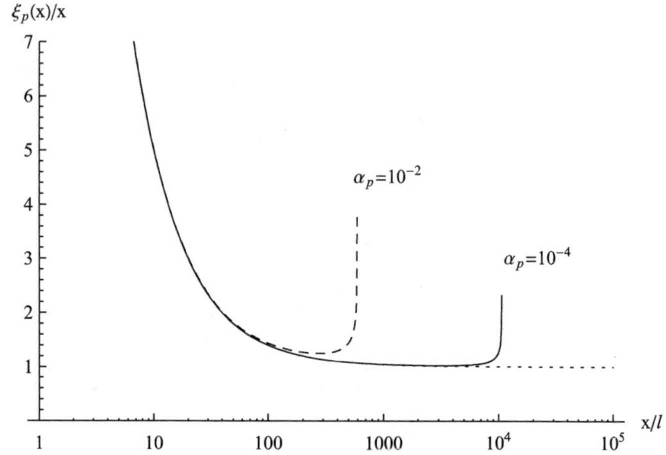

Figure 7.3. $\xi_p(x)/x$ given by (7.76) versus $\xi_p(x)/x = \alpha_p + 1 + \beta_p/x$, (dotted), for $\alpha_p = 10^{-4}$ (full), $\alpha_p = 10^{-2}$ (dashed), $\ell' = \ell$, $\beta_p = 40\ell$, as functions of x/ℓ.

The shift D is thus unobservable, so the freedom of the scale is reduced to $u(x) \mapsto Cu(x)$. The relevant Abel equations are consistent with each other,

$$\tilde{u}(\tilde{\xi}_p(x)/\ell') - \tilde{u}(x/\ell) = \tilde{\omega}_p, \tag{7.79}$$

$$u(\tilde{\xi}_p(x)/\ell') - u(x/\ell) = \tilde{\omega}_p/C. \tag{7.80}$$

It would be, of course, premature to claim that the sensory scale $u(x) = C \tanh x$ is *the* scale employed by realistic psychophysical systems. Still, it has sufficiently many appealing qualitative features to motivate its formal studies.

Chapter 8

Conclusion

Learning never exhausts the mind.
Leonardo da Vinci

Thus, we have demonstrated that, on the one hand, there is a rich and insightful theory of prearithmetics and arithmetics, which unifies and provides a common foundation for many useful mathematical structures such as rings, fields, semirings, linear algebras, ordered rings, ordered fields, topological rings, topological fields, normed rings, normed algebras, tropical and thermodynamic semirings, normed field Ω-groups, Ω-rings, lattices and Boolean algebras. Naturally, all varied utilizations of these structures are also applications of prearithmetics exhibiting their usefulness and efficacy.

On the other hand, prearithmetics and arithmetics have diverse utilizations in new areas of mathematics such as non-Newtonian calculus (Grossman and Katz, 1972; Pap, 1993; Czachor, 2016), idempotent analysis (Maslov, 1987; Maslov and Samborskii, 1992; Maslov and Kolokoltsov, 1997), tropical analysis (Litvinov, 2007; Speyer and Sturmfels, 2009), non-commutative and non-associative pseudo-analysis (Pap and Vivona, 2000; Pap and Štajner-Papuga, 2001), non-Newtonian functional analysis and idempotent functional analysis (Litvinov *et al.*, 2001). As a matter of fact, prearithmetics and non-Diophantine arithmetics form unassailable foundation for all these areas.

Besides, as we have demonstrated in this book, non-Diophantine arithmetics have diverse applications in physics (cf. also Czachor, 2016; 2017; Czachor and Posiewnik, 2016), psychology (cf. also Czachor, 2017a) and theory of fractals (cf. also Aerts *et al.*, 2016; 2016a; 2018). In addition, non-Diophantine arithmetics also found numerous applications:

- In decision making, dynamical systems, differential equations, chaos theory, economics, marketing, finance, fractal geometry, image analysis and electrical engineering as the base of the non-Newtonian calculus (Grossman and Katz, 1972; Grossman *et al.*, 1980; Pap, 1993; Mesiar and Rybařik, 1993; Pap and Vivona, 2000; Pap and Štajner-Papuga, 2001; Mora *et al.*, 2012; Filip and Piatecki, 2014);
- In optimization theory, optimal design of computer systems and media, optimal organization of data processing, dynamic programming, computer science, discrete mathematics, and mathematical logic as the base of the idempotent analysis and tropical analysis (Maslov, 1987; Maslov and Samborskii, 1992; Maslov and Kolokoltsov, 1997; Litvinov, 2007; Speyer and Sturmfels, 2009).

All these applications open new approaches to solving many other problems in the mentioned fields.

Writing about applications, it is necessary to emphasize that the discovery of non-Diophantine arithmetics demonstrated that there is no one universal arithmetic and it is crucial to find an appropriate arithmetic for each specific situation to correctly apply mathematics and efficiently solve emerging problems.

To conclude, we formulate some open problems, which, in turn, open new directions for the future research.

In Chapter 1, we discussed functions of language, specifying three basic functions — communicative or informative, expressive and instructive. Mathematics in general and arithmetic in particular are also specific languages. According to Galileo, mathematics is the language of science. That is why speaking about mathematics and arithmetic as a language only their informative function has been taken into account.

Problem 1. Explore expressive and instructive functions of mathematics and arithmetic.

It is demonstrated that mathematics as a whole has three characteristic realities: mental, structural and material realities (Burgin, 2018). Numbers as the basic structures of mathematics have their presence in all these realities (cf. Chapter 1).

Problem 2. Explore interaction between structural, mental and material reality of arithmetic.

We have proved that in an abstract prearithmetic $\boldsymbol{A} = (A; +, \circ, \leq)$ with linear order $\leq$, subtraction and monotone, commutative and associative addition, the operation *subtraction* is monotone from the left and antitone from the right.

Problem 3. Find if the statement remains true without assumption that the order is linear.

Different algebraic systems, such as rings, fields, linear algebras, and lattices, form special classes of abstract prearithmetics.

Problem 4. Select/define other useful/important classes of abstract prearithmetics and study structural properties of systems from these classes.

For instance, there are advanced structural theories of rings, fields, linear algebras, and lattices (cf., for example, Allenby, 1991; Faith, 2004; González, 1994; Grätzer, 1996; Rowen, 1988; Pierce, 1982). It would be interesting to do similar studies in the context of prearithmetics and non-Diophantine arithmetics. Namely, we have the following problem.

Problem 5. Build a structural theory of abstract prearithmetics.

There are different operations with universal algebras: unions, intersections, Cartesian products, free products, etc. (cf., for example, Kurosh, 1967).

Problem 6. Define and study operations with abstract prearithmetics.

Free universal algebras, such as free groups or free rings, play important role in algebra (cf., for example, Kurosh, 1967).

Problem 7. Define and study free arithmetics and prearithmetics.

In partially ordered algebraic systems, such as ordered groups or ordered rings, operations are monotone and there is an extended theory of these systems (Fuchs, 1963).

Problem 8. Study arithmetics and prearithmetics with monotone addition and/or multiplication.

It is proved (Proposition 2.3.26) that existence of a monomorphism from one abstract prearithmetic with the total order into another abstract prearithmetic implies projectivity of the first one with respect to the second one. This brings us to the following problem.

Problem 9. Does existence of a homomorphism from one abstract prearithmetic into another abstract prearithmetic implies weak projectivity of the first one with respect to the second one?

It is proved (Theorem 2.2.2) that inverse weak projectivity preserves additive factoring property in abstract prearithmetics. This brings us to the following problem.

Problem 10. Does inverse weak projectivity preserve strong additive factoring property in abstract prearithmetics?

It is proved (Theorem 2.2.4) that inverse weak projectivity preserves multiplicative factoring property in abstract prearithmetics. This brings us to the following problem.

Problem 11. Does inverse weak projectivity preserve strong multiplicative factoring property in abstract prearithmetics?

It is proved (Proposition 2.7.28) that there are W-prearithmetics in which addition coincides with the operation of taking the maximum.

Problem 12. Are there W-prearithmetics in which addition coincides with the operation of taking the minimum?

It is proved (Proposition 2.7.30) that there are W-prearithmetics in which multiplication coincides with the operation of taking the maximum.

Problem 13. Are there W-prearithmetics in which multiplication coincides with the operation of taking the minimum?

It is proved (Proposition 2.5.24) that in exact projectivity, projector and coprojector preserve subprearithmetics.

Problem 14. Find conditions when in weak projectivity (projectivity), projector or/and coprojector preserves subprearithmetics.

It is proved (Theorem 2.2.1) that abstract prearithmetics are objects of the category **APAWP**, in which morphisms are weak projectivity relations. This brings us to the following problem.

Problem 15. Study properties of the category **APAWP**, objects of which are abstract prearithmetics and morphisms are weak projectivity relations.

It is proved (Theorems 2.2.2–2.2.5) that additive and multiplicative factoring properties are preserved by inverse weak projectivity.

Problem 16. Find under what conditions inverse weak projectivity preserves strong additive and multiplicative factoring properties.

It is proved (Theorem 3.1.1) that partially extended abstract prearithmetics are objects of the category **PEAPAWP**, in which morphisms are weak projectivity relations. This brings us to the following problem.

Problem 17. Study properties of the category **PEAPAWP**, objects of which are partially extended abstract prearithmetics and morphisms are weak projectivity relations.

It is proved (Theorem 3.2.1) that wholly extended abstract prearithmetics are objects of the category **WEAPAWP**, in which morphisms are weak projectivity relations. This brings us to the following problem.

Problem 18. Study properties of the category **WEAPAWP**, objects of which are wholly extended abstract prearithmetics and morphisms are weak projectivity relations.

It is proved (Theorem 2.3.1) that abstract prearithmetics are objects of the category **APAP**, in which morphisms are projectivity relations. This brings us to the following problem.

Problem 19. Study properties of the category **APAP**, objects of which are abstract prearithmetics and morphisms are projectivity relations.

It is proved (Theorem 2.4.3) that abstract prearithmetics are objects of the category **APAEP**, in which morphisms are exact projectivity relations. This brings us to the following problem.

Problem 20. Study properties of the category **APAEP**, objects of which are abstract prearithmetics and morphisms are exact projectivity relations.

It is proved (Theorem 2.5.1) that a subprearithmetic of an abstract prearithmetic $\boldsymbol{A}$ is projective with respect to $\boldsymbol{A}$.

Problem 21. Find if an abstract prearithmetic $\boldsymbol{A}_2$, which is projective with respect to its subprearithmetic $\boldsymbol{A}_1$, is exactly projective with respect to $\boldsymbol{A}_1$.

In Section 2.5, we introduced and explored properties of partial subarithmetics. It is natural to do the same for universal algebras with several operations. This brings us to the following problem.

Problem 22. Study partial subalgebras of universal algebras with several operations such as rings, groups with operators or lattices.

It is demonstrated that various systems of abstract prearithmetics form a category. This brings us to the following problem.

Problem 23. Study properties of the category **PEAPAWP** (category **WEAPAWP**), objects of which are partially (wholly) extended abstract prearithmetics and morphisms are weak projectivity relations.

In the theory of non-Diophantine arithmetics in general and in this book in particular, we study weak projectivity relations between abstract prearithmetics of different types. However, this concept is also essential for universal algebras such as groups or vector spaces. This brings us to the following problem.

Problem 24. Study weak projectivity relations for universal algebras in general and some of their special classes such as groups or vector spaces in particular.

Categories with projectivity relations as morphisms are also important.

Problem 25. Study properties of the category **PEAPAP** (category **WEAPAWP**), objects of which are partially (wholly) extended abstract prearithmetics and morphisms are projectivity relations.

In the theory of non-Diophantine arithmetics in general and in this book in particular, we study projectivity relations between abstract prearithmetics of different types. However, this concept is also essential for universal algebras such as groups or vector spaces. This brings us to the following problem.

Problem 26. Study projectivity relations for universal algebras in general and some of their special classes such as groups or vector spaces in particular.

Categories with exact projectivity relations as morphisms are also important.

Problem 27. Study properties of the category **PEAPAEP** (category **WEAPAEP**), objects of which are partially (wholly) extended abstract prearithmetics and morphisms are exact projectivity relations.

In the theory of non-Diophantine arithmetics in general and in this book in particular, we study exact projectivity relations between abstract prearithmetics of different types. However, this concept is also essential for universal algebras such as groups or vector spaces. This brings us to the following problem.

Problem 28. Study exact projectivity relations for universal algebras in general and some of their special classes such as groups or vector spaces in particular.

It is demonstrated that various systems of non-Diophantine arithmetics form a category. This brings us to the following problems.

Problem 29. Study properties of the category **NDWP**, objects of which are non-Diophantine arithmetics and morphisms are weak projectivity relations.

We have a similar problem for projectivity relations.

Problem 30. Study properties of the category **NDP**, objects of which are non-Diophantine arithmetics and morphisms are projectivity relations.

It is also necessary to note that originally non-Diophantine arithmetics are defined in a constructive way by a generative schema (Burgin, 1977; 1997). At the same time, there are axiomatic descriptions of the Diophantine arithmetics, which are discussed in Section 2.11 of this book. Thus, we come to the following problem.

Problem 31. Develop logical description of non-Diophantine arithmetics and prearithmetics giving their axiomatic characterizations.

Elements of higher arithmetic (number theory) for prearithmetics are elaborated in Sections 2.1 and 2.3 of this book.

Problem 32. Further develop higher arithmetic (number theory) for different types of non-Diophantine arithmetics.

As we saw, there are many functional direct projective arithmetics of natural numbers, which are essentially different from the Diophantine arithmetic $\boldsymbol{N}$. This brings us to the following problem.

Problem 33. Find conditions when a functional direct projective arithmetic (prearithmetic) is isomorphic to the Diophantine arithmetic $\boldsymbol{N}$.

As we saw, there are many functional dual projective arithmetics of whole numbers, which are essentially different from the Diophantine arithmetic $\boldsymbol{W}$. This brings us to the following problem.

Problem 34. Find conditions when a functional dual projective arithmetic (prearithmetic) is isomorphic to the Diophantine arithmetic $\boldsymbol{W}$.

We considered categories of prearithmetics and arithmetics, in which morphisms are weak projectivity, projectivity and exact projectivity relations. However, the popular approach for algebraic categories is based on homomorphisms of algebraic structures (cf., for example, Bucur and Deleanu, 1968; Herrlich and Strecker, 1973). This brings us to the following problem.

Problem 35. Study homomorphisms of non-Diophantine arithmetics.

Topology plays important role in mathematics and its applications. That is why many mathematical structures have topology. For instance, the arithmetic $\boldsymbol{R}$ of real numbers has the natural topology (cf., for example, Kelly, 1955). It is possible to consider abstract arithmetics and prearithmetics as well as quasilinear spaces and non-Grassmannian linear spaces with topological structures, in which operations become continuous functions. This brings us to the following problem.

Problem 36. Study topological arithmetics and prearithmetics.

Many important topological structures are determined by metrics and pseudometrics. For instance, Euclidean spaces are metric vector spaces.

Problem 37. Study metric and pseudometric arithmetics, prearithmetics, quasilinear spaces and non-Grassmannian linear spaces.

In many important cases, metrics are determined by norms while pseudometrics are determined by seminorms. For instance, normed vector spaces and algebras play important role in physics and economics.

Problem 38. Study normed and seminormed arithmetics and prearithmetics.

Hypermetrics form an important generalization of metrics (Burgin, 2017d).

Problem 39. Study hypermetric and hyperpseudometric arithmetics and prearithmetics.

Hypernorms form an important generalization of norms (Burgin, 2017d).

Problem 40. Study hypernormed and hyperseminormed arithmetics and prearithmetics.

We built the very beginning of non-Diophantine number theory in Sections 2.1 and 2.3.

Problem 41. Continue developing non-Diophantine number theory finding when the classical results are true and when they are invalid.

We built the very beginning of the theory of non-Grassmannian linear spaces and algebras in Section 3.5. Naturally, there are still many open problems in this area. Here arc some of them.

Problem 42. Study operators in quasilinear and non-Grassmannian linear spaces and algebras.

We have described some basic properties of the inner structures of quasilinear and non-Grassmannian linear spaces.

Problem 43. Build a structural theory of quasilinear and non-Grassmannian linear spaces and algebras.

We demonstrated that it is possible to build categories of quasilinear and non-Grassmannian linear spaces using homomorphisms as morphisms or using some type of projectivity relations as morphisms.

Problem 44. Study categories of quasilinear and non-Grassmannian linear spaces and algebras.

Topological vector spaces play an important role in functional analysis (Burgin, 2017d). Standard techniques allow defining topology or a norm in quasilinear and non-Grassmannian linear spaces.

Problem 45. Study topological and normed quasilinear and non-Grassmannian linear spaces and algebras.

Rényi entropies studied in Section 6.7 are obtained from Kolmogorov–Nagumo averages (Kolmogorov, 1930; Nagumo, 1930; de Finetti, 1931) combined with the standard Newtonian calculus. In essence, it is possible to treat such averages as non-Diophantine projections of the conventional addition of numbers. These projections are carried out by exact projectivity studied in Section 2.4. At the same time, frequency interpretation of probability is based on averages of the results of experiments (Burgin, 2015a).

Problem 46. Explore the properties and applications of information theory based on a consistent treatment of Kolmogorov–Nagumo averaging as an approach to a non-Diophantine probability theory.

There is also a related problem directly concerning probability.

Problem 47. Are quantum probabilities an example of a non-Diophantine probability theory, applicable to small numbers?

Studies of non-Diophantine arithmetics and their applications bring us to problems related to dynamical systems, fractals and differential equations used in physics. In particular, the conventional calculus is successfully used in dynamical system theory and its applications in biology. Thus, it is naturally to consider the following problem.

Problem 48. Are models based on non-Newtonian calculus relevant for population dynamics? Are such models applicable to systems where perception of change is more important than the change itself? What about non-Newtonian game theory and its biological applications?

For F^{α}-calculus studied in Parvate and Gangal (2005, 2011) and Satin and Gangal, 2016, one discusses fractional derivatives which are conjugate (by means of a bijection) to the ordinary Newtonian derivatives.

Problem 49. Are all the other forms of fractional calculus examples of a non-Newtonian calculus discussed in Chapter 4?

Fractal non-Newtonian calculus is an alternative to the more traditional approaches to calculus on fractals. Certain problems, such as Fourier analysis, are in a non-Newtonian formalism almost trivial, as opposed to the difficulties one encounters in the more standard approaches.

Problem 50. Study optimization problems in fractal systems.

For instance, the problem of finding a minimum of a function with a fractal domain can be addressed in different versions of the calculus.

Problem 51. Are the results always equivalent? If not, what are the differences?

Cantor, Sierpiński, and Koch sets are easy to represent as carriers of some arithmetics since the corresponding bijections are easy to find.

Problem 52. Is the same true for other well-known fractals such as the Mandelbrot of Julia sets?

In this book, applications of non-Diophantine arithmetics to physics and psychology are described. Thus, the general problem is the following one.

Problem 53. Further develop and extend applications of non-Diophantine arithmetics to physics and psychology.

There are also more specific problems in physics related to the utilized arithmetic.

In general relativity, a geometry of space–time is determined by means of a physical law (Einstein–Hilbert variational principle, Einstein equations).

Problem 54. Is there a physical principle that determines the choice of arithmetic? Can it apply to both physics and psychophysics?

Dark energy is still an enigma in physics. Non-Diophantine arithmetics open a new approach to study this mysterious phenomenon.

Problem 55. Are sets of Lebesgue measure zero and cardinality of the continuum (equipped with non-Diophantine arithmetic and non-Newtonian calculus) candidates for mathematical models of a Dark Universe (dark matter, dark energy, etc.)? Is dark energy an indication of a non-Newtonian calculus whose manifestations become observable at large space–time scales?

A non-Diophantine arithmetic based on a non-trivial bijection $f_A : A \to R$ in general requires dimensionless arguments x in $f_A(x)$. Physical quantities are not dimensionless. Dimensionless quantities can be obtained for the price of introducing fundamental physical units, for example a fundamental length. However, fundamental units do exist in Nature (Planck constant, velocity of light, charge of the electron, etc.).

Problem 56. Is it an indication of a fundamental and non-trivial f_A?

Bijections $f_A : A \to R$ are always continuous in topologies induced from R, but in typical examples they are discontinuous in metric topologies of A. A canonical example is provided by $A = R^3$. So, points in A that are close to each other when analyzed form the point of view of the induced topology may be separated by large distances if considered in terms of the metric of A. In consequence, physics, which is local at the non-Newtonian level, becomes non-local when analyzed in Newtonian terms. This phenomenon has striking similarity to the problem of non-locality of hidden variables in quantum mechanics.

Problem 57. Is it just a similarity, or perhaps there are fundamental non-Newtonian/non-Diophantine principles behind it?

Length of any fragment of a Koch curve is infinite, but non-Diophantine arithmetic turns it into a finite number. Techniques of extracting finite experimental predictions from infinite theoretical results are used in quantum and classical field theory in various forms (regularization and renormalization).

Problem 58. Are renormalization methods examples of a non-Newtonian calculus? Can one invent renormalization methods based on non-Newtonian calculus? Is quantum field theory so ill defined mathematically because it is implicitly based on the Newtonian but non-physical calculus?

Many different generalizations of quantum mechanics can be found in the literature, in particular, those that are based on some kinds of a nonlinear Schrödinger equation. Such generalizations are known to possess interpretational difficulties. However, quantum mechanics or quantum field theory based on non-Newtonian calculus has never been seriously treated as a candidate for a generalized quantum theory. Considering such a theory, we come to the following problem.

Problem 59. Can one invent a quantum experiment, which can directly test whether quantum mechanics is based on a non-Newtonian calculus?

"Physics is geometry". This popular slogan means that n-tuples of real numbers one encounters in various physical models are in fact coordinates on n-dimensional manifolds.

Problem 60. Is it possible that "physics is actually arithmetic"?

If the answer is positive, then the n-tuples may correspond to n images of carriers A of some abstract arithmetic under bijections $f_A : A \to R$. The fundamental object is then not a manifold but an arithmetic.

The bijection $f_A : A \to R$ maps in general a higher (or lower) dimensional set A into a one-dimensional set R. The dimension of A becomes "masked" by the one-dimensionality of the image $f_A(A)$. In particular, a four-dimensional-space-time may be an image of an arbitrarily-dimensional A, a carrier of some arithmetic.

Problem 61. Is there a relation between such higher/lower dimensional spaces and the problem of dimensionality of space–time one encounters in string theory and Kaluza–Klein theories?

As we know, quantum physics is essentially different from the classical physics of macro-bodies.

Problem 62. Is the transition between micro-physics and macro-physics an example of a change of arithmetic in transition from small numbers to large numbers?

Let us consider some problems in psychology related to utilizations of arithmetic. For instance, from a non-Diophantine perspective, one can say that many psychophysical theories employ a non-Diophantine subtraction.

Problem 63. What about other arithmetical operations? Can they possess a psychophysical interpretation?

Non-Newtonian signal analysis is basically unexplored, although non-linear filtering and cepstral transforms posses certain properties one would expect as an influence of non-Diophantine arithmetics and a non-Newtonian calculus (Oppenheim *et al.*, 1968; 1968a; Oppenheim and Schafer, 1975; Childers *et al.*, 1977). This brings us to the following problem.

Problem 64. Study applications of non-Diophantine arithmetics and non-Newtonian calculi to signal analysis.

Appendix: Notations and Basic Definitions

Mathematics is the art of giving the same name to different things.
Henri Poincare (1854–1912)

To make the book easier for the reader, notations and definitions of the conventional mathematical concepts and structures used in this book are included in the appendix.

A.1. General Concepts and Structures

$N = \{1, 2, 3, \ldots\}$ is the set of all natural numbers.
$\boldsymbol{N}$ is the arithmetic of all natural numbers.
ω is the sequence of all natural numbers.
$W = \{0, 1, 2, 3, \ldots\}$ is the set of all whole numbers.
$\boldsymbol{W}$ is the arithmetic of all whole numbers.
Z is the set of all integer numbers.
$\boldsymbol{Z}$ is the arithmetic of all integer numbers.
Q is the set of all rational numbers.
$\boldsymbol{Q}$ is the arithmetic of all rational numbers.
Q^+ is the set of all non-negative rational numbers.
$\boldsymbol{Q}^+$ is the arithmetic of all non-negative rational numbers.
Q^{++} is the set of all positive rational numbers.
$\boldsymbol{Q}^{++}$ is the arithmetic of all positive real numbers.
R is the set of all real numbers.

$\boldsymbol{R}$ is the arithmetic of all real numbers.
R^+ is the set of all non-negative real numbers.
$\boldsymbol{R}^+$ is the arithmetic of all non-negative real numbers.
R^{++} is the set of all positive real numbers.
$\boldsymbol{R}^{++}$ is the arithmetic of all positive real numbers.
R^- is the set of all non-positive real numbers.
R^{--} is the set of all negative real numbers.
$[a, b] = \{x; a \leq x \leq b, x \in R\}$ is a closed interval in R.
$(a, b) = \{x; a < x < b, x \in R\}$ is an open interval in R.
$[a, b) = \{x; a \leq x < b, x \in R\}$ and $(a, b] = \{x; a < x \leq b, x \in R\}$ are semi-open intervals in R.
There are also infinite intervals:
$[a, \infty) = \{x; a \leq x, x \in R\}$.
$(-\infty, b] = \{x; x \leq b, x \in R\}$.
$(a, \infty) = \{x; a < x, x \in R\}$.
$(-\infty, b) = \{x; x < b, x \in R\}$.
C is the set of all complex numbers.
$\boldsymbol{C}$ is the arithmetic of all complex numbers.
Ø is the *empty set*, i.e., the set that has no elements.
Functions are denoted by small letters $f, g, h, \ldots$.
Operators and linear transformations are denoted by capital letters $A, B, C, F, G, H, \ldots$.
Matrices are denoted by capital letters $M, K, H, \ldots$.
Elements are denoted by small letters $a, b, c, x, y, z, \ldots$.
Sets are denoted by capital letters $X, Y, Z, A, B, C, \ldots$.
Variables are denoted by small letters $x, y, z, \ldots$.
Algebras, e.g., groups, are denoted by capital letters $A, B, C, G, H, \ldots$.
Arithmetics and prearithmetics are denoted by bold capital letters $\boldsymbol{A}$, $\boldsymbol{B}$, $\boldsymbol{C}$, ...
Sets, on which arithmetical operations are defined, are denoted by capital letters $A, B, C, \ldots$.
If a is a real number, then $|a|$ or $||a||$ denotes the absolute value or modulus of a.
If a is a complex number, then $||a||$ or $|a|$ denotes the magnitude or absolute value or modulus of a.
If b is a vector from the n-dimensional space R^n, then $||a||$ denotes its modulus.

Zero as a number is denoted by 0, while zero as an element of a vector space is denoted by **0**.

In general, a sequence of elements a_i is denoted either by $\{a_i; i = 1, 2, 3, \ldots\}$ or by $\{a_i; i \in \omega\}$ or by $(a_i)_{i\in\omega}$.

$F(\boldsymbol{R},\boldsymbol{R})$ is the space of all real functions.

$C(\boldsymbol{R},\boldsymbol{R})$ is the space of all continuous real functions.

$F[a, b]$ is the space of all real functions defined in the interval $[a, b]$.

$C[a, b]$ is the space of all continuous real functions defined in the interval $[a, b]$.

If X is a set (class), then $r \in X$ means that r belongs to X or r is a member of X.

Two sets (classes) X and Y are called *equipotent* if there is a one-to-one correspondence between X and Y.

If X and Y are sets (classes), then $Y \subseteq X$ means that Y is a *subset* (subclass) of X, i.e., Y is a set such that all elements of Y belong to X, and X is a *superset* of Y. A subset is *proper* if it does coincide with the whole set.

The *union* $Y \cup X$ of two sets (classes) Y and X is the set (class) that consists of all elements from Y and from X. The union $Y \cup X$ is called disjoint if $Y \cap X = \emptyset$.

The *intersection* $Y \cap X$ of two sets (classes) Y and X is the set (class) that consists of all elements that belong both to Y and to X.

The *union* $\bigcup_{i\in I} X_i$ of sets (classes) X_i is the set (class) that consists of all elements from all sets (classes) $X_i, i \in I$.

The *intersection* $\bigcap_{i\in I} X_i$ of sets (classes) X_i is the set (class) that consists of all elements that belong to each set (class) $X_i, i \in I$.

The *difference* $Y\backslash X$ of two sets (classes) Y and X is the set (class) that consists of all elements that belong to Y but do not belong to X.

If a set (class) X is a subset of a set (class) Y, i.e., $X \subseteq Y$, then the difference $Y\backslash X$ is called the *complement* of the set (class) X in the set (class) Y and is denoted by $C_Y X$.

The *symmetric difference* $Y\Delta X$ of two sets (classes) Y and X is equal to $(Y\backslash X) \cup (Y\backslash X)$.

If X is a set, then 2^X is the *power set* of X, which consists of all subsets of X. The *power set* of X is also denoted by $\mathrm{P}X$.

If X and Y are sets (classes), then $X \times Y = \{(x, y); x \in X, y \in Y\}$ is the *Cartesian product* of X and Y, in other words, $X \times Y$ is the set (class) of all pairs (x, y), in which x belongs to X and y belongs to Y.

A mapping (function) f from a set X into a set Y is denoted by $f\colon X \to Y$ and a mapping of an element x into an element y is denoted by $f\colon x \mapsto y$.

Y^X denotes the set of all mappings from X into Y.

$X^n = \underbrace{X \times X \times \cdots \times X \times X}_{n}$

Elements of the set X^n have the form $(x_1, x_2, \ldots, x_n)$ with all $x_i \in X$ and are called n-tuples, or simply, tuples.

One of the most fundamental structures of mathematics is a *function.* However, functions are special kinds of binary relations between two sets, which are defined below.

A *binary relation* T between sets X and Y, also called a *correspondence* from X to Y, is a subset of the Cartesian product $X \times Y$. The set X is called the *domain* of $T(X = \mathrm{Dom}(T))$ and Y is called the *codomain* of $T(Y = \mathrm{Codom}(T))$. The *range* of the relation T is $\mathrm{Rg}(T) = \{y; \exists x \in X((x, y) \in T)\}$. The *domain of definition* also called the *definability domain* of the relation T is $\mathrm{DDom}(T) = \{x; \exists y \in Y((x, y) \in T)\}$. If $(x, y) \in T$, then one says that the elements x and y are in the relation T, and one also writes $T(x, y)$.

The image $T(x)$ of an element x from X is the set $\{y; (x, y) \in T\}$ and the coimage $T^{-1}(y)$ of an element y from Y is the set $\{x; (x, y) \in T\}$.

The *graph* of binary relation T between sets of real numbers is the set of points in the two dimensional vector space (a plane), the coordinates of which satisfy this relation.

Binary relations are also called *multivalued functions* (*mappings* or *maps*).

$BR(X, Y)$ is the set of all binary relations between sets X and Y.

Taking binary relations $T \subseteq X \times Y$ and $R \subseteq Y \times Z$, it is possible to build a new binary relation $RT \subseteq X \times Z$ that is called the (*sequential*) *composition* or *superposition* of binary relations T and R and is defined as

$$R \circ T = \{(x, z); x \in X, z \in Z;\ \text{where } (x, y) \in T$$

$$\text{and } (y, z) \in R \text{ for some } y \in Y\}.$$

A *preorder* (also called *quasiorder*) on a set (class) X is a binary relation Q on X that satisfies the following axioms:

O1. Q is *reflexive*, i.e., $x\mathrm{Q}x$ for all x from X.

O2. Q is *transitive*, i.e., $x\mathrm{Q}y$ and $y\mathrm{Q}z$ imply $x\mathrm{Q}z$ for all $x, y, z \in X$.

A preorder can be *partial* or *total* when for all $x, y \in X$, we have either $x\mathrm{Q}y$ or $y\mathrm{Q}x$.

A *partial order* is a preorder that satisfies the following additional axiom:

O3. Q is *antisymmetric*, i.e., $x\mathrm{Q}y$ and $y\mathrm{Q}x$ imply $x = y$ for all $x, y \in X$.

A *strict* also called *sharp partial order* is a preorder that is not reflexive, is transitive and satisfies the following additional axiom:

O4. Q is *asymmetric*, i.e., only one relation $x\mathrm{Q}y$ or $y\mathrm{Q}x$ is true for all $x, y \in X$.

A *linear* or *total order* is a strict partial order that satisfies the following additional axiom:

O5. We have either $x\mathrm{Q}y$ or $y\mathrm{Q}x$ for all $x, y \in X$.

A set (class) X is *well-ordered* if there is a partial order on X such that any of its non-empty subsets has the least element. Such a partial order is called *well-ordering.*

An *equivalence* on a set (class) X is a binary relation Q on X that is reflexive, transitive and satisfies the following additional axiom:

O6. Q is *symmetric*, i.e., $x\mathrm{Q}y$ implies $y\mathrm{Q}x$ for all x and y from X.

If we have an equivalence σ on a set X, this set is a disjoint union of classes of the equivalence σ where each class consists of equivalent elements from X and there are no equivalent elements in different classes. The set of these equivalence classes is called the quotient set of $X$and it is denoted by X/σ.

A *tolerance relation* is a binary relation that is reflexive and symmetric.

A *function* (also called a *mapping* or *map* or *total function* or *total mapping* or *everywhere defined function*) f from X to Y is defined as a binary relation between sets X and Y in which there are no elements from X which are corresponded to more than one element from Y and to any element from X, some element from Y is corresponded. Traditionally, the function f is also denoted by $f\colon X \to Y$ or by $f(x)$. In the latter formula, x is a variable that takes values in X. The *support*, or *carrier*, of a numerical function f is the closure of the set where $f(x) \neq 0$. Usually the element $f(a)$ is called the *image* of the element a and denotes the value of f on the element a from X. The *coimage* $f^{-1}(y)$ of an element y from Y is the set $\{x; f(x) = y\}$.

A partial *function* f from X to Y is defined as a binary relation between sets X and Y in which there are no elements from X which are corresponded to more than one element from Y.

However, the traditional definition does not include all kinds of functions and their representations.

There are three basic forms of function representation (definition):

1. (The *set-theoretical*, e.g., *table, representation*) A function f is given as a subset R_f of the Cartesian product $X \times Y$ such that the first element if each pair from R_f uniquely defines the second element in this pair, e.g., in a form of a table or of a list of pairs (x, y) where the first element x is taken from X, while the second element y is the image $f(x)$ of the first one. The set R_f is called the *graph* of the function f. When X and Y are sets of points in a geometrical space, e.g., their elements are real numbers, the graph of the function f is called the *geometrical graph* of f.
2. (The *analytic representation*) A function f is described by a formula, i.e., a relevant expression in a mathematical language, e.g., $f(x) = \sin(e^{x+\cos x})$.
3. (The *algorithmic representation*) A function f is given as an algorithm that computes $f(x)$ given x.

$f(x) \equiv a$ means that the function $f(x)$ is equal to a at all points where$f(x)$ is defined.

A function (mapping) f from X to Y is an *injection* if the equality $f(x) = f(y)$ implies the equality $x = y$ for any elements x and y from X, i.e., different elements from X are mapped into different elements from Y.

A function (mapping) f from X to Y is a *projection* also called *surjection* if for any y from Y there is x from X such that $f(x) = y$.

A function (mapping) f from X to Y is a *bijection* if it is both a projection and injection.

A function (mapping) f from X to Y is an *inclusion* if the equality $f(x) = x$ holds for any element x from X.

Two important concepts of mathematics are the domain and range of a function. However, there is some ambiguity for the first of them. Namely, there are two distinct meanings in current mathematical usage for this concept. In the majority of mathematical areas, including the calculus and analysis, the term "domain of f" is used for the set of all values x such that $f(x)$ is defined. However, some mathematicians (in particular, category theorists), consider the domain of a function $f\colon X \to Y$ to be X, irrespective of whether $f(x)$ is defined for all x in X. To eliminate this ambiguity, we suggest the following terminology consistent with the current practice in mathematics.

If f is a function from X into Y, then the set X is called the *domain* of f (it is denoted by Dom f) and Y is called the *codomain* of T (it is denoted by Codom f). The *range* Rg f of the function f is the set of all elements from Y assigned by f to, at least, one element from X, or formally, Rg $f = \{y; \exists x \in X(f(x) = y)\}$. The *domain of definition* also called the *definability domain*, DDom f, of the function f is the set of all elements from X that related by f to, at least, one element from Y is or formally, DDom $f = \{x; \exists y \in Y(f(x) = y)\}$. Thus, for a partial function f, its domain of definition DDom f is the set of all elements for which $f(x)$ is defined.

$F(X, Y)$ is the set of all mappings (functions) from a set X into a set Y.

$FT(X, Y)$ is the set of all total mappings (total functions) from a set X into a set Y.

$PF(X, Y)$ is the set of all partial mappings (partial functions) from X into Y. Naturally, $F(X, Y) \subseteq PF(X, Y)$.

Taking two mappings (functions) $f\colon X \to Y$ and $g\colon Y \to Z$, it is possible to build a new mapping (function) $gf\colon X \to Z$ that is called the (*sequential*) *composition* or *superposition* of mappings (functions) f and g and defined by the rule $gf(x) = g(f(x))$ for all x from X.

For any set $S, \chi_S(x)$ is its *characteristic function*, also called the *set indicator function*, if $\chi_S(x)$ is equal to 1 when $x \in S$ and is equal to 0 when $x \notin S$, and $\mathrm{C}_S(x)$ is its partial characteristic function if $\mathrm{C}_S(x)$ is equal to 1 when $x \in S$ and is undefined when $x \notin S$.

If $f\colon X \to Y$ is a function and $Z \subseteq X$, then the restriction $f|_Z$ of f on Z is the function defined only for elements from Z and $f|_Z(z) = f(z)$ for all elements z from Z.

Sequential composition or superposition of binary relations defines sequential composition or superposition of functions, i.e., if $f\colon X \to Y$ and $g\colon Y \to Z$ are functions (mappings), then the mapping (function) $g \circ f\colon X \to Z$ is called the (*sequential*) *composition* or *superposition* of functions (mappings) f and g and defined by the rule $(g \circ f)(x) = g(f(x))$ for all x from$X$.

A real function $f\colon \boldsymbol{R} \to \boldsymbol{R}$ is *bounded* if there is a real number c such that $|f(x)| < c$ for all elements x from $\boldsymbol{R}$.

A real function $f\colon \boldsymbol{R} \to \boldsymbol{R}$ is *increasing* if the inequality $a \leq b$ implies the inequality $f(a) \leq f(b)$.

A real function $f\colon \boldsymbol{R} \to \boldsymbol{R}$ is *strictly increasing* if the inequality $a < b$ implies the inequality $f(a) < f(b)$.

For real functions, it is possible to define *arithmetical compositions*:

$$(g + f)(x) = f(x) + g(x),$$

$$(g \cdot f)(x) = f(x) \cdot g(x).$$

A real function f is called a *contraction* or *contraction function* if there exists a real number k such that $0 \leq k \leq 1$ and for all real numbers x and y, we have

$$|f(x) - f(y)| < k|x - y|.$$

A real function f is called *periodic* if there exists a real number $k > 0$ such that for all real numbers x, we have $f(x) = f(x + k)$.

The *ceiling function* $\lceil x \rceil$ returns the smallest integer that is greater than or equal to x, e.g., $\lceil 4.5 \rceil = 5$.

The *floor function* $\lfloor x \rfloor$ returns the largest integer that is less than or equal to x, e.g., $\lfloor 4.5 \rfloor = 4$.

The *round function* $[x]$ returns the nearest integer to x, while when there are two such integers the larger of them is taken as the value of the function, e.g., $[4.5] = 5$ and $[4.8] = 5$.

A function f from a partially ordered set X to a partially ordered set X is called *monotone* (*antitone*) if $x \leq y$ implies $f(x) \leq f(y)$ $(f(y) \leq f(x))$ for any elements x and y from X.

If U is a correspondence of a set X to a set Y (a binary relation between X and Y), i.e. $U \subseteq X \times Y$, then $U(x) = \{y \in Y; (x, y) \in U\}$ and $U^{-1}(y) = \{x \in X; (x, y) \in U\}$.

An n-ary relation R in a set X is a subset of the nth power of X, i.e., $R \subseteq X^n$. If $(a_1, a_2, \ldots, a_n) \in R$, then one says that the elements $a_1, a_2, \ldots, a_n$ from X are in relation R.

Named sets as the most encompassing and fundamental mathematical construction include ordinary sets and all their generalizations, such as fuzzy sets and multisets, providing unified foundations for the whole mathematics (Burgin, 2004b). Functions, mappings, operations, relations, graphs, multigraphs, operators, fiber bundles, morphisms, functors, enumerations and many other mathematical structures are named sets. Moreover, all mathematical structures are built of named sets (Burgin, 2011).

A *named set* (also called a *fundamental triad*) has the following graphic representation (Burgin, 1990; 1995; 2011):

$$\text{Entity 1} \xrightarrow{\text{connection}} \text{Entity 2} \qquad (1)$$

or

$$\text{Essence 1} \xrightarrow{\text{correspondence}} \text{Essence 2} \qquad (2)$$

In the fundamental triad (named set) (1) or (2), Entity 1 (Essence 1) is called the *support*, the Entity 2 (Essence 2) is called the *reflector* (also called the *set* or *component of names*) and the connection (correspondence) between Entity 1 (Essence 1) and Entity 2 (Essence 2) is called the *reflection* (also called the *naming correspondence*) of the fundamental triad (1) (respectively, (2)).

In the symbolic form, a *named set* (*fundamental triad*) $\mathbf{X}$ is a triad (X,f,I) where X is the *support* of $\mathbf{X}$ and is denoted by $\mathrm{S}(\mathbf{X})$, I is the *component of names* (also called *set of names* or *reflector*) of $\mathbf{X}$ and is denoted by $\mathrm{N}(\mathbf{X})$, and f is the *naming correspondence* (also called *reflection*) of the named set $\mathbf{X}$ and is denoted by $\mathrm{n}(\mathbf{X})$. The most popular type of named sets is a named set $\mathbf{X} = (X,f,I)$ in which X and I are sets and f consists of connections between their elements. When these connections are set-theoretical, i.e., each connection is represented by a pair (x, a) where x is an element from X and a is its name from I, we have a *set-theoretical named set*, which is binary relation. Even before the concept of a fundamental triad was introduced, Bourbaki in their fundamental monograph (Bourbaki, 1960) had also represented binary relations in the form of a triad (named set).

A.2. Logical Concepts and Structures

If P and Q are two statements, then $P \rightarrow Q$ means that P implies Q and $P \leftrightarrow Q$ means that P is equivalent to Q.

Logical operations:

- *negation* is denoted by $\neg$ or by $\sim$,
- *conjunction* also called logical "and" is denoted by $\wedge$ or by & or by $\cdot$,

- *disjunction* also called logical "or" is denoted by $\vee$,
- *implication* is denoted by $\rightarrow$ or by $\Rightarrow$ or by $\supset$,
- *equivalence* is denoted by $\leftrightarrow$ or by $\equiv$ or by $\Leftrightarrow$.

The logical symbol $\forall$ is called the *universal quantifier* and means "for any".

The logical symbol $\exists$ is called the *existential quantifier* and means "there exists".

Logical formulas and operations allow mathematicians and logicians to work both with finite and infinite systems. However, to do this in a more constructive form when the number of elements in the system is very big or infinite, mathematicians use the Principle of Induction.

Descriptive Principle of Induction. Given an infinite (very big) system $R = \{x_1, x_2, \ldots, x_n, \ldots\}$ of elements enumerated by natural numbers and a predicate P defined for these elements, if it is proved that $P(x_1)$ is true and assuming for an arbitrary number n, that all $P(x_1), (x_2), \ldots, P(x_{n-1})$ are true, it is also proved that $P(x_n)$ is true, then $P(x)$ is true for all elements from R.

Constructive Principle of Induction. If an infinite (very big) system $R = \{x_1, x_2, \ldots, x_n, \ldots\}$ of elements enumerated by natural numbers is described by some property represented by a predicate P defined for elements of R, then if there is an algorithm (constructive method) A that builds x_1 and assuming that for an arbitrary number n, all $x_1, x_2, \ldots, x_{n-1}$ are built, A can also build x_n. Then A can (potentially) build the whole R, i.e., R exists potentially.

A.3. Topological Concepts and Structures

A *topology* in a set X is a system $\mathrm{O}(X)$ of subsets of X that are called *open subsets* and satisfy the following axioms:

T1. $X \in O(X)$ and $\emptyset \in \mathrm{O}(X)$.
T2. For all subsets A and B of X, if $A, B \in \mathrm{O}(X)$, then $A \cap B \in \mathrm{O}(X)$.
T3. For all subsets A_i of X $(i \in I)$, if all $A_i \in \mathrm{O}(X)$, then $\bigcup_{i \in I} A_i \in \mathrm{O}(X)$.

A set X with a topology in it is called a *topological space*.

Topology in a set can be also defined by a system of neighborhoods of points from this set. In this case, a set is *open* in this topology if it contains a standard neighborhood of each of its points. For instance, if a is a real number and $t \in \boldsymbol{R}^{++}$, then an open interval $\mathrm{O}_t a = \{x \in \boldsymbol{R}; a-t < x < a+t\}$ is a standard neighborhood of a.

To define a topology, a system of sets has to satisfy the following *neighborhood axioms* (Kuratowski, 1966).

NB1. Any neighborhood of a point $x \in X$ contains this point.
NB2. For any two neighborhoods $\mathrm{O}_1 x$ and $\mathrm{O}_2 x$ of a point $x \in X$, there is a neighborhood $\mathrm{O}x$ of x that is a subset of the intersection $\mathrm{O}_1 x \cap \mathrm{O}_2 x$.
NB3. For any neighborhood $\mathrm{O}x$ of a point $x \in X$ and a point $y \in \mathrm{O}x$, there is a neighborhood $\mathrm{O}y$ of y that is a subset of $\mathrm{O}x$.

One more way to define topology in a set is to use the *closure operation* (Kuratowski 1966).

A topology σ in a set X is *stronger* than topology τ in the same set X if any open in the topology τ set is also open in the topology σ.

If X is a subset of a topological space, then $\mathrm{Cl}(X)$ denotes the *closure* of the set X.

A *cluster point* of a set A in a topological space X is a point any neighborhood of which contains infinitely many elements from A.

A topological space X can satisfy the following axioms (Kelly, 1957):

$\mathbf{T_0}$ (the *Kolmogorov Axiom*). $\forall x, y \in X(\exists Ox(y \notin Ox) \vee \exists Oy(x \notin Oy))$.

In other words, for every pair of points a and b there exists an open set U in O such that at least one of the following statements is true: (1) a belongs to U and b does not belong to U, and (2) b belongs to U and a does not belongs to U.

$\mathbf{T_1}$ (the *Alexandroff Axiom*). $\forall x,y \in X \exists\ Ox\ \exists Oy\ (x \notin Oy\ \&\ y \notin Ox)$.

In other words, for every pair of points a and b there exists an open set U such that U contains a but not b. To say that a space is T_1 is equivalent to saying that sets consisting of a single point are closed.

$\mathbf{T_2}$ (the *Hausdorff Axiom*). $\forall x,y \in X \exists Ox\ \exists Oy\ (Ox \cap Oy = \emptyset)$.

In other words, for every pair of points a and b there exist disjoint open sets which separately contain a and b. In this case, open sets *separate* points.

Here Ox, Oy are some neighborhoods of x and y, respectively.

A topological space, which satisfies the axiom $\mathbf{T}_i$, is called a $\mathbf{T}_i$*-space*. Each axiom $\mathbf{T_{i+1}}$ is stronger than axiom $\mathbf{T_i}$. $\mathbf{T}_0$*-spaces* are also called the *Kolmogorov spaces*. $\mathbf{T}_1$*-spaces* are also called the *Fréchet spaces*. $\mathbf{T}_2$-spaces are also called the *Hausdorff spaces* (Kelly 1957).

There are also $\mathbf{T}_3$-*spaces* or *regular spaces*, in which for every point a and closed set B there exist disjoint open sets which separately contain a and B. That is, any two closed sets are separated. Many authors require that $\mathbf{T}_3$-spaces also be $\mathbf{T}_0$-spaces, since with this added condition, they are also $\mathbf{T}_2$-spaces (Alexandroff, 1961).

There are also $\mathbf{T}_4$-*spaces* or *normal spaces*, in which for every pair of closed sets A and B there exist disjoint open sets which separately contain A and B. That is, points and closed sets are separated. Many authors require that $\mathbf{T}_4$-spaces also satisfy Axiom $\mathbf{T}_1$ (Alexandroff, 1961).

If X is a set in a topological space Z, then the *border* of X is equal to the intersection of X with the closure of the complement of X in Z.

For instance, the border of the interval (a, b) is equal to the set $\{a, b\}$ and the border of the interval $[a, b]$ is equal to the set $\{a, b\}$.

A mapping f: $X \to Y$ is called *continuous* if the inverse image of any open set in Y is an open set in X.

A mapping f: $X \to Y$ is called *sequentially continuous* if the image of any converging sequence in X is a converging sequence in Y.

Another topological construction is a metric space. Let X be some set.

A mapping $\mathbf{d}$: $X \times X \to \boldsymbol{R}^+$ is called a *metric* (or a *distance function*) if it satisfies the following axioms:

M1. $\mathbf{d}(x, y) = 0$ if and only if $x = y$.
M2. $\mathbf{d}(x, y) = \mathbf{d}(y,x)$ for all $x, y \in X$.
M3. $\mathbf{d}(x, y) \leq \mathbf{d}(x, z) + \mathbf{d}(z, y)$ for all $x, y, z \in X$.

A set X with a metric $\mathbf{d}$ is called a *metric space.*

The real number $\mathbf{d}(x,y)$ is called the *distance* between x and y in the metric space X.

If X and Y are metric spaces, then a mapping f: $X \to Y$ is called an *isometry* if it preserves the distance between points, i.e., $\mathbf{d}(x, y) = \mathbf{d}(f(x), f(y))$ for any $x, y \in X$.

It is possible to find other structures from topology and their properties, for example, in (Kelly, 1957; Kuratowski, 1966; 1968).

A *fiber bundle* $\boldsymbol{B}$ (also called fibre bundle) is a triad (E, p, B) where the topological space E is called the *total space* or simply, *space* of the fiber bundle $\boldsymbol{B}$, the topological space B the *base space* or simply, *base* of the fiber bundle $\boldsymbol{B}$, and p is a topological projection of E onto B such that every point in the base space has a neighborhood U such that $p^{-1}(b) = F$ for all points b from B and $p^{-1}(U)$ is homeomorphic to the Cartesian product

$U \times F$. The topological space F is called the *fiber* of the fiber bundle $\boldsymbol{B}$. Informally, a fiber bundle is a topological space which looks locally like a product space $U \times F$.

Fiber bundles are special cases of topological bundles. A *topological bundle* $\boldsymbol{D}$ is a triad (named set) (E, p, B) where E and B are topological spaces and p is a topological (i.e., continuous) projection of E onto B.

It is possible to find introductory information on topological manifolds, fiber and topological bundles, and other topological structures in (Husemöller, 1994; Lee, 2000).

A.4. Algebraic Concepts and Structures

An *algebraic system* is a structure $A = (X, \Omega, R)$ that consists of: a non-empty set X called the *carrier* or the *underlying set* of A and elements of which are called the elements of A; a family Ω of algebraic operations, which are mappings $\omega_i : X^{ni} \to X(i \in I)$; and a family R of relations $r_j \subseteq X^{mj}(j \in J)$ defined on X.

The non-negative integers n_i and m_j are called the *arities* of the respective operations ω_i and relations r_j. For instance, addition and subtraction of numbers are binary operations. A nullary operation is selection of a specific element in the universal algebra. When ω is an operation from Ω with arity n, then the image $\omega(a_1, a_2, \ldots, a_n)$ of the element $(a_1, a_2, \ldots, a_n)$ from X^n under the mapping $\omega : X^n \to X$ is called the *value* of the operation ω for elements $a_1, a_2, \ldots, a_n$.

In a general case, some operations from Ω may be partial mappings. For instance, division in such a universal algebra (algebraic system) as a field is not defined for 0, i.e., it is impossible to divide by 0.

The operations from Ω and the relations from R are called *basic* or *primitive*. Using basic operations, it is possible to build many different derivative operations in the algebra A. The pair of families ($\{n_i; i \in I\}$; $\{m_j; j \in J\}$) is called the *type* of the algebraic system A. Two algebraic systems A and A' have the *same type* if $I = I', J = J'$, and $n_i = n'_i$ and $m_j = m'_j$ for all $i \in I, j \in J$. Usually it is assumed that algebraic systems of the same type have the same operations and relations. For instance, group-oids have the same operation, which is called multiplication.

An algebraic system A is called finite if the set X is *finite* and it is called of finite type if the set $I \cup J$ is finite.

An algebraic system $A = (X, \Omega, R)$ is called a *universal algebra* or simply, an *algebra* if the set R of its basic relations is empty.

An algebraic system $A = (X, \Omega, R)$ is called a *model* (in logic) or a *relational system* if the set Ω of basic operations is empty.

A mapping f of universal algebras having the same type is called a *homomorphism* if it preserves all operations. For instance, a homomorphism of vector spaces is a linear mapping (operator).

An injective homomorphism is called a *monomorphism.*

A surjective homomorphism is called an *epimorphism.*

A bijective homomorphism is called an *isomorphism.*

A mapping f of a universal algebra $A = (X, \Omega)$ into a universal algebra $B = (Y, \Omega)$ is called a *homomorphism with respect to operations* from a subset Φ of Ω if it preserves all operations from Φ. For instance, a linear mapping (operator) of linear algebras over a field F is a homomorphism with respect to addition and multiplication by the elements from the field F.

An injective homomorphism with respect to operations from Φ is called a *monomorphism with respect to operations* from Φ.

A surjective homomorphism with respect to operations from Φ is called an *epimorphism with respect to operations* from Φ.

A bijective homomorphism with respect to operations from Φ is called an *isomorphism with respect to operations* from Φ.

A *semigroup* S is a set with an associative binary operation.

An *abelian semigroup* A is a semigroup in which the binary operation is commutative.

A *monoid* S is a semigroup with an identity (unit) element.

A monoid in which all elements have inverse elements is called a *group.* It is possible to consider a group as a universal algebra with one binary operation usually called multiplication, one unary operation, which assigns the inverse to each element, and one nullary operation, which specifies the identity (unit) element.

A *group* G is a set with one operation:

multiplication: $K \times K \to K$ denoted by xy where x and y belong to K.

This operation satisfies the following axioms:

1. *Multiplication is associative*:

For all x, y, z from K, we have $x + (y + z) = (x + y) + z$.

2. *Multiplication has an identity element*:

There exists an element 1 from G, called the *one*, such that $1x = x1 = x$ for all x from K.

3. *Each element has the inverse element*:

For any x from K, there exists an element z from K such that $xz = 1$.

An *abelian group* A is a group in which the binary operation is commutative.

A *semiring* K is a set with two operations:

addition: $K \times K \to K$ denoted by $x + y$ where x and y belong to K;

multiplication: $K \times K \to K$ denoted by xy where x and y belong to K.

These operations satisfy the following axioms:

1. *Addition is associative*:

For all x, y, z from K, we have $x + (y + z) = (x + y) + z$.

2. *Addition is commutative*:

For all x, y from K, we have $x + y = y + x$.

3. *Addition has an identity element*:

There exists an element 0 from K, called the *zero*, such that $x + 0 = x$ for all x from K.

4. *Multiplication is distributive over addition*:
For all elements x, y, z from K, we have

$$x(y + z) = xy + xz$$

and

$$(x + y)z = xz + yz$$

A *ring* K is a set with two operations:

addition: $K \times K \to K$ denoted by $x + y$ where x and y belong to K;

multiplication: $K \times K \to K$ denoted by xy where x and y belong to K.

These operations satisfy the following axioms:

4. *Addition is associative*:

For all x, y, z from K, we have $x + (y + z) = (x + y) + z$.

5. *Addition is commutative*:

For all x, y from K, we have $x + y = y + x$.

6. *Addition has an identity element*: There exists an element 0 from K, called the *zero*, such that $x + 0 = x$ for all x from K.
7. *Each element has the opposite element*:

For any x from K, there exists an element z from K, called the *opposite* of x, such that $x + z = 0$.

8. *Multiplication is distributive over addition*:

For all elements x, y, z from K, we have

$$x(y + z) = xy + xz$$

and

$$(x + y)z = xz + yz$$

Sets $F(\boldsymbol{R})$, $C(\boldsymbol{R})$, $F[a, b]$ and $C[a, b]$ are rings.

A commutative ring with non-zero multiplicative cancellation is called a *domain*.

A *left module* M over a ring K has two operations:

addition: $M \times M \to M$ denoted by $x + y$ where x and y belong to M;

multiplication: $K \times M \to M$ denoted by ax where x belongs to M and a belongs to K.

These operations satisfy the following axioms:

1. *Addition is associative*:

For all x, y, z from M, we have $x + (y + z) = (x + y) + z$.

2. *Addition is commutative*:

For all x, y from M, we have $x + y = y + x$.

3. *Addition has an identity element*:

There exists an element 0 from M, called the *zero*, such that $x + 0 = x$ for all x from M.

4. *Each element has the opposite element*:

For any x from M, there exists an element z from M, called the *opposite* of x, such that $x + z = 0$.

multiplication is distributive over addition in M:

For all elements x, from K and y, z from M, we have

$$x(y + z) = xy + xz$$

multiplication is distributive over addition in K:

For all elements x, from M and a, b from K, we have

$$(a + b)x = ax + bx$$

In a *right module* M, elements from M are multiplied from the right by elements from M.

Sets $F(\boldsymbol{R})$ and $F[a, b]$ of all functions in $\boldsymbol{R}$ and in $[a, b]$, correspondingly, are modules over rings $C(\boldsymbol{R})$ and $C[a, b]$ of all continuous functions in $\boldsymbol{R}$ and in $[a, b]$, respectively.

A *field* $\boldsymbol{F}$ is a ring in which:

1. *Multiplication is associative*:

For all x, y, z from $\boldsymbol{F}$, we have $x(yz) = (xy)z$.

2. *Multiplication is commutative*:

For all x, y from $\boldsymbol{F}$, we have $xy = yx$.

3. *Multiplication has an identity element*:
There exists an element e from $\boldsymbol{F}$, such that $xe = ex = x$ for all x from $\boldsymbol{F}$.
4. *Multiplication has inverse element*:

For any x from $\boldsymbol{F}$, there exists an element z from $\boldsymbol{F}$, called the *multiplicative inverse* of x, such that $xz = zx = e$.

Examples of fields are all real numbers $\boldsymbol{R}$ and all complex numbers $\boldsymbol{C}$.

A *linear space* also called a *vector space* or a *linear vector space* L over the field $\boldsymbol{F}$ has two operations:

addition: $L \times L \to L$ denoted by $x + y$ where x and y belong to L;

scalar multiplication: $\boldsymbol{F} \times L \to L$ denoted by ax where $a \in \boldsymbol{F}$ and $x \in L$. These operations satisfy the following axioms:

1. Addition is associative:

   For all x, y, z from L, we have $x + (y + z) = (x + y) + z$.

2. Addition is commutative:

   For all x, y from L, we have $x + y = y + x$.

3. Addition has an identity element:
   There exists an element $\mathbf{0}$ from L, called the *zero vector*, such that $x + \mathbf{0} = x$ for all x from L.
4. *Each element has the opposite element*:
   For any x from L, there exists an element z from L, called the *opposite* of x, such that $x + z = \mathbf{0}$.
5. Scalar multiplication is distributive over addition in L:

   For all elements a from $\boldsymbol{F}$ and vectors y, z from L, we have

$$a(y + z) = ay + az.$$

6. Scalar multiplication is distributive over addition in $\mathbf{F}$:

   For all element elements a, b from $\boldsymbol{F}$ and any vector $\boldsymbol{y}$ from L, we have

$$(a + b)y = ay + by$$

7. Scalar multiplication is compatible with multiplication in $\mathbf{F}$:

   For all elements a, b from $\boldsymbol{F}$ and any vector y from L, we have

$$a(by) = (ab)y.$$

8. The identity element 1 from the field $\mathbf{F}$ also is an identity element for scalar multiplication:

   For all vectors $\boldsymbol{x}$ from L, we have $1\boldsymbol{x} = \boldsymbol{x}$.

The sets of functions $F(\boldsymbol{R})$, $C(\boldsymbol{R})$, $F[a,b]$ and $C[a,b]$ are vector spaces over the field $\boldsymbol{R}$.

A subset M of a vector space L is called a *linear* (*vector*) *subspace* of L if M is closed with respect to addition and scalar multiplication.

An expression $\sum_{i=1}^{n} a_i \boldsymbol{x}_i$ where a_i are elements from $\boldsymbol{F}$ and $\boldsymbol{x}_1, \boldsymbol{x}_2, \ldots, \boldsymbol{x}_n$ are vectors from L is called a *linear combination* of vectors $\boldsymbol{x}_1, \boldsymbol{x}_2, \ldots, \boldsymbol{x}_n$.

Vectors $\boldsymbol{x}_1, \boldsymbol{x}_2, \ldots, \boldsymbol{x}_n$ from L are called *linearly dependent* in L if there is an equality $\sum_{i=1}^{n} a_i \boldsymbol{x}_i = 0$ where a_i are elements from $\boldsymbol{F}$ and not all of them are equal to 0. When there are no such an equality, vectors $\boldsymbol{x}_1, \boldsymbol{x}_2, \ldots, \boldsymbol{x}_n$ are called *linearly independent.*

If L is a vector space, then the linear subspace $L\{\boldsymbol{x}_1, \boldsymbol{x}_2, \ldots, \boldsymbol{x}_n\}$generated by vectors $\boldsymbol{x}_1, \boldsymbol{x}_2, \ldots, \boldsymbol{x}_n$ from L consists of all linear combinations of vectors $\boldsymbol{x}_1, \boldsymbol{x}_2, \ldots, \boldsymbol{x}_n$.

Mappings of vector spaces are also called *operators.* A mapping of a linear space over a field F into F is called a *functional.*

A system B of linearly independent vectors from L is called a *basis* of L if any element $\boldsymbol{x}$ from L is equal to a sum $\sum_{i=1}^{n} a_i \boldsymbol{x}_i$ where n is some natural number, $\boldsymbol{x}_i$ are elements from B and a_i are elements from $\boldsymbol{F}$.

The number of elements in a basis is called the *dimension* of the vector space L. It is proved that all bases of the same space have the same number of elements. The number of elements in a basis is called the *dimension* of the space L. It is proved that for each vector space L, its dimension is defined in a unique way.

If X are Y are real vector spaces, then a mapping $f\colon X \to Y$ is called *linear* if $f(c \cdot u + d \cdot v) = c \cdot f(u) + d \cdot f(v)$ for any elements u and v from X and any real numbers c and d.

$\mathrm{L}(X, Y)$ is the set of all linear mappings from X into Y. $\mathrm{PL}(X, Y)$ is the set of all partial linear mappings from X into Y. Naturally, $\mathrm{L}(X, Y) \subseteq \mathrm{PL}(X, Y)$.

The space $\boldsymbol{R}$ is a one-dimensional vector (linear) space over itself. The space $\boldsymbol{R}^n$ is an n-dimensional vector (linear) space over $\boldsymbol{R}$.

A linear space A over $\boldsymbol{R}$ is called a *linear algebra* over $\boldsymbol{R}$ if a binary operation called multiplication is also defined in A and this operation satisfies the following additional axioms:

1. *Multiplication is distributive over addition* in A:

For all elements x, y and z from A, we have

$$x(y + z) = xy + xz.$$

2. *Multiplication is distributive over addition* in $\boldsymbol{R}$:

For all elements x from M and y, z from $\boldsymbol{R}$, we have

$$(z + y)x = zx + yx.$$

Sets $F(\boldsymbol{R})$, $C(\boldsymbol{R})$, $F[a,b]$ and $C[a,b]$ are linear algebras over the field $\boldsymbol{R}$.

Note that any linear algebra is also a ring and thus, it is possible to consider modules over linear algebras.

It is possible to find other structures from algebra and their properties, for example, in (Kurosh, 1963; Van der Varden, 1971; Rotman, 1996).

There are two approaches to the mathematical structure called a *category*. One approach treats categories in the framework of the general set-theoretical mathematics. Another approach establishes categories independently of sets and uses them as a foundation of mathematics different from set theory. It is possible to build the whole mathematics in the framework of categories. For instance, such a basic concept as a binary relation is frequently studied in categories (cf., for example, Burgin, 1970). Topos allows one to reconstruct set theory as a subtheory of category theory (cf., for example, Goldblatt, 1979). According to the first approach, we have the following definition of a category.

A *category* **C** consists of two collections Ob **C**, the objects of **C**, and Mor **C**, the morphisms of **C** that satisfy the following three axioms:

A1. For every pair A, B of objects, there is a set $\mathrm{Mor}_{\mathbf{C}}(A, B)$, also denoted by $\mathrm{H}_{\mathbf{C}}(A, B)$ or $\mathrm{Hom}_{\mathbf{C}}(A, B)$, elements of which are called morphisms from A to B in **C**. When f is a morphism from A to B, it is denoted by f: $A \to B$.

A2. For every three objects A, B and C from Ob **C**, there is a binary partial operation, which is a partial function from pairs of morphisms that belong to the direct product $\mathrm{Mor}_{\mathbf{C}}(A, B)\times \mathrm{Mor}_{\mathbf{C}}(B, C)$ to morphisms in $\mathrm{Mor}_{\mathbf{C}}(A, C)$. In other words, when f: $A \to B$ and g: $B \to C$, there is a morphism $g \circ f$: $A \to C$ called the composition of morphisms g and f in **C**. This composition is associative, that is, if f: $A \to B, g$: $B \to C$ and h: $C \to D$, then $h \circ (g \circ f) = (h \circ g) \circ f$.

A3. For every object A, there is a morphism 1_A in $\mathrm{Mor}_C(A, A)$, called the identity on A, for which if f: $A \to B$, then $1_B \circ f = f$ and $f \circ 1_A = f$.

Examples of categories:

1. The category of sets **SET**: objects are arbitrary sets and morphisms are mappings of these sets.
2. The category of groups **GRP**: objects are arbitrary groups and morphisms are homomorphisms of these groups.

3. The category of topological spaces **TOP**: objects are arbitrary topological spaces and morphisms are continuous mappings of these topological spaces.

A category **K** is a *subcategory* of a category **C** if:

(a) All objects of **K** are objects of **C**.
(b) All morphisms of **K** are morphisms of **C**.
(c) For any two objects A and B from **K**, the set $\mathrm{Mor}_{\mathbf{K}}(A, B)$ is a subset of the set $\mathrm{Mor}_{\mathbf{C}}(A, B)$.

A subcategory **K** of a category **C** is called *wide* if **K** and **C** have the same objects.

Mapping of categories that preserve their structure are called functors. There are functors of two types: covariant functors and contravariant functors.

A *covariant functor* **F: C → K**, also called a *functor*, from a category **C** to a category **K** is a mapping that is stratified into two related mappings $F_{\mathrm{Ob}\mathbf{C}}$: Ob**C**→ Ob **K** and $F_{Mor\mathbf{C}}$: Mor **C**→ Mor**K**, i.e., F_{ObC} associates an object $F(A)$ from the category **K** to each object A from the category **C** and $F_{\mathrm{Mor}\mathbf{C}}$ associates a morphism $F(f)$: $F(B) \to F(A)$ from the category **K** to each morphism f: $A \to B$ from the category **C**. In addition, F satisfies the following two conditions:

1. $F(1_A) = 1_{F(A)}$ for every object A from the category **C**;
2. $F(f \circ g) = F(f) \circ F(g)$ for all morphisms$f$ and g from the category **C** when their composition $f \circ g$ exists.

That is, functors preserve identity morphisms and composition of morphisms.

A *contravariant functor* **F: C → K** from a category **C** to a category **K** consists of two mappings $F_{\mathrm{Ob}\mathbf{C}}$: Ob**C**→ Ob**K** and $F_{\mathrm{Mor}\mathbf{C}}$: Mor**C**→ Mor**K**, i.e., F_{ObC} associates an object $F(A)$ from the category **K** to each object A from the category **C** and $F_{\mathrm{Mor}\mathbf{C}}$ associates a morphism $F(f)$: $F(A) \to F(B)$ from the category **K** to each morphism f: $A \to B$ from the category **C**, that satisfy the following two conditions:

1. $F(1_A) = 1_{F(A)}$ for every object A from the category **C**;
2. $F(f \circ g) = F(g) \circ F(f)$ for all morphisms$f$and g from the category **C** when their composition $f \circ g$ exists.

It is possible to define a contravariant functor as a covariant functor on the dual category $\mathbf{C}^{\text{op}}$.

A functor from a category to itself is called an *endofunctor*.

There is an approach to the definition of a category in which a category $\mathbf{C}$ consists only of the collection Mor $\mathbf{C}$ of the morphisms (also called *arrows*) of $\mathbf{C}$ with corresponding axioms. Objects of $\mathbf{C}$ are associated with identity morphisms 1_A. It is possible to do this because 1_A is unique in each set $\text{Mor}_{\mathbf{C}}(A, A)$, and uniquely identifies the object A. In any case, morphism is the central concept in a category. But what is a morphism?

If $f: A \to B$ is a morphism, then it is a one-to-one named set $(\{A\}, f, \{B\})$. Thus, the main object of a category is a named set, and categories are built from these named sets. Besides, a construction or separation of a category begins with separation of all elements into two sets and calling all elements from one of these sets by the name "objects" and elements from the other sets by the name "morphisms". In such a way, two named sets appear. In addition, composition of morphisms, as any algebraic operation, is also represented by a named set $(\text{Mor}_{\mathbf{C}}(A, B) \times \text{Mor}_{\mathbf{C}}(B, C), \circ, \text{Mor}_{\mathbf{C}}(A, C))$. This shows that the informal notion of a named set is prior both to categories and sets. As a result, we come to the conclusion that any category is built of different named sets. Moreover, functors between categories, which are structured mappings of categories (Herrlich and Strecker, 1973), are morphisms of those named sets.

It is possible to read more about categories, functors and their properties, for example, in (Goldblatt 1979; Herrlich and Strecker, 1973).

Bibliography

1 + 1 = 3, Urban Dictionary, https://www.urbandictionary.com/define.php?term=1%2B1%3D3.

2+2 = 5, Cambridge Dictionary, http://dictionary.cambridge.org/us/dictionary/english/2-2-5.

Abbott, L. F. and Wise, M. B. (1981) Dimension of a quantum-mechanical path, *Am. J. Phys.* v. 49, 37.

Aberth, O. (1968) Analysis in the Computable Number Field, *J. Assoc. Comput. Mach. (JACM)*, v. 15, No. 2, pp. 276–299.

Abian, A. (1965) *The Theory of Sets and Transfinite Arithmetic*, W.B. Saunders Company, Philadelphia/London.

Ackermann, W. (1924) Begréndung des "tertium non datur" mittels der Hilbertschen Theorie der Widerspruchsfreiheit, *Math. Ann.*, v. 93, pp. 1–36.

Ackermann, W. (1940) Zur Widerspruchsfreiheit der Zahlentheorie, *Math. Ann.*, v. 117, pp. 162–194.

Aczél, J. (1966) *Lectures on Functional Equations and Their Applications*, Mathematics in Science and Engineering, v. 19, Academic Press, New York.

Aerts, D., Czachor, M. and Kuna, M. (2016) Crystallization of space: Space-time fractals from fractal arithmetic, *Chaos, Solitons Fractals*, v. 83, pp. 201–211.

Aerts, D., Czachor, M. and Kuna, M. (2016a) Fourier transforms on Cantor sets: A study in non-Diophantine arithmetic and calculus, *Chaos, Solitons Fractals*, v. 91, pp. 461–468.

Aerts, D., Czachor, M. and Kuna, M. (2018) Simple fractal calculus from fractal arithmetic, *Rep. Math. Phys.*, v. 81, pp. 357–370.

Agahi, H., Ouyang, Y., Mesiar, R., Pap, E. and Štrboja, M. (2011) Hölder and Minkowski type inequalities for pseudo-integral, *App. Math. Comput.*, v. 217, pp. 8630–8639.

Agrawal, A. and Jaffe, J. (2000) The post merger performance puzzle, *Adv. Mergers Acquisitions*, v. 1, pp. 119–156.

Agrillo, C. and Bisazza, A. (2018) Understanding the origin of number sense: A review of fish studies, *Philos. Trans. R. Soc. Lond.*, B *Biol. Sci.*, v. 373, 20160511.

Agrillo, C. and Bisazza, A. (2014) Spontaneous versus trained numerical abilities. A comparison between the two main tools to study numerical competence in non-human animals, *J. Neurosci. Methods*, v. 234, pp. 82–91.

Agrillo, C., Dadda, M. and Bisazza, A. (2006) Quantity discrimination in female mosquitofish, *Animal Cognition*, v. 10, No. 1, pp. 63–70.

Aigner, M. (1979) *Combinatorial Theory*, Springer-Verlag, New York/Berlin.

Alexandroff, P. S. (1977) *Introduction to Set Theory and General Topology*, Nauka, Moscow (in Russian).

Allenby, R. B. J. T. (1991) *Rings, Fields and Groups*, Edward Arnold, London.

Alp, K. O. (2010) A comparison of sign and symbol (their contents and boundaries), *Semiotica*, v. 182, No.1/4, pp. 1–13.

Amalric, M. and Dehaene, S. (2016) Origins of the brain networks for advanced mathematics in expert mathematicians, *Proc. Natl. Acad. Sci. USA*, v. 113, No.18, pp. 4909–4917.

Amat Plata, S. (2005) Mentimos a nuuestros hijos cuuando les decimos quue $1+1$ son 2? *Rev. Euureka*, pp. 33–37.

Ambjørn, J., Jurkiewicz, J. and Loll, R. (2005) Reconstructing the Universe, *Phys. Rev.* D, v. 72, 064014.

Amelino-Camelia, G. (2002) Relativity in spacetimes with short-distance structure governed by an observer-independent (Planckian) length scale, *Int. J. Mod. Phys.* D, v. 11, 35.

Anderson, R. L. (2004) It adds up after all: Kant's philosophy of Arithmetic in light of the traditional logic, *Philos. Phenomenolog. Res.*, v. 69, No. 3, pp. 501–540.

Andres, M., Michaux, N. and Pesenti, M. (2012) Common substrate for mental arithmetic and finger representation in the parietal cortex, *Neuroimage*, v. 62, pp. 1520–1528.

Andrews, U., Bonik, G., Chen, J. P., Martin, R. W. and Teplyaev, A. (2015) *Wave equation on one-dimensional fractals with spectral decimation and the complex dynamics of polynomials*, Preprint in Mathematical Physics [math-ph], arXiv:1505.05855.

Anglin, W. S. and Lambek, J. (1995) Plato and Aristotle on mathematics, in *The Heritage of Thales*, Undergraduate Texts in Mathematics (Readings in Mathematics), Springer, New York, NY, pp. 67–69.

Aniszewska, D. (2007) Multiplicative Runge–Kutta methods, *Nonlinear Dyn.*, v. 50, No. 1–2, pp. 265–272.

Annas, J. (1975) Aristotle, number and time, *Philos. Quart.*, v.25, No. 99, pp. 97–113.

Ansari, D. (2016) The neural roots of mathematical expertise, *Proc. Natl. Acad. Sci. USA*, v. 113, No.18, pp. 4887–4889.

Apostol, T. M. (1967) Calculus, v. 1: *One-Variable Calculus with an Introduction to Linear Algebra*, John Wiley & Sons, New York.

Aragona, J. and Juriaans, S. O. (2001) Some structural properties of the topological ring of Colombeau's generalized numbers, *Comm. Alg.*, v. 29, No. 5, pp. 2201–2230.

Archibald, J. (2014) *One Plus One Equals One: Symbiosis and the Evolution of Complex Life*, Oxford University Press, New York.

Argand, J.-R. (1806) *Essai sur une manière de représenter des quantités imaginaires dans les constructions géométriques*, Chez Mme Vve Blanc, Paris.

Archimedes (1880) *Archimedis Opera Omnia cum Commentariis Eutocii*, J. L. Heiberg, Ed., B. G. Teubner, Leipzig.

Aristotle (1984) *The Complete Works of Aristotle*, Princeton University Press, Princeton.

Arithmetic (2016) *New World Encyclopedia.*

Arsalidou, M. and Taylor, M.J. (2011) Is $2+2=4$? Meta-analyses of brain areas needed for numbers and calculations, *Neuroimage*, v. 54, pp. 2382–2393.

Artin, M. (1991) *Algebra*, Prentice-Hall, Englewood Cliffs, NJ.

Arzt, P. (2014) *Measure theoretic trigonometric functions*, Preprint arXiv:1405.4693.

Ashcraft, M. H. (1992) Cognitive arithmetic: A review of data and theory, *Cognition*, v. 44, pp. 75–106.

Ashcraft, M. H. (1995) Cognitive psychology and simple arithmetic: A review and summary of new directions, *Math. Cogn.*, v. 1, pp. 3–34.

Asher, M. (2002) *Mathematics Elsewhere: An Exploration of Ideas Across Cultures*, Princeton University Press, Princeton/Oxfordshire.

Ashkenazi, S. (2008) Basic numerical processing in left intraparietal sulcus (IPS) acalculia, *Cortex*, v. 44, No. 4, pp. 439–448.

Atkinson, K. (1989) *An Introduction to Numerical Analysis*, Wiley, New York.

Atkinson, R. L., Atkinson, R. C., Smith E. E. and Bem, D. J. (1990) *Introduction to Psychology*, Harcourt Brace Jovanovich, Inc., San Diego.

Avigad, J. (2015) *Mathematics and language*, Preprint in Mathematics History and Overview (math.HO), arXiv:1505.07238.

Aydin, K., Ucar, A., Oguz, K. K., Okur, O. O., Agayev, A., Unal, Z., Yilmaz, S. and Ozturk, C. (2007) Increased gray matter density in the parietal cortex of mathematicians: a voxel-based morphometry study, *Am. J. Neuroradiol.* (*AJNR*), v. 28, No. 10, pp. 1859–1864.

Baccheli, B. (1986) Representation of continuous associative functions, *Stochastica: Rev. Mate. Aplicada*, v. 10, No. 1, pp. 13–28.

Bachem, A. (1952) Weber's law in physics and arithmetic, *Am. J. Psychol.*, v. 65, No. 1, pp. 106–107.

Badets, A. and Pesenti, M. (2010) Creating number semantics through finger movement perception, *Cognition*, v. 115, pp. 46–53.

Badiou, A. (1990/2008) *Number and Numbers*, (Mackay, R. Trans.), Polity, Cambridge, UK/Oxford.

Bagla, J. S., Yadav, J. and Seshadri, T. R. (2008) Fractal dimensions of a weakly clustered distribution and the scale of homogeneity, *Mon. Not. R. Astron. Soc.*, v. 390, p. 829.

Bair, J., Błaszczyk, P., Ely, R., Henry, V., Kanovei, V., Katz, K. U., Katz, M. G., Kutateladze, S. S., McGaffey, T., Reeder, P., Schaps, D.M., Sherry, D. and Shnider, S. (2017) Interpreting the infinitesimal mathematics of Leibniz and Euler, *J. General Philos. Sci.*, v. 48, No. 2, pp. 195–238.

Baird, J. C. (1975) Psychophysical study of numbers (IV): Generalized preferred state theory, *Psychol. Res.*, v. 38, pp. 175–187.

Baird, J. C. (1975a) Psychophysical study of numbers (V): Preferred state theory of matching functions, *Psych. Res.*, v. 38, pp. 188–207.

Baird, J. C. (1997) *Sensation and Judgment: Complementarity Theory of Psychophysics*, Lawrence Erlbaum Associates, Mahwah.

Baird, J. C. and Noma, E. (1975) Psychophysical study of numbers (I): Generation of numerical responses, *Psychol. Res.*, v. 37, pp. 281–297.

Baird, J. C. and Noma, E. (1978) *Fundamentals of Scaling and Psychophysics*, Wiley, New York.

Balankin, A. S. (2014) *Toward the mechanics of fractal materials: Mechanics of continuum with fractal metric*, Preprint in physics [cond-mat.mtrl-sci], arXiv:1409.5829.

Balcazar, J. L., Diaz, J. and Gabarro, J. (1988) *Structural Complexity*, Springer-Verlag.

Bali, N. P. (2005) *Golden Differential Calculus*, Laxmi Publication, New Delhi, India.

Balian, R. and Schaeffer, R. (1988) Galaxies — fractal dimensions, counts in cells, and correlations, *Astrophys. J.*, v. 335, L43.

Banks, V. P. and Coleman, M. J. (1981) Two subjective scales of number, *Perception Psychophys.*, v. 29, No. 2, pp. 95–105.

Banks, W. P. and Hill, D. K. (1974) The apparent magnitude of number scaled by random production, *J. Exp. Psychol.*, v. 102, pp. 353–376.

Barker, S. (1967) Number, in *Encyclopedia of Philosophy* (Edwards, P., Ed.), Macmillan Publishing Company, New York, v. 5, pp. 526–530.

Barlow, M. T. (1998) *Diffusions on Fractals*, Lecture Notes in Mathematics, v. 1690, Springer, Berlin.

Barner, D. and Bachrach, A. (2010) Inference and exact numerical representation in early language development, *Cogn. Psychol.*, v. 60, pp. 40–62.

Bartle, R. G. (2001) *A Modern Theory of Integration*, American Mathematical Society, Providence, RI.

Barton, B. (2008) *The Language of Mathematics: Telling Mathematical Tales*, Springer, New York.

Bashirov, A. and Riza, M. (2011) On complex multiplicative differentiation, *TWMS J. Appl. Eng. Math.*, v. 1, No. 1, pp. 75–85.

Bashirov, A. E., Kurpınar, E. M. and Özyapıcı, A. (2008) Multiplicative calculus and its applications, *J. Math. Anal. Appl.*, v. 337, No. 1, pp. 36–48.

Bashirov, A. E., Mısırlı, E., Tandoğdu, Y. and Özyapıcı, A. (2011) On modeling with multiplicative differential equations, *Appl. Math.:J. Chinese Univ.*, v. 26, No. 4, pp. 425–438.

Beck, C. and Schlögl, F. (1993) *Thermodynamics of Chaotic Systems: An Introduction*, Cambridge University Press, Cambridge.

Beck, A., Bleicher, M. N. and Crowe, D. W. (2000) *Excursions into Mathematics*, A K Peters, Natick, MA.

Beckenbach, E. F. (1956) *Modern Mathematics for the Engineer*, McGraw-Hill, New York/Toronto/London.

Beechler, D. *How to Create* "1+1 = 3" *Marketing Campaigns*, 2013, http://www.marketingcloud.com/blog/how-to-create-1-1-3-marketing-campaigns.

Bell, E. T. (1937) *Men of Mathematics*, Simon & Schuster, New York.

Bell, J. L. (2008) *A Primer of Infinitesimal Analysis*, Cambridge University Press, Cambridge.

Bellavitis, G. (1835) Saggio di applicazioni di un nuovo metodo di geometria analitica (Calcolo delle equipollenze), *Ann. Sci. Regno Lombaro-Veneto, Padova*, v. 5, pp. 244–259.

Belna, J.-P. (1996) *La notion de nombre chez Dedekind, Cantor, Frege: Théories, conceptions, et philosophie*, Librairie Philosophique Vrin, Paris.

Benacerraf, P. (1965) What numbers could not be, *Philos. Rev.*, v. 74, pp. 47–73.

Benci, V. and Di Nasso, M. (2005) A purely algebraic characterization of the hyperreal numbers, *Proc. Am. Math. Soc.*, v. 133, pp. 2501–2505.

Benedetti, D. (2009) Fractal properties of quantum spacetime, *Phys. Rev. Lett.*, v. 102, 111303.

Benioff, P. (2001) The representation of natural numbers in quantum mechanics, *Phys. Rev.*, v. A63, 032305.

Benioff, P. (2001a) Efficient implementation and the product state representation of numbers, *Phys. Rev.*, v. 64A, 052310.

Benioff, P. (2001b) The representation of numbers in quantum mechanics, *Algorithmica*, v. 34, pp. 529–559.

Benioff, P. (2002) Towards a coherent theory of physics and mathematics, *Found. Phys.*, v. 32, pp. 989–1029.

Benioff, P. (2005) Towards a coherent theory of physics and mathematics: The theory-experiment connection, *Found. Phys.*, v. 35, pp. 1825–1856.

Benioff, P. (2006) Complex rational numbers in quantum mechanics, *Int. J. Mod. Phys.* B, v. 20, pp. 1730–1741.

Benioff, P. (2007) A Representation of real and complex numbers in quantum theory, *Int. J. Pure Appl. Math.*, v. 39, pp. 297–339.

Benioff, P. (2009) A possible approach to inclusion of space and time in frame fields of quantum representations of real and complex numbers, *Adv. Math. Phy.*, ID 452738.

Benioff, P. (2011) New gauge field from extension of spacetime parallel transport of vector spaces to the underlying number systems, *Int. J. Theor. Phys.*, v. 50, 1887.

Benioff, P. (2012) Local availability of mathematics and number scaling: Effects on quantum physics, in *Quantum Information and Computation* X (Donkor, E., Pirich, A. and Brandt, H. Eds.) Proceedings of SPIE, v. 8400, SPIE, Bellingham, WA, 84000T, also arXiv:1205.0200.

Benioff, P. (2012a) Effects on quantum physics of the local availability of mathematics and space time dependent scaling factors for number systems, in *Advances in Quantum Theory* (Cotaescu, I. I., Ed.), InTech, also arXiv:1110.1388.

Benioff, P. (2013) Representations of each number type that differ by scale factors, *Adv. Pure Math.*, v. 3, pp. 394–404.

Benioff, P. (2015) Fiber bundle description of number scaling in gauge theory and geometry, *Quantum Studies: Math. Found.*, v. 2, pp. 289–313.

Benioff, P. (2016) Space and time dependent scaling of numbers in mathematical structures: Effects on physical and geometric quantities, *Quantum Inform. Process.*, v. 15, No. 3, pp. 1081–1102, also arXiv:1508.01732.

Benioff, P. (2016a) Effects of a scalar scaling field on quantum mechanics, *Quantum Inform. Process.*, v. 15, No. 7, pp. 3005–3034.

Benvenuti, P., Mesiar, R. and Vivona, D. (2002) Monotone set functions-based integrals, in *Handbook of Measure Theory*, v. II, (Pap, E., Ed.), Elsevier, North-Holland, pp. 1329–1379.

Berch, D. B. (2005) Making sense of number sense: implications for children with mathematical disabilities, *J. Learning Disabil.*, v. 38, No. 4, pp. 333–339.

Berčić, B. (2005) Zašto 2 + 2 = 4? *Filozofska Istrazivanja*, v. 25, No. 4, pp. 945–961.

Berkeley, G. (1707) *Arithmetica absque algebra aut Euclide demonstrata and Miscellanea Mathematica*, London, A. & J. Churchill, Dublin, J. Pepyat.

Berman, K. and Knight, J. (2008) *Financial Intelligence for Entrepreneurs: What You Really Need to Know About the Numbers*, Harvard Business Press, Brighton, MA.

Bernays, P. (1935) Sur le platonisme dans les mathematiques, *L'Enseignement math.*, v. 34, pp. 52–69.

Bernays, P. (1951) Über das Inductionsschema in der recursiven Zahlentheorie, in *Kontrolliertes Denken, Festschrift für Wilhelm Britzelmayr*, Freiburg/München, pp. 10–17.

Bernays, P. (1967) David Hilbert, in *The Encyclopedia of Philosophy*, v. 3, Macmillan Publishing Company and The Free Press, New York, pp. 496–504.

Berry, M. (1979) Distribution of modes in fractal resonators, in *Structural Stability in Physics* (W. Guttinger and H. Elkheimer, Eds.), Springer, Berlin, pp. 51–53.

Berteletti, I. and Booth, J. R. (2015) Perceiving fingers in single-digit arithmetic problems, *Front. Psychol.*, v. 6, pp. 226–241.

Beutelspacher, A. (2016) *Numbers: Histories, Mysteries, Theories*, Dover, New York.

Biagioli, F. (2016) *Space, Number, and Geometry from Helmholtz to Cassirer*, Springer, New York.

Bigelow, J. (1988) *The Reality of Numbers: A Physicalist's Philosophy of Mathematics*, Oxford University Press, New York.

Binney, J. J., Dowrick, N. J., Fisher, A. J. and Newman, M. E. J. (1993) *The Theory of Critical Phenomena*, Oxford University Press, New York.

Bird, J. (1993) *Higher Engineering Mathematics*, Newnes, Oxford & Boston.

Birget, J.-C., Margolis, S., Meakin, J. and Sapir, M.V. (Eds.) (2000) *Algorithmic Problems in Groups and Semigroups*, Birkhäuser Verlag AG, Boston.

Birkhoff, G. (1938) On product integration, *J. Math. Phys.*, v. 16, pp. 104–132.

Birkhoff, G. and Bartee, T. C. (1967) *Modern Applied Algebra*, McGraw–Hill, New York.

Biro D. and Matsuzawa, T. (2001) Use of numerical symbols by the chimpanzee (Pan troglodytes): Cardinals, ordinals, and the introduction of zero, *Anim. Cogn.*, v. 4, pp. 193–199.

Black, M. (1951) Review: Gottlob Frege, The foundations of arithmetic. A logico-mathematical enquiry into the concept of number, *J. Symbolic Logic*, v. 16, No. 1, p. 67.

Blass, A., Di Nasso, M. and Forti, M. (2011) *Quasi-selective ultrafilters and asymptotic numerosities*, Preprint (ArXiv:1011.2089v2).

Blehman, I. I., Myshkis, A. D. and Panovko, Ya. G. (1983) *Mechanics and Applied Logic*, Nauka, Moscow (in Russian).

Blok, W. and Pigozzi, D. (1989) *Algebraizable Logics*, Memoires of the American Mathematical Society, v. 396.

Blyth, D. (2000) Platonic number in the parmenides and metaphysics XIII, *Int. J. Philos. Studies*, v. 8, No. 1, pp. 23–45.

Boccuto, A. and Candeloro, D. (2011) Differential calculus in Riesz spaces and applications to g-calculus, mediterranean *J. Math.*, v. 8, No. 3, pp. 315–329.

Boissoneault, L. (2017) *How Humans Invented Numbers — And How Numbers Reshaped Our World*: Anthropologist Caleb Everett explores the subject in his new book, *Numbers and the Making Of Us*, smithsonian.com, https://www.smithsonianmag.com/innovation/how-humans-invented-numbersand-how-numbers-reshaped-our-world-180962485/.

Boksic, B. (2017) Getting to $1 + 1 = 3$: The 3 types of relationships in your life, https://www.goalcast.com/2017/06/26/getting-113-3-types-relationships-life/.

Bondecka-Krzykowska, I. (2004) Strukturalizm jako alternatywa dla platonizmu w filozofii matematyki, *Filozofia Nauki*, No. 1.

Boole, G. (1860) *Calculus of Finite Differences*, Chelsea Publishing Company.

Boole, G. (1880) *A Treatise on the Calculus of Finite Differences*, Macmillan., London.

Boolos, G. (1987) The consistency of Frege's Foundations of Arithmetic, in *On Being and Saying*, Essays in Honor of Richard Cartwright, MIT Press, Cambridge, pp. 3–20.

Boolos, G. (1990) The standard of equality of numbers, in *Meaning and Method*, Essays in Honor of Hilary Putnam, Cambridge University Press, Cambridge, pp. 261–277.

Borel, E. (1903) Contribution a l'analyse arithmetique du continu, *J. Math. Appl.*, 5e série, tome IX, pp. 329–375.

Borel, E. (1956) *Probabilité et Certitude*, Presses Universitaires de France, Paris.

Borges, E. (2004) A possible deformed algebra and calculus inspired in nonextensive thermo-statistics, *Physica A*, v. 340, pp. 95–101.

Boruah, K. and Hazarika, B. (2016) *Some basic properties of G-Calculus and its applications in numerical analysis*, Preprint in General Mathematics [math.GM], arXiv:1607.07749.

Boscarino, G. (2018) An interpretation of Plato's idea and Plato's criticism of Parmenides according to Peano's ideography, *Athens J. Humanities Arts*, v. 5, No. 1, pp. 7–28.

Boumans, M. (2005) *How Economists Model the World into Numbers*, Routledge, New York.

Bourbaki, N. (1948) L'architecture des mathématiques, Legrands courants de la pensée mathématiques, *Cahiers Sud*, pp. 35–47.

Bourbaki, N. (1950) The architecture of mathematics, *Amer. Math. Monthly*, v. 57, pp. 221–232.

Bourbaki, N. (1957) *Structures*, Hermann, Paris.

Bourbaki, N. (1960) *Theorie des Ensembles*, Hermann, Paris (English translation: Bourbaki, N. *Theory of Sets*, Hermann, Paris, 1968).

Boyer, C. B. (1985) *A History of Mathematics*, Princeton University Press, Princeton.

Brannon, E. M. and Terrace, H. S. (1998) Ordering of the numerosities 1 to 9 by monkeys, *Science*, v. 282, pp. 746–749.

Braβel, B., Fischer, S. and Huch, F. (2007) Declaring numbers, in *Workshop on Functional and (Constraint) Logic Programming*, pp. 23–36.

Brekke, L. and Freund, P. (1993) p-Adic numbers in physics, *Phys. Rep*, v. 231, pp. 1–66.

Breukelaar, J. W. C. and Dalrymple-Alford, J. C. (1998) Timing ability and numerical competence in rats, *J. Exp. Psychol. Anim. Behav. Process.*, v. 24, pp. 84–97.

Broadbent, T. A. A. (1971) The higher arithmetic, *Nature*, v. 229 No. 6, pp. 187–188.

Brockman, J. (1997) *What Kind of Thing is a Number*? A talk with Reuben Hersh, *Edge*, https://www.edge.org/conversation/reuben_hersh-what-kind-of-thing-is-a-number.

Brodsky, A. E., Rogers Senuta, K., Weiss, C. L. A. Marx, C. M., Loomis, C., Arteaga, S. S., Moore, H., Benhorin, R. and Castagnera-Fletcher, A. (2004) When one plus one equals three: The role of relationships and context in community, *Amer. J. Community Psychol.*, v. 33, No. 3–4, pp. 229–241.

Brown, J. R. (2017) Proofs and guarantees, *Math. Intelligencer*, v. 39, No. 4, pp. 47–50.

Brown, N. (2015) *One plus one equals three: Combining email + Facebook ads to reach more customers*, February 17th, 2015, https://socialmediaweek.org/blog/2015/02/one-plus-one-equals-three-combining-email-facebook-ads-reach-customers/.

Brown, P. and Lauder, H. (2001) Human capital, social capital and collective intelligence, in *Social Capital: Critical Perspectives*, Oxford University Press, Oxford, pp. 226–242.

Buchanan, W. (1974) Nominal and ordinal bivariate statistics: The practitioner's view, *Amer. J. Political Sci.*, v. 18, No. 3, pp. 625–646.

Bucur, I. and Deleanu, A. (1968) *Introduction to the Theory of Categories and Functors*, Wiley, London.

Bueti, D., and Walsh, V. (2009) The parietal cortex and the representation of time, space, number and other magnitudes, *Philos. Trans. R. Soc.*, B 364, pp. 1831–1840.

Buium, A. (2005) *Arithmetic Differential Equations*, Mathematical Surveys and Monographs, v. 118, American Mathematical Society.

Burgess, J. P. and Rosen, G. (1997) *A Subject With No Object: Strategies for Nominalistic Interpretation of Mathematics*, Clarendon Press, Oxford.

Burgin, M. (1970) Permutational products of linear Ω-algebras, *Soviet Math. Izvestiya*, 1970, v. 4, No. 3, pp. 977–999 (translated from Russian).

Burgin, M. (1972) Free quotients of free linear Ω-algebras, *Mat. Zametki*, 1972, v. 11, No. 5, pp. 537–544 (translated from Russian).

Burgin, M. (1977) Non-classical models of natural numbers, *Russian Math. Sur.*, v. 32, No. 6, pp. 209–210 (in Russian).

Burgin, M. (1980) Dual arithmetics, in *Abstracts Presented to the American Mathematical Society*, v.1, No. 6.

Burgin, M. (1982) Products of operators in a multidimensional structured model of systems, *Math. Social Sci.*, No. 2, pp. 335–343.

Burgin, M. (1985) Abstract theory of properties, in *Non-classical Logics*, Institute of Philosophy, Moscow, pp. 109–118 (in Russian).

Burgin, M. (1986) Quantifiers in the theory of properties, in *Nonstandard Semantics of Non-classical Logics*, Institute of Philosophy, Moscow, pp. 99–107 (in Russian).

Burgin, M. (1989) Numbers as properties, in *Abstracts of Papers Presented to the American Mathematical Society*, v. 10, No. 1.

Burgin, M. (1989a) *Named Sets, General Theory of Properties, and Logic*, Institute of Philosophy, Kiev (in Russian).

Burgin, M. (1990) Theory of named sets as a foundational basis for mathematics, in *Structures in Mathematical Theories*, San Sebastian, Spain, pp. 417–420.

Burgin, M. (1990a) Abstract theory of properties and sociological scaling, in *Expert Evaluation in Sociological Studies*, Kiev, pp. 243–264 (in Russian).

Burgin, M. (1990b) Hypermeasures and hyperintegration, *Notices Nat. Acad. Sci. Ukraine*, No. 6, pp. 10–13 (in Russian and Ukrainian).

Burgin, M. (1991) Logical methods in artificial intelligent systems, *Vestnik Comput. Soc.*, No. 2, pp. 66–78 (in Russian).

Burgin, M. (1992) Infinite in finite or metaphysics and dialectics of scientific abstractions, *Philos. Sociological Thought*, No. 8, pp. 21–32 (in Russian and Ukrainian).

Burgin, M. (1992a) Algebraic structures of multicardinal numbers, in *Problems of Group Theory and Homological Algebra*, Yaroslavl, pp. 3–20 (in Russian).

Burgin, M. (1994) Is it possible that mathematics gives new knowledge about reality, *Philos. Sociological Thought*, No. 1, pp. 240–249 (in Russian and Ukrainian).

Burgin, M. (1995) Named sets as a basic tool in epistemology, *Epistemologia*, v. XVIII, pp. 87–110.

Burgin, M. (1997) *Non-Diophantine Arithmetics or What Number is* 2 + 2, Ukrainian Academy of Information Sciences, Kiev (in Russian, English summary).

Burgin, M. (1997a) Mathematical theory of technology, in *Methodological and Theoretical Problems of Mathematics and Informatics*, Ukrainian Academy of Information Sciences, Kiev, pp. 91–100 (in Russian).

Burgin, M. (1997b) *Fundamental Structures of Knowledge and Information*, Ukrainian Academy of Information Sciences, Kiev (in Russian).

Burgin, M. (1997c) Scientific foundations of structuralism, in *Language and Culture*, v.1, Kiev, pp. 24–25 (in Russian).

Burgin, M. (1997d) Logical varieties and covarieties, in *Methodological and Theoretical Problems of Mathematics and Information and Computer Sciences*, Ukrainian Academy of Information Sciences, Kiev, pp. 18–34 (in Russian).

Burgin, M. (1998) *On the Nature and Essence of Mathematics*, Ukrainian Academy of Information Sciences, Kiev (in Russian).

Burgin, M. (1998a) Finite and infinite, in *On the Nature and Essence of Mathematics*, Appendix, Kiev, pp. 97–108 (in Russian).

Burgin, M. (2001) *Diophantine and Non-Diophantine Arithmetics: Operations with Numbers in Science and Everyday Life*, LANL, Preprint Mathematics GM/0108149.

Burgin, M. (2001a) *How we Count or is it Possible that Two Times Two is not Equal to Four*, Science Direct Working Paper No S1574-0358(04)70635-8, http://www.sciencedirect.com/preprintarchive.

Burgin, M. (2002) Theory of hypernumbers and extrafunctions: Functional spaces and differentiation, *Discrete Dynam. Nature Soc.*, v. 7, No. 3, pp. 201–212.

Burgin, M. (2003) Levels of system functioning description: From algorithm to program to technology, in *Proceedings of the Business and Industry Simulation Symposium*, Society for Modeling and Simulation International, Orlando, FL, pp. 3–7.

Burgin, M. (2003a) Nonlinear phenomena in spaces of algorithms, *Int. J. Comput. Math.*, v. 80, No. 12, pp. 1449–1476.

Burgin, M. (2004) Hyperfunctionals and generalized distributions, in *Stochastic Processes and Functional Analysis*, A Dekker Series of Lecture Notes in Pure and Applied Mathematics, v. 238, pp. 81–119.

Burgin, M. (2004a) Logical tools for program integration and interoperability, in *Proceedings of the IASTED International Conference on Software Engineering and Applications*, MIT, Cambridge, pp. 743–748.

Burgin, M. (2004b) *Unified Foundations of Mathematics*, Preprint in Mathematics, arxiv:math.LO/0403186.

Burgin, M. (2005) *Super-recursive Algorithms*, Springer, New York.

Burgin, M. (2005a) Hypermeasures in general spaces, *Int. J. Pure Appl. Math.*, v. 24, pp. 299–323.

Burgin, M. (2006) Nonuniform operations on named sets, in 5th *Annual International Conference on Statistics, Mathematics and Related Fields, 2006 Conference Proceedings*, Honolulu, Hawaii, pp. 245–271.

Burgin, M. (2006a) Operational and program schemas, in *Proceedings of the* 15th *International Conference on Software Engineering and Data Engineering* (*SEDE-2006*), ISCA, Los Angeles, California, pp. 74–78.

Burgin, M. (2007) Elements of non-Diophantine arithmetics, in 6th *Annual International Conference on Statistics, Mathematics and Related Fields, 2007 Conference Proceedings*, Honolulu, Hawaii, January, pp. 190–203.

Burgin, M. (2007a) Universality, reducibility, and completeness, *Lect. Notes Comput. Sci.*, v. 4664, pp. 24–38.

Burgin, M. (2008) *Neoclassical Analysis: Calculus Closer to the Real World*, Nova Science Publishers, New York.

Burgin, M. (2008a) Hyperintegration approach to the Feynman integral, *Integration: Math. Theory Appl.*, v. 1, No. 1, pp. 59–104.

Burgin, M. (2009) Mathematical theory of information technology, in *Proceedings of the 8th WSEAS International Conference on Data Networks, Communications, Computers* (*DNCOCO'09*), Baltimore, Maryland, USA, November, pp. 42–47.

Burgin, M. (2009a) Structures in mathematics and beyond, in *Proceedings of the 8th Annual International Conference on Statistics, Mathematics and Related Fields*, Honolulu, Hawaii, pp. 449–469.

Burgin, M. (2010) *Introduction to Projective Arithmetics*, Preprint in Mathematics, arxiv:math.GM/1010.3287.

Burgin, M. (2010a) Integration in bundles with a hyperspace base: indefinite integration, *Integration: Math. Theory Appl.*, v. 2, pp. 395–435.

Burgin, M. (2010b) *Theory of Information: Fundamentality, Diversity and Unification*, World Scientific, New York.

Burgin, M. (2010c) *Measuring Power of Algorithms, Computer Programs, and Information Automata*, Nova Science Publishers, New York.

Burgin, M. (2010d) Algorithmic complexity of computational problems, *Int. J. Comput. Inform. Technol.*, v. 2, No. 1, pp. 149–187.

Burgin, M. (2011) *Theory of Named Sets*, Mathematics Research Developments, Nova Science Publishers, New York.

Burgin, M. (2011a) *Differentiation in Bundles with a Hyperspace Base*, Preprint in Mathematics, math.CA/1112.3421.

Burgin, M. (2012) *Hypernumbers and Extrafunctions: Extending the Classical Calculus*, Springer, New York.

Burgin, M. (2012a) *Structural Reality*, Nova Science Publishers, New York.

Burgin, M. (2012b) Integration in bundles with a hyperspace base: definite integration, *Integration: Math. Theory Appl.*, v. 3, No. 1, pp. 1–54.

Burgin, M. (2012c) Fuzzy continuous functions in discrete spaces, *Ann Fuzzy Sets, Fuzzy Logic Fuzzy Syst.*, v. 1, No. 4, pp. 231–252.

Burgin, M. (2013) Named sets and integration of structures, in *Topics in Integration Research*, Nova Science Publishers, New York, pp. 55–98.

Burgin, M. (2015) Operations with extrafunctions and integration in bundles with a hyperspace base, in *Functional Analysis and Probability*, Chapter 1, Nova Science Publishers, New York, pp. 3–76.

Burgin, M. (2015a) Picturesque diversity of probability, in *Functional Analysis and Probability*, Nova Science Publishers, New York, 2015, pp. 301–354.

Burgin, M. (2016) Probability theory in relational structures, *J. Adv. Res. Appl. Math. Statist.*, v. 1, No. 3 & 4, pp. 19–29.

Burgin, M. (2016a) *Theory of Knowledge: Structures and Processes*, World Scientific, New York.

Burgin, M. (2017) *Functional Algebra and Hypercalculus in Infinite Dimensions: Hyperintegrals, Hyperfunctionals and Hyperderivatives*, Nova Science Publishers, New York.

Burgin, M. (2017a) Bidirectional named sets as structural models of interpersonal communication, *Proceedings*, v. 1, No. 3, 58.

Burgin, M. (2017b) Ideas of Plato in the context of contemporary science and mathematics, *Athens J. Humanities Arts*, v. 4, No. 3, pp. 161–182.

Burgin, M. (2017c) *Mathematical knowledge and the role of an observer: Ontological and epistemological aspects*, Preprint in Mathematics History and Overview (math.HO), arXiv: 1709.06884.

Burgin, M. (2017d) *Semitopological Vector Spaces: Hypernorms, Hyperseminorms and Operators*, Apple Academic Press, Toronto.

Burgin, M. (2018) *Mathematics as an Interconnected Whole*, Preprint in General Mathematics, viXra: 1801.0135.

Burgin, M. (2018a) Mathematical analysis of the concept equality, *Res. Rep. Math.*, v. 2, No. 2, 10.4172/RRM.1000e102.

Burgin, M. (2018b) Triadic structures in interpersonal communication, *Information*, v. 9, No. 11, 283.

Burgin, M. (2018c) Introduction to Non-Diophantine Number Theory, *Theory Appl. Math. Comput. Sci.*, v. 8, No. 2, pp. 91–134.

Burgin, M. Information-Oriented Analysis of Discovery and Invention in Mathematics, in *Philosophy and Methodology of Information: The Study of Information in the Transdisciplinary Perspective*, World Scientific, New York/London/Singapore, 2019, pp. 171–199.

Burgin, M. (2019) On Weak Projectivity in Arithmetic, *European Journal of Pure and Applied Mathematics*, v. 12, No. 4, pp. 1787–1810.

Burgin, M. S. and Baranovich, T. M. (1975) Linear Ω-algebras, *Russian Math. Surv.*, v. 30, No. 4, pp. 61–106 (translated from Russian).

Burgin, M. and Karasik, A. (1976) Operators of multidimensional structured model of parallel computations, *Automation Remote Control*, v. 37, No. 8, pp. 1295–1300.

Burgin, M. and Meissner, G. (2017) $1 + 1 = 3$: Synergy arithmetic in economics, *Appl. Math.*, v. 8, No. 2, pp. 133–144.

Burgin, M. and Rocchi, P. *Basic Concepts of the Theory of Ample Probability*, UPI Journal of Mathematics and Biostatistics (UPI-JMB), v. 2, No. 1, 2019, 28 p.

Burgin, M. and Milov, Yu. (1998) Grammatical aspects of language in the context of the existential triad concept, in *On the Nature and Essence of Mathematics*, Ukrainian Academy of Information Sciences, Kiev, pp. 136–142.

Burgin, M. and de Vey Mestdagh, C. N. J. (2011) The representation of inconsistent knowledge in advanced knowledge based systems, *Lect. Notes Comput. Sci., Knowledge-Based Intelligent Inform. Eng. Syst.*, v. 6882, pp. 524–537.

Burgin, M. and de Vey Mestdagh, C. N. J. (2015) Consistent structuring of inconsistent knowledge, *J. Intelligent Inform. Syst.*, v. 45, No. 1, pp. 5–28.

Burgin, M. and Schumann, J. (2006) Three levels of the symbolosphere, *Semiotica*, v. 160, No. 1/4, pp. 185–202.

Burr, S. A. (Ed.) (1992) The unreasonable effectiveness of number theory, *Proc. Symposia Appl. Math.*, Orono, Maine, August 1991, v. 46.

Burr, D. and Ross, J. (2008) A visual sense of number, *Curr. Biol.*, v. 18, pp. 425–428.

Burton, D. M. (1997) *The History of Mathematics*, McGraw-Hill, New York.

Busch, P., Grabowski, M. and Lahti, P. J. (1995) *Operational Quantum Physics*, Springer, Berlin.

Buss, S. R. (2013) The computational power of bounded arithmetic from the predicative viewpoint, in *New Computational Paradigms: Changing Conceptions of What is Computable*, Elsevier, Amsterdam, pp. 213–222.

Bussmann, J. B. (2013) One plus one equals three (or more ...): combining the assessment of movement behavior and subjective states in everyday life, *Frontiers Psychol.*, 16 May 2013, https://doi.org/10.3389/fpsyg.2013.00216.

Butterworth, B. (2005) The development of arithmetical abilities, *J. Child. Psychol. Psychiatry*, v. 46, pp. 3–18.

Byers, W. (2007) *How Mathematicians Think: Using Ambiguity, Contradiction, and Paradox to Create Mathematics*, Princeton University Press, Princeton, NJ.

Cain, F. (2015) Venus, the morning star and evening star, *Universe Today*, 2008/2015, https://www.universetoday.com/22570/venus-the-morning-star/.

Çakır, Z. (2013) Spaces of continuous and bounded functions over the field of geometric complex numbers, *J. Ineq. Appl.*, v. 2013, Article 363.

Cakmak, A. F. and Basar, F. (2012) Some new results on sequence spaces with respect to non-Newtonian calculus, *J. Ineq. Appl.*, v. 228, pp. 1–17.

Çakmak, A. F. and Başar, F. (2012a) On the classical sequence spaces and non-Newtonian calculus, *J. Ineq. Appl.*, v. 2012, Article ID 932734, 12 p.

Çakmak, A. F. and Başar, F. (2014) Certain spaces of functions over the field of non-Newtonian complex numbers, *Abstract Appl. Anal.*, v. 2014, Article ID 236124, 12 p.

Calcagni, G. (2010) Fractal universe and quantum gravity, *Phys. Rev. Lett.*, v. 104, 251301.

Calcagni, G. (2012) Geometry of fractional spaces, *Adv. Theor. Math. Phys.*, v. 16, 549.

Calcagni, G. (2012a) Geometry and field theory in multi-fractional spacetime, *J. High Energ. Phys.*, v. 1201, 065.

Calcagni, G. (2013) Multi-scale gravity and cosmology, *J. Cosmol. Astro. Phys.*, v. 1312, 041.

Calcagni, G., Oriti, D. and Thürigen, J. (2015) Dimensional flow in discrete quantum geometries, *Phys. Rev.* D, v. 91, 084047.

Cambridge Business English Dictionary, Cambridge University Press, London, 2011.

Cantlon, J. F. (2012) Math, monkeys, and the developing brain, *Proc. Natl. Acad. Soc. USA*, v. 109 (Supplement 1), pp. 10725–10732.

Cantlon, J. F. and Brannon, E. M. (2006) Shared system for ordering small and large numbers in monkeys and humans, *Psychol. Sci.*, v. 17, pp. 401–406.

Cantlon, J. F. and Brannon, E. M. (2007) Basic math in monkeys and college students, *PLoS Biol.*, v. 5:e328.

Cantlon, J. F. and Brannon, E. M. (2007a) How much does number matter to a monkey (Macaca mulatta)? *J. Exp. Psychol. Anim. Behav. Process*, v. 33, pp. 32–41.

Cantlon, J. F., Libertus, M. E., Pinel, P., Dehaene, S., Brannon, E. M. and Pelphrey, K. A. (2009) The neural development of an abstract concept of number, *J. Cognitive Neurosci.*, v. 21, No. 11, pp. 2217–2229.

Cantor, G. (1872) Über die Ausdehnung eines Satzes aus der Theorie der trigonometrischen Reihen, *Math. Ann.*, v. 5, No. 1, 123–132.

Cantor, G. (1874) Über eine Eigenschaft des Inbegriffes aller reelen algebraischen Zahlen, *J. Reine Angew. Math.*, b. 77, s. 258–262.

Cantor, G. (1878) Ein Beitrag zur Mannigfaltigkeitslehre, *J. Reine Angew. Math.*, b. 84, s. 242–258.

Cantor, G. (1883) *Grundlagen einer allgemeinen Mannigfaltigkeitslehre: Ein mathematisch-philosophischer Versuch in der Lehre des Unendlichen*, Teubner, Leipzig.

Cantor, G. (1895/1897) Beiträge zur Begrűndung der transfiniten Mengenlehre, *Math. Ann.*, b. 46, s. 481–512; b. 49, s. 207–246.

Cantor, G. (1932) *Gesammelte Abhandlungen Mathematischen und Philosophischen Inhalts*, Springer, Berlin.

Cantù, P. (2018) The epistemological question of the applicability of mathematics, *J. History Anal. Philos.*, v. 6, No. 3, pp. 96–114.

Cardano, G. (1545) *Artis magnae, sive de regulis algebraicis (known as Ars magna)*, Johannes Petreius, Nuremberg.

Cardano, G. (1577) *Practica arithmetice et mensurandi singularis*, Castellioneus, J.A., Milan.

Carey S. (1998) Knowledge of number: Its evolution and ontogeny, *Science*, v. 282, pp. 641–642.

Carey, S. (2001) Cognitive foundations of arithmetic: Evolution and ontogenesis, *Mind Lang.*, v. 16, pp. 37–55.

Carroll, L. (2009) *Euclid and His Modern Rivals*, Barnes and Noble, New York.

Carroll, M. L. (2001) The natural chain of binary arithmetic operations and generalized derivatives, arXiv:math/0112050 [math.HO].

Carroll, P. and Mui, C. (2009) *Billion Dollar Lessons: What You Can Learn from the Most Inexcusable Business Failures*, Portfolio.

Carruth, P. W. (1942) Arithmetic of ordinals with applications to the theory of ordered abelian groups, *Bull. Amer. Math. Soc*, v. 48, pp. 262–271.

Carter, J. (2008) Structuralism as a Philosophy of Mathematical Practice, *Synthese*, v. 163, No. 2, pp. 119–131.

Cartwright, S. and Schoenberg, R. (2006) 30 years of mergers and acquisitions research, *British J. Manag.*, v. 15, No. 51, pp. 51–55.

Cartwrite, J. H. E. and Piro, O. (1992) The dynamics of Runge–Kutt methods, *Int. J. Bifurcation Chaos*, v. 2, No. 3, pp. 427–450.

Cassani, O. and Conway, J. H. (2018) Numbering, *Math. Intell.*, v. 40, No.1, pp. 91–92.

Cassels, J. W. S. (1957) *An Introduction to Diophantine Approximation*, Cambridge Tracts in Mathematics and Mathematical Physics, v. 45, Cambridge University Press, Cambridge.

Castaneda, H. (1959) Arithmetic and reality, *Australasian J. Philos.*, v. 37, No. 2, pp. 91–107.

Cayley, A. (1843) Chapters in the analytical geometry of (n) dimensions, *Cambridge Math. J.*, v. 4, pp. 119–127.

Cayley, A. (1845) On Jacobi's elliptic functions and on quaternions, *Philosophical Magazine*, v. 26, pp. 208–211.

Chacón-Cardona, C. A. and Casas-Miranda, R. A. (2012) Millennium simulation dark matter haloes: Multifractal and lacunarity analysis and the transition to homogeneity, *Mon. Not. R. Astron. Soc.*, v. 427, 2613.

Chan, J., Ngai, S.-M. and Teplyaev, A. (2015) One-dimensional wave equations defined by fractal Laplacians, *Anal. Math.*, v. 127, pp. 219–246.

Chester, R. (1915) *Algebra of al-Khowarizmi,* Macmillan, London.

Cheyne, C. (1997) Getting in touch with numbers: Intuition and mathematical platonism, *Philos. Phenomenol. Res.*, v. 57, No. 1, pp. 111–125.

Chihara, C. (1982) A Gödelian thesis regarding mathematical objects: Do they exist? And can we perceive them? *Philos. Rev.*, v. 91, pp. 211–227.

Childers, D. G., Skinner, D. P. and Kemarait, R. C. (1977) The cepstrum: a guide to processing, *Proc. IEEE*, v. 65, No. 10, pp. 1428–1443.

Chin, G. and Culotta, E. (2014) What the numbers tell us, *Science*, v. 344, No. 6186, pp. 818–821.

Chrisman, N. R. (1998) Rethinking levels of measurement for cartography, *Cartography Geographic Inform. Sci.*, v. 25, No. 4, pp. 231–242.

Chrisomalis, S. (2010) *Numerical Notation: A Comparative History*, Cambridge University Press, Cambridge.

Christie, J. A. (1865) *The Constructive Arithmetic*, Virtue Brothers and Co., London.

Chung, W. and Hassanabadi, H. (2019) Deformed classical mechanics with α-deformed translation symmetry and anomalous diffusion, *Mod. Phys. Lett. B*, v. 33, p. 950368.

Chung, W. S. and Hounkonnou, M. N. (2020) Deformed special relativity based on α-deformed binary operations, arXiv:2005.11155 [physics.gen-ph].

Church, A. (1932) A set of postulates for the foundation of logic, *Ann. Math.*, (2), v. 33, No. 2, pp. 346–366.

Church, A. (1936) An unsolvable problem of elementary number theory, *Amer. J. Math.*, v. 58, pp. 345–363.

Church, A. (1956) *Introduction to Mathematical Logic*, Princeton University Press, Princeton.

Ciesielski, Z. (1995) Fractal functions and Schauder bases, *Panoramas Math.*, v. 34, pp. 47–54.

Clark, P. (1995) *Hellman Geoffrey. Mathematics without numbers. Towards a modal-structural interpretation*, Clarendon Press, Oxford University Press, Oxford and New York 1989, xi + 154 pp [Book Review], *J. Symbolic Logic*, v. 60, No. 4, pp. 1310–1312.

Clarke-Doane, J. (2008) Multiple reductions revisited, *Philos. Math.*, v. 16, No. 2, pp. 244–255.

Clawson, C. C. (1994) *The Mathematical Traveler: Exploring the Grand History of Numbers (Language of Science)*, Plenum Press, New York.

Cleveland, A. (2008) Circadian math: one plus one doesn't always equal two, *RPI News*, June 6, 2008, https://news.rpi.edu/luwakkey/2456.

Clifford, W. K. (1873) Preliminary sketch of bi-quaternions, *Proc. London Math. Soc.*, v. 4, pp. 381–395.

Clifford, A. H. (1954) Naturally totally ordered commutative semigroups, *Amer. J. Math.*, v. 76, pp. 631–646.

Close, F. (2011) *The Infinity Puzzle: Quantum Field Theory and the Hunt for an Orderly Universe*, Basic Books, New York.

Cohen, L. W. and Ehrlich, G. (1977) *The Structure of the Real Number System*, Robert E. Krieger P.C., Huntington, NY.

Cohen, G., Gaubert, S. and Quadrat, J. P. (1999) Max-plus algebra and system theory: Where we are and where to go now, *Ann. Rev. Control*, v. 23, pp. 207–219.

Cohen Kadosh, R. and Dowker, A. (Eds.) (2015) *Oxford Library of Psychology. The Oxford Handbook of Numerical Cognition*, Oxford University Press, New York.

Cohen Kadosh, R. and Walsh, V. (2009) Numerical representation in the parietal lobes: Abstract or not abstract? *Behav. Brain Sci.*, v. 32 (discussion 328–373), pp. 313–328.

Cohen Kadosh, R., Cohen Kadosh, K., Kaas, A., Henik, A. and Goebel, R. (2007) Notation-dependent and -independent representations of numbers in the parietal lobes, *Neuron*, v. 53, pp. 307–314.

Cohn, P. M. (1965) *Universal Algebra*, Harper & Row, New York.

Coleman, P. and Pietronero, L. (1992) The fractal structure of the universe, *Phys. Rep.*, v. 213, 311.

Collins, J. C. (1984) *Renormalization*, Cambridge University Press, Cambridge.

Conant, L. L. (1896) *The Number Concept, Its Origin and Development*, Macmillan, New York.

Condry, K. F. and Spelke, E. S. (2008) The development of language and abstract concepts: the case of natural number, *J. Exp. Psychol.*, v. 137, pp. 22–28.

Conway, J. H. (1976) *On Numbers and Games*, Academic Press, London.

Conway, J. H. (1994) The Surreals and the Reals, in *Real Numbers, Generalizations of the Reals, and Theories of Continua*, Springer, Netherlands, pp. 93–103.

Conway, J. H. and Guy, R. K. (1996) *The Book of Numbers*, Springer-Verlag, New York.

Córdova-Lepe, F. (2006) The multiplicative derivative as a measure of elasticity in economics, *TMAT Rev. Latinoamericana Ciencias Ingeniería*, v. 2, No. 3.

Corfield, D. (2003) *Towards a Philosophy of Real Mathematics*, Cambridge University Press, Cambridge.

Corry, L. (1996) *Modern Algebra and the Rise Mathematical Structures*, Birkhäuser, Basel.

Courant, R. and Robbins, H. (1960) *What is Mathematics*, Oxford university Press, London.

Covey, S. R. (2004) *The 7 Habits of Highly Effective People: Personal Workbook*, Touchstone.

Crandall, R. and Pomerance, C. (2001) *Prime Numbers: A Computational Perspective*, Springer, New York.

Crollen, V., Grade, S., Pesenti, M. and Dormal, V. (2013) A common metric magnitude system for the perception and production of numerosity, length, and duration, *Frontiers Psychol.*, v. 4, Article 449.

Crump, T. (1990) *The anthropology of numbers*, Cambridge University Press, Cambridge.

Cummins, D. D. (1991) Childrens's interpretations of arithmetic word problems, *Cogn. Instr.*, v. 8, No. 3, pp. 261–289.

Cuninghame-Green, R. A. (1995) Minimax algebra and applications, in *Advances in Imaging and Electron Physics*, v. 90, Academic Press, New York, pp. 1–121.

Curry, H. B. (1941) A formalization of recursive arithmetic, *Amer. J. Math*, v. 63, pp. 263–283.

Czachor, M. (2016) Relativity of arithmetic as a fundamental symmetry of physics, *Quantum Stud.: Math. Found.*, v. 3, pp. 123–133.

Czachor, M. (2017) If gravity is geometry, is dark energy just arithmetic? *Int. J. Theoret. Phys.*, v. 56, pp. 1364–1381.

Czachor, M. (2017a) Information processing and Fechner's problem as a choice of arithmetic, in *Information Studies and the Quest for Transdisciplinarity: Unity through Diversity*, World Scientific, New York, pp. 363–372.

Czachor, M. (2019) Waves along fractal coastlines: From fractal arithmetic to wave equations, *Acta Phys. Polon.* B, v. 50, No. 4, pp. 813–831 (also: Preprint in Mathematics, [math.DS] 2017b; arXiv:1707.06225).

Czachor, M. (2020) Non-Newtonian mathematics instead of non-Newtonian physics: Dark matter and dark energy from a mismatch of arithmetics, *Found. Sci.* (2020) — in print; arXiv:1911.10903 [physics.gen-ph].

Czachor, M. (2020) A loophole of all 'loophole-free' Bell-type theorems, *Found. Sci.* (2020), DOI: 10.1007/s10699-020-09666-0; arXiv:1710.06126v2 [quant-ph].

Czachor, M. (2020) Arithmetic loophole in Bell's theorem: An overlooked threat for entangled-state quantum cryptography, arXiv:2004.04097 [physics.gen-ph].

Czachor, M. and Posiewnik, A. (2016) Wavepacket of the Universe and its spreading, *Int. J. Theor. Phys.*, v. 55, pp. 2001–2011.

Dacke, M., and Srinivasan, M. V. (2008) Evidence for counting in insects, *Anim. Cogn.*, v. 11, pp. 683–689.

Dantzig, T. (1930/2007) *Number: The Language of Science*, Macmillan, New York.

Darrigol, O. (2003) Number and measure: Hermann von Helmholtz at the crossroads of mathematics, physics, and psychology, *Studies History Philos. Sci. Part A*, v. 34, No. 3, pp. 515–573.

Davenport, C. M. (1978) *An Extension of the Complex Calculus to Four Real Dimensions, with an Application to Special Relativity*, University of Tennessee, Knoxville, TN.

Davenport, H. (1992) *The Higher Arithmetic: An Introduction to the Theory of Numbers*, Cambridge University Press, Cambridge.

Davies, E. B. (2003) *Science in the Looking Glass: What Do Scientists Really Know?* Oxford University Press, New York.

Davis, P. J. (1972) Fidelity in mathematical discourse: Is one and one really two? *Amer. Math. Monthly*, pp. 252–263.

Davis, P. J. (2000) Four thousand — or possibly thirty-seven thousand — years of mathematics, *SIAM News*, v. 33, No. 9, p. 6.

Davis, P. J. (2003) Is mathematics a unified whole, *SIAM News*, v. 36, No. 3, p. 6.

Davis, P. J. and Hersh, R. (1986) *The Mathematical Experience*, Houghton Mifflin Co., Boston, MA.

Davis, P. J. and Hersh, R. (1987) *Descartes' Dream: The World According to Mathematics*, Houghton Mifflin Co., Boston, MA.

De Cruz, H. (2008) An extended mind perspective on natural number representation, *Philosop. Psychol.*, v. 21, pp. 475–490.

de Finetti, B. (1931) Sulconcetto di media, *Giornale dell' Instituto, Italiano degli Attuarii*, v. 2, pp. 369–396.

De Villiers, M. (1923) *The Numeral-Words, Their Origin, Meaning, History and Lesson*, Witherby, London.

Deco, G. and Rolls, E. T. (2006) Decision-making and Weber's law: A neurophysiological model, *Eur. J. Neurosci.*, v. 24, pp. 901–916.

Decock, L. (2008) The conceptual basis of numerical abilities: one-to-one correspondence versus the successor relation, *Philosop. Psychol.*, v. 21, pp. 459–473.

Dedekind, R. (1872) *Stetigkeit und irrational Zahlen*, Friedr. Viemeg & Sohn, Braunschweig.

Dedekind, R. (1888) *Was sind und was sollen die Zahlen*? Vieweg, Braunschweig.

Dedekind, R. (1890) Letter to Keferstein, in *From Frege to Gödel: A Source Book in Mathematical Logic*, 1879–1931, pp. 98–103.

Dedekind R. (1901) *Essays on the Theory of Numbers*, Open Court Publishing Company, Chicago.

Dehaene, S. (1992) Varieties of numerical abilities, *Cognition*, v. 44, No. 1–2, pp. 1–42.

Dehaene, S. (1997) *The Number Sense: How the Mind Creates Mathematics*, Oxford University Press, New York.

Dehaene, S. (1997a) What are numbers, really? A cerebral basis for number sense, *Edge*, http://www.edge.org/3rd_culture/dehaene/index.html.

Dehaene, S. (2001) Precis of the number sense, *Mind Lang.*, v. 16, pp. 16–36.

Dehaene, S. (2002) Single-neuron arithmetic, *Science*, v. 297, pp. 1652–1653.

Dehaene, S. (2003) The neural basis of the Weber–Fechner law: A logarithmic mental number line, *Trends in Cognitive Sciences*, v. 7, pp. 145–147.

Dehaene, S. (2009) Origins of mathematical intuitions: the case of arithmetic, *Ann. N. Y. Acad. Sci.*, v. 1156, pp. 232–259.

Dehaene, S. and Brannon, E. (Eds.) (2011) *Space, Time and Number in the Brain: Searching for the Foundations of Mathematical Thought*, Attention and performance Series, Academic Press, Amsterdam.

Dehaene, S. and Changeux, J. P. (1993) Development of elementary numerical abilities: A neuronal model, *J. Cogn. Neurosci.*, v. 5, pp. 390–407.

Dehaene, S. and Cohen, L. (1994) Dissociable mechanisms of subitizing and counting: Neuropsychological evidence from simultanagnosic patients, *J. Exp. Psychol.: Human Perception Perform.*, v. 20, pp. 958–975.

Dehaene, S. and Cohen, L. (1995) Towards an anatomical and functional model of number processing, *Math. Cognition*, v. 1, pp. 83–120.

Dehaene, S., Dehaene-Lambertz, G. and Cohen, L. (1998) Abstract representations of numbers in the animal and human brain, *Trends. Neurosci.*, v. 21, pp. 355–361.

Dehaene, S., Izard, V., Spelke, E. and Pica, P. (2008) Log or linear? Distinct intuitions of the number scale in Western and Amazonian indigene cultures, *Science*, v. 320, pp. 1217–1220.

Dehaene, S., Molko, N., Cohen, L. and Wilson, A.J. (2004) Arithmetic and the brain, *Current Opinion Neurobiol.*, v. 14, pp. 218–224.

Dehaene, S., Piazza, M., Pinel, P. and Cohen, L. (2003) Three parietal circuits for number processing, *Cogn. Neuropsychol.*, v. 20, pp. 487–506.

Deligne, P. and Morgan, J. W. (1999) Notes on supersymmetry, in *Quantum Fields and Strings: A Course for Mathematicians*, v. 1, American Mathematical Society, pp. 41–97.

Dembart, L. Book review: Should we count on mathematics? *The Los Angeles Times*, Tuesday, November 11, 1986, p. 92.

Demopoulos, W. (1998) The philosophical basis of our knowledge of number, *Noûs*, v. 32 pp. 481–503.

Demopoulos, W. (2000) On the origin and status of our conception of number, *Notre Dame J. Formal Logic*, v. 41, No. 3, pp. 210–226.

Derboven, J. (2011) One plus one equals three: Eye-tracking and semiotics as complementary methods in HCI, in *CCID2: The Second International Symposium on Culture, Creativity, and Interaction Design*, Newcastle, UK, https://ccid2.files.wordpress.com/2011/02/one-plus-one-equals-three_derboven.pdf.

Descartes, R. (1637/2006) *A Discourse on the Method of Correctly Conducting One's Reason and Seeking Truth in the Sciences* (Transl. I. Maclean), Oxford University Press, New York.

Devito, J. A. (1994) *Human Communication: The Basic Course*, Harper Collins, New York.

DeWitt, B. (1984) *Supermanifolds*, Cambridge University Press, Cambridge.

Di Giorgio, N. (2010) *Non-Standard Models of Arithmetic: A Philosophical and Historical perspective*, MSc Thesis, Universiteit van Amsterdam.

Dickson, L. E. (1919/2005) *History of the Theory of Numbers*, v. I: *Divisibility and primality*, Dover Publications, New York.

Dickson, L. E. (1932) *History of the Theory of Numbers*, Carnegie Institute of Washington, Washington.

Diez, J. A. (1997) A hundred years of numbers. An historical introduction to measurement theory 1887–1990, Pt. 1, *Studies History Philos. Sci.*, v. 28, No. 1, pp. 167–185.

Diez, J. A. (1997a) A hundred years of numbers. An historical introduction to measurement theory 1887–1990, Pt. 2, *Studies History Philos. Sci.*, v. 28, No. 2, pp. 237–265.

Dijkman, J. G., van Haeringen, H. and De Lange, S. J. (1983) Fuzzy Numbers, *J. Math. Anal. Appl.*, v. 92, pp. 301–341.

Dijksterhuis, E. J. (1987) *Archimedes, With a New Bibliographic Essay by Wilbur R. Knorr*, Princeton University Press, New Jersey.

Dijkstra, E. W. (1982) *Why Numbering Should Start at Zero*, https://www.cs.utexas.edu/users/EWD/transcriptions/EWD08xx/EWD831.html.

Diophantus (1974) *Arithmetic* (translated by Veselovskii, I.N.), Nauka, Moscow (in Russian).

Dirac, P. A. M. (1978) *Directions in Physics*, John Wilcy & Sons, New York, NY.

Dixon, G. M. (1994) *Division Algebras: Octonions, Quaternions, Complex Numbers and the Algebraic Design of Physics*, Kluwer Academic Publishers, Dordrecht.

Dollard, J. D. and Friedman, C. N. (1977) Product integrals and the Schrödinger equation, *J. Math. Phys.*, v. 18, No. 8, pp. 1598–1607.

Dollard, J. D. and Friedman, C. N. (1979) *Product Integration, with Applications to Differential Equations*, Addison-Wesley, Reading, MA.

Domański, Z. and Błaszak, M. (2017) *Deformation quantization with minimal length*, Preprint in Mathematical Physics [math-ph] (arXiv:1706.00980).

Donaldson, T. M. E. (2014) If there were no Numbers, what would you Think? *Thought: J. Philos.*, v. 3, No. 4, pp. 283–287.

Drake, F. R. (1974) *Set Theory: An Introduction to Large Cardinals*, Studies in Logic and the Foundations of Mathematics, v. 76, Elsevier Science Ltd., Amsterdam.

Du Bois-Reymond, P. (1870/1871) Sur la grandeur relative des infinis des fonctions, *Ann. Mat. Pura Appli.*, v. 4, pp. 338–353.
Du Bois-Reymond, P. (1875) Über asymptotische Werthe, infinitäre Approximationen und infinitäre Auflösung von Gleichungen, *Math. Ann.*, v. 8, pp. 363–414.
Du Bois-Reymond, P. (1877) Über die Paradoxen des Infinitärcalcüls, *Math. Ann.*, v. 11, pp. 149–167.
Du Bois-Reymond, P. (1882) *Die allgemeine Functionentheorie I: Metaphysik und Theorie der mathematischen Grundbegriffe: Grösse, Grenze, Argument und Function*, Verlag der H. Laupp'schen Buchhandlung, Tübingen.
Duff, M. J. (2004) *Comment on time-variation of fundamental constants*, Preprint in High Energy Physics (arXiv: hep-th/0208093v3).
Dummett, M. (1975) Wang's paradox, *Synthese*, v. 30, No. 3/4, pp. 301–324.
Dummett, M. (1991) *Frege: Philosophy of Mathematics*, Harvard University, Cambridge, MA.
Durkheim, E. (1984) *The Division of Labor in Society*, The Free Press, New York.
Duyar, C., Sagır, B. and Ogur, O. (2015) Some basic topological properties on non-Newtonian real line, *British J. Math. & Comput. Sci.*, v. 9, No. 4, pp. 300–307.
Dzhafarov, E. N. and Colonius, H. (2011) The Fechnerian idea, *Ameri. J. Psychol.*, 124, 127–140.
Edgar, G. (2008) *Measure, Topology, and Fractal Geometry*, Springer, New York.
Eger, E., Michel, V., Thirion, B., Amadon, A., Dehaene, S. and Kleinschmidt, A. (2009) Deciphering cortical number coding from human brain activity patterns, *Curr. Biol.*, v. 19, pp. 1608–1615.
Ehrlich, P. (1992) Universally extending arithmetic continua, in *Le Labyrinthe du Continu*, Colloque de Cerisy, Springer-Verlag, Paris, pp. 168–178.
Ehrlich, P. (1994) All numbers great and small, in *Real Numbers, Generalizations of the Reals, and Theories of Continua*, Kluwer Academic Publishers, Dordrecht, pp. 239–258.
Ehrlich, P. (Ed) (1994a) *Real Numbers, Generalizations of the Reals, and Theories of Continua*, Kluwer Academic Publishers. Dordrecht.
Ehrlich, P. (1995) Hahn's "Über die Nichtarchimedischen Grössensysteme" and the origins of the modern theory of magnitudes and numbers to measure them, in *From Dedekind to Gödel: Essays on the Development of the Foundations of Mathematics* (J. Hintikka, Ed.), Kluwer Academic Publishers, pp. 165–213.
Ehrlich, P. (2001) Number systems with simplicity hierarchies: A generalization of Conway's theory of surreal numbers, *J. Symbolic Logic*, v. 66, pp. 1231–1258.
Ehrlich, P. (2006) The rise of non-Archimedean mathematics and the roots of a misconception. I. The emergence of non-Archimedean systems of magnitudes, *Arch. Hist. Exact Sci.*, v. 60, No. 1, pp. 1–121.
Ehrlich, P. (2011) Conway names, the simplicity hierarchy and the surreal number tree, *J. Logic Anal.*, v. 3, No. 1, pp. 1–26.
Ehrlich, P. (2012) The absolute arithmetic, continuum and the unification of all numbers great and small, *Bull. Symbolic Logic*, v. 18, pp. 1–45.

Eisler, H. (1963) Magnitude scales, category scales, and Fechnerian integration, *Psychol. Rev.*, v. 70, pp. 243–253.

Ekman, G. (1964) Is the power law a special case of Fechner's law? *Perceptual Motor Skills*, v. 19, 730.

Ekman, G. and Hosman, B. (1965) Note on subjective scales of number, *Perceptual Motor Skills*, v. 21, pp. 101–102.

Enge, E. (2017) *SEO and social*: 1 + 1 = 3, SearchEngineLand, https://searchengineland.com/seo-social-1-1-3-271978.

Enriques, F. (1911) Sui numeri non archimedei e su alcune loro interpretazioni, *Boll. Math. Soc. Italiana Mat. IIIa*, pp. 87–105.

Epstein, M. and Śniatycki, J. (2006) Fractal mechanics, *Physica D*, v. 220, pp. 54–68.

Ershov, Yu. L. (1999) Theory of numberings, in *Handbook of Computability Theory*, Studies Logic Foundation. Mathematics, v. 140, North-Holland, Amsterdam, pp. 473–503.

Euclid (1956) *The Thirteen Books of Euclid's Elements, with Introduction and Commentary* by T. L. Heath, Dover, New York.

Evans, M. G. (1955) Aristotle, Newton, and the theory of continuous magnitude, *J. History of Ideas*, v. 16, No. 4, pp. 548–557.

Everett, C. (2017) *Numbers and the Making of Us: Counting and the Course of Human Cultures*, Harvard University Press.

Eves, H. (1990) *An Introduction to the History of Mathematics*, Saunders College Publishing, Philadelphia.

Faith, C. (2004) *Rings and Things and a Fine Array of Twentieth Century Associative Algebra* (Mathematical Surveys & Monographs), American Mathematical Society.

Falconer, K. (2003) *Fractal Geometry: Mathematical Foundations and Applications*, Wiley, New York.

Falmagne, J. C. (1971) The generalized Fechner problem and discrimination, *J. Math. Psychol.*, v. 8, pp. 22–43.

Falmagne, J. C. (1985) *Elements of Psychophysical Theory*, Oxford University Press.

Falmagne, J.-C. and Doble, C. (2015) *On Meaningful Scientific Laws*, Springer, New York.

Farmelo, G. (2019) *The Universe Speaks in Numbers: How Modern Math Reveals Nature's Deepest Secrets*, Basic Books, New York.

Fechner, G. T. (1860) *Elemente der Psychophysik*, Breitkopf und Hartel, Leipzig.

Feigenson, L., Dehaene, S. and Spelke, E. (2004) Core systems of number, *Trends. Cogn. Sci.*, v. 8, pp. 307–314.

Feferman, S. (1960) Arithmetization of metamathematics in a general setting, *Fund. Math.*, v. 49, pp. 35–92.

Feferman, S. (1974) *The Number Systems: Foundations of Algebra and Analysis*, Addison-Wesley, Reading, MA.

Feferman, S. and Hellman, G. (1995) Foundations of predicative arithmetic, *Philos. Logic*, v. 24, pp. 1–17.

Feferman, S. and Strahm, T. (2010) Unfolding finitist arithmetic, *Rev. Symbolic Logic*, v. 3, No. 4, pp. 665–689.

Felka, K. (2014) Number words and reference to numbers, *Philos. Studies*, v. 168, No. 1, pp. 261–282.

Fellows, M. R., Gaspers, S. and Rosamond, F. A. (2010) *Parameterizing by the Number of Numbers*, Preprint in Data Structures and Algorithms (cs.DS) (arXiv:1007.2021).

Ferreirós, J. (2005) Dedekind, R. (1888) and Peano, G. (1889), Booklets on the foundations of arithmetic, in *Landmark Writings in Western Mathematics, 1640–1940* (Grattan-Guinness, I. Ed.), Elsevier, Amsterdam, Chapter 47, pp. 613–626.

Ferreirós, J. (2007) The rise of pure mathematics, as arithmetic in Gauss, in *The Shaping of Arithmetic after C.F. Gauss's Disquisitiones Arithmeticae*, (Goldstein, C. Schappacher, N. and Schwermer, J. Eds.), Springer, Berlin, pp. 235–268.

Fetterman J. G. (1993) Numerosity discrimination: both time and number matter, *J. Exp. Psychol. Anim. Behav. Process.*, v. 19, pp. 149–164.

Fetters, M. D. and Freshwater, D. (2015) The 1 + 1 = 3 integration challenge, *J. Mixed Methods Res.* (*JMMR*), v. 9, No. 2, pp. 115–117.

Feynman, R. P. (1948) Space-time approach to non-relativistic quantum mechanics, *Rev. Mod. Phys.*, v. 20, pp. 367–387.

Feynman, R. P. and Hibbs, A. R. (1965) *Quantum Mechanics and Path Integrals*, McGraw-Hill, New York.

Fias, W. (2001) Two routes for the processing of verbal numbers: Evidence from the SNARC effect, *Psychol. Res.*, v. 65, pp. 250–259.

Fias, W., Brysbaert, M., Geypens, F. and d'Ydewalle, G. (1996) The importance of magnitude information in numerical processing: Evidence from the SNARC effect, *Math. Cogn.*, v. 2, No. 1, pp. 95–110.

Field, H. (1980) *Science Without Numbers*: *The Defense of Nominalism*, Princeton Legacy Library, Princeton University Press, Princeton.

Filip, D. A. and Piatecki, C. (2014) A non-Newtonian examination of the theory of exogenous economic growth, *Math. Aetherna*, v. 4, No. 2, pp. 101–117.

Fine, H. B. (1891) *The Number System of Algebra Treated Theoretically and Historically*. Leach, Shewell & Sanborn, Boston.

Fisher, G. (1994) Veronese's Non-Archimedean Linear Continuum, in *Real Numbers, Generalizations of the Reals, and Theories of Continua*, Springer, The Netherlands, pp. 107–145.

Fitting, M. (2015) Intensional Logic, in *The Stanford Encyclopedia of Philosophy* (E. N. Zalta, Ed.), https://plato.stanford.edu/archives/sum2015/entries/logic-intensional/.

Flannery, S. and Flannery, D. (2000) *In Code*: *A Mathematical Journey*, Profile Books, London.

Flegenheimer, M. When the calculator says 1 + 1 = 4, *The New York Times*, April 13, 2012.

Flegg, G. (2002) *Numbers: Their History and Meaning*, Dover Publications, Mineola, NY.

Florack, L. (2012) Regularization of positive definite matrix fields based on multiplicative calculus, in *Scale Space and Variational Methods in Computer Vision*, Lecture Notes in Computer Science, v. 6667, Springer, pp. 786–796.

Florack, L. and van Assen, H. (2012) Multiplicative calculus in biomedical image analysis, *J. Math. Imaging Vis.*, v. 42, no. 1, 64–75.

Flynn, M. J. and Oberman, S. S. (2001) *Advanced Computer Arithmetic Design*, Wiley, New York.

Fraenkel, A. A., Bar-Hillel, Y., and Levy, A. (1973) *Foundations of Set Theory*, North-Holland P.C., Amsterdam.

Frame, A. and Meredith, P. (2008) One plus one equals three: Legal hybridity in Aotearoa/New Zealand, in *Hybrid Identities*, pp. 313–332.

Franklin, J. (2014) *An Aristotelian Realist Philosophy of Mathematics: Mathematics as the Science of Quantity and Structure*, Palgrave Macmillan.

Frege, G. (1879) *Begriffsschrift, eine der arithmetischen nachgebildete Formalsprache des reinen Denkens*, L. Nebert, Halle.

Frege, G. (1884) *Die Grundlagen der Arithmetik: Eine logisch-mathematische Untersuchung über den Begriff der Zahl*, Verlage Wilhelm Koebner, Breslau.

Frege, G. (1892) Über Sinn und Bedutung. *Zeit. Philos. Philosophische Kritik*, v. 100, pp. 25–50.

Frege, G. (1893) *Grundgesetze der Arithmetic, Begriffsschriftlich abgeleitet*, Bd I, Verlag Hermann Pohle, Jena.

Frege, G. (1903) *Grundgesetze der Arithmetik*, Begriffsschriftlich abgeleitet, Bd. II, Verlag Hermann Pohle, Jena.

Frege, G. (1984) *Collected Papers on Mathematics, Logic, and Philosophy*, Blackwell, Oxford.

Frege, G. (1988) Begriffsschrift, eine der arithmetischen Nachgebildete Formelsprache des reinen Denkens, in *Begriffsschrift und andere Aufsätze*, Georg Olms Verlag, Hildesheim.

Freiberg, U. (2003) A survey on measure geometric Laplacians on Cantor like sets, *Arabian J. Sci. Eng.*, v. 28, pp. 189–198.

Freiberg, U. and Seifert, C. (2015) Dirichlet forms for singular diffusion in higher dimensions, *J. Evolution Equations*, v. 15, pp. 869–878.

Freiberg, U. and Zähle, M. (2002) Harmonic calculus on fractals: A measure geometric approach (I), *Potential Analysis*, v. 16, pp. 265–277.

Friedberg, R. (1968) *An Adventurer's Guide to Number Theory*, McGraw-Hill, New York.

Friedrich, B. (2019) *Multiplikative Euklidische Vektorräume als Grundlage für das Rechnen mit positiv-reellen Größen*, Logos Verlag, Berlin.

Friedrich, B. and Heil, P. (2017) Onset-duration matching of acoustic stimuli revisited: Conventional arithmetic vs. proposed geometric measures of accuracy and precision, *Front. Psychol.* 7: 2013. doi: 10.3389/fpsyg.2016.02013.

Frieman, J. A., Turner, M. S. and Huterer, D. (2008) Dark energy and the accelerating Universe, *Annual Rev. Astron. Astrophys.*, v. 46, pp. 385–432.

Fuchs, L. (1963) *Partially Ordered Algebraic System*, Pergamon Press, Oxford.

Fujita, T. (1990) Some asymptotic estimates of transition probability densities for generalized diffusion processes with self-similar speed measures, *Publ. Res. Inst. Math. Sci.*, v. 26, pp. 819–840.

Fujita, T. (1987) A fractional dimension, self-similarity and a generalized diffusion operator, Probabilistic Methods in Mathematical Physics, *Proceedings of Taniguchi International Symposium* (Ito, K. and Ikeda, N. Eds.), Kinokuniya.

Gaite, J. (2007) Halos and voids in a multifractal model of cosmic structure, *Astrophys. J.*, v. 658, No. 11.

Galileo (1638) *Discorsi e dimostrazioni matematiche, intorno à due nuove scienze*, v. 213, Leida, Appresso gli Elsevirii (Louis Elsevier, Leiden) (Mathematical discourses and demonstrations, relating to Two New Sciences, English translation by Henry Crew and Alfonso de Salvio, 1914).

Gallistel, C. R. (1988) Counting versus subitizing versus the sense of number. (Commentary on Davis and Pérusse: Animal counting), *Behav. Brain Sci.*, v. 11, pp. 585–586.

Gallistel, C. R. and Gelman, R. (1990) The what and how of counting, *Cognition*, v. 44, pp. 43–74.

Gallistell, R. C. and Gellman, R. (2000) Non-verbal numerical cognition: From reals to integers, *Trends Cogn. Sci.*, v. 4, pp. 59–65.

Galovich, S. (1989) *Introduction to Mathematical Structures*, Harcourt Brace Jovanovich, San Diego.

Gardner, M. (2005) Review of science in the looking glass: What do scientists really know? By E. Brian Davies (Oxford University Press, 2003), *Notices Amer. Math. Soc.*, v. 52, No. 11, http://www.ams.org/notices/200511/rev-gardner.pdf.

Gasking, D. A. T. (1940) Mathematics and the world, *Australasian Journal of Philosophy*, v. 18, No. 2, pp. 97–116.

Gates, Jr., S. J., Grisaru, M.T., Roček, M. and Siegel, W. (1983) *Superspace or One Thousand and One Lessons in Supersymmetry*, Benjamins Cumming Publishing.

Gauss, C. F. (1801) *Disquisitiones Arithmeticae*, Fleischer, Leipzig.

Gauss, C. F. (1808) Theorematis arithmetici demonstratio nova, *Comment. Soc. Regiae Sci. Göttingen*, XVI, 69; Werke II, pp. 1–8.

Gauss, C. F. (1831) Theoria residuorum biquadraticorum: Commentatio secunda, *Göttingische Gelehrte Anzeigen*, pp. 625–638.

Gauthier, Y. (2000) The internal consistency of arithmetic with infinite descent, *Modern Logic*, v. 8, No. 1–2, pp. 47–87.

Gel'fand, I. M., Raikov, D. and Shilov, G. E. (1964) *Commutative Normed Rings*, Chelsea P.C., New York.

Gelman, R. and Butterworth, B. (2005) Number and language: How are they related? *Trends Cogn. Sci.*, v. 9, pp. 6–10.

Gentzen, G. (1936) Die Widerspruchsfreiheit der reinen Zahlentheorie, *Math. Ann.*, v. 112, pp. 493–565.

Gentzen, G. (1938) Neue Fassung des Widerspruchsfreiheitsbeweises für die reine Zahlentheorie, *Forschungen Logik Grundl. exakten Wissens.*, v. 4, pp. 19–44.

Gentzen, G. (1943) Beweisbarkeit und Unbeweisbarkeit der Anfangsfällen der transfiniten Induktion in der reinen Zahlentheorie, *Math. Ann.*, v. 120, pp. 140–161.

Gershaw, D. A. (2015) *Two Plus Two Equals Four, But Not Always*, http://virgil.azwestern.edu/~dag/lol/TwoPlusTwo.html.

Gersten, R. and Chard, D. (1999) Number sense: Rethinking arithmetic instruction for students with mathematical disabilities, *J. Special Education*, v. 33, No. 1 pp. 18–28.

Ghosh, S. (2014) Spontaneous generation of a crystalline ground state in a higher derivative theory, *Physica A*, v. 407, 245.

Giaquinto, M. (2001) Knowing numbers, *J. Philos.*, v. 98, pp. 5–18.

Giaquinto, M. (2001a) What cognitive systems underlie arithmetical abilities? *Mind Lang.*, v. 16, pp. 56–68.

Gibbon, J. (1977) Scalar expectancy theory and Weber's Law in animal timing, *Psychol. Rev.*, v. 84, pp. 279–335.

Gill, R. D. and Johansen, S. (1990) A survey of product-integration with a view toward application in survival analysis, *Ann. Statist.*, v. 18, pp. 1501–1555.

Gillies, D. A. (1982) *Frege, Dedekind, and Peano on the Foundations of Arithmetic*, Van Gorcum, Assen.

Gioia, A. (2001) *Number Theory, An Introduction*, Dover Publications, New York.

Girard, P. R. (1984) The quaternion group and modern physics, *Eur. J. Phys.*, v. 5, pp. 25–32.

Glaser, A. (1981) *History of Binary and Other Non-decimal Numeration*, Tomash Publishers.

Gleyzal, A. (1937) Transfinite real numbers, *Proc. Natl. Acade. Sci.*, v. 23, pp. 581–587.

Glyn, A. (2017) *One plus one equals three — the power of Data Combinations*, Luciad, 30 Nov 2017, http://www.luciad.com/blog/one-plus-one-equals-three-the-power-of-data-combinations.

Gödel, K. (1931/1932) Über formal unentscheidbare Sätze der Principia Mathematica und verwandter Systeme I, *Monatsh. Math. Phys.*, v. 38, No. 1, pp. 173–198.

Gödel, K. (1934) Besprechung von Über die Unmöglichkeit einer vollstandigen Charakterisierung der Zahlenreihe mittels eines endlichen Axiomensystems, *Z. Math. Grenzgebiete*, v. 2, No. 3.

Golan, J. S. (2003) *Semirings and Affine Equations over Them*, Springer Science & Business Media, New York.

Goldblatt, R. (1984) *Topoi: The Categorical analysis of Logic*, North-Holland P.C., Amsterdam.

Goldman, J. R. (1997) *The Queen of Mathematics: A Historically Motivated Guide to Number Theory*, A K Peters/CRC Press, New York.

Goldman, J. G. (2010) What are the Origins of (Large) Number Representation? *Scientific American*, August 17, 2010, https://blogs.scientificamerican.com/thoughtful-animal/what-are-the-origins-of-large-number-representation/.

Goldstein, S. (1987) Random walks and diffusions on fractals, Percolation theory and ergodic theory of infinite particle systems, *IMA Vol. Math. Appl.*, v. 8, pp. 121–129.

Golmankhaneh, A. K. and Baleanu D. (2015) About Schrödinger equation on fractals curves imbedding in R^3, *Int. J. Theor. Physics*, v. 54, No. 4, pp. 1275–1282.

Golmankhaneh, A. K. (2017) On the calculus of parametrized fractal curves, *Turk. J. Phys.*, v. 41, pp. 418–425.

Gómez-Torrente, M. (2015) On the Essence and Identity of Numbers, *Theoria: Rev. Teoría, Historia y Fund. Ciencia*, v. 30, No. 3, pp. 317–329.

Gondran, M. and Minoux, M. (1979) *Graphes et Algorithmes*, Editions Eyrolles, Paris.

Gontar, V. (1993) New theoretical approach for physicochemical reactions dynamics with chaotic behavior, in *Chaos in Chemistry and Biochemistry*, World Scientific, London, pp. 225–247.

Gontar, V. (1997) Theoretical foundation for the discrete dynamics of physicochemical systems: Chaos, self-organization, time and space in complex systems, *Discrete Dynam. Nature Soc.*, v. 1, No. 1, pp. 31–43.

Gontar, V. and Ilin, I. (1991) New mathematical model of physicochemical dynamics, *Contrib. Plasma Physics*, 31, No. 6, pp. 681–690.

Gonthier, G. (2008) Formal proof — the four-color theorem, *Notices Amer. Math. Soc.*, v. 55, No. 11, pp. 1382–1393.

González, S. (Ed.) (1994) *Non-associative Algebra and Its Applications*, Springer, New York.

Goodman, N. (1968) *Languages of Art: An Approach to a Theory of Symbols*, Bobbs-Merrill, Indianapolis.

Goodman, N. D. (1979) Mathematics as an objective science, *Amer. Math. Monthly*, v. 88, pp. 540–551.

Goodstein, R. L. (1954) Logic-free formalizations of recursive arithmetic, *Mat. Scand.*, v. 2, pp. 247–261.

Gordon, P. A. (2004) Numerical cognition without words: Evidence from Amazonia, *Science*, v. 306, pp. 496–499.

Gottlieb, A. (2013). '$1 + 1 = 3$': The synergy between the NEW key technologies, Next-generation Enterprise WANs, *Network World*, July 15, 2013, http://www.networkworld.com/article/2224950/cisco-subnet/-1---1---3---the-synergy-between-the-new-key-technologies.html.

Grabisch, M., Marichal, J.-L., Mesiar, R. and Pap, E. (2009) *Aggregation Functions*, Cambridge University Press, New York.

Grant, M. and Johnston, C. (2013) $1 + 1 = 3$: *CMO & CIO Collaboration Best Practices That Drive Growth*, Canadian Marketing Association, Don Mills, Canada.

Grassmann, H. (1844) *Die lineale Ausdehnungslehre, ein neuer Zweig der Mathematik, dargestellt und durch Anwendungen auf die übrigen Zweige der Mathematik, wie auch die Statistik, Mechanik, die Lehre vom Magnetismus und die Krystallonomie erläutert*, Wiegand, Leipzig.

Grassmann, H. (1861) *Lehrbuch der Arithmetik für höhere Lehranstalten*, bd 1, Enslin, Berlin.

Grassmann, H. (1862) *Die Ausdehnungslehre.* Vollständig und in strenger Form *bearbietet*, Enslin, Berlin.

Grassmann, H. (1894–1911) *Gesammelte Mathematische und Physikalische Werke*, 3 vols., (Engel, F. Ed.) B.G. Teubner, Leipzig.

Grassmann, R. (1872) *Die Formenlehre oder Mathematik*, Georg Olms Verlagsbuchhandlung, Hildesheim.

Grassmann, R. (1872) *Die Begriffslehre oder Logik.* Zweites Buch der Formenlehre oder Mathematik, Verlag von R. Graßmann, Stettin.

Grassmann, R. (1891) Die Zahlenlehre oder Arithmetik — streng wissenschaftlich in strenger Formelentwicklung, Verlag von R. Graßmann, Stettin.

Grattan-Guinness, I. (1998) *The Norton History of the Mathematical Sciences: The Rainbow of Mathematics*, W.W. Norton, New York.

Grätzer, G. (1996) *General Lattice Theory*, Birkhäuser, Boston.

Gray, J. (1979) Non-Euclidean geometry: A Re-interpretation, *Historia Mathematics*, v. 6, pp. 236–258.

Gray, J. (2008) *Plato's Ghost: The Modernist Transformation of Mathematics*, Princeton University Press, Princeton.

Gregg, D. G. (2010) Designing for collective intelligence, *Communications of the ACM*, v.53, No.4, pp. 134–138.

Gregory, F. H. (1996) *Arithmetic and Reality: A Development of Popper's Ideas*, Working Paper Series No. WP96/01, Dept. of Information Systems, City University of Hong Kong, Hong Kong.

Griffin, M. (2008) Looking behind the symbol: Mythic algebra, numbers, and the illusion of linear sequence, *Semiotica*, v. 171, pp. 1–13.

Griffiths, P. (1990) *Equality in Language: Aspects of the Theory of Linguistic Equality*, Durham theses, Durham University, Available at Durham E-Theses.

Grossman, J. (1981) *Meta-Calculus: Differential and Integral*, Archimedes Foundation, Rockport, MA.

Grossman, M. (1979) *The First Non-linear System of Differential and Integral Calculus*, Mathco, Rockport, MA.

Grossman, M. (1979) An introduction to non-Newtonian calculus, *Int. J. Math. Education Sci. Technol.*, v. 10, No. 4, pp. 525–528.

Grossman, M. (1983) Bigeometric Calculus: A System with Scale-Free Derivative, Archimedes Foundation, Rockport.

Grossman, M. (1988) Calculus and discontinuous phenomena, *Int. J. Math. Education Sci. Technol.*, v. 19, No. 5, pp. 777–779.

Grossman, M. and Katz, R. (1972) *Non-Newtonian Calculus*, Lee Press, Pigeon Cove, MA.

Grossman, M. and Katz, R. (1984) Isomorphic calculi, *Int. J. Math. Education Sci. Technol.*, v. 15, No. 2, pp. 253–263.

Grossman, M. and Katz, R. (1986) A new approach to means of two positive numbers, *Int. J. Math. Education Sci. Technol.*, v. 17, No. 2, pp. 205–208.

Grossman, J., Grossman, M. and Katz, R. (1980) *The First Systems of Weighted Differential and Integral Calculus*, Archimedes Foundation, Rockport, MA.
Grossman, J., Grossman, M. and Katz, R. (1983) *Averages: A New Approach*, Archimedes Foundation, Rockport, MA.
Grossman, J., Grossman, M. and Katz, R. (1987) Which growth rate?, *Int. J. Math. Education Sci. Technol.*, v. 18, No. 1, pp. 151–154.
Grotheer, M., Herrmann, K.-H. and Kovaćs, G. (2016) Neuroimaging evidence of a bilateral representation for visually presented numbers, *J. Neurosci.*, v. 36, No. 1, pp. 88–97.
Gouvêa, F. (1997) Q. *p-Adic Numbers: An Introduction*, Springer-Verlag, New York.
Guenther, R. A. (1983) Product integrals and sum integrals, *Int. J. Math. Education Sci. Technol.*, v. 14, pp. 243–249.
Gullberg, J. (1997) *Mathematics: From the Birth of Numbers*, W. W. Norton, New York.
Gwiazda J. (2010) *Infinite Numbers Are Large Finite Numbers*, https://philarchive.org/rec/GWIINA?all_versions=1.
Gwiazda, J. (2012) On infinite number and distance, *Constructivist Found.*, v. 7, No. 2, pp. 126–130.
Hahn, H. (1907) Über die nichtarchimedischen Grössensysteme, *Sitzungsberichte der Kaiserlichen Akademie der Wissenschaften*, Wien, *Mathematisch — Naturwissenschaftliche Klasse*, v. 116 (Abteilung IIa), pp. 601–655.
Halberda, J. and Feigenson, L. (2008) Set representations required for the acquisition of the "natural number" concept, *Behavioral Brain Sci.*, v. 31, pp. 655–656.
Halberda, J., Mazzocco, M. M. M. and Feigenson, L. (2008) Individual differences in non-verbal number acuity correlate with maths achievement, *Nature*, v. 455, pp. 665–668.
Hale, B. (2000) Reals by abstractio, in *Philos. Math.*, v. 8, pp. 100–123.
Hale, B. (2002) Real numbers, quantities and measurement, *Philos. Math.*, v. 10, pp. 304–323.
Hale, B. (2004) Real numbers and set theory — extending the neo-Fregean programme beyond arithmetic, *Synthese*, v. 147, pp. 21–41.
Halsey, T. C., Jensen, M. H., Kadanoff, L. P., Procaccia, I. and Shraiman, B. I. (1986) Fractal measures and their singularities: The characterization of strange sets, *Phys. Rev. A*, v. 33, p. 1141.
Hamilton, W. R. (1833) *Introductory Lecture on Astronomy*, Dublin University Review and Quarterly Magazine, v. I, Trinity College.
Hamilton, W. R. (1843) On Quaternions; or on a new System of Imaginaries in Algebra, in *A letter to John T. Graves*, Dated October 17.
Hamilton, W. R. (1844) On quaternions, or on a new system of imaginaries in algebra. *Philos. Mag.*, v. 25, No. 3, pp. 489–495.
Hamilton (1848) Note, by Sir W. R. Hamilton, respecting the researches of John T. Graves, Esq., *Trans. Royal Irish Acad.*, v. 21, pp. 338–341.
Hamilton, W. R. (1866) *Elements of Quaternions*, Longmans, Green, London.

Hankel, H. (1867) Vorlesungenüber die complexen Zahlen und ihre Functionen, Voss, Leipzig.

Hanna, R. (2002) Mathematics for humans: Kant's philosophy of arithmetic revisited, *European J. Philos.*, v. 10, No. 3, pp. 328–352.

Hardy, G. H. (1910) Orders of Infinity, The "Infinitärcalcül" of Paul Du Bois-Reymond, Cambridge University Press, Cambridge.

Hardy, G. H., Littlewood, J. E. and Pólya, G. (1934) *Inequalities*, Cambridge University Press, Cambridge.

Hartle, J. B. (2003) *Gravity: An Introduction to Einstein's General Relativity*, Pearson, Boston.

Hartman, S. and Mikusinski, J. (1961) *The Theory of Lebesgue Measure and Integration*, Pergamon Press, Oxford.

Hartnett, P. M. (1992) The development of mathematical insight: From one, two, three to infinity, *Dissertation Abstracts Int.*, v. 52, p. 3921.

Harvey, B. M., Fracasso, A., Petridou, N. and Dumoulin, S. O. (2015) Topographic representations of object size and relationships with numerosity reveal generalized quantity processing in human parietal cortex, *Proc. Natl Acad. Sci. USA*, v. 112, No. 44, pp. 13525–13530.

Hauser, M., Tsao, F., Garcia, P. and Spelke, E. (2003) Evolutionary foundations of number: spontaneous representation of numerical magnitudes by cotton-top tamarins. *Proc. Royal Soc. B: Biol. Sci.*, v. 270, No. 1523, pp. 1441–1446.

Hayes, B. (2009) The higher arithmetic, *Amer. Scientist*, v. 97, No. 5, pp. 364–367.

Heaton, H. (1898) Infinity, the infinitesimal, and zero, *Amer. Math. Monthly*, v. 5, No. 10, pp. 224–226.

Heck, R. (1993) The development of arithmetic in Frege's Grundgesetze der Arithmetik, *J. Symbolic Logic*, v. 58, pp. 579–601.

Heck, R. G. (1995) Definition by induction in Frege's Grundgesetze der Arithmetik, in *Frege's Philosophy of Mathematics*, Harvard University Press, Cambridge, MA, pp. 295–333.

Heck, R. G., Jr. (1999) Grundgesetze der Arithmetik I §10, *Philos. Math.*, v. 7 pp. 258–292.

Heck, R. (2000) Cardinality, counting, and equinumerosity, *Notre Dame J. Formal Logic*, v. 41, No. 3, pp. 187–209.

Heidelberger, M. (1993) *Nature from Within: Gustav Theodore Fechner and His Psychophysical Worldview* (C. Klohr, trans.), University of Pittsburgh Press, Pittsburgh.

Heidelberger, M. (1993a) Fechner's impact for measurement theory, commentary on D.J. Murray, "A perspective for viewing the history of psychophysics", *Behavioral Brain Sci.*, v. 16, No. 1, pp. 146–148.

Heis, J. (2015) Arithmetic and number in the philosophy of symbolic forms, in *The Philosophy of Ernst Cassirer: A Novel Assessment*, pp. 123–140.

Heisenberg, W. (1998) *Philosophie — Le manuscrit de* 1942, Seuil, Paris.

Hellman, G. (1989) *Mathematics Without Numbers*, Oxford University Press, New York.

Hellman, G. (2001) Three varieties of mathematical structuralism, *Philos. Math.*, v. 9, pp. 184–211.

Helmut, W. C. (2010) Frege, the complex numbers, and the identity of indiscernibles, *Logique Anal.*, v. 53, No. 209, pp. 51–60.

Hentschel, H. G. E. and Procaccia, I. (1983) The infinite number of generalized dimensions of fractals and strange attractors, *Physica D*, v. 8, 435.

Herbert, N. (1987) *Quantum Reality: Beyond the New Physics*, Anchor Books, New York.

Herbrand, J. (1931) Sur la non-contradiction de l'arithmétiqué, *J. Reine Angew. Math.*, v. 166, pp. 1–8.

Herrmann, E., Hernández-Lloreda, M. V., Call, J., Hare, B. and Tomasello, M. (2010) The structure of individual differences in the cognitive abilities of children and chimpanzees, *Psychol Sci.*, v. 21, No. 1, pp. S102–10.

Herrlich, H. and Strecker, G. E. (1973) *Category Theory*, Allyn and Bacon Inc., Boston.

Hersh, R. (1999) *What is Mathematics, Really?* Oxford University Press, New York.

Hewitt, E. and Hewitt, R. E. (1979) The Gibbs-Wilbraham phenomenon: An episode in Fourier analysis, *Arch. History of Exact Sci.*, v. 21, No. 2, pp. 129–160.

Higgins, P. M. (2008) *Number Story: From Counting to Cryptography*, Copernicus, London.

Hilbert, D. (1899) *Die Grundlagen der Geometrie*, Festschrift zur Feier der Enthüllung des Gauss-Weber Denkmals in Göttingen, Teubner, Leipzig.

Hilbert, D. (1900) Über der Zahlbegriff, *Jaresbericht der Deutschen Mathematiker-Vereinigung*, bd. 8.

Hilbert, D. (1902) Mathematical problems, *Bull Amer. Math. Soc.*, v. 8, No. 10, pp. 437–479.

Hilbert, D. and Bernays, P. (1968) *Grundlagen der Mathematik*, I, Springer-Verlag, Berlin.

Hill, C. O. (2010) Husserl on axiomatization and arithmetic, in *Phenomenology and Mathematics*, Springer, New York.

Hinz, M., Lancia, M. R., Teplyaev, A. and Vernole, P. (2016) *Fractal snowflake domain diffusion with boundary and interior drifts*, Preprint, arXiv:1605.06785 [math.AP].

Hirsch, M. W. (1994) *Differential Topology*, Springer, New York.

Hjelmslev, J. (1950) Eudoxus' axiom and Archimedes' lemma, *Centaurus*, v. 1, pp. 2–11.

Hlodovskii, I. (1959) A new proof of the consistency of arithmetic, *Uspehi Mat. Nauk*, v. 14, No. 6, pp. 105–140 (in Russian).

Hodes, H. (1984) Logicism and the ontological commitments of arithmetic, *J. Philos.*, v. 81, pp. 123–149.

Hölder, O. (1901) Die Axiome der Quantität und die Lehre vom Mass, *Berichteüber die Verhandlungen der Königlich Sachsischen Gesellschaft der Wissenschaften zu Leipzig, Mathematische-Physicke Klasse*, v. 53, pp. 1–64.

Hölder, O. (1996) The axioms of quantity and the theory of measurement, *J. Math. Psycho.* 40, pp. 235–252.

(2013) *Homotopy Type Theory: Univalent Foundations of Mathematics*, The Univalent Foundations Program, Institute for Advanced Study.

Hopcroft, J. E., Motwani, R., and Ullman, J. D. (2007) *Introduction to Automata Theory, Languages, and Computation*, Addison-Wesley, Boston.

Huizing, C., Kuiper, R. and Verhoeff, T. (2012) Generalizations of Rice's theorem, applicable to executable and non-executable formalisms, in *Turing-100, The Alan Turing Centenary*, EPiC Series, v. 10, pp. 168–180.

Hurford, J. (1987) *Language and Number: The Emergence of a Cognitive System*, Blackwell, Oxford.

Husemöller, D. (1994) *Fibre Bundles*, Springer Verlag, Berlin.

Husserl, E. (1983) *Studien zur Arithmetik und Geometrie*, Husserliana, v. XXI, M. Nijhoff, The Hague.

Husserl, E. (1887) On the concept of number, in *Husserl: Shorter Works* (P. Mc Cormick and F. Elliston Eds), University of Notre Dame Press, Notre Dame, 1981, pp. 92–120.

Husserl, E. (1891) *Philosophie der Arithmetik*, Pfeffer, Halle.

Imaeda, K. and Imaeda, M. (2000) Sedenions: algebra and analysis, *Appl. Math. Comput.*, v. 115, No. 2, pp. 77–88.

Inönü, E. and Wigner, E. P. (1953) On the Contraction of Groups and Their Representations, *Proc. Natl. Acad. Sci.*, v. 39, No. 6, pp. 510–524.

Iverson, K. E. (1962) *A Programming Language*, Wiley, New York.

Izard, V., Pica, P., Spelke, E. S. and Dehaene, S. (2008) Understanding exact numbers, *Philos. Psychol.*, v. 21, No. 4, pp. 491–505.

Izhakian, Z. (2005) *Tropical arithmetic & algebra of tropical matrices*, Preprint in Mathematics, arXiv: 0505458.

Jackson, B. B. (2013) Defusing easy Arguments for numbers, *Linguistics Philos.*, v. 36, No. 6, pp. 447–461.

Jackson, J. D. (1975) *Classical Electrodynamics*, Wiley, New York.

Jay, C. B. (1989) A note on natural numbers objects in monoidal categories, *Studia Logica*, v. 48, No. 3, pp. 389–393.

Jech, T. (2002) *Set theory*, Springer, New York.

Jizba, P. and Scardigli, F. (2012) The emergence of special and doubly special relativity, *Phys. Rev.* D, v. 86, 025029.

Jizba, P. and Scardigli, F. (2013) Special relativity induced by granular space, *Eur. Phys. J. C*, v. 73, 2491.

Jizba, P. and Korbel, J. (2020) When Shannon and Khinchin meet Shore and Johnson: equivalence of information theory and statistical inference axiomatics, *Phys. Rev. E*, v. 101, p. 042126.

Jonsson, A. (1998) Wavelets on fractals and Besov spaces, *J. Fourier Anal. Appl.*, v. 4, pp. 329–340.

Jørgensen, K. F. (2005) *Kant's Schematism and the Foundations of Mathematics*, PhD thesis, Roskilde University, Roskilde.

Jørgensen, K. F. (2005a) *Kant and the Natural numbers*, https://www.scribd.com/document/116697225/Kant-and-the-Natural-Numbers.

Jorgensen, P. E. T. (2006) Analysis and Probability: Wavelets, Signals, Fractals, Springer, New York.

Jorgensen, P. E. T. and Pedersen, S. (1998) Dense analytic subspaces in fractal L^2 spaces, *J. Anal. Math.*, v. 75, pp. 185–228.

Jude, B. (2014) *Synergy* - 1 + 1 = 3, Aug 20, 2014, https://www.linkedin.com/pulse/20140820054514-115081853-synergy-1-1-3.

Jung, C. G. (1969) *The Structure and Dynamics of the Psyche*, Princeton University Press, Princeton.

Kalderon, M. E. (1996) What numbers could be (and, hence, necessarily are), *Philos. Math.*, v. 4, No. 3, pp. 238–255.

Kang, J., Wu, J., Smerieri, A. and Feng, J. (2010) Weber's law implies neural discharge more regular thana Poisson process, *European J. Neurosci.*, v. 31, pp. 1006–1018.

Kant, I. *Inaugural Dissertation*, (Translated by J. Handyside), The Open Court, Chicago, 1770/1929.

Kant, I. (1786/2004) *Metaphysical Foundations of Natural Science*, Cambridge University Press, Cambridge.

Kant, I. (1787/1958) *The Critique of Pure Reason*, (Translated by N. Kemp Smith), Macmillan, & Co, London.

Kaniadikis, G. (2001) Non-linear kinetics underlying generalized statistics, *Physica A*, v. 296, pp. 405–425.

Kaniadikis, G. and Scarfone, A. M. (2002) A new one parameter deformation of the exponential function, *Physica A*, v. 305, pp. 69–75.

Kaniadikis, G. (2002) Statistical mechanics in the context of special relativity, *Phys. Rev. E*, v. 66, p. 056125.

Kaniadikis, G. (2005) Statistical mechanics in the context of special relativity (II), *Phys. Rev. E*, v. 72, p. 036108.

Kaniadikis, G. (2011) Power-law tailed statistical distributions and Lorentz transformations, *Phys. Lett. A*, v. 375, pp. 356–359.

Kaniadikis, G. (2012) Physical origin of the power-law tailed statistical distributions, *Mod. Phys. Lett. B*, v. 26, p. 1250061.

Kantor, I. L. and Solodovnikov, A. S. (1989) *Hypercomplex Numbers: An Elementary Introduction to Algebras*, Springer-Verlag, New York.

Karpinski, L. C. (1925) *The History of Arithmetic*, Rand McNally & Company, Chicago.

Katz, E. (1957) The two-step flow of communication, *Public Opinion Quarterly*, v. 21, pp. 61–78.

Katz, E. (2017) What is tropical geometry? *Notices Amer. Math. Soc.*, v. 64, No. 4, pp. 380–382.

Kauffman, L. H. (1995) Arithmetic in the form, *Cybernet. Syst.*, v. 26, pp. 1–57.

Kauffman, L. H. (1999) What is a number? *Cybernet. Syst.*, v. 30, No. 2, pp. 113–130.

Kaufmann, R. M. and Yeomans, C. (2017) Math by pure thinking: R first and the divergence of measures in Hegel's philosophy of mathematics, *European J. Philos.*, v. 25, No. 4, pp. 985–1020.

Kaye, R. (1991) *Models of Peano Arithmetic*, Oxford Logic Guides, Oxford University Press, New York.

Kaye, R. (1993) Using Herbrand-type theorems to separate strong fragments of arithmetic, in *Arithmetic, Proof Theory, and Computational Complexity*, Oxford Logic Guides, v. 23, Oxford University Press, New York, pp. 238–246.

Kelly, J. L. (1955) *General Topology*, Van Nostrand Co., Princeton.

Kesseböhmer, M., Samuel, T., and Weyer, H. (2014) *A note on measure-geometric Laplacians*, Preprint in Mathematics, arXiv:1411.2491.

Khrennikov, A. (1997) *Non-Archimedean Analysis: Quantum Paradoxes, Dynamical Systems and Biological Models*, Kluwer Academic Publishers, Dordrecht.

Kigami, J. (1989) A harmonic calculus on the Sierpinski spaces, *Japan J. Appl. Math.*, v. 6, pp. 259–290.

Kigami, J. (2001) *Analysis on Fractals*, Cambridge Tracts in Mathematics, v. 143, Cambridge University Press, Cambridge.

Kim, J. (2014) Euclid strikes back at Frege, *Philos. Quart.*, v. 64 (254), pp. 20–38.

Kirby, A. (2000) Water arithmetic doesn't adds, *BBC News*, http://news.bbc.co.uk/hi/english/sci/tech/newsid_671000/671800.stm.

Kiselyov, O., Byrd, W. E., Friedman, D. P. and Shan, C.-C. (2008) Pure, declarative, and constructive arithmetic relations (declarative pearl), in *Proceedings of the 9th International Conference on Functional and Logic Programming (FLOPS'08)*, pp. 64–80.

Kleene, S. C. (1956) Representation of events in nerve sets and finite automata, in *Automata Studies*, Princeton University Press, Princeton, pp. 3–40.

Kleene, S. C. (2002) *Mathematical Logic*, Courier Dover Publications, New York.

Klees, E. (2006) *One Plus One Equals Three — Pairing Man/Woman Strengths: Role Models of Teamwork* (The Role Models of Human Values Series, v. 1) Cameo Press, New York.

Klement, E. P., Mesiar, R. and Pap, E. (2000) *Triangular Norms*, Springer-Science+Business Media, B.V., New York.

Kline M. (1967) *Mathematics for Nonmathematicians*, Dover Publications, New York.

Kline, M. (1980) *Mathematics: The Loss of Certainty*, Oxford University Press, New York.

Kline, M. (1990) *Mathematical Thought from Ancient to Modern Times*, Oxford University Press, New York.

Klotz, I. M. (1995) Number mysticism in scientific thinking, *Math. Intelligencer*, v. 17, pp. 43–51.

Knapp, T. R. (2009) *Percentages: The most useful statistics ever invented*, http://www.statlit.org/pdf/2009Knapp-Percentages.pdf.

Kneusel, R. T. (2015) *Numbers and Computers*, Springer International Publishing, Switzerland.

Knight, J., Case, J. and Berman K. (2006) Financial Intelligence: A Manager's Guide to Knowing What the Numbers Really Mean, Harvard Business Review Press.

Knuth, D. E. (1974) *Surreal Numbers: How Two Ex-students Turned on to Pure Mathematics and Found Total Happiness*, Addison-Wesley, Reading, MA.

Knuth D. E. (1976) Mathematics and computer science: Coping with finiteness, *Science*, v. 194, No. 4271, pp. 1235–1242.

Knuth, D. (1997) *The Art of Computer Programming*, v. 2: *Seminumerical Algorithms*, Addison-Wesley, Reading, MA.

Koblitz, N. (1977) *p-Adic Numbers, p-Adic Analysis, and Zeta-Functions*, Springer-Verlag, New York.

Koeplinger, J. and Shuster, J. A. (2012) "$1 + 1 = 2$" *A step in the wrong direction*? http://fqxi.org/community/forum/topic/1449.

Kolmogorov, A. N. (1930) Sur la notion de la moyenne, *Atti Accad. Naz. Lincei*, v. 12, pp. 388–391.

Kolmogorov, A. N. (1961) Automata and life, *Tekhnika Molodezhi*, No. 10, pp. 16–19; No. 11, pp. 30–33 (in Russian).

Kolokoltsov, V. N. and Maslov, V. P. (1997) *Idempotent Analysis and Its Applications*, Kluwer Academic Publishers, Dordrecht.

Kolwankar, K. M. and Gangal, A. D. (1998) Local fractional Fokker-Planck equation, *Phys. Rev. Lett.*, v. 80, 214.

Kossak, R. and Schmerl, J. H. (2006) *The Structure of Models of Peano Arithmetic*, Clarendon Press, Oxford.

Kozen, D. (1997) *Automata and Computability*, Springer-Verlag, New York.

Krantz, S. G. (2003) *Calculus Demystified*, McGraw-Hill, New York.

Krantz, D. H. (1971) Integration of Just-Noticeable Differences, *J. Math. Psychol.*, v. 8, pp. 591–599.

Krantz, D. H., Luce, R. D, Suppes, P., and Tversky, A. (1971) *Foundations of Measurement*, v. 1, Academic Press, San Diego, CA.

Krause, E. F. (1987) *Taxicab Geometry*, Dover Publications, New York.

Krause, G. M. (1981) *A strengthening of Ling's theorem on representation of associative functions*, Ph.D. Thesis in Mathematics, Illinois Institute of Technology.

Kreisel, G. (1965) *Mathematical Logic*, Lectures on Modern Mathematics, John Wiley & Sons, Inc., New York.

Kress, S. (2015) *Synergy: When one plus one equals three*, http://www.summitteambuilding.com/synergy-when-one-plus-one-equals-three.

Kroiss, M., Fischer, U. and Schultz, J. (2009) When one plus one equals three: Biochemistry and bioinformatics combine to answer complex questions, *Fly*, v. 3, No. 3, pp. 212–214.

Kronecker, L. (1887) Über den Zahlbegriff, Reworked and expanded version in *Werke*, v. 3 (Hensel, K. Ed.) Teubner, Leipzig, pp. 251–274.

Krysztofiak, W. (2008) Modalna arytmetyka indeksowanych liczb naturalnych: możliwe światy liczb, *Przegl̆ăd Filozoficzny — Nowa Seria*, v. 17, pp. 79–107 (in Polish).

Krysztofiak, W. (2012) Indexed Natural Numbers in Mind: A Formal Model of the Basic Mature Number Competence, *Axiomathes*, v. 22, No. 4, pp. 433–456.

Kuczma, M., Choczewski, B. and Ger, R. (1990) *Iterative Functional Equations*, Cambridge University Press, Cambridge.

Kuich, W. (1986) *Semirings, Automata, Languages*, Springer-Verlag, Berlin.

Kulisch, U. W. (1982) Computer arithmetic and programming languages, in *APL'82 Proceedings of the International Conference on APL*, Heidelberg, Germany, pp. 176–182.

Kulsariyeva, A. T. and Zhumashova, Z. (2015) Numbers as cultural significant, *Procedia — Social Behavioral Sci.*, v. 191, pp. 1660–1664.

Kul'vetsas, L. L. (1989) The status of the concept of quantity in physics theory and H Helmholtz's book "Zählen und Messen", in *Studies in the History of Physics and Mechanics,* Nauka, Moscow, pp. 170–186 (in Russian).

Kuratowski, K. (1966) *Topology,* v. 1, Academic Press, Warszawa.

Kuratowski, K. (1968) *Topology,* v. 2, Academic Press, Warszawa.

Kuratowski, K. and Mostowski, A. (1967) *Set Theory,* North Holland P.C., Amsterdam.

Kurosh, A. G. (1963) *Lectures on General Algebra,* Chelsea P. C., New York.

Kusuoka, S. (1987) A diffusion process on a fractal, probabilistic methods in mathematical physics, in *Proceedings of Taniguchi International Symposium* (Ito, K. and Ikeda, N. Eds.), pp. 251–271, Kinokuniya.

Kusuoka, S. (1989) Dirichlet forms on fractals and products of random matrices, *Publ. Res. Inst. Math. Sci.*, v. 25, pp. 659–680.

Kutter, E. F., Bostroem, J., Elger, C. E., Mormann, F. and Nieder, A. (2018) Single Neurons in the Human Brain Encode Numbers, *Neuron,* 20 September 2018; DOI: https://doi.org/10.1016/j.neuron.2018.08.036.

Ladyman, J. (1998) What is structural realism? *Studies History Philos. Sci.*, v. 29, No. 3, pp. 409–424.

Lakoff, G. and Núñez, R. (1997) The metaphorical structure of mathematics: Sketching out cognitive foundations for a mind-based mathematics, in *Mathematical Reasoning: Analogies, Metaphors, and Images* (English, L., Ed), Erlbaum, Mahwah, NJ, pp. 267–280.

Lakoff, G. and Núñez, R. (2000) *Where Mathematics Comes From: How the Embodied Mind Brings Mathematics Into Being,* Basic Books, New York.

Lamb, R. (2010) *How math works,* https://science.howstuffworks.com/math-concepts/math1.htm.

Lancia, M. R. (2002) A transmission problem with a fractal interface, *Zeit. Anal. Anwendungen,* v. 21, pp. 113–133.

Landau, E. (1966) *Foundations of Analysis: The Arithmetic of Whole, Rational, Irrational and Complex Numbers,* A Supplement to Textbooks on the Differential and Integral Calculus, Chelsea, New York.

Landau, E. (1999) *Elementary Number Theory,* American Mathematical Society, RI.

Lang, M. (2014) One plus one equals three: multi-line fiber lasers for nonlinear microscopy, *Optik&Photonik,* v. 9, No. 4, pp. 53–56.

Lapidus, M. L. (1991) Fractal drum, inverse spectral problems for elliptic operators and a partial resolution of the Weyl–Berry conjecture, *Trans. Amer. Math. Soc.*, v. 325, pp. 465–529.

Lapidus, M. L. and Pang, M. M. H. (1995) Eigenfunctions of the Koch snowflake domain, *Commun. Math. Phys.*, v. 172, pp. 359–376.

Lapidus, M. L., Radunović, G. and Žubrinić, D. (2017) *Fractal Zeta Functions and Fractal Drums: Higher-Dimensional Theory of Complex Dimensions*, Springer Monographs in Mathematics, Springer, New York.

Lapidus, M. L. and van Frankenhuijsen, M. (2013) *Fractal Geometry, Complex Dimensions and Zeta Functions: Geometry and Spectra of Fractal Strings*, Springer Monographs in Mathematics, Springer, New York.

Lasswell, H. (1948) The structure and function of communication in society, in *The Communication of Ideas*, (Bryson, Lymon; Ed.) Institute for Religious and Social Studies, New York, pp. 37–51.

Laugwitz, D. (1961) Anwendungen unendlichkleiner Zahlen I. Zur Theorie der Distributionen, *J. Reine Angew. Math.*, v. 207, pp. 53–60.

Laugwitz, D. (1961a) Anwendungen unendlichkleiner Zahlen II. Ein Zugang zur Operatorenrechnung von Mikusinski Distributionen, *J. Reine Angew. Math.*, v. 208, pp. 22–34.

Law, S. (2012) A brief history of numbers and counting, Pt 1: Mathematics advanced with civilization, *Deseret News*, https://www.deseretnews.com/article/865560110/A-brief-history-of-numbers-and-counting-Part-1-Mathematics-advanced-with-civilization.html.

Law, S. (2012a) A brief history of numbers and counting, Pt 2: Indian invention of zero was huge in development of math, *Deseret News*, https://www.deseretnews.com/article/865560133/A-brief-history-of-numbers-and-counting-Part-2-Indian-invention-of-zero-was-huge-in-development-of.html.

Lawrence, C. (2011) Making $1+1=3$: Improving sedation through drug synergy, *Gastrointestinal Endoscopy*, February 2011.

Le Corre, M. and Carey, S. (2007) One, two, three, four, nothing more: an investigation of the conceptual sources of the verbal counting principles, *Cognition*, v. 105, pp. 395–438.

Lea, R. (2016) Why one plus one equals three in big analytics, *Forbes*, May 27, 2016, https://www.forbes.com/sites/teradata/2016/05/27/why-one-plus-one-equals-three-in-big-analytics/#1aa2070056d8.

Lee, J. M. (2000) *Introduction to Topological Manifolds*, Graduate Texts in Mathematics 202, Springer, New York.

Leibniz, G. (1703) Explication de l'arithmétique binaire, Que se sert seul caractéres 0 & 1; avec des Remarques sur son utilité, & sur ce qu'elle donne le sens des anciennes figures Chinoises de Fohy, *Mem. Acad. Roy. Sci.*, pp. 85–93.

Leibovich, T. and Ansari, D. (2016) The symbol-grounding problem in numerical cognition: A review of theory, evidence, and outstanding questions, *Can. J. Exp. Psychol.*, v. 70, No. 1, pp. 12–23.

Lemaire, P., Abdi, H. and Fayol, M. (1996) The role of working memory resources in simple cognitive arithmetic, *Euro. J. Cogn. Psychol.*, v. 8, No. 1, pp. 73–103.

Leng, M. (2018) Does $2+3=5$? In defence of a near absurdity, *Math Intelligencer*, v. 40, pp. 14–17.

Leuthesser, L., Kohli, C. and Suri, R. (2003) 2+2=5? A framework for using co-branding to leverage a brand, *J. Brand Manag.*, v. 11, No. 1, pp. 35–47.

Li, M. and Vitanyi, P. (1997) *An Introduction to Kolmogorov Complexity and its Applications*, Springer-Verlag, New York.

Li, J. and Ostoja-Starzewski, M. (2009) Fractal solids, product measures and fractional wave equations, *Proc. Roy. Soc. Lond.* A, v. 465, pp. 2521–2536.

Ling, C.-H. (1995) Representation of associative functions, *Publicationes Mathematicae*, v. 12, pp. 189–212.

Lipshitz, L. (1979) Diophantine correct models of arithmetic, *Proc. Amer. Math. Soc.*, v. 73, No. 1, pp. 107–108.

Littlewood, J. E. (1953) *Miscellany*, Methuen, London.

Litvinov, G. L. (2007) The Maslov dequantization, idempotent and tropical mathematics: A brief introduction, *J. Math. Sci.* (*New York*), v. 140, No. 3, pp. 426–444.

Livanova, A. (1969) *Three Destinies — Comprehension of the World*, Znaniye, Moscow (in Russian).

Livesey, D. A. (1974) The importance of numerical algorithms for solving economic optimization problems, *Int. J. Sys. Sci.*, v. 5, No. 5, pp. 435–451.

Livio, M. (2009) *Is God a Mathematician*? Simon & Schuster, New York.

Lorenzen, P. (1951) Algebraische und logistische Untersuchungen über freie Verbände, *J. Symbolic Logic*, v. 16, pp. 81–106.

Losev, A. F. (1994) *Myth, Number, Essence*, Mysl', Moscow (in Russian).

Löwenheim, L. (1915) Über Möglichkeiten im Relativkalkül, *Math. Annalen*, v. 76, No. 4, pp. 447–470.

Luce, R. D. (1964) The mathematics used in mathematical psychology, *Amer. Math. Monthly*, v. 71, pp. 364–378.

Luce, R. D. (1987) Measurement structures with Archimedean ordered translation groups, *Order*, v. 4, pp. 165–189.

Luce, R. D. (2002) A psychophysical theory of intensity proportions, joint presentations, and matches, *Psychol. Rev.*, v. 109, No. 3, pp. 520–532.

Luce, R. D. and Edwards, W. (1958) The derivation of subjective scales from just noticeable differences, *Psychol. Rev.*, v. 65, pp. 222–237.

Luce, R. D., Bush, R. R. and Galanter, E. (1963) *Handbook of Mathematical Psychology*, v. 1, Wiley, New York.

Mac Lane, S. and Birkhoff, G. (1999) *Algebra*, v. 330 of AMS Chelsea Publishing Series, American Mathematical Society.

MacNeal, E. (1995) *Mathsemantics, Making Numbers Talk Sense*, Penguin Books, England.

Magueijo, J. (2003) New varying speed of light theories, *Rep. Prog. Phys.*, v. 66, 2025.

Mandelbrot, B. (1967) How long is the coast of Britain? Statistical self-similarity and fractional dimension, *Science*, v. 155, pp. 636–638.

Mandelbrot, B. (1983) *The Fractal Geometry of Nature*, Macmillan, New York.

Mane, R. (1952) Evolution of mutuality: one plus one equals three; formula characterizing mutuality, *La Revue Du Praticien*, v. 2, No. 5, pp. 302–304.

Manin, Y. I. (1991) *Course in Mathematical Logic*, Springer-Verlag, New York.

Manin, Y. (2007) *Mathematical knowledge: internal, social and cultural aspects*, Preprint in Mathematics History and Overview, arXiv:math/0703427 [math.HO].

Marichal, J.-L. (2009) *Aggregation functions for decision-making*, Preprint math.ST, arXiv: 0901.4232.

Marie, K. L. (2007) One Plus One Equals Three: Joint-Use Libraries in Urban Areas — The Ultimate Form of Library Cooperation, *Library Leadership & Management* (LL&M), v. 21, No. 1.

Marcolli, M. and Thorngren, R. (2011) *Thermodynamic semirings*, Preprint in Quantum Algebra, arXiv:1108.2874.

Markov, A. A. and Nagornii, N.M. (1988) *The Theory of Algorithms*, Springer, New York.

Marks, L. E. (1974) *Sensory Processes: The New Psychophysics*, Academic Press, London.

Marks, M. L. and Mirvis, P. H. (2010) Joining forces: Making one plus one equal three in mergers, in *Acquisitions, and Alliances*, Jossey-Bass.

Martin, G. (1985) *Arithmetic and Combinatorics: Kant and his Contemporaries*, Southern Illinois University Press, Carbondale and Edwardsville.

Martinez, A. A. (2006) *Negative Math: How Mathematical Rules Can Be Positively Bent*, Princeton University Press, Princeton.

Maslov, V. P. (1987) Asymptotic Methods for Solving Pseudo-Differential Equations, Nauka, Moscow (in Russian).

Maslov, V. P. and Samborskii, S. N. (Eds.) (1992) *Idempotent Analysis*, American Mathematical Society, New York.

Matejko, A. A. and Ansari, D. (2015) Drawing connections between white matter and numerical and mathematical cognition: A literature review, *Neurosci. Biobehav. Rev.*, v. 48, pp. 35–52.

Matson, J. (2009) The origin of zero, *Scientific American*, August 21, 2009, http://www.scientificamerican.com/article.cfm?id=history-of-zero.

Matsushita, S. (1951) On the foundations of orders in groups, *J. Inst. Polytechn. Osaka City Univ.*, A, v. 2, pp. 19–22.

Mattessich, R. (1998) From accounting to negative numbers: A signal contribution of medieval India to mathematics, *Accounting Historians J.*, v. 25, No. 2, pp. 129–145.

Maugin, G. A. (2017) Non-classical continuum mechanics, in *A Dictionary Advanced Structured Materials*, v. 51, Springer Nature, Singapore.

Mazur, B. (2008) When is one thing equal to some other thing? in *Proof and Other Dilemmas: Mathematics and Philosophy*, (Gold, B. and Simons, R. Eds.), Mathematical Association of America, Washington, DC, pp. 221–243.

Mazur, B. (2008a) Mathematical Platonism and its opposites, *European Math. Soc. Newsletter*, No. 68, pp. 19–21.

McCall, S. (2014) *The Consistency of Arithmetic and Other Essays*, Oxford University Press, New York.

McComb, K., Packer, C. and Pusey, A. (1994) Roaring and numerical assessment in contests between groups of female lions, Panthera leo, *Animal Behav.*, v. 47, pp. 379–387.

McCulloch, W. S. (1961) What is a number that a man may know it, and a man, that he may know a number? (the Ninth Alfred Korzybski Memorial Lecture), *General Semantics Bull.*, No. 26/27, Institute of General Semantics, pp. 7–18.

McLarty, C. (1993) Numbers can be just what they have to, *Nous*, v. 27, pp. 487–498.

McLeish, J. (1994) *The Story of Numbers: How Mathematics Has Shaped Civilization*, Fawcett Columbine, New York.

Meck W. H. and Church R. M. (1983) A mode control model of counting and timing processes, *J. Exp. Psychol. Anim. Behav. Process.*, v. 9, pp. 320–334.

Meiert, J. O. 1 + 1 = 3: Explaining busyness and background noise on websites, in *On Web Development: Articles 2005–2015*, Kindle Edition, 2015.

Meinong, A. (1882) Zur Relationstheorie [On the theory of relations], *Sitzungsberichte der Phil.-Hist. Classe d. K. Akademie der Wissenschaften zu Wien*, v. 101, pp. 573–752.

Mendell, H. Aristotle and Mathematics, in *The Stanford Encyclopedia of Philosophy* (Spring 2017 Edition), Edward N. Zalta (ed.), https://plato.stanford.edu/archives/spr2017/entries/aristotle-mathematics/.

Menger, K. (1942) Statistical metrics, *Proc. Natl. Acad. Sci. USA*, v. 8, pp. 535–537.

Menninger, K. (1992) *Number Words and Number Symbols: A Cultural History of Numbers*, Dover Publications, New York.

Menon, V. (2010) Developmental cognitive neuroscience of arithmetic: implications for learning and education. *ZDM: Int. J. Math. Education*, v. 42, No. 6, pp. 515–525.

Menon, V. (2015) Arithmetic in the child and adult brain, in *Oxford Library of Psychology. The Oxford Handbook of Numerical Cognition*, Oxford University Press, New York, pp. 502–530.

Merritt, D. J., Rugani, R. and Brannon, E. M. (2009) Empty sets as part of the numerical continuum: Conceptual precursors to the zero concept in rhesus monkeys, *J. Exp. Psychol. Gen.*, v. 138, pp. 258–269.

Mesiar, R. and Rybařik, J. (1993) Pseudo-arithmetical operations, *Tatra Mountains Math. Publ.*, v. 2, pp. 185–192.

Meyer, R. K. (1976) Relevant arithmetic, *Bull. Section Logic Polish Acad. Sci.*, v. 5, pp. 133–137.

Meyer, R. K. and Mortensen, C. (1984) Inconsistent models for relevant arithmetics, *J. Symbolic Logic*, v. 49, pp. 917–929.

Mikhalkin, G. (2004) Amoebas of algebraic varieties and tropical geometry, in *Different Faces of Geometry*, International Mathematical Series, v. 3, Kluwer Academic/Plenum Publishers, New York, NY, pp. 257–300.

Miller, G. A. (1956) The magical number seven, plus or minus two: Some limits on our capacity for processing information, *Psychol. Rev.*, v. 63, pp. 81–97.

Miller, G. R. (1966) *Speech Communication: A Behavioral Approach*, Bobbs-Merril, Indianapolis, IN.

Miller, J. (1982) *Numbers in Presence and Absence*, M. Nijhoff, The Hague.

Miller, M. (1999) *Time, Clocks and Causality*, Quackgrass Press, Calgary (electronic edition: http://www.quackgrass.com/home.html).

Minsky, M. (1967) *Computation: Finite and Infinite Machines*, Prentice-Hall, Englewood Cliffs, NJ.

Misner, C. W., Thorne, K. S. and Wheeler, J. A. (1973) *Gravitation*, W.H. Freeman and Company, New York.

Möbius, A. F. (1827) *Der Barycentrische Calcul*, Johan Ambrosius Barth, Leipzig.

Modesto, L. (2009) Fractal spacetime from the area spectrum, *Class. Quantum Grav.*, v. 26, 242002.

Modesto, L. and Nicolini, P. (2010) Spectral dimension of a quantum universe, *Phys. Rev. D*, v. 81, 104040.

Molitor, D., Ott, N. and Strichartz, R. (2015) Using Peano curves to construct Laplacians on fractals, *Fractals*, v. 23, 1550048.

Mollin, R. A. (2005) *Codes: The Guide to Secrecy from Ancient to Modern Times*, Discrete Mathematics and its Applications, Chapman & Hall/CRC, Boca Raton, FL.

Moltmann, F. (2013) Reference to numbers in natural language, *Philos. Studies*, v. 162, No. 3, pp. 499–536.

Moltmann, F. (2016) The number of planets, a number-referring term? in *Abstractionism: Essays in Philosophy of Mathematics*, Oxford University Press, Oxford, pp. 113–129.

Monk, J. D. (1976) *Mathematical Logic*, Graduate Texts in Mathematics, Springer-Verlag, Berlin.

Moore, R. E. (1966) *Interval Analysis*, Prentice-Hall, New York.

Mora, M., Córdova-Lepe, F. and Del-Vall, R. (2012) A non-Newtonian gradient for contour detection in images with multiplicative noise, *Pattern Recogn. Lett.*, v. 33, pp. 1245–1256.

Morgan, A. (1836) *The Connexion of Number and Magnitude: An Attempt to Explain the Fifth Book of Euclid*, Taylor and Walton, London.

Morita, K. (2017) *Theory of Reversible Computing*, Springer, Japan KK.

Morris, M. *Lessons in Math and Marriage: One Plus One Equals ONE — How to be Better Together*, Marriage Dynamics Institute, https://marriagedynamics.com/tips-growing-healthy-marriage-recognizing-rooting-selfishness/.

Mortensen, C. (1995) *Inconsistent Mathematics*, Kluwer Mathematics and Its Applications Series, Kluwer, Dordrecht.

Mortensen, C. (2000) Prospects for inconsistency, in *Frontiers of Paraconsistent Logic*, (Batens, D. *et al.*, Eds.), Research Studies Press, London, pp. 203–208.

Mosteller, F. (1977) *Data Analysis and Regression: A Second Course in Statistics*, Addison-Wesley, Reading, MA.

Mumford, D. (2008) Why I am a Platonist, *EMS Newsletter*, No. 12, pp. 27–29.

Murawski, R. (1998) Undefinability of truth. The problem of the priority: Tarski vs. Gödel, *History Philos. Logic*, v. 19, pp. 153–160.

Murawski, R. (2001) On proofs of the consistency of arithmetic, *Studies Logic, Grammar Rhetoric*, v. 4, No. 17, pp. 41–50.

Murphy, M. and Miller, M. (2010) Making 1 + 1 = 3, *FTI J.*, http://www.ftijournal.com/article/making-1-1-3.

Nagumo, M. (1930) Übereine Klasse der Mittelwerte, *Jap. J. Math.*, v. 7, pp. 71–79.

Naimark, M. A. (1959) *Normed Rings*, P. Noordhoff N. V., Groningen.

Naimark, M. A. (1972) *Normed Algebras*, Wolters-Noordhof Publishing Co., Groningen.

Napier, J. (1614) *Mirifici Logarithmorum Canonis Descriptio* [*The Description of the Wonderful Rule of Logarithms*], Andrew Hart, Edinburgh, Scotland (in Latin).

Napier, J. (1990) *Rabdology*, (Richardson, W.F., Transl.), MIT Press.

Narens, L. (1980) A qualitative treatment of Weber's law, *J. Math. Psychol.*, v. 13, pp. 88–91.

Narens, L. (1981) On the scales of measurement, *J. Math. Psychol.*, v. 24, No. 3, pp. 249–275.

Narens, L. (1981a) A general theory of ratio scalability with remarks about the measurement-theoretic concept of meaningfulness, *Theory Decision*, v. 13, pp. 1–70.

Narens, L. (1985) *Abstract Measurement Theory*, MIT Press, Cambridge, MA.

Narens, L. (1988) Meaningfulness and the Erlanger program of Felix Klein. *Math. Inform. Sci. Humaines*, v. 101, 61–72.

Narens, L. and Mausfeld, R. (1992) On the relationship of the psychological and the physical in psychophysics, *Psychol. Rev.*, v. 99, No. 3, pp. 467–479.

National Research Council (2001) Number: What is there to know? in *Adding It Up: Helping Children Learn Mathematics*, The National Academies Press, Washington, DC.

Naudts, J. (2002) Deformed exponentials and logarithms in generalized thermostatistics, *Physica A*, v. 316, pp. 323–334.

Naudts, J. (2011) *Generalised Thermostatistics*, Springer, London.

Nelson, E. (1986) *Predicative Arithmetic*, Mathematical Notes, v. 32, Princeton University Press, Princeton, NJ.

Neugebauer (1969) *The Exact Sciences in Antiquity*, Courier Dover Publications, New York.

Nevanlinna, R. (1939) Le Theoreme de Picard–Borel et la theorie des fonctions meromorphes, Gauthier-Villars, Paris.

Newman, J. R. (Ed.) (1956) *The World of Mathematics*, Simon and Schuster, New York.

Newton, I. (1728/1967) Universal arithmetic: Or, a treatise of arithmetical composition and resolution, in *The Mathematical Works of Isaac Newton* (D.T. Whiteside, Ed.), Johnson Reprint Corp., New York, v. 2, pp. 3–134.

Newton, I. (1736) *The Method of Fluxions and Infinite Series: With Its Application to the Geometry of Curve-lines*, Translated from the Author's Latin Original Not Yet Made Publick, Henry Woodfall, London.

Nguen, N. T. (2008) Inconsistency of knowledge and collective intelligence, *Cybernet. Syst.*, v. 39, No. 6, pp. 542–562.

Nicolini, P. and E. Spallucci, E. (2011) Un-spectral dimension and quantum spacetime phases, *Phys. Lett. B*, v. 695, p. 290.

Nicomachus of Gerasa (1926) *Introduction to Arithmetic*, (M. L. D'Ooge, Trs.), Macmillan, New York.

Nieder, A. (2005) Counting on neurons: the neurobiology of numerical competence, *Nat. Rev. Neurosci.*, v. 6, pp. 177–190.

Nieder, A. (2016) The neuronal code for number, *Nature*, v. 17, pp. 366–382.

Nieder, A., Freedman D. J. and Miller E. K. (2002) Representation of the quantity of visual items in the primate prefrontal cortex, *Science*, v. 297, pp. 1708–1711.

Nieder, A. and Miller, E. K. (2004) A parieto-frontal network for visual numerical information in the monkey, *Proc. Natl. Acad. Sci. USA*, v. 101, pp. 7457–7462.

Nieder, A. and Dehaene, S. (2009) Representation of number in the brain, *Annual Rev. Neurosci.*, v. 32, pp. 185–208.

Nieuwmeijer, C. (2013) 1 + 1 = 3: *The positive effects of the synergy between musician and classroom teacher on young children's free musical play*, Dissertation, Roehampton University, London.

Nivanen, L., Le Mehaute, A. and Wang, Q. A. (2003) Generalized algebra within a nonextensive statistics, *Rep. Math. Phys*, v. 52, pp. 437–444.

Noma, E. and Baird, J. C. (1975) Psychophysical study of numbers: II. Theoretical models of number generation, *Psychol. Res.*, v. 38, pp. 81–95.

North, J. (2009) The "structure" of physics: A case study, *J. Philos.*, v. 106, pp. 57–88.

Norwich, K. H. (1993) *Information, Sensation, and Perception*, Academic Press, San Diego.

Nottale, L. and Schneider, J. (1984) Fractals and nonstandard analysis, *J. Math. Phys.*, v. 25, p. 1296.

Nottale, L. (1993) *Fractal Space-Time and Microphysics: Towards a Theory of Scale Relativity*, World Scientific, Singapore.

Nottale, L. (2011) *Scale Relativity and Fractal Space-Time*, Imperial College Press, London.

Number (2016) *New World Encyclopedia.*

Núñez, R. (2009) Numbers and arithmetic: Neither Hardwired nor out there, *Biol. Theory*, v. 4, No. 1, pp. 68–83.

Núñez, R. and Lakoff, G. (1998) What did Weierstrass really define? The cognitive structure of natural and ε–δ continuity, *Math. Cognition*, v. 4, No. 2, pp. 85–101.

Núñez, R. and Lakoff, G. (2005) The cognitive foundations of mathematics: The role of conceptual metaphor, in *Handbook of Mathematical Cognition* (Campbell, J., Ed.), Psychology Press, New York, pp. 109–124.

Oakley, B. (2014) *A Mind For Numbers: How to Excel at Math and Science* (*Even If You Flunked Algebra*), Tarcher Perigee, New York.

O'Connor, J. J. and Robertson, E. F. Jaina mathematics, *MacTutor History of Mathematics*, http://www-history.mcs.st-andrews.ac.uk/HistTopics/Jaina_mathematics.html.

O'Connor, J. J. and Robertson, E. F. Nine chapters on the mathematical art, *MacTutor History of Mathematics*, http://www-groups.dcs.st-and.ac.uk/history/HistTopics/Nine_chapters.html.

Oppenheim, A. V., Schafer, R. W. (1975) *Digital Signal Processing*, Prentice-Hall.

Oppenheim, A. V., Schafer, R. W. and Stockham, T. G. (1968) Nonlinear filtering of multiplied and convolved signals, *IEEE Trans. Audio Electroacoustics*, v. AU-16, No. 3, pp. 437–466.

Oppenheim, A. V., Schafer, R. W. and Stockham, T. G. (1968a) Nonlinear filtering of multiplied and convolved signals, *Proc. IEEE*, v. 56, pp. 1264–1291.

Ord, G. N. (1983) Fractal space-time: a geometric analogue of relativistic quantum mechanics, *J. Phys. A: Math. Gen.*, v. 16, p. 1869.

Ore, O. (1935) On the foundation of abstract algebra, I, *Ann. of Math.*, v. 36, pp. 406–437.

Ore, O. (1948) *Number Theory and its History*, McGraw-Hill, New York.

Orsi, R. (2007) When $2 + 2 = 5$, *The American Scholar*, March 1 2007, https://theamericanscholar.org/when-2-2-5/#.W7GLXWhKjIU.

Ostoja-Starzewski, M., Li, J., Joumaa, H. and Demmie, P. N. (2013) From fractal media to continuum mechanics, *Z. Angew. Math. Mech.*, v. 93, pp. 1–29.

Otte, M. (1990) Arithmetic and geometry: some remarks on the concept of complementarity, *Stud. Philos. Ed.*, v. 10, pp. 37–62.

Ozyapıcı, A. and Bilgehan, B. (2016) Finite product representation via multiplicative calculus and its applications to exponential signal processing, *Numer. Algorithms*, v. 71, No. 2, pp. 475–489.

Palmer, C. and Stavrinou, P. N. (2004) Equations of motion in a non-integer-dimensional space, *J. Phys.* A: *Math. Gen.*, v. 37, pp. 6987–7003.

Pandit, S. N. N. (1961) A new matrix calculus, *SIAM J. Appl. Math.* 9, pp. 632–639.

Pap, E. (1993) g-calculus, Univ. u Novom Sadu, *Zb. Rad. Prirod.-Mat. Fak. Ser. Mat.*, v. 23, No. 1, pp. 145–150.

Pap, E. (2008) Generalized real analysis and its applications, *Int. J. Approx. Reasoning*, v. 47, pp. 368–386.

Pap, E., Štrboja, M. and Rudas, I. (2014) Pseudo-L^p space and convergence, *Fuzzy Sets Syst.*, v. 238, pp. 113–128.

Pap, E., Takaci, D. and Takaci, A. (2002) The g-operational calculus, *Int. J. Uncertainty, Fuzziness and Knowledge-Based Syst.*, v. 10, Supplement, pp. 75–88.

Parhami, B. (2002) Number representation and computer arithmetic, in *Encyclopedia of Information Systems*, Academic Press.

Parhami, B. (2010) *Computer Arithmetic*: *Algorithms and Hardware Designs*, Oxford University Press, New York.

Parikh, R. (1971) Existence and feasibility in arithmetic, *J. Symbolic Logic*, v. 36, No. 3, pp. 494–508.

Parsons, C. (1964) Infinity and Kant's conception of the 'possibility of experience', *Philos. Rev.*, v. 73, No. 2, pp. 182–197.

Parsons, C. (1969) Kant's Philosophy of Arithmetic, in *Philosophy, Science and Method: Essays in Honor of Ernest Nagel* (S. Morgenbesser, P. Suppes, and M. White, Eds.), St. Martin's Press, New York.

Parsons, C. (1984) Arithmetic and the categories, *Topoi*, v. 3, No. 2, pp. 109–121.

Parsons, C. (1990) The structuralist view of mathematical objects, *Synthese*, v. 84, pp. 303–346.

Parsons, T. (2000) *Indeterminate Identity: Metaphysics and Semantics*, Clarendon Press, Oxford.

Parvate, A. and Gangal, A. D. (2005) Fractal differential equations and fractal-time dynamical systems, *Pramana*, v. 64, No. 3, 389.

Parvate, A. and Gangal, A. D. (2009) Calculus on fractal subsets of real line — I, *Fractals*, v. 17, No. 1, pp. 53–81.

Parvate, A. and Gangal, A. D. (2009) Calculus on fractal subsets of real line — II: Conjugacy with integer order calculus, Pune University Preprint.

Parvate, A. and Gangal, A. D. (2011) Calculus on fractal subsets of real line: (II) Conjugacy with ordinary calculus, *Fractals*, v. 19, No. 3, pp. 271–290.

Parvate, A., Satin, S. and Gangal, A. D. (2011) Calculus on fractal sets in R^n, *Fractals*, v. 19, No. 1, pp. 15–27.

Pascoe, M. One plus one makes more than two: Our overlooked immigration benefits, *The Sydney Morning Herald*, 16 November 2017.

Peano, G. (1888) *Calcolo Geometrico secondo l'Ausdehnungslehre di H. Grassmann preceduto dalle Operazioni della Logica Deduttiva*, Fratelli Bocca Editori, Bocca, Turin.

Peano, G. (1889) *Arithmetices Principia Nova Method Exposita*, Bocca, Torino.

Peano, G. (1890) Sur une courbe, qui remplit toute une aire plane, *Math. Ann.*, v. 36, No. 1, pp. 157–160.

Peano, G. (1908) *Formulario Mathematico*, Bocca, Torino.

Peano, G. (1973) *Selected Works of Giuseppe Peano*, (Kennedy, H. C. Ed.), Allen & Unwin, London.

Pegg, D. (2014). *These are 25 Famous Numbers and Why They are Important*, https://list25.com/25-famous-numbers-and-why-they-are-important/.

Peirce, C. S. (1881) On the logic of number, *Amer. J. Math.*, v. 4, pp. 85–95.

Peirce, C. S. (1931–1935) *Collected Papers*, v. 1–6, Cambridge University Press, Cambridge.

Penrose, R. (1972) On the nature of quantum geometry, in *Magic Without Magic*, (J. Klauder, Ed), Freeman, San Francisco, pp. 333–354.

Penrose, R. and Rindler, W. (1984) *Spinors and Space-Time. Volume 1: Two-Spinor Calculus and Relativistic Fields*, Cambridge University Press, Cambridge.

Perez Velazquez, J. L. (2005) Brain, behavior and mathematics: Are we using the right approaches? *Physica D*, v. 212, pp. 161–182.

Pesenti, M., Zago L., Crivello, F., Mellet, E., Samson, D., Duroux, B., Seron, X., Mazoyer, B. and Tzourio-Mazoyer, N. (2001) Mental calculation in a prodigy is sustained by right prefrontal and medial temporal areas, *Nat. Neurosci.*, v. 4, No. 1, pp. 103–107.

Peters, L. and De Smedt, B. (2018) Arithmetic in the developing brain: A review of brain imaging studies, *Developmental Cognitive Neurosci.*, v. 30, pp. 265–279.

Phillips, J. (2016) When one plus one equals three, *Wellness Universe*, April 10, 2016 http://blog.thewellnessuniverse.com/when-one-plus-one-equals-three.

Piaget, J. (1952) *The Child's Conception of Number*, Norton, New York.

Piaget, J. (1964) Mother structures and the notion of number, in *Cognitive Studies and Curriculum Development*, School of Education, Cornell University, pp. 33–39.

Piazza, M. and Dehaene, S. (2004) From number neurons to mental arithmetic: The cognitive neuroscience of number sense, in *The Cognitive Neurosciences* (M. S. Gazzaniga; Ed.), MIT Press, Cambridge, MA, pp. 865–875.

Pica, P., Lemer, C., Izard, V. and Dehaene, S. (2004) Exact and approximate arithmetic in an Amazonian indigene group, *Science*, v. 306, pp. 499–503.

Pierce, R. C. (1977) A brief history of logarithm, *Two-Year College Math. J.*, v. 8, No. 1, pp. 22–26.

Pierce, R. S. (1982) *Associative Algebras*, Springer-Verlag, New York.

Pierce, D. (2011) *Numbers*, Preprint in Logic (math.LO), arXiv:1104.5311.

Pietronero, L. (1987) The fractal structure of the universe: correlations of galaxies and clusters and the average mass density, *Physica* A, v. 144, p. 257.

Pin, J.-E. (1998) Tropical semirings, in Idempotency, Gunawardena, J. (Ed.). Publications of the Newton Institute, v. 11. Cambridge University Press. pp. 50–69.

Plato (1961) *The Collected Dialogues of Plato*, Princeton University Press, Princeton.

Plotinus (2018) *The Enneads*, Cambridge University Press, Cambridge.

Podlubny, I. (1998) *Fractional Differential Equations*, Academic Press, New York.

Poincaré, H. (1902) *La Science et l'hypothèse*, Flammarion, Paris.

Pollack, C. and Ashby, N. C. (2018) Where arithmetic and phonology meet: The meta-analytic convergence of arithmetic and phonological processing in the brain, *Developmental Cognitive Neurosci.*, v. 30, pp. 251–264.

Poonen, B. and Tschinkel, Y. (Eds.) (2004) *Arithmetic of Higher-Dimensional Algebraic Varieties*, Birkhäuser, Boston.

Popper, K. R. (1959) *The Logic of Scientific Discovery*, Basic Books, New York.

Popper, K. R. (1972) *The Logic of Scientific Discovery*, Hutchinson, London.

Popper, K. R. (1974) Replies to my critics, in *The Philosophy of Karl Popper* (P.A. Schilpp, Ed.), Open Court, La Salle, IL, pp. 949–1180.

Porter, T. M. (1995) *Trust in Numbers: The Pursuit of Objectivity in Science and Public Life*, Princeton University Press, Princeton, New Jersey.

Potter, M. (2002) *Reason's Nearest Kin: Philosophies of Arithmetic from Kant to Carnap*, Oxford University Press, Oxford.

Prado, J., Noveck, I. A. and van der Henst, J. B. (2010) Overlapping and distinct neural representations of numbers and verbal transitive series, *Cereb. Cortex*, v. 20, No. 3, pp. 720–729.

Presburger, M. (1929) Über die Vollstaendigkeit eines gewissen Systems der Arithmetik ganzer Zahlen, in welchem die Addition als einzige

Operation hervortritt, *C.R.I Congr. Math. Pays Slaves*, Warszawa, pp. 92–101.

Price, G. B. (1991) *An Introduction to Multicomplex Spaces and Functions*, Marcel-Dekker.

Price, G. R. and Ansari, D. (2011) Symbol processing in the left angular gyrus: Evidence from passive perception of digits, *Neuroimage*, v. 57, No. 3, pp. 1205–1211.

Price, G. R. Mazzocco, M. M. M. and Ansari, D. (2013) Why mental arithmetic counts: Brain activation during single digit arithmetic predicts high school math scores, *J. Neurosci.*, v. 33, No. 1, pp. 156–163.

Priest, G. (1994) Is arithmetic consistent? *Mind*, v. 105, No. 420, pp. 649–659.

Priest, G. (1994a) What could the least inconsistent number be? *Logique Anal.*, v. 37, pp. 3–12.

Priest, G. (1997) Inconsistent models for arithmetic: I, Finite models, *J. Philos. Logic*, v. 26, pp. 223–235.

Priest, G. (1998) Number, in *Routledge Encyclopedia of Philosophy*, v. 7, Routledge, London, pp. 47–54.

Priest, G. (2000) Inconsistent models for arithmetic: II, The general case, *J. Symbolic Logic*, v. 65, pp. 1519–1529.

Priest, G. (2003) On alternative geometries, arithmetics, and logics: A tribute to Łukasiewicz, *Studia Logica*, v. 74, No. 3, pp. 441–468.

Putnam, H. (1960) Minds and machines, in *Dimensions of Mind*, pp. 148–179.

Pycior, H. M. (2011) *Symbols, Impossible Numbers, and Geometric Entanglements: British Algebra through the Commentaries on Newton's Universal Arithmetick*, Cambridge University Press, Cambridge.

Radu, M. (2003) A debate about the axiomatization of arithmetic: Otto Hölder against Robert Graßmann, *Historia Math.*, v. 30, No. 3, pp. 341–377.

Rasch, G. (1934) Zur theorie und anwendung des produktintegrals, *J. Reine Angew. Math.*, v. 191, pp. 65–119.

Rashevsky, P. K. (1973) On the Dogma of the natural numbers, *Russian Math. Surv.*, v. 28, No. 4, pp. 243–246.

Rasiowa, H. and Sikorski, R. (1963) *The Mathematics of Metamathematics*, Panswowe Wydawnictwo Naukowe, Warszawa.

Rechter, O. (2006) The View from 1763: Kant on the Arithmetical Method Before Intuition", in *Intuition and the Axiomatic Method*, (Carson, E. and Huber, R. Eds.), Springer, Dordrecht.

Reck, E. (2017) Dedekind's contributions to the foundations of mathematics, *The Stanford Encyclopedia of Philosophy* (Winter 2017 Edition), Zalta, E. N. (Ed.), https://plato.stanford.edu/archives/win2017/entries/dedekind-foundations/.

Recorde, R. (1557) *Arithmetike: containyng the extraction of Rootes: The Coßike practise, with the rule of Equation and the woorkes of Surde Nombers*, Jhon Kyngstone, London.

Reid, C. (2006) *From Zero to Infinity: What Makes Numbers Interesting*, A K Peters, Ltd., Natick, MA.

Rényi, A. (1960) Some fundamental questions of information theory, MTA III.

Resnick, M. (1997) *Mathematics as a Science of Structures*, Oxford University Press, Oxford.

Resnik, M. D. (1999) *Mathematics as a Science of Patterns*, Clarendon Press, Oxford.

Reves, G. E. (1951) Outline of the history of arithmetic, *School Sci. Math.*, v. 51, No. 8, pp. 611–617.

Richards, M. (1967) *The BCPL Reference Manual*, Massachusetts Institute of Technology, Boston.

Riedell, W. E., Pikul, J. L. and Carpenter-Boggs, L. (2002) *One Plus One Equals Three: The Synergistic Effects Of Crop Rotation On Soil Fertility And Plant Nutrition*, UNL's Institutional Repository, https://digitalcommons.unl.edu/usdaarsfacpub/1061/.

Ries, A. (2014) *In the marketing world, one plus one equals three-fourths*, http://adage.com/article/al-ries/marketing-world-equals-fourths/295251/.

Riley, M. S., Greeno, J. G. and Heller, J. I. (1983) Development of children's problem-solving ability in arithmetic, in *The Development of Mathematical Thinking*, Academic, Orlando, FL, pp. 153–196.

Rips, L. J. (2013) How many is a zillion? Sources of number distortion, *J. Exp. Psychol.: Learning, Memory, Cognition*, v. 39, pp. 1257–1264.

Rips, L. J. (2015) Beliefs about the nature of numbers, in *Mathematics, Substance and Surmise: Views on the Meaning and Ontology of Mathematics*, (Davis, E. and Davis, P. Eds.), Springer, Berlin, pp. 321–345.

Ritchie, J. (2014) 1+1=3 — *How Partner Marketing Defies the Laws of Math*, 2014, https://blog.marketo.com/2014/08/113-how-partner-marketing-defies-the-laws-of-math.html.

Roberts, B. W. (2011) Group structural realism, *British J. Philos. Sci.*, v. 62. No. 1, pp. 47–69.

Roberts, W. A. (1995) Simultaneous numerical and temporal processing in the pigeon, *Curr. Dir. Psychol. Sci.*, v. 4, pp. 47–51.

Roberts, W. A. and Boisvert, M. J. (1998) Using the peak procedure to measure timing and counting processes in pigeons, *J. Exp. Psychol. Anim. Behav. Process.*, v. 24, pp. 416–430.

Robinson, A. (1961) Non-standard analysis, *Indagationes Math.* v. 23, pp. 432–440.

Robinson, A. (1966) *Non-Standard Analysis*, Studies of Logic and Foundations of Mathematics, North-Holland, New York.

Robinson, A. (1967) Nonstandard arithmetic, *Bull. Amer. Math. Soc.*, v. 73, No. 6, pp. 818–843.

Rogers, H. (1958) Gödel numberings of partial recursive functions, *J. Symbolic Logic*, v. 23, pp. 331–341.

Rogers, L. The history of negative numbers, in *Enriching Mathematics*, http://Enrich.maths.org/5961.

Roitman, J. D., Brannon E. M. and Platt M. L. (2007) Monotonic coding of numerosity in macaque lateral intraparietal area, *PLoS. Biol.*, v. 5, e208; 10.1371/journal.pbio.0050208.

Rose, H. E. (1961) On the consistency and undecidability of recursive arithmetic, *Zeit. Math. Logik Grundlagen der Math.*, v. 7, pp. 124–135.

Rosen, G. (2006) What are numbers? *Philosophy Talk*, March 14, 2006, http://www.philosophytalk.org/pastShows/Number.html.

Rosen, G. (2011) The reality of mathematical objects, in *Meaning in Mathematics* (Polkinghorne, J. Ed.), Oxford University Press, Oxford, pp. 113–132.

Rosinger, E. E. (2008) *On the safe use of inconsistent mathematics*, Preprint in Mathematics, math. GM, arXiv.org, 0811.2405v2.

Rosner, R. S. (1965) The power law and subjective scales of number, *Perceptual Motor Skills*, v. 21, p. 42.

Ross, K. A. (1996) *Elementary Analysis: The Theory of Calculus*, Springer-Verlag, New York.

Roth, P. (1608) *Arithmetica Philosophica, Oder schöne wolgegründete Uberauß Kunstliche Rechnung der Coß oder Algebrae: In drey unterschiedliche Theil getheilt*, Lantzenberger, Nürnberg.

Rotman, B. (1996) Counting information: A note on physicalized numbers, *Minds and Machines*, v. 6, No. 2, pp. 229–238.

Rotman, B. (1997) The truth about counting, *Sciences*, No. 11, pp. 34–39.

Rotman, B. (2003) Will the digital computer transform classical mathematics? *Phil. Trans. R. Soc. Lond.* A, v. 361, pp. 1675–1690.

Rotman, J. (1996) *A First Course of Abstract Algebra*, Prentice-Hall, Upper Saddle River, NJ.

Roux, A. (2007) Un plus un égale trois: La perte de l'objet primaire comme condition de l'apparition du symbole. *Rev. Française Psychanalyse*, v. 71, No. 1, pp. 153–168.

Rowen, L. H. (1988) *Ring Theory*, Academic Press, Boston, MA.

Rozeboom, W. W. (1966) Scaling theory and the nature of measurement, *Synthese*, v. 16, No. 2, pp. 170–233.

Rucker, R. (1987) *Mind Tools: The Five Levels of Mathematical Reality*, Houghton Mifflin Co., Boston.

Russell, B. (1903) *Principles of Mathematics*, Allen and Unwin, London.

Ryle, G., Lewy, C. and Popper, K. R. (1946) Symposium: Why are the calculuses of logic and arithmetic applicable to reality? *Proceedings of the Aristotelian Society*, Supplementary volumes, v. 20, Logic and Reality, pp. 20–60.

Ryll-Nardzawski, C. (1952) The role of the axiom of induction in the elementary arithmetic, *Fund. Math.*, v. 39, pp. 239–263.

Salmon, N. (2008) Numbers versus nominalists, *Analysis*, v. 68, No. 3, pp. 177–182.

Satin, S. and Gangal, A. D. (2016) Langevin equation on fractal curves, *Fractals*, v. 24, No. 3, 1650028.

Sawamura, H., Shima, K. and Tanji, J. (2002) Numerical representation for action in the parietal cortex of the monkey, *Nature*, v. 415, pp. 918–922.

Sawamura, H., Shima, K. and Tanji, J. (2010) Deficits in action selection based on numerical information after inactivation of the posterior parietal cortex in monkeys, *J. Neurophysiol.*, v. 104, pp. 902–910.

Sayward, C. (2002) A conversation about numbers and knowledge, *Amer. Philos. Quarterly*, v. 39, No. 3, pp. 275–287.

Sazonov, V. Yu. (2002) On feasible numbers, in *Logic and Computational Complexity, Lecture Notes in Computer Science*, v. 960, Springer, New York, pp. 30–51.

Schirn, M. (1995) Axiom V and Hume's principle in Frege's foundational project, *Dialogos*, v. 66, pp. 7–20.

Schneider, B., Parker, S., Ostrosky, D., Stein, D. and Kanow, G. (1974) A scale for the psychological magnitude of numbers, *Perception Psychophys.*, v.16, pp. 43–46.

Schlote, K.-H. (1996) Hermann Guenther Grassmann and the theory of hypercomplex number systems, *Boston Studies Philos. Sci.*, v. 187, pp. 165–174.

Schröder, E. (1873) *Lehrbuch der Arithmetik und Algebra für Lehrer und Studirende*, Teubner, Leipzig.

Schroeder, M. R. (1985) Number theory and the real world, *Math. Intelligencer*, v. 7, No. 4, pp. 18–26.

Schroeder, M. R. (1988) The unreasonable effectiveness of number theory in science and communication (1987 Rayleigh Lecture), *IEEE ASSP Magazine*, v.5, No. 1, pp. 5–12.

Schütte, K. (1951) Beweistheoretische Erfassung der unendlichen Induktion in der Zahlentheorie, *Math. Ann.*, v. 122, pp. 369–389.

Schütte, K. (1977) *Proof Theory*, Springer-Verlag, Berlin.

Schwartz, L. (1950/1951) *Théorie des Distributions*, v. I–II, Hermann, Paris.

Schweizer, B. and Sklar, A. (1960) Statistical metric spaces, *Pacific J. Math.*, v. 10, pp. 313–334.

Segre, C. (1892) Le rappresentazioni real idelle forme complesse e glientii per algebrici, *Math. Ann.*, v. 40, pp. 413–467.

Segre, M. (1994) Peano's Axioms in their Historical Context, *Arch. History Exact Sci.*, v. 48, No. 3/4, pp. 201–342.

Sella, F., Sader, E., Lolliot, S. and Cohen Kadosh, R. (2016) Basic and advanced numerical performances relate to mathematical expertise but are fully mediated by visuospatial skills, *J. Exp. Psychol. Learn. Mem. Cogn.*, v. 42, No. 9, pp. 1458–1472.

Sellars, W. (1962) Naming and saying, *Philos. Sci.*, v. 29, pp. 7–26.

Semadeni, Z. (1982) *Schauder Bases in Banach Spaces of Continuous Functions*, Lecture Notes in Mathematics, v. 918, Springer, Berlin.

Shabel, L. (2016) Kant's philosophy of mathematics, *The Stanford Encyclopedia of Philosophy* (Spring 2016 Edition), Zalta, E. N. (Ed.), https://plato.stanford.edu/archives/spr2016/entries/kant-mathematics/.

Shannon, C. (1948) A mathematical theory of communication, *Bell Syst. Techn. J.*, v. 27, pp. 379–423.

Shannon, C. E. (1993) *Collected Papers*, (Sloane, N. J. A. and Wyner, A. D. Eds.), IEEE Press, New York.

Shapiro, S. (1983) Mathematics and reality, *Philos. Sci.*, v. 50, pp. 523–548.

Shapiro, S. (1996) Space, number, and structure: A tale of two debates, *Philos. Math.*, v. 4, No. 3, pp. 148–173.

Shapiro, S. (1997) *Philosophy of Mathematics: Structure and Ontology*, Oxford University Press, New York.

Sharma, G. (2011) Color imaging arithmetic: Physics $\cup$ math $>$ physics $+$ math, in *CCIW 2011*, Lecture Notes in Computer Science, v. 6626, Springer, pp. 31–46.

Shields, P. (1997) Peirce's Axiomatization of Arithmetic, in *Studies in the Logic of Charles Sanders Peirce*, (Houser, N., Roberts, D. D. and Van Evra, J. Eds.), Indiana University Press, Bloomington, pp. 43–52.

Shields, P. (2012) *Peirce on the Logic of Number*, Docent Press.

Shiozawa, Y. (2015) International trade theory and exotic algebras, *Evolutionary Inst. Econ. Rev.*, v. 12, pp. 177–212.

Shirley, J. W. (1951) Binary numeration before Leibniz, *Amer. J. Phy.*, v. 19, No. 8, pp. 452–454.

Shoenfield, J. R. (2001) *Mathematical Logic*, Association for Symbolic Logic, K Peters, Ltd., Natick, MA.

Shulman, M. (2016) *Homotopy Type Theory: A synthetic approach to higher equalities*, Preprint, arXiv:1601.05035v3.

Shum, J., Hermes, D., Foster, B. L., Dastjerdi, M., Rangarajan, V., Winawer, J., Miller, K. J. and Parvizi, J. (2013) A brain area for visual numerals, *J. Neurosci.*, v. 33, No. 16, pp. 6709–6715.

Sicha, J. F. (1970) Counting and the natural numbers, *Philos. Sci.*, v. 37, No. 3, pp. 405–416.

Šikić, Z. (1996) What are numbers? *Int. Studies Philos. Sci.*, v. 10, No. 2, pp. 159–171.

Silver, R. A. (2010) Neuronal arithmetic, *Nat. Rev. Neuroscience*, v. 11, No.7, pp. 474–489.

Simpson, S. G. (2009) *Subsystems of Second Order Arithmetic*, Cambridge University Press, Cambridge.

Sipser, M. (1997) *Introduction to the Theory of Computation*, PWS Publishing Co., Boston.

Sitomer, M. and Sitomer, H. (1976) *How Did Numbers Begin?* Crowell, New York.

Skolem, T. (1923/1976) Begründung der elementaren Arithmetik durch die rekurrierende Denkweise ohne Anwendung scheinbarer Veränderlichen mit unendlichem Ausdehnungsbereich, *Skrifter utgit av Videnskapsselskapet i Kristiania*. I, *Matematisk-naturvidenskabelig klasse*, v. 6, pp. 1–38 (in English: The foundations of elementary arithmetic established by means of the recursive mode of thought, without the use of apparent variables ranging over infinite domains , in *From Frege to Gödel* (Jean van Heijenoort, Ed.), Harvard University Press, pp. 306–333).

Skolem, T. (1934) Über die Nicht-charakterisierbarkeit der Zahlen reihe mittels endlich oder abzählbar unendlich vieler Aussagen mit ausschließlich Zahlenvariablen, *Fund. Math.*, v. 23, No. 1, pp. 150–161.

Skolem, T. (1955) Peano's axioms and models of arithmetic, in *Mathematical Interpretation of Formal Systems*, North-Holland Publishing Company, Amsterdam, pp. 1–14.

Slaveva-Griffin, S. (2009) *Plotinus on Number*, Oxford University Press, Oxford.

Slavík, A. (2007) *Product Integration, Its History and Applications*, History of Mathematics, v. 29, Matfyzpress, Prague.

Smith, D. E. (1963) *Mathematics*, Cooper Square Publishers, New York.

Smeltzer, D. (1959) *Man and Number*, Emerson Books, New York.

Smorynski, C. (1984) Lectures on nonstandard models of arithmetic: Commemorating Giuseppe Peano, in *Logic Colloquium'82, Proceedings of the Colloquium Held in Florence*, 23–28 August, 1982, Elsevier Science Publication Co., Amsterdam, pp. 1–70.

Śniatycki, J. (2013) *Differential Geometry of Singular Spaces and Reduction of Symmetry*, Cambridge University Press, Cambridge.

Soderstrand, M. A., Jenkins, W. K., Jullien, G. A. and Taylor, F. J. (Eds.) (1986) *Residue Number System Arithmetic*, IEEE Press.

Sondheimer, E. and Rogerson, A. (1981) *Numbers and Infinity: A Historical Account of Mathematical* Concepts, Dover Books on Mathematics, Dover Publications, New York.

Sonin, A. A. (2001) *The Physical Basis of Dimensional Analysis*, MIT, Cambridge.

Sonnen, M. (2013) Merger math: Can 1 + 1 = 3? *Investment News*, Dec 11, 2013, https://www.investmentnews.com/article/20131211/FREE/131219982/merger-math-can-1-1-3.

Spelke E. and Barth, H. (2003) The construction of large number representations in adults, *Cognition*, v. 86, No. 3, pp. 201–221.

Spelke, E. S. and Tsivkin, S. (2001) Language and number: A bilingual training study, *Cognition*, v. 78, No. 1, pp. 45–88.

Speyer, D. and Sturmfels, B. (2009) Tropical mathematics, *Math. Magazine*, v. 82, No. 3, pp. 163–173.

Spiegel, M. R. (1971) *Calculus of Finite Differences and Difference Equations*, McGraw-Hill Education, New York.

Stähle, K. (1931) *Die Zahlenmystik bei Philon von Alexandreia*, Teubner, Tuebingen.

Stanley, D. (1999) A multiplicative calculus, *Primus*, v. 9, No. 4, pp. 310–326.

Steen, S. W. P. (1972) *Mathematical Logic with Special Reference to the Natural Numbers*, Cambridge University Press, Cambridge.

Steenrod, N. (1951) *The Topology of Fibre Bundles*, Princeton University Press.

Stein, H. (1990) Eudoxos and Dedekind: on the ancient Greek theory of ratios and its relation to modern mathematics, *Synthese*, v. 84, pp. 163–211.

Stevens, S. S. (1935) The operational definition of psychological concepts, *Psychol. Rev.*, v. 42, pp. 517–527.

Stevens, S. S. (1946) On the theory of scales of measurement, *Science*, v. 103, No. 2684, pp. 677–680.

Stevens, S. S. (1951) Mathematics, measurement and psychophysics, in *Handbook of experimental psychology*, Wiley, New York, pp. 1–49.

Stevens, S. S. (1975) *Psychophyscis: Introduction to its Perceptual, Neuronal, and Social Prospects*, Wiley, New York.

Stevin, S. (1585) *De Thiende*, Chriftoffel Plantijn, Leyden (in Flemish).

Stevin, S. (1958) *The principal works of Simon Stevin*, C. V. Swets & Zeitlinger, Amsterdam.

Sthananga Sutra, *Wikipedia*, https://en.wikipedia.org/wiki/Sthananga_Sutra.

Stillinger, F. H. (1977) Axiomatic basis for spaces with noninteger dimension, *J. Math. Phys.*, v. 18, pp. 1224–1234.

Stolz, O. (1881) B. Bolzano's Bedeutung in der Geschichte der Infinitesimalrechnung, *Math. Ann.*, v. 18, pp. 255–279.

Stolz, O. (1882) Zur Geometrie der Alten, insbesondere über ein Axiom des Archimedes, *Berichtedes Naturwissenschaftlich-Medizinischen Vereines in Innsbruck* 12, pp. 74–89.

Stolz, O. (1883) Zur Geometrie der Alten, insbesondere über ein Axiom des Archimedes, *Math. Ann.* v. 22, pp. 504–519.

Stolz, O. (1885) *Vorlesungen über Allgemeine Arithmetik, Erster Theil: Allgemeines und Arithhmetik der Reelen Zahlen*, Teubner, Leipzig.

Stolz, O. (1886) *Vorlesungen über Allgemeine Arithmetik, Zweiter Theil: Arithhmetik der Complexen Zahlen*, Teubner, Leipzig.

Stolz, O. (1888) Ueber zwei Arten von unendlich kleinen und von unendlich grossen Grössen, *Math. Ann.* v. 31, pp. 601–604.

Stolz, O. (1891) Ueber das axiom des archimedes, *Math. Ann.* v. 39, pp. 107–112.

Stolz, O. and Gmeiner, J. A. (1902) *Theoretische Arithmetik*, Teubner, Leipzig.

Stolz, O. and Gmeiner, J. A. (1915) *Theoretische Arithmetik, Abteilung II: Die Lehren Von Den Reelen Und Von Den Komplexen Zahlen*, Zweite Auflage, Teubner, Leipzig.

Štrboja, M., Pap, E. and Mihailović, B. (2018) Discrete bipolar pseudo-integrals, *Inform. Sci.*, v. 468, pp. 72–88.

Štrboja, M., Pap, E. and Mihailović, B. (2019) Transformation of the pseudo-integral and related convergence theorems, *Fuzzy Sets Syst.*, v. 355, pp. 67–82.

Strichartz, R. S. (2006) *Differential Equations on Fractals*, Princeton University Press, Princeton.

Struik, D. J. (1987) *A Concise History of Mathematics*, Dover Publications, New York.

Stewart, I. (2018) Number symbolism, *Encyclopedia Britannica.*

Sulkowski, G. M. and Hauser, M. D. (2001) Can rhesus monkeys spontaneously subtract? *Cognition*, v. 79, pp. 239–262.

Sunehag, P. (2007) On a connection between entropy, extensive measurement and memoryless characterization arXiv:0710.4179 [physics.data-an].

Suroweicki, J. (2004) *The Wisdom of Crowds: Why the Many Are Smarter Than the Few and How Collective Wisdom Shapes Business*, Economies, Societies and Nations, Little & Brown, Boston.

Sutherland, D. (2006) Kant on arithmetic, algebra, and the theory of proportions, *Journal of the History of Philosophy*, v. 44, No. 4, pp. 533–558.

Sutherland, D. (2017) Kant's conception of number, *Philos. Rev.*, v. 126, No. 2, pp. 147–190.

Svozil, K. (1987) Quantum field theory on fractal spacetime: a new regularization method, *J. Phys. A*, v. 20, pp. 3861–3875.

Swartzlander, E. E., Jr. (1990) *Computer Arithmetic*, v. I & II, IEEE Computer Society Press.

Swift, J. D. (1956) Diophantus of Alexandria, *Amer. Math. Monthly*, v. 63, pp. 163–170.

Sylos Labini, F., Montuori, M. and Pietronero, L. (1998) Scale-invariance of galaxy clustering, *Phys. Rep.*, v. 293, p. 61.

Synergy (2011) Cambridge Business English Dictionary, Cambridge University Press, London.

Szuba, T. (2001) *Computational Collective Intelligence*, Wiley, New York.

Tait, W. W. (1996) Frege versus Cantor and Dedekind: On the concept of number, in *Frege: Importance and Legacy* (M. Schirn, Ed.), de Gruyter, Berlin, pp. 70–113.

Tall, D. O. (2004) The three worlds of mathematics, *For the Learning of Math.*, v. 23, No. 3, pp. 29–33.

Tall, D. O. (2004a) Thinking Through Three Worlds of Mathematics, in *Proceedings of the 28th Conference of the International Group for the Psychology of Mathematics Education*, Bergen, Norway, v. 4, pp. 281–288.

Tall, D. (2008) The transition to formal thinking in mathematics, *Math. Education Res. J.*, v. 20 No. 2, pp. 5–24.

Tang, Y., Zhang, W., Chen, K., Feng, S., Ji, Y., Shen, J., Reiman, E.M. and Liu, Y. (2006) Arithmetic processing in the brain shaped by cultures, *Proc. Natl. Acad. Sci. USA*, v. 103, No. 28, pp. 10775–10780.

Tappenden, J. (1995) Geometry and generality in Frege's philosophy of arithmetic, *Synthese*, v. 102, pp. 319–361.

Tarasov, V. E. (2005) Continuous medium model for fractal media, *Phys. Lett. A*, v. 336, pp. 167–174.

Tarasov, V. E. (2014) Anisotropic fractal media by vector calculus in non-integer dimensional space, *J. Math. Phys.*, v 55, 083510.

Tarasov, V. E. (2016) Acoustic waves in fractal media: Non-integer dimensional spaces approach, *Wave Motion*, v. 63, pp. 18–22.

Tarski, A. (1936) Der Wahrheitsbegriff in den formalisierten Sprachen, *Studia Philosophica*, v. 1, pp. 261–405.

Tarski, A. and Vaught, R. L. (1957) Arithmetical extensions of relational systems, *Compos. Math.*, v. 13, pp. 81–102.

Taylor, R. and Wiles A. (1995) Ring theoretic properties of certain Hecke algebras, *Annals of Mathematics*, v. 141, No. 3, pp. 553–572.

Tegmark, M. (2014) *Our Mathematical Universe: My Quest for the Ultimate Nature of Reality*, Alfred A. Knopf, New York.

Teplyaev, A. (2000) Gradients on fractals, *J. Funct. Anal.*, v. 174, pp. 128–154.

(2014) *The Business Book*: *Big Ideas Simply Explained*, Books, London, UK.

Thomson, J. (2001) Comments on Professor Benacerraf's Paper, in *Zeno's Paradoxes*, (Salmon, W.C. Ed.) Hackett, Indianapolis, IN.

Thompson, R. F., Mayers, K. S., Robertson, R. T. and Patterso, C. J. (1970) Number coding in association cortex of the cat, *Science*, v. 168, No. 3928, pp. 271–273.

Tolpygo, A. (1997) Finite Infinity, in *Methodological and Theoretical Problems of Mathematics and Information Sciences*, Kiev, Ukrainian Academy of Information Sciences, pp. 35–44 (in Russian).

Trabacca, A., Moro, G., Gennaro, L. and Russo, L. (2012) When one plus one equals three: the ICF perspective of health and disability in the third millennium, *European J. Phys. Rehabilitation Med.*, v. 48, No. 4, pp. 709–710.

Trick L. M. and Pylyshyn Z. W. (1994) Why are small and large numbers enumerated differently? A limited-capacity preattentive stage in vision, *Psychol. Rev.*, v. 101, pp. 80–102.

Troelstra, A. S. (Ed) (1973) *Metamathematical Investigation of Intuitionistic Arithmetic and Analysis*, Springer, New York.

Trott, D. (2015) *One Plus One Equals Three: A Masterclass in Creative Thinking*, Macmillan Publishing Company, New York.

Trudeau, R. J. (1987) *The Non-Euclidean Revolution*, Birkhäuser, Boston.

Tubbs, R. (2009) *What is a Number?: Mathematical Concepts and Their Origins*, Johns Hopkins University Press, Baltimore, MD.

Tufte, E. R. (1990) *Envisioning Information*, Graphics Press, Cheshire, CT.

Turing, A. (1936) On computable numbers with an application to the Entscheidungs-problem, *Proc. Lond. Math. Soc.*, Ser.2, v. 42, pp. 230–265.

Urbaniak, R. (2012) Numbers and propositions versus nominalists: yellow cards for Salmon & Soames, *Erkenntnis*, v. 77, No. 3, pp. 381–397.

Uzer, A. (2010) Multiplicative type complex calculus as an alternative to the classical calculus, *Comput. Math. Appl.*, v. 60, No. 10, pp. 2725–2737.

Van Bendegem, J. P. (1994/1996) Strict finitism as a viable alternative in the foundations of mathematics, *Logique Anal.*, v. 37, No. 145, pp. 23–40.

van Dantzig, D. (1956) Is $10^{10^{10}}$ a finite number? *Dialectica*, No. 9.

Van de Voorde, N. (2017) One plus one equals three? The electoral effect of multiple office-holding in national and local elections, in *ECPR General Conference* 2017, https://biblio.ugent.be/publication/8534419.

Van der Waerden, B. L. (1971) *Algebra*, Springer-Verlag, Berlin.

Varadarajan, V. S. (2002) Some remarks on arithmetic physics, *J. Statist. Planning Inference*, v. 103, No. 1–2, pp. 3–13.

Varadarajan, V. S. (2004) Arithmetic quantum physics: Why, what, and whether, *Tr. Mat. Inst. Steklova*, v. 245, pp. 273–280.

Varadarajan, V. S. (2004a) *Supersymmetry for Mathematicians: An Introduction*, Courant Lecture Notes in Mathematics, v. 11. American Mathematical Society.

Varley, R. A., Klessinger, N. J., Romanowski, C. A. and Siegal, M. (2005) Agrammatic but numerate, *Proc. Natl. Acad. Sci. USA*, v. 102, No. 9, pp. 3519–3524.

Vasas, V. and Chittka, L. (2018) Insect-inspired sequential inspection strategy enables an artificial network of four neurons to estimate numerosity, *Science*, v. 11, pp. 85–92.

Verguts, T. and Fias, W. (2004) Representation of number in animals and humans: A neural model, *J. Cogn. Neurosci.*, v. 16, pp. 1493–1504.

Verguts, T. and Fias, W. (2008) Symbolic and nonsymbolic pathways of number processing, *Philos. Psychol.*, v. 21, pp. 539–554.

Vernaeve, H. (2010) Ideals in the ring of Colombeau generalized numbers, *Comm. Alg.*, v. 38, No. 6, pp. 2199–2228.

Veronese, G. (1889) Il continuo rettilineo e l'assioma V di Archimede, *Memorie della Reale Accademia dei Lincei, Atti della Classe di Scienze Naturali, Fisiche e Matematiche*, v. 6, No. 4, pp. 603–624.

Volovich, I. V. (1987) *Number theory as the ultimate physical theory*, Preprint CERN-TH 87, pp. 4781–4786.

Volterra, V. (1887) Sui fondamenti della teoria delle equazioni differenziali lineari, *Memorie della Società Italiana della Scienze*, v. 3, p. VI.

Volterra, V. (1887a) Sulle equazioni differenziali lineari, *Rendiconti dell' Academia dei Lincei*, v. 3, pp. 393–396.

von Helmholtz, H. (1887) *Zahlen und Messen*, Philosophische Aufsatze, Fues's Verlag, Leipzig, pp. 17–52 (Translated by C. L. Bryan, "*Counting and Measuring*", Van Nostrand, 1930).

von Koch, H. (1904) Sur une courbe continue sans tangente, obtenue par une construction géométrique élémentaire, *Arkiv Mat.*, v. 1, pp. 681–704.

von Neumann, J. (1927) Zur Hilbertschen Beweistheorie, *Math. Zeit.*, v. 26, pp. 1–46.

von Neumann, J. (1955) *Mathematical Foundations of Quantum Mechanics*, Princeton University Press, Princeton, NJ.

von Neumann, J. (1961) *Collected Works*, v. 1, *Logic, Theory of Sets and Quantum Mechanics*, Pergamon Press, New York.

von Plato J. (2017) Husserl and Grassmann, in *Essays on Husserl's Logic and Philosophy of Mathematics* (Centrone S. Ed.), Synthese Library (Studies in Epistemology, Logic, Methodology, and Philosophy of Science), v. 384, Springer, Dordrecht.

Vorobjev, N. N. (1963) The extremal matrix algebra, *Soviet Math. Dokl.,* v. 4, pp. 1220–1223.

Vrandečić, D., Krötzsch, M., Rudolph, S. and Lösch, U. (2010) Leveraging Non-Lexical Knowledge for the Linked Open Data Web, *The Fifth RAFT'2010, The Yearly Bilingual Publication on Nonchalant Research*, v. 5, No. 1, pp. 18–27.

Walsh, S. (2010) *Arithmetical Knowledge and Arithmetical Definability: Four Studies*, Dissertation, University of Notre Dame, Notre Dame, IN.

Walsh, S. (2012) Comparing Peano arithmetic, Basic Law V, and Hume's Principle, *Ann. Pure Appl. Logic*, v. 163, No. 11, pp. 1679–1709.

Walsh, V. (2003) A theory of magnitude: Common cortical metrics of time, space and quantity, *Trends Cogn. Sci.*, v. 7, pp. 483–488.

Walsh, V. (2015) A theory of magnitude: the parts that sum to number, in *The Oxford Handbook of Numerical Cognition* (Cohen Kadosh R. and Dowker A., Eds.), Oxford University Press, Oxford, pp. 552–565.

Wang, H. (1956) Arithmetic translations of axiom systems, *J. Symbolic Logic*, v. 21, No. 4, pp. 402-403.

Wang, H. (1957) The axiomatization of arithmetic, *J. Symbolic Logic*, v. 22, No. 2, pp. 145–158.

Wang, L., Uhrig, L., Jarraya, B. and Dehaene, S. (2015) Representation of numerical and sequential patterns in Macaque and human brains, *Curr. Biol.*, v. 25, No. 15, pp. 1966–1974.

Wapnick, K. (2013) *When 2 + 2 = 5: Reflecting Love in a Loveless World*, Foundation for A Course in Miracles, Henderson, NV.

Waszkiewicz, W. (1971) The notions of isomorphism and identity for many-valued relational structures, *Studia Logica*, v. 27, pp. 93–98.

Watzlawick, P, Beavin, J. H. and Jackson, D. D. (1967) *Pragmatics of Human Communication: A Study of Interactional Patterns, Pathologies, and Paradoxes*, W W Norton, New York.

Wedderburn, J. (1908) On hypercomplex numbers, *Proc. London Math. Soc.*, v. 6, pp. 77–118.

Weil, A. (1984) *Number Theory: An Approach Through History from Hammurapi to Legendre*, Birkhäuser, Boston.

Weiss, A. (2005) The Power of Collective Intelligence, netWorker — Beyond file-sharing, *Collective Intell.*, v. 9, No. 3, pp. 16–24.

Weissmann, S. M., Hollingsworth, S. R. and Baird, J. C. (1975) Psychophysical study of numbers (III): Methodological applications, *Psychol. Res.*, v. 38, pp. 97–115.

Wells, D. G. (1998) *The Penguin Dictionary of Curious and Interesting Numbers*, Penguin Books, London, UK.

Wellman H. M. and Miller K. F. (1986) Thinking about nothing: developmental concepts of zero, *Br. J. Dev. Psychol.*, v. 4, pp. 31–42.

Weyl, H. (1927) *Philosophie der Mathematik und Naturwissenschaft*, Druck und Verlag Von R. Oldenbourg, München und Berlin.

Weyl, H. (1985) Axiomatic versus constructive procedures in mathematics, *Math. Intelligencer*, v. 7, No. 4, pp. 10–17, 38.

Whalen, J., Gallistel, C. R. and Gelman, R. (1999) Nonverbal counting in humans: the psychophysics of number representation, *Psychol. Sci.*, v. 10, pp. 130–137.

White, M. (1999) Incommensurables and incomparables: On the conceptual status and the philosophical use of hyperreal numbers, *Notre Dame Journal of Formal Logic*, v. 40, pp. 420–446.

Whitney, H. (1933) Characteristic functions and the algebra of logic, *Annals of Mathematics*, v. 34, pp. 405–414.

Wiener, N. (1933) *The Fourier Integral and Certain of Its Applications*, Cambridge University Press, Cambridge.

Wiles, A. (1995) Modular elliptic curves and Fermat's last theorem, *Annals of Math.*, v. 142, No. 3, pp. 443–551.

Wilson, A. M. (1996) *The Infinite in Finite*, Oxford University Press, Oxford.

Wittgenstein, L. (1921) *Tractatus Logico-Philosophicus*, Routledge and Kegan Paul, London.

Wittgenstein, L. (1953) *Philosophical Investigations*, Blackwell, Oxford.

Wolpert, D. H. and Tumer, K. (2000) An introduction to collective intelligence, in *Handbook of Agent Technology*, AAAI/MIT Press.

Worrall, J. (1989) Structural realism: The best of both worlds? *Dialectica*, v. 43, pp. 99–124.

Wright, C. (1983) *Frege's Conception of Numbers as Objects*, Aberdeen University Press, Aberdeen.

Wynn, K. (1990) Children's understanding of counting, *Cognition*, v. 36, pp. 155–193.

Wynn, K. (1992) Children's acquisition of number words and the counting system, *Cogn. Psychol.*, v. 24, pp. 220–251.

Xu, F. and Spelke, E. S. (2000) Large number discrimination in 6-month-old infants, *Cognition*, v. 74, pp. B1–B11.

Xu, F., Spelke, E. S. and Goddard, S. (2005) Number sense in human infants, *Dev. Sci.*, v. 8, pp. 88–101.

Yalcin, N., Celik, E. and Gokdogan, A. (2016) Multiplicative Laplace transform and its applications, *Optik*, v. 127, No. 20, pp. 9984–9995.

Yanovskaya, S. A. (1963) Are the problems known as Zeno's aporia solved in contemporary science? in *Problems of Logic*, Nauka, Moscow, pp. 116–136 (in Russian).

Yellot, J. I. Jr. (1977) The relationship between Luce's choice axiom, thurstone's theory of comparative judgement, and the double exponential distribution, *J. Math. Psych.*, v. 15, pp. 109–144.

Yershov, Yu. A. and Palyutin, E. A. (1979) *Mathematical Logic*, Nauka, Moscow (in Russian).

Yesenin-Volpin, A. S. (1960) On the grounding of set theory, in *Application of Logic in Science and Technology*, Nauka, Moscow, pp. 22-118 (in Russian).

Yesenin-Volpin, A. S. (1970) The ultra-intuitionistic criticism and the antitraditional program for foundations of mathematics, in *Intuitionism and Proof Theory*, North–Holland, Amsterdam, pp. 3-45.

Zähle, M. (2005) Harmonic calculus on fractals: A measure geometric approach (II), *Trans. Amer. Math. Soc.*, v. 357, pp. 3407–3423.

Žarnić, B. (1999) Mathematical Platonism: from objects to patterns, *Synthesis Philosophica* No. 27–28, pp. 53–64.

Zeldovich, Ya. B., Ruzmaikin, A. A. and Sokoloff, D. D. (1990) *The Almighty Chance*, World Scientific Lecture Notes, v. 20, World Scientific, Singapore.

Zermelo, E. (1908) Untersuchungen über die Grundlagen der Mengenlehre, *Math. Annalen*, v. 65, pp. 261–281.

Zimmermann, U. (1981) Linear and combinatorial optimization in ordered algebraic structures, *Ann. Discrete Math.*, v. 10, pp. 1–380.

Zimmermann, H.-J. (1996) *Fuzzy Set Theory – and its Applications*, 3rd Edition, Kluwer, Boston.

Author Index

Subject Index

B

C

D

E

G

H

I

J

K

L

O

P

Q

R

S

T

U

V

9 789811 214301
IPW0136017